The Elements

Name	Symbol	Atomic Number	Relative Atomic Mass
Actinium	Ac	89	227.028
Aluminum	Al	13	26.9815
Americium	Am	95	(243)
Antimony	Sb	51	121.75
Argon	Ar	18	39.948
Arsenic	As	33	74.9216
Astatine	At	85	(210)
Barium	Ba	56	137.327
Berkelium	Bk	97	(247)
Beryllium	Be	4	9.01218
Bismuth	Bi	83	208.980
Boron	B	5	10.811
Bromine	Br	35	79.904
Cadmium	Cd	48	112.411
Calcium	Ca	20	40.078
Californium	Cf	98	(251)
Carbon	C	6	12.011
Cerium	Ce	58	140.115
Cesium	Cs	55	132.905
Chlorine	Cl	17	35.4527
Chromium	Cr	24	51.9961
Cobalt	Co	27	58.9332
Copper	Cu	29	63.546
Curium	Cm	96	(247)
Dysprosium	Dy	66	162.50
Einsteinium	Es	99	(252)
Erbium	Er	68	167.26
Europium	Eu	63	151.965
Fermium	Fm	100	(257)
Fluorine	F	9	18.9984
Francium	Fr	87	(223)
Gadolinium	Gd	64	157.25
Gallium	Ga	31	69.723
Germanium	Ge	32	72.61
Gold	Au	79	196.967
Hafnium	Hf	72	178.49
Helium	He	2	4.00260
Holmium	Ho	67	164.930
Hydrogen	H	1	1.00794
Indium	In	49	114.818
Iodine	I	53	126.904
Iridium	Ir	77	192.22
Iron	Fe	26	55.847
Krypton	Kr	36	83.80
Lanthanum	La	57	138.906
Lawrencium	Lr	103	(260)
Lead	Pb	82	207.2
Lithium	Li	3	6.941
Lutetium	Lu	71	174.967
Magnesium	Mg	12	24.3050
Manganese	Mn	25	54.9381
Mendelevium	Md	101	(258)

Name	Symbol	Atomic Number	Relative Atomic Mass
Mercury	Hg	80	200.59
Molybdenum	Mo	42	95.94
Neodymium	Nd	60	144.24
Neon	Ne	10	20.1797
Neptunium	Np	93	237.048
Nickel	Ni	28	58.69
Niobium	Nb	41	92.9064
Nitrogen	N	7	14.0067
Nobelium	No	102	(259)
Osmium	Os	76	190.23
Oxygen	O	8	15.9994
Palladium	Pd	46	106.42
Phosphorus	P	15	30.9738
Platinum	Pt	78	195.08
Plutonium	Pu	94	(244)
Polonium	Po	84	(209)
Potassium	K	19	39.0983
Praseodymium	Pr	59	140.908
Promethium	Pm	61	(145)
Protactinium	Pa	91	231.036
Radium	Ra	88	226.025
Radon	Rn	86	(222)
Rhenium	Re	75	186.207
Rhodium	Rh	45	102.906
Rubidium	Rb	37	85.4678
Ruthenium	Ru	44	101.07
Samarium	Sm	62	150.36
Scandium	Sc	21	44.9559
Selenium	Se	34	78.96
Silicon	Si	14	28.0855
Silver	Ag	47	107.868
Sodium	Na	11	22.9898
Strontium	Sr	38	87.62
Sulfur	S	16	32.066
Tantalum	Ta	73	180.948
Technetium	Tc	43	(98)
Tellurium	Te	52	127.60
Terbium	Tb	65	158.925
Thallium	Tl	81	204.383
Thorium	Th	90	232.038
Thulium	Tm	69	168.934
Tin	Sn	50	118.710
Titanium	Ti	22	47.88
Tungsten	W	74	183.85
Uranium	U	92	238.029
Vanadium	V	23	50.9415
Xenon	[illegible]	[illegible]	[illegible]31.29
Ytterbium	[illegible]	[illegible]	[illegible]73.04
Yttrium	[illegible]	[illegible]	[illegible]88.9059
Zinc	[illegible]	[illegible]	[illegible]65.39
Zirconium	[illegible]	[illegible]	[illegible]91.224

Atomic masses in this table are relative to carbon-12 and limited to six significant figures, although some atomic masses are known more precisely. For certain radioactive elements the numbers listed (in parentheses) are the mass numbers of the most stable isotopes.

GENERAL CHEMISTRY

GENERAL CHEMISTRY

John W. Hill
University of Wisconsin–River Falls

Ralph H. Petrucci
California State University, San Bernardino

Prentice Hall
Upper Saddle River, New Jersey 07458

Library of Congress Cataloging-in-Publication Data
Hill, John William.
General chemistry / John W. Hill, Ralph H. Petrucci.
p. cm.
Includes index.
ISBN 0-02-354481-3
1. Chemistry. I. Petrucci, Ralph H. II. Title.
QD33.H654 1996
540—dc20 95-3374
CIP

Acquisitions Editor: *Ben Roberts*
Editor in Chief: *Paul F. Corey*
Editorial Director: *Tim Bozik*
Assistant Vice President of Production and Manufacturing: *David W. Riccardi*
Executive Managing Editor: *Kathleen Schiaparelli*
Development Editor: *Alan MacDonell*
Senior Project Manager: *Elisabeth Belfer*
Assistant Managing Editors: *Margaret Antonini and Shari Toron*
Marketing Manager: *Linda Taft*
Marketing Assistant: *Amy Reed*
Manufacturing Manager: *Trudy Pisciotti*
Creative Director: *Paula Maylahn*
Art Director: *Heather Scott*
Interior Design: *Electronic Publishing Services Inc. and Initial Graphics*
Cover Designer: *Tom Nery*
Front Cover Photography: *Fundamental Photographs*
Back Cover Photography: *Carey Van Loon*
Editorial Assistants: *Ashley Scattergood and Nancy Bauer*
Art Studios: *Network Graphics and BioGrafx*
Copyediting and Text Composition: *Electronic Publishing Services Inc.*

Printed in the United States of America

10 9 8 7 6 5 4 3 2

ISBN 0-02-354481-3

Prentice-Hall International (UK) Limited, *London*
Prentice-Hall of Australia Pty. Limited, *Sydney*
Prentice-Hall Canada Inc., *Toronto*
Prentice-Hall Hispanoamericana, S.A., *Mexico*
Prentice-Hall of India Private Limited, *New Delhi*
Prentice-Hall of Japan, Inc., *Tokyo*
Simon & Schuster Asia Pte. Ltd., *Singapore*
Editora Prentice-Hall do Brasil, Ltda., *Rio de Janeiro*

Brief Contents

Contents

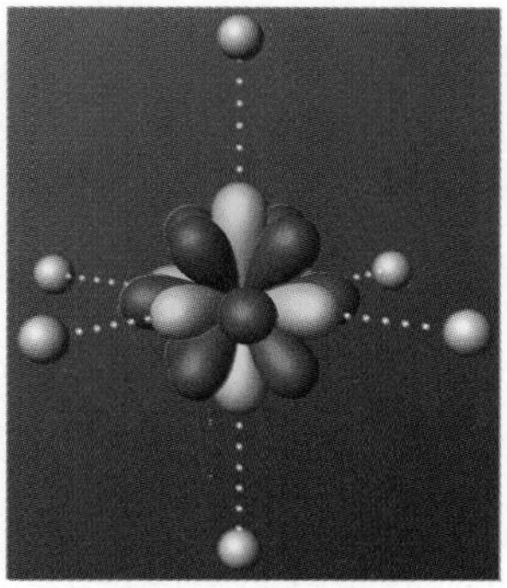

Featured Applications and Essays

Preface

You and your classmates come to this course with a variety of backgrounds and interests. Most of you plan to be scientists or professionals in a medical field or engineering. A knowledge of chemistry is essential to an understanding of fields that range from cell biology to the materials science of computer chips. Indeed, the chemical properties and principles you learn in this course will pervade almost every aspect of your private and professional lives. In this text we have tried to provide you with principles and applications of chemistry that will help you in your professional practice, but that will enrich your everyday life as well.

You may have studied chemistry before. If so, you may find much in the early chapters that will be a review of that earlier course. However, you are likely to find that the pace is more rapid here. If you have little or no background in chemistry, the first few chapters will provide essential concepts that underlie much of the remainder of the course.

Problem Solving

Problem-solving skills and the ability to think critically are essential for success in today's world. For every type of problem, we provide carefully worked out examples to guide you in solving similar problems, and we follow these examples with practice problems you can use to check your understanding. It is not enough, however, just to be able to plug numbers into an equation and get an answer. We hope to help you develop judgment about whether or not the answer is *reasonable*; we do this through worked out **estimation examples** followed by **estimation exercises**. You also need to develop some insight into the chemical concepts on which problems are based. We provide guided **conceptual examples**, followed by **conceptual exercises**, to foster this process. Throughout, we use drawings, computer graphics, and photographs to help you visualize chemical phenomena at both the microscopic (molecular) and macroscopic (visible) levels. In summary, we have striven to combine our collective experience in teaching and in writing chemistry textbooks for various audiences to provide a modern treatment of the subject that strikes a balance between the principles that give meaning to chemistry and the applications that make it come alive.

Organization

A major goal of ours in writing this text was to provide a truly general course that integrates all the major areas of chemistry. Physical principles, inorganic compounds, and analytical techniques are addressed throughout. Some simple organic chemistry is introduced in Chapter 2 and used thereafter in acid-base, oxidation-reduction, and other areas to illustrate chemical principles. Biochemistry is introduced in Chapter 5 (carbohydrates and fats as fuels for our bodies) and used frequently in following chapters where appropriate.

We have integrated descriptive chemistry throughout all the chapters, but five of them—Chapters 8, 14, 21, 22, and 23—are specifically descriptive. Even in these chapters, though, we relate most of the properties of substances to principles learned earlier. Other chapters are largely built upon a theoretical framework, but we illustrate principles throughout with concrete examples.

The early placement of the *s*-block elements in Chapter 8 allows us to use the relatively simple descriptive chemistry of these elements to illustrate some of the principles learned in the first seven chapters. It also allows us to demonstrate the need for additional principles that are developed in following chapters: atomic and molecular structure, liquids, solids, solutions, and so on. In the next descriptive chapter (14), we use these further principles to describe some of the chemistry of the gases that make up Earth's atmosphere and some of the important substances that are manufactured from atmospheric gases. Chapter 14, in turn, raises questions about reaction rates, equilibrium, and other topics that are then discussed in the six following chapters. Three more descriptive chapters—on the *p*-block elements, the *d*-block elements, and complex ions and coordination compounds—complete the body of the text. In these last three chapters, we apply many of the principles learned in the preceding chapters.

We have included a number of special interest features throughout the book. Most of these focus on a practical topic related to the subject matter of the chapter. The topics are varied, ranging from air batteries, to divers' bends and Henry's Law, to the way a molecule's shape affects the action of a drug. All of these features are intended for enjoyment and enlightenment and should not require the sort of intense study that is necessary for much of the textual material.

Also provided with this textbook — as a separate optional bound volume — are six selected topics, following the same format as the text chapters:

A Metallurgy
B Bonding in Metals and Semiconductors
C Spectroscopy: Determination of Structure
D Nuclear Chemistry
E Polymers
F Nucleic Acids: Molecules of Heredity

For those who wish to use them, these topics can be taken up at appropriate times in conjunction with other chapters. For example, Topic E might be used after Chapter 9, or Topic B after Chapter 10. Alternatively, the selected topics can be considered as a group following Chapter 23.

Exercises

In addition to the many examples and exercises in the body of each chapter, we have three kinds of end-of-chapter exercises.

- Review Questions are intended to provide a qualitative measure of your understanding of the chapter.
- The Problems are arranged by topic; they test your mastery of the problem-solving techniques introduced in the chapter.

- The Additional Problems are not grouped by type. Some are intended to be a bit more challenging; they often require a synthesis of ideas from more than one chapter. Others, however, are not any more difficult than those arranged by topic. Rather, they pursue an idea further than is done in the text or they introduce new ideas.

Some of the end-of-chapter exercises are estimation exercises or conceptual exercises, echoing similar types of exercises within each chapter. We have chosen not to label these, in order to give students experience recognizing different kinds of problems as well as solving them.

Acknowledgments

JWH would like to thank his colleagues at the University of Wisconsin-River Falls for so many ideas that made their way into his previous texts—some of which appear in this one—and to Kathy Sumter, Program Aide in the Department of Chemistry. He is especially indebted to Ina Hill and Cynthia Hill for library research, typing, and unfailing support throughout this project, and to Mike Davis for his encouragement and for sharing his love of learning.

RHP would like to thank his colleagues at California State University, San Bernardino, who have offered many helpful suggestions for previous texts that carry over into this one; Robert K. Wismer of Millersville University, who is ever ready to offer advice on the broad range of issues that crop up in the preparation of a new text; and Ruth Petrucci, who has freed him from most other tasks so that his full attention could be devoted to this one.

Both of us would like to thank our students, who have challenged us to be better teachers, and the reviewers of this and our other books, who have challenged us to be better writers.

List of Reviewers

Donald Baird	Florida Atlantic University
Marvin E. Bunch	Corning Community College
Roger Bunting	Illinois State University
Julia Burdge	University of Akron
Jorge Castillo	Chicago State University
A. Wallace Cordes	University of Arkansas
James Crosthwaite	University of North Carolina
John DeKorte	Glendale Community College
Marian Douglas	University of Arkansas
Kirt Dreyer	Bemidji State University
Michael Eastman	Northern Arizona University
Grover Everett	University of Kansas
Raymond Fort	University of Maine
Robin L. Garrell	University of California, Los Angeles
Elsie Gross	Hillsborough Community College
Anthony Guzzo	University of Wyoming
Alton Hassell	Baylor University
Sherman Henzel	Monroe Community College
John Hershberger	North Dakota State University
Bruce Hoffman	Lewis & Clark Community College
Paul W. W. Hunter	Michigan State University

Roger Hurdlik	Olive-Harvey College
Michael A. Janusa	Nicholls State University
Dale Johnson	University of Arkansas
Philip Kinsey	University of Evansville
Donald Kleinfelter	University of Tennessee
Roy A. Lacey	State University of New York
James Long	University of Oregon
C. Luehrs	Michigan Technological University
Terry McCreary	Murray State University
Paul O'Brien	West Valley College
Peter Sheridan	Colgate University
Charles Trapp	University of Louisville
Garth Welch	Weber State University
Robert K. Wismer	Millersville University
Don Roach	Miami-Dade Community College
E. Alan Sadorski	Ohio Northern University
Catherine Shea	Mission College
David White	Beekman Institute

Supplements

For the Student

Student Study Guide by Dixie Goss of Hunter College provides further learning material, including summaries, self tests, practice examples, and learning goals.

Solutions Manual by Alton Hassell of Baylor University contains worked-out solutions to all in-chapter, end-of-chapter, review, conceptual, and estimation exercises and problems.

Selected Solutions Manual by Alton Hassell contains worked-out solutions to over half of the text's problems. The answers to these problems also appear in Appendix F of the text (Answers to Selected Problems).

Math Review Toolkit by Gary Long of Virginia Polytechnic Institute provides a chapter-by-chapter review of the mathematics used throughout the book; a guide to preparing for a career in chemistry; and a review of some of the special writing requirements often needed in the general chemistry course, focusing particularly on the laboratory notebook.

Chemistry Explorer Software This interactive simulation software program is based on worked problems and examples from within the chapters. It allows students to manipulate variables and physical parameters in performing experiments to observe how the changing variables affect the results of experiments. It also provides data analysis tools, such as spreadsheets and graphs, and is available in both Macintosh and Windows.

General Chemistry CD-ROM is an interactive CD-ROM designed to augment and enhance the traditional learning experiences of lecture, lab, and text. For a number of the more difficult topics that lend themselves to the visual and interactive capabilities of multimedia, this package provides an alternative, in-depth, dynamic learning environment. Here the student will be able to interact with and visualize chemical concepts in ways that are not possible with traditional learning programs.

The New York Times/Prentice Hall Themes of the Times

The New York Times and Prentice Hall are sponsoring *Themes of the Times*, a program designed to enhance student access to current information of relevance in the classroom.

Through this program, the core subject matter provided in the text is supplemented by a collection of time-sensitive articles from one of the world's most distinguished newspapers. These articles demonstrate the vital, ongoing connection between what is learned in the classroom and what is happening in the world around us.

To enjoy the wealth of information of *The New York Times* daily, a reduced subscription rate is available. For information, call toll free: 1-800-631-1222.

Prentice Hall and *The New York Times* are proud to cosponsor *Themes of the Times*. We hope it will make the reading of both textbooks and newspapers a more dynamic, involving process.

For the Professor

Instructor's Resource Manual prepared by Robert K. Wismer of Millersville University provides lecture outlines, additional resources, problem outlines for conceptual and estimation exercises, and information on how to correlate the multimedia and video products with the textbook.

Transparencies. Over 250 full-color transparencies chosen from the text put principles into visual perspective and save time in preparing lectures.

Test Item File prepared by Ronald J. Wikholm of the University of Connecticut offers a wide variety of questions written specifically for this textbook.

Prentice Hall Custom Test. This software program is based on the powerful testing technology developed by Engineering Software Associates, Inc. Available for Windows, Macintosh, and DOS, the Custom Test allows you to create and tailor the exam to your own needs. With the Online Testing option, you can also administer exams online, automatically transferring data for evaluation. A comprehensive desk reference guide is included, along with online assistance.

Media and Multimedia Products for the Professor

Prentice Hall Chemistry Laserdisc. A visual encyclopedia that combines animation, demonstrations, applications, and still images and art.

Prentice Hall Multimedia Presentation Manager. A lecture media manager software program designed to help you integrate the power of Prentice Hall multimedia products into your classroom with minimal start-up time. It allows you to sequence video demonstrations, animation, computer simulations, and any other applications resident on your hard drive.

Chemistry Explorer Simulation Software. Interactive simulation software based on selected worked examples and problems from within this text. Allows you to manipulate variables and physical parameters in experiments to demonstrate how the changes affect results.

Prentice Hall Video Demonstration Library. An archive of video demonstrations, including some that would be difficult, dangerous, or expensive for routine classroom use.

Multimedia Development Tools for the Professor
(available for purchase only)

MACROMEDIA

Director Academic. An educational adaptation of Macromedia's Director 4.0. It enables you to create professional-quality multimedia products using graphics, animations, video, and sound.

MACROMEDIA

Authorware Academic. An educational adaptation of Macromedia's Authorware Professional 2.0, the industry standard authoring tool for creating a variety of interactive applications. This icon-based authoring system allows you to create curriculum applications, from simple lecture presentations to complex student tutorials. Includes twenty-five models (pre-defined templates) whose content can be replaced with your own. For further information or purchase of the above products, contact your local representative or the Prentice Hall Multimedia Group at 1-800-887-9998.

About the Authors

John W. Hill

John Hill is currently Professor of Chemistry at the University of Wisconsin–River Falls. As a specialist in chemical education, Professor Hill has over 40 publications in refereed journals and is the author of several books, including *Chemistry and Life: An Introduction to General, Organic and Biological Chemistry* and the market-defining *Chemistry for Changing Times*. He has also been the recipient of several awards for excellence in teaching, and he is active in the American Chemical Society.

> "Chemical properties and principles pervade almost every aspect of our private and professional lives. We have therefore tried in this text to provide principles and applications of chemistry that will help students to be more proficient in their professional practice and also enrich their everyday lives."
>
> — *John Hill*

Ralph H. Petrucci

Ralph Petrucci received his B.S. in Chemistry from Union College and his Ph.D. from the University of Wisconsin–Madison. Following several years of teaching, research, consulting, and directing NSF Institutes for Secondary School Science Teachers at Case Western Reserve University, Professor Petrucci joined the teaching and planning staff of California State University, San Bernardino. Professor Petrucci is the author of several books, including *General Chemistry: Principles and Modern Applications*, now in its sixth edition with William S. Harwood.

> "Our purpose in writing this new general chemistry textbook is to combine our experiences in teaching and writing textbooks for various audiences into a modern treatment of the subject that is truly general, that is, encompassing and integrating all the major areas of chemistry."
>
> — *Ralph Petrucci*

False-color image of a cluster of gold atoms (yellow, orange, and red) on a surface of carbon atoms (green) obtained by scanning tunnelling microscopy. The hypothesis that all matter is composed of atoms is over 2000 years old.

1

Chemistry: Matter and Measurement

CONTENTS

People in the industrialized nations have a higher standard of living than the human race has ever known: more nutritious food, better health, more wealth—and much of this is due to chemistry. Chemistry enables us to design all sorts of materials: drugs to fight disease; pesticides to protect our health and crops; fertilizers to grow abundant food; fuels for transportation; fibers to provide comfort and variety in clothes; building materials for affordable housing; plastics to package foods, replace worn-out body joints, and stop bullets; sports equipment to enrich our leisure time; and much more.

Chemistry also helps us understand the nature of our environment and of ourselves. It provides essential information as we address the most fundamental question of all: What is the nature of life? The theories of chemistry illuminate our understanding of the material world from tiny atoms to giant galaxies.

1.1 Chemistry: Principles and Applications

The principles of chemistry that we consider in this book have many useful applications. In some cases, exploring the practical side of chemistry has stimulated the discovery of new principles. In chemistry, theory and applications are intertwined. To illustrate, let's consider some issues concerning chlorine, one of the most familiar chemical elements.

Chlorine is a pale green gas at room temperature. It is not found free in nature, but only combined with other elements, such as sodium in sodium chloride—common table salt. Seawater is 3% salt, and blood serum is about 0.8% salt. Salt, and hence chlorine, is essential to life.

Chlorine was discovered by Karl Scheele in 1774, but it did not become an important commercial chemical until later in the nineteenth century when an inexpensive method was developed for preparing it from salt. The method, called electrolysis, is based on principles that relate chemistry and electricity.

Today the chemical industry turns out some 10,000 chlorine compounds, ranging from polyvinyl chloride (PVC) for pipes to compounds used as bleaches, disinfectants, solvents, flame retardants, pesticides, and drugs. In some cases,

Chlorine gas.

as in the manufacture of titanium, a space-age metal, chlorine plays an important role but is not found in the final product.

Cholera, typhoid fever, and dysentery are dangerous diseases transmitted through impure drinking water. In 1900, for example, there were 35,000 deaths from typhoid in the United States. Water-borne diseases are all but eliminated when water supplies are treated with chlorine to kill pathogenic (disease-causing) organisms. Water supplies in the United States have been treated with chlorine since the 1920s.

Using chlorine to disinfect water does have a drawback. It converts some dissolved organic compounds into organic chlorine compounds that are suspected carcinogens (cancer-causing substances). However, the levels at which these chemicals are found are usually minute, often measured only in parts per billion. In fact, if it were not for the application of chemical principles to new methods of chemical analysis, it is unlikely that anyone would know these substances were there.

Trace amounts of toxic compounds called dioxins are produced by the combustion of chlorine-containing plastics and by the use of chlorine as a bleach in the pulp and paper industry. Dioxins harm fish and wildlife and perhaps even humans. Another class of chlorine compounds that cause environmental concern is the chlorofluorocarbons (CFCs), used in refrigerators and air conditioners and in making foamed plastics. Chlorine atoms released from CFCs in the stratosphere deplete the ozone that protects life on Earth from harmful ultraviolet radiation. In this case, the study of chemical reactions led to a prediction of the ozone problem even before the problem had been detected and confirmed.

Scientists and other people are now rethinking many of the uses of chlorine. PCBs (polychlorinated biphenyls), once used in printing inks and electrical transformers, and DDT, a once common insecticide, are no longer used. The CFCs are now being replaced as refrigerants by substances less damaging to the ozone layer, but these replacements are more costly and less efficient. Two chlorine compounds, chlorine dioxide and sodium chlorite, are increasingly being used as bleaches for paper and other products. Neither produces the organic chlorine compounds that are formed when chlorine itself is used. Ozone, a form of the element oxygen, is used in place of chlorine in water treatment in some cities. It is as

Karl W. Scheele (1742–1786) made many chemical discoveries, but he rarely received proper credit because he was not the first to report them. In addition to chlorine, he was involved in the discovery of the elements manganese, barium, molybdenum, tungsten, nitrogen, and oxygen.

Laboratory technicians routinely check water samples for various contaminants, including chlorine-containing substances such as the solvents chloroform and carbon tetrachloride and the pesticides chlordane and lindane.

Halons, compounds that contain bromine as well as chlorine, have long been used to extinguish fires, particularly those on aircraft. Halons, like CFCs, have been implicated in the depletion of the ozone layer. To date no substitutes have been found that have all the desirable properties of the halons: essentially nontoxic, nonconducting, noncorrosive, and leaving no residue.

effective as chlorine, but it does not produce chlorinated byproducts or impart a "chemical" taste to the water. However, unlike chlorine, ozone does not provide a residual disinfectant action in the treated water. Overall, the cost of replacing chlorine in the economy has been estimated at over $100 billion a year in the United States, and there are some important uses of chlorine and its compounds for which no suitable replacements are known.

Chemical knowledge was used to make the chlorine products that have provided so many benefits; chemical knowledge played a large role in revealing the negative side of these products; and even more chemical knowledge will be required in the search for suitable substitutes. The applications of chemistry are much like the science itself: undergoing constant change.

1.2 Getting Started: Some Basic Ideas

For the most part, we will introduce new terms as we need them. To get started, though, we do need a little basic vocabulary. You probably already know something about the ideas and terms in this section, but let's try to establish a common understanding.

Chemistry is a study of the composition, structure, and properties of matter and of the changes that occur in matter. What is matter? It is the stuff things are made of. One way to think about matter is that particles or objects of matter occupy space, and no two objects can occupy the same space at the same time. Wood, sand, people, water, and air are all examples of matter. Light is not matter; it is a form of energy. The main concern of chemists is with the tiny, submicroscopic building blocks of matter known as atoms and molecules. **Atoms** are the smallest units that we associate with the chemical behavior of matter and **molecules** are larger units made up of groupings of atoms. Ultimately, samples of matter are what they are because of the atoms or molecules that form them.

Figure 1.1 Properties of matter. Copper (left) and ethyl alcohol (right) are easily distinguished by their properties. Copper is a solid; ethyl alcohol is a liquid. Copper is opaque and has a red-brown color. Ethyl alcohol is transparent and colorless. Also, ethyl alcohol burns and copper does not.

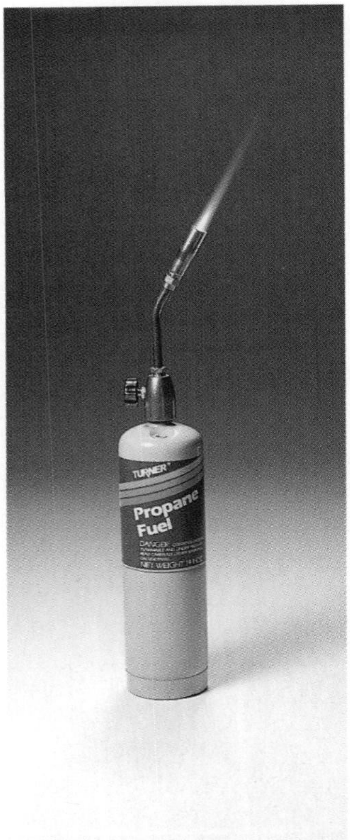

Figure 1.2 Physical and chemical changes. Propane fuel is stored in the tank as a liquid under pressure. When the tank valve is opened, the liquid vaporizes, a physical change. Propane gas mixes with air and burns, a chemical change. The products of the combustion are carbon dioxide and water.

Properties

To distinguish between samples of matter, we can compare their *properties* (see Figure 1.1). When we observe a **physical property**, such as color, odor, or hardness, the atomic or molecular building blocks of a sample do not change. When we observe a **chemical property** at work, matter is undergoing a *chemical* change—the original substance is replaced by one or more new substances. The burning of sulfur in air demonstrates a chemical property. Sulfur, made up of one type of atom, and oxygen from air, made up of another type, combine to form sulfur dioxide, made up of molecules that have sulfur and oxygen atoms in the ratio 1:2.

When ice melts, solid water is changed to liquid water. The process of melting is a physical change, and the temperature at which it occurs—the melting point—is a physical property. Water is still water, whether liquid or solid. Although it may be difficult at times to decide whether a change is physical or chemical, the answer depends on what happens to the composition of the matter involved. **Composition** refers to what elements are present and what their relative proportions are. With but few exceptions, a chemical change results in a change in composition; a physical change does not. Figure 1.2 illustrates a situation involving both physical and chemical changes.

Example 1.1

Which of the following are chemical changes and which are physical changes?

a. Your hair is trimmed by a barber.
b. Bacteria convert milk to yogurt.
c. Water boils.
d. Water is broken down into hydrogen gas and oxygen gas.

Solution

a. Physical change; the hair is not changed by clipping.
b. Chemical change; yogurt and milk differ in composition.
c. Physical change; liquid water and invisible water vapor formed when it boils have the same composition. Water merely changes from a liquid to a gas.
d. Chemical change; new substances, hydrogen and oxygen, are formed.

Exercise 1.1

Which of the following are chemical changes and which are physical changes?

a. Liquid acetic acid freezes to a solid in a cold room.

b. A strip of magnesium metal burns brightly in air to form a white powder called magnesium oxide.
c. A dull knife blade is sharpened by hand with a grindstone.

Classifying Matter

Figure 1.3 shows several ways that chemists classify matter. A **substance** has a definite, or fixed, composition that does not vary from one sample to another. In contrast, the composition of a **mixture** may be variable. Pure gold (24-karat gold) consists of just one type of atom; it is a substance. All samples of water are made up of molecules consisting of two hydrogen atoms and one oxygen atom; water is a substance. On the other hand, a saline solution—a solution of salt in water—is a mixture. The proportions of salt and water can vary from sample to sample.

Any given saline solution is a **homogeneous mixture**: It has the same composition and properties—the same "saltiness"—throughout the solution. By contrast, a **heterogeneous mixture** varies in composition and/or properties from one part of the mixture to another. An ice–water mixture is heterogeneous. Although the composition is the same, depending on whether we examine a piece of ice or the water on which it floats, the physical properties are different. In a sand–water mixture, both the composition and properties vary within the mixture.

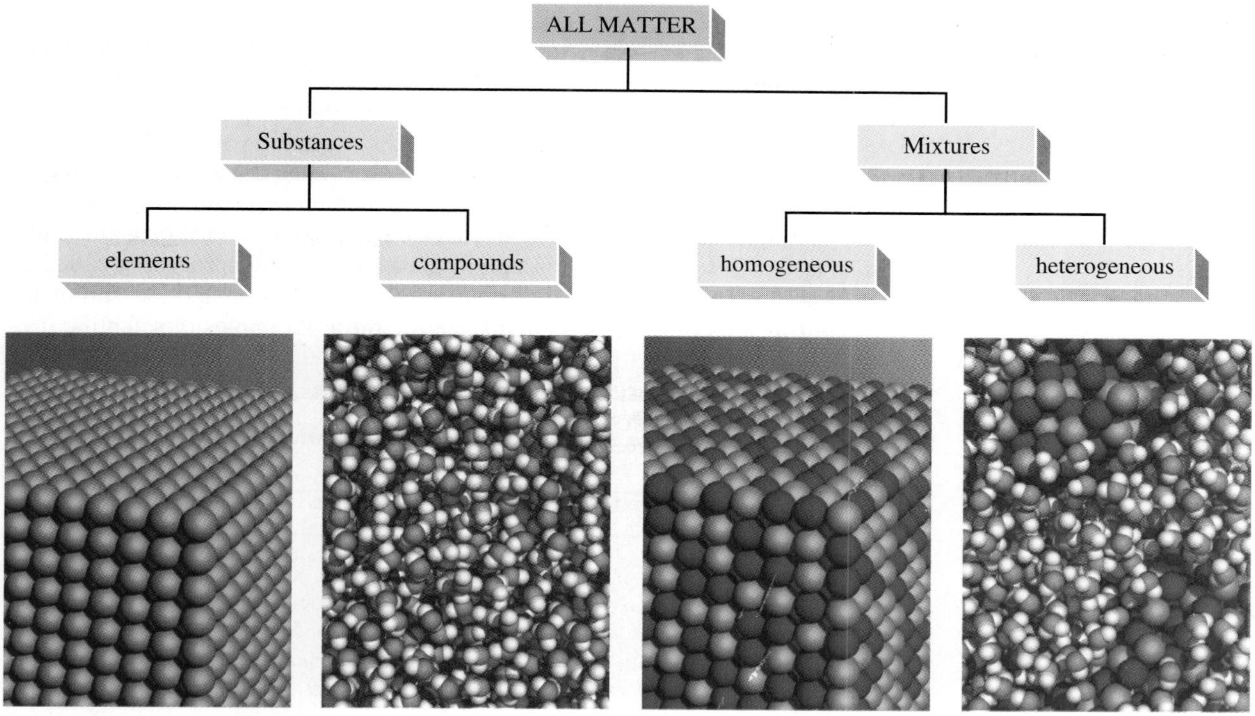

Figure 1.3 A scheme for classifying matter. The "molecular-level" views are of gold—an *element*; water—a *compound*; "12-carat" gold—a *homogeneous mixture* of silver and gold; and a *heterogeneous mixture* of particles of "12-carat" gold in water.

All substances are either elements or compounds. An **element** is composed entirely of a single type of atom. Because bulk matter cannot be made exclusively from any particles smaller than atoms, elements are the simplest of all substances. At the present time 111 elements are known, but many are quite rare. Among the familiar elements are oxygen, nitrogen, carbon, sulfur, iron, copper, silver, and gold. A **compound** is made up of atoms of two or more elements, with the different kinds of atoms combined in fixed ratios. In contrast to the limited number of elements, the possible number of compounds is essentially limitless. Currently, over twelve million compounds have been recorded in the chemical literature. Water, carbon dioxide, sodium chloride (table salt), sucrose (table sugar), and iron oxide (rust) are all compounds.

The names and symbols of all the elements are listed inside the front cover of this book.

Because elements and compounds are so fundamental to our study, we find it useful to refer to them by symbols. A **chemical symbol** is a one- or two-letter designation derived from the name of an element. Most symbols are based on English names; a few are based on the Latin name of the element or one of its compounds (see Table 1.1). For a two-letter symbol, the first letter is capitalized and the second is always lowercase. (It makes a difference. For example, Co is the symbol for the element cobalt; CO represents the poisonous compound carbon monoxide.) Compounds are designated by combinations of chemical symbols called *formulas*. Writing chemical formulas is somewhat more difficult than writing symbols, and we will postpone doing this to the next chapter.

The Scientific Method

Chemists and other scientists use certain terms to describe the way in which they conduct their studies. We will briefly consider some of these terms, but the key idea to keep in mind is that scientific knowledge is *testable*, *reproducible*, *explanatory*, *predictive*, and *tentative*.

Scientists often begin a study by making observations and then formulating a hypothesis. A **hypothesis** is a tentative explanation or prediction concerning some phenomenon. It may be just an educated guess, but it must be a guess that can be tested. Testing is done through a carefully controlled procedure called an **experiment**. The facts obtained through careful observation and measurements made during experiments are called scientific **data**. Examples of scientific data

TABLE 1.1 **Some Elements with Symbols Derived from Latin Names**

Usual Name	Latin Name	Symbols
Copper	cuprum	Cu
Gold	aurum	Au
Iron	ferrum	Fe
Lead	plumbum	Pb
Mercury	hydrargyrum	Hg
Potassium	kalium	K
Silver	argentum	Ag
Sodium	natrium	Na
Tin	stannum	Sn

are the melting point of iron (1535 °C) and the speed of light (2.99792458×10^8 meters per second). Further experiments may refine these data to some degree, but the basic facts have been verified repeatedly; they are *reproducible*.

Patterns in large amounts of scientific data can sometimes be summarized in brief statements called **scientific laws**. Many of these laws can be stated mathematically. Scientists often use scientific models—tangible items or pictures—to represent invisible processes and explain complicated phenomena. For example, the invisible particles (atoms and molecules) of a gas can be visualized as billiard balls, marbles, or as dots or circles on paper (Figure 1.4). The ultimate goal of scientists is to formulate theories. A **theory** is a framework for organizing scientific knowledge that provides explanations of observed natural phenomena and predictions that can be tested by further experiments.

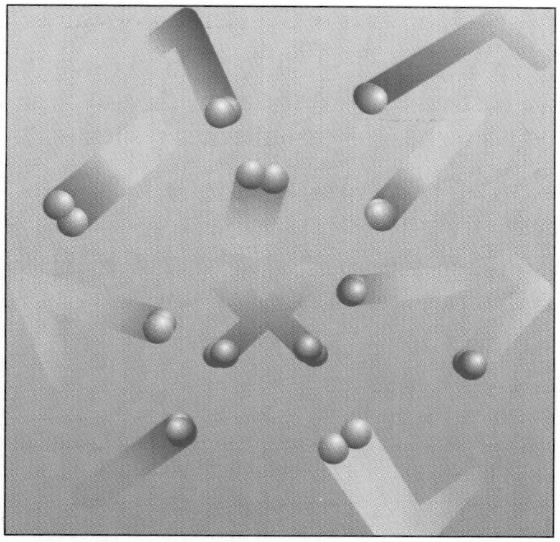

Figure 1.4 A scientific model of a gas. Chlorine, pictured on page 4, is made up of chlorine molecules; each molecule, in turn, is a combination of two chlorine atoms. Chlorine molecules, like all gas molecules, are in constant, random, chaotic motion and undergo frequent collisions with each other and with the container walls. This model is used in Chapter 4 to explain several properties of gases. For now, we can see that the model explains how it is possible to force a second gas into a container of gas. Molecules of one gas fit into the empty spaces among molecules of the other gas.

Contrary to some popular notions, scientific knowledge is not absolute. No theory or hypothesis can ever be proved completely true; it can only be disproved. The most promising hypothesis can be destroyed by one stubborn fact. The body of scientific knowledge is growing, changing, and never final. Old concepts, or even old "facts," are discarded when new tools, new questions, and new techniques reveal new data and generate new concepts.

1.3 Scientific Measurements

In order to test a hypothesis, a scientist must gather data by measurement. And before the hypothesis is accepted, other scientists must reproduce the measured data. Data gathering and checking are much easier to accomplish if all scientists agree to use a common system of measurement. The system agreed upon since 1960 is the International System of Units (SI), a modernized version of the metric system established in France in 1791.

All measured quantities can be expressed in terms of the seven base units listed in Table 1.2. We will make use of the first six in this text and introduce the first four now.

The standard of length in the original metric system, the meter, was taken to be 1/10,000,000th of the distance from the equator to the North Pole. Other standards were related to the meter.

TABLE 1.2 **The Seven SI Base Units**

Physical Quantity	Name of Unit	Symbol of Unit
Mass	kilogram	kg
Length	meter[a]	m
Time	second	s
Temperature	kelvin	K
Amount of substance	mole	mol
Electric current	ampere	A
Luminous intensity	candela	cd

[a] Spelled *metre* in English-speaking countries other than the United States.

Length

The SI base unit of length is the **meter (m)**, a unit about 10% longer than the yard. Units larger and smaller than the base unit are expressed by the use of prefixes (see Table 1.3). To measure lengths much larger than the meter, such as distances along a highway, we often use the kilometer (km).

$$1 \text{ km} = 1000 \text{ m} = 10^3 \text{ m}$$

In the laboratory, lengths smaller than the meter are often most convenient, for example, the centimeter (cm) and the millimeter (mm).

$$1 \text{ cm} = 0.01 \text{ m} = 10^{-2} \text{ m}$$

$$1 \text{ mm} = 0.001 \text{ m} = 10^{-3} \text{ m}$$

On the submicroscopic scale the micrometer (μm), the nanometer (nm), and the picometer (pm) are all used.

$$1\ \mu\text{m} = 10^{-6} \text{ m} \qquad 1 \text{ nm} = 10^{-9} \text{ m} \qquad 1 \text{ pm} = 10^{-12} \text{ m}$$

TABLE 1.3 **Some Common SI Prefixes**

Multiple	Prefix
10^9	giga (G)
10^6	mega (M)
10^3	kilo (k)
10^{-1}	deci (d)
10^{-2}	centi (c)
10^{-3}	milli (m)
10^{-6}	micro (μ)[a]
10^{-9}	nano (n)
10^{-12}	pico (p)

[a] The Greek letter μ (spelled "mu" and pronounced "mew").

There are no SI base units for area and volume, but their units can be obtained from the base unit of length. The SI unit of area is the square meter (m^2), although for laboratory work we often find it more convenient to work with square centimeters (cm^2).

$$1 \text{ cm}^2 = (10^{-2} \text{ m})^2 = 10^{-4} \text{ m}^2$$

A square centimeter is easy to picture—it's about the area of a button on a touch-tone telephone.

The SI unit of volume is the cubic meter (m^3), but Figure 1.5 pictures two units that are more likely to be used in the laboratory: the cubic centimeter (cm^3) and the cubic decimeter (dm^3).

$$1 \text{ cm}^3 = (10^{-2} \text{ m})^3 = 10^{-6} \text{ m}^3$$

$$1 \text{ dm}^3 = (10^{-1} \text{ m})^3 = 10^{-3} \text{ m}^3$$

Although it is not an SI unit, the old metric unit *liter* is also commonly used. A **liter (L)**, which is slightly larger than a quart, is the same as one cubic decimeter, or 1000 cubic centimeters.

$$1 \text{ L} = 1 \text{ dm}^3 = 1000 \text{ cm}^3$$

The milliliter (mL) is the same as a cubic centimeter: 1 mL = 1 cm^3.

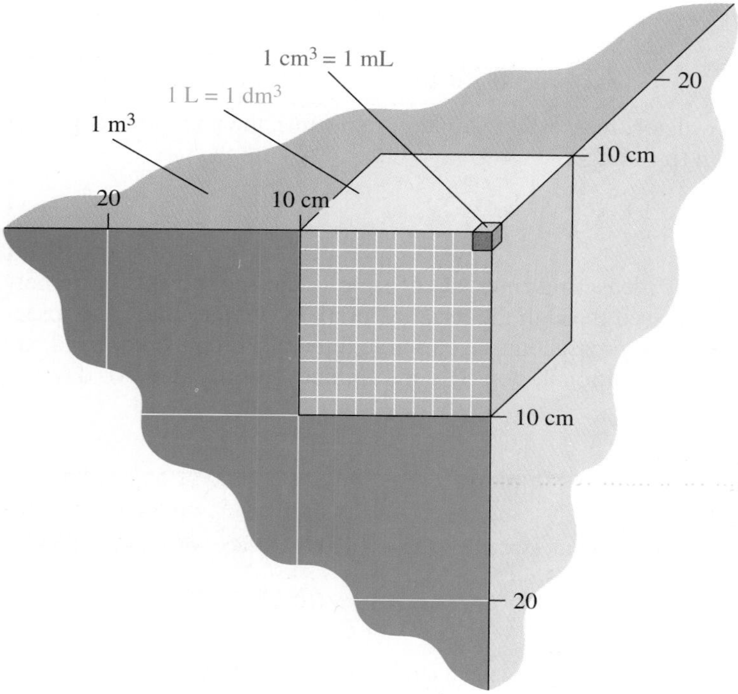

Figure 1.5 Some volume units compared. The largest volume, shown in part, is the SI standard—1 cubic meter (m^3). A cube with a length of 10 cm (1 dm) on edge (in blue) has a volume of 1000 cm^3 (1 dm^3) and is called 1 liter (1 L). The smallest cube is 1 cm on edge (red) and has a volume of 1 cm^3 = 1 mL.

Mass

Mass measures how much matter there is in an object. Mass is usually measured by weighing (Figure 1.6). The SI base quantity of mass is the **kilogram (kg)** (about 2.2 lb). This base quantity is unusual in that it already has a prefix. A more convenient mass unit for most laboratory work is the gram (g).

$$1\text{ kg} = 1000\text{ g} = 10^3\text{ g}$$

Figure 1.6 Measuring mass by weighing. Weight is the force of gravity on an object. This force is proportional to the mass.

In the balance shown here, the force of gravity on the object is counterbalanced by a magnetic force, and the magnitude of this force is registered as a mass in the digital readout of the balance. The portion of the metallic cylinder shown has a mass of 999.0 g, and the very thin section, 1.0 g. Their combined mass is 1000.0 g = 1.0000 kg.

The milligram (mg) is a suitable unit for small quantities of materials, such as drug dosages.

$$1 \text{ mg} = 10^{-3} \text{ g}$$

Chemists can now detect masses in the microgram (μm), the nanogram (ng), and even the picogram (pg) range.

Time

The SI base unit for measuring intervals of time is the **second (s)**. Extremely short times are expressed through the usual SI prefixes: *milli*seconds, *micro*seconds, *nano*seconds, and *pico*seconds. Long time intervals, on the other hand, are usually expressed in traditional, non-SI units: minute (min), hour (h), day (d), and year (y).

Example 1.2

Convert each of the following measurements to a unit that replaces the power of ten by a prefix.

a. 3.22×10^{-6} s **b.** 9.56×10^{-3} m **c.** 1.07×10^{3} g

Solution

We merely need to match each exponential term (for example, 10^{-6}) with the appropriate prefix from Table 1.3.

a. 10^{-6} corresponds to the prefix *micro* (μ); 3.22 μs.
b. 10^{-3} corresponds to the prefix *milli*; 9.56 mm.
c. 10^{3} corresponds to the prefix *kilo*; 1.07 kg.

Exercise 1.2

Convert each of the following measurements to a unit that replaces the power of ten by a prefix.

a. 7.42×10^{-3} s **b.** 5.41×10^{-6} m
c. 1.19×10^{-3} g **d.** 5.98×10^{3} m

Temperature

Temperature is difficult to define. We can say that it is a measure of "hotness," but that isn't very precise. From common experience we know that if two objects at different temperatures are brought together, heat flows from the warmer to the colder object. For example, if you touch a hot test tube, heat will flow from the tube to your hand. If the tube is hot enough, your hand will be burned. The temperature of the warmer object drops and that of the colder object increases, until finally the two objects are at the same temperature. Temperature is therefore a property that tells us in what direction heat will flow. Later in the text we will present some additional, perhaps more satisfying, ideas about temperature.

The SI base unit of temperature is the **kelvin (K)**. We'll define the Kelvin temperature when we first encounter a situation where we need it. For most routine laboratory work we can use a more familiar temperature scale: the Celsius scale. On this temperature scale, the freezing point of water is 0 °C and the boiling point is 100 °C. The interval between these two reference points is divided into 100 equal parts, each a degree Celsius. Another temperature scale, not used in scientific work and probably unfamiliar to most people in the world, is the Fahrenheit scale. This temperature scale is still widely used in the United States, both in everyday life and in industry and commerce. The Celsius and Fahrenheit scales are compared in Figure 1.7. The relationship between the two is

$$t_F = 1.8\,t_C + 32 \qquad \text{or} \qquad t_C = (t_F - 32)/1.8$$

You will not often need to convert between Celsius and Fahrenheit temperature, but Example 1.3 illustrates a practical situation where this would be necessary.

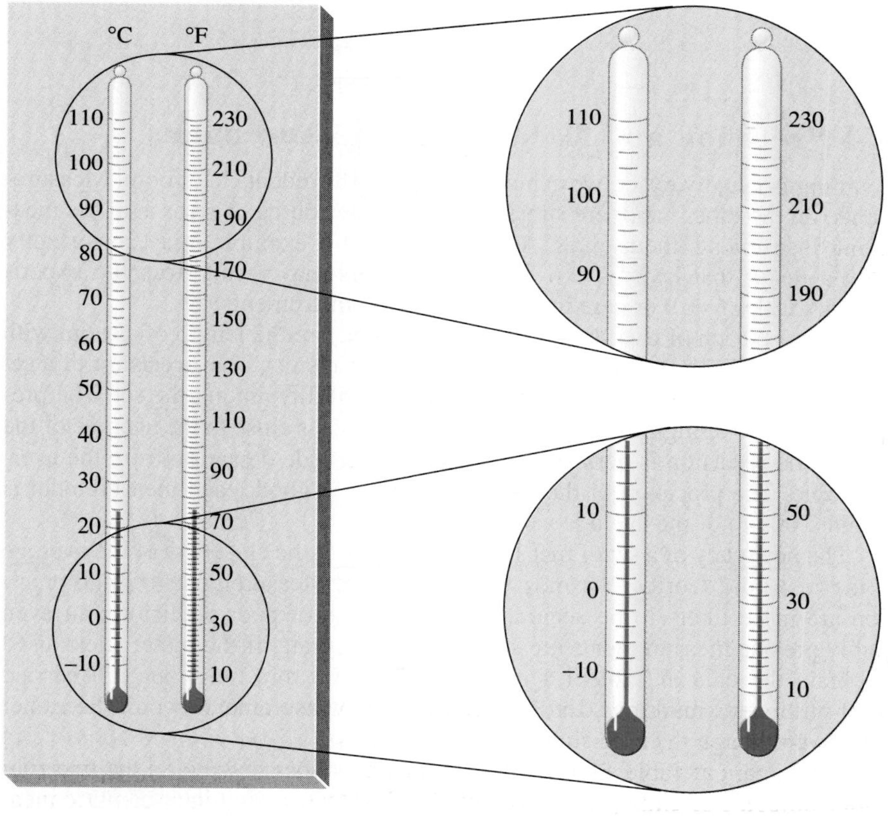

Figure 1.7 The Celsius and Fahrenheit temperature scales compared. The thermometer on the left is marked in degrees Celsius and the one on the right, in degrees Fahrenheit. The freezing point of water is at 0 °C and 32 °F; the boiling point is at 100 °C and 212 °F. Note that for an interval of 10 °C the corresponding interval on the Fahrenheit scale is 18 °F.

Example 1.3

A parasite that causes trichinosis is sometimes found in undercooked pork. It is killed when the meat is cooked to 66 °C. Assume you have only a Fahrenheit meat thermometer, and determine the minimum Fahrenheit temperature to which the pork should be heated when it is being cooked.

Solution

The pork must be heated to a temperature of 66 °C or higher. We must substitute $t_C = 66$ in the following equation and solve for t_F. This will give us the minimum Fahrenheit temperature.

$$t_F = 1.8\, t_C + 32$$

$$t_F = (1.8 \times 66) + 32 = 151 \text{ °F}$$

Exercise 1.3

Carry out the following temperature conversions.

a. 85.0 °C to degrees Fahrenheit
b. −12.2 °C to degrees Fahrenheit
c. 355 °F to degrees Celsius
d. −20.8 °F to degrees Celsius

1.4 Precision and Accuracy in Measurements

Counting is usually exact: we can count exactly 18 students in a room. Measurements, on the other hand, are subject to error. One source of error is in the measuring instruments themselves. A thermometer, for example, may consistently yield a result that is 2 °C too low. Other errors may result from the experimenter's lack of skill or care in using measuring instruments.

Suppose each of five students is asked to measure the length of a room with a meter stick. Table 1.4 presents a possible set of results. The **precision** of a set of measurements refers to the degree of reproducibility among the set. The precision is good (or high) if each of the measurements is close to the average of the series. The precision is poor (or low) if there is a wide deviation from the average value. The precision of the data in Table 1.4 is good; each measurement is within 0.005 m of the average value.

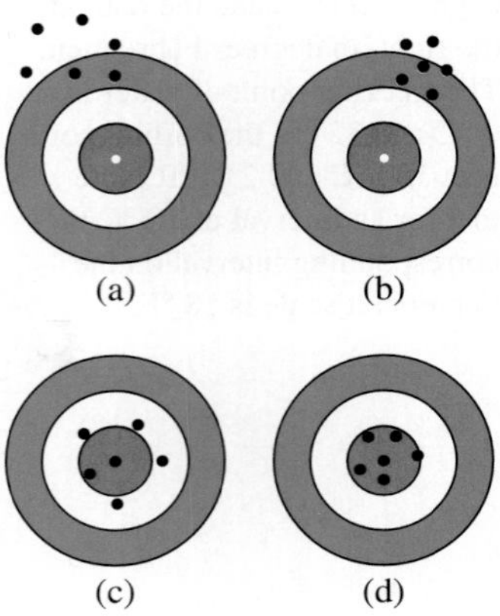

Comparing precision and accuracy. (a) Measurements of *low* accuracy and *low* precision are scattered and off-center; (b) those with *low* accuracy and *high* precision form a tight off-center cluster; (c) those with *high* accuracy and *low* precision are evenly distributed but distant from the center; and (d) those with *high* accuracy and *high* precision are bunched in the center of the target.

The **accuracy** of a set of measurements refers to the closeness of the average of the set to the "correct" or most probable value. Measurements of high precision are more likely to be accurate than are those of poor precision, but even highly precise measurements are sometimes inaccurate. If the meter sticks used to obtain the data in Table 1.4 were actually 1005 mm long, but still carried 1000-millimeter markings, the accuracy of the measurements would be rather poor, even though the precision remained high.

Look again at Table 1.4. Notice that the five students agree on the first four digits (14.15); they differ only in the fifth digit. The last digit in a scientific measurement is usually regarded as uncertain; and all digits known with certainty, plus one of uncertain value, are called **significant figures**. The measurements in Table 1.4 have five significant figures. In other words, we are quite sure that the length of the room is between 14.15 m and 14.16 m. Our best estimate, including the uncertain digit, is the average value: 14.155 m.

It is easy to establish that 14.155 has five significant figures; we simply count the number of digits. In any measurement that is properly reported, all *nonzero* digits are significant. *Zeros* present problems because they can be used in two ways: to indicate a measured value or to position a decimal point.

- Zeros between two other digits are significant.
  Examples: 1107 (four significant figures); 50.002 (five).
- Zeros that precede the first nonzero digit are *not* significant.
  Examples: 0.000163 (three significant figures); 0.06801 (four).
- Zeros at the end of a number are significant if they are to the *right* of the decimal point.
  Examples: 0.2000 (four significant figures); 0.050120 (five).
- Zeros at the end of a number may or may not be significant if the number is written *without* a decimal point.
  Example: 400.
  We do not know whether this number was measured to the nearest unit, ten, or hundred. To avoid this confusion, we use exponential notation (see Appendix A). In exponential notation, 400 would be recorded as 4×10^2 or 4.0×10^2 or 4.00×10^2 to indicate one, two, or three significant figures, respectively.

TABLE 1.4 A Set of Measurements of Length of a Room

Student	Length, m
1	14.157
2	14.150
3	14.153
4	14.159
5	14.156
Average:	14.155

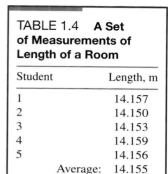

The concept of significant figures applies only to *measurements*—quantities subject to error. It does not apply to a quantity that is *counted*, such as the 6 faces of a cube or 12 items in a dozen. It also does not apply to *defined* quantities, such as 1 km = 1000 m. In this context, the numbers 6, 12, and 1000 are not limited to one, two, and three significant figures. In effect, each has an unlimited number of significant figures (6.000..., 12.000..., 1000.000...) or, more properly, each is an *exact* value.

Significant Figures in Calculations: Multiplication and Division

Not only must we recognize the relationship between the precision of a measurement and the number of significant figures used to express it, but we have to obey this fundamental rule: *A calculated quantity may be no more precise than the data used in the calculation.*

In multiplication and division this leads to the practical rule that an answer should have no more significant figures than the factor with the *fewest* significant figures.

The "answer" on this electronic calculator is 59.7402, suggesting six significant figures, but the rule for multiplication tells us there can only be three: 59.7.

Example 1.4

What is the area, in square meters, of a room that is 12.42 m long and 4.81 m wide?

Solution

The length of the room is expressed to *four* significant figures and the width to three. By whatever method we use to carry out the multiplication, we are limited to three significant figures in our answer.

$$12.42 \text{ m} \times 4.81 \text{ m} = 59.7 \text{ m}^2$$

Example 1.5

For a laboratory experiment, a teacher wants to divide all of a 453.6-g bottle of sulfur among the 21 members of her class. What mass of sulfur, in grams, will each student receive? (1 lb = 453.6 g.)

Solution

Here we need to recognize that the number "21" is a counted number. It is not subject to significant figure rules. The answer should carry four significant figures, the same as in 453.6 g.

$$\frac{453.6 \text{ g}}{21} = 21.60 \text{ g}$$

Exercise 1.4

Perform the indicated operations and give answers with the proper number of significant figures.

a. 73 m × 1.340 m × 0.41 m

b. 0.137 cm × 1.43 cm

c. 3.132 cm × 5.4 cm × 5.4 cm

d. $\dfrac{51.79 \text{ m}}{4.6 \text{ s}}$

e. $\dfrac{456.1 \text{ mi}}{7.13 \text{ h}}$

f. $\dfrac{305.5 \text{ mi}}{14.7 \text{ gal}}$

Significant Figures in Calculations: Addition and Subtraction

In addition or subtraction the concern is not with the number of significant figures but with digits to the right of the decimal point. If the quantities being added or subtracted have varying numbers of digits to the right of the decimal point, find the one with the *fewest* such digits. The result of the addition or subtraction should contain the same number of digits to the right of *its* decimal point. The idea is this: If you are adding several lengths, and one of them is measured only to the nearest *centi*meter, the total length cannot be stated to the nearest *milli*meter, no matter how precise the other measurements are.

In addition and subtraction it is often necessary to round off numbers. The usual procedure is to round off the final result, using the following rules:

- If the leftmost digit dropped (in red) is 0, 1, 2, 3, or 4, leave the remaining final digit *unchanged.*
  Example: 369.648 rounds to 369.6 if we want one decimal place.
- If the leftmost digit dropped is 5, 6, 7, 8, or 9, *increase* the final digit by *one.*
  Example: 538.768 rounds to 538.77 if we want two decimal places; 538.768 rounds to 538.8 if we want one decimal place.

This rule is easy to follow when using a hand-held calculator.

We apply these ideas in Example 1.6, and illustrate another point as well: In a calculation involving several steps, round off only the final result.

Example 1.6

Perform the following calculation and round off the answer to the correct number of digits.

$$49.146 + 72.13 - 9.1434 = ?$$

Solution

In this calculation we must add two numbers and, from their sum, subtract a third. We do this in two ways below. In both cases, we express the answer to two decimal places, the number of decimal places in "72.13."

(a)

$$\begin{array}{r} 49.146 \\ +72.13 \\ \hline 121.276 \end{array} = \begin{array}{r} 121.28 \\ -\ 9.1434 \\ \hline 112.1366 \end{array} = 112.14$$

(b)

$$\begin{array}{r} 49.146 \\ +72.13 \\ \hline 121.276 \\ -\ 9.1434 \\ \hline 112.1326 \end{array} = 112.13$$

The preferred method is (b), where we do *not* round off the intermediate result: 121.276. Note that if we use a hand-held calculator, there is no need even to write down or otherwise take note of the intermediate result.

Exercise 1.5

Perform the indicated operations and give answers with the proper number of digits.

a. 48.2 m + 3.82 m + 48.4394 m
b. 148 g + 2.39 g + 0.0124 g
c. 451 g − 15.46 g − 20.3 g
d. 15.436 L + 5.3 L − 6.24 L − 8.177 L

1.5 A Problem-Solving Method

Throughout this text we present chemistry problems. To solve a chemistry problem means to use basic principles to answer a question about a new situation. Often, a chemistry problem is stated in a way that requires calculations and a *quantitative* or numerical result. Here we describe a useful method for doing such calculations, and in the next section we present some additional ideas on problem solving.

The Unit-Conversion Method

Imagine that an American orders a leather belt from a foreign country. The customer knows her waist length in common units but must give it in metric units. Suppose her waist is 26 in. What is this length in centimeters? We can answer this practical question by the same method that works for many scientific calculations, the *unit-conversion method.* This is the basic idea:

When we measure something in one unit and then convert this to another unit, we must ensure that the measured quantity is not changed in any fundamental way. A 26-in. waist is a fixed length, whether it is expressed in inches, feet, centimeters, or millimeters. The key is to recognize that when we multiply a quantity by "1" its value is unchanged. A **conversion factor**, then, is a ratio of terms, a fraction, that is equivalent to the number one. When the known quantity is multiplied by a conversion factor, the unit used to express the known quantity cancels out and is replaced by the desired unit. To put the matter another way, the general approach is

$$\text{Desired quantity and unit} = \text{given quantity and unit} \times \text{conversion factors}$$

To convert between inches and centimeters, we need the relationship between these two units listed in Table 1.5.

$$1 \text{ in.} = 2.54 \text{ cm}$$

We can divide both sides of this equation by 1 in., or we can divide both sides by 2.54 cm. In either case the resulting ratio is a conversion factor equal to "1."

$$\frac{2.54 \text{ cm}}{1 \text{ in.}} = 1 \qquad \frac{1 \text{ in.}}{2.54 \text{ cm}} = 1$$

Now to return to our original question. The measured quantity is 26 in., and the appropriate conversion factor must have the desired unit, cm, in the *numerator* and the unit to be canceled, in., in the *denominator*.

$$? \text{ cm} = 26 \text{ in.} \times \frac{2.54 \text{ cm}}{1 \text{ in.}} = 66 \text{ cm}$$

It's not hard to see why the other conversion factor won't work. It gives an absurd answer.

$$26 \text{ in.} \times \frac{1 \text{ in.}}{2.54 \text{ cm}} = 10 \frac{\text{in.}^2}{\text{cm}}$$

Example 1.7 differs from what we've just illustrated in one important respect: we need more than one conversion factor.

TABLE 1.5 Some Conversions Between Common (U.S.) and Metric Units

Metric		Common
Mass		
1 kg	=	2.205 lb
453.6 g	=	1 lb
28.35 g	=	1 ounce (oz)
Length		
1 m	=	39.37 in.
1 km	=	0.6214 mi
2.54 cm	=	1 in.[a]
Volume		
1 L	=	1.057 qt
3.785 L	=	1 gal
29.57 mL	=	1 fluid ounce (fl oz)

[a] The U.S. inch is defined as exactly 2.54 cm. The other equivalencies are rounded off.

Example 1.7

What is the length, in millimeters, of a 1.25-ft rod?

Solution

Table 1.5 does not have a relationship between ft and mm, but we can achieve the conversion in a three-step approach. First, we use the fact that 1 ft = 12 in. To convert from in. to m we use data from Table 1.5. Finally, we convert from m to mm with our knowledge of prefixes.

$$? \text{ mm} = 1.25 \text{ ft} \times \frac{12 \text{ in.}}{1 \text{ ft}} \times \frac{1 \text{ m}}{39.37 \text{ in.}} \times \frac{1000 \text{ mm}}{1 \text{ m}} = 381 \text{ mm}$$

desired quantity and unit; given quantity and unit; conversion factors; answer

Exercise 1.6

Carry out the following conversions.

a. 76.3 mm to meters
b. 0.0856 kg to grams
c. 0.927 lb to ounces (1 lb = 16 oz)
d. 415 in. to yards (1 yd = 3 ft; 1 ft = 12 in.)
e. 3.00 L to fluid ounces

Notice that in Example 1.7 we used *four* significant figures in the factor 1 m/39.37 in., even though we knew that our answer could not be stated with more than three significant figures. Usually, we will express equivalencies, numerical constants, and well-established quantities with more significant figures than the least precisely known measured quantity. By doing this we ensure that the precision of the result is dictated only by the measured quantity that is least precisely known. Note also in Example 1.7 that the "12" in one factor and the "1000" in another are *exact* values.

In Example 1.8 we find that *two* units in the original measured quantity must be replaced. We solve the problem in two ways and compare the two approaches.

Example 1.8

A sprinter runs a 100.0-m dash in 11.00 s. What is her speed in kilometers per hour (km/h)?

Solution

First let's identify the measured quantity. It is a speed, expressed by the ratio

$$\frac{100.0\ \text{m}}{11.00\ \text{s}}$$

Our task is to convert this speed to one in km/h. We must convert from meters to kilometers in the numerator and from seconds to hours in the denominator. We need this set of equivalent values to formulate conversion factors:

$$1\ \text{km} = 1000\ \text{m} \qquad 1\ \text{min} = 60\ \text{s} \qquad 1\ \text{h} = 60\ \text{min}$$

In one approach we arrange all the necessary conversion factors in a single straight-line setup that yields a final answer with the desired units.

$$?\ \frac{\text{km}}{\text{h}} = \frac{100.0\ \cancel{\text{m}}}{11.00\ \cancel{\text{s}}} \times \frac{1\ \text{km}}{1000\ \cancel{\text{m}}} \times \frac{60\ \cancel{\text{s}}}{1\ \cancel{\text{min}}} \times \frac{60\ \cancel{\text{min}}}{1\ \text{h}} = 32.73\ \text{km/h}$$

In an alternate approach, we do the conversions in the numerator and in the denominator separately and then combine the results.

$$\text{Numerator:} \quad 100.0\ \cancel{\text{m}} \times \frac{1\ \text{km}}{1000\ \cancel{\text{m}}} = 0.1000\ \text{km}$$

$$\text{Denominator:} \quad 11.00\ \cancel{\text{s}} \times \frac{1\ \cancel{\text{min}}}{60\ \cancel{\text{s}}} \times \frac{1\ \text{h}}{60\ \cancel{\text{min}}} = 3.056 \times 10^{-3}$$

$$\text{Division:} \quad \frac{0.1000\ \text{km}}{3.056 \times 10^{-3}\ \text{h}} = 32.72\ \text{km/h}$$

Exercise 1.7

Carry out the following conversions.

a. 90.0 km/h to m/s **b.** 1.39 ft/s to km/h **c.** 4.17 g/s to kg/h

You may wonder which of the two methods illustrated in Example 1.8 to use. In general, the first approach is preferred. Do an unbroken series of calculations to obtain a single result; round off and record the result. In the second approach we rounded off and wrote down an intermediate result (3.0555556×10^{-3} to 3.056×10^{-3}) as well as the final answer. Notice that the answer obtained by rounding off twice, 32.72 km/h, differed slightly from that in which only the final result was rounded off, 32.73 km/h. With practice, you should be able to do most calculations in a single setup. On the other hand, there may be cases where you will more clearly visualize how to do a problem if you break it down into individual steps. Either way, remember that the overall goal of problem solving is to devise a sound strategy that works.

Some conversions make repeated use of the same conversion factor. To convert square inches (in.^{2}) to square meters (m^2), we have to cancel "in." twice and replace it by "m" twice. We need to use the relationship "1 m = 39.37 in." twice—that is,

$$\frac{1\ \text{m}}{39.37\ \text{in.}} \times \frac{1\ \text{m}}{39.37\ \text{in.}} \qquad \text{or} \qquad \left[\frac{1\ \text{m}}{39.37\ \text{in.}}\right]^2$$

Example 1.9

A box has a volume of 482.2 in.^{3}. What is its volume in cm^3?

Solution

Here we need a conversion factor based on the relationship 1 in. = 2.54 cm, and we must use it three times.

$$?\ \text{cm}^3 = 482.2\ \text{in.}^3 \times \left[\frac{2.54\ \text{cm}}{1\ \text{in.}}\right]^3 = 7902\ \text{cm}^3$$

Exercise 1.8

Carry out the following conversions.

a. 476 cm^2 to square inches **b.** 124 ft^3 to cubic meters **c.** 15.8 lb/in.^{2} to kg/m^2

Density: A Physical Property and Conversion Factor

People often say that iron is "heavy" and aluminum is "light." Do they mean that an iron skillet weighs more than an aluminum extension ladder? Of course not. What they mean is that iron is more *dense* than aluminum. If we compare the masses of *equal volumes* of the two metals, the iron indeed weighs more. **Density (*d*)** is the mass per unit volume of a substance, defined by the equation

$$d = \frac{m}{V}$$

The SI unit of density is kilograms per cubic meter (kg/m^3 or kg·m^{-3}), but densities are more often measured in grams per cubic centimeter (g/cm^3) or grams per milliliter (g/mL). The density of aluminum is 2.70 g/cm^3, and that of iron is 7.87 g/cm^3.

Mass does not vary with temperature, but volume does, and so does density. At around room temperature we can generally assume the density of water to be 1.00 g/mL.

Ethylene glycol (the principal ingredient in many antifreeze solutions), water, and ethanol are all colorless liquids. They can be distinguished by their densities. At 20 °C, the density of ethylene glycol is 1.114 g/mL, that of water is 0.998 g/mL, and that of ethanol is 0.789 g/mL. If you combine two liquids that are *immiscible* (that do not mix to form solutions), the liquid with lower density will float on top of the liquid of higher density (Figure 1.8).

Example 1.10

The mass of 325 mL of the liquid methanol (wood alcohol) is found to be 257 g. What is the density of methanol?

Solution

We know the mass and volume of a sample of the liquid, and so we can apply the density equation.

$$d = \frac{m}{V} = \frac{257\ \text{g}}{325\ \text{mL}} = 0.791\ \text{g/mL}$$

Exercise 1.9

Calculate the density of liquid mercury, in g/mL, if a 76.0-lb sample is found to occupy a volume of 2.54 L.

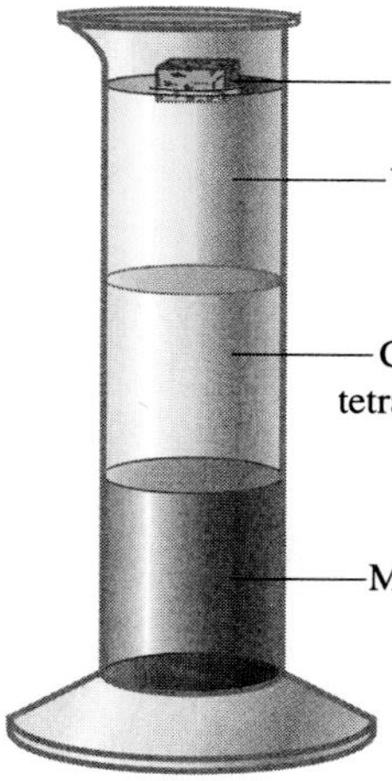

Figure 1.8 A comparison of some densities. Liquid carbon tetrachloride (d = 1.59 g/mL) floats on liquid mercury (d = 13.6 g/mL). Water (d = 1.00 g/mL), which does not mix with carbon tetrachloride, floats on it. Finally, a cork (d = 0.22 g/cm³) floats on the water.

In Example 1.10 we found that the density of methanol is 0.791 g/mL. Another way to describe this density is through the equation

$$1\ \text{mL methanol} = 0.791\ \text{g methanol}$$

This then allows us to formulate two conversion factors between the mass and volume of methanol.

$$\text{(a)}\quad \frac{0.791\ \text{g methanol}}{1\ \text{mL methanol}} \quad \text{and} \quad \text{(b)}\quad \frac{1\ \text{mL methanol}}{0.791\ \text{g methanol}}$$

To convert a volume to a mass we use density as a conversion factor, that is, factor (a) above. To convert from mass to volume, we use the inverse of the density, factor (b).

Example 1.11

What is the mass, in kilograms, of methanol in the 15.5-gal fuel tank of an automobile modified to run on methanol?

Solution

Before we can use conversion factor (a) above, we need to do conversions from gallons to milliliters. Then we apply the factor needed to make the final conversion from grams to kilograms.

$$?\ \text{kg} = 15.5\ \text{gal} \times \frac{3.785\ \text{L}}{1\ \text{gal}} \times \frac{1000\ \text{mL}}{1\ \text{L}} \times \frac{0.791\ \text{g}}{1\ \text{mL}} \times \frac{1\ \text{kg}}{1000\ \text{g}} = 46.4\ \text{kg}$$

Exercise 1.10

An experiment calls for 8.65 g of carbon tetrachloride (d = 1.59 g/mL). What volume would you use?

Exercise 1.11

What volume of methanol (d = 0.791 g/mL) has the same mass as 10.00 gal of gasoline (d = 0.660 g/mL)?

Fat Floats: Body Density and Fitness

How fit are you physically? If you are an athlete, or if you simply want to get in better physical condition through exercise, your first step might well be a fitness evaluation. One part of the evaluation is the determination of percent body fat, and this is best done by measuring body density. The density of human fat is 0.903 g/mL, whereas that of water is 1.00 g/mL. If placed in water, fat floats. The higher the proportion of body fat in a person, the more buoyant the person is in water. Body density is determined from body mass (weight) and volume. Body volume is determined by weighing a person when the person is submerged in water (Figure 1.9). Once body density is determined, the percent body fat is estimated from a graph or from a table of compiled data. The average percent body fat is about 16% for an adult male and 25% for an adult female. Male athletes in superb condition will have less than 7% body fat and females, less than 12% body fat.

Figure 1.9 Determining body volume. A person's mass is easily measured but what about a person's volume? When submerged in water, a person displaces his or her own volume of water. The difference in a person's weight in air and when submerged in water corresponds to the mass of displaced water. This mass, divided by the density of water, yields the volume of displaced water and hence, with an appropriate correction for the volume of air in the lungs, the person's volume.

1.6 Further Remarks on Problem Solving

We have just considered a method of solving *quantitative* chemistry problems. At times, though, we need only a *qualitative* answer: a statement, a picture, symbols, or a diagram to represent what is expected. Occasionally, the answer may be a simple "yes" or "no," although considerable thought may be required to get to this answer.

Regardless of the type of problem, it often helps to visualize what is required, perhaps by drawing a diagram or sketch. The next action is vital: outline *a* strategy, a step-by-step approach to solving the problem. We emphasize the *a* because often more than one strategy will work. In general, any logical scheme you choose will be valid as long as it leads to an appropriate conclusion. You demonstrate your mastery of the underlying principles by devising a problem-solving strategy that works.

At times you may devise a strategy that was not anticipated by the person who posed the problem. In doing so, you show insight into the principles involved.

To assist you in developing a broad range of problem-solving skills, from time to time we will present two types of examples and exercises that differ somewhat from those introduced earlier. We close this chapter by illustrating both types.

Estimation Examples and Exercises

A calculator will always give you an answer, but the answer may not be correct. You may have punched a wrong number or performed the wrong function. If you learn to make *estimations* of answers, you at least will know whether your answer is a reasonable one. And, at times, an approximate answer is all that is possible because there are not enough data to make a precise calculation. Besides, the ability to estimate answers is important in everyday life as well as in chemistry.

Keep in mind as you consider the following illustrations that the idea behind an estimation is *not* to do a detailed calculation. Only a very rough calculation is required, or, at times, no calculation at all.

Estimation Example 1.1

Which of the following is a reasonable height for a female of average height?

1.68 cm   16.8 cm   1.68 m   16.8 m

Solution

If you remember that a centimeter is just under half an inch, you can see that a person's height must be more than a few centimeters. If you also remember that a meter is just over 3 ft, you can see that a person cannot be several meters tall. The only reasonable answer is 1.68 m.

Estimation Example 1.2

Which of the following is an approximate room temperature on the Celsius scale?

5 °C   20 °C   50 °C   70 °C

Solution

Two easy points to remember are (1) the freezing point of water is 0 °C = 32 °F, and (2) there are about two Fahrenheit degrees for every degree Celsius (1.8, to be exact). The first value, 5 °C, is less than 10 °F above the freezing point of 32 °F; it is rather far below room temperature. The second answer, 20 °C, is about 40 °F above the freezing point, which makes it about 70 °F; this is a typical room temperature. We can stop here. The other two temperatures are obviously much above room temperature.

Estimation Exercise 1.1

Which of the following is a reasonable mass for a male of average size?

70 g   700 g   70 kg   700 kg

Estimation Exercise 1.2

Which of the following is a reasonable estimate of the mass of a 20-qt pail of water?

5 kg   10 kg   15 kg   20 kg

(*Hint*: What's the approximate relationship between a quart and a liter?)

Conceptual Examples and Exercises

A typical illustrative example or exercise in this text is designed to illustrate a limited idea—the topic under immediate discussion (significant figures, unit conversions, density, . . .). However, many practical problems require us to draw upon several concepts to arrive at an answer. These problems may be either qualitative or quantitative; their solutions may require calculations or simply estimates.

Conceptual Example 1.1

Could we use the 5.15-kg object pictured in Figure 1.10 to mark the location of a submerged pipeline in a lake? That is, would the object float?

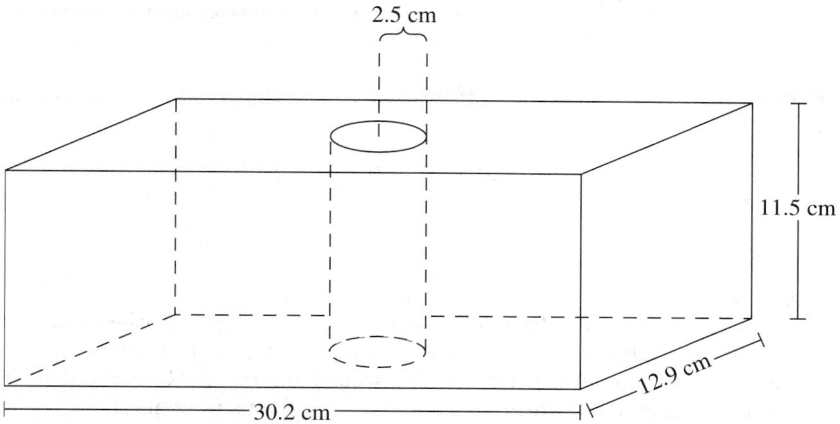

Figure 1.10 Will this 5.15-kg object float on water?

Solution

Figure 1.9 suggests that an object will float on a liquid if its density is *less* than that of the liquid. To determine the density of the object in question, we need to determine its volume. (We are given its mass.) Let's try a simple approach. Let's *estimate* the volume of the object, forgetting for the moment that a cylindrical hole is bored through it.

$$V = 30.2 \text{ cm} \times 12.9 \text{ cm} \times 11.5 \text{ cm} = 4.48 \times 10^3 \text{ cm}^3$$

Now, let's estimate the density of the object.

$$d = \frac{m}{V} = \frac{5.15 \text{ kg} \times 1000 \text{ g/kg}}{4.48 \times 10^3 \text{ cm}^3} = 1.15 \text{ g/cm}^3$$

Because this density is *greater* than that of water (1.00 g/mL), we conclude that *the object will not float.*

How about the cylindrical hole? The actual volume of material in the block is *less* than 4.48×10^3 cm^3 because of the material missing from the hole. This means that the true density is *greater than* 1.15 g/cm^3, and our conclusion is still that the object will not float. Can you see that we could not have reached this conclusion so readily if our rough estimate of the density had been less than about 1 g/cm^3? In that case we would have had to take into account the volume of the "missing" material.

Conceptual Exercise 1.1

Show that if we plugged each end of the cylindrical hole in the object in Figure 1.10, it still would not float on water.

Conceptual Exercise 1.2

Show that if we drilled out and plugged three more holes like the one in Figure 1.10, the object would float on water. (*Hint*: The volume of a cylinder of height h and radius r is $\pi r^2 h$.)

SUMMARY

The science of chemistry deals with the composition and physical and chemical properties of matter. Characteristics of a science are that it is testable, reproducible, explanatory, predictive, and tentative. The scientific method involves making observations, forming hypotheses, doing experiments, gathering data, and formulating laws and theories.

Matter is made up of atoms and molecules. All matter can be divided into two categories: substances and mixtures. Substances are further subdivided into elements and compounds, and mixtures into homogeneous and heterogeneous mixtures.

The SI system has seven base units, four of which are introduced in this chapter. Some measurements, such as length in meters, are directly in terms of base units. However, many measurements are expressed as multiples or submultiples of a base unit, such as length in kilometers or millimeters. In this chapter, temperature measurements on the Celsius scale are emphasized. The SI base unit of temperature, the kelvin, will be discussed later.

To indicate its precision, a measured quantity must be expressed with the proper number of significant figures. Furthermore, special attention must be paid to the concept of significant figures in reporting calculated quantities. Calculations themselves frequently can be done by the unit-conversion method. Also useful in problem solving are techniques for estimating answers.

KEY TERMS

(See Glossary for definitions of these terms.)

accuracy (1.4)
atom (1.2)
chemical property (1.2)
chemical symbol (1.2)
chemistry (1.2)
composition (1.2)
compound (1.2)
conversion factor (1.5)
data (1.2)
density (d) (1.5)
element (1.2)
experiment (1.2)
heterogeneous mixture (1.2)
homogeneous mixture (1.2)
hypothesis (1.2)
kelvin (K) (1.3)
kilogram (kg) (1.3)
liter (L) (1.3)
mass (1.3)
meter (m) (1.3)
mixture (1.2)
molecule (1.2)
physical property (1.2)
precision (1.4)
scientific law (1.2)
second (s) (1.3)
significant figure (1.4)
substance (1.2)
theory (1.2)

REVIEW QUESTIONS

1. What is matter?

2. Which of the following are examples of matter?
a. iron **b.** the human body
c. air **d.** gasoline
e. love **f.** an idea

3. State some distinguishing characteristics of science. Which characteristic(s) best serves to distinguish science from other disciplines?

4. What is a hypothesis? How are hypotheses tested?

5. What are scientific data and what characterizes good data?

6. What is a scientific law? How does it differ from a legislative law?

7. What is a theory? What must one be able to do with a theory? Can a theory be proved?

8. Two samples are weighed under identical conditions in a laboratory. Sample A weighs 112 g and sample B weighs 224 g. Does sample B have twice the mass of sample A? Explain your answer.

9. A formerly chubby person completes a successful diet. Has the person's weight changed? Has the person's mass changed? Explain.

10. Sample A, which is on the moon, has exactly the same mass as sample B on Earth. Do the two samples weigh the same? Explain your answer. (*Hint*: See Figure 1.6.)

11. How do physical and chemical properties differ?

12. How do physical and chemical changes differ?

13. Which of the following describes a chemical change, and which, a physical change?
a. Sheep are sheared and the wool is spun into yarn.
b. A lawn grows thicker after being fertilized and watered.
c. Milk turns sour when left out of the refrigerator for many hours.

d. Silkworms feed on mulberry leaves and produce silk.
e. An overgrown lawn is manicured by mowing it with a lawn mower.

14. Which of the following represent elements and which do not? Explain.
a. C **b.** CO **c.** Cl
d. $CaCl_2$ **e.** Na **f.** KI

15. Which of the following are substances and which are mixtures?
a. helium gas used to fill a balloon
b. maple syrup collected from a maple tree
c. vinegar made from wine
d. salt used to de-ice roads

16. Which of the following mixtures are homogeneous and which are heterogeneous?
a. high-octane gasoline
b. iced tea
c. Italian salad dressing
d. white wine

17. What are the names and symbols of the SI base units for length, mass, time, and temperature?

18. Express the SI units for area, volume, and density in terms of SI base units.

19. What is the difference between the precision and the accuracy of a set of measurements?

20. How do the number of significant figures in a measured quantity relate to the precision of the measurement? to the accuracy?

21. If the numerical value of a measured quantity includes zeros, are these zeros significant? Explain.

22. What is a conversion factor? How must the numerator and denominator of a conversion factor be related?

PROBLEMS

A word of advice: You cannot learn to work problems by reading them or watching your teacher work them, any more than you could become a piano player solely by reading about pianistic skills or attending a performance. As you work through these problems, you will find many opportunities to solidify your understanding of the ideas of the chapter, and also to practice your estimation skills and your ability to synthesize concepts. Plan to work through a large number of these practice problems.

Measurement and Unit Conversion

23. Change each of the following measurements to one in which the unit has an appropriate SI prefix.
a. 4.54×10^{-9} g **b.** 3.76×10^{3} m
c. 6.34×10^{-6} g

24. Change each of the following measurements to one in which the unit has an appropriate SI prefix.
a. 1.09×10^{-9} g **b.** 9.01×10^{-3} s
c. 7.77×10^{-12} s

25. Carry out the following temperature conversions.
a. 23.5 °C to °F **b.** 98.6 °F to °C
c. 212 °C to °F

26. Carry out the following temperature conversions.
a. 173.9 °F to °C **b.** −98.0 °C to °F
c. −10.0 °F to °C

27. Make the following temperature conversions.
a. The high temperature record for the continent of Africa is 136 °F, recorded at Azizia, Libya, in 1922. What is this temperature in degrees Celsius?
b. In the Martian winter, the temperature at the poles drops to −120 °C, freezing water vapor and carbon dioxide from the atmosphere. What is this temperature in degrees Fahrenheit?

28. Answer the following temperature scale questions.
a. A candy recipe calls for heating a sugar mixture to the "soft ball" stage (234 to 240 °F). Can a laboratory thermometer with a range of −10 to 110 °C be used for this measurement?
b. At what temperature do the Fahrenheit and Celsius readings have the same numerical value? Can they have the same value at more than one temperature?

29. Carry out the following conversions.
a. 50.0 km to meters **b.** 546 mm to meters
c. 98.5 kg to grams **d.** 47.9 mL to liters
e. 578 μs to ms **f.** 237 mm to cm

30. Carry out the following conversions.
a. 87.6 μg to kg **b.** 1.00 h to μs
c. 0.0962 km/min to m/s **d.** 55 mi/h to km/min
e. 87.4 cm^2 to mm^2 **f.** 46.4 m^3 to liters

31. Arrange the following in order of increasing length (shortest first): (1) a 1.21-m chain, (2) a 75-in. rope, (3) a 3-ft 5-in. rattlesnake, (4) a yardstick.

32. Arrange the following in order of increasing mass (lightest first): (1) a 5-lb bag of potatoes, (2) a 1.65-kg cabbage, (3) 2500 g sugar.

Significant Figures

33. How many significant figures are there in each of the following measured quantities?
a. 8008 m **b.** 0.00075 s
c. 0.049300 g **d.** 6.02×10^{5} m
e. 4.200×10^{5} s **f.** 0.1050 °C

Density - g/cm³
Mass - g
Volume - cm³ (cc) OR mL

34. How many significant figures are there in each of the following measured quantities?
a. 4051 m **b.** 0.0169 s
c. 0.0430 g **d.** 5.00×10^9 m
e. 1.60×10^{-9} s **f.** 0.0150 °C

35. Express each of the following measured quantities in exponential notation. Assume all the zeros in parts **a** and **b** are significant.
a. 2800 m **b.** 9000 s
c. 0.00090 cm **d.** 20.00 s

36. Express each of the following measured quantities in exponential notation. Assume all the zeros in parts **a** and **b** are significant.
a. 80,000 m **b.** 8900 s
c. 0.09000 cm **d.** 300.0 s

37. Perform the indicated operations and give answers with the proper number of significant figures.
a. 48.2 m + 3.82 m + 48.4394 m
b. 151 g + 2.39 g + 0.0124 g
c. 100.53 cm − 46.1 cm
d. 451 g − 15.46 g
e. 15.44 mL − 9.1 mL + 105 mL
f. 12.52 cm + 5.1 cm − 3.18 cm − 12.02 cm

38. Perform the indicated operations and give answers in the indicated unit and with the proper number of significant figures. (*Note*: You must convert all quantities to a common unit.)
a. 13.25 cm + 26 mm − 7.8 cm + 0.186 m (in cm)
b. 48.834 g + 717 mg − 0.166 g + 1.0251 kg (in kg)

39. Perform the indicated operations and give answers with the proper number of significant figures.
a. $73.0 \times 1.340 \times 0.41 = ?$
b. $265.02 \times 0.000581 \times 12.18 = ?$
c. $\dfrac{33.58 \times 1.007}{0.00705} = ?$
d. $\dfrac{22.61 \times 0.0587}{135 \times 28} = ?$

40. Perform the indicated operations and give answers with the proper number of significant figures.
a. $\dfrac{418.7 \times 31.8}{19.27 - 1.1} = ?$
b. $\dfrac{33.62 + 12.2 - 8.36}{26.4 \times 12.13} = ?$
c. $\dfrac{2.023 - (1.8 \times 10^{-3})}{1.05 \times 10^4} = ?$
d. $\dfrac{(4.6 \times 10^3) + (2.2 \times 10^2)}{3.11 \times 10^4 \times 7.12 \times 10^{-2}} = ?$

Density

41. A 25.0-mL sample of liquid bromine has a mass of 78.0 g. What is the density of the bromine?

42. What is the density of a salt solution if 50.0 mL has a mass of 57.0 g? (cm³)

43. Some metal chips having a volume of 3.29 cm³ are placed on a piece of paper and weighed. The combined mass is found to be 18.43 g. The paper itself weighs 1.2140 g. Calculate the density of the metal to the proper number of significant figures.

44. A glass container weighs 48.462 g. A sample of 4.00 mL of antifreeze solution is added, and the container plus the antifreeze weigh 54.51 g. Calculate the density of the antifreeze solution to the proper number of significant figures.

45. A rectangular block of lead is 1.20 cm × 2.41 cm × 1.80 cm and has a mass of 59.01 g. Calculate the density of lead.

46. A rectangular block of gold-colored material measures 3.00 cm × 1.25 cm × 1.50 cm and has a mass of 28.12 g. Can the material be gold? The density of gold is 19.3 g/cm³.

47. What is the mass, in grams, of 30.0 mL of grenadine (a syrup made from pomegranate juice), which has a density of 1.32 g/mL?

48. What is the mass, in kg, of 2.75 L of the liquid glycerol, which has a density of 1.26 g/mL?

49. What is the volume of a 898-kg piece of cast iron (d = 7.76 g/cm³)? If the iron is formed into a cylindrical bar with a base area of 1.50 cm², how long is the bar?

50. What is the volume of 5.79 mg of gold (d = 19.3 g/cm³)? If the gold is hammered into a gold leaf of uniform thickness with an area of 44.6 cm², what is the thickness of the gold leaf?

51. A box with a base 0.80 m on a side and a height of 1.20 m is filled with 3.2 kg of expanded polystyrene packing material. What is the bulk density, in g/cm³, of the packing material? (The bulk density includes the air between the pieces of polystyrene foam.)

52. Hylon VII, a starch-based substitute for polystyrene packing material, has a bulk density of 12.8 kg/m³. What mass, in grams, of the material is needed to fill a volume of 2.00 ft³?

53. Which of the following items would be most

difficult to lift onto the back of a pickup truck: (1) a 100-lb bag of potatoes, (2) a 15-gal plastic bottle filled with water, (3) a 3.0-L flask filled with mercury (d = 13.6 g/cm^3)?

54. Liquid mercury (d = 13.6 g/cm^3) is commonly shipped in iron flasks that contain 76 lb of mercury. Will one of these flasks fit inside a wooden box that is 4.0 in. × 4.0 in. × 8.0 in.?

ADDITIONAL PROBLEMS

55. If the meter stick used in the measurements in Table 1.4 was actually 1.005 m long, how large would the error have been in measuring the length of the room?

56. In its nonstop, round-the-world trip, the aircraft *Voyager* traveled 25,102 mi in 9 days, 3 min, and 44 s. To the maximum number of significant figures permitted, calculate the average speed of *Voyager* in mi/h.

57. The *furlong* is a unit used in horse racing, and the units *chain* and *link* are used in surveying land. There are 8 furlongs in 1 mi, 10 chains in 1 furlong, and 100 links in 1 chain. Calculate the length of 1 link, in inches, to three significant figures. (1 mi = 5280 ft; 1 ft = 12 in.)

58. In the United States land area is commonly measured in acres: 640 acres = 1 mi^2. In most of the rest of the world land area is measured in hectares: 1 hectare = 1 hm^2 [1 hectometer (hm) = 100 m]. Which is the larger area, the acre or the hectare? Write a conversion factor that relates the acre and the hectare. (1 mi = 5280 ft; 1 m = 39.37 in.)

59. In scientific work, densities usually are expressed in g/cm^3. In engineering work, the unit lb/ft^3 is often used. The density of water at 20 °C is 0.998 g/cm^3. What is its density in lb/ft^3?

60. A square of aluminum foil (d = 2.70 g/cm^3) is 5.10 cm on a side and has a mass of 1.762 g. Calculate the thickness of the foil, in mm.

61. An empty 3.00-L bottle weighs 1.70 kg. Filled with a homemade wine, the bottle weighs 4.72 kg. The wine contains 11.0% ethyl alcohol by mass. How many ounces of ethyl alcohol are present in a drink that contains 275 mL of this wine?

62. Aerogels consist of a solid framework with most of their volume occupied by air. Silica (silicon dioxide) powder has a density of 2.2 g/cm^3. Silica can be expanded into an aerogel with a bulk density of 0.015 g/cm^3. Calculate the volume of silica aerogel that can be made from 125 cm^3 of silica powder.

63. Calculate the total surface area of the aerogel described in Problem 62, if there is 470 m^2 of surface per gram of silica in the aerogel.

64. Dust from air that is not significantly polluted is deposited ("dustfall") at a typical rate of 10 tons/mi^2 per month. What is this dustfall expressed in mg/m^2 per hour?

65. Refer to the object pictured in Figure 1.10. It can be made to float by enlarging the cylindrical hole and plugging both ends of the hole. What is the *minimum* diameter hole that must be drilled and plugged if the object is to float? Why must the actual diameter of the hole be somewhat larger than this calculated value?

66. The hexagonal nut pictured is 7.0 mm on edge, 6.0 mm thick, and has a 7.0-mm diameter hole. The density of the metal used in the nuts is 7.87 g/cm^3. How many of these nuts are present in a 1.00-lb package?

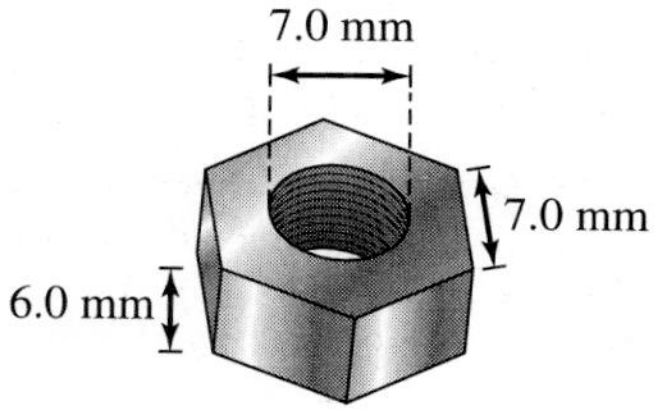

67. The two vessels shown are completely filled with water. A brass cube 2.0 cm on an edge is gently placed on the water in the vessel on the left, and a rectangular block of cork, 5.0 cm × 4.0 cm × 2.0 cm, on the water in the vessel on the right. The density of brass is 8.40 g/cm^3 and that of cork is 0.22 g/cm^3. From which vessel will the greater volume of water overflow?

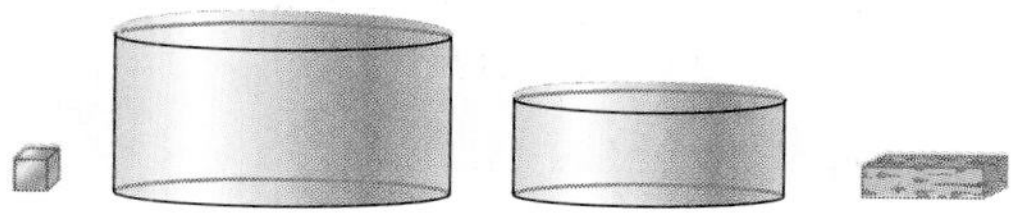

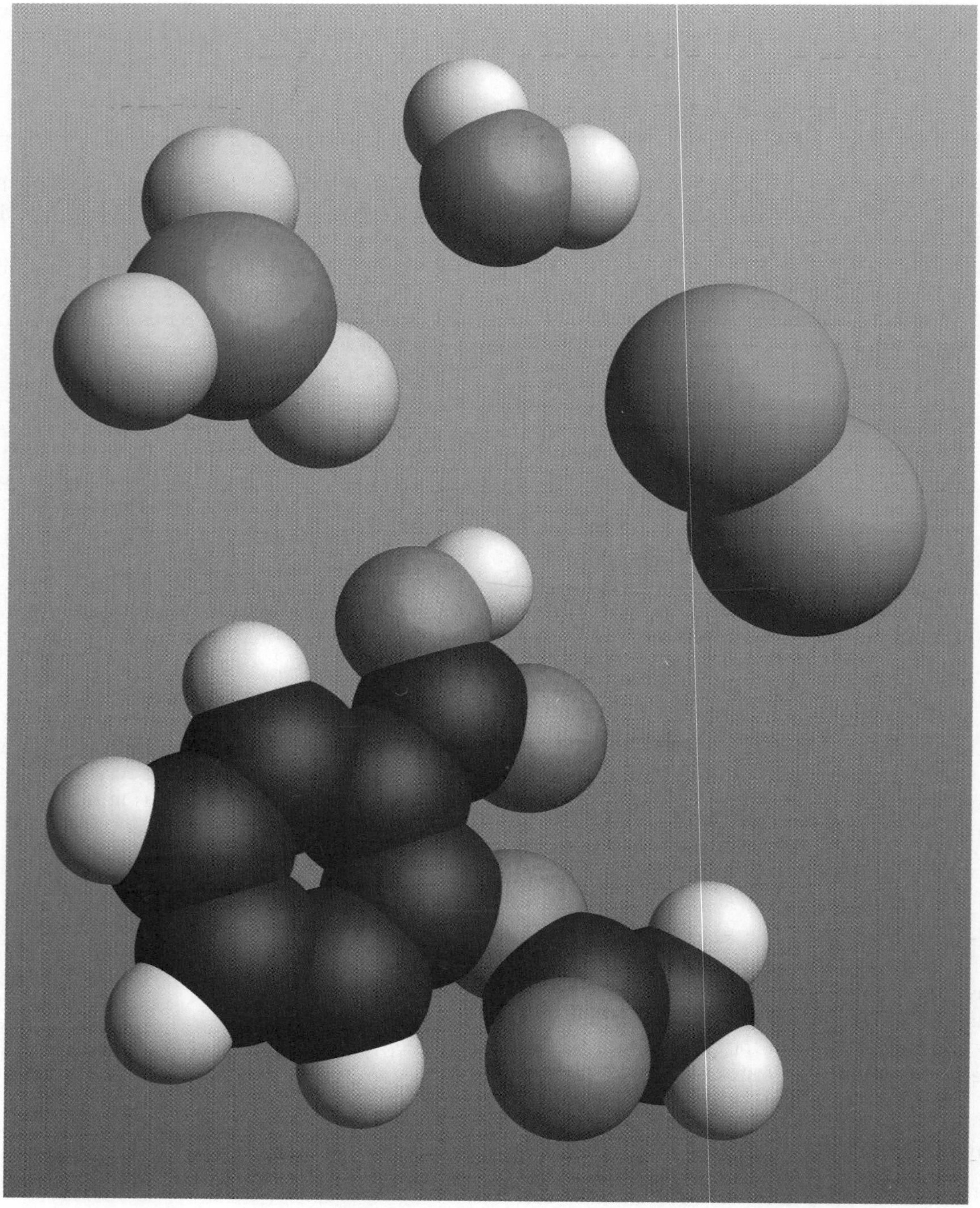

Molecules are often represented by space-filling models. In clockwise fashion from the top are models of water, H_2O; chlorine, Cl_2; acetylsalicylic acid (aspirin), $CH_3COOC_6H_4COOH$; and boron trifluoride, BF_3.

2

Laws, Symbols, and Formulas: The Language of Chemistry

CONTENTS

The language that chemists use has two facets. First, in writing and talking *about* chemistry, chemists use a special vocabulary: atoms, ions, molecules, conservation of mass, definite proportions, saturated hydrocarbons, ... Second, they use a special symbolic chemical language.

Chemical symbols and formulas of elements and compounds, like K, Mg^{2+}, H_2O, and C_5H_{12}, make up the "alphabet and words" of chemistry, as we see in this chapter. In the next chapter we will look at the "sentences" of chemistry: chemical equations to represent chemical reactions.

2.1 Laws of Chemical Combination

We have learned that data obtained by experiment can often be summarized into natural laws. Scientific theories are then formulated to explain the natural laws. In this section we consider the discovery of two natural laws that John Dalton subsequently set out to explain with his atomic theory (Section 2.2).

Lavoisier: The Law of Conservation of Mass

Modern chemistry dates from the eighteenth century, when scientists began to make *quantitative* observations. Antoine Lavoisier, through measurements precise to about 0.0001 g, found that total mass was unchanged in a chemical reaction. For example, when he decomposed the red oxide of mercury, the mass of the products (mercury metal and oxygen gas) was exactly the same as the mass of the mercury oxide he started with.

Lavoisier summarized his findings through the **law of conservation of mass**, one form of which states that

The total mass of the products of a reaction is always equal to the total mass of the reactants (starting materials) consumed in the reaction.

An alternative statement of this law is: Matter is neither created nor destroyed during a chemical change.

Antoine Lavoisier (1743–1794) used the return on investments in a much-hated private tax-collecting agency of King Louis XVI to equip a private laboratory and finance his research. As a result he earned the enmity of leaders of the French Revolution and was beheaded on a guillotine in 1794. Little did his executioners realize that their victim would later become known as the father of modern chemistry.

This law is the basis for many of the chemical calculations that we will perform in Chapter 3.

The law of conservation of mass is not just a matter of academic interest. It states that we cannot create materials from nothing. New materials can be made only by changing the way atoms are combined. Nor can we get rid of wastes by the destruction of matter. Chemistry offers alternatives, however. Through chemical reactions we can change some kinds of potentially hazardous wastes to less harmful forms. Such transformations of matter are what chemistry is all about.

Proust: The Law of Definite Proportions

By the end of the eighteenth century, Lavoisier and other scientists had succeeded in decomposing many compounds into their elements. One of these scientists was Joseph Proust (1754–1826), who, through painstaking quantitative studies, established the **law of definite proportions**.

A given compound always contains its constituent elements in certain fixed proportions by mass.

Particularly important were Proust's investigations of basic copper carbonate, described in Figure 2.1.

Figure 2.1 The law of definite proportions. Proust based the law of definite proportions on a substance that we now know to be basic copper carbonate. Whether a sample of this material is synthesized in the laboratory (left) or obtained as the mineral *malachite* (right), it has the same composition: 57.48% copper, 5.43% carbon, 0.91% hydrogen, and 36.18% oxygen, by mass.

Not only does a compound have a fixed composition, it also has fixed properties. At 20 °C, 100.0 g of pure water always dissolves 35.9 g of sodium chloride (salt)—no more, no less. Under normal atmospheric pressure, water always freezes at 0 °C and boils at 100 °C. The properties of chemical compounds depend on their composition; they are not a matter of chance.

2.2 John Dalton and the Atomic Theory of Matter

In 1803, John Dalton proposed a model to explain the laws of chemical combination that we have just described. As he developed his model, Dalton uncovered a *third* law that his theory would have to explain, the **law of multiple proportions.** This law recognizes that a given set of elements may combine in more than one set of proportions, giving rise to different compounds. Moreover, we find that

> *When these different compounds are compared, the masses of one element that combine with a fixed mass of the second are in the ratio of small whole numbers.*

To see how the law of multiple proportions works, let's consider two compounds of oxygen and carbon. Oxygen combines with carbon in the ratio 8.0 g of oxygen to 3.0 g of carbon in carbon dioxide, the familiar gas produced in respiration and in the burning of fuels. But oxygen and carbon can also combine in the ratio 4.0 g of oxygen to 3.0 g of carbon in carbon monoxide, a poisonous gas formed when a fuel is burned in limited air. The masses of oxygen (8.0 g and 4.0 g) that combine with a fixed mass of carbon (3 g) are in a small whole-number ratio, 8.0:4.0 = 2:1 (see Figure 2.2).

John Dalton (1766–1844) was not a good experimenter (perhaps because he was color-blind), but he skillfully used the results of others in formulating his atomic theory.

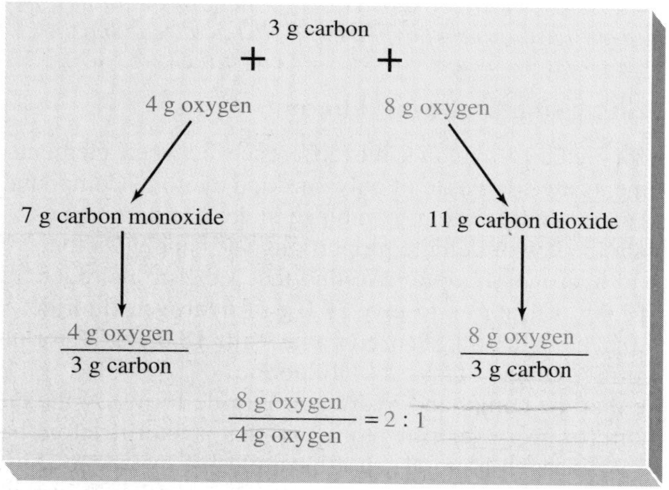

Figure 2.2 **The law of multiple proportions demonstrated.** The oxygen-to-carbon mass ratio in carbon dioxide is twice that found in carbon monoxide.

Example 2.1

A nitrogen–oxygen compound is found to have an oxygen-to-nitrogen mass ratio of 1.142—that is, 1.142 g of oxygen for every 1.000 g of nitrogen. Which of the following oxygen-to-nitrogen mass ratio(s) is (are) possible for a different nitrogen–oxygen compound(s)?

a. 0.571 **b.** 1.000 **c.** 2.285 **d.** 2.500

Solution

The given compound has 1.142 g of oxygen and 1.000 g of nitrogen. Response (c) has 2.285 g of oxygen for the same 1.000 g of nitrogen. The ratio of the masses of oxygen, 2.285:1.142, is almost exactly 2:1. Response **c** is a possibility. So is response **a**. Here the ratio is 0.571:1.142 = 0.500 = 1:2. Responses **b** and **d** are not possibilities. They yield ratios of 1.000:1.142 = 0.8757 and 2.500:1.142 = 2.189, respectively. Neither of these can be expressed as the ratio of small whole numbers.

Exercise 2.1

From the information given in Example 2.1, determine which of the following oxygen-to-nitrogen mass ratio(s) is (are) also possible for a nitrogen–oxygen compound(s).

a. 0.612 **b.** 1.250 **c.** 1.713 **d.** 2.856

Dalton's Atomic Theory

The atomic model Dalton developed to explain the laws of chemical combination is based on four ideas.

- *All matter is composed of extremely small, indivisible particles called atoms.*
- *All atoms of a given element are alike in mass and other properties, but atoms of one element differ from the atoms of every other element.*
- *Compounds are formed when atoms of different elements unite in fixed proportions.* [The numbers of each kind of atom form a simple ratio, such as one atom of A to one of B (AB), two atoms of A to one of B (A_2B), etc.]
- *A chemical reaction involves a rearrangement of the atoms. No atoms are created, destroyed, or broken apart in a chemical reaction.*

Explanations Using Dalton's Theory

Dalton's theory clearly explains the difference between elements and compounds. Elements are composed of only one kind of atom. Compounds are made up of two or more kinds of atoms combined in definite proportions.

To explain the law of definite proportions, let's apply Dalton's reasoning to the compound hydrogen fluoride. We find that 1.0 g of hydrogen always combines with 19.0 g of fluorine to give 20.0 g of hydrogen fluoride. Why? Why shouldn't 1.0 g of hydrogen also combine with 18.0 g of fluorine? Or with 20.0 g of fluorine? Or any other mass of fluorine?

Suppose that the compound hydrogen fluoride forms by the union of one hydrogen atom for every fluorine atom, and that a fluorine atom has 19 times the mass of a hydrogen atom. Then all samples of hydrogen fluoride would have to exhibit a fluorine-to-hydrogen mass ratio of 19:1. If we tried to combine 1.0 g hydrogen with 20.0 g fluorine, we would end up with 20.0 g of hydrogen fluoride (1.0 g H + 19.0 g F) and one leftover gram of fluorine. An atomic model offers an easy explanation of the law of definite proportions. Figure 2.3 gives interpretations of both this law and the law of conservation of mass in terms of Dalton's atomic theory.

Dalton's theory also nicely explains the law of multiple proportions. In carbon dioxide, two oxygen atoms are combined with each carbon atom (CO_2); in carbon monoxide there is one oxygen atom per carbon atom (CO). The ratio of the numbers of oxygen atoms per carbon between the two compounds is 2:1. Because all oxygen atoms have the same mass, this 2:1 atom ratio is also the mass ratio, just as described in Figure 2.2.

Along these same lines, the first compound described in Example 2.1 is NO, and the others are (a) N_2O and (c) NO_2.

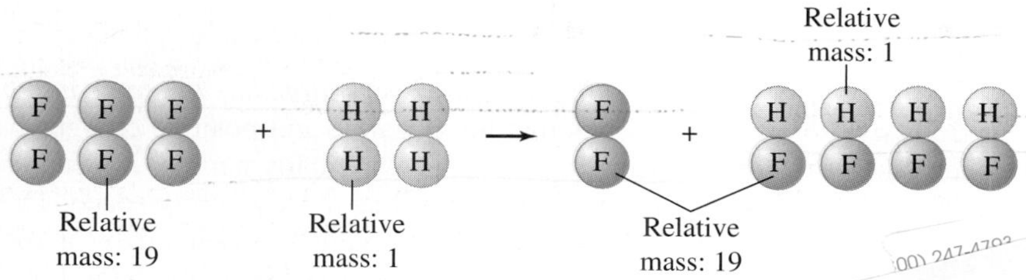

Figure 2.3 **Dalton's theory and the laws of definite proportion... tion of mass.** Hydrogen, fluorine, and hydrogen fluoride are all gaseous... their atoms joined in pairs in molecules. Since a fluorine atom has 19 times t... hydrogen atom, nineteen-twentieths or 95% of the mass of hydrogen fluoride... one-twentieth or 5% is hydrogen, regardless of the size of the sample. The law... proportions is illustrated.

Even though six atoms of fluorine are available, only four of them combi... gen atoms to form hydrogen fluoride. After the reaction the numbers of hydrog... rine atoms are four and six, respectively, just as before the reaction. The law of c... of mass is illustrated.

2.3 The Divisible Atom

Dalton's concept of indivisible atoms lasted for nearly all of the nineteenth c... tury. Like most scientific theories, however, Dalton's required modification i... light of later discoveries. About 100 years ago, experiments (described mo... fully in Chapter 6) began to reveal that atoms are made up of smaller parts. C... the dozens of smaller-than-atomic particles now known, three are of special importance to chemists.

Subatomic Particles

The **proton** has a mass that for now we will call a relative mass of 1. The proton also carries one fundamental unit of positive electric charge, 1+. The **neutron** is a particle with a mass just slightly larger than that of the proton, but it has no electric charge. The third particle, the **electron**, has a mass that is 1/1836 (0.0005447) of a proton's mass. An electron has the same amount of charge as a proton, but it is a negative charge, that is, 1− . Protons, neutrons, and electrons are *fundamental* particles. That is, all protons are alike, all neutrons are alike, and all electrons are alike. These three subatomic particles and some of their properties are listed in Table 2.1.

Electric charge, electricity, and other basic physical quantities are described in Appendix B.

TABLE 2.1 Three Subatomic Particles

Particle	Symbol	Approximate Relative Mass	Relative Charge	Location in Atom
Proton	p^+	1	1+	nucleus
Neutron	n	1	0	nucleus
Electron	e^-	0.000545	1−	outside nucleus

Protons and neutrons are densely packed into a tiny, massive, positively charged core of the atom known as the *nucleus*. The extremely lightweight electrons are widely dispersed around the nucleus. *Every atom has the same number of electrons as protons*. This means that the negative and positive charges balance each other out, and the atom as a whole is electrically neutral: it has no net charge. To picture an atom, we should think of it as mostly empty space. Think of it as something like this: If an entire atom were represented by a room 5 m × 5 m × 5 m, the nucleus would be only about as big as the period at the end of this sentence. The electrons would be of a comparable size.

Dalton believed that the *mass* of an atom determines the element. We now know that it is not the mass but the *number of protons* in the nucleus that determines the kind of atom and therefore the identity of an element. The **atomic number (*Z*)** is the number of protons in the nucleus of an atom of the element. An atom with two protons has $Z = 2$; it is an atom of helium. An atom with 92 protons has $Z = 92$ and is an atom of uranium. A list of all the known elements, with their symbols and atomic numbers, is located inside the front cover of this book.

Isotopes

Dalton thought that all atoms of a given element have the same mass. This isn't quite so. Any two atoms of an element have the same number of protons (and electrons as well), but they may have different numbers of neutrons. For example, there are three kinds of hydrogen atoms. The most abundant, sometimes called protium, has a single proton and no neutrons in its nucleus. About one in every 6700 hydrogen atoms, however, has a neutron as well as a proton. The mass of this kind of hydrogen atom, called deuterium, is about twice that of protium. A third, very rare form of hydrogen atom, called tritium, has two neutrons and one proton in the nucleus. A tritium atom has about three times the mass of the protium atom.

The names deuterium and tritium are commonly used, but protium is not.

Because the masses of electrons are negligible compared to those of protons and neutrons, for practical purposes the mass of an atom is determined by the numbers of protons and neutrons present. We can refer to the mass of an atom through its **mass number (*A*)**, the sum of its numbers of protons and neutrons. Atoms that have a given number of protons but varying numbers of neutrons are called **isotopes**. Isotopes have the same atomic number but different mass numbers. The three isotopes of hydrogen all have the atomic number 1, but their mass numbers are 1, 2, and 3, respectively. Only the hydrogen isotopes have special names. Other isotopes are usually identified by the name of the element followed by the mass number, such as carbon-14, cobalt-60, and uranium-235.

Some elements have the same mass number for all their naturally occurring atoms, such as fluorine-19, sodium-23, and phosphorus-31. Most, however, occur in two or more isotopic forms; tin has the greatest number: ten. For all but a few elements, the naturally occurring isotopes always occur in certain precise proportions. In chlorine from natural sources, for example, 75.77% of the atoms are chlorine-35 and 24.23% are chlorine-37.

Chemical symbols for isotopes are commonly written in the form

$${}^{A}_{Z}\mathrm{E}$$

with A as the mass number and Z as the atomic number of the element E. For the two naturally occurring isotopes of chlorine we can write

$$^{35}_{17}\text{Cl} \quad \text{and} \quad ^{37}_{17}\text{Cl}$$

indicating mass numbers of 35 and 37 for chlorine-35 and chlorine-37, respectively. Because the atomic number of an element is implied through its chemical symbol, sometimes simplified forms such as ^{35}Cl and ^{37}Cl are used. The number of neutrons in an atom is easily calculated from the values of A and Z.

$$\text{Number of neutrons} = A - Z$$

Example 2.2

How many neutrons are there in the nucleus of a ^{238}U atom?

Solution

Even though the atomic number is not shown, we know that it must be 92. That's all that it can be for uranium.

$$\text{Number of neutrons} = A - Z = 238 - 92 = 146$$

Exercise 2.2

What is the mass number of an isotope of tin that has 66 neutrons?

2.4 Atomic Weights

Because he had no means of isolating and weighing individual atoms, Dalton set the hydrogen atom as a standard and assigned it a mass of 1. He then assigned to each of the other known atoms a mass based on this standard and called it the *atomic weight* of the element. Consider how Dalton would have dealt with the element fluorine if he had been aware of its existence. He would have *measured* the combining *mass* ratio of fluorine to hydrogen in hydrogen fluoride and found it to be 19.0 g F:1.0 g H. He would have *assumed* that fluorine and hydrogen atoms combine in the ratio, 1:1. So, if a given number of fluorine atoms weighed 19 times as much as the same number of hydrogen atoms, one fluorine atom must weigh 19 times as much as one hydrogen atom, the standard. He would have assigned fluorine an atomic weight of 19.0.

Using this approach, Dalton immediately got off to a bad start. First, he assumed that water, the only hydrogen–oxygen compound known at the time, had a 1:1 ratio of hydrogen to oxygen atoms. Next, he found the best available value of the combining ratio by mass: 1 g H:7 g O. From these two bits of information he assigned oxygen the atomic weight of 7. You are probably already aware of where Dalton went wrong. As we now know, the actual combining ratios are 2 H atoms to 1 O atom and 1 g H:8 g O. Therefore, as we see from Figure 2.4, an atom of oxygen must have *16* times the mass of an atom of hydrogen.

During most of the nineteenth and well into the twentieth century, oxygen was used as the atomic weight standard. A problem arose, however, when isotopes were discovered. Chemists assigned an atomic weight of exactly 16 to the naturally occurring *mixture* of oxygen isotopes, but physicists assigned this

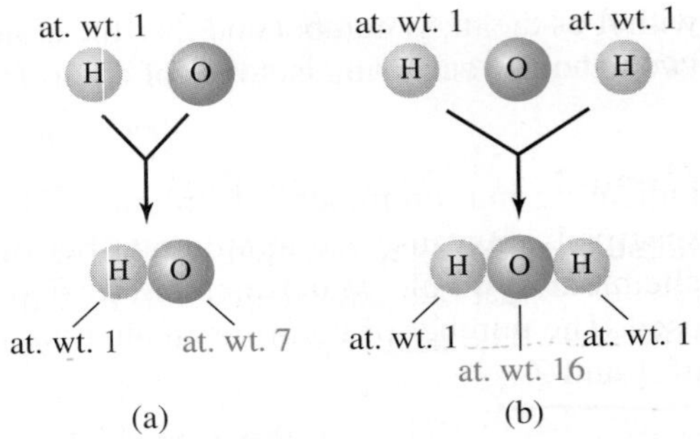

Figure 2.4 Dalton's atomic weight problem. (a) Dalton assumed a combining ratio of hydrogen to oxygen atoms of 1:1. Data at the time suggested that the mass ratio of hydrogen to oxygen in water was 1:7. If the atomic weight of hydrogen was taken as one, that of oxygen had to be seven. (b) Modern data indicate that the combining ratio of hydrogen to oxygen atoms is 2:1, and that the mass ratio is 1:8 (or 2:16). If the atomic weight of hydrogen is taken to be one, that of oxygen must be 16.

value to the *single* isotope oxygen-16. The chemist's and physicist's atomic weight scales differed by a factor of 1.000275.

In the late 1950s, by an international agreement, scientists agreed to adopt as the atomic weight standard the *pure* isotope carbon-12, which was assigned a mass of *exactly* 12 atomic mass units (12 u). Based on this standard, we can define an **atomic mass unit** (abbreviated amu and having the unit, u) as exactly one-twelfth the mass of a carbon-12 atom. In more familiar units of mass, 1 u = 1.66054×10^{-24} g.

Naturally occurring carbon consists of a *mixture* of two isotopes. The much more abundant isotope is ^{12}C. The other is ^{13}C, with a mass of 13.00335 u. When carbon atoms participate in a chemical reaction, even though a mixture of isotopes is involved, it's more convenient to think in terms of a hypothetical, "average" atom. This "average" carbon atom would have a mass somewhere between 12.00000 u and 13.00335 u. However, this mass must be "weighted" toward the mass of the more abundant carbon-12 isotope. The weighted average of the masses of the naturally occurring carbon atoms is 12.011 u. The **atomic weight*** of an element, then, is the weighted average of the masses of the naturally occurring atoms of that element.

Looking Ahead

Atomic weights are discussed further in Section 6.5. There we describe mass spectrometry, the experimental method used to determine the masses of atoms. We will also show how to calculate a weighted-average atomic weight.

2.5 The Periodic Table: An Organizational Concept for Chemistry

Dmitri Ivanovich Mendeleev (1834–1907) invented the periodic table while trying to systematize the properties of the elements for presentation in a chemistry textbook. His highly influential text lasted for 13 editions— five of them after his death.

After the time of Lavoisier, Dalton, and their contemporaries, chemists discovered elements at a rapid rate. By 1830, there were 55 known elements, all with different properties and no apparent pattern among them. Chemists were in great need of a way to organize the growing mass of chemical data by arranging the elements in a meaningful manner. The first successful arrangement, called a periodic table, was published by the Russian chemist Dmitri Mendeleev in 1869. Mendeleev's table and modern refinements of it are so important that we devote most of a later chapter to this topic. Our need for now is

*The term *atomic mass* is more appropriate than atomic weight. However, the term atomic weight is so widely used by chemists that it is likely to remain in service for many years.

just to consider a few ideas that will help us in naming and writing formulas of chemical compounds.

Mendeleev's Periodic Table

Mendeleev arranged the elements according to increasing atomic weights in a tabular format that placed elements with similar properties in the same vertical columns of the table, a format called a **periodic table**. Because organization by properties was paramount, a few elements had to be placed "out of order," rather than strictly in order of increasing atomic weight. For example, Mendeleev placed tellurium (at. wt. 127.6) ahead of iodine (at. wt. 126.9) so that tellurium would be in the same column as sulfur and selenium, elements that closely resemble it. The notion that elements with similar properties must fall in the same group also led Mendeleev to leave gaps in his table. Instead of seeing these gaps as defects, he boldly predicted the existence of undiscovered elements. Furthermore, because this table was based on patterns of *properties*, he was able to predict some properties of the missing elements. Thus, he left a blank space for the undiscovered element that he called "eka-silicon" and used its location between silicon and tin to predict its properties (Figure 2.5). Table 2.2 shows just how accurate his predictions were when compared to the properties of the actual element, germanium, discovered 15 years later.

In subsequent years, other gaps were filled in by the discovery of more new elements. The last gap was filled by the discovery of technetium in 1937. It was the *predictive* nature of Mendeleev's periodic table that led to its wide acceptance as a great scientific accomplishment.

Reihen	Gruppe I R^2O	Gruppe II RO	Gruppe III R^2O^3	Gruppe IV RH^4 RO^2
1	H = 1			
2	Li = 7	Be = 9,4	B = 11	C = 12
3	Na = 23	Mg = 24	Al = 27,3	Si = 28
4	K = 39	Ca = 40	-- = 44	Ti = 48
5	(Cu = 63)	Zn = 65	-- = 68	-- = 72
6	Rb = 85	Sr = 87	?Yt = 88	Zr = 90
7	(Ag = 108)	Cd = 112	In = 113	Sn = 118

Figure 2.5 Reproduction of a portion of Mendeleev's periodic table. Mendeleev arranged the elements into eight main groups (Gruppe) and twelve rows (Reihen). The formulas R^2O, RO, … , are those of the element oxides, such as Li_2O, MgO, … ; the formulas RH^4, RH^3, … are those of the element hydrides, such as CH_4, NH_3, … . Note the blank spaces (in color) corresponding to undiscovered elements whose atomic weights Mendeleev predicted.

The Modern Periodic Table

The modern periodic table shown inside the front cover contains 111 elements. Each entry in the table gives the chemical symbol, atomic number (Z), and

TABLE 2.2 Properties of Germanium: Predicted and Observed

Property	Predicted: Eka-silicon[a] (1871)	Observed: Germanium (1886)
Atomic weight	72	72.6
Density, g/cm^3	5.5	5.47
Color	dirty gray	grayish white
Density of oxide, g/cm^3	EsO_2: 4.7	GeO_2: 4.703
Boiling point of chloride	$EsCl_4$: below 100 °C	$GeCl_4$: 86 °C
Density of chloride, g/cm^3	$EsCl_4$: 1.9	$GeCl_4$: 1.887

[a] The term "eka" is derived from Sanskrit and means "first." Literally, eka-silicon means, first comes silicon (and then comes the unknown element).

atomic weight for an element. Note that the elements are placed in order of increasing *atomic number*, a more fundamental property than atomic weight.

The periodic table is divided into groups and periods. *Groups* (or families) are the vertical columns of elements having similar properties. *Periods* are the horizontal rows of elements. The periods vary in length from two (the first period) to 32 (the sixth and seventh periods). In order to keep the maximum width of the table at 18 members, 14-member series are extracted from the sixth and seventh series and listed as "footnotes" at the bottom of the table. The sixth period footnote is called the *lanthanide* series and the seventh period footnote, the *actinide* series.

Elements are also divided into two main classes by a heavy, stepped, diagonal line. Those to the left of the line, except for hydrogen, are **metals**, elements that have a characteristic luster and are generally good conductors of heat and electricity. Most metals are *malleable*, which means that they can be hammered into thin sheets or foil. Also, most metals are *ductile*; they can be drawn into wires. Except for mercury, a liquid, all the metals are solid at room temperature.

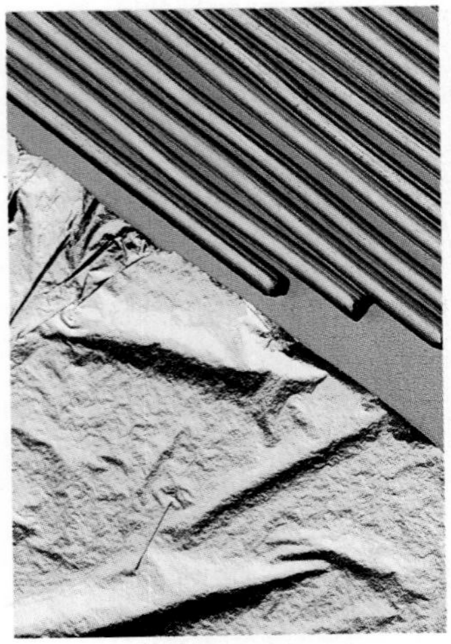

The metals copper (top) and gold (bottom) are ductile and malleable.

Elements to the right of the stepped line are **nonmetals**, elements that lack metallic properties. For example, nonmetals generally are poor conductors of heat and electricity. At room temperature, several nonmetals, such as oxygen, nitrogen, fluorine, and chlorine, are gases. Others, such as carbon, sulfur, phosphorus, and iodine, are brittle solids. Bromine is the only nonmetal that is a liquid at room temperature.

Looking Ahead

Among the topics that we will consider in Chapter 7 are

- the theoretical basis of the periodic table;
- why the periods are not all of the same length;
- why different systems are used in numbering the groups;
- why hydrogen, a nonmetal, appears in Group 1A, a group of metals;
- alternate definitions of metals and nonmetals based on the structures of their atoms;
- the idea that some elements bordering the zigzag line that separates metals and nonmetals (called metalloids) have some properties like metals and some like nonmetals.

2.6 Chemical Formulas: The "Words" of Chemistry

The chemical *symbols* that we introduced in Chapter 1 are like the letters in an alphabet of chemistry. To describe compounds, these symbols are combined into chemical *formulas*, much as letters are combined to make words. A *chemical formula* indicates the composition of a compound. It uses symbols to denote the elements present and subscripts to indicate the relative numbers of atoms of each element. For example, the formula B_2O_3 signifies that the compound boron oxide contains two boron atoms for every three oxygen atoms.

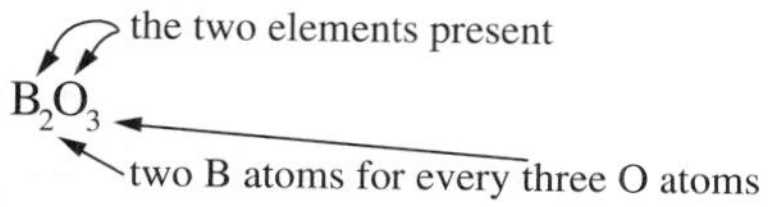

Ammonia has nitrogen and hydrogen atoms in the ratio 1 to 3, but the formula is written as NH_3, *not* N_1H_3. The lack of a subscript number in a formula signifies that the subscript "1" is intended.

Chemical substances are classified in several ways. One scheme uses two major categories: molecular and ionic. Another divides chemical compounds into inorganic and organic. And still another uses the categories acids, bases, and salts. In the next few sections we outline systems for assigning names and formulas to chemical compounds. Sometimes a compound has more than one name, depending on how it is being classified. Different names are used to convey specific information about the compound.

2.7 Molecules and Molecular Substances

A **molecule** is an electrically neutral group of two or more atoms held together in a definite spatial arrangement by forces called covalent bonds. A *molecular* substance has molecules as its smallest characteristic entities; the component atoms of the molecule, considered separately, do not have the attributes that characterize the molecular substance.

Covalent bonds are discussed in Chapters 9 and 10. In this section we concentrate on the atoms present and not on the forces between them.

An **empirical formula** is the simplest formula that can be written for a compound. It lists the elements present and indicates the smallest integral ratio in which their atoms are combined. The empirical formula CH_2O indicates that the elements C, H, and O are present in the atom ratios 1:2:1, respectively. However, several different compounds have this same empirical formula, for example, acetic acid (found in vinegar) and glucose (blood sugar). A **molecular formula** shows the difference by giving the symbol and the exact number of each kind of atom in a molecule. The molecular formula of acetic acid is $C_2H_4O_2$ and that of glucose is $C_6H_{12}O_6$. Notice that in both of these molecular formulas the atom ratios are the same as in the empirical formula, 1:2:1, for C, H, and O, respectively. Water (H_2O), ammonia (NH_3), and carbon dioxide (CO_2) are familiar substances whose molecular formulas are the same as their empirical formulas.

Pure hydrogen in nature does not exist as individual H atoms, but rather in pairs of atoms joined into the molecules, H_2. These are called *diatomic* (two-atom) molecules. Other common elements that form diatomic molecules are nitrogen, N_2, and oxygen, O_2. The Group 7A elements, the halogens, are also diatomic: F_2, Cl_2, Br_2, I_2. A few elements exist as *polyatomic* (many-atom) molecules. Well-known examples are phosphorus, P_4, and sulfur, S_8. The noble gases of Group 8A exist in *monatomic* (one-atom) form. Thus we represent helium, neon, and argon simply as He, Ne, and Ar.

The atoms in a molecule are not arranged at random. Instead, they are attached in a definite order. In water molecules the order is always H—O—H, never H—H—O. A **structural formula** is a chemical formula that shows how the atoms are attached to one another. The structural formulas of ammonia, methane, and acetic acid are shown below.

The formula CH_3COOH tells us much more about the acetic acid molecule than does the ordinary molecular formula, $C_2H_4O_2$. We will say more about acetic acid in Section 2.11.

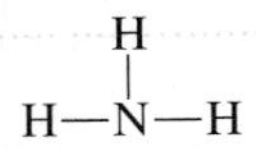
Ammonia (NH_3)

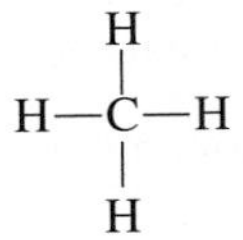
Methane (CH_4)

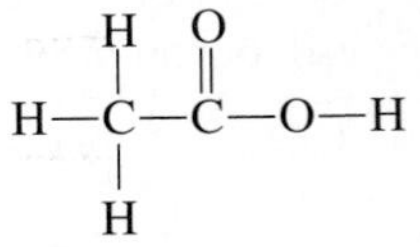
Acetic acid (CH_3COOH)

The lines in structural formulas represent the bonds between atoms. A single line represents a single bond; a double line, a double bond; and a triple line, a triple bond. We'll discuss these different bond types in Chapters 9 and 10, but for now simply look upon a double bond as being stronger than a single bond, and a triple bond, stronger still.

Molecules have definite shapes, but these are often difficult to represent on paper. Shapes are best represented by molecular models. Figure 2.6 shows two types of models of the acetic acid molecule. The *ball-and-stick* model shows the spatial arrangements of the bonds in a way that can only be implied by the structural formula. The *space-filling* model shows that atoms in a molecule occupy space and that they are in actual contact with one another. We'll have much more to say about molecular models in Chapter 10.

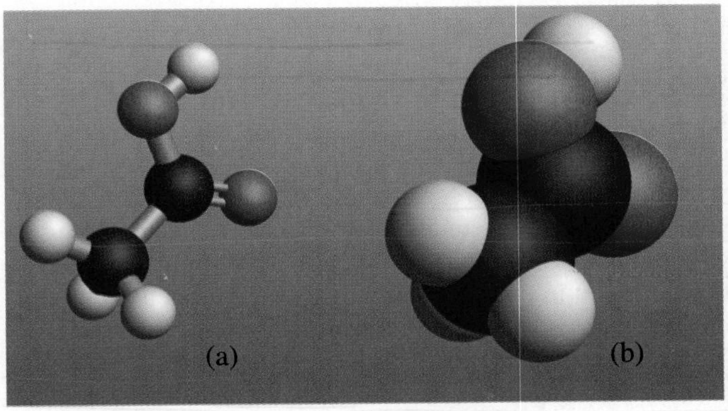

Figure 2.6 Two types of molecular models. The ball-and-stick model of acetic acid (a) and the space-filling model (b) conform to the structural formula of the acetic acid molecule

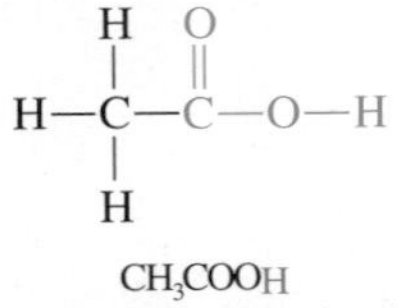
CH_3COOH
Acetic acid or ethanoic acid

but are more informative about its three-dimensional shape.

Naming Binary Molecular Compounds

A *binary* molecular compound is composed of two elements. A number of these compounds have widely used common names, such as water, ammonia, and methane. For others we use the prefixes *mono-*, *di-*, *tri-*, *tetra-*, and so on, to indicate the number of atoms of each element in the molecule. To name the two elements, leave the first element name unchanged, but give the second element name an *-ide* ending. For example, P_4S_3 is called *tetra*phosphorus *tri*sulf*ide*. Prefixes for up to ten atoms and examples of their use are shown in Table 2.3. Note that in these examples the prefix *mono-* is treated in a special way. We do not use

TABLE 2.3 **Naming Binary Molecular Compounds**

Number of Atoms	Prefix	Examples[a]	
1	mono	NO	nitrogen monoxide
2	di	NO_2	nitrogen dioxide
3	tri	N_2O_3	dinitrogen trioxide
4	tetra	N_2O_4	dinitrogen tetroxide
5	penta	N_2O_5	dinitrogen pentoxide
6	hexa	SF_6	sulfur hexafluoride
7	hepta	S_2O_7	disulfur heptoxide
8	octa		
9	nona		
10	deca		

[a] When the prefix ends in "a" or "o" and the element name begins with "a" or "o," the final vowel of the prefix is dropped for ease of pronunciation. For example, nitrogen *mon*oxide and not nitrogen *monoo*xide, and dinitrogen *tetr*oxide, not dinitrogen *tetra*oxide. However, PI_3 is phosphorus *tri*iodide, not phosphorus *tr*iodide.

it for the first-named element, but we do for the second, as in the name carbon monoxide for CO, *not* monocarbon monoxide.

Example 2.3

What is the name for SCl_2? for BF_3?

Solution

SCl_2 has one sulfur atom and two chlorine atoms; it is sulfur dichloride. BF_3 has one boron atom and three fluorine atoms; it is boron trifluoride.

Example 2.4

Give the formula for carbon tetrachloride.

Solution

The name indicates a molecule with one carbon atom and four chlorine atoms. The formula is CCl_4.

Exercise 2.3

What is the name for S_4N_2?

Exercise 2.4

Give the formula for tetraphosphorus decoxide.

ionic compound - metals form with non-metals.

2.8 Ions and Ionic Substances

Most molecular compounds, and the molecules of which they are composed, are formed by a combination of two or more nonmetals. When metal atoms combine with nonmetal atoms, they usually form an *ionic* compound. As we'll see in this section, the smallest characteristic entity of an ionic compound is not a molecule but a *formula unit* made up of *ions*.

An isolated atom has an equal number of protons and electrons and is electrically neutral. But in some chemical reactions, an atom or a grouping of bonded atoms may lose or gain one or more electrons, acquire a net electric charge, and become an **ion**. If electrons are *lost*, there is an excess of protons over electrons and the ion has a *positive* charge. If electrons are *gained*, there is an excess of electrons over protons and the ion has a *negative* charge. *Simple* ions are formed when a single atom loses or gains one or more electrons. *Polyatomic* ions are formed when the loss or gain of electrons occurs in a grouping of bonded atoms. Positively charged ions are called **cations** and negatively charged ions, **anions**. In general, metal atoms form cations and nonmetal atoms form anions.

Simple Ions

To some extent we can use the periodic table to predict the charges on certain simple ions (Figure 2.7). For metal atoms in the A groups, in most cases the number of electrons given up is equal to the periodic table group number. Thus, the metal atoms of Group 1A give up one electron to form cations with charges of 1+; Group 2A atoms give up two electrons to form cations with 2+ charges; and aluminum, the most important metal in Group 3A, forms cations with 3+ charges.

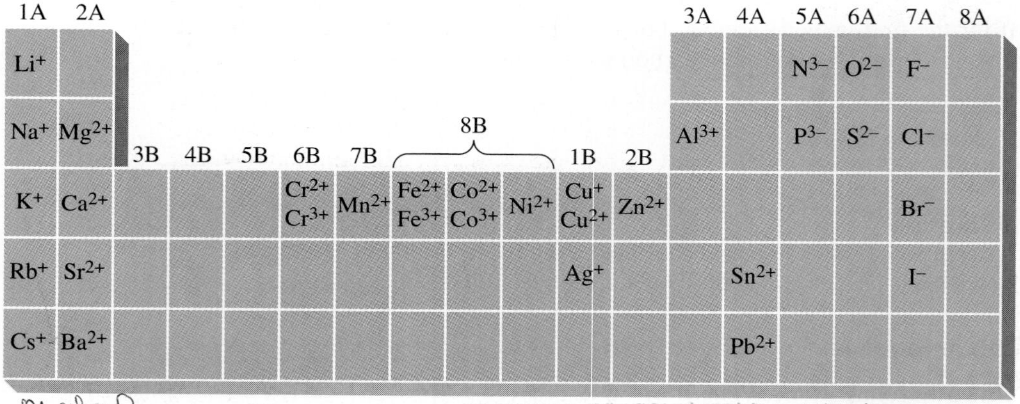

Figure 2.7 Symbols and periodic table locations of some common, simple ions. The word "ion" is part of each name; for example, Li^+ is lithium ion and Cl^- is chloride ion. In general, (a) the metals of Groups 1A and 2A and aluminum have just one cation, which carries a positive charge equal in magnitude to the A-group number; (b) the metals of the Bgroups have two or more cations of different charges, though in some cases only one of these is commonly encountered; and (c) the nonmetals of Groups 7A and 6A, nitrogen, and phosphorus form anions with a charge equal to "the group number minus eight."

The names of simple cations are derived from those of the parent elements by adding the word *ion*: lithium ion, sodium ion, magnesium ion, and so on. The charge is written as an Arabic numeral followed by a "+" sign. For a 1+ charge, however, the 1 is usually omitted and a simple + sign is used. We write Mg^{2+} and Al^{3+}, but we write Na^+ rather than Na^{1+}.

When combining with metal atoms, nonmetal atoms generally gain electrons to form anions with a charge equal to "the periodic table group number minus eight." The nonmetal atoms of Group 7A gain one electron to form anions with

What Is a Low-Sodium Diet?

Some familiar foods with a high Na^+ content.

People with high blood pressure are usually advised to follow a low-sodium diet. Just what does that mean? Surely they are not being advised to reduce their consumption of sodium itself. Sodium is an extremely reactive metal that reacts violently with moisture; eating it wouldn't be safe. The concern is really with sodium *ions*, Na^+. The average adult in the United States consumes too much Na^+, most of it from sodium chloride, common table salt. It is not uncommon for some individuals to eat 6 or 7 g of sodium chloride a day, most of it in prepared foods. Many snack foods, such as potato chips, pretzels, and corn chips are especially high in salt. The American Heart Association recommends that adults limit their salt intake to about 3 g per day.

We cannot overemphasize the difference between ions and the atoms from which they are made. A metal atom and its cation are as different as a whole peach (atom) and a peach pit (ion). The names and symbols look a lot alike, but the species themselves are quite different. Unfortunately, the situation is confused when people talk about diets with too much "sodium" or of taking "calcium" for healthy teeth and bones. They really mean sodium *ions* and calcium *ions*. Scientists try to be more precise in their terminology.

the charge 7 − 8 = 1−, such as F^- and Cl^-. Group 6A atoms gain two electrons to form anions such as O^{2-} and S^{2-}. Nitrogen, in Group 5A, forms the anion N^{3-}. The names of simple anions are derived from those of their parent elements by using an *-ide* ending and adding the word *ion.* When a chlor*ine* atom gains one electron, it becomes a chlor*ide* ion. An ox*ygen* atom, by gaining two electrons, becomes an ox*ide* ion.

There is no simple way to use the periodic table to determine the most likely charge on ions formed by B-group elements. In a few cases the magnitude of the charge is equal to the group number, but in most cases it is not. Moreover, you will notice from Figure 2.7 that B-group elements may form ions with different charges, such as Fe^{2+} and Fe^{3+}. In such cases chemists use Roman numerals to indicate the charge. Thus, the ions *iron(II)* and *iron(III)* are Fe^{2+} and Fe^{3+}, respectively.

An older system of naming ions is occasionally used, based on the Latin stems, and "ous" and "ic" endings are used. The "ous" indicates the lower of two possible charges and "ic," the higher charge: Fe^{2+} is *ferrous* ion and Fe^{3+} is *ferric* ion.

Formulas and Names for Binary Ionic Compounds

Binary (two-element) ionic compounds are made up of simple cations and simple anions. These combinations of ions must be electrically neutral; their net charge can be neither positive nor negative. This requirement gives us a basis for writing formulas of ionic compounds.

Consider the formula for aluminum oxide. The symbols of the ions are Al^{3+} and O^{2-}. We cannot represent the compound just by combining the symbols for the ions; we would get an entity with a net positive charge: AlO^{+}. The simplest combination of ions that produces an electrically neutral unit, called a **formula unit** of the compound, must have *two* Al^{3+} ions and *three* O^{2-} ions, giving the formula Al_2O_3. Note that in this unit the net charge is

$$2(3+) + 3(2-) = +6 - 6 = 0$$

Perhaps you are wondering why we called Al_2O_3 aluminum oxide instead of *di*aluminum *tri*oxide? The answer is that we don't use prefixes if we don't need them, and we don't need them for ionic compounds. If we know the charges on the cations and anions, we can always figure out their correct proportions to produce an electrically neutral formula unit. Then we name the formula unit in terms of the cations and anions present, regardless of their relative numbers. You should generally have no difficulty in writing names and formulas of ionic compounds.

Example 2.5

Determine the formulas for (**a**) calcium chloride and (**b**) magnesium oxide.

Solution

a. First write the symbols for the ions in the order cation followed by anion: Ca^{2+} and Cl^{-}. The simplest combination of these ions that will give an electrically neutral formula unit is *one* Ca^{2+} ion for every *two* Cl^{-} ions. The formula is $CaCl_2$.

$$Ca^{2+} + 2\ Cl^{-} = CaCl_2$$

b. The ions are Mg^{2+} and O^{2-}. The simplest ratio for an electrically neutral formula unit is 1:1. The formula of the compound is MgO.

$$Mg^{2+} + O^{2-} = MgO$$

Exercise 2.5

Give the formula for each of the following ionic compounds.

a. aluminum fluoride **b.** potassium sulfide
c. calcium nitride **d.** lithium oxide

Example 2.6

What are the names of (**a**) MgS and (**b**) $FeCl_3$?

Solution

a. MgS is made up of Mg^{2+} and S^{2-} ions. Its name is magnesium sulfide.
b. The ions are Fe^{3+} and Cl^{-}. How do we know that the iron is present as Fe^{3+} and not Fe^{2+}? Since there are three Cl^{-} ions, the cation must have a charge of 3+, that is, $3 + 3(1-) = 3 - 3 = 0$. The compound is iron(III) chloride.

Exercise 2.6

Give names for the following compounds.

a. $CaBr_2$ **b.** Li_2S **c.** $FeBr_2$ **d.** CuI

Polyatomic Ions

Atoms and simple ions remain unchanged in chemical reactions (although an atom may be converted into an ion or vice versa). This is not always true of polyatomic ions, but often it is the case. For example, in the reaction of water solutions of NaOH and $MgCl_2$, the products are a water solution of NaCl and solid $Mg(OH)_2$. The simple ions Na^+, Mg^{2+}, and Cl^-, and the polyatomic anion OH^- all emerge unchanged. All that occurs is that Mg^{2+} exchanges partners: OH^- replaces Cl^-.

Table 2.4 gives a list of some of the more common polyatomic ions. Notice that polyatomic anions are more common than cations, that the suffixes *-ite* and *-ate* are frequently used, and that the prefixes *hypo-* and *per-* are occasionally used. The *-ite* anion has one less oxygen than the *-ate* anion. Where more than two polyatomic anions of the same elements exist, the prefix *hypo-* represents one less oxygen atom than the *-ite* anion, and the prefix *per-* represents one more oxygen atom than the *-ate* anion. In summary, the scheme used is

Increasing number of oxygen atoms ⟶

hypo___ite ___ite ___ate per___ate

For example, the oxo anions of chlorine are: ClO^-, hypochlorite ion; ClO_2^-, chlorite ion; ClO_3^-, chlorate ion; and ClO_4^-, perchlorate ion.

If a polyatomic anion has hydrogen as a third element, its presence is indicated in the name: HPO_4^{2-} is the *hydrogen* phosphate ion and $H_2PO_4^-$ is the *dihydrogen* phosphate ion.

You can write formulas and names for compounds containing polyatomic ions by combining information from Figure 2.7 and Table 2.4 on page 48.

Example 2.7

What are the formulas of the following?

a. sodium sulfite **b.** ammonium sulfate

Solution

a. Notice that you can identify the sulfite ion without memorizing all the ions in Table 2.4. If you remember the name and formula of one of the sulfur–oxygen polyatomic anions, you can deduce the names of others. Suppose you remember that sulf*ate* is SO_4^{2-}. The *-ite* anion has one less oxygen, 3 instead of 4; it is SO_3^{2-}. A formula unit of sodium sulfite has Na^+ and SO_3^{2-} in the ratio 2:1. The formula is Na_2SO_3.

b. Ammonium ion is NH_4^+ and sulfate ion is SO_4^{2-}. A formula unit has two NH_4^+ ions and one SO_4^{2-}. To represent the two NH_4^+ ions we need to use parentheses, that is, $(NH_4)_2$. The formula of ammonium sulfate is $(NH_4)_2SO_4$.

Example 2.8

What are the names for the following compounds?

a. NaCN **b.** KH_2PO_4

Solution

a. The ions are Na^+ and CN^-. The name is sodium cyanide.

b. The ions are K^+ and $H_2PO_4^-$. The name is potassium dihydrogen phosphate.

Exercise 2.7

Give names or formulas of the following compounds.

a. ammonium nitrate **b.** $FePO_4$ **c.** sodium hypochlorite **d.** $KHCO_3$

TABLE 2.4 **Some Common Polyatomic Ions**

Name	Formula	Typical Compound
Cations		
Ammonium ion	NH_4^+	NH_4Cl
Hydronium ion	H_3O^+	[a]
Anions		
Acetate ion	[b]$C_2H_3O_2^-$	$NaC_2H_3O_2$
Carbonate ion	CO_3^{2-}	Li_2CO_3
Hydrogen carbonate ion (or bicarbonate ion)[c]	HCO_3^-	$NaHCO_3$
Hypochlorite ion	ClO^-	$Ca(ClO)_2$
Chlorite ion	ClO_2^-	$NaClO_2$
Chlorate ion	ClO_3^-	$NaClO_3$
Perchlorate ion	ClO_4^-	$Mg(ClO_4)_2$
Chromate ion	CrO_4^{2-}	K_2CrO_4
Dichromate ion	$Cr_2O_7^{2-}$	$(NH_4)_2Cr_2O_7$
Cyanate ion	OCN^-	KOCN
Thiocyanate ion[d]	SCN^-	KSCN
Cyanide ion	CN^-	KCN
Hydroxide ion	OH^-	NaOH
Nitrite ion	NO_2^-	$NaNO_2$
Nitrate ion	NO_3^-	$NaNO_3$
Oxalate ion	$C_2O_4^{2-}$	CaC_2O_4
Permanganate ion	MnO_4^-	$KMnO_4$
Phosphate ion	PO_4^{3-}	Na_3PO_4
Hydrogen phosphate ion	HPO_4^{2-}	Na_2HPO_4
Dihydrogen phosphate ion	$H_2PO_4^-$	NaH_2PO_4
Sulfite ion	SO_3^{2-}	Na_2SO_3
Hydrogen sulfite ion (or bisulfite ion)[c]	HSO_3^-	$NaHSO_3$
Sulfate ion	SO_4^{2-}	Na_2SO_4
Hydrogen sulfate ion (or bisulfate ion)[c]	HSO_4^-	$NaHSO_4$
Thiosulfate ion	$S_2O_3^{2-}$	$Na_2S_2O_3$

[a] In water solution, H^+ associates itself with water molecules and is generally represented as H_3O^+. There are no common compounds containing H_3O^+.
[b] The acetate ion is also represented as CH_3COO^-.
[c] The prefix "bi" means that the ion contains a replaceable H atom. This should not be confused with the prefix "di," which means two (used to represent a doubling of a simpler unit).
[d] The prefix "thio" means that a sulfur atom has replaced an oxygen atom.

Hydrates

If you scan the labels in a chemical storeroom, you may find some that don't match what we have written. For example, a bottle of calcium chloride may carry the label $CaCl_2 \cdot 6H_2O$ instead of $CaCl_2$. This label indicates that the substance is a **hydrate**, called calcium chloride *hexa*hydrate. In this substance each $CaCl_2$ formula unit has with it *six* water molecules, as well as one Ca^{2+} ion and two Cl^- ions. A few other examples are barium chloride *di*hydrate, $BaCl_2 \cdot 2H_2O$, lithium perchlorate *tri*hydrate, $LiClO_4 \cdot 3H_2O$, and magnesium carbonate *penta*hydrate, $MgCO_3 \cdot 5H_2O$.

A hydrate may form from an anhydrous (nonhydrated) material by exposure to atmospheric water vapor or, more commonly, when the ionic compound is crystallized from a water solution. Although hydrates are common, many ionic compounds do not form them, and there is no need to try to learn which ones do. You only need to recognize a hydrate formula when you see one.

Copper sulfate pentahydrate, $CuSO_4 \cdot 5H_2O$

2.9 Acids, Bases, and Salts

All of us practice a complicated chemistry, from digesting food and shedding tears to taking an antacid or baking bread. Central to much of this chemistry are two special kinds of compounds called acids and bases. We eat them and drink them and our bodies produce them. Some common acids and bases used around the home are shown in Figure 2.8.

(a) (b)

Figure 2.8 Some common acids and bases. (a) Some common acids: toilet bowl cleaner, vinegar, aspirin, tomato and fruit juices, and so on. (b) Some common bases: drain cleaners, oven cleaner, quinine water, and so on.

Historically, acids and bases have been classified according to some distinctive properties. *Acids* are substances that, when dissolved in water,

- Taste sour, if diluted with enough water to be tasted safely.
- Produce a prickling or stinging sensation on the skin.
- Turn the color of the indicator dye litmus from blue to red.
- React with many metals, such as magnesium, zinc, and iron, to produce ionic compounds and hydrogen gas.
- React with bases, thereby losing their acidic properties.

Bases are substances that, when dissolved in water,

- Taste bitter, if diluted with enough water to be tasted safely.*
- Feel slippery or soapy on the skin.
- Turn the color of the indicator dye litmus from red to blue.
- React with acids, thereby losing their basic properties.

Acids and Bases: The Arrhenius Concept

In 1887, the Swedish chemist Svante Arrhenius proposed that an **acid** is a molecular substance that breaks up in water solution into H^+ ions and anions. The acid is said to ionize. He viewed a **base** as a substance that produces OH^- in water solution, either because it contains OH^-, such as an ionic hydroxide, or because it can ionize to produce OH^-. He proposed that the essential reaction between an acid and a base, *neutralization*, is the combination of H^+ and OH^- to form water. The cation originally associated with the OH^- and the anion associated with the H^+ give rise to an ionic compound, a **salt**.

Like many scientific theories, the Arrhenius theory has been supplanted by a better one based on newer data. For example, we now know that simple hydrogen ions, H^+, do not exist in water solutions. We study this more modern acid–base theory in Chapter 17, but the Arrhenius theory will help us to identify acids and bases and to write names and formulas. That's all we will attempt for now.

Naming Acids and Bases

We can use the following scheme to relate the names of acids and the anions produced when the acids ionize.

Acid name		Anion name
hydro___ ic acid	⟶	___ide
hypo___ ous acid	⟶	hypo ___ ite
___ ous acid	⟶	___ ite
___ ic acid	⟶	___ ate
per___ ic acid	⟶	per___ ate

The number of H atoms in the molecular acid is the same as the number of H^+ ions required to balance the electric charge of the anion. That is, one H atom in HCl, two H atoms in H_2SO_4, and three H atoms in H_3PO_4. The scheme for naming acids is illustrated by the examples in Table 2.5.

Foremost among the ionic bases are those that contain the Group 1A and 2A cations. They include enough OH^- ions to balance the charge of the accompanying cation, and they are named like other ionic compounds:

$$NaOH = \text{sodium hydroxide}$$
$$KOH = \text{potassium hydroxide}$$
$$Ca(OH)_2 = \text{calcium hydroxide}$$

*According to the biblical account, the Israelites, in their journey from Egypt to Canaan, came upon the bitter waters at Marah (Exod. 15:23). Although the writers may not have meant to, they thus recorded for posterity the existence of a base.

TABLE 2.5 **Names of Some Common Acids and Their Salts**

Formula of Acid	Name of Acid	Sodium Salt Formula	Sodium Salt Name
HCl	*hydro*chlor*ic* acid	$NaCl$	sodium chlor*ide*
$HClO$	*hypo*chlor*ous* acid	$NaClO$	sodium *hypo*chlor*ite*
$HClO_2$	chlor*ous* acid	$NaClO_2$	sodium chlor*ite*
$HClO_3$	chlor*ic* acid	$NaClO_3$	sodium chlor*ate*
$HClO_4$	*per*chlor*ic* acid	$NaClO_4$	sodium *per*chlor*ate*
HNO_2	nitr*ous* acid	$NaNO_2$	sodium nitr*ite*
HNO_3	nitr*ic* acid	$NaNO_3$	sodium nitr*ate*
$HC_2H_3O_2$[a]	acet*ic* acid	$NaC_2H_3O_2$	sodium acet*ate*
H_2S	*hydro*sulfur*ic* acid	Na_2S	sodium sulf*ide*
H_2SO_3[b]	sulfur*ous* acid	Na_2SO_3	sodium sulf*ite*
H_2SO_4[b]	sulfur*ic* acid	Na_2SO_4	sodium sulf*ate*
H_3PO_4[b]	phosphor*ic* acid	Na_3PO_4	sodium phosph*ate*

[a] Acetic acid is also represented as CH_3COOH (see Section 2.11) and sodium acetate as CH_3COONa.
[b] Table 2.4 also lists salts of these acids in which only one of the H atoms is replaced by Na (one or two in the case of H_3PO_4).

Most bases are molecular substances and do not contain hydroxide ion. They produce it by reacting with water. Molecular bases are discussed in Chapters 12 and 17, and the only one we will encounter until then is ammonia, NH_3.

2.10 Organic Compounds

In this section and the next we introduce a category of compounds that constitute over 95% of the 12 million or so known compounds. All of these compounds, called *organic* compounds, contain carbon atoms; almost all contain hydrogen atoms; and many also contain other atoms, such as O, N, and S. Most organic compounds are molecular; only a few are ionic. Some also fit into other categories, such as acid, base, and salt.

Roughly speaking, *inorganic* compounds are those containing elements other than carbon.

The distinctive feature of organic compounds is that the carbon atoms can join together into chains or rings to form a backbone to which other atoms are attached. This makes possible an almost limitless number of different structures and, therefore, different compounds. Writing names and formulas of organic compounds is a challenging task that we will undertake only to a limited extent. Often, we can name an organic compound in two ways: through a common or trivial name or with a systematic name. For example, glycerol and glycerine are two common (trivial) names for a byproduct of soap manufacture with the formula $C_3H_8O_3$ that is used in lotions and cosmetic creams. Figure 2.9 shows that the systematic name of this glycerol molecule is related to its structure. In our discussion of organic compounds we'll use mostly common or trivial names, but you will find a brief survey of the systematic nomenclature of organic chemistry in Appendix C.

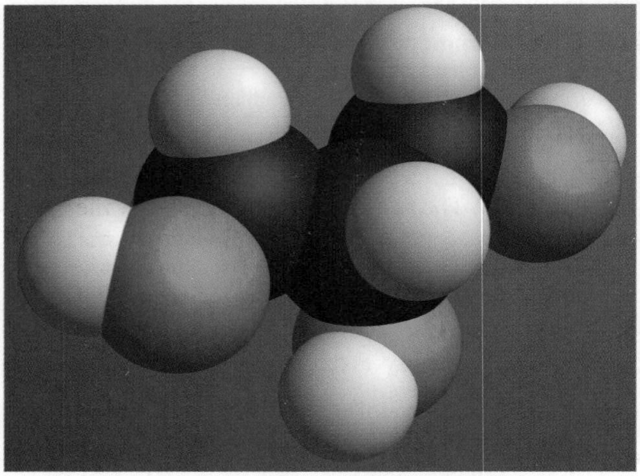

Figure 2.9 The glycerol molecule. The systematic name of glycerol is 1,2,3-propanetriol. In relation to the molecular model, *propane* signifies a chain of three carbon atoms. The ending *-ol* indicates the presence of —OH groups. *Triol* means that there are three of these groups, and the numbers "1,2,3" signify that they are attached to the first, second, and third carbon atoms of propane. The system of nomenclature that leads to this name is outlined in Appendix C. (Note that two of the H atoms are "hidden" in this model.)

Alkanes: Saturated Hydrocarbons

The simplest organic compounds contain only carbon and hydrogen atoms; they are *hydrocarbons*. There are several types of hydrocarbons, and in this section we'll discuss the kind known as alkanes. The **alkanes** are said to be *saturated* hydrocarbons because their molecules contain the maximum number of hydrogen atoms for the carbon atoms present; the molecules are "saturated" with H atoms.

The simplest alkane is methane, the principal component of natural gas. In methane four hydrogen atoms are attached to a central carbon atom.

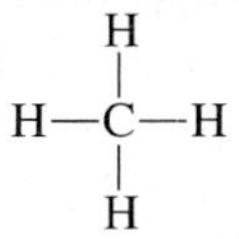

Methane (CH_4)

The next member of the series is ethane, a minor component of natural gas. In this molecule two carbon atoms are joined together and three hydrogen atoms are attached to each carbon atom.

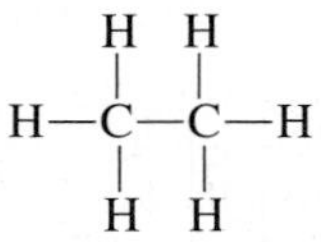

Ethane (C_2H_6)

Propane, the familiar "bottled gas" used as a fuel in portable torches, gas grills, and stoves, is the third member of the series.

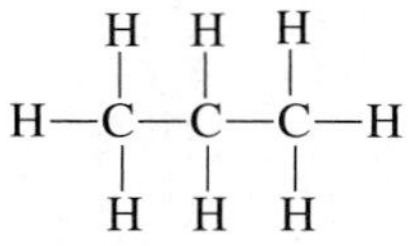

Propane (C_3H_8)

Perhaps the pattern is becoming apparent: Each member of the alkane series differs from the preceding member by a $—CH_2$ unit, one carbon atom and two hydrogen atoms. The general formula for an alkane is

$$C_nH_{2n+2}$$

The first four alkanes are given trivial names and the rest are named in the systematic fashion outlined in Table 2.6. The *-ane* ending indicates that the hydrocarbon belongs to the alk*ane* family, and the prefixes indicate the number of carbon atoms in the chain. Thus, C_5H_{12} is *pent*ane and C_6H_{14} is *hex*ane.

Another pattern in the alkane series is that each carbon atom always forms *four* bonds and each hydrogen atom, *one* bond. By counting the number of bonds to each C and H atom we can quickly decide whether a given structural formula is plausible.

When we attempt to write the structural formula of the fourth member of the alkane series we find that there are *two* possible structures with the formula C_4H_{10}.

TABLE 2.6
Word Stems Indicating the Number of Carbon Atoms in Organic Molecules

Stem	Number of C Atoms
meth-	1
eth-	2
prop-	3
but-	4
pent-	5
hex-	6
hept-	7
oct-	8
non-	9
dec-	10

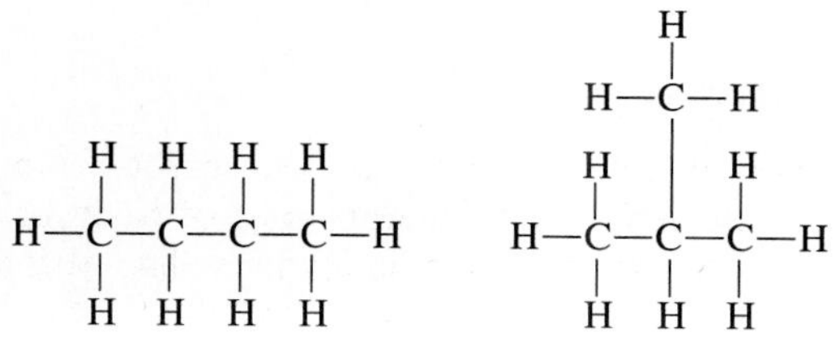

Butane (C_4H_{10}) Isobutane (C_4H_{10})

Compounds with the same molecular formula but different structural formulas are called **isomers.** Isomers are distinctly different compounds. Although both butane and isobutane are gases at room temperature, liquid butane boils at about 0 °C, whereas the boiling point of isobutane is –12 °C. Differences in structure of isomers are brought out clearly by models (Figure 2.10), but we can also suggest these differences on paper. We can use the structural formulas shown above, or we can use *condensed* structural formulas. These are formulas, often written on a single line, that bring out the main structural features of molecules.

$CH_3CH_2CH_2CH_3$ (Butane)

$CH_3CH(CH_3)CH_3$, drawn with the CH_3 above the central C, or $CH_3CH(CH_3)CH_3$ (Isobutane)

The number of isomers increases rapidly with the number of carbon atoms: three pentanes, five hexanes, nine heptanes, eighteen octanes, and so on.

There are, for example, more than 4 billion possible isomers of $C_{30}H_{62}$, very few of which have been isolated or synthesized.

A series of compounds whose formulas and structures vary in a regular manner also have properties that vary in a predictable manner, a principle called *homology*. For example, the boiling points of the straight-chain alkanes increase by about 30 °C per $—CH_2$ group: butane, –0.5 °C; pentane, 36.1 °C; hexane, 68.7 °C, ... Homology helps us in our study of organic chemistry in much the same way that the periodic table provides an organizing principle for the chemistry of the elements. Instead of studying the properties of a bewildering array of individual organic compounds, organic chemists study a few members of a homologous series and deduce the properties of the others.

Figure 2.10 Ball-and-stick models. Butane (left); isobutane (right).

Estimation Exercise 2.1

Estimate the boiling point of the gasoline hydrocarbon *octane*.

Counting Isomers

At first glance it may be difficult to tell whether structures represent isomers or are merely different representations of the same compound. Consider the question of how many different structures one can write for the alkane C_5H_{12}. The following structures, in which H atoms are omitted for simplicity, are all the same. (Think of the carbon atoms as being hinged.)

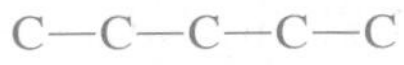
(1) Pentane

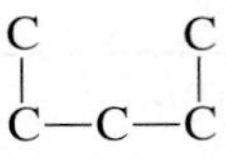
(1a)

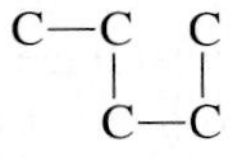
(1b)

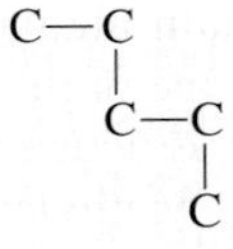
(1c)

Now consider some possibilities that have four carbon atoms in a chain, with a one-carbon branch.

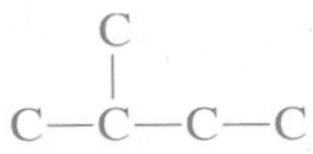
(2) Isopentane

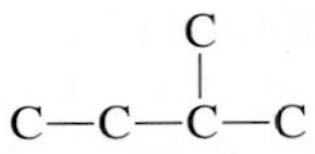
(2a)

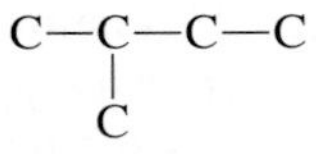
(2b)

These structures are also identical. Structure (2a), if flopped from left to right, and (2b), if turned top to bottom, are both the same as (2).

Finally, consider a three-carbon chain with two one-carbon branches. There is only one possibility.

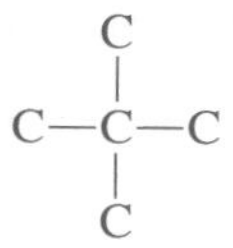

(3) Neopentane

Other structures that we might consider will be identical to one of the three structures shown in color. There are only three isomeric C_5H_{12} alkanes.

Although the prefixes iso and neo (new) enable us to name the C_5H_{12} isomers, we need a more systematic approach when large numbers of isomers are possible. This system is outlined in Appendix C.

Exercise 2.8

Draw structures of the *five* C_6H_{14} alkanes. (*Hint*: First draw a straight-chain structure, then a structure(s) with a one-carbon branch, and so on.)

Ring structures in alkanes are indicated by the prefix *cyclo*. The smallest number of carbon atoms in an alkane ring structure is three, cyclopropane. Another common cyclic alkane is one based on a six-carbon-atom ring and called cyclohexane. Of the two representations shown below, the second is much preferred; it's easier to write.

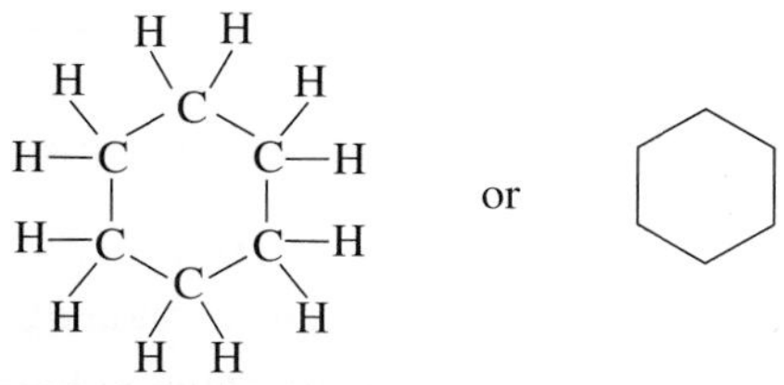

Cyclohexane

The generic formula for straight- and branched-chain alkanes is C_nH_{2n+2}, but for alkane ring structures it is C_nH_{2n}. Think of the carbon atoms at each end of a straight-chain alkane joining together to form a ring. Each of these C atoms must shed one of its H atoms in order to do this.

Cyclic structures can also have other atoms or groupings of atoms attached to the ring. Methylcyclopropane consists of a three-carbon ring, with a $-CH_3$ (methyl) group attached.

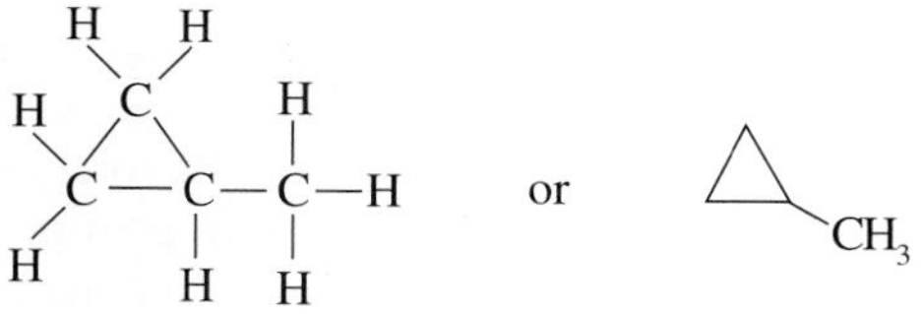

Methylcyclopropane

CONCEPTUAL EXAMPLE 2.1

Are the following pairs of molecules isomers or not? Explain.

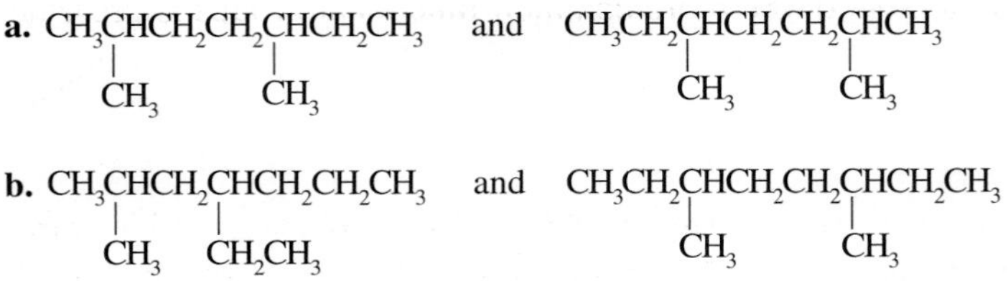

Solution

a. The two molecules have the same molecular formula, C_9H_{20}. However, if the second structure is flopped from left to right, we can see that it is identical to the first. Each molecule has a CH_3 group on both the second and fifth atoms of the carbon chain, starting from the end nearest the first branch. The structures do not represent isomers; they are two representations of the same substance.

b. The two molecules have the same molecular formula, $C_{10}H_{22}$. The first has seven C atoms in a chain and two branches, one with one C atom and the other with two C atoms. The second molecule has eight C atoms in a chain and two branches. However, each branch has only one C atom. The two molecules have the same molecular formula ($C_{10}H_{22}$) but different structures; they are isomers.

Conceptual Exercise 2.1

Are the following pairs of molecules isomers or not? Explain.

a. $CH_3CH_2CH_2CH_2CH(CH_3)_2$ and

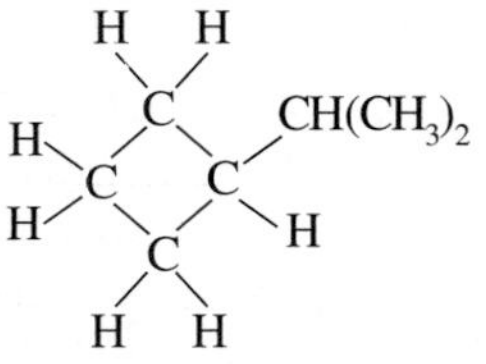

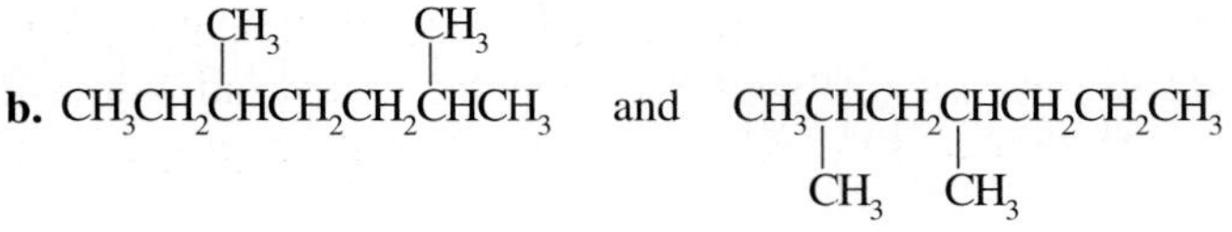

2.11 Organic Functional Groups

In addition to homology, another important organizing principle in organic chemistry is that of functional groups. A **functional group** is an atom or group of atoms attached to a hydrocarbon chain that confers characteristic properties to the molecule as a whole. Simple organic molecules are often composed of two parts: a functional group where most of the reactions of the molecule occur, and a hydrocarbon chain that is relatively unreactive. Molecules having the same functional group generally have similar properties. We consider two functional groups here, and from time to time in the text we will introduce others. For ready reference, a table of the more common functional groups is presented in Appendix C.

Alcohols

The functional group common to **alcohols** is the *hydroxyl* group, —OH. The simplest alcohol has the —OH group substituted for one of the H atoms in methane.

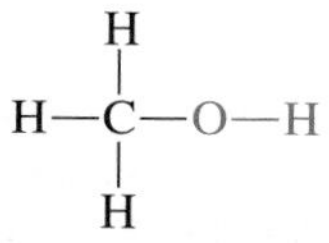

CH_3OH
Methyl alcohol or methanol

Both names suggest a relationship to methane. In the common name, methyl alcohol, the word *alcohol* indicates the family name. In the systematic name, methanol, the ending *-ol* indicates that the compound is an alcoh*ol.*

Methanol is widely used as a solvent, and because it burns more cleanly than gasoline, methanol is used in some fleet vehicles in smoggy areas. Methanol is fairly toxic. It causes blindness and even death when as little as 30 mL is ingested.

The next higher alcohol is based on the two-carbon alkane, ethane.

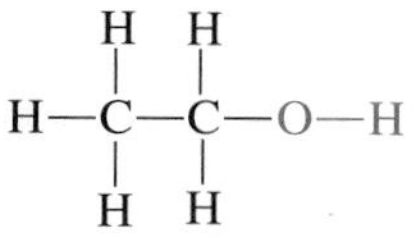

CH_3CH_2OH
Ethyl alcohol or ethanol

Ethanol is the alcohol found in alcoholic beverages. It is considerably less toxic than methanol. (The lethal dose of ethanol, when rapidly ingested, is about 500 mL.) Like methanol, ethanol is widely used as a solvent. It is also used as a gasoline additive and, in some areas (Brazil, for example), as a gasoline substitute.

There are *two* different three-carbon alcohols, based on propane. They are isomers.

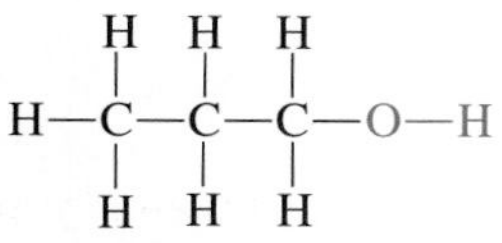

$CH_3CH_2CH_2OH$
Propyl alcohol
1-Propanol

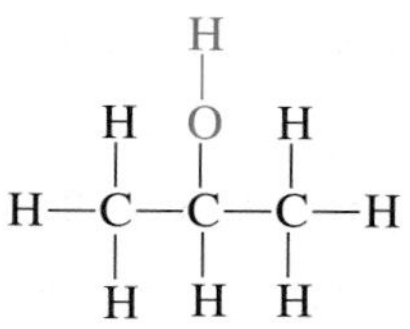

$CH_3CHOHCH_3$
Isopropyl alcohol
2-Propanol

Alkanes: Our Principal Fuels

The modern world runs mainly on alkanes. Petroleum is a complex mixture of hydrocarbons, mostly alkanes, but—in the form of a thick, sticky, smelly liquid—as it comes from the ground it is of limited use. To better suit our needs, petroleum is boiled in a distilling column and separated into portions, called *fractions*, having different boiling point ranges (see Figure 2.11). The lighter molecules, those with one to four carbon atoms, come off the top of the column. The next fraction contains, for the most part, molecules having from five to twelve carbon atoms. Molecules in succeeding fractions contain still more carbon atoms, until finally a residue is obtained in which molecules have twenty or more carbon atoms.

Natural gas is about 80% methane, 10% ethane, and 10% a mixture of higher alkanes. Natural gas is used as a cooking fuel (and in Bunsen burners). It is the cleanest of the fossil fuels because it contains the smallest quantity of sulfur compounds. Propane and the butanes are familiar fuels. Although they are gases at ordinary temperatures and pressures, they liquefy under high pressure. In liquid form they are known as liquefied petroleum gas (LPG). Gasoline, like the petroleum from which it is derived, is a mixture of hydrocarbons. The typical alkanes in gasoline have formulas ranging from C_5H_{12} to $C_{12}H_{26}$. There are also small amounts of other kinds of hydrocarbons present, and even some sulfur- and nitrogen-containing compounds.

Because gasoline is generally the petroleum fraction in greatest demand, the higher boiling fractions are often in excess supply. These can be converted to gasoline by heating them in the absence of air. This process, called *cracking*, breaks the big molecules apart. Not only does cracking convert higher alkanes into those in the gasoline range, but it affords a variety of useful byproduct chemicals. From these, chemists can synthesize a remarkable array of substances: plastics, painkillers, antibiotics, stimulants, depressants, and detergents, to name just a few. Future shortages of petroleum could mean a great deal more than just scarce, high-priced gasoline.

Catalytic cracking unit in a petroleum refinery. Catalytic cracking breaks down large hydrocarbon molecules into smaller molecules in the gasoline range, thereby increasing the yield of gasoline from crude oil.

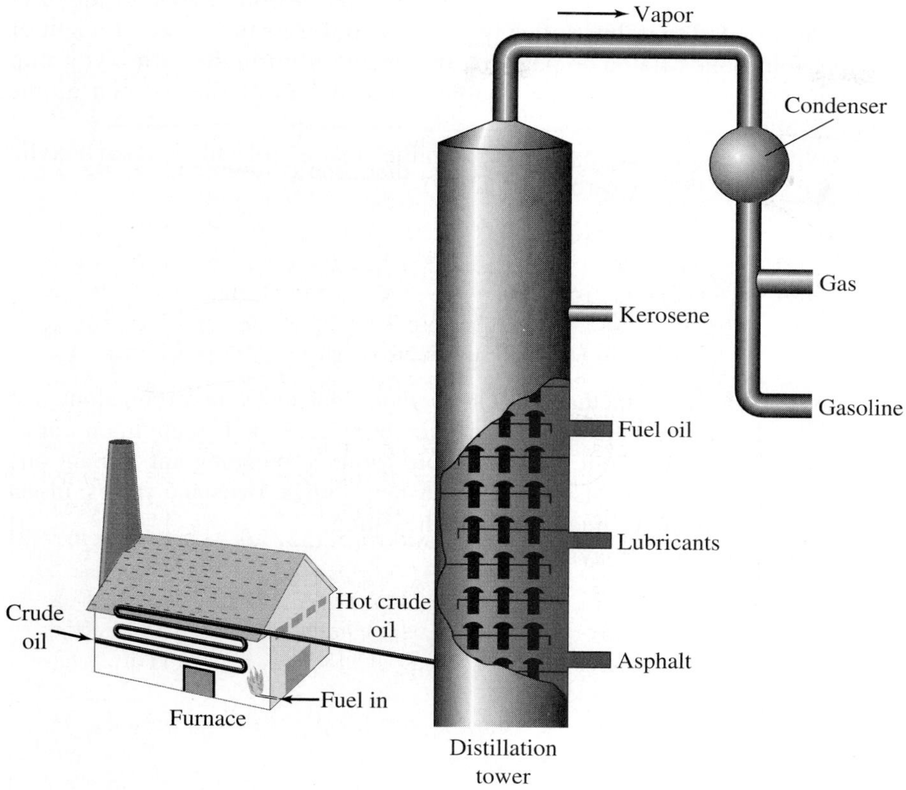

Figure 2.11 Petroleum distillation and gasoline production. Crude oil is vaporized and the fractional distillation column separates the components of the vapor according to their boiling points. The lower boiling components come off at the top of the column, and higher boiling components come off lower in the column. A nonvolatile residue collects at the bottom.

In the systematic names, the prefix "1-" indicates that the —OH group is on the first or end carbon atom, and the prefix "2-," that the —OH group is on the second carbon from the end. Rubbing alcohol is a water solution containing 70% isopropyl alcohol.

It is important to note that even though alcohols contain the group —OH, they are not bases in the Arrhenius sense. The hydroxyl group is not present as OH^- (as in an ionic base), nor is OH^- produced when an alcohol dissolves in water.

Carboxylic Acids

The functional group that most commonly confers acidic properties to an organic substance is the *carboxyl* group:

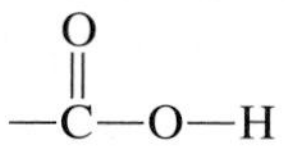

The carbon atom in a carboxyl group forms a *double* bond to one of the oxygen atoms bonded to it, but notice that it still conforms to the requirement of forming four bonds stated on page 53. In condensed form the carboxyl group is often written —COOH; the double bond to one of the oxygen atoms is understood.

The presence of a carboxyl group signifies that a molecule is a **carboxylic acid**. The simplest of the carboxylic acids is

It is the hydrogen of the carboxyl group that becomes an H^+ ion in water solution; hydrogen atoms attached to carbon atoms do *not* break off as H^+. Hydrogen atoms that come off as H^+ are called *acidic hydrogens* or *ionizable hydrogens.*

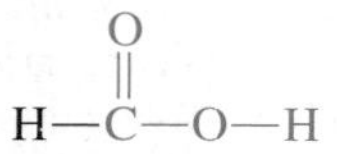

HCOOH

Formic acid or methanoic acid

In the systematic name, methanoic acid, *methan-* indicates one carbon atom, and *-oic acid* tells us that the compound is a carboxylic acid. The common name, formic acid, is derived from the Latin word *formica*, meaning ant. An ant bite stings because the ant injects formic acid when it bites. Bees and wasps, insect relatives of ants, inject formic acid when they sting.

The two-carbon carboxylic acid is

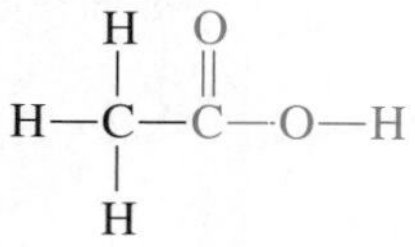

CH_3COOH

Acetic acid or ethanoic acid

We can distinguish between butyric acid and isobutyric acid in a rather unusual way—by odor. Butyric acid is one of the most foul-smelling substances known. The odor of isobutyric acid is not as unpleasant.

Acetic acid can be made by the fermentation of cider or wine, producing the familiar product vinegar, which is about 4 to 10% acetic acid. Acetic acid is probably the most frequently used organic acid in chemical laboratories.

The three-carbon acid is propionic (propanoic) acid, and there are two isomers of the four-carbon acid: butyric and isobutyric acid. As you might expect, the number of possible isomers goes up rapidly as the chain length increases.

SUMMARY

The basic laws of chemical combination are the laws of conservation of mass, definite proportions, and multiple proportions. Each played an important role in Dalton's development of the atomic theory.

The three components of atoms of most concern to chemists are protons, neutrons, and electrons. Protons and neutrons make up the nucleus, and their combined number is the mass number, *A*, of the atom. The number of protons is the atomic number, *Z*. Electrons, found outside the nucleus, neutralize the positive charge associated with the protons. All atoms of an element have the same atomic number but may have different mass numbers, giving rise to isotopes.

A chemical formula indicates the relative numbers of atoms of each type in a compound. An empirical formula is the simplest that can be written, and a molecular formula reflects the actual composition of a molecule. A structural formula describes the arrangement of atoms within a molecule. A system of naming simple binary molecular compounds uses prefixes to denote the numbers of atoms of each element and an *-ide* ending for the second named element.

Ions are formed by the loss or gain of electrons by single atoms or groups of atoms. Positive ions are known as cations and negative ions as anions. Formulas of ionic compounds are based on an electrically neutral combination of cations and anions called a formula unit. Assistance in formula writing is provided by the periodic table, an arrangement of the elements by atomic number that places elements with similar properties into vertical columns called groups or families. Some monatomic cations have Roman numerals in their names; monatomic anions have an *ide* ending; and among polyatomic anions the prefixes *hypo-* and *per-* and the endings *-ite* and *-ate* are commonly found.

Many compounds are acids, bases, or salts. By the simplest acid–base theory, the Arrhenius theory, an acid produces H^+ in water solution and a base produces OH^-. The products of a neutralization reaction between acid and base solutions are water and an ionic compound called a salt.

Among the vast number of organic compounds, some contain just the elements carbon and hydrogen. Alkane hydrocarbons have carbon atoms joined together by single bonds into chains or rings, and hydrogen atoms attached to the carbon atoms. Alkanes with more than three carbon atoms can exist as isomers, molecules having the same formula but different structures and properties.

Functional groups confer distinctive properties to an organic molecule when they are substituted for hydrogen atoms in a hydrocarbon. Alcohols feature the hydroxyl group, —OH, and carboxylic acids, the carboxyl group,

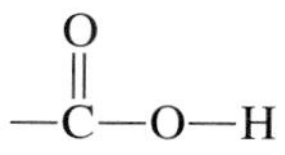

Appendix C contains a table of functional groups and a discussion of the nomenclature of organic compounds.

KEY TERMS

acid (2.9)
alcohol (2.11)
alkane (2.10)
anion (2.8)
atomic mass unit (2.4)
atomic number (Z) (2.3)
atomic weight (2.4)
base (2.9)
carboxylic acid (2.11)
cation (2.8)
electron (2.3)
empirical formula (2.7)
formula unit (2.8)
functional group (2.11)
hydrate (2.8)
ion (2.8)
isomer (2.10)
isotope (2.3)
law of conservation of mass (2.1)
law of definite proportions (2.1)
law of multiple proportions (2.2)
mass number (A) (2.3)
metal (2.5)
molecular formula (2.7)
molecule (2.7)
neutron (2.3)
nonmetal (2.5)
periodic table (2.5)
proton (2.3)
salt (2.9)
structural formula (2.7)

REVIEW QUESTIONS

1. A balloon filled with helium floats near the ceiling. After several days, the balloon is deflated and on the floor. Have the helium atoms been destroyed? If so, how? If not, where are they?

2. Sugar consists of carbon, hydrogen, and oxygen atoms. When a cube of sugar is burned in a crucible, nothing remains in the vessel. Have the carbon, hydrogen, and oxygen atoms of the sugar been destroyed? If so, how? If not, where are they?

3. Polychlorinated biphenyls (PCBs) are environmental contaminants composed of carbon, hydrogen, and chlorine atoms. A news report states that a new method of disposal of PCBs converts them completely to carbon dioxide and water. The necessary oxygen atoms come from oxygen in the air. Criticize the report in light of the law of conservation of mass.

4. Heptane is always composed of 84.0% carbon and 16.0% hydrogen by mass. What law does this observation illustrate?

5. When 3.00 g of carbon is burned in 8.00 g of oxygen, 11.00 g of carbon dioxide is formed. When 3.00 g of carbon is burned in 50.00 g of oxygen,
a. what is the total mass of substances present after the reaction?
b. what mass of carbon dioxide is formed?
c. what law(s) is (are) illustrated by this reaction?

6. Sulfur and oxygen form two compounds. The mass ratio of oxygen to sulfur in compound A is 1.0:1.0 and that in compound B is 1.5:1.0. What is the ratio of oxygen to sulfur in compound B as compared to that in compound A? Express the ratio in the smallest whole numbers. What law do these observations illustrate?

7. A photographic flash bulb weighing 0.750 g contains magnesium and air. The flash produces magnesium oxide. After flashing, the bulb weighs 0.750 g. What law does this observation illustrate?

8. Jan Baptista van Helmont (1579–1644), a Flemish alchemist, planted a young willow tree in a weighed bucket of soil. After five years he found that the tree had gained 75 kg in mass, yet the soil had lost only 0.057 kg. He had added only water to the system, so he concluded that the substance of the tree had come from water. Criticize his conclusion.

9. Outline the main points of Dalton's atomic theory.

10. To the nearest atomic mass unit, an atom of calcium has a mass of 40 u and an atom of vanadium has a mass of 50 u. Are these findings in agreement with Dalton's atomic theory? Explain your answer.

11. To the nearest atomic mass unit, an atom of calcium has a mass of 40 u and an atom of potassium has a mass of 40 u. Are these findings in agreement with Dalton's atomic theory? Explain your answer.

12. To the nearest atomic mass unit, one atom of calcium has a mass of 40 u and another calcium atom has a mass of 44 u. Are these findings in agreement with Dalton's atomic theory? Explain your answer.

13. Use Dalton's atomic theory, together with an illustrative example, to explain
a. the law of conservation of mass,
b. the law of definite proportions,
c. the law of multiple proportions.

14. Give the distinguishing characteristics of the proton, the neutron, and the electron.

15. What is the atomic nucleus? What subatomic particles are found in the nucleus?

16. What are isotopes, and what is meant by the mass number of an isotope?

17. Which of the following pairs of symbols represent isotopes?
a. ${}^{70}_{33}E$ and ${}^{70}_{34}E$ **b.** ${}^{57}_{28}E$ and ${}^{66}_{28}E$
c. ${}^{186}_{74}E$ and ${}^{186}_{74}E$ **d.** ${}^{7}_{3}E$ and ${}^{3}_{2}E$
e. ${}^{22}_{11}E$ and ${}^{11}_{6}E$

18. Use the symbolism ${}^{A}_{Z}E$ to represent each of the following atoms. You may refer to the periodic table.
a. boron-8 **b.** carbon-14
c. uranium-235 **d.** cobalt-60

19. Explain what tabulated atomic weight values, such as those found inside the front cover, represent.

20. List some characteristic properties of metals, and indicate where these elements are located in the periodic table.

21. Explain why a chemist calls the compound $MgCl_2$ magnesium chloride and not magnesium dichloride.

22. Describe two ways in which a chemist might distinguish between $FeCl_2$ and $FeCl_3$ through the names assigned to them. Which is the preferred way?

23. Of the names listed below, which refer to actual substances that you might find in containers on a storeroom shelf? What do the other names represent?
a. magnesium **b.** hydroxyl **c.** chloride
d. ammonia **e.** ammonium **f.** ethane

24. A substance has the molecular formula $C_4H_8O_2$.
a. What is the *empirical* formula of this substance?
b. What additional information about the substance would be revealed by its *structural* formula?
c. Can you write the structural formula from the molecular formula given? Explain.

25. According to the Arrhenius theory, all acids have one element in common. What is that element? Are all compounds containing that element acids? Explain.

26. Suggest some ways in which you might determine whether a particular water solution contains an acid or a base.

27. Can a substance whose molecular formula is C_3H_4 be an *alkane*? Explain.

28. Are hexane and cyclohexane isomers? Explain.

29. For which of the following types of compounds can you characterize the compound from the *molecular* formula alone. For which must you have the *structural* formula?

a. an organic compound
b. a hydrocarbon
c. an alcohol
d. an alkane
e. a carboxylic acid

30. Explain the difference in meaning between each pair of terms.

a. a group and a period of the periodic table
b. an ion and an ionic substance
c. an acid and a salt
d. an isomer and an isotope

PROBLEMS

Laws of Chemical Combination

31. A student heats 1.000 g of zinc with 0.200 g of sulfur. She obtains 0.608 g of zinc sulfide and recovers 0.592 g of unreacted zinc. Show by calculation whether her results can be used to confirm the law of conservation of mass.

32. A student heats 0.5585 g of iron with 0.3550 g of sulfur. He obtains 0.8792 g of iron(II) sulfide and recovers 0.0344 g of unreacted sulfur. Show by calculation whether his results can be used to confirm the law of conservation of mass.

33. A colorless liquid is thought to be a pure compound. Analyses of three samples of the material gives the following results.

	Mass of sample	Mass of carbon	Mass of hydrogen
Sample 1	1.000 g	0.625 g	0.0419 g
Sample 2	1.549 g	0.968 g	0.0649 g
Sample 3	0.988 g	0.618 g	0.0414 g

Could the material be a pure compound?

34. Azulene, a blue solid, is thought to be a pure compound. Analyses of three samples of the material give the following results.

	Mass of sample	Mass of carbon	Mass of hydrogen
Sample 1	1.000 g	0.937 g	0.0629 g
Sample 2	0.244 g	0.229 g	0.0153 g
Sample 3	0.100 g	0.094 g	0.0063 g

Could the material be a pure compound?

The Atomic Nucleus

35. The table below describes four atoms:

	Atom A	Atom B	Atom C	Atom D
Number of protons	10	11	11	10
Number of neutrons	11	10	11	10
Number of electrons	10	11	11	10

Are atoms A and B isotopes? A and C? A and D? B and C?

36. Give mass numbers of the atoms in Problem 35.

37. Indicate how many electrons and how many protons there are in a neutral atom of each of these elements. You may use the periodic table.

a. calcium **b.** sodium **c.** fluorine
d. argon **e.** beryllium

38. Indicate how many electrons and how many protons there are in a neutral atom of each of these elements. You may use the periodic table.

a. nitrogen **b.** iron
c. cadmium **d.** uranium

39. Indicate the number of protons and the number of neutrons in atoms of the following isotopes.

a. ^{62}Zn **b.** ^{241}Pu
c. ^{99}Tc **d.** ^{99}Mo

40. Indicate the number of protons and the number of neutrons in atoms of the following isotopes.

a. ^{11}B **b.** ^{154}Sm
c. ^{81}Kr **d.** ^{121}Te

41. Indicate the group and period in which each of the following elements is found. Classify each as a metal or nonmetal. You may use the periodic table.
 a. C **b.** Ca **c.** Cd
 d. Cl **e.** B **f.** Ba
 g. Bi **h.** Br

42. Indicate the group and period in which each of the following elements is found. Classify each as a metal or nonmetal. You may use the periodic table.
 a. S **b.** Sn **c.** Sm
 d. Sr **e.** Ta **f.** Tc
 g. Ti **h.** Tl

43. Identify the element in each of the following descriptions. You may use the periodic table.
 a. Group 3A, period 4 **b.** Group 1B, period 4
 c. Group 7A, period 5

44. Identify the element in each of the following descriptions. You may use the periodic table.
 a. Group 4A, nonmetal **b.** Group 7B, period 5
 c. Group 1A, period 2

Chemical Formulas: Elements

45. Give symbols and also, where appropriate, molecular formulas for the following elements.
 a. helium **b.** oxygen **c.** chlorine
 d. phosphorus

46. Give symbols and also, where appropriate, molecular formulas for the following elements.
 a. hydrogen **b.** nitrogen **c.** bromine
 d. sulfur

Chemical Formulas: Binary Molecular Compounds

47. Give formulas for the following binary molecular compounds.
 a. dinitrogen monoxide
 b. tetraphosphorus trisulfide
 c. phosphorus pentachloride
 d. sulfur hexafluoride

48. Give formulas for the following binary molecular compounds.
 a. oxygen difluoride
 b. dinitrogen pentoxide
 c. phosphorus tribromide
 d. tetrasulfur tetranitride

49. Name the following binary molecular compounds.
 a. CS_2 **b.** N_2S_4
 c. PF_3 **d.** S_2F_{10}

50. Name the following binary molecular compounds.
 a. CBr_4 **b.** Cl_2O_7
 c. P_4S_{10} **d.** I_2O_5

Chemical Symbols and Formulas: Ions

51. Name the following ions.
 a. Na^+ **b.** Mg^{2+} **c.** Al^{3+}
 d. Cl^- **e.** O^{2-} **f.** N^{3-}

52. Name the following ions.
 a. K^+ **b.** Ca^{2+} **c.** Zn^{2+}
 d. Br^- **e.** Li^+ **f.** S^{2-}

53. Name the following ions.
 a. Fe^{3+} **b.** Cu^{2+} **c.** Ag^+

54. Name the following ions.
 a. Fe^{2+} **b.** Cu^+ **c.** I^-

55. Give symbols for the following ions.
 a. bromide ion **b.** calcium ion
 c. potassium ion **d.** iron(II) ion
 e. sodium ion

56. Give symbols for the following ions.
 a. aluminum ion **b.** oxide ion
 c. copper(II) ion **d.** nitride ion

57. Name the following ions.
 a. CO_3^{2-} **b.** HPO_4^{2-}
 c. MnO_4^- **d.** OH^-

58. Name the following ions.
 a. NO_3^- **b.** SO_4^{2-}
 c. $H_2PO_4^-$ **d.** HCO_3^-

59. Give formulas for the following ions:
 a. ammonium ion **b.** hydrogen sulfate ion
 c. cyanide ion **d.** nitrite ion

60. Give formulas for the following ions.
 a. phosphate ion **b.** hydrogen carbonate ion
 c. dichromate ion **d.** oxalate ion

Chemical Formulas: Ionic Compounds

61. Name the following ionic compounds.
 a. NaBr **b.** $FeCl_3$ **c.** LiI
 d. Na_2O **e.** K_2S **f.** CuBr
 g. KCl **h.** $MgBr_2$ **i.** CaS
 j. $FeCl_2$ **k.** Al_2O_3

62. Name the following ionic compounds.
  a. KNO_2 b. LiCN c. $NaClO_2$
  d. $NaNO_3$ e. $KMnO_4$ f. $CaSO_4$
  g. NaH_2PO_4 h. $(NH_4)_3PO_4$ i. $Al(NO_3)_3$
  j. NH_4NO_3

63. Give formulas for the following ionic compounds.
  a. magnesium sulfate
  b. sodium hydrogen carbonate
  c. potassium nitrate
  d. calcium monohydrogen phosphate
  e. calcium chlorite
  f. calcium carbonate
  g. lithium hydrogen sulfate
  h. magnesium cyanide
  i. potassium dihydrogen phosphate
  j. sodium hypochlorite

64. Give formulas for the following ionic compounds.
  a. iron(II) phosphate
  b. potassium dichromate
  c. copper(I) iodide
  d. ammonium nitrite
  e. potassium perchlorate
  f. iron(III) oxalate
  g. sodium permanganate
  h. copper(II) bromide
  i. zinc monohydrogen phosphate
  j. sodium chlorate

65. Name the following ionic compounds.
  a. $NaHSO_4$ b. $Al(OH)_3$ c. Na_2CO_3
  d. $KHCO_3$ e. NH_4NO_2

66. Name the following ionic compounds.
  a. $Ca(ClO)_2$ b. Li_2CO_3 c. $Na_2Cr_2O_7$
  d. $Ca(H_2PO_4)_2$ e. $(NH_4)_2C_2O_4$

Acids and Bases

67. Give the formulas for the following acids and bases.
  a. hydrochloric acid
  b. sulfuric acid
  c. carbonic acid
  d. hydrocyanic acid
  e. lithium hydroxide
  f. magnesium hydroxide

68. Give the formulas for the following acids and bases.
  a. nitric acid b. sulfurous acid
  c. phosphoric acid d. hydrosulfuric acid
  e. calcium hydroxide f. potassium hydroxide

69. Name the following acids and bases.
  a. NaOH b. H_3PO_4 c. HNO_3
  d. H_2SO_3 e. $Ca(OH)_2$ f. H_2S

70. Name the following acids and bases.
  a. HCl b. H_2SO_4 c. LiOH
  d. H_2CO_3 e. $Mg(OH)_2$ f. HCN

Organic Compounds: Formulas, Structures, and Functional Groups

71. Give a structure for each of the following organic compounds.
  a. pentane b. butanoic acid
  c. propyl alcohol d. cyclobutane
  e. isobutane f. acetic acid

72. Give a structure for each of the following organic compounds.
  a. hexane b. pentanoic acid
  c. isopropyl alcohol d. cyclopentane
  e. 1-butanol f. propionic acid

73. Write structural formulas for these two acids mentioned in the chapter: butyric acid and isobutyric acid.

74. Write a plausible structure(s) for a saturated hydrocarbon with the molecular formula C_4H_8.

75. Each of the formulas below belongs to one of the following classes of substances: straight-chain alkane, branched-chain alkane, cyclic alkane, hydrocarbon, alcohol, carboxylic acid, inorganic compound. Match each formula with the term that is *most specific*.
  a. $CH_3(CH_2)_6CH_3$ b. $CH_3CH_2CHOHCH_3$
  c. C_5H_{10} d. C_2H_2
  e. $CH_3(CH_2)_6COOH$ f. Na_2CO_3

76. Each of the formulas below belongs to one of the following classes of substances: straight-chain alkane, branched-chain alkane, cyclic alkane, hydrocarbon, alcohol, carboxylic acid, inorganic compound. Match each formula with the term that is *most specific*.
  a. $CH_3(CH_2)_3CH(CH_3)CH_2CH_3$ (with CH_3 branch on the CH) b. CH_3CH_2COOH
  c. $KHCO_3$ d. $CH_3(CH_2)_3CH(OH)CH_2CH_3$ (with OH on the CH)
  e. C_6H_6 f. HCOOH
  g. $CH_3(CH_2)_4CH_3$ h. C_8H_{16}

77. Which of the following formulas applies to the molecular model pictured?

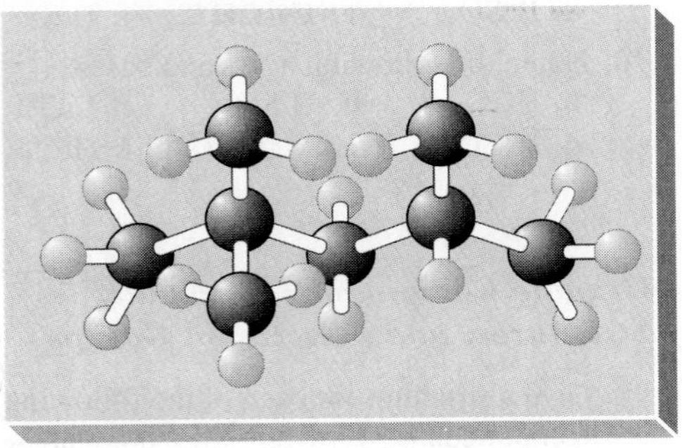

(a) $CH_3(CH_2)_6CH_3$

(b) $CH_3\overset{CH_3}{\underset{CH_3}{C}}CH_2\underset{CH_3}{CH}CH_3$

(c) $CH_3\overset{CH_3}{\underset{CH_3}{C}}—\underset{CH_3}{CH}CH_3$

(d) $CH_3\overset{CH_3}{\underset{CH_3}{C}}—CH_2—\overset{CH_3}{\underset{CH_3}{C}}CH_3$

78. Write a structural formula for the molecular model of the alcohol pictured.

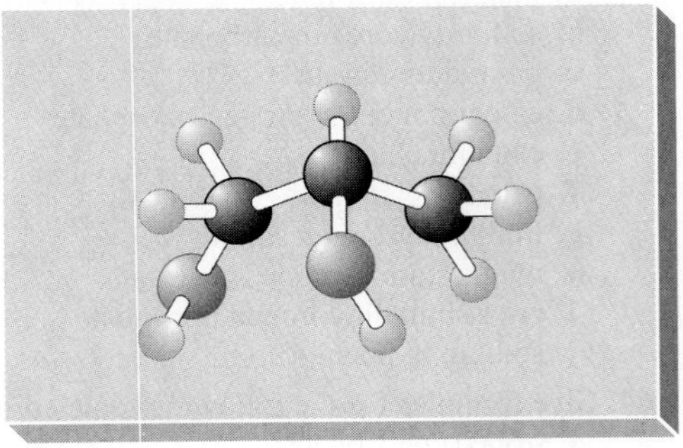

ADDITIONAL PROBLEMS

79. Show that the following experiment is consistent with the law of conservation of mass (within the limits of experimental error): A 10.00-g sample of calcium carbonate was dissolved in 100.0 mL of hydrochloric acid solution (d = 1.148 g/mL). The products were 120.40 g of solution (a mixture of hydrochloric acid and calcium chloride) and 2.22 L of carbon dioxide gas (d = 0.0019769 g/mL).

80. When 3.06 g of hydrogen was allowed to react with an excess of oxygen, 27.35 g of water was obtained. In a second experiment the electrolysis of a sample of water produced 1.45 g of hydrogen and 11.51 g of oxygen. Are these results consistent with the law of definite proportions? Demonstrate why or why not.

81. In one experiment 0.312 g of sulfur was completely burned, producing 0.623 g of sulfur dioxide as the sole product. In a second experiment 0.842 g of sulfur dioxide was obtained. What mass of sulfur was burned in the second experiment?

82. William Prout hypothesized in 1815 that all other atoms are built up of hydrogen atoms. If so, all elements should have integral (whole number) atomic weights relative to an atomic weight of 1 for hydrogen. However, the hypothesis appeared inconsistent with atomic weights such as 24.3 for magnesium and 35.5 for chlorine. Based on modern knowledge, Prout's hypothesis seems more reasonable. Explain why.

83. Mercury and oxygen form two compounds. One contains 96.2% mercury, by mass, and the other, 92.6%. Show that these data conform to the law of multiple proportions.

84. One oxide of iron is found to consist of 22.36% oxygen by mass. Which of the following represents a possible mass percent oxygen of a second iron oxide? Explain.

(a) 27.64% (b) 44.72%
(c) 53.00% (c) 66.00%

85. Identify the isotope for which the numbers of protons, neutrons, and electrons in its atoms are equal and their *total* is 60.

86. Identify the isotope for which the mass number of its atoms is 234 and its atoms have 60.0% more neutrons than protons.

87. The following name and formula suggest compounds that do not exist. Explain why they do not.

a. isoethanol **b.** $CH_3CH_2C(CH_3)_3CH_2CH_3$

88. Use the rules for naming organic compounds from Appendix C to tell what is wrong with each of these proposed names of organic compounds, and replace each by a correct name.

a. 2-ethylbutane **b.** 4-methylhexane
c. methyl hydroxide **d.** hydrocarboxylic acid

89. Are the following pairs of molecules isomers or not? Explain.

a. 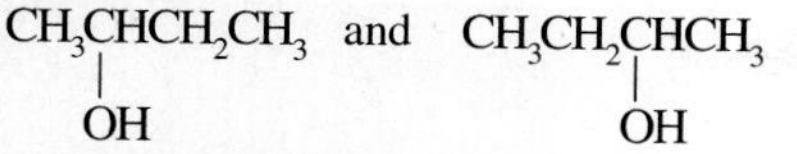

b. $CH_3(CH_2)_5CH(CH_3)_2$ and $CH_3(CH_2)_4CH(CH_3)CH_2CH_3$

c. $CH_3(CH_2)_4CH(CH_3)_2$ and $(CH_3)_2CHCH_2CH_2CH_3$

d. $CH_3CH_2C(CH_3)_2(CH_2)_3CH_3$ and $C_2H_5C(CH_3)_2C_4H_9$

90. Write the general formula for

a. a straight-chain or branched-chain alcohol having one —OH group per molecule.

b. a straight-chain or branched-chain carboxylic acid having one —COOH group per molecule.

c. a cyclic alcohol having two —OH groups per molecule.

91. Use the rules for naming organic compounds described in Appendix C to propose likely names for the following compounds.

a. 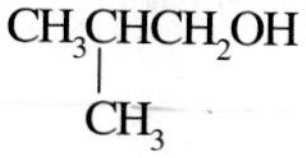

b. 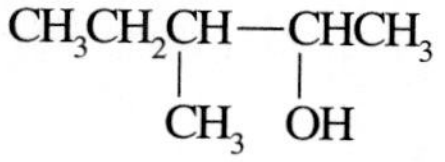

c. 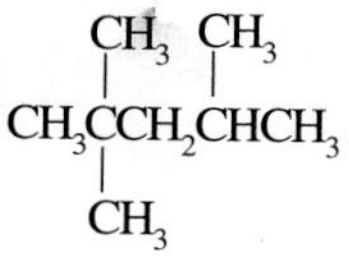

d. 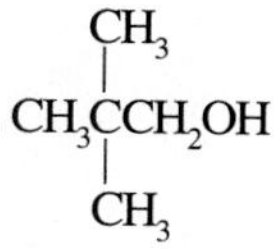

92. Similar to the previous problem, write the structural formulas implied by the following names.

a. 2-ethyl-1-butanol

b. 2-hydroxyoctanoic acid

c. 1,2,6-hexanetriol

d. hydroxypropanedioic acid

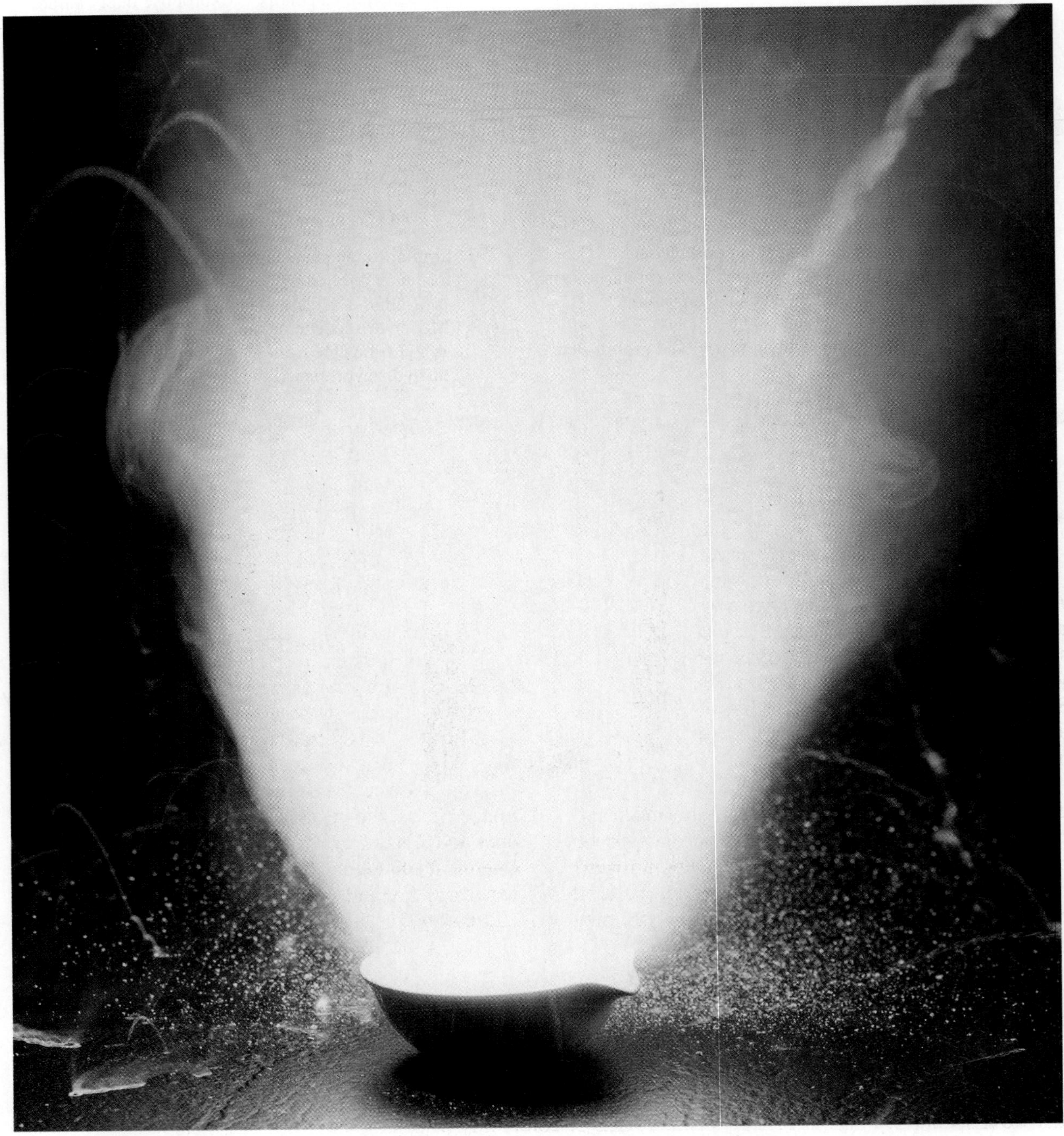

This vigorous reaction between powdered zinc and powdered sulfur is described in the chapter (page 92).

3

Stoichiometry: Chemical Calculations

CONTENTS

You can learn many important physical and chemical properties just by reading about them: Ammonium nitrate is a substance used both as a fertilizer and as a high explosive. Chlorine is a poisonous, greenish yellow gas used to disinfect water. Sodium is a soft, silvery metal that reacts violently with water to produce hydrogen gas.

However, chemists often face questions that require *quantitative* answers: How much ammonium nitrate should be used to provide an avocado tree with the desired amount of the element nitrogen? How much chlorine gas is required to establish a level of two parts per million of chlorine in a swimming pool? How much of a base such as soda ash (Na_2CO_3) is required to neutralize the acidity resulting from the reaction of the chlorine with the pool water? How much hydrogen gas is produced when 1 kg of sodium metal reacts with water, and how much heat is released when this hydrogen gas is burned?

Quantitative answers require mathematics, and indeed some chemical questions require quite sophisticated mathematics. However, chemical questions like the ones above require only arithmetic or simple algebra. Chemical equations and calculations based on chemical formulas and chemical equations, called **stoichiometry**, are the subjects of this chapter.

3.1 Molecular Weights and Formula Weights

As we learned in Chapter 2, each element has a characteristic atomic weight. Because chemical compounds are made up of two or more elements, the masses that we associate with molecules or formula units of compounds are combinations of atomic weights. Atomic weights, molecular weights, and formula weights all enter into the calculations we do in this chapter.

Molecular Weights

We use the term "average" when speaking of the mass of an individual molecule because two molecules of a compound may have different isotopes of one or more of their constituent elements.

For a molecular substance, the **molecular weight** is the average mass of a molecule of a substance relative to that of a carbon-12 atom. More simply, it is the sum of the masses of the atoms represented in a molecular formula. For example, because the formula O_2 specifies two O atoms per molecule of oxygen, the molecular weight of oxygen (O_2) is twice the atomic weight of oxygen.

$$2 \times \text{atomic weight of O} = 2 \times 15.9994\ \text{u} = 31.9988\ \text{u}$$

The molecular weight of carbon dioxide (CO_2) is the sum of the atomic weight of carbon and twice the atomic weight of oxygen.

$$\begin{aligned} 1 \times \text{atomic weight of C} &= 1 \times 12.011\ \text{u} = 12.011\ \text{u} \\ 2 \times \text{atomic weight of O} &= 2 \times 15.9994\ \text{u} = \underline{31.9988\ \text{u}} \\ \text{Molecular weight of } CO_2 &= 44.010\ \text{u} \end{aligned}$$

Example 3.1

Calculate the molecular weight of sulfur hexafluoride, SF_6, a gas used as an electric insulator in high-voltage equipment.

Solution

We think about the problem in this way: Add the atomic weight of sulfur to six times the atomic weight of fluorine. However, if we use an electronic calculator, we need only write down the final answer, 146.056 u. That is, we have no need to record the numbers 32.066 and 113.9904.

$$\begin{aligned} 1 \times \text{atomic weight of S} &= 1 \times 32.066\ \text{u} = 32.066\ \text{u} \\ 6 \times \text{atomic weight of F} &= 6 \times 18.9984\ \text{u} = \underline{113.9904\ \text{u}} \\ \text{Molecular weight of } SF_6 &= 146.056\ \text{u} \end{aligned}$$

Exercise 3.1

Calculate, to five significant figures, the molecular weight of each of the following compounds.

a. $C_6H_{12}Br_2$ **b.** $C_2H_2Cl_2O_2$ **c.** $H_5P_3O_{10}$ **d.** $C_6H_4Cl_2O_2S$

Formula Weights

The term "molecular weight" is not always appropriate. As we learned in Chapter 2, we use the term *formula unit* for ionic compounds and other substances in which individual molecules do not exist. **Formula weight** is the average mass of a formula unit relative to that of a carbon-12 atom. In short, the formula weight is the sum of the masses of the atoms or ions represented by the formula.

Example 3.2

Calculate the formula weight of ammonium sulfate, a fertilizer commonly used by home gardeners.

Solution

Before we can determine a formula weight, we must find the correct chemical formula. Ammonium sulfate is an ionic compound made up of ammonium ions, NH_4^+, and sulfate ions, SO_4^{2-}; its formula is $(NH_4)_2SO_4$. In the summation below, remember that everything within the parentheses must be multiplied by 2.

$$\begin{aligned}
2 \times \text{atomic weight of N} &= 2 \times 14.0067\ \text{u} = 28.0134\ \text{u} \\
8 \times \text{atomic weight of H} &= 8 \times 1.00794\ \text{u} = 8.06352\ \text{u} \\
1 \times \text{atomic weight of S} &= 1 \times 32.066\ \text{u} = 32.066\ \text{u} \\
4 \times \text{atomic weight of O} &= 4 \times 15.9994\ \text{u} = 63.9976\ \text{u} \\
\text{Formula weight of } (NH_4)_2SO_4 &= 132.141\ \text{u}
\end{aligned}$$

Exercise 3.2

Calculate the formula weight of each of the following compounds.

a. K_2SbF_5 **b.** barium bromate **c.** iron(III) phosphate **d.** $NaB(C_6H_5)_4$

3.2 The Concept of the Mole and Avogadro's Number (6.022×10^{23})

When a material rich in carbon, such as coal or charcoal, is burned, the carbon combines with oxygen to produce carbon monoxide and/or carbon dioxide as products. If the available quantity of oxygen is limited, the product is rich in carbon monoxide—a poisonous gas. With abundant oxygen, carbon is converted completely to carbon dioxide. How can we determine the minimum quantity of oxygen required to ensure that carbon is completely converted to carbon dioxide?

From the formula of carbon dioxide we see that each CO_2 molecule consists of one C atom and two O atoms. In Chapter 2 we learned that oxygen is *diatomic*—that is, made up of O_2 molecules. Thus, we need at least as many O_2 molecules as we have C atoms. But any actual sample of carbon contains so many atoms that we can't count them. How can we measure a sample of oxygen gas that will contain this same number of O_2 molecules or more?

Even a sample of carbon as small as a pencil-mark period at the end of a sentence contains about 10^{18} C atoms, that is, about 1,000,000,000,000,000,000 C atoms.

We can deal with questions of this sort through a quantity called the *mole*. A **mole (mol)** is an amount of substance that contains as many elementary units as there are atoms in exactly 12 g of the carbon-12 isotope. For chemical species, such as atoms, molecules, ions, or formula units, the elementary units are those described by the symbol or formula. The actual number of elementary units in a mole is called **Avogadro's number, N_A**, named for Amedeo Avogadro, who had the first clear idea of the significance of this unit. The mole is such an important quantity that it is the SI base unit for the amount of a substance.

Now, to complete the discussion of how much oxygen is required to convert carbon to carbon dioxide, let's reason along the following lines.

- The mass of a carbon-12 atom is defined to be *exactly* 12 u.
- The mass of an "average" carbon atom (that is, accounting for the relative abundances of ^{12}C and ^{13}C) is 12.011 u.
- The mass of an "average" oxygen atom, relative to that of a carbon-12 atom, is 15.9994 u.
- The mass of an "average" oxygen molecule is $2 \times 15.9994\ \text{u} = 31.9988\ \text{u}$.
- The mass of Avogadro's number (N_A) of ^{12}C atoms is *exactly* 12 g; the mass of N_A "average" C atoms is 12.011 g; and the mass of N_A "average" O_2 molecules is 31.9988 g.

To have the same number of O_2 molecules as of C atoms, we can choose Avogadro's number (N_A) of each. Then, according to the law of conservation of mass, the reaction of 12.011 g of C with 31.9988 g of O_2 produces 44.010 g of CO_2, with neither carbon nor oxygen left over. To ensure the complete combustion of carbon to carbon dioxide, the ratio of mass of oxygen to mass of carbon should equal or exceed 31.9988 g O_2/12.011 g C = 2.6641. Some of the ideas discussed here are illustrated in Figure 3.1.

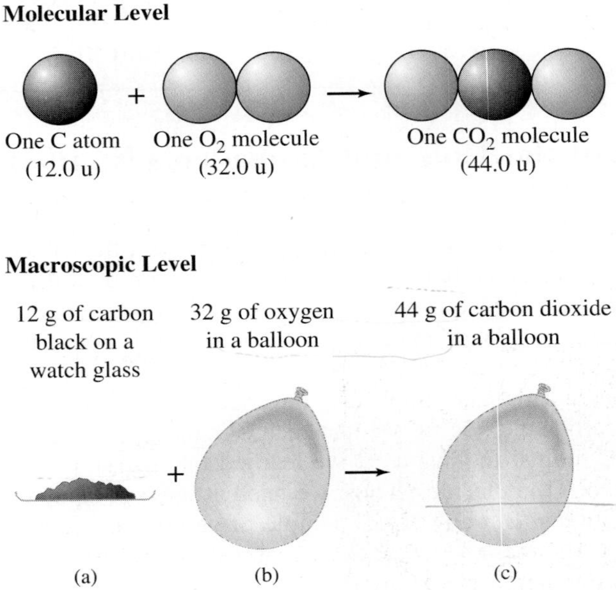

Figure 3.1 Visualizing the reaction of carbon and oxygen to form carbon dioxide. Although chemical reactions occur between small numbers of atoms and molecules at the molecular level (top), extremely large numbers are involved at the macroscopic level (bottom). The artist's sketch suggests Avogadro's number of (a) C atoms (12.0 g), (b) O_2 molecules (32.0 g), and (c) CO_2 molecules (44.0 g).

Notice how we have used the *concept* of Avogadro's number without requiring the actual number itself. Later in the text we'll indicate how the numerical value of N_A is established, but we might as well state it now.

$$N_A = 6.022137 \times 10^{23}\ \text{mol}^{-1}$$

Depending on the number of significant figures needed in a calculation, we can usually round off Avogadro's number to 6.022×10^{23} mol^{-1}, or even 6.02×10^{23} mol^{-1}. The unit "mol^{-1}" can be read "per mole"; it signifies that the entities we are counting are those present in one mole.

Avogadro's number is unimaginably large. If you had 6.02×10^{23} dollars, you could spend a billion dollars a second for your entire lifetime and still have used only 0.001% of your money. If carbon atoms were the size of peas, 6.02×10^{23} of them would cover the entire surface of our planet to a depth of 15 m.

We could speak of a mole of eggs, but the term would not be very useful. All the hens' eggs laid since the first chicken wouldn't come close to totaling one mole.

3.3 More on the Mole

We buy eggs by the dozen (12 eggs), pencils by the gross (144 pencils), and paper by the ream (500 sheets). As we have seen, chemists also use a term to refer to a particular number of things—the *mole*.

The definition of the mole includes a reference to Avogadro's number of "elementary units." The elementary units must be specified by a chemical symbol or formula. They may be atoms, such as C, O, or Pu, or they may be molecules, such as O_2, CO_2, or even $C_{46}H_{65}N_{15}O_{12}S_2$ (vasopressin, a hormone). They may even be units that we cannot gather up by themselves and put in a bottle, such as

$$1 \text{ mol } Mg^{2+} = 6.022 \times 10^{23} \; Mg^{2+} \text{ ions}$$

We can obtain 1 mol Mg^{2+} by measuring out 1 mol $MgCl_2$. However, because 1 mol $MgCl_2$ consists of 1 mol Mg^{2+} *and* 2 mol Cl^-, we cannot measure the Mg^{2+} separately. To obtain 6.022×10^{23} Mg^{2+} ions we have to take $2 \times 6.022 \times 10^{23}$ Cl^- ions at the same time.

Molar Mass

A dozen is the same *number*, whether we have a dozen oranges or a dozen watermelons. However, a dozen oranges and a dozen watermelons do not have the same *mass*. Similarly, a mole of magnesium and a mole of iron contain the same numbers of atoms but have *different* masses. Figure 3.2 is a photograph of 1 mole of each of several chemical substances.

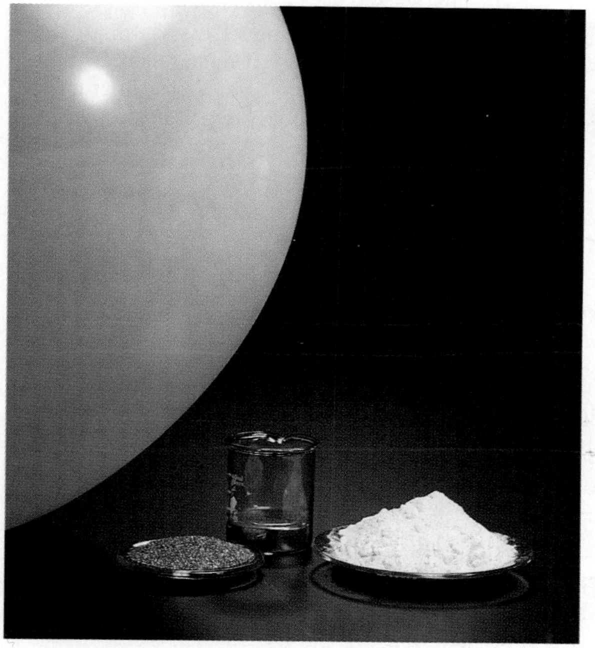

Figure 3.2 One mole each of four different elements. The watch glasses contain one mole of copper atoms (left) and one mole of sulfur atoms (right). The beaker contains one mole of liquid mercury, and the balloon contains one mole of helium gas.

The **molar mass** of a substance is the mass of 1 mol of that substance. The molar mass is numerically equal to the atomic weight, molecular weight, or formula weight, but it is expressed in the units *g/mol*.

The atomic weight of sodium is 22.99 u; its molar mass is 22.99 g/mol. The molecular weight of carbon dioxide is 44.01 u; its molar mass is 44.01 g/mol. The formula weight of magnesium chloride is 95.21 u; its molar mass is 95.21 g/mol.

We can use these facts, together with the basic definition of the number of elementary units in a mole, to write the following relationships.

$$1 \text{ mol Na} = 22.99 \text{ g Na} = 6.022 \times 10^{23} \text{ Na atoms}$$
$$1 \text{ mol } CO_2 = 44.01 \text{ g } CO_2 = 6.022 \times 10^{23}\ CO_2 \text{ molecules}$$
$$1 \text{ mol } MgCl_2 = 95.21 \text{ g } MgCl_2 = 6.022 \times 10^{23}\ MgCl_2 \text{ formula units}$$

In turn, these relationships supply the conversion factors that we need in order to make conversions among mass in grams, amount in moles, and number of elementary units, as illustrated in the following examples.

Example 3.3

Calculate the mass, in grams, of 0.250 mol Na.

Solution

Sodium has an atomic weight of 22.99 u and a molar mass of 22.99 g Na/mol Na. Therefore,

$$? \text{ g Na} = 0.250 \text{ mol Na} \times \frac{22.99 \text{ g Na}}{1 \text{ mol Na}} = 5.75 \text{ g Na}$$

Example 3.4

Calculate the number of moles of CO_2 present in a 225-g sample of the gas.

Solution

In this case we need the molar mass of a molecular substance, CO_2. We first determine its molecular weight, 12.011 u + 2 × 15.9994 u = 44.01 u, and then its molar mass, 44.01 g CO_2/mol CO_2. Also, we see that in converting from a mass in grams to an amount in moles we must use the *inverse* of the molar mass to get the proper cancellation of units.

$$? \text{ mol } CO_2 = 225 \text{ g } CO_2 \times \frac{1 \text{ mol } CO_2}{44.01 \text{ g } CO_2} = 5.11 \text{ mol } CO_2$$

Example 3.5

How many Cl^- ions are present in 1.38 g $MgCl_2$?

Solution

First we need the formula weight of $MgCl_2$ and then the molar mass. Once we have established the amount of $MgCl_2$ in moles, we can use Avogadro's number to determine the number of formula units (f.u.) of $MgCl_2$. Finally, we can use a factor (shown in red) that establishes the number of Cl^- ions per formula unit.

$$? \text{ Cl}^- \text{ ions} = 1.38 \text{ g } MgCl_2 \times \frac{1 \text{ mol } MgCl_2}{95.21 \text{ g } MgCl_2} \times \frac{6.022 \times 10^{23} \text{ f.u.}}{1 \text{ mol } MgCl_2} \times \frac{2 \text{ Cl}^- \text{ ions}}{1 \text{ f.u.}}$$
$$= 1.75 \times 10^{22} \text{ Cl}^- \text{ ions}$$

Example 3.5 illustrates the common practice of expressing molar mass and Avogadro's number with one more significant figure than the least precisely known quantity (1.38 g $MgCl_2$). This ensures that the precision of the calculated result is limited only by the least precisely known quantity.

Example 3.6

What is the mass of a sodium atom, in grams?

Solution

Because 1 mole of atoms has a mass equal to the molar mass and contains Avogadro's number of atoms, we can write

$$? \text{ g/Na atom} = \frac{22.99 \text{ g}}{1 \text{ mol Na}} \times \frac{1 \text{ mol Na}}{6.0221 \times 10^{23} \text{ Na atoms}}$$
$$= 3.818 \times 10^{-23} \text{ g/Na atom}$$

We can ask for the mass of a sodium atom, rather than the *average* mass, because there is only one naturally occurring isotope, sodium-23.

Exercise 3.3

Calculate the amount, in moles, of each of the following.

a. 3.71 g Fe **b.** 76.0 mg of phosphoric acid

c. 1.65×10^3 kg C_4H_{10} **d.** 1.99 g of potassium chromate

Exercise 3.4

Calculate the mass, in grams, of each of the following.

a. 55.5 mol H_2O **b.** 0.0102 mol $C_4H_{10}O$

c. 2.45×10^{-4} mol of ethane **d.** 2.13 mol of nitric acid

Exercise 3.5

Calculate the number of (**a**) atoms in 6.17 g Ca; (**b**) molecules in 0.0100 g l (**c**) molecules in 18.5 g of butane, C_4H_{10}; (**d**) *atoms* in 215 g of sucrose ($C_{12}H_{22}O_{11}$).

Exercise 3.6

Calculate the average mass in grams of (**a**) a helium atom; (**b**) a bismuth atom; (**c**) a carbon tetrachloride (CCl_4) molecule; (**d**) a propane molecule.

Estimation Exercise 3.1

The mass of 1.0×10^{23} magnesium atoms is about

2.4×10^{-23} g 4.0×10^{-23} g 2.4 g 4.0 g

Estimation Exercise 3.2

Which of the following is a reasonable value for the number of atoms in 1.00 g of helium?

0.25 1.51 1.51×10^{-23} 1.51×10^{23}

Estimation Exercise 3.3

Which of the following is a reasonable value for the mass of a calcium atom?

6.7×10^{-23} g 40.1 g 6.7×10^{-23} u 6.7×10^{23} u

3.4 Mass Percent Composition from Chemical Formulas

Sometimes we need to know how much of the mass of a compound is taken up by each individual element. When we determine the molar mass of a compound from its formula, we can keep track of the contribution that each element makes to the total mass. With this information we can then establish a ratio of the mass of each element to that of the compound as a whole; this gives us the fractional composition of the compound, by mass. If we multiply these fractions by 100%, we obtain the mass percent composition of the compound.

Consider these three compounds, all of which are used as fertilizers: ammonium nitrate, NH_4NO_3; ammonium sulfate, $(NH_4)_2SO_4$; and urea, $CO(NH_2)_2$. If the three compounds were each available at the same cost per pound, which would you choose as the cheapest source of nitrogen, an essential plant nutrient? You could make your choice most simply by determining the mass percent nitrogen in each compound and choosing the one with the greatest percentage of nitrogen. The method outlined in the preceding paragraph is applied to NH_4NO_3 in Example 3.7.

"5-10-5" Fertilizer: What Is It?

A common "5-10-5" fertilizer.

Most farmers buy *complete fertilizers*, which, despite the name, usually contain only three main nutrients: nitrogen, phosphorus, and potassium. There are usually three numbers found on a fertilizer bag. The first number represents the mass percent N; the second, the percent P_2O_5; and the third, the percent K_2O. So, "5-10-5" means that a fertilizer contains 5% N, 10% P_2O_5, and 5% K_2O. The rest is inert material.

A 100-lb bag of this fertilizer contains 5 lb of nitrogen, N. The actual mass of the nitrogen-containing compound, of course, depends on its formula, for example, whether it is NH_4NO_3 (ammonium nitrate) or $CO(NH_2)_2$ (urea). But what about the K_2O and P_2O_5? There are no compounds with these formulas in the fertilizer. These percentages based on oxides are a holdover from the way compositions were reported in the early days of analytical chemistry, but we can convert them to actual % P and % K without difficulty. By the method of Example 3.7, we can show that P_2O_5 is 43.64% P and that K_2O is 83.01% K. Thus, 10% P_2O_5 is the same as 4.4% P (10% × 0.4364), and 5% K_2O is about 4.2% K (5% × 0.8301). The 100-lb bag of fertilizer contains 4.4 lb P and 4.2 lb K.

Phosphorus can be supplied by several compounds; $Ca(H_2PO_4)_2$ and $(NH_4)_2HPO_4$ are common ones. The potassium is nearly always supplied as KCl, although any potassium salt will furnish the needed K^+ ion.

Fertilizers greatly increase the production of food and fiber, but they also can cause problems. The fertilizers must be water soluble to be used by plants. When it rains, the nutrients from the fertilizers are washed into streams and lakes, where they stimulate blooms of algae. These chemicals, particularly the nitrates, also penetrate the groundwater. In some areas the nitrates present in well water have reached levels that are toxic to infants.

Example 3.7

Calculate the mass percent of each element in ammonium nitrate.

Solution

First, determine the molar mass of NH_4NO_3.

$$\begin{aligned}
\text{Formula wt.} &= 2 \times \text{at.wt. N} + 4 \times \text{at.wt. H} + 3 \times \text{at.wt. O} \\
&= (2 \times 14.01)\text{ u} + (4 \times 1.01)\text{ u} + (3 \times 16.00)\text{ u} \\
&= 28.02\text{ u} + 4.04\text{ u} + 48.00\text{ u} = 80.06\text{ u} \\
\text{Molar mass} &= 80.06\text{ g/mol } NH_4NO_3
\end{aligned}$$

Then, for one mole of compound, determine mass ratios and percents.

$$\% \text{ N} = \frac{28.02 \text{ g N}}{80.06 \text{ g NH}_4\text{NO}_3} \times 100\% = 35.00\%$$

$$\% \text{ H} = \frac{4.04 \text{ g H}}{80.06 \text{ g NH}_4\text{NO}_3} \times 100\% = 5.05\%$$

$$\% \text{ O} = \frac{48.00 \text{ g O}}{80.06 \text{ g NH}_4\text{NO}_3} \times 100\% = 59.96\%$$

An alternate way to determine % O is to note that % O = 100.00% − 35.00% N − 5.05% H = 59.95% O.

To check, add the percentages to ensure that they add up to 100.00% (Here, they actually total 100.01% due to rounding; that is close enough.)

Exercise 3.7

Calculate the mass percent of each element in (**a**) ammonium sulfate and (**b**) urea, $CO(NH_2)_2$. Which compound has the greatest mass percent nitrogen: ammonium nitrate, ammonium sulfate, or urea?

We can set up a conversion factor based on mass percent and use it to determine the mass of an element in any sample of a compound. For example, using the result from Example 3.7 that ammonium nitrate is 35.00% N by mass, we can find the mass of nitrogen in 46.34 g NH_4NO_3 by writing

$$? \text{ g N} = 46.34 \text{ g NH}_4\text{NO}_3 \times \frac{35.00 \text{ g N}}{100.00 \text{ g NH}_4\text{NO}_3} = 16.22 \text{ g N}$$

However, Example 3.8 illustrates a simpler approach. It involves formulating factors to convert from grams of NH_4NO_3 to grams of N from the chemical formula, without first evaluating the mass percent N. Similar conversion factors are compared in Estimation Example 3.1, where no detailed calculations are required.

Example 3.8

What mass of nitrogen, in grams, is present in 46.34 g ammonium nitrate? (Assume you do not know that the % N in NH_4NO_3 is 35.00%.)

Solution

The central factor in the conversion (shown in red) is based on the chemical formula NH_4NO_3. The other factors are based on the molar masses of ammonium nitrate and nitrogen.

$$? \text{ g N} = 46.34 \text{ g NH}_4\text{NO}_3 \times \frac{1 \text{ mol NH}_4\text{NO}_3}{80.06 \text{ g NH}_4\text{NO}_3} \times \frac{2 \text{ mol N}}{1 \text{ mol NH}_4\text{NO}_3} \times \frac{14.01 \text{ g N}}{1 \text{ mol N}}$$

$$= 16.22 \text{ g N}$$

Exercise 3.8

People with hypertension (high blood pressure) are advised to limit the amount of sodium (actually sodium ion, Na^+) in their diets. Calculate the mass, in mg, of Na^+ in 5.00 g of sodium hydrogen carbonate (sodium bicarbonate).

ESTIMATION EXAMPLE 3.1

Which of these compounds contains the greatest mass of sulfur per gram of compound: barium sulfate, lithium sulfate, sodium sulfate, or lead sulfate?

Solution

To make this comparison we need the formulas of the compounds, which we can get from their names.

$$BaSO_4 \qquad Li_2SO_4 \qquad Na_2SO_4 \qquad PbSO_4$$

The compound with the greatest mass of sulfur per gram of compound is the one with the greatest % S by mass. From the formulas we see that in one mole of each compound there is one mole of sulfur, 32.066 g S. The compound with the greatest % S is the one with the *smallest* formula weight. Each formula unit has one SO_4^{2-} ion, so all we have to do is compare some atomic weights: that of barium to twice that of lithium, and so on. With just a glance at an atomic weight table we see the answer must be lithium sulfate, Li_2SO_4.

Estimation Exercise 3.4

Which of these compounds has the greatest percent nitrogen by mass: ammonium sulfate, ammonium nitrate, or ammonium iodide?

3.5 Chemical Formulas from Mass Percent Composition

In the preceding section we described how the mass percent composition can be obtained from the formula of a compound. Often, however, what chemists do is *deduce* the chemical formula of an unknown compound from its mass percent composition. Mass percent compositions of compounds can generally be measured with considerable precision.

Determining Empirical Formulas

We have learned that the *empirical* formula is the simplest formula that we can write for a compound. It uses the smallest integers possible for its subscripts. When we determine the mass percent composition of a compound by experiment, we specify the constituent elements on a *mass* basis. To get the empirical formula, we must specify the elements on a *mole* basis.

We could base our calculation on a sample of any size, but the task is made simpler if we choose 100.00 g. When we do this, the masses of the elements in the sample are numerically equal to their mass percentages. The following four-step procedure is illustrated in the examples. The first three steps are suggested by Figure 3.3.

Convert from mass to moles:

Step 1. Convert percentages to masses.

Step 2. Convert masses to amounts in moles.

Create a tentative formula:

Step 3. Attempt to get integers as subscripts by dividing each of the numbers of moles from step 2 by the smallest one.

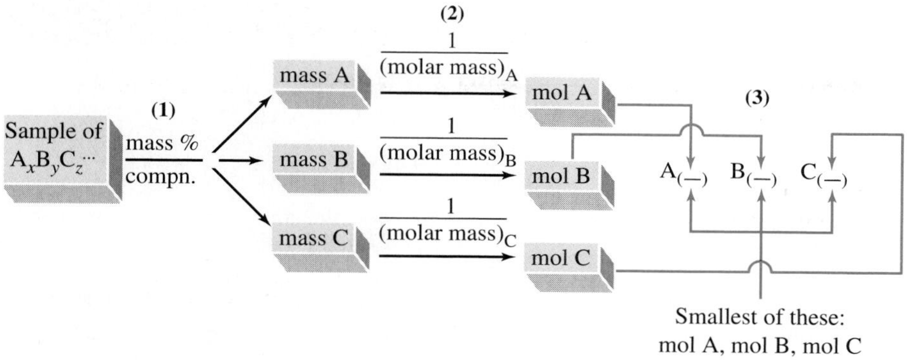

Figure 3.3 Determining the empirical formula from the mass percent composition of the compound $A_xB_yC_z\ldots$ The first three steps of the procedure outlined in the text are suggested in this diagram, leading to a tentative formula in step (3). The remainder of the procedure, illustrated through Examples 3.9 and 3.10, involves converting the subscripts in the tentative formula to integral numbers.

Obtain an empirical formula:

Step 4: If the subscripts obtained after step 3 are decimals, multiply each subscript by a common factor to convert them all to integers.

Example 3.9

Phenol, a general disinfectant, is 76.57% C, 6.43% H, and 17.00% O. Determine its empirical formula.

Solution

Step 1. A 100.00-g sample of phenol contains 76.57 g C, 6.43 g H, and 17.00 g O.
Step 2. Convert the masses of C, H, and O to numbers of moles.

$$76.57 \text{ g C} \times \frac{1 \text{ mol C}}{12.011 \text{ g C}} = 6.375 \text{ mol C}$$

$$6.43 \text{ g H} \times \frac{1 \text{ mol H}}{1.008 \text{ g H}} = 6.38 \text{ mol H}$$

$$17.00 \text{ g O} \times \frac{1 \text{ mol O}}{15.999 \text{ g O}} = 1.063 \text{ mol O}$$

Step 3. Divide each of the quantities above by the smallest, and use the results as subscripts in a tentative formula.

$$C_{\frac{6.375}{1.063}}H_{\frac{6.38}{1.063}}O_{\frac{1.063}{1.063}} \longrightarrow C_{5.997}H_{6.00}O_{1.000} \longrightarrow C_6H_6O$$

Step 4 is not required, because all the subscripts in Step 3 are integers. The empirical formula of phenol is C_6H_6O.

In step 3 of Example 3.9 we rounded off the subscript 5.997 to the integer 6. Rounding off subscripts that are within one or two hundredths of an integral value is generally within the limits of uncertainty of the calculation.

Example 3.10

Diethylene glycol, used as an antifreeze, is 45.27% C, 9.50% H, and 45.23% O. Determine its empirical formula.

Solution

Step 1. A 100.00-g sample of diethylene glycol contains 45.27 g C, 9.50 g H, and 45.23 g O.

Step 2. Convert the masses of C, H, and O to numbers of moles.

$$45.27 \text{ g C} \times \frac{1 \text{ mol C}}{12.011 \text{ g C}} = 3.769 \text{ mol C}$$

$$9.50 \text{ g H} \times \frac{1 \text{ mol H}}{1.008 \text{ g H}} = 9.42 \text{ mol H}$$

$$45.23 \text{ g O} \times \frac{1 \text{ mol O}}{15.999 \text{ g O}} = 2.827 \text{ mol O}$$

Step 3. Divide each of the quantities above by the smallest, and use the results as subscripts in a tentative formula.

$$C_{\frac{3.769}{2.827}}H_{\frac{9.42}{2.827}}O_{\frac{2.827}{2.827}} \longrightarrow C_{1.333}H_{3.33}O_{1.000}$$

Some decimal and fractional equivalents that you will find useful in solving these problems are 0.500 = 1/2; 0.333 = 1/3; 0.250 = 1/4, 0.200 = 1/5, and so on.

Step 4. Multiply each of the subscripts from step 3 by a common factor to convert them all to integers. By recognizing that 1.333 = 4/3 and 3.33 = 10/3, we can see that the common factor we need is three.

$$C_{(1.333 \times 3)}H_{(3.33 \times 3)}O_{(1.000 \times 3)} \longrightarrow C_4H_{10}O_3$$

Exercise 3.9

Mebutamate, a diuretic "water pill" used to treat hypertension, is 51.70% C, 8.68% H, 12.06% N, and 27.55% O. Determine its empirical formula.

Exercise 3.10

Trinitrotoluene (TNT), an explosive, is 37.01% C, 2.22% H, 18.50% N, and 42.27% O. Determine its empirical formula.

Relating Molecular Formulas to Empirical Formulas

The subscript integers in the molecular formula of a molecular substance are either the same as those in its empirical formula or they are common "multiples" of them (see Section 2.7). For example, the subscripts in the molecular formula of acetylene, C_2H_2, are twice those in the empirical formula, CH. In benzene, C_6H_6, the multiplier to convert empirical-formula subscripts to those of the molecular formula is six. To establish the molecular formula of a compound we must know both the empirical formula and the molecular weight. We have just learned how to calculate an empirical formula from mass percent composition, and in later chapters we will consider several experimental methods of determining molecular weights.

Example 3.11

The empirical formula of hydroquinone, a chemical used in photography, is C_3H_3O, and its molecular weight is 110 u. What is its molecular formula?

Solution

The formula weight based on the empirical formula is 55.0 u. If all the subscripts are multiplied by two (110/55.0 = 2), the formula weight becomes equal to the molecular weight (2 × 55.0 u = 110 u). The molecular formula is 2 × C_3H_3O or $C_6H_6O_2$.

Exercise 3.11

Ethylene (molecular weight: 28.0 u), cyclohexane (84.0 u), and 1-pentene (70.0 u) all have the empirical formula CH_2. Give the molecular formula of each compound.

3.6 Elemental Analysis: Experimental Determination of Carbon, Hydrogen, and Oxygen

A chemist has just synthesized a new compound in an organic chemistry research laboratory. Most likely he or she has some idea of the composition and structure of the compound. To publish this discovery in the chemical literature or to obtain a patent covering this work the chemist must first present proof of the identity of the compound. One of the first steps in that proof is likely to be an elemental analysis establishing the mass percent composition of the compound. The methods of analysis are quite varied, depending on the elements present. We limit our discussion to the analysis of compounds that contain only the elements carbon, hydrogen, and oxygen.

As we have noted previously, over 95% of all known compounds are carbon compounds; they are organic substances. Nearly all of these organic compounds also contain hydrogen, and many of them have oxygen atoms as well.

In the apparatus pictured in Figure 3.4, a weighed sample of a compound in a combustion "boat" is burned in a stream of oxygen gas, producing carbon dioxide and water. We can represent the combustion as

$$C_xH_yO_z + O_2 \longrightarrow x\,CO_2 + y/2\,H_2O$$

In the combustion each mole of carbon in the compound yields 1 mol CO_2. That is, the x mol C in 1 mol $C_xH_yO_z$ appears in x mol CO_2. Each mole of hydrogen in the compound yields $\frac{1}{2}$ mol H_2O. That is, y mol H in 1 mol $C_xH_yO_z$ appears in $y/2$ mol H_2O. The mass percentages of carbon and hydrogen in the original sample can therefore be related to the masses of carbon dioxide and water obtained. Because oxygen atoms in the CO_2 and H_2O come from *both* the compound *and* the oxygen gas in which the compound is burned, the mass percent of oxygen in the compound must be determined indirectly, by subtracting the mass percentages of C and H from 100.00%.

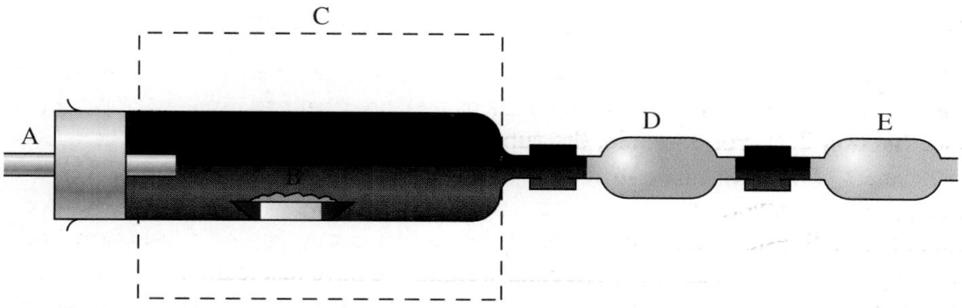

Figure 3.4 Apparatus for combustion analysis. Oxygen gas (A) streams over the sample under analysis (B) in a combustion tube enclosed within a high-temperature furnace (C). Water vapor produced in the combustion is absorbed by magnesium perchlorate (D) and carbon dioxide by sodium hydroxide (E). The masses of these products are determined by the differences in mass of the absorbers D and E after and before the combustion.

The specific conversions required to establish the mass percent composition of a compound from combustion analysis data are

$$\text{g } CO_2 \longrightarrow \text{mol } CO_2 \longrightarrow \text{mol C} \longrightarrow \text{g C} \longrightarrow \text{\% C in orig. sample}$$
$$\text{g } H_2O \longrightarrow \text{mol } H_2O \longrightarrow \text{mol H} \longrightarrow \text{g H} \longrightarrow \text{\% H in orig. sample}$$

These calculations and a typical application of combustion analysis are illustrated in Example 3.12.

Example 3.12

A 0.1000-g sample of a compound of C, H, and O is burned in oxygen to yield 0.1953 g CO_2 and 0.1000 g H_2O. Its molecular weight had previously been found to be 90 u.

a. Calculate the mass percent composition of the compound.
b. What is the empirical formula of the compound?
c. What is its molecular formula?

Solution

a. We do the conversions outlined on page 81, first to calculate the mass of carbon in the CO_2.

$$? \text{ g C} = 0.1953 \text{ g } CO_2 \times \frac{1 \text{ mol } CO_2}{44.010 \text{ g } CO_2} \times \frac{1 \text{ mol C}}{1 \text{ mol } CO_2} \times \frac{12.011 \text{ g C}}{1 \text{ mol C}}$$
$$= 0.05330 \text{ g C}$$

We then use this mass to calculate the mass percent of carbon in the compound.

$$?\% \text{ C} = \frac{0.05330 \text{ g C}}{0.1000 \text{ g sample}} \times 100\% = 53.30\% \text{ C}$$

Now we do the conversions to calculate the mass of hydrogen in the H_2O.

$$? \text{ g H} = 0.1000 \text{ g } H_2O \times \frac{1 \text{ mol } H_2O}{18.015 \text{ g } H_2O} \times \frac{2 \text{ mol H}}{1 \text{ mol } H_2O} \times \frac{1.0079 \text{ g H}}{1 \text{ mol H}}$$
$$= 0.01119 \text{ g H}$$

We then use this mass to calculate the mass percent of hydrogen in the compound.

$$\% \text{ H} = \frac{0.01119 \text{ g H}}{0.1000 \text{ g sample}} \times 100\% = 11.19\% \text{ H}$$

Finally, we calculate the mass percent of oxygen by difference

$$\% \text{ O} = 100.00 - 53.30\% - 11.19\% = 35.51\% \text{ O}$$

b. Here, we can again apply the method of Examples 3.9 and 3.10.
Step 1. A 100.00-g sample of the compound contains 53.30 g C, 11.19 g H, and 35.51 g O.
Step 2. Convert the masses of C, H, and O to numbers of moles.

$$53.30 \text{ g C} \times \frac{1 \text{ mol C}}{12.011 \text{ g C}} = 4.438 \text{ mol C}$$

$$11.19 \text{ g H} \times \frac{1 \text{ mol H}}{1.0079 \text{ g H}} = 11.10 \text{ mol H}$$

$$35.51 \text{ g O} \times \frac{1 \text{ mol O}}{15.999 \text{ g O}} = 2.220 \text{ mol O}$$

Step 3. Divide each of the quantities above by the smallest, and use the results as subscripts in a tentative formula.

$$C_{\frac{4.438}{2.220}}H_{\frac{11.10}{2.220}}O_{\frac{2.220}{2.220}} \longrightarrow C_{1.999}H_{5.000}O_{1.000} \longrightarrow C_2H_5O$$

c. Because the experimentally determined molecular weight (90 u) is almost exactly twice the empirical formula weight of C_2H_5O (45.1 u), the true molecular formula must be $C_4H_{10}O_2$.

Exercise 3.12

A 0.3629-g sample of tetrahydrocannabinol, a compound of C, H, and O and the principal active component of marijuana, is burned in oxygen to yield 1.067 g of carbon dioxide and 0.3120 g of water.

a. Calculate its mass percent composition.

b. Calculate its empirical formula.

3.7 Writing and Balancing Chemical Equations

As we have just seen, quite a few chemical calculations are based on chemical formulas. When we deal with chemical reactions, though, the range of calculations becomes even more extensive, and for these calculations, we must use chemical equations.

A **chemical equation** is a shorthand way of describing a chemical reaction, using symbols and formulas to represent the elements and compounds involved. For the reaction of carbon and oxygen to produce carbon dioxide we can write

$$C + O_2 \longrightarrow CO_2$$

The plus sign (+) indicates that carbon and oxygen combine in some way, and the arrow ($\longrightarrow$), usually read as "yields," points to the result of that combination. Substances on the left of the arrow are called the **reactants** or starting materials, and those on the right, the **products** of the reaction.

For some applications it is necessary to indicate the physical states of the reactants and products. We can do this through the symbols

(g) = gas; (l) = liquid; (s) = solid; (aq) = aqueous (water) solution

which are attached to the formulas of reactants and products. For example,

$$C(s) + O_2(g) \longrightarrow CO_2(g)$$

Chemical equations can be interpreted on the atomic and molecular levels, as in the statement, "One atom of carbon (C) reacts with one molecule of oxygen (O_2) to produce one molecule of carbon dioxide (CO_2)." In the laboratory we must work at the macroscopic level: "1 mol (12 g) of carbon reacts with 1 mol (32 g) of oxygen to produce 1 mol (44 g) of carbon dioxide." We will stress the macroscopic interpretation in the next section.

We can't represent all chemical reactions as simply as that between carbon and oxygen to form carbon dioxide. For the reaction of hydrogen and oxygen to form water, we might first write

$$H_2(g) + O_2(g) \longrightarrow H_2O(l) \quad \textit{(not balanced)}$$

However, this representation is not consistent with the law of conservation of mass. Two oxygen atoms are shown among the reactants, as O_2, and only one among the products, in H_2O; the equation is not balanced. For the equation to represent the chemical event correctly, it must be *balanced.* First, to balance oxygen atoms we place the coefficient "2" in front of the formula for water.

$$H_2(g) + O_2(g) \longrightarrow 2\,H_2O(l) \quad \textit{(oxygen balanced, hydrogen not balanced)}$$

This coefficient means that two molecules of water are involved, and therefore two oxygen atoms. A coefficient preceding a formula multiplies everything in the formula, and a coefficient of "1" is understood when no other number appears.

In the above equation, the coefficient 2 not only increases the number of oxygen atoms on the right to two, but it also increases the number of hydrogen atoms to four. The equation is still not balanced. To balance the numbers of hydrogen atoms, we place the coefficient "2" in front of H_2 on the left.

$$2\,H_2(g) + O_2(g) \longrightarrow 2\,H_2O(l) \quad \textit{(balanced)}$$

Now there are enough hydrogen atoms on the left. In fact, there are four hydrogen atoms and two oxygen atoms on each side of the equation. The law of conservation of mass is obeyed. Figure 3.5 illustrates two common pitfalls as well as the correct method of balancing equations.

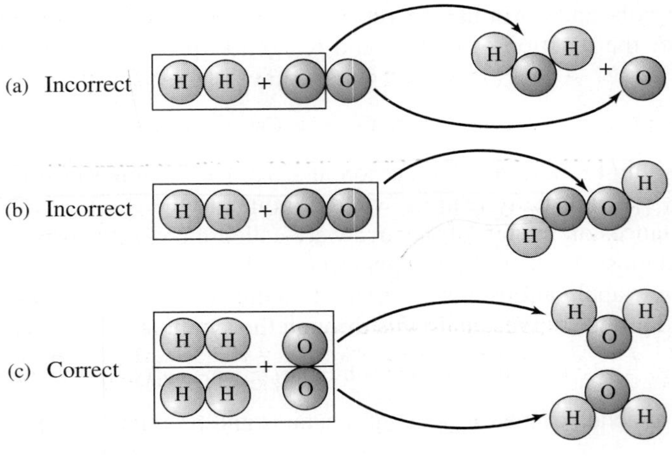

Figure 3.5 Balancing a chemical equation: the reaction of hydrogen and oxygen to form water. (a) *Incorrect*: There is no evidence of the presence of atomic oxygen (O) as a product. *Extraneous formulas cannot be introduced for the purpose of balancing an equation.* (b) *Incorrect*: The product of the reaction is water, H_2O, not hydrogen peroxide, H_2O_2, *A formula cannot be changed for the purpose of balancing an equation.* (c) *Correct*: An equation can be balanced only through the use of *correct formulas and coefficients.*

Although simple equations can be balanced by inspection (trial and error), there are strategies that can be used to balance them. For example,

- If an element occurs in just one substance on each side of the equation, try balancing that element *first*.
- If a reactant or product exists as the *free* element, try balancing that element *last*.

Perhaps the most important step in any strategy is to check an equation to ensure that it is indeed balanced: For each element, the same number of atoms of the element must appear on each side of the equation. Atoms are conserved in chemical reactions.

Example 3.13

Balance the following equation.

$$Fe + O_2 \longrightarrow Fe_2O_3$$

Solution

Begin by balancing the oxygen atoms. The least common multiple of two and three is six. We need *three* molecules of O_2 and *two* formula units of Fe_2O_3.

$$Fe + 3\,O_2 \longrightarrow 2\,Fe_2O_3 \quad (\textit{not balanced})$$

We now have four atoms of iron on the right side. We can get four on the left by placing the coefficient 4 in front of Fe.

$$4\,Fe + 3\,O_2 \longrightarrow 2\,Fe_2O_3 \quad (\textit{balanced})$$

Example 3.14

Balance the following equation.

$$C_2H_6 + O_2 \longrightarrow CO_2 + H_2O$$

Solution

Oxygen appears as the free element on the left. Let's leave oxygen for last and balance the other two elements first. To balance carbon, we place the coefficient 2 in front of CO_2.

$$C_2H_6 + O_2 \longrightarrow 2\,CO_2 + H_2O \quad (\textit{not balanced})$$

To balance hydrogen, we need the coefficient 3 before H_2O.

$$C_2H_6 + O_2 \longrightarrow 2\,CO_2 + 3\,H_2O \quad (\textit{still not balanced})$$

Now, if we count oxygen atoms, we find 2 on the left and 7 on the right. We can get 7 on each side by using the *fractional* coefficient $\frac{7}{2}$ on the left ($\frac{7}{2} \times 2 = 7$).

$$C_2H_6 + \tfrac{7}{2}\,O_2 \longrightarrow 2\,CO_2 + 3\,H_2O \quad (\textit{balanced})$$

There are some applications where we prefer equations with fractional coefficients, but generally we will write equations only with integral (whole number) coefficients. To do so here, we multiply each coefficient by 2. That is,

$$2 \times (C_2H_6 + \tfrac{7}{2}\,O_2 \longrightarrow 2\,CO_2 + 3\,H_2O)$$

leads to

$$2\,C_2H_6 + 7\,O_2 \longrightarrow 4\,CO_2 + 6\,H_2O \quad (\textit{balanced})$$

(To check that the equation is balanced, we count four C atoms, 12 H atoms, and 14 O atoms on each side.)

Example 3.15

Balance the following equation.

$$H_2SO_4 + NaCN \longrightarrow HCN + Na_2SO_4$$

Solution

Notice that the SO_4 and CN groups remain unchanged in the reaction. For purposes of balancing the equation, we can treat each group as a whole rather than breaking it down into its constituent atoms. To balance hydrogen, we place a 2 before HCN.

$$H_2SO_4 + NaCN \longrightarrow 2\,HCN + Na_2SO_4 \quad (\textit{not balanced})$$

To balance sodium, atoms, we put a 2 in front of the NaCN.

$$H_2SO_4 + 2\,NaCN \longrightarrow 2\,HCN + Na_2SO_4 \quad (\textit{balanced})$$

Note that in the final balanced equation we find one SO_4 group and two CN groups on each side of the equation.

Exercise 3.13

Balance the following equations.

a. $Mg + B_2O_3 \longrightarrow B + MgO$

b. $NO_2 + H_2O \longrightarrow HNO_3 + NO$

c. $H_2 + Fe_2O_3 \longrightarrow Fe + H_2O$

d. $CaO + P_4O_{10} \longrightarrow Ca_3(PO_4)_2$

e. $C_5H_{12} + O_2 \longrightarrow CO_2 + H_2O$

f. $C_4H_{10} + O_2 \longrightarrow CO_2 + H_2O$

CONCEPTUAL EXAMPLE 3.1

Write as complete a chemical equation as you can for the combustion of liquid 1,4-butanediol in an abundant supply of oxygen gas.

Solution

The key to writing this equation is to identify 1,4-butanediol, and this requires us to use several ideas from Chapter 2. A portion of the name signifies a relationship to the four-carbon alkane *butane*, C_4H_{10}. The *-ol* ending indicates that the liquid is an alcohol, and thus contains the group —OH. The prefix "di" tells us that there are *two* —OH groups. In alcohols, —OH groups substitute for H atoms in a hydrocarbon molecule. 1,4-Butanediol has the molecular formula $C_4H_8(OH)_2$. As we learned in Section 3.6, the complete combustion of a compound of C, H, and O yields carbon dioxide gas and water as products.

$$C_4H_8(OH)_2(l) + O_2(g) \longrightarrow CO_2(g) + H_2O(l) \quad \textit{(not balanced)}$$

First, we balance the C and H atoms. Notice that to count H atoms on the left we must combine the H atoms from "C_4H_8" and "$(OH)_2$," for a total of 10.

$$C_4H_8(OH)_2(l) + O_2(g) \longrightarrow 4\,CO_2(g) + 5\,H_2O(l) \quad \textit{(not balanced)}$$

When balancing O atoms, notice that two O atoms are associated with "$(OH)_2$" and the rest with O_2(g) molecules.

$$C_4H_8(OH)_2(l) + \tfrac{11}{2}\,O_2(g) \longrightarrow 4\,CO_2(g) + 5\,H_2O(l) \quad \textit{(balanced)}$$

Finally, to remove the fractional coefficient multiply all coefficients by two.

$$2\,C_4H_8(OH)_2(l) + 11\,O_2(g) \longrightarrow 8\,CO_2(g) + 10\,H_2O(l) \quad \textit{(balanced)}$$

Conceptual Exercise 3.1

Electricity is produced by an automotive lead storage battery as a result of the reaction of lead metal, solid lead(IV) oxide, and an aqueous solution of sulfuric acid to produce solid lead(II) sulfate and liquid water. Write a complete balanced equation to represent this reaction.

Looking Ahead

Some equations of a type called oxidation–reduction are difficult to balance by inspection. We will develop a method for these equations in Chapter 12. A more important activity than simply balancing an equation is *predicting* whether a chemical reaction will occur and then writing an equation to represent it. We will learn to do this in several chapters later in the text.

3.8 Stoichiometry of Chemical Reactions

Earlier in this chapter we stressed mole and mass relationships based on chemical *formulas*. Whether making medicines, obtaining metals from their ores, studying the combustion of a rocket fuel, synthesizing new compounds, or testing hypotheses, chemists also need to consider mole and mass relationships in chemical reactions. These relationships are derived from chemical *equations*.

Consider the combustion of propane in air to form carbon dioxide and water.

$$C_3H_8 + 5\ O_2 \longrightarrow 3\ CO_2 + 4\ H_2O$$

We can interpret the coefficients in this equation to mean *one* molecule of C_3H_8, *five* molecules of O_2, and so on. Also, we can "scale up" these quantities by multiplying each of them by Avogadro's number (N_A). This leads to statements such as

1 mol C_3H_8 reacts with 5 mol O_2;
3 mol CO_2 is produced for every 1 mol C_3H_8 that reacts;
4 mol H_2O is produced for every 3 mol CO_2 produced;

and so on. Moreover, we can turn these statements into conversion factors known as stoichiometric factors. A **stoichiometric factor** relates the amounts of any two substances involved in a chemical reaction, on a *mole* basis. In the examples that follow stoichiometric factors are shown in red.

Example 3.16

When 0.105 mol propane is burned in a rich supply of oxygen, how many moles of oxygen are consumed.

$$C_3H_8 + 5\ O_2 \longrightarrow 3\ CO_2 + 4\ H_2O$$

Solution

The equation tells us that 5 mol O_2 is required to burn 1 mol C_3H_8. This gives us the stoichiometric factor 5 mol O_2/1 mol C_3H_8 to use as a conversion factor in the calculation.

$$?\ \text{mol } O_2 = 0.105\ \text{mol } C_3H_8 \times \frac{5\ \text{mol } O_2}{1\ \text{mol } C_3H_8} = 0.525\ \text{mol } O_2$$

Exercise 3.14

For the combustion of propane in Example 3.16:

a. How many moles of carbon dioxide are formed when 0.529 mol C_3H_8 is burned?
b. How many moles of water are produced when 76.2 mol C_3H_8 is burned?
c. How many moles of carbon dioxide are produced when 1.010 mol O_2 is consumed?

Although the mole is essential in calculations based on chemical equations, we cannot measure out molar amounts directly. We have to relate them to quantities that we can measure: mass in grams or kilograms, volume in milliliters or liters, and so on. If the quantities of substances are expressed through their masses, we can follow the four-step approach outlined below.

Step 1. Write a balanced equation for the reaction.
Step 2. Convert grams of the substance for which information is given, the *given* substance, to moles of that substance using its molar mass. (The given substance may be either a reactant or a product.)

Step 3. Obtain a stoichiometric factor from the balanced equation to convert from moles of the given substance to moles of the substance about which information is sought, the *desired* substance.

Step 4. Use its molar mass to convert moles of the desired substance to grams of that substance.

We illustrate this four-step approach in Figure 3.6 and Example 3.17, where we also show how the steps can easily be combined into the preferred single-setup method.

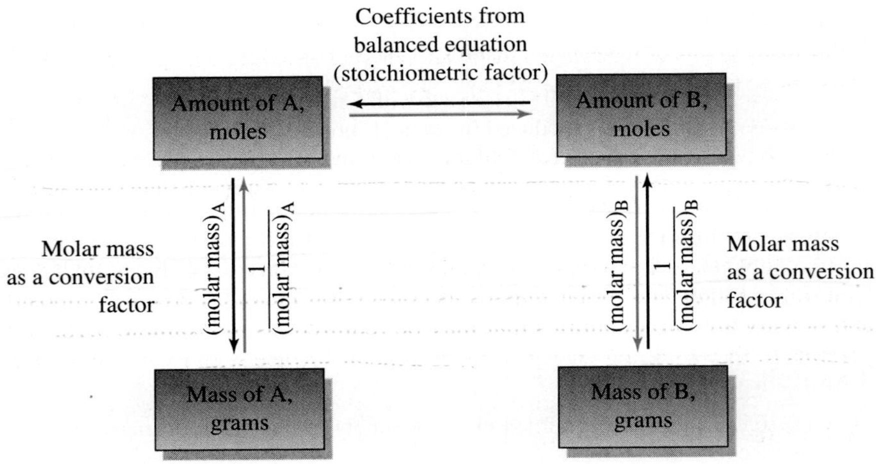

Figure 3.6 Relating masses of a reactant (A) and product (B) through a chemical equation. The required unit conversions are from grams to moles of one substance, to moles of a second substance, and then to grams of the second substance, by the routes traced by the arrows. The required conversion factors are molar masses and a stoichiometric factor from the balanced equation.

Example 3.17

Ammonia, NH_3, a common fertilizer, is made by causing hydrogen and nitrogen to react at a high temperature and pressure. How many grams of ammonia can be made from 60.0 g of hydrogen?

Solution

Step 1. We first write a chemical equation

$$N_2 + H_2 \longrightarrow NH_3 \quad \textit{(not balanced)}$$

and balance it.

$$N_2 + 3\,H_2 \longrightarrow 2\,NH_3 \quad \textit{(balanced)}$$

Step 2. We convert the mass of the given substance, hydrogen, to an amount in moles.

$$?\ \text{mol } H_2 = 60.0\ \text{g } H_2 \times \frac{1\ \text{mol } H_2}{2.016\ \text{g } H_2} = 29.8\ \text{mol } H_2$$

Step 3. We use coefficients from the balanced equation to establish the stoichiometric factor that relates the amount of ammonia to that of hydrogen.

$$?\ \text{mol } NH_2 = 29.8\ \text{mol } H_2 \times \frac{2\ \text{mol } NH_3}{3\ \text{mol } H_2} = 19.9\ \text{mol } NH_3$$

Step 4. We convert from moles of ammonia to grams of ammonia.

$$? \text{ g NH}_3 = 19.9 \text{ mol NH}_3 \times \frac{17.03 \text{ g NH}_3}{1 \text{ mol NH}_3} = 339 \text{ g NH}_3$$

As is usually the case, all of the steps outlined above can be combined into a single setup.

$$? \text{ g NH}_3 = 60.0 \text{ g H}_2 \times \frac{1 \text{ mol H}_2}{2.016 \text{ g H}_2} \times \frac{2 \text{ mol NH}_3}{3 \text{ mol H}_2} \times \frac{17.03 \text{ g NH}_3}{1 \text{ mol NH}_3}$$

$$= 338 \text{ g NH}_3$$

Exercise 3.15

How many grams of magnesium metal are required to reduce 83.6 g of titanium(IV) chloride to titanium metal? Magnesium chloride is the other product.

Exercise 3.16

The decomposition of potassium chlorate produces potassium chloride and oxygen gas. How many grams of oxygen can be made from 2.47 g of potassium chlorate?

A calculation based on a chemical equation often requires more than a stoichiometric factor and molar masses as conversion factors. Percent composition and density are two quantities that may be required, as in Example 3.18, which pertains to the "pickling" or cleaning of a metal surface with hydrochloric acid.

Example 3.18

A sheet of iron with a surface area of 525 cm^2 is covered with a coating of rust that has an average thickness of 0.0021 cm. What mass of a hydrochloric acid solution that is 15% HCl by mass is required to clean the surface of the metal by dissolving the rust? Assume that the rust is iron(III) oxide and that its density is 5.2 g/cm^3.

$$\text{Fe}_2\text{O}_3\text{(s)} + 6 \text{ HCl(aq)} \longrightarrow 2 \text{ FeCl}_3\text{(aq)} + 3 \text{ H}_2\text{O(l)}$$

Solution

As suggested by Figure 3.7, we begin by calculating the volume of Fe_2O_3 and its mass. This is followed by the usual three steps to produce a result in grams of HCl. The final conversion from grams of HCl to grams of HCl solution requires a factor that is the inverse of percent composition. All these steps can be easily combined into a single setup.

$$? \text{ g HCl soln} = 525 \text{ cm}^2 \times 0.0021 \text{ cm} \times \frac{5.2 \text{ g Fe}_2\text{O}_3}{1 \text{ cm}^3} \times \frac{1 \text{ mol Fe}_2\text{O}_3}{159.7 \text{ g Fe}_2\text{O}_3}$$

$$\times \frac{6 \text{ mol HCl}}{1 \text{ mol Fe}_2\text{O}_3} \times \frac{36.46 \text{ g HCl}}{1 \text{ mol HCl}} \times \frac{100 \text{ g HCl soln}}{15 \text{ g HCl}}$$

$$= 52 \text{ g HCl soln}$$

Exercise 3.17

How many milliliters of liquid water should be produced by the combustion in oxygen of 775 mL of liquid octane? Assume that the volumes of both liquids are measured at 20 °C, where the densities are 0.7025 g/mL for octane and 0.9982 g/mL for water. (*Hint:* First write a balanced equation.)

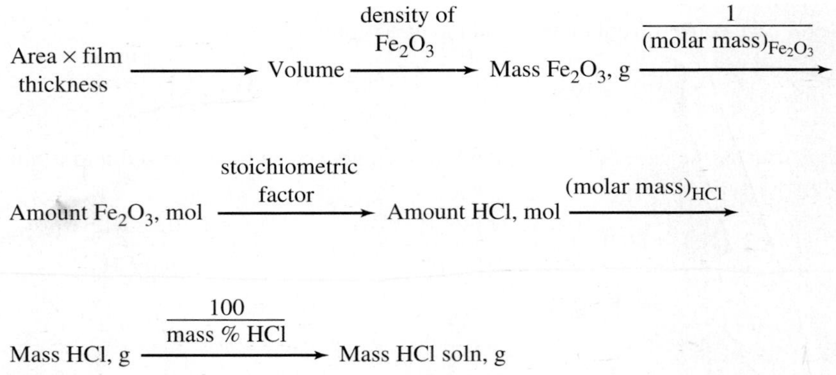

Figure 3.7 Outline of a stoichiometric calculation: Example 3.18 visualized. The quantities arrived at in each step of the calculation are joined by arrows. The required conversion factors are written above the arrows.

3.9 Limiting Reactants

When we mix reactants in mole ratios according to the coefficients in a balanced equation, we say that the reactants are in **stoichiometric proportions**. All initial reactants are totally consumed when the reaction goes to completion. However, we carry out most reactions with a limited amount of one reactant and excess quantities of the others. We do this to minimize the extent to which a reverse reaction can occur and thus maximize the yield of products.

Sometimes the products of a reaction react to re-form the original reactants. Such reactions are *reversible*. We discuss reversible reactions in Chapter 16.

The reactant that is completely consumed limits the amounts of products formed and is called the **limiting reactant** or **limiting reagent**. (Reagent is a general term for a chemical. When used in a chemical reaction, a reagent is called a reactant.) In combustion analysis (recall Figure 3.4) the amounts of CO_2 and H_2O produced depend only on the amount of substance that is burned, the limiting reactant, and not on the amount of oxygen gas present, a reactant in excess.

The situation is much like the task illustrated in Figure 3.8, that of packaging a snack meal for airline passengers. Each package consists of a sandwich, two cookies, and an orange; that is, the "stoichiometric" proportions are 1:2:1. If we have 100 sandwiches, 200 cookies, and 100 oranges, we can prepare 100 snack meals and have nothing left over. But from 98 sandwiches, 202 cookies, and 102 oranges we can assemble only 98 packages. The sandwiches are the "limiting reactant," and the cookies and oranges are reactants in excess. From 105 sandwiches, 202 cookies, and 107 oranges we can assemble 101 packaged meals. Here the cookies are the "limiting reactant." In effect, what we do is determine how much of the product can be made from each "reactant" assuming we have enough of the others. We then pick the *smallest* answer.

Figure 3.8 An analogy to determining a limiting reactant. To package the snack meal shown on the right requires sandwiches, cookies, and oranges in the ratio 1:2:1. For the situation shown, cookies are the limiting reactant.

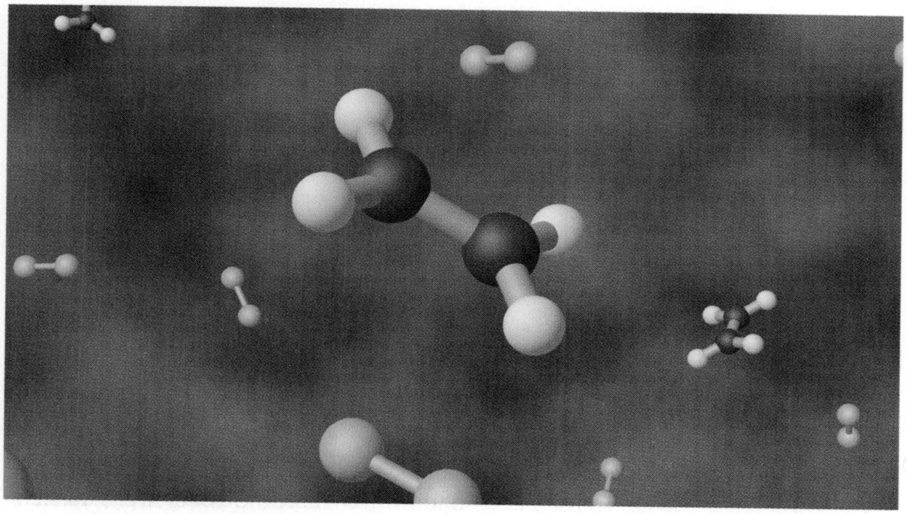

The reaction of ethylene and bromine. Ethylene and bromine react in a 1:1 mole ratio.

$$C_2H_4(g) + Br_2(g) \longrightarrow C_2H_4Br_2(g)$$

Here, bromine molecules outnumber ethylene molecules. Bromine is present in excess and ethylene is the limiting reactant.

Example 3.19

Potassium iodide is used as a dietary supplement to prevent goiter, an iodine deficiency disease. It is prepared by the reaction of hydrogen iodide and potassium hydrogen carbonate.

$$HI + KHCO_3 \longrightarrow KI + H_2O + CO_2$$

a. How many grams of KI can be made by mixing 481 g HI and 318 g $KHCO_3$?
b. Which reactant is in excess, and how many grams of it remain after the reaction?

Solution

a. First we calculate the number of moles of KI that could be produced from each reactant, assuming that one to be the limiting reactant.

$$? \text{ mol KI} = 481 \text{ g HI} \times \frac{1 \text{ mol HI}}{127.9 \text{ g HI}} \times \frac{1 \text{ mol KI}}{1 \text{ mol HI}} = 3.76 \text{ mol KI}$$

$$? \text{ mol KI} = 318 \text{ g KHCO}_3 \times \frac{1 \text{ mol KHCO}_3}{100.1 \text{ g KHCO}_3} \times \frac{1 \text{ mol KI}}{1 \text{ mol KHCO}_3} = 3.18 \text{ mol KI}$$

$KHCO_3$ is the limiting reactant, and the mass of the product is

$$? \text{ g KI} = 3.18 \text{ mol KI} \times \frac{166.0 \text{ g KI}}{1 \text{ mol KI}} = 528 \text{ g KI}$$

b. Having found that the amount of product is 3.18 mol KI, we can now calculate how much HI must have been consumed.

$$? \text{ g HI} = 3.18 \text{ mol KI} \times \frac{1 \text{ mol HI}}{1 \text{ mol KI}} \times \frac{127.9 \text{ g HI}}{1 \text{ mol HI}} = 407 \text{ g HI}$$

The mass of HI present in excess is

$$481 \text{ g HI}_{\text{initially}} - 407 \text{ g HI}_{\text{consumed}} = 74 \text{ g HI}_{\text{in excess}}$$

Exercise 3.18

Hydrogen sulfide is a foul-smelling gas that was extensively used at one time to precipitate (separate from solution as solids) metal sulfides in a laboratory scheme to identify certain cations in aqueous solutions. One way to produce the gas is by the reaction of iron(II) sulfide with hydrochloric acid.

$$FeS(s) + 2\ HCl(aq) \longrightarrow FeCl_2(aq) + H_2S(g)$$

If 10.2 g HCl is added to 13.2 g FeS, how many grams of H_2S can be formed?

3.10 Percent Yield

The calculated amount of product in a reaction is called the **theoretical yield** of the reaction. In Example 3.19, the calculated quantity is 528 g; that is, 528 g is the theoretical yield of potassium iodide. If you carried out the reaction described in the example and determined the quantity of product by weighing it—the **actual yield**—you would likely obtain *less* than the theoretical yield of 528 g. Actual yields of chemical reactions are often less than the calculated or theoretical yields for a variety of reasons (see Figure 3.9). The starting materials may not be pure, meaning that the actual quantities of reactants are less than what is weighed out. Some of the product may be left behind during the process of separating it from excess reactants. Side reactions may occur in addition to the main reaction, converting some of the original reactants into products other than the desired one. In many cases, a reverse reaction may prevent complete conversion of the limiting reactant to product.

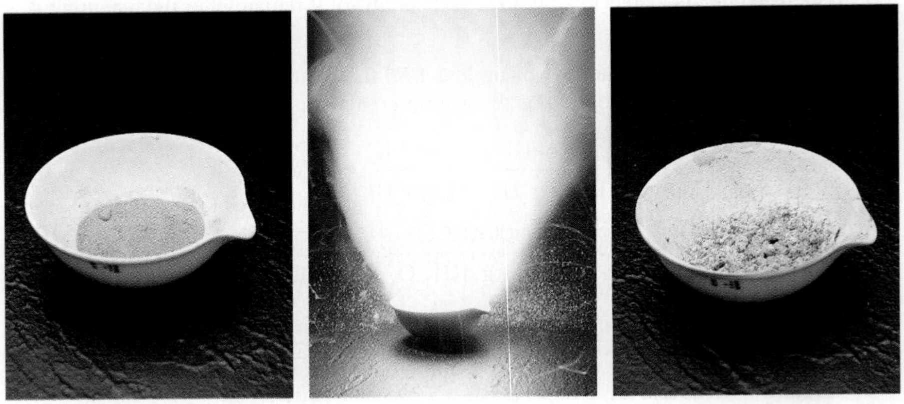

(a) (b) (c)

Figure 3.9 A reaction with a less-than-100% yield.

$$8\ Zn(s) + S_8(s) \longrightarrow 8\ ZnS(s)$$

The actual yield of ZnS(s) obtained in (c) is less than that calculated for the starting mixture in (a) for several reasons:

- Neither the powdered zinc nor the powdered sulfur is pure, and as suggested in (b)
- Zinc can combine with $O_2(g)$ in air to produce ZnO(s),
- Some of the sulfur burns in air to produce SO_2 (g),
- Some of the product escapes from the reaction mixture as lumps of material and as a fine dust.

Yields usually are expressed as percentages. The **percent yield** is the ratio of the actual yield to the theoretical yield times 100%.

$$\text{Percent yield} = \frac{\text{actual yield}}{\text{theoretical yield}} \times 100\%$$

If the actual yield of KI in Example 3.19 had been 496 g, the percent yield would have been

$$\text{Percent yield} = \frac{496\ \text{g}}{528\ \text{g}} \times 100\% = 93.9\%$$

Yields in Organic Reactions

Yield calculations are important in synthesis reactions, especially those in organic chemistry. Consider, for example, the reaction by which ethyl alcohol is converted to diethyl ether, a common laboratory solvent once widely used as an anesthetic. The reaction is carried out in the presence of sulfuric acid, a fact that is indicated by writing the formula H_2SO_4 above the yield arrow.

$$2\,CH_3CH_2OH \xrightarrow{H_2SO_4} \underset{\text{Diethyl ether}}{CH_3CH_2OCH_2CH_3} + H_2O$$

The reaction appears to be straightforward, but there are several complications. There is an important side reaction: some of the ethyl alcohol is converted to ethylene, a hydrocarbon with a double bond (described in Chapter 9).

$$CH_3CH_2OH \longrightarrow \underset{\text{Ethylene}}{CH_2{=}CH_2} + H_2O$$

Any ethanol that is converted to ethylene obviously cannot also form diethyl ether. The yield of diethyl ether is diminished.

There are practical problems also. For example, the diethyl ether is purified by distilling it from the reaction mixture, and some will always remain in the distillation glassware. Also, some of the ethanol may distill with the ether, effectively removing it as a reactant. Even under the best of conditions, yields above 80 to 85% are difficult to achieve. Chemists often have to settle for 50%—and sometimes even less than that.

Example 3.20

Ethyl acetate is a solvent used as fingernail polish remover. How many grams of acetic acid are needed to prepare 252 g of ethyl acetate if the expected percent yield is only 85.0%? Assume that the other reactant, ethyl alcohol, is present in excess. The equation for the reaction, carried out in the presence of H_2SO_4, is

$$\underset{\text{Acetic acid}}{CH_3COOH} + \underset{\text{Ethanol}}{HOCH_2CH_3} \xrightarrow{H_2SO_4} \underset{\text{Ethyl acetate}}{CH_3COOCH_2CH_3} + HOH$$

Solution

First, determine the theoretical yield of this reaction.

$$\text{Percent yield} = \frac{\text{actual yield}}{\text{theoretical yield}} \times 100\%$$

$$85.0\% = \frac{252 \text{ g ethyl acetate}}{\text{theoretical yield}} \times 100\%$$

$$\text{Theoretical yield} = 252 \text{ g ethyl acetate} \times \frac{100}{85.0} = 296 \text{ g ethyl acetate}$$

If the percent yield is only 85.0%, we need a theoretical yield of 296 g of ethyl acetate to get an actual yield of 252 g. The remainder of the calculation is to determine the mass of acetic acid required to produce 296 g of ethyl acetate.

$$? \text{ g } CH_3COOH = 296 \text{ g } CH_3COOCH_2CH_3 \times \frac{1 \text{ mol } CH_3COOCH_2CH_3}{88.11 \text{ g } CH_3COOCH_2CH_3}$$

$$\times \frac{1 \text{ mol } CH_3COOH}{1 \text{ mol } CH_3COOCH_2CH_3} \times \frac{60.05 \text{ g } CH_3COOH}{1 \text{ mol } CH_3COOH}$$

$$= 202 \text{ g } CH_3COOH$$

Exercise 3.19

Isopentyl acetate is the main component of banana flavoring. Calculate the theoretical yield of isopentyl acetate that can be obtained from 20.0 g of isopentyl alcohol and 25.0 g of acetic acid.

$$\underset{\text{Acetic acid}}{CH_3COOH} + \underset{\text{Isopentyl alcohol}}{HOCH_2CH_2CH(CH_3)_2} \longrightarrow \underset{\text{Isopentyl acetate}}{CH_3COOCH_2CH_2CH(CH_3)_2} + HOH$$

If the percent yield of the reaction is 90.0%, what is the actual yield of isopentyl acetate?

Exercise 3.20

How many grams of isopentyl alcohol are needed to make 433 g of isopentyl acetate in the reaction described in Exercise 3.19 if the expected yield is 85.0%? Assume the acetic acid is in excess.

3.11 Solutions and Solution Stoichiometry

Because reactions between solids usually are extremely slow, chemists often carry out reactions in solution. Most of the reactions in our bodies also occur in aqueous solution.

As we learned in Chapter 1, a solution is a homogeneous mixture of two or more substances. A solution of sugar in water does not consist of tiny particles of solid sugar dispersed among droplets of liquid water. Rather, individual sugar molecules are randomly distributed among water molecules in a uniform liquid medium. A solution is *homogeneous*; the composition and physical and chemical properties are identical in all portions of a solution.

The components of a solution are the **solute**(s), the substance(s) being dissolved, and the **solvent**, the substance doing the dissolving. The solutes are usually the components present in smaller amounts, and the solvent is usually present in the greatest amount. There are many solvents: Hexane dissolves grease. Ethanol dissolves many drugs. Isopentyl acetate, a component of banana oil, is a solvent for the glue used in making model airplanes. Water is no doubt the most familiar solvent, dissolving as it does many common substances such as sugar, salt, and ethanol. We limit our discussion here to *aqueous* solutions, those in which water is the solvent.

The *concentration* of a solution refers to the quantity of a solute in a given quantity of solvent or of solution. A *dilute* solution is one that contains relatively little solute in lots of solvent. A *concentrated* solution contains a relatively large amount of solute in a given quantity of solvent. These terms are imprecise and can be used only subjectively. For example, a dilute sugar solution tastes faintly sweet and a concentrated solution has a sickeningly sweet taste, but the terms "dilute" and "concentrated" do not in themselves tell us the precise proportions of sugar and water.

For commercially available acids and bases, the term *concentrated* generally signifies the highest concentration commonly available. These concentrations are usually expressed as percentages. Commercial concentrated hydrochloric acid is about 38% HCl by mass; the rest is water. Commercial concentrated sulfuric acid contains 93 to 98% H_2SO_4 by mass, and, again, the rest is water.

Molar Concentration

There are several ways to describe the concentration of a solution. The method chosen often depends on how the solution is to be used. Two facts dictate the choice most often made by chemists.

- Substances enter into chemical reactions according to certain *molar* ratios.
- Volumes of solutions are more convenient to measure than their masses.

The concentration unit that chemists use most is *molarity*. The **molarity (M)** or **molar concentration** is the amount of solute, in moles, per liter of solution.

Recall from Chapter 1 that the liter is the same as a cubic decimeter: 1 L = 1 dm³. The derived SI unit for molarity is moles per cubic decimeter, mol/dm³. Although the liter isn't an official SI unit, the unit mol/L is still widely used.

$$\text{Molarity (M)} = \frac{\text{moles of solute}}{\text{liters of solution}}$$

The molarity of a solution made by dissolving 3.50 mol NaCl in enough water to produce 2.00 L of solution is

$$\text{Molarity} = \frac{\text{moles of solute}}{\text{liters of solution}} = \frac{3.50 \text{ mol NaCl}}{2.00 \text{ L solution}} = 1.75 \text{ M NaCl}$$

We read "1.75 M NaCl" as "1.75 molar NaCl."

It is important to understand that molarity signifies moles of solute per liter of *solution*, not per liter of solvent. Thus, to make 1 L of 0.01000 M $KMnO_4$ we weigh out 1.580 g $KMnO_4$ and add it to a 1.000-L volumetric flask partially filled with water. After the solid has completely dissolved, we add more water to bring the volume up to the mark that indicates a solution volume of 1.000 L. This procedure is outlined in Figure 3.10. Of course, we can use a volumetric flask of any volume and a proportionate quantity of solute, as shown in Example 3.21.

Example 3.21

What is the molarity of a solution in which 333 g of potassium hydrogen carbonate is dissolved in enough water to make 10.0 L of solution?

Solution

First, prepare the setup that would convert from mass of $KHCO_3$ to the number of moles of $KHCO_3$.

$$333 \text{ g } KHCO_3 \times \frac{1 \text{ mole } KHCO_3}{100.1 \text{ g } KHCO_3}$$

Now, without solving this expression, use it as the numerator in the defining equation for molarity. The solution volume, 10.0 L, is the denominator.

$$\text{Molarity} = \frac{333 \text{ g KHCO}_3 \times \dfrac{1 \text{ mole KHCO}_3}{100.1 \text{ g KHCO}_3}}{10.0 \text{ L solution}} = 0.333 \text{ M KHCO}_3$$

Exercise 3.21

Calculate the molarity of each of the following solutions.

a. 18.0 mol H_2SO_4 in 2.00 L of solution
b. 3.00 mol KI in 2.39 L of solution
c. 0.206 mol HF in 752 mL of solution
d. 0.522 g HCl in 0.592 L of solution
e. 4.98 g $C_6H_{12}O_6$ in 224 mL of solution
f. 10.5 g C_2H_5OH in 24.7 mL of solution

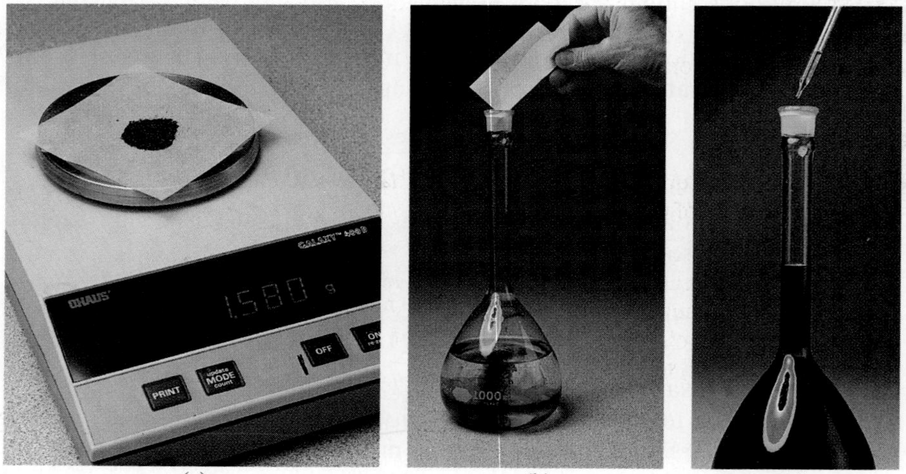

(a) (b) (c)

Figure 3.10 Preparation of 0.01000 M $KMnO_4$. (a) The balance is set to zero (tared) when just the weighing paper is present. The sample of $KMnO_4$ weighs 1.580 g (0.01000 mole). (b) The $KMnO_4$ is dissolved in water in the partially filled 1.000-L volumetric flask. (c) More water is added and the solution is thoroughly mixed. Finally, the flask is filled to the mark by adding a small quantity of water one drop at a time.

Estimation Exercise 3.5

You have available 10.0 g of each of the following solutes.

$CO(NH_2)_2$ $C_6H_{12}O_6$ CH_3CH_2OH NH_3

When used to prepare 100.0 mL of solution, which one yields the solution of greatest molarity?

Frequently we need to know the mass of solute required to prepare a given volume of solution of a specified molarity. In such calculations we can use molarity as a conversion factor between moles of solute and liters of solution. Thus, in Example 3.22, "6.67 M NaOH" means 6.67 mol NaOH per liter of solution, expressed as the conversion factor 6.67 mol NaOH/1 L soln.

Example 3.22

How many grams of NaOH are required to prepare 0.500 L of 6.67 M NaOH?

Solution

The required conversions are L soln ⟶ mol NaOH ⟶ g NaOH, and we can do them in a single setup.

$$? \text{ g NaOH} = 0.500 \text{ L soln} \times \frac{6.67 \text{ mol NaOH}}{1 \text{ L soln}} \times \frac{40.01 \text{ g NaOH}}{1 \text{ mol NaOH}}$$

$$= 133 \text{ g NaOH}$$

Exercise 3.22

How many grams of potassium hydroxide are required to prepare each of the following solutions?

a. 2.00 L of 6.00 M KOH
b. 100.0 mL of 1.00 M KOH
c. 10.0 mL of 0.100 M KOH
d. 33.0 mL of 2.50 M KOH

Labels on bottles of stock solutions of acids and bases often indicate concentrations only in percent solute by mass. Sometimes we need to calculate the molarity of such a solution, but, as illustrated in Example 3.23, to do so we need to know the solution density as well.

Example 3.23

A stock bottle of aqueous ammonia indicates that the solution is 28.0% NH_3 by mass and has a density of 0.898 g/mL. Calculate the molarity of the solution.

Solution

The most convenient volume of solution on which to base the calculation is 1.00 L. To determine the number of moles of NH_3 in this 1.00 L of solution, we need to convert (1) L soln to mL soln; (2) mL soln to g soln (density); (3) g soln to g NH_3 (mass percent); (4) g NH_3 to mol NH_3.

$$? \text{ mol NH}_3 = 1.00 \text{ L soln} \times \overset{(1)}{\frac{1000 \text{ mL soln}}{1 \text{ L soln}}} \times \overset{(2)}{\frac{0.898 \text{ g soln}}{1 \text{ mL soln}}} \times \overset{(3)}{\frac{28.0 \text{ g NH}_3}{100 \text{ g soln}}}$$

$$\times \overset{(4)}{\frac{1 \text{ mol NH}_3}{17.03 \text{ g NH}_3}} = 14.8 \text{ mol NH}_3$$

Do you see why we chose 1.00 L of solution? This made the number of moles of NH_3 the same as the solution molarity and simplified the calculation.

Finally, calculate the molarity.

$$\text{Molarity} = \frac{\text{moles of NH}_3}{\text{liters of soln}} = \frac{14.8 \text{ mol NH}_3}{1.00 \text{ L soln}} = 14.8 \text{ M NH}_3$$

Exercise 3.23

A stock bottle of aqueous formic acid indicates that the solution is 90.0% HCOOH by mass and has a density of 1.20 g/mL. Calculate the molarity of the solution.

Dilution of Solutions

Quite often in the laboratory we need to make a more dilute solution from a more concentrated one. This process, called **dilution**, is a common procedure for making solutions of the desired concentration. Concentrated solutions, often ones that are commercially available, are kept in stock in a storeroom. The principle of dilution is that

> *addition of solvent does not change the amount of solute in a solution but does change the solution concentration.*

We apply this principle in Example 3.24.

Example 3.24

How many milliliters of a 2.00 M $CuSO_4$ stock solution would you use to prepare 0.250 L of 0.400 M $CuSO_4$?

Solution

The key to this problem lies in noting that all the solute in the unknown volume of the stock solution appears in the 0.250 L of 0.400 M $CuSO_4$. First, let's calculate the amount of solute in question.

$$? \text{ mol } CuSO_4 = 0.250 \text{ L} \times \frac{0.400 \text{ mol } CuSO_4}{1 \text{ L}} = 0.100 \text{ mol } CuSO_4$$

Now we can answer the question: "What volume of 2.00 M $CuSO_4$ contains 0.100 mol $CuSO_4$?" In doing so we will have answered the original question.

$$? \text{ mL} = 0.100 \text{ mol } CuSO_4 \times \frac{1 \text{ L}}{2.00 \text{ mol } CuSO_4} \times \frac{1000 \text{ mL}}{1 \text{ L}} = 50.0 \text{ mL}$$

Of course we could have done all of this in a single setup.

$$? \text{ mL} = 0.250 \text{ L} \times \frac{0.400 \text{ mol } CuSO_4}{1 \text{ L}} \times \frac{1 \text{ L}}{2.00 \text{ mol } CuSO_4} \times \frac{1000 \text{ mL}}{1 \text{ L}}$$

$$= 50.0 \text{ mL}$$

Either way, we conclude that to make the dilute solution, measure out 50.0 mL of 2.00 M $CuSO_4$ and add enough water to make 0.250 L of solution, as illustrated in Figure 3.11.

Exercise 3.24

How many milliliters of a 10.15 M NaOH stock solution would you use to prepare 15.0 L of 0.315 M NaOH?

Visualizing the dilution of a solution. When pure water is added to an aqueous solution of methanol (left), a more dilute solution is produced (right). However, the diluted solution contains the same number of CH_3OH molecules as the original solution.

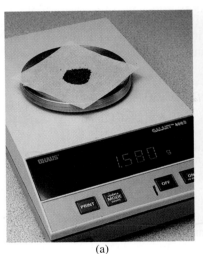

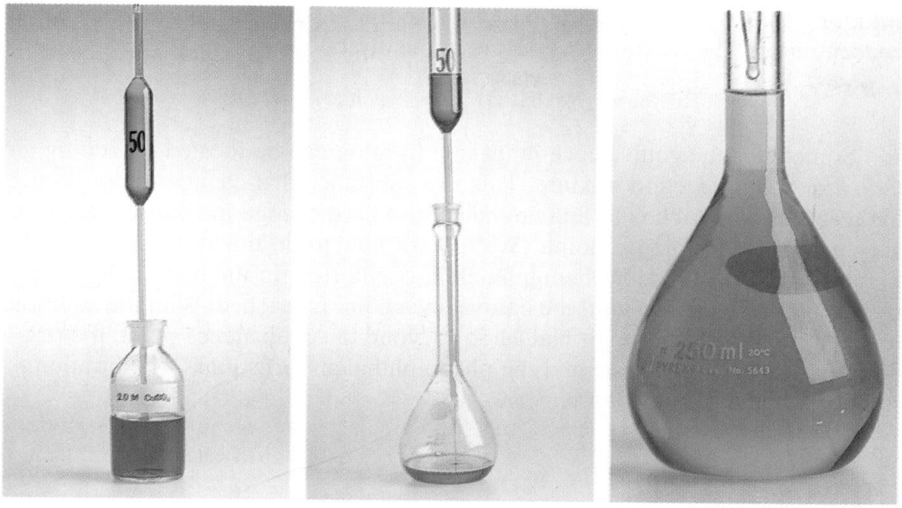

(a) (b) (c)

Figure 3.11 Dilution of a copper(II) sulfate solution: Example 3.24 illustrated. (a) The pipet is being filled with 50.0 mL of 2.00 M $CuSO_4$. The amount of $CuSO_4$ in the filled pipet will be 0.100 mole. (b) The 50.0 mL of 2.00 M $CuSO_4$ is transferred to a 250.0-mL volumetric flask; water is added and the solution is thoroughly mixed. (c) Finally, the flask is filled to the mark by adding the remaining water dropwise.

Solutions in Chemical Reactions

Molarity gives us a way to obtain a given number of moles of a solute by measuring out a calculated volume of solution. Such a volume measurement is often made in a **buret**, a graduated, long glass tube constructed to deliver precise volumes of solution through a stopcock valve. A buret is the chief instrument used in a **titration**, a procedure in which two reactants in solution are made to react in their stoichiometric proportions.

Consider, for example, a typical acid–base titration: A measured volume of an acid solution of *unknown* concentration is transferred to a flask. Then, a solution of a base of *known* concentration is added carefully from a buret until the reaction of the acid with the base is just complete. This is called the *equivalence point* of the titration. Figure 3.12 illustrates the titration of a hydrochloric acid

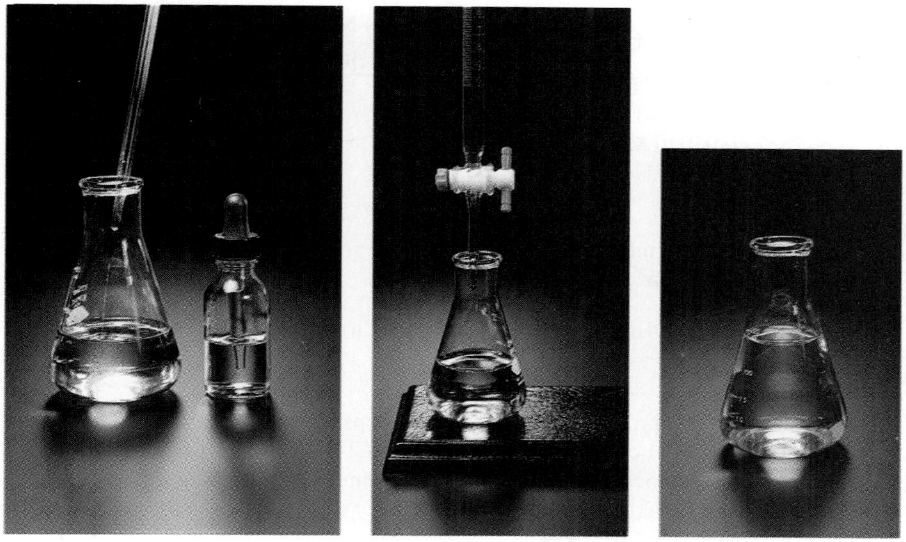

(a) (b) (c)

Figure 3.12 The technique of titration. (a) A precisely measured volume of HCl(aq) solution of unknown concentration is added to a quantity of water in a small flask. A few drops of phenolphthalein indicator solution are also added. (b) NaOH(aq) of known concentration is slowly added from a buret. (c) As long as the acid is in excess, the solution remains colorless. When the acid has just been neutralized, an additional drop of NaOH(aq) makes the solution slightly basic; the indicator turns to a light pink color. This is taken to be the equivalence point of the titration.

solution of unknown concentration with a sodium hydroxide solution of known concentration. The reaction that occurs in the titration is

$$HCl(aq) + NaOH(aq) \longrightarrow NaCl(aq) + H_2O(l)$$

Sometimes the equivalence point in a titration can be located by adding an *indicator* to the reaction mixture. This is a substance that changes color during the reaction. The trick is to find an indicator whose change in color comes at or very near the equivalence point. (We'll learn how to do this in Chapter 17.) In Figure 3.12, the indicator phenolphthalein is colorless in the hydrochloric acid solution and remains so until the equivalence point is reached. With the addition of just a fraction of a drop of NaOH(aq) beyond the equivalence point, the reaction mixture becomes basic and the phenolphthalein turns pink. The titration is stopped when the pink color appears.

Example 3.25

A flask contains 25.00 mL of $H_2SO_4(aq)$ of unknown concentration. Titration of the sample requires 25.20 mL of 0.1000 M NaOH(aq) for neutralization. What is the molarity of the sulfuric acid solution?

$$2\ NaOH(aq) + H_2SO_4(aq) \longrightarrow Na_2SO_4(aq) + 2\ H_2O(l)$$

Solution

We can do this problem in three separate steps, or we can combine the steps into a single setup. In either case we must think in these terms: (1) convert the quantity of NaOH used in the titration to moles; (2) determine the number of moles of H_2SO_4 in the solution of unknown concentration; and (3) use the defining equation for molarity. We number the first two conversions in the setup below. This setup also assumes that we can do two of the conversions in our heads:

$$25.20\ mL = 0.02520\ L \qquad \text{and} \qquad 25.00\ mL = 0.02500\ L$$

$$?\ M = \frac{0.02520\ \text{L NaOH(aq)} \times \overset{(1)}{\dfrac{0.1000\ \text{mol NaOH}}{1\ \text{L NaOH(aq)}}} \times \overset{(2)}{\dfrac{1\ \text{mol}\ H_2SO_4}{2\ \text{mol NaOH}}}}{0.02500\ L}$$

$$= 0.05040\ M\ H_2SO_4$$

Example 3.26

A flask contains 20.00 mL of 0.1030 M HCl. What volume of 0.2010 M NaOH must be added to just neutralize the acid?

$$NaOH(aq) + HCl(aq) \longrightarrow NaCl(aq) + H_2O(l)$$

Solution

Again, we can do this problem in three separate steps, or in a single setup. This time our thoughts are to (1) convert the quantity of HCl used in the titration to moles; (2) determine the number of moles of NaOH consumed in the titration; and (3) use the molarity of NaOH in a conversion factor that relates the number of moles of NaOH to the volume of solution.

$$? \text{ L NaOH(aq)} = 0.02000 \text{ L HCl(aq)} \times \overset{(1)}{\frac{0.1030 \text{ mol HCl}}{1 \text{ L HCl(aq)}}} \times \overset{(2)}{\frac{1 \text{ mol NaOH}}{1 \text{ mol HCl}}}$$

$$\times \overset{(3)}{\frac{1 \text{ L NaOH(aq)}}{0.2010 \text{ mol NaOH}}} = 0.01025 \text{ L NaOH(aq)}$$

The titration requires 10.25 mL (0.01025 L) of the NaOH(aq).

Exercise 3.25

A flask contains 20.00 mL of KOH(aq) of unknown concentration. Titration of the solution requires 15.62 mL of 0.1104 M H_2SO_4(aq).

$$2 \text{ KOH(aq)} + \text{H}_2\text{SO}_4\text{(aq)} \longrightarrow \text{K}_2\text{SO}_4\text{(aq)} + 2 \text{ H}_2\text{O}$$

Calculate the molarity of the KOH(aq).

Exercise 3.26

How many milliliters of 0.100 M $AgNO_3$(aq) are required to react completely with 750.0 mL of 0.0250 M Na_2CrO_4(aq)?

$$2 \text{ AgNO}_3\text{(aq)} + \text{Na}_2\text{CrO}_4\text{(aq)} \longrightarrow \text{Ag}_2\text{CrO}_4\text{(s)} + 2 \text{ NaNO}_3\text{(aq)}$$

SUMMARY

Molecular and formula weights relate to the masses of molecules and formula units. Molecular weight applies to molecular compounds, but only formula weight is appropriate for ionic compounds.

A mole is an amount of substance containing a number of elementary units equal to the number of atoms in exactly 12 g of carbon-12. This number, Avogadro's number, is $N_A = 6.022 \times 10^{23}$. The mass, in grams, of one mole of substance, called the molar mass, is numerically equal to an atomic, molecular, or formula weight.

Formulas and molar masses can be used to calculate the mass percent compositions of compounds. An *empirical* formula can be established from the mass percent composition of a compound. To establish a *molecular* formula, though, the molecular weight must also be known. The mass percents of carbon, hydrogen, and oxygen in organic compounds can be determined by combustion analysis.

A chemical equation uses symbols and formulas for the elements and/or compounds involved in a reaction. Numerical coefficients are used in the equation to reflect that a chemical reaction obeys the law of conservation of mass.

Calculations concerning reactions use factors based on coefficients in the balanced equation. Also required are molar masses and often other factors, such as densities. In some reactions, all but one reactant, called the limiting reactant, are present in excess amounts. The limiting reactant determines the amounts of products. The calculated quantity of a product is the theoretical yield of a reaction. The quantity obtained is often less and is commonly expressed as a percentage of the theoretical yield, known as the percent yield.

The molarity of a solution is the number of moles of solute per liter of solution. Common calculations include relating an amount of solute with solution volume and molarity and finding the volume of one solution required to produce another of lesser molarity. Problems involving chemical reactions in solutions use molarity as a conversion factor, in addition to other conversion factors.

KEY TERMS

actual yield (3.10)
Avogadro's number (N_A) (3.2)
buret (3.11)
chemical equation (3.7)
dilution (3.11)
formula weight (3.1)
limiting reactant (reagent) (3.9)

molar concentration (3.11)
molarity (M) (3.11)
molar mass (3.3)
mole (mol) (3.2)
molecular weight (3.1)
percent yield (3.10)
product (3.7)
reactant (3.7)
solute (3.11)
solvent (3.11)
stoichiometry (Introduction)
stoichiometric factor (3.8)
stoichiometric proportions (3.9)
theoretical yield (3.10)
titration (3.11)

REVIEW QUESTIONS

1. Explain the difference between the *atomic weight* of oxygen and the *molecular weight* of oxygen. Explain how each is determined from data in the periodic table.
2. What is Avogadro's number and how is it related to the quantity called one mole?
3. How many oxygen molecules and how many oxygen atoms are in 1.00 mol O_2?
4. How many calcium ions and how many chloride ions are in 1.00 mol $CaCl_2$?
5. What is the molecular weight, and what is the molar mass, of carbon dioxide? Explain how each is determined from the formula, CO_2.
6. Describe how the mass percent composition of a compound is established from its formula.
7. Describe how the empirical formula of a compound is determined from its mass percent composition.
8. What are the empirical formulas of the compounds with the following molecular formulas?
   a. H_2O_2 **b.** C_8H_{16} **c.** $C_{10}H_8$ **d.** $C_6H_{16}O$
9. Describe how the empirical formula of a compound that contains carbon, hydrogen, and oxygen is determined by combustion analysis.
10. What is the purpose of balancing a chemical equation?
11. Explain the meaning of the equation
$$CH_4 + 2\,O_2 \longrightarrow CO_2 + 2\,H_2O$$
at the molecular level. Interpret the equation in terms of moles. State the mass relationships conveyed by the equation.
12. Translate the following chemical equations into words.
   a. $2\,H_2(g) + O_2(g) \longrightarrow 2\,H_2O(l)$
   b. $2\,KClO_3(s) \longrightarrow 2\,KCl(s) + 3\,O_2(g)$
   c. $2\,Al(s) + 6\,HCl(aq) \longrightarrow 2\,AlCl_3(aq) + 3\,H_2(g)$
13. What is meant by the limiting reactant in a chemical reaction?
14. Why are the actual yields of products often less than the theoretical yields? Can actual yields ever be greater than theoretical yields?
15. Define each of the following terms.
   a. solution **b.** solvent **c.** solute
16. Define each of the following terms.
   a. molarity
   b. concentrated solution
   c. dilute solution
17. Is the volume of a solution changed by dilution? Is the concentration? Is the number of moles of solute?
18. Describe how you would determine the concentration of a solution of an acid by using a solution of a base of known concentration.

PROBLEMS

Molecular Weights and Formula Weights

19. Calculate the molecular weight or formula weight of each of the following.
   a. C_6H_5Br **b.** H_3PO_4
   c. $K_2Cr_2O_7$ **d.** $Al_2(SO_4)_3 \cdot 18H_2O$
20. Calculate the molecular weight or formula weight of each of the following.
   a. $(NH_4)_3PO_4$ **b.** $Na_2S_2O_3$
   c. $C_2H_5NO_2$ **d.** $Fe(NO_3)_3 \cdot 9H_2O$

Avogadro's Number and Molar Masses

21. Calculate the average mass, in grams, of an atom of
   a. silicon **b.** copper **c.** rhodium
22. Calculate the average mass, in grams, of an atom of
   a. cobalt **b.** strontium **c.** lead
23. Calculate the mass, in grams, of each of the following.
   a. 0.00500 mol MnO_2
   b. 1.12 mol CaH_2
   c. 0.250 mol $C_6H_{12}O_6$

24. Calculate the mass, in grams, of each of the following.
 a. 4.61 mol $AlCl_3$
 b. 0.615 mol of chromium(III) oxide
 c. 0.158 mol of iodine pentafluoride

25. Calculate the amount, in moles, of each of the following.
 a. 98.6 g HNO_3 b. 9.45 g CBr_4
 c. 9.11 g $FeSO_4$ d. 11.8 g $Pb(NO_3)_2$

26. Calculate the amount, in moles, of each of the following.
 a. 16.3 g of sulfur hexafluoride
 b. 25.4 g of lead(II) acetate
 c. 35.6 g of iron(III) chloride heptahydrate
 d. 75.3 g of cobalt(II) chlorate hexahydrate

Percent Composition and Empirical Formulas

27. Calculate the mass percent of each element in each of the following compounds.
 a. $BaSiO_3$ b. $C_6H_5NO_2$
 c. $Mg(HCO_3)_2$ d. $Al(BrO_3)_3 \cdot 9H_2O$

28. Calculate the mass percent of each element in each of the following compounds.
 a. C_3H_8O b. $(NH_4)_2SO_4$
 c. $C_6H_8N_2$ d. $Ba(ClO_4)_2 \cdot 3H_2O$

29. The empirical formula of *para*-dichlorobenzene, used as a moth repellent, is C_3H_2Cl. The molecular weight of the compound is 147 u. What is the molecular formula?

30. The empirical formula of apigenin, a yellow dye for wool, is C_3H_2O. The molecular weight of the compound is 270 u. What is the molecular formula?

31. Chloroform, an organic solvent, is 10.05% C, 0.84% H, and 89.10% Cl by mass. Determine its empirical formula.

32. Urea, used as a fertilizer and in the manufacture of plastics, is 20.00% C, 6.71% H, 46.65% N, and 26.64% O by mass. Determine its empirical formula.

33. Resorcinol, used in manufacturing resins, drugs, and other products, is 65.44% C, 5.49% H, and 29.06% O by mass. Determine its empirical formula. The molecular weight of resorcinol is 110 u. What is its molecular formula?

34. Sodium tetrathionate is formed when sodium thiosulfate reacts with iodine.

$$Na_2S_2O_3 + I_2 \longrightarrow Na_xS_yO_z + NaI \quad (\textit{not balanced})$$

Its composition is 17.01% Na, 47.46% S, and 35.52% O by mass. Determine its empirical formula. The formula weight of sodium tetrathionate is 270 u. What is its actual formula?

35. An 0.1204-g sample of a carboxylic acid is burned in oxygen to yield 0.2147 g CO_2 and 0.0884 g H_2O. Calculate the mass percentages of carbon, hydrogen, and oxygen in the compound.

36. An 0.0989-g sample of an alcohol is burned in oxygen to yield 0.2160 g CO_2 and 0.1194 g H_2O. Calculate the mass percentages of carbon, hydrogen, and oxygen in the compound.

Chemical Equations

37. Balance the following equations.
 a. $Cl_2O_5 + H_2O \longrightarrow HClO_3$
 b. $V_2O_5 + H_2 \longrightarrow V_2O_3 + H_2O$
 c. $Al + O_2 \longrightarrow Al_2O_3$
 d. $C_4H_{10} + O_2 \longrightarrow CO_2 + H_2O$
 e. $Sn + NaOH \longrightarrow Na_2SnO_2 + H_2$
 f. $PCl_5 + H_2O \longrightarrow H_3PO_4 + HCl$
 g. $CH_3OH + O_2 \longrightarrow CO_2 + H_2O$
 h. $Zn(OH)_2 + H_3PO_4 \longrightarrow Zn_3(PO_4)_2 + H_2O$

38. Balance the following equations.
 a. $TiCl_4 + H_2O \longrightarrow TiO_2 + HCl$
 b. $WO_3 + H_2 \longrightarrow W + H_2O$
 c. $C_5H_{12} + O_2 \longrightarrow CO_2 + H_2O$
 d. $Al_4C_3 + H_2O \longrightarrow Al(OH)_3 + CH_4$
 e. $Al_2(SO_4)_3 + NaOH \longrightarrow Al(OH)_3 + Na_2SO_4$
 f. $Ca_3P_2 + H_2O \longrightarrow Ca(OH)_2 + PH_3$
 g. $Cl_2O_7 + H_2O \longrightarrow HClO_4$
 h. $MnO_2 + HCl \longrightarrow MnCl_2 + Cl_2 + H_2O$

39. Write balanced equations to represent the following.
 a. the reaction of solid magnesium and gaseous oxygen to form solid magnesium oxide
 b. the decomposition of solid ammonium nitrate into dinitrogen monoxide gas and gaseous water
 c. the combustion of liquid 2-butanol to produce gaseous carbon dioxide and liquid water
 d. the reaction of aluminum metal with aqueous hydrochloric acid to form an aqueous solution of aluminum chloride and hydrogen gas

40. a. The decomposition by heating of solid potassium chlorate yields solid potassium chloride and oxygen gas as products. Write a balanced equation for this reaction.
 b. The reaction of magnesium nitride with water produces magnesium hydroxide and ammonia as products. Write a balanced equation for this reaction.

Stoichiometry of Chemical Reactions

41. Consider the equation for the combustion of octane in oxygen.

$$2\ C_8H_{18}(l) + 25\ O_2(g) \longrightarrow 16\ CO_2(g) + 18\ H_2O(l)$$

a. How many moles of CO_2 are produced when 2.0×10^{10} mol C_8H_{18} is burned?
b. How many moles of oxygen are required to burn 4.4×10^{10} mol C_8H_{18}?

42. Use the equation in Problem 41 to determine
a. How many moles of H_2O are produced when 2.0×10^8 mol C_8H_{18} is burned?
b. How many moles of CO_2 are produced when 4.0×10^8 mol O_2 is consumed?

43. Given the equation

$$N_2(g) + H_2(g) \longrightarrow NH_3(g) \quad \textit{(not balanced)}$$

a. How many grams of ammonia can be made from 440 g H_2?
b. How many grams of hydrogen are needed to react completely with 892 g N_2?

44. What mass of oxygen, in grams, can be prepared from 24.0 g H_2O_2?

$$H_2O_2 \longrightarrow H_2O + O_2 \quad \textit{(not balanced)}$$

45. Toluene and nitric acid are used in the production of trinitrotoluene (TNT), an explosive.

p43

$$\underset{\text{Toluene}}{C_7H_8} + HNO_3 \longrightarrow \underset{\text{TNT}}{C_7H_5N_3O_6} + H_2O \quad \textit{(not balanced)}$$

a. What mass of nitric acid is required to react with 454 g C_7H_8?
b. Calculate the mass of TNT that can be made from 829 g C_7H_8.

46. How many kilograms of quicklime (calcium oxide) can be made when 4.72×10^9 g of limestone (calcium carbonate) is decomposed by heating?

$$CaCO_3(s) \longrightarrow CaO(s) + CO_2(g)$$

47. How many grams of nitric acid can be made from 971 g of ammonia?

$$NH_3 + O_2 \longrightarrow HNO_3 + H_2O \quad \textit{(not balanced)}$$

48. In an oxyacetylene welding torch, acetylene (C_2H_2) burns in pure oxygen with a very hot flame.

$$C_2H_2 + O_2 \longrightarrow CO_2 + H_2O \quad \textit{(not balanced)}$$

How many grams of oxygen are required to react with 52.0 g C_2H_2?

49. Kerosene is a mixture of hydrocarbons used in domestic heating and as a jet fuel.

$$C_{14}H_{30}(l) + O_2(g) \longrightarrow CO_2(g) + H_2O(l) \quad \textit{(not balanced)}$$

a. How many grams of CO_2 are produced by the combustion of 1.00 gal of kerosene? Assume that kerosene can be represented as $C_{14}H_{30}$ and that it has a density of 0.763 g/mL. (1 gal = 3.785 L.)
b. How many milliliters of kerosene must be burned to produce 1.00 kg $CO_2(g)$?

50. Ordinary chalkboard chalk is a solid mixture with limestone ($CaCO_3$) and gypsum ($CaSO_4$) as its principal ingredients. The limestone dissolves in dilute HCl(aq), but the gypsum does not.

$$CaCO_3(s) + HCl(aq) \longrightarrow CaCl_2(aq) + CO_2(g) + H_2O \quad \textit{(not balanced)}$$

a. If a 5.05-g piece of chalk that is 72.0% $CaCO_3$ is dissolved in excess HCl(aq), what mass of $CO_2(g)$ will be produced?
b. What is the mass percent $CaCO_3$ present if a 4.38-g piece of chalk yields 1.31 g CO_2 when it reacts with excess HCl(aq)?

Limiting Reactant and Yield Calculations

51. Lithium hydroxide absorbs carbon dioxide to form lithium carbonate and water.

p44

$$2\ LiOH + CO_2 \longrightarrow Li_2CO_3 + H_2O$$

If a reaction vessel contains 0.150 mol LiOH and 0.080 mol CO_2, which compound is the limiting reactant? How many moles of Li_2CO_3 can be produced?

52. Boron trifluoride reacts with water to produce boric acid (H_3BO_3) and fluoroboric acid (HBF_4).

$$4\ BF_3 + 3\ H_2O \longrightarrow H_3BO_3 + 3\ HBF_4$$

If a reaction vessel contains 0.496 mol BF_3 and 0.313 mol H_2O, which compound is the limiting reactant? How many moles of HBF_4 can be produced?

53. Propane, C_3H_8, burns in oxygen to produce carbon dioxide and water. If a reaction vessel contains 4.81 g C_3H_8 and 16.4 g O_2, which compound is the limiting reactant? What is the maximum number of grams of carbon dioxide that can be produced?

54. Butane, C_4H_{10}, burns in oxygen to produce carbon dioxide and water. If a reaction vessel contains 8.37 g C_4H_{10} and 31.9 g O_2, which compound is the limiting reactant? What is the maximum number of grams of carbon dioxide that can be produced?

55. Calculate the theoretical yield of ZnS, in grams, that can be made from 0.488 g Zn and 0.503 g S_8.

$$8\ Zn + S_8 \longrightarrow 8\ ZnS$$

If the actual yield is 0.606 g ZnS, what is the percent yield?

56. Calculate the theoretical yield of C_2H_5Cl, in grams, that can be made from 11.3 g of ethanol and 3.48 g PCl_3.

$$3\ C_2H_5OH + PCl_3 \longrightarrow 3\ C_2H_5Cl + H_3PO_3$$

If the actual yield is 1.24 g C_2H_5Cl, what is the percent yield?

57. **a.** A student prepares ammonium bicarbonate by the reaction

$$NH_3 + CO_2 + H_2O \longrightarrow NH_4HCO_3$$

She uses 14.8 g NH_3 and 41.3 g CO_2. Water is present in excess. What is her actual yield of ammonium bicarbonate if she obtains a 74.7% yield in the reaction?

b. A student needs 625 g of zinc sulfide, a white pigment, for an art project. He can synthesize it using the reaction

$$Na_2S(aq) + Zn(NO_3)_2(aq) \longrightarrow ZnS(s) + 2\ NaNO_3(aq)$$

How many grams of zinc nitrate will he need if he can make the zinc sulfide in 85.0% yield? Assume that he has plenty of sodium sulfide.

58. Only one of the outcomes suggested by the illustration is possible for the reaction

$$2\ Hg(l) + O_2(g) \longrightarrow 2\ HgO(s)$$

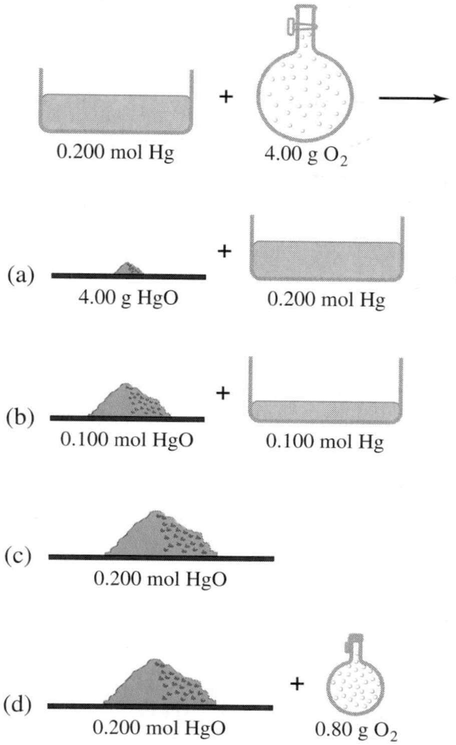

Indicate which one represents the substances present after the reaction. Explain why the other outcomes are not possible for the given initial conditions.

Solution Stoichiometry

59. Calculate the molarity of each of the following solutions.
a. 6.00 mol HCl in 2.50 L of solution
b. 0.00700 mol Li_2CO_3 in 10.0 mL of solution

60. Calculate the molarity of each of the following solutions.
a. 2.50 mol H_2SO_4 in 5.00 L of solution
b. 0.200 mol C_2H_5OH in 18.4 mL of solution

61. Calculate the molarity of each of the following solutions.
a. 8.90 g H_2SO_4 in 100.0 mL of solution
b. 439 g $C_6H_{12}O_6$ in 1.25 L of solution

62. Calculate the molarity of each of the following solutions.
a. 44.3 g KOH in 125 mL of solution
b. 2.46 g $H_2C_2O_4$ in 750.0 mL of solution

63. How many grams of solute are needed to prepare each of the following solutions?
a. 2.00 L of 1.00 M NaOH
b. 10.0 mL of 4.25 M $C_6H_{12}O_6$

64. How many grams of solute are needed to prepare each of the following solutions?
a. 250 mL of 2.50 M $K_2Cr_2O_7$
b. 20.0 mL of 0.0100 M $KMnO_4$

65. How many milliliters of 6.00 M NaOH are required to contain 1.25 mol NaOH?

66. How many milliliters of 2.50 M NaOH are required to contain 1.05 mol NaOH?

67. How many milliliters of 0.0250 M $KMnO_4$ must one take to get 8.10 g $KMnO_4$?

68. How many milliliters of 4.25 M $C_6H_{12}O_6$ must one take to get 205 g $C_6H_{12}O_6$?

69. How many milliliters of 12.0 M HCl are required to make 2.00 L of 1.00 M HCl?

70. How many milliliters of 8.89 M HBr are required to make 2.00 L of 1.00 M HBr?

71. If 25.00 mL of 1.04 M Na_2CO_3 is diluted to 0.500 L, what is the molarity of Na_2CO_3 in the diluted solution?

72. If 50.00 mL of 19.1 M NaOH is diluted to 2.00 L, what is the molarity of NaOH in the diluted solution?

73. Which of the following is the approximate molarity of a solution obtained by mixing 0.100 L, of 0.100 M NH_3 and 0.200 L of 0.200 M NH_3?
(a) 0.13 M NH_3 (b) 0.15 M NH_3
(c) 0.17 M NH_3 (d) 0.30 M NH_3

74. The solution in the vial in the illustration, when diluted with water to exactly 1.000 L, produces 0.1000 M HCl. Which of the following is a plausible approximate concentration of the HCl solution in the vial?

(a) 0.5 M HCl (b) 2.0 M HCl
(c) 7.5 M HCl (d) 12.0 M HCl

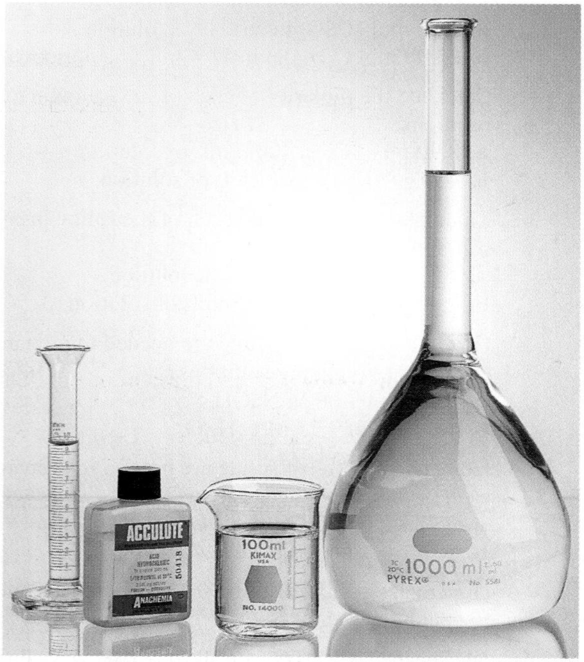

75. A stock bottle of nitric acid indicates that the solution is 67.0% HNO_3 by mass and has a density of 1.40 g/mL. Calculate the molarity of the solution.

76. A stock bottle of potassium hydroxide solution indicates that the solution is 50.0% KOH by mass and has a density of 1.52 g/mL. Calculate the molarity of the solution.

77. Calculate the molarity of an HCl solution if 20.00 mL of it requires 33.22 mL of 0.1503 M NaOH for its neutralization.

$$HCl(aq) + NaOH(aq) \longrightarrow NaCl(aq) + H_2O(l)$$

78. Calculate the molarity of an HNO_3 solution if 30.00 mL of it requires 18.34 mL of 0.1044 M KOH for its neutralization.

$$HNO_3(aq) + KOH(aq) \longrightarrow KNO_3(aq) + H_2O(l)$$

79. Calculate the molarity of a $Ca(OH)_2$ solution if 18.50 mL of it requires 28.27 mL of 0.01025 M HCl for its neutralization.

$$Ca(OH)_2(aq) + 2\ HCl(aq) \longrightarrow CaCl_2(aq) + 2\ H_2O(l)$$

80. Calculate the molarity of a $H_2C_2O_4$ solution if 12.50 mL of it requires 25.72 mL of 0.0995 M NaOH for its neutralization.

$$H_2C_2O_4(aq) + 2\ NaOH(aq) \longrightarrow Na_2C_2O_4(aq) + 2\ H_2O(l)$$

81. How many milliliters of 0.1000 M H_2SO_4 are required to react with 10.32 mL of 0.4042 M $NaHCO_3$?

$$H_2SO_4(aq) + 2\ NaHCO_3(aq) \longrightarrow Na_2SO_4(aq) + 2\ H_2O(l) + 2\ CO_2(g)$$

82. How many milliliters of 0.1104 M H_2SO_4 are required to react with 30.07 mL of 0.08872 M $Ba(OH)_2$?

$$H_2SO_4(aq) + Ba(OH)_2(aq) \longrightarrow BaSO_4(s) + 2\ H_2O(l)$$

83. Suppose you need about 80 mL of 0.1000 M $AgNO_3$ to use in some titrations. You have available about 150 mL of 0.04000 M $AgNO_3$ and also about 1.2 g of solid $AgNO_3$. Assume that you have available standard laboratory equipment such as an analytical balance, 10.00-mL and 25.00-mL pipets, 100.0-mL and 250.0-mL volumetric flasks, and so on. Describe how you would prepare the desired $AgNO_3$ solution, including actual masses or volumes required.

84. Two sucrose solutions—125 mL of 1.50 M $C_{12}H_{22}O_{11}$ and 275 mL of 1.25 M $C_{12}H_{22}O_{11}$—are mixed. Assuming the solution volumes are additive, what is the molarity of $C_{12}H_{22}O_{11}$ in the final solution?

ADDITIONAL PROBLEMS

85. Calcium tablets for use as dietary supplements are available in the form of several different compounds. Calculate the mass of each required to furnish 875 mg Ca^{2+}.

a. calcium carbonate, $CaCO_3$
b. calcium lactate, $Ca(C_3H_5O_3)_2$
c. calcium gluconate, $Ca(C_6H_{11}O_7)_2$
d. calcium citrate, $Ca_3(C_6H_5O_7)_2$

86. Iron, as Fe^{2+}, is an essential nutrient. Pregnant women often take 325-mg ferrous sulfate ($FeSO_4$) tablets as a dietary supplement. Yet iron tablets are the leading cause of poisoning deaths in children. As little as 550 mg Fe^{2+} can be fatal to a 22-lb child. How many 325-mg tablets would it take to constitute a lethal dose to a 22-lb child?

87. Chlorophyll, found in plant cells and essential to the process of photosynthesis, contains 2.72% Mg by mass. Assuming one magnesium atom per chlorophyll molecule, calculate the molecular weight of chlorophyll.

88. A 2.55-mL sample of heptane, C_7H_{16}, is burned in an excess of oxygen gas in a carbon–hydrogen analyzer. What masses of CO_2 and H_2O should be obtained? The density of heptane is 0.684 g/mL.

89. When burned in oxygen in combustion analysis, a 0.1888-g sample of a hydrocarbon produced 0.6260 g CO_2 and 0.1602 g H_2O. The molecular weight of the compound is 106 u. Calculate (**a**) the mass percent composition, (**b**) the empirical formula, and (**c**) the molecular formula of the hydrocarbon.

90. Dimethylhydrazine is a compound of C, H, and N used in rocket fuels. When burned completely in oxygen, a 0.312-g sample yields 0.458 g CO_2 and 0.374 g H_2O. The nitrogen content of a separate 0.525-g sample is converted to 0.244 g N_2. What is the empirical formula of dimethylhydrazine?

91. The combustion in oxygen of 1.525 g of a compound of carbon, hydrogen, and oxygen derived from an alkane yields 3.047 g CO_2 and 1.247 g H_2O. The molecular weight of this compound is 88.1 u. Draw a plausible structural formula for this compound. Is there more than one possibility? Explain.

92. At temperatures above 300 °C, silver oxide, Ag_2O, decomposes to give metallic silver and oxygen gas. A 2.95-g sample of *impure* silver oxide yields 0.183 g of oxygen. Assuming that $Ag_2O(s)$ is the only source of oxygen, what is the mass percent of Ag_2O in the sample?

93. A piece of sheet aluminum that measures 12.3 cm × 14.3 cm × 2.2 mm reacts with an excess of hydrochloric acid. What mass of hydrogen is produced? The density of aluminum is 2.70 g/cm³.

$$Al(s) + HCl(aq) \longrightarrow AlCl_3(aq) + H_2(g) \quad \textit{(not balanced)}$$

94. At 400 °C, hydrogen gas is passed over iron(III) oxide. Water vapor is formed together with a black residue—a compound that is 72.3% Fe and 27.7% O by mass. Write a balanced chemical equation for this reaction.

95. A coal-burning power plant burns 228 trainloads of Western subbituminous coal per year. Each train is comprised of 115 cars, and each car carries 90.5 metric tons of coal (1 metric ton = 1000 kg). If the coal is 64.3% carbon, how many metric tons of carbon dioxide are produced by the plant each year?

96. The following is a side reaction in the manufacture of rayon fibers from wood pulp.

$$3\,CS_2 + 6\,NaOH \longrightarrow 2\,Na_2CS_3 + Na_2CO_3 + 3\,H_2O$$

How many grams of Na_2CS_3 are produced in the reaction of 88.0 mL of liquid CS_2 (d = 1.26 g/mL) and 3.12 mol NaOH?

97. A laboratory manual calls for 13.0 g of butanol, 21.6 g of sodium bromide, and 33.8 g H_2SO_4 as reactants in this reaction.

$$C_4H_9OH + NaBr + H_2SO_4 \longrightarrow C_4H_9Br + NaHSO_4 + H_2O$$

A student following these directions obtains 16.8 g of butyl bromide (C_4H_9Br). What are the theoretical yield and the percent yield of this reaction?

98. How many milliliters of a concentrated hydrochloric acid solution (36.0% HCl by mass; d = 1.18 g/mL) are required to produce 15.0 L of 0.225 M HCl?

99. How many grams of sodium metal must react with 250.0 mL of water to produce a solution that is 0.315 M NaOH? (Assume the final solution volume is 250.0 mL.)

$$2\,Na(s) + 2\,H_2O(l) \longrightarrow 2\,NaOH(aq) + H_2(g)$$

100. For the reaction

$$Ca(OH)_2(s) + 2\,HCl(aq) \longrightarrow CaCl_2(aq) + 2\,H_2O(l)$$

How many kilograms of $Ca(OH)_2$ are needed to neutralize 325 L of an HCl solution that is 30.12% HCl by mass and has a density of 1.15 g/mL?

101. A sample of battery acid is to be analyzed for its sulfuric acid content. A 1.00-mL sample weighs 1.239 g. This 1.00-mL sample is diluted to 250.0 mL with water, and 10.00 mL of this diluted acid requires 32.44 mL of 0.00986 M NaOH for its titration. What is the mass percent H_2SO_4 in the battery acid?

$$H_2SO_4(aq) + 2\,NaOH(aq) \longrightarrow Na_2SO_4(aq) + 2\,H_2O(l)$$

Some of these helium-filled balloons, launched at Kiruna, Sweden, in 1990 in a study of atmospheric ozone, reached an altitude of 28 km.

4

Gases

CONTENTS

Earth's life-support system relies on a thin blanket of gases—air. Air is so insubstantial that it is difficult to think of it as matter. But air is matter; it is matter in the gaseous state. Gases have mass and occupy space.

We may have a more intuitive feeling for liquids and solids, but their behavior is more difficult to explain than the behavior of gases. Several simple natural laws apply to gases, and among the most successful of scientific theories is one dealing with gases. In this chapter we study gas laws and the kinetic-molecular theory of gases. In a later chapter (Chapter 15) we will find that the kinetic-molecular theory not only explains the behavior of gases, but also provides important insights into how molecules enter into chemical reactions.

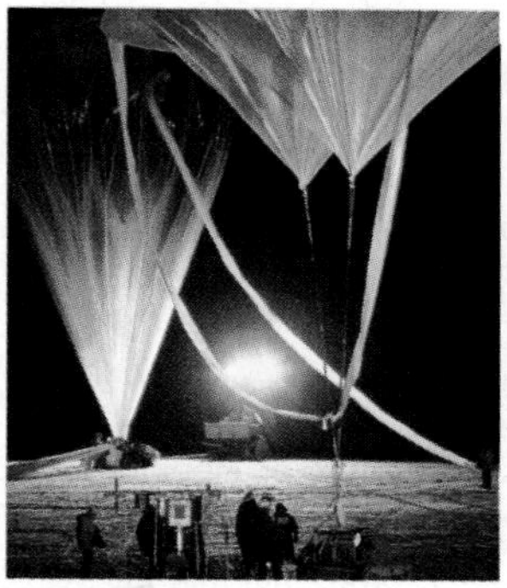

4.1 Gases: What Are They Like?

Gases are composed of widely separated molecules in constant, random motion. Gases flow readily and occupy the entire volume of their container, regardless of its shape. If we release ammonia in one part of a room, its odor is soon detectable throughout the room.

Unlike liquids and solids, in which molecules or atoms are already quite close to one another, gases are readily compressed by the application of pressure. It is possible to compress enough air for an hour or more of breathing into a small portable tank for underwater diving. When air is compressed, the molecules are closer together, but they are still far apart compared to those of a liquid or a solid.

A **vapor** is the gaseous state of a substance that under ordinary conditions exists mainly as a liquid (for example, water or ethanol) or a solid (for example, *para*-dichlorobenzene, used as a moth repellent and restroom deodorizer).

At room temperature (25 °C) nearly all ionic substances are solids, whereas most molecular substances with low molar masses are gases or liquids that are easily vaporized.

Table 4.1 lists several common gases. All of these gases and many others are commercially available in tanks, either as highly compressed gases or as liquids that vaporize when the pressure on them is released. Perhaps the most familiar and important gases are those found mixed together in ordinary air—N_2, O_2, Ar, and CO_2. So important are they that we will devote an entire later chapter (Chapter 14) to these and other atmospheric gases.

Several gases are among the top industrial chemicals and are frequently shipped as liquids in tanker trucks. Among all industrial chemicals produced in the United States, typical annual rankings, by mass, are nitrogen, No. 2; oxygen, No. 3; ethylene, No. 4; ammonia, No. 6; and chlorine, No. 10. The tanker truck here is being filled with liquefied natural gas.

TABLE 4.1 **Some Common Gases**[a]

Substance	Formula	Typical Use(s)
Acetylene	C_2H_2	fuel for welding of metals
Ammonia	NH_3	fertilizer; manufacture of plastics
Argon	Ar	filling gas for specialized light bulbs
Butane	C_4H_{10}	fuel for heating (LPG)
Carbon dioxide	CO_2	carbonation of beverages
Carbon monoxide	CO	obtaining metals from ores
Chlorine	Cl_2	disinfectant; bleach
Ethylene	C_2H_4	manufacture of plastics
Helium	He	lifting gas for balloons
Hydrogen	H_2	chemical reagent
Hydrogen sulfide	H_2S	chemical reagent
Methane	CH_4	fuel; manufacture of hydrogen
Nitrogen	N_2	manufacture of ammonia
Nitrous oxide	N_2O	anesthetic
Oxygen	O_2	support of combustion; respiration
Propane	C_3H_8	fuel for heating (LPG)
Sulfur dioxide	SO_2	preservative, disinfectant, bleach
Sulfur trioxide	SO_3	maufacture of sulfuric acid

[a] All of these substances are gases at room temperature (about 25 °C) and ordinary pressures (about 1 atm), but they can be converted to liquids and solids by a sufficient lowering of the temperature and/or increase of the pressure.

4.2 The Kinetic-Molecular Theory: An Introduction

In the name "kinetic-molecular" theory, "kinetic" conveys the idea of motion and "molecular" that the motion is that of molecules. We will use the term "molecules" for the particles of a gas, even those that are monatomic.

The **kinetic-molecular theory** was developed in the mid-nineteenth century to provide a model for gases at the *molecular* level and to explain their observed physical properties. The theory treats gases as collections of particles in rapid, random motion. The particles of argon gas, for example, are individual atoms (Ar); in nitrogen gas they are diatomic molecules (N_2).

The molecules of a gas are in such rapid motion that they seem to resist the force of gravity. They do not fall and collect at the bottom of a container, as do the molecules in a liquid. Gases fill their containers completely. *Filled* does not mean that the gas molecules are tightly packed, but rather that they are distributed uniformly throughout the entire volume of a container. In fact, because the distances between molecules are generally much greater than the dimensions of the molecules themselves, the typical gas is mostly empty space.

The movement of gaseous molecules through three-dimensional space, their *translational* motion, is random. That is, the speed and direction of motion of any given molecule are at all times unpredictable. The speed can range over a wide set of values, and all directions of motion are equally probable. A molecule of a gas moves along a straight-line path unless it strikes another molecule of the gas or hits the container wall. Then it bounces off at an angle and travels along a new straight-line path until its next collision (see Figure 4.1). Some molecules lose energy and are slowed down by collisions, but others gain energy and are speeded up. There is neither a gain nor a loss of energy in collisions; therefore the average energy for all the molecules in a gas is not changed by the collisions. Because of this conservation of kinetic energy, collisions of gas molecules are said to be *elastic*.

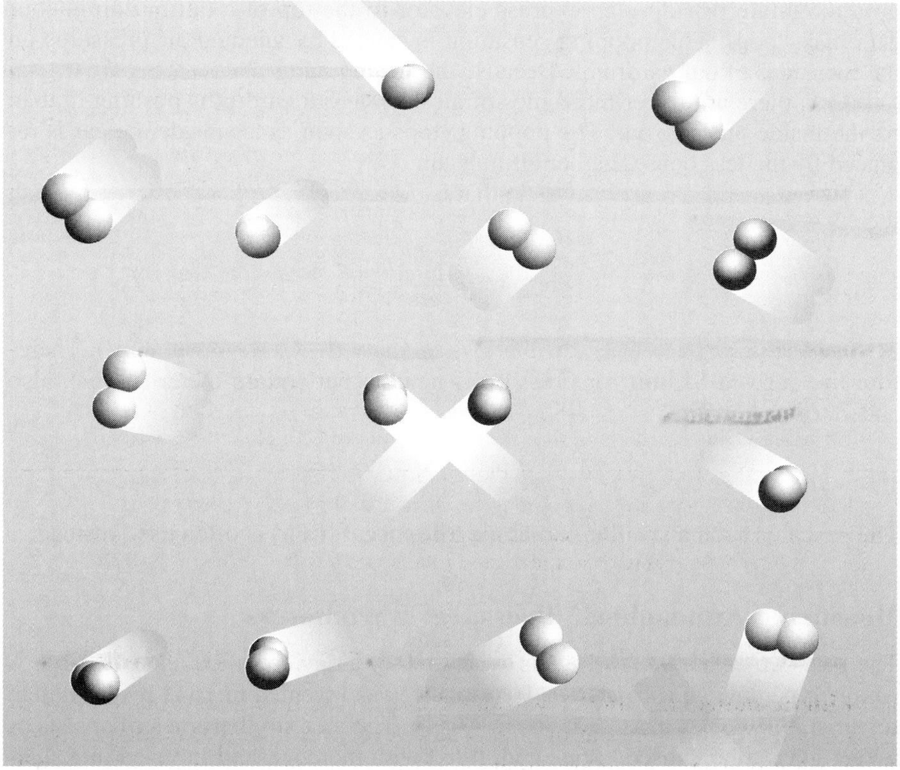

Figure 4.1 Visualizing molecular motion in a gas. Molecules of a gas are in constant, random, straight-line motion and undergo elastic collisions with each other and with their container walls.

The kinetic-molecular theory explains what we are measuring when we measure temperature. According to the theory, temperature is related to the translational kinetic energies of the molecules in a sample of gas. The higher the average translational kinetic energy (that is, the faster, on average, the molecules are moving), the higher the temperature of the gas. On average, molecules of a cold sample of a gas are moving more slowly than those of a hot sample of the same gas. Temperature reflects the *average* translational

Temp goes up, kinetic energy increases, more motion

kinetic energy of the molecules of a gas because the molecules, moving at different speeds, have different kinetic energies.

The kinetic-molecular theory also explains the origin of gas pressure, an important gas property explored in the next section. Consider a gas-filled balloon: When a molecule of the gas strikes the wall of the balloon, it gives the wall a little push. When we measure gas pressure, we assess the effect of huge numbers of these "molecular pushes." We will return to a more quantitative discussion of the kinetic-molecular theory in Section 4.11, after we have studied some of the natural laws describing gas behavior.

4.3 Gas Pressure

Although gas molecules do not settle under the force of gravity in containers of ordinary size, they tend to do so in that largest of containers—the atmosphere. The density of air decreases with increased height above Earth's surface.

Molecules of air are constantly bouncing off our skin; but they are so tiny that we don't feel their individual impacts. In fact, we usually are not aware of their collective impacts. However, when we increase our altitude rapidly—by driving up a mountain or riding an express elevator to the top of a tall building—our ears may "pop." This popping sensation is caused by unequal air pressures on the two sides of our eardrums. Because the density of air decreases at the higher altitudes, there are fewer molecules of air outside our eardrums pushing in than on the inside pushing out. The popping stops as soon as air inside the ear is replaced by the less dense, higher altitude air.

Pressure is force per unit area, that is, force divided by the area over which the force is exerted.

$$\text{Pressure} = \frac{\text{force}}{\text{area}} = \frac{F}{A}$$

For a review of fundamental physical quantities, such as the newton, see Appendix B.

In SI, force is expressed in *newtons (N)* and area in square meters (m^2). Therefore the derived SI unit for pressure is newton per square meter (N/m^2), also called a **pascal (Pa)**.

$$1 \text{ Pa} = 1 \text{ N/m}^2$$

The pascal is such a small unit that the **kilopascal (kPa)** is often used instead.

Measuring Atmospheric Pressure: Barometers

The pressure of the atmosphere is measured by a device called a **barometer**. A simple type, known as a mercury barometer, was invented in 1643 by the Italian scientist Evangelista Torricelli (1608–1647). The mercury barometer consists of a long glass tube, closed at one end, filled with mercury, and inverted in a shallow dish that also contains mercury (Figure 4.2).

Suppose the tube is 1 m long. Some of the mercury in the tube drains into the dish, but *not all* of it. The mercury drains out only until the pressure exerted by the mercury remaining in the tube exactly balances the pressure exerted by the atmosphere on the surface of the mercury in the dish. The mercury in the tube tries to push its way out under the influence of gravity, and the air pressure tries to push the mercury back in. At some point these two opposing tendencies reach a balance.

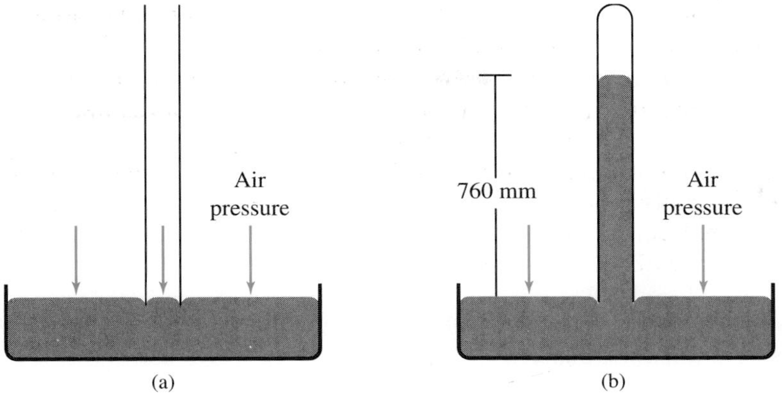

Figure 4.2 Measurement of air pressure with a mercury barometer. (a) The mercury levels are equal inside and outside the *open-end* tube, since the tube is open to the atmosphere and filled with air. (b) A column of mercury 760 mm high is maintained in the *closed-end* tube. The space above the mercury is devoid of air and contains only a trace of mercury vapor.

Mercury is a dense liquid. On average, at sea level, a column of mercury about 760 mm high will balance the push of a column of air many kilometers high. The pressure that is exerted by a column of mercury that is *exactly* 760 mm high is called **1 atmosphere (atm)**.* The pressure unit **1 millimeter of mercury (mmHg)** is often called a **torr** (after Torricelli). That is,

$$1 \text{ atm} = 760 \text{ mmHg} = 760 \text{ torr}$$

These are the pressure units that we will use for the most part in this text.

The relationship of the atmosphere unit to SI units is

$$1 \text{ atm} = 101{,}325 \text{ Pa} = 101.325 \text{ kPa}$$

For approximate work, it is helpful to remember that 1 atm is about 100 kPa.

Several other pressure units are widely used. Weather reports in the United States often include atmospheric pressure in *millibars (mb)* or *inches of mercury (in.Hg)*.

$$1 \text{ atm} = 29.921 \text{ in.Hg} = 1.01325 \text{ bar} = 1013.25 \text{ mb}$$

Engineers generally use *pounds per square inch* (*lb/in.*^{2}) for practical applications like steam pressure in boilers and turbines.

$$1 \text{ atm} = 14.696 \text{ lb/in.}^2$$

Respiratory therapists use the unit *centimeters of water*, a useful scale for measuring small differences in pressure.

$$1 \text{ mmHg} = 13.6 \text{ mmH}_2\text{O} = 1.36 \text{ cmH}_2\text{O}$$

*For the pressure to be exactly 1 atm, the 760-mm column of mercury must be at 0 °C (d = 13.59508 g/cm^3) and at a location where the acceleration due to gravity (g) is 9.80665 m/s^2.

Example 4.1

A Canadian weather report gives the atmospheric pressure as 100.2 kPa. What is the pressure in torr?

Solution

Because 1 atm = 760 torr = 101.325 kPa, use 760 torr/101.325 kPa as a conversion factor.

$$? \text{ torr} = 100.2 \text{ kPa} \times \frac{760 \text{ torr}}{101.325 \text{ kPa}} = 751.6 \text{ torr}$$

Exercise 4.1

Carry out the following conversions.

a. 722 torr to mmHg **b.** 98.2 kPa to torr
c. 29.95 in.Hg to torr **d.** 768 torr to atm

Measuring Other Gas Pressures: Manometers

A mercury barometer is fine for measuring the pressure of the atmosphere, but we often need to know the pressure of a gas in a closed container. Because it is not usually practical to place a mercury barometer inside a flask, scientists developed a device, called a **manometer**, to measure gas pressure in a closed container.

Figure 4.3 illustrates the operation of an open-end manometer. Mercury is usually chosen as the liquid in a manometer because its great density (13.595 g/cm³) permits the measurement of large pressure differences in columns of reasonable height. But other liquids can be used as well. For example, if the differ-

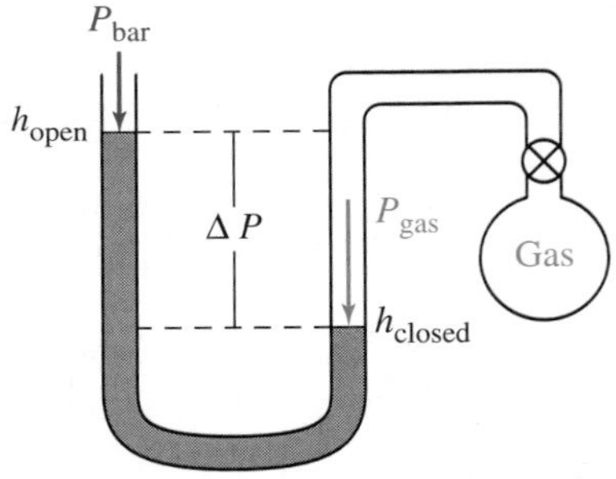

(a) Gas pressure greater than barometric pressure

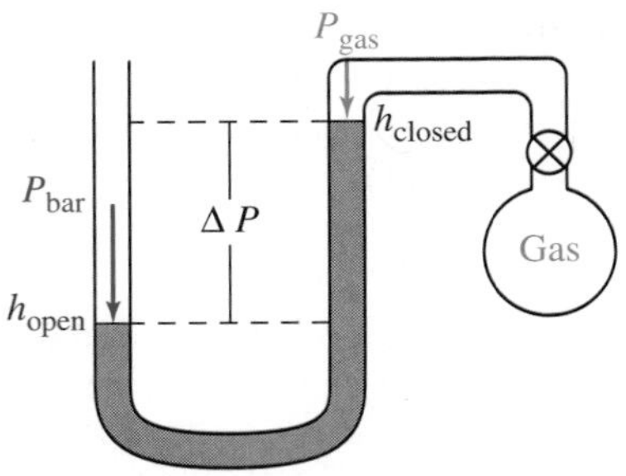

(b) Gas pressure less than barometric pressure

Figure 4.3 **Measuring gas pressure with an open-end manometer.** ΔP, the difference between barometric pressure, P_{bar}, and the pressure of the gas, P_{gas}, is given by the difference in the mercury levels in the open and closed arms of the manometer.

$$\Delta P = \Delta h = h_{open} - h_{closed}$$

Treat Δh as a *positive* quantity, and
(a) If P_{gas} is greater than P_{bar},

$$P_{gas} = P_{bar} + \Delta h$$

(b) If P_{gas} is less than P_{bar},

$$P_{gas} = P_{bar} - \Delta h$$

ence between P_{gas} and P_{bar} is quite small, we might choose a liquid of low density and limited ability to vaporize (see Problem 35).

The general expression that relates the pressure exerted by a column of liquid to the density of the liquid and height of the column is

$$P = g \cdot d \cdot h$$

where P is the pressure, g is the acceleration due to gravity (9.807 m/s²), d is the density of the liquid, and h is the height of the column. If density is expressed in kg/m³ and height in m, the pressure is in N/m² (Pa). It is more common, however, just to express the pressure exerted by a liquid through the height of a liquid column, as in the following examples.

Example 4.2

Calculate the height of a column of water that would exert the same pressure as a column of mercury that is 760 mm high. (Use densities of 13.6 g/cm³ for mercury and 1.00 g/cm³ for water.)

Solution

The pressure of the mercury column is given by

$$P_{Hg} = g \cdot d_{Hg} \cdot h_{Hg}$$

and that of the water column by

$$P_{H_2O} = g \cdot d_{H_2O} \cdot h_{H_2O}$$

The two pressures must be equal, so

$$\cancel{g} \cdot d_{Hg} \cdot h_{Hg} = \cancel{g} \cdot d_{H_2O} \cdot h_{H_2O}$$

Canceling g and substituting numerical values, we have

$$13.6 \text{ g/cm}^3 \times 760 \text{ mm} = 1.00 \text{ g/cm}^3 \times h_{H_2O}$$

$$h_{H_2O} = \frac{13.6 \text{ g/cm}^3 \times 760 \text{ mm}}{1.00 \text{ g/cm}^3} = 10{,}300 \text{ mm} = 10.3 \text{ m}$$

To measure normal atmospheric pressures, a water-filled barometer would have to be over 10 m tall—as tall as a three-story building!

Example 4.3

Elephant seals dive to depths as great as 1250 m. What pressure, in atm, is exerted by water at that depth? (Use densities of 13.6 g/cm³ for mercury and 1.00 g/cm³ for water.)

Solution

First, calculate the pressure exerted by a column of *mercury* 1250 m high.

$$? \text{ atm} = 1250 \text{ m} \times \frac{1000 \text{ mm}}{1 \text{ m}} \times \frac{1 \text{ atm}}{760 \text{ mm}} = 1640 \text{ atm}$$

Now use the equations $P_{Hg} = g \cdot d_{Hg} \cdot h_{Hg}$ and $P_{H_2O} = g \cdot d_{H_2O} \cdot h_{H_2O}$. Divide the second equation by the first to obtain the ratio

$$\frac{P_{H_2O}}{P_{Hg}} = \frac{g \cdot d_{H_2O} \cdot h_{H_2O}}{g \cdot d_{Hg} \cdot h_{Hg}}$$

and solve for P_{H_2O}, noting that $h_{H_2O} = h_{Hg}$ and that $P_{Hg} = 1640$ atm.

$$P_{H_2O} = P_{Hg} \times \frac{d_{H_2O}}{d_{Hg}} = 1640 \text{ atm} \times \frac{1.00 \text{ g/cm}^3}{13.6 \text{ g/cm}^3} = 121 \text{ atm}$$

We could have solved the problem directly by using the answer to Example 4.2.

$$? \text{ atm} = 1250 \text{ mH}_2\text{O} \times \frac{1 \text{ atm}}{10.3 \text{ mH}_2\text{O}} = 121 \text{ atm}$$

Data such as 10.3 mH_2O = 1 atm are not always readily available, however.

Exercise 4.2

Calculate the height of a column of carbon tetrachloride, CCl_4, that would exert the same pressure as a column of mercury that is 760 mm high. The density of CCl_4 is 1.59 g/cm³ and that of Hg is 13.6 g/cm³.

Exercise 4.3

A diver reaches a depth of 30.0 m. What pressure, in atm, is exerted by water at that depth?

CONCEPTUAL EXAMPLE 4.1

Without doing calculations, arrange the sketches in Figure 4.4 so that the pressures denoted in color are in *increasing* order.

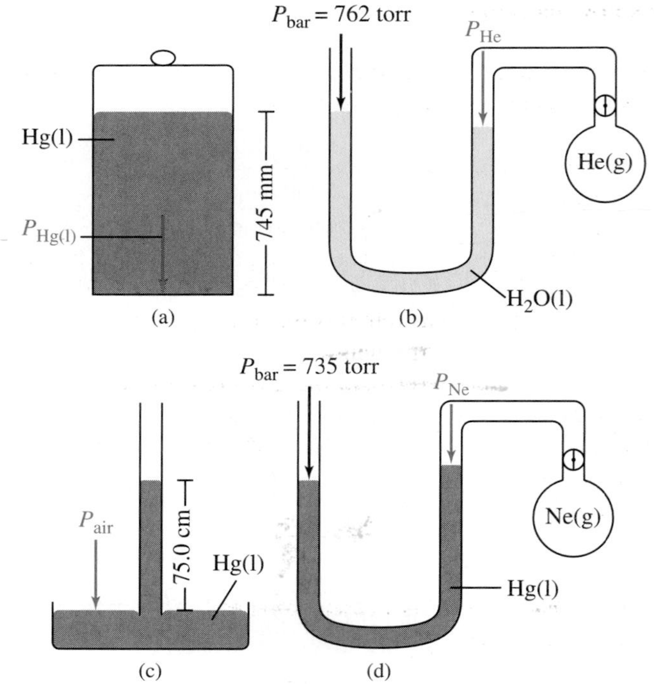

Figure 4.4 Conceptual Example 4.1 illustrated.

Solution

A good way to begin is to determine which of the pressures is (are) less than 1 atm, and which is (are) greater. In (c), P_{air} is 75.0 cmHg = 750 mmHg; this is slightly less than 1 atm. In (d), P_{Ne} is slightly less than P_{bar}, which is 735 torr. Thus P_{air} is greater than P_{Ne}. In (a), although there appears to be a lot more mercury than in the barometer in (c), the pressure exerted by the liquid column depends only on the liquid density and the height. Thus, $P_{Hg(l)}$ = 745 mmHg; this is intermediate to the first two pressures. Now we can say that $P_{Ne} < P_{Hg(l)} < P_{air}$. Finally, only in (b) is the pressure greater than 1 atm. This is because $P_{He} > P_{bar}$, and P_{bar} is itself greater than 1 atm. The order of increasing pressure is

$$P_{Ne} < P_{Hg(l)} < P_{air} < P_{He}$$
$$(d) < (a) < (c) < (b)$$

4.4 Boyle's Law: The Pressure–Volume Relationship

Four variables are used to describe a sample of a gas in calculations: amount of gas in moles (n), volume (V), temperature (T), and pressure (P). These variables are related by several gas laws that provide qualitative and quantitative descriptions of how one of the variables will change as a second variable is changed, while the remaining two variables are held constant.

The first of these simple gas laws, discovered by the Irish chemist Robert Boyle in 1662, concerns the relationship between the pressure and volume of a gas. **Boyle's law** states that

For a given amount of gas at a constant temperature, the volume of the gas varies inversely *with its pressure.*

That is, when the pressure increases, the volume decreases; when the pressure decreases, the volume increases.

Think of gases as pictured in the kinetic-molecular theory (Figure 4.5). A gas exerts a particular pressure because the gas molecules bounce against the container walls with a certain frequency. If the volume of the container is expanded while the amount of gas remains fixed, the number of molecules per unit volume of gas decreases. The frequency with which molecules strike a unit area of the container walls decreases, and the gas pressure decreases. Thus, as the volume of a gas is increased, its pressure decreases.

Mathematically, for a given amount of gas at a constant temperature, Boyle's law is written

$$V \propto \frac{1}{P}$$

where the symbol $\propto$ means "is proportional to." This may be changed to an equation by inserting a proportionality constant, a.

$$V = \frac{a}{P}$$

Robert Boyle (1627–1691) dabbled in alchemy, but his experiments on air and his textbook, *The Sceptical Chemist*, helped to replace alchemy with modern chemistry.

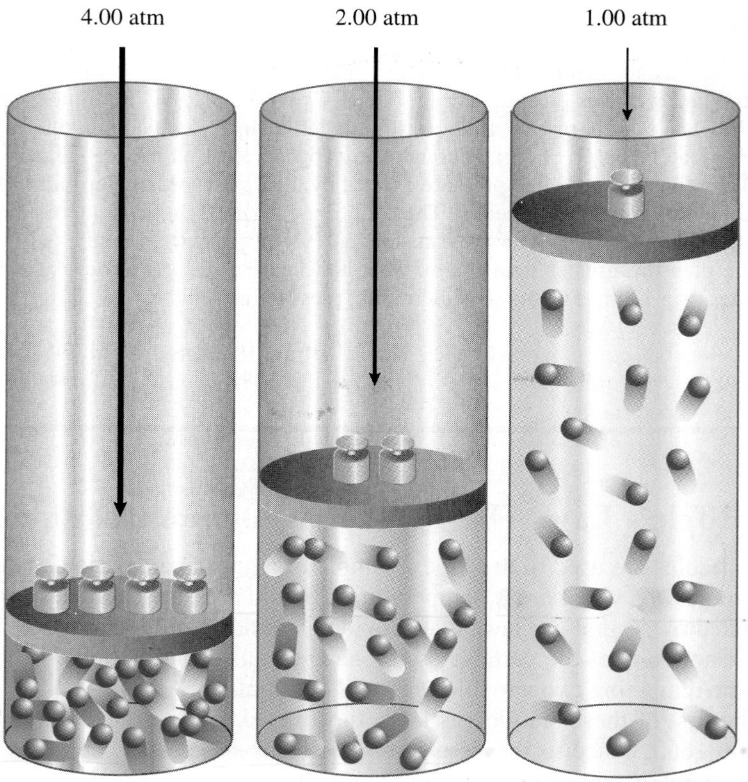

Figure 4.5 Boyle's law: a kinetic theory view. As the pressure is successively reduced from 4.00 atm to 2.00 atm and then to 1.00 atm, the volume doubles and then doubles again.

Multiplying both sides of the equation by P, we get

$$PV = a$$

Another way to state Boyle's law, then, is that for a given amount of gas at a constant temperature, the product of the pressure and volume is a constant. This is an elegant and precise, if somewhat abstract, way of summarizing a lot of experimental data. If the product $P \times V$ is to be constant, then if V increases P must decrease, and vice versa. This relationship is demonstrated in the pressure–volume graph in Figure 4.6.

Gases are usually stored under high pressure, even though they will be used at atmospheric pressure. This allows a large amount of gas to be stored in a small volume.

Boyle's law has a number of practical applications perhaps best illustrated by some examples. Note that in these applications any units can be used for pressure and volume, as long as the same units are used throughout a calculation.

Example 4.4

A cylinder of oxygen has a volume of 2.00 L. The pressure of the gas is 1360 lb/in.^{2} at 20 °C. What volume will the oxygen occupy at standard atmospheric pressure (14.7 lb/in.^{2}), assuming there is no temperature change?

Solution

The mathematical expression of Boyle's law tells us that

$$P_{\text{initial}} \times V_{\text{initial}} = a = P_{\text{final}} \times V_{\text{final}}$$

We can solve this expression for V_{final} and substitute the given values for the other terms.

$$V_{\text{final}} = V_{\text{initial}} \times \frac{P_{\text{initial}}}{P_{\text{final}}} = 2.00\ \text{L} \times \frac{1360\ \text{lb/in.}^2}{14.7\ \text{lb/in.}^2} = 185\ \text{L}$$

Because the final pressure in Example 4.4 is *less than* the initial pressure, we expect the final volume to be *larger than* the original volume; and it is.

Example 4.5

A weather balloon is partially filled with helium. On the ground, where the atmospheric pressure is 748 torr, the volume of the balloon is 10.0 m³. When the balloon reaches an altitude of 5300 m, its volume is found to be 20.2 m³. Assuming that the temperature remains constant, what must be the prevailing air pressure at 5300 m?

Solution

Again, we use the expression

$$P_{\text{initial}} \times V_{\text{initial}} = a = P_{\text{final}} \times V_{\text{final}}$$

but this time we solve for P_{final}.

$$P_{\text{final}} = P_{\text{initial}} \times \frac{V_{\text{initial}}}{V_{\text{final}}} = 748\ \text{torr} \times \frac{10.0\ \text{m}^3}{20.2\ \text{m}^3} = 370\ \text{torr}$$

Exercise 4.4

A sample of air occupies 73.3 mL at 98.7 kPa and 0 °C. What volume will the air occupy at 4.02 atm and 0 °C? (*Hint*: Recall that the initial and final pressure must be expressed in the same unit.)

Exercise 4.5

A sample of helium occupies 535 mL at 988 torr and 25 °C. If the sample is transferred to a 1.05-L flask at 25 °C, what will be the gas pressure in the flask?

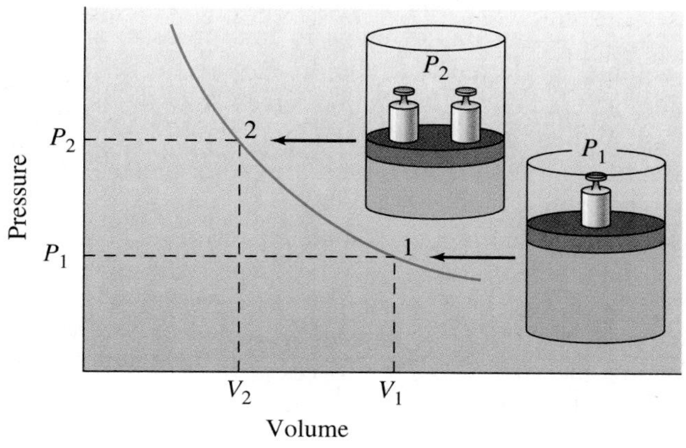

Figure 4.6 Boyle's law: a graphical representation. As the pressure of a gas is increased, its volume decreases. When the pressure is doubled ($P_2 = 2 \times P_1$), the volume of a gas decreases to one-half of its original value ($V_2 = \frac{1}{2} V_1$). The pressure–volume product is a constant ($PV = a$).

Boyle's Law and Breathing

The pressure–volume relationship helps to explain the mechanics of breathing. When we breathe in (inspire), the diaphragm is lowered and the chest wall is expanded, increasing the volume of the chest cavity (Figure 4.7). According to Boyle's law, the pressure inside the cavity must decrease. Outside air enters the lungs because it is at a higher pressure than the air in the chest cavity. When we breathe out (expire), the diaphragm rises and the chest wall contracts, decreasing the volume of the chest cavity. The pressure is increased, and some air is forced out.

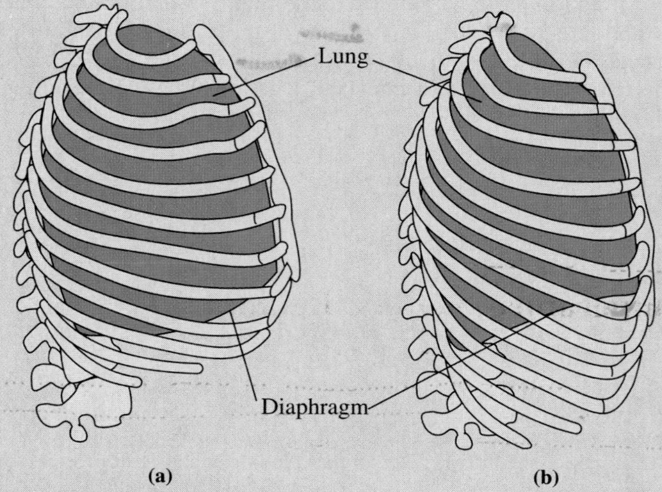

Figure 4.7 The mechanics of breathing. (a) Inspiration. The diaphragm is pulled down, and the rib cage is lifted up and out, increasing the volume of the chest cavity. (b) Expiration. The diaphragm is relaxed; the rib cage is down; and the volume of the chest cavity decreases.

During normal inspiration, the pressure inside the lungs drops about 3 torr below atmospheric pressure. During expiration, the internal pressure is about 3 torr above atmospheric pressure. About one-half liter of air is moved in and out of the lungs in this process, and this normal breathing volume is referred to as the *tidal volume*. The *vital capacity* is the maximum volume of air that can be forced from the lungs and ranges from 3 to 7 L, depending on the individual. A pressure inside the lungs 100 torr greater than the external pressure is not unusual during such a maximum expiration.

The lungs are never emptied completely, however. The space around the lungs is maintained at a slightly lower pressure than are the lungs themselves, causing the lungs to be kept partially inflated by the higher pressure within them. If a lung, the diaphragm, or the chest wall is punctured, allowing the two pressures to equalize, the lung will collapse. Sometimes a damaged lung will be collapsed intentionally to give it time to heal. Closing the opening reinflates the lung.

ESTIMATION EXAMPLE 4.1

A gas is enclosed in a cylinder fitted with a piston. The volume of the gas is 2.00 L at 398 torr. The piston is moved to increase the gas pressure to 5.15 atm. Which of the following is a reasonable value for the volume of the gas at the greater pressure?

0.20 L 0.40 L 1.00 L 16.0 L

Solution

The initial pressure (398 torr) is about 0.5 atm. The pressure increase is about tenfold. The volume should drop to about one-tenth of the initial value. We estimate a volume of 0.20 L. (The calculated value is 0.203 L.)

Estimation Exercise 4.1

A gas is enclosed in a 10.2-L tank at 1208 torr. Which of the following is a reasonable value for the pressure when the gas is transferred to a 30.0-L tank?

0.40 atm 25 lb/in.^{2} 400 mmHg 3600 torr

4.5 Charles's Law: The Temperature–Volume Relationship

In 1787, the French physicist Jacques Charles (1746–1823) studied the relationship between volume and temperature of gases. He found that when a fixed mass of gas is cooled at constant pressure, its volume decreases. When the gas is heated, its volume increases. Temperature and volume vary directly; that is, they rise or fall together. But this law requires a bit more thought. If a quantity of gas that occupies 1.00 L is heated from 100 °C to 200 °C at constant pressure, the volume does not double but only increases to about 1.27 L. The relationship between temperature and volume is not as tidy as it may seem on first impression.

Zero pressure or zero volume really means zero—no pressure or volume to be measured. Zero degrees Celsius (0 °C) means only a temperature the same as the freezing point of water. This zero point is arbitrarily set, much like mean sea level is the arbitrary zero used to measure altitudes on Earth. Temperatures below 0 °C are often encountered, as are altitudes below sea level.

Charles noted that for each Celsius degree rise in temperature, the volume of a gas expands by 1/273 of its volume at 0 °C. For each Celsius degree drop in temperature, the volume of a gas decreases by 1/273 of its value at 0 °C. If we plot volume against temperature we get a straight line (Figure 4.8). We can *extrapolate* the line beyond the range of measured temperatures to the temperature at which the volume of the gas appears to become zero.

Before a gas ever reaches this temperature, however, it liquefies—and then the liquid freezes—so this is an exercise for the imagination.

This temperature is −273.15 °C. In 1848, William Thomson (Lord Kelvin) made this temperature the **absolute zero** on a temperature scale, now called the **Kelvin scale**. The unit of temperature on this scale is the **kelvin (K)**, which is made to be equal to a degree of Celsius temperature. This means that

$$T(\text{K}) = T(°\text{C}) + 273.15$$

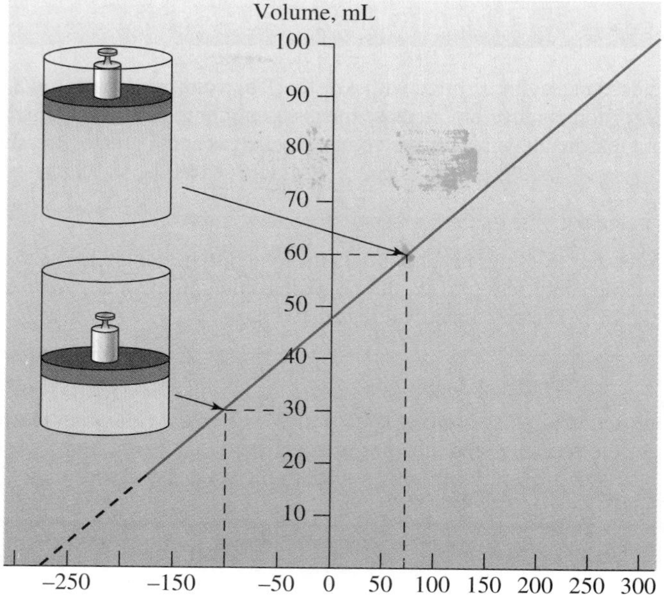

Figure 4.8 Charles's law: gas volume as a function of temperature. When the gas shown has been cooled to about 70 °C, its volume is 60 mL. In the temperature interval from about 70 °C to −100 °C, the volume drops to 30 mL. The volume continues to fall in a linear fashion as the temperature is lowered. The extrapolated line intersects the temperature axis (corresponding to a volume of zero) at about −270 °C.

A modern statement of **Charles's law** is:

The volume of a fixed amount of a gas at a constant pressure is directly proportional to its Kelvin temperature.

Mathematically, this relationship can be expressed as

$$V \propto T \qquad \text{and} \qquad V = bT$$

The kinetic-molecular model easily accounts for the relationship between gas volume and temperature. When we heat a gas we supply the gas molecules with energy and they begin to move faster. These speedier molecules strike the walls of their container harder and more often. For the pressure to stay the same, the volume of the container must increase so that the increased molecular motion will be distributed over a greater space. In this way, the pressure exerted by the faster molecules in the larger volume (high temperature) is the same as that of the slower moving molecules in the smaller volume (low temperature). Figure 4.9 illustrates the dramatic change in gas volume that occurs over an extreme range of temperatures.

Example 4.6

A balloon indoors, where the temperature is 27 °C, has a volume of 2.00 L. What will its volume be outdoors, where the temperature is −23 °C? (Assume no change in the gas pressure.)

Solution

First, convert both temperatures to the Kelvin scale, using the relationship

$$T(\text{K}) = T(°\text{C}) + 273.15$$

The initial temperature is

$$T_{\text{initial}} = 27 + 273.15 = 300\ \text{K}$$

and the final temperature is

$$T_{\text{final}} = -23 + 273.15 = 250 \text{ K}$$

Now apply Charles's law in the form

$$\frac{V_{\text{final}}}{T_{\text{final}}} = b = \frac{V_{\text{initial}}}{T_{\text{initial}}}$$

Solving for V_{final} leads to the expression

$$V_{\text{final}} = V_{\text{initial}} \times \frac{T_{\text{final}}}{T_{\text{initial}}} = 2.00 \text{ L} \times \frac{250 \text{ K}}{300 \text{ K}} = 1.67 \text{ L}$$

Because the final temperature is *less than* the initial temperature, we expect the final volume to be *smaller than* the initial volume; and it is.

Example 4.7

At what temperature, in kelvins and degrees Celsius, should the balloon described in Example 4.6 be held to have a volume of 2.25 L? (Assume a constant pressure.)

Solution

In this case Charles's law takes the form

$$T_{\text{final}} = T_{\text{initial}} \times \frac{V_{\text{final}}}{V_{\text{initial}}} = 300 \text{ K} \times \frac{2.25 \text{ L}}{2.00 \text{ L}} = 338 \text{ K}$$

$$T(°\text{C}) = T(\text{K}) - 273.15 = 338 - 273.15 = 65 \text{ °C}$$

Because the final volume is *larger than* the initial volume, we expect the final temperature to be *greater than* the initial temperature; and it is.

Exercise 4.6

A sample of hydrogen occupies 692 L at 602 °C. If the pressure is held constant, what volume will the gas occupy after cooling to 23 °C?

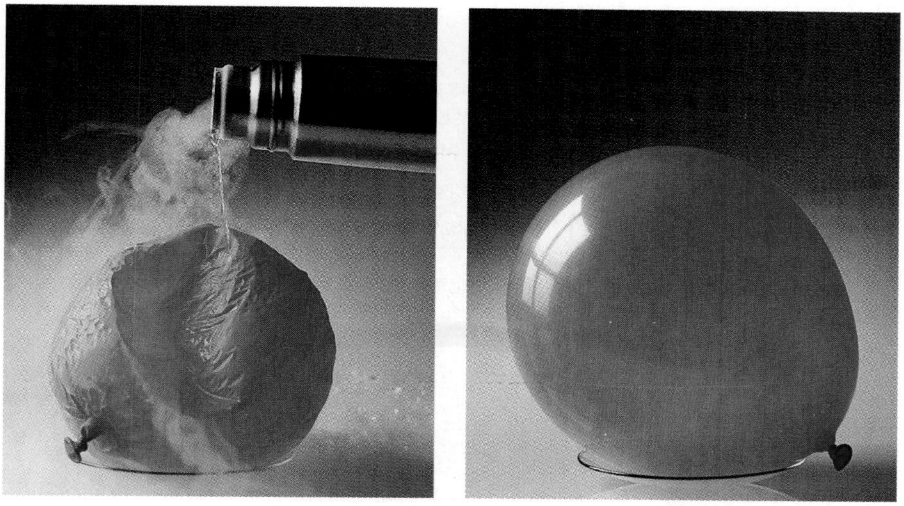

Figure 4.9 Charles's law: a dramatic illustration. (a) Liquid nitrogen (boiling point −196 °C) cools the air in the balloon to a temperature far below room temperature—the balloon collapses. (b) As the balloon warms to room temperature, it reinflates. The air in the balloon regains its original room-temperature volume.

Estimation Exercise 4.2

A sample of nitrogen occupies a volume of 2.50 L at −120 °C and 1 atm pressure. To which of the following approximate temperatures should the gas be brought in order to double its volume while maintaining a constant pressure?

−240 °C −60 °C −12 °C 30 °C

Amedeo Avogadro (1776–1856) did not live to see his ideas accepted by the scientific community. This acceptance came in 1860 at an international conference at which Stanislao Cannizzaro (1826–1910) effectively communicated Avogadro's ideas from five decades before.

4.6 Avogadro's Law: The Amount–Volume Relationship

As we will describe in Section 4.9, Amedeo Avogadro proposed an important hypothesis in 1811 to explain some observations about the ratios in which volumes of gases combine in chemical reactions. His hypothesis was that equal numbers of molecules of different gases compared at the same temperature and pressure occupy equal volumes. For now, let's restate Avogadro's hypothesis in the form generally called **Avogadro's law**.

At a fixed temperature and pressure, the volume of a gas is directly proportional to the amount of gas (that is, to the number of molecules of gas or to the number of moles of gas, n).

If we double the number of moles of gas at a fixed T and P, the volume of the gas doubles. Because the mass of a gas is proportional to the number of moles, doubling the *mass* of a gas also doubles its volume. Mathematically, we can state Avogadro's law as

$$V \propto n \qquad \text{or} \qquad V = cn \quad \text{(where } c \text{ is a constant)}$$

When we use Avogadro's law to compare different gases, the gases must be at the same temperature and pressure. A convenient temperature/pressure combination for such comparisons is 0 °C (273.15 K) and 1 atm (760 torr), known as **standard conditions of temperature and pressure (STP)**.

Molar Volume of a Gas

Suppose that in comparing different gases we use STP as the fixed temperature and pressure and Avogadro's number as the number of molecules present. Avogadro's hypothesis states that under these conditions, 1 mol (6.022×10^{23} molecules) of *all* gases should occupy the same volume. By experiment, this **molar volume of a gas** at STP is found to be

$$22.428\ \text{L}\ H_2;\quad 22.404\ \text{L}\ N_2;\quad 22.394\ \text{L}\ O_2;\quad 22.360\ \text{L}\ CH_4;\quad \text{and so on}$$

To *three* significant figures, we can state that

$$1\ \text{mol gas} = 22.4\ \text{L gas (at STP)}$$

Figure 4.10 pictures a volume of 22.4 L and relates it to some familiar objects. At STP, the 22.4-L container would hold 28.0 g N_2, 32.0 g O_2, or 44.0 g CO_2.

If a gas is *not* at STP, relating the amount of gas and its volume requires a different calculation. We will consider one way to do this in Section 4.8.

Example 4.8

Calculate the volume occupied by 4.11 kg of methane gas at STP.

Solution

We must convert the mass of gas to an amount in moles, and then use the molar volume relationship as a conversion factor to go from the amount of gas to its volume at STP. We can do all of this in a single setup.

$$? \text{ L } CH_4 = 4.11 \text{ kg } CH_4 \times \frac{1000 \text{ g } CH_4}{1 \text{ kg } CH_4} \times \frac{1 \text{ mol } CH_4}{16.04 \text{ g } CH_4} \times \frac{22.4 \text{ L } CH_4}{1 \text{ mol } CH_4}$$

$$= 5.74 \times 10^3 \text{ L } CH_4$$

Exercise 4.7

Solid carbon dioxide, called "dry ice," is useful in maintaining frozen foods because it vaporizes to $CO_2(g)$ rather than melting to a liquid. How many kilograms of "dry ice" can be produced from 5.00×10^3 L $CO_2(g)$ measured at STP?

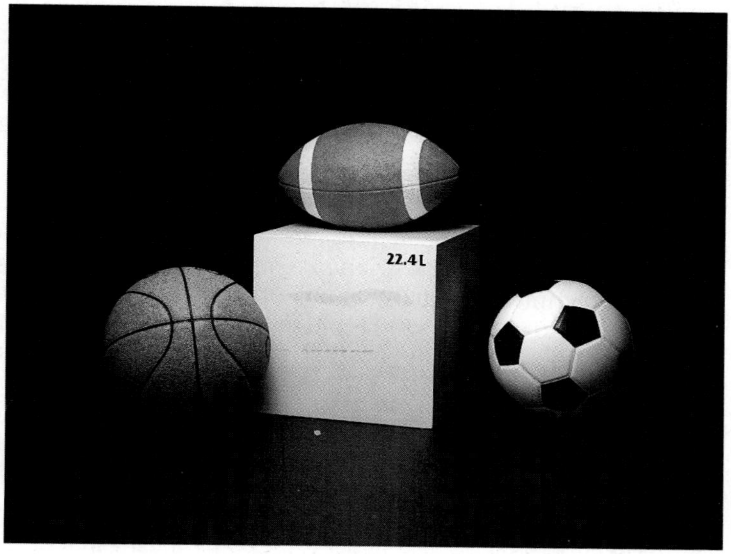

Figure 4.10 Molar volume of a gas visualized. The wooden cube has the same volume as one mole of gas at STP: 22.4 L. By contrast, the volume of the basketball is 7.5 L, the soccer ball, 6.0 L, and the football, 4.4 L.

4.7 The Combined Gas Law

From the three simple gas laws,

$$V = \frac{a}{P}; \quad V = bT; \quad \text{and} \quad V = cn$$

it seems reasonable that the volume of a gas (V) should be *directly* proportional to the Kelvin temperature (T) and to the amount of gas (n), and *inversely* proportional to the pressure (P), and that happens to be the case. That is,

$$V \propto \frac{nT}{P}$$

Or, expressed as an equation rather than a proportionality,

$$\frac{PV}{nT} = \text{constant}$$

This *combined* gas law is most useful in situations where a gas is described under an *initial* set of conditions, the gas is subjected to a change, and a question is asked about the *final* set of conditions. In these cases we write,

$$\begin{array}{ccc} \text{Initial} & & \text{Final} \\ \dfrac{P_1V_1}{n_1T_1} & = \text{constant} = & \dfrac{P_2V_2}{n_2T_2} \end{array}$$

Often we can simplify this expression because one or more of the gas properties is held constant during the change. (Boyle's law and Charles's law are just such simplifications.) We simply eliminate any constant terms. For example, consider a *fixed amount* of gas, $n_1 = n_2$, in a *fixed volume*, $V_1 = V_2$. The combined gas equation

$$\frac{P_1 \cancel{V_1}}{\cancel{n_1} T_1} = \frac{P_2 \cancel{V_2}}{\cancel{n_2} T_2}$$

simplifies to

$$\frac{P_1}{T_1} = \frac{P_2}{T_2}$$

This equation shows that the pressure of a fixed amount of gas in a constant volume is proportional to its Kelvin temperature. This relationship is a simple gas law sometimes called *Amontons's* law. This law certainly seems reasonable from the standpoint of the kinetic-molecular theory, as is illustrated in Figure 4.11.

Guillaume Amontons, a French physicist, was the first to use mercury as the fluid in a thermometer.

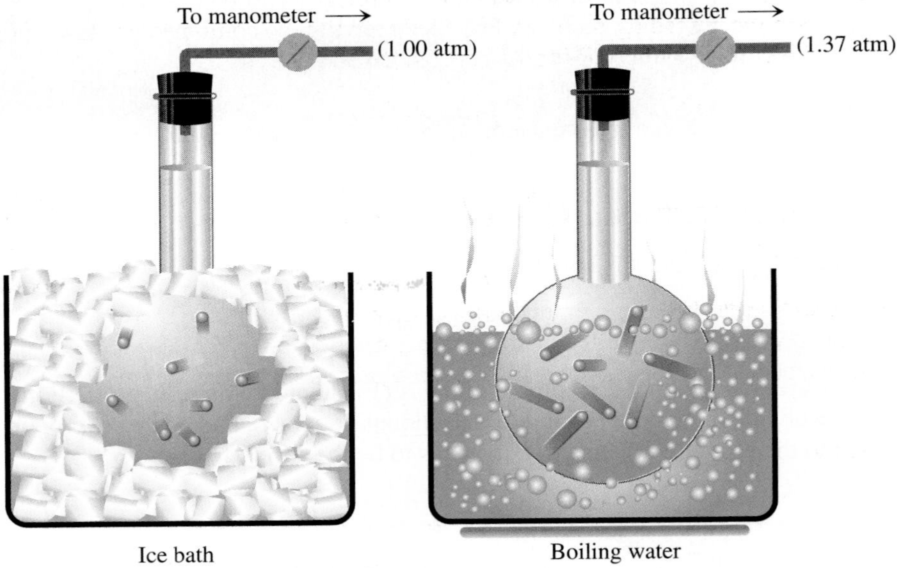

Figure 4.11 The effect of temperature on the pressure of a fixed amount of gas in a constant volume: a kinetic theory interpretation. The amount of gas and its volume are the same in either case, but if the gas in the ice bath (0 °C) exerts a pressure of 1 atm, the gas in the boiling water bath (100 °C) exerts a pressure of 1.37 atm. The frequency and the force of the molecular collisions with the container walls are greater at the higher temperature.

Example 4.9

The flask pictured in Figure 4.11 contains O_2(g), first at STP and then at 100 °C. What is the pressure at 100 °C?

Solution

We start with the expression

$$\frac{P_1}{T_1} = \frac{P_2}{T_2}$$

where the subscripts "1" represent the initial condition (STP) and "2," the final condition. Then we solve it for the final pressure, P_2.

$$P_2 = P_1 \times \frac{T_2}{T_1} = 1.00 \text{ atm} \times \frac{(100 + 273) \text{ K}}{273 \text{ K}} = 1.37 \text{ atm}$$

Note that in this calculation we need to know that the amount and volume of $O_2(g)$ remain constant, but we don't need to know their actual values.

Exercise 4.8

Aerosol containers often carry the warning that they should not be heated. Suppose such a container was filled with a gas at 2.5 atm and 22 °C and that the container may rupture if the pressure exceeds 8.0 atm. At what temperature is that rupture likely to occur?

Exercise 4.9

In the manner we used to derive the simple gas law relating pressure and temperature (Amontons's law), derive an expression relating the pressure and amount of gas when both the temperature and volume are held constant. Rationalize your result in terms of the kinetic-molecular theory, and draw a sketch in the manner of Figure 4.11.

4.8 The Ideal Gas Law

We will find the single most useful equation for gases to be one that combines ideas from the preceding sections. First, let's rewrite the combined gas law from page 125 with the symbol R for the constant term.

$$\frac{PV}{nT} = \text{constant} = R$$

Then, let's substitute data into the left side of the equation that will allow us to obtain the value of R. The data we can use for this purpose are those associated with the molar volume of a gas at STP, but what's the "best" value for the molar volume? It's the value found for an *ideal gas*. An **ideal gas** is a gas that *strictly* obeys all the simple gas laws, and for an ideal gas the molar volume at STP is 22.4141 L. We can obtain the three different values for R listed in Table 4.2 depending on the units we use to express pressure. If we choose pressure in atmospheres,

$$R = \frac{PV}{nT} = \frac{1 \text{ atm} \times 22.4141 \text{ L}}{1 \text{ mol} \times 273.15 \text{ K}} = 0.082058 \frac{\text{L} \cdot \text{atm}}{\text{mol} \cdot \text{K}}$$

TABLE 4.2 **Units for the Universal Gas Constant, *R***

R has the value	When
$0.082058 \text{ L} \cdot \text{atm} \cdot \text{mol}^{-1} \cdot \text{K}^{-1}$	*P* is in atm
$62.364 \text{ L} \cdot \text{torr} \cdot \text{mol}^{-1} \cdot \text{K}^{-1}$	*P* is in torr
$8.3145 \text{ J} \cdot \text{mol}^{-1} \cdot \text{K}^{-1}$	*P* is in Pa; *V* is in m^3

R is called the **universal gas constant**, or simply the *ideal gas constant*. Its value is often rounded off to three significant figures and the units in the denominator are written with negative exponents, that is, $R = 0.0821\ \text{L} \cdot \text{atm} \cdot \text{mol}^{-1} \cdot \text{K}^{-1}$.

The equation we have been describing is generally written in the form

$$PV = nRT$$

and called the **ideal gas law** or the **ideal gas equation**.

Real gases only approach the behavior described by the ideal gas equation (as we shall see in Section 4.11), but under suitable conditions most real gases fit the equation well enough to make it quite useful. Specifically, we can use it to calculate any one of the four quantities—*P*, *V*, *n*, or *T*—if the other three are known. As indicated in the following examples, we begin each calculation by solving the ideal gas equation for the unknown variable, and we also make certain to express quantities in the units needed to match the units of *R*: for example, L (volume), atm (pressure), mol (amount of gas), and K (temperature) to match $R = 0.0821\ \text{L} \cdot \text{atm} \cdot \text{mol}^{-1} \cdot \text{K}^{-1}$.

Example 4.10

What is the pressure exerted by 0.508 mol O_2 in a 15.0-L container at 303 K?

Solution

Solve the ideal gas equation, $PV = nRT$, for *P*; then substitute the data given.

$$P = \frac{nRT}{V} = \frac{0.508\ \text{mol} \times 0.0821\ \text{L} \cdot \text{atm} \times 303\ \text{K}}{15.0\ \text{L} \qquad \text{mol} \cdot \text{K}} = 0.842\ \text{atm}$$

Example 4.11

What is the volume occupied by 16.0 g ethane gas (C_2H_6) at 720 torr and 18 °C?

Solution

Express the amount of gas, temperature, and pressure in the appropriate units.

$$n = 16.0\ \text{g}\ C_2H_6 \times \frac{1\ \text{mol}\ C_2H_6}{30.07\ \text{g}\ C_2H_6} = 0.532\ \text{mol}\ C_2H_6$$

$$T = (273.15 + 18)\ \text{K} = 291\ \text{K}$$

$$P = 720\ \text{torr} \times \frac{1\ \text{atm}}{760\ \text{torr}} = 0.947\ \text{atm}$$

Then, substitute into the ideal gas equation to obtain

$$V = \frac{nRT}{P} = \frac{0.532\ \text{mol} \times 0.0821\ \text{L} \cdot \text{atm} \times 291\ \text{K}}{0.947\ \text{atm} \qquad \text{mol} \cdot \text{K}} = 13.4\ \text{L}$$

An alternative approach is to leave pressure in torr and to use $R = 62.4\ \text{L} \cdot \text{torr} \cdot \text{mol}^{-1} \cdot \text{K}^{-1}$.

Exercise 4.10

How many moles of nitrogen, N_2, are there in a sample that occupies 35.0 L at a pressure of 3.15 atm and a temperature of 852 K?

Exercise 4.11

What is the temperature, in degrees Celsius, at which 15.0 g O_2 will exert a pressure of 785 torr in a volume of 5.00 L?

Exercise 4.12

How many grams are there in a sample of N_2 that occupies 525 mL at a pressure of 546 torr and a temperature of 35 °C?

Molecular Weight Determination

We can use the ideal gas equation to determine the amount of a gas, in moles, by measuring the volume, pressure, and temperature of a fixed quantity of the gas.

$$n = \frac{PV}{RT}$$

If we also measure the mass of this fixed quantity of gas, we can establish the molar mass of the gas by relating the mass, *m*, and amount of gas, *n*.

$$\text{Molar mass} = \frac{m}{n}$$

As we have previously seen, the molar mass and molecular weight are numerically equal; for example, the molar mass of CO_2 = 44.01 g and the molecular weight of CO_2 is 44.01 u.

The method of molecular weight determination suggested here works for any gas or easily vaporized liquid, as illustrated in Example 4.12.

Example 4.12

Calculate the molecular weight of a gas if 0.550 g of the gas occupies 0.200 L at 0.968 atm and 289 K.

Solution

First, calculate the amount of gas, in moles, from the ideal gas equation.

$$n = \frac{PV}{RT} = \frac{0.968\ \text{atm} \times 0.200\ \text{L}}{0.0821\ \text{L} \cdot \text{atm} \cdot \text{mol}^{-1} \cdot \text{K}^{-1} \times 289\ \text{K}} = 0.00816\ \text{mol}$$

Now use the known mass and the calculated number of moles to determine the mass per mole.

$$\text{Molar mass} = \frac{0.550\ \text{g}}{0.00816\ \text{mol}} = 67.4\ \text{g/mol}$$

The molar mass of the gas is 67.4 g/mol. The molecular weight, therefore, is 67.4 u.

An alternative approach to that illustrated in Example 4.12 is to combine the two equations

$$n = \frac{PV}{RT} \quad \text{and} \quad \text{molar mass} = \frac{m}{n}$$

into this variation of the ideal gas equation

$$n = \frac{m}{\text{molar mass}} = \frac{PV}{RT};\ \frac{m}{PV} = \frac{\text{molar mass}}{RT};\ \text{and molar mass} = \frac{mRT}{PV}$$

before substituting any of the gas data. This is the method used in Example 4.13.

Example 4.13

Calculate the molecular weight of a liquid which, when vaporized at 100 °C and 755 mmHg, yields 185 mL of vapor with a mass of 0.523 g.

Solution

First, change all quantities to units that match those of the value of *R* that is used.

$$P = 755 \text{ mmHg} \times \frac{1 \text{ atm}}{760 \text{ mmHg}} = 0.993 \text{ atm}$$

$$V = 185 \text{ mL} \times \frac{1 \text{ L}}{1000 \text{ mL}} = 0.185 \text{ L}$$

$$T = (273 + 100) \text{ K} = 373 \text{ K}$$

Now, substitute into the equation

$$\text{Molar mass} = \frac{mRT}{PV} = \frac{0.523 \text{ g} \times 0.821 \text{ L} \cdot \text{atm} \cdot \text{mol}^{-1} \cdot \text{K}^{-1} \times 373 \text{ K}}{0.993 \text{ atm} \times 0.185 \text{ L}}$$
$$= 87.2 \text{ g/mol}$$

The molar mass of the gas is 87.2 g/mol. The molecular weight, therefore, is 87.2 u.

Exercise 4.13

Calculate the molar mass of a gas if 0.440 g occupies 179 mL at 741 mmHg and 86 °C.

Exercise 4.14

Calculate the molecular weight of a liquid which, when vaporized at 98 °C and 715 mmHg, yields 121 mL of vapor with a mass of 0.471 g.

We can establish the *molecular* formula of an unknown gaseous carbon–hydrogen–oxygen compound by (1) doing a combustion analysis (recall Example 3.12) to establish the *empirical* formula and (2) doing a molecular weight determination as described in this section. This combination of calculations is required in Exercise 4.15.

Exercise 4.15

Diacetyl, a substance contributing to the characteristic flavor and aroma of butter, consists of 55.80% C, 7.03% H, and 37.17% O. In the gaseous state at 100 °C and 747 mmHg, a 0.3060-g sample occupies a volume of 111 mL. Establish the molecular formula of diacetyl.

Gas Densities

Gases are much less dense than liquids and solids, and we generally express their densities in g/L rather than g/mL. To establish the density of $O_2(g)$ at STP, we simply divide the molar mass of O_2 by its molar volume at STP.

$$d = \frac{32.00 \text{ g } O_2}{22.4 \text{ L}} = 1.43 \text{ g } O_2/\text{L}$$

However, the volume of a given mass of gas depends both on temperature and pressure. Because volume appears in the definition of density, gas density also depends on these factors. We must therefore always specify the temperature and pressure at which a value for a gas density applies. These relationships can best be seen in another rearrangement of the ideal gas equation.

$$PV = nRT; \qquad PV = \frac{mRT}{\text{molar mass}}$$

which rearranges to

$$\frac{m}{V} = \frac{\text{molar mass} \times P}{RT} = d$$

Thus, we see that the density of a gas is *directly* proportional to its molar mass and pressure and *inversely* proportional to its Kelvin temperature.

Can you rationalize this statement about gas densities in terms of the kinetic-molecular theory? (See Review Question 22.)

Example 4.14

Calculate the density of methane gas (CH_4), in grams per liter, at 25 °C and 0.978 atm.

Solution

We could do this as a two-step problem in which we (1) determine the number of moles of CH_4 in 1.00 L of the gas at the stated temperature and pressure and (2) determine the mass of this gas. To do so, however, is really the same as solving the gas density equation derived above.

$$d = \frac{\text{molar mass} \times P}{RT} = \frac{16.04 \text{ g} \cdot \text{mol}^{-1} \times 0.978 \text{ atm}}{0.0821 \text{ L} \cdot \text{atm} \cdot \text{mol}^{-1} \cdot \text{K}^{-1} \times 298 \text{ K}}$$

$$= 0.641 \text{ g } CH_4/\text{L}$$

We can also use the equation

$$d = \frac{\text{molar mass} \times P}{RT}$$

to calculate temperatures or pressures, as illustrated in Example 4.15 and Exercise 4.17.

Example 4.15

Under what pressure must O_2(g) be maintained at 25 °C to have a density of 1.50 g/L?

Solution

Here we simply solve the gas density equation for the unknown pressure.

$$P = \frac{dRT}{\text{molar mass}} = \frac{1.50 \text{ g} \cdot \text{L}^{-1} \times 0.0821 \text{ L} \cdot \text{atm} \cdot \text{mol}^{-1} \cdot \text{K}^{-1} \times 298 \text{ K}}{32.00 \text{ g} \cdot \text{mol}^{-1}}$$

$$= 1.15 \text{ atm}$$

To ensure that you have used a correct variation of the ideal gas equation, you should use the appropriate cancellation of units as a check. Here, all units cancel except for the desired "atm."

Exercise 4.16

Calculate the density of ethane gas (C_2H_6), in grams per liter, at 15 °C and 748 torr.

Exercise 4.17

To what temperature must propane gas (C_3H_8) at 785 torr be heated to have a density of 1.51 g/L?

Connections: Balloons, the Gas Laws, and Chemistry

Boyle developed his pressure–volume law in 1662. Over a century elapsed before the next simple gas law was formulated. The experiments that led to this law were performed with the help of an eighteenth-century innovation: human flight using balloons.

In June 1782 two French brothers, Joseph Michel and Jacques Étienne Montgolfier, launched the first such balloon. They lit a fire under an opening in a large bag. The less dense hot air allowed the balloon to rise slowly through the denser, cooler atmosphere. By August of that year, Jacques Charles filled a balloon with hydrogen gas, discovered only 16 years before, in 1766, by Henry Cavendish. Charles made hydrogen on a scale never before attempted by reacting about 500 kg of iron with acid.

$$Fe(s) + 2\,HCl(aq) \longrightarrow FeCl_2(aq) + H_2(g)$$

At any given temperature and pressure, hydrogen is only one-fourteenth as dense as air, which means that each kilogram of hydrogen can carry aloft a payload of nearly 13 kg. In December 1782 Charles and a companion flew for 25 km from Paris in a balloon filled with hydrogen. They landed in a small village where they were attacked by frightened farmers who tore the balloon with pitchforks.

By 1804 Joseph Gay-Lussac had ascended to an altitude of 7 km and brought back samples of the rarefied air at that altitude. Within just a few years, scientists had used balloons to learn a great deal about the nature of gases, about Earth's atmosphere, and about weather.

A contemporary drawing depicting the first ascent of the Montgolfier brothers in a hot-air balloon in 1782.

Estimation Exercise 4.3

Which of the following gases has the greatest density? H_2 at 100 °C and 745 torr; He at STP; CH_4 at −10 °C and 765 torr.

4.9 Gases in Reaction Stoichiometry

The French scientist Joseph Gay-Lussac (1778–1850) was a pioneer in the study of reactions between gases. His results led to Avogadro's hypothesis, which in turn was the starting point of the modern treatment of stoichiometry that we explored in detail in Chapter 3.

The Law of Combining Volumes

In 1809, Gay-Lussac published an experimental result known as the law of **combining volumes.**

> *When gases measured at the same temperature and pressure are allowed to react, the volumes of gaseous reactants and products are in small whole-number ratios.*

For example, at 100 °C two volumes of hydrogen unite with one volume of oxygen to produce two volumes of steam (water vapor), as suggested by Figure 4.12. The combining ratio is 2:1:2.

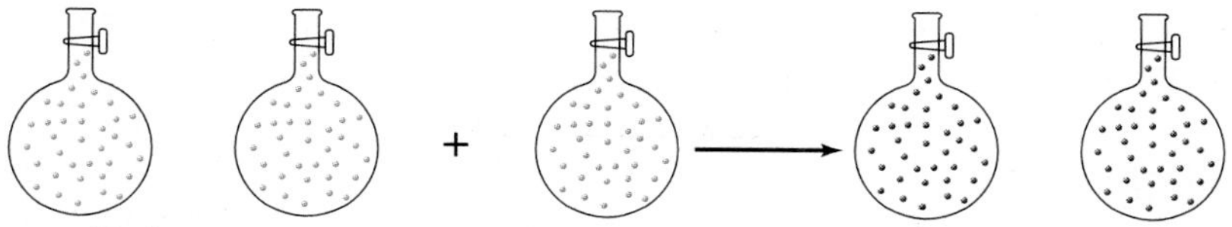

Figure 4.12 Gay-Lussac's law of combining volumes.

From what we know today, the law of combining volumes is not hard to explain. The chemical equation for the reaction of hydrogen and oxygen is

$$2\ H_2(g) + O_2(g) \longrightarrow 2\ H_2O(g)$$

and the coefficients mean that *two* molecules of H_2 and *one* molecule of O_2 react for every *two* molecules of H_2O produced. Because the combining ratio by numbers of molecules, 2:1:2, is the same as the combining ratio of the gas volumes, it must be that equal volumes of gases at the same temperature and pressure contain identical numbers of molecules.

But consider how difficult it was to arrive at this conclusion in Dalton's time. Here is how Dalton described the reaction of hydrogen and oxygen to form steam.

$$H(g) + O(g) \longrightarrow OH(g)$$

His expectation, if the "equal volumes–equal numbers" hypothesis were correct, was a combining ratio by volumes of 1:1:1. Even if he had known the true formula of water, H_2O, he would have written

$$2\ H(g) + O(g) \longrightarrow H_2O(g)$$

This formulation would have led to the combining ratio by volume of 2:1:1, still not what was observed.

The genius of Avogadro was not just in the "equal volumes–equal numbers" hypothesis. He also proposed that gases may exist in *molecular* form. By postulating the molecules H_2 and O_2, Avogadro was able to explain the reaction in the same way that we do today.

$$2\,H_2(g) + O_2(g) \longrightarrow 2\,H_2O(g)$$

Figure 4.13 suggests Avogadro's line of reasoning.

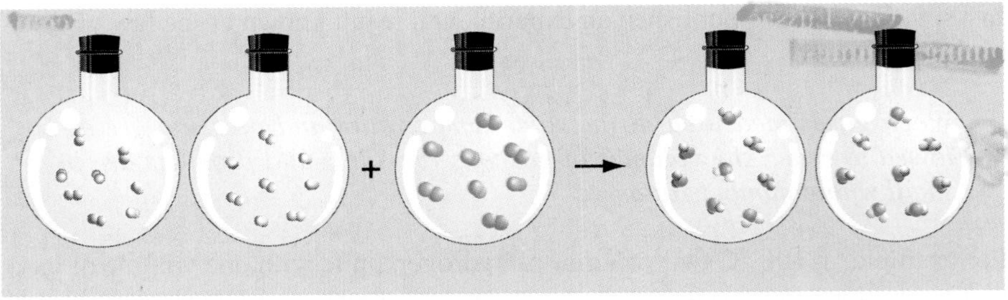

Figure 4.13 Avogadro's explanation of Gay-Lussac's law of combining volumes. When the gases are measured at the same temperature and pressure, each of the identical flasks contains the same number of molecules.

Example 4.16 is a straightforward application of the law of combining volumes. Exercise 4.18 demonstrates that, even if solids or liquids are involved in a reaction, the law still applies, as long as both the substances in a stoichiometric factor are *gases*.

Example 4.16

What volume of oxygen is required to burn 0.556 L of propane, if both gases are measured at the same temperature and pressure?

$$C_3H_8(g) + 5\,O_2(g) \longrightarrow 3\,CO_2(g) + 4\,H_2O(g)$$

Solution

The equation indicates that *five* volumes of $O_2(g)$ are required for every volume of $C_3H_8(g)$. Thus,

$$?\,L\,O_2(g) = 0.556\,L\,C_3H_8(g) \times \frac{5\,L\,O_2(g)}{1\,L\,C_3H_8(g)} = 2.78\,L\,O_2(g)$$

Exercise 4.18

What volume of methane must decompose to produce 10.0 L of hydrogen in the following reaction, if the two gases are compared at the same temperature and pressure?

$$CH_4(g) \longrightarrow C(s) + 2\,H_2(g)$$

The Ideal Gas Equation in Reaction Stoichiometry

We can substitute ratios of gas volumes for mole ratios *only* for *gases* at *the same temperature and pressure.* Often the amount of a gaseous reactant or product needs to be related to that of a *solid* or *liquid.* In these cases, we must work with mole ratios, but the ideal gas equation allows us to relate moles of gas to other gas properties, as shown in Example 4.17 and Exercises 4.19 and 4.20.

Example 4.17

In the chemical reaction used in automotive air-bag safety systems, $N_2(g)$ is produced by the high-temperature decomposition of sodium azide, NaN_3.

$$2\ NaN_3(s) \longrightarrow 2\ Na(l) + 3\ N_2(g)$$

What volume of $N_2(g)$, measured at 25 °C and 745 torr, is produced by the decomposition of 62.5 g NaN_3?

Solution

First, determine the amount of $N_2(g)$ in moles.

$$?\ \text{mol } N_2 = 62.5\ \text{g } NaN_3 \times \frac{1\ \text{mol } NaN_3}{65.01\ \text{g } NaN_3} \times \frac{3\ \text{mol } N_2}{2\ \text{mol } NaN_3} = 1.44\ \text{mol } N_2$$

Then, use the ideal gas equation to determine the volume of this amount of $N_2(g)$ under the stated conditions.

$$V = \frac{nRT}{P} = \frac{1.44\ \text{mol} \times 0.0821\ \text{L} \cdot \text{atm} \cdot \text{mol}^{-1} \cdot \text{K}^{-1} \times (273 + 25)\ \text{K}}{(745/760)\ \text{atm}}$$

$$= 35.9\ \text{L}$$

Although we have solved this problem in two steps, there is no need to re-enter the intermediate result (1.44 mol N_2) in your calculator. The number displayed at the end of the first step simply becomes the first entry of the second step.

Exercise 4.19

The manufacture of quicklime (CaO) for use in the construction industry is accomplished by decomposing limestone ($CaCO_3$) by heating.

$$CaCO_3(s) \longrightarrow CaO(s) + CO_2(g)$$

How many liters of $CO_2(g)$ at 825 °C and 754 torr are produced in the decomposition of 45.8 kg $CaCO_3(s)$?

Exercise 4.20

How many kilograms of CaO(s) are formed in the decomposition of $CaCO_3(s)$ if the volume of $CO_2(g)$ obtained is 1.25×10^4 L at 825 °C and 733 torr?

4.10 Mixtures of Gases: Dalton's Law of Partial Pressures

Early experimenters did not make a strong distinction between air and other gases. Carbon dioxide was first called "fixed air"; oxygen, "dephlogisticated air"; and hydrogen, "flammable air." We now know that air is a mixture, but fortunately the ideal gas equation applies to mixtures of gases about as well as it does to individual gases. Let's begin by considering Dalton's views on gaseous mixtures.

Partial Pressures and Mole Fractions

Dalton is most renowned for his atomic theory, but he had wide-ranging interests, one of which was meteorology. He performed several experiments on atmospheric gases in an attempt to understand weather. In fact, he formulated his atomic theory to explain the results of those experiments. In one experiment, he found that if he added water vapor at a certain pressure to dry air, the pressure exerted by the air increased by an amount equal to the pressure of the water vapor. Based on this and other experiments, Dalton concluded that each gas in a mixture behaves independently of the other gases. **Dalton's law of partial pres-**

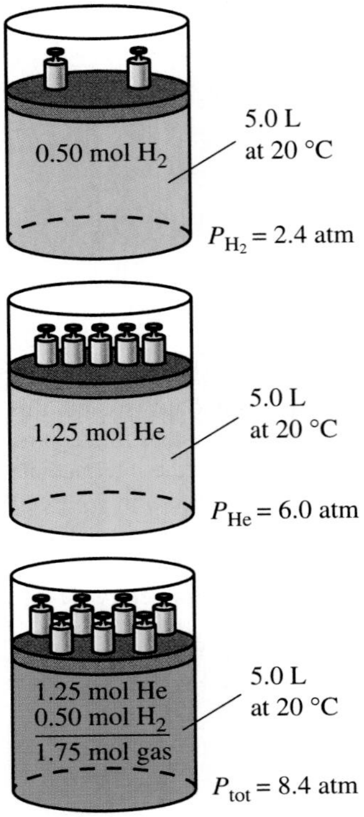

Figure 4.14 Dalton's law of partial pressures illustrated. Each gas expands to fill the container and exerts a pressure that is easily calculated with the ideal gas equation. The total pressure of the mixture is the sum of the partial pressures of the individual gases.

sures states that in a mixture of gases each gas expands to fill the container and exerts its own pressure; this pressure is called a **partial pressure**. *The total pressure of the mixture is equal to the sum of the partial pressures exerted by the separate gases* (Figure 4.14).

Mathematically, Dalton's law of partial pressures is simply

$$P_{total} = P_1 + P_2 + P_3 + \cdots$$

where the terms on the right side refer to the partial pressures of gases 1, 2, 3, and so on. The partial pressure of each gas follows the ideal gas law. For gas 1,

$$P_1 = \frac{n_1 RT}{V}$$

where n_1 is the number of moles of gas 1. Similarly, for gas 2,

$$P_2 = \frac{n_2 RT}{V}$$

where n_2 is the number of moles of gas 2, and so on.

Example 4.18 illustrates the idea that a component (N_2) of a mixture of gases (air) expands to fill its container and exerts a distinctive partial pressure.

Example 4.18

A 1.00-L sample of dry air at 25 °C contains 0.0319 mol N_2, 0.00856 mol O_2, 0.000381 mol Ar, and 0.00002 mol CO_2. Calculate the partial pressure of $N_2(g)$ in the mixture.

Solution

From the ideal gas equation,

$$P_{N_2} = \frac{n_{N_2} RT}{V}$$

$$= \frac{0.0319 \text{ mol} \times 0.0821 \text{ L} \cdot \text{atm} \cdot \text{mol}^{-1} \cdot \text{K}^{-1} \times 298 \text{ K}}{1.00 \text{ L}} = 0.780 \text{ atm}$$

Exercise 4.21

Calculate the partial pressure of each of the other components of the air sample in Example 4.18. What is the total pressure exerted by all the gases in the sample?

Exercise 4.22

What is the pressure exerted by a mixture of 4.05 g N_2, 3.15 g H_2, and 6.05 g He when confined to a 6.10-L container at 25 °C?

We can obtain another useful expression from the equations we wrote earlier. Let's designate one of the components of a gaseous mixture as "1", and then take the *ratio* of the partial pressure of component "1" to the total gas pressure.

$$\frac{P_1}{P_{total}} = \frac{P_1}{P_1 + P_2 + \cdots}$$

$$\frac{P_1}{P_{total}} = \frac{n_1 RT/V}{n_1 RT/V + n_2 RT/V + \cdots} = \frac{n_1}{n_1 + n_2 + \cdots} = \frac{n_1}{n_{total}}$$

We can write similar expressions for each of the other components in the gaseous mixture.

The ratio

$$\frac{n_1}{n_1 + n_2 + \cdots} = \frac{n_1}{n_{\text{total}}}$$

has a special name. It is called the *mole fraction* of component "1" in the mixture. The mole fraction represents the fraction of all the molecules in a mixture that are of a given type. The sum of the mole fractions of all the components in a gaseous mixture is one. That is,

$$\frac{n_1}{n_1 + n_2 + \cdots} + \frac{n_2}{n_1 + n_2 + \cdots} + \cdots = \frac{n_1 + n_2 + \cdots}{n_1 + n_2 + \cdots} = 1$$

As with other fractional parts of the whole, we can multiply mole fractions by 100% to obtain *mole percents*.

The significance of expressions such as

$$\frac{P_1}{P_{\text{total}}} = \frac{n_1}{n_{\text{total}}} = \text{mole fraction of component "1"}$$

is that they allow us to relate the partial pressures of the components of a gaseous mixture to the total gas pressure. That is,

$$P_1 = \text{mol fraction of component "1"} \times P_{\text{total}}$$

The compositions of gaseous mixtures are often given in percent by volume, which is actually a mole percent. In Example 4.19, we calculate the partial pressures of the components of air from its volume percent composition.

Example 4.19

The main components of dry air, by volume, are N_2, 78.08%; O_2, 20.95%; Ar, 0.93%; and CO_2, 0.04%. What are the partial pressures of each of the four gases in a sample of air at 1.000 atm?

Solution

Volume percent for an ideal gas mixture is the same as mole percent, and from the mole percents we can write the mole fractions. Each partial pressure is given by the expression

$$\text{Partial pressure} = \text{mole fraction} \times \text{total pressure}$$

Thus,

$$P_{N_2} = 0.7808 \times 1.000 \text{ atm} = 0.7808 \text{ atm}$$

$$P_{O_2} = 0.2095 \times 1.000 \text{ atm} = 0.2095 \text{ atm}$$

$$P_{Ar} = 0.0093 \times 1.000 \text{ atm} = 0.0093 \text{ atm}$$

$$P_{CO_2} = 0.0004 \times 1.000 \text{ atm} = 0.0004 \text{ atm}$$

Exercise 4.23

A sample of expired air (air that has been exhaled) is composed, by volume, of the following main components: N_2, 74.1%; O_2, 15.0%; H_2O, 6.0%; Ar, 0.9%; and CO_2, 4.0%. What are the partial pressures of each of the five gases in the expired air at 37 °C and 1.000 atm?

Exercise 4.24

A mixture of gases has methane at a partial pressure of 505 torr, ethane at 201 torr, propane at 43 torr, and butane at 11.2 torr. Calculate the mole fraction of each component. (*Hint*: What is the total pressure?)

CONCEPTUAL EXAMPLE 4.2

Describe what we would need to do to get from condition (a) to condition (b) in Figure 4.15.

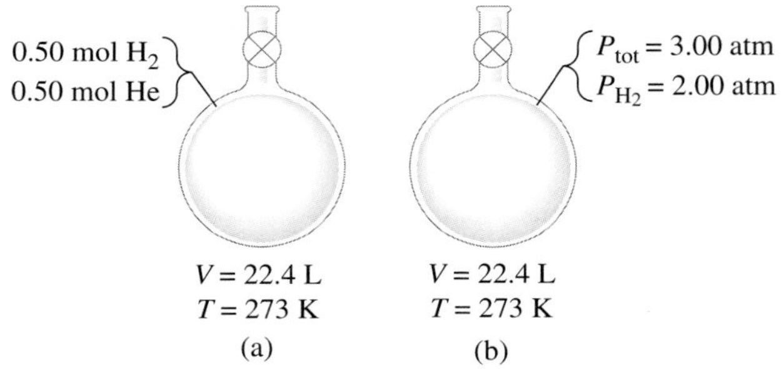

Figure 4.15 Conceptual Example 4.2 illustrated.

Solution

The key here is to note that in (a) 1.00 mol of gas occupies a volume of 22.4 L at 273 K (0 °C); P_{total} = 1.00 atm. Furthermore, because the mole fraction of each gas is 0.50, $P_{H_2} = P_{He}$ = 0.50 atm. The principal change between (a) and (b) is that the pressure *triples*, to 3.00 atm. In (b), P_{H_2} = 2.00 atm. To produce a fourfold increase in the partial pressure of H_2—that is, from 0.50 atm to 2.00 atm—we must increase the amount of H_2, from 0.50 mol to 2.00 mol. The remaining partial pressure(s) in (b) is 1.00 atm. This is a doubling of the partial pressure of helium in (a). To produce this increase in partial pressure we must add an additional 0.50 mol of gas; it can be He or it can be any other gas, *other than* hydrogen.

Conceptual Exercise 4.1

Why can't the increase in gas pressure between (a) and (b) in Figure 4.15 be achieved simply by adding hydrogen gas? That is, why must a second gas also be added?

Conceptual Exercise 4.2

Prepare a sketch to represent the conditions in Figure 4.15 at the molecular level.

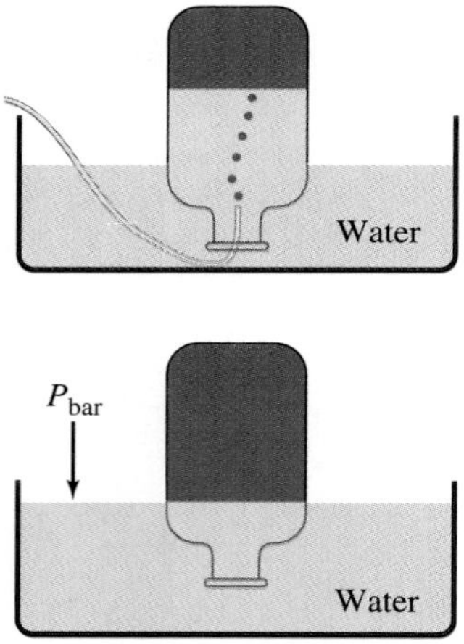

Figure 4.16 Collection of a gas over water. To make the total pressure of the gaseous mixture in the bottle equal to the barometric pressure, it is necessary to adjust the position of the bottle so that the water levels inside and outside the bottle are equal. When this is done, the partial pressure of the gas, $P_{gas} = P_{bar} - P_{H_2O}$. P_{bar} is the barometer reading and P_{H_2O}, the vapor pressure of the water, is obtained from tabulated data such as Table 4.3.

Collection of Gases Over Water

Gases such as oxygen, nitrogen, and hydrogen are only slightly soluble in water. They are often collected over water by the technique of displacement (Figure 4.16).

As a gas is being collected by displacement, water vapor produced by evaporation gathers in the collection vessel as well. The total pressure in the collection vessel is that of the gas *plus* that of water vapor.

The partial pressure of the water vapor, known as the *vapor pressure* of the water, depends only on the temperature of the water. The warmer the water is, the higher is the vapor pressure (Table 4.3). To find the partial pressure of the gas being collected, we need only subtract the vapor pressure of the water from the total pressure within the collection vessel, as illustrated in Example 4.20.

TABLE 4.3
Vapor Pressure of Water as a Function of Temperature

Temp., °C	Pressure, mmHg
15	12.8
16	13.6
17	14.5
18	15.5
19	16.5
20	17.5
21	18.7
22	19.8
23	21.1
24	22.4
25	23.8
30	31.8
40	55.3

Example 4.20

Oxygen is collected over water at 21 °C. The total pressure inside the collection jar is 741 torr. What is the pressure due to the oxygen alone?

Solution

From Table 4.3 we find that the vapor pressure of water at 21 °C is 18.7 torr. Because the total pressure is equal to 741 torr, we have

$$P_{\text{total}} = P_{O_2} + P_{H_2O}$$

$$741 \text{ torr} = P_{O_2} + 18.7 \text{ torr}$$

$$P_{O_2} = 741 \text{ torr} - 18.7 \text{ torr} = 722 \text{ torr}$$

Example 4.21

Hydrogen, produced by the following reaction, is collected over water at 23 °C and 742 torr barometric pressure.

$$2\,Al(s) + 6\,HCl(aq) \longrightarrow 2\,AlCl_3(aq) + 3\,H_2(g)$$

What is the volume of the "wet" gas that will be collected in the reaction of 1.50 g Al(s) with excess HCl(aq)?

Solution

The "wet" gas is the mixture of hydrogen and water vapor. However, because the two gases are found in the same collection vessel, each expands throughout the entire volume, and we need to calculate the volume of only one of them. We can find the pressure of the water vapor in Table 4.3—21.1 torr—but this and the temperature are not enough data with which to calculate the volume of the water vapor. For hydrogen, we know the temperature, 23 °C, and we can get the partial pressure:

$$P_{H_2} = 742 \text{ torr} - 21.1 \text{ torr} = 721 \text{ torr}$$

Moreover, we have a way of determining the number of moles of hydrogen from the stoichiometry of the reaction.

$$? \text{ mol } H_2 = 1.50 \text{ g Al} \times \frac{1 \text{ mol Al}}{26.98 \text{ g Al}} \times \frac{3 \text{ mol } H_2}{2 \text{ mol Al}} = 0.0834 \text{ mol } H_2$$

So, now we can use the ideal gas equation to calculate the volume of hydrogen, and, therefore, that of the "wet" gas.

$$V = \frac{nRT}{P} = \frac{0.0834 \text{ mol} \times 0.0821 \text{ L} \cdot \text{atm} \cdot \text{mol}^{-1} \cdot \text{K}^{-1} \times (273 + 23) \text{ K}}{(721/760) \text{ atm}}$$

$$= 2.14 \text{ L}$$

Exercise 4.25

What is the total volume of gas, in liters, that will be obtained if 1.28 g N_2 is collected over water at 21 °C and a barometric pressure of 696 torr?

Exercise 4.26

Hydrogen gas is collected over water at 18 °C. The total pressure inside the collection jar is set at the barometric pressure of 738 torr. If the volume of the gas is 246 mL, what mass of hydrogen is collected? What is the mass of the "wet" gas? (*Hint*: What is the mass of water vapor?)

Exercise 4.27

A sample of $KClO_3$ was decomposed to potassium chloride and oxygen by heating. The evolved O_2 was collected over water at 21 °C and a barometric pressure of 746 mmHg. A 155-mL volume of the gaseous mixture was obtained. What mass of $KClO_3$ was decomposed? (*Hint*: First write a balanced equation for the reaction.)

4.11 The Kinetic-Molecular Theory: Some Quantitative Aspects

Let us now return to an examination of the kinetic-molecular theory introduced in Section 4.2, this time in a more quantitative fashion. Recall that a theory is invented to explain data but that successful theories also allow for predictions. To begin, let's restate the principal assumptions of the kinetic-molecular theory.

1. A gas is comprised of molecules that are in constant, random straight-line motion.
2. Molecules of a gas are far apart—a gas is mostly empty space.
3. There are no forces between molecules except during the instant of collision. Each molecule acts independently of all the others; except for collisions, it is unaffected by their presence.
4. Individual molecules may gain or lose energy as a result of collisions; however, in a collection of molecules at constant temperature, the total energy remains constant.

With these assumptions it is possible to calculate the pressure exerted by a collection of molecules when confined to a given volume at a constant temperature. The derivation is too complex for us to consider in detail, but the final result is the equation

$$P = \frac{1}{3} \cdot \frac{N}{V} \cdot m\overline{u^2}$$

where V is the volume containing N molecules having mass m, and $\overline{u^2}$ is the *average* of the *squares* of their speeds. (The bar above a term indicates an average value.) P is the pressure exerted by the gas. The important conclusions of the kinetic-molecular theory stem from this equation.

The Kinetic-Molecular Theory and Temperature

The kinetic-molecular theory gives us an important insight into the meaning of temperature. Suppose we modify the basic equation by replacing the fraction $\frac{1}{3}$ by the product $\frac{2}{3} \times \frac{1}{2}$.

$$PV = \tfrac{1}{3}N \cdot m\overline{u^2} = \tfrac{2}{3}N(\tfrac{1}{2}m\overline{u^2})$$

We do this to be able to isolate the term $\frac{1}{2}m\overline{u^2}$, which represents the average translational kinetic energy, $\overline{KE}$, of the gas molecules (see Appendix B for a discussion of kinetic energy). Now, let's assume we have one mole of gas, which means N will be Avogadro's number, N_A. We then use the ideal gas equation for one mole of gas: $PV = RT$. This allows us to substitute RT for PV and to write

$$RT = \frac{2}{3}N_A\overline{KE}$$

Finally, we can solve for $\overline{KE}$.

$$\overline{KE} = \frac{3}{2} \cdot \frac{R}{N_A} \cdot T$$

Because $\frac{3}{2}$, R, and N_A are all fixed quantities, this equation is equivalent to the expression

$$\overline{KE} = \text{constant} \times T$$

That is,

The average translational kinetic energy of the molecules of a gas, $\overline{KE}$, is directly proportional to the Kelvin temperature.

Thus, *any* gas at a given temperature has the same average translational kinetic energy as any *other* gas at the same temperature.

When we heat a gas, we increase the average translational kinetic energy of the gas molecules and the temperature increases. If we were to cool a gas to 0 K, the average translational kinetic energy of the molecules would drop to zero. This, then, is the kinetic-molecular interpretation of absolute zero temperature: At 0 K, translational molecular motion ceases; molecules stop moving around. (Translational molecular motion is movement of an entire molecule in three-dimensional space.)

Absolute zero temperature is unattainable, but recent attempts have produced temperatures of only a few nanokelvins (nK).

Molecular Speeds: Faster than a Speeding Bullet

Let's explore the notion of average molecular speeds that is required in the equations of the kinetic-molecular theory. The average speed $\bar{u}$ is defined as

$$\bar{u} = \frac{\text{sum of the speeds of all the molecules}}{\text{total number of molecules}}$$

$$= \frac{u_1 + u_2 + \cdots + u_n}{N}$$

For most applications, however, the root-mean-square speed is a more significant expression of molecular speed than is the average speed. The **root-mean-square speed, u_{rms}** is the *square root* of the *average* of the *squares* of the molecular speeds. That is,

$$u_{rms} = \sqrt{\overline{u^2}} = \sqrt{\frac{u_1^2 + u_2^2 + \cdots + u_n^2}{N}}$$

We can't measure speeds of individual molecules, so we can't calculate average speeds with this equation. One of the triumphs of the kinetic-molecular theory, however, is that it gives us mathematical equations for calculating average speeds. The equation for root-mean-square speed is

$$u_{rms} = \sqrt{\overline{u^2}} = \sqrt{\frac{3RT}{\text{molar mass}}}$$

Typical u_{rms} speeds of molecules are high. For hydrogen gas at 25 °C, for example, $u_{rms} = 1.92 \times 10^3$ m/s. This is about 4300 mi/h, or twice as fast as the speed of a bullet fired from an M-16 rifle.

To determine u_{rms} in meters per second with this equation requires that we express the gas constant, R, as 8.3145 joules (J)* per mole per kelvin ($J \cdot mol^{-1} \cdot K^{-1}$) and the molar mass in kilograms per mole. But, for now, let's just see how we can apply this equation in a qualitative way.

First, we see that the root-mean-square speed is *inversely* proportional to the *square root* of the molar mass and *directly* proportional to the *square root* of the Kelvin temperature. Thus, the *lower* the molar mass and the *higher* the temperature, the faster a molecule moves. Figure 4.17 illustrates the effect of molar mass on molecular speeds, and Figure 4.18 illustrates the effect of temperature on molecular speeds.

Our interest in molecular speeds is primarily in being able to make comparisons, as illustrated in Estimation Example 4.2.

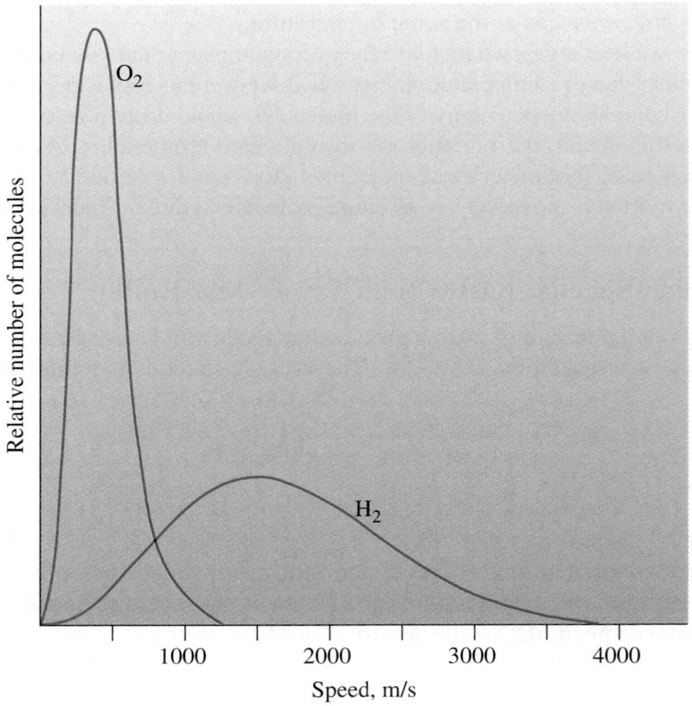

Figure 4.17 Distribution of molecular speeds: effect of molar mass. The relative numbers of molecules having given speeds are plotted against the speed. The peaks of the curves represent the most probable speeds for molecules of the gas (average speeds are proportional to most probable speeds). The gas with the heavier molecules (higher molar mass) has the lower molecular speeds. At 0 °C the most probable speeds for O_2 and H_2 molecules are 377 and 1500 m/s, respectively.

*The *joule* is the basic unit of work and energy in SI. It is defined in Section 5.1.

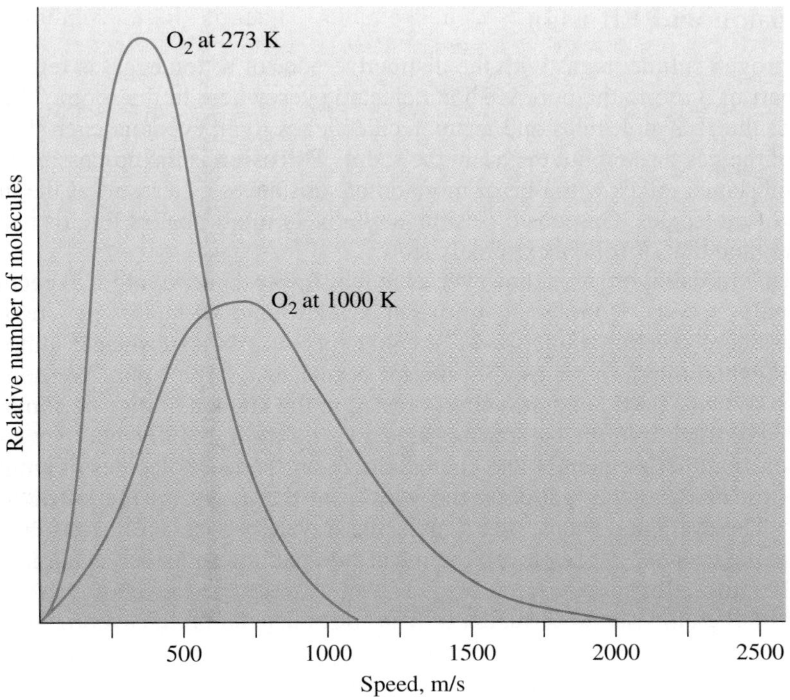

Figure 4.18 Distribution of molecular speeds: effect of temperature. The distribution curve is broadened and the molecular speeds shift to higher values as the temperature is increased.

ESTIMATION EXAMPLE 4.2

If u_{rms} of H_2 molecules at 0 °C is 1838 m/s, which of the following is a likely value for u_{rms} of O_2 molecules at 0 °C?

(a) 115 m/s (b) 460 m/s (c) 1838 m/s (d) 7352 m/s (e) 29,400 m/s

Solution

At first glance, response (c) looks good—that the H_2 and O_2 molecules travel at the same speed. But this is wrong. Molecules at the same temperature do *not* have equal speeds. They have equal average translational kinetic energies: $\frac{1}{2}m\overline{u^2}$. O_2 molecules are *heavier* than H_2 molecules and should move *more slowly*. This fact eliminates the responses (d) and (e), as well as (c).

The O_2-to-H_2 molar mass ratio is $32/2 = 16$. Response (a), 115 m/s, is one-sixteenth of 1838 m/s; but this is also *incorrect* for u_{rms} for O_2. The molecular speeds are inversely related as the *square root* of the molar mass ratio: $\sqrt{16} = 4$. The u_{rms} speed of O_2 molecules is *one-quarter* of that for hydrogen; it is 460 m/s. The correct response is (b).

Estimation Exercise 4.4

At which temperature(s) listed below will u_{rms} of O_2 be *greater* than u_{rms} of H_2 at 0 °C; that is, greater than 1838 m/s?

(a) 1000 K (b) 2000 K (c) 3000 K (d) 4000 K (e) 5000 K

Diffusion and Effusion

If hydrogen sulfide, a gas with the distinctive odor of rotten eggs, is released in one part of a room, the odor is soon detected everywhere in the room. The motion of the H_2S molecules and air molecules leads to an eventual even distribution of the gas throughout the air in the room. **Diffusion** is the process by which one substance mixes with one or more other substances as a result of the movement of molecules. Gaseous diffusion is relatively rapid; that of liquids is much slower; and that of solids, extremely slow.

The diffusion of gases, however, is much slower than we might expect from molecular speeds. If molecules move at several hundred meters per second—equivalent to speeding bullets—we should expect to smell substances almost instantly when they are released in the far corner of a large room. We don't, of course. When a bottle of ammonia is opened in the kitchen, it may be some time before we smell it in the bathroom. Molecules do move rapidly, but they collide with one another frequently and change direction. Those molecules of ammonia travel tortuous, zigzag paths on the way from the mouth of the bottle to our noses. The average distance that a molecule travels between collisions is called the *mean free path.* At sea level, the mean free path of an air molecule is about 6×10^{-8} m.

Although the rate of diffusion of a gas does depend on the average molecular speed of the gas molecules, frequent collisions make precise calculations quite complicated. Let's look at a somewhat simpler process for some insight into the process.

Effusion is the process suggested by Figure 4.19, in which a gas escapes from its container through a tiny hole (called an orifice). Because lighter molecules have greater speeds, we should expect them to escape more quickly through an orifice than heavier molecules. Although the situation is actually more complicated, we can roughly relate the rates at which gases effuse to their root-mean-square speeds. This expression for two different gases whose molar masses are M_1 and M_2,

$$\frac{\text{rate}_1}{\text{rate}_2} = \frac{\sqrt{3RT/M_1}}{\sqrt{3RT/M_2}} = \sqrt{\frac{M_2}{M_1}}$$

is a kinetic-molecular description of a nineteenth century law proposed by Thomas Graham (1805–1869) and known as *Graham's law of effusion.*

At a given temperature, the rates of effusion of gas molecules are inversely proportional to the square roots of their molar masses.

In Example 4.22 we compare the rates or speeds of effusion of two different gases. In Example 4.23, instead of working with rates of effusion we work with effusion *times*. Effusion times and rates are inversely related. That is,

$$\text{Effusion time} \propto \frac{1}{\text{effusion rate}}$$

This relationship should seem reasonable. It's like saying that a car traveling at 60 mi/h makes a 100-mi journey in half the time of a car traveling at 30 mi/hr.

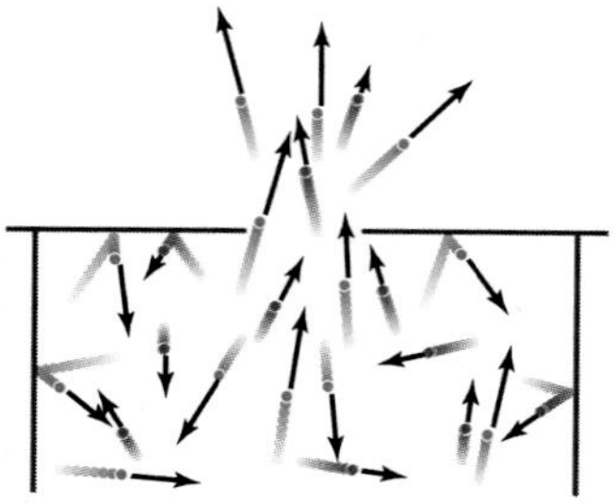

Figure 4.19 Effusion of gases through an orifice. Average speeds of the two different types of molecules are suggested by the lengths of the arrows. The faster molecules (shown in red) effuse more rapidly.

Separation of Uranium Isotopes

Naturally occurring uranium is a mixture of isotopes, with only about 0.7% being the fissionable uranium-235 isotope. To make fuel for a nuclear power plant or material for a nuclear bomb, the ^{235}U must be separated from the more abundant ^{238}U. During World War II, the United States undertook a massive program, called the Manhattan Project, to develop a nuclear bomb. Project scientists developed a technique for separating the uranium isotopes by gaseous diffusion, a method still used today to produce ^{235}U-enriched fuels for nuclear power plants.

One of the few uranium compounds that can be converted to a gas at moderate temperatures is uranium hexafluoride, UF_6. When high-pressure $UF_6(g)$ is forced through a porous barrier, molecules containing the lighter ^{235}U isotope move through a tiny bit faster than those containing ^{238}U. The portion of $UF_6(g)$ that has passed through the barrier has a slightly higher ratio of ^{235}U to ^{238}U than the original gas; it is said to be "enriched" in ^{235}U. The more times the process is repeated, the greater the concentration of ^{235}U. Uranium must be enriched to 3% or 4% ^{235}U for use in typical nuclear power plants, and to about 90% to make a nuclear bomb.

Example 4.22

Calculate the relative rates of effusion of hydrogen and helium under the same conditions of temperature and pressure.

Solution

$$\frac{r_{H_2}}{r_{He}} = \sqrt{\frac{M_{He}}{M_{H_2}}} = \sqrt{\frac{4.00}{2.02}} = 1.41$$

Hydrogen, the lighter gas, effuses about 1.41 times as fast as helium.

Exercise 4.28

Calculate the relative rates of effusion of N_2 and Ar under the same conditions of temperature and pressure.

Example 4.23

A sample of Ar(g) escapes through a tiny hole in 77.3 s. An unknown gas escapes under the same conditions in 97.6 s. Calculate the molar mass of the unknown gas.

Solution

First, let's get the appropriate form for an expression relating effusion times and molar masses. We can do this by noting that

$$\frac{(\text{effusion time})_{\text{unk}}}{(\text{effusion time})_{\text{Ar}}} = \frac{1/(\text{effusion rate})_{\text{unk}}}{1/(\text{effusion rate})_{\text{Ar}}} = \frac{(\text{effusion rate})_{\text{Ar}}}{(\text{effusion rate})_{\text{unk}}}$$

$$\frac{97.6\text{ s}}{77.3\text{ s}} = \sqrt{\frac{M_{\text{unk}}}{M_{\text{Ar}}}} = \sqrt{\frac{M_{\text{unk}}}{39.95}} = 1.26$$

Simple reasoning helps us see that this is a sensible answer. Because the unknown gas effuses more slowly than does Ar, its molar mass must be greater than that of Ar.

If we square both sides of the final equation and solve for M_{unk}, we get

$$M_{unk} = 39.95 \times (1.26)^2$$

The molar mass of the unknown gas is

$$M_{unk} = 63.4 \text{ g/mol}$$

Exercise 4.29

A sample of $N_2(g)$ effuses from a tiny opening in 57 s. An unknown gas escapes under the same conditions in 83 s. Calculate the molar mass of the unknown gas.

Exercise 4.30

A sample of $O_2(g)$ effuses from an orifice in 123 s. How long should it take a similar sample of methane, $CH_4(g)$, to effuse under the same conditions?

4.12 Real Gases

In both the kinetic-molecular theory and the ideal gas law, we assume that there are no attractive forces between the molecules of a gas and that the molecules themselves have no volume. No real gas strictly meets these conditions.

We can account for deviations from ideal gas behavior by reexamining the assumptions of the kinetic-molecular theory. It is all right to assume minimal forces of attraction between molecules when molecules are far apart. However, at high pressures molecules are more closely spaced, and at low temperatures they pass by one another more slowly. Therefore we must reckon with the forces of attraction. We must also rethink the idea that molecules always have negligible volume because when they are closely spaced, molecules occupy a much greater percentage of the total gas volume than they would otherwise. Equations describing real gases must account for these factors. One equation that is reasonably successful in doing so is the *van der Waals equation*.

$$\left(P + \frac{n^2a}{V^2}\right)(V - nb) = nRT$$

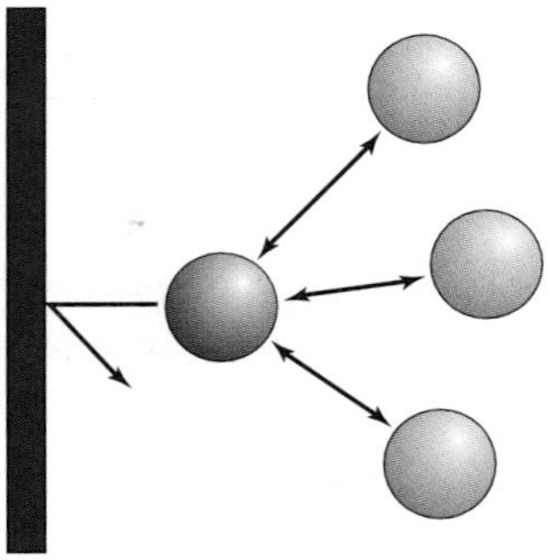

Figure 4.20 Intermolecular forces of attraction.
Attractive forces of the orange molecules for the purple molecule cause the purple molecule to exert less force when it collides with the wall than it would if these attractions did not exist.

Here, we use the equation to help us to see how the ideal gas equation, $PV = nRT$, would have to be "fixed up" to work better for a real gas. (Problem 104b involves a calculation using the van der Waals equation.) For example, because of intermolecular forces, the measured pressure of a real gas is less than what we would expect (Figure 4.20). The term n^2a/V^2 is related to intermolecular forces of attraction and is *added* to the measured pressure to bring it closer to an "ideal" value. The term nb is related to the volume occupied by the molecules themselves. It is *subtracted* from the measured volume so as to more closely represent the volume of space in a gas, the *free* volume.

For a real gas to approach ideal gas behavior, the term n^2a/V^2 should be small compared to P, and nb should be small compared to V. These conditions are most likely to be met when a small amount of gas (n) is found in a large volume (V) of gas, or, generally speaking, when a gas is at a *high temperature* and a *low pressure*. At room temperature or above, and at pressures less than a few atmospheres, most gases obey the ideal gas equation reasonably well.

SUMMARY

Gases consist of widely separated molecules, moving constantly and randomly throughout their containers. Molecular collisions with the container walls create a gas pressure. Atmospheric pressure can be measured with a barometer, and other gas pressures, with a manometer. The SI unit of pressure is the pascal (Pa), but more commonly used are the standard atmosphere (atm), the millimeter of mercury (mmHg), and the torr: 1 atm = 760 mmHg = 760 torr.

For a given amount of gas at a constant temperature, the volume and pressure are *inversely* proportional: $V = a/P$. For a given amount of gas at a constant pressure, the volume and temperature are *directly* proportional: $V = bT$. The temperature in this relationship is absolute temperature, measured on the Kelvin scale, with 0 K = −273.15 °C. At a fixed temperature and pressure, the volume and amount of a gas are *directly* proportional: $V = cn$. At 0 °C and 1 atm pressure (STP), the molar volume of a gas is about 22.4 L. The three simple gas laws can be consolidated into the combined gas law.

Gases that obey the simple gas laws also conform to the ideal gas law, $PV = nRT$. The ideal gas equation can be used in place of the simple or combined gas laws, especially in molecular weight and gas density determinations. Of particular importance is its use in stoichiometric calculations for reactions involving gases.

Each gas in a mixture expands to fill the container and exerts its own *partial* pressure. Dalton's law of partial pressures states that the total pressure in a gaseous mixture is the sum of the partial pressures. A gas collected over water is a mixture. It contains water vapor at a partial pressure equal to the vapor pressure of water. Stoichiometric calculations involving gases collected over water use stoichiometric factors from the balanced equation, Dalton's law of partial pressures, and the ideal gas equation.

The postulates of the kinetic-molecular theory lead to an equation for gas pressure and a relationship between Kelvin temperature and the average translational kinetic energy of gas molecules. The theory also provides a basis for describing the diffusion and effusion of gases.

Real gases are most likely to exhibit ideal behavior at high temperatures and low pressures. When the ideal gas equation fails, it often can be replaced by another equation of state, such as that of van der Waals.

KEY TERMS

absolute zero (4.5)
atmosphere (atm) (4.3)
Avogadro's law (4.6)
barometer (4.3)
Boyle's law (4.4)
Charles's law (4.5)
Dalton's law of partial pressures (4.10)
diffusion (4.11)
effusion (4.11)
ideal gas (4.8)
ideal gas equation (4.8)
ideal gas law (4.8)
kelvin (K) (4.5)
Kelvin scale (4.5)
kilopascal (kPa) (4.3)
kinetic-molecular theory of gases (4.2, 4.11)
law of combining volumes (4.9)
manometer (4.3)
millimeter of mercury (mmHg) (4.3)
molar volume of a gas (4.6)
partial pressure (4.10)
pascal (Pa) (4.3)
pressure (4.3)
root-mean-square speed (u_{rms}) (4.11)
standard conditions of temperature and pressure (STP) (4.6)
torr (4.3)
universal gas constant (*R*) (4.8)

REVIEW QUESTIONS

1. Describe what a gas is like at the molecular level.
2. Why don't the molecules of a gas settle to the bottom of their container?
3. Give a kinetic-molecular explanation of the origin of gas pressure.
4. What does a mercury barometer measure? How does it work?

5. Why is mercury (rather than water or another liquid) used as the fluid in barometers?

6. How does a manometer differ from a barometer? How is a manometer used to measure gas pressures?

7. State Boyle's law in words and as a mathematical equation.

8. Use the kinetic-molecular theory to explain Boyle's law.

9. What is the advantage of storing gases under high pressure (for example, oxygen used in respiration therapy)?

10. State Charles's law in words and in the form of a mathematical equation.

11. Use the kinetic-molecular theory to explain Charles's law.

12. Why must an absolute temperature scale rather than the Celsius scale be used for Charles's law calculations?

13. How is the Kelvin scale related to the Celsius temperature scale?

14. What effect will the following changes have on the volume of a fixed amount of gas?
  a. an increase in pressure at constant temperature
  b. a decrease in temperature at constant pressure
  c. a decrease in pressure coupled with an increase in temperature

15. What effect will the following changes have on the pressure of a fixed amount of a gas?
  a. an increase in temperature at constant volume
  b. a decrease in volume at constant temperature
  c. an increase in temperature coupled with a decrease in volume

16. According to the kinetic-molecular theory, (**a**) what change in temperature is occurring if the molecules of a gas begin to move more slowly, on average? (**b**) what change in pressure occurs when the walls of the container are struck less often by molecules of the gas?

17. Container A has twice the volume but holds twice as many gas molecules as container B at the same temperature. Use the kinetic-molecular theory to compare the pressures in the two containers.

18. For each of the following, indicate in which container the gas would be expected to have the higher density.
  a. Containers A and B have the same volume and are at the same temperature, but the gas in A is at a higher pressure.
  b. Containers A and B are at the same pressure and temperature, but the volume of A is greater than that of B.
  c. Containers A and B are at the same pressure and volume, but the gas in A is at a higher temperature.

19. What are the standard conditions of temperature and pressure for gases? Why is it useful to define such conditions?

20. What is meant by the *molar volume* of a gas? What is the value of the molar volume of a gas at STP?

21. State the ideal gas law in words and in the form of a mathematical equation.

22. Use the kinetic-molecular theory to explain the dependence of gas density on molar mass, temperature, and pressure referred to on page 131.

23. State Dalton's law of partial pressures in words and in the form of a mathematical equation.

24. Describe how Dalton's law of partial pressures is applied to gases collected over water.

25. State the definition of mole fraction in the form of a mathematical equation, and then restate it in terms of partial pressures of gases.

26. Interpret what we mean by temperature in terms of the kinetic-molecular theory.

27. What is meant by u_{rms}, the root-mean-square speed of molecules of a gas? What properties of a gas determine the value of u_{rms}?

28. State Graham's law of effusion, and indicate whether $O_2(g)$ or $N_2(g)$ has the greater effusion rate when the gases are compared under identical conditions.

29. Describe the primary factors that cause a gas not to conform to the ideal gas equation.

30. Under what conditions do the properties of real gases differ substantially from those of ideal gases?

PROBLEMS

Assume that the simple and ideal gas laws apply in all cases, unless the problem states otherwise.

Pressure

31. Calculate the height of a mercury column that exerts a pressure of 4.36 atm.

32. Calculate the height of a mercury column that exerts a pressure of 213 torr.

33. Carry out the following conversions between pressure units.
 a. 0.985 atm to mmHg **b.** 849 torr to atm
 c. 721 torr to atm

34. Carry out the following conversions between pressure units.
 a. 4.00 atm to mmHg **b.** 642 torr to kPa
 c. 105.7 kPa to torr

35. A manometer is filled with an oil that has a density of 0.798 g/mL. Calculate the gas pressure measured by the manometer (**a**) if barometric pressure is 755 mmHg and the oil column on the side of the gas flask, h_{closed}, is 44 mm lower than on the open side, h_{open}; (**b**) if barometric pressure is 735 torr and the oil column on the side of the gas flask, h_{closed}, is 22.3 cm higher than on the open side, h_{open}.

36. **a.** Calculate the heights of columns of carbon tetrabromide (d = 3.42 g/mL) that exert the same pressures as a column of mercury (d = 13.6 g/mL) that is 760 mm high and a column of water (d = 1.00 g/mL) that is 25.5 cm high.
 b. City fire codes specify the water pressure that must be maintained at the end of a water line—for example, 35.5 lb/in.^{2}. Calculate the height of a water column that exerts a pressure of 35.5 lb/in.^{2}. (*Hint*: 1 atm = 14.7 lb/in.^{2}.)

Boyle's Law

37. A sample of helium occupies 521 mL at 1572 torr. Assume that the temperature is held constant, and determine (**a**) the volume of the helium at 752 torr; (**b**) the volume of the helium at 3.55 atm; (**c**) the pressure, in torr, if the volume is changed to 315 mL; (**d**) the pressure, in atm, if the volume is changed to 2.75 L.

38. A decompression chamber used by deep-sea divers has a volume of 10.3 m^3 and operates at an internal pressure of 4.50 atm. How many cubic meters would the air in the chamber occupy if it were at normal atmospheric pressure, assuming no temperature change?

39. A novel energy storage system involves storing air under high pressure. (Energy is released when the air is allowed to expand.) How many cubic feet of air, measured at standard atmospheric pressure of 14.7 pounds per square inch (psi), can be compressed into a 19-million-ft^3 underground cavern at a pressure of 1070 psi?

40. Oxygen used in respiratory therapy is stored at room temperature under a pressure of 1.50×10^2 atm in gas cylinders with a volume of 60.0 L.
 a. What volume would the gas occupy at a pressure of 750.0 torr? Assume no temperature change.
 b. If the oxygen flow to the patient is adjusted to 8.00 L per minute, at room temperature and 750.0 torr, how long will the tank of gas last?

41. A gas is confined in an upright cylinder closed off by a piston that has a diameter of 15.0 cm. The volume of the gas is 1.20 L at standard atmospheric pressure. What total pressure must be exerted on the piston to cause it to move down 5.25 cm?

42. The pressure within a 2.25-L balloon is 1.10 atm. If the volume of the balloon increases to 7.05 L, what will be the final pressure within the balloon, if the temperature does not change?

Charles's Law

43. A gas at a temperature of 100 °C occupies a volume of 154 mL. What will the volume be at a temperature of 10 °C, assuming no change in pressure?

44. A balloon is filled with helium. Its volume is 5.90 L at 26 °C. What will be its volume at −78 °C, assuming no pressure change.

45. A 567-mL sample of a gas at 305 °C and 1.20 atm is cooled at constant pressure until its volume becomes 425 mL. What is the new gas temperature?

46. A sample of gas at STP is to be heated at constant pressure until its volume triples. What is the new gas temperature?

Avogadro's Law and Molar Volume at STP

47. Which of the following gas samples at STP contains the greatest number of molecules?
a. 5.0 g H_2 **b.** 50 L SF_6
c. 1.0×10^{24} molecules of CO_2

48. a. How many molecules are present in 475 mL of $CO_2(g)$ at STP?
b. What is the volume occupied by 3.50×10^{24} $SO_2(g)$ molecules at STP?

49. What is the mass of 498 L of neon gas at STP?

50. What is the volume occupied by 0.837 g of xenon gas at STP?

The Combined Gas Law

51. A sealed can with an internal pressure of 721 torr at 25 °C is thrown into an incinerator operating at 755 °C. What will be the pressure inside the heated can, assuming the container remains intact during incineration?

52. A fixed amount of He exerts a pressure of 775 mmHg in a 1.05-L container at 26 °C. To what value must the temperature be changed to reduce the gas pressure to 725 mmHg? Assume the volume of gas remains constant.

53. Suppose we wish to contain a 1.00-mol sample of gas at 1.00 atm pressure and 25 °C. What volume container would we need?

54. At 25 °C, the pressure in a gas cylinder containing 8.00 mol O_2 is 5.05 atm. To maintain both pressure and volume constant, how many moles of $O_2(g)$ must be released when the temperature is raised to 235 °C?

55. If a fixed amount of gas occupies 2.53 m³ at a temperature of −15 °C and 191 torr, what volume will it occupy at 25 °C and 1142 torr?

56. In terms of pressure (P), volume (V), Kelvin temperature (T), and amount of gas (n), and in the manner of Figures 4.6 and 4.8, sketch a graph of each of the following.
a. V as a function of P, with T and n held constant
b. n as a function of P, with T and V held constant
c. T as a function of P, with V and n held constant
d. n as a function of T, with P and V held constant

The Ideal Gas Law

57. Calculate the volume, **(a)** in liters, of 1.12 mol $H_2S(g)$ at 62 °C and 1.38 atm, and **(b)** in mL, of 6.00×10^{-3} mol of a gas at 31 °C and 661 mmHg.

58. Calculate the volume, **(a)** in mL, of 3.45 mg $O_2(g)$ at 24 °C and 775 mmHg, and **(b)** in m³, of 6.92 kg $SF_6(g)$ at 100.0 °C and 743 torr.

59. Calculate the pressure, **(a)** in atm, of 4.64 mol CO(g) in a 3.96-L tank at 29 °C, and **(b)** in mmHg, of 0.0108 mol $CH_4(g)$ in a 0.265-L flask at 37 °C.

60. Calculate the pressure, **(a)** in mmHg, of 1.42 g CO_2 in a 735-mL flask at −12 °C, and **(b)** in kPa, of 35.5 g $N_2(g)$ in a 5.35-L flask at 0 °C.

61. Calculate the mass, **(a)** in grams, of Kr(g) in 2.22 L at 698 torr and 45 °C, and **(b)** in milligrams, of CO(g) in 7.45 mL at 784 torr and 36 °C.

62. a. If the gas present in 4.65 L at STP is changed to a temperature of 15 °C and a pressure of 756 torr, what will be the new volume?
b. What volume will 498 mL of a fixed amount of gas, measured at 27 °C and 722 torr, occupy at STP?

63. The interior volume of the Hubert H. Humphrey Metrodome in Minneapolis is 1.70×10^{10} L. The Teflon-coated fiberglass roof is supported by air pressure provided by 20 huge electric fans. How many moles of air are present in the dome if the pressure is 1.02 atm at 18 °C?

64. A hyperbaric chamber is an enclosure containing oxygen at higher-than-normal pressure used in the treatment of certain heart and circulatory conditions. What volume of $O_2(g)$, from a cylinder at 25 °C and 151 atm, is required to fill a hyperbaric chamber with a volume of 4.20×10^3 L to a pressure of 3.00 atm at 17 °C?

Molecular Weight Determinations

65. Calculate the molecular weight of a gas if 0.549 g occupies 211 mL at 747 mmHg and 24 °C.

66. Calculate the molecular weight of a gas if 0.233 g occupies 334 mL at 721 torr and 28 °C.

67. Calculate the molecular weight of a liquid that, when vaporized at 98 °C and 756 torr, gave 125 mL of vapor with a mass of 0.625 g.

68. Calculate the molecular weight of a liquid that, when vaporized at 99 °C and 716 torr, gave 225 mL of vapor with a mass of 0.773 g.

Gas Densities

69. Calculate the density, in g/L, of each of the following.
   a. $CO(g)$ at STP
   b. $Ar(g)$ at 1.26 atm and 325 °C

70. Calculate the density, in g/L, of each of the following.
   a. $AsH_3(g)$ at STP
   b. $N_2(g)$ at 715 torr and 98 °C

71. At what temperature will $O_2(g)$ at 0.982-atm pressure have a density of 1.05 g/L?

72. At what pressure will $N_2(g)$ have a density of 0.985 g/L at 25 °C?

73. What must be the molar mass of a gas found to have a density of 2.57 g/L at 25 °C and 745 torr?

74. Which of the following gases is (are) *more* dense than $O_2(g)$ at STP?
   a. N_2 at STP
   b. CO at 0 °C and 1.5 atm
   c. SO_2 at 300 °C and 1 atm
   d. H_2 at 25 °C and 10 atm

Stoichiometry of Gaseous Reactions

75. How many liters of $SO_3(g)$ can be produced by the reaction of 1.15 L $SO_2(g)$ and 0.65 L $O_2(g)$, if all three gases are measured at the same temperature and pressure?

$$2\ SO_2(g) + O_2(g) \longrightarrow 2\ SO_3(g)$$

76. What minimum volume of $O_2(g)$ is required to convert 3.06 L $CO(g)$ to $CO_2(g)$, if all three gases are measured at the same temperature and pressure?

$$CO(g) + \tfrac{1}{2}\ O_2(g) \longrightarrow CO_2(g)$$

77. How many liters of $CO_2(g)$, measured at 22.5 °C and 743.5 mmHg, are produced in the decomposition of one metric ton (1.00×10^3 kg) of limestone?

$$CaCO_3(s) \longrightarrow CaO(s) + CO_2(g)$$

78. What volume of $O_2(g)$, measured at 22 °C and 763 torr, is consumed in the combustion of 7.50 L of $C_2H_6(g)$, measured at STP?

$$2\ C_2H_6(g) + 7\ O_2(g) \longrightarrow 4\ CO_2(g) + 6\ H_2O(l)$$

79. How many milligrams of magnesium metal must react with excess HCl(aq) to produce 28.50 mL of $H_2(g)$, measured at 26 °C and 758 torr?

$$Mg(s) + 2\ HCl(aq) \longrightarrow MgCl_2(aq) + H_2(g)$$

80. A 100.0-g sample of aqueous hydrogen peroxide solution decomposes over time, producing 2.17 L $O_2(g)$ at 25 °C and 755 torr.

$$2\ H_2O_2(aq) \longrightarrow 2\ H_2O(l) + O_2(g)$$

What must have been the mass percent H_2O_2 in the solution?

Mixtures of Gases

81. Calculate the mole fraction of each gas in the following mixtures.
   a. 0.354 g Ar, 0.0521 g Ne, and 0.00419 g Kr
   b. 1.98 g N_2, 0.390 g O_2, and 0.0201 g Ar

82. A gas sample has 76.8 mole percent N_2, 20.1 mole percent O_2, and 3.1 mol percent CO_2. If the total pressure is 762 mmHg, what are the partial pressures of the three gases?

83. A sample of intestinal gas was collected and found to consist of 44% CO_2, 38% H_2, 17% N_2, 1.3% O_2, and 0.003% CH_4, by volume. (The percentages do not total 100% because of rounding.) What is the partial pressure of each gas, if the total pressure in the intestine is 818 torr? (*Hint*: Recall that volume percent is the same as mole percent for ideal gas mixtures.)

84. Mixtures of helium and oxygen are used in scuba diving. What are the partial pressures of the two gases in a mixture of 1.96 g He and 60.8 g O_2 confined in a 5.00-L tank at 25 °C?

85. Oxygen is collected over water at 30 °C and a barometric pressure of 742 torr. What is the partial pressure of $O_2(g)$ in the container?

86. An oxygen–helium gas sample, collected over water at 23 °C, exerts a total pressure of 758 torr. Calculate the mole fraction of water vapor in the sample.

87. *Elodea* is a green plant that carries out photosynthesis under water.

$$6\ CO_2(g) + 6\ H_2O(l) \longrightarrow C_6H_{12}O_6(aq) + 6\ O_2(g)$$

In an experiment some *Elodea* produce 122 mL of $O_2(g)$, collected over water at 743 torr and 21 °C. What mass of oxygen is produced? What mass of glucose ($C_6H_{12}O_6$) is produced concurrently?

88. A 9.90-g sample of potassium chlorate is decomposed by heating, yielding potassium chloride and oxygen. The $O_2(g)$ is collected over water at 18 °C. What are the partial pressures of the two gases (oxygen and water vapor) when contained in a 1.00-L container at 22 °C? What is the total pressure? (*Hint*: Write an equation for the chemical reaction.)

Kinetic-Molecular Theory

89. A gaseous mixture with equal numbers of molecules of H_2 and He is allowed to effuse through an orifice for a certain period of time. Which of the conditions pictured (a, b, c, or d) is most likely to result?

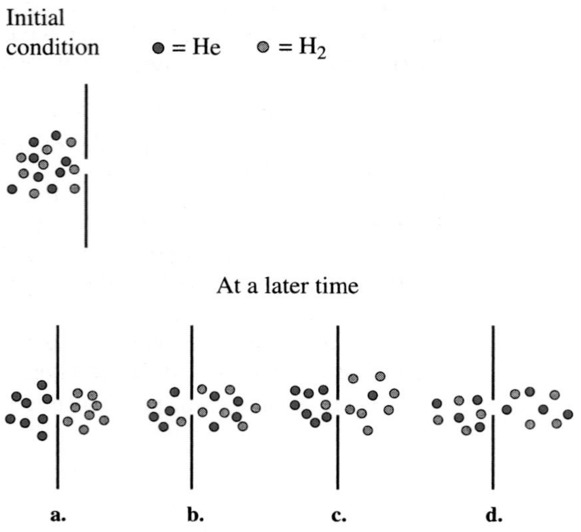

90. Calculate the relative rates of effusion of the following pairs of gases under the same conditions of temperature and pressure.
   a. He and CF_4 **b.** CH_4 and C_4H_{10}

91. It takes 44 s for a sample of $N_2(g)$ to effuse through a tiny orifice. Determine the molecular weights of gases whose effusion times, under exactly the same conditions, are (**a**) 75 s and (**b**) 42 s.

92. At a certain temperature, the root-mean-square speed of CH_4 molecules is 1610 km/h. What is the root-mean-square speed of CO_2 molecules at the same temperature?

ADDITIONAL PROBLEMS

93. What mass of Ne(g) should be added to 1.83 L $O_2(g)$ at 18 °C and 735 mmHg to raise the total gas pressure to 3.00 atm? (Assume that the volume and temperature are held constant.)

94. Stephen Malaker of Cryodynamics, Inc., has developed a refrigerator that uses compressed helium as a refrigerant gas. A typical system uses 5.00 in.^{3} of He compressed to 195 lb/in.^{2} at 20 °C. How many grams of helium are needed for one refrigerator?

95. In an attempt to verify Avogadro's hypothesis, small quantities of several different gases were weighed in 100.0-mL syringes. Masses were determined within an experimental error of 1% on an analytical balance. The following masses were obtained: 0.0080 g H_2, 0.1112 g N_2, 0.1281 g O_2, 0.1770 g CO_2, 0.2320 g C_4H_{10}, and 0.4824 g CCl_2F_2. Within 1%, are these results consistent with Avogadro's hypothesis? Explain.

96. The density of sulfur vapor at 445 °C and 755 mmHg is 4.33 g/L. What is the molecular formula of sulfur vapor?

97. A gaseous hydrocarbon is 85.63% C and 14.37% H by mass. It has a density of 1.69 g/L at 24 °C and 743 mmHg. What is the molecular formula of this hydrocarbon?

98. A 2.135-g sample of a gaseous chlorofluorocarbon (a type of gas implicated in the destruction of the ozone layer) occupies a volume of 315.5 mL at 739.2 mmHg and 26.1 °C. Analysis of the compound shows it to be 14.05% C, 41.48% Cl, and 44.46% F, by mass. What is the molecular formula of this compound?

99. The gaseous hydrocarbon known as 1,3-butadiene is used to make synthetic rubber. The following measurements were made to determine its molecular weight: A glass container weighed 45.0143 g when evacuated, 192.8273 g when filled with Freon-113, a liquid with a density of 1.576 g/mL, and 45.2217 g when filled with butadiene at 751.2 mmHg and 21.48 °C. What is the molecular weight of 1,3-butadiene?

100. Calculate the volume of $H_2(g)$ required to react with 15.0 L CO(g) in the reaction

$$3\ CO(g) + 7\ H_2(g) \longrightarrow C_3H_8 + 3\ H_2O(l)$$

(**a**) if both gases are measured at STP; (**b**) if the CO(g) is measured at STP, and $H_2(g)$ at 22 °C and 745 mmHg; (**c**) if both gases are measured at 22 °C and 745 mmHg; (**d**) if the CO(g) is measured at 25 °C and 757 mmHg, and the H_2(g) at 22 °C and 745 mmHg.

101. Use C_8H_{18} as the "formula" of gasoline and 0.71 g/mL for its density. If a car gets 31.2 mi/gal (1 gal = 3.785 L), what volume of CO_2(g), measured at 28 °C and 732 mmHg, is produced in a trip of 265 mi? Assume complete combustion of the gasoline.

102. Use the definitions of $\bar{u}$ and u_{rms} on page 141 to calculate $\bar{u}$ and u_{rms} for a group of six particles with the speeds: 9.83×10^3, 9.05×10^3, 8.33×10^3, 6.48×10^3, 3.67×10^3, and 1.75×10^3 m/s, respectively.

103. Calculate the root-mean-square speed of SO_2 molecules at 27 °C. (*Hint*: Use $R = 8.3145\ J \cdot mol^{-1} \cdot K^{-1}$ and $1\ J = 1\ kg \cdot m^2 \cdot s^{-2}$.)

104. Calculate the pressure exerted by 1.00 mol CO_2(g) when it is confined to a volume of 2.50 L at 298 K by using (**a**) the ideal gas equation and (**b**) the van der Waals equation (page 146). For CO_2, the van der Waals constants are $a = 3.59\ L^2 \cdot atm \cdot mol^{-2}$ and $b = 0.0427\ L \cdot mol^{-1}$. (**c**) Compare the two results, and comment on the reason(s) for the difference between them.

105. A person at rest breathes about 80 mL of air per second. Assuming this air is at 25 °C and 755 mmHg, how long would it take to breathe "a mole of air"?

106. Typically, when a person coughs, he or she first inhales about 2.0 L of air at 1.0 atm and 25 °C. The epiglottis and the vocal chords then shut, trapping the air in the lungs, where it is warmed to 37 °C and compressed to a volume of about 1.7 L by the action of the diaphragm and chest muscles. The sudden opening of the epiglottis and vocal chords releases this air explosively. Just prior to this release, what is the approximate pressure of the gas inside the lungs?

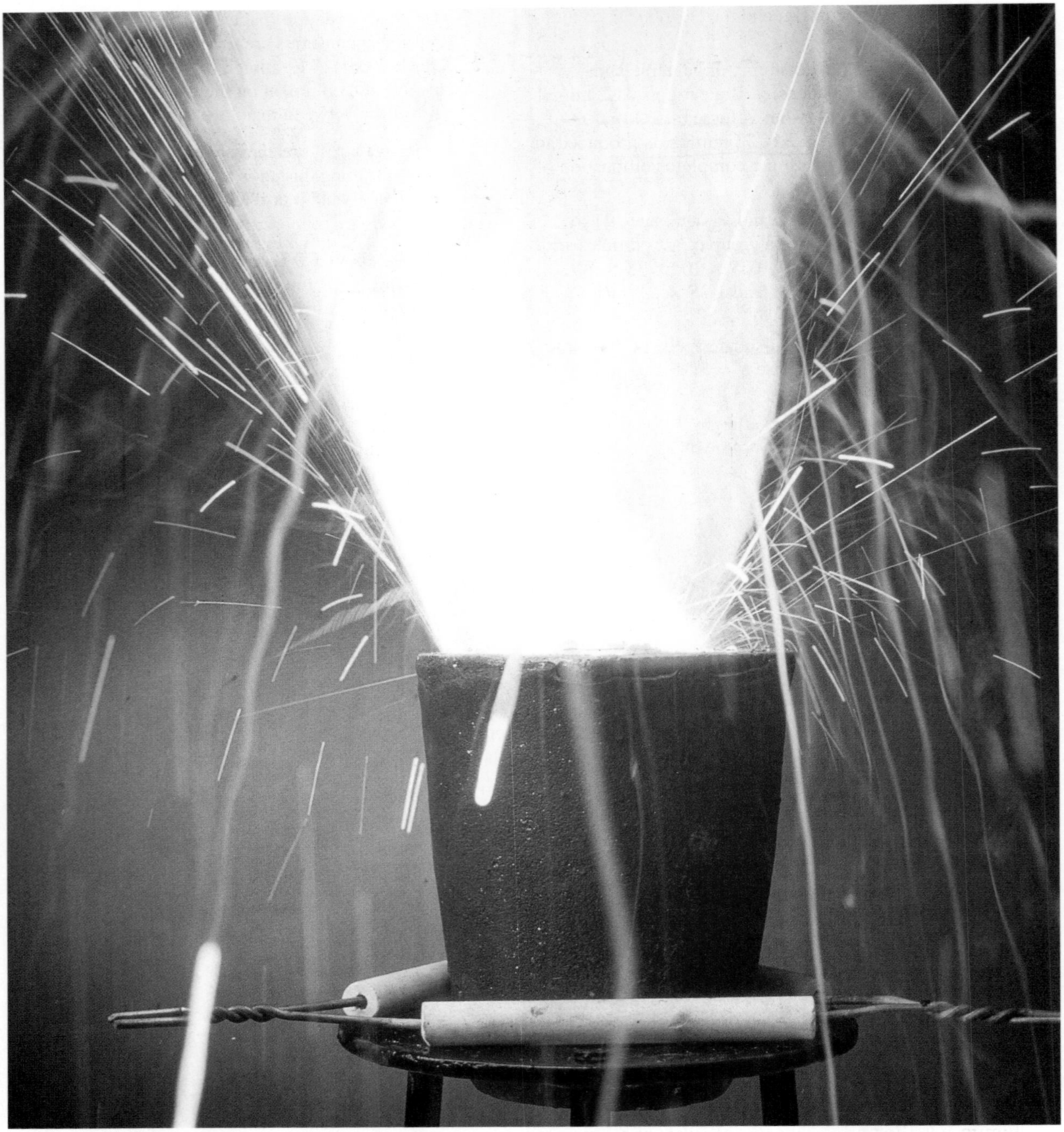

The thermite reaction—the reaction between powdered aluminum metal and iron(III) oxide—yields aluminum oxide, metallic iron, and a quantity of heat sufficient to melt the iron. Thermochemistry deals with the heat effects accompanying chemical reactions.

Exercises 1-4, 6-17

5

Thermochemistry

CONTENTS

Many chemical reactions are carried out just for the associated energy changes, not to obtain products. Consider the combustion of methane, the principal component of natural gas.

$$CH_4(g) + 2\ O_2(g) \longrightarrow CO_2(g) + 2\ H_2O(l)$$

We use methane for cooking and to warm our homes because of the heat that is released when methane burns. We continue to use it even though one of its products, $CO_2(g)$, contributes to global warming. We use it despite the fact that other undesirable substances, such as carbon monoxide and oxides of nitrogen, usually accompany its combustion. In fact, most of the energy that sustains modern industrial society comes from burning fossil fuels: natural gas, coal, and petroleum products such as fuel oil and gasoline. All of these fuels yield undesirable products in addition to heat.

In this chapter, after first examining a few fundamental concepts, we will look at some quantitative relationships involving chemical reactions and energy changes. We conclude the chapter with a discussion of the foods that sustain life and of the fuels that enrich it.

5.1 Energy

Most of what we do in this chapter requires us to deal with the concept of energy and energy changes. We will briefly survey some basic ideas about energy here, but for more details on these matters, you should refer to Appendix B.

Energy is the capacity to do work, and it exists in two basic forms: potential energy and kinetic energy. **Potential energy** is energy due to condition, position, or composition; it is the energy associated with forces of attraction or repulsion between objects. As we learned in our discussion of the kinetic-molecular theory in Chapter 4, **kinetic energy** is energy of motion.

Figure 5.1 illustrates these two types of energy. Water at the top of a dam has potential energy due to its gravitational attraction to Earth's center. The water has the capacity to do work, but as long it remains behind the dam it does none. When the water is allowed to flow through a pipe to a lower level, some of

Figure 5.1 Energy conversion. Potential energy is converted to kinetic energy (and then to electricity) as water falls 480 feet through pipes that connect the reservoir to the electric power station at Shasta Dam in California.

its potential energy is converted to kinetic energy. Water rushing through a pipe can be made to turn the blades of a turbine (a water wheel), which in turn can rotate a coil of wire in an electrical generator, producing electricity. The net result is that some of the potential energy originally stored in the water is converted to electrical work.

The amount of work required to move an object is the product of the force applied and the distance the object is moved.

$$\text{Work} = \text{force} \times \text{distance}$$

The work done by a force of 1 newton (N) acting over a distance of 1 meter is called **1 joule (J)**. In terms of the SI base quantities of mass, length, and time,

$$1\ \text{J} = 1\ \text{N} \times 1\ \text{m} = 1\ \text{kg} \cdot \text{m}^2 \cdot \text{s}^{-2}$$

This definition doesn't help us to visualize whether 1 joule is a lot of energy or a little. We can get a better idea by recalling the definition of kinetic energy:

$$\text{KE} = \tfrac{1}{2} mu^2$$

If we express mass (m) in kilograms and speed (u) in meters per second, the units of kinetic energy are

$$\text{kg} \times (\text{m/s})^2 = \text{kg} \cdot \text{m}^2 \cdot \text{s}^{-2}$$

A 2-kg object moving at a speed of 1 m/s has the kinetic energy

$$\text{KE} = \tfrac{1}{2} \times 2\ \text{kg} \times (1\ \text{m/s})^2 = 1\ \text{kg} \cdot \text{m}^2 \cdot \text{s}^{-2}$$

A bowling ball rolling slowly down a bowling alley has a kinetic energy of a few joules.

If this object were brought to a rest, perhaps by hitting a wall, it would do 1 joule of work. The joule is the derived SI unit of both energy and work.

In many ways, the joule is a small energy unit. A 60-watt light bulb, for example, uses 60 J of energy per second. In thermochemistry the kilojoule (kJ) is commonly used: 1 kJ = 1000 J. Chemists have also used the calorie and kilocalorie as units of energy.

$$1 \text{ cal} = 4.184 \text{ J} \qquad 1 \text{ kcal} = 1000 \text{ cal}$$

We can use the bouncing ball in Figure 5.2 to illustrate something more about the nature of work and energy. First, we have to do work to raise the ball to the starting position. (We have to apply a force to overcome the force of gravity.) This work of lifting is "stored" in the ball as potential energy.

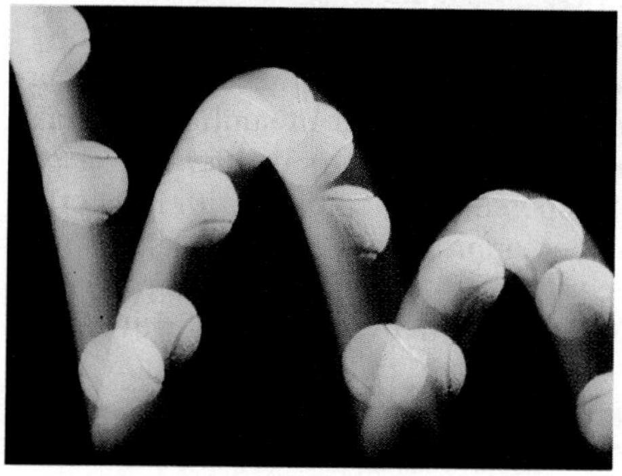

Figure 5.2 Potential energy and kinetic energy. A bouncing ball illustrates the interconversion of potential energy and kinetic energy.

When we release the ball, it falls as it is pulled toward Earth's center by the force of gravity. During this fall, potential energy is converted to kinetic energy, and the kinetic energy reaches a maximum when the ball strikes the surface. As the ball rebounds, its kinetic energy decreases (the ball slows down) and its potential energy increases (the ball rises).

If its collisions with the immovable surface were perfectly *elastic*, the ball would always rebound with the same energy that it had before the collision. At every point on its path the *sum* of the potential and kinetic energies would be constant, and the ball would go on bouncing forever. Experience tells us that a real ball behaves otherwise.

In a real situation, the maximum height on each bounce is less than on the previous one, and the ball eventually comes to rest on the surface. The potential energy invested in lifting the ball is lost, but it does not appear as kinetic energy of the ball. Instead, it appears as additional kinetic energy in the atoms and molecules that make up the ball, the surface, and the surrounding air. The temperatures of the ball and its surroundings increase slightly.

This illustration suggests our need to explore more fully a few basic ideas in the next section.

5.2 Thermochemistry: Some Basic Terms

Thermochemistry is the study of energy changes associated with chemical reactions or physical processes, especially those that involve heat. When we speak of an energy change as heat, we imply that one thing gives off the heat and something else absorbs it. We need terms to describe these two "things."

Thermochemistry is a limited study within the broader field of thermodynamics. We examine other aspects of thermodynamics in Chapters 19 and 20.

System and Surroundings

We define the **system** as that part of the universe that we are studying. The system may be a solution in a beaker, a gas in a cylinder, or a block of frozen spinach in a dish. The system may be as complicated as a polluted lake or even the entire atmosphere. The **surroundings** are the rest of the universe, although we typically can limit our concern only to the parts of the universe in the vicinity of the system (specifically, those parts that interact with the system). *Interactions* refer to the exchange of energy and/or matter between a system and its surroundings. The surroundings of a solution in a beaker, for example, might be the beaker itself, a hot plate on which the beaker is placed, and that part of the surrounding air with which the solution effectively exchanges heat.

There are three kinds of systems. An *open* system interacts readily with its surroundings, exchanging matter and energy, as does, for example, a "steaming" hot cup of coffee. A *closed* system exchanges energy but not matter with the surroundings. A capped bottle of water and tea bags placed in sunlight to make "sun" tea is a closed system. An *isolated* system exchanges neither matter nor energy with its surroundings. A tightly stoppered thermos flask with hot coffee is, at least over a short period of time, a good approximation of an isolated system. In our study of thermochemistry we will try to keep systems simple, often isolating them from the surroundings. We know from experience, however, that we cannot isolate a system completely.

Energy exchanges between a system and its surroundings can occur in the form of *heat* or in several other forms collectively referred to as *work*. These energy exchanges affect the total amount of energy contained *within* a system, the **internal energy** (E). The components of internal energy that interest us here are *thermal energy*, energy associated with random molecular motion, and *chemical energy*, energy associated with chemical bonds and intermolecular forces.

Heat

Heat (q) is an energy transfer into or out of a system that is caused by a temperature difference between a system and its surroundings. Heat always passes spontaneously from a region of higher temperature to a region of lower temperature. Set a hot apple pie on the counter and heat passes from the pie to the countertop and the air around the pie. They get warmer as the pie cools. When a system and its surroundings reach the same temperature, heat transfer ceases. The system and its surroundings are in *thermal equilibrium*.

Let's consider a system consisting of methane gas burning in oxygen. This is the principal reaction that takes place in a typical laboratory Bunsen burner (see Figure 5.3). The methane and oxygen gases and the reaction products comprise the system. The burner and everything else around it make up the surroundings. When the methane burns, chemical energy is converted to thermal energy and the temperature of the system rises. Because the temperature of the system is now above that of the surroundings, heat is transferred from the system to the surroundings. A process in which heat energy is transferred from a system to its surroundings is an *exothermic* process. The combustion of methane is an exothermic chemical reaction. An example of an exothermic physical change is the dissolving of sodium hydroxide (lye) in water. The thermal energy

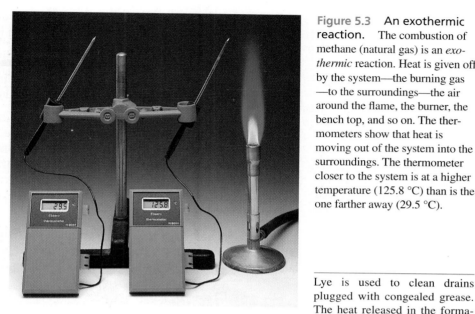

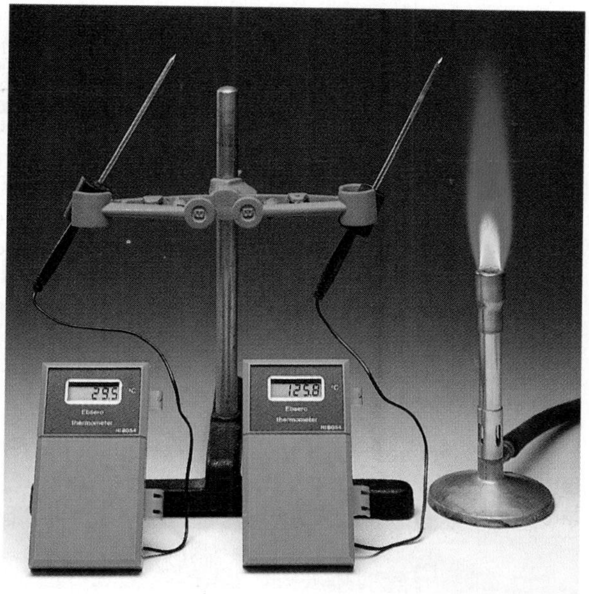

Figure 5.3 An exothermic reaction. The combustion of methane (natural gas) is an *exothermic* reaction. Heat is given off by the system—the burning gas—to the surroundings—the air around the flame, the burner, the bench top, and so on. The thermometers show that heat is moving out of the system into the surroundings. The thermometer closer to the system is at a higher temperature (125.8 °C) than is the one farther away (29.5 °C).

Lye is used to clean drains plugged with congealed grease. The heat released in the formation of NaOH(aq) may melt the grease, allowing it to be flushed from the pipe.

of the solution increases so rapidly that the solution gets quite hot, even though it continuously gives up heat to the surroundings.

Next, consider a system composed of water, initially in the form of a block of ice. When the ice is placed in surroundings above 0 °C, heat passes from the surroundings into the ice (see Figure 5.4). The heat absorbed by the system is used to melt the ice, which remains at 0 °C while melting occurs. A process, like the melting of ice, in which heat energy is transferred from the surroundings into a system is an *endothermic* process.

Figure 5.4 An endothermic reaction. As indicated by the thermometers, heat passes from the warmer surroundings (19.1 °C and 8.4 °C) into the colder system (the block of ice at 0.0 °C). The heat that enters the system is used to convert ice to liquid water: $H_2O(s) \longrightarrow H_2O(l)$, an *endothermic* process. Only after all the ice has melted will the temperature of the system (now liquid water) and its surroundings become equal.

Work

Work (*w*), like heat, is an energy transfer into or out of a system. Work, though, can be of several types. The most common type of work in chemical reactions is *pressure–volume work*—the work of compression or expansion of gases.

In Figure 5.5, a quantity of gas is confined in a cylinder by a freely moving piston. When one of the two objects having a mass m is removed from the piston, the remaining object is raised through the distance h. To see that work is done in this case, recall the defining equation for pressure: $P = F/A$. Thus, the product of the pressure P and the area A over which it is exerted is the force F. And the product of this force F and the distance h represents a quantity of work w.

$$\text{Work } (w) = \text{force } (F) \times \text{distance } (h) = P \times A \times h$$

The product of the cross-sectional area A of the cylinder and the height h represents the *change* in volume of the gas ΔV. The magnitude of work associated with the expanding gas at constant pressure, then, is

$$\text{Work } (w) = P\Delta V$$

The Greek letter *delta*, Δ, represents a *change* in some quantity. The change is always formulated as a *final* value minus an *initial* value. That is, $\Delta V = V_{\text{final}} - V_{\text{initial}}$.

Figure 5.5 not only suggests the origin of the term "pressure–volume" work, but it suggests a more general idea about work. It is something that we can relate, directly or indirectly, to the lifting or lowering of weights.

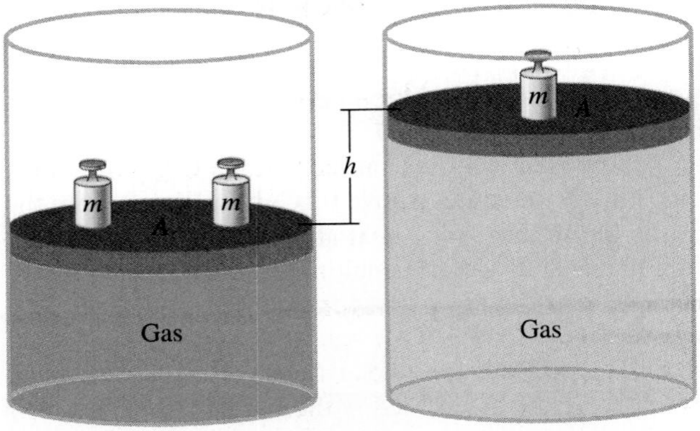

Figure 5.5 Pressure–volume work. The gas expands when one of the objects of a mass m is removed. The remaining object is lifted through the distance h. The gas does work.

Our discussion of heat and work is not yet complete because, in addition to having a magnitude, each quantity of heat and work must also be given a positive or negative sign. We consider this requirement, together with the relationship between heat and work that is one of the underlying principles of thermodynamics, in the next section.

5.3 First Law of Thermodynamics: Internal Energy (*E*)

One of the most fundamental laws of all natural science is the **law of conservation of energy**, also known as the **first law of thermodynamics**:

In a physical or chemical change, energy is neither created nor destroyed.

The first law of thermodynamics says that the internal energy (E) of an *isolated* system is constant. Or, if a system interacts with its surroundings by exchanging heat (q) or work (w), the exchange must take place in such a way that no energy is created or destroyed. These ideas can be expressed through the mathematical equation

$$\Delta E = q + w$$

To conform to the first law of thermodynamics, this equation requires us to follow certain conventions.

- Energy *entering* a system carries a *positive* sign: If heat is *absorbed* by the system, $q > 0$. If work is done *on* a system, $w > 0$.
- Energy *leaving* a system carries a *negative* sign: If heat is *given off* by a system $q < 0$. If work is done *by* a system, $w < 0$.

Can you see by this convention that the work of expansion in Figure 5.5 is a *negative* quantity?

As illustrated through Example 5.1 and Exercise 5.1, ΔE is *positive* if more energy enters a system than leaves; ΔE is *negative* if more energy leaves a system than enters.

Example 5.1

A gas does 135 J of work while expanding, and at the same time it absorbs 156 J of heat. What is the change in internal energy?

Solution

We are given the quantities w and q and seek ΔE. The key is to assign the correct signs to the values of w and q, as indicated in parentheses below. Note specifically that heat is *absorbed* and work is done *by* the system.

$$\Delta E = q + w = (+156\text{ J}) + (-135\text{ J}) = +21\text{ J}$$

Exercise 5.1

In a process in which 89 J of work is done on a system, 567 J of heat is given off. What is ΔE of the system?

We can conceive of what internal energy E is—the total of all the forms of energy contained within a system—but we cannot actually measure it for any real system. Fortunately, E has a property that enables us to use it successfully: E is a *function of state*. When we specify the temperature, pressure, and the kinds and amounts of substances in a system, we define the *state* of the system. Any property having a unique value when the state of a system is specified is called a **state function**.

To illustrate, let's consider the mountain-climbing analogy in Figure 5.6. Think of the base of the mountain as representing an initial state of a system,

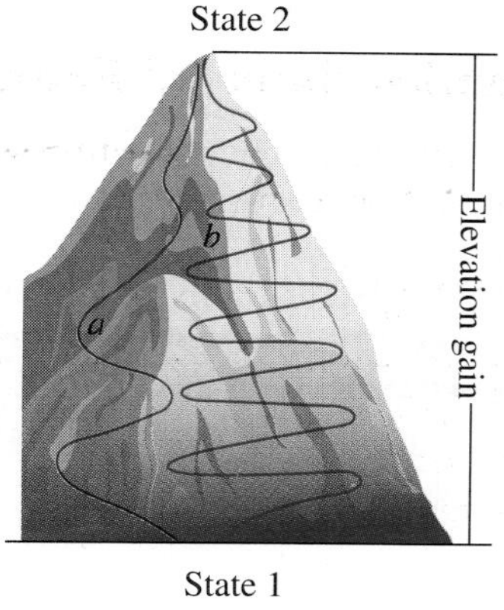

Figure 5.6 Mountain-climbing analogy to a function of a state. Path a is shorter than path b, so a shorter distance is walked along path a. The gain in elevation, however, is independent of the path chosen. E of a system is also independent of the path by which a change occurs. Internal energy, E, is a state function.

state 1, and the summit as a final state, *state 2*. Climbing the mountain is like a thermodynamic process. The *elevation gain*, the difference in elevation between the base and the summit of the mountain, is analogous to the change in internal energy, ΔE. We can measure the elevation gain without knowing the actual elevation of either the base or the summit of the mountain. Similarly, we can measure the *change* in internal energy without knowing the actual internal energy of either the initial or the final state of a system. Furthermore, the elevation gain has a unique value, regardless of the route we follow to climb the mountain. So too is ΔE independent of the particular way in which a process is carried out to get from state 1 to state 2. Finally, the distance traveled in climbing the mountain depends on the path chosen. The distance traveled is *not* a function of state. Heat and work in a thermodynamic process are analogous to the distance traveled. They also depend on the path chosen and *are not* state functions. For example, the amount of work performed by the expanding gas in Figure 5.5 would be *different* if we removed the mass m in little bits and pieces rather than in one big lump, even though the starting and final conditions of the gas would be the same.

We will consider this alternative method of gas expansion in Chapter 19.

An important characteristic of a function of state is that it returns to its initial value when a process is reversed. In our analogy, a trip to the top of a mountain and back down results in a total elevation gain of *zero*. Similarly, when a system is returned to its initial state, the total change in internal energy is *zero* and E returns to its initial value. That is,

$$\text{State 1 } (E_1) \xrightarrow{\Delta E} \text{state 2 } (E_2) \xrightarrow{-\Delta E} \text{state 1 } (E_1)$$

5.4 Heats of Reaction and Enthalpy Change, ΔH

In thermochemistry, we think of a chemical reaction as a process in which a system changes from an initial state (the reactants) to a final state (the products). In an **exothermic reaction**, chemical energy is converted to thermal energy. In an isolated system, this increase in thermal energy produces a temperature increase; in a non-isolated system, it is in the form of heat *given off* to the surroundings: $q < 0$—a *negative* quantity. In an **endothermic reaction**, thermal energy is converted to chemical energy. In an isolated system, this loss of thermal energy appears as a temperature decrease; in a non-isolated system, heat is *absorbed* from the surroundings: $q > 0$—a *positive* quantity.

The **heat of reaction (*q*)** is defined as the quantity of heat exchanged between a system and its surroundings for a reaction at constant temperature. In actual practice, reactions usually are not conducted at a constant temperature. Consider, for example, how rapidly the temperature may rise in a combustion reaction. This definition simply requires that an assessment of the quantity of heat exchanged between a system and its surroundings include a provision for returning the products of a reaction to the same temperature as that of the reactants.

We will consider, in Chapter 20, how chemical reactions can be made to produce electrical work.

A reaction can produce work in several different forms, but often the only work involved is pressure–volume work. This is the work associated with the expansion or compression of gaseous reactants and products (recall Figure 5.5).

In Section 5.3, we learned that in a chemical reaction the values of ΔE, q, and w are related through the expression

$$\Delta E = q + w$$

and that ΔE has a unique value for a given reaction. The values of q and w, on the other hand, depend on the method by which the reaction is carried out. Two methods are generally used.

If the reaction is carried out in a system of *constant volume*, no pressure–volume work is possible: ΔV is zero, which in turn makes $P\Delta V = 0$. Then, if no other forms of work are permitted,

$$\Delta E = q_V$$

That is, $w = 0$ and the observed heat of reaction, q_V, is equal to ΔE for the reaction. (The subscript "V" denotes that the reaction volume is held constant.)

If a reaction is carried out at *constant pressure*, if the only work permitted is pressure–volume work, and if we express this as work done *by* the system (that is, $w = -P\Delta V$), we can write the first law of thermodynamics as

$$\Delta E = q + w = q_P - P\Delta V$$

and rearrange it to

$$q_P = \Delta E + P\Delta V$$

To summarize, the heat of reaction at constant volume, q_V, is equal to ΔE, and the heat of reaction at constant pressure, q_P, is equal to $\Delta E + P\Delta V$.

The vast majority of chemical reactions are carried out at constant pressure—that is, in open vessels that are exposed to the atmosphere under the prevailing barometric pressure. It is quite convenient to have a thermodynamic function whose change in a chemical reaction is exactly equal to the heat of reaction at constant pressure, q_P. A function called *enthalpy* has been designed for this purpose. **Enthalpy (H)** is the sum of the internal energy and the pressure–volume product of a system.

$$H = E + PV$$

The **enthalpy change (ΔH)** for a process carried out at constant temperature and pressure and with work limited to pressure–volume work is

$$\Delta H = \Delta E + P\Delta V$$

We have previously shown that $\Delta E + P\Delta V = q_P$. When we make this substitution into the right side of the above equation, we obtain the desired result:

$$\Delta H = q_P$$

Some Properties of Enthalpy

Enthalpy, H, is especially useful because of certain properties that it shares with internal energy, E, and a few other thermodynamic functions.

1. **Enthalpy is an *extensive* property.** That is, the enthalpy of a system depends on the quantities of substances present. Even though we do not know the absolute enthalpy of 1.00 mol CO_2, we *do* know that the enthalpy of 2.00 mol CO_2 is exactly twice as much.

2. **Enthalpy is a function of state.** The enthalpy of a system depends only on its present state or condition, and *not* on the path or route by which it got there. Because E, P, and V are all functions of state and H is a function only of these variables (that is, $H = E + PV$), then H is also a state function.
3. **Enthalpy changes have unique values.** Because the enthalpy in each of two states of a system has a unique value, the difference in enthalpy between the two states—the enthalpy *change*, ΔH—also has a unique value. And this change in enthalpy is equal to the heat of reaction at constant pressure: $\Delta H = q_P$.

Now let us examine some practical consequences of these statements about enthalpy and enthalpy change.

Representing ΔH for a Chemical Reaction

First we can expand our symbolic description of a chemical reaction, as shown below for the combustion of methane under certain conditions at 25 °C.

The value of ΔH varies somewhat depending on the pressures of the gaseous reactants and products. We will say more in Section 5.7 about the usual conditions for which ΔH values are reported.

$$CH_4(g) + 2\,O_2(g) \longrightarrow CO_2(g) + 2\,H_2O(l) \qquad \Delta H = -890.3\text{ kJ}$$

The enthalpy change in the combustion of 1 mol $CH_4(g)$ under these conditions is −890.3 kJ. Because $\Delta H = q_P$, we see that the combustion of methane at constant pressure is an *exothermic* reaction.

The equation representing the decomposition of mercury(II) oxide shows this to be an *endothermic* reaction.

$$2\,HgO(s) \longrightarrow 2\,Hg(l) + O_2(g) \qquad \Delta H = +181.66\text{ kJ}$$

The decomposition of *two* moles of HgO(s) requires the absorption of 181.66 kJ of heat in a constant-pressure reaction at 25 °C. Because enthalpies and enthalpy changes are extensive properties, the decomposition of *one* mole of HgO(s) is accompanied by exactly one-half the enthalpy change noted above. That is,

$$HgO(s) \longrightarrow Hg(l) + \tfrac{1}{2}\,O_2(g) \qquad \Delta H = \tfrac{1}{2} \times 181.66 = +90.83\text{ kJ}$$

We can represent enthalpy change through a chemical equation, but we can also do so through a graphical representation known as an **enthalpy diagram**, as illustrated in Figure 5.7.

If a process is carried out first in one direction and then brought back to its initial state, the total enthalpy change must be *zero*. This is because enthalpies are state functions, and the initial and final states in such a process (called a *cyclic* process) are the same. This means that ΔH changes sign when a process is reversed (see Figure 5.8). Thus, for the *formation* of one mole of HgO(s) from its elements at 25 °C,

$$Hg(l) + \tfrac{1}{2}\,O_2(g) \longrightarrow HgO(s) \qquad \Delta H = -90.83\text{ kJ}$$

In Example 5.2 we write a new equation based on only a fraction ($\tfrac{1}{8}$) of the amounts represented in the given equation, and in Example 5.3 we must reverse an equation and then reduce the amounts by $\tfrac{1}{2}$.

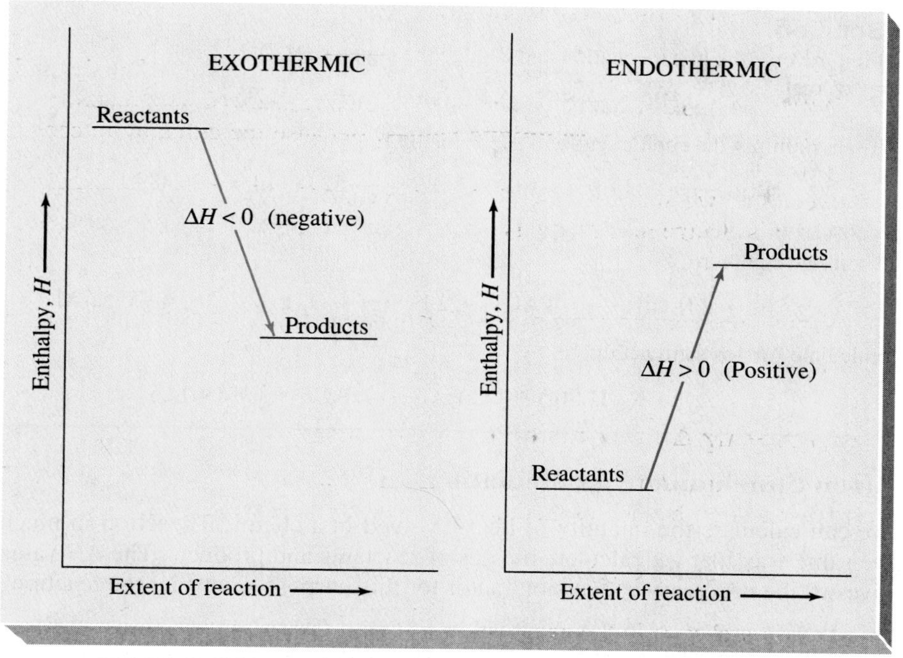

Figure 5.7 Enthalpy diagrams. No numbers are shown on the enthalpy axis because absolute values of enthalpy cannot be measured. For an *exothermic* reaction (red), the products have a lower enthalpy than the reactants and ΔH is *negative* ($\Delta H < 0$). For an *endothermic* reaction (blue), the products have a higher enthalpy than the reactants and ΔH is *positive* ($\Delta H > 0$).

Example 5.2

Use the following equation

$$S_8(s) + 4\,O_2(g) + 8\,Cl_2(g) \longrightarrow 8\,SOCl_2(l) \qquad \Delta H = -1694\text{ kJ}$$

to calculate ΔH for the reaction

$$\tfrac{1}{8}\,S_8(s) + \tfrac{1}{2}\,O_2(g) + Cl_2(g) \longrightarrow SOCl_2(l)$$

Solution

Because the coefficients have been multiplied by $\frac{1}{8}$, we must also multiply ΔH by $\frac{1}{8}$.

$$\tfrac{1}{8}\,S_8(s) + \tfrac{1}{2}\,O_2(g) + Cl_2(g) \longrightarrow SOCl_2(l) \qquad \Delta H = \tfrac{1}{8} \times (-1694) = -211.8\text{ kJ}$$

Exercise 5.2

Use the equation

$$3\,O_2(g) \longrightarrow 2\,O_3(g) \qquad \Delta H = +285.4\text{ kJ}$$

to calculate ΔH for the reaction

$$\tfrac{3}{2}\,O_2(g) \longrightarrow O_3(g)$$

Example 5.3

Given the equation

$$H_2(g) + I_2(s) \longrightarrow 2\,HI(g) \qquad \Delta H = +52.96\text{ kJ}$$

calculate ΔH for the reaction

$$HI(g) \longrightarrow \tfrac{1}{2}\,H_2(g) + \tfrac{1}{2}\,I_2(s)$$

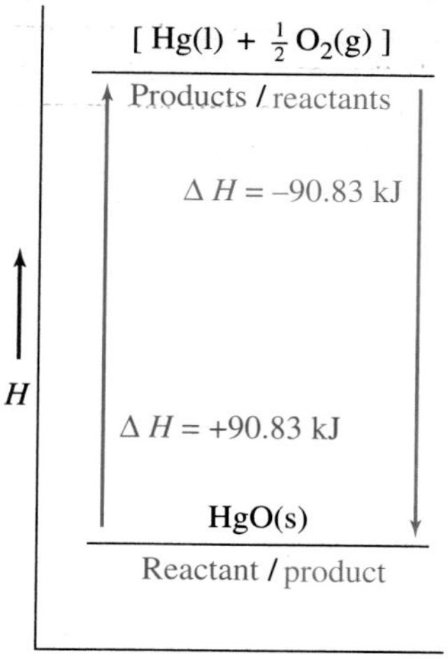

Figure 5.8 Reversing a chemical reaction. The *forward* reaction, the dissociation at 25 °C of one mole of HgO(s) into its elements, is accompanied by an increase in enthalpy of 90.83 kJ; $\Delta H = +90.83$ kJ. (The forward reaction is shown in blue.) In the *reverse* reaction, one mole of HgO(s) is formed from its elements. This reaction is accompanied by a *decrease* in enthalpy of 90.83 kJ; $\Delta H = -90.83$ kJ. (The reverse reaction is shown in red.) When a reaction is reversed, the magnitude of ΔH remains the same but its sign changes.

Solution

First, reverse the given equation and change the sign of ΔH.

$$2\,HI(g) \longrightarrow H_2(g) + I_2(s) \qquad \Delta H = -52.96 \text{ kJ}$$

Then, multiply the coefficients and ΔH value by $\frac{1}{2}$.

$$HI(g) \longrightarrow \tfrac{1}{2}H_2(g) + \tfrac{1}{2}I_2(s) \qquad \Delta H = \tfrac{1}{2}(-52.96) \text{ kJ} = -26.48 \text{ kJ}$$

Exercise 5.3

From the equation

$$2\,Ag_2S(s) + 2\,H_2O(l) \longrightarrow 4\,Ag(s) + 2\,H_2S(g) + O_2(g) \qquad \Delta H = +595.5 \text{ kJ}$$

calculate ΔH for the reaction

$$Ag(s) + \tfrac{1}{2}H_2S(g) + \tfrac{1}{4}O_2(g) \longrightarrow \tfrac{1}{2}Ag_2S + \tfrac{1}{2}H_2O(l)$$

ΔH in Stoichiometric Calculations

We can calculate the quantity of heat involved in a chemical reaction in much the same way that we calculate masses of reactants and products. The ΔH value gives us the necessary conversion factor for this purpose. Consider the reaction

$$H_2(g) + Cl_2(g) \longrightarrow 2\,HCl(g) \qquad \Delta H = -184.6 \text{ kJ}$$

for which we can write conversion factors such as

$$\frac{-184.6 \text{ kJ}}{1 \text{ mol } H_2} \qquad \frac{-184.6 \text{ kJ}}{1 \text{ mol } Cl_2} \qquad \frac{-184.6 \text{ kJ}}{2 \text{ mol HCl}}$$

One of these conversion factors is used in Example 5.4, and another in Exercise 5.4.

Perhaps more typical is the situation presented in Exercise 5.5. There a conversion factor is used to relate the amount of a reactant [$CH_4(g)$] to a quantity of heat through a ΔH value. Then the amount of the gaseous reactant is converted to a gas volume at a given temperature and pressure by using the ideal gas equation.

Example 5.4

What is the enthalpy change associated with the formation of 5.67 mol HCl(g) in the reaction?

$$H_2(g) + Cl_2(g) \longrightarrow 2\,HCl(g) \qquad \Delta H = -184.6 \text{ kJ}$$

Solution

The conversion factor we need from the chemical equation is shown in color below.

$$? \text{ kJ} = 5.67 \text{ mol HCl} \times \frac{-184.6 \text{ kJ}}{2 \text{ mol HCl}} = -523 \text{ kJ}$$

The formation of 5.67 mol HCl in this *exothermic* reaction results in the release of 523 kJ of heat to the surroundings.

Exercise 5.4

What is the enthalpy change when 12.8 g $H_2(g)$ reacts with excess $Cl_2(g)$ to form HCl(g)?

$$H_2(g) + Cl_2(g) \longrightarrow 2\,HCl(g) \qquad \Delta H = -184.6 \text{ kJ}$$

Exercise 5.5

What volume of $CH_4(g)$, measured at 25 °C and 745 torr, must be burned in excess oxygen to release 1.00×10^6 kJ of heat to the surroundings?

$$CH_4(g) + 2\,O_2(g) \longrightarrow CO_2(g) + 2\,H_2O(l) \quad \Delta H = -890.3 \text{ kJ}$$

5.5 Calorimetry: Measuring Quantities of Heat

We have seen that ΔH for a chemical reaction is the heat of reaction at constant pressure, q_P, and we have learned how to incorporate ΔH values into chemical equations. But how do we measure heats of reaction? The key is in measuring quantities of heat.

The measurement of a quantity of heat is called *calorimetry*, and the device in which the measurement is made is called a **calorimeter**. The basic principle of calorimetry is the *law of conservation of energy*, which requires that whatever heat is *lost* by a system must be *gained* by its surroundings (or vice versa). The key to successful calorimetry is the ability to measure quantities of heat accurately.

In order to use a calorimeter, we must be familiar with the concepts of heat capacity and specific heat. We begin by introducing and illustrating these terms.

Heat Capacity

Different substances have differing capacities to store energy that is absorbed as heat. The *heat capacity* (*C*) of a system is the quantity of heat required to change the temperature of the system by 1 °C (or by 1 K). The quantity of heat q associated with a change in temperature of a system from an initial temperature T_i to a final temperature T_f is given by

$$q = C(T_f - T_i)$$

q- quantity of heat

If we represent the change in temperature as

$$\Delta T = T_f - T_i$$

then we have

$$q = C \times \Delta T$$

The heat capacity of a system is given by

$$C = \frac{q}{\Delta T}$$

and the units of heat capacity are joules per degree Celsius (J/°C) or joules per kelvin (J/K). Because a change in temperature of one kelvin is equal to a change of one Celsius degree, heat capacity has the same value whichever of these units is used.

A temperature—T—has different values on the Kelvin and Celsius scales, but a temperature *difference*—ΔT—is the same on the two scales.

Example 5.5

Calculate the heat capacity of an aluminum block if a temperature rise from 22 °C to 145 °C requires the absorption of 629 J of heat.

Solution

$$C = \frac{q}{\Delta T} = \frac{629 \text{ J}}{145 - 22 \text{ °C}} = \frac{629 \text{ J}}{123 \text{ °C}} = 5.11 \text{ J/°C}$$

Exercise 5.6

Calculate the heat capacity of a sample of brake fluid if a temperature rise from 15 °C to 100 °C requires the absorption of 911 J of heat.

Specific Heats

The heat capacity of a system depends on the quantity and type(s) of matter in the system. A large block of aluminum has a higher heat capacity than does a small piece of the metal. Generally we must relate heat capacity to the amount of material in question. **Molar heat capacity** is the heat capacity of one mole of a substance. **Specific heat** is the heat capacity of a one-gram sample. That is, it is the quantity of heat required to change the temperature of *one gram* of the substance by 1 K (or by 1 Celsius degree). We get the specific heat when we divide the heat capacity of a substance by its mass.

$$\text{Specific heat} = \frac{\text{heat capacity}}{\text{mass}} = \frac{C}{m}$$

Because the heat capacity itself is $C = q/\Delta T$,

$$\text{Specific heat} = \frac{q}{m \times \Delta T}$$

If the aluminum block described in Example 5.5 has a mass of 5.7 g, then the specific heat of aluminum is

$$\text{Specific heat} = \frac{629 \text{ J}}{5.7 \text{ g} \times 123 \text{ °C}} = 0.90 \text{ J} \cdot \text{g}^{-1} \cdot \text{°C}^{-1}$$

The molar heat capacity of aluminum is

$$\text{Specific heat} \times \text{molar mass} = 0.90 \text{ J} \cdot \text{g}^{-1} \cdot \text{°C}^{-1} \times 27.0 \text{ g} \cdot \text{mol}^{-1} = 24 \text{ J} \cdot \text{mol}^{-1} \cdot \text{°C}^{-1}$$

The specific heats of several familiar substances are given in Table 5.1.

TABLE 5.1 **Specific Heats of Several Substances at 25 °C**

Substance	Specific Heat, $J \cdot g^{-1} \cdot °C^{-1}$
Aluminum (Al)	0.902
Copper (Cu)	0.385
Ethanol (C_2H_5OH)	2.46
Iron (Fe)	0.449
Lead (Pb)	0.128
Mercury (Hg)	0.139
Silver (Ag)	0.235
Sulfur (S)	0.706
Water (H_2O)	4.182

A typical calorimetric calculation requires relating a quantity of heat, a temperature change, and the mass and specific heat of a substance. For this purpose we can rearrange the equation for specific heat into the form

$$q = \text{mass} \times \text{specific heat} \times \Delta T$$

Note that because $\Delta T = (T_f - T_i)$, a temperature increase in the system means that T_f is greater than T_i. ΔT is *positive* and q is *positive*, indicating that heat is *gained* by the system. A temperature decrease means that T_f is smaller than T_i. ΔT is *negative* and q is *negative*, indicating that heat is *lost* by the system.

The specific heat of a substance varies with temperature, and some of the calculations we do here yield only approximate answers. For example, the specific heat of silver, which is 0.235 J · g^{-1} · K^{-1} at 298 K (25 °C), increases steadily to a value of 0.278 J · g^{-1} · K^{-1} at 1000 K. The most important specific heat value that we will use in calculations is that of water. At about room temperature the specific heat of water is 4.18 J · g^{-1} · °C^{-1}, and over the temperature range from 0 °C to 100 °C it remains within 1% of this value.

The specific heat of water at 15 °C is 4.184 J · g^{-1} · °C. At one time the quantity of heat required to raise the temperature of one gram of water from 14.5 °C to 15.5 °C was defined as one **calorie (cal)**. As we saw in Section 5.1, the relationship between the calorie and the joule now is simply defined as 1 cal = 4.184 J.

Example 5.6

How much heat, in joules and kilojoules, does it take to raise the temperature of 225 g of water from 25.0 °C to 100.0 °C?

Solution

The specific heat of water is 4.18 J · g^{-1} · °C^{-1}. The temperature change is (100.0 – 25.0) °C = 75.0 °C, and the quantity of water to be heated is 225 g.

$$q = m \times \text{specific heat} \times \Delta T$$

$$q = 225\ \text{g H}_2\text{O} \times \frac{4.18\ \text{J}}{\text{g H}_2\text{O} \cdot {}^\circ\text{C}} \times (100.0 - 25.0)\ {}^\circ\text{C}$$

$$= 7.05 \times 10^4\ \text{J} = 70.5\ \text{kJ}$$

Exercise 5.7

How much heat, in calories and kilocalories, does it take to raise the temperature of 814 g of water from 18.0 °C to 100.0 °C?

Example 5.7

How many kilograms of water can be heated from 5.5 °C to 55.0 °C by 9.09 × 10^{10} J of heat?

Solution

Solve the basic calorimetry equation for mass. Then substitute the given quantities.

$$\text{Mass H}_2\text{O} = \frac{q}{\text{specific heat} \times \Delta T}$$

$$= \frac{9.09 \times 10^{10}\ \text{J}}{4.18\ \text{J} \cdot \text{g}^{-1} \cdot {}^\circ\text{C}^{-1} \times (55.0 - 5.5)\ {}^\circ\text{C}}$$

$$= 4.39 \times 10^8\ \text{g} = 4.39 \times 10^5\ \text{kg}$$

Example 5.8

What will be the final temperature if 25.0 J of heat is removed from a 5.00-g silver ring that is initially at 37.0 °C? Use 0.235 $J \cdot g^{-1} \cdot °C^{-1}$ as the specific heat of silver.

Solution

Because heat is removed from the silver, q is negative, that is, $q = -25.0$ J. The final temperature should be *lower* than the initial temperature. We solve the calorimetry equation for the temperature change ($T_f - T_i$), which will be a *negative* quantity. Then we solve for the final temperature, T_f.

$$\Delta T = (T_f - T_i) = \frac{q}{\text{mass} \times \text{specific heat}} = \frac{-25.0\ \text{J}}{5.00\ \text{g} \times 0.235\ \text{J} \cdot \text{g}^{-1} \cdot °\text{C}^{-1}}$$

$$= -21.3\ °\text{C}$$

$$T_f = T_i - 21.3\ °\text{C} = 37.0\ °\text{C} - 21.3\ °\text{C} = 15.7\ °\text{C}$$

Exercise 5.8

A 454-g block of lead is at an initial temperature of 22.5 °C. What will be the temperature of the lead after 4.22 kJ of heat is added? Use 0.13 $J \cdot g^{-1} \cdot °C^{-1}$ as the specific heat of lead.

Exercise 5.9

How many grams of copper can be heated from 22.5 °C to 35.0 °C by a quantity of heat sufficient to raise the temperature of 145 g H_2O through this same interval? (*Hint*: Refer to Table 5.1.)

Estimation Exercise 5.1

Which of these metals will be raised to the highest temperature when 1.00 kJ of heat is absorbed by a 100.0-g sample at 22 °C: aluminum, iron, or silver? (*Hint*: Use data from Table 5.1.)

Figure 5.9 A simple calorimeter made from Styrofoam cups. The inner cup is closed off with a cork stopper through which a thermometer and stirrer are inserted into the calorimeter. The outer cup provides additional thermal insulation from the surroundings.

Measuring Specific Heats

The simple calorimeter pictured in Figure 5.9 is made from Styrofoam cups. It is quite suitable for determining the specific heats of substances that are insoluble in water.

A measured mass of the substance is heated to a given temperature and dropped into the calorimeter, which contains a fixed mass of water at a known temperature. Heat is transferred from the hot substance into the cooler water, causing a temperature rise in the water. We assume that all the heat lost by the substance is gained by the water in the cup—that is, that the heat gained by the Styrofoam itself in negligible. From the mass and temperature change of the water, we determine the quantity of heat transferred. From this quantity of heat and the mass and temperature change of the substance, we determine its specific heat. The method is illustrated in Example 5.9.

Example 5.9

A 15.5-g sample of a metal alloy is heated to 98.9 °C and then dropped into 25.0 g of water in a calorimeter. The temperature of the water rises from 22.5 °C to 25.7 °C. Calculate the specific heat of the alloy.

Solution

First, calculate the quantity of heat absorbed by the water.

$$q_{H_2O} = 25.0 \text{ g} \times \frac{4.18 \text{ J}}{\text{g} \cdot {}^\circ\text{C}} \times (25.7 - 22.5)\ {}^\circ\text{C} = 334 \text{ J}$$

The alloy must lose 334 J of heat, meaning that

$$q_{\text{alloy}} = -334 \text{ J}$$

Finally, calculate the specific heat of the alloy.

$$\text{Specific heat}_{\text{alloy}} = \frac{q_{\text{alloy}}}{\text{mass} \times \Delta T} = \frac{-334 \text{ J}}{15.5 \text{ g} \times (25.7 - 98.9\ {}^\circ\text{C})}$$

$$= 0.29 \text{ J} \cdot \text{g}^{-1} \cdot {}^\circ\text{C}^{-1}$$

You may recognize that by the significant figure rules we are not justified in carrying three significant figures in the intermediate result: 334 J. This intermediate result, which remains stored in an electronic calculator, is simply the numerator in the final setup. We need to round off only once, in the final answer.

Exercise 5.10

A 23.9-g sample of iridium is heated to 89.7 °C and then dropped into 20.0 g of water in a calorimeter. The temperature of the water rises from 20.1 °C to 22.6 °C. Calculate the specific heat of iridium.

CONCEPTUAL EXAMPLE 5.1

Which of the following is a likely approximate final temperature when 100 g of iron at 100 °C is added to 100 g of water in a Styrofoam-cup calorimeter at 20 °C?

20 °C 30 °C 60 °C 70 °C

Solution

We need to see that the temperature change produced in an object is directly proportional to the amount of heat involved and inversely proportional to the mass of the object and its specific heat. That is, we need the expression

$$\Delta T = \frac{q}{\text{mass} \times \text{specific heat}}$$

Because the masses of the iron and the water are the same, as are the quantities of heat, q, we can simply compare the specific heats of iron and water. From Table 5.1 we see that they are about 0.45 $\text{J} \cdot \text{g}^{-1} \cdot {}^\circ\text{C}^{-1}$ for iron and 4.18 $\text{J} \cdot \text{g}^{-1} \cdot {}^\circ\text{C}^{-1}$ for water. The drop in temperature of the iron must be much greater than the rise in temperature of the water. The final temperature must certainly be *below* 60 °C, the average of 20 °C and 100 °C. Because the water does absorb some heat, the final temperature must be greater than 20 °C. The only possible approximate final temperature of those given is 30 °C.

See Problem 77 for a calculation of the final temperature.

Conceptual Exercise 5.1

What is the approximate final temperature if 200 mL of water at 80 °C is added to 100 mL of water at 20 °C?

Measuring Enthalpy Changes for Chemical Reactions

Certain chemical reactions, particularly those involving aqueous solutions, can be carried out and the heat of reaction measured in the simple calorimeter of Figure 5.9. This kind of experiment is slightly different from the specific heat determination just considered. In the previous case, we treated the metal alloy as the system and the water as the surroundings. In this case, let's treat the contents of the cup as an *isolated* system. That is, chemical energy is converted to thermal energy (or vice versa), but we assume that no heat is exchanged with the surroundings outside the cup. The contents of the cup either warm up or cool down.

However, our definition of a heat of reaction requires that the products of a reaction be at the same temperature as the reactants. We need to calculate the quantity of heat that *would have been transferred* to keep the temperature from changing if the system had been allowed to interact with the surroundings. This quantity of heat is simply the *negative* of the change in thermal energy observed within the system. And this, in turn, is just the product of the heat capacity of the system and the temperature change (product of the mass, specific heat, and temperature change). If we represent the change in thermal energy in the calorimeter as q_{calorim}, this means that the heat of reaction—let's call it q_{rxn}—is

$$q_{\text{rxn}} = -q_{\text{calorim}}$$

Notice how this treatment of the calorimetry data is consistent with other thermochemical ideas: For example, in an exothermic reaction in an *isolated* system, the temperature rises, corresponding to a *positive* value of q_{calorim}. The heat of reaction, on the other hand, is *negative*, just as it should be for an *exothermic* reaction—heat given off by a system to its surroundings. (In an endothermic reaction, q_{calorim} is negative and q_{rxn} is positive.) Finally, because reactions carried out in the calorimeter of Figure 5.9 are under constant atmospheric pressure, $q_{\text{rxn}} = q_P = \Delta H$, and

$$\Delta H = -q_{\text{calorim}}$$

In Example 5.10 we calculate the enthalpy change for an acid–base reaction in the manner outlined above.

Remember that 0.250 M HCl, read as "0.250 molar HCl," means HCl at 0.250 mole per liter.

Example 5.10

A 50.0-mL sample of 0.250 M HCl at 19.50 °C is added to 50.0 mL of 0.250 M NaOH, also at 19.50 °C, in a Styrofoam-cup calorimeter. Upon mixing, the solution temperature rises to 21.21 °C. Calculate the enthalpy change, ΔH, for the reaction

$$HCl(aq) + NaOH(aq) \longrightarrow NaCl(aq) + H_2O(l)$$

Solution

We must make several simplifying assumptions: (1) The specific heat and density of the product solution, NaCl(aq), are the same as those for pure water; which is a good assumption as long as the solution is dilute; (2) the final volume is 100.0 mL (that is, there is no volume change on mixing); (3) no energy escapes from the calorimeter (an isolated system); and (4) the warming of any part of the calorimeter assembly other than the NaCl(aq) involves an insignificant amount of heat.

The increase in thermal energy occurring within the calorimeter (the system) is

$$q_{\text{calorim}} = \text{mass} \times \text{specific heat} \times \Delta T$$

$$= 100.0 \text{ mL} \times \frac{1.00 \text{ g}}{\text{mL}} \times \frac{4.18 \text{ J}}{\text{g} \cdot {}^\circ\text{C}} \times (21.21 - 19.50)\,{}^\circ\text{C} = 715 \text{ J}$$

The enthalpy change is

$$\Delta H = q_P = -q_{\text{calorim}} = -715 \text{ J}$$

But this enthalpy change is not for the formation of the 1.00 mol H_2O implied in the chemical equation. The amount actually formed in the reaction is

$$? \text{ mol } H_2O = 50.0 \text{ mL soln} \times \frac{0.250 \text{ mol HCl}}{1000 \text{ mL soln}} \times \frac{1 \text{ mol } H_2O}{1 \text{ mol HCl}} = 0.0125 \text{ mol } H_2O$$

For the formation of 1.00 mol H_2O,

$$\Delta H = 1.00 \text{ mol } H_2O \times \frac{-715 \text{ J}}{0.0125 \text{ mol } H_2O} = -57{,}200 \text{ J} = -57.2 \text{ kJ}$$

Exercise 5.11

A volume of 100.0 mL of 0.500 M HBr at 20.29 °C is added to 100.0 mL of 0.500 M KOH, also at 20.29 °C, in a Styrofoam-cup calorimeter. Upon mixing, the temperature rises to 23.65 °C. Calculate the enthalpy change, ΔH, for the reaction.

$$HBr(aq) + KOH(aq) \longrightarrow KBr(aq) + H_2O(l)$$

Bomb Calorimetry: Reactions at Constant Volume

Reactions carried out in Styrofoam-cup calorimeters, like reactions in most laboratory vessels, are open to the atmosphere, that is, under constant pressure. (The lid on the calorimeter is there to minimize heat loss; it is not airtight.) We cannot carry out reactions involving gases in these calorimeters.

For reactions involving gases, such as combustion reactions, we need to use a device like the one illustrated in Figure 5.10—a *bomb calorimeter*. This calorimeter confines the reactants and products to a constant volume. The strong steel walls keep the container intact even if the contents react explosively in an exothermic reaction.

A small weighed sample is placed in a metal cup in the bomb, and the bomb is assembled and filled with oxygen at a pressure of about 30 atm. The bomb is then placed in an insulated container filled with a known quantity of water. The reaction is initiated with an electric ignition coil, and the contents of the insulated container are allowed to come to thermal equilibrium at a temperature somewhat higher than the initial temperature.

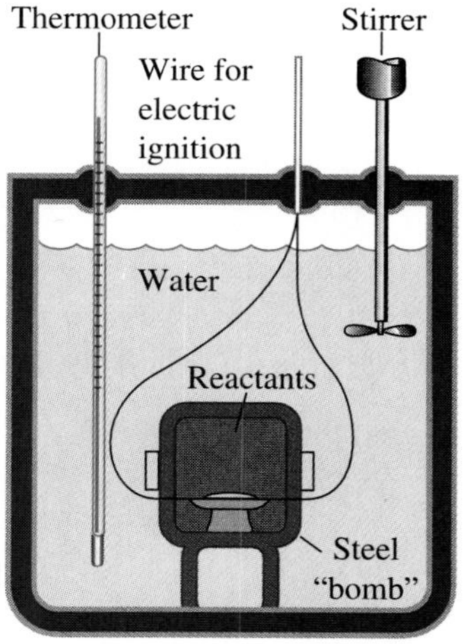

Figure 5.10 A bomb calorimeter. The sample to be burned is placed in a small cup in the bomb, which is then filled with oxygen. The sample is ignited by an electric current. The heat of the reaction is determined from the temperature rise in the water that surrounds the bomb. The steel bomb confines the reactants and products to a constant volume.

From the temperature rise and the heat capacity of the calorimeter (established in a separate experiment), the increase in thermal energy of the calorimeter contents, $q_{calorim}$, can be established. That is,

$$q_{calorim} = \text{heat capacity of calorimeter} \times \Delta T$$

The contents of the calorimeter are treated as an *isolated* system; and, as was the case with the Styrofoam-cup calorimeter in Example 5.10,

$$q_{rxn} = -q_{calorim}$$

The pressure within the bomb in which the reaction occurs does not necessarily remain constant, but the *volume* does. As a result, the quantity measured is q_V; and, as we demonstrated on page 163, $q_V = \Delta E$. Thus, for a reaction carried out in a bomb calorimeter,

$$\Delta E = q_V = q_{rxn} = -q_{calorim}$$

A heat of reaction measured in a Styrofoam-cup calorimeter is $q_P = \Delta H$. In a bomb calorimeter, the heat of reaction is $q_V = \Delta E$.

Even though we measure heats of combustion in bomb calorimeters, we usually carry out combustion reactions in the open atmosphere, such as in a natural gas heater or a propane torch. Thus, the more useful property of combustion reactions is ΔH (the heat of reaction at constant pressure), not ΔE (the heat of reaction at constant volume). Although it is not difficult to calculate a ΔH value from a ΔE value, we will not do so here.

What we can say, though, is that in many cases ΔH and ΔE are nearly equal because the $P\Delta V$ term is relatively small. The principal exceptions are reactions that involve a change in the number of moles of gas. If we treat the gases as ideal, $P\Delta V = \Delta n_g RT$, where $\Delta n_g = n_{\text{gaseous products}} - n_{\text{gaseous reactants}}$. If $\Delta n_g = 0$, the difference between ΔH and ΔE is typically negligible. If the number of moles of gas changes in a reaction, $\Delta n_g \neq 0$ and the difference between ΔH and ΔE may occasionally be significant.

Example 5.11 and Exercise 5.12 illustrate cases where ΔH and ΔE are nearly identical. Exercise 5.13 describes an experiment to establish the heat capacity of a bomb calorimeter, a necessary preliminary in bomb calorimetry.

Example 5.11

In a preliminary experiment, the heat capacity of a bomb calorimeter assembly is found to be 6.52 kJ/°C. In a second experiment, a 0.250-g sample of diamond (carbon) is placed in the bomb with an excess of oxygen. The water, bomb, and other contents of the calorimeter are in thermal equilibrium at 20.00 °C. The diamond is ignited and burned, and the water temperature rises to 21.26 °C. Calculate ΔH for the reaction

$$\text{C(diamond)} + O_2\text{(g)} \longrightarrow CO_2\text{(g)} \qquad \Delta H = ?$$

Solution

First, we calculate $q_{calorim}$ from the heat capacity of the calorimeter and the temperature change.

$$q_{calorim} = 6.52 \text{ kJ} \cdot {}^\circ\text{C}^{-1} \times (21.26 - 20.00)\ {}^\circ\text{C} = 8.22 \text{ kJ}$$

Then, we establish q_{rxn}.

$$q_{rxn} = -q_{calorim} = -8.22 \text{ kJ}$$

To get ΔH we need the heat released by the combustion of one mole (12.01 g) of diamond.

$$\Delta H = 12.01\ \text{g} \times \frac{-8.22\ \text{kJ}}{0.250\ \text{g}} = -395\ \text{kJ}$$

$$\text{C(diamond)} + O_2(g) \longrightarrow CO_2(g) \qquad \Delta H = -395\ \text{kJ}$$

Exercise 5.12

A 0.480-g sample of graphite (another form of carbon) is burned with an excess of $O_2(g)$ in a bomb calorimeter having a heat capacity of 5.15 kJ/°C. The calorimeter temperature rises from 25.00 °C to 28.05 °C. Calculate ΔH for the reaction

$$\text{C(graphite)} + O_2(g) \longrightarrow CO_2(g) \qquad \Delta H = ?$$

Exercise 5.13

A 0.8082-g sample of glucose ($C_6H_{12}O_6$) is burned in a bomb calorimeter assembly and the temperature is noted to rise from 25.11 °C to 27.21 °C. Determine the heat capacity of the bomb calorimeter assembly.

$$C_6H_{12}O_6(s) + 6\ O_2(g) \longrightarrow 6\ CO_2(g) + 6\ H_2O(l) \qquad \Delta H = -2803\ \text{kJ}$$

The result in Exercise 5.12 is slightly different from that in Example 5.11 because the initial states of the systems differ [C(graphite) in one case and C(diamond) in the other].

5.6 Hess's Law of Constant Heat Summation

We have previously learned how to work with ΔH values for chemical reactions and how to establish ΔH values experimentally through calorimetry. In this section we find that we can add chemical equations and their enthalpy changes just as we can add algebraic equations. Why is this important to us? Because it gives us a way to *calculate* heats of reaction for certain chemical reactions that do not lend themselves to accurate direct measurements. We can even calculate heats of reaction for hypothetical reactions (that is, for reactions that have never been observed or may not actually occur at all).

The principle necessary for doing these things was first stated by Germain Hess (1802–1850), a Swiss chemist who worked in Russia. **By Hess's law** the heat of a reaction is constant, whether the reaction is carried out directly in one step or indirectly through a number of steps. If a chemical equation can be expressed as the sum of other equations, the ΔH value for that equation is the sum of the ΔH values for the other equations. Consider, for example, the reaction

$$(1)\ \ 2\ \text{C(graphite)} + O_2(g) \longrightarrow 2\ CO(g) \qquad \Delta H = ?$$

We cannot measure ΔH directly because it is difficult to get the reaction to stop at CO(g). The CO(g) is likely to react further with $O_2(g)$ to form $CO_2(g)$:

$$(2)\ \ 2\ CO(g) + O_2(g) \longrightarrow 2\ CO_2(g) \qquad \Delta H = -566.0\ \text{kJ}$$

Also, C(graphite) may react with O_2 to form CO_2 directly:

$$(3)\ \ \text{C(graphite)} + O_2(g) \longrightarrow CO_2(g) \qquad \Delta H = -393.5\ \text{kJ}$$

Attempts to measure the heat of reaction (1) lead to ambiguous results; reactions (1), (2), and (3) all occur at the same time. However, we can measure the enthalpy changes for reactions (2) and (3). By combining equations (2) and (3) in an appropriate manner, we can get equation (1) *and its ΔH value*. We must

- *reverse* the direction of equation (2) and change the sign of its ΔH;
- *double* equation (3) and its ΔH; and
- add together the results of (a) and (b).

–Eq (2):	$2\,CO_2(g) \longrightarrow 2\,CO(g) + O_2(g)$	$\Delta H = (-1)(-566.0\text{ kJ}) = +566.0\text{ kJ}$
2 × Eq (3):	$2\,C(\text{graphite}) + 2O_2 \longrightarrow 2\,CO_2$	$\Delta H = 2\,(-393.5\text{ kJ}) = -787.0\text{ kJ}$
Eq (1):	$2\,C(\text{graphite}) + O_2(g) \longrightarrow 2\,CO(g)$	$\Delta H = -221.0\text{ kJ}$

Notice that the two $CO_2(g)$ on each side cancel and that two $O_2(g)$ on the left and one $O_2(g)$ on the right leave a net of one $O_2(g)$ on the left when the equations are added.

Hess's law is based on the fact that enthalpy is a function of state. As shown by the enthalpy diagram of Figure 5.11, the enthalpy change is the same by any route we take to get from a given initial state to a given final state.

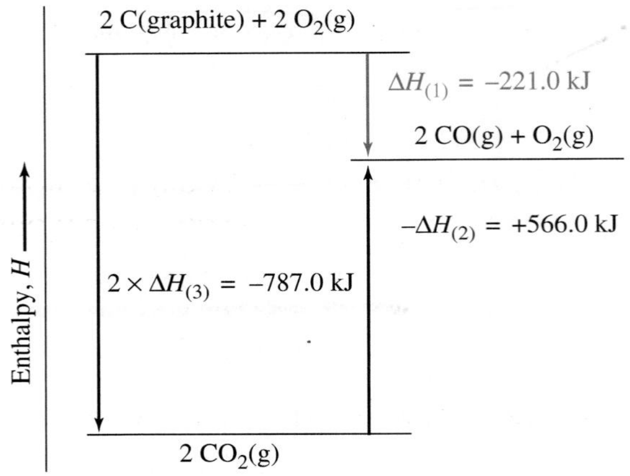

Figure 5.11 Hess's law illustrated through an enthalpy diagram. Whether the reaction occurs through a single step (red arrow) or in two steps (black arrows), the enthalpy change is $\Delta H = -221.0$ kJ for the net reaction:

$2\,C(\text{graphite}) + O_2(g) \longrightarrow 2\,CO(g)$

Just as we cannot burn carbon to produce CO(g) exclusively, we cannot burn a hydrocarbon to yield only CO(g) and $H_2O(l)$. Example 5.12 illustrates how we can use Hess's law to calculate ΔH in a *hypothetical* combustion in which CO(g) and no $CO_2(g)$ is formed. To do this, we use the heat of combustion of the hydrocarbon to form $CO_2(g)$ and $H_2O(l)$, which we can measure in a bomb calorimeter.

Example 5.12

Calculate the enthalpy change for the reaction

$$2\,CH_4(g) + 3\,O_2(g) \longrightarrow 2\,CO(g) + 4\,H_2O(l) \qquad \Delta H = ?$$

from the following thermochemical data.

$$CH_4(g) + 2\,O_2(g) \longrightarrow CO_2(g) + 2\,H_2O(l) \qquad \Delta H = -890.3\text{ kJ}$$

$$2\,CO(g) + O_2(g) \longrightarrow 2\,CO_2(g) \qquad \Delta H = -566.0\text{ kJ}$$

Solution

Because we want *two* $CH_4(g)$ on the left, we *double* the first equation and its ΔH value. We need *two* CO(g) on the right, and this suggests that we need to reverse the second equation and change the sign of its ΔH. Then we add the two equations and their ΔH values.

$$2\,CH_4(g) + 4\,O_2(g) \longrightarrow 2\,CO_2(g) + 4\,H_2O(l) \qquad \Delta H = 2(-890.3\text{ kJ}) = -1781\text{ kJ}$$
$$2\,CO_2(g) \longrightarrow 2\,CO(g) + O_2(g) \qquad \Delta H = +566.0\text{ kJ}$$
$$2\,CH_4(g) + 3\,O_2(g) \longrightarrow 2\,CO(g) + 4\,H_2O(l) \qquad \Delta H = -1215\text{ kJ}$$

Exercise 5.14

Calculate the enthalpy change for the reaction

$$2\,C(\text{graphite}) + 2\,H_2(g) \longrightarrow C_2H_4(g) \qquad \Delta H = ?$$

given the following data.

$$C(\text{graphite}) + O_2(g) \longrightarrow CO_2(g) \qquad \Delta H = -393.5\text{ kJ}$$
$$C_2H_4(g) + 3\,O_2(g) \longrightarrow 2\,CO_2(g) + 2\,H_2O(l) \qquad \Delta H = -1410.9\text{ kJ}$$
$$2\,H_2(g) + O_2(g) \longrightarrow 2\,H_2O(l) \qquad \Delta H = -571.6\text{ kJ}$$

Exercise 5.15

Calculate the enthalpy change for the reaction

$$C_2H_4(g) + H_2(g) \longrightarrow C_2H_6(g) \qquad \Delta H = ?$$

given the following additional data.

$$C_2H_4(g) + 3\,O_2(g) \longrightarrow 2\,CO_2(g) + 2\,H_2O(l) \qquad \Delta H = -1410.9\text{ kJ}$$
$$C_2H_6(g) + \tfrac{7}{2}\,O_2(g) \longrightarrow 2\,CO_2(g) + 3\,H_2O(l) \qquad \Delta H = -1559.7\text{ kJ}$$

5.7 Standard Enthalpies of Formation

Enthalpy changes for some reactions can be determined by calorimetry. Other enthalpy changes can be *calculated* with Hess's law, using ΔH values that can be measured. It would be nice if we could calculate ΔH values directly, using the relationship

$$\Delta H = H_{\text{products}} - H_{\text{reactants}}$$

We can't do this because we can't obtain *absolute* values of enthalpies. On the other hand, though, isn't this situation similar to that of determining the gain in elevation in climbing a mountain? We can express the elevation gain in a climb as

$$\text{Elevation gain } (\Delta h) = \text{elevation}_{\text{top}} - \text{elevation}_{\text{base}}$$

Yet we don't have *absolute* values for the elevations either. We use instead the elevations *relative* to mean sea level, as illustrated in Figure 5.12.

Chemists have devised a similar scale of *relative* enthalpies called *enthalpies of formation* (see Figure 5.13). To understand how this scale works, we need first to describe so-called *standard states* for substances.

The **standard state** of a solid or liquid substance is the pure element or compound at 1 atm pressure and at the temperature of interest. For a gaseous substance the standard state is the (hypothetical) pure gas behaving as an ideal gas at 1 atm pressure and the temperature of interest. The **standard enthalpy of reaction** is the enthalpy *change* for a reaction in which the reactants in their

Most of the ΔH values given to this point in the chapter have been $\Delta H°$ values.

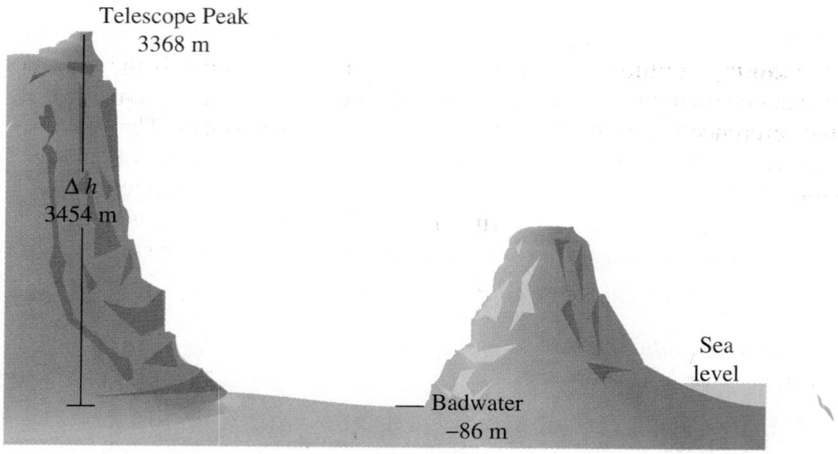

Figure 5.12 Determining an elevation gain: an analogy to ΔH_f°. Relative to a mean sea level that is *arbitrarily* set at zero, the elevation of Badwater is −86 m (86 meters below sea level) and that of Telescope Peak on the other side of Death Valley is 3368 m. The elevation gain in climbing from Badwater to the top of Telescope Peak is

$$\begin{aligned}\text{Elevation gain } (\Delta h) &= \text{elevation}_{\text{Telescope Peak}} - \text{elevation}_{\text{Badwater}} \\ &= 3368 \text{ m} - (-86 \text{ m}) \\ &= 3454 \text{ m}\end{aligned}$$

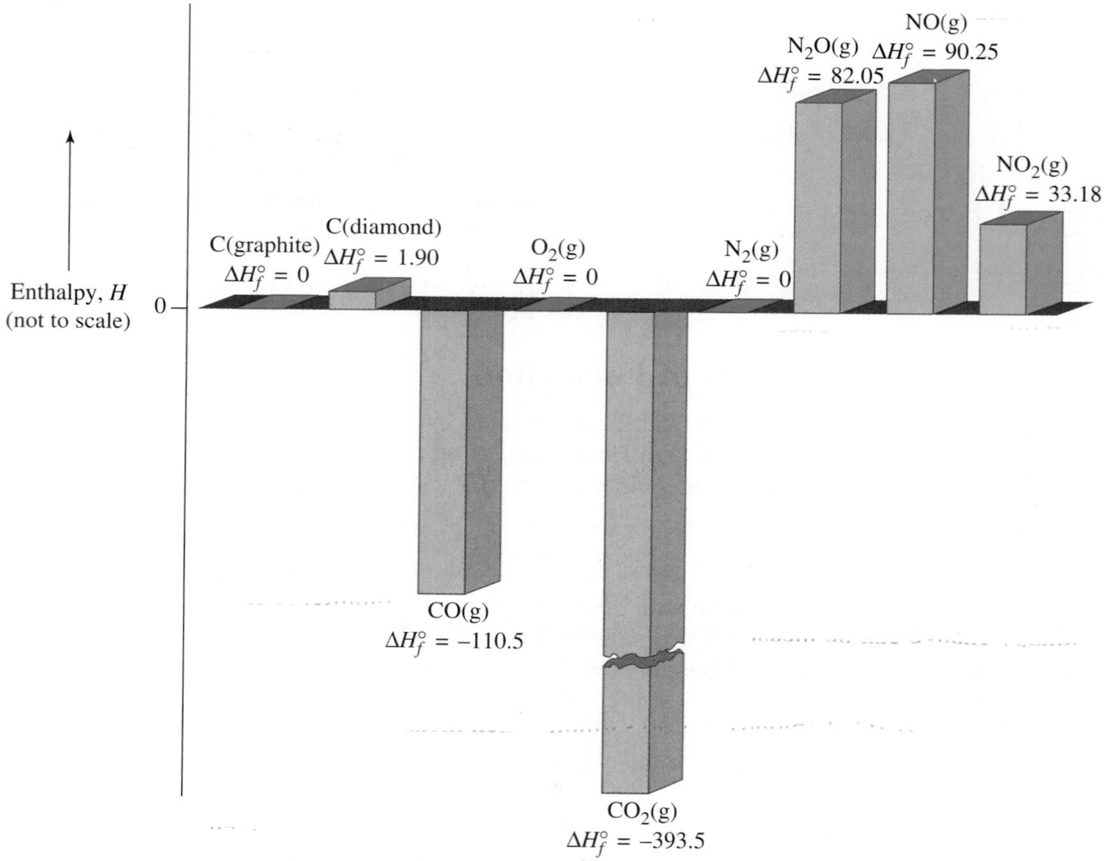

Figure 5.13 Enthalpies of formation of some oxides of carbon and nitrogen at 298 K. Values are given in kJ/mol. O_2(g), N_2(g), and graphite, the most stable form of carbon at 298 K and 1 atm, have the values $\Delta H_f^\circ = 0$. Diamond has a slightly positive ΔH_f°. The ΔH_f° values for the oxides of nitrogen are all positive. The reactions in which these compounds are formed from their elements are *endothermic*. The oxides of carbon have negative values of ΔH_f°; their formation reactions are *exothermic*.

standard states yield products in their standard states; it is denoted with a superscript degree symbol—that is, $\Delta H°$.

The **standard enthalpy of formation ($\Delta H_f°$)** of a substance is the enthalpy *change* that occurs in the formation of 1 mol of the substance in its standard state from the reference forms of its elements in their standard states. The reference forms of the elements are their most stable forms at the given temperature and 1 atm pressure. The superscript degree sign labels the enthalpy change as a standard enthalpy change, and the subscript "f" refers to the reaction in which a substance is *formed* from its elements. Because the formation of the most stable form of an element from itself is really no change at all, its $\Delta H_f°$ value = 0. That is,

The standard enthalpy of formation of a pure element in its most stable form is 0.

The standard enthalpy of formation, which is actually an enthalpy *change*, is often called the standard heat of formation, or more simply, the *heat of formation*.

Extensive tables of standard enthalpies of formation have been compiled, usually given at 298 K (25 °C) and expressed in kJ/mol. You will find a short list in Table 5.2 and a longer tabulation in Appendix D.

TABLE 5.2 Some Standard Enthalpies of Formation at 25 °C

Substance	$\Delta H_f°$, kJ/mol	Substance	$\Delta H_f°$, kJ/mol
$CO(g)$	−110.5	$HCl(g)$	−92.31
$CO_2(g)$	−393.5	$HF(g)$	−271.1
$CH_4(g)$	−74.81	$HI(g)$	26.48
$C_2H_2(g)$	226.7	$H_2O(g)$	−241.8
$C_2H_4(g)$	52.26	$H_2O(l)$	−285.8
$C_2H_6(g)$	−84.68	$NH_3(g)$	−46.11
$C_3H_8(g)$	−103.8	$NO(g)$	90.25
$C_4H_{10}(g)$	−125.7	$N_2O(g)$	82.05
$C_6H_6(l)$	48.99	$NO_2(g)$	33.18
$CH_3OH(l)$	−238.7	$N_2O_4(g)$	9.16
$CH_3CH_2OH(l)$	−277.7	$SO_2(g)$	−296.8
$HBr(g)$	−36.40	$SO_3(g)$	−395.7

What do we mean by the "most stable form" of an element? Two forms of solid carbon are readily attainable at 25 °C, diamond and graphite. A small enthalpy change can be measured when diamond is converted to graphite, so diamond and graphite cannot both be assigned an enthalpy of formation of zero. Graphite is the more stable form at 1 atm pressure, and so its $\Delta H_f° = 0$. For diamond, $\Delta H_f°$(C, diamond) = 1.90 kJ/mol, which means that

$$\text{C(graphite)} \longrightarrow \text{C(diamond)} \qquad \Delta H° = \Delta H_f°(\text{diamond}) = +1.90 \text{ kJ}$$

One way in which the graphite–diamond difference shows up is in the enthalpy changes for their combustion reactions. The enthalpy of combustion of graphite is equal to the enthalpy of formation of $CO_2(g)$; the enthalpy of combustion of diamond is not.

$$\text{C(graphite)} + O_2(g) \longrightarrow CO_2(g) \qquad \Delta H° = \Delta H_f°[CO_2(g)] = -393.5 \text{ kJ}$$

$$\text{C(diamond)} + O_2(g) \longrightarrow CO_2(g) \qquad \Delta H° = -395.4 \text{ kJ}$$

Can you see how to get the enthalpy of formation of diamond by applying Hess's law to these two equations?

The enthalpy diagram in Figure 5.13 should help you to picture enthalpies of formation. The enthalpies of formation of CO(g) and CO_2(g) are −110.5 kJ/mol and −393.5 kJ/mol, respectively. Their formation from the elements C(graphite) and O_2(g) are *exothermic* reactions; the conversion of graphite to diamond is an *endothermic* process.

Now let's explore how standard enthalpies of formation are used to calculate standard enthalpy changes of chemical reactions. Suppose we need to know the standard enthalpy change for the formation of dinitrogen tetroxide from nitrogen dioxide at 25 °C.

$$2\,NO_2(g) \longrightarrow N_2O_4(g) \qquad \Delta H° = ?$$

From Table 5.2 we can find the enthalpies of formation of these two gases.

$$(1)\ \tfrac{1}{2}N_2(g) + O_2(g) \longrightarrow NO_2(g) \qquad \Delta H° = \Delta H_f°[NO_2(g)] = 33.18\text{ kJ}$$

$$(2)\ N_2(g) + 2\,O_2(g) \longrightarrow N_2O_4(g) \qquad \Delta H° = \Delta H_f°[N_2O_4(g)] = 9.16\text{ kJ}$$

With Hess's law we can combine these two equations and arrive at the equation of interest to us. To do this we must first *reverse* and then *double* equation (1).

$$-2 \times \text{Eq (1)}:\ 2\,NO_2(g) \longrightarrow N_2(g) + 2\,O_2(g) \qquad \Delta H° = -2\,\Delta H_f°[NO_2(g)] = -2 \times 33.18 = -66.36\text{ kJ}$$

$$\text{Eq (2)}:\ N_2(g) + 2\,O_2(g) \longrightarrow N_2O_4(g) \qquad \Delta H° = \Delta H_f°[N_2O_4(g)] = 9.16\text{ kJ}$$

$$2\,NO_2(g) \longrightarrow N_2O_4(g) \qquad \Delta H° = -66.36 + 9.16 = -57.20\text{ kJ}$$

If we look carefully at this calculation, we see that it amounts to

$$\Delta H° = \Delta H_f°[N_2O_4(g)] - 2 \times \Delta H_f°[NO_2(g)] = 9.16\text{ kJ} - (2 \times 33.18)\text{ kJ} = -57.20\text{ kJ}$$

This is just a special case of a more general expression that allows us to relate a standard enthalpy of reaction and standard enthalpies of formation, *without* formally going through Hess's law.

$$\Delta H° = \sum \nu_p\,\Delta H_f°(\text{products}) - \sum \nu_r\,\Delta H_f°(\text{reactants})$$

The symbol Σ means *the sum of* and ν_p and ν_r are the coefficients of the substances as they appear in the balanced chemical equation.

Example 5.13 shows how this equation is used to calculate a standard enthalpy of reaction, $\Delta H°$, from tabulated standard enthalpies of formation, $\Delta H_f°$. Example 5.14 illustrates the reverse: how to establish an unknown standard enthalpy of formation from a measured standard enthalpy of reaction. With organic compounds, the measured $\Delta H°$ is often the enthalpy of combustion, $\Delta H°_{comb}$.

Example 5.13

Synthesis gas is a carbon monoxide–hydrogen mixture used to synthesize a variety of organic compounds, such as methanol. One reaction for producing synthesis gas is

$$3\,CH_4(g) + 2\,H_2O(l) + CO_2(g) \longrightarrow 4\,CO(g) + 8\,H_2(g) \qquad \Delta H° = ?$$

Use standard enthalpies of formation from Table 5.2 to calculate the standard enthalpy change for this reaction.

Solution

A convenient way to do this kind of calculation is to list under the formula of each substance in an equation the product of its coefficient and its ΔH_f° value.

$$\begin{array}{ccccc} 3\,CH_4(g) & +\ 2\,H_2O(l) & +\ CO_2(g) \longrightarrow & 4\,CO(g) & +\ 8\,H_2(g) \\ 3 \times (-74.81) & 2 \times (-285.8) & -393.5 & 4 \times (-110.5) & 8 \times 0 \end{array}$$

$$\begin{aligned} \Delta H^\circ &= \sum \nu_p\, \Delta H_f^\circ(\text{products}) - \sum \nu_r\, \Delta H_f^\circ(\text{reactants}) \\ &= (-4 \times 110.5) - [(-3 \times 74.81) + (-2 \times 285.8) + (-393.5)] \\ &= -442.0 - (-224.4 - 571.6 - 393.5) \\ &= -442.0 - (-1189.5) \\ &= -442.0 + 1189.5 = +747.5\ \text{kJ} \end{aligned}$$

A net chemical equation simply notes the initial and final conditions, but often there are rather specific requirements for bringing about a chemical reaction. This reaction will not occur just by passing methane and carbon dioxide gases into water.

Example 5.14

The combustion of isopropyl alcohol, common rubbing alcohol, is represented by the equation

$$2\,(CH_3)_2CHOH(l) + 9\,O_2(g) \longrightarrow 6\,CO_2(g) + 8\,H_2O(l) \qquad \Delta H^\circ = -4011\ \text{kJ}$$

Use this equation and data from Table 5.2 to establish the standard enthalpy of formation for isopropyl alcohol.

Solution

We can proceed as in Example 5.13, but here we must solve for an unknown standard enthalpy of formation, ΔH_f°.

$$\begin{array}{ccccc} 2\,(CH_3)_2CHOH(l) & +\ 9\,O_2(g) \longrightarrow & 6\,CO_2(g) & +\ 8\,H_2O(l) & \Delta H^\circ = -4011\ \text{kJ} \\ 2 \times \Delta H_f^\circ & 9 \times 0 & 6 \times (-393.5) & 8 \times (-285.8) & \end{array}$$

$$\begin{aligned} \Delta H^\circ &= \sum \nu_p\, \Delta H_f^\circ(\text{products}) - \sum \nu_r\, \Delta H_f^\circ(\text{reactants}) \\ -4011 &= [(-6 \times 393.5) + (-8 \times 285.8)] - 2 \times \Delta H_f^\circ \\ -4011 &= -2361 - 2286 - 2 \times \Delta H_f^\circ \\ \Delta H_f^\circ &= (-2361 - 2286 + 4011)/2 = -318\ \text{kJ/mol}\ (CH_3)_2CHOH \end{aligned}$$

Exercise 5.16

Ethylene, derived from petroleum, is used to make ethanol for use as a fuel or solvent. The chief reaction used is

$$C_2H_4(g) + H_2O(l) \longrightarrow CH_3CH_2OH(l)$$

Use data from Table 5.2 to calculate ΔH° for this reaction.

Exercise 5.17

The combustion of thiophene, a compound used in the manufacture of pharmaceuticals, is represented by the equation

$$C_4H_4S(l) + 6\,O_2(g) \longrightarrow 4\,CO_2(g) + 2\,H_2O(l) + SO_2(g) \qquad \Delta H^\circ = -2523\ \text{kJ}$$

Use this equation and data from Table 5.2 to establish the standard enthalpy of formation for thiophene.

CONCEPTUAL EXAMPLE 5.2

Which of these two substances should yield the greatest quantity of heat upon complete combustion, on a per mole basis: ethane, $C_2H_6(g)$, or ethanol, $CH_3CH_2OH(l)$? Use relevant standard enthalpies of formation from Table 5.2.

Solution

We could, of course, calculate the two heats of combustion in the manner of Example 5.13 and compare them. But the question is much more simple if we recognize this fact: Each combustion has precisely the same final state and enthalpy, that is, 2 mol $CO_2(g)$ + 3 mol $H_2O(l)$. The initial state is one mole of the compound and enough oxygen to complete the combustion. Because $\Delta H_f^\circ[O_2(g)] = 0$, the relative enthalpy of the initial state is simply that of one mole of the compound. The combustion reaction that gives the most negative value of ΔH°_{comb} is the one that starts with the compound having the higher (less negative) enthalpy of formation. The greater heat of combustion, that is, the more negative ΔH°_{comb}, is obtained with ethane. The sketch in Figure 5.14 should help you to visualize this conclusion.

Conceptual Exercise 5.2

Without doing a detailed calculation, determine which alcohol gives off the most heat upon combustion, on a per mole basis: $CH_3OH(l)$ or $CH_3CH_2OH(l)$? (Use data from Table 5.2.)

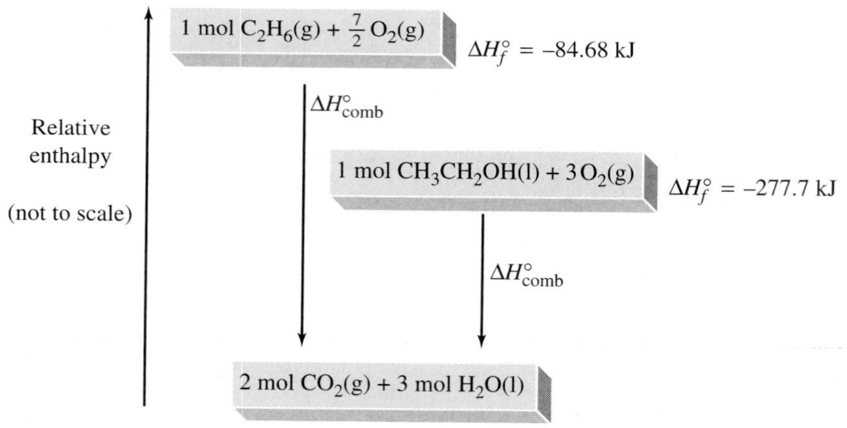

Figure 5.14 Conceptual Example 5.2 illustrated.

Looking Ahead

A reaction that occurs simply by bringing the reactants together under the appropriate conditions is said to be *spontaneous*. Most exothermic reactions are spontaneous. In these reactions enthalpy decreases. The tendency for the enthalpy of a system to be lowered is a powerful driving force, much like the decrease in the potential energy of water as it flows downhill. However, there is more to the matter. Some endothermic processes, like the dissolving of many solids in liquids, also occur spontaneously, even though they involve an enthalpy increase. We need to consider a second factor, called entropy, before we can fully answer the question of why some processes are spontaneous and others are not. We will do this in Chapter 19.

In describing spontaneous change we will also have to distinguish between the *tendency* for a reaction to occur and the *rate* at which a reaction occurs. Some reactions that should occur according to thermodynamics may occur too slowly to be observed. We will discuss rates of chemical reactions in Chapter 15.

5.8 Combustion and Respiration: Fuels and Foods

Two types of exothermic reactions are of utmost importance: (1) The combustion of fuels provides the energy that sustains our modern way of life. (2) The foods we eat are "burned" in the respiratory process to give us the energy to stay alive and perform our many activities.

Fossil Fuels: Coal, Natural Gas, and Petroleum

A *fuel* is a substance that burns with the release of heat. In much of the world today, wood is the principal fuel for cooking food and keeping people warm, just as it was in ancient times. The modern industrial world, however, is powered mainly by the fossil fuels coal, petroleum, and natural gas. These materials were formed over a period of millions of years from organic matter that became buried and compressed under mud and water. We use these fossil fuels today at a rate 50,000 times as fast as they were formed.

Coal is a complex organic material that also contains minerals. Several varieties of coal are listed in Table 5.3. The fuel value of coal lies mainly in its carbon content. Complete combustion of carbon produces carbon dioxide.

$$C(s) + O_2(g) \longrightarrow CO_2(g)$$

If the carbon in coal were graphite (it isn't), the combustion reaction would release 393.5 kJ of heat per mole, or

$$\frac{-393.5 \text{ kJ}}{1 \text{ mol C}} \times \frac{1 \text{ mol C}}{12.011 \text{ g C}} = -32.76 \text{ kJ/g C}$$

TABLE 5.3 **Approximate Composition (in Mass Percent) of Some Typical Coal Samples**

Grade of Coal	Carbon	Hydrogen	Oxygen	Nitrogen	Sulfur	Ash
Lignite (brown coal)	41.2	6.9	45.6	0.7	0.4	5.2
Subbituminous coal	53.9	6.9	33.4	1.0	0.5	4.3
Bituminous coal	76.7	4.9	6.2	1.5	0.8	9.9
Semianthracite coal	78.3	3.7	4.0	1.7	2.2	10.1
Anthracite coal	79.8	3.4	6.2	1.0	0.6	9.0

Burning fossil fuels has caused an increase in the concentration of carbon dioxide gas in Earth's atmosphere. This increase has contributed to a concern, called global warming, that we will discuss in Chapter 14.

High-grade coal yields about 30 kJ per gram. The burning of coal also produces carbon monoxide and soot (unburned carbon) from incomplete combustion. Mineral matter is left unburned as ashes and enters the atmosphere as fly ash. Sulfur oxides are formed from sulfur compounds in the coal, and nitrogen oxides are produced from air by the reaction of nitrogen and oxygen at the high temperatures at which coal burns.

Natural gas is mainly methane (CH_4), with some ethane (C_2H_6), propane (C_3H_8), and butane (C_4H_{10}). Complete combustion of methane gives carbon dioxide and water.

$$CH_4(g) + 2\,O_2(g) \longrightarrow CO_2(g) + 2\,H_2O(l)$$

The enthalpy change in the combustion of methane (the heat of combustion) is −890.3 kJ/mol or −55.50 kJ/g. Natural gas has a heat of combustion on a mass basis that is almost twice as great as that of coal. With far fewer impurities than coal, natural gas also burns more cleanly than coal. Some carbon monoxide, soot, and nitrogen oxides are formed, but there are only traces of sulfur oxides and no ash.

As we learned in Section 2.10, *petroleum* is an exceedingly complex mixture of hydrocarbons. Gasoline is the petroleum fraction most in demand. Like petroleum itself, gasoline is a mixture of hydrocarbons. Typical formulas range from C_5H_{12} to $C_{12}H_{26}$, with octane, C_8H_{18}, often used as a representative gasoline molecule. Complete combustion of octane yields carbon dioxide and water.

Isooctane, $(CH_3)_3CCH_2CH(CH_3)_2$, an excellent automotive fuel, is assigned an octane number of 100. Heptane, $CH_3(CH_2)_5CH_3$, is a poor fuel with an octane number of 0. Gasolines are rated against these two standards. For example, a gasoline with an octane number of 87 is equivalent to a mixture of 87% isooctane and 13% heptane.

$$C_8H_{18}(l) + \tfrac{25}{2}\,O_2(g) \longrightarrow 8\,CO_2(g) + 9\,H_2O(l) \qquad \Delta H^\circ = -5450\text{ kJ}$$

Petroleum products such as gasoline burn more cleanly than coal, although not as cleanly as natural gas.

Estimation Exercise 5.2

Heats of combustion of several alkanes are listed in Table 5.4. Calculate the heat of combustion of each, in kilojoules per mol C. Use these data to make a graph of the heat of combustion per mol C for the first eight alkanes (CH_4 through C_8H_{18}). Estimate the heat of combustion of decane ($C_{10}H_{22}$) and dodecane ($C_{12}H_{26}$).

TABLE 5.4 Heats of Combustion of Some Alkanes

Alkane		
Name	Formula	ΔH°, kJ/mol
Methane	$CH_4(g)$	−890
Ethane	$C_2H_6(g)$	−1560
Propane	$C_3H_8(g)$	−2220
Butane	$C_4H_{10}(g)$	−2879
Pentane	$C_5H_{12}(l)$	−3536
Hexane	$C_6H_{14}(l)$	−4163
Heptane	$C_7H_{16}(l)$	−4811
Octane	$C_8H_{18}(l)$	−5450

Foods: Fuels for the Body

Our bodies require food for the growth and repair of tissue and as a source of energy. The three principal classes of foods are carbohydrates, fats, and proteins. The body's preferred fuel is carbohydrates (starches and sugars). During digestion, starches and sugars are converted to a simple sugar: glucose, $C_6H_{12}O_6$.

If glucose is burned in a calorimeter, its heat of combustion at 25 °C is found to be −2803 kJ/mol or −15.56 kJ/g.

$$C_6H_{12}O_6(s) + 6\ O_2(g) \longrightarrow 6\ CO_2(g) + 6\ H_2O(l) \qquad \Delta H° = -2803\ kJ$$

When "burned" in the body, glucose produces the same products and yields about the same quantity of energy. It isn't exactly the same because the reactions occur at body temperature, about 37 °C, rather than 25 °C. It isn't exactly burned either. Rather, it is metabolized through a sequence of many reaction steps. Still, Hess's law applies. The enthalpy change is the same.

Fats are formed by the reaction of glycerol (a three-carbon alcohol with an —OH group on each carbon atom) with carboxylic acids known as fatty acids. Fats belong to a class of organic compounds called *esters*. A typical fat, glyceryl trilaurate, has the molecular formula $C_{39}H_{74}O_6$. Complete combustion gives carbon dioxide and water.

Fatty acids and esters are discussed further in Section 9.10.

$$C_{39}H_{74}O_6(s) + 54.5\ O_2(g) \longrightarrow 39\ CO_2(g) + 37\ H_2O(g) \qquad \Delta H° = -2.39 \times 10^4\ kJ$$

The heat of combustion is -2.39×10^4 kJ/mol $C_{39}H_{74}O_6$ or −37.4 kJ/g $C_{39}H_{74}O_6$. Note that, per gram, this heat of combustion is more than twice that of glucose.

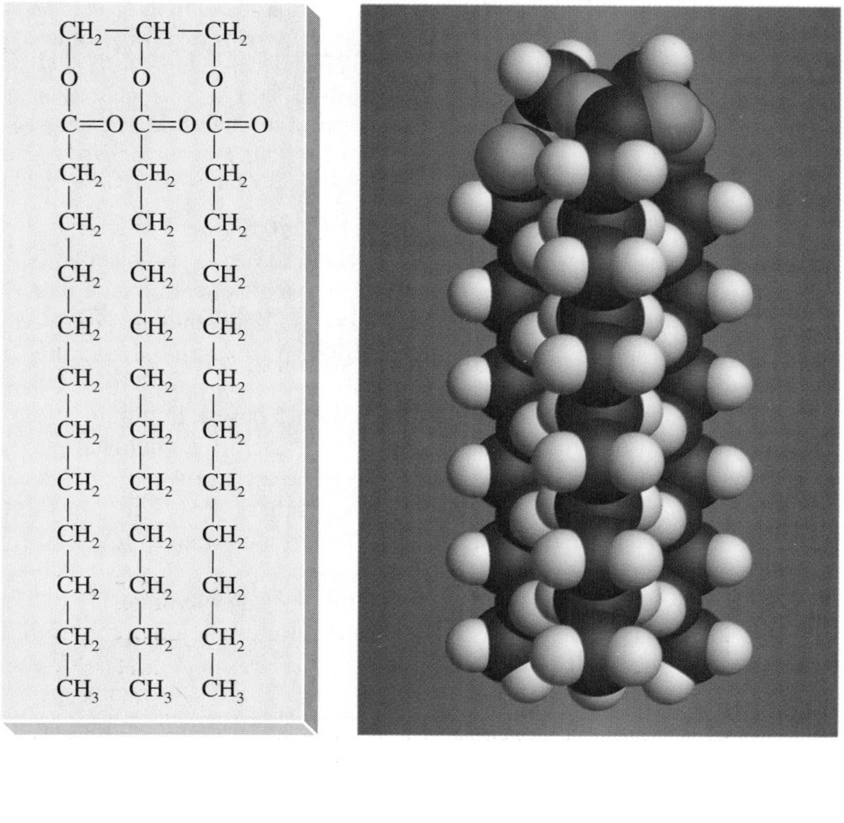

Glyceryl trilaurate, a typical fat. Glycerol provides the three-carbon backbone, and lauric acid molecules form three tails.

The Food Calorie

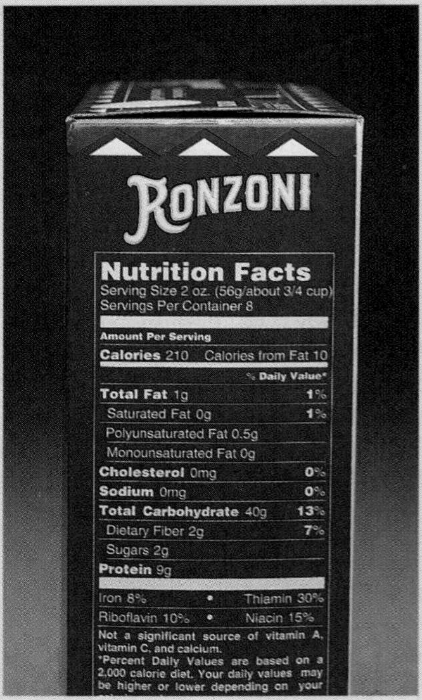

A nutritional label from a box of pasta. Can you verify, per serving, the total calories and the calories from fat shown on the label? Use other information in the label and in the text.

In nutrition the energy value of foods is measured in kilocalories.

$$1 \text{ kcal} = 1000 \text{ cal} = 4.184 \text{ kJ}$$

The energy value of glucose is 3.72 kcal/g, and that of glyceryl trilaurate is 8.94 kcal/g. In everyday life, people generally refer to the kilocalorie as the "Calorie."

$$1 \text{ kcal} = 1 \text{ Cal} = 1000 \text{ cal}$$

The capital C is used to distinguish the food calorie (a kilocalorie) from the calorie used in science. The term "large calorie" has also been used to designate the food calorie. People are often not careful to make a clear distinction between the two different meanings of "calorie," and we generally need to decide what is intended by the context in which the word is used.

In general, the energy value of carbohydrates and proteins is about 4.0 Cal/g and that of fats is more than twice as high, about 9.0 Cal/g.

Our daily requirement for protein is about 0.8 g per kilogram of body mass. A person who weighs 60 kg (132 lb) therefore requires about 48 g of protein a day. A 3-oz serving of lean meat, such as beef, chicken, or fish, supplies approximately 20 g of protein. Protein deficiency is rare in developed countries. The average American diet, for example, has more than twice the required quantity of protein.

Our bodies require some fat. Males should have a minimum of about 3% by mass of body fat; females about 10%. Dietary guidelines specify a maximum of 30% of calories from fats. The typical American diet has 34% of calories from fat; some have as much as 45%. Most of this fat comes from meats such as hamburgers, from foods fried in fats, and from snack foods such as potato chips and cookies.

There is no set dietary recommendation for carbohydrates, but at least 65%, and perhaps as much as 80%, of our calories should come from this food category.

This is true in general—fats yield over twice as much energy per gram as carbohydrates. Fats are the body's preferred material for energy storage. More energy can be packed into a given mass of fat than in the same mass of carbohydrate. The body can store at most about 500 g of carbohydrate as glycogen, a form of starch. The ability to store fat appears to be almost without limit.

Proteins are used mainly for structural materials such as muscle, connective tissue, skin, and hair, and for enzymes that regulate the many chemical reactions that occur in living cells. Proteins eaten in excess of the body's need for growth, repair, and replacement can be used for energy. Proteins have about the same energy value as carbohydrates.

SUMMARY

Thermochemistry concerns energy changes in physical processes or chemical reactions. Some of its basic ideas are the notion of a system and its surroundings; the concepts of kinetic energy, potential energy, thermal energy, and chemical energy; and the distinction between two types of energy exchanges, heat (q) and work (w).

The law of conservation of energy or the first law of thermodynamics is expressed as $\Delta E = q + w$. Internal energy (E) is a function of state; it has a unique value once the state or condition of a system is defined. The heat exchanged between a reaction mixture and its surroundings—the heat of reaction—is not a unique quantity. If a reaction is carried out at constant volume, $q_V = \Delta E$. At a constant pressure and with work limited to pressure–volume work, the heat of reaction is equal to the change in enthalpy: $q_P = \Delta H$. Enthalpy (H) is related to internal energy ($H = E + PV$) and, like internal energy, it is a function of state.

A reaction in which enthalpy decreases is *exothermic*. In an *endothermic* reaction enthalpy increases. The enthalpy change (ΔH) provides a conversion factor between the amount of a reactant or product and the quantity of heat lost or gained in the reaction.

A calorimeter is used to measure quantities of heat. The actual measurements are of masses and temperature changes. Also required are heat capacities and/or specific heats. Two common types of calorimeters are the bomb calorimeter, which yields $q_V = \Delta E$, and the "coffee cup" calorimeter, which yields $q_P = \Delta H$.

Because internal energy and enthalpy are state functions, the values of ΔE and ΔH do not depend on how a reaction is carried out (e.g., in one step or several). This makes it possible, through Hess's law, to evaluate ΔE or ΔH for one reaction from a knowledge of their values for other reactions.

The concepts of the standard state, a standard enthalpy change, $\Delta H°$, and a standard enthalpy of formation (heat of formation), $\Delta H_f°$, are important in thermochemical calculations, The standard enthalpy of formation of an element has a value of zero, and the standard enthalpy of formation of a compound is the standard enthalpy change of a reaction in which the compound is formed from its elements. Standard enthalpies of formation (usually listed at 25 °C) can be used to *calculate* standard enthalpies of reaction.

Some practical applications of thermochemistry deal with the heats of combustion of fossil fuels—coal, natural gas, and petroleum—and the energy content of foods: carbohydrates, fats, and proteins.

KEY TERMS

calorie (cal) (5.5)
calorimeter (5.5)
endothermic (5.4)
enthalpy (H) (5.4)
enthalpy change (ΔH) (5.4)
enthalpy diagram (5.4)
exothermic (5.4)
first law of thermodynamics (5.3)
heat (q) (5.2)
heat of reaction (5.4)
Hess's law (5.6)
internal energy (E) (5.2)
joule (J) (5.1)
kinetic energy (5.1)

law of conservation of energy (5.3)
molar heat capacity (5.5)
potential energy (5.1)
specific heat (5.5)
standard enthalpy of formation (ΔH_f°) (5.7)
standard enthalpy of reaction (ΔH°) (5.7)
standard state (5.7)
state function (5.3)
surroundings (5.2)
system (5.2)
thermochemistry (5.2)
work (5.2)

REVIEW QUESTIONS

1. What is a thermodynamic system? What are the surroundings?
2. Describe the three types of thermodynamic systems and give an example of each.
3. You are studying a sample of copper(II) sulfate solution in a beaker on a ring stand. The beaker is being heated by the flame of a Bunsen burner. What is a logical choice to describe as the system? What type of system is it? Describe the surroundings for your choice of system.
4. Describe the transfer of heat between a system and its surroundings. What is meant by thermal equilibrium?
5. Can a system do work and absorb heat at the same time? Can it do so while maintaining a constant internal energy?
6. Can a system do work and give off heat at the same time? Can it do so while maintaining a constant internal energy?
7. Define the term exothermic reaction. Give an example, and indicate the sign of ΔH.
8. Define the term endothermic reaction. Give an example, and indicate the sign of ΔH.
9. Why must we indicate the physical state of all reactants and products when calculating the enthalpy change for a reaction?
10. How is the enthalpy of a system related to its internal energy?
11. Under what conditions is the heat of reaction equal to the change in internal energy in a reaction?
12. Under what conditions is the heat of reaction equal to the change in enthalpy in a reaction?
13. Upon heating under constant pressure, solid calcium carbonate decomposes to solid calcium oxide and carbon dioxide gas. In one experiment, 17.8 kJ of heat is added to the system to decompose 0.100 mol $CaCO_3$. Is the reaction exothermic or endothermic? What is the value of ΔH for the reaction of one mole of $CaCO_3$?
14. Given that

$$P_4(s) + 6\,Cl_2(g) \longrightarrow 4\,PCl_3(g) \quad \Delta H = -1148 \text{ kJ}$$

what is the value of ΔH for the following reaction?

$$\tfrac{1}{4}\,P_4(s) + \tfrac{3}{2}\,Cl_2(g) \longrightarrow PCl_3(g)$$

15. Given that

$$N_2(g) + 3\,H_2(g) \longrightarrow 2\,NH_3(g) \quad \Delta H = -92.22 \text{ kJ}$$

what is the value of ΔH for the following reaction?

$$NH_3(g) \longrightarrow \tfrac{1}{2}\,N_2(g) + \tfrac{3}{2}\,H_2(g)$$

16. What is calorimetry? How does a calorimeter work?
17. Define heat capacity. Give some typical units that can be used to express it.
18. Define specific heat. Give some typical units that can be used to express it.
19. Describe how the specific heat of a substance can be determined.
20. Describe how enthalpy changes for chemical reactions are determined in a Styrofoam-cup calorimeter.
21. Describe how a bomb calorimeter works. What kind of reactions are carried out in bomb calorimeters?
22. State Hess's law of constant heat summation.
23. What do we mean when we say a substance is in its standard state?
24. What is meant by the standard molar enthalpy of formation of a substance?
25. Write the equation for a reaction which has as its heat of reaction the standard molar enthalpy of formation of $Fe_2O_3(s)$.
26. Consider the reactions at 25 °C represented by the following equations. For which of these is the enthalpy of reaction equal to a standard molar enthalpy of formation? Explain your reasoning in each case.
    a. $2\,Cu(s) + O_2(g) \longrightarrow 2\,CuO(s) \quad \Delta H^\circ = -314.6 \text{ kJ}$
    b. $Cu(s) + \tfrac{1}{2}\,O_2(g) \longrightarrow CuO(s) \quad \Delta H^\circ = -157.3 \text{ kJ}$
    c. $CO(g) + \tfrac{1}{2}\,O_2(g) \longrightarrow CO_2(g) \quad \Delta H^\circ = -283.0 \text{ kJ}$
    d. $\tfrac{1}{2}\,Br_2(g) + \tfrac{1}{2}\,Cl_2(g) \longrightarrow BrCl(g) \quad \Delta H^\circ = -16.31 \text{ kJ}$
27. Describe how a standard enthalpy of reaction can be calculated from tabulated standard enthalpies of formation, and how an unknown standard enthalpy of formation can be determined from a measured

standard enthalpy of reaction and tabulated standard enthalpies of formation.

28. What is a fuel? List the three principal fossil fuels and some advantages of gaseous and liquid fuels over solid fuels.

29. How is the food Calorie related to the calorie used in science?

30. List the three major classes of foods. Which type has the highest energy value per gram?

PROBLEMS

First Law of Thermodynamics

31. What is the change in internal energy of a system that *absorbs* 455 J of heat and does 325 J of work?

32. What is the change in internal energy of a system that has 625 J of work done on it and *gives off* 515 J of heat?

33. How much work, in joules, must be involved in a process in which a system *gives off* 58 cal of heat, if the internal energy of the system is to remain unchanged?

34. If ΔE for a system is 217 J in a process in which the system absorbs 185 cal of heat, how much work, in joules, is involved?

Enthalpy Changes

35. At high temperatures, water is decomposed to hydrogen and oxygen.

$$2\ H_2O(g) \longrightarrow 2\ H_2(g) + O_2(g)$$

Decomposition of 10.0 g H_2O at constant pressure requires that 134 kJ of heat be absorbed by the system. Is the reaction endothermic or exothermic? What is the value of q for the reaction, per mole of water? Is the value of q equal to ΔE or ΔH? Explain.

36. Combustion of 0.144 g of sucrose (table sugar, $C_{12}H_{22}O_{11}$) results in the release of 2.38 kJ of heat.

$$C_{12}H_{22}O_{11}(s) + 12\ O_2(g) \longrightarrow 12\ CO_2(g) + 11\ H_2O(l)$$

Is the reaction endothermic or exothermic? What is the value of q for the reaction, per mole of sucrose? How would you expect the values of ΔE and ΔH to compare for this reaction? Explain.

37. When 0.0500 mol of solid calcium carbonate is heated in air, it decomposes to solid calcium oxide and carbon dioxide gas; 8.90 kJ of heat is absorbed. Write a chemical equation for the decomposition of one mole of calcium carbonate, including the physical states of all the substances and the value of ΔH.

38. When 0.200 mol of ethane gas is burned in excess oxygen, 312 kJ of heat is evolved. Write a chemical equation for the combustion of one mole of ethane, including the physical states of all the substances and the value of ΔH. (*Hint*: What are the products of the reaction and in what forms do they appear at 25 °C and 1 atm pressure?)

39. Given the reaction

$$3\ Fe_2O_3(s) + CO(g) \longrightarrow 2\ Fe_3O_4(s) + CO_2(g) \qquad \Delta H° = -46\ \text{kJ}$$

Calculate $\Delta H°$ for the following reactions.

a. $Fe_2O_3(s) + \frac{1}{3}\ CO(g) \longrightarrow \frac{2}{3}\ Fe_3O_4(s) + \frac{1}{3}\ CO_2(g)$

b. $Fe_3O_4(s) + \frac{1}{2}\ CO_2(g) \longrightarrow \frac{3}{2}\ Fe_2O_3(s) + \frac{1}{2}\ CO(g)$

40. Given the reaction

$$2\ Na_2O_2(s) + 2\ H_2O(l) \longrightarrow 4\ NaOH(s) + O_2(g) \qquad \Delta H° = -109\ \text{kJ}$$

Calculate $\Delta H°$ for the following reactions.

a. $Na_2O_2(s) + H_2O(l) \longrightarrow 2\ NaOH(s) + \frac{1}{2}\ O_2(g)$

b. $NaOH(s) + \frac{1}{4}\ O_2(g) \longrightarrow \frac{1}{2}\ Na_2O_2(s) + \frac{1}{2}\ H_2O(l)$

Calculation of Heats of Reaction

41. Calcium oxide (lime) reacts with water to form calcium hydroxide (slaked lime).

$$CaO(s) + H_2O(l) \longrightarrow Ca(OH)_2(s) \qquad \Delta H° = -65.2\ \text{kJ}$$

How much heat, in joules, is evolved when 0.333 mol $Ca(OH)_2(s)$ is formed?

42. Calcium carbide (CaC_2) can be made by heating calcium oxide (lime) with carbon (charcoal).

$$CaO(s) + 3\ C(s) \longrightarrow CaC_2(s) + CO(g) \qquad \Delta H° = +464.8\ \text{kJ}$$

How much heat, in joules, is absorbed in a reaction in which 2.79 mol C(s) is consumed under these conditions?

43. Iron(III) oxide is reduced to iron metal by reaction with carbon monoxide.

$$Fe_2O_3(s) + 3\ CO(g) \longrightarrow 2\ Fe(s) + 3\ CO_2(g) \qquad \Delta H° = -24.8\ \text{kJ}$$

How many kilograms of CO are consumed when 429 kJ of heat is released under these conditions?

44. Calcium carbide (CaC_2) reacts with water to form acetylene (C_2H_2), a gas used as a fuel in welding.

$$CaC_2(s) + 2\ H_2O(l) \longrightarrow C_2H_2(g) + Ca(OH)_2(s) \qquad \Delta H° = -128.0\ kJ$$

How many kilograms of CaC_2 are consumed in a reaction that releases 3.64×10^4 kJ of heat under these conditions?

Calorimetry: Heat Capacity and Specific Heat

(Use data from Table 5.1, as necessary.)

45. Calculate the heat capacity of a piece of iron if a temperature rise from 18 °C to 45 °C requires 112 J of heat.

46. Calculate the heat capacity of a sample of radiator coolant if a temperature rise from –5 °C to 142 °C requires 932 J of heat.

47. How much heat, in kilojoules, is required to raise the temperature of **(a)** 20.0 g of water from 20.0 °C to 96.0 °C; **(b)** 120.0 g of ethanol from –10.5 °C to 44.5 °C?

48. How much heat, in kilojoules, is released when the temperature of **(a)** 47.0 g of water drops from 45.4 °C to 10.0 °C; **(b)** when 209 g of iron drops from 400.0 °C to 22.6 °C?

49. A 48.7-g block of lead initially at 27.0 °C absorbs 93.5 J of heat. What is the final temperature of the lead?

50. A 454-g iron block initially at 16 °C absorbs 63.9 kJ of heat. What is the final temperature of the iron?

51. A 10.25-g sample of a metal alloy is heated to 99.10 °C and is then quickly dropped into 20.0 g of water in a calorimeter. The water temperature rises from 18.51 °C to 22.03 °C. Calculate the specific heat of the alloy.

52. A 2.05-g sample of a metal alloy wire is heated to 98.88 °C. It is then quickly dropped into 28.0 g of water in a calorimeter. The water temperature rises from 19.73 °C to 21.23 °C. Calculate the specific heat of the alloy.

53. A 1.35-kg piece of iron (sp. heat = $0.449\ J \cdot g^{-1} \cdot °C^{-1}$) is quickly dropped into 0.817 kg of water, and the water temperature rises from 23.3 °C to 39.6 °C. What must have been the initial temperature of the iron?

54. A piece of stainless steel (sp. heat = $0.50\ J \cdot g^{-1} \cdot °C^{-1}$) is taken from an oven at 178 °C and quickly immersed in 225 mL of water at 25.9 °C. The water temperature rises to 42.4 °C. What is the mass of the piece of steel? How precise is this method of mass determination? Explain.

Calorimetry: Measuring Heats of Reaction

55. A 500.0-mL sample of 0.500 M NaOH at 20.00 °C is mixed with an equal volume of 0.500 M HCl at the same temperature in a Styrofoam-cup calorimeter. The reaction

$$HCl(aq) + NaOH(aq) \longrightarrow NaCl(aq) + H_2O(l)$$

takes place and the temperature rises to 23.21 °C. Calculate the enthalpy change for the reaction. (You may make the same assumptions as in Example 5.10.)

56. A 65.0-mL sample of 0.600 M HI at 18.46 °C is mixed with 84.0 mL of a solution containing excess potassium hydroxide at 18.46 °C in a Styrofoam-cup calorimeter. The reaction

$$HI(aq) + KOH(aq) \longrightarrow KI(aq) + H_2O(l)$$

takes place and the temperature rises to 21.96 °C. Calculate the enthalpy change for the reaction. (You may make the same assumptions as in Example 5.10.)

57. A 0.309-g sample of coal reacts with excess oxygen in a bomb calorimeter. The heat capacity of the calorimeter is 4.62 kJ/°C. The water and bomb are initially in thermal equilibrium at 20.45 °C. The coal is ignited and burns, causing the water temperature to rise to 22.28 °C. Calculate the heat of reaction, in kilojoules per gram of coal. Is this heat of reaction equal to ΔH for the reaction? If not, is it much different from ΔH? Explain.

58. A 0.196-g sample of gasoline reacts with excess oxygen in a bomb calorimeter. The heat capacity of the calorimeter is 5.01 kJ/°C. The water and bomb are initially in thermal equilibrium at 22.75 °C. The gasoline is ignited and burns, causing the water temperature to rise to 24.50 °C. Calculate the heat of reaction, in kilojoules per gram of gasoline. Is this heat of reaction equal to ΔH for the reaction? If not, is it much different from ΔH? Explain.

59. A pure substance of known heat of combustion can be used to determine the heat capacity of a bomb calorimeter. When 2.00 g of sucrose is burned in a particular calorimeter, the temperature rises from 22.83 °C to 25.67 °C. What is the heat capacity of the calorimeter? The heat of combustion of sucrose (q_V) is –16.5 kJ/g.

60. Benzoic acid (C_6H_5COOH) is sometimes used as a standard to determine the heat capacity of a bomb calorimeter. When 1.22 g C_6H_5COOH is burned in a calorimeter that is being calibrated, the temperature rises from 21.13 °C to 22.93 °C. What is the heat capacity of the calorimeter? The heat of combustion of benzoic acid (q_V) is −26.42 kJ/g.

Hess's Law of Constant Heat Summation

61. Use the following equations

$$N_2(g) + 2\,O_2(g) \longrightarrow N_2O_4(g) \qquad \Delta H° = +9.2\text{ kJ}$$
$$N_2(g) + 2\,O_2(g) \longrightarrow 2\,NO_2(g) \qquad \Delta H° = +33.2\text{ kJ}$$

to calculate the enthalpy change for the reaction

$$2\,NO_2(g) \longrightarrow N_2O_4(g) \qquad \Delta H° = ?$$

62. Use the following equations

$$C(\text{graphite}) + O_2(g) \longrightarrow CO_2(g) \qquad \Delta H° = -393.5\text{ kJ}$$
$$2\,CO(g) + O_2(g) \longrightarrow 2\,CO_2(g) \qquad \Delta H° = -566.0\text{ kJ}$$

to calculate the enthalpy change for the reaction

$$2\,C(\text{graphite}) + O_2(g) \longrightarrow 2\,CO(g) \qquad \Delta H° = ?$$

63. Use the following equations

$$C_3H_8(g) + 5\,O_2(g) \longrightarrow 3\,CO_2(g) + 4\,H_2O(l) \qquad \Delta H° = -2219.9\text{ kJ}$$
$$CO(g) + \tfrac{1}{2}\,O_2(g) \longrightarrow CO_2(g) \qquad \Delta H° = -283.0\text{ kJ}$$

to calculate the enthalpy change for the reaction

$$C_3H_8(g) + \tfrac{7}{2}\,O_2(g) \longrightarrow 3\,CO(g) + 4\,H_2O(l) \qquad \Delta H° = ?$$

64. Use the following equations

$$N_2H_4(l) + O_2(g) \longrightarrow N_2(g) + 2\,H_2O(l) \qquad \Delta H° = -622.2\text{ kJ}$$
$$H_2(g) + \tfrac{1}{2}\,O_2(g) \longrightarrow H_2O(l) \qquad \Delta H° = -285.8\text{ kJ}$$
$$H_2(g) + O_2(g) \longrightarrow H_2O_2(l) \qquad \Delta H° = -187.8\text{ kJ}$$

to calculate the enthalpy change for the reaction

$$N_2H_4(l) + 2\,H_2O_2(l) \longrightarrow N_2(g) + 4\,H_2O(l) \qquad \Delta H° = ?$$

Standard Enthalpies of Formation

65. Use standard enthalpies of formation from Appendix D to calculate the standard enthalpy change for each of the following reactions.
a. $NH_3(g) + HCl(g) \longrightarrow NH_4Cl(s)$
b. $NH_3(g) + HNO_3(l) \longrightarrow NH_4NO_3(s)$
c. $MgCl_2(s) + Ca(s) \longrightarrow Mg(s) + CaCl_2(s)$
d. $FeO(s) + CO(g) \longrightarrow Fe(s) + CO_2(g)$

66. Use standard enthalpies of formation from Appendix D to calculate the standard enthalpy change for each of the following reactions.
a. $Cl_2(g) + I_2(s) \longrightarrow 2\,ICl(g)$
b. $NO(g) + O_3(g) \longrightarrow NO_2(g) + O_2(g)$
c. $Mg(s) + 2\,HCl(g) \longrightarrow MgCl_2(s) + H_2(g)$
d. $3\,C_2H_2(g) \longrightarrow C_6H_6(l)$
e. $2\,CH_3CH_2OH(l) + O_2(g) \longrightarrow 2\,CH_3CHO(g) + 2\,H_2O(l)$

67. Use the following equation and data from Appendix D to calculate the enthalpy of formation, per mole, of ZnS(s).

$$2\,ZnS(s) + 3\,O_2(g) \longrightarrow 2\,ZnO(s) + 2\,SO_2(g) \qquad \Delta H° = -878.2\text{ kJ}$$

68. Use the following equation and data from Appendix D to calculate the enthalpy of formation, per mole, of sucrose, $C_{12}H_{22}O_{11}(s)$.

$$C_{12}H_{22}O_{11}(s) + 12\,O_2(g) \longrightarrow 12\,CO_2(g) + 11\,H_2O(l) \qquad \Delta H° = -5.65 \times 10^3\text{ kJ}$$

69. When it undergoes complete combustion in oxygen, a 1.050-g sample of the industrial solvent diethylene glycol, $C_4H_{10}O_3$, gives off 23.50 kJ of heat to the surroundings. Calculate the standard enthalpy of formation of diethylene glycol. Assume that the initial reactants and the products of the combustion are at 25 °C and 1 atm pressure. (*Hint*: Write an equation for the reaction and use data from Table 5.2.)

70. When it undergoes complete combustion in excess oxygen, a 658.0-mg sample of adipic acid, $HOOC(CH_2)_4COOH$, a substance used in the manufacture of nylon, gives off 12.63 kJ of heat to the surroundings. Calculate the standard enthalpy of formation of adipic acid. Assume that the initial reactants and the products of the combustion are at 25 °C and 1 atm pressure. (*Hint*: Write an equation for the reaction and use data from Table 5.2.)

Fuels and Foods

71. Without doing detailed calculations, determine which of the alkanes listed in Table 5.4 has the greatest heat of combustion **(a)** per mole of alkane; **(b)** per gram of alkane; **(c)** per gram of carbon content; **(d)** per gram of $CO_2(g)$ produced.

72. A person on a 2500-Cal low-fat diet attempts to maintain a fat intake of no more than 20% of food

Calories from fat. What percent of the daily fat allowance is represented by a tablespoon of peanut butter (15.0 g and 50.1% fat)? Recall that fat has a food value of about 9.0 Cal/g.

73. Verify the claim by sugar manufacturers that a teaspoon (about 4.8 g) of sugar (sucrose, $C_{12}H_{22}O_{11}$) "contains only 19 Cal." Use the heat of combustion from Problem 68, and recall the definition of a food Calorie.

ADDITIONAL PROBLEMS

74. Which of the following quantities of energy is the largest: (a) the kinetic energy of a hydrogen molecule at STP; (b) the kinetic energy of a 1.0-g "BB" shot traveling with a speed of 100 m/s; (c) the heat required to raise the temperature of 10 mL of water from 20 °C to 21 °C.

75. The internal energy of a fixed quantity of an ideal gas depends only on its temperature. A sample of an ideal gas is allowed to expand at a *constant* temperature.

a. Does the gas do work?
b. Is any heat exchanged with the surroundings?
c. What is ΔE for the gas?

76. A 225-mL sample of water at 35.4 °C is added to 334 mL of water at 20.7 °C in a thermally insulated container. What will be the final water temperature?

77. Calculate the final temperature for the situation described in Conceptual Example 5.1 on page 171.

78. A British thermal unit (Btu) is defined as the quantity of heat required to change the temperature of 1 lb of water by 1 °F. Assuming the specific heat of water to be independent of temperature, how much heat is required to raise the temperature of the water in a 30-gal water heater from 66 °F to 145 °F: **(a)** in Btu; **(b)** in kcal; **(c)** in kJ?

79. Out of concern that the greenhouse effect (Chapter 14) will alter Earth's climate, it is desirable to calculate the amount of carbon dioxide produced by various fuels per unit of energy. For each of the following fuels, calculate the mass of CO_2 produced per kJ of heat released: hydrogen ($\Delta H^\circ_{comb} = -286$ kJ/mol); methane (−890 kJ/mol); methanol (−726 kJ/mol); octane (−5450 kJ/mol).

80. The combustion of methane is represented by the equation

$$CH_4(g) + 2\,O_2(g) \longrightarrow CO_2(g) + 2\,H_2O(l) \qquad \Delta H = -890.3\text{ kJ}$$

a. What mass of $CH_4(g)$ must be burned to give off 1.00×10^5 kJ of heat?
b. What quantity of heat is given off by the combustion of 105 L $CH_4(g)$, measured at 23 °C and 746 mmHg?

81. The complete combustion of octane in oxygen is represented by the equation

$$C_8H_{18}(l) + 12.5\,O_2(g) \longrightarrow 8\,CO_2(g) + 9\,H_2O(l) \qquad \Delta H^\circ = -5.45 \times 10^3\text{ kJ}$$

How much heat is liberated per gallon of $C_8H_{18}(l)$ burned? [1 gal = 3.785 L; density of $C_8H_{18}(l)$ = 0.703 g/mL.]

82. The combustion of hydrogen–oxygen mixtures is used to produce very high temperatures (about 2500 °C) needed for certain types of welding. Consider the combustion reaction to be

$$H_2(g) + \tfrac{1}{2}O_2(g) \longrightarrow H_2O(g) \qquad \Delta H^\circ = -241.8\text{ kJ}$$

How much heat is evolved when a 100.0-g mixture containing 15.5% $H_2(g)$, by mass, is burned? (*Hint*: What is the limiting reactant?)

83. The thermite reaction is highly exothermic.

$$Fe_2O_3(s) + 2\,Al(s) \longrightarrow Al_2O_3(s) + 2\,Fe(s) \qquad \Delta H^\circ = -852\text{ kJ}$$

The reaction is started in a room-temperature (25 °C) mixture of 1.00 mol $Fe_2O_3(s)$ and 2.00 mol Al(s). The liberated heat is retained within the products, whose combined specific heat over a broad temperature range is about 0.8 $J \cdot g^{-1} \cdot {}^\circ C^{-1}$. Show that the quantity of heat liberated is sufficient to raise the temperature of the products to the melting point of iron (1530 °C).

84. The composition of a particular natural gas, expressed on a mole fraction basis, is CH_4, 0.830; C_2H_6, 0.112; C_3H_8, 0.058. A 215-L sample of this natural gas, measured at 24.5 °C and 744 mmHg, is burned in an excess of oxygen. How much heat is evolved in the combustion? (*Hint*: What are the heats of combustion of the individual gases?)

85. A 1.108-g sample of naphthalene, $C_{10}H_8(s)$, is burned in a bomb calorimeter assembly and a temperature increase of 5.92 °C is noted. When a 1.351-g sample of thymol, $C_{10}H_{14}O(s)$ (a preservative and mold and mildew inhibitor), is burned in the same calorimeter assembly, the temperature increase is 6.74 °C. If the heat of combustion of naphthalene is −5153.5 kJ/mol $C_{10}H_8$,

what is the heat of combustion of thymol, in kJ/mol $C_{10}H_{14}O$?

86. A 1.148-g sample of benzoic acid is burned in an excess of oxygen in a bomb calorimeter. The temperature of the water rises from 24.96 °C to 30.25 °C. The heat of combustion of benzoic acid is −26.42 kJ/g. In a second experiment, a 0.895-g powdered coal sample is burned in the same calorimeter assembly. The temperature of the water rises from 24.98 °C to 29.73 °C. What mass of this coal, in kg, would have to be burned to liberate 1.00×10^9 kJ of heat?

87. Care must be taken in preparing solutions of solutes that liberate heat on dissolving. The heat of solution of NaOH is −42 kJ/mol NaOH. What will be the approximate maximum temperature reached in the preparation of 500.0 mL of 6.0 M NaOH from NaOH(s) and water?

88. The heat of neutralization of HCl(aq) by NaOH(aq) is −55.90 kJ/mol H_2O produced. If 50.00 mL of 1.16 M NaOH is added to 25.00 mL of 1.79 M HCl, with both solutions originally at 25.15 °C, what will be the final solution temperature? Assume a density of 1.00 g/mL and a specific heat of 4.18 J • g^{-1} • °C^{-1} for all solutions. (*Hint*: What is the chemical reaction that occurs? What is the limiting reactant?)

89. Calculate the standard enthalpy change for the reaction

$$BrCl(g) \longrightarrow Br(g) + Cl(g) \qquad \Delta H° = ?$$

by using the following data:

$$Br_2(l) \longrightarrow Br_2(g) \qquad \Delta H° = +30.91\text{ kJ}$$
$$Br_2(g) \longrightarrow 2\,Br(g) \qquad \Delta H° = +192.9\text{ kJ}$$
$$Cl_2(g) \longrightarrow 2\,Cl(g) \qquad \Delta H° = +243.4\text{ kJ}$$
$$Br_2(l) + Cl_2(g) \longrightarrow 2\,BrCl(g) \qquad \Delta H° = +29.2\text{ kJ}$$

90. Substitute natural gas (SNG) is a gaseous mixture containing $CH_4(g)$ that can be used as a fuel. One reaction for the production of SNG is

$$4\,CO(g) + 8\,H_2(g) \longrightarrow 3\,CH_4(g) + CO_2(g) + 2\,H_2O(l) \qquad \Delta H° = ?$$

Show how the following data, as necessary, can be used to determine $\Delta H°$ for this SNG reaction.

$$C(\text{graphite}) + \tfrac{1}{2}O_2(g) \longrightarrow CO(g) \qquad \Delta H° = -110.5\text{ kJ}$$
$$CO(g) + \tfrac{1}{2}O_2(g) \longrightarrow CO_2(g) \qquad \Delta H° = -283.0\text{ kJ}$$
$$H_2(g) + \tfrac{1}{2}O_2(g) \longrightarrow H_2O(l) \qquad \Delta H° = -285.8\text{ kJ}$$
$$C(\text{graphite}) + 2\,H_2(g) \longrightarrow CH_4(g) \qquad \Delta H° = -74.81\text{ kJ}$$

Compare this result with that obtained by using enthalpy of formation data directly on the SNG reaction.

91. Which of the following combustion reactions would you expect to liberate the greater amount of heat? What is the difference, in kJ/mol $C_{12}H_{26}$, in the enthalpies of combustion of these two reactions?

(a) $2\,C_{12}H_{26}(l) + 37\,O_2(g) \longrightarrow 24\,CO_2(g) + 26\,H_2O(l)$

(b) $2\,C_{12}H_{26}(l) + 37\,O_2(g) \longrightarrow 24\,CO_2(g) + 26\,H_2O(g)$

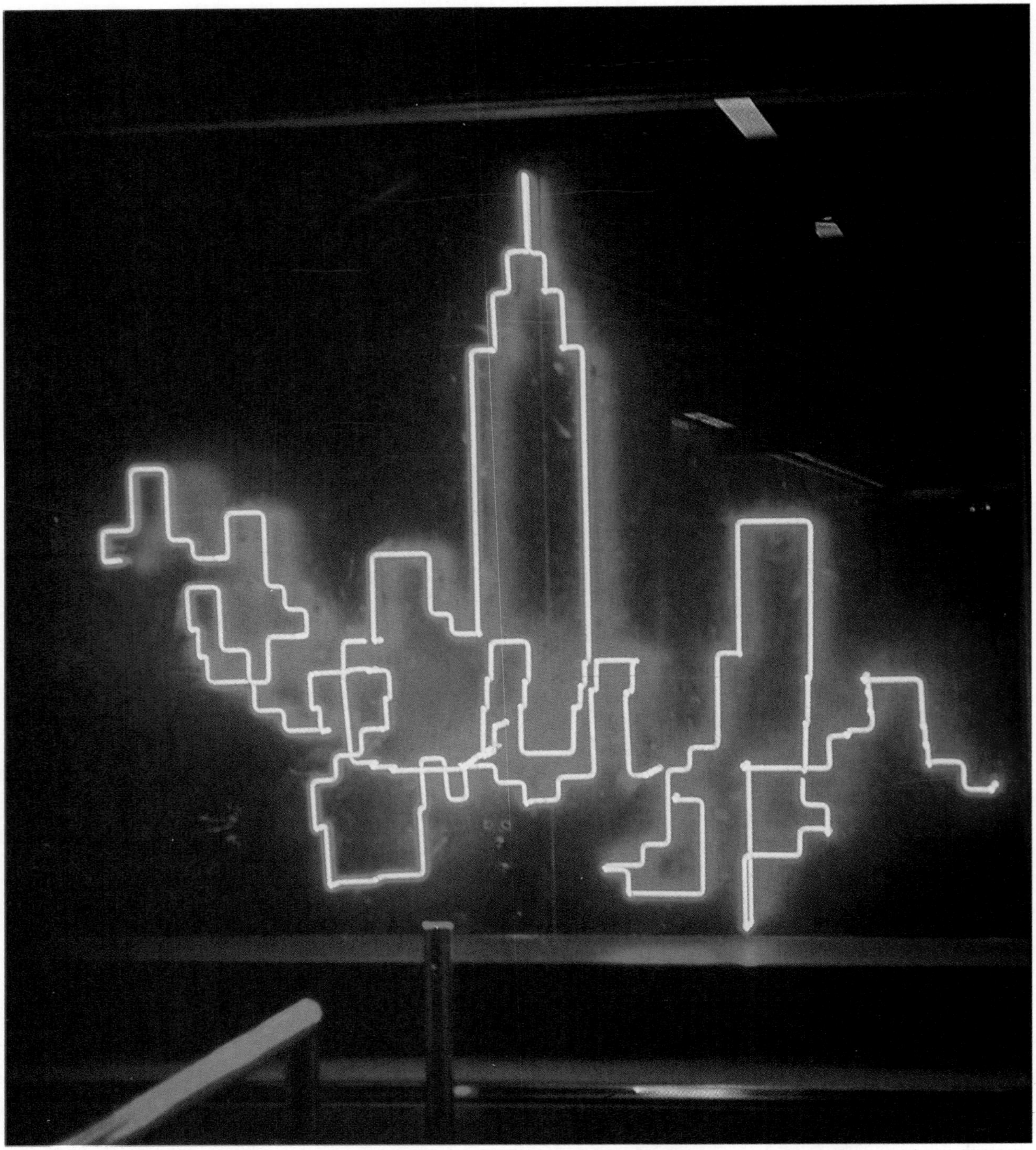

The red light emitted by neon signs and this neon sculpture arises from changes that occur in neon atoms when an electric discharge is passed through the gas. The study of light emission stimulated the development of theories of atomic structure.

6

Atomic Structure

CONTENTS

Some discoveries in chemistry have had a great impact on other disciplines. For example, chemists have determined the structures of proteins and other biochemical substances, and this knowledge has enabled biologists to study life processes at the molecular level. By the same token, discoveries in other sciences have had dramatic effects on chemistry. Sometimes discoveries in two or more areas come together at the right time and in a way that changes forever how we look at the world. In this chapter we resume the story of the atom begun in Chapter 2 by considering developments in physics that have profoundly changed the science of chemistry.

6.1 The Electron

In Section 2.3 we identified the three subatomic particles of primary interest to chemists—protons, neutrons, and electrons—and described them briefly. Here we show how knowledge about electrons has been greatly expanded through a small number of critical experiments.

Static electricity, such as that produced by running a comb through your hair or walking across a carpet in rubber-soled shoes, has been known since ancient times. Current electricity, the familiar kind that flows through wires, was discovered early in the nineteenth century. In 1800, Alessandro Volta (1745–1827) invented a device called a *voltaic pile* that converted chemical energy into electricity. In principle, his invention was much like a modern electric battery. If the electrodes (terminals) of a battery are connected by a wire, electricity flows through the wire. The electric current is sustained by chemical reactions inside the battery. Volta's invention quickly found its way into many areas of science and everyday life.

You may find it helpful at this time to read the portions of Appendix B that deal with electricity and magnetism.

Humphry Davy (1778–1829) promoted nitrous oxide ("laughing gas") as a chemical anesthetic, invented a safety lamp for coal miners, and proposed that hydrogen is the common element in all acids.

Electrolysis

Soon after Volta's invention, Humphry Davy, a British chemist, built a powerful battery that he used to pass electricity through molten ionic compounds. He quickly isolated several elements—potassium and sodium in 1807 and magnesium, strontium, barium, and calcium in 1808. The science of *electrochemistry* was born.

Davy's protégé, Michael Faraday, greatly extended this new science and defined many of the terms we still use today. Figure 6.1 illustrates *electrolysis*, the decomposition of a compound by electricity. The compound, called an *electrolyte*, is in the molten state or in a liquid solution. Electric current is carried to and from the liquid by metal strips or carbon rods called *electrodes*. Faraday hypothesized that electric current is carried through an electrolyte by charged particles called *ions*. In electrolysis, *anions*, having negative charges, are attracted to the positively charged electrode called the *anode*. *Cations* are positively charged, so they are attracted to the negatively charged electrode called the *cathode*.

Faraday also attempted to pass electricity through air and other gases at very low pressures (see Figure 6.2). Although he did discover some new phenomena, Faraday was not generally successful in this line of research. The vacuum pumps he used to lower gas pressures were not good enough. Later in the century, better vacuum pumps and more powerful sources of electricity led to a reopening of the study of electric discharges through gases, and with some dramatic results.

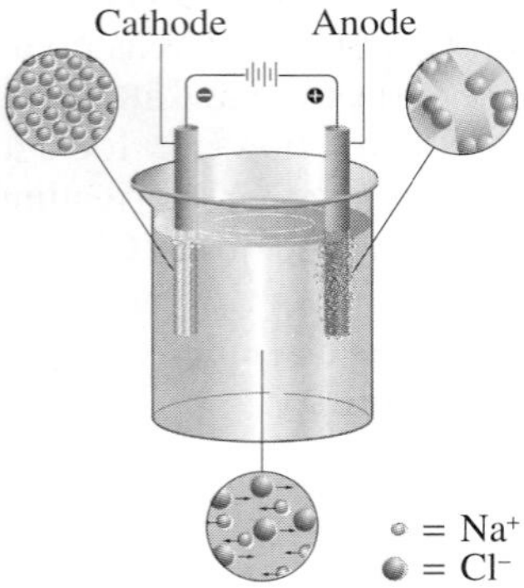

Figure 6.1 Electrolysis of molten sodium chloride. When electric current is passed through molten NaCl (an electrolyte), Na^+ ions (cations) are attracted to the negative electrode—the *cathode*. At the same time, Cl^- ions (anions) are attracted to the positive electrode—the *anode*. At the cathode, Na^+ ions pick up electrons to become atoms of liquid sodium metal, which floats on the molten sodium chloride. At the anode, Cl^- ions give up electrons to become chlorine atoms, which join to form chlorine molecules (Cl_2) that escape as a gas.

Cathode Rays

In 1879, the English scientist William Crookes (1832–1919) experimented with gas discharge tubes like the one pictured in Figure 6.3. Electric current in the tube was concentrated into a narrow beam as it passed through a metal slit. The current then passed along a zinc sulfide coated screen that gave off a green fluorescence, thus making the path of the electric current visible. The beam, which

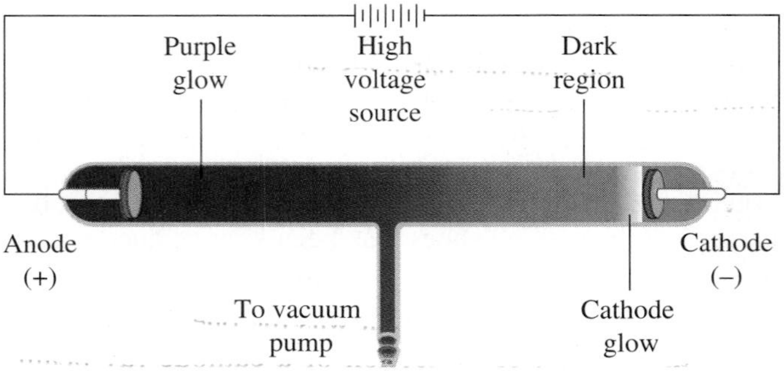

Figure 6.2 Electric discharge in an evacuated tube. The glass tube is evacuated by connection to a vacuum pump. As the air is pumped out, a purple glow is produced in the tube. The glow extends from the anode almost to the cathode. The cathode also glows, but there is a dark region between the cathode and the column of purple light. As the evacuation continues, the dark region replaces the purple glow throughout the tube while the glass itself emits a phosphorescent glow.

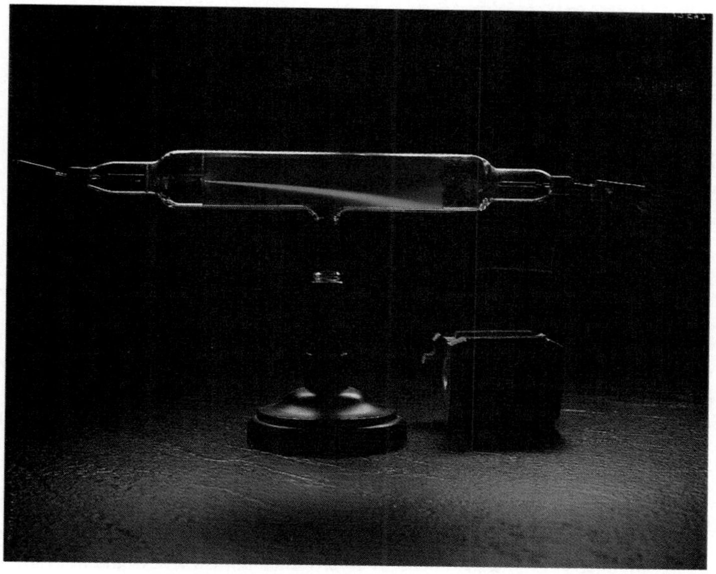

Figure 6.3 Cathode rays and their deflection in a magnetic field. Cathode rays are invisible. We detect them only through their interaction with a fluorescent material. Here the rays cause a green fluorescence as they strike the screen coated with zinc sulfide. The beam of cathode rays begins at the cathode on the left, and it is deflected as it enters the field of the magnet, which is located slightly behind the anode. The deflection we observe is that expected of negatively charged particles.

originated at the cathode and traveled to the anode, became known as a **cathode ray**. Crookes observed that cathode rays normally travel in straight lines, but are deflected when a magnet is brought nearby.

A television picture tube is actually a cathode ray tube (CRT). Cathode rays are deflected by varying magnetic fields to the appropriate spots on a fluorescent material coating the screen, thereby creating an image.

Mass-to-Charge Ratio of Cathode Rays: Thomson's Experiment

Considerable speculation arose as to the nature of cathode rays. In general, British scientists believed the rays to be beams of electrified particles. Most German scientists held that the rays were more likely to consist of a form of energy much like visible light. The answer came in the only way that scientific answers can come: from experiments. The crucial ones were performed by the English physicist Joseph John Thomson (1856–1940) in 1897. By studying the behavior of cathode rays in electric and magnetic fields, Thomson established unequivocally that the rays consist of negatively charged particles. Moreover, his experiments showed that the particles were identical, regardless of the materials from which the electrodes were made or the type of gas in the tube. Thomson concluded that these negatively charged particles were constituents of every kind of atom. We now call these particles *electrons*, a term that had been coined by the Irish physicist George Stoney in 1891 to describe the smallest unit of negative electric charge. Cathode rays are beams of electrons.

Thompson: Cathode Rays
Electrons are the negatively charged particles in them

Perhaps Thomson's most significant experiment was the one described in Figure 6.4. By measuring the amount of deflection of a cathode ray beam in electric and magnetic fields of known strengths, Thomson was able to calculate the ratio of the mass of an electron to its charge. If we denote the mass of the electron as m_e and its electrical charge as e, the mass-to-charge ratio is m_e/e. This ratio has a value of -5.686×10^{-12} kilograms per coulomb (kg/C). Thomson was awarded the Nobel Prize in physics in 1906 for his studies on the nature of the electron.

$M_e/e = -5.686 \times 10^{-12}$ Kg/C

The coulomb (C) is the SI unit of electric charge.

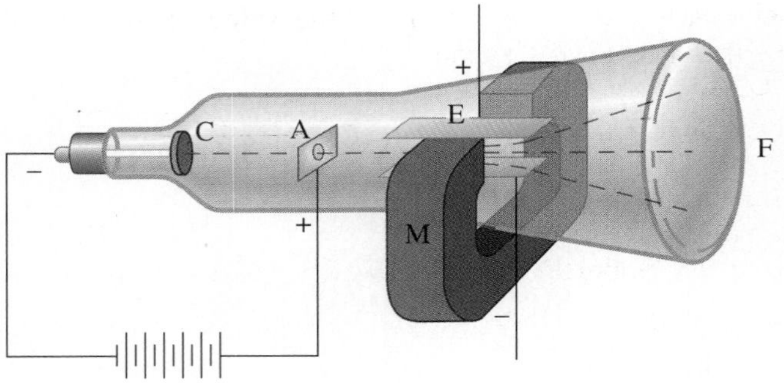

Figure 6.4 **Thomson's apparatus for determining the mass-to-charge ratio, m_e/e, of cathode rays.** Electrons pass from the cathode, C, to the anode, A, but the anode is perforated to allow a narrow beam of cathode rays to pass through. There is an electric field between the electrically charged condenser plates, E, and a magnetic field between the poles of the magnet, M. Since they consist of negatively charged particles, cathode rays are attracted upward toward the positive plate of E. Because of their motions and charge, the rays are bent downward by the magnetic field. If the forces on the particles exerted by the electric and magnetic fields are exactly counterbalanced by appropriate adjustment of the field strengths, the cathode ray beam will strike the fluorescent screen, F, undeflected. From the strengths of the two fields and other data, one can obtain a value for m_e/e.

Electron Charge: Millikan's Oil-Drop Experiment

Robert Millikan (1868–1953) was a professor at the University of Chicago at the time of his famous experiments on the electronic charge (1909–1913). In 1921 he moved to the California Institute of Technology and developed an interest in a new type of radiation found to enter Earth's atmosphere from outer space, for which he coined the term "cosmic radiation."

In 1909, Robert A. Millikan, an American physicist, determined the charge on the electron by observing the behavior of electrically charged oil drops in an electric field. A diagram and explanation of Millikan's apparatus are given in Figure 6.5. Based on hundreds of careful experiments, Millikan established the

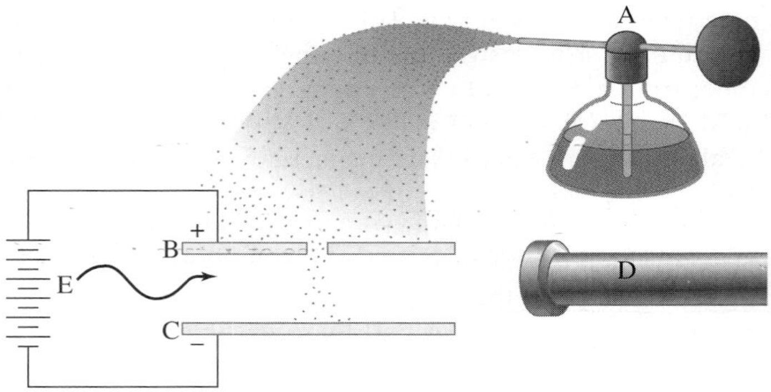

Figure 6.5 **Millikan's oil drop experiment.** Oil droplets from the atomizer (A) enter the apparatus through a tiny hole in the top plate of an electrical condenser. The droplets are observed through a telescope with a micrometer eyepiece (D). Ions are produced by ionizing radiation such as X-rays (E). Some of the oil droplets acquire an electric charge by attaching ions to their surface as they fall between the electrically charged plates B and C.

By measuring the velocity of a falling droplet, both with an electric field present and with no field, Millikan obtained the data needed to establish the magnitude of the charge, q, on the droplet. His conclusion, based on the measurement of many droplets, was that q is always an integral multiple of a particular fundamental charge, called the electronic charge e. That is, $q = ne$, where $n = 1, 2, 3$, and so on.

charge on an electron (e) as -1.602×10^{-19} C. From this value and the value for m_e/e, we can calculate the mass of an electron.

$$m_e = \frac{m_e}{e} \times e = -5.686 \times 10^{-12}\ \text{kg/C} \times \frac{-1.602 \times 10^{-19}\ \text{C}}{1\ \text{electron}}$$

$$= 9.109 \times 10^{-31}\ \text{kg/electron}$$

Millikan was awarded the Nobel Prize in physics in 1923 for his oil-drop experiment.

The mass of an electron is so small that it takes more than 1800 electrons to equal the mass of one hydrogen atom, the lightest atom known. This is why, in Chapter 2, we were able to neglect the mass of an electron compared to those of a proton and a neutron.

6.2 X-rays and Radioactivity: Serendipity in Science

Many scientific discoveries are the result of serendipity—they are fortunate findings made accidentally. But it is no accident that most of these discoveries are made by scientists. Only a trained observer is likely to notice and grasp the significance of unexpected phenomena and to follow up the observation with crucial experiments. We consider here two prime examples of serendipity that were offshoots of cathode ray research.

Wilhelm Roentgen (1845–1923). Within a year of Roentgen's discovery over one thousand papers were published on X-rays. He was awarded the first Nobel Prize in physics in 1901.

The Discovery of X-rays

It has been reported that William Crookes complained to his supplier that photographic film stored in his laboratory was frequently fogged. If this was the case, Crookes uncharacteristically did not follow up on this accidental observation and so missed out on the discovery of X-rays.

In 1895, Wilhelm Roentgen, while conducting experiments with a Crookes cathode ray tube covered with black paper, noticed that a material elsewhere in the laboratory emitted a fluorescent glow. He hypothesized that the radiation responsible for the fluorescence was coming from the cathode ray tube but that it was not ordinary light. He referred to this unknown radiation as **X-rays**. In a matter of a few weeks he achieved a fairly complete characterization of X-rays, and almost immediately after his discovery was announced, the first medical X-rays were taken.

Fluorescence is the emission of light by a substance that itself is being struck by another form of radiant energy, such as the emission of visible light during irradiation with X-rays or ultraviolet light.

Roentgen found that the impact of cathode rays on a dense metal anode (called a *target*) provided an especially effective source of X-rays. Figure 6.6 is a diagram of a modern X-ray tube.

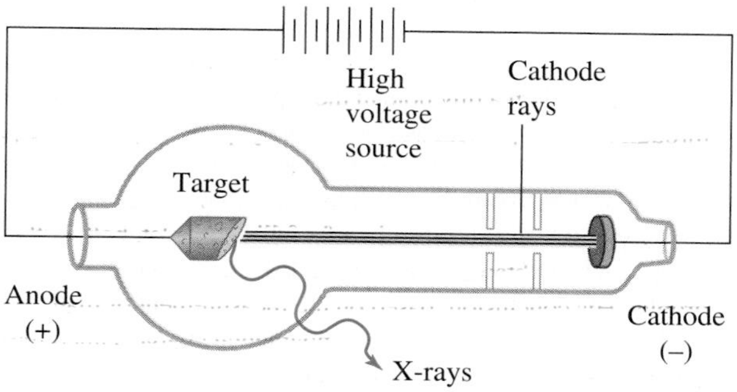

Figure 6.6 The production of X-rays.

Marie Sklodowska Curie (1867–1934) was born in Poland and went as a young woman to Paris to work for her doctorate degree in mathematics and physics. There she met and married Pierre Curie (1859–1906), a French physicist of some note. Pierre Curie was killed in a trafffic accident in 1906. Madame Curie died in 1934 of pernicious anemia, perhaps brought on by long exposure to the radiation with which she worked.

The Discovery of Radioactivity

In Roentgen's original experiments, the impact of cathode rays on glass caused both fluorescence of the glass and the emission of X-rays. In 1896, Antoine Henri Becquerel (1852–1908), a French physicist long interested in fluorescence, asked himself the questions: Is the production of X-rays associated with fluorescence? If so, do materials that fluoresce in ordinary light also produce X-rays? To answer these questions, Becquerel wrapped photographic film with black paper, placed a few crystals of a fluorescent material on the paper, and placed the entire assembly in strong sunlight. Sure enough, fogging of the film occurred under the crystals, meaning that some type of radiation—but not ordinary light—had penetrated the black paper. Becquerel believed this radiation to be X-rays.

In the course of these studies, on several cloudy days when exposures to sunlight were not possible, Becquerel kept busy by preparing samples and placing them in a drawer. To his surprise, the photographic film was fogged even though the fluorescent crystals had not been exposed to sunlight. Rather than just discard the film and start over, Becquerel followed up his accidental observation with experiments to determine what had fogged the film. He showed that it was not fluorescence, but rather radiation emitted continuously by uranium, one of the elements in the crystals, that was responsible for fogging the film.

Right away other scientists began to study this new radiation. Marie Sklodowska Curie called it radioactivity. **Radioactivity** is the spontaneous emission of radiation from the unstable nuclei of certain isotopes. Marie Curie, her husband Pierre, and Antoine Becquerel won the Nobel Prize in physics in 1903 for their studies of radioactivity, and for her continued work, Marie Curie was awarded the Nobel Prize in chemistry in 1911.

Three Types of Radiation

Shortly after the discovery of radioactivity, three types of radiation were identified in the emanations from radioactive substances. One type, called **alpha** (α) **particles**, consists of particles that have a mass about four times that of a hydrogen atom and a charge twice the magnitude of an electron, but positive rather than negative. An alpha particle is now known to be a doubly ionized helium atom, that is, He^{2+}. A second type of radiation was shown to consist of negatively charged particles identical to cathode rays. That is, these particles, called **beta** (β) **particles**, are electrons, but they come from inside the nucleus The third type of radiation, called **gamma** (γ) **rays**, is a form of electromagnetic radiation, much like the X-rays used in medical work, but of even higher energy. Like X-rays, but unlike alpha and beta radiation, gamma rays are a form of energy and not a form of matter. The three types of radiation are illustrated in Figure 6.7.

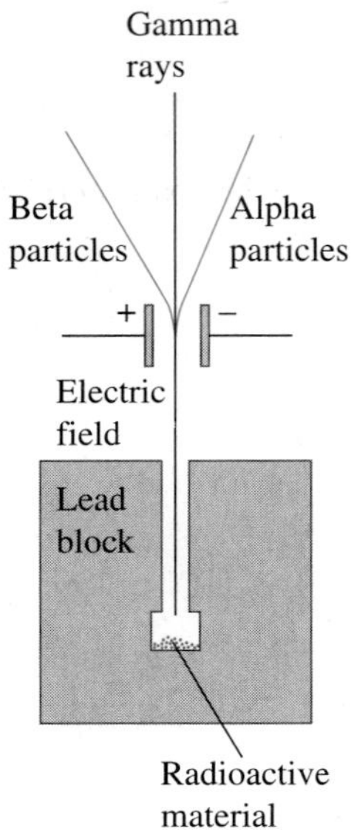

Figure 6.7 Three types of radiation from radioactive materials. The radioactive material is enclosed in a lead block. All the radiation except that passing through the opening is absorbed by the lead. When the radiation is passed through an electric field, it splits into three beams. One beam is attracted to the negative plate; it is composed of positively charged alpha (α) particles. A second beam is attracted to the positive plate; it is a beam of beta (β) particles—electrons. A third beam, called gamma (γ) rays, is undeflected. There is no charge associated with these rays. Because of their greater momentum (mass times velocity), α particles are deflected to a smaller extent than β particles. (Similar results are obtained if a magnetic field is used instead of an electric field.)

6.3 The Nuclear Model of the Atom: Rutherford's Experiment

Ernest Rutherford, a New Zealander who spent his career in Canada and Great Britain, was a pioneer in the investigation of radioactivity. In 1909, Rutherford and his assistant, Hans Geiger (1882–1945), began a series of researches to study the structure of the atom by using alpha-particle beams as probes. The prevailing atomic model at the time was J. J. Thomson's "plum pudding" model. The atom was pictured as a diffuse cloud of positive electric charge with electrons embedded in the cloud, much like plums in a popular pudding dessert of the time. Based on this model Rutherford expected that most alpha particles would pass through atoms undeflected, but an alpha particle that came close to an electron should be deflected to some extent. By measuring the deflections of the alpha particles, he hoped to gain information about the distribution of electrons in an atom.

Ernest Rutherford (1871–1937) grew up on a farm in New Zealand before entering Cambridge University to study under J. J. Thomson. Although he was a physicist, for his pioneering work in atomic structure Rutherford was awarded the Nobel prize in chemistry in 1908.

In the experiment diagrammed in Figure 6.8, Geiger and Ernest Marsden, an undergraduate student, bombarded a very thin foil of gold with alpha particles from a radioactive source. They found that most of the particles went right through the foil undeflected or deflected only slightly, just as Rutherford had expected. However, a few particles were deflected sharply, and once in a while an alpha particle would bounce right back in the

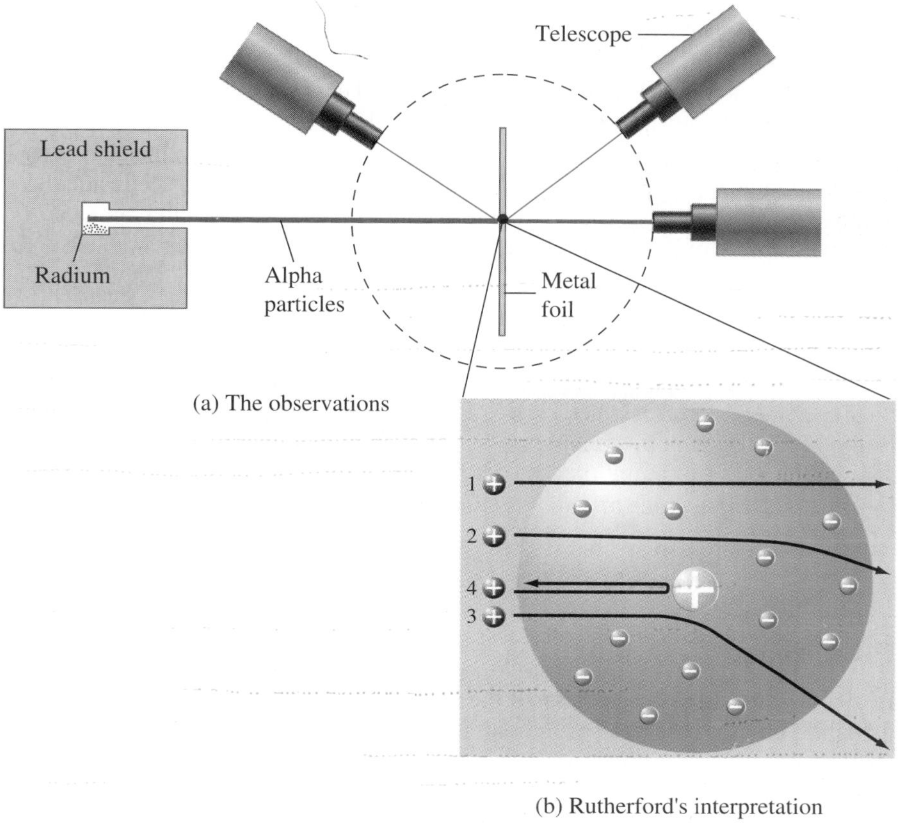

Figure 6.8 The scattering of alpha particles by thin metal foil. (a) The observations. (1) Most of the α particles pass through the foil undeflected. (2) Some α particles are deflected slightly as they penetrate the foil. (3) A few (about 1 in 20,000) are greatly deflected. (4) A similar small number do not appear to penetrate the foil at all, but rather are reflected back toward the source. (b) Rutherford's interpretation. If the atoms of the foil have a massive, positively charged nucleus and light electrons outside the nucleus, one can explain how: (1) an alpha particle passes through the atom undeflected because it encounters no particles (a fate shared by most of the particles); (2) an alpha particle is deflected slightly as it passes near an electron; (3) an alpha particle is deflected significantly by passing quite close to an atomic nucleus; and (4) an alpha particle bounces back toward the source as a result of approaching a nucleus head on.

Commenting on this observation years later, Rutherford said, "It is about as incredible as if you had fired a 15-inch shell at a piece of tissue paper and it came back and hit you."

To picture Rutherford's model, visualize the atom as a giant indoor football stadium. A pea at the middle of the volume represents the nucleus. A few mosquitoes, flitting here and there throughout the stadium, represent the electrons.

direction from which it had come. This observation came as a complete surprise to Rutherford.

To explain the Geiger–Marsden experiment, Rutherford concluded that all the positive charge of an atom is concentrated at the center of the atom in a tiny core called the *nucleus*. When an alpha particle, which is positively charged, approaches a positively charged nucleus, it is strongly repelled and therefore sharply deflected. Because only a few alpha particles were deflected in the Geiger–Marsden experiment, Rutherford concluded that the nucleus must occupy only a tiny fraction of the volume of an atom. Most of the particles passed right through because, except for the tiny nucleus and a small number of electrons outside, an atom is mostly empty space.

6.4 Protons and Neutrons: Positive Particles and Missing Mass

In this section we first examine some positively charged particles produced when cathode rays pass through gases. Then we consider a powerful technique, mass spectrometry, that grew out of Thomson's experiments. Mass spectrometry gives us a way to obtain highly accurate atomic weights.

Positive Rays: Goldstein's Experiment

In 1886, Eugen Goldstein (1850–1930) performed experiments with gas discharge tubes with perforated cathodes (Figure 6.9). He found that while cathode rays were formed at the cathode and sped off toward the anode as usual, positive ions were also formed in the residual gas in the tube. These ions moved in the opposite direction, toward the cathode, and some passed through the holes in the cathode. A study of the deflection of these beams of positive ions, called *positive rays*, in a magnetic field indicated that they had varying masses that depended on the residual gas in the tube. The lightest particles were formed when hydrogen was the remnant gas; these particles were later shown to have a mass about 1836 times that of an electron.

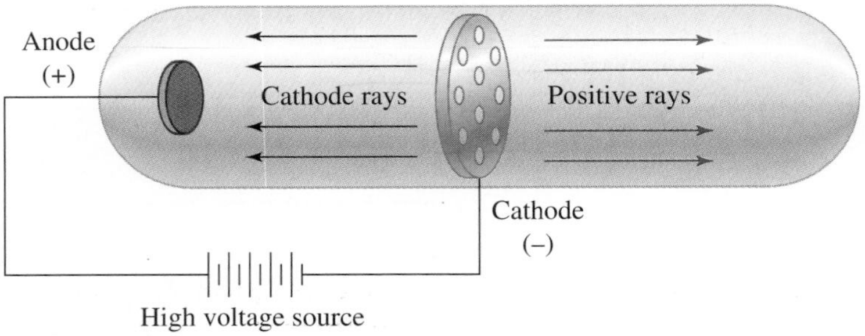

Figure 6.9 The production of positive rays. Cathode rays (black) stream toward the anode (+). They collide with residual gas atoms and knock electrons from the atoms, producing positively charged ions. These ions are attracted to the cathode (–), but some pass through the holes in the cathode and appear as a stream of positive particles (red) on the other side. These beams of positive ions are called *positive rays.*

Protons and Neutrons

In 1914, Rutherford suggested that the smallest positive ray particle was the fundamental unit of positive charge in all nuclei. He called this particle, which has a charge equal in magnitude but opposite in sign to that of an electron, a *proton*. Rutherford proposed that protons constitute the positively charged matter in the nuclei of all atoms.

However, except for the lightest hydrogen isotope, protium (^1_1H), atoms have more mass than is indicated by the number of protons. For example, a helium nucleus, with *two* protons, has a mass *four* times that of hydrogen. If all the mass came from the protons, a helium atom would have only twice the mass of a hydrogen atom. The reason for this "excess" mass puzzled scientists for several years. One hypothesis was that the atomic nucleus also contained electrically neutral fundamental particles.

In 1919, Rutherford demonstrated the existence of protons as nuclear particles by bombarding nitrogen nuclei with alpha particles. Protons were ejected in the process.

In the 1920s and early 1930s, alpha particles were used as projectiles to bombard a variety of materials. Bombardment of beryllium atoms produced a strange, highly penetrating form of radiation. In 1932, James Chadwick (1891–1972) showed that this radiation was best explained as a beam of neutral particles. These particles, called *neutrons*, were found to have about the same mass as protons but no electric charge. This discovery finally provided an explanation for the mysterious excess mass: A helium atom has two protons and *two* neutrons. Because protons and neutrons have roughly the same mass (and electrons have almost no mass), the helium atom should have about four times the mass of the hydrogen atom.

Mass Spectrometry

A **mass spectrometer** is a device that separates positive ions according to their mass-to-charge ratios. A simple diagram of a mass spectrometer is shown and described in Figure 6.10. The essence of the method is to produce a mixture of

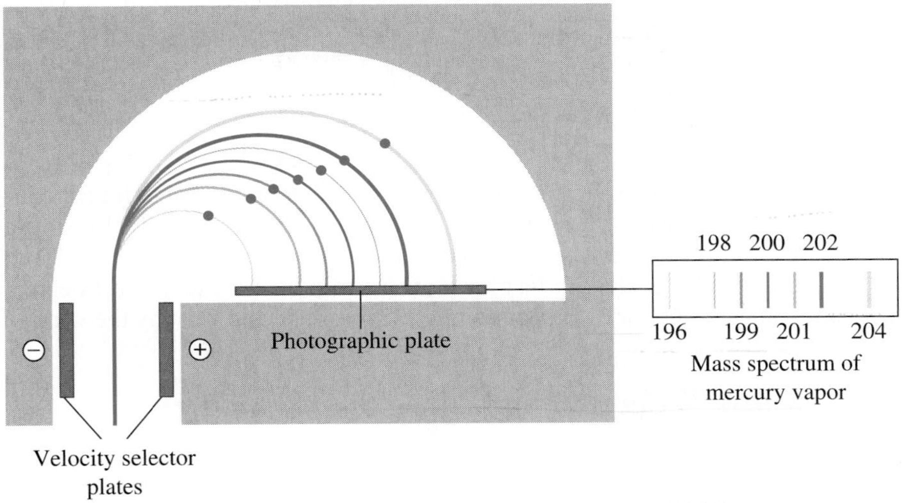

Figure 6.10 A mass spectrometer. A gaseous sample is ionized by bombarding it with electrons in the lower part of the apparatus (not shown). The positive ions thus formed pass through an electric field in which they are accelerated to a particular velocity. The ions then pass through a narrow slit into a curved chamber. A magnetic field is applied perpendicular to the beam of ions (perpendicular to the page), causing a circular motion; all the ions with the same mass-to-charge ratio are deflected into the same circular path. (The usual ionic charge is 1+, so that the mass-to-charge ratio is generally the same as the mass.) Ions with different masses strike the detector (here a photographic plate) in different regions. The more ions of a given mass, the greater is the response of the detector (depth of the deposit on the photographic plate).

positive ions in an ionization chamber. Electric and magnetic fields are used to sort the mixture into groups of ions that are deflected into a set of curved paths through the spectrometer. All ions with the same mass-to-charge ratio travel the same path. An ion detector is used to collect the ions. Ions with different mass-to-charge ratios are collected at different places in the apparatus. The output of the detector indicates the masses of ions on a scale of atomic mass units. The relative abundances of the ions are indicated by the intensities of the detector signals. Generally these relative values are printed as a bar graph like the one shown in Figure 6.11.

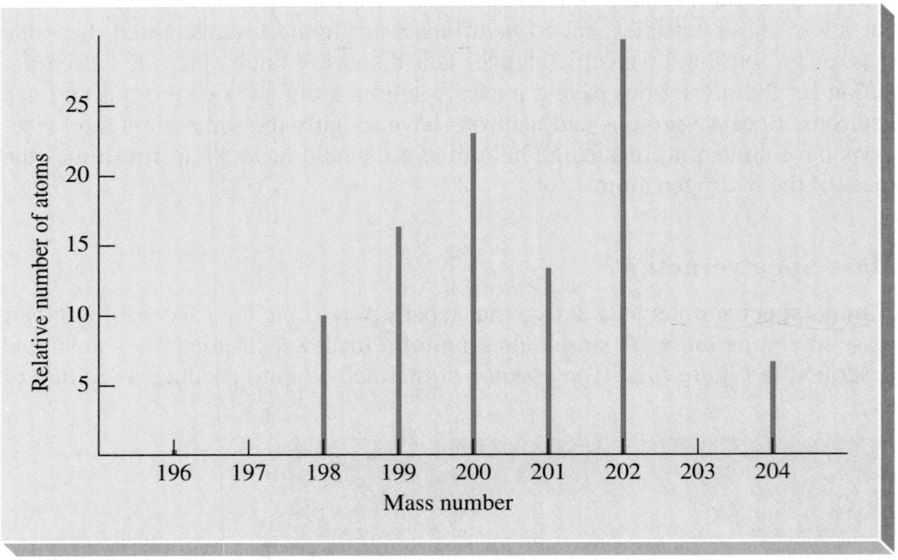

Figure 6.11 Mass spectrum for mercury. The response of the ion detector in Figure 6.10 (depth of deposits on the photographic plate) has been converted to a scale of relative numbers of atoms. The percent natural abundances for the mercury isotopes are ^{196}Hg, 0.146%; ^{198}Hg, 10.02%; ^{199}Hg, 16.84%; ^{200}Hg, 23.13%; ^{201}Hg, 13.22%; ^{202}Hg, 29.80%; ^{204}Hg, 6.85%.

As an analogy to the separation of ions by their mass-to-charge ratios, consider three balls thrown at the same speed in a straight-line direction with a strong crosswind blowing. The first, a tennis ball, will be blown farthest off course. The second, a heavier baseball, will be blown off its intended path to a lesser degree. The third, a heavy shot used in shot putting, will be deflected least by the wind.

6.5 Atomic Weights from Mass Spectrometry

Nuclide is a term used to signify an atomic species having a particular atomic number and mass number, such as $^{12}_{6}C$. Where two or more nuclides of an element exist, such as $^{12}_{6}C$ and $^{13}_{6}C$, we refer to these as *isotopes*. As we learned in Section 2.4, the atomic weight of an element is an average of the atomic masses of the isotopes of the element, weighted in proportion to their fractional abundances in nature.

Some elements have only one naturally occurring nuclide. Fluorine, for example, consists only of the nuclide fluorine-19; it is a *mononuclidic* element. The atomic weight of a mononuclidic element is the same as the atomic mass of the one nuclide. For elements having naturally occurring isotopes, each isotope contributes to the atomic weight a quantity equal to the product of its atomic mass and its fractional abundance. The atomic masses and fractional abundances

This method of determining atomic weights works quite well because, with few exceptions, the fractional abundances of the isotopes of an element are the same from one sample to another.

are both established by mass spectrometry, and the weighted-average atomic weight is calculated as illustrated in Example 6.1.

Example 6.1

Chlorine exists naturally as two isotopes: chlorine-35, with a mass of 34.96885 u and a fractional abundance of 0.75771, and chlorine-37, with a mass of 36.96590 u and a fractional abundance of 0.24229. Calculate the atomic weight of chlorine.

Abundances are often given as percentages, for example, 75.771% chlorine-35 and 24.229% chlorine-37. A fractional abundance = percent abundance/100.

Solution

The atomic weight must be closer to 35 u than to 37 u because there are more chlorine-35 atoms than chlorine-37 atoms. To obtain the right average, we must "weight" the chlorine-35 isotope more heavily than the chlorine-37 isotope. It is this "weighted average" that we obtain when we determine the contribution of each isotope and add them together.

$$\text{Contribution of } ^{35}\text{Cl} = 0.75771 \times 34.96885 \text{ u} = 26.496 \text{ u}$$
$$\text{Contribution of } ^{37}\text{Cl} = 0.24229 \times 36.96590 \text{ u} = 8.9565 \text{ u}$$
$$\text{At. wt. of chlorine} = 26.496 \text{ u} + 8.957 \text{ u} = 35.453 \text{ u}$$

Exercise 6.1

Carbon exists as the isotopes carbon-12, with a fractional abundance of 0.9890 and a mass of exactly 12 u, and carbon-13, with a fractional abundance of 0.0110 and a mass of 13.00335 u. Calculate the atomic weight of carbon.

Exercise 6.2

Neon exists as three isotopes with isotopic masses and *percent* abundances as follows: neon-20, 19.99244 u, 90.51%; neon-21, 20.99395 u, 0.27%; neon-22, 21.99138 u, 9.22%. Calculate the atomic weight of neon.

Estimation Exercise 6.1

Indium has an atomic weight of 114.82 u and consists of *two* principal isotopes. One of the isotopes has a mass of 112.9043 u. Which of the following is likely to be the second isotope? Explain.

^{111}In ^{112}In ^{114}In ^{115}In

6.6 Light in Nature: The Nature of Light

Space is vast and dark. Here and there a star lights a small segment of the void. Our sun is one such star; it lights Earth and the other parts of its solar system. Our eyes perceive sunlight to be "white," but under proper conditions, white light can be separated into a rainbow of colors. Later in the chapter we will see that to describe the arrangements of electrons outside the nuclei of atoms, we need to analyze the light emitted by energetic atoms. To perform this analysis we need to know something about the physical nature of light.

The Wave Nature of Light

Waves are common in nature. Waves of water break upon the shores of oceans and lakes. A concentric pattern of waves is formed when a pebble is tossed into a pond. Earthquakes send waves through Earth's crust. The song "America the Beautiful" refers to "amber waves of grain." Sound waves carry music to our ears. With microwaves we can thaw a TV dinner or heat a cup of soup. But just what is a wave?

A **wave** is a progressive, repeating disturbance that spreads from a point of origin to more distant points. There is little movement of material present in the path of the disturbance. It is rather like a whispered message ("disturbance") passed along by seated people ("little movement") from one end of a row ("point of origin") to the other ("distant point"). Two people holding a long string can start wave motion in the string if one of them jerks an end of the string up and down (Figure 6.12).

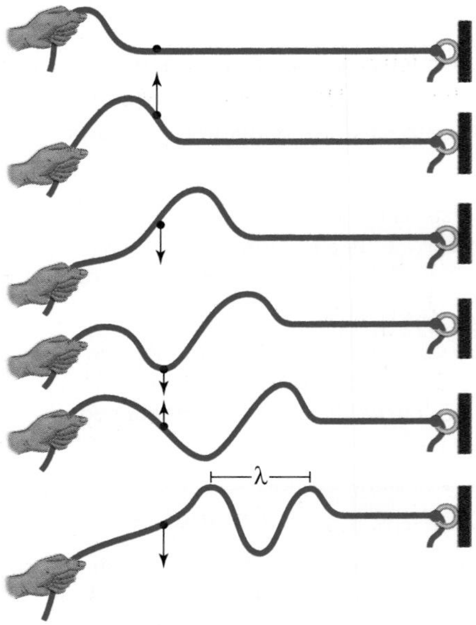

Figure 6.12 The simplest wave motion: a traveling wave in a string. Imagine a taut, infinitely long string. Up and down hand motion (top to bottom) causes waves to pass along the string from left to right. The up and down motion of a typical point (dot) on the string is also shown. This one-directional moving wave is called a traveling wave.

Electromagnetic waves originate in the movement of charged particles—for example, when an electron moves with respect to the nucleus of an atom. This movement produces oscillations in electric and magnetic fields that are propagated over distances. Unlike water waves and sound waves, electromagnetic waves require no medium for their propagation. This quality of electromagnetic radiation makes possible the transmission through empty space of some of the sun's energy to Earth.

Electromagnetic radiation can be described by its wavelength and frequency (Figure 6.13). The **wavelength** is the distance between any two identical points in consecutive cycles; peaks or crests are chosen for convenience. Wavelength is denoted by the Greek letter λ (lambda). A common SI unit of wavelength, especially for visible light, is the nanometer (1 nm = 10^{-9} m). A non-SI unit that has long been used in scientific work and continues to be to some extent is the angstrom unit, Å (1 Å = 10^{-10} m).

The **frequency** of a wave is the number of cycles of the wave (the number of wavelengths) that pass through a point in a unit of time. Frequency is denoted

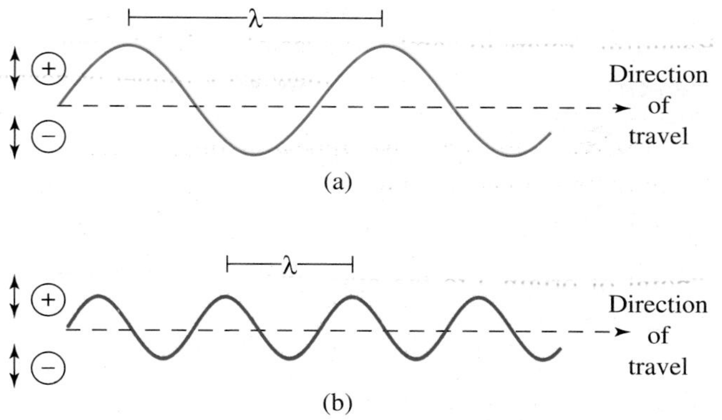

Figure 6.13 **An electromagnetic wave.** Although greatly oversimplified, this representation suggests that electromagnetic radiation originates from the motions of electrically charged particles and shows one characteristic of the radiation: the wavelength λ. The wave in (a) has a *longer* wavelength than the one in (b).

by the Greek letter ν (nu, pronounced the same as "new"). The SI unit of frequency is the *hertz* (Hz). A hertz is one cycle per second, but generally the word "cycle" is understood. We simply use 1/s or s^{-1}.

$$1 \text{ Hz} = 1 \text{ s}^{-1}$$

Imagine yourself on a raft at sea. The raft bobs up and down as each wave passes. The number of times you bob up and down per unit time is the frequency of the ocean waves.

Consider again the standing wave in a string shown in Figure 6.12. If ν cycles of the wave pass through a point on the string each second, and if the length of each cycle is the wavelength, λ, then the total distance the wave impulse travels in one second is

$$\nu \cdot \lambda = c$$

Because c is a distance per unit time, it is a speed—the speed of the wave.

In a vacuum, light travels at a *constant* speed of 2.99792458×10^8 m/s (often rounded to 3.00×10^8 m/s). In air its speed is slightly less, but in calculations we will generally assume that light waves are traveling in a vacuum. Because the speed of light in a medium is constant, wavelength and frequency are inversely related. The longer the wavelength, the lower is the frequency, or the shorter the wavelength, the greater is the frequency.

The speed of light is the fastest speed attainable. A beam of light would travel a distance equivalent to that from London to San Francisco in about 0.03 s, and from Earth to Moon in 1.28 s.

Example 6.2

Calculate the wavelength, in nanometers, of ultraviolet radiation that has a frequency of 1.22×10^{15} s^{-1}.

Solution

We can rearrange the basic expression, $\nu \cdot \lambda = c$, and substitute values for ν and c.

$$\lambda = \frac{c}{\nu} = \frac{3.00 \times 10^8 \text{ m} \cdot \text{s}^{-1}}{1.22 \times 10^{15} \text{ s}^{-1}} = 2.46 \times 10^{-7} \text{ m}$$

Expressed in nanometers, the wavelength is

$$2.46 \times 10^{-7} \text{ m} \times \frac{1 \text{ nm}}{1 \times 10^{-9} \text{ m}} = 246 \text{ nm}$$

Example 6.3

Calculate the frequency, in s^{-1}, of an X-ray that has a wavelength of 8.21 nm.

Solution

First convert the wavelength to meters (m):

$$8.21 \text{ nm} \times \frac{1 \times 10^{-9} \text{ m}}{1 \text{ nm}} = 8.21 \times 10^{-9} \text{ m}$$

Then solve the basic expression for frequency, ν,

$$\nu = \frac{c}{\lambda}$$

and substitute the known data.

$$\nu = \frac{3.00 \times 10^{8} \text{ m} \cdot \text{s}^{-1}}{8.21 \times 10^{-9} \text{ m}} = 3.65 \times 10^{16} \text{ s}^{-1}$$

Exercise 6.3

Calculate the wavelength, in nanometers, of infrared radiation that has a frequency of 9.76×10^{13} Hz.

Exercise 6.4

Calculate the frequency, in hertz, of a microwave that has a wavelength of 1.07 mm.

Electromagnetic Spectrum → Range of wavelengths + frequencies that range from short-wavelength, high frequency gamma rays to long-wavelength low frequency radio waves. - invisible.

The Electromagnetic Spectrum

Types of electromagnetic radiation range from short-wavelength, high-frequency gamma rays to long-wavelength, low-frequency radio waves. This range of wavelengths and frequencies is known as the **electromagnetic spectrum** (Figure 6.14).

The electromagnetic spectrum is largely *invisible* to the human eye; we see only a tiny portion somewhere near the middle of the range. The visible spectrum ranges from about 750 nm, where red light blends into the infrared, to about 400 nm, where violet fades into the ultraviolet. Our senses can detect some radiation other than visible. We can feel the warmth due to infrared radiation, perhaps as it comes through the glass of a car window, even on a cold winter day. Sunburned skin is a sign that we have been exposed too long to ultraviolet radiation, although we may not have been aware of it at the time. X-rays and gamma rays also can burn us if we are overexposed, and they can cause much more serious damage because of their high energies. The different forms of electromagnetic radiation can be detected by the appropriate devices: radios detect radio waves, night-vision goggles respond to infrared light, our eyes and ordinary photographic film are sensitive to visible light, and so on.

Materials vary in their abilities to absorb or transmit different parts of the spectrum. Our bodies are opaque to visible light but largely transparent to X-rays. Ordinary window glass obviously is transparent to visible light; it also transmits some infrared energy but effectively blocks most ultraviolet radiation. (You can't get a suntan in sunlight coming through window glass.) We frequently make use of the selective properties of materials to shield us from the rays we don't want to absorb and pass on those that we do.

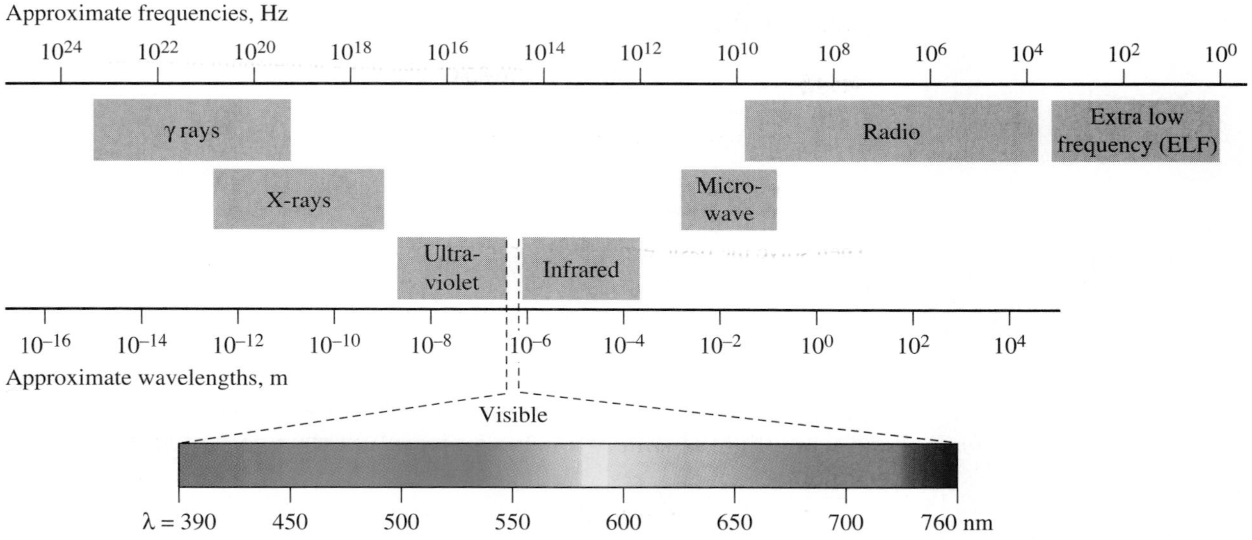

Figure 6.14 The electromagnetic spectrum. The visible region, which extends from red at the longest wavelength to violet at the shortest wavelength, is only a small portion of the entire spectrum. The approximate wavelength and frequency ranges of some other forms of electromagnetic radiation are also indicated.

Continuous and Line Spectra

When white light from an incandescent lamp is passed through a slit and a prism, it is separated into a *continuous spectrum*, a spreading out of the various components of white light into an unbroken "rainbow" of colors (Figure 6.15).

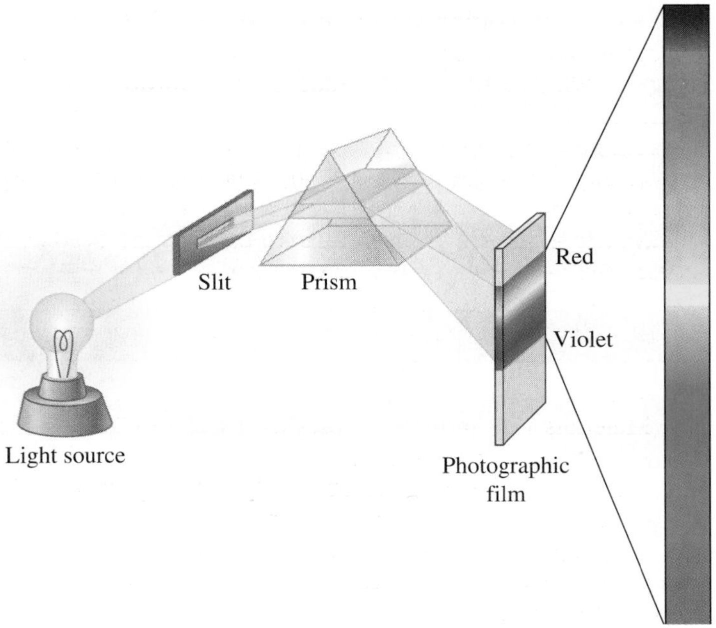

Figure 6.15 The spectrum of ordinary ("white") light. Red light is bent the least and violet light the most when "white" light is passed through a glass prism. The other colors of the visible spectrum are found between the red and the violet.

(a)

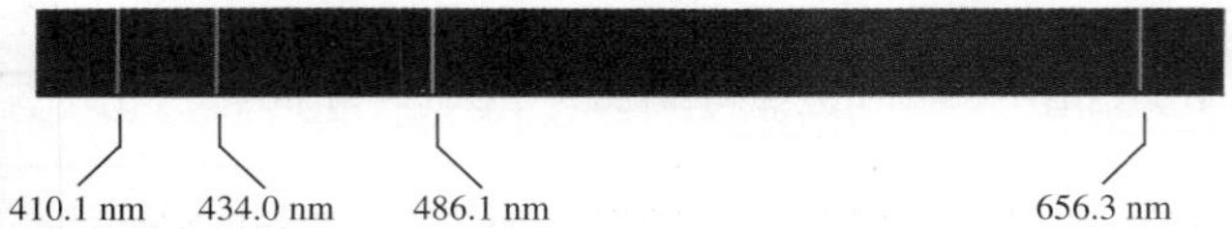

(b)

Figure 6.16 The visible spectrum of hydrogen. (a) Light from a hydrogen lamp as it appears to the unaided eye. (b) The Balmer series of the hydrogen spectrum. The Balmer series consists of four lines in the visible region and a much larger number of closely spaced lines in the near ultraviolet that are invisible to the eye.

Sunlight passing through raindrops produces the same effect. The different colors of light correspond to different wavelengths and frequencies. A blue light has a shorter wavelength (higher frequency) than does a red one. All visible wavelengths are present in the continuous visible spectrum, and white light is simply a combination of all the wavelength components or colors.

Now consider light from a hydrogen lamp—that is, from a gas discharge tube containing hydrogen. If light from that lamp is passed through a slit and then through a prism, only a few images of the slit are observed, appearing as narrow colored lines separated by dark regions (Figure 6.16); this is a *discontinuous* spectrum. Each line corresponds to electromagnetic radiation of a specific frequency and wavelength. The pattern of lines produced by the light emitted by an element is called an **emission spectrum** or a *line spectrum*. The line spectrum of an element is characteristic of that element and can be used to identify it (see Figure 6.17). However, we cannot see many of the lines in a line spectrum because they appear in the ultraviolet or infrared regions of the electromagnetic spectrum.

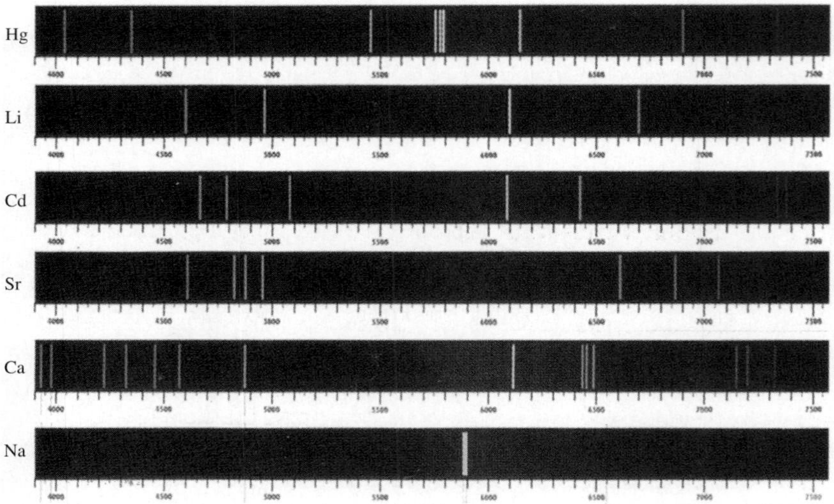

Figure 6.17 Line spectra of selected elements. Atomic spectra of several elements are shown here. Wavelengths are given in angstrom units (1 Å = 10^{-10} m). Each element has its own distinctive spectrum which can be used to identify the element. In addition to their practical use in analyzing matter, atomic spectra have led to many of the ideas concerning atomic structure.

Fireworks, Flame Tests, and Emission Spectroscopy

An analytical technique called emission spectroscopy is based on an analysis of the light emitted when an element is strongly heated or energized in an electric spark or a gas discharge tube such as the helium lamp in Figure 6.18. The emitted light is dispersed into individual wavelength components. A photograph or other record of this light is called the *emission spectrum* of the element.

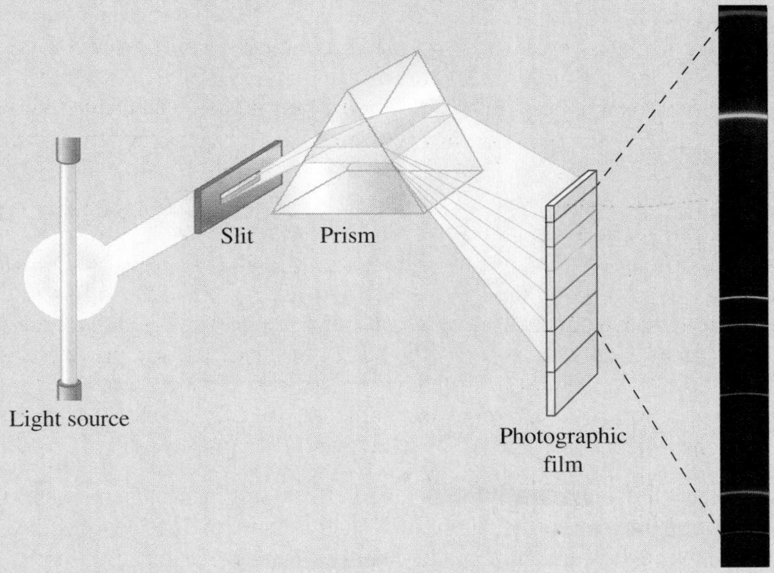

Figure 6.18 Production of an emission spectrum. The light source is a helium lamp. The visible spectrum of helium consists of six colored lines that can be seen with the unaided eye. In *emission spectroscopy* an element is identified by the *positions* of its spectral lines. These positions correspond to specific frequencies and wavelengths. The quantity of the element is determined from the *intensities* of the lines.

Every element has a unique emission spectrum. That spectrum is sort of an "atomic fingerprint," which can be used to identify the element. The emission spectrum of helium in Figure 6.18 was first observed during a solar eclipse in 1868. Twenty-seven years later helium was discovered on Earth, and its spectrum proved to be identical to that of helium on the sun.

Atoms of certain elements—Li, Na, K, Rb, Cs, Ca, Sr, and Ba, for example—can be induced to emit light simply by heating the elements or their compounds in a gas flame. The flame takes on a distinctive color determined by the particular element, and the colored flame is referred to as a *flame test* for the element. The emission spectrum of sodium, for example, is dominated by two closely spaced, exceptionally bright lines in the yellow region of the spectrum (see again Figure 6.17). The flame test for sodium is a yellow flame color that appears when sodium metal or sodium compounds are heated. Figure 6.19a shows this color as emitted by a sodium vapor lamp, and Figure 6.19b shows the same color when a piece of glass tubing is heated in a gas flame. (The raw materials used to make ordinary soda-lime glass are sodium and calcium carbonates and silicon dioxide.)

Figure 6.19 **The emission spectrum of sodium.** (a) A sodium vapor lamp. (b) The sodium flame color.

In the early days of spectroscopy, there was a need for a steady, clean-burning gas flame. Robert Bunsen (1811–1899), a pioneer chemist-spectroscopist, developed a burner to produce such a flame in 1855; to this day we continue to use the familiar laboratory "Bunsen burner." Using a Bunsen burner and a spectroscope, Bunsen and his coworker Gustav Kirchhoff (1824–1887) discovered a new element in 1860 and named it *cesium* for its blue flame color (Latin *caesius*, sky blue). In 1861 they discovered another new element through its characteristic red flame color, *rubidium* (L. *rubidius*, deepest red).

Perhaps the most familiar application of flame colors is in fireworks displays. Fireworks originated in ancient China, long before any of the developments described here, but the process for creating fireworks is unchanged. In modern fireworks, brilliant reds are produced by strontium compounds [for example, $SrCO_3$, $Sr(NO_3)_2$, or $SrSO_4$]. Barium compounds yield bright green colors; sodium compounds yield yellow; copper compounds produce a greenish blue; and powdered magnesium and aluminum produce bright white light when they burn. A brilliant blue is a highly desired color, but no common compound is known that produces it; cesium compounds are too expensive for general use.

Although flame colors excited interest through fireworks and proved useful in chemical analyses, no one in the nineteenth century was able to explain them. Why, for example, is the dominant flame color from sodium compounds always the same shade of yellow—and never red, green, or blue—regardless of the compound used? We will discover why later in the chapter.

The brilliant colors in fireworks displays are characteristic of the elements present, in this photograph probably aluminum or magnesium metal (white), strontium (red), and sodium (yellow).

All solids emit electromagnetic radiation at all temps, mostly infrared radiation (invisible)

6.7 Photons: Energy by the Quantum

Toward the end of the nineteenth century, a great body of knowledge, now called *classical physics,* included all the basic laws of mechanics, thermodynamics, electricity, magnetism, and light. Scientists seemed poised to answer, in principle at least, all remaining scientific questions. Only a few natural phenomena stubbornly refused to reveal their secrets to the powerful tools of classical physics. The existence of line spectra was one such phenomenon. Let's consider another.

All solids emit some electromagnetic radiation at all temperatures, but in most cases this is invisible infrared radiation. This is the type of radiation, emitted by Earth's surface and trapped by CO_2 and certain other gases, that is associated with global warming (Chapter 14). It is also the radiant heat by which objects can be seen through night-vision goggles.

At high temperatures, the wavelengths of much of the radiation emitted by solids is in the visible range; we can see it. At about 750 °C, for example, a solid emits a considerable amount of red light (think of a "red hot poker"). As the temperature rises further, more light in the yellow and blue portions of the spectrum blends with the red, until at about 1200 °C the solid glows white (hence, the term "white hot"). This radiation, which depends on the temperature of a solid and not on the particular elements present, is called *black-body radiation*.

Radiation associated with line spectra is emitted by *gaseous* atoms and depends on the particular element. Black-body radiation is emitted by *solids* and does not depend on the identity of the solid.

Planck's Quantum Hypothesis

According to classical physics, the atoms of a solid vibrate about fixed points. As the temperature increases, the atoms vibrate more vigorously. Black-body radiation represents the release of some of the energy of a system of vibrating atoms as electromagnetic radiation. The failure of classical physics was that it could not account for details of the observed black-body spectrum, such as the distribution of this radiation among different frequencies (colors) and the variation of this distribution with temperature. We now know that this failure stemmed from a prevailing assumption in classical physics that a particle within a system of particles could be endowed with energy in any amount, without preset limits.

In 1900, Max Planck (1858–1947) developed a startling theory to explain all aspects of black-body radiation. The essential points of his theory are these.

- The energy of a vibrating atom in a system of such atoms may have only a specific set of values, not just any value.
- Electromagnetic radiation emitted by the vibrating atoms can correspond only to the exact *difference* between two allowable energy values.
- The smallest quantity of energy, E, that can be emitted as electromagnetic radiation, referred to as a **quantum** of energy, is given by the expression

$$E = h\nu$$

where ν is the frequency of the radiated light, and h is a numerical constant, now called **Planck's constant**, which has a value of

$$h = 6.626 \times 10^{-34}\ \text{J} \cdot \text{s}$$

- Light from a given emitter can be emitted only as a single quantum, the energy of which must be an *exact integral multiple* of a simplest quantum, that is, $h\nu$, $2h\nu$, $3h\nu$, and so on, but not a value like $1.5h\nu$ or $3.06h\nu$.

Even though we seldom give much thought to it, we occasionally encounter situations in everyday life where things are quantized. Think of a vending machine that accepts only nickels, dimes, and quarters. Items in the machine can have prices at five-cent intervals (for example, \$0.50, \$0.55, \$0.60, and so on). A purchase price of \$0.57 is not possible.

Scientists in Planck's time, including Planck himself, had difficulty accepting quantum theory. To them it was strange. Albert Einstein and Niels Bohr soon changed all that by successfully applying Planck's theory to other areas in which classical physics had failed. Planck was awarded the Nobel Prize in physics in 1918 for his part in changing forever the way we view the world.

The Photoelectric Effect: Einstein and Photons

In 1905, Albert Einstein (1879–1955) extended Planck's quantum theory and used it to explain the phenomenon known as the *photoelectric effect*. When a beam of light ("photo") shines on certain surfaces, particularly those of certain metals, a beam of electrons ("electric") is produced.

The photoelectric effect cannot be explained by classical physics. Classical physics would predict that the kinetic energies of the electrons leaving the surface should depend on the *brightness* of the light. No such dependence exists; on the contrary, the kinetic energies depend on the *frequency* (color) of the light (Figure 6.20). Feeble blue light produces photoelectrons with higher energies than does bright red light. Moreover, if the frequency of the light falls below a certain minimum value, called the *threshold frequency*, the photoelectric effect is not observed at all.

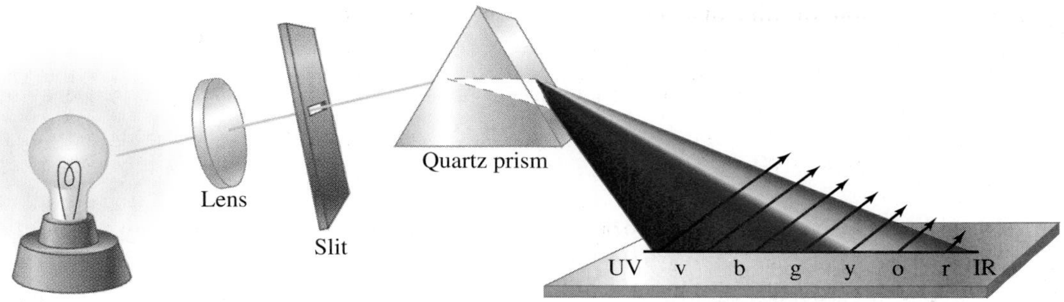

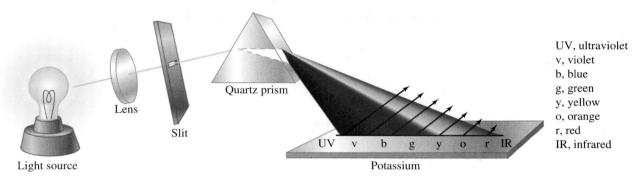

Figure 6.20 The photoelectric effect and the frequency of light. A beam of "white" light is dispersed into its wavelength components by a quartz prism and falls on a metal sample (here potassium). Light of the highest frequencies (ultraviolet and violet) produces the most energetic photoelectrons (longest arrows). Light of lower frequencies (for example, yellow) results in less energetic photoelectrons (shorter arrows). Light with a frequency lower than $4.23 \times 10^{14}\ s^{-1}$ (710 nm) produces no photoelectric effect at all on potassium, regardless of how bright (intense) the light is.

Einstein suggested that electromagnetic radiation itself is quantized; that light can be thought of as little packets of energy that we now call **photons**. He proposed that the energy of a photon of light of a particular frequency is given by

$$\text{Energy of photon} = E = h\nu$$

Einstein assumed that an electron is knocked from a surface atom when that atom absorbs a single photon of light. Obviously, the photon must have enough energy to overcome the attractive forces that normally hold the electron in the surface atom. More energetic photons will also do the job; any excess energy appears as kinetic energy in the ejected electron. But even repeated hits by photons with too little energy will not dislodge an electron. Einstein won the Nobel Prize in physics in 1921 for his explanation of the photoelectric effect.

To understand the photoelectric effect, consider the analogy of a truck stuck in the mud. A stream of people can come by, and each of them individually can give a push without getting the truck unstuck. A small garden tractor also would have less than the threshold energy necessary to dislodge the truck. A farm tractor, on the other hand, would be able to overcome the attractive forces of the mud and get the truck out. Larger tractors also could do the job. Some could even exert more pull than necessary, giving the freed truck some kinetic energy.

We can use the relationships

$$E = h\nu \quad \text{and} \quad c = \nu\lambda$$

to calculate the energy of a photon of light. These calculations, illustrated in the following examples and exercises, show how the energy associated with electromagnetic radiation is related to the portion of the electromagnetic spectrum. Ultraviolet radiation is more energetic than visible light. Violet light is more energetic than red, which in turn is more energetic than infrared. To stress that the expression $E = h\nu$ gives the energy of a single photon of light, we express Planck's constant as $h = 6.626 \times 10^{-34}\ \text{J} \cdot \text{s} \cdot \text{photon}^{-1}$.

Example 6.4

Calculate the energy, in joules, of a photon of violet light that has a frequency of $6.15 \times 10^{14}\ \text{s}^{-1}$.

Solution

$$\begin{aligned} E &= h\nu \\ &= 6.626 \times 10^{-34}\ \text{J} \cdot \text{s} \cdot \text{photon}^{-1} \times 6.15 \times 10^{14}\ \text{s}^{-1} \\ &= 4.07 \times 10^{-19}\ \text{J/photon} \end{aligned}$$

Exercise 6.5

Calculate the energy, in joules per photon, of microwave radiation that has a frequency of $2.89 \times 10^{10}\ \text{s}^{-1}$.

Example 6.5

A laser produces red light of wavelength 632.8 nm. Calculate the energy, in kilojoules, of *one mole* of photons of this red light.

Solution

A three-step approach is to determine (1) the frequency of the light using the expression $\nu = c/\lambda$; (2) the energy per photon with Planck's equation: $E = h\nu$; and (3) the energy per mole of photons by multiplying the result of the second step by Avogadro's number. Another approach that avoids having to write intermediate results is first to combine the two relevant expressions

$$E = h\nu = \frac{hc}{\lambda}$$

and to multiply this by Avogadro's number. That is,

$$E = \frac{6.626 \times 10^{-34}\ \text{J} \cdot \text{s} \times 3.00 \times 10^{8}\ \text{m} \cdot \text{s}^{-1}}{\text{photon}\left(632.8\ \text{nm} \times \dfrac{1 \times 10^{-9}\ \text{m}}{1\ \text{nm}}\right)} \times \frac{6.022 \times 10^{23}\ \text{photons}}{1\ \text{mol photons}}$$

$$= 1.89 \times 10^{5}\ \text{J/mol photons} = 189\ \text{kJ/mol photons}$$

Exercise 6.6

Calculate the wavelength in the far ultraviolet that corresponds to an energy of 1609 kJ/mol photons.

Estimation Exercise 6.2

Use Figure 6.14 and the result of Example 6.5 to estimate the portion of the electromagnetic spectrum where you would expect to find radiation with photons that have an energy content of 100 kJ/mol.

6.8 Bohr's Hydrogen Atom: A Planetary Model

Now back to our study of atomic structure. Classical physics requires the negatively charged electrons to be in motion around the positive nucleus of an atom; otherwise they would be attracted into it. But according to classical physics, electromagnetic radiation arises from the relative motion of charged particles (recall Figure 6.13). Electron motion would be accompanied by the *continuous* emission of light. With the loss of energy as light, electrons would be drawn closer to the nucleus and eventually spiral into it. Such an atom would therefore be *unstable*. Classical physics failed to explain atomic structure, just as it failed to explain emission spectra.

In 1913, Niels Bohr, a Danish physicist, combined ideas from classical physics and the new quantum theory to explain the structure of the hydrogen atom. In doing so, he was also able to explain the spectrum of light emitted by hydrogen atoms in gas discharge tubes.

Basing his work on that of Planck and Einstein, Bohr made the revolutionary assumption that certain properties of the electron in a hydrogen atom—including energy—can have only certain specified values. That is to say, these properties are *quantized*. Having made this basic assumption, Bohr was then able to use classical physics to calculate properties of the hydrogen atom. In particular, he derived an equation for the electron energy (E_n). Each specified energy value (E_1, E_2, E_3, . . .) is called an **energy level** of the atom, and the only allowable values are

$$E_n = \frac{-B}{n^2}$$

where n is an *integer* (that is, $n = 1, 2, 3, \ldots$), and B is a constant based on quantities such as Planck's constant and the mass and charge of an electron: $B = 2.179 \times 10^{-18}$ J. The parts of the hydrogen atom stay together (that is, the atom is stable) because the electron and nucleus attract one another. Energies associated with forces of attraction are taken to be *negative*, and this accounts for the negative sign in the energy level equation.

Bohr imagined the electron to be orbiting about the nucleus much as planets orbit the sun. Different energy levels correspond to different orbits, and only a discrete set of energy levels or orbits is possible (Figure 6.21).

Niels Bohr (1885–1962) proposed his theory of the hydrogen atom early in his career. Later, he directed the Institute of Theoretical Physics in Copenhagen, a center of attraction for theorectical physicists in the 1920s and 1930s. He worked on the atomic bomb project in World War II, but after the war became one of the strongest proponents of peaceful uses of atomic energy.

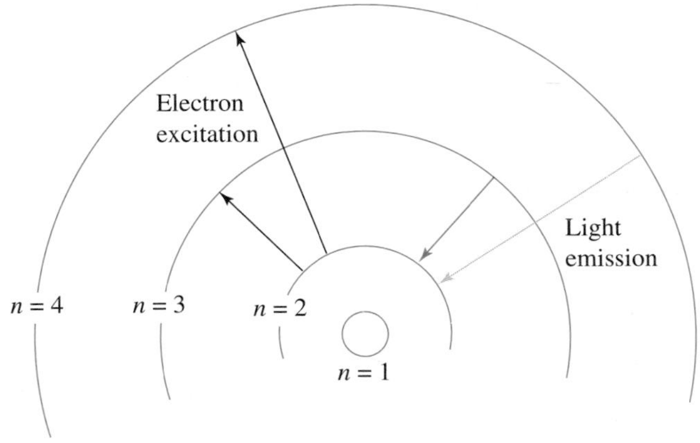

Figure 6.21 The Bohr model of the hydrogen atom. A portion of the hydrogen atom model is shown. The nucleus is at the center of the atom and the electron is in one of a set of discrete orbits $n = 1, 2, 3, \ldots$. When the atom is excited, the electron moves to a higher orbit (black arrows). Transitions in which an electron falls to a lower level are accompanied by the emission of light. Two spectral lines are shown in the approximate colors in which they appear in the Balmer series.

Example 6.6

Calculate the energy of an electron in the second energy level of a hydrogen atom.

Solution

Use the Bohr equation for the hydrogen atom, with $B = 2.179 \times 10^{-18}$ J and $n = 2$.

$$E_2 = \frac{-B}{n^2} = \frac{-2.179 \times 10^{-18}\ \text{J}}{2^2}$$

$$= \frac{-2.179 \times 10^{-18}\ \text{J}}{4} = -5.448 \times 10^{-19}\ \text{J}$$

Exercise 6.7

Calculate the energy of an electron in the level $n = 6$ of a hydrogen atom.

Bohr's Explanation of Line Spectra

Generally, we are more concerned with *differences* in energy levels than in the absolute value of an energy level. The change in energy (ΔE) when an electron moves from an initial energy level E_i to a final energy level E_f is

$$\Delta E = E_f - E_i$$

Or, if we use the Bohr energy-level equation as well,

$$\Delta E = \frac{-B}{n_f^2} - \frac{-B}{n_i^2} = B\left(\frac{1}{n_i^2} - \frac{1}{n_f^2}\right)$$

If $n_f > n_i$, the electron *absorbs* a quantum of energy and moves from the energy level n_i to the higher level n_f, and ΔE is positive. This energy absorption can occur, for example, in a gas discharge tube. If $n_f < n_i$, the electron drops from a higher energy level n_i to a lower energy level n_f, and a quantum of energy is *emitted* as a photon of light. In this case, ΔE is negative, corresponding to a loss of energy by the atom. In all transitions, the electron moves in a single "jump" from one level to another; there is no stopping between levels. Every transition in which energy is emitted produces a spectral line, and the collection of these lines is the observed emission spectrum.

Consider the analogy of a person on a ladder. The person can stand on the first rung, the second rung, the third rung, and so on, but is unable to stand between rungs. As the person goes from one rung to another, the potential energy (energy due to position) changes by definite amounts or quanta. For an electron, its total energy (both potential and kinetic) changes as it moves from one energy level to another.

Example 6.7

Calculate the energy change, in joules, that occurs when an electron falls from the $n_i = 5$ to the $n_f = 3$ energy level in a hydrogen atom.

Solution

We substitute the indicated energy levels and the value of B into the equation

$$\Delta E = B\left(\frac{1}{n_i^2} - \frac{1}{n_f^2}\right)$$

$$\Delta E = 2.179 \times 10^{-18}\left(\frac{1}{5^2} - \frac{1}{3^2}\right) = 2.179 \times 10^{-18}\left(\frac{1}{25} - \frac{1}{9}\right)$$

$$= 2.179 \times 10^{-18}\,(0.04000 - 0.1111) = -1.550 \times 10^{-19}\ \text{J}$$

The negative sign indicates that the atom has given up energy as a photon of light; by doing so it has become more stable.

Exercise 6.8

Calculate the energy change that occurs when an electron is raised from the $n_i = 2$ to the $n_f = 4$ energy level of a hydrogen atom.

We can easily calculate the frequency and wavelength of the photons released when an electron drops from one energy level to a lower one. Simply calculate the energy change, as in Example 6.7, and then use Planck's equation to determine the frequency of the light corresponding to this energy change. Finally, if necessary, use the relationship $c = \nu\lambda$ to calculate the wavelength of the light.

Example 6.8

Calculate the frequency, in s^{-1}, of the radiation released by the electron energy-level change described in Example 6.7.

Solution

An energy per photon of 1.550×10^{-19} J corresponds to radiation with frequency

$$\nu = \frac{E}{h} = \frac{1.550 \times 10^{-19}\ \text{J}}{6.626 \times 10^{-34}\ \text{J} \cdot \text{s}} = 2.339 \times 10^{14}\ \text{s}^{-1}$$

Exercise 6.9

Calculate the wavelength, in nanometers, that corresponds to the radiation released by the electron energy-level change from $n_i = 5$ to $n_f = 2$ in a hydrogen atom.

The Line Spectrum of Hydrogen

The emission spectrum of hydrogen consists of several series of lines. The series most frequently encountered are those in the visible, the ultraviolet, and the near-infrared portions of the electromagnetic spectrum. The electron transitions between energy levels that result in these spectral series are shown in Figure 6.22. The series in which each transition terminates at $n = 1$ appears

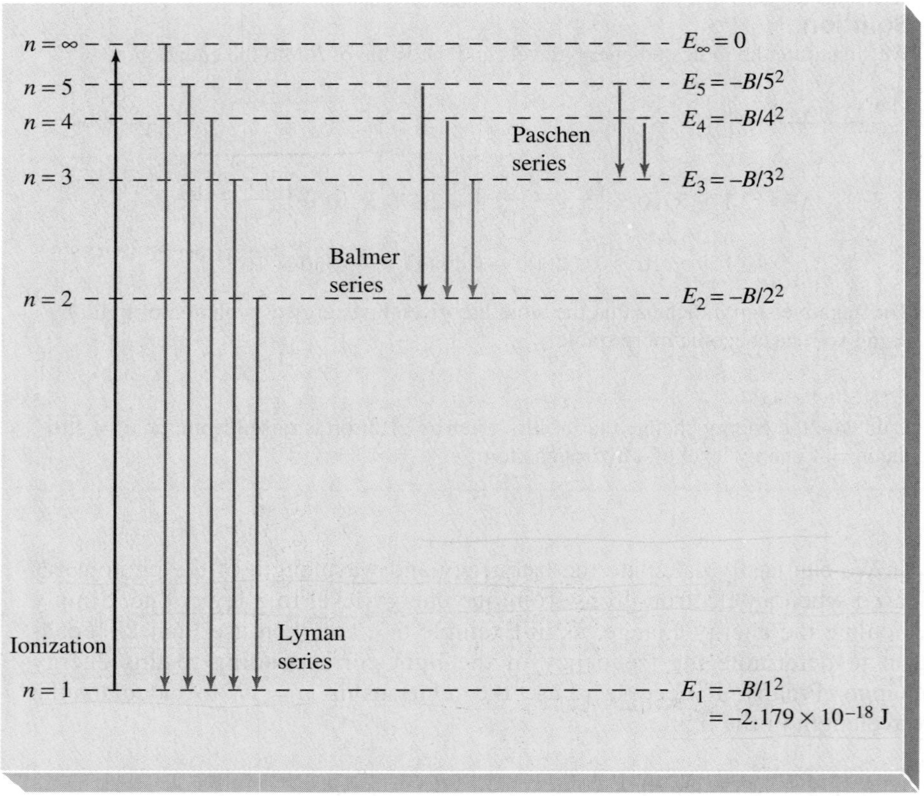

Figure 6.22 Energy levels and spectral lines for hydrogen. This diagram (not to scale) shows that

- The zero of energy corresponds to the completely ionized atom ($n = \infty$).
- Ionization of the normal hydrogen atom requires moving the electron from the level $n = 1$ to $n = \infty$, a process that requires 2.179×10^{-18} J of energy.
- Electron transitions from higher levels to the $n = 2$ level produce the visible lines in the Balmer series (three of the four lines are shown). Electron transitions to the $n = 1$ level produce lines in the ultraviolet, and transitions to the $n = 3$ level yield lines in the infrared. Each series is named for the scientist who discovered or characterized it.

in the ultraviolet. Four of the spectral lines in the series in which electrons end up at the $n = 2$ level are in the visible spectrum (recall Figure 6.16); the rest of the series is in the ultraviolet. The third series, in which transitions terminate at $n = 3$ (as in Example 6.7), appears in the infrared.

Ground States and Excited States

The electron in the hydrogen atom is usually in the lowest energy level. Normally, electrons remain in their lowest possible energy levels (orbits closest to the nucleus). Atoms whose electrons are in this situation are said to be in their **ground state**. When an electric discharge, a flame, or some other source supplies the energy to promote an electron from the lowest possible level to a higher level, the atom is said to be in an **excited state**. An atom in an excited state eventually emits energy as photons as the electron drops back down to one of the lower levels and ultimately reaches the ground state.

Bohr's theory established the important idea of energy levels in atoms and was a spectacular success in explaining the line spectrum of hydrogen. Bohr was awarded the Nobel Prize in physics in 1922 for this work. However, Bohr's theory does not work for atoms with several electrons, nor does it explain other details of atomic structure.

Atomic Absorption Spectroscopy

On page 211, we described emission spectroscopy. In that approach, light emitted by excited atoms is dispersed into line spectra and these spectra are analyzed. Now we know how the atoms become excited: by absorbing quanta with just the right energy to promote ground-state atoms to an excited state.

An effective way of raising ground-state atoms to an excited state is to strike the atoms in a high-temperature gaseous sample with a beam of light that has the same frequency that the excited atoms would themselves emit. Ground-state atoms *absorb* photons from the light beam, and the beam leaves the high-temperature region with a reduced intensity. The more atoms present of the type that can absorb the photons, the lower will be the intensity of the transmitted light.

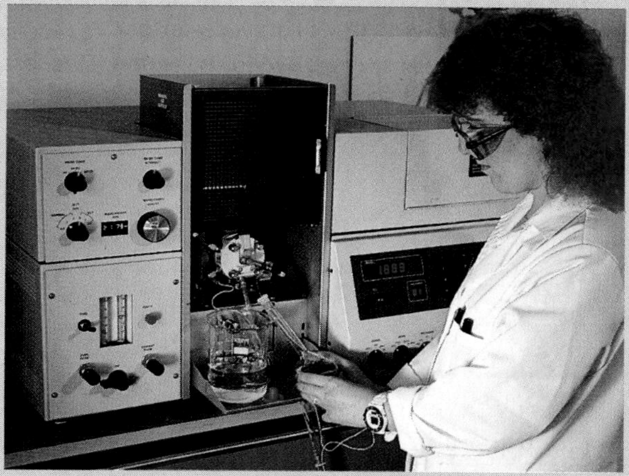

Atomic absorption spectrophotometry. The sample to be analyzed is vaporized in the flame, and the light-absorbing ability of the vaporized atoms is measured.

Suppose, for example, that we were continuously to vary the wavelength of a light beam passing through a sample of gaseous sodium. When we hit the wavelength 589.0 nm, an unusually strong *absorption* of light would occur. This is the same wavelength as one of the two bright yellow lines in the sodium emission spectrum. In *atomic absorption spectroscopy*, we analyze light that is absorbed rather than light emitted.

Over 70 elements can be quantitatively determined by atomic absorption spectroscopy, even in amounts as low as a few parts per billion in some cases. Atomic absorption spectroscopy is routinely used in the food industry to test for metals such as calcium, copper, and iron; in environmental analysis for metals such as cadmium, lead, and mercury; and in astronomy to study the chemical makeup of the stars and the atmospheres of the planets.

CONCEPTUAL EXAMPLE 6.1

Which of the four electronic transitions shown in Figure 6.23 produces the shortest wavelength spectral line in the hydrogen emission spectrum?

Solution

We could calculate the energy difference between each pair of energy levels, convert this energy difference to the frequency and wavelength of a photon of light (as in Examples 6.7 and 6.8), and choose the shortest wavelength. But we don't need to do all of this.

First, we should recognize that transition (a), even though it involves the greatest span of energy levels, requires energy *absorption*, not light emission. We can eliminate (a). The other three transitions do involve light emission. Let's compare the energy changes in these transitions, recognizing that the larger the energy change, the *shorter* the wavelength of the spectral line produced. Transitions (b) and (d) both terminate at the same level, $n = 2$, but the drop in energy in (b) is greater than in (d). Transition (c) is comparable to transition (d) but terminates at a higher energy level, $n = 3$. The energy change for transition (c) is less than that for transition (d). In conclusion, the shortest wavelength spectral line of the three is that produced by transition (b).

Conceptual Exercise 6.1

With a *minimum* of calculation, determine which of the following electronic transitions in a hydrogen atom requires the greatest amount of energy to be absorbed: (**a**) from $n = 1$ to $n = 2$; (**b**) from $n = 3$ to $n = \infty$; (**c**) from $n = 4$ to $n = 1$; (**d**) from $n = 2$ to $n = 3$.

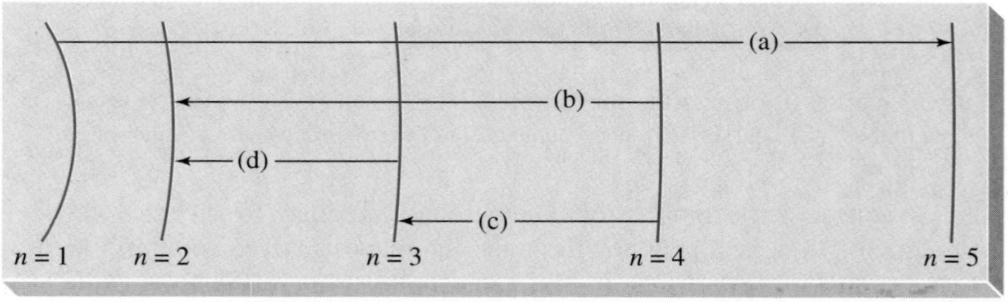

Figure 6.23 Several possible electronic transitions in a hydrogen atom.

6.9 Wave Mechanics: Matter as Waves

When passed through a prism, light is dispersed (bent and spread out) into a spectrum, like the scattering of water waves around obstacles in their path. In the photoelectric effect, light acts as if it consists of a stream of particles or packets of energy (photons). It is as if light has two natures—wave and particle. It reveals one or another of its natures depending on the type of observation we make.

We usually regard matter as consisting of particles. Is it possible that matter could, under the proper circumstances, behave as waves? Such speculation led Louis de Broglie (1892–1987), a French physicist, to propose a startling new theory in 1923. De Broglie's theory, in turn, led to a new mathematical description of atoms that still has broad application in modern chemistry.

De Broglie's Equation

De Broglie proposed that a particle with a mass m moving at a speed u will have some wave nature, consistent with a wavelength given by

$$\lambda = \frac{h}{m \cdot u}$$

The symbol h, once again, is Planck's constant.

Even large objects presumably have wave properties, but their associated wavelengths are so short that they cannot be observed. For example, a 1000-kg car moving at 100 km/h has an associated wavelength of 2.39×10^{-38} m, far shorter than that of the shortest wavelength (most energetic) gamma rays (see Figure 6.14 for comparison). Subatomic particles, on the other hand, have wave properties that can be readily observed.

Example 6.9

Calculate the wavelength, in meters and nanometers, of an electron moving at a speed of 2.74×10^6 m/s. The mass of an electron is 9.11×10^{-31} kg. Recall that 1 J = 1 kg $\cdot$ m^2 $\cdot$ s^{-2}.

Solution

$$\lambda = \frac{h}{m \cdot u} = \frac{6.626 \times 10^{-34}\ \text{kg} \cdot \text{m}^2 \cdot \text{s}^{-2} \cdot \text{s}}{9.11 \times 10^{-31}\ \text{kg} \times 2.74 \times 10^6\ \text{m} \cdot \text{s}^{-1}}$$

$$= 2.65 \times 10^{-10}\ \text{m}$$

The wavelength of the electron is 2.65×10^{-10} m or 0.265 nm.

Exercise 6.10

Calculate the wavelength, in nanometers, of a proton moving at a speed of 3.79×10^3 m/s. The mass of a proton is 1.67×10^{-27} kg.

De Broglie's predictions were verified six years later, and his work soon led to the development of the electron microscope. This instrument, which makes use of the wave nature of electrons, is now a standard piece of equipment in many scientific laboratories. It can produce images of objects only a few hundred picometers in size (1 pm = 10^{-12} m). De Broglie was awarded the Nobel Prize in physics in 1929 for his work on the wave nature of matter.

The smallest objects visible through a microscope are those having dimensions of the same order of magnitude as the wavelength of the radiation—about 1000 nm for visible light microscopes and 1 nm or less for electron microscopes.

Wave Functions

The sophisticated mathematical description of atomic structure based on the wave properties of subatomic particles is called **wave mechanics** or **quantum mechanics**. A model of the hydrogen atom based on the wave nature of the electron was developed in the late 1920s, principally by the Austrian physicist Erwin

Erwin Schrödinger (1887–1961) was at the University of Zurich when he promulgated his wave equation (1926) and then moved to the University of Berlin in 1928 to succeed Max Planck. Although he was not Jewish, Schrödinger left Nazi Germany in 1933, the same year he was awarded the Nobel Prize in physics. In his later years Schrödinger's interest turned to biology, expressed in his famous 1944 book, *What Is Life?*

Werner Heisenberg (1901–1976) devised an approach to atomic structure while on a North Sea vacation in 1927. His approach, called matrix mechanics, was purely mathematical but proved to be equivalent to Schrödinger's wave equation. Later in that year he deduced his famous uncertainty principle. Unlike Schrödinger, Heisenberg remained in Nazi Germany during World War II, directing an unsuccessful atomic bomb project.

Schrödinger. Mathematical equations describing the nature of electron waves in atoms are fundamental to the modern picture of the atom. The wave equations that are acceptable solutions to the Schrödinger equation are called *wave functions.* To obtain one of these acceptable solutions, we must assign integral values, called **quantum numbers**, to three quantities in the wave equation, much as an integral value of n is required in the Bohr equation for the hydrogen atom (see Figure 6.24).

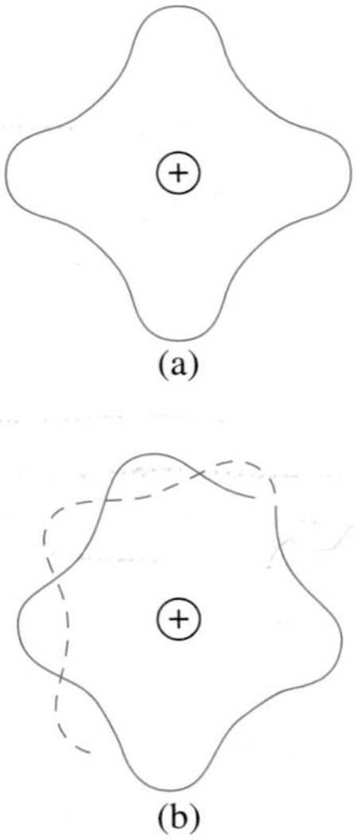

Figure 6.24 **The electron as a matter wave.** These patterns are two-dimensional cross sections of a much more complicated three-dimensional wave. The wave pattern in (a), called a standing wave, is an acceptable representation because it has an integral number of wavelengths (four) around the nucleus. The pattern in (b) is unacceptable because the number of wavelengths is nonintegral (4.5). The crests in one part of the wave in (b) overlap the troughs in another part, the wave cancels itself, and there is no wave at all. Quantum numbers, which are integers, restrict solutions of a wave equation to acceptable ones; that is to ones like (a) rather than (b).

In contrast to the precise planetary orbits of the Bohr atom, wave mechanics provides a less certain picture of the hydrogen atom. Instead of speaking of the exact location of the electron, we can only say something about the probability of the electron being found in certain regions of the atom. Or, if we adopt the view that the electron is just a cloud of negative electric charge, we can only speak of the charge densities in various parts of the atom. This "fuzzier" atomic model is the only type acceptable according to an important principle of science established in 1927 by Werner Heisenberg.

Heisenberg's **uncertainty principle** states, roughly, that we can't simultaneously know exactly where a tiny particle is and exactly how it is moving. One way to visualize problems associated with the principle is that when we try to make measurements on extremely small particles, the act of measuring interferes with the particle we are measuring. That is, we produce changes in the one measured quantity just in the act of measuring some other quantity. We can focus on a particle's position, but if we do, we lose our sense of its momentum. (Momentum is the product of the particle's mass and velocity.) If we focus on the particle's momentum, we lose our sense of its exact position. Heisenberg created an equation that permits a calculation of the combined degree of uncertainty that exists in the simultaneous measurements of an object's position and momentum.

According to Heisenberg's principle, if the mass of a particle is large, the uncertainties in position and momentum are extremely small compared to the values of those quantities. We can, therefore, readily and accurately describe the orbit of Earth about the sun or of the smallest artificial satellite about Earth. The mass of an electron, on the other hand, is so small that the uncertainties of its position and momentum are significant. In fact, the uncertainty in its position is of about the same size as the atom itself. We can, however, use quantum mechanics to determine the chance or *probability* of finding an electron in a certain region of an atom.

Bohr's theory fails with respect to the uncertainty principle because it tells us more than we can know with certainty. It gives us a precise location for the electron in a hydrogen atom—the particular orbit—and it enables us to make a precise calculation of the electron speed in this orbit. It is interesting to note, however, that the "most probable" distance from the nucleus of an electron in the first energy level of a hydrogen atom calculated from wave mechanics—52.9 pm—is the same as the radius of the first orbit calculated with Bohr's theory.

A new theory replaces an old one when it explains something the old one can't. Moreover, the new theory must be able to answer all the questions that the old theory can—and answer them better.

Each wave function that is a solution to the Schrödinger equation is denoted by the Greek letter ψ (psi, pronounced "sigh") and represents an energy state of the atom. Max Born (1882–1970) was the first to propose an interpretation of wave functions in terms of probabilities: The square of the wave function (ψ^2) gives the probability of finding the electron in a particular volume of space in the atom.

Suppose we were to divide the space around the nucleus of an atom into tiny volumes and calculate the probability of finding an electron in each of these volumes. Now suppose we were to plot those electron probabilities as dots on a map. The result would be analogous to a map of population density for a country; people are most likely to be found where the dots are thickest and least likely to be found where there are fewest dots. We will consider such plots in detail in the next section, but for now consider this general idea:

The plot of electron probability for the first energy level shows that the electron is likely to be found close to the nucleus. In the second energy level, the electron would most likely be found farther from the nucleus. In the third energy level, the most probable distance would be farther still, and so on. We do not know exactly where the electron is in any case, but we do know where it is most likely to be found.

6.10 Quantum Numbers and Atomic Orbitals

As we have noted, finding acceptable wave functions to satisfy Schrödinger's wave equation for the hydrogen atom requires assigning specific integral values called quantum numbers to three parameters in the equation. A given set of these three quantum numbers yields a wave function for an electron called an **atomic orbital**. Atomic orbitals, although they are only mathematical expressions, allow us to visualize an entire three-dimensional region in an atom in which there is a significant probability of finding electrons. Consequently, we tend to think of orbitals in terms of these geometric regions rather than just as mathematical expressions. Because orbitals are related to quantum numbers, let's first say something more about quantum numbers.

Quantum Numbers

Here we describe the three quantum numbers and their possible values. Specifying a set of values for the three quantum numbers defines a specific atomic orbital. Table 6.1 summarizes the discussion that follows, and you may find it helpful to refer to the table from time to time.

TABLE 6.1 Electronic Shells, Orbitals and Quantum Numbers

Principal Shell	1st		2nd					3rd							
$n =$	1	2	2	2	2	3	3	3	3	3	3	3	3	3
$l =$	0	0	1	1	1	0	1	1	1	2	2	2	2	2
$m_l =$	0	0	−1	0	+1	3	−1	0	+1	−2	−1	0	+1	+2
Subshell and orbital designation	1*s*	2*s*	2*p*	2*p*	2*p*	3*s*	3*p*	3*p*	3*p*	3*d*	3*d*	3*d*	3*d*	3*d*
Number of orbitals in the subshell	1	1		3		1		3				5		

1. The first or **principal quantum number (n)** designates the main or *principal energy level*. This is the most important quantum number because values of the other two depend on the value assigned to n. The value of n is a *positive integer* (whole number).

$$n = 1, 2, 3, 4, 5, 6, \ldots$$

The quantum number n is analogous to that specified by Bohr in his planetary model. As n increases, the electron spends more of its time farther from the nucleus. The orbital becomes larger and the electron in it is higher in energy. Orbitals with the same value of n are said to be in the same *principal shell* or **principal level**.

2. The second or **angular momentum quantum number (l)** designates the particular energy **sublevel** or *subshell* within a principal energy level. It can have positive integral values from zero to one less than the value of n.

$$l = 0, 1, 2, \ldots, (n - 1)$$

The value of the l quantum number also determines the *type* of orbital. The four types most commonly encountered are

Value of l:	0	1	2	3
Energy sublevel or orbital designation:	*s*	*p*	*d*	*f*

Among the chief differences for these orbital types are the geometric shapes of the regions of space they describe. Also note that the number of sublevels or different kinds of orbitals in a principal energy level or shell is equal to the principal quantum number n. For example, the third principal level (with $n = 3$) has three sublevels and three kinds of orbitals: *s*, *p*, and *d*, corresponding to l values of 0, 1, and 2, respectively.

3. The third or **magnetic quantum number (m_l)** designates a particular orbital within a given sublevel. It can have any integral value ranging from

$-l$ to $+l$. The value of m_l distinguishes among the orbitals with given n and l values. For $l = 0$, m_l must also be zero, and there is only one orbital of the s type. But for $l = 1$, m_l can have values of -1, 0, or $+1$, and there are *three* orbitals of the p type. An s subshell consists of a single s orbital; a p subshell is made up of *three* p orbitals; a d subshell is made up of *five* d orbitals ($m_l = -2, -1, 0, +1, +2$); and an f subshell is made up of *seven* f orbitals ($m_l = -3, -2, -1, 0, +1, +2, +3$).

orbitals - ½ of subshell capacity.

Example 6.10 tests your understanding of relationships among the quantum numbers. Example 6.11 asks you to use this understanding to answer questions about principal shells, subshells, and atomic orbitals.

Example 6.10

Considering the limitations on values for the various quantum numbers, state whether an electron can have each of the following sets. If a set is not possible, state why it is not.

a. $n = 2, l = 1, m_l = -1$
b. $n = 1, l = 1, m_l = +1$
c. $n = 7, l = 3, m_l = +3$
d. $n = 3, l = 1, m_l = -3$

Solution

a. All the quantum numbers are allowed values.
b. Not possible. The value of l must be less than n.
c. All the quantum numbers are allowed values.
d. Not possible. The value of m_l must be in the range $-l$ to $+l$ (in this case, -1 to $+1$).

Exercise 6.11

Consider the limitations on values for the various quantum numbers, and state whether an electron can have each of the following sets. If a set is not possible, state why it is not.

a. $n = 2, l = 1, m_l = -2$
b. $n = 3, l = 2, m_l = +2$;
c. $n = 4, l = 3, m_l = +3$
d. $n = 5, l = 2, m_l = +3$

Example 6.11

Consider the relationship among quantum numbers and orbitals, subshells, and principal shells to answer the following: **(a)** How many orbitals are there in the $4d$ subshell? **(b)** What is the first principal shell in which f orbitals can be found? **(c)** Can an atom have a $2d$ subshell? **(d)** Does a hydrogen atom have a $3p$ subshell?

Solution

a. The d subshell corresponds to a value of $l = 2$. With $l = 2$ there are five possible values of m_l: $-2, -1, 0, +1$, and $+2$. There are *five d* orbitals in the $4d$ subshell.
b. The f orbital corresponds to a value of $l = 3$. Since the maximum value of l is $n - 1$, the allowable values of n are 4, 5, 6, . . . The first shell to contain f orbitals is the fourth principal shell ($n = 4$).
c. For a d subshell, $l = 2$. The maximum value of l is $n - 1$, so l cannot be 2 if $n = 2$. There cannot be a $2d$ subshell.
d. Yes, although the electron in a hydrogen atom is usually in the $1s$ orbital, it can be excited to one of the higher energy states, such as one of the $3p$ orbitals.

Exercise 6.12

Consider the relationship among quantum numbers and orbitals, subshells, and principal shells to answer the following: **(a)** How many orbitals are there in the $5p$ subshell? **(b)** What is the *total* number of orbitals in the principal shell $n = 4$? **(c)** What subshell of the hydrogen atom consists of a total of seven orbitals?

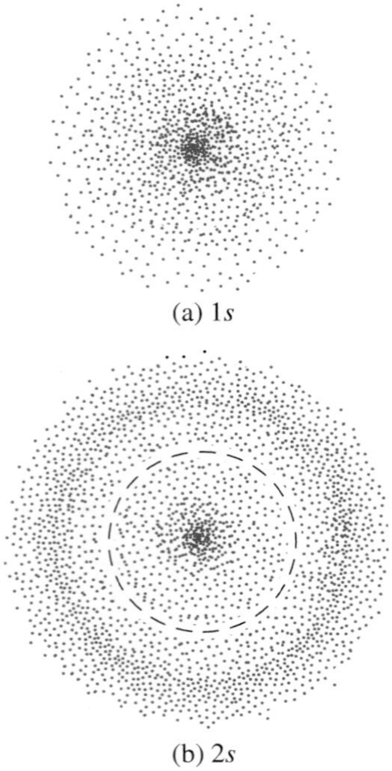

Figure 6.25 **The 1*s* and 2*s* orbitals.** (a) A pattern of dots representing electron probabilities for the 1*s* orbital in a plane with the nucleus at the center. Closely spaced dots indicate a greater electron density. (b) The pattern of dots for the 2*s* orbital covers a larger region than that in a 1*s* orbital. The broken circle indicates a region (called a *node*) in which there is no probability of finding an electron. The most probable location of the electron is in a spherical shell found in the second ring of dots.

The Geometric Shapes of Atomic Orbitals

What do the regions of high electron probabilities described by atomic orbitals look like? We can't see these regions, of course, but the mathematical forms of the orbitals can be translated into geometric pictures. Suppose you had a high-speed movie camera that could photograph electrons (there is no such thing, but we are just supposing) and you left the shutter open while an electron zipped about the nucleus. When you developed the picture, you would have a record of where the electron had been. (Doing the same thing with an electric fan would give a picture in which the blades of the fan looked like a solid disk of material. The blades move so rapidly that their photographic image is blurred.)

The wave mechanical picture of an electron in an *s* orbital looks like a fuzzy ball (that is, the orbital has *spherical* symmetry). The size of the orbital and the specific electron probability distribution depend on the value of n. Figure 6.25 suggests that the greatest probability of finding an electron in a tiny unit of volume in a 1*s* orbital is at the nucleus itself. A more meaningful quantity, however, is the probability of finding an electron somewhere in a spherical shell surrounding the nucleus. This requires adding up the probabilities for all the volume units at a given distance from the nucleus. When this is done, the region of maximum probability is a spherical shell of radius 52.9 pm, a distance that is identical to the first Bohr orbit of hydrogen. Of course, the wave–particle duality of an electron allows us to think of the electron as something like a cloud of negative electric charge rather than as a particle. In this case we can substitute electron charge density for electron probability. Thus, the spherical shell at 52.9 pm is the region of highest electron charge density in the 1*s* orbital.

The 2*s* orbital has two regions of high probability; both are spherical. The one near the nucleus is separated from the outer region by a spherical *node*—a spherical shell in which the electron probability is zero. The 2*s* electron is more likely to be found in the outer region.

Strictly speaking, atomic orbitals must have an infinite size to account for an electron with 100% certainty. Quantum mechanics states that there is at least a tiny probability that any electron can be anywhere in the universe. For convenience in representing orbitals, however, an outline drawing is used to indicate a certain probability (say, a 90% probability) that the electron is located within the volume so outlined (see Figure 6.25). Alternatively, we can say that the volume contains a certain percentage of the electron's charge (such as 90%).

The second principal energy level consists of four different orbitals. One, having $l = 0$, is the 2*s* orbital just described. The other three orbitals of the $n = 2$ level have $l = 1$. They describe dumbbell-shaped regions (Figure 6.26);

Figure 6.26 **The three 2*p* orbitals.** The region of high electron probabilities are dumbbell-shaped and oriented along the *x*,*y*, and *z* axes, respectively. Each of the orbitals has a *nodal plane*, a planar region of zero electron probability that passes through the nucleus of the atom.

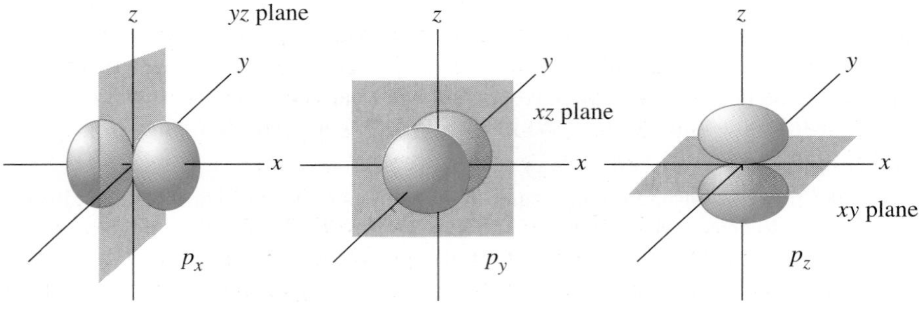

two lobes lie along a line with the nucleus at the center between the lobes. In a 2*p* orbital there is a planar node between the two "halves" of the orbital. To distinguish among the different 2*p* orbitals they are sometimes referred to as the $2p_x$, $2p_y$, and $2p_z$ orbitals because they are perpendicular to one another and can be drawn along the *x*, *y*, and *z* coordinate axes (see Figure 6.26). Higher level *p* orbitals (that is, 3*p*, 4*p*, . . .) have similar overall shapes, but they have spherical and planar nodes between regions of higher electron probability; their sizes depend on the value of *n*.

The third principal energy level is divided into three sublevels: one 3*s* orbital, three 3*p* orbitals, and five 3*d* orbitals. The *d* orbitals are pictured in Figure 6.27.

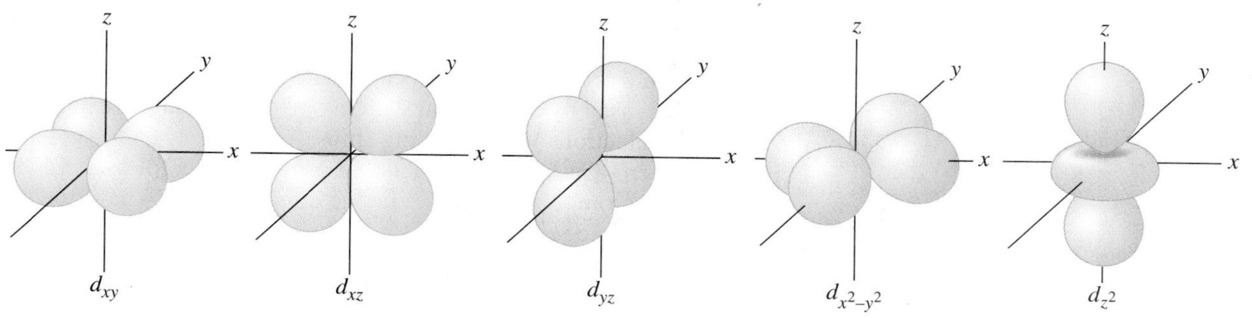

Figure 6.27 The five *d* orbitals. The designations *xy*, *xz*, *yz*, and so on are determined by the directional characteristics of certain combinations of the orbitals with the allowed values of m_l ($l = 2$) for all *d* orbitals.

The fourth principal energy level is divided into four sublevels or subshells. There are a 4*s* subshell with one orbital, a 4*p* subshell with three orbitals, a 4*d* subshell with five orbitals, and a 4*f* subshell with seven orbitals. The *f* orbitals have complex shapes. Because *f* orbitals are seldom involved in chemical interactions, we will not consider them further in this text.

6.11 Electron Spin: A Fourth Quantum Number

The quantum numbers *n*, *l*, and m_l arise from Schrödinger's treatment of the electron in a hydrogen atom as a matter wave; by extension, this concept can be applied to electrons in other atoms as well. Although these three quantum numbers fully characterize the orbitals in an atom, there is a need for a fourth quantum number to describe the electrons that occupy these orbitals. It was proposed in 1925 by Samuel Goudsmit and George Uhlenbeck to explain some of the finer features of atomic emission spectra. This fourth quantum number is called the **electron spin quantum number (m_s)**. The two possible values of the spin quantum number are $+\frac{1}{2}$ (also represented by the arrow, ↑) and $-\frac{1}{2}$ (↓).

The name, electron spin quantum number, suggests that electrons have a spinning motion. However, there is no way to attach a precise physical reality to electron spin. All that we know is that *experiments* have shown that electrons endow atoms with properties that can be explained as if the electron were spinning.

In 1921 Otto Stern and Walter Gerlach performed the experiment suggested by Figure 6.28. When they shot a beam of gaseous silver atoms into a powerful

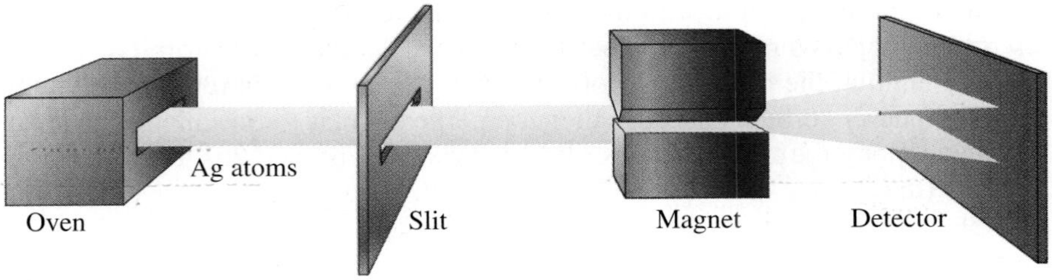

Figure 6.28. The Stern-Gerlach experiment: demonstration of electron spin. Silver atoms vaporized in the oven are shaped into a beam by the slit, and the beam is passed through a nonuniform magnetic field. The beam splits in two. (The beam of atoms would not experience a force if the magnetic field were uniform. The field strength must be greater in certain directions than others.)

magnetic field, the beam was split in two by the field, indicating that the atoms themselves act like small magnets.

It has long been known that a moving electric charge produces a magnetic field about itself, so a spinning electron should have an associated magnetic field. The magnetic field produced by an electron with $m_s = +\frac{1}{2}$ is in the opposite direction to the field caused by an electron with $m_s = -\frac{1}{2}$. The magnetic effect is canceled for a pair of electrons with opposing spins.

To explain the Stern–Gerlach experiment we can reason as follows: Of the 47 electrons in a silver atom, 23 have a spin in one direction and 24 have the opposite spin. We can consider, then, that 46 of the electrons have their spins "paired." The direction of deflection of a given silver atom depends on the type of spin on the 47th electron. In a collection of silver atoms, there is an equal chance that the "odd" electron will have a spin of $+\frac{1}{2}$ (↑) or $-\frac{1}{2}$ (↓). The beam of silver atoms is thus split into two beams. A beam of hydrogen atoms would be similarly split in a magnetic field, indicating that half the electrons have a spin of $+\frac{1}{2}$ and half, $-\frac{1}{2}$.

Looking Ahead

We have now described the four quantum numbers that are used to characterize electrons in atoms. In the next chapter we will see how these quantum numbers, together with a few simple rules, allow us to establish which orbitals in atoms are occupied by electrons. And we will discover a close relationship between the electronic structures of atoms and the periodic table. In Chapters 8 and 9 we will see how the electronic structures of atoms in turn determine the nature of chemical bonds between atoms and the structures of molecules.

SUMMARY

Cathode rays (electrons) are produced when electricity passes through evacuated tubes. X-rays form when cathode rays strike matter.

Radioactivity is the emission of radiation by unstable atomic nuclei, and the most common types of radiation are alpha (α) particles, beta (β) particles, and gamma (γ) rays. Alpha particles are helium nuclei; beta particles are electrons; and gamma rays are high-energy electromagnetic radiation similar to X-rays.

Rutherford's atomic model is that of a very small positively charged nucleus and extranuclear electrons. The nucleus consists of protons and neutrons and contains practically all the mass of an atom. Atomic masses and relative abundances of the isotopes of an element can be established by mass spectrometry. The atomic weight of the element is the weighted average of these atomic masses.

Electromagnetic radiation is the transmission of electric and magnetic fields as a wave motion. The waves are characterized by their frequencies (ν), wavelengths (λ), and constant velocity in a medium: $c = \nu \cdot \lambda$. A light source that emits an essentially unbroken series of wavelength components has a continuous spectrum. Only a discrete set of wavelength components is present in the emission spectrum of an element.

Einstein's explanation of the photoelectric effect views light as packets of energy called photons. The energy of a photon (E) is given by the expression $E = h \cdot \nu$; where h is Planck's constant.

Bohr's theory requires the electron in a hydrogen atom to be in one of a discrete set of energy levels. The fall of an electron from a higher to a lower energy level releases a discrete amount of energy as a photon of light with a characteristic frequency. Bohr's theory accounts for the observed atomic spectrum of hydrogen.

The electron in a hydrogen atom can be viewed as a matter wave enveloping the nucleus. The matter wave is represented by a wave equation, and solutions of the wave equation are wave functions. Each wave function is characterized by the value of three quantum numbers: the principal quantum number, n; the angular momentum quantum number, l; and the magnetic quantum number, m_l. Wave functions with acceptable values of the three are called atomic orbitals. An orbital describes a region in an atom that has a high probability of containing an electron or a high electronic charge density. Orbitals with the same value of n are in the same principal energy level or principal shell. Those with the same value of n and of l are in the same sublevel or subshell. The shapes associated with orbitals depend on the value of l. Thus, the s orbital ($l = 0$) is spherical and the p orbital ($l = 1$) is dumbbell shaped.

The n, l, and m_l quantum numbers define an orbital, but a fourth quantum number is also required to characterize an electron in an orbital—the spin quantum number, m_s. This quantum number may have either of two values: $+\frac{1}{2}$ or $-\frac{1}{2}$.

KEY TERMS

alpha particle (α) (6.2)
angular momentum quantum number (l) (6.10)
atomic orbital (6.10)
beta particle (β) (6.2)
cathode ray (6.1)
electromagnetic spectrum (6.6)
electromagnetic wave (6.6)
electron spin quantum number (m_s) (6.11)
emission (line) spectrum (6.6)
energy level (6.8)
excited state (6.8)
frequency (ν) (6.6)
gamma ray (γ) (6.2)
ground state (6.8)
magnetic quantum number (m_l) (6.10)
mass spectrometer (6.4)
nuclide (6.5)
photon (6.7)
Planck's constant (h) (6.7)
principal level (shell) (6.10)
principal quantum number (n) (6.10)
quantum (6.7)
quantum (wave) mechanics (6.9)
quantum numbers (6.9)
radioactivity (6.2)
sublevel (subshell) (6.10)
uncertainty principle (6.9)
wave (6.6)
wavelength (λ) (6.6)
X-ray (6.2)

REVIEW QUESTIONS

1. What is a cathode ray? How was the nature of cathode rays determined?
2. Describe J. J. Thomson's experiment that determined the mass-to-charge ratio of the electron.
3. Describe Millikan's oil-drop experiment. How did Millikan determine the charge on an electron?
4. What is radioactivity? How was it discovered?
5. What are alpha particles? What are beta particles? What are gamma rays? How was the nature of each established?
6. How do gamma rays differ from X-rays? How are the two kinds of rays similar?
7. How did the discovery of radioactivity cause Dalton's atomic theory to be modified?
8. J. J. Thomson proposed the "plum-pudding" model in which electrons (plums) are embedded in a smear

of positive charge (pudding). How did Rutherford's interpretation of the Geiger–Marsden gold-foil experiments contradict Thomson's model?

9. What is the nucleus? Describe the nuclear atom.

10. How does a mass spectrometer work?

11. Sketch a bar-graph mass spectrum of a sample of silicon, which has three isotopes with the following masses and fractional abundances. (*Hint*: Recall Figure 6.11.)

Isotope	Mass, u	Fractional abundance
Silicon-28	27.977	0.9223
Silicon-29	28.976	0.0467
Silicon-30	29.974	0.0310

12. What is meant by the wave–particle duality of light?

13. Describe Planck's explanation of the light given off by a hot solid at different temperatures (black-body radiation).

14. Describe the photoelectric effect. How did Einstein explain it?

15. How was Rutherford's nuclear model of the atom refined by Bohr?

16. When light is emitted by an atom, what change has occurred within the atom?

17. When light is absorbed by an atom, what change has occurred within the atom?

18. Which atom absorbed more energy: one in which an electron moved from the second energy level to the third level or an otherwise identical atom in which an electron moved from the first to the third energy level?

19. What is an atomic orbital? What information does it give about the electronic structure of an atom?

20. What limitations on our knowledge of the electronic structure of atoms are required by Heisenberg's uncertainty principle?

21. Describe the three quantum numbers that arise from the solution of Schrödinger's wave equation and list their possible values.

22. Give the subshell notation for each of the following sets of quantum numbers.
a. $n = 3$ and $l = 2$
b. $n = 2$ and $l = 0$
c. $n = 4$ and $l = 1$
d. $n = 4$ and $l = 3$

23. How many sublevels are there in an energy level with each of the following values of n?
a. $n = 3$ **b.** $n = 2$ **c.** $n = 4$

24. What is the geometric shape described by an s orbital. How does a $2s$ orbital differ from a $1s$ orbital?

25. What is the geometric shape described by a p orbital?

26. Describe the property called electron spin and an experiment that can be explained in terms of electron spin.

PROBLEMS

Subatomic Particles

27. The mass-to-charge ratio of the proton is found to be 1.044×10^{-8} kg/C. The charge on the proton is 1.602×10^{-19} C. Calculate the mass of the proton.

28. The mass-to-charge ratio of the positron (a product of the radioactive decay of certain isotopes) is 5.686×10^{-12} kg/C. The charge on the positron is 1.602×10^{-19} C. Calculate the mass of the positron. Does any other particle have this same mass?

29. An oil-drop experiment results in the following charges on oil droplets: 6.4×10^{-19} C, 3.2×10^{-19} C, and 4.8×10^{-19} C. What value for the electronic charge can be deduced from this experiment?

30. An oil-drop experiment results in the following charges on oil droplets: 9.6×10^{-19} C, 3.2×10^{-19} C, and 16.0×10^{-19} C. What possible value of the electronic charge can be deduced from this experiment? Why might this value of the electronic charge be different from that deduced in Problem 29? Can the results of both experiments be correct? Explain.

31. Determine approximate mass-to-charge ratios for the following ions.
a. $^{80}Br^{-}$ **b** $^{18}O^{2-}$ **c.** $^{40}Ar^{+}$

Why are these values only approximate? (*Hint:* What are the masses of these ions?)

Atomic Weights from Mass Spectrometry

32. Gallium in nature consists of two isotopes: gallium-69, with a mass of 68.926 u and a fractional abundance of 0.601, and gallium-71, with a mass of 70.925 u and a fractional abundance of 0.399. Calculate the atomic weight of gallium.

33. Europium in nature consists of two isotopes: europium-151, with a mass of 150.92 u and a fractional abundance of 0.478, and europium-153, with a mass of 152.92 u and a fractional abundance of 0.522. Calculate the atomic weight of europium.

34. Neon in nature consists of the following isotopes.

Isotope	Atomic mass, u	Percent abundance
Neon-20	19.9924	90.51
Neon-21	20.9940	0.27
Neon-22	21.9914	9.22

Calculate the atomic weight of neon.

35. Strontium in nature consists of the following isotopes.

Isotope	Atomic mass, u	Percent abundance
Strontium-84	83.913	0.56
Strontium-86	85.909	9.86
Strontium-87	86.909	7.00
Strontium-88	87.906	82.58

Calculate the atomic weight of strontium.

36. The isotope potassium-40 is a relatively rare, radioactive isotope with an atomic mass of 39.974 u. Given that the atomic weight of potassium is 39.0983 u, what can you infer about other possible isotopes of potassium?

Electromagnetic Radiation

37. When the *Voyager* space probe radioed information back from Neptune to Earth, the electromagnetic signals had to travel about 4.4×10^9 km. How much time, in seconds, did it take for these signals to reach Earth?

38. The satellite Luna is located 3.78×10^5 km from the planet Terra. How much time, in seconds, does it take for a radio signal to travel from Luna to Terra?

39. Radio station XOOX operates at a frequency of 992 kHz on the AM band. What is the wavelength, in meters, of the radio waves?

40. Radio station ZAAZ operates at a frequency of 88.7 MHz on the FM band. What is the wavelength, in meters, of the radio waves?

41. Gadolinium-153, used to determine bone density in the diagnosis of osteoporosis, emits gamma radiation with a frequency of 9.74×10^{18} Hz. What is the wavelength, in nanometers, of this gamma radiation?

42. Nuclear magnetic resonance (NMR) spectrometers use radio frequency waves to obtain information from which the structures of molecules can be deduced. One kind of NMR operates at a frequency of 200 MHz. What is the wavelength, in meters, of this radiation?

43. What is the frequency, in s^{-1}, of green light that has a wavelength of 539 nm?

44. What is the frequency, in s^{-1}, of light that has a wavelength of 650 nm? What color is the light? (*Hint*: Refer to Figure 6.14.)

45. John P. Bagby of Hughes Aircraft claims to have found gravitational vibrations with frequencies of 2.67 h^{-1} and 4.00 h^{-1}. What is the wavelength, in meters, of these gravity waves?

Photons and the Photoelectric Effect

46. Calculate the energy, in joules, of a photon of violet light that has a frequency of $7.42 \times 10^{14}\ s^{-1}$.

47. Calculate the energy, in joules, of a photon of red light that has a frequency of $3.73 \times 10^{14}\ s^{-1}$.

48. The laser in a compact disc player uses light with a wavelength of 780 nm. Calculate the energy of this radiation, in joules per photon and in kilojoules per mole of photons.

49. A microwave oven operates with radiation having a wavelength of 1.20 cm. Calculate the energy of this radiation, in joules per photon and in kilojoules per mole of photons.

50. The minimum energy needed to knock an electron from an iron atom is 7.21×10^{-19} J. What is the maximum wavelength of light, in nanometers, that will show a photoelectric effect with iron?

51. The minimum energy needed to knock an electron from a lithium atom is 4.65×10^{-19} J. What is the maximum wavelength of light, in nanometers, that will show a photoelectric effect with lithium?

52. How many photons of yellow light of 625-nm wavelength are equivalent to 1.00 J of energy?

53. What is the wavelength, in nanometers, of light that has an energy content of 487 kJ/mol? In what portion of the electromagnetic spectrum will this light be found?

The Bohr Model of the Hydrogen Atom

54. What is the energy of the photon emitted when the electron in a hydrogen atom drops from the $n = 5$ level to (**a**) the $n = 1$ level and (**b**) the $n = 4$ level?

55. Calculate the frequency, in s^{-1} of the electromagnetic radiation emitted by a hydrogen atom when its electron drops from (**a**) the $n = 3$ level to the $n = 2$ level and (**b**) from the $n = 4$ level to the $n = 1$ level.

56. Calculate the wavelength, in nanometers, of the electromagnetic radiation emitted by a hydrogen atom when its electron drops from (**a**) the $n = 5$ level to the $n = 2$ level and (**b**) from the $n = 3$ level to the $n = 1$ level.

57. Use the Bohr equation to calculate the energy of an electron in each of the first seven energy levels of a hydrogen atom. What is the energy of an electron when $n = \infty$?

58. Can the electron in a hydrogen atom be at the energy level -1.09×10^{-19} J? $+1.09 \times 10^{-19}$ J? Explain why each is possible or not possible?

Matter as Waves

59. Calculate the wavelength, in nanometers, associated with a proton traveling at a speed of 2.55×10^6 m/s. Use 1.67×10^{-27} kg as the proton mass.

60. Calculate the wavelength, in nanometers, associated with an electron traveling 90.0% of the speed of light.

61. Electrons have an associated wavelength of 84.4 nm. At what speed, in meters per second, are the electrons traveling?

62. At what speed, in meters per second, must electrons travel to have an associated wavelength of 174 pm?

Quantum Numbers and Atomic Orbitals

63. How many orbitals are there of the 4*f* type?

64. Is there a 3*d* subshell in the hydrogen atom? Explain.

65. What is the lowest-numbered principal shell in which *p* orbitals are found?

66. What is the lowest-numbered principal shell in which *f* orbitals are found?

67. If $n = 5$, what are the possible values of l? If $l = 3$, what are the possible values of m_l?

68. If $n = 4$, what are the possible values of l? If $l = 1$, what are the possible values of m_l?

69. Indicate whether each of the following is a permissible set of quantum numbers. If the set is not permissible, state why it is not.
a. $n = 2, l = 1, m_l = +1$
b. $n = 3, l = 3, m_l = -3$
c. $n = 3, l = 2, m_l = -2$
d. $n = 0, l = 0, m_l = 0$

70. Indicate whether each of the following is a permissible set of quantum numbers. If the set is not permissible, state why it is not.
a. $n = 3, l = 1, m_l = +2$
b. $n = 4, l = 3, m_l = -3$
c. $n = 3, l = 2, m_l = +2$
d. $n = 5, l = 0, m_l = 0$

71. Consider the electronic structure of an atom: (**a**) What are the n, l, and m_l quantum numbers corresponding to the 3*s* orbital? (**b**) List all the possible quantum number values for an orbital in the 5*f* subshell. (**c**) In what specific subshell will an electron having the quantum numbers $n = 3$, $l = 1$, and $m_l = -1$ be found?

72. Consider the electronic structure of an atom: (**a**) What are the n, l, and m_l quantum numbers corresponding to the 3*p* subshell? (**b**) List all the possible quantum number values for an orbital in the 4*d* subshell. (**c**) In what specific subshell will an electron having the quantum numbers $n = 4$, $l = 2$, and $m_l = 0$ be found?

ADDITIONAL PROBLEMS

73. Lutetium occurs in nature as two isotopes: lutetium-175, with a mass of 174.941 u, and lutetium-176, with a mass of 175.943 u. Use these data and the atomic weight of lutetium to calculate the fractional abundances of the lutetium isotopes.

74. There are three naturally occurring isotopes of magnesium. The masses and percent natural abundances of two of the isotopes are 24.98584 u (10.00%) and 25.98259 u (11.01%). Use these data and the atomic weight of magnesium to determine the mass of the *third* magnesium isotope.

75. We sometimes refer to the *period* of a wave, that is, the time required for one cycle. Household electricity has a frequency of 60 s^{-1}. What is the period of this electricity? How are period and frequency related?

76. The photoelectric *work function* of an element is the energy required for an electron to escape a solid surface. What is the minimum wavelength of light that will eject an electron from (**a**) selenium, with a work function of 9.4×10^{-19} J, and (**b**) cesium, with a work function of 3.42×10^{-19} J?

77. One photon of ultraviolet light can dislodge an electron from the surface of a metal. Two photons of red light, with the same total energy as the one photon of ultraviolet, do not produce any photoelectrons. Explain why this is so.

78. Iodine molecules (I_2) can be dissociated into separate iodine atoms by light of wavelengths less than 499.5 nm. What is the minimum energy, in joules, required to separate an iodine molecule into iodine atoms? What is the minimum energy, in kilojoules, required to dissociate 1 mol of I_2 molecules into I atoms?

79. The energy difference between the first and second energy levels of the Bohr hydrogen atom is 3/4 *B*. Which of the following represents the energy difference between the first and third energy levels? Explain your reasoning.

a. $3/2 \times (3/4\,B)$ **b.** $2/3 \times (3/4\,B)$

c. $32/27 \times (3/4\,B)$ **d.** $(3/4\,B) + (4/9\,B)$

80. Show that there should be only *four* lines in the visible spectrum of hydrogen and at the wavelengths indicated in Figure 6.16.

81. What are the maximum and minimum wavelengths of lines in the Lyman series of the hydrogen spectrum (recall Figure 6.22)?

82. From what energy level must an electron fall to produce a line at 1094 nm in the Paschen series of the hydrogen spectrum (recall Figure 6.22)?

83. Between which two energy levels of a hydrogen atom must an electron fall to account for a spectral line at 486.1 nm?

84. Is it possible that a line in the Lyman series occurs at 150.5 nm? Explain.

85. Show that if an electron in a hydrogen atom drops from the level $n = 3$ to $n = 2$, followed by a drop from $n = 2$ to $n = 1$, the total energy emitted is the same as if the electron had fallen directly from the level $n = 3$ to $n = 1$.

86. The absorption spectrum of sunlight shows a pair of dark lines at 589.0 nm and 589.6 nm, whereas the emission spectrum of sodium has bright yellow lines at these same wavelengths. Explain how the dark lines might arise.

87. Calculate the wavelength, in meters, associated with a bullet with a mass of 25.0 g traveling with a speed of 110 m/s.

88. Einstein's equation for the equivalence of matter and energy is $E = mc^2$. Use this equation, together with Planck's equation, to derive a relationship similar to de Broglie's equation.

89. Which must have the greater velocity to produce matter waves of the same wavelength (for example, 1 nm), protons or electrons? Explain your reasoning.

90. Radio signals from *Voyager 1* spacecraft on its trip to Jupiter in the late 1970s were broadcast at a frequency of $8.4 \times 10^9\ s^{-1}$. These signals were intercepted on Earth by an antenna capable of detecting 4×10^{-21} watt (1 watt = 1 J/s). At a minimum, approximately how many photons per second of electromagnetic radiation was the antenna capable of intercepting?

91. A typical microwave oven uses microwave radiation with a frequency of $2.45 \times 10^9\ s^{-1}$. How many moles of photons of this radiation are required to raise the temperature of 345 g water from 26.5 °C to 99.8 °C? (Assume a specific heat of $4.18\ J \cdot g^{-1} \cdot {}^{\circ}C^{-1}$ for water.)

92. Einstein was never able to accept the Heisenberg uncertainty principle. He stated, "God does not play dice with the Universe." What do you suppose Einstein meant by this remark?

93. In reply to Einstein's remark in Problem 92, Niels Bohr is supposed to have said, "Albert, stop telling God what to do." What do you suppose that Bohr meant by this remark?

94. The greatest probability of finding the electron in a small volume element of the 1*s* orbital of the hydrogen atom is at the nucleus. Yet, the most probable distance of the electron from the nucleus is 52.9 pm. How can you reconcile these two statements?

Music consists of a series of tones that build octave after octave. Similarly, elements have properties that recur period after period. An early attempt to organize elements systematically was called Newlands's law of octaves.

7

Electron Configurations and the Periodic Table

CONTENTS

In Chapter 2 we described how, in 1869, Dmitri Mendeleev arranged elements with similar properties into groups, in a format called the periodic table. We learned that to make his classification scheme work, Mendeleev shrewdly corrected some atomic weight values and left gaps in his table for undiscovered elements. Mendeleev's approach to the periodic table was *empirical*; he did what was necessary to make his classification scheme conform to the observed facts. In this chapter we will learn that what Mendeleev did also makes perfect sense from a *theoretical* standpoint. As a new aspect of atomic structure, we will learn how electrons are distributed among the regions of an atom described by atomic orbitals. This description is called the *electron configuration* of the atom. Our focus will be on the intimate relationship between the electron configurations of atoms and properties of the elements that vary periodically.

7.1 Multielectron Atoms

Schrödinger's wave-mechanics treatment of atomic structure described in Chapter 6 applies to the hydrogen atom or to other species consisting of a positively charged nucleus and a *single* electron; that is, H, He^{+}, Li^{2+}, ... But most atoms have more than one electron. Wave mechanics quickly gets extremely complicated as we move on to *multielectron* species.

The difficulty with wave equations for multielectron atoms is that in addition to the terms that relate to the attractions between electrons and the nucleus, we must also include terms that account for the mutual *repulsions* between electrons. We cannot state the nature of these repulsions exactly because exact positions of the electrons are not known. The best we can do is to approximate the behavior of electrons one at a time in an environment consisting of the nucleus and the remaining electrons. Fortunately, this approach to wave mechanics yields orbitals that are of the same types as in the hydrogen atom, so-called *hydrogenlike* orbitals.

The chief differences between orbitals of a hydrogen atom and the equivalent orbitals in multielectron atoms concern their energies. The differences, suggested by Figure 7.1 and described below, are of two types.

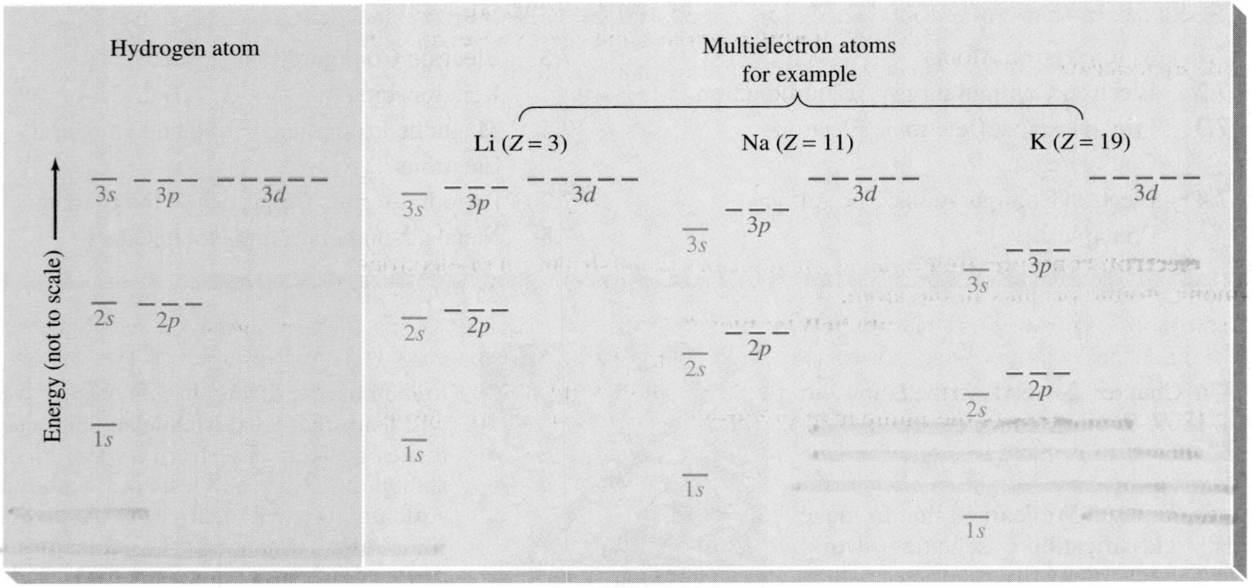

Figure 7.1 Orbital energy diagrams. Energy levels for the sublevels of the first three principal levels are shown for a hydrogen atom (left) and three typical atoms with more than one electron (right). Each multielectron atom has its own energy-level diagram. Note that in a hydrogen atom all the orbitals in a principal shell are degenerate (have the same energy). In multielectron atoms the orbitals within a sublevel are degenerate, but there is a difference in energy between the sublevels of a principal level. Note also that orbital energies decrease with increasing atomic number.

1. The number of electrons in an atom, the number of protons in the nucleus, and the positive charge on the nucleus all increase as the atomic number increases. Thus, with an increase in atomic number comes a stronger attractive force between the nucleus and any given electron and a lowering of the energy of the orbital in which that electron is found. For example, as shown in Figure 7.1, the 1*s* orbital in a sodium atom ($Z = 11$) is at a lower energy than is the 1*s* orbital in a lithium atom ($Z = 3$), which in turn is at a lower energy than is the 1*s* orbital in a hydrogen atom ($Z = 1$).
2. Electrons in orbitals that are concentrated between any particular electron and the atomic nucleus tend to *screen* or *shield* that electron from the full attractive force of the nucleus. Just how effective these inner-shell electrons are in this shielding depends on the type of orbital in which the particular shielded electron is found. An electron in an *s* orbital spends more time

close to the nucleus than does one in a p orbital in the same principal shell. The s electron is therefore not shielded as effectively as is the p electron; it is attracted more strongly to the nucleus and is thus at a lower energy than is the p electron. Thus, for example, we note in Figure 7.1 that the $3s$ sublevel is lower in energy than is the $3p$ sublevel. Similarly, the $3p$ sublevel is lower in energy than is the $3d$ sublevel for all multielectron atoms.

There is no further splitting of energy levels within a subshell of an isolated multielectron atom—all three of the $3p$ orbitals are at the same energy level, as are all five of the $3d$ orbitals. Orbitals at the same energy level are said to be **degenerate**. In the hydrogen atom all orbitals in all subshells in a principal shell are degenerate. In a multielectron atom only orbitals in the same subshell are degenerate.

7.2 Electron Configurations: An Introduction

The **electron configuration** of an atom describes the distribution of electrons among atomic orbitals in the atom. Two general methods are used to denote electron configurations. The **subshell (sublevel) notation** uses numbers to designate the principal energy levels or principal shells and the letters s, p, d, and f to identify the sublevels or subshells. A superscript number following the letter indicates the number of electrons in the designated subshell. The notation

This *"spdf" notation* originated from terms used to describe spectral lines: *s*harp, *p*rincipal, *d*iffuse, and *f*undamental.

$$1s^22s^22p^3$$

therefore indicates an atom with two electrons in the $1s$ subshell, two in the $2s$ subshell, and three in the $2p$ subshell. The atom having this electron configuration has an atomic number of seven. It is an atom of the element *nitrogen*.

The notation we have just written leaves an unanswered question: How are the three $2p$ electrons distributed among the three orbitals in the $2p$ subshell? We can show this through the *expanded spdf notation*

$$1s^22s^22p_x^12p_y^12p_z^1$$

which shows that each of the three $2p$ orbitals contains a single electron. Another way to answer the question is to represent the electron configuration through an **orbital diagram**, in which boxes are used to indicate orbitals within subshells and arrows to represent electrons in these orbitals. The directions of the arrows represent the directions of the electron spins. The orbital diagram for nitrogen

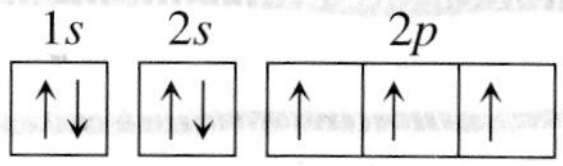

indicates two electrons of opposite or opposing spins in the $1s$ subshell, and two more with opposing spins in the $2s$ subshell. Electrons with opposing spins are said to be paired. Each of the orbitals of the $2p$ subshell has one electron, and in the most stable arrangement all three electrons have spins in the same direction; they are said to have parallel spins.

It does not matter whether a pair of electrons with opposing spins is represented as

[↑↓] or [↓↑]

although we generally write

[↑↓]

Similarly, unpaired electrons with parallel spins can be represented either as,

[↑] [↑] [↑]

or

[↓] [↓] [↓]

although we generally write

[↑] [↑] [↑]

Example 7.1

Interpret the electron configuration represented by each notation.

a. $1s^22s^22p^63s^1$

b.

1s	2s	2p		
↑↓	↑↓	↑↓	↑	↑

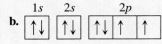

Solution

a. This *spdf* notation indicates two electrons in the 1*s* subshell, two in the 2*s* subshell, six in the 2*p* subshell, and one in the 3*s* subshell. This is the electron configuration of sodium ($Z = 11$).

b. This orbital diagram indicates two electrons in the 1*s* subshell, two in the 2*s* subshell, and four in the 2*p* subshell. Of the four electrons in the 2*p* subshell, two electrons with opposing spins are found in one of the 2*p* orbitals, and the remaining two 2*p* orbitals are singly occupied by electrons with parallel spins. This is the electron configuration of oxygen ($Z = 8$).

Exercise 7.1

Interpret the following electron configurations.

a. $1s^22s^22p^63s^23p^63d^{10}4s^2$

b.

1s	2s	2p			3s	3p		
↑↓	↑↓	↑↓	↑↓	↑↓	↑↓	↑	↑	

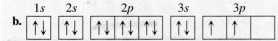

Several questions arise when we look carefully at the electron configurations we have written: Why are there never more than two electrons in an atomic orbital? When there are two electrons in an orbital, why do these always have opposing spins? Why are orbitals occupied singly before pairing of electrons occurs? Why do the electrons in singly occupied orbitals have parallel spins?

7.3 Principles That Determine Electron Configurations

To answer the questions just posed and to lay the groundwork for *predicting* electron configurations of the elements, we need to describe three basic principles that govern the distribution of electrons among atomic orbitals.

1. *Electrons occupy orbitals of the lowest energy available.*

For a given a multielectron atom, the order in which orbitals of the first three principal shells fill is suggested by Figure 7.1. That is, the 1*s* orbital is at a lower energy than the 2*s*; it fills before the 2*s*; the three orbitals in the 2*p* subshell fill before the 3*s* orbital; the three orbitals in the 3*p* subshell fill before the

3*d* subshell, and so on. The observed order in which the subshells of atoms are filled with electrons is usually

1*s*, 2*s*, 2*p*, 3*s*, 3*p*, 4*s*, 3*d*, 4*p*, 5*s*, 4*d*, 5*p*, 6*s*, 4*f*, 5*d*, 6*p*, 7*s*, 5*f*, 6*d*, 7*p*.

In Section 7.5 we will consider an especially simple way of establishing this order of filling by relating it to the periodic table.

The energies of certain subshells at higher quantum numbers are so similar in value that the order of increasing orbital energies and the order of filling of subshells may not correspond in some cases because electrons in various orbitals interact in ways that are sometimes hard to predict. For example, a given 3*d* subshell may be at a slightly lower energy than the corresponding 4*s* subshell; nevertheless, the 4*s* subshell fills first. The order of filling of subshells is based not just on orbital energies viewed in isolation but on what produces the lowest energy state for the atom as a whole when all electrons are in place. In the final analysis, the order of filling of the subshells has been determined *experimentally*, through spectroscopic and magnetic studies. These experiments have established essentially the order described here, with a few exceptions that we mention later.

2. *No two electrons in the same atom may have all four quantum numbers alike.*

This is a statement, known as the **Pauli exclusion principle**, made by Wolfgang Pauli (1900–1958) in 1926 to explain the complex features of the emission spectra of atoms in a magnetic field. The important consequence of this principle for electron configurations is this: Because electrons in a given orbital must have the same values of n, l, and m_l (for example, $n = 3$, $l = 0$, $m_l = 0$ in the 3*s* orbital), they must have different values of m_s. Only two values of m_s are possible: $+\frac{1}{2}$ and $-\frac{1}{2}$. We conclude that

An atomic orbital can accommodate only two electrons and these electrons must have opposing spins.

Each principal shell consists of a given number of subshells, and each subshell in turn contains a given number of orbitals. Pauli's exclusion principle therefore limits the number of electrons that can be found in individual orbitals, subshells, and principal shells. These limitations are set forth in Table 7.1.

TABLE 7.1 **Maximum Capacities of Subshells and Principal Shells**

$n =$	1	2		3			4				. . .	n
$l =$	0	0	1	0	1	2	0	1	2	3		
Subshell designation	*s*	*s*	*p*	*s*	*p*	*d*	*s*	*p*	*d*	*f*		
Orbitals in subshell	1	1	3	1	3	5	1	3	5	7		
Subshell capacity	2	2	6	2	6	10	2	6	10	14		
Principal shell capacity	2	8		18			32				. . .	$2n^2$

3. *When entering orbitals of identical energy, electrons initially occupy them singly and with the same spins, that is, with parallel spins.*

This declaration, known as **Hund's rule**, results in an atom having as many unpaired electrons as possible. Hund's rule can be rationalized as follows: Two electrons, because they carry identical charges and therefore repel each other, tend not to occupy the same region of space. As a result, electrons go into orbitals separately as long as empty orbitals of the appropriate energy are available. This is rather like the inclination of bus passengers to sit singly. This gives each person maximum seat space. Only when all seats have one passenger do passengers begin to sit two to a seat.

It is because of Hund's rule that we wrote the ground-state electron configuration of nitrogen as

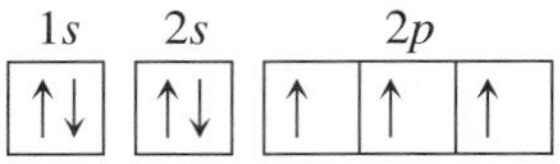

rather than

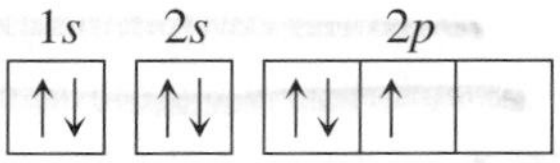

(*incorrect*)

The reason why electrons in singly occupied orbitals have *parallel* spins is not so easily explained. However, both experiment and theory indicate that an electron configuration in which all unpaired electrons have parallel spins represents a lower energy state of an atom than any other electron configuration we can write for that atom. This fact warns us, for example, *not* to write the ground-state electron configuration of nitrogen as

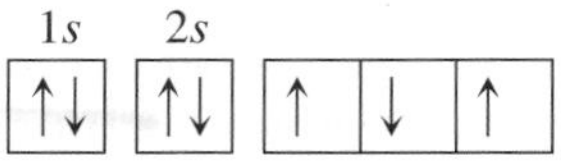

(*incorrect*)

7.4 Electron Configurations: The Aufbau Principle

Aufbau is a German word that means "building up."

To predict probable ground-state electron configurations of the elements, we will use a combination of the rules stated in the preceding section and something called the *aufbau principle*. The **aufbau principle** is a conceptual process in which we think about building up an atom from the one that precedes it in atomic number. We do this by adding a proton (and some neutrons) to the nucleus and one electron to an atomic orbital. That is, we think about building a helium atom from a hydrogen atom, a lithium atom from a helium atom, and so on. In this process we focus on the particular atomic orbital into which the added electron goes to produce a *ground-state* atom, an atom in its lowest energy state.

An *excited-state* electron configuration has one or more electrons in a higher energy orbital or spin state than in the ground state. We will deal with such excited-state electron configurations in Chapter 10.

Let's illustrate the aufbau principle by "building up" a few atoms. A hydrogen atom, with atomic number 1, has only one electron. That single electron goes into the orbital with the lowest possible energy, the 1*s* orbital.

$$(Z = 1) \text{ H } 1s^1$$

The electron we add in moving from hydrogen to helium also goes into the 1*s* orbital.

$$(Z = 2) \text{ He } 1s^2$$

In the lithium atom, the first two electrons are in the 1*s* orbital, just as in the helium atom. The Pauli exclusion principle tells us that we *cannot* put the third electron into the 1*s* orbital. The first principal shell is filled, and the added electron must go into the 2*s* orbital, the vacant orbital next lowest in energy. The second principal shell begins to fill.

$$(Z = 3) \text{ Li } 1s^22s^1$$

In the beryllium atom, the added electron also goes into the 2*s* orbital.

$$(Z = 4) \text{ Be } 1s^22s^2$$

With the boron atom, the Pauli exclusion principle dictates that the added electron enter a 2*p* orbital.

$$(Z = 5) \text{ B } 1s^22s^22p^1$$

The 2*p* subshell fills in the succession of atoms: carbon, nitrogen, oxygen, fluorine, and neon. In these atoms, Hund's rule requires that electrons enter 2*p* orbitals singly (and with parallel spins) before electrons pair up. This is a fact that is not explicitly shown in the usual *spdf* notation but is brought out clearly in orbital diagrams.

			1*s*	2*s*	2*p*		
(Z = 6)	C	$1s^22s^22p^2$	↑↓	↑↓	↑	↑	
(Z = 7)	N	$1s^22s^22p^3$	↑↓	↑↓	↑	↑	↑
(Z = 8)	O	$1s^22s^22p^4$	↑↓	↑↓	↑↓	↑	↑
(Z = 9)	F	$1s^22s^22p^5$	↑↓	↑↓	↑↓	↑↓	↑
(Z = 10)	Ne	$1s^22s^22p^6$	↑↓	↑↓	↑↓	↑↓	↑↓

We often use a noble-gas-core abbreviated electron configuration in which the portion that corresponds to that of a noble gas is represented by a bracketed chemical symbol. Thus, [He] represents the configuration $1s^2$; so the electron configuration of lithium, which is $1s^22s^1$, can be written instead as [He]$2s^1$. Likewise, [He]$2s^22p^3$ represents the electron configuration of nitrogen: $1s^22s^22p^3$.

The electron configuration of neon has both the first and second principal shells filled to their maximum capacities (recall Table 7.1). In moving from neon to sodium, the added electron must go into the *s* orbital of the third principal shell, that is, the 3*s* orbital.

$$(Z = 11) \quad \text{Na} \quad 1s^2 2s^2 2p^6 3s^1 \quad or \quad [\text{Ne}]3s^1$$

Other atoms with electrons entering the third principal shell are featured in Example 7.2 and Exercise 7.2, where we illustrate the following general procedure for writing electron configurations.

1. Find the atomic number for the element (use the periodic table).
2. Place that number of electrons in various subshells according to the order of filling of orbitals outlined on page 241.
3. Do not exceed the maximum number of electrons for any orbital or for any subshell; two electrons in the same orbital must have opposite spin (Pauli exclusion principle).
4. In orbital diagrams, be sure that when a subshell is only partly filled, orbitals in that subshell, where possible, are singly occupied by electrons having parallel spins.

Example 7.2

Write out the electron configuration for sulfur, using both the *spdf* notation and an orbital diagram.

Solution

The sulfur atom, with atomic number 16, has 16 electrons, and we place them into the lowest energy subshells available. *Two* go into the 1*s* subshell, *two* into the 2*s* subshell, *six* into the 2*p* subshell, and *two* into the 3*s* subshell. That leaves four electrons to be placed in the 3*p* subshell. According to Hund's rule, two electrons with opposing spins will occupy one of the 3*p* orbitals, and the remaining two electrons will

			1s	2s	2p	3s	3p
(Z = 16)	S	$1s^2 2s^2 2p^6 3s^2 3p^4$	[↑↓]	[↑↓]	[↑↓ | ↑↓ | ↑↓]	[↑↓]	[↑↓ | ↑ | ↑]

have parallel spins and singly occupy the other 3*p* orbitals.
The noble-gas-core abbreviated electron configuration is

				3s	3p
(Z = 16)	S	$[\text{Ne}]3s^2 3p^4$	[Ne]	[↑↓]	[↑↓ | ↑ | ↑]

Exercise 7.2

Write out the electron configuration for phosphorus, using the *spdf* notation, the noble-gas-core abbreviated electron configuration, and an orbital diagram.

The electron configurations predicted by the aufbau principle follow a regular pattern through argon, in which the 3*p* subshell fills.

$$(Z = 18) \quad [\text{Ar}]1s^2 2s^2 2p^6 3s^2 3p^6 \quad or \quad [\text{Ne}]3s^2 3p^6$$

Now we have to be careful. In proceeding to potassium (Z = 19), we must be guided by the order of filling of subshells, rather than by Table 7.1, which states that the maximum capacity of the third principal shell is 18. That is, the 19th electron goes into the 4*s* subshell, not the 3*d*.

$$(Z = 19) \quad \text{K} \quad 1s^22s^22p^63s^23p^64s^1 \quad or \quad [\text{Ar}]4s^1$$

In calcium, the 20th electron pairs up with the 19th electron in the 4*s* orbital.

$$(Z = 20) \quad \text{Ca} \quad 1s^22s^22p^63s^23p^64s^2 \quad or \quad [\text{Ar}]4s^2$$

Representative and Transition Elements

The first 20 elements are all of a type called **main-group** or **representative elements**—elements in which the orbitals being filled in the aufbau process are either *s* or *p* orbitals of the outermost shell (the shell with electrons having the highest principal quantum number). Scandium ($Z = 21$) is the first of the **transition elements.** In these elements, the subshell being filled in the aufbau process is an inner one—that is, the principal quantum number of its electrons is smaller than that of the electrons in the outermost shell.

The electron configuration of scandium is shown below in two commonly used representations.

$$\text{(a) Sc} \quad [\text{Ar}]3d^14s^2 \qquad \text{(b) Sc} \quad [\text{Ar}]4s^23d^1$$

Method (a) groups together all subshells within the same principal shell and arranges the principal shells according to increasing principal quantum number. Method (b) arranges the subshells in the apparent order in which they fill. In this text we will use method (a).

Electron configurations of the elements from scandium through zinc, known as the first transition series, are summarized in Table 7.2. Following zinc, we return to a series of representative elements. The third principal shell having been filled to its maximum capacity of 18 in zinc, electrons now follow the aufbau order, entering the lowest energy subshell available, the 4*p* subshell. The six atoms in which this subshell fills range from

$$(Z = 31) \quad \text{Ga} \quad [\text{Ar}]3d^{10}4s^24p^1$$

to

$$(Z = 36) \quad \text{Kr} \quad [\text{Ar}]3d^{10}4s^24p^6.$$

Exceptions to the Aufbau Principle

Examine the electron configurations of chromium and copper in Table 7.2. The expected configurations, those based on the aufbau principle, are not the ones observed through the emission spectra and magnetic properties of the elements.

	Expected	*Observed*
Cr ($Z = 24$)	$[\text{Ar}]3d^44s^2$	$[\text{Ar}]3d^54s^1$
Cu ($Z = 29$)	$[\text{Ar}]3d^94s^2$	$[\text{Ar}]3d^{10}4s^1$

The reason for these exceptions to the aufbau principle are not completely understood, but it seems that the half-filled 3*d* subshell of chromium ($3d^5$) and

TABLE 7.2 **Electron Configurations of the First Transition Series Elements**

		3d					4s	
Sc	[Ar]	↑					↑↓	$[Ar]3d^14s^2$
Ti	[Ar]	↑	↑				↑↓	$[Ar]3d^24s^2$
V	[Ar]	↑	↑	↑			↑↓	$[Ar]3d^34s^2$
Cr	[Ar]	↑	↑	↑	↑	↑	↑	$[Ar]3d^54s^1$
Mn	[Ar]	↑	↑	↑	↑	↑	↑↓	$[Ar]3d^54s^2$
Fe	[Ar]	↑↓	↑	↑	↑	↑	↑↓	$[Ar]3d^64s^2$
Co	[Ar]	↑↓	↑↓	↑	↑	↑	↑↓	$[Ar]3d^74s^2$
Ni	[Ar]	↑↓	↑↓	↑↓	↑	↑	↑↓	$[Ar]3d^84s^2$
Cu	[Ar]	↑↓	↑↓	↑↓	↑↓	↑↓	↑	$[Ar]3d^{10}4s^1$
Zn	[Ar]	↑↓	↑↓	↑↓	↑↓	↑↓	↑↓	$[Ar]3d^{10}4s^2$

the filled 3*d* subshell of copper ($3d^{10}$) lend a special stability to the observed electron configurations. Apparently, having a half-filled 4*s* subshell and a half-filled 3*d* subshell gives a lower energy state for a Cr atom than having a filled 4*s* subshell. Actually, because there is little difference between the 4*s* and 3*d* orbital energies, the expected and observed electron configurations are quite close in energy.

At higher principal quantum numbers, the energy difference between certain subshells is even smaller than that between the 3*d* and 4*s* subshells. As a result, there are still more exceptions to the aufbau principle among the heavier transition elements.

7.5 Electron Configurations: Periodic Relationships

We applied the aufbau process to the first 36 elements in the preceding section. We could extend it to all the elements, but let's consider instead a simpler way of working with electron configurations. As we shall see both in this chapter and in the next, electron configurations correlate closely with the groups and periods of the periodic table, and they help to explain the similar properties of elements within a group.

If we write electron configurations for the elements arranged as in their periodic-table format, as in Figure 7.2 on pages 248 and 249, we quickly spot some distinctive patterns. With some exceptions, the number of electrons in the outer shell (principal shell containing electrons with the highest quantum number) is the same for each element within a group of the periodic table. The most common pattern is that

for A-group elements (also called main-group or representative elements) the number of outer-shell electrons is the same as the periodic-table group number.

Thus, except for helium, which has only two electrons, all the noble gases (Group 8A) have *eight* outer-shell electrons. These electrons have the configuration ns^2np^6, where n is the principal quantum number for the outer shell of electrons (that is, $2s^22p^6$ for neon, $3s^23p^6$ for argon, $4s^24p^6$ for krypton, and so on).

Each element in Group 1A (the alkali metals) has a *single* electron in an s orbital of the outermost principal shell: $2s^1$ for Li, $3s^1$ for Na, $4s^1$ for K, or, in general, ns^1. Similarly, the Group 2A elements have the outer-shell electron configuration ns^2, that is, with the two outermost electrons in an s orbital. As a further example, the seven electrons of highest principal quantum number in the Group 7A elements are in the configuration ns^2np^5.

The correlation between an A-group number and the number of outer-shell electrons is a useful one, but it is important to note that this correlation does *not* extend to the B-group elements, the *transition* elements. Atoms of the transition elements typically have only *two* outer-shell electrons (ns^2); some have only one (ns^1).

Palladium ($Z = 46$) is unique among the transition elements in having no electrons in its outermost subshell (the $5s$).

Look once more at Figure 7.2 and at the periodic table inside the front cover and notice this fundamental relationship.

The period number is the same as the principal quantum number, n, *of the electrons in the outer shell.*

All elements in the fourth period, for example, have one or more electrons with $n = 4$ and none with a higher value of n. The period begins with K, which has the electron configuration $[Ar]4s^1$, and ends with Kr ($[Ar]3d^{10}4s^24p^6$). The next subshell to fill after $4p$ is $5s$, so the element Rb, with the electron configuration $[Kr]5s^1$, begins the fifth period.

Each period of the periodic table begins with a Group 1A element and ends with a noble gas element. The periods differ in length because the number of subshells that must fill to get from the electron configuration ns^1 to ns^2np^6 differs. In the two-member first period, we go directly from $1s^1$ to $1s^2$ (there is no p subshell in the first principal shell). In the eight-member second and third periods, the subshells that must fill are the $2s$ and $2p$ and the $3s$ and $3p$, respectively. The fourth and fifth periods have 18 members because the subshells that must fill are the $4s$, $3d$, and $4p$ and the $5s$, $4d$, and $5p$, respectively. Finally, the 32-member sixth and seventh periods require the filling of the $6s$, $4f$, $5d$, and $6p$ and the $7s$, $5f$, $6d$, and $7p$ subshells, respectively.

Example 7.3

Write out the subshell notation for the outer-shell electrons of strontium and arsenic.

Solution

Strontium is in Group 2A and the fifth period of the periodic table. Its outer-shell electron configuration is $5s^2$. Arsenic is in Group 5A and the fourth period. Its five outer-shell electrons have the configuration $4s^24p^3$.

Exercise 7.3

Write out the subshell notation for the outer-shell electrons of gallium and tellurium.

1A	2A	3B	4B	5B	6B	7B	8B	
1 **H** $1s^1$								
3 **Li** $2s^1$	4 **Be** $2s^2$							
11 **Na** $3s^1$	12 **Mg** $3s^2$							
19 **K** $4s^1$	20 **Ca** $4s^2$	21 **Sc** $3d^14s^2$	22 **Ti** $3d^24s^2$	23 **V** $3d^34s^2$	24 **Cr** $3d^54s^1$	25 **Mn** $3d^54s^2$	26 **Fe** $3d^64s^2$	27 **Co** $3d^74s^2$
37 **Rb** $5s^1$	38 **Sr** $5s^2$	39 **Y** $4d^15s^2$	40 **Zr** $4d^25s^2$	41 **Nb** $4d^45s^1$	42 **Mo** $4d^55s^1$	43 **Tc** $4d^55s^2$	44 **Ru** $4d^75s^1$	45 **Rh** $4d^85s^1$
55 **Cs** $6s^1$	56 **Ba** $6s^2$	57 ***La** $5d^16s^2$	72 **Hf** $4f^{14}5d^26s^2$	73 **Ta** $4f^{14}5d^36s^2$	74 **W** $4f^{14}5d^46s^2$	75 **Re** $4f^{14}5d^56s^2$	76 **Os** $4f^{14}5d^66s^2$	77 **Ir** $4f^{14}5d^76s^2$
87 **Fr** $7s^1$	88 **Ra** $7s^2$	89 **†Ac** $6d^17s^2$	104 $5f^{14}6d^27s^2$	105 $5f^{14}6d^37s^2$	106 $5f^{14}6d^47s^2$			

*	58 **Ce** $4f^26s^2$	59 **Pr** $4f^36s^2$	60 **Nd** $4f^46s^2$	61 **Pm** $4f^56s^2$	62 **Sm** $4f^66s^2$
†	90 **Th** $6d^27s^2$	91 **Pa** $5f^26d^17s^2$	92 **U** $5f^36d^17s^2$	93 **Np** $5f^46d^17s^2$	94 **Pu** $5f^67s^2$

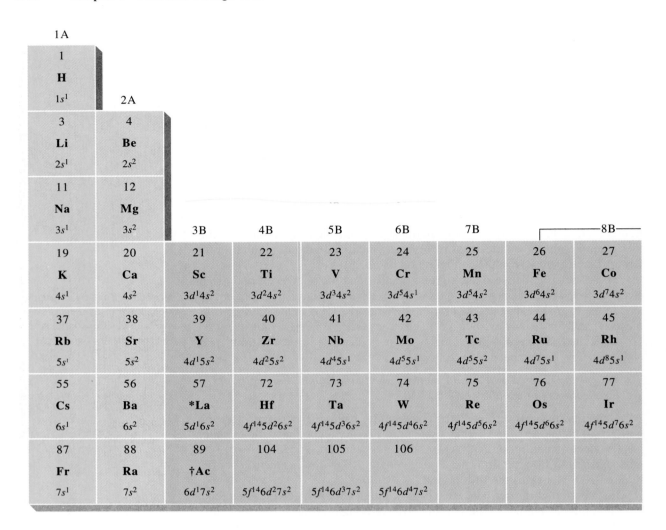

Figure 7.2 **Electron configurations and the periodic table.** Each configuration has as its core the electron configuration of the noble gas of the preceding period, that is [He] for the second period, [Ne] for the third period, [Ar] for the fourth period, and so on.

	1B	2B	3A	4A	5A	6A	7A	8A
								2 **He** $1s^2$
			5 **B** $2s^22p^1$	6 **C** $2s^22p^2$	7 **N** $2s^22p^3$	8 **O** $2s^22p^4$	9 **F** $2s^22p^5$	10 **Ne** $2s^22p^6$
			13 **Al** $3s^23p^1$	14 **Si** $3s^23p^2$	15 **P** $3s^23p^3$	16 **S** $3s^23p^4$	17 **Cl** $3s^23p^5$	18 **Ar** $3s^23p^6$
28 **Ni** $3d^84s^2$	29 **Cu** $3d^{10}4s^1$	30 **Zn** $3d^{10}4s^2$	31 **Ga** $3d^{10}4s^24p^1$	32 **Ge** $3d^{10}4s^24p^2$	33 **As** $3d^{10}4s^24p^3$	34 **Se** $3d^{10}4s^24p^4$	35 **Br** $3d^{10}4s^24p^5$	36 **Kr** $3d^{10}4s^24p^6$
46 **Pd** $4d^{10}$	47 **Ag** $4d^{10}5s^1$	48 **Cd** $4d^{10}5s^2$	49 **In** $4d^{10}5s^25p^1$	50 **Sn** $4d^{10}5s^25p^2$	51 **Sb** $4d^{10}5s^25p^3$	52 **Te** $4d^{10}5s^25p^4$	53 **I** $4d^{10}5s^25p^5$	54 **Xe** $4d^{10}5s^25p^6$
78 **Pt** $4f^{14}5d^96s^1$	79 **Au** $4f^{14}5d^{10}6s^1$	80 **Hg** $4f^{14}5d^{10}6s^2$	81 **Tl** $4f^{14}5d^{10}6s^26p^1$	82 **Pb** $4f^{14}5d^{10}6s^26p^2$	83 **Bi** $4f^{14}5d^{10}6s^26p^3$	84 **Po** $4f^{14}5d^{10}6s^26p^4$	85 **At** $4f^{14}5d^{10}6s^26p^5$	86 **Rn** $4f^{14}5d^{10}6s^26p^6$

63 **Eu** $4f^76s^2$	64 **Gd** $4f^75d^16s^2$	65 **Tb** $4f^96s^2$	66 **Dy** $4f^{10}6s^2$	67 **Ho** $4f^{11}6s^2$	68 **Er** $4f^{12}6s^2$	69 **Tm** $4f^{13}6s^2$	70 **Yb** $4f^{14}6s^2$	71 **Lu** $4f^{14}5d^16s^2$
95 **Am** $5f^77s^2$	96 **Cm** $5f^76d^17s^2$	97 **Bk** $5f^97s^2$	98 **Cf** $5f^{10}7s^2$	99 **Es** $5f^{11}7s^2$	100 **Fm** $5f^{12}7s^2$	101 **Md** $5f^{13}7s^2$	102 **No** $5f^{14}7s^2$	103 **Lr** $5f^{14}6d^17s^2$

Using the Periodic Table to Predict Electron Configurations

We don't have to memorize an order of filling of orbitals or use an orbital energy diagram (Figure 7.1) to write probable electron configurations. We can deduce configurations directly from the periodic table. All we need to know is which subshells fill in different regions of the periodic table. To assist in this, we refer to the four blocks of elements shown in Figure 7.3.

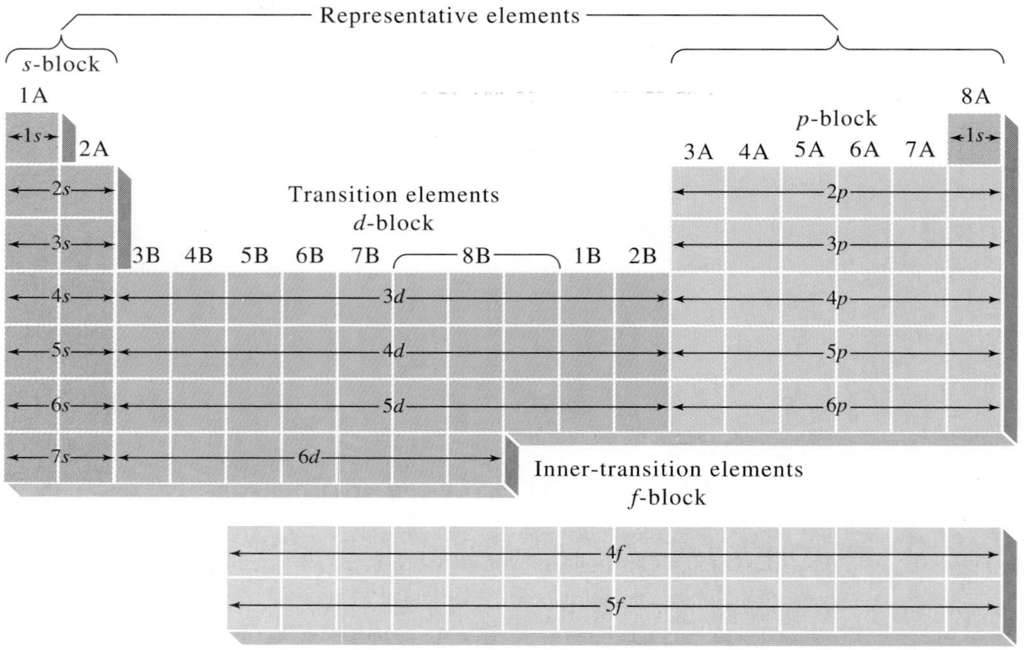

Figure 7.3 The periodic table and the order of filling of subshells. Read through this periodic table starting from the upper left and you will discover the same order of filling of subshells as shown on page 241.

***s*-block:** The *ns* subshell (the *s* subshell of the outer shell) fills by the aufbau process. These are *main-group* or *representative* elements.

***p*-block:** The *np* subshell (the *p* subshell of the outer shell) fills. These are also *main group* or *representative* elements.

***d*-block:** The $(n - 1)d$ subshell (the *d* subshell of the next-to-outermost shell) fills. These are *transition* elements found in the main body of the periodic table.

***f*-block:** The $(n - 2)f$ subshell (the *f* subshell of the second-from-outermost shell) fills. To keep the periodic table at a maximum width of 18 members, these elements are placed below the main body of the table. The 4*f* subshell fills in the **lanthanide** series, and the 5*f* subshell fills in the **actinide** series.

Because they fall within series of *d*-block elements, the lanthanides and actinides are sometimes called the *inner-transition* elements. The lanthanide series follows lanthanum ($Z = 57$) in the periodic table, and the actinide series follows actinium ($Z = 89$).

Example 7.4

Give the complete ground-state electron configuration of a strontium atom, both in the expanded *spdf* notation and in the noble-gas-core abbreviated notation.

Solution

We proceed through Figure 7.3 until we reach Sr ($Z = 38$). The 1*s* subshell fills as we go through the first period. Now, as we continue through the second period (Li through Ne), the 2*s* and 2*p* subshells fill. Moving through the third period, the 3*s* and 3*p* subshells fill. In the fourth period, the 4*s*, 3*d*, and 4*p* subshells fill in succession. In the fifth period, we find Sr in Group 2A, indicating two electrons in the 5*s* orbital. The electron configuration, in the order of filling, is

$$1s^22s^22p^63s^23p^64s^23d^{10}4p^65s^2$$

or, grouping the subshells into principal shells, we have

$$1s^22s^22p^63s^23p^63d^{10}4s^24p^65s^2$$

The noble-gas-core abbreviated notation is $[Kr]5s^2$.

Exercise 7.4

Refer only to the periodic table and give the complete ground-state electron configuration, in the expanded *spdf* notation and in the noble-gas-core abbreviated notation, for each of the following.

a. germanium **b.** zinc **c.** titanium **d.** iodine

Valence Electrons and Core Electrons

In some of the discussions that follow, we will find it convenient to use the term *valence* electrons. **Valence electrons** are the electrons in the outermost occupied principal shell—that is, they are the electrons with the highest principal quantum number, *n*. We refer to the electrons in inner shells as **core electrons**. Thus, in a calcium atom, with the electron configuration $[Ar]4s^2$, the two 4*s* electrons are valence electrons and those in the [Ar] configuration are core electrons. Bromine, with the electron configuration $[Ar]3d^{10}4s^24p^5$, has seven valence electrons, two 4*s* and five 4*p*; the core electrons are in the configuration $[Ar]3d^{10}$.

Another view of valence electrons is that they are the electrons that can be involved when atoms participate in chemical reactions. From this standpoint some of the inner-shell electrons of the transition elements might also be considered as valence electrons. However, we will limit our use of the terms valence and core electrons to that outlined in the preceding paragraph.

7.6 Magnetic Properties: Paired and Unpaired Electrons

In Chapter 6 we introduced a property called electron spin. This property causes each electron in an atom to create a magnetic field; that is, each electron acts like a tiny magnet. For a *pair* of electrons with opposite spins, the magnetic fields cancel one another, and the pair shows no resultant magnetism. Most substances

Unsettled Issues Concerning the Periodic Table

The periodic table has been around for well over a century, and in that time many variations of the table have been proposed. Yet, in all that time, there still is no single table that has gained acceptance by all chemists.

A particular state of confusion results from differing use of the letters A and B. In the United States, A is used to designate the representative elements and B, the transition elements. In Europe, the groups to the left of Fe/Ru/Os have typically been labeled as A groups, and those to the right of Ni/Pd/Pt, as B groups.

In order to eliminate confusion over the use of A and B, the International Union of Pure and Applied Chemistry has recommended that the groups simply be numbered from 1 to 18. There are advantages to each system. The A/B system used in this text, for example, is helpful in that the A-group numbers are equal to the numbers of outer-shell electrons in atoms of the representative elements.

There is also some difficulty in the proper placement of hydrogen. Although it is a member of Group 1A, hydrogen is *not* an alkali metal; it isn't even a metal. Most periodic tables put hydrogen in Group 1A because it has the electron configuration $1s^1$. But it could also be placed in Group 7A because, like the halogen atoms, it is one electron short of a noble gas electron configuration. Some periodic tables place it in *both* groups, and some place it in neither, but alone at the top of the table in the center.

Although chemists disagree over just what form of the periodic table is *most* useful, there is no disagreement over the great utility of the table.

exhibit no magnetic properties until they are placed in a magnetic field. Then two types of behavior are possible.

- If the atoms or molecules of a substance have *unpaired* electrons, the substance is **paramagnetic** and is *attracted* to an external magnetic field. Paramagnetism is associated with unpaired electrons.
- If all the electrons in the atoms or molecules of a substance are *paired*, the substance is **diamagnetic** and is *weakly repelled* by an external magnetic field.

The magnetic properties of a substance can be determined rather simply by weighing the substance in the absence and in the presence of a magnetic field, as indicated in Figure 7.4.

Looking Ahead

Iron, cobalt, and nickel atoms are all paramagnetic, but these metals and certain metal alloys exhibit a magnetic effect that is far stronger than that attributable to paramagnetism alone. This special type of magnetism is called *ferromagnetism* and is described in Section 22.3.

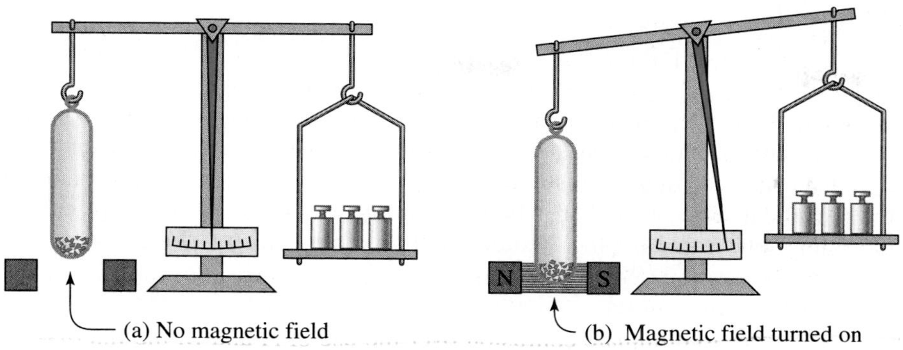

Figure 7.4 Paramagnetism illustrated. (a) A sample is weighed in the absence of a magnetic field. (b) When the field is turned on, the balanced condition is upset. The sample appears to gain weight. This is because it is now subjected to two attractive forces: the force of gravity *and* the force exerted by the interaction of the magnetic field and the unpaired electrons.

Example 7.5

A sample of chlorine gas is found to be diamagnetic. Can this gaseous sample be made up of individual Cl atoms?

Solution

We can deduce the electron configuration of Cl from its position in the periodic table and represent it by an orbital diagram.

		1*s*	2*s*	2*p*	3*s*	3*p*
($Z = 17$)	Cl	↑↓	↑↓	↑↓ ↑↓ ↑↓	↑↓	↑↓ ↑↓ ↑

The diagram shows one unpaired electron per atom. We would expect atomic chlorine (Cl) to be paramagnetic. Because the gas is diamagnetic, it cannot be comprised of individual Cl atoms. (In fact, chlorine gas consists of diatomic molecules, Cl_2. In Chapter 9 we will see that all electrons in the Cl_2 molecule are paired.)

Exercise 7.5

Which of the following elements, in atomic form, are expected to exhibit paramagnetism?

a. potassium **b.** mercury **c.** barium
d. gallium **e.** sulfur **f.** lead

Of course, we can establish that chlorine gas is not Cl(g) by using the ideal gas law.

$$PV = \left(\frac{m}{\text{molar mass}}\right) RT$$

By measuring the volume of a known mass of chlorine at a given temperature and pressure, we would find that the molecular weight is 71 u. Chlorine gas is Cl_2(g).

7.7 Periodic Atomic Properties of the Elements

Mendeleev developed his original periodic table by noting the periodic variation in properties such as density, melting point, and boiling point when the elements were arranged according to increasing atomic weight. His arrangement also produced groups whose elements formed oxides, hydrides, and other compounds with similar formulas.

Periodic relationships can be summarized by a general statement called the periodic law. In its modern form, the **periodic law** states that certain sets of physical and chemical properties recur at regular intervals (periodically) when the elements are arranged according to increasing atomic number.

Some physical properties, such as thermal and electrical conductivity, density, and hardness, are displayed only by *bulk* matter, that is, by large aggregations of atoms. Other properties, called *atomic properties*, are associated with individual atoms. In this section we examine three periodic atomic properties: atomic radii, ionization energies, and electron affinities.

Atomic Radii

We cannot measure the exact size of an isolated atom because its outermost electrons have a chance of being found at relatively large distances from the nucleus. What we can measure is the distance between the nuclei of two adjacent atoms, and we can derive a property called the **atomic radius** from this distance. Internuclear distances are not unique, however; they depend on the particular environment in which the atoms are found. Thus, there can be more than one value for the "atomic" radius of an element. We consider two specific cases next.

The **covalent radius** of an atom is one-half the distance between the nuclei of two like atoms joined into a molecule. Consider, for example, the I_2 molecule pictured in Figure 7.5. The distance between the two iodine nuclei is 266 pm. The covalent radius of iodine is one-half of this distance, or 133 pm.* The **metallic radius** is one-half the distance between the nuclei of adjacent atoms in a solid metal. In this text, when we use the single expression "atomic radius," we mean *covalent* radius for *nonmetals* and *metallic* radius for *metals*.

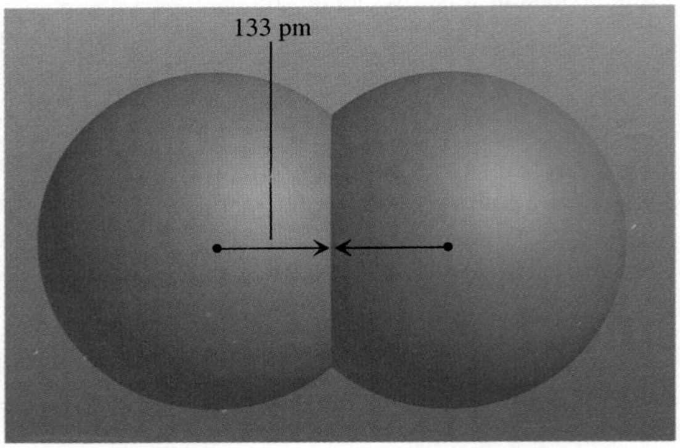

Figure 7.5 Atomic radius represented through the covalent radius of iodine. The covalent radius is one-half the distance between the centers of two I atoms in the molecule I_2.

Atomic radii often are measured in *angstroms* (Å), a non-SI unit.

$$1\ \text{Å} = 10^{-10}\ \text{m} = 100\ \text{pm}$$

However, in this book we will use picometers (pm).

$$1\ \text{pm} = 10^{-12}\ \text{m}$$

Periodic properties of the elements are clearly illustrated by the atomic radii tabulated in Figure 7.6 and shown graphically in Figure 7.7. To explain the trends shown in these figures, let us think of an atomic radius as the distance from the nucleus to the outer-shell (valence) electrons. Any factor that causes this distance to increase makes for a larger atomic radius.

*The value of the covalent radius depends on whether the bond between the two atoms is single, double, or triple. The covalent radius for iodine is a single-bond covalent radius.

1 H 37																	2 He —
3 Li 152	4 Be 111											5 B 80	6 C 77	7 N 74	8 O 74	9 F 71	10 Ne —
11 Na 186	12 Mg 160											13 Al 143	14 Si 118	15 P 110	16 S 103	17 Cl 99	18 Ar —
19 K 227	20 Ca 197	21 Sc 161	22 Ti 145	23 V 131	24 Cr 125	25 Mn 137	26 Fe 124	27 Co 125	28 Ni 125	29 Cu 128	30 Zn 133	31 Ga 122	32 Ge 123	33 As 125	34 Se 116	35 Br 114	36 Kr —
37 Rb 248	38 Sr 215	39 Y 178	40 Zr 159	41 Nb 143	42 Mo 136	43 Tc 135	44 Ru 133	45 Rh 135	46 Pd 138	47 Ag 145	48 Cd 149	49 In 163	50 Sn 141	51 Sb 145	52 Te 143	53 I 133	54 Xe —
55 Cs 265	56 Ba 217	57 La 187	72 Hf 156	73 Ta 143	74 W 137	75 Re 137	76 Os 134	77 Ir 136	78 Pt 139	79 Au 144	80 Hg 150	81 Tl 170	82 Pb 175	83 Bi 155	84 Po —	85 At —	86 Rn —
85 Fr —	86 Ra —																

Figure 7.6 **Atomic radii and the periodic table.** The data listed are metallic radii for metals and covalent radii for nonmetals, expressed in picometers (1 pm = 10^{-12} m = 0.01 Å). Data are not listed for the noble gases because it is difficult to measure their covalent radii (only Kr and Xe compounds are known).

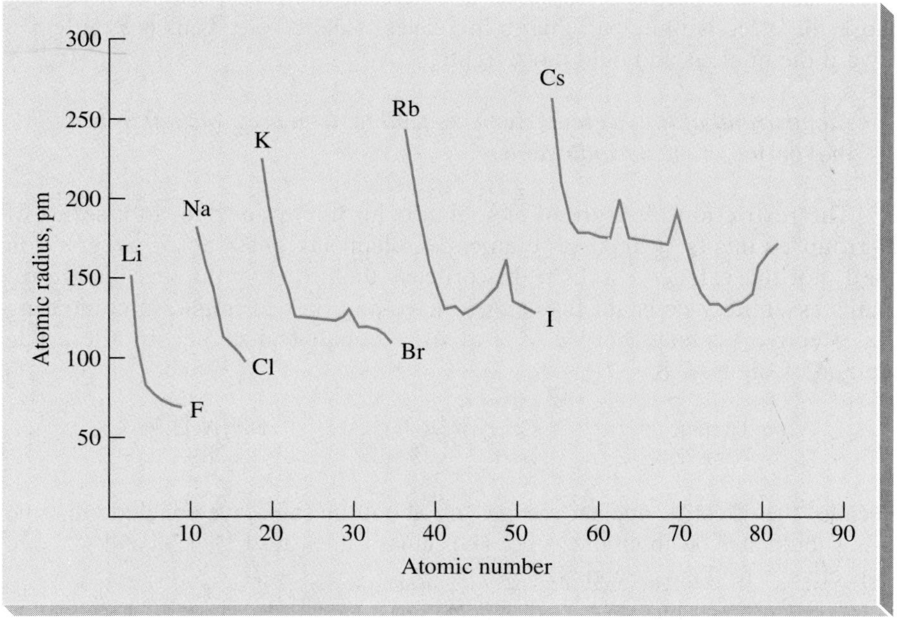

Figure 7.7 **Atomic radii of the elements.** The values shown here, in picometers (pm), are metallic radii for metals and covalent radii for nonmetals. Explanations have been suggested for the small peaks in the middle of some periods and for other irregularities, but these are beyond the scope of this discussion.

Within a vertical *group* of the periodic table, each succeeding member has one more principal shell occupied by electrons, that is, a shell with a principal quantum number n that is one unit higher than that of its immediate predecessor $(n - 1)$; the outer electrons are farther from the nucleus.

Atomic radii increase from top to bottom within a group *of the periodic table.*

To describe how atomic radii vary within a *period* of the periodic table, it is helpful to employ a new concept. The **effective nuclear charge (Z_{eff})** acting on an electron is the actual nuclear charge less the screening effect of other electrons in the atom.

First, as an oversimplification, consider the sodium atom pictured in Figure 7.8. *If* the $3s$ valence electron were at all times completely outside the region in which the ten electrons of the neon core ($1s^2 2s^2 2p^6$) are found, the $3s$ electron would be perfectly screened or shielded from the positively charged nucleus; it would experience an attraction to a net positive charge of only $+11 - 10 = +1$. The corresponding situation for a magnesium atom would be that of two $3s$ electrons outside the neon core and a net positive charge of $+12 - 10 = +2$ acting on each $3s$ electron. In a similar fashion we would find that the net positive charge acting on the valence electrons would progressively increase across the third period.

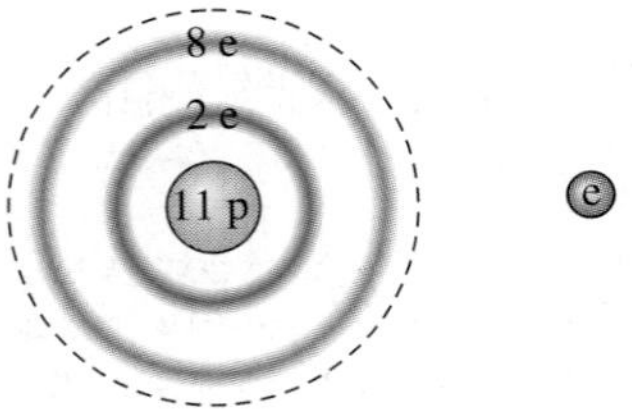

Figure 7.8 Shielding effect and effective nuclear charge, illustrated for a sodium atom. The charge on the nucleus is +11. If the nucleus and the two inner electronic shells (contained within the broken circle) acted as a unit, the effective nuclear charge would be +1. Shielding by the inner electrons is not perfect, however, and the effective nuclear charge is somewhat greater than +1.

The limitation of this simple description is that it fails to take into account that (1) the valence electrons do partially occupy the same region as the core electrons, (2) not all core electrons are equally effective in shielding valence electrons, and (3) to some extent, valence electrons shield one another. But whether we make crude or precise estimates of effective nuclear charges, we arrive at the same conclusion: The effective nuclear charge increases from left to right in a period of representative elements in the periodic table. Because the effective nuclear charge increases, valence electrons are pulled in toward the nucleus and held more tightly.

Atomic radii of the A-group elements tend to decrease from left to right in a period of the periodic table.

The restriction to A-group elements is an important one. In a series of B-group elements (transition elements), electrons enter an *inner* electron shell, not the valence shell. In this process the effective nuclear charge remains essentially constant instead of increasing. For example, in comparing the effective nuclear charges, Z_{eff}, of iron, cobalt and nickel, by the crude method of Figure 7.8,

Fe: $[Ar]3d^6 4s^2$	Co: $[Ar]3d^7 4s^2$	Ni: $[Ar]3d^8 4s^2$
$Z_{eff} \approx 26 - 24 \approx 2$	$Z_{eff} \approx 27 - 25 \approx 2$	$Z_{eff} \approx 28 - 26 \approx 2$

Because the effective nuclear charges are about the same, we conclude that the radii should also be about the same. The actual values are 124, 124, and 125 pm, respectively, for Fe, Co, and Ni.

Figure 7.7 clearly illustrates the difference in the variation of atomic radii with atomic number between A-group and B-group elements.

Example 7.6

Arrange each set of elements in order of increasing atomic radius, that is, from smallest to largest.

a. Mg, Si, S **b.** As, N, P **c.** As, Se, Sb

Solution

a. All three elements are A-group elements in the same period (third). Atomic radii within a period decrease from left to right. The order of increasing radius is:

S (smallest) < Si < Mg (largest)

b. All three elements are in the same group (5A). Atomic radii increase from top to bottom. The order of increasing radius is

N (smallest) < P < As (largest)

c. Of this set, As and Se are in the same period (fourth); As is to the left and therefore larger than Se. Because Sb is below As in the same group (5A), it is larger than As. The overall order is

Se (smallest) < As < Sb (largest)

Exercise 7.6

Arrange each set of elements in order of increasing atomic radius.

a. Be, F, N **b.** Ba, Be, Ca **c.** Cl, F, S **d.** Ca, K, Mg

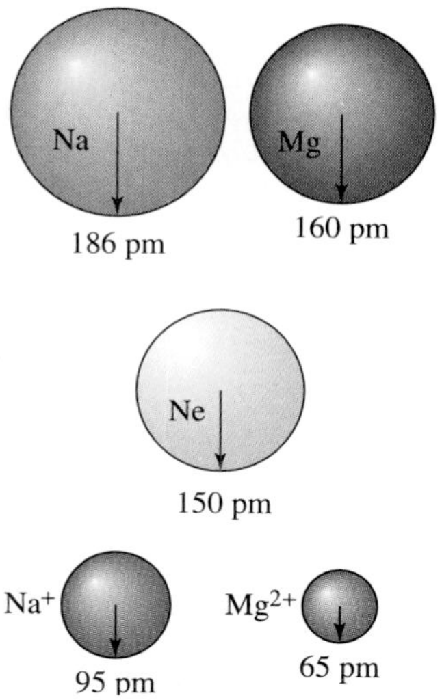

Figure 7.9 A comparison of atomic and cationic sizes. Metallic radii are shown for Na and Mg, and ionic radii for Na^+ and Mg^{2+}. The radius for Ne is for a nonbonded atom; it is called the van der Waals radius.

Ionic Radii

When atoms of metals react, they often lose all their valence (outer-shell) electrons. With one electronic shell fewer, metal ions are smaller than the atoms from which they come. Also, because the nuclear charge is now greater than the number of electrons, the nucleus attracts the remaining electrons more strongly and holds them closer in a cation (positive ion) than in the corresponding atom.

Cations are smaller *than the atoms from which they are formed.*

Figure 7.9 compares five species: a sodium atom (Na), a magnesium atom (Mg), a sodium ion (Na^+), a magnesium ion (Mg^{2+}), and a neon atom (Ne). The Mg atom is smaller than the Na atom, and the ions are smaller than the corresponding atoms. Na^+, Mg^{2+}, and Ne are *isoelectronic*; they all have the same number (10) of electrons. They also have identical electron configurations ($1s^22s^22p^6$). Ne has a nuclear charge of +10. Because it has a nuclear charge of +11, Na^+ is *smaller* than Ne. Because its nuclear charge is +12, Mg^{2+} is smaller still.

When a nonmetal atom gains an electron to form a negative ion (anion), the nuclear charge remains constant while repulsions among the electrons increase. The electrons spread out more and the size increases (Figure 7.10). By the same reasoning, an anion with a charge of 2− is somewhat larger than an anion with the same electron configuration but a charge of only 1−.

Anions are larger *than the atoms from which they are formed.*

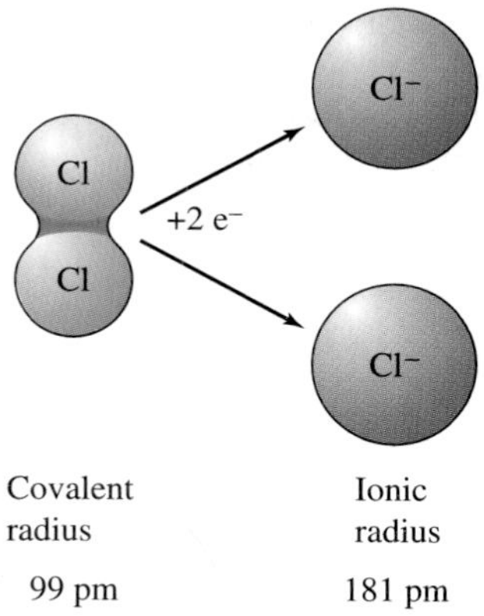

Figure 7.10 Atomic (covalent) and anionic radii compared. The two chlorine atoms in a Cl_2 molecule gain one electron each to form two chloride ions (Cl^-).

In summary, for a series of ions having the same electron configuration, the greater the nuclear charge, the smaller is the ion. This generalization is illustrated in Figure 7.11, which lists the radii of a number of common cations and anions.

Li^+ 60	Be^{2+} 31											B^{3+} 20	C —	N^{3-} 171	O^{2-} 140	F^- 136
Na^+ 95	Mg^{2+} 65											Al^{3+} 50	Si —	P^{3-} 212	S^{2-} 184	Cl^- 181
K^+ 133	Ca^{2+} 99	Sc^{3+} 81	Ti^{2+} 80	V^{2+} 88	Cr^{2+} 83	Mn^{2+} 80	Fe^{2+} 74	Co^{2+} 72	Ni^{2+} 69	Cu^{2+} 72	Zn^{2+} 74				Se^{2-} 198	Br^- 195
Rb^+ 148	Sr^{2+} 113														Te^{2-} 221	I^- 216
Cs^+ 169	Ba^{2+} 135															

Figure 7.11 **Some representative ionic radii in picometers.** Many of the elements form more than one ion. Different ions of the same element differ in size (Fe^{2+} is larger than Fe^{3+}; Why?). The data listed here are only meant to be representative.

Example 7.7

Arrange the following species in order of increasing radius: S^{2-}, Ca^{2+}, Mg^{2+}, K^+, Se^{2-}.

Solution

Three of these—S^{2-}, K^+, and Ca^{2+}—are isoelectronic and have the same electron configuration: $1s^22s^22p^63s^23p^6$. According to the generalization stated above, their comparative radii should be $Ca^{2+} < K^+ < S^{2-}$. Because it has the same charge but fewer electron shells than Ca^{2+}, we expect the Mg^{2+} ion to be smaller than Ca^{2+}. To get a feel for the size of the Se^{2-} ion, note that it is in the same group as S^{2-}. Because Se^{2-} has the same charge but more electron shells than S^{2-}, we expect Se^{2-} to be larger than S^{2-}. We conclude that the overall order is

$$Mg^{2+} < Ca^{2+} < K^+ < S^{2-} < Se^{2-}$$

Exercise 7.7

Arrange the following species in order of increasing radius: Br^-, Rb^+, Se^{2-}, Sr^{2+}, Y^{3+}.

Ionization Energy

A common characteristic of metals is a tendency to lose valence electrons *when they undergo chemical reactions*. However, isolated atoms do not eject electrons spontaneously. Work must be done to remove an electron from an atom, and the amount of work is very much dependent on the size of the atom.

The **ionization energy** is the energy required to remove an electron from a ground-state atom (or ion) in the gaseous state. Atoms with more than one electron can ionize in successive steps. Consider the boron atom, which has five electrons—two in an inner core ($1s^2$) and three valence electrons ($2s^22p^1$). The five steps and their ionization energies, I_1 through I_5, are

$$B(g) \longrightarrow B^+(g) + e^- \qquad I_1 = 801 \text{ kJ/mol}$$

$$B^+(g) \longrightarrow B^{2+}(g) + e^- \qquad I_2 = 2427 \text{ kJ/mol}$$

$$B^{2+}(g) \longrightarrow B^{3+}(g) + e^- \qquad I_3 = 3660 \text{ kJ/mol}$$

$$B^{3+}(g) \longrightarrow B^{4+}(g) + e^- \qquad I_4 = 25{,}025 \text{ kJ/mol}$$

$$B^{4+}(g) \longrightarrow B^{5+}(g) + e^- \qquad I_5 = 32{,}822 \text{ kJ/mol}$$

The first electron to be removed is from the highest energy level ($2p$). It is by far the easiest to remove. The energy required to remove this first electron is called the *first* ionization energy, I_1. Note that the *second* ionization energy, I_2, is about three times as great as the first. Why is this so? The first electron is removed from the $2p$ orbital of a *neutral* B atom. The second electron is removed from the $2s$ orbital of a B^+ *ion*. The second ionization energy is larger than the first, partly because the $2s$ orbital is at a lower energy than the $2p$, but mostly because the second electron must be stripped from a positive ion, to which it is strongly attracted. Removal of the third electron must occur from a more highly charged ion, B^{2+}, making I_3 larger still.

Compared to the first three, the fourth and fifth ionization energies, I_4 and I_5, of boron are extremely large. Note that the first three electrons are *valence* electrons and the last two are *inner core* electrons, characterized by a lower value of n. The large difference in ionization energy for these two types of electrons suggests a special stability for inner-core electron configurations. For the representative elements, removing a core electron takes *much* more work than removing a valence electron. This is consistent with the observation that only valence electrons are associated with the chemical reactivity of the representative elements.

Table 7.3 lists some ionization energies for representative elements, and Figure 7.12 is a graph of first ionization energy versus atomic number. From these two sources we can infer several useful generalizations.

- The first ionization energy of an atom is its lowest. Compare I_1 and I_2 values for the members of Group 2A in Table 7.3.
- A large increase in ionization energy occurs between the removal of the last valence electron and the removal of the first core electron. Compare I_1 and I_2 values for the members of Group 1A in Table 7.3.
- Ionization energies *decrease* down a group in the periodic table—that is, from lower to higher atomic numbers. Compare I_1 values for different members of Group 1A and of Group 2A of Table 7.3. Also, notice the gradual decrease in the minimum point of the graph of Figure 7.12 each time a minimum recurs.
- In general, ionization energies *increase* in going from left to right through a period in the periodic table. Compare the I_1 values across the top row of Table 7.3. In Figure 7.12 this is seen in the steady rise in I_1 values (with some notable exceptions) from the minima for Group 1A (alkali metal) elements to the maxima for Group 8A (noble gas) elements.

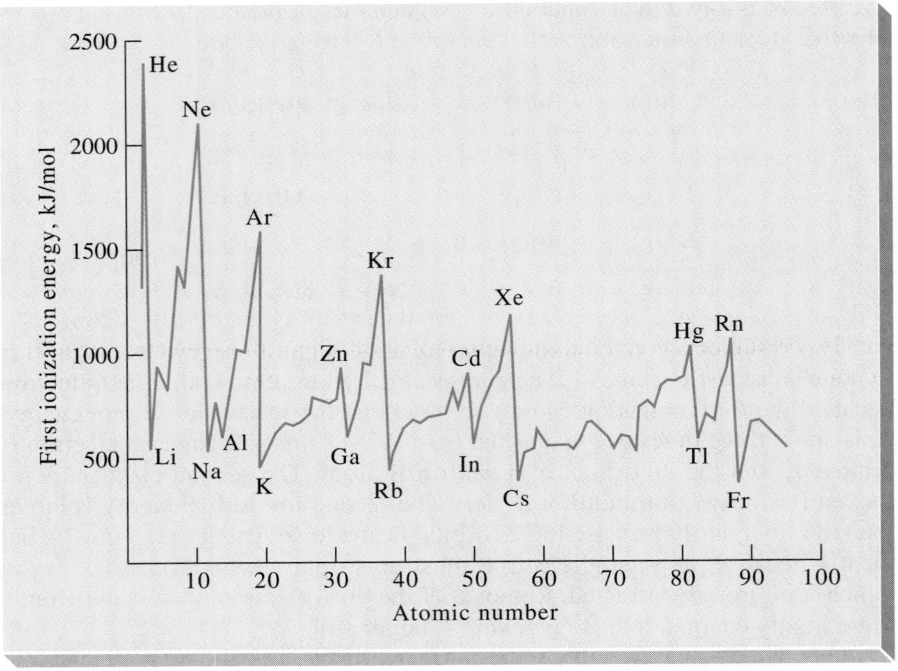

Figure 7.12 First ionization energy as a function of atomic number.

Even though other factors contribute to these patterns, we can explain these general trends most easily in terms of atomic sizes. The greater the distance between the atomic nucleus and the electron to be removed, the less tightly that electron is held to the nucleus and the more readily ionization occurs. As we have already seen, the general trends in atomic radii are an increase down a group and a decrease across a period of the periodic table.

TABLE 7.3 Some Selected Ionization Energies, kJ/mol

	1A	2A	3A	4A	5A	6A	7A	8A
	Li	Be	B	C	N	O	F	Ne
I_1	520	900	801	1086	1402	1314	1681	2081
I_2	7298	1757						
	Na	Mg						
I_1	496	738						
I_2	4562	1451						
	K	Ca						
I_1	419	590						
I_2	3051	1145						
	Rb	Sr						
I_1	403	550						
I_2	2633	1064						
	Cs	Ba						
I_1	376	503						
I_2	2230	965						

It is also possible to explain the irregularities in the trends across periods. Note, for example, that boron, with $Z = 5$ and the electron configuration $1s^2 2s^2 2p^1$, has a smaller first ionization energy (801 kJ/mol) than does beryllium ($Z = 4$), which has the electron configuration $1s^2 2s^2$ and $I_1 = 900$ kJ/mol. The $2p$ electron of boron is at a higher energy than the $2s$ electrons of beryllium and is therefore easier to remove. This kind of discontinuity occurs generally in proceeding from a Group 2A or Group 2B element to a Group 3A element.

To explain the slight irregularities that occur between Groups 5A and 6A let's consider I_1 values for nitrogen ($[He]2s^2 2p_x^1 p_y^1 p_z^1$) and oxygen ($[He]2s^2 2p_x^2 p_y^1 p_z^1$). In $2p$ orbitals, the repulsion between the paired electrons in the $2p_x$ orbital of oxygen makes the removal of one of those electrons easier to accomplish than the removal of an unpaired electron from a half-occupied $2p$ orbital of nitrogen. The observed I_1 values are 1314 kJ/mol for oxygen and 1401 kJ/mol for nitrogen.

Example 7.8

Arrange each set of elements in order of increasing first ionization energy.

a. Mg, Si, S **b.** As, N, P **c.** As, Ge, P

Solution

a. All three elements are in the same period. Within a period, I_1 *increases* from left to right as the atoms become smaller. The order of increasing I_1 is

Mg (lowest) < Si < S (highest)

b. All three elements are in the same group. Within a group, I_1 *decreases* from top to bottom as atoms become larger. The order of increasing I_1 is

As (lowest) < P < N (highest)

c. Of this set, As and Ge are in the same period; Ge is to the left of As and has a lower I_1. Because it is below P in the same group, As has a lower I_1 than P. The order of increasing I_1 is

Ge (lowest) < As < P (highest)

Exercise 7.8

Arrange each set of elements in order of increasing first ionization energy.

a. Be, F, N **b.** Ba, Be, Ca **c.** P, F, S **d.** Ca, K, Mg

Electron Affinity

Ionization energy relates to the process of forming a gaseous *positive* ion from a gaseous atom. The corresponding atomic property for the formation of a gaseous *negative* ion is **electron affinity**, the energy change that occurs when an electron is added to a gaseous atom.

An electron approaching a neutral atom experiences an attraction for the positively charged nucleus. Repulsion of the incoming electron by electrons already present in the atom tends to offset this attraction. Still, in many cases the incoming electron is absorbed by the atom and energy is evolved, as in the process

$$F(g) + e^- \longrightarrow F^-(g) \qquad EA = -328 \text{ kJ/mol}$$

When a fluorine atom gains an electron, energy is given off. The process is *exothermic*, and the electron affinity is therefore a *negative* quantity.

Table 7.4 lists several electron affinity values, but we see fewer clear-cut trends and more irregularities than in the Table 7.3 listing of ionization energies. Some of the data suggest a rough correlation between electron affinity and atomic size: The smaller an atom, the more negative is its electron affinity. We might reason that the smaller the atom, the closer an added electron can approach the atomic nucleus and the more strongly it is attracted to the nucleus. This certainly seems to be the case for the Group 1A elements, for Group 6A from S to Po, and for Group 7A from Cl to At. The first row elements present some problems, though. The electron affinity of O is not as negative as that of S, nor is that of F as negative as that of Cl. Here it may be that electron repulsions in the small compact atoms keep the added electron from being as tightly bound as we might expect.

TABLE 7.4 **Some Selected Electron Affinities, kJ/mol**

1A	2A	3A	4A	5A	6A	7A	8A
Li –60	Be —	B –27	C –154	N –7	O –141	F –328	Ne —
Na –53					S –200	Cl –349	
K –48					Se –195	Br –325	
Rb –47					Te –190	I –295	
Cs –46					Po –183	At –270	

In most cases in Table 7.4, the added electron goes into an energy sublevel that is already partly filled. For the Groups 2A and 8A atoms, however, the added electron would be required to enter a significantly higher energy level, the *np* level for the Group 2A atoms and the *s* level of the next principal level for the Group 8A atoms. In these cases a stable anion does not form.

Just as we described the stepwise loss of multiple electrons in the formation of positive ions, we can speak of the stepwise addition of electrons in anion formation. And we can write a separate electron affinity for each step. The situation for the oxide ion is

$$O(g) + e^- \longrightarrow O^-(g) \qquad EA_1 = -141 \text{ kJ/mol}$$

$$O^-(g) + e^- \longrightarrow O^{2-}(g) \qquad EA_2 = +744 \text{ kJ/mol}$$

It's not hard to see why EA_2 is a positive quantity. Here an electron approaches an ion with a net charge of –1. It is strongly repelled, and work must be done to force the extra electron onto the $O^-(g)$ ion. In fact, we only expect the O^{2-} ion to form in situations where the large energy expense to create it can be offset by other energetically favorable processes. This happens in the formation of ionic oxides, as we shall see in Section 9.3.

ESTIMATION EXAMPLE 7.1

Which of the values given is a reasonable estimate of EA_2 for this process?

$$S^-(g) + e^- \longrightarrow S^{2-}(g) \qquad EA_2 = ?$$

−200 kJ/mol +450 kJ/mol +800 kJ/mol +1200 kJ/mol

Solution

Because of the repulsion between the $S^-(g)$ anion and the electron it is to gain, the energy change must be *positive* (the process is endothermic). This eliminates −200 kJ/mol as a possibility. The key to choosing among the other three values lies in the value of EA_2 for oxygen, +744 kJ/mol (page 262). Because of its *smaller* size, we should expect the $O^-(g)$ anion to exert a stronger repulsion for the incoming electron than would the $S^-(g)$ anion. The EA_2 value for $S^-(g)$ should be less than +744 kJ/mol. The only reasonable estimate is +450 kJ/mol.

Estimation Exercise 7.1

Based on the result of Estimation Example 7.1, which of the values given is a reasonable estimate of EA_2 for this process?

$$Se^-(g) + e^- \longrightarrow Se^{2-}(g) \qquad EA_2 = ?$$

−390 kJ/mol +400 kJ/mol +460 kJ/mol +500 kJ/mol

7.8 Metals, Nonmetals, and Noble Gases

In Section 2.5 we distinguished between metals and nonmetals in terms of general appearance and bulk physical properties. We are now in a position to consider metals and nonmetals in relation to electron configurations and positions in the periodic table.

Generally speaking, **metal** atoms have small numbers of electrons in their valence shells. Except for hydrogen and helium, all *s*-block elements are metals. All *d*- and *f*-block elements are metals. A few of the *p*-block elements also are metals. **Nonmetal** atoms generally have larger numbers of electrons in their valence shell than do metals; except for the special case of hydrogen, nonmetals are all *p*-block elements. The metals and nonmetals are separated by a heavy, stepped diagonal line in the periodic table. Most of the elements along this line have the physical appearance of metals but some nonmetallic properties as well; these "borderline" elements are sometimes called **metalloids**. The gaseous nonmetal elements in the column at the extreme right of the periodic table are often singled out as a special group, the **noble gases**. Each period of the periodic table ends with a noble gas element. Figure 7.13, which is in the format of the periodic table, summarizes this information.

Perhaps more interesting than the electron configurations of the metals and nonmetals are the configurations of the *ions* that they tend to form in ionic compounds. The formation of *anions* from nonmetal atoms is easiest to describe. When forming an ionic compound, a nonmetal atom can alter its

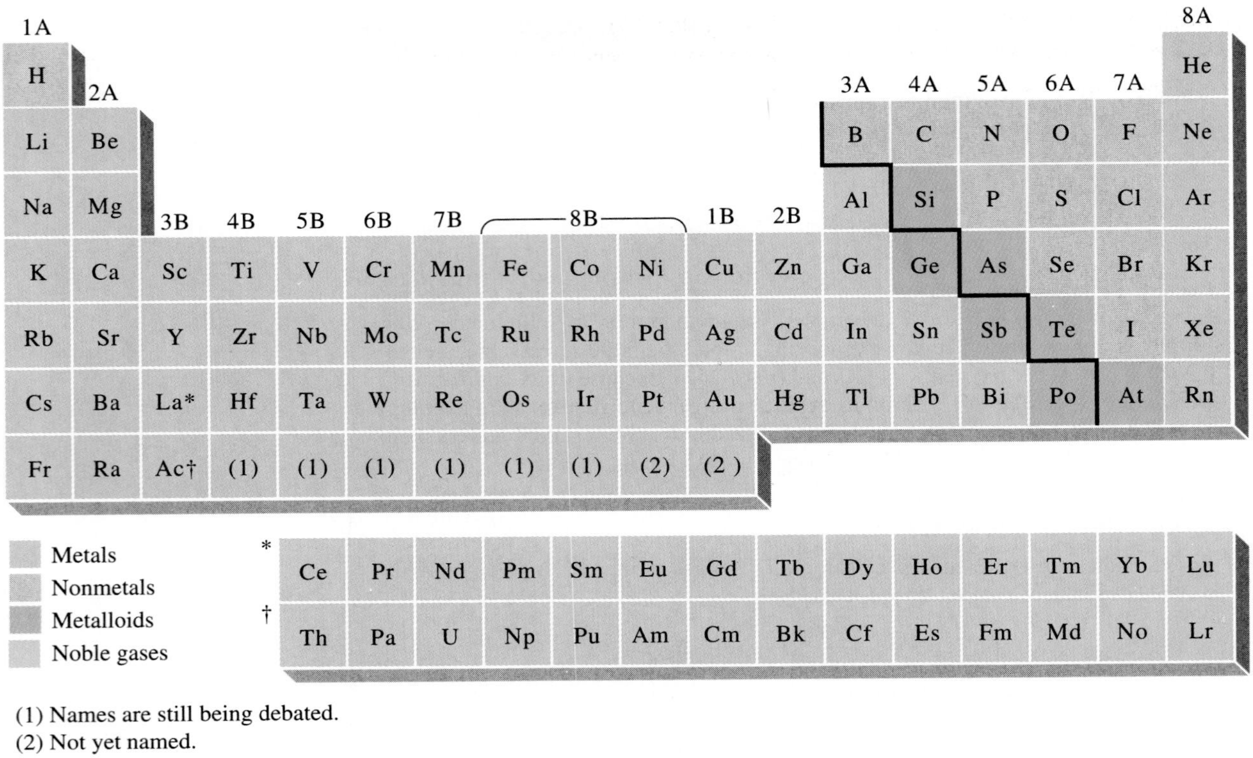
1A
8A
H
2A
He
3A 4A 5A 6A 7A
Li Be
B C N O F Ne
Na Mg
3B 4B 5B 6B 7B 8B 1B 2B
Al Si P S Cl Ar
K Ca Sc Ti V Cr Mn Fe Co Ni Cu Zn Ga Ge As Se Br Kr
Rb Sr Y Zr Nb Mo Tc Ru Rh Pd Ag Cd In Sn Sb Te I Xe
Cs Ba La* Hf Ta W Re Os Ir Pt Au Hg Tl Pb Bi Po At Rn
Fr Ra Ac† (1) (1) (1) (1) (1) (1) (2) (2)
*
Ce Pr Nd Pm Sm Eu Gd Tb Dy Ho Er Tm Yb Lu
†
Th Pa U Np Pu Am Cm Bk Cf Es Fm Md No Lr
Metals
Nonmetals
Metalloids
Noble gases
(1) Names are still being debated.
(2) Not yet named.

Figure 7.13 Metals, nonmetals, metalloids, and noble gases.

electron configuration by gaining a small number of electrons (usually 1 or 2, sometimes 3) to complete its valence shell and acquire the electron configuration of a noble gas. For example,

$$\text{Br}\ ([\text{Ar}]3d^{10}4s^24p^5) + e^- \longrightarrow \text{Br}^-([\text{Kr}])$$

$$\text{S}\ ([\text{Ne}]3s^23p^4) + 2\ e^- \longrightarrow \text{S}^{2-}([\text{Ar}])$$

$$\text{N}\ ([\text{He}]2s^22p^3) + 3\ e^- \longrightarrow \text{N}^{3-}([\text{Ne}])$$

When forming an ionic compound, a metal atom loses a small number of electrons. For representative metals, in some cases all the valence electrons are lost and the resulting electron configuration is that of a noble gas atom. For example,

$$\text{Na}\ ([\text{Ne}]3s^1) \longrightarrow \text{Na}^+\ ([\text{Ne}]) + e^-$$

$$\text{Mg}\ ([\text{Ne}]3s^2) \longrightarrow \text{Mg}^{2+}\ ([\text{Ne}]) + 2\ e^-$$

$$\text{Al}\ ([\text{Ne}]3s^23p^1) \longrightarrow \text{Al}^{3+}\ ([\text{Ne}]) + 3\ e^-$$

However, in other cases the electron configuration of the cation is not that of a noble gas.

$$\text{Ga}\ ([\text{Ar}]3d^{10}4s^24p^1) \longrightarrow \text{Ga}^{3+}\ ([\text{Ar}]3d^{10}) + 3\ e^-$$

$$\text{Sn}\ ([\text{Kr}]4d^{10}5s^25p^2) \longrightarrow \text{Sn}^{2+}\ ([\text{Kr}]4d^{10}5s^2) + 2\ e^-$$

Many transition metal atoms tend to lose not only their valence-shell electrons, but one or more inner-shell electrons as well. Moreover, depending on the number of electrons lost, the same atom may be capable of forming more than one ion. And, in general, transition metal ions do not have noble gas electron configurations.

$$Fe\ ([Ar]3d^6 4s^2) \longrightarrow Fe^{2+}\ ([Ar]3d^6) + 2\ e^-$$

$$Fe\ ([Ar]3d^6 4s^2) \longrightarrow Fe^{3+}\ ([Ar]3d^5) + 3\ e^-$$

The possible types of electron configurations of cations are summarized in Table 7.5.

TABLE 7.5 **Electron Configurations of Some Metal Ions**

"Noble Gas"			"Pseudo Noble Gas"[a]		"18 + 2"[b]	"Various"	
Li^+	Be^{2+}	Al^{3+}	Cu^+	Zn^{2+}	In^+	Cr^{2+}	$[Ar]3d^4$
Na^+	Mg^{2+}		Ag^+	Cd^{2+}	Tl^+	Cr^{3+}	$[Ar]3d^3$
K^+	Ca^{2+}		Au^+	Hg^{2+}	Sn^{2+}	Mn^{2+}	$[Ar]3d^5$
Rb^+	Sr^{2+}				Pb^{2+}	Mn^{3+}	$[Ar]3d^4$
Cs^+	Ba^{2+}				Sb^{3+}	Fe^{2+}	$[Ar]3d^6$
					Bi^{3+}	Fe^{3+}	$[Ar]3d^5$
						Co^{2+}	$[Ar]3d^7$
						Co^{3+}	$[Ar]3d^6$
						Ni^{2+}	$[Ar]3d^8$

[a] In the "pseudo noble gas" configuration all valence electrons are lost and the remaining $(n-1)$ shell has 18 electrons in the configuration $(n-1)s^2(n-1)p^6(n-1)d^{10}$.
[b] In the "18 + 2" configuration, $(n-1)s^2(n-1)p^6(n-1)d^{10}ns^2$, two valence electrons remain.

Metallic character is closely related to ionization energy, which in turn is closely related to atomic radius. The easier it is to remove electrons from an atom, the more metallic the element is. Ease of removal of electrons corresponds to low ionization energy and large atomic radius. As a result,

> *Metallic character* increases *from top to bottom in a group and* decreases *from left to right in a period of the periodic table.*

The more readily an atom gains an electron, the more nonmetallic is the character of the atom. A strong tendency to gain electrons corresponds to a large negative electron affinity, a property found among the smaller atoms. In general, we can say that

> *Nonmetallic character* decreases *from top to bottom in a group and* increases *from left to right in a period of the periodic table.*

From the first generalization we can identify the alkali metals (Group 1A) as a highly metallic group, and from the second generalization, the halogens

(Group 7A) as a highly nonmetallic group. In the middle of the periodic table we see intermediate behavior. Group 4A has carbon, a nonmetal, at the top and two metals, tin and lead, at the bottom. In between are two metalloids, silicon and germanium. Figure 7.14 summarizes trends in atomic properties and metallic/nonmetallic behavior in the periodic table.

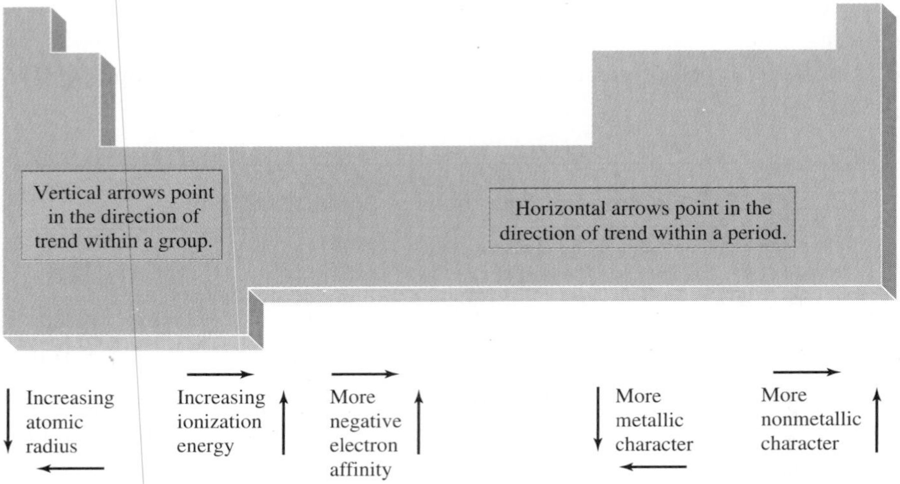

Figure 7.14 Atomic properties—a summary of trends in the periodic table. Metallic character refers generally to the ability to lose electrons and nonmetallic behavior, to gain electrons. Ionization energy refers to I_1.

Example 7.9

In each set, indicate which is the more metallic element.

a. Ba, Ca **b.** Sb, Sn **c.** Ge, S

Solution

a. Ba is below Ca in Group 2a. The Ba atom is larger than the Ca atom. We expect its first and second ionization energies to be lower than those of Ca. As a result, Ba is more metallic than Ca.

b. Sn is to the left of Sb in the fifth period. We expect it to be a larger atom with lower ionization energies than those of Sb. Sn is more metallic than Sb.

c. Ge is to the left of S and in the following period. Because of its larger atoms, we expect Ge to be more metallic than S. (Actually, Ge is a metalloid and S is a nonmetal.)

Exercise 7.9

In each set, indicate which is the more nonmetallic element.

a. O, P **b.** As, S **c.** P, F

CONCEPTUAL EXAMPLE 7.1

Without reference to tables or figures in the text, write, in the proper position in the blank periodic table of Figure 7.15, (**a**) the atomic number of the element that has the electron configuration $4s^2 4p^6 4d^5 5s^1$ for its outermost and next-to-outermost shells and (**b**) the symbol of the most metallic of the fifth-period *p*-block elements.

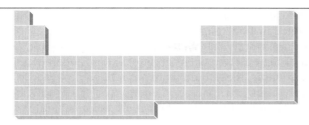

Figure 7.15

Solution

a. The portion $4d^55s^1$ of the electron configuration identifies this as a transition element in the d-block (the $4d$ subshell is being filled) and in the fifth period ($n = 5$ is the highest principal quantum number). The underlying noble gas electron configuration is that of krypton ($Z = 36$): $1s^22s^22p^63s^23p^63d^{10}4s^24p^6$. The complete noble-gas-core abbreviated electron configuration is [Kr] $4d^55s^1$, signifying this to be the element with the atomic number $Z = 42$ (molybdenum). It is the sixth element in the fifth period (in Group 6B).

b. In general, the more metallic elements are those with large atomic radii and low ionization energies. These elements tend to be found toward the *left* of their respective periods. The *fifth*-period element in the p-block that is farthest to the left is indium. We expect it to be the most metallic of the fifth period p-block elements. (The measured I_1 value of In is 558 kJ/mol, that of Sn is 709 kJ/mol, and that of Sb is 834 kJ/mol.)

Conceptual Exercise 7.1

Without reference to tables or figures in the text, write, in the proper position in the blank periodic table of Figure 7.15, the following information.

a. The atomic number of the element having as the electron configuration of its two outermost shells $4s^24p^64d^{10}5s^25p^4$.

b. The atomic number of the largest atom in the first transition series.

c. The atomic numbers of the d-block elements of the fifth period having the lowest and highest I_1 values.

d. The symbol of the most *nonmetallic* element of Group 5A.

The Noble Gases

In the last decade of the nineteenth century, a group of elements was discovered that made up an entirely new family, one completely unexpected by Mendeleev and his contemporaries. Nonetheless, this new group, called the noble gases, fit neatly between the highly active nonmetals of Group 7A and the very reactive alkali metals (Group 1A). In the usual form of the periodic table, the noble gases are placed to the far right as Group 8A, in the column immediately following Group 7A.

The six noble gases are helium, neon, argon, krypton, xenon, and radon. All are found to some extent in the atmosphere. Argon is relatively abundant, making up nearly 1% of the atmosphere by volume. Xenon, on the other hand, is quite rare, accounting for only 91 parts per billion (that is, having a mole fraction of 0.000000091) of the atmosphere. Radioactive radon, produced by the radioactive decay of heavy elements such as uranium, seeps from the ground. It makes up only an insignificant portion of the atmosphere, but it can be a health problem when concentrated in a building with poor ventilation.

The noble gases rarely enter into chemical reactions. This lack of reactivity is a reflection of their electron configurations, ionization energies and electron affinities. For example, the helium atom has a filled first energy level ($1s^2$), and an exceptionally high first ionization energy (2372 kJ/mol). Helium loses an electron only with great difficulty. The helium atom has little or no affinity for an additional electron because this would have to be accommodated in the $2s$ orbital, which is at a much higher energy than the filled $1s$ orbital. Helium therefore does not form an anion. The other noble gases, with the valence-shell electron configuration ns^2np^6, show a similar aversion both to losing and gaining an electron.

Because of their lack of reactivity, noble gas atoms do not tend to combine with other atoms, even of their own kind. Therefore the noble gases occur naturally only in elemental form and only as monatomic species. The noble gases were once called the "inert gases," but since 1962 a few compounds of krypton and xenon have been prepared. As yet, no compounds have been made of the lighter noble gases, helium, neon, and argon. While this family of elements is no longer called "inert," its nobility is unquestioned. More than any other family in the periodic table, the noble gases avoid interactions with other elements.

Looking Ahead

Electron configurations, the periodic table, and atomic properties such as atomic radii and ionization energies all enter into the discussion of the *s*-block elements in the next chapter. Ionization energies and electron affinities figure prominently in the treatment of ionic compounds in Chapter 9. Ideas about unpaired and paired electrons are important to the discussion of covalent bonds in Chapters 9 and 10.

SUMMARY

The wave-mechanical treatment of the hydrogen atom can be extended to multielectron atoms, but with this essential difference: Principal energy levels are (a) lower than those of the hydrogen atom and (b) split, that is, having different energies for the different subshells.

Electron configuration refers to the distribution of electrons among orbitals in an atom. Introduced here are the subshell (or *"spdf"*) notation and the orbital diagram. Key ideas required to write a probable electron configuration are that (a) electrons tend to occupy the lowest energy orbitals available, (b) no two electrons can have all four quantum numbers alike, and (c) where possible, electrons occupy orbitals singly rather than in pairs.

The aufbau principle describes a hypothetical process of building up one atom from the atom of preceding atomic number. With this principle and the ideas cited above, it is possible to predict probable electron configurations for many of the elements. In the aufbau process, electrons are added to the *s* or *p* subshell of highest principal quantum number in the representative or main-group elements. In transition elements, electrons go into the *d* subshell of the next-to-outermost shell, and in the inner-transition elements, into the *f* subshell of the second-from-outermost principal shell.

Elements with similar valence-shell electron configurations fall in the same group of the periodic table. For A-group elements, the group number corresponds to the number of electrons in the principal shell of highest quantum number. The period number is the same as the highest numbered principal shell containing electrons (the outer shell). The division of the periodic table into *s*, *p*, *d*, and *f* blocks greatly assists in the assignment of probable electron configurations.

An atom with all electrons paired is diamagnetic, and an atom with one or more unpaired electrons is paramagnetic. Experimentally determined magnetic properties can be used to verify electron configurations.

Certain atomic properties vary periodically when atoms are considered in terms of increasing atomic number. The properties and trends considered in this chapter are those of atomic radius, ionic radius, ionization energy, and electron affinity. Values of these atomic properties strongly influence physical and chemical properties of the elements.

The elements can be categorized according to the tendency for their atoms to lose electrons (metals) or to gain electrons (nonmetals). Each of these categories, together with those labeled metalloids and noble gases, is found in its own general region of the periodic table.

KEY TERMS

actinide (7.5)
atomic radius (7.7)
aufbau principle (7.4)
core electrons (7.5)
covalent radius (7.7)
d-block (7.5)
degenerate (7.1)
diamagnetic (7.6)
effective nuclear charge (Z_{eff}) (7.7)
electron affinity (7.7)
electron configuration (7.2)
f-block (7.5)
Hund's rule (7.3)
ionization energy (7.7)
lanthanide (7.5)
metal (7.8)
metallic radius (7.7)
metalloid (7.8)
noble gas (7.8)
nonmetal (7.8)
orbital diagram (7.2)
paramagnetic (7.6)
Pauli exclusion principle (7.3)
p-block (7.5)
periodic law (7.7)
representative (main-group) element (7.4)
s-block (7.5)
subshell (sublevel) notation (7.2)
transition element (7.4)
valence electrons (7.5)

REVIEW QUESTIONS

1. How do the orbitals of multielectron atoms differ from those of hydrogen?

2. What is meant by the term degenerate orbitals?

3. Which of the following hydrogen orbitals have identical energies? How would your answer change for an atom other than hydrogen?
a. 1*s*, 2*s*, 2*p* **b.** 3*s*, 3*p*, 3*d* **c.** $3p_x$, $3p_y$, $3p_z$

4. State Pauli's exclusion principle. How is it applied to the electron configurations of atoms?

5. State what is meant by a ground-state and an excited-state electron configuration.

6. State Hund's rule. Describe how it helps us to select the correct electron configuration for the carbon atom.

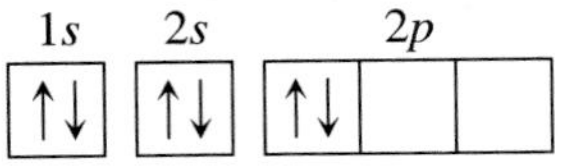

or

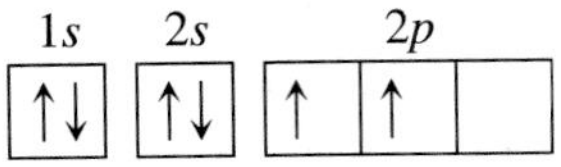

7. Tell in words what is meant by each of the following electron configuration notations. What element corresponds to each configuration?
a. $1s^22s^22p^5$ **b.** $1s^22s^22p^63s^23p^63d^94s^2$
c. $1s^22s^22p_x^12p_y^12p_z^1$ **d.** $[Ne]3s^1$

e. [Ar] 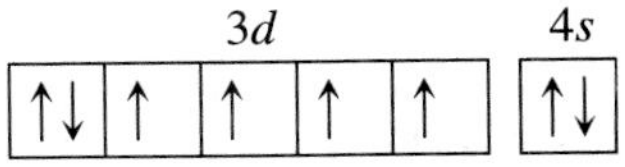

8. What is meant by the aufbau process? How is it used?

9. What is the configuration of the electrons in the outermost shell in the Group 3A elements? in Group 5A?

10. What subshell(s) is(are) being filled in each of the following regions of the periodic table?
a. Groups 1A and 2A
b. Groups 3A through 7A
c. the transition elements
d. the lanthanides and actinides

11. What similarity in electron configuration is shared by lithium, sodium, and potassium? by beryllium, magnesium, and calcium?

12. How many electrons are in the outermost principal shell of an atom of each of the following?
a. C **b.** Ne **c.** F **d.** Al **e.** Mg?

13. Referring only to the periodic table inside the front cover, indicate what similarity in electron configuration is shared by fluorine and chlorine, and by carbon and silicon. What is the difference in the electron configurations of each pair? What is the difference in the electron configurations of oxygen and fluorine?

14. Do you think it possible that someone might discover (**a**) a new element with atomic number 113; (**b**) a new element that would fit between magnesium and aluminum in the periodic table? Explain.

15. Give the periodic table period number and group number for the element whose atoms have the electron configuration
a. $1s^22s^22p^6$ **b.** $1s^22s^22p^63s^23p^2$
c. $1s^22s^22p^63s^1$ **d.** $1s^22s^2$
e. $1s^22s^22p^3$ **f.** $1s^22s^22p^63s^23p^1$

16. What is the configuration of the electrons in the outermost shell in Group 4A? in Group 6A?

17. How many electrons are described in the subshell notation $2p^6$? What is the general shape of the orbitals described in the notation? How many orbitals are included in the notation?

18. What is the maximum number of electrons that go into a *d* orbital? a *d* subshell?

19. Which member of each set of orbitals is higher in energy?
a. $2s$ or $3s$ **b.** $3s$ or $2p$
c. $2s$ or $2p$ **d.** $3s$ or $3p$

20. Write out the *spdf* notation for the electrons in the highest principal energy level for each of the following elements.
a. Cs **b.** Se **c.** In.

21. Write out the complete *spdf* notation for each of the following elements. Identify the core electrons and the valence electrons in each.
a. Si **b.** Rb **c.** Br

22. What are the differences in electron configurations that distinguish (**a**) representative and transition elements; (**b**) paramagnetic and diamagnetic elements?

23. Consider monatomic vapors of each of the following elements. Which are paramagnetic and which are diamagnetic?
a. Ba **b.** Hg **c.** Cs
d. Se **e.** Xe **f.** Sn

24. State the periodic law in its modern form.

25. Explain why the several periods of the periodic table do not all contain the same number of elements.

26. Based on the relationship between electron configurations and the periodic table, give the number of (**a**) outer-shell electrons in an atom of Bi; (**b**) electrons in the *fourth* principal shell of Au; (**c**) elements whose atoms have five outer-shell electrons; (**d**) unpaired electrons in an atom of Se; (**e**) transition elements in the fifth period.

27. On the basis of the periodic table and rules for electron configurations, indicate the number of (**a**) $3p$ electrons in an atom of P; (**b**) $4s$ electrons in an atom of Cs; (**c**) $4d$ electrons in an atom of Se; (**d**) $4f$ electrons in an atom of Bi; (**e**) unpaired electrons in an atom of Ga; (**f**) elements in Group 5A of the periodic table; (**g**) elements in the sixth period of the periodic table.

28. Explain why the sizes of atoms do not simply increase uniformly with increasing atomic number.

29. Explain why the difference in atomic radius between the elements $Z = 11$ (Na; 186 pm) and $Z = 12$ (Mg; 160 pm) is so large, whereas between $Z = 24$ (Cr; 125 pm) and $Z = 25$ (Mn; 124 pm) this difference is negligible.

30. In what location of the periodic table would you expect to find the two or three elements having the largest atoms? Explain.

31. Why are cations smaller than the atoms from which they are formed, whereas anions are larger?

32. Why is it that isoelectronic ions in the same electron configuration do not have the same ionic radii?

33. Arrange each of the following pairs of elements in order of increasing atomic radius.
a. Cl or S **b.** Al or Mg
c. As or Ge **d.** Ca or K.

34. Describe the trend in successive ionization energies as electrons are removed one at a time from an aluminum atom. Why is there a big jump between I_3 and I_4?

35. What is the general trend in first ionization energies within a period? Within a group? Explain each trend.

36. Why does sulfur have a lower first ionization energy than phosphorus?

37. Arrange each of the following sets of elements in order of increasing first ionization energy.
a. Ca, Mg, Sr **b.** Cl, P, S
c. As, Ge, Sn **d.** Br, Cl, Se

38. Which group of elements has electron affinities with the largest negative values? Explain why.

39. Which of the main groups of elements do not form stable negative ions? Use electron configurations and electron affinities to explain this behavior.

40. Silicon has an electron affinity of –134 kJ/mol. The electron affinity of phosphorus is –72 kJ/mol. Give a plausible reason for this difference.

41. Lithium has an electron affinity of –60 kJ/mol. That of boron is –27 kJ/mol. Give a plausible reason for this difference.

PROBLEMS

Pauli Exclusion Principle and Hund's Rule

42. Use the Pauli exclusion principle and Hund's rule to determine which of the following orbital diagrams are possible for a ground-state electron configuration and which are not. If the orbital diagram is not allowed, state why it is not.

	1s	2s	2p		
a.	↑↓	↑↓	↑↓	↑↑	↑↓
b.	↑↓	↑↓	↑↓	↑	↑↓↑
c.	↑↓	↑↓	↑	↑	↑
d.	↓↑	↓↑	↓	↓	↓
e.	↑↓	↑↓	↑	↓	↑

	1s	2s	2p			3s
f.	↑↓	↑↓	↑↓	↑↓	↑↓	↑↓

43. Use the Pauli exclusion principle and Hund's rule to determine which of the following orbital diagrams are possible for a ground-state electron configuration and which are not. If the orbital diagram is not allowed, state why it is not.

	1s	2s	2p			3s
a.	↓↑	↓↑	↓↑	↓↑	↓↑	↓
b.	↑↓	↑↑	↑↓	↑	↑↓↑	↑↓
c.	↑↓	↑↓	↑↓	↑	↑↓↑	↑↓

	1s	2s	2p			3s	3p		
d.	↓↑	↓↑	↓↑	↓↑	↓↑	↓↑	↑	↓	↑
e.	↑↓	↑↓	↑↓	↑↓	↑↓	↑↓	↑↓	↑	
f.	↑↓	↑↓	↑↓	↑↓	↑↓	↑↓	↑	↑	↑

44. None of the following electron configurations is reasonable for a ground-state atom. In each case, explain why.

a. $1s^2 2s^2 3s^2$ **b.** $1s^2 2s^2 2p^2 3s^1$
c. $1s^2 2s^2 2p^6 2d^5$

45. None of the following electron configurations is reasonable for a ground-state atom. In each case, explain why.

a. $1s^2 2s^2 2p^6 3s^1 3p^1$ **b.** $1s^2 2s^2 2p^6 3s^2 3p^6 3d^1$
c. $1s^2 2s^2 2p^6 3s^2 3p^6 3d^8 4s^2 4p^1$

46. Explain the principle(s) or rule(s) that each of the following electron configurations violates.

a. $1s^2 2s^6 3s^2$ **b.** $1s^2 2s^2 2p^7 3s^1$
c. $1s^2 2s^2 2p^6 2d^3$

47. Explain the principle(s) or rule(s) that each of the following electron configurations violates.

a. $[Ar]2d^{10}$ **b.** $[Ar]3f^3 4s^2$
c. $[Kr]4d^{10} 4f^{14} 5s^2$

Electron Configurations: The Aufbau Principle

48. Using expanded *spdf* notation and referring only to the periodic table inside the front cover, write out the ground-state electron configuration of each of the following.

a. Al **b.** Cl **c.** Na
d. B **e.** He **f.** O
g. C **h.** Li **i.** Si

49. Using expanded *spdf* notation and referring only to the periodic table inside the front cover, write out the ground-state electron configuration of each of the following.

a. Ar **b.** H **c.** Ne
d. Be **e.** K **f.** P
g. Ca **h.** Mg **i.** Br

50. Using a noble-gas-core abbreviated *spdf* notation and referring only to the periodic table inside the front cover, write out the ground-state electron configuration for each of the following.

a. Ba **b.** Rb **c.** As
d. F **e.** Se **f.** Sn

51. Using a noble-gas-core abbreviated *spdf* notation and referring only to the periodic table inside the front cover, write out the ground-state electron configuration for each of the following.

a. Ga **b.** Te **c.** I
d. Cs **e.** Sb **f.** Sr

52. Give orbital diagrams for the ground-state electron configuration of each of the following.

a. C **b.** O **c.** K
d. Al **e.** S **f.** Mg

53. Give orbital diagrams for the ground-state electron configuration of each of the following.

a. N **b.** B **c.** Si
d. Ca **e.** Cl **f.** Sc

54. Give the orbital diagram for the electrons beyond the xenon core of the hafnium (Hf) atom. Relate this electron configuration to the position of hafnium in the periodic table.

55. Give the orbital diagram for the electrons beyond the xenon core of the mercury (Hg) atom. Relate this electron configuration to the position of mercury in the periodic table.

Periodic Properties

56. Arrange each set of elements in order of increasing atomic radius and explain the basis for this order.

a. Al, Mg, Na **b.** Ca, Mg, Sr

57. Arrange each set of elements in order of increasing atomic radius and explain the basis for this order.

a. Ca, Rb, Sr **b.** Al, C, Si

58. Arrange each set of elements in order of increasing first ionization energy and explain the basis for this order.

a. K, Na, Rb **b.** P, S, Si

59. Arrange each set of elements in order of increasing first ionization energy and explain the basis for this order.

a. Ca, K, Mg **b.** Br, I, Te

60. Arrange the elements in each set in Problem 56 in order of decreasing metallic character (most metallic first) and explain the basis for this order.

61. Arrange the elements in each set in Problem 57 in order of decreasing metallic character (most metallic first) and explain the basis for this order.

Periodic Law

62. One of the first periodic properties studied was that of *atomic volume*, the atomic weight of an element divided by its density in the solid state. Draw a graph to show that atomic volume is a periodic property of the following elements. Densities are in g/cm^3. Na, 0.971; Mg, 1.74; Al, 2.70; Si, 2.33; P, 2.20; S, 2.07; Cl, 2.03; Ar, 1.66; K, 0.862; Ca, 1.55; Sc, 2.99; Cr, 7.19; Co, 8.90; Zn, 7.13; Ga, 5.91; As, 4.70; Br, 4.05; Kr, 2.82; Rb, 1.53; Sr, 2.54. To what atomic property described in this chapter does the atomic volume seem most closely related? Explain.

63. The following melting points are in °C. Draw a graph to show that melting point is a periodic property of these elements. Al, 660; Ar, −189; Be, 1278; B, 2300; C, 3350; Cl, −101; F, −220; Li, 179; Mg, 651; Ne, −249; N, −210; O, −218; P, 590; Si, 1410; Na, 98; S, 119. For this purpose, does it matter whether melting points are expressed on the Celsius scale or the Kelvin scale? Explain.

ADDITIONAL PROBLEMS

64. Without referring to any tables or listing in the text, mark an appropriate location for each of the following in the blank periodic table provided: (**a**) the fourth period noble gas; (**b**) a fifth period element whose atoms have three unpaired electrons; (**c**) the *d*-block element having one 3*d* electron; (**d**) a *p*-block element that is a metalloid; (**e**) a metal that forms the oxide M_2O_3.

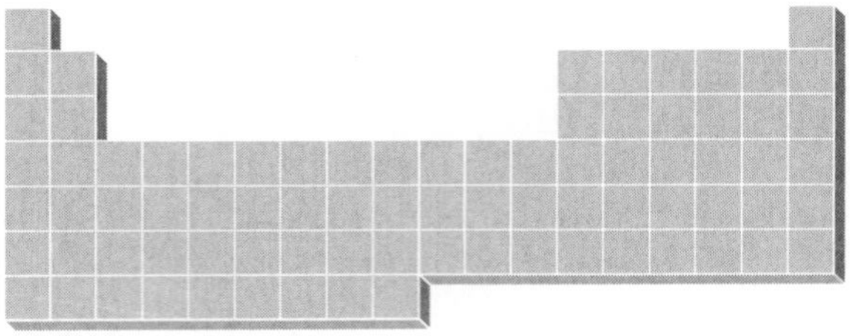

65. Propose a probable electron configuration for the unknown and as yet undiscovered element with $Z = 114$.

66. Use ideas presented in this chapter to indicate (**a**) three metals that you would expect to exhibit the photoelectric effect with visible light, and three that you would not; (**b**) the approximate first ionization energy of fermium ($Z = 100$); (**c**) the approximate atomic radius of francium.

67. Arrange the following ionization energies in the most probable order of increasing value and explain your reasoning: I_1 for B, I_1 for Cs, I_2 for In, I_2 for Sr, I_2 for Xe, I_3 for Ca.

68. Explain why the subshells within a given principal shell of a hydrogen atom all have the same energy, whereas in multielectron atoms they do not.

69. The production of gaseous chloride ions from chlorine molecules can be considered to be a two-step process in which the first step is

$$Cl_2(g) \longrightarrow 2\,Cl(g) \qquad \Delta H = +242.8\text{ kJ}$$

What is the second step? Is the overall process endothermic or exothermic?

70. Use ionization energies and electron affinities to determine whether the following reaction is endothermic or exothermic.

$$Mg(g) + 2\,Cl(g) \longrightarrow Mg^{2+}(g) + 2\,Cl^-(g)$$

71. When sodium chloride is strongly heated in a flame, the flame takes on the yellow color associated with the emission spectrum of sodium atoms. The reaction that occurs is

$$Na^+(g) + Cl^-(g) \longrightarrow Na(g) + Cl(g)$$

Show that this reaction is *exothermic*.

72. Show how you might use the atomic weight and density of a solid metallic element, together with Avogadro's number, to make a rough estimate of the volume of an atom of the metal. How would you estimate the metallic radius from this volume? Why is this estimate not exact, even if the atomic weight, the density, and Avogadro's number are all known with considerable accuracy? Is the estimated atomic radius likely to be too high or too low? Explain. [*Hint:* Use sodium as a specific example ($d = 0.971\text{ g/cm}^3$), and compare the estimated metallic radius with that shown in Figure 7.9.]

73. In this text ionization energies are given in the unit kJ/mol. Another way of expressing these quantities is in terms of a *single* atom rather than a mole of atoms, through the unit electron volts per atom, eV/atom. Use physical constants and other data from the appendices and inside the back cover to show that 1 eV/atom = 96.49 kJ/mol.

A characteristic reaction of most of the *s*-block elements is their reaction with water to produce hydrogen gas. If the reaction is sufficiently vigorous, as here with potassium metal and water, the hydrogen will ignite.

8

The *s*-Block Elements: An Introduction to Descriptive Chemistry

CONTENTS

The study of chemistry includes a consideration of both practical information—often called *descriptive chemistry*—and theoretical concepts. In its earliest stages, chemistry was largely descriptive; texts were mostly descriptions of the manufacture, reactions, and uses of substances, and of their physical and chemical properties. As their science matured, chemists developed important theories to explain the growing body of factual information. Today, the practical and theoretical are interwoven into a powerful science that illuminates our understanding of the modern world. The two facets are inseparable; the theoretical without the factual is sterile, and the factual without a theoretical framework overwhelms us with detail. In this chapter, we consider the descriptive chemistry of the *s*-block elements—those that have valence electrons in an *s* subshell only. These elements include hydrogen, the alkali metals (Group 1A), and the alkaline earth metals (Group 2A). We will also see additional applications of reaction stoichiometry, atomic properties, and relationships based upon the periodic table.

8.1 Hydrogen

A hydrogen atom has only one proton in its nucleus, and in the ground state its only electron is in the $1s^1$ configuration. In nature, hydrogen occurs as diatomic molecules, H_2. A gas at room temperature, hydrogen liquefies at 20.39 K and solidifies at 13.98 K; it has an extraordinarily short liquid range.

As the element with the simplest atoms, hydrogen has been used as a model in the development of theories about the structure and behavior of matter (Figure 8.1). But for all its theoretical significance, hydrogen also has many practical uses, as we learn in this section.

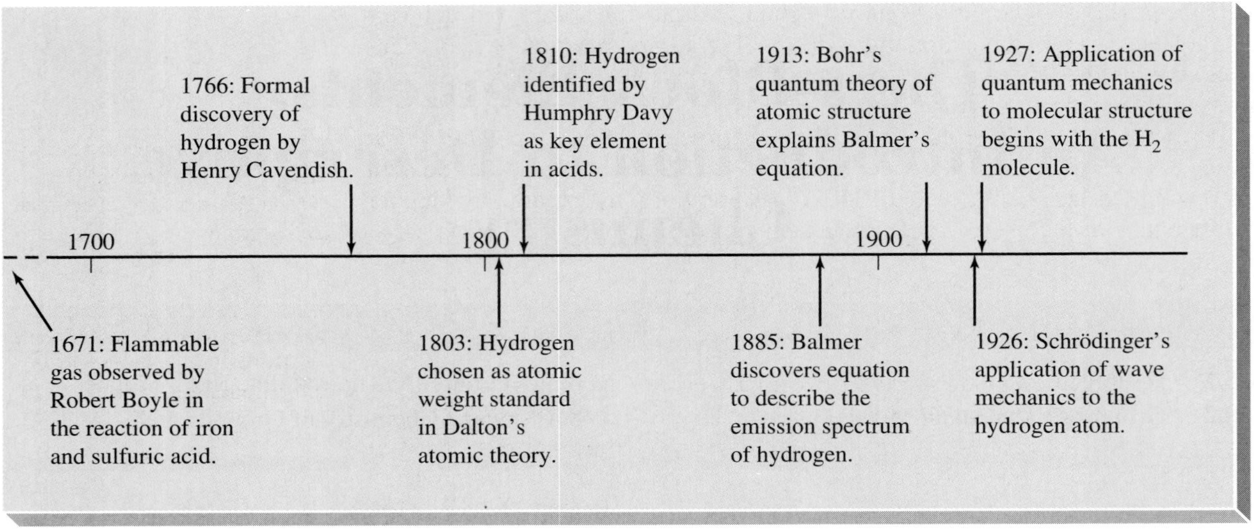

Figure 8.1 Some theoretical developments linked to the element hydrogen.

Optical image of the emission nebula NGC 6357, located about 5500 light years from Earth. It is a cloud of hydrogen (60 light years or 3.5×10^{14} mi across) ionized and lit up by young, hot, blue stars embedded in it. NGC 6357 does not show a blue component in its emitted light because clouds of dust situated between the nebula and Earth absorb the blue light.

A *catalyst* is a substance that speeds up a reaction without itself being consumed in the reaction. Catalysts are discussed in some detail in Chapter 15.

Occurrence and Preparation

By mass, hydrogen makes up only about 0.9% of Earth's crust. Nevertheless, hydrogen, the lightest element, is quite abundant. In numbers of *atoms* in Earth's crust, hydrogen ranks third (15.1%), after oxygen (53.3%) and silicon (15.9%). If we look beyond our home planet, it has been estimated that hydrogen atoms make up 89% of the atoms on the sun, and that 85 to 95% of the atoms in the atmospheres of the outer planets (Jupiter, Uranus, Saturn, and Neptune) are hydrogen atoms. In the universe as a whole, about 90% of all atoms are hydrogen—and the rest are mainly helium.

Only trace amounts of free hydrogen (H_2) are found on Earth; hydrogen must be obtained from its compounds. Although hydrogen occurs in more compounds than does any other element, only a few of these compounds are economically viable sources of the element. Water (H_2O), the most abundant hydrogen-containing compound, is usually the first choice. Elemental hydrogen can be extracted from water by reaction with *carbon* (as coal or coke), with *carbon monoxide*, and with *hydrocarbons*—particularly methane (natural gas). The first of these reactions is carried out at high temperatures,

$$\textit{Water gas reaction:}\quad \mathrm{C(s) + H_2O(g) \xrightarrow{1000\ ^\circ C} CO(g) + H_2(g)}$$

The other two reactions require a catalyst as well.

$$\textit{Water gas shift reaction:}\quad \mathrm{CO(g) + H_2O(g) \xrightarrow[1000\ ^\circ C]{catalyst} CO_2(g) + H_2(g)}$$

$$\textit{Reforming of methane:}\quad \mathrm{CH_4(g) + H_2O(g) \xrightarrow[1000\ ^\circ C]{catalyst} CO(g) + 3\,H_2(g)}$$

The first reaction produces "water gas," a combustible mixture that consists mainly of CO and H_2. The mixture also contains some CO_2 produced by the second reaction. Water gas was an important fuel gas for homes and industry before the advent of the natural gas industry. The second reaction is used to increase the yield of H_2 when that gas is the desired product. The third reaction is now the principal source of hydrogen.

Hydrogen is an important byproduct of petroleum refining operations. For example, in the process of *catalytic reforming*, an alkane hydrocarbon (Section 2.10) of low octane rating (page 184) is converted to a hydrocarbon of higher octane rating, as in the conversion of hexane to benzene.

Reforming refers to the restructuring of a hydrocarbon: CH_4 to CO and H_2, or hexane to benzene and hydrogen.

$$\underset{\text{Hexane}}{C_6H_{14}} \xrightarrow{\text{catalyst}} \underset{\text{Benzene}}{C_6H_6} + 4\,H_2$$

Benzene belongs to a class of compounds, called *aromatic* hydrocarbons, discussed in Section 9.10.

The most direct method of producing hydrogen (and oxygen as well) is the simple decomposition of water. Some compounds can be decomposed into their elements simply by heating strongly enough to break chemical bonds, but even at 2000 °C water is only about 1% decomposed. When decomposition by heating is not feasible, chemists often use *electrolysis*—decomposition by electric current (see Figure 8.2).

$$2\,H_2O(l) \xrightarrow{\text{electrolysis}} 2\,H_2(g) + O_2(g)$$

The electrolysis of water is expensive. About 0.1 kilowatt hour (kWh) of electrical energy is required to produce 1 mol $H_2(g)$; it takes as much energy to obtain just 2 g of hydrogen by the electrolysis of water as to operate a 100-watt electric light bulb for 1 h. Electrolysis of water is economically feasible only where hydroelectric power is cheap (as in Canada and Norway).

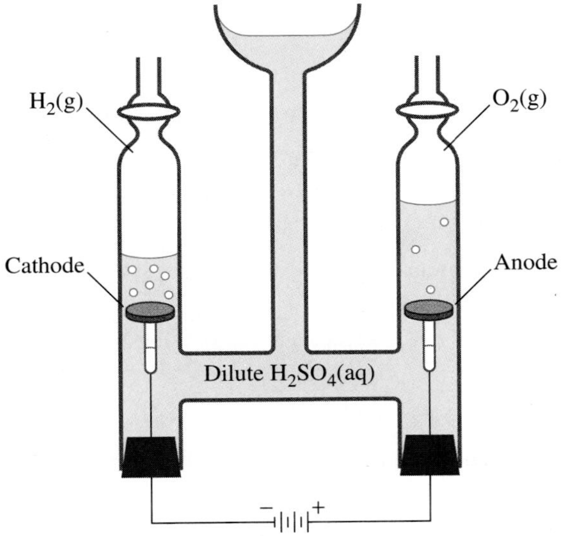

Figure 8.2 The electrolysis of water. Electric current is passed through water containing a small quantity of sulfuric acid to make it an electrical conductor. Hydrogen gas forms at the negative electrode (cathode) and oxygen gas at the positive electrode (anode). The volume of hydrogen produced is exactly twice that of oxygen. This result is consistent with the chemical equation, Avogadro's hypothesis (Section 4.6), and the law of combining volumes (Section 4.9). We will consider the details of electrolysis reactions in Chapter 20.

In the laboratory we often can prepare chemicals by methods that are not economically feasible for commercial production, such as the preparation of hydrogen and oxygen by the electrolysis of water. A common laboratory method for preparing small quantities of hydrogen involves the reaction of certain metals with an aqueous solution of a strong acid such as HCl(aq). The actual reactants are metal atoms and hydrogen ions in the acid solution, as shown in this representative equation.

$$Zn(s) + 2\ H^+(aq) \longrightarrow Zn^{2+}(aq) + H_2(g)$$

An equation written to show individual ions, rather than ionic compounds themselves, is called an *ionic equation*.

The most active of the metals, those of Group 1A and the heavier members of Group 2A, displace $H_2(g)$ even from pure water, where the concentration of H^+ is very low. Here H_2O is the reactant, rather than H^+.

$$2\ M(s) + 2\ H_2O(l) \longrightarrow 2\ MOH(aq) + H_2(g) \quad \text{(where M is any Group 1A metal)}$$

$$M(s) + 2\ H_2O(l) \longrightarrow M(OH)_2(aq) + H_2(g) \quad \text{(where M is Ca, Sr, Ba, or Ra)}$$

(a)

(b)

Figure 8.3 Displacment of $H_2(g)$ from water by active metals. (a) The reaction of the Group 1A metal potassium with water is so exothermic that the temperature is raised to the point that the liberated $H_2(g)$ ignites; it combines with O_2 from air to form water. Note that potassium, being less dense than water, floats. (b) The reaction of the Group 2A metal calcium with water proceeds less vigorously than that of potassium in (a). Note that calcium is more dense than water. Also, phenolphthalein indicator present in the water changes to a pink color, signaling the buildup of $OH^-(aq)$.

Some alkali metals (Group 1A) react violently with water, but the alkaline earth metals (Group 2A) react more slowly (Figure 8.3). Because it is relatively cheap and not too reactive, calcium is often the metal chosen.

Example 8.1, in which we review equation writing and reaction stoichiometry, describes another method of producing hydrogen.

Example 8.1

Hydrogen is sometimes made by the reaction of depleted uranium with water and acid. When the acid is $HNO_3(aq)$, the other product is uranyl nitrate, $UO_2(NO_3)_2$. (Depleted uranium is spent "fuel" from a nuclear reactor; most of the uranium-235 is gone, leaving mainly uranium-238.)

a. Write a balanced equation for the reaction.

b. What mass of $H_2(g)$ can be made from 1.00 kg of U(s)?

Solution

a. From the description of the reaction we first write an unbalanced equation.

$$U(s) + H_2O(l) + HNO_3(aq) \longrightarrow UO_2(NO_3)_2(aq) + H_2(g) \quad \textit{(not balanced)}$$

The U atoms are balanced, and we next balance "NO_3," which we can treat as a unit.

$$U(s) + H_2O(l) + 2\ HNO_3(aq) \longrightarrow UO_2(NO_3)_2(aq) + H_2(g) \quad \textit{(not balanced)}$$

Now we can balance the remaining O atoms by placing the coefficient "2" in front of H_2O on the left.

$$U(s) + 2\ H_2O(l) + 2\ HNO_3(aq) \longrightarrow UO_2(NO_3)_2(aq) + H_2(g) \quad \textit{(not balanced)}$$

Finally, we can balance H atoms.

$$U(s) + 2\ H_2O(l) + 2\ HNO_3(aq) \longrightarrow UO_2(NO_3)_2(aq) + 3\ H_2(g) \quad \textit{(balanced)}$$

b. To calculate the mass of $H_2(g)$ produced in the reaction of 1.00 kg (1000 g) U with an excess of H_2O and $HNO_3(aq)$, we proceed in the fashion introduced in Section 3.8.

$$?\ g\ H_2 = 1000\ g\ U \times \frac{1\ mol\ U}{238\ g\ U} \times \frac{3\ mol\ H_2}{1\ mol\ U} \times \frac{2.016\ g\ H_2}{1\ mol\ H_2} = 25.4\ g\ H_2$$

Exercise 8.1

How does the mass of $H_2(g)$ calculated in Example 8.1 compare with that obtainable from 1.00 kg of Zn(s)? (*Hint*: What reaction would you use?)

Binary Compounds of Hydrogen

Hydrogen reacts with other nonmetals to form binary (two-element) molecular compounds such as hydrogen chloride (HCl) and ammonia (NH_3).

$$H_2(g) + Cl_2(g) \longrightarrow 2\,HCl(g)$$
$$3\,H_2(g) + N_2(g) \longrightarrow 2\,NH_3(g)$$

In aqueous solution, hydrogen chloride is the familiar strong acid, hydrochloric acid [HCl(aq)]. A solution of ammonia in water, [NH_3(aq)], is our most common weak base solution.

Among the most important molecular hydrogen compounds are those of carbon and hydrogen—the *hydrocarbons*.

Hydrogen reacts with the most active metals to form *ionic hydrides*. In these compounds hydrogen exists as the *hydride* ion, H^-. The formation reactions for NaH(s) and CaH_2(s) are represented by the equations

$$Na(s) + \tfrac{1}{2}H_2(g) \longrightarrow NaH(s) \qquad \Delta H_f^\circ = -56.3\text{ kJ}$$
$$Ca(s) + H_2(g) \longrightarrow CaH_2(s) \qquad \Delta H_f^\circ = -186\text{ kJ}$$

We will learn how to calculate ΔH_f° values of ionic compounds from atomic properties in the next chapter.

The electron affinity of hydrogen is −72.8 kJ/mol. This means that the addition of an electron to the half-filled 1*s* orbital of a gaseous hydrogen atom is energetically favorable. Now consider the loss of an electron by a gaseous Na atom, which is governed by the ionization energy of Na: +496 kJ/mol. Formation of the ion Na^+ from a gaseous Na atom is *not* energetically favorable. However, when these processes occur together and the resulting ions cluster into a solid ionic compound, the overall change is energetically favorable; the formation reaction for NaH(s) is *exothermic*.

Ionic hydrides react with water to liberate H_2(g) and are sometimes used as a source of hydrogen for weather observation balloons.

$$CaH_2(s) + 2\,H_2O(l) \longrightarrow Ca^{2+}(aq) + 2\,OH^-(aq) + 2\,H_2(g)$$

Hydrides of the transition elements retain some metallic properties, such as electrical conductivity, but many are *nonstoichiometric*—the ratio of H atoms to metal atoms is variable, not a fixed number. For example, the formula of titanium hydride can vary from $TiH_{1.8}$ to TiH_2. We can picture H atoms in such compounds as occupying the voids or holes among metal atoms in the solid metal—much like fitting cherries into the holes among oranges in a box. Because some holes fill while others do not, the ratio of H to metal atoms is variable.

Nonstoichiometric compounds contradict Dalton's assumption that atoms combine only in small whole number ratios (Section 2.2). His idea still works well, though, for simple molecular compounds and for most ionic compounds of main group elements.

Estimation Exercise 8.1

Which of the following reactions produces the greatest mass of H_2 for a given mass of the limiting reactant: (**a**) Zn(s) with excess HCl(aq), (**b**) Na(s) with excess H_2O, (**c**) CaH_2(s) with excess H_2O? (*Hint:* Refer to the chemical equations for these reactions.)

The greenhouse effect that may result in global warming is discussed in Chapter 14.

A Possible Future: The Hydrogen Economy

The prospect of dwindling supplies and higher prices for coal, natural gas, and petroleum and the threat of global warming mean that we may need a substitute for fossil fuels in future centuries. Hydrogen is often mentioned as a candidate.

Hydrogen is an attractive fuel for several reasons. One is the fact that an automobile engine burning hydrogen is 25 to 50% more energy-efficient than is an engine burning gasoline. Because the only significant combustion product is H_2O, a hydrogen engine's exhaust is far lower in pollutants than the exhaust of a gasoline engine. The heat of combustion per gram of liquid hydrogen is more than twice that of jet fuel. Aircraft using liquid hydrogen could fly much farther than those using conventional jet fuel.

In addition to taking the place of gasoline and jet fuels for transportation, hydrogen could also replace natural gas for heating homes and other buildings. Because it is an excellent reducing agent, hydrogen could largely replace carbon in metallurgy. Finally, with hydrogen readily available, the cost of producing ammonia and products derived from it would remain low. If we adopt the widespread use of hydrogen, as outlined here, major changes in our way of life would follow, producing what is called a *hydrogen economy*.

Hydrogen looks good for a future fuel, but tough problems must be solved before we adopt a hydrogen economy. There are no hydrogen mines, nor can we pump it from the ground by drilling holes. A hydrogen economy requires a cheap way of making hydrogen and storing it. The most likely future source of hydrogen will be water. The decomposition of water might be accomplished as the net reaction of a series of thermochemical reactions, or by electrolysis if this can be done cheaply enough.

Because of the large volumes involved, storing hydrogen as a gas is not feasible. Hydrogen, when liquefied, occupies a much smaller volume, but then it must be maintained at extremely low temperatures (below −240 °C). And, in

Uses of Hydrogen

The largest single use of hydrogen, accounting for nearly half of the gas produced, is in the manufacture of ammonia (NH_3). Ammonia, in turn, is used in the manufacture of fertilizers, plastics, and explosives. The second most important use of hydrogen is in the petrochemical industry. For example, hydrogen is used to convert benzene to cyclohexane, a cyclic hydrocarbon used as an intermediate in the production of nylon.

$$\underset{\text{Benzene}}{C_6H_6} + 3\,H_2 \xrightarrow{\text{catalyst}} \underset{\text{Cyclohexane}}{C_6H_{12}}$$

In reactions of this sort, called **hydrogenation reactions**, hydrogen atoms are added to double or triple bonds in other molecules. Hydrogenation, for example, converts unsaturated hydrocarbons (those with double and triple bonds) to satu-

The 1937 fire aboard the dirigible *Hindenburg* as it docked at Lakehurst, New Jersey, after crossing the Atlantic Ocean spelled the end of projected air transportation by hydrogen-filled airships. Any future hydrogen economy must contend with the flammability of hydrogen and the explosive nature of hydrogen–oxygen mixtures.

either case, hydrogen must be kept from contact with oxygen (air), with which it forms explosive mixtures. One promising possibility is to dissolve the hydrogen gas in a metal, such as an iron–titanium alloy. The gas can be released from the metal by mild heating. In a hydrogen-burning automobile engine, for example, the required heat might come from the engine exhaust.

Planning for a possible hydrogen economy is a source of fascinating challenges for chemists and other natural and social scientists as well.

rated hydrocarbons (alkanes). Hydrogenation converts oleic acid, a liquid, into stearic acid, a solid.

$$\underset{\text{Oleic acid}}{CH_3(CH_2)_7CH{=}CH(CH_2)_7COOH(l)} + H_2(g) \xrightarrow{\text{catalyst}} \underset{\text{Stearic acid}}{CH_3(CH_2)_7CH_2CH_2(CH_2)_7COOH(s)}$$

This is the type of reaction used to convert food oils (from corn or soybeans, for example) into solid or semisolid fats (margarine or vegetable shortenings such as Crisco).

Another chemical manufacturing process that uses hydrogen is the synthesis of methyl alcohol (methanol), an industrial solvent and a raw material for making other organic compounds.

$$CO(g) + 2\,H_2(g) \xrightarrow{\text{catalyst}} CH_3OH(l)$$

Although a definition of reduction based on the removal of oxygen atoms (or the addition of hydrogen atoms) works well in many cases, especially in biochemistry, a more general definition is given and applied in Chapter 12.

In metallurgy, a process called *reduction* is used to extract pure metals from metal oxides. The reactant responsible for removing the oxygen, called a *reducing agent*, is typically carbon (in the form of charcoal, coke, or coal), but in more difficult cases, hydrogen is frequently used. An example is the preparation of tungsten.

$$WO_3(s) + 3\,H_2(g) \xrightarrow{850\ °C} W(s) + 3\,H_2O(g)$$

Liquid hydrogen is used as a rocket fuel. The space shuttle rocket engines, for example, use the reaction of hydrogen and oxygen.

$$H_2(g) + \tfrac{1}{2}O_2(g) \longrightarrow H_2O(l) \quad \Delta H° = -285.8\ kJ$$

Both reactants are stored as liquids. The fuel tank holds 1.5×10^6 L of liquid hydrogen. The oxygen tank carries 5.4×10^5 L of liquid oxygen. During "lift off," these propellants power the shuttle's main engines for about 8.5 min; liquid hydrogen is consumed at a rate of nearly 3000 L/s.

When hydrogen is passed through an electric arc, H_2 molecules absorb energy and dissociate into H atoms. The H atoms then recombine into H_2 molecules, in a highly exothermic process. This recombination, coupled with the combustion of H_2, creates an extremely high flame temperature.

$$\begin{array}{ll} 2\,H(g) \longrightarrow H_2(g) & \Delta H° = -436\ kJ \\ H_2(g) + \tfrac{1}{2}O_2(g) \longrightarrow H_2O(g) & \Delta H° = -241.8\ kJ \\ \hline 2\,H(g) + \tfrac{1}{2}O_2(g) \longrightarrow H_2O(g) & \Delta H° = -678\ kJ \end{array}$$

An oxyhydrogen welding torch readily cuts through steel and can be used to melt tungsten, which has a melting point of (3400 °C).

8.2 Group 1A: The Alkali Metals

As noted on page 252, even though hydrogen has the electron configuration $1s^2$ and is often placed in Group 1A of the periodic table, it is *not* and alkali metal.

The six metals that make up Group 1A—lithium, sodium, potassium, rubidium, cesium, and francium—have the valence-shell electron configuration ns^1. Because their atoms lose their single valence electrons so readily to form 1+ ions, these elements are the most reactive of the metals. This reactivity is due to low ionization energies that in turn are mainly the result of large atomic radii (see Table 8.1).

Sodium, a soft metal, can be cut with a knife. Note that this active metal is covered with a thick oxide coating.

TABLE 8.1 **Atomic Properties of the Alkali Metals**

	Electron Configuration	Atomic Radius,[a] pm	Ionization Energy (I_1), kJ/mol
Li	$[He]2s^1$	152	520.2
Na	$[Ne]3s^1$	186	495.8
K	$[Ar]4s^1$	227	418.8
Rb	$[Kr]5s^1$	248	403.0
Cs	$[Xe]6s^1$	265	375.7
Fr	$[Rn]7s^1$	ca. 270	ca. 375

[a] Values of atomic radii are for metallic radii.

The energy required to promote the valence-shell electron of an alkali metal atom to a higher energy level can be supplied by an ordinary flame. The return of excited-state atoms to the ground state is accompanied by the emission of visible light. The frequency of this light depends on the separation of energy levels in the atoms (Section 6.6); as a result, each of the alkali metals exhibits a characteristic flame color (Figure 8.4).

In the elemental form, the alkali metals are soft solids with low melting points. On a hot day, in fact, cesium would be a liquid; it melts at 29 °C. Alkali metals have low densities, the lowest of all the metals. Three of the alkali metals—Li, Na, and K—have densities low enough that they float on water as they react. These physical properties are related principally to the large atomic radii of the Group 1A metals. When freshly cut, the alkali metals are bright and shiny, but they tarnish quickly as they react with oxygen in the atmosphere.

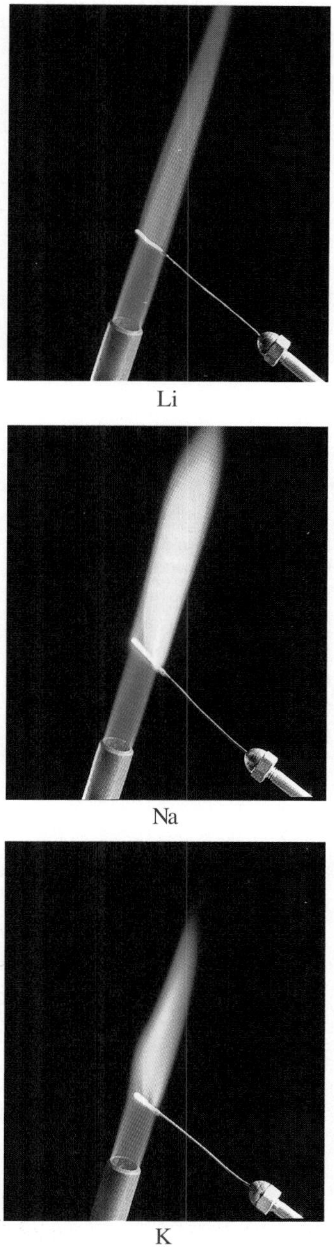

Figure 8.4 Flame colors of some alkali metals.

Occurrences and Uses

The natural abundances of sodium and potassium in Earth's solid crust are 2.27% and 1.84% by mass, respectively; the other alkali metals are much scarcer, ranging from 78 to 18 to 2.6 parts per million (ppm) of Rb, Li, and Cs, respectively. Francium is exceptionally rare. It is formed by the radioactive decay of heavier elements, and there is probably no more than 15 g of francium in the top 1 km of Earth's crust. For this reason, few of francium's properties have been determined.

The alkali metals have a few important uses. *Liquid* sodium is used as a heat transfer medium in some types of nuclear reactors. Liquid sodium is especially good for this purpose because it has

- A higher specific heat than most liquid metals—large quantities of heat per gram are required to change its temperature.
- Good thermal conductivity—heat is readily conveyed through the liquid and delivered to other media.
- A low density and low viscosity—liquid sodium is easy to pump.
- A low vapor pressure—the tendency of liquid sodium to vaporize is limited, even at the high temperatures (550 °C) used in these applications.

Small quantities of sodium, as sodium vapor, are used in lamps for outdoor lighting. Perhaps the most important use of sodium metal is as a reducing agent in producing refractory (high melting point) metals, such as titanium, zirconium, and hafnium.

$$MCl_4 + 4\,Na \longrightarrow M + 4\,NaCl \quad (\text{where } M = Ti, Zr, \text{ or } Hf)$$

General use of potassium is limited to a few applications in which sodium, the cheaper metal, cannot be used. One such application is its combustion to form potassium superoxide, KO_2, for use in life-support systems. (Sodium superoxide can be prepared only under very carefully controlled conditions.) The function of the KO_2 is both to absorb CO_2 and to produce O_2.

$$4\,KO_2(s) + 2\,CO_2(g) \longrightarrow 2\,K_2CO_3(s) + 3\,O_2(g)$$

A possible future use of lithium is the production of the hydrogen isotope tritium (^3H) for use in nuclear fusion reactors.

Lithium is used in lightweight electrical batteries of the type found in clocks and watches, hearing aids, and heart pacemakers. When added in small quantities, lithium gives high-temperature strength to aluminum metal and ductility to magnesium metal. When alloyed with silver, it is used for brazing (welding together) metals.

Preparation of the Alkali Metals

Brines are aqueous solutions of ionic compounds; they usually contain NaCl, and often also have KCl, $MgCl_2$, and various others.

The principal compounds from which sodium and potassium metals are extracted are NaCl and KCl, whether mined as solids or isolated from natural brines. Lithium is extracted primarily from the mineral spodumene, $LiAl(SiO_3)_2$.

To change an alkali metal ion to an alkali metal atom, the ion must be forced to take on an electron. This is a process called *reduction*. Reduction of alkali metal ions is not easy to accomplish through chemical reactions, but reduction can be achieved rather easily by the electrolysis of a molten salt, usually the chloride. In the electrolysis of molten NaCl, an electric current takes electrons away from Cl^- ions—a process called *oxidation*—and forces them onto Na^+ ions.

Oxidation is a loss of electrons, and reduction is a gain of electrons; oxidation and reduction always occur together. These terms are defined more precisely and discussed in more detail in Chapter 12.

Oxidation: $$2\,Cl^- \longrightarrow Cl_2(g) + 2\,e^-$$

Reduction: $$2\,Na^+ + 2\,e^- \longrightarrow 2\,Na(l)$$

Net ionic equation: $$2\,NaCl(l) \xrightarrow{\text{electrolysis}} 2\,Na(l) + Cl_2(g)$$

Lithium can be similarly prepared from LiCl.

Potassium metal was the first to be prepared by electrolysis; Humphry Davy electrolyzed molten KOH in 1807. The usual industrial method involves a chemical reaction between sodium metal and potassium chloride.

$$KCl(l) + Na(l) \xrightarrow{850\ ^\circ C} NaCl(l) + K(g)$$

In this reaction, Na atoms lose their valence electrons and K^+ ions acquire them. The resulting gaseous potassium escapes from the reaction mixture and is converted, first to a liquid and then to a solid, by cooling. Were it not for the fact that potassium is a gas whereas sodium is a liquid at 850 °C, the method would not work. The reverse reaction would be favored instead because K atoms actually lose electrons more readily than do Na atoms (they have a lower ionization energy).

Some Reactions of Li, Na, and K

The chemistry of the Group 1A metals reflects the relative ease of removal of the ns^1 electrons to form the metal ions. The alkali metals (M) react directly with elements of Group 7A, the halogens (X_2), to form ionic binary halides (MX). As we saw in Section 8.1, they also react with hydrogen to form ionic hydrides (MH). The reaction of $O_2(g)$ with the metals produces various products. Table 8.2 summarizes some of the reaction chemistry of the Group 1A metals.

Lithium, sodium, and potassium are such active metals (as are also rubidium and cesium) that they will displace $H_2(g)$ from water itself, as well as from acid solutions. In their reactions with water, as we saw in Section 8.1, they produce hydrogen gas and aqueous solutions of ionic hydroxides, MOH.

$$2\,M(s) + 2\,H_2O(l) \longrightarrow 2\,MOH(aq) + H_2(g)$$

TABLE 8.2 Some Typical Reactions of the Alkali Metals, M

With halogens (Group 7A), X_2:

$2\,M(s) + X_2 \longrightarrow 2\,MX(s)$ (e.g., LiF, NaCl, KBr, CsI)

With hydrogen, H_2:

$2\,M(s) + H_2(g) \longrightarrow 2\,MH(s)$ (e.g., LiH, NaH)

With excess oxygen, O_2[a]

$2\,Li(s) + O_2(g) \longrightarrow Li_2O(s)$ (plus some Li_2O_2)

$2\,Na(s) + O_2(g) \longrightarrow Na_2O_2(s)$ (plus some Na_2O)

$M(s) + O_2(g) \longrightarrow MO_2(s)$ (where M = K, Rb, and Cs)

With water, H_2O

$2\,M(s) + 2\,H_2O(l) \longrightarrow 2\,MOH(aq) + H_2(g)$

[a] Li_2O is a *normal* oxide; Na_2O_2 is a *peroxide;* MO_2 is a *superoxide.*

Because Group 1A hydroxides are *basic* or *alkaline*, the Group 1A metals have long been called the "alkali" metals.

Some Important Compounds of Li, Na, and K

More important than the free (uncombined) metals are the alkali metal compounds, especially those of sodium. The compounds are relatively inexpensive, and almost all are quite soluble in water. Alkali metal compounds therefore are good sources of alkali metal cations and accompanying anions in aqueous solutions. Some uses of a few common lithium, sodium, and potassium compounds are listed in Table 8.3. Compounds of rubidium and cesium are rare and quite expensive; they have few uses.

TABLE 8.3 Some Uses of a Few Alkali Metal Compounds

Li_2CO_3	Preparation of other lithium compounds. Also used in ceramic and porcelain glazes, luminescent paints, varnishes, and dyes; in medicine as an antidepressant.
NaCl	Manufacture of sodium, chlorine, sodium hydroxide, sodium sulfate, hydrochloric acid, and other chemicals.
NaOH	Manufacture of aluminum, cellophane, rayon, soaps and detergents, and salts of various acids. Also used in textile processing and petroleum refining.
Na_2CO_3	Manufacture of glass, soaps and detergents, and other sodium chemicals. Also used for water treatment.
Na_2SO_4	Manufacture of paper and glass. Also used in textile processing, solar heat storage, and as a filler in synthetic detergents.
KCl	Preparation of other potassium salts. Also used in fertilizers, as a food additive, and as a salt substitute.
KNO_3	Manufacture of fertilizers and explosives. Also used in curing foods and as an oxidizer in gunpowder and rocket propellants.

Sodium chloride (salt) is mined, pumped from brine wells, and, in some seaside locations, harvested from seawater.

Among the most important lithium compounds are the carbonate, halides, hydroxide, and hydride. Lithium carbonate is the usual starting material for making other lithium compounds. For example, lithium hydroxide, LiOH, is prepared by the reaction

$$Li_2CO_3(aq) + Ca(OH)_2(aq) \longrightarrow CaCO_3(s) + 2\ LiOH(aq)$$

Solid $CaCO_3$ is removed by filtration, leaving an aqueous solution of LiOH. The LiOH can be obtained as a solid by evaporation of the water from the solution.

An interesting use of lithium hydroxide is to remove $CO_2(g)$ from expired air in confined quarters such as submarines and space vehicles.

$$2\ LiOH(s) + CO_2(g) \longrightarrow Li_2CO_3(s) + H_2O$$

LiOH is preferred over other ionic hydroxides because, per mole of CO_2 removed, a smaller mass of LiOH is required than of other, cheaper hydroxides. Similarly, in its reaction with water, LiH produces a greater volume of $H_2(g)$ per unit mass of hydride than do other, cheaper hydrides.

With an annual production in the United States of about 50 million tons, sodium chloride is easily the most important sodium compound. In fact, it is the most used of all minerals for the production of chemicals. Table 8.4 lists several of these NaCl-based chemicals and their estimated annual production in the United States. Figure 8.5 is a diagram that shows how sodium chloride can be converted to several other compounds. In this figure, the principal reactants required for the conversions are written with the arrow ($\longrightarrow$). A Δ symbol is used if a reaction mixture needs to be heated. Also, some of the reactions yield products in addition to those noted in the diagram. Example 8.2 and Exercise 8.2 illustrate the use of Figure 8.5.

TABLE 8.4 **Some Chemicals Produced from NaCl**

Chemical	U.S. Production, 1994, billions of pounds
Sodium hydroxide, NaOH	25.83
Chlorine, Cl_2	24.20
Sodium carbonate, Na_2CO_3	20.56[a]
Hydrochloric acid, HCl	6.71[b]
Sodium sulfate, Na_2SO_4	1.02

[a] Includes Na_2CO_3 from natural sources.
[b] Most HCl is produced as a by-product of the chlorination of hydrocarbons; for example, in the chlorination of methane:

$$CH_4(g) + Cl_2(g) \longrightarrow CH_3Cl(g) + HCl(g)$$

Example 8.2

Write chemical equations for the Leblanc process—the conversion of NaCl to Na_2CO_3 by way of Na_2SO_4.

Solution

Figure 8.5 describes the necessary transformations in an abbreviated form. To translate this information into chemical equations, we need only identify the reactants in each reaction and predict plausible products. For example, to convert NaCl(s) to Na_2SO_4(s) requires reaction with H_2SO_4(concd aq). The other likely product is HCl(g).

$$2\ NaCl(s) + H_2SO_4(concd\ aq) \longrightarrow Na_2SO_4(s) + 2\ HCl(g)$$

The reaction of Na_2SO_4(s) with carbon yields Na_2S(s), with CO(g) as another plausible product. [CO_2(g) might seem as plausible as CO(g) at this point; we choose CO(g) because it is the actual product.]

$$Na_2SO_4(s) + 4\ C(s) \longrightarrow Na_2S(s) + 4\ CO(g)$$

The final step is the reaction of Na_2S(s) with $CaCO_3$(s), yielding Na_2CO_3(s) and CaS(s).

$$Na_2S(s) + CaCO_3(s) \xrightarrow{\Delta} Na_2CO_3(s) + CaS(s)$$

Water-soluble Na_2CO_3 is extracted from the mixed solid.

Exercise 8.2

Write plausible chemical equations for each of the following conversions outlined in Figure 8.5.

a. NaCl to NaH **b.** NaCl to NaOCl

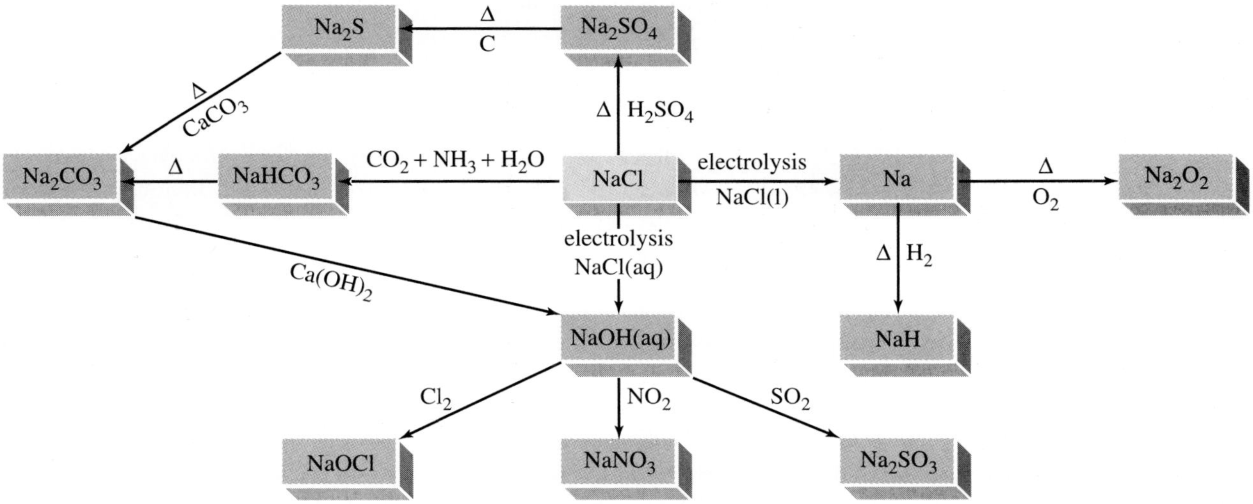

Figure 8.5 **Preparation of Sodium Compounds from NaCl.** The methods of preparation suggested by this diagram are not necessarily the preferred industrial methods.

Sodium hydroxide is produced by the electrolysis of concentrated NaCl(aq). From the ionic form of the electrolysis equation we see that Na^+ is unchanged, Cl^- is converted to Cl_2(g), and H_2O is changed to H_2(g) and OH^-.

$$2\ Na^+(aq) + 2\ Cl^-(aq) + 2\ H_2O \xrightarrow{\text{electrolysis}} 2\ Na^+(aq) + 2\ OH^-(aq) + H_2(g) + Cl_2(g)$$

The gases are collected separately. If *solid* NaOH is desired (rather than an aqueous solution), water is evaporated from the solution after the electrolysis is completed.

Sodium Carbonate: The Economy and the Environment

In 1775, in order to avoid dependence on foreign sources of naturally occurring Na_2CO_3, the French government offered a prize to anyone who could devise a way to synthesize Na_2CO_3 from NaCl. Nicholas Leblanc responded with the process suggested by Figure 8.5 and described in Example 8.2. The Na_2CO_3 was needed for the reaction

$$Ca(OH)_2(s) + Na_2CO_3(aq) \longrightarrow CaCO_3(s) + 2\ NaOH(aq)$$

This reaction converts a base that is only moderately soluble, $Ca(OH)_2$, to one that is highly soluble, NaOH. The NaOH was needed primarily for the manufacture of soap (described later in the chapter).

By the end of the nineteenth century the Leblanc process had been almost completely displaced by an alternate method devised by the Belgian chemist Ernest Solvay (1838–1922). The key step in the Solvay process, outlined in Figure 8.6 (page 290), involves the reaction of $NH_3(g)$ and $CO_2(g)$ with concentrated NaCl(aq).

$$NaCl(aq) + NH_3(g) + CO_2(g) + H_2O \longrightarrow NaHCO_3(s) + NH_4Cl(aq)$$

Solid sodium hydrogen carbonate precipitates from the reaction mixture. Sodium carbonate is then produced by heating the solid sodium hydrogen carbonate.

$$2\ NaHCO_3(s) \xrightarrow{\Delta} Na_2CO_3(s) + H_2O(g) + CO_2(g)$$

The Solvay process was a great engineering success because it used raw materials efficiently by recycling several of them. For example, limestone ($CaCO_3$) is heated to produce $CO_2(g)$ and CaO(s). The $CO_2(g)$ is a

Sodium sulfate is obtained partly from natural sources and partly through the following reaction, introduced by J. R. Glauber more than 300 years ago.

$$H_2SO_4(\text{concd aq}) + 2\ NaCl(s) \xrightarrow{\Delta} Na_2SO_4(s) + 2\ HCl(g)$$

A *volatile* substance is one that vaporizes readily. A *nonvolatile* substance has little tendency to vaporize.

The principle of this reaction is that a volatile acid (HCl) is produced by heating one of its salts (NaCl) with a nonvolatile acid (H_2SO_4). Other acids can be produced by similar reactions. The major use of sodium sulfate (Na_2SO_4) is in the paper industry (about 70% of the annual U.S. consumption). In making kraft paper—the kind used in grocery bags—undesirable lignin is removed from wood by treating the wood with an alkaline solution of sodium sulfide (Na_2S). The Na_2S is produced by the reaction of Na_2SO_4 with carbon. About 100 lb of Na_2SO_4 is used in the production of one ton of paper.

Potassium compounds have some uses similar to their sodium counterparts (for example, K_2CO_3 in glass and ceramics), but their most important use by far is in fertilizers, which account for 95% of commercial use of potassium compounds. Potassium is one of the three main nutrients required by plants (nitrogen and phosphorus are the other two). KCl is commonly used as a fertilizer because this is the form in which most potassium is obtained from natural sources.

Glass bottles being manufactured. The principal current use of sodium carbonate is in glassmaking.

principal reactant in the main reaction. CaO is converted to $Ca(OH)_2$ by treatment with water, and $Ca(OH)_2$ is used to convert NH_4Cl, a main reaction byproduct, to NH_3. The NH_3 is then recycled into the production of ammoniated brine.

The only ultimate by-product of the Solvay process is $CaCl_2$. The demand for $CaCl_2$ is limited, however, and in the past much of the $CaCl_2$ was dumped, generally into lakes or streams. Environmental regulations no longer permit this. Partly for this reason, and partly because of the discovery of abundant natural sources of Na_2CO_3 in Wyoming, the Solvay process is now obsolete in the United States. It continues in use elsewhere in the world, however.

The case of sodium carbonate emphasizes the importance of economic and environmental factors in establishing a kind of natural progression for typical industrial chemical processes—advent, modification, and decline.

Diagonal Relationships: The Special Case of Lithium

The first member of a group of the periodic table often differs in some significant ways from the other group members. The eccentric member of Group 1A is lithium, and the main reason for this is the high charge density of Li^+ compared to those of the other alkali metal ions. The charge density is the ratio of the ionic charge to the ionic radius.

These are some of the ways in which lithium and its compounds differ from the other alkali metals:

- Lithium carbonate, fluoride, hydroxide, and phosphate are much less soluble in water than are corresponding salts of the other alkali metals.
- Lithium forms a nitride (Li_3N), the only one of the alkali metals to do so.
- When it burns in air, lithium forms a normal oxide (Li_2O) rather than a peroxide (M_2O_2) or a superoxide (MO_2).
- Lithium carbonate and lithium hydroxide decompose on heating to form lithium oxide, while the carbonates and hydroxides of the other Group 1A metals are thermally stable.

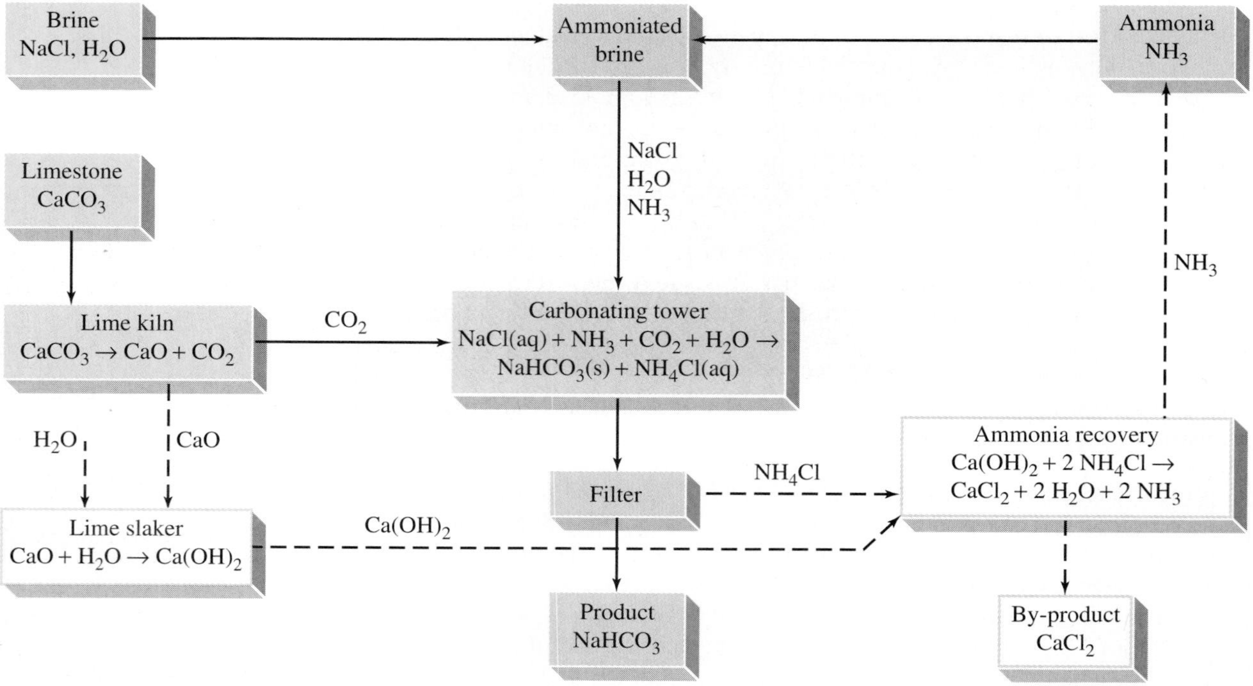

Figure 8.6 **The Solvay process.** The colored boxes and solid arrows outline the steps essential to the main reaction and product. The other boxes and broken-line arrows indicate the recycling steps and the by-product.

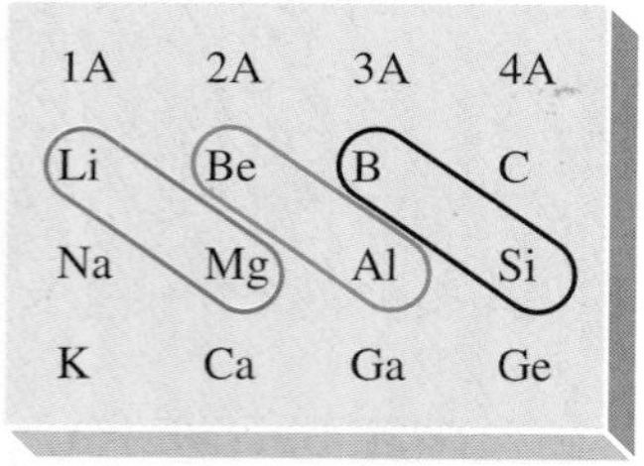

Figure 8.7 **The diagonal relationship in the periodic table.** The elements in each encircled pair have a number of similar properties.

Note that the names for the diseases associated with Na^+ and K^+ are derived from the Latin names for the elements sodium (*natrium*) and potassium (*kalium*).

In some of these properties, lithium and its compounds resemble magnesium and its compounds. This is an example of the **diagonal relationship**, depicted in Figure 8.7. The similarity between Li and Mg probably results from the roughly equal sizes of the Li and Mg atoms and of the Li^+ and Mg^{2+} ions. The diagonal relationship between beryllium and aluminum is described in Section 8.3.

The Alkali Metals and Living Organisms

Living organisms require both Na^+ and K^+ ions. These ions are involved in the regulation and control of body fluids. Too high a concentration of Na^+, a condition called *hypernatremia*, leads to edema (swelling of tissues due to excess fluid retention), thirst, and lessened urine production. (The involvement of excess Na^+ in hypertension was mentioned on page 45.) *Hyponatremia*, too little Na^+, is rare; in fact, most of us eat more salt (NaCl) than we need. Hyponatremia is characterized by a low level of body fluid and by diarrhea, anxiety, and circulatory failure. Too much K^+, called *hyperkalemia*, is characterized by irritability, nausea, decreased urine production, and cardiac arrest. Too little K^+, called *hypokalemia*, results in lethargy, muscle pain and weakness, and a failure of nerve impulses. Ordinary salt (NaCl), which is naturally present in most foods and is often added as table salt, supplies the body with Na^+ and also with chloride ions, which are necessary for the production of stomach acid, HCl(aq). Bananas and orange juice are good sources of K^+.

Potassium ion is an essential nutrient for plants. It is generally abundant and is readily available to plants, except in soil depleted by high-yield agriculture. The usual form of potassium in commercial fertilizers is potassium chloride, KCl.

Because most alkali metal compounds are water soluble, many acidic drugs are administered in the form of their sodium or potassium salts. For example, "free" penicillin G (benzylpenicillic acid) is only sparingly soluble in water, whereas potassium penicillin G is freely soluble in water.

Lithium carbonate is used in medicine to level out the dangerous manic "highs" that occur in manic-depressive psychoses. Some practitioners also recommend lithium carbonate for the depression stage of the cycle. It appears to act by affecting the transport of chemical substances across cell membranes in the brain.

Sodium sulfate decahydrate, $Na_2SO_4 \cdot 10H_2O$, was one of the first synthetic chemicals used in medicine. Still known today as *Glauber's salt*, it acts as a cathartic (a substance that purges the bowels). Glauber's synthesis of this salt helped turn the focus of alchemists from attempts to turn base metals into gold to the synthesis of medicines.

Ca

Sr

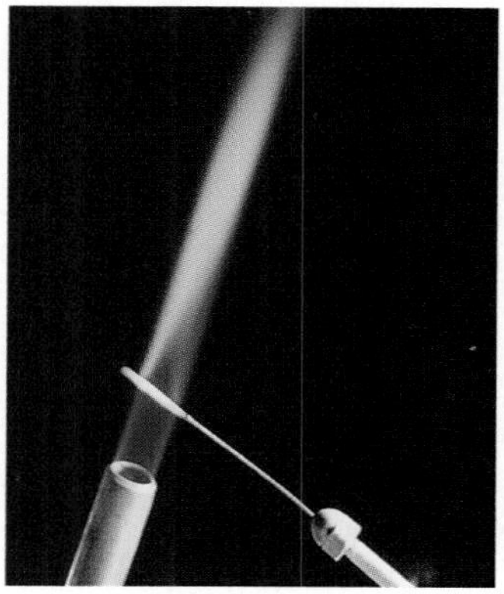

Ba

Figure 8.8 Flame colors of some alkaline earth metals.

8.3 Group 2A: The Alkaline Earth Metals

The six elements making up Group 2A—beryllium, magnesium, calcium, strontium, barium, radium—have the valence-shell electron configuration ns^2. The Group 2A elements are metals, though as suggested by Table 8.5 and the diagonal relationship, beryllium is different enough to warrant separate consideration. The following generalizations apply quite well to the Group 2A metals *other than beryllium*.

- The metals are fairly soft and of low density.
- The metals are reactive; their atoms show a tendency to give up two electrons to form 2+ ions.
- The Group 2A metal compounds are almost exclusively ionic.

A property that can be used to detect the presence of Ca, Sr, and Ba is that they produce characteristic flame colors (Figure 8.8).

TABLE 8.5 **Some Properties of the Alkaline Earth Metals**

	Electron Configuration	Atomic Radius,[a] pm	Ionization Energies, kJ/mol I_1	I_2	Density, g/cm³	Melting Point, °C
Be	$[He]2s^2$	111	899	1757	1.85	1278
Mg	$[Ne]3s^2$	160	738	1451	1.74	649
Ca	$[Ar]4s^2$	197	590	1145	1.54	839
Sr	$[Kr]5s^2$	215	550	1064	2.63	769
Ba	$[Xe]6s^2$	217	503	965	3.65	725
Ra	$[Rn]7s^2$	223	509	979	5.50	700

[a] Values of atomic radii are for metallic radii.

The Group 2A oxides, MO, and hydroxides, $M(OH)_2$, are basic or *alkaline*, though none is soluble in water to any great extent. In the early days of chemistry, the term "earth" was used to describe substances that are insoluble or only slightly soluble in water and that are not decomposed by heating. These properties together account for the group family name: the *alkaline earth* metals.

Many familiar gemstones, including aquamarine and emerald, are based on the mineral beryl.

Sources and Uses of the Alkaline Earth Metals

An important natural source of beryllium and its compounds is the mineral beryl, $3BeO \cdot Al_2O_3 \cdot 6SiO_2$. Beryllium alloys have many applications. For example, an alloy of copper with about 2% Be is used in springs, clips, and electrical contacts because it is able to withstand metal fatigue. Other Be alloys have structural uses where light weight is important. The small Be atom has little stopping power for X-rays or neutrons; this makes beryllium useful for "windows" in X-ray tubes and for various components in nuclear reactors. Beryllium is quite toxic, a drawback to other possible applications.

The free element can be prepared from beryllium fluoride by the reaction described in Example 8.3.

Example 8.3

Beryllium is prepared by the reaction of BeF_2 with magnesium metal at about 1000 °C. The other product is MgF_2. Write a balanced chemical equation for the reaction, including the physical state of each substance.

Solution

First, translate the words into chemical symbols and formulas.

$$BeF_2 + Mg \xrightarrow{1000\ °C} Be + MgF_2$$

Then, use a handbook or other reference to look up the melting point of each substance, and, if necessary, the boiling point or sublimation temperature. (A substance *sublimes* when it goes directly from a solid to a gas without ever becoming a liquid.)

	Melting Point, °C	Boiling Point, °C
BeF_2	sublimes at 800	no normal bp
Mg	649	1090
Be	1278	not needed
MgF_2	1261	not needed

Because BeF_2(s) sublimes to a gas at 800 °C, it is also a gas at 1000 °C. Mg is a liquid under these conditions; it melts below 1000 °C but doesn't boil until 1090 °C. Be and MgF_2 are both solids; they melt above 1000 °C. The complete equation is

$$BeF_2(g) + Mg(l) \xrightarrow{1000\ °C} Be(s) + MgF_2(s)$$

Exercise 8.3

Barium can be prepared by the reaction of BaO with aluminum metal at about 1800 °C. The other product is Al_2O_3. Write a balanced chemical equation for the reaction, including the physical state of each substance.

Magnesium has a lower density than any other structural metal (except beryllium, which is used mainly in alloys with copper and aluminum) and is valued for its light weight. Aircraft parts, for example, are manufactured from magnesium alloyed with aluminum and other metals. Magnesium is used in a number of metallurgical processes, such as the production of beryllium (recall Example 8.3) and titanium.

$$TiCl_4 + 2\ Mg \longrightarrow Ti + 2\ MgCl_2$$

Magnesium is also used in batteries, fireworks, flash photography, and in the synthesis of important reagents for organic chemical reactions.

The free metal is generally prepared by the electrolysis of a molten magnesium salt, typically magnesium chloride in the Dow process described below.

Calcium is used to reduce the oxides or fluorides of other, less common metals (Sc, W, Th, U, Pu, and most of the lanthanides) to the free metals. For example,

$$UO_2 + 2\ Ca \longrightarrow U + 2\ CaO$$

$$2\ ScF_3 + 3\ Ca \longrightarrow 2\ Sc + 3\ CaF_2$$

Calcium is also used in the manufacture of batteries and in forming alloys with aluminum, silicon, and lead.

Metallic calcium is generally obtained by the electrolysis of a molten calcium salt, such as calcium chloride. The free metals strontium, barium, and radium are not widely used, but they too can be prepared by the electrolysis of an appropriate molten salt.

The Dow Process for the Production of Magnesium

Seawater is a solution containing many different ions. There is a lot of Na^+ and Cl^-, of course, from the "salt" in the water. But there is also about 1.3 g Mg^{2+} per kilogram of seawater, and Mg^{2+} is the only common cation in seawater that forms an insoluble hydroxide.

In the Dow process, calcium carbonate (limestone or sea shells) is decomposed to CaO(s) and CO_2(g) by heating. Treatment of CaO with water produces $Ca(OH)_2$, the source of OH^- for the precipitation of $Mg(OH)_2$(s).

$$Mg^{2+}(aq) + 2\ OH^-(aq) \longrightarrow Mg(OH)_2(s)$$

The precipitated $Mg(OH)_2$(s) is washed, filtered, and dissolved in HCl(aq).

$$Mg(OH)_2(s) + 2\ HCl(aq) \longrightarrow MgCl_2(aq) + 2\ H_2O(l)$$

The resulting concentrated $MgCl_2$(aq) is evaporated to dryness, and the $MgCl_2$ is melted and electrolyzed, yielding pure Mg metal and Cl_2(g).

$$Mg^{2+} \quad + 2\,Cl^- \xrightarrow{\text{electrolysis}} Mg(l) + Cl_2(g)$$

The Cl_2(g) is converted to HCl and recycled. The electrolysis of $MgCl_2$(l) is pictured in Figure 8.9.

Of all industrial operations, electrolyses and those that require heating materials to high temperatures consume the most energy. The Dow process involves both. The production of 1 kg Mg requires about 300 MJ of energy (300 MJ = 300×10^6 J). At best, the process might be modified to reduce this requirement to perhaps 200 MJ. By contrast, it takes only about 7 MJ of energy to melt and recast 1 kg of recycled Mg. Recycling can be especially cost effective in the production and use of materials like magnesium.

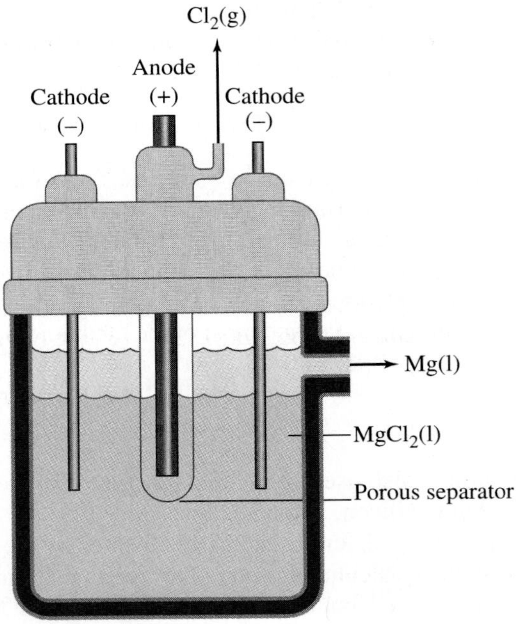

Figure 8.9 **The electrolysis of $MgCl_2$(l).** The positive electrode (anode), where Cl_2(g) is formed, is made of carbon (graphite). The negative electrodes (cathode), where Mg(l) is formed, are made of iron. The Mg(l) floats on the $MgCl_2$(l) and is removed through a tap.

Reactions of the Alkaline Earth Metals

To illustrate the trends in physical and chemical properties found in Group 2A of the periodic table, let's consider the reaction of the metals with water.

$$M(s) + 2\,H_2O \longrightarrow M(OH)_2 + H_2(g)$$

If M = Be, no reaction;
= Mg, slow reaction with steam;
= Ca, slow reaction with cold water;
= Sr, reaction more rapid than with Ca;
= Ba, reaction more rapid than with Sr.

The slow reaction of calcium with water was illustrated in Figure 8.3. In the case of magnesium, an impervious film of $Mg(OH)_2$ covers the surface and immediately stops the reaction. Magnesium does react with steam, however, but MgO is formed rather than $Mg(OH)_2$.

$$Mg(s) + H_2O(g) \xrightarrow{\Delta} MgO(s) + H_2(g)$$

Beryllium fails to react with either cold water or steam.

Another trend is found in the water solubilities of the hydroxides. The solubilities increase with increasing atomic number within the family, although none of the hydroxides is particularly soluble (Table 8.6).

Beryllium reveals itself as the eccentric member of the group because it does not react with water. Another difference is the tendency to form molecular rather than ionic compounds. According to the diagonal relationship illustrated in Figure 8.7, we even expect beryllium to resemble a member of Group 3A—aluminum. For example, aluminum also forms many molecular compounds. Much of the characteristic behavior of beryllium can be attributed to the small size and high ionization energy of the Be atom.

TABLE 8.6	**Solubility of $M(OH)_2$ at 20 °C**
$M(OH)_2$	Solubility, mol/L
$Mg(OH)_2$	0.0002
$Ca(OH)_2$	0.021
$Sr(OH)_2$	0.066
$Ba(OH)_2$	0.23

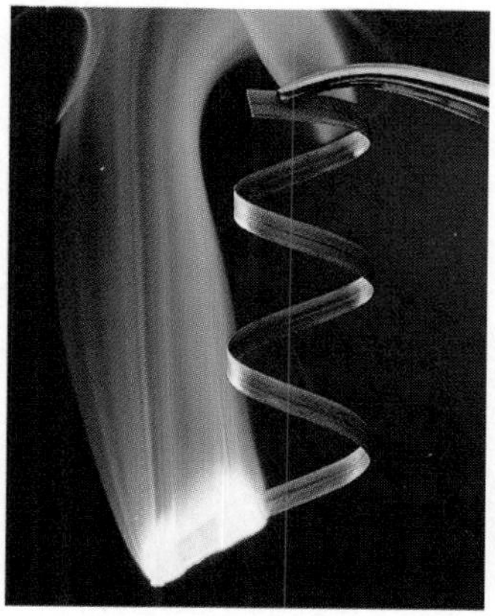

Magnesium burns in air by combining with oxygen gas to produce a brilliant white light and a "smoke" of magnesium oxide.

Some typical reactions of the alkaline earth metals, illustrated below for magnesium, are with the halogens (X_2), with oxygen, and with nitrogen.

$$Mg + X_2 \longrightarrow MgX_2 \quad \text{(where X = F, Cl, Br, I)}$$

$$2\,Mg + O_2 \longrightarrow 2\,MgO$$

$$3\,Mg + N_2 \longrightarrow Mg_3N_2$$

CONCEPTUAL EXAMPLE 8.1

The equations above show that magnesium metal reacts with both oxygen and nitrogen. When a 0.267-g sample of Mg is heated in air, 0.420 g of product is formed. All the Mg appears in the product. **(a)** What mass, in grams, would have been obtained if the product were pure MgO? **(b)** What mass, in grams, would have been obtained if the product were pure Mg_3N_2? **(c)** Could the product be a mixture of MgO and Mg_3N_2?

Solution

a. If the sole product were MgO, the mass of the product would be

$$? \text{ g MgO} = 0.267 \text{ g MgO} \times \frac{1 \text{ mol Mg}}{24.31 \text{ g Mg}} \times \frac{2 \text{ mol MgO}}{2 \text{ mol Mg}} \times \frac{40.32 \text{ g MgO}}{1 \text{ mol MgO}}$$

$$= 0.443 \text{ g MgO}$$

b. If the sole product were Mg_3N_2, the mass of the product would be

$$? \text{ g } Mg_3N_2 = 0.267 \text{ g Mg} \times \frac{1 \text{ mol Mg}}{24.31 \text{ g Mg}} \times \frac{1 \text{ mol } Mg_3N_2}{3 \text{ mol Mg}} \times \frac{101.0 \text{ g } Mg_3N_2}{1 \text{ mol } Mg_3N_2}$$

$$= 0.370 \text{ g } Mg_3N_2$$

c. Since the actual mass of product obtained (0.420 g) is between the values calculated for the pure products, it could well be a mixture of the two.

Conceptual Exercise 8.1

Assuming that the result of the reaction in Conceptual Example 8.1 is indeed a mixture of MgO and Mg_3N_2, calculate the mass percent of MgO in the product.

Important Compounds of Magnesium and Calcium

Several magnesium compounds occur naturally, either in mineral form or in brine solutions. These include the carbonate, chloride, hydroxide, and sulfate. Other magnesium compounds can be prepared from these. A few important compounds and some of their uses are listed in Table 8.7.

Limestone is a naturally occurring form of calcium carbonate, containing some clay and other impurities. Calcium carbonate is the most widely used calcium compound, in part because it is relatively easy to make nearly all other calcium compounds from it. We examine the special role of limestone in the chemical industry in the next section.

TABLE 8.7 **Some Important Magnesium Compounds**

$MgCO_3$	Manufacture of refractory bricks; glass, inks, rubber reinforcing agent; dentrifices, cosmetics, antacids and laxatives.
$MgCl_2$	Manufacture of magnesium metal; manufacture of textiles and paper; fireproofing agents, cements; refrigeration brine.
MgO	Refractories (furnace linings); ceramics syntheses; cements; SO_2 removal from stack gases.
$MgSO_4$	Fireproofing; textile manufacturing; ceramics; fertilizers; cosmetics; dietary supplements.

Limestone: Building Stone and Chemical Raw Material

The decomposition (calcination) of limestone is carried out in a long rotary kiln, producing quicklime, CaO, and carbon dioxide. In a cement kiln, the limestone is mixed with clay and sand to produce the complex mixture of calcium silicates and aluminates known as Portland cement.

Some limestone is shaped and used as a building stone. However, most limestone is used to manufacture other building materials. *Portland cement* is a complex mixture of calcium and aluminum silicates obtained by heating limestone, clay, and sand in a high-temperature rotary kiln. When the cement is mixed with sand, gravel, and water, it solidifies into the familiar material *concrete*. Ordinary *soda-lime glass* (used to make bottles and windows, for example) is a mixture of sodium and calcium silicates formed by heating together limestone, sand, and sodium carbonate.

Limestone is used in the metallurgy of iron and steel as a *flux*—a material producing an easily liquefied calcium silicate mixture, called *slag*, that carries away impurities from molten metals. And, through chemical reactions producing CaO and $Ca(OH)_2$, limestone is the basis of a large part of the inorganic chemical industry.

In practical applications using limestone, often the first step is its decomposition by heating, a process called *calcination*.

$$\textit{Calcination:}\quad CaCO_3(s) \xrightarrow{\Delta} CaO(s) + CO_2(g)$$

Calcination is carried out in a high-temperature kiln (about 1000 °C) with continuous removal of $CO_2(g)$ to promote the forward reaction. The product formed, CaO(s), is called *lime* or *quicklime*. The reaction of quicklime with water, a process called *hydration*, produces $Ca(OH)_2$, known as *slaked lime*.

$$\textit{Hydration:}\quad CaO(s) + H_2O(l) \longrightarrow Ca(OH)_2(s)$$

If $CO_2(g)$ is bubbled through a suspension of $Ca(OH)_2(s)$ in water, a process called *carbonation* takes place, and $CaCO_3(s)$ is formed once more.

$$\textit{Carbonation:} \quad Ca(OH)_2(s) + CO_2(g) \longrightarrow CaCO_3(s) + H_2O(l)$$

The three steps just described can be combined and used to prepare chemically pure $CaCO_3(s)$ from limestone, an impure material. This chemically pure $CaCO_3(s)$, called *precipitated* calcium carbonate, is extensively used as a filler to provide bulk to such materials as paint, plastics, printing inks, and rubber. It is also used in toothpastes, food, cosmetics, and antacids and other pharmaceuticals. Added to paper, calcium carbonate makes the paper bright, opaque, smooth, and capable of absorbing ink well.

Quicklime and slaked lime are the cheapest and most widely used bases. Because they are cheap, they are usually the first choice among bases in any application where unwanted acids must be neutralized. Thus, lime is used to neutralize acidic soils of lawns, gardens, and farmland, and to treat excess acidity in lakes.

$$CaO(s) + 2\,H^+(aq) \longrightarrow Ca^{2+}(aq) + H_2O(l)$$

Another application of quicklime is in air-pollution control. When coal is burned in an electric power plant, sulfur-containing impurities in the coal are converted to sulfur dioxide gas, a major culprit in the formation of acid rain. When powdered limestone is mixed with powdered coal before combustion, the limestone decomposes to $CaO(s)$, which then reacts with $SO_2(g)$ that would otherwise escape.

$$CaO(s) + SO_2(g) \longrightarrow CaSO_3(s)$$

By reaction with oxygen, the calcium sulfite is converted to calcium sulfate (gypsum), which has a number of uses (page 298). Quicklime is also used in treating wastewater effluents and sewage.

Slaked lime, $Ca(OH)_2$, is the cheapest commercial base and is used in all applications where high water solubility is not essential. Slaked lime is also used in the manufacture of other alkalis and bleaching powder, in sugar refining, in tanning hides, and in water softening.

A mixture of slaked lime, sand, and water is the familiar mortar used in bricklaying. In the initial setting of the mortar, bricks absorb excess water, which is then lost through evaporation. In the final setting, $CO_2(g)$ from air reacts with $Ca(OH)_2$ and converts it back to $CaCO_3$.

$$Ca(OH)_2(s) + CO_2(g) \longrightarrow CaCO_3(s) + H_2O(g)$$

Hydrates

The mineral gypsum has the formula $CaSO_4 \cdot 2H_2O$. Recall that a compound that incorporates water molecules into its fundamental solid structure is called a *hydrate* (Section 2.8). In gypsum, *two* water molecules are present for every formula unit of $CaSO_4$ in the solid; the chemical name of gypsum is calcium sulfate *dihydrate*.

Another hydrate of calcium sulfate has one water molecule for every *two* formula units of $CaSO_4$. We could write its formula as $2CaSO_4 \cdot H_2O$, but more commonly we write $CaSO_4 \cdot \frac{1}{2}H_2O$ and call the compound calcium sulfate *hemihydrate*. This hydrate is commonly called "plaster of Paris." It is obtained by heating gypsum.

$$\underset{\text{Gypsum}}{CaSO_4 \cdot 2H_2O(s)} \longrightarrow \underset{\text{Plaster of Paris}}{CaSO_4 \cdot \tfrac{1}{2}H_2O(s)} + \tfrac{3}{2}H_2O(g)$$

When mixed with water, plaster of Paris reverts to gypsum, and in doing so it expands slightly. A mixture of plaster of Paris and water is used to make castings where sharp details of an object must be retained, as in dental work and jewelry making. The most important use is in making gypsum board, which has largely supplanted plaster in the construction industry.

Hydrate formation occurs infrequently among alkali metal compounds, but it is commonly found in alkaline earth metal compounds. Typical hydrates, for example, are $MX_2 \cdot 6H_2O$ (where M = Mg, Ca, or Sr and X = Cl or Br).

Estimation Exercise 8.2

Which of the hydrates just described by the formula $MX_2 \cdot 6H_2O$ has the greatest mass percent of water?

The Alkaline Earth Metals and Living Organisms

Magnesium and calcium are essential to all living organisms. Magnesium ions, for example, are a part of the chlorophyll molecule. Chlorophyll is the catalyst essential to photosynthesis, the process by which plants convert carbon dioxide, water, and sunlight into sugars. Both magnesium and calcium are essential for proper functioning of the nerves that control muscles.

Calcium ions are necessary for the proper development of bones and teeth. For this reason growing children are usually encouraged to drink milk, a rich source of calcium. Adults also require calcium, which is necessary for clotting of blood, maintenance of a regular heartbeat, and—especially in older women—prevention of osteoporosis, a condition in which the bones become porous, brittle, and easily broken.

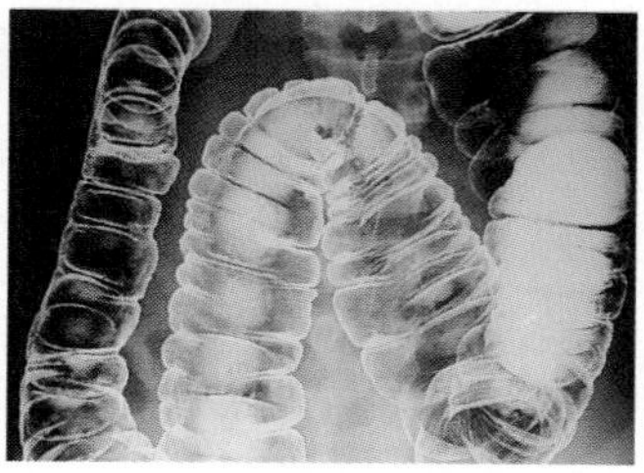

The gastrointestinal tract is rendered visible through the X-ray-absorbing ability of a barium sulfate coating on the tract walls.

Strontium is not essential to living organisms, but it is of interest because of its chemical similarity to calcium. Strontium can follow some of the same pathways in living organisms as does calcium. Because of this, the body easily ingests and absorbs the dangerously radioactive isotope strontium-90, a product of the fallout from the nuclear fission that occurs in nuclear explosions.

Barium also has no known function in living organisms; in fact the Ba^{2+} ion is toxic. Despite this fact, water suspensions of $BaSO_4(s)$ have been used in X-ray imaging of the gastrointestinal tract: a barium "milkshake" for the upper tract or a "barium enema" for the lower tract. Barium atoms and ions are good absorbers of X-rays and make the tract visible in an X-ray photograph. Because it is insoluble, $BaSO_4(s)$ is eliminated by the body with no significant absorption of Ba^{2+} ion.

8.4 Some Chemistry of Groundwater

Carbonates, especially $CaCO_3$, are involved in a number of natural phenomena. One process begins when rainwater dissolves atmospheric $CO_2(g)$. The $CO_2(g)$ reacts with the water to form carbonic acid, H_2CO_3.

$$CO_2(g) + H_2O(l) \longrightarrow H_2CO_3(aq)$$

As rainwater charged with CO_2 (and possibly other acids) seeps through limestone, *insoluble* $CaCO_3$ is converted to *soluble* calcium hydrogen carbonate, $Ca(HCO_3)_2$.

$$CaCO_3(s) + H_2O + CO_2 \longrightarrow Ca(HCO_3)_2(aq)$$

Over time, this dissolving action can produce a large cavity in a limestone bed—a limestone cave. This reaction is reversible, however. Evaporation of a solution of $Ca(HCO_3)_2$ leads to a loss of both water and CO_2, and $Ca(HCO_3)_2$ is converted back to $CaCO_3(s)$.

$$Ca(HCO_3)_2(aq) \longrightarrow CaCO_3(s) + H_2O(g) + CO_2(g)$$

This process occurs very slowly, but over a period of many years, as $Ca(HCO_3)_2(aq)$ drips from the ceiling of a cave, it is converted to icicle-like deposits of $CaCO_3(s)$ called *stalactites*. When some of the dripping $Ca(HCO_3)_2(aq)$ hits the floor of the cave, decomposition occurs there and limestone ($CaCO_3$) deposits build up from the floor in formations called *stalagmites*. Some of the stalactites and stalagmites grow together into limestone columns (see Figure 8.10).

Figure 8.10 Stalactites, stalagmites, and columns in a limestone cavern.

Stalactites hang from the ceiling; stalagmites rise from the ground.

Hard Water

We have just seen how natural rainwater can become infused with calcium hydrogen carbonate. **Hard water** is groundwater that contains significant concentrations of ions from natural sources, principally Ca^{2+}, Mg^{2+}, and sometimes Fe^{2+}, along with associated anions. If the primary anion is the hydrogen carbonate ion, HCO_3^-, the water is said to be **temporary hard water**. If the predominant anions are other than HCO_3^-, for example, Cl^- or SO_4^{2-}, the water is said to be **permanent hard water**.

Recall that the HCO_3^- ion is also called bicarbonate ion. Temporary hardness is sometimes called bicarbonate hardness.

When temporary hard water is heated, bicarbonate ions decompose.

$$2\,HCO_3^-(aq) \xrightarrow{\Delta} CO_3^{2-}(aq) + H_2O(l) + CO_2(g)$$

The CO_3^{2-} formed in this way reacts with Ca^{2+}, Mg^{2+}, and Fe^{2+} cations to form solid carbonates.

$$M^{2+}(aq) + CO_3^{2-}(aq) \longrightarrow MCO_3(s)$$

The mixed precipitate of $CaCO_3$, $MgCO_3$, and $FeCO_3$, together with any other undissolved solids, is commonly called *boiler scale*. Formation of boiler scale is a serious problem associated with hard water. A boiler clogged with boiler scale heats unevenly, and overheating in parts of the boiler can lead to an explosion. Hard water also interferes with the action of soap (pages 302–304). For many of its uses, then, hard water must be softened.

Water Softening

Water softening refers to the removal of objectionable cations and anions in either temporary or permanent hard water. Temporary hard water can be softened by boiling, but this produces boiler scale. A better way is to treat the water with a base; this converts HCO_3^- to CO_3^{2-}. The CO_3^{2-} then combines with M^{2+} ions to form precipitates. These metal carbonates are thus removed from the water, but remain as a gritty solid that can cause problems in washing machines and other devices if not removed by filtration or other methods.

$$HCO_3^-(aq) + OH^-(aq) \longrightarrow CO_3^{2-}(aq) + H_2O(l)$$

$$CO_3^{2-}(aq) + M^{2+}(aq) \longrightarrow MCO_3(s)$$

In this method of water softening, the usual source of $OH^-(aq)$ is slaked lime, $Ca(OH)_2$.

Permanent hard water cannot be softened by boiling. Addition of washing soda (Na_2CO_3) softens permanent hard water by precipitating cations such as Ca^{2+}, Mg^{2+}, and Fe^{2+} as carbonates, but salts such as NaCl and Na_2SO_4 remain in solution.

Ion Exchange

One of the best ways to soften water is through **ion exchange**, that is, to exchange the undesirable ions in hard water for ions that are less objectionable. The ion-exchange medium may be a synthetic resin or a natural sodium aluminosilicate called a *zeolite*. In either case, the ion-exchange materials consist of macromolecular (polymer) particles. In contact with water, these materials ionize to produce two types of ions:

A *macromolecular* substance is composed of large molecules. *Polymers* are macromolecular substances formed by joining many simpler units, called *monomers* (see Section 9.10).

1. *Fixed* ions that remain attached to the polymer surface.
2. Free or mobile *counter* ions. The counter ions of the ion-exchange resin exchange places with the undesirable ions in hard water when a sample of the water passes through the medium.

In the ion-exchange resin pictured in Figure 8.11, the fixed ions R are negatively charged and the counter ions are cations. In a fully charged resin, all the counter ions at the beginning of the process are Na^+. When hard water is passed through, the more highly charged Ca^{2+}, Mg^{2+}, and Fe^{2+} ions in the water attach to the negatively charged surfaces of the ion-exchange resin and displace the Na^+ ions as counter ions. The Na^+ ions do not interfere with the action of soap when the softened water is used for cleaning. To regenerate the resin, concentrated NaCl(aq) is passed through the bed. In *high* concentration, through their sheer numbers, the Na^+ ions are able to dislodge the 2+ cations and restore the resin to its original condition. The ion-exchange material has an indefinite lifetime. To represent the ion-exchange process through simple chemical equations, we can write

$$Na_2Z + M^{2+}(aq) \longrightarrow MZ + 2\,Na^+(aq) \quad \text{(where Z represents a zeolite)}$$

$$Na_2R + M^{2+}(aq) \longrightarrow MR + 2\,Na^+(aq) \quad \text{(where R represents a synthetic resin)}$$

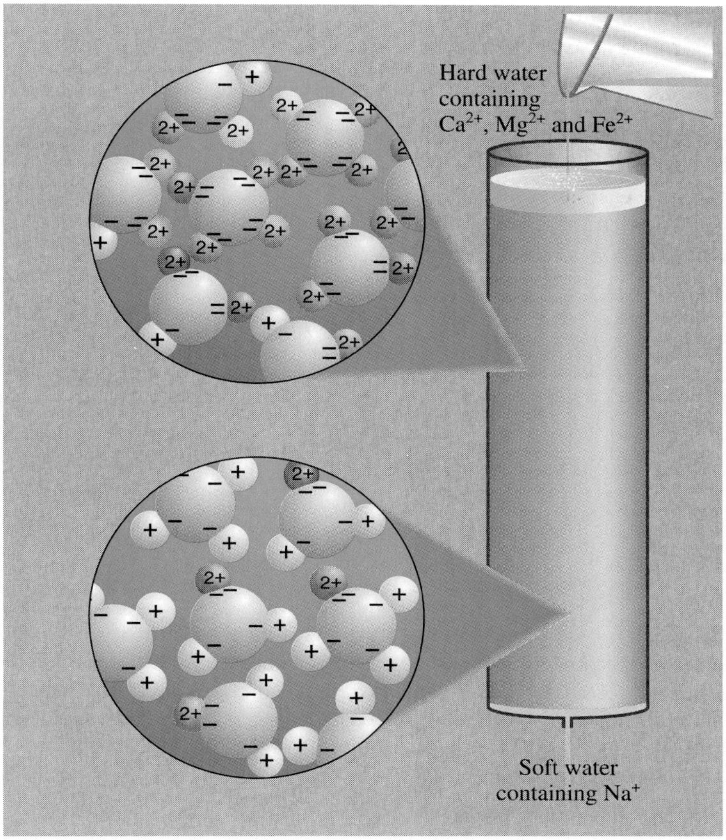

Figure 8.11 Water softening by ion exchange. Multiply charged cations (gray)—Ca^{2+}. Mg^{2+}, Fe^{2+}—replace Na^+ (orange) as counter ions to the negatively charged surfaces of the ion exchange particles (top). By the time the water has reached the bottom of the column, all the water has been softened. The resin there still has mainly Na^+ counter ions.

Suppose that instead of NaCl we use concentrated HCl(aq) to recharge an ion-exchange resin. This leaves H^+ as the counter ions. Now, when a hard water sample is passed through the resin, the exchange reaction is

$$H_2R + M^{2+}(aq) \longrightarrow MR + 2\,H^+(aq)$$

The water becomes acidic, and there are two consequences of this fact.

1. We can determine the amount of acid in a solution by titration (Section 3.11). We can therefore titrate the H^+ eluted (washed) from an ion-exchange column, and in this way establish the level of hardness in the water (see Problem 62).
2. Following replacement of all cations in a water sample by H^+, we can pass the somewhat acidic water through an *anion*-exchange resin in which OH^- replaces all the anions. The H^+ and OH^- neutralize each other.

$$H^+(aq) + OH^-(aq) \longrightarrow H_2O(l)$$

As a result of this two-stage ion-exchange process the water is essentially freed of all its ions. The product is called **deionized water**. We use deionized water rather than ordinary tap water in the laboratory because ions present in tap water may interfere with chemical reactions, for example, by forming unwanted precipitates (see Figure 8.12).

Figure 8.12 Deionized water in the chemical laboratory. When a solution of $AgNO_3$ is prepared in tap water, a white precipitate forms (left). Most likely this precipitate is AgCl(s), formed by the reaction of Ag^+(aq) with traces of Cl^-(aq) in the water. No precipitate forms when $AgNO_3$ is dissolved in *deionized* or *distilled* water (right).

Soaps and Detergents

Soaps are salts of *fatty acids*, long-chain carboxylic acids derived from fats. Soaps are formed in the reaction of fats with a base such as NaOH(aq). Sodium palmitate, a typical soap, is a salt of the 16-carbon palmitic acid. It can be represented as

$$\underset{\text{Sodium palmitate (a soap)}}{CH_3(CH_2)_{14}COO^- Na^+} \quad \text{or} \quad RCOO^- Na^+ \ [\text{where } R = CH_3(CH_2)_{14}]$$

Except when very dilute, alkali metal soaps do not form true solutions in water. Rather, they form colloidal dispersions (see Section 13.10).

The function of a soap as a cleaning agent is to disperse grease and oil films into microscopic droplets. The droplets detach themselves from the surfaces being cleaned, become suspended in water, and are removed by rinsing. As suggested by Figure 8.13 on page 304, at an oil–water interface soap molecules orient themselves with their hydrocarbon residues, R—, dissolved in the oil and their ionic ends, $—COO^- Na^+$, in the water. The soap emulsifies or "solubilizes" the oil.

The alkali metal soaps are water soluble, but soaps having an M^{2+} cation are not. This means that if an alkali metal soap is used in hard water, a calcium, magnesium, or iron soap precipitates. This precipitate is the major component of the familiar "bathtub ring." For example,

$$\underset{\text{Sodium soap}}{2\,RCOO^- Na^+} + Ca^{2+} \longrightarrow \underset{\text{Calcium soap ("bathtub ring")}}{Ca(RCOO)_2} + 2\,Na^+(aq)$$

Cholestyramine: Lowering Cholesterol Through Ion-Exchange

High levels of blood cholesterol pose an increased risk of heart disease. Lowering these levels reduces the risk. Cholestyramine (Questran), a medicine used to lower the concentration of blood cholesterol, is actually an ion-exchange resin.

The resin has positive sites fixed to a polymer surface, R^+, and Cl^- ions as counter ions; it is an *anion*-exchange resin. The resin is taken internally, but the polymer molecules are too large to be absorbed from the gastrointestinal tract, making cholestyramine one of the safest drugs for treating high blood-cholesterol levels.

Bile acids, which we can represent as HA, ionize to form H^+ and A^- anions. The bile anions are soap-like structures that aid in the emulsification and subsequent digestion of fats. In action in the intestines, the ion-exchange resin exchanges chloride ions for the bile anions. That is,

$$R^+Cl^-(s) + A^-(aq) \longrightarrow RA(s) + Cl^-(aq)$$

where R is the resin.

One fate of cholesterol in the body is its conversion to bile acids. In effect, removal of bile acids through the ion-exchange process constitutes a removal of some of the body's stored cholesterol. The liver compensates for the reduced level of bile acids by converting more stored cholesterol to bile acids. In addition, bile acids are required for the absorption of cholesterol from the digestive tract into the bloodstream, so lowering the level of bile acids also decreases the intestinal absorption of new cholesterol. In all, treatment with cholestyramine results in about a 20% reduction in blood-cholesterol levels and a commensurate decrease in the risk of heart disease.

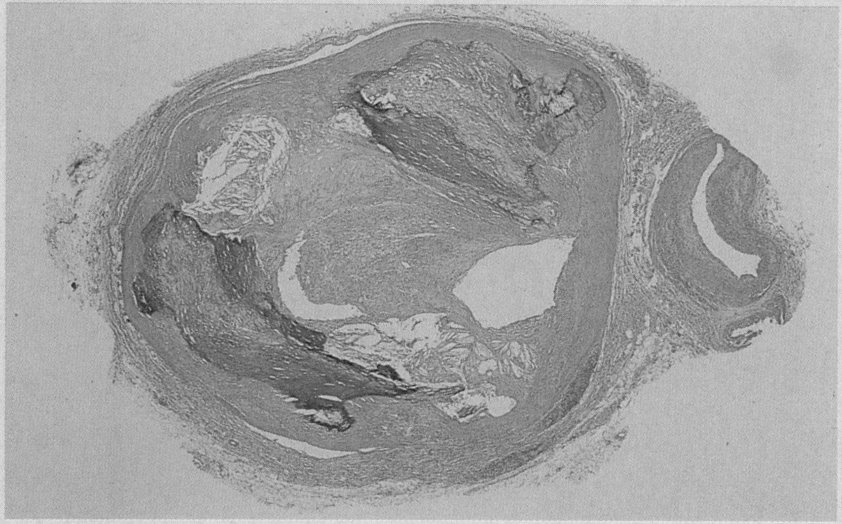

A human coronary artery blocked by cholesterol deposits.

Earlier detergents, called *alkylbenzene sulfonates (ABS)*, had highly branched alkyl groups. ABS detergents were *not* biodegradable.

Before a soap can function well in laundry water, part of it must be used up to precipitate all the M^{2+} ions present. In other words, the soap must first soften the water before it will work. But this leaves an objectionable dirty grayish film of precipitated calcium soap on clothes. An alternative, of course, is to soften water by one of the methods previously described before using it with soap. Still another alternative is to use a synthetic detergent in place of a soap.

A synthetic detergent functions in much the same way as a soap, but it does not form precipitates with Ca^{2+} and other highly charged cations. The detergent can therefore be used in hard water. The detergent type represented by the following structural formula is called a *linear alkylbenzene sulfonate* (LAS).

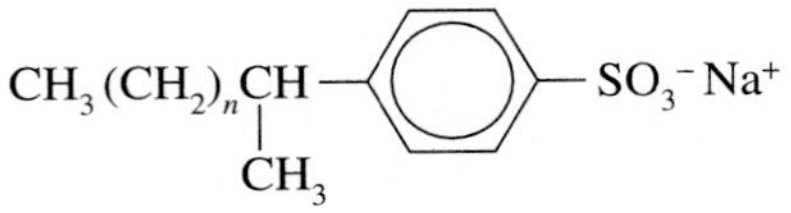

(where *n* is usually 9 to 13)

LAS detergents, like soaps, are biodegradable. Bacteria break them down to carbon dioxide, water, and inorganic ions such as sulfate ion.

Figure 8.13 Cleaning action of a soap visualized. (a) Sodium palmitate, $CH_3(CH_2)_{14}COO^- Na^+$, a typical soap. (b) A microscopic view of soap action. In an oil droplet suspended in water, the hydrocarbon residues (R groups) of soap molecules are immersed in the oil and the ionic ends extend into the water. Attractive forces between water molecules and the ionic ends cause the oil droplet to be "solubilized."

SUMMARY

The main sources of free hydrogen are the electrolysis of water, the water gas reactions, and the reforming of hydrocarbons. Small quantities of the gas can be produced by the reaction of very active metals (Group 1A and Ca, Sr, or Ba) with water or less active metals (for example, Zn and Fe) with mineral acids. Among the uses of hydrogen are (a) the synthesis of ammonia, (b) the hydrogenation of oils, (c) as a reducing agent in metallurgy, and (d) as a fuel.

Of all the elements, the Group 1A metals (the alkali metals) have the largest atomic radii and lowest ionization energies; they also have low densities and low melting points. They form ionic solids with nonmetals and react with water to produce ionic hydroxides and hydrogen gas. In some of its physical and chemical behavior, lithium resembles magnesium. This resemblance is called a *diagonal relationship*. Sodium is formed in the electrolysis of molten sodium chloride. Sodium hydroxide, chlorine, and hydrogen form in the electrolysis of aqueous sodium chloride. Other chemicals produced from sodium chloride include sodium carbonate, sodium sulfate, and hydrochloric acid. The vast majority of Group 1A compounds are water soluble. Sodium and potassium ions are essential to living organisms, and many Group 1A compounds have medicinal value.

The Group 2A metals (the alkaline earth metals) have smaller atomic radii and greater ionization energies, densities, and melting points than do the Group 1A metals. Beryllium is an exception to general trends found in Group 2A. Calcium and magnesium are found in many minerals, for example, in limestone ($CaCO_3$) and dolomite ($CaCO_3 \cdot MgCO_3$). The principal method used to obtain the metals is by electrolysis of the molten chloride. The heavier Group 2A metals react with water to liberate hydrogen; magnesium does so with steam; but beryllium does not react with water. Among the important Group 2A compounds are the carbonates, chlorides, hydroxides, oxides, and sulfates of calcium and magnesium. Many alkaline earth compounds are insoluble or only slightly soluble in water. Also, many occur as *hydrates*, as in $CaCl_2 \cdot 6H_2O$. Magnesium and calcium are essential to all living organisms.

The slight dissolution of minerals (for example, $CaCO_3$, $CaSO_4$) in acidified rainwater introduces ions into groundwater. This dissolving action is responsible for natural caverns, limestone formations, and "hardness" in water. Hard water is objectionable because it yields mineral deposits when heated and interferes with the cleaning action of soaps. Water can be "softened" by chemical reactions that remove the offending ions Ca^{2+}, Mg^{2+}, and Fe^{2+}. Another method that can be used is ion exchange, in which offending ions are replaced by innocuous ones. Ion exchange can also be used to eliminate virtually all ions from water, yielding *deionized* water. Another alternative that permits the use of hard water for laundry purposes is to substitute detergents for soaps.

KEY TERMS

deionized water (8.4)
diagonal relationship (8.2)
hard water (8.4)
hydrogenation reaction (8.1)
ion exchange (8.4)
permanent hard water (8.4)
soap (8.4)
temporary hard water (8.4)

REVIEW QUESTIONS

1. Which is the most abundant element in Earth's crust? In the sun? In the universe as a whole?
2. Which is greater, the percent abundance of hydrogen in Earth's crust on a mass basis or on the basis of numbers of atoms? Explain.
3. In Earth's crust, which is the most abundant alkali metal? Which is the most abundant alkaline earth metal?
4. Name several naturally occurring forms in which alkali metal and alkaline earth metal compounds are found.

5. Which of the alkali metal ions is (are) essential to living organisms? Which of the alkaline earth metal ions?

6. Which of the Group 1A elements react with cold water to produce $H_2(g)$? Which of the Group 2A elements?

7. Name some elements that form *ionic* hydrides and some that form binary molecular hydrogen compounds. Cite some of the differences between these two types of compounds.

8. What is a hydrogenation reaction? Give a commercially important example.

9. Explain the function of the metallurgical process known as *reduction*.

10. What is meant by the term *calcination* (for example, the calcination of limestone)?

11. What is meant by the *diagonal relationship* among elements? To which elements does it apply?

12. Describe several ways in which (**a**) lithium differs from the other Group 1A elements; (**b**) beryllium differs from the other Group 2A elements.

13. Describe several ways in which (**a**) lithium resembles magnesium; (**b**) beryllium resembles aluminum.

14. Describe how flame tests can be used to detect the presence of certain elements. List several of these elements and their flame colors.

15. What is the principal substance produced in each of the following processes?
a. the Dow process **b.** the Solvay process

16. What is the primary chemical substance comprising each of the following materials?
a. limestone **b.** quicklime
c. slaked lime **d.** gypsum
e. Glauber's salt

17. What are the principal raw materials required for the manufacture of each of the following materials?
a. Portland cement **b.** soda–lime glass
c. mortar **d.** plaster of Paris

18. Describe how each of the following is used in medicine.
a. lithium carbonate **b.** barium sulfate

19. Explain the difference between *temporary* and *permanent* hard water.

20. Describe some ways in which hard water can be softened.

21. What is "boiler scale"? How is it formed?

22. Describe how limestone caverns and the distinctive features found in them—stalactites, stalagmites, and columns—are formed.

23. Describe how deionized water can be prepared from ordinary tap water through the use of ion-exchange resins.

24. What are the basic structural features of each of the following?
a. a soap **b.** a detergent

25. How do soaps and detergents function; that is, how do they emulsify oil or grease?

PROBLEMS

Nomenclature

26. Supply a name or formula for each of the following.
a. Li_2CO_3
b. magnesium nitride
c. calcium bromide hexahydrate
d. $KHSO_4$

27. Supply a name or formula for each of the following.
a. LiH
b. magnesium iodide
c. barium hydroxide octahydrate
d. $Ca(HCO_3)_2$

Chemical Equations

28. Complete the following equations representing the reactions of substances with water.
a. $Sr(s) + H_2O \longrightarrow$
b. $LiH(s) + H_2O \longrightarrow$
c. $BaO(s) + H_2O \longrightarrow$
d. $C(s) + H_2O(g) \xrightarrow{\Delta}$

29. Complete the following equations representing the reactions of substances with acids.
   a. $Ca(s) + HCl(aq) \longrightarrow$
   b. $CaO(s) + HCl(aq) \longrightarrow$
   c. $NaF(s) + H_2SO_4(\text{concd aq}) \xrightarrow{\Delta}$

30. Write chemical equations to represent each of the following.
   a. the displacement of $H_2(g)$ from HCl(aq) by Al(s)
   b. the *reforming* of ethane gas (C_2H_6) with steam
   c. the complete hydrogenation of methylacetylene, $CH_3C{\equiv}CH$
   d. the reduction of $MnO_2(s)$ to Mn(s) with $H_2(g)$

31. Write chemical equations to represent each of the following.
   a. the reaction of lithium metal with chlorine gas
   b. the reaction of potassium metal with water
   c. the reaction of cesium metal with liquid bromine
   d. the combustion of potassium metal to form potassium superoxide

32. Write chemical equations to represent each of the following.
   a. the reaction of BeF_2 with metallic sodium to produce metallic beryllium
   b. the reaction of calcium metal with dilute acetic acid, $CH_3COOH(aq)$
   c. the reaction of plutonium(IV) oxide with calcium to produce metallic plutonium
   d. the calcination of dolomite, a mixed calcium magnesium carbonate ($CaCO_3 \cdot MgCO_3$)

33. Write equations for the reactions you would expect to occur when (**a**) $MgCO_3(s)$ is heated to a high temperature; (**b**) $CaCl_2(l)$ is electrolyzed; (**c**) Ca(s) is added to cold dilute HCl(aq); (**d**) an excess of slaked lime is added to an aqueous solution of sulfuric acid.

34. Write an equation for the reaction that you would expect to occur when (**a**) $CaCO_3(s)$ is heated to a high temperature; (**b**) $MgCl_2(l)$ is electrolyzed; (**c**) Ba(s) is added to cold dilute HCl(aq).

35. Write an equation for the reaction that you would expect to occur when (**a**) $Mg(OH)_2(s)$ is added to HCl(aq); (**b**) $CO_2(g)$ is bubbled into KOH(aq); (**c**) KCl(s) is heated with concentrated $H_2SO_4(aq)$.

36. Write equations to show how each of the following substances can be used in the preparation of $H_2(g)$.
   a. $H_2O(l)$
   b. HI(aq)
   c. Mg(s)
   d. $CH_4(g)$

   Use other reactants as necessary—water, acids or bases, metals, etc.

37. Write plausible equations to show how you would convert (**a**) NaCl(s) to $Na_2SO_4(s)$; (**b**) $Mg(HCO_3)_2(aq)$ to MgO(s).

38. Write plausible equations to show how you would convert (**a**) $BaCO_3(s)$ to $Ba(OH)_2(s)$; (**b**) NaCl(s) to $Na_2O_2(s)$.

39. Write chemical equations to represent (**a**) the recovery of NH_3 in the Solvay process; (**b**) the production of an alkaline solution for the precipitation of $Mg(OH)_2(s)$ in the Dow process.

Reaction Stoichiometry

(Hint: Write chemical equations, as necessary.)

40. How many grams of $CaH_2(s)$ are required to generate sufficient $H_2(g)$ to fill a 126-L weather observation balloon at 746 mmHg and 15 °C?

41. How many liters of $H_2(g)$ at 22 °C and 15.5 atm pressure, are required to convert 175 g of propylene ($CH_3CH{=}CH_2$) to propane (C_3H_8)? What mass of propane is obtained?

42. How many liters of 6.0 M HCl are required to neutralize 55.6 kg $Ca(OH)_2(s)$?

43. Assume that the magnesium content of seawater is 1270 g Mg^{2+}/ton seawater and that the density of seawater is 1.03 g/mL. What minimum volume of seawater, in liters, must be used in the Dow process to obtain 1.00 kg of magnesium? Why is the actual volume required greater than this calculated minimum volume?

44. In the Solvay process, for every 1.00 kg $NaHCO_3(s)$ produced in the main reaction, how many kilograms of $CaCl_2$ are obtained as a by-product of the ammonia recovery step?

45. How many cubic meters of $CO_2(g)$, at 748 mmHg and 22 °C, would be produced in the calcination of 1.00×10^3 kg of the mineral dolomite ($CaCO_3 \cdot MgCO_3$)?

Hard Water

46. Describe how *temporary* hard water can be softened by the addition of $NH_3(aq)$. (*Hint*: What ions are present in the water? What chemical reactions occur?)

47. With reference to Problem 46, do you think that *permanent* hard water can be softened with $NH_3(aq)$? Explain.

48. A particular water sample has a hardness expressed as 115 parts per million (ppm) HCO_3^-. Assuming that Ca^{2+} is the only cation present, how many milligrams of "boiler scale" would you expect to deposit when 725 mL of the water is boiled for a period of time? (*Hint*: Think of parts per million as g $HCO_3^-/10^6$ g water ≈ g $HCO_3^-/10^3$ L water.)

49. A water sample has a hardness of 126 ppm HCO_3^-. How many kilograms of $Ca(OH)_2$ are required to soften 1.50×10^7 L of the water? (*Hint*: Refer to Problem 48 and the equations on page 300.)

50. Explain why it is necessary to know what cations are present in the hard water sample of Problem 48 but not in that of Problem 49.

51. Why is the quantity of soap required to wash clothes in hard water greater than that used in soft water?

52. Describe how a water sample containing the ions Fe^{3+}, Ca^{2+}, Cl^-, and SO_4^{2-} can be *deionized* by the use of ion-exchange resins.

53. Write chemical equations to represent the preparation of deionized water from the following solutions:
a. $Ca(HCO_3)_2(aq)$
b. $NaCl(aq)$
c. $NaOH(aq)$.

ADDITIONAL PROBLEMS

54. For the following groupings of substances, select (**a**) the most metallic of K, Be, and Ca; (**b**) the hardest of the solids Na(s), K(s), and Mg(s); (**c**) the least water soluble of Li_2CO_3, Na_2CO_3, and $CaCO_3$.

55. Write chemical equations for the following reactions referred to on page 289.
a. the formation of lithium oxide
b. the reaction of lithium with nitrogen gas to form lithium nitride
c. the calcination of lithium carbonate

56. The use of potassium superoxide in life-support systems is described on page 283. Write a balanced equation to show how sodium peroxide can be used instead.

57. In a manner similar to that shown in Figure 8.5 for NaCl, construct a diagram to show how each of the following compounds can be prepared from $Ca(OH)_2$.
a. $CaCO_3$ **b.** $CaCl_2$
c. $CaSO_4$ **d.** $CaHPO_4$

58. Chemetall GmbH of Frankfurt am Main, Germany, is the world's leading producer of cesium products. Assuming the availability of water, common reagents (acids, bases, salts), and simple laboratory equipment, give a practical method that could be used to prepare each of the following substances from cesium chloride (CsCl).
a. Cs metal **b.** CsOH
c. $CsNO_3$ **d.** Cs_2SO_4
e. $CsHCO_3$ **f.** Cs_2CO_3

59. A particular water sample has a hardness of 117 parts per million (ppm) SO_4^{2-}. The cations present are Ca^{2+} ions.
a. Show how this water can be softened with Na_2CO_3 (washing soda).
b. What mass of Na_2CO_3, in grams, is required to soften 162 L of this water?

60. How many grams of the soap sodium palmitate are consumed in softening 5.00 L of a hard water that has a hardness of 135 ppm HCO_3^-? Assume that the only cations in the water are Ca^{2+} ions. (*Hint*: Recall the discussion on page 304, and see the hint in Problem 48.)

61. A sample of water whose hardness is expressed as 185 ppm Ca^{2+} is passed through an ion-exchange column and the Ca^{2+} is replaced by Na^+. What is the molarity of Na^+ in the water so treated? (*Hint*: Base your calculation on a 1000-L sample, which weighs 1.00×10^6 g.)

62. A 100.0-mL sample of hard water is passed through a column of the ion-exchange resin H_2R. The water coming off the column requires 15.17 mL of 0.02650 M NaOH for its titration. What is the hardness of the water, expressed as ppm Ca^{2+}?

63. A wastewater stream is 1.38 M in HCl(aq). Calculate how many kilograms of (**a**) CaO(s) and (**b**) NaOH(s) are required to neutralize 985 L of the acidic solution.

64. The standard enthalpy of formation of $Ca(OH)_2$ is −986.1 kJ/mol, that of H_2O is −285.8 kJ/mol, and that of CaO is −635.1 kJ/mol. Calculate the heat of reaction for the conversion of CaO to $Ca(OH)_2$.

65. When lime is stored exposed to air, it slowly loses the properties of CaO. What reaction occurs?

66. When a 5.00-g sample of impure lime containing some $CaCO_3$ is dissolved in HCl(aq), 143 mL of a gas, measured at 746 mmHg and 22 °C, was liberated. Calculate the percent of $CaCO_3$ in the sample.

67. Determine whether liquid hydrogen or jet fuel has the greater heat of combustion (**a**) on a mass basis and (**b**) on a volume basis. Jet fuel is a mixture of hydrocarbons, but assume that $C_{12}H_{26}$ is a representative formula. The densities of $H_2(l)$ and $C_{12}H_{26}(l)$ are 0.0708 g/mL and 0.749 g/mL, respectively. (*Hint*: Recall Estimation Exercise 5.2, page 184.)

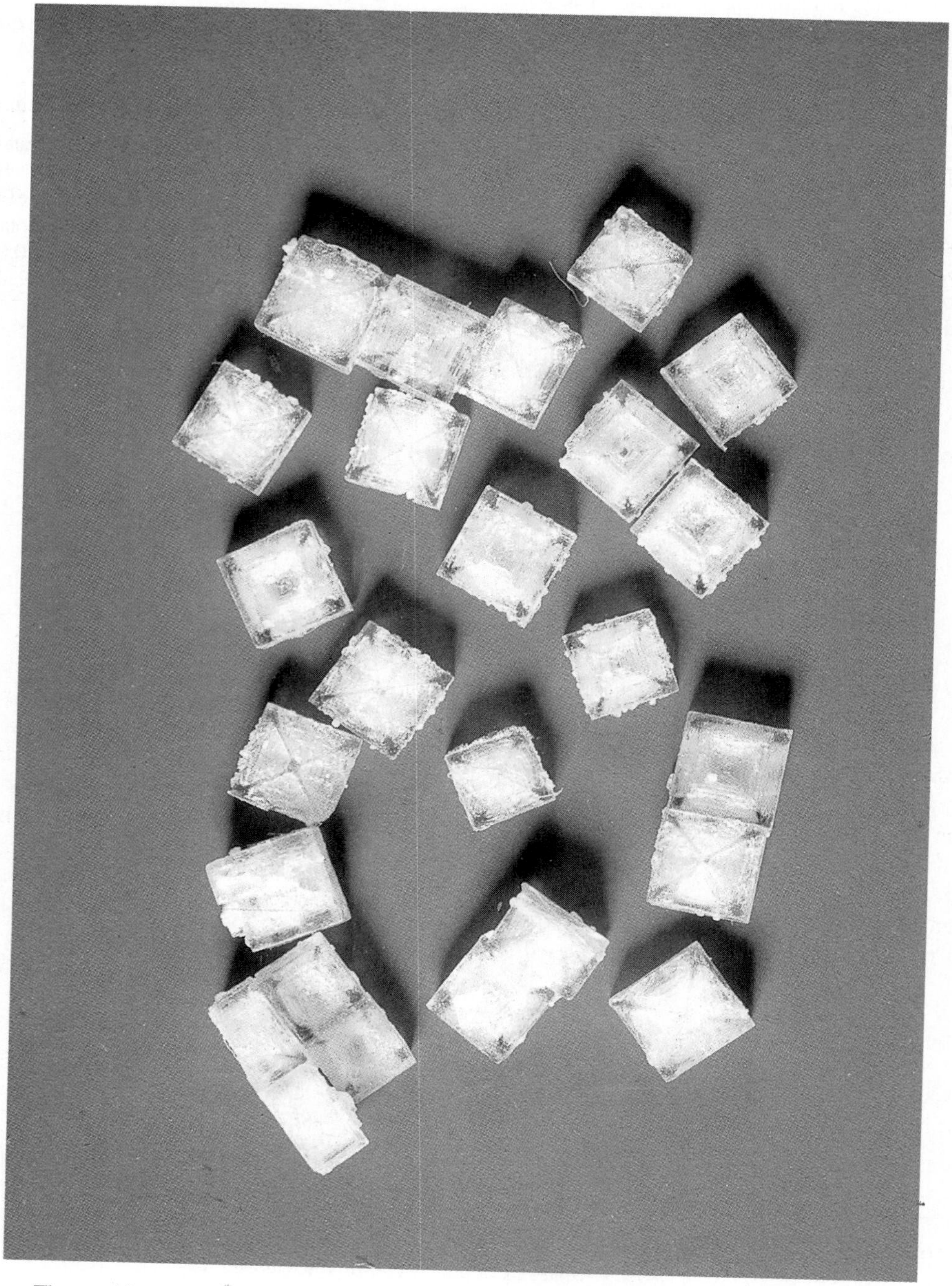

These cubic crystals of sodium chloride—common table salt—reflect the arrangement of sodium ions (Na^+) and chloride ions (Cl^-) at the microscopic level. The bonds that hold these ions in this particular arrangement are discussed on page 313.

9

Chemical Bonds

CONTENTS

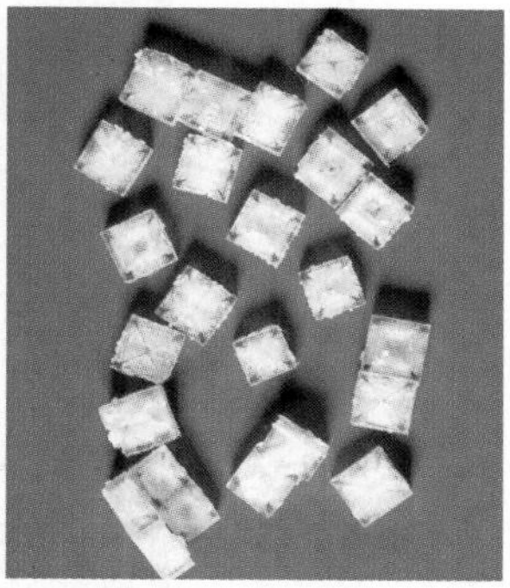

The forces that hold atoms together in molecules and those that keep ions in place in solid ionic compounds are called **chemical bonds**. Nearly all the properties of a material are determined by the nature of its chemical bonds. The concepts described in this and the two chapters that follow can be used to explain some of life's most fundamental questions, and a host of practical questions as well. For example, concepts of chemical bonding can be used to explain

- Why carbon atoms are the key atoms in most of the molecules of living things—fats, proteins, carbohydrates, and nucleic acids.
- Why the action of most drugs depends on the shape of the drug molecule, and how this shape is determined by the types of chemical bonds in the molecule.
- Why carbon monoxide and oxygen both can form bonds to hemoglobin, but why carbon monoxide kills and oxygen sustains life.
- How the energy required for maintaining a heartbeat, breathing, and other physiological functions is stored in chemical bonds.

9.1 The Ionic Bond: A Preview

Historically, the lack of reactivity of the noble gases has been attributed to their electron configurations. Noble gas atoms have high ionization energies, which means that only with great difficulty do they lose electrons to form cations. And noble gas atoms have no affinity for additional electrons: they do not form energetically stable anions. After the discovery of the noble gases, chemists began to wonder if other atoms combine in such a way as to acquire electron configurations like those of the noble gases. One way atoms might do this is to give up or acquire one or more electrons. And that is precisely what certain kinds of atoms do.

The Reaction of Sodium with Chlorine

As we saw in Chapter 8, sodium is a soft, silvery metal that reacts violently with water. Chlorine is a yellow-green gas made up of the diatomic molecules Cl_2. It is highly reactive and quite toxic if inhaled even in moderate amounts. When sodium and chlorine react (Figure 9.1), they form a stable white crystalline solid—sodium chloride, common table salt.

Figure 9.1 The reaction of sodium and chlorine. Sodium metal and chlorine gas provide striking visual evidence of their reaction to produce the ionic substance sodium chloride.

Let's look at the electron configurations of sodium and chlorine atoms for a possible interpretation of the reaction between them. A sodium atom, Na, by *losing* an electron, forms a sodium ion, Na^+, which has the same electron configuration as the noble gas neon.

$$\begin{array}{lccc} & \text{Na} & \longrightarrow & \text{Na}^+ + \text{e}^- \\ \text{Electron configuration:} & [\text{Ne}]3s^1 & & [\text{Ne}] \end{array}$$

A chlorine atom, Cl, by *gaining* an electron, forms a chloride ion, Cl^-, which has the same electron configuration as the noble gas argon.

$$\begin{array}{lccc} & \text{Cl} & + \text{e}^- \longrightarrow & \text{Cl}^- \\ \text{Electron configuration:} & [\text{Ne}]3s^23p^5 & & [\text{Ar}] \end{array}$$

These two processes occur together in the reaction in Figure 9.1. That is, the sodium atoms lose electrons and the chlorine atoms gain them.

$$\begin{array}{lccccc} & Na & + & Cl & \longrightarrow & Na^+ + Cl^- \\ \text{Electron configuration:} & [Ne]3s^1 & & [Ne]3s^23p^5 & & [Ne] \quad [Ar] \end{array}$$

Let us emphasize that in giving up an electron a sodium atom does not become a neon atom. The sodium ion and the neon atom do have the same electron configuration. However, the sodium ion has 11 protons in its nucleus and a charge of 1+, whereas the neon atom has 10 protons in its nucleus and is electrically neutral. Similarly, a chlorine atom does not become an argon atom. The two have the same electron configuration, but their nuclei differ, and it is the number of protons in the nucleus—the atomic number—that defines an element.

Chlorine gas is composed of Cl_2 molecules. In the reaction, each atom of the molecule receives an electron from a sodium atom. Two sodium ions and two chloride ions are formed.

$$Cl_2 + 2\,Na \longrightarrow 2\,Cl^- + 2\,Na^+$$

Ionic Bonds

The two ions formed in the reaction between a sodium atom and a chlorine atom have opposite charges. They are strongly attracted to one another. However, the attractive force of a given sodium ion is not limited to one chloride ion. Each sodium ion most strongly attracts (and is attracted by) six neighboring chloride ions; it also attracts distant chloride ions, but more weakly. Similarly, each chloride ion most strongly attracts (and is attracted by) six neighboring sodium ions, and attracts distant sodium ions more weakly. Moreover, ions of like charge repel one another. Although the attractive and repulsive forces between ions counteract one another to some extent, the net effect of all the interactions within a cluster of oppositely charged ions of an ionic compound is one of attraction. The net attractive forces that hold positive and negative ions together in a crystalline solid are called **ionic bonds** (see Figure 9.2).

A **crystal** is a structure characterized by plane surfaces, sharp edges, and a regular geometric shape. These features exist because the crystal has a distinctive repeating pattern of its constituent particles—atoms, ions, or molecules—at the microscopic level.

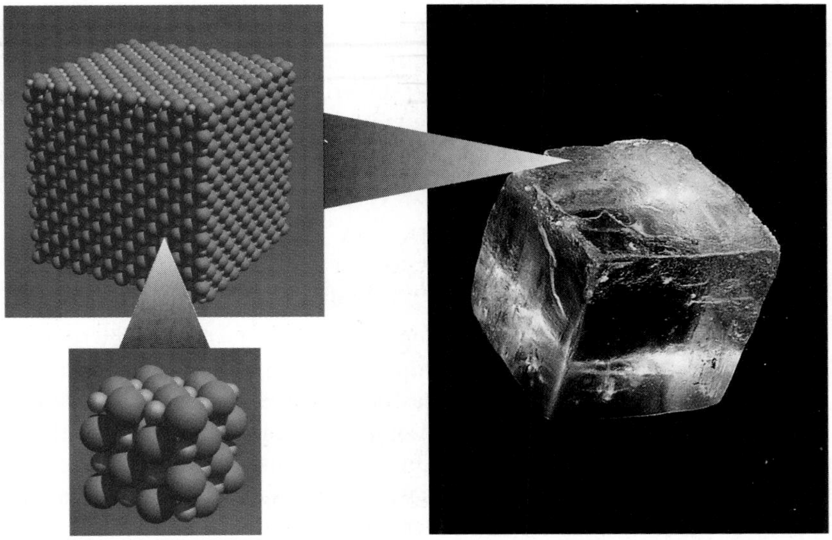

Figure 9.2 Formation of a crystal of sodium chloride. Each Na^+ ion (small sphere) is surrounded by six Cl^- ions (large spheres). In turn, each Cl^- ion is surrounded by six Na^+ ions. This arrangement repeats itself many, many times over, ultimately resulting in a crystal of sodium chloride (right).

9.2 Lewis (Electron-Dot) Symbols

The core electron configurations of sodium and chlorine atoms do not change when these atoms form ions. This fact underlies a simple method of representing electron configurations in chemical bonds. We let the chemical

Gilbert Newton Lewis (1875–1946) was one of the foremost American chemists of the first half of the twentieth century. In addition to his pioneer work in describing chemical bonding through the symbolism named after him, Lewis was a driving force in the introduction of thermodynamics into the mainstream of chemistry, and he made important contributions to acid–base theory.

symbol represent the core or kernel of an atom: the nucleus plus inner-shell electrons. Then we represent the valence (outer) electrons by placing dots around the symbol. To illustrate,

	Na	$\longrightarrow$	$Na^+ + e^-$
Electron configuration:	[Ne] $3s^1$		[Ne]

reduces to

$$\text{Na}\cdot \longrightarrow \text{Na}^+ + e^-$$

and

	Cl	$+ e^- \longrightarrow$	Cl^-
Electron configuration:	[Ne] $3s^2 3p^5$		[Ne] $3s^2 3p^6$

becomes

$$:\dot{\underset{\cdot\cdot}{\text{Cl}}}: + e^- \longrightarrow :\ddot{\underset{\cdot\cdot}{\text{Cl}}}:^-$$

The reaction of sodium and chlorine can be represented as

$$\text{Na}\cdot + :\dot{\underset{\cdot\cdot}{\text{Cl}}}: \longrightarrow \text{Na}^+ + :\ddot{\underset{\cdot\cdot}{\text{Cl}}}:^-$$

Representations like these, in which the symbol of an element stands for the core of the atom and dots stand for its valence electrons, are called **Lewis symbols** or **electron-dot symbols**.

Lewis Symbols and the Periodic Table

Lewis symbols are generally written only for the main group or representative elements. This is easy to do because the number of valence electrons for these elements is equal to the group number. In writing Lewis symbols, we will follow the practice initiated by Lewis himself. We will place the first four bold dots on the four sides of the chemical symbol, and then we will pair them up as we add the next four. For example,

1A	2A	3A	4A	5A	6A	7A	8A
Li·	$\underset{\cdot}{\text{Be}}\cdot$	$\cdot\underset{\cdot}{\text{B}}\cdot$	$\cdot\dot{\underset{\cdot}{\text{C}}}\cdot$	$\cdot\dot{\underset{\cdot}{\text{N}}}:$	$\cdot\dot{\underset{\cdot\cdot}{\text{O}}}:$	$:\dot{\underset{\cdot\cdot}{\text{F}}}:$	$:\ddot{\underset{\cdot\cdot}{\text{Ne}}}:$

It makes no difference on which side of the symbol we begin the process of adding dots. Here we have placed them around the symbol in a clockwise direction, starting at "3 o'clock."

Notice that for elements in Groups 2A, 3A, and 4A, the Lewis symbols do *not* reflect the electron pairing we would expect in ground-state electron configurations—that is, that Be actually has no unpaired electrons, B has one, and C has two. However, electron spin and other fine points about electronic structure had not yet been proposed when Lewis formulated his theory. We will write Lewis symbols as Lewis did, and, as we will see in Section 10.3, in a number of cases Lewis symbols lead to a better prediction of chemical bonding than do ground-state electron configurations.

- electron spin
- electron pairing
- ground-state elec. config.

Example 9.1

Give Lewis symbols for magnesium, sulfur, and phosphorus.

Solution

Magnesium is in Group 2A, sulfur is in Group 6A, and phosphorus is in Group 5A. The Lewis symbols are therefore

$$\mathrm{Mg}\!: \qquad \cdot\ddot{\mathrm{S}}\!: \qquad \cdot\dot{\mathrm{P}}\!:$$

Exercise 9.1

Give Lewis symbols for each of the following atoms.

a. Ar **b.** Ca **c.** Br
d. As **e.** K **f.** Se

Using Lewis Symbols: More Ionic Compounds

As we might expect, atoms of potassium, a metal in the same family as sodium, react with bromine, a reddish brown liquid nonmetal in the same family as chlorine. The reaction yields potassium bromide, KBr.

$$\mathrm{K}\cdot + :\dot{\underset{\cdot\cdot}{\mathrm{Br}}}: \longrightarrow \mathrm{K}^+ + :\ddot{\underset{\cdot\cdot}{\mathrm{Br}}}:^-$$

Magnesium, a Group 2A metal, reacts with oxygen, a Group 6A element, to form a stable white crystalline solid—magnesium oxide, MgO.

$$\mathrm{Mg}\!: + \cdot\dot{\underset{\cdot\cdot}{\mathrm{O}}}: \longrightarrow \mathrm{Mg}^{2+} + :\ddot{\underset{\cdot\cdot}{\mathrm{O}}}:^{2-}$$

A magnesium atom must give up two electrons and an oxygen atom must gain two electrons for each to acquire the electron configuration of the noble gas neon.

An atom like oxygen, which needs two electrons to complete a noble gas configuration, may react with lithium atoms, which have only one electron to give. In this case, *two* atoms of lithium are needed for each oxygen atom. The product is lithium oxide, Li_2O.

$$\begin{matrix}\mathrm{Li}\cdot \\ \mathrm{Li}\cdot\end{matrix} + \cdot\dot{\underset{\cdot\cdot}{\mathrm{O}}}: \longrightarrow \begin{matrix}\mathrm{Li}^+ \\ \mathrm{Li}^+\end{matrix} + :\ddot{\underset{\cdot\cdot}{\mathrm{O}}}:^{2-} \quad \text{or} \quad 2\,\mathrm{Li}\cdot + \cdot\dot{\underset{\cdot\cdot}{\mathrm{O}}}: \longrightarrow 2\,\mathrm{Li}^+ + :\ddot{\underset{\cdot\cdot}{\mathrm{O}}}:^{2-}$$

In this reaction, each lithium atom achieves the helium electron configuration. Oxygen again attains the neon electron configuration.

Generally speaking, metallic elements in Groups 1A, 2A, and 3A of the periodic table react with nonmetallic elements in Groups 5A, 6A, and 7A to form stable crystalline ionic solids. The ions so formed have noble gas electron configurations.

The Octet Rule

The ns^2np^6 electron configuration of the valence shell of all the noble gas atoms except helium is commonly called an *octet* of electrons. In forming chemical bonds, atoms often gain, lose, or share electrons until they have achieved an outer shell that contains an octet of electrons (eight electrons). This is called the **octet rule**. The rule applies to most ionic bonds between representative metal atoms and nonmetal atoms. In lithium compounds, such as Li_2O, the Li^+ ion has just two valence-shell electrons in the electron configuration of helium, $1s^2$. We might say that Li^+ has an outer-shell *duet.* In lithium hydride, LiH, both Li^+ and H^- have an outer-shell duet.

There are some important facts to keep in mind about the octet rule, though. First, as we will see in the next section, atoms lose or gain electrons not to satisfy the octet rule but to reach a lower energy state in an ionic compound. It is in reaching this lower energy state that they often follow the octet rule as well. The principal exceptions to the octet rule are found in ionic compounds in which the cations do not acquire noble gas electron configurations (recall Table 7.5). Thus, the cations do not follow the octet rule in $FeCl_3$, CuO, or PbI_2.

Example 9.2

Use Lewis symbols to show the transfer of electrons from magnesium atoms to nitrogen atoms to form ions with noble gas electron configurations.

Solution

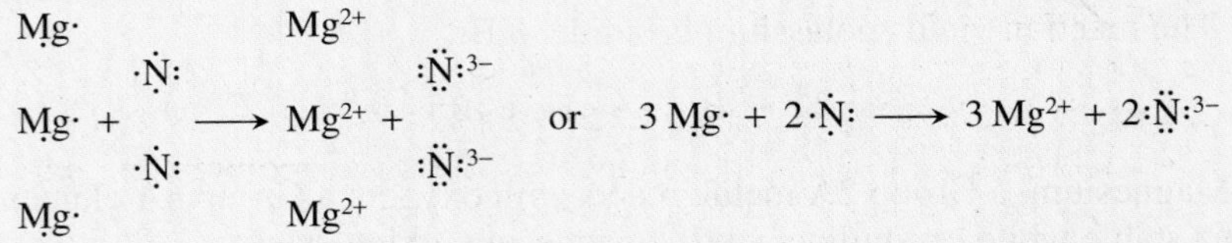

Each of three Mg atoms gives up two electrons (a total of six), and each of two N atoms acquires three (a total of six). Notice that the total positive and negative charges on the products are equal (6+ and 6–). Magnesium reacts with nitrogen to yield magnesium nitride, Mg_3N_2.

Exercise 9.2

Use Lewis symbols to show the transfer of electrons from aluminum to oxygen atoms to form ions with noble gas electron configurations. What is the formula and name of the product?

9.3 Energy Changes in Ionic Compound Formation

If an ionic crystal is to be stable with respect to the elements from which it is formed, there must be a net decrease in energy in its formation. In other words, heat must be released when the ionic crystal is formed from its elements. Let's assess the situation for sodium chloride: The energy needed to remove the outer electron from a gaseous sodium atom is its *first ionization energy.* The energy released when an electron is added to a gaseous chlorine atom is the *electron affinity* of chlorine.

$$Na(g) \longrightarrow Na^+(g) + e^- \qquad \Delta H = I_1 = +495 \text{ kJ/mol}$$

$$Cl(g) + e^- \longrightarrow Cl^-(g) \qquad \Delta H = EA = -349 \text{ kJ/mol}$$

The net energy change for the transfer of an electron from an isolated sodium atom to a lone chlorine atom is 495 kJ/mol + (–349) kJ/mol = +146 kJ/mol.

Based on this calculation alone, the simultaneous formation of an isolated sodium cation and an isolated chloride anion from gaseous atoms is not energetically favorable. We know, however, that sodium chloride is a stable substance. What is wrong with our analysis? It is simply that the formation of a crystal of NaCl, as we have previously noted, is more complicated than just an electron transfer from an Na atom to a Cl atom. We must consider the effect of a very large number of Na^+ and Cl^- ions coming together to form a crystal of NaCl, and we must begin with solid sodium metal and molecules of chlorine gas (Cl_2), rather than with separate gaseous Na and Cl atoms. We should emphasize the *enthalpy of formation* of NaCl(s)—that is, the enthalpy change accompanying the reaction

$$Na(s) + \tfrac{1}{2}Cl_2(g) \longrightarrow NaCl(s) \qquad \Delta H_f^\circ = ?$$

We can consider the enthalpy of formation of NaCl to be the net enthalpy change of a five-step process that begins with Na(s) and Cl_2(g) and ends with NaCl(s). From Hess's law, ΔH_{net} is the sum of the ΔH values of the individual steps (Section 5.6). This general approach to describing the energetics of ionic compound formation is known as the *Born–Haber cycle*. We apply it to sodium chloride in the steps outlined below and illustrate it in Figure 9.3.

1. Solid sodium is converted to sodium vapor. The enthalpy change for this process, ΔH_1, is the *enthalpy of sublimation* of sodium, determined experimentally to be +108 kJ/mol.
2. Cl_2 molecules are dissociated into Cl atoms. The enthalpy change for this process is the *bond dissociation energy* for Cl_2, found by experiment to be +243 kJ/mol Cl_2. To produce one mole of Cl atoms requires one-half mole of Cl_2 and, therefore, 122 kJ. This is the value for ΔH_2.
3. One mole of Na atoms are ionized to Na^+ ions. This requires energy equivalent to the first ionization energy: $\Delta H_3 = I_1 = +495$ kJ/mol.
4. One mole of Cl^- ions are formed from Cl atoms. This liberates energy in an amount equal to the electron affinity: $\Delta H_4 = EA = -349$ kJ/mol.

Sublimation is the conversion of a solid directly to a gas, without going through a liquid phase. Dry ice (solid carbon dioxide) sublimes.

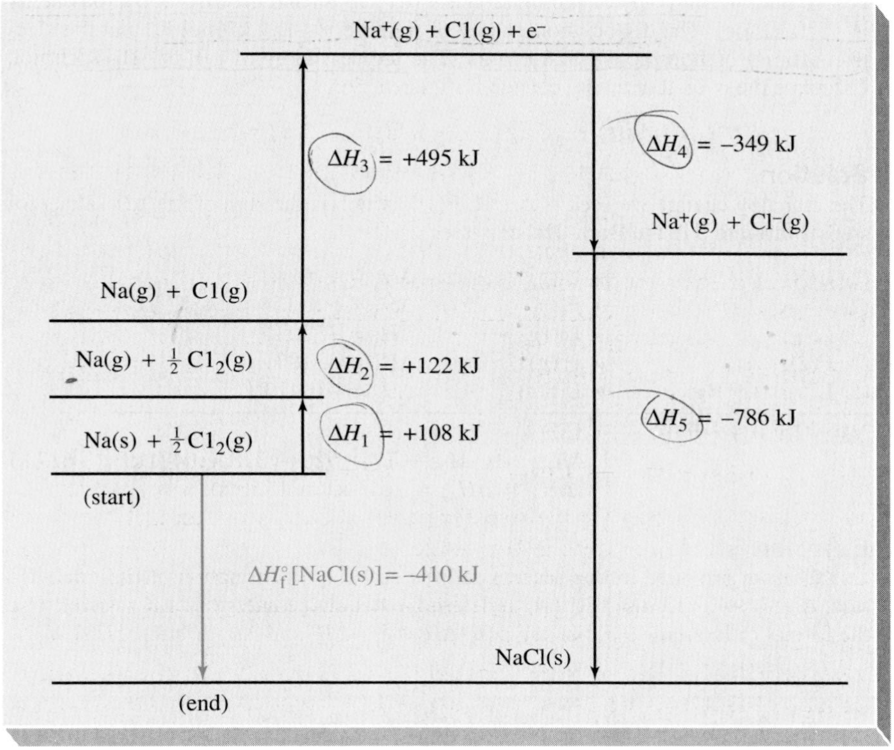

Figure 9.3 Born–Haber cycle for 1 mol of sodium chloride. The starting point and the five steps of the cycle are shown. (ΔH values are not to scale.) The sum of the five enthalpies gives ΔH_f°. The equivalent one-step reaction for the formation of NaCl(s) directly from Na(s) and $\tfrac{1}{2}Cl_2$(g) is shown in color.

5. One mole of NaCl(s) is formed from the gaseous ions. The enthalpy change for the formation of one mole of an ionic solid from its separated gaseous ions is called the **lattice energy.** For NaCl, the lattice energy is $\Delta H_5 = -786$ kJ/mol.

The equations for these five steps are written below, together with their ΔH values. When we sum the equations (canceling out terms that appear on both sides of the summation) we see that the net equation is for the formation of one mole of NaCl(s) from its elements. The sum of the ΔH values gives ΔH_f° of NaCl(s).

$$
\begin{array}{llll}
(1)\ \mathrm{Na(s)} & \longrightarrow \mathrm{Na(g)} & \Delta H_1 & = +108\ \mathrm{kJ} \\
(2)\ \tfrac{1}{2}\mathrm{Cl_2(g)} & \longrightarrow \mathrm{Cl(g)} & \Delta H_2 & = +122\ \mathrm{kJ} \\
(3)\ \mathrm{Na(g)} & \longrightarrow \mathrm{Na^+(g)} + e^- & \Delta H_3 & = +495\ \mathrm{kJ} \\
(4)\ \mathrm{Cl(g)} + e^- & \longrightarrow \mathrm{Cl^-(g)} & \Delta H_4 & = -349\ \mathrm{kJ} \\
(5)\ \mathrm{Na^+(g)} + \mathrm{Cl^-(g)} & \longrightarrow \mathrm{NaCl(s)} & \Delta H_5 & = -786\ \mathrm{kJ} \\
\hline
\textit{Net}:\ \mathrm{Na(s)} + \tfrac{1}{2}\mathrm{Cl_2(g)} & \longrightarrow \mathrm{NaCl(s)} & \Delta H_{\mathrm{net}} & = (108 + 122 + 495 - 349 - 786)\ \mathrm{kJ} \\
 & & & = -410\ \mathrm{kJ} \\
 & & \Delta H_f^\circ & = \Delta H_{\mathrm{net}} = -410\ \mathrm{kJ/mol\ NaCl(s)}
\end{array}
$$

The negative value of ΔH_f° indicates that the formation of crystalline NaCl(s) from the elements in their standard states is energetically favorable, as we thought it should be. Note that the major factor driving this process is the large negative value of the lattice energy, ΔH_5.

Example 9.3

For lithium, the enthalpy of sublimation is +161 kJ/mol and the first ionization energy is +520 kJ/mol. The dissociation energy of fluorine is +154 kJ/mol F_2, and the electron affinity of fluorine is −328 kJ/mol. The lattice energy of LiF is −1047 kJ/mol. Calculate the overall enthalpy change for the reaction

$$\mathrm{Li(s)} + \tfrac{1}{2}\mathrm{F_2(g)} \longrightarrow \mathrm{LiF(s)} \qquad \Delta H_f^\circ = ?$$

Solution

The enthalpy change we seek is ΔH_f°[LiF(s)], which is the sum of the ΔH values for steps 1 through 5 in the Born–Haber cycle.

$$
\begin{array}{llll}
(1)\ \mathrm{Li(s)} & \longrightarrow \mathrm{Li(g)} & \Delta H_1 & = +161\ \mathrm{kJ} \\
(2)\ \tfrac{1}{2}\mathrm{F_2(g)} & \longrightarrow \mathrm{F(g)} & \Delta H_2 & = \tfrac{1}{2} \times 154 = +77\ \mathrm{kJ} \\
(3)\ \mathrm{Li(g)} & \longrightarrow \mathrm{Li^+(g)} + e^- & \Delta H_3 & = +520\ \mathrm{kJ} \\
(4)\ \mathrm{F(g)} + e^- & \longrightarrow \mathrm{F^-(g)} & \Delta H_4 & = -328\ \mathrm{kJ} \\
(5)\ \mathrm{Li^+(g)} + \mathrm{F^-(g)} & \longrightarrow \mathrm{LiF(s)} & \Delta H_5 & = -1047\ \mathrm{kJ} \\
\hline
\textit{Net}:\ \mathrm{Li(s)} + \tfrac{1}{2}\mathrm{F_2(g)} & \longrightarrow \mathrm{LiF(s)} & & \\
 & \Delta H_{\mathrm{net}} & = (+161 + 77 + 520 - 328 - 1047)\ \mathrm{kJ} = -617\ \mathrm{kJ} & \\
 & \Delta H_f^\circ & = \Delta H_{\mathrm{net}} = -617\ \mathrm{kJ/mol\ LiF(s)} &
\end{array}
$$

Exercise 9.3

Use the data provided in this section and the enthalpy of formation of lithium chloride, $\Delta H_f^\circ = -409$ kJ/mol LiCl(s), to determine the lattice energy of LiCl. (*Hint*: Use the format of Example 9.3, but solve for ΔH_5 using ΔH_f° and ΔH_1 through ΔH_4.)

9.4 Covalent Bonds: Shared Electron Pairs

electron affinity- energy change when an e⁻ is added to an atom in the gaseous state.

We have stated several times that hydrogen gas is made up of H_2 molecules. This means that the energy state of two H atoms joined by a chemical bond must be lower than that of two isolated H atoms. But the bond cannot be ionic. One hydrogen atom can hardly accept an electron from another because all hydrogen atoms have an equal electron affinity. Just as important is the fact that hydrogen atoms have a high ionization energy (I_1 = 1312 kJ/mol); they lose electrons only with great difficulty. Lewis proposed that in cases like this the chemical bond is a pair of electrons *shared* between the bonded atoms; it is called a **covalent bond**.

H· + ·H ⟶ H:H (covalent bond (shared pair of electrons))

The term *covalent* was proposed by Irving Langmuir (1881–1957), another American chemist who contributed significantly to our understanding of chemical bonding.

A representation of covalent bonding through Lewis symbols and shared electron pairs is called a **Lewis structure**. Notice that if we "double count" the shared electrons, each H atom appears to have two electrons in its valence shell. This is the electron configuration of helium.

Consider next the case of chlorine, which also exists in the diatomic form, Cl_2. Chlorine atoms, too, are joined by a covalent bond.

:C̤̈l· + ·C̤̈l: ⟶ :C̤̈l:C̤̈l:

Because the shared electrons once again "count" for both atoms, each chlorine atom in the chlorine molecule has *eight* valence electrons, an arrangement like that of the noble gas argon. The chlorine atoms follow the *octet* rule.

The shared pairs of electrons in a molecule are called **bonding pairs**. The other electron pairs that stay with one atom and are not shared are called *nonbonding pairs* or **lone pairs**. Bonding pairs (:) and nonbonding or lone pairs (:) in the Cl_2 molecule are shown below. Also illustrated is the practice of representing a bonding pair of electrons by a dash (—).

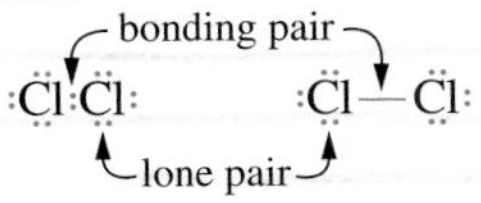

9.5 Polar Covalent Bonds: Unequal Sharing of Electron Pairs

So far, we have seen two different ways in which atoms can combine. Metals and nonmetals combine by the transfer of one or more electrons from the metal atoms to the nonmetal atoms to form ionic bonds. Atoms that are identical combine by sharing pairs of electrons to form covalent bonds. Now let's consider bond formation between atoms that are different, but not different enough to form ionic bonds.

What we show for HCl is not a chemical equation in the usual sense, but a way of describing the covalent bond and rationalizing the formula HCl. The chemical equation for the reaction may be represented as

$$H_2(g) + Cl_2(g) \longrightarrow 2\,HCl(g)$$

or

$$\tfrac{1}{2}H_2(g) + \tfrac{1}{2}Cl_2(g) \longrightarrow HCl(g)$$

Hydrogen Chloride

Both hydrogen and chlorine atoms need an electron to achieve a noble gas electron configuration, which they can do by sharing a pair of electrons between them.

H· + ·C̤̈l: ⟶ H:C̤̈l: or H—C̤̈l:

You might wonder why hydrogen and chlorine molecules react at all. Haven't we just explained (in Section 9.4) that the diatomic molecules themselves are more energetically stable than the isolated atoms? But there is stable, and there is more stable. Given an opportunity, a chlorine atom will bond with a hydrogen atom rather than with another chlorine atom. Why? Because two hydrogen chloride molecules are in a lower energy state (a more stable state) than are one hydrogen molecule and one chlorine molecule. We will say something more about the energetics of covalent bond formation in Section 9.7.

Electronegativity

In the hydrogen chloride molecule, atoms that are unlike (H and Cl) share a pair of electrons. Sharing does not necessarily mean sharing equally though. Chlorine atoms have a greater attraction for a shared pair of electrons than do hydrogen atoms, as reflected by a property known as *electronegativity*. We have already examined two atomic properties that relate to electron attraction—that is, ionization energy and electron affinity. The greater the ionization energy of an atom, the more inclined the atom is to retain its electrons, and the more negative its electron affinity, the more inclined it is to acquire an additional electron. However, these properties apply only to isolated gaseous atoms, and not directly to atoms in molecules. **Electronegativity**, which is related to ionization energy and electron affinity, is a measure of the tendency of an atom to attract bonding electrons to itself when it is in a molecule.

The greater the electronegativity of an atom in a molecule, the more strongly it attracts the electrons in a covalent bond.

The electronegativity of a chlorine atom is greater than that of a hydrogen atom.

Atoms of the elements in the upper right of the periodic table—small, nonmetal atoms—attract bonding electrons most strongly; they have the greatest electronegativities. Atoms of the elements to the left side of the table—large, metal atoms—have a weaker hold on electrons; they have the smallest electronegativities. On an electronegativity scale devised by Linus Pauling, the most nonmetallic, and hence most electronegative element, fluorine, is assigned a value of 4.0. Typical active metals have electronegativities of about 1.0 or less.

Within a period, elements become more electronegative from left to right (Figure 9.4). In the second period the trend is completely regular, increasing by about 0.5 per element as we move from lithium at the far left to fluorine at the far right. In other periods the trend is in the same direction but less regular.

Within a group, electronegativity decreases from top to bottom. Chlorine is less electronegative than fluorine and sulfur is less electronegative than oxygen. A comparison of electronegativities is not quite so straightforward when considering two elements that are neither in the same period nor in the same group. As suggested in Figure 9.5, generally the element above or to the right (or both) is more electronegative than one below or to the left (or both).

Electronegativity values provide the rationale for some periodic tables that locate hydrogen at the top and in the middle of the periodic table, rather than placing it in either Group 1A or 7A (page 252). Its electronegativity (2.1) is

Linus Pauling (1901–1994) extended Lewis's theory through quantum mechanics. Pauling's work was summarized in the 1939 text *The Nature of the Chemical Bond*, and he was awarded the Nobel Prize in chemistry in 1954. Pauling also won the Nobel Peace Prize in 1962 for his fight to control nuclear weapons that was influential in the establishment of the 1963 nuclear test ban treaty. Later in life, Pauling became interested in the medical value of megadoses of vitamins, particularly vitamin C.

1A	2A	3B	4B	5B	6B	7B	8B	8B	8B	1B	2B	3A	4A	5A	6A	7A
H 2.1																
Li 1.0	Be 1.5											B 2.0	C 2.5	N 3.0	O 3.5	F 4.0
Na 0.9	Mg 1.2											Al 1.5	Si 1.8	P 2.1	S 2.5	Cl 3.0
K 0.8	Ca 1.0	Sc 1.3	Ti 1.5	V 1.6	Cr 1.6	Mn 1.5	Fe 1.8	Co 1.8	Ni 1.8	Cu 1.9	Zn 1.7	Ga 1.6	Ge 1.8	As 2.0	Se 2.4	Br 2.8
Rb 0.8	Sr 1.0	Y 1.2	Zr 1.4	Nb 1.6	Mo 1.8	Tc 1.9	Ru 2.2	Rh 2.2	Pd 2.2	Ag 1.9	Cd 1.7	In 1.7	Sn 1.8	Sb 1.9	Te 2.1	I 2.5
Cs 0.7	Ba 0.9	La* 1.1	Hf 1.3	Ta 1.5	W 1.7	Re 1.9	Os 2.2	Ir 2.2	Pt 2.2	Au 2.4	Hg 1.9	Tl 1.8	Pb 1.8	Bi 1.9	Po 2.0	At 2.2
Fr 0.7	Ra 0.9	Ac† 1.1														

below 1.0 — 1.0–1.4 — 1.5–1.9 — 2.0–2.4 — 2.5–2.9 — 3.0–4.0

*Lanthanides: 1.1–1.3
†Actinides: 1.3–1.5

Figure 9.4 Pauling's electronegativities of the elements. Values are from L. Pauling, *The Nature of the Chemical Bond,* 3rd edition, Cornell University, Ithaca, NY, 1960, p. 93. (Some values have been modified by later investigators.)

Because noble gas compounds are limited to a small number of compounds of Kr and Xe, we have not included electronegativities for the Group 8A elements in Figure 9.4.

comparable to those of boron (2.0) and carbon (2.5), considerably higher than that of lithium (1.0), and much less than that of fluorine (4.0).

Example 9.4

Refer only to the periodic table inside the front cover and arrange the following sets of atoms in order of increasing electronegativity, that is, from the lowest to highest electronegativity value.

a. Cl, Mg, Si **b.** As, N, Sb **c.** As, Se, Sb

Solution

a. The order of increasing electronegativity within a period (the third, in this case) is from left to right: Mg < Si < Cl.

b. The order of increasing electronegativity within a group (5A) is from bottom to top: Sb < As < N.

c. As and Se are in the same period; Se is more electronegative than As. As and Sb are in the same group; As is more electronegative than Sb. The order of increasing electronegativity is Sb < As < Se.

Exercise 9.4

Refer only to the periodic table inside the front cover and arrange the following sets of atoms in order of increasing electronegativity.

a. Ba, Be, Ca **b.** Ga, Ge, Se **c.** Cl, S, Te **d.** Bi, P, S

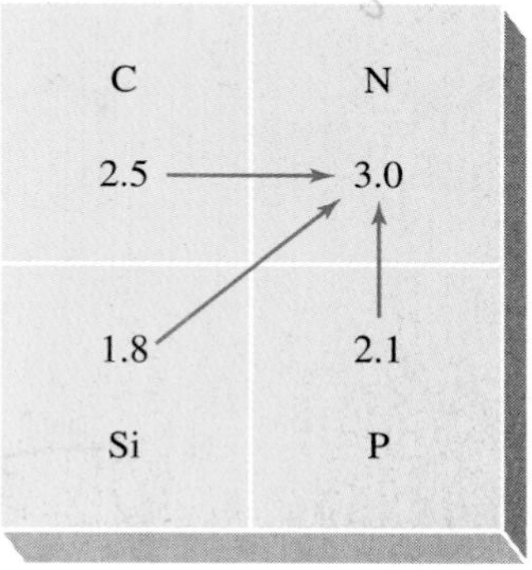

Figure 9.5 Electronegativities in relation to position in the periodic table. In general, electronegativities increase in the directions of the colored arrows.

Electronegativity Difference and Bond Type

In a covalent bond between two atoms with equal electronegativities, there is an equal sharing of an electron pair, and the electrons are not drawn any closer to one atom than to the other. A bond of this type is said to be **nonpolar**. The H—H and Cl—Cl bonds are nonpolar. In a covalent bond between atoms of different electronegativities, there is an unequal sharing of an electron pair, and the electrons are drawn closer to the atom of higher electronegativity. Such a bond is said to be **polar**. The H—Cl bond is a polar bond. With large differences in electronegativity, such as those that exist between some metal and nonmetal atoms, complete electron transfer occurs from the metal to the nonmetal and the bond is essentially ionic. Figure 9.6 illustrates that there is no distinct boundary between covalent and ionic bonds. Some bonds are clearly nonpolar covalent, some are essentially 100% ionic, but many others have an intermediate character—they are polar covalent.

Note that it is the extent of the electronegativity *difference*, not the actual electronegativities of the bonded atoms, that determines the extent of polar character in a bond.

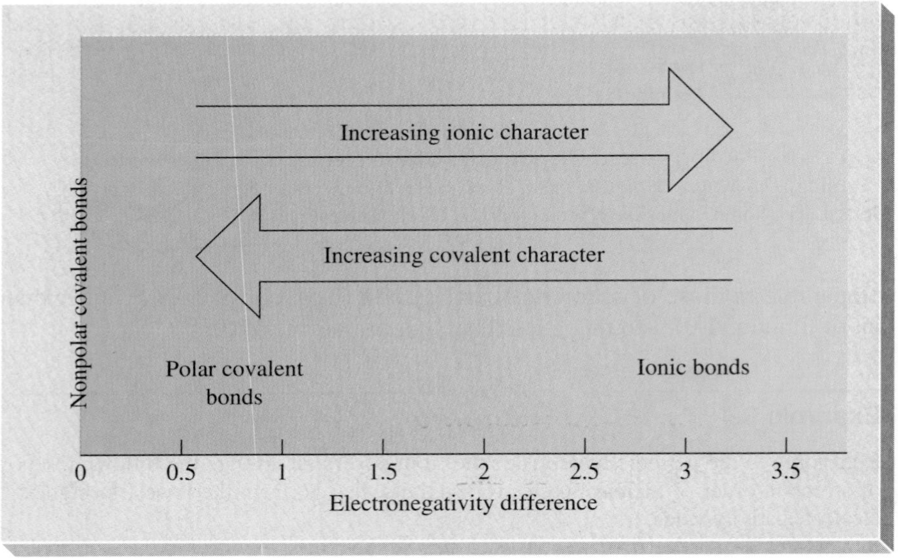

Figure 9.6 Electronegativity and bond type.

Example 9.5

Use electronegativity differences to arrange the following bonds in order of increasing polarity: Br—Cl, Cl—Cl, Cl—F, H—Cl, I—Cl.

Solution

The electronegativity differences are

Br—Cl	Cl—Cl	Cl—F	H—Cl	I—Cl
0.2	0.0	1.0	0.9	0.5

The order of increasing polarity is: Cl—Cl < Br—Cl < I—Cl < H—Cl < Cl—F.

Exercise 9.5

Use electronegativity differences to arrange the following bonds in order of increasing polarity: C—Cl, C—H, C—Mg, C—O, C—S.

Electron-Charge Cloud Pictures of Covalent Bonds

Let's picture the electron-pair bond between two atoms as a cloud of negative electric charge that encompasses both atoms. (Modern quantum theory enables us to do this, as we will discuss in Chapter 10). In a *nonpolar* covalent bond such as H—H, the electron cloud density is greatest between the two hydrogen nuclei but is otherwise uniformly distributed (Figure 9.7). In a *polar* covalent bond such as H—Cl, the greatest electron density is still found between the two nuclei, but the cloud is strongly displaced toward the more electronegative chlorine.

There are two methods, both widely used, to indicate the polar nature of a bond. In the representation

$$\overset{\delta+}{H}-\overset{\delta-}{Cl}$$

the δ+ and δ– (read "delta plus" and "delta minus") signify that one end (H) is partially positive and the other end (Cl), partially negative. The term *partial* charge signifies something less than the full charges of the ions that would result from complete electron transfer. In the other method,

$$\overset{+\!\!\longrightarrow}{H-Cl}$$

the cross-based arrow indicates the direction of displacement of negative charge, from the less to the more electronegative atom. Note that the "+" of the δ+ and "+" on the tail of the cross-based arrow are always associated with the *positive* end of the bond—the less electronegative atom.

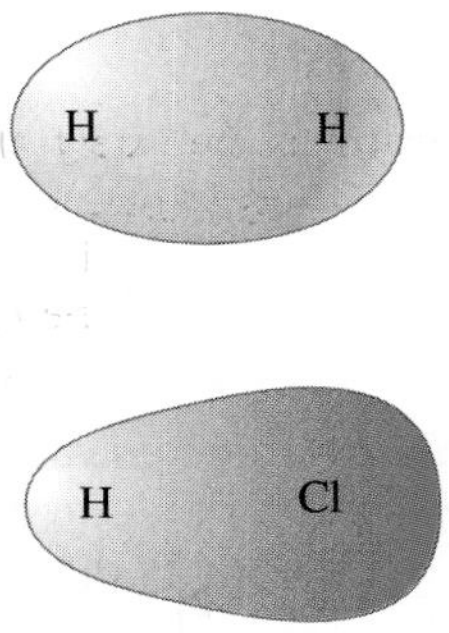

Figure 9.7 Nonpolar and polar covalent bonds. In H—H there is an even distribution of electron charge density between the atoms. In H—Cl electron charge density is displaced toward the Cl atom.

9.6 Additional Examples of Covalent Bonding

In our discussion of covalent bonding to this point, we have considered only molecules with one bond between two atoms in diatomic molecules such as H_2, Cl_2, and HCl. Whether the bonds are nonpolar, as in H_2 and Cl_2, or polar, as in HCl, we write Lewis structures that conform with the octet rule. An additional rule is useful when writing Lewis structures for *polyatomic* molecules: Nonmetallic atoms, particularly those of the second period, often form a number of covalent bonds equal to *eight minus the group number*. Thus, oxygen (Group 6A) should form two bonds; nitrogen (Group 5A), three bonds; and carbon (Group 4A), four bonds. The following three examples illustrate this rule.

Water

An oxygen atom shares two pairs of electrons in order to acquire an octet in its valence shell, but a hydrogen atom shares only one pair. As a result, an oxygen atom must bond with two hydrogen atoms.

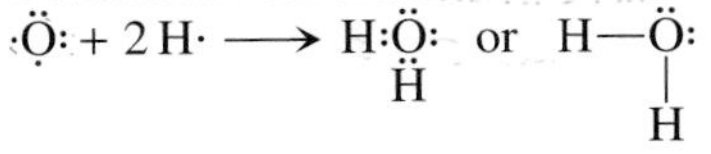

Consequently, the oxygen atom acquires the electron configuration of neon, and the hydrogen atoms, that of helium.

We have chosen to represent the water molecule with the hydrogen atoms arranged at an angle rather than on a straight line with the oxygen atom. This arrangement conforms to the bent shape of the water molecule (see Figure 9.8) that has been determined *by experiment*. You are not expected to predict molecular shapes yet, so you need not concern yourself with the different possible arrangements of Lewis symbols in Lewis structures. In fact, there is no requirement that Lewis structures must suggest the correct molecular geometry, so writing the structure as H—$\ddot{\underset{..}{\text{O}}}$—H is acceptable.

Figure 9.8 **Three polyatomic molecules.** Space-filling models of H_2O, NH_3, and CH_4 (left to right) give an idea of the shapes of the electron-charge clouds in these molecules. The molecular formulas can be deduced by writing Lewis structures, but Lewis structures alone do not permit us to predict the shapes of these molecules. Molecular shapes are discussed in Section 10.1.

Ammonia

A nitrogen atom has five valence electrons. It can assume the neon electron configuration by sharing three pairs of electrons with *three* hydrogen atoms to form ammonia (see Figure 9.8).

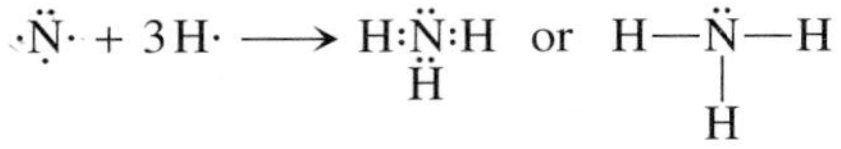

Methane

A carbon atom has four valence electrons. It can assume the neon configuration by sharing electron pairs with four hydrogen atoms to form methane (see Figure 9.8).

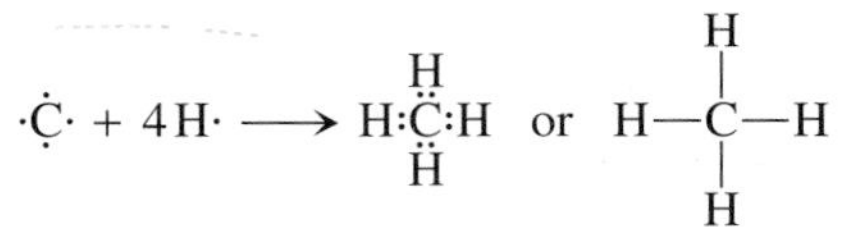

Coordinate Covalent Bonds

In all cases we have examined so far, covalent bonds are formed through the contribution of one electron each by the atoms involved in an electron-pair bond. In some cases, though, one atom provides *both* electrons of the shared pair. A bond formed in this way is called a **coordinate covalent bond**.

In a water molecule, the oxygen atom shares electrons with two hydrogen atoms, and it has two lone pairs of electrons. When an acid (a substance capable of donating hydrogen ions, H^+) is added to water, in some of the water molecules a lone pair of electrons on an oxygen atom accepts an H^+ and forms a *third* bond to a hydrogen. In a case like this, the positive charge of the hydrogen ion is transferred to the product, H_3O^+. That is, the positive charge is associated with the polyatomic ion as a whole and not with any particular atom.

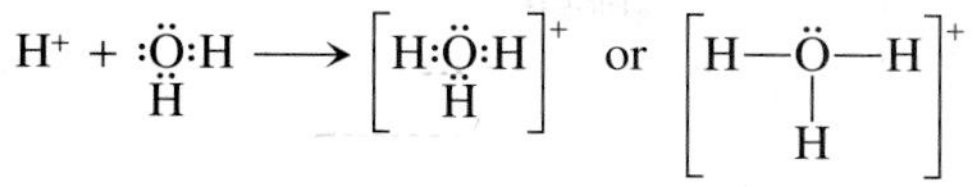

The H_3O^+ ion is called the *hydronium ion*. We will describe it further in Chapter 17, but until then, we will continue to represent acidic solutions as $H^+(aq)$.

$$H^+ + \overset{\displaystyle .}{\underset{\displaystyle H}{:\ddot{O}:H}} \longrightarrow \left[\overset{\displaystyle }{\underset{\displaystyle H}{H:\ddot{O}:H}} \right]^+ \quad or \quad \left[\overset{\displaystyle }{\underset{\displaystyle H}{H-\overset{\displaystyle }{\underset{\displaystyle |}{\ddot{O}}}-H}} \right]^+$$

In a similar manner, the nitrogen atom of ammonia can form a fourth bond to a hydrogen ion, H^+, through its lone pair of electrons to give the ammonium ion, NH_4^+.

$$H^+ + \overset{\displaystyle H}{\underset{\displaystyle H}{:\ddot{N}:H}} \longrightarrow \left[\overset{\displaystyle H}{\underset{\displaystyle H}{H:\ddot{N}:H}} \right]^+ \quad or \quad \left[\overset{\displaystyle H}{\underset{\displaystyle H}{H-\overset{\displaystyle |}{\underset{\displaystyle |}{N}}-H}} \right]^+$$

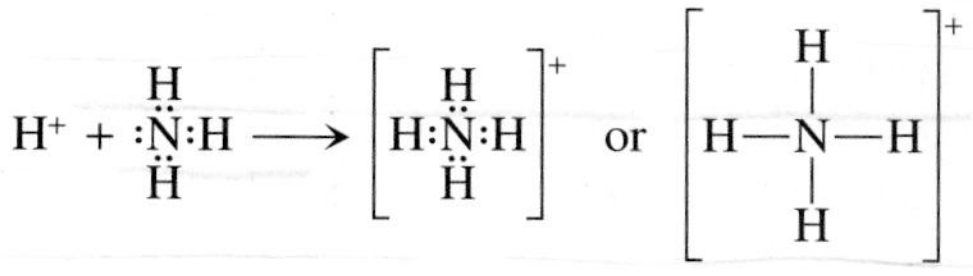

Once it is formed, the coordinate covalent bond is really indistinguishable from other covalent bonds. All bonding pairs of electrons are the same, and the term coordinate covalent refers only to how the bond is formed.

Multiple Covalent Bonds

Applying what we have learned so far, our first attempt at a Lewis structure for carbon dioxide, CO_2, might be

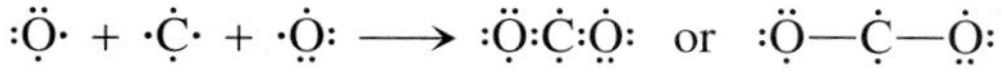

$$:\ddot{O}\cdot + \cdot\dot{C}\cdot + \cdot\ddot{O}: \longrightarrow :\ddot{O}:\ddot{C}:\ddot{O}: \quad or \quad :\ddot{O}-\dot{C}-\ddot{O}: \; (not\ correct)$$

But this structure is unsatisfactory. None of the atoms has acquired a valence-shell octet. By shifting the four unpaired electrons into the regions between the C and O atoms, however, we obtain a satisfactory structure in which each atom does have a valence-shell octet.

$$:\ddot{O}\frown\ddot{C}\frown\ddot{O}: \longrightarrow :\ddot{O}=C=\ddot{O}:$$

Each O atom is joined to the C atom by a **double bond**, a covalent linkage in which two atoms share *two* pairs of electrons between them.

Our first attempt at a Lewis structure for the nitrogen molecule, N_2, looks equally bad.

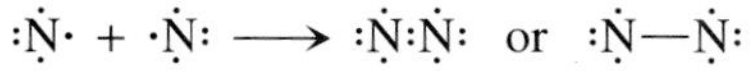

$$:\dot{N}\cdot + \cdot\dot{N}: \longrightarrow :\dot{N}:\dot{N}: \quad or \quad :\dot{N}-\dot{N}: \; (not\ correct)$$

Again we can produce an octet for each nitrogen atom by shifting all the unpaired electrons into the region between the two N atoms.

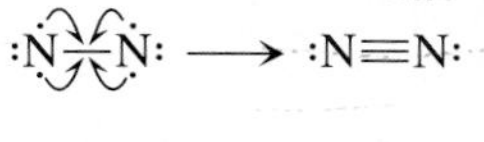

These electron manipulations may seem like sleight of hand, but we'll consider a more systematic approach to writing Lewis structures later in the chapter.

$$:\dot{N}\frown\dot{N}: \longrightarrow :N\equiv N:$$

In doing so we obtain a **triple bond**, a covalent linkage in which two atoms share *three* pairs of electrons between them.

Altogether, then, the possibilities for covalent bonds are a **single bond**—the sharing of one pair of electrons between the bonded atoms—and a **multiple bond**—the sharing of two pairs of electrons (to form a double bond) or three pairs of electrons (to form a triple bond). The electron pairs can also be represented by dashes, a single dash for a single bond, two parallel dashes for a double bond, and three for a triple bond.

H—Cl	O=C=O	N≡N
single bond	double bonds	triple bond
	multiple bonds	

The Importance of Experimental Evidence

We can write what seems to be a good Lewis structure for the O_2 molecule using the same approach to covalent bonding that we used for CO_2 and N_2. Doing so, we quickly arrive at a structure with a double bond.

> Interestingly, the strength of the O_2 bond indicates that it is a double bond, but no simple Lewis structure adequately describes all aspects of the bonding in the O_2 molecule. We present a method of describing the molecule that does conform to all the experimental evidence in Section 10.5.

$$:\dot{\underset{..}{O}}\cdot + \cdot\dot{\underset{..}{O}}: \longrightarrow :\underset{..}{O}=\underset{..}{O}: \quad (\textit{not correct})$$

This Lewis structure conforms to the octet rule, but it is *not* consistent with an important property of the O_2 molecule. Oxygen is *paramagnetic*; the O_2 molecule must contain unpaired electrons. It is important to note, then, that no matter how plausible a Lewis structure may be, we cannot accept it as the true structure unless it conforms to all the available *experimental* evidence.

Different types of experimental evidence can be used to establish the validity of a Lewis structure, and among the most important are bond lengths and bond energies, described in the next section.

9.7 Bond Lengths and Bond Energies

The electron charge density associated with shared electron pairs is concentrated in the region between the nuclei of the bonded atoms. The greater the attractive forces between the positively charged nuclei and the bonding electrons, the more tightly the atoms are joined. Thus, we can think of shared electrons as the glue that binds atoms together in covalent molecules. Although bond strengths vary for different combinations of atoms and for different numbers of shared electrons, there is a general pattern for a bond between two particular atoms: *the more electrons in the bond, the more tightly the atoms are held together.* We use the term **bond order** to indicate whether a covalent bond is *single* (bond order = 1), *double* (bond order = 2), or *triple* (bond order = 3). Often we can figure out bond orders by writing plausible Lewis structures, but also, if we have experimental evidence that is pertinent to bond order, we should use it when we write a Lewis structure.

Two important properties of bonds are related to the bond order: bond length and bond energy. **Bond length** is the distance between the nuclei of two atoms joined by a covalent bond. For a particular pair of atoms, a single bond has a certain bond length that usually varies only slightly when the bond appears in different molecules. A double bond between the same two atoms is shorter than a single bond (the atoms are more tightly bound). A triple bond is shorter still. This relationship is clearly shown in Table 9.1. For example, compare the three different bond lengths for nitrogen-to-nitrogen single, double, and triple bonds.

TABLE 9.1 Some Representative Bond Lengths and Bond Energies

Bond	Bond Length, pm	Bond Energy, kJ/mol	Bond	Bond Length, pm	Bond Energy, kJ/mol
H—H	74	436	C—O	143	360
H—C	110	414	C═O	120	736[a]
H—N	100	389	C—Cl	178	339
H—O	97	464	N—N	145	163
H—S	132	368	N═N	123	418
H—F	92	565	N≡N	110	946
H—Cl	127	431	N—O	136	222
H—Br	141	364	N═O	120	590
H—I	161	297	O—O	145	142
C—C	154	347	O═O	121	498
C═C	134	611	F—F	143	159
C≡C	120	837	Cl—Cl	199	243
C—N	147	305	Br—Br	228	193
C═N	128	615	I—I	266	151
C≡N	116	891			

[a] The value for the C—O bond in CO_2 is considerably different: 799 kJ/mol.

Recall that we introduced a particular kind of atomic radius—the *single covalent radius*—in Section 7.7. Now you can appreciate its meaning: Because it is one-half the distance between the nuclei of identical atoms when they are joined by a *single* covalent bond, the single covalent radius is *one-half the single covalent bond length.* If you look once more at Figure 7.5, you will see that the single covalent radius of an iodine atom is about 133 pm. That value is one-half the I—I bond length given in Table 9.1, that is, $\frac{1}{2} \times 266$ pm. Furthermore, even for *unlike* atoms, as a *rough* generalization we can say that

The length of the covalent bond joining two atoms is the sum of the covalent radii of the two atoms.

At times, experimentally determined bond lengths can help us to choose the best Lewis structure for a molecule. But also, as illustrated in Estimation Example 9.1, we can estimate a bond length from a plausible Lewis structure.

ESTIMATION EXAMPLE 9.1

Estimate the indicated bond lengths.

a. The nitrogen-to-nitrogen bond in N_2H_4.

b. The bromine-to-chlorine bond in BrCl.

Solution

a. Table 9.1 lists three N-to-N bond lengths. Which do we use? First draw a plausible Lewis structure for the molecule to see whether the N-to-N bond is single, double, or triple; and then select the appropriate value from Table 9.1.

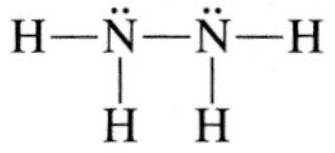

The N-to-N bond is a single bond, and so we expect its bond length to be 145 pm.

b. The Br-to-Cl bond length is not given in Table 9.1. Here we can use the approximation that the covalent bond length is the sum of the covalent radii of the atoms. To get the Lewis structure of BrCl, we imagine substituting one Br atom for one Cl atom in the Lewis structure of Cl_2.

$$:\ddot{\underset{..}{Br}}—\ddot{\underset{..}{Cl}}:$$

The BrCl molecule therefore contains a Br—Cl single bond. The length of the Br—Cl bond is approximately one-half the Cl—Cl bond length *plus* one-half the Br—Br bond length. From Table 9.1 we get $(\frac{1}{2} \times 199) + (\frac{1}{2} \times 228) = 214$ pm.

Estimation Exercise 9.1

Estimate the indicated bond lengths.

a. The nitrogen-to-nitrogen bond in CH_3NNCH_3. (*Hint*: Draw a plausible Lewis structure first.)

b. The oxygen-to-fluorine bond in OF_2.

Bond Energy

Energy must be *absorbed* to break a covalent bond. We define **bond dissociation energy (*D*)** as the quantity of energy required to break one mole of covalent bonds between atoms, with both the reactants and the products *in the gas phase.* These energies are generally expressed in kilojoules per mole of bonds (kJ/mol) and compiled in tables such as Table 9.1.

Bond dissociation energies are easy to understand for *diatomic* molecules, because there is only one bond (be it single, double, or triple) per molecule. We can represent bond dissociation energy (or bond dissociation enthalpy) as an enthalpy change or a heat of reaction. The enthalpy change for the reverse reaction, in which a bond is formed, is the *negative* of the bond dissociation energy. For example,

Bond breaking: $Cl_2(g) \longrightarrow 2\,Cl(g) \qquad \Delta H = D(Cl—Cl) = +243$ kJ/mol
Bond forming: $2\,Cl(g) \longrightarrow Cl_2(g) \qquad \Delta H = -D(Cl—Cl) = -243$ kJ/mol

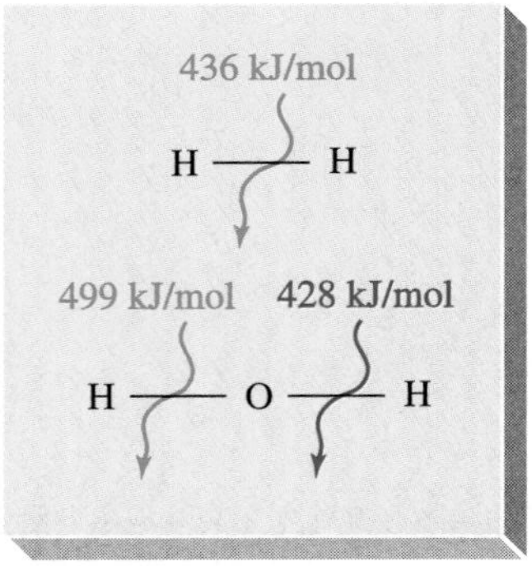

Figure 9.9 Some bond energies compared. The same quantity of energy (436 kJ/mol) is required to break all H—H bonds. In water more energy is require to break the first O—H bond (499 kJ/mol) than to break the second (428 kJ/mol). The value of the O—H bond energy in Table 9.1 is an average value based on water and other compounds that contain the O—H bond.

With a polyatomic molecule, such as H_2O, the situation is different. As shown in Figure 9.9, the energy required to dissociate one mole of H atoms by breaking one O—H bond per molecule

$$H—OH(g) \longrightarrow H(g) + OH(g) \qquad \Delta H = +499 \text{ kJ/mol}$$

is different from the energy required to dissociate a second mole of H atoms by breaking the remaining O—H bond in OH(g).

$$O—H(g) \longrightarrow H(g) + O(g) \qquad \Delta H = +428 \text{ kJ/mol}$$

The value for the O—H bond energy listed in Table 9.1 is the average of these two.

As we can readily see from Table 9.1, there is a relationship between bond order and bond dissociation energy. A double bond between a given pair of atoms has a higher bond dissociation energy than does a single bond between the same atoms, and a triple bond has a still higher bond dissociation energy. What is not apparent from Table 9.1 is that the bond dissociation energy of a given bond depends on the *environment* of that bond; that is, on the other atoms in the molecule that are near the bond. The bond dissociation energy of the O—H bond is therefore somewhat different in H—O—H from that in H—O—O—H and in H_3C—O—H. For many purposes, it is customary to use *average* values. An **average bond energy** is the average of the bond dissociation energies for a number of different molecules containing the particular bond. In Table 9.1, the values for bonds found in stable diatomic molecules, such as H—H, H—Cl, and Cl—Cl, are bond dissociation energies. For other bonds, such as H—C, H—N, and H—O, the values are average bond energies. In the remainder of this discussion we will use only average bond energies.

Calculations Involving Bond Energies

For a reaction *in the gas phase*, we can imagine a process in which all the bonds in the reactant molecules are broken and a completely new set of bonds is formed in the product molecules. The sum of the enthalpy changes for breaking the old bonds and forming the new bonds yields the enthalpy change, ΔH, for the reaction. That is,

$$\Delta H = \Delta H_{\text{bonds broken}} + \Delta H_{\text{bonds formed}}$$

If in this expression we use *average* bond energies (BE), then

$$\Delta H_{\text{bonds broken}} = \text{BE(reactants)} \qquad \text{and} \qquad \Delta H_{\text{bonds formed}} = -\text{BE(products)}$$

Estimation Example 9.2 illustrates the use of these expressions.

ESTIMATION EXAMPLE 9.2

Use average bond energies from Table 9.1 to estimate the enthalpy of formation of ammonia. Compare the result with the tabulated value of $\Delta H_f^\circ[NH_3(g)]$.

Solution

As shown in Figure 9.10, we can write Lewis structures, and from these structures we can determine what bonds are broken and what bonds are formed.

ΔH for bond breakage:

$$1 \text{ mol N}\equiv\text{N bonds} = 946 \text{ kJ}$$

$$3 \text{ mol H—H bonds} = 3 \times 436 = 1308 \text{ kJ}$$

$$\text{Sum for bonds broken} = 2254 \text{ kJ}$$

ΔH for bond formation:

$$6 \text{ mol H—N bonds} = 6 \times (-389) = -2334 \text{ kJ}$$

$$\text{Enthalpy change for the reaction, } \Delta H = 2254 \text{ kJ} - 2334 \text{ kJ} = -80 \text{ kJ}$$

The reaction in Figure 9.10 involves the formation of *two* moles of $NH_3(g)$: $2\Delta H_f^\circ[NH_3(g)] = -80$ kJ.

$$\Delta H_f^\circ[NH_3(g)] = -40 \text{ kJ/mol } NH_3(g)$$

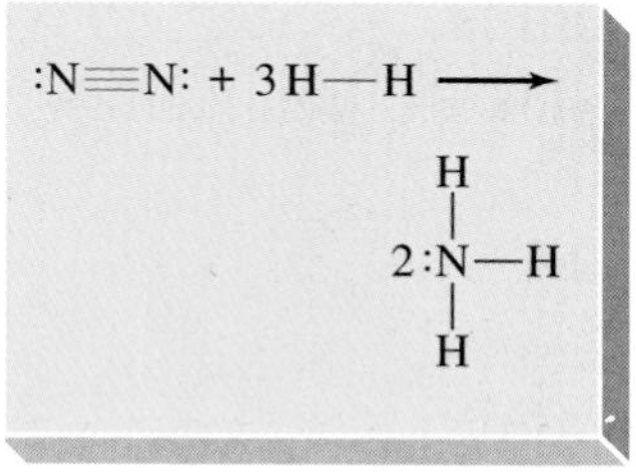

Figure 9.10 Bond breakage and formation in the ammonia synthesis reaction. The energies of the highlighted bonds are used in the estimation of $\Delta H_f^\circ[NH_3(g)]$ in Estimation Example 9.2.

This estimation from average bond energies is reasonably close to the tabulated value of $\Delta H_f^\circ[NH_3(g)]$, which is –46.11 kJ/mol NH_3.

Estimation Exercise 9.2

Estimate ΔH for the reaction

$$C_2H_6(g) + Cl_2(g) \longrightarrow C_2H_5Cl(g) + HCl(g)$$

The result of Estimation Example 9.2 clearly suggests that enthalpy changes calculated from average bond energies are neither as precise nor as accurate as those calculated from standard enthalpies of formation (Section 5.7). Average bond energies are most useful in applications where the necessary thermodynamic data are not available, or where rough estimates will suffice.

9.8 Strategies for Writing Lewis Structures

We are now armed with several important concepts of chemical bonding: the octet rule, multiple bonds, coordinate covalent bonds, electronegativity, bond lengths, and bond energies. Those concepts, together with one or two others introduced in this section, provide us with a strategy for writing a broad range of Lewis structures.

Whenever possible, a Lewis structure should be checked to see if it is consistent with *experimental evidence*. Nevertheless, the strategy we introduce here gives us a reasonable assurance that a Lewis structure is *plausible*, even if we lack experimental data.

Skeletal Structures

Just by applying the octet rule, we know that the arrangement of atoms in an H_2O molecule is HOH, not HHO. In many cases, however, we may be less certain about the proper arrangement of atoms in the skeletal structure. The **skeletal structure** of a polyatomic ion or molecule indicates the order in which the atoms are attached to one another. It usually consists of one or more central atoms and at least two terminal atoms. A **central atom** is bonded to two or more atoms in the structure, and a **terminal atom** is bonded only to one other atom. In writing a skeletal structure, we apply the idea that every atom must be connected to the rest of the structure by at least one bond. We do this by joining the bonded atoms by single dashes and making no attempt to account for all the valence electrons. The skeletal structure is *not* a Lewis structure; it is rather the *first step* in deducing a plausible Lewis structure. In the absence of experimental evidence, the following observations help us to devise likely skeletal structures.

1. Hydrogen atoms are nearly always terminal atoms; they form only one bond. This is a consequence of the fact that a bonded hydrogen atom can have only two electrons in its valence shell. Thus, in ethane, the two

carbon atoms are central atoms, and the six hydrogen atoms are terminal atoms.

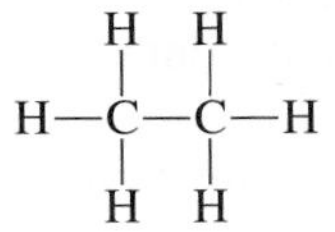

Exceptions to this observation are found in boron–hydrogen compounds, such as B_2H_6. In B_2H_6, two of the H atoms are bonded to both B atoms while the other four H atoms are terminal atoms. The very unusual bonding in B_2H_6 and related compounds is considered in Chapter 21.

2. In polyatomic molecules and ions the central atom(s) usually has the lowest electronegativity, and the terminal atoms have higher electronegativities. (The reasons for this will become clear later in this section.) Hydrogen is an exception; it is always a terminal atom, even when bonded to a more electronegative atom. Electronegativity (EN) values are indicated in the skeletal structure below for phosgene ($COCl_2$).

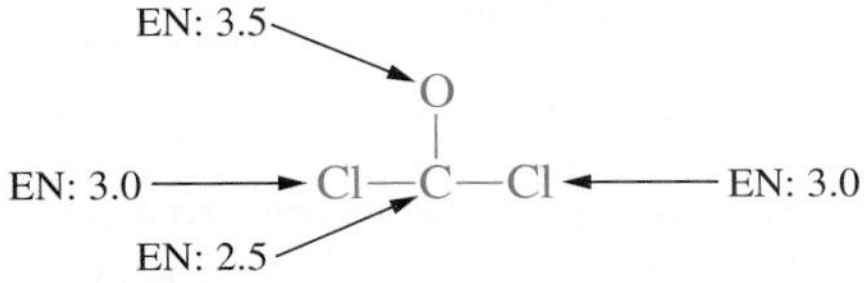

3. In oxoacids, hydrogen atoms are usually bonded to oxygen atoms.

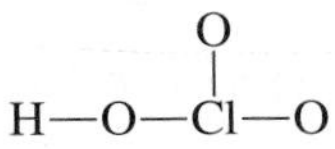

Chloric acid ($HOClO_2$)

4. With the major exception of carbon compounds, in which long chains of carbon atoms are common, polyatomic molecules and ions usually form compact structures.

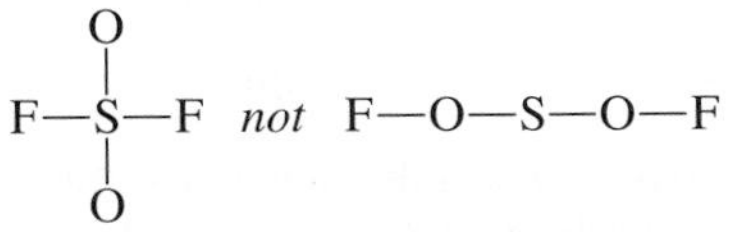

Sulfuryl fluoride (SO_2F_2)

5. Occasionally, the application of these rules fails to identify a particular skeletal structure as the most plausible. In these cases additional information may be necessary.

A Method for Writing Lewis Structures

After choosing a skeletal structure for a polyatomic molecule or ion, follow these five steps.

1. Determine the total number of valence electrons. This is the number of electrons that must appear in the final Lewis structure. The total number for a molecule is the sum of the valence electrons for each atom. For a polyatomic anion, which has one or more extra electrons, add one electron for each unit of negative charge. For a polyatomic cation, which is missing one or more electrons, subtract one electron for each unit of positive charge. Ex-

amples: N_2O_4 has $(2 \times 5) + (4 \times 6) = 34$ valence electrons; NO_3^- has $5 + (3 \times 6) + 1 = 24$ valence electrons; and NH_4^+ has $5 + (4 \times 1) - 1 = 8$ valence electrons.

2. Write the skeletal structure, and connect bonded atoms with an electron-pair bond (a dash).
3. Place electron pairs around terminal atoms so that each (except hydrogen) has an octet.
4. Assign any remaining electrons as lone pairs around the central atom(s).
5. If at this point a central atom has fewer than eight electrons, a multiple bond(s) is likely. Move one or more lone pairs of electrons from a terminal atom(s) to a region between it and the central atom to form a double or triple bond.

Example 9.6

Write the Lewis structure of nitrogen trifluoride, NF_3.

Solution

1. The total number of valence electrons is $5 + (3 \times 7) = 26$.
2. The skeletal structure has nitrogen—the less electronegative atom—as the central atom (item 2 on page 331).

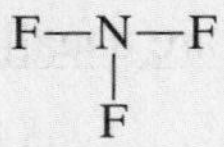

3. Complete the octets of the fluorine atoms by placing three lone pairs on each.

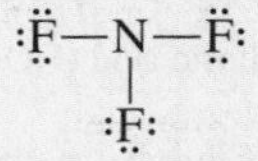

4. We have now assigned 24 of the 26 valence electrons. Place the remaining two as a lone pair on the N atom.

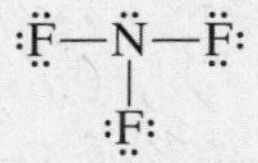

Each atom now has an octet. This is the Lewis structure of NF_3. Step 5 of the general method (dealing with multiple bonds) is not needed.

Exercise 9.6

Write the Lewis structure of ethyl chloride, C_2H_5Cl.

Example 9.7

Write a plausible Lewis structure for the poisonous gas phosgene, $COCl_2$, used in the manufacture of plastics.

Solution

1. The total number of valence electrons is $4 + 6 + (2 \times 7) = 24$.
2. The electronegativities of the elements are C (2.5), O (3.5), and Cl (3.0). We expect the *least* electronegative atom—carbon—to be the central atom. The expected skeletal structure is

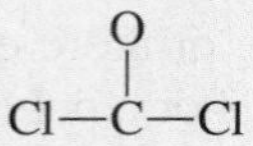

3. Complete the octets of the oxygen and chlorine atoms by placing three lone pairs on each

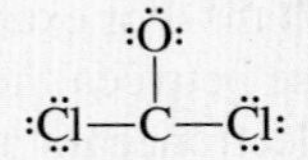

4. We have now assigned all 24 valence electrons. There are none available to complete the octet of the central C atom.
5. To complete the octet of the C atom, shift the indicated lone-pair electrons from the O atom into the bonding region to form a C-to-O double bond.

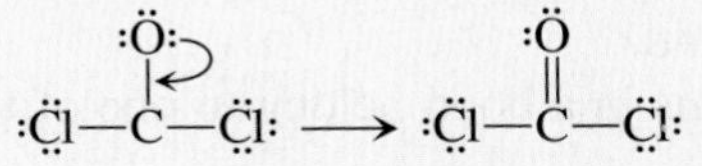

Exercise 9.7

Write a plausible Lewis structure for carbonyl sulfide, COS.

Note that the way in which a formula is commonly written may not correspond to the skeletal structure. The formula $OCCl_2$ is more revealing of the skeletal structure of phosgene than is $COCl_2$.

Double bonds form most readily to C, N, O, and S atoms (see page 342). This is one reason that we place the double bond between the C and O atoms and not between the C and Cl atoms. We will discover another reason in Example 9.9.

Example 9.8

Write a plausible Lewis structure for the chlorate ion, ClO_3^-.

Solution

1. The total number of valence electrons is $7 + (3 \times 6) + 1 = 26$. (Note that the electron gained to form this polyatomic anion is counted as one of the valence electrons.)
2. The electronegativities of the elements are Cl (3.0) and O (3.5). We expect the *less* electronegative chlorine atom to be the central atom. The expected skeletal structure is

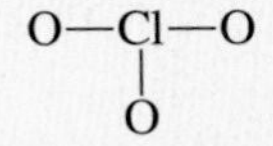

3. Complete the octets of the oxygen atoms by placing three lone pairs on each.

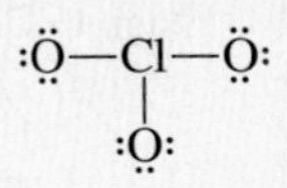

4. We have now assigned 24 of the 26 valence electrons. Place the remaining two electrons as a lone pair on the central Cl atom.

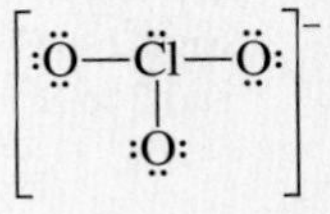

Each atom now has an octet. This is a plausible Lewis structure of ClO_3^-. As in Example 9.6, Step 5 of the general method (dealing with multiple bonds) is not needed.

Exercise 9.8

Write a plausible Lewis structure for the phosphonium ion, PH_4^+. (*Hint*: Note that this is a cation.)

This Lewis structure of ClO_3^- is entirely plausible according to the rules we've considered to this point. In Section 9.9, we'll consider an alternate Lewis structure based on some additional ideas.

Formal Charge

A concept known as *formal charge* can help us to write some Lewis structures that might otherwise present problems. For example, to write a Lewis structure for carbon disulfide we have to choose between the skeletal structures CSS and SCS. Because both C and S have an electronegativity of 2.5, the rule on choosing the central atom based on electronegativity doesn't apply.

Suppose that in addition to requiring valence shell octets on each of the atoms in a Lewis structure, we also assign the valence electrons to particular atoms. We do this as follows.

1. All unshared (lone pair) electrons are the sole property of the atom on which they are found.
2. Half of the electrons in a bond belong to one of the bonding atoms, half to the other.
3. After assigning all the valence electrons, determine whether any of the atoms has a formal charge.

Formal charge is the number of valence electrons in a free neutral atom minus the number of electrons assigned to that atom in a Lewis structure. We can express the formal charge (FC) of a bonded atom in this way:

$$\text{FC} = \text{no. valence electrons in free atom} - \tfrac{1}{2}\,[\text{no. bonding (shared) electrons}] - (\text{no. lone pair electrons})$$

However, you may find the method outlined in Figure 9.11 to be the simplest approach.

Where formal charges are found, they can be designated as shown below for structure (b) in Figure 9.11.

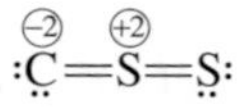

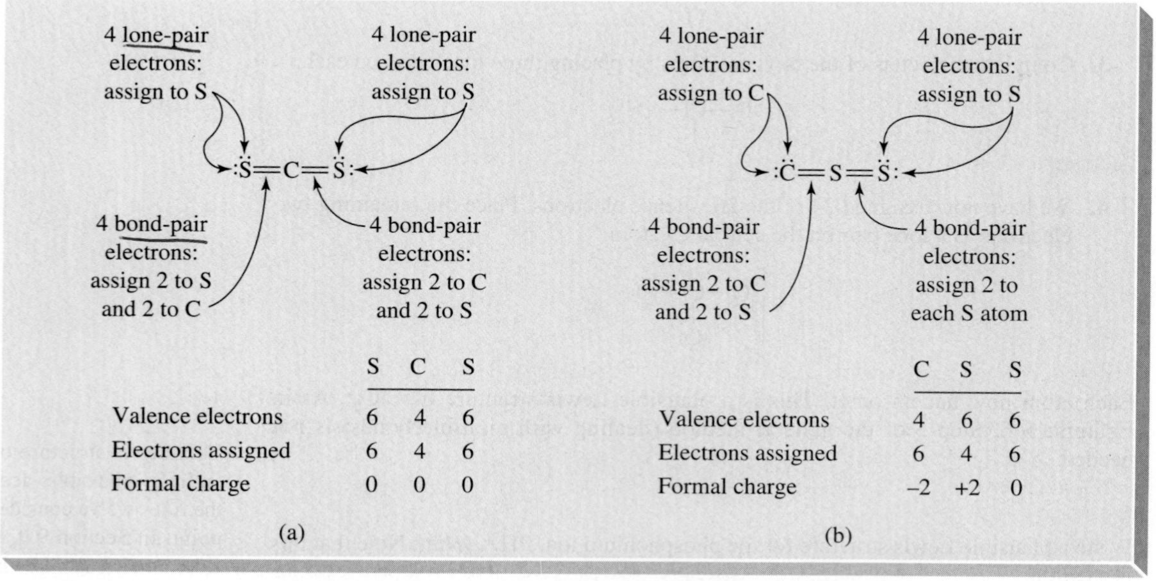

Figure 9.11 The concept of formal charge illustrated. Lewis structure (a) is more plausible than (b) because it has no formal charges.

Formal charges are *hypothetical*; the individual atoms in a covalent molecule do not carry these charges in actuality. We use small encircled numbers for formal charges to distinguish them from actual charges.

If all the covalent bonds in a Lewis structure involve equal contributions of electrons by the bonded atoms, there are no formal charges. In other words, formal charges arise only if coordinate covalent bonds are involved. *In applying the formal charge concept, the basic idea is to keep formal charges to a minimum.* We can achieve this goal with a few simple rules.

- Usually, the most plausible Lewis structure is one with no formal charges (that is, with formal charges of zero on all atoms).
- Where formal charges are required, they should be as small as possible, and negative formal charges should appear on the most electronegative atoms.
- Adjacent atoms in a structure should not carry formal charges of the same sign.
- Formal charges on the atoms in a Lewis structure must total to *zero* for a neutral molecule and to the net charge for a polyatomic ion.

When we apply the formal charge concept to the $COCl_2$ molecule (Example 9.9), we confirm that the most likely Lewis structure has the least electronegative atom as its central atom and that the double bond in the structure is C-to-O and not C-to-Cl.

Example 9.9

In Example 9.7 we wrote structure (a) below for the molecule $COCl_2$. Show that structure (a) is more plausible than (b) or (c).

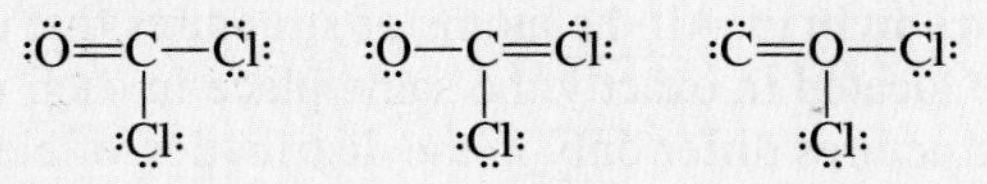

(a) (b) (c)

Solution

Assign formal charges to the atoms in each structure in the manner outlined in Figure 9.11. Then evaluate the results in accordance with the rules stated above. The formal charges in each structure are indicated below.

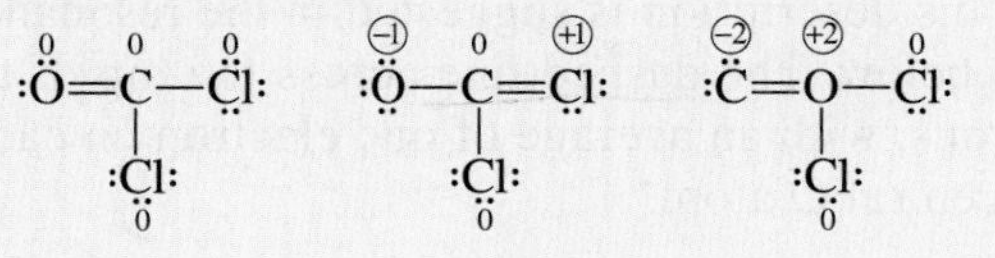

(a) (b) (c)

Structure (a) is most plausible because it has no formal charges. Structure (b), with its C-to-Cl double bond, is less plausible; it has formal charges. Structure (c) is the least plausible because the central O atom has the highest rather than the lowest electronegativity and because it has a positive formal charge. In structure (c) the negative formal charge appears on the *least* electronegative atom.

Exercise 9.9

Nitrosyl chloride, NOCl, is present in *aqua regia*, a mixture of concentrated nitric and hydrochloric acids capable of dissolving gold. Write the best Lewis structure that you can for NOCl.

Resonance: Delocalized Bonding

When we apply the general strategy for Lewis structures to the ozone molecule, O_3, we quickly get the structure

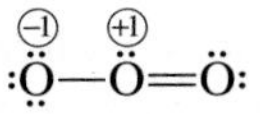

This structure implies that one of the O-to-O bonds should be a single bond and the other a double bond. However, experimental data show that the two O-to-O bonds are identical, each with a length of 128 pm. The bonds are intermediate between a single and a double bond, and we cannot write a single Lewis structure consistent with this fact. The best we can do is to write some plausible Lewis structures and try to "average" these in our minds into a composite or hybrid.

(I) (II)

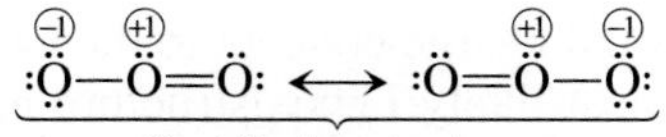

Contributing structures

Consider this analogy: A mule is a hybrid of a horse and a donkey. This does not mean that a mule is half-horse and half-donkey, nor does it mean that a mule is a horse half of the time and a donkey the rest. The mule is a distinctive animal that has some of the characteristics of a horse and some of a donkey.

This description of O_3 is called **resonance**, a circumstance in which two or more plausible Lewis structures can be written and the true structure is a composite or hybrid of them. The different structures used to represent the molecule or ion are called **contributing structures** or **resonance structures**, and we write them linked by double-headed arrows. The actual species that exists is called a **resonance hybrid**. All the atoms of structures that contribute to a resonance hybrid are located in exactly the same place in each resonance structure. The resonance structures differ only in the distribution of electrons.

In molecules such as H_2O, NH_3, and CH_4, in which resonance is not involved, we describe bonding electron pairs as existing in fairly well defined regions between two atoms. These electrons are *localized*. In O_3, to produce O-to-O bonds that are intermediate between single and double bonds, we need to think of some of the electrons in the resonance hybrid as being *delocalized*. Delocalized electrons are bonding electrons that are spread out over several atoms. This description is suggested in the resonance hybrid structure of ozone shown below. The dashed line across the top of the structure represents four electrons, with an average of one electron on each terminal O atom and two on the central O atom.

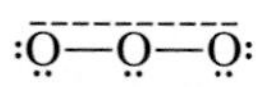

Resonance hybrid

Whenever you are able to write two or more plausible Lewis structures based on the same skeletal structure, you should suspect that the true Lewis structure is a resonance hybrid. Generally, you will be asked to write one or more of these resonance structures but not to depict the resonance hybrid.

Example 9.10

Write three equivalent structures for the SO_3 molecule that obey the octet rule. Describe the resonance hybrid of these three structures.

Solution

1. There are 6 + (3 × 6) = 24 valence electrons.
2. The skeletal structure is

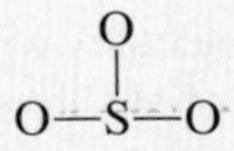

3. Place three lone pairs of electrons on each O atom.

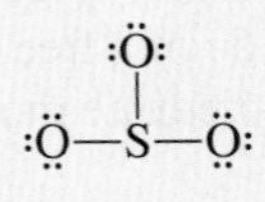

4. We have assigned all 24 electrons.
5. The central S atom has only six valence electrons shown. Move a lone pair of electrons from a terminal O atom to form a double bond to the central S atom. Because the double bond can go to any one of the three O atoms, we get three structures that differ only in the position of the double bond.

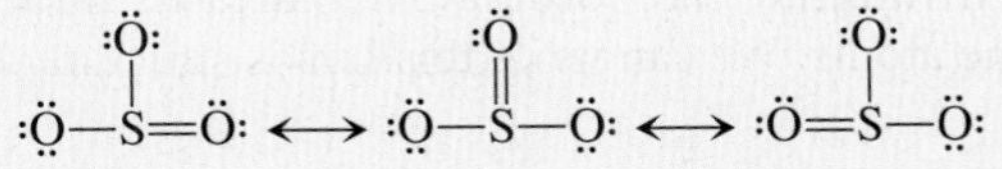

The SO_3 molecule is a resonance hybrid of these three contributing structures. The three sulfur-to-oxygen bonds are identical; each is intermediate between a single and a double bond.

Exercise 9.10

Write three equivalent Lewis structures for the nitrate ion, NO_3^-. Describe its resonance hybrid structure.

9.9 Molecules That Don't Follow the Octet Rule

Most molecules made of atoms of the main group elements can be satisfactorily represented by Lewis structures that follow the octet rule. There are exceptions, however, and these fall into three categories. Each type is readily identified by some structural characteristic.

Odd-Electron Molecules

In all Lewis structures we have written to this point we have dealt with electrons in *pairs*, either as bonding pairs or as lone pairs. In Lewis structures with an *odd* number of valence electrons, it is not possible for all the valence electrons to be in pairs, nor is it possible to satisfy the octet rule for all the atoms. Examples are nitrogen monoxide, NO, with 5 + 6 = 11 valence electrons and nitrogen dioxide, NO_2, with 17 valence electrons.

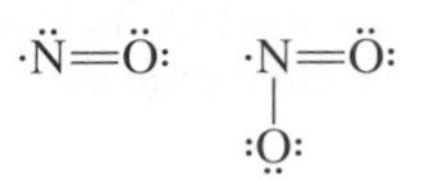

These oxides of nitrogen are major components of smog. Chlorine dioxide, ClO_2, whose molecules have 19 valence electrons, is made in multi-ton quantities and is used for bleaching flour and paper.

There are only a few stable molecules with odd numbers of electrons. Most of the other species with odd numbers of electrons are unstable fragments of molecules called **free radicals.** Generally, free radicals are highly reactive and have only a fleeting existence as intermediates in chemical reactions. An important free radical in the atmosphere is the *hydroxyl* radical (·OH). Often, a bold dot is used to represent the unpaired electron(s) characteristic of a free radical, as in this equation for the reaction of a hydroxyl radical with a methane molecule to produce a *methyl* radical.

$$\cdot OH(g) + CH_4(g) \longrightarrow \cdot CH_3(g) + H_2O(g)$$

Incomplete Octets

Boron atoms have three valence electrons; fluorine atoms have seven. In the molecule boron trifluoride, the boron atom shares its three valence electrons with three fluorine atoms. We can write this Lewis structure

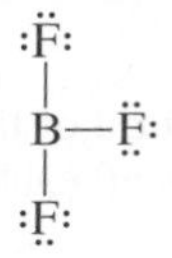

which accounts for all 24 valence electrons but does not complete the octet of the central B atom. Nevertheless, we should note that this structure is quite acceptable according to the formal charge rules—all atoms have a formal charge of 0. We can also write a Lewis structure that conforms to the octet rule through the use of a B-to-F double bond.

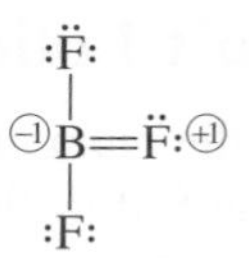

There are actually three equivalent structures with B-to-F double bonds. However, there might be some initial objection to these structures because they place positive formal charges on the more electronegative fluorine atoms. Nevertheless, experimental evidence suggests that the best representation of BF_3 is a resonance hybrid of four contributing structures.

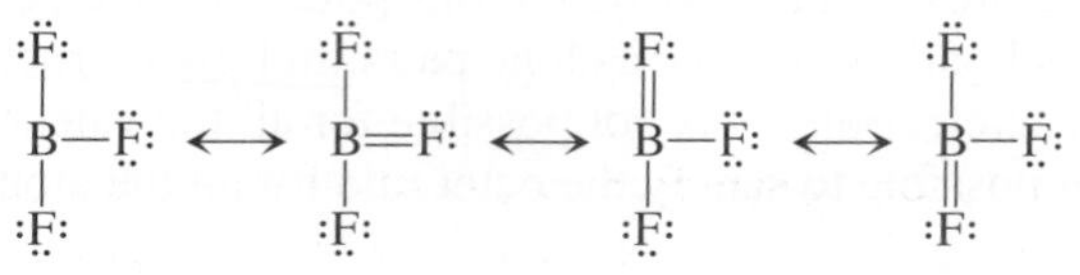

(I) (II) (III) (IV)

The high chemical reactivity of BF_3 suggests that the central boron atom is electron deficient, as in structure (I). For example, BF_3 readily reacts to form a coordinate covalent bond with a lone pair of electrons, such as those provided by a fluoride ion.

The primary uses of BF_3, an important industrial chemical, are based, not on its boron or fluorine content, but on properties related to its electron structure.

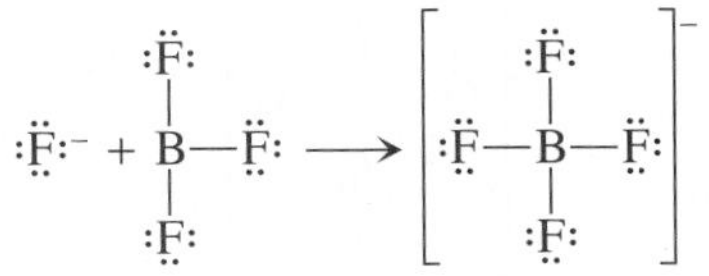

On the other hand, the B-to-F bond length in BF_3 is much shorter than in the single-bonded BF_4^- ion (130 pm compared to 145 pm). This suggests that some double bonding actually does occur in BF_3.

In conclusion, we might say that BF_3 probably exists as a resonance hybrid of the four contributing structures shown above, with perhaps the dominant structure being the one with an incomplete octet on the boron atom. This conclusion adds an important dimension to our description of resonance: It is not necessary that the contributing structures to a resonance hybrid all be equivalent (as they are in O_3 and SO_3). Also, as revealed through experimental evidence, one or more resonance structures may contribute more heavily to the resonance hybrid structure than do others.

Expanded Octets

The second-period elements carbon, nitrogen, oxygen, and fluorine nearly always obey the octet rule, with odd-electron molecules as obvious exceptions. The valence electron shell of the second-period elements holds a maximum of eight electrons ($2s^22p^6$). Although the third period ends with argon ($3s^23p^6$), the third principal shell can hold up to 18 electrons ($3s^23p^63d^{10}$). Apparently, third-period elements can depart from the octet rule by having more than eight electrons in their valence shells. These so-called **expanded octets** are employed in the following structures.

Phosphorus pentachloride   Sulfur hexafluoride

Although many of their compounds still follow the octet rule, nonmetal atoms in the third period and beyond are not limited to an octet. And sometimes Lewis structures based on expanded octets are in better agreement with experimental data than are those based on the octet rule. Consider the case of chlorine monoxide, Cl_2O, and perchlorate ion, ClO_4^-. A Lewis structure of Cl_2O with O as the central atom has valence shell octets for all three atoms and no formal charges.

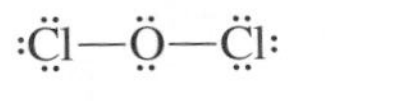

Let's assume that this satisfactory Lewis structure for Cl_2O describes the normal Cl—O single bond. Experimental evidence indicates that the Cl—O bonds in Cl_2O have a bond length of 170 pm.

The Lewis structure for the perchlorate ion shown below conforms to the octet rule but has formal charges. The positive formal charge on the Cl atom is particularly high.

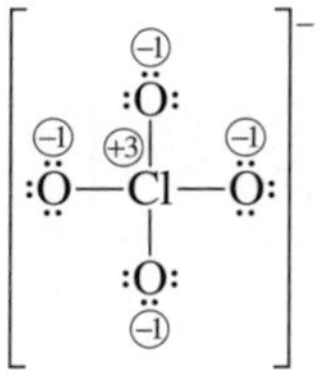

Similarly, the ClO_3^- ion of Example 9.8 can also be described as a resonance hybrid with some of the contributing structures, such as

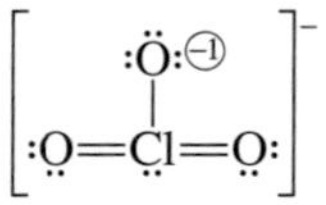

having expanded octets and reduced formal charges

Experimental evidence indicates that the O—Cl bond lengths in this ion are 144 pm. These shorter bonds suggest that the O—Cl bond has some multiple bond character. With the concept of expanded octets, we can write a number of Lewis structures having double bonds and reduced formal charges. The true structure is a resonance hybrid of many structures, not all of which are equivalent or equally important. One example of a possible contributing structure is

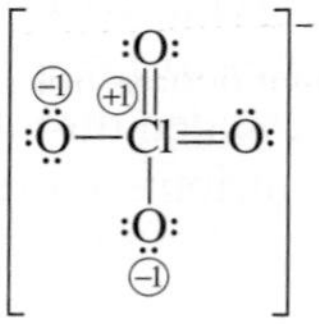

Example 9.11

Write the Lewis structure for bromine pentafluoride, BrF_5.

Solution

1. There are 7 + (5 × 7) = 42 valence electrons.
2. The skeletal structure is

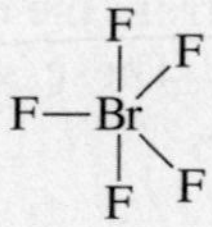

3. Complete the octet of each F atom with three lone pairs of electrons.

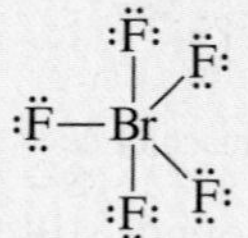

4. We have assigned 40 electrons. Two remain to be placed. They can be shown as a lone pair on the bromine atom. In this representation, the Br atom acquires an expanded octet of 12 valence electrons.

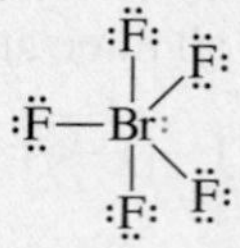

(Notice that this structure has no formal charges.)

Exercise 9.11

Write the Lewis structure for chlorine trifluoride, ClF_3.

CONCEPTUAL EXAMPLE 9.1

Indicate the error involved in representing a molecule by each of the following Lewis structures. Replace each by a more acceptable structure(s).

a. :C≡N: **b.** 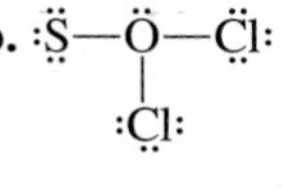 **c.** :Ö—S̈=Ö:

Solution

a. This structure conforms to the octet rule, but let's back up to the first step in the general strategy for writing Lewis structures—assessing the total number of valence electrons. This number is 4 (from C) + 5 (from N) = 9, not the 10 electrons shown. The structure shown here is not for a molecular species but for the cyanide *ion*, CN^-. It should be written

$$[:C{\equiv}N:]^-$$

b. If we assign formal charges to the atoms in this structure, we find them to be −1 for S and +1 for O. The less electronegative S atom should be the central atom, not the O atom. Furthermore, formal charges can be eliminated if we expand the octet of the S atom to a total of 10, as in the Lewis structure

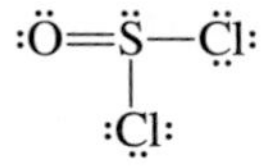

c. The Lewis structure has the correct skeletal structure and number of valence electrons, and each atom satisfies the octet rule. It is only one of two possible equivalent *resonance* structures. The other equivalent structure is

:Ö=S̈—Ö:

Notice that each of these structures has a formal charge of +1 on S and −1 on the singly bonded oxygen. Still another possible resonance structure, one that has no formal charges, is based on an expanded octet for the sulfur atom.

:Ö=S̈=Ö:

Conceptual Exercise 9.1

Only one of the following Lewis structures is correct. Identify that one and indicate the error(s) in the others.

a. chlorine dioxide, :Ö—C̈l—Ö:

b. hydrogen peroxide, H—Ö—Ö—H

c. dinitrogen difluoride, :F̈—N̈—N̈—F̈:

9.10 Unsaturated Hydrocarbons

Some of the most representative examples of covalent bonding are found in organic compounds. In Chapter 2 we learned to write structural formulas of alkane hydrocarbons and compounds derived from them. We can now see that the Lewis structure of an alkane is the same as its structural formula, when we represent electron pairs by dashes.

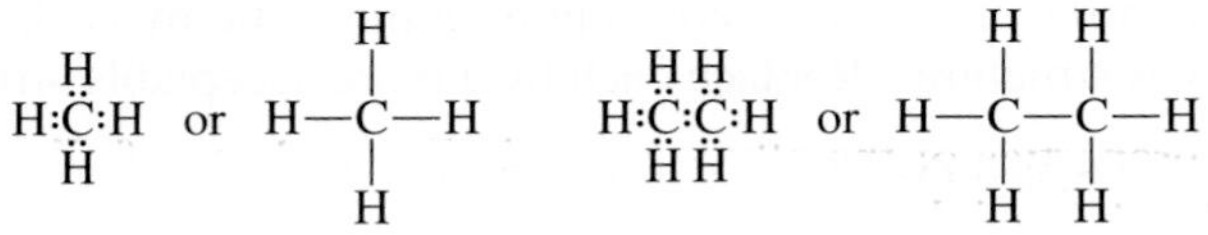

Methane Ethane

In the alkanes all carbon-to-carbon and carbon-to-hydrogen bonds are single bonds, and hydrocarbon molecules having only single bonds are called *saturated* hydrocarbons.

Multiple bonds are common in organic compounds. Collectively, hydrocarbons with double or triple bonds between carbon atoms are called **unsaturated hydrocarbons**. Carbon atoms can also form double bonds with nitrogen, oxygen, and sulfur atoms, and a triple bond with nitrogen atoms.

The Alkenes

The **alkenes** (note the *-ene* ending) are hydrocarbons whose molecules contain carbon-to-carbon double bonds. *Simple* alkenes have just one double bond in the molecule and the general formula C_nH_{2n}. Hydrocarbons with this formula constitute a homologous series parallel to the straight-chain and branched-chain alkanes, C_nH_{2n+2}. The simplest alkene is *ethene*, C_2H_4. We can see that this molecule must have a carbon-to-carbon double bond when we construct its Lewis structure from Lewis symbols.

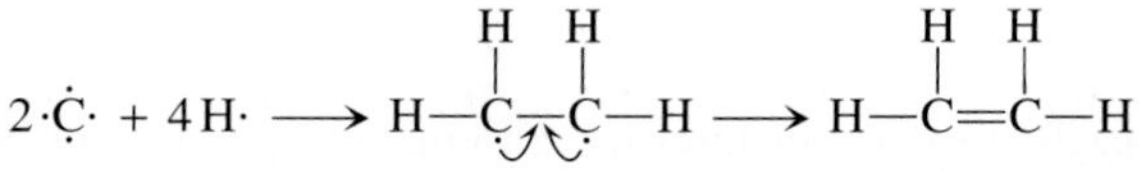

We can also represent this molecule with a structural formula or a condensed structural formula. The structural formula shows that the double bond is shared by the two carbon atoms and does not involve the hydrogen atoms. The condensed structural formula does not make this point quite as obvious, but it is the easier formula to set in type, requires less space on a printed page, and is consequently the most widely used.

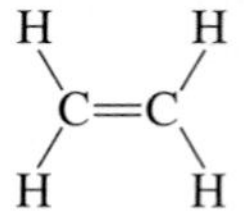

$CH_2{=}CH_2$ or $H_2C{=}CH_2$

Structural formula Condensed structural formula

Figure 9.12 Molecular model of ethene (ethylene).

A space-filling molecular model of ethene is shown in Figure 9.12.

Many simple alkenes have common, nonsystematic names. Ethene, the two-carbon alkene, is most often called *ethylene*. The three-carbon alkene, propene, C_3H_6, is frequently called *propylene*. Four different compounds—isomers—have the molecular formula C_4H_8. Whenever large numbers of isomers occur, common names are more difficult to deal with. Chemists resort to the IUPAC system described in Appendix C.

Alkenes and Alkanes Compared

The physical properties of the alkenes are similar to those of alkanes of corresponding molecular weight, as suggested by the following data.

	Ethene, C_2H_4	Ethane, C_2H_6
mp:	−169 °C	−172 °C
bp:	−104 °C	−88 °C

	Propene, C_3H_6	Propane, C_3H_8
mp:	−185 °C	−188 °C
bp:	−48 °C	−42 °C

Also, like alkanes (and all other hydrocarbons), the alkenes burn—that is, they undergo combustion. On the other hand, because they possess a double bond, alkenes are much more chemically reactive than alkanes.

In a typical *alkane* reaction, a hydrogen atom is replaced by a new incoming group, a type of reaction known as a **substitution reaction**. In the reaction of ethane with chlorine in sunlight, for example, a chlorine atom substitutes for a hydrogen atom.

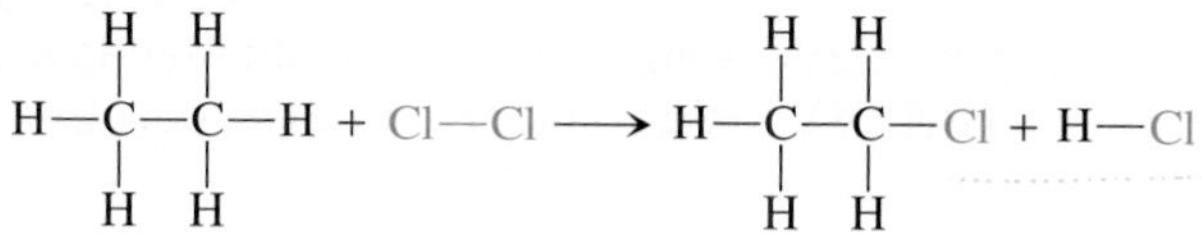

Ethane Ethyl chloride

Two molecules, ethane and chlorine, react to form two product molecules, ethyl chloride and hydrogen chloride.

The typical reactions of alkenes are **addition reactions**, reactions in which two reactant molecules combine or "add together" to give a single product molecule. In addition reactions, one of the electron pairs in a multiple bond is uncoupled and used to attach additional atoms or groups to the molecule. Here is the reaction of chlorine with ethylene to yield 1,2-dichloroethane.

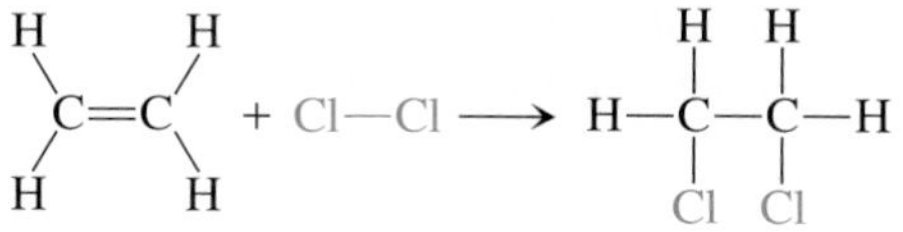

Ethene 1,2-Dichloroethane

Note that the product molecule no longer has a double bond; it is no longer an alkene but rather a halogenated alkane.

Fats and Oils: Hydrogenation

Chemically, fats are called **triglycerides** because a molecule of fat is composed of three fatty acids joined to a molecule of glycerol, a *trihydroxy alcohol*, $CH_2OHCHOHCH_2OH$. One classification of triglycerides is made on the basis of their physical states at room temperature. In general, a triglyceride is called a *fat* if it is a solid at 25 °C, and an *oil* if it is a liquid at that temperature. An important factor contributing to these differences in melting points is the degree of *un*saturation—that is, the number of carbon-to-carbon double bonds—of the constituent fatty acids.

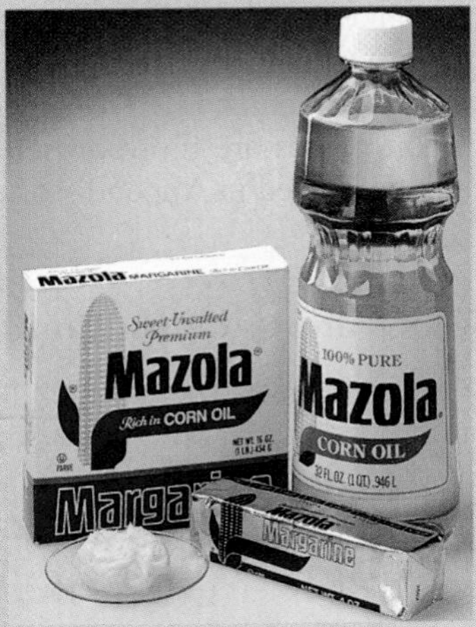

Unsaturated fatty acids make up about 85% of the mass of corn oil. To convert the oil to a solid (margarine), some of the unsaturated fatty acids must be converted to saturated fatty acids. This is accomplished through hydrogenation.

Saturated fatty acids contain no carbon-to-carbon double bonds; stearic acid, $C_{17}H_{35}COOH$, is a typical example.

$$CH_3CH_2CH_2CH_2CH_2CH_2CH_2CH_2CH_2CH_2CH_2CH_2CH_2CH_2CH_2CH_2CH_2\overset{\overset{\displaystyle O}{\|}}{C}{-}OH$$

Stearic acid (saturated)

Monounsaturated fatty acids contain one C-to-C double bond per molecule; oleic acid, $C_{17}H_{33}COOH$, is a common one.

$$CH_3CH_2CH_2CH_2CH_2CH_2CH_2CH_2CH{=}CHCH_2CH_2CH_2CH_2CH_2CH_2CH_2\overset{\overset{\displaystyle O}{\|}}{C}{-}OH$$

Oleic acid (monounsaturated)

Polyunsaturated fatty acids are those that have two or more C-to-C double bonds per molecule. Linolenic acid, $C_{17}H_{29}COOH$, with three such bonds, is a typical one.

Bromine, Br_2, forms brownish red solutions in carbon tetrachloride, CCl_4. When such a solution is added to an alkene, the color disappears because the bromine reacts with the alkene to form a colorless dibromoalkane.

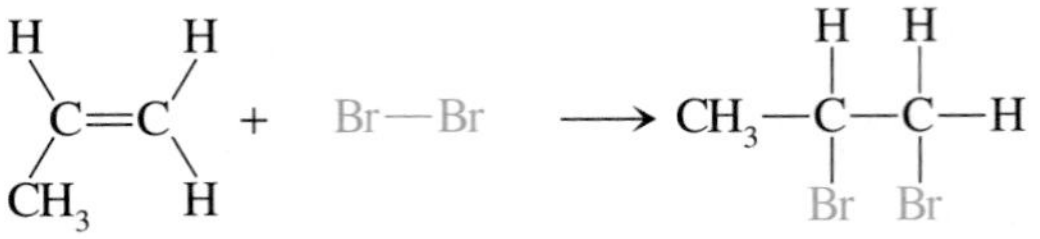

Propylene (colorless) + Bromine (brownish red) ⟶ 1,2-Dibromopropane (colorless)

$$CH_3CH_2CH{=}CHCH_2CH{=}CHCH_2CH{=}CHCH_2CH_2CH_2CH_2CH_2CH_2CH_2\overset{\overset{\large O}{\|}}{C}{-}OH$$

Linolenic acid (polyunsaturated)

Saturated fats contain a high proportion of saturated fatty acids; the fat molecules have relatively few double bonds. *Polyunsaturated fats* (oils) incorporate mainly unsaturated fatty acids; these fat molecules have many double bonds. Triglycerides obtained from animal sources are usually solids, whereas those of plant origin are generally oils. Therefore, we speak of *animal fats* and *vegetable oils*. Coconut and palm oils, which are highly saturated, and fish oils, which are relatively unsaturated, are notable exceptions to the general rule.

Saturated fats have been implicated, along with cholesterol, in one type of arteriosclerosis (hardening of the arteries). There is a strong correlation that has led to a concern over the relative amounts of saturated and unsaturated fats in our diets. Advertisers who recommend that you buy corn oil margarine (prepared from relatively unsaturated vegetable oil) rather than butter (a relatively saturated animal fat) highlight this concern.

Unsaturated oils can be converted to more saturated ones by hydrogenation.

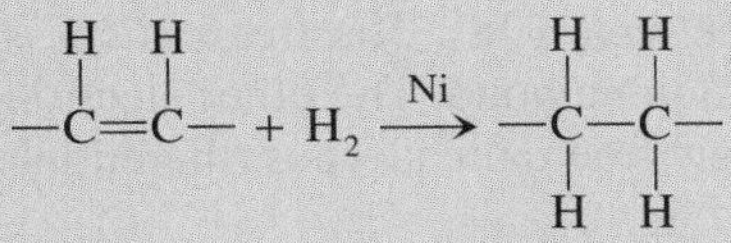

Margarine, a butter substitute, and vegetable shortening, a lard substitute, consist of vegetable oils that have been partially hydrogenated. (If all the bonds were hydrogenated, the product would become hard and brittle like tallow.) If reaction conditions are properly controlled, it is possible to prepare a fat with a desirable physical consistency (soft and pliable). In this manner, inexpensive and abundant vegetable oils (cottonseed, corn, soybean) are converted into oleomargarine and cooking fats (Crisco, for example). Of course, the consumer would get much greater unsaturation by using the oils directly, but most people would rather spread margarine than pour oil on their toast.

The decolorization of bromine solutions is frequently used as a simple test for the presence of alkenes.

A commercially important addition reaction is that of hydrogen, H_2, a reaction called *hydrogenation*. A catalyst is required for the reaction, usually nickel, platinum, or palladium.

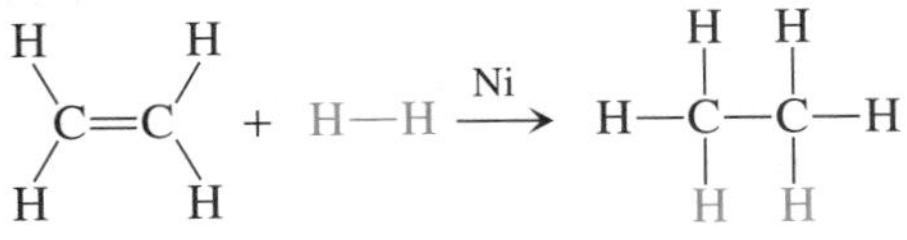

Ethene Ethane

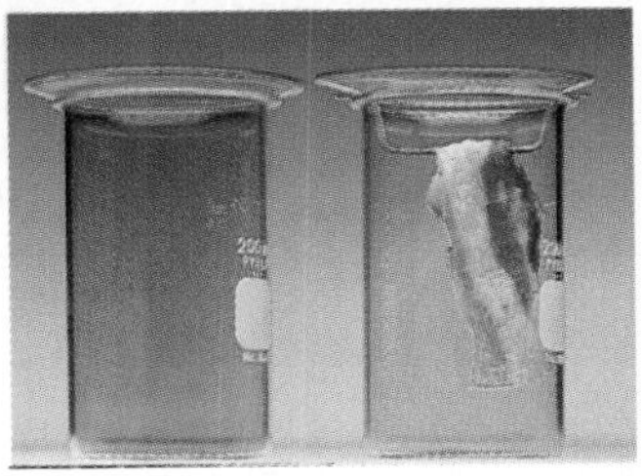

A strip of bacon (right) removes bromine vapor from a beaker (left). An addition reaction occurs between bromine molecules and unsaturated fatty acids in the bacon.

The product of this reaction is an alkane with the same carbon skeleton as the original alkene. An unsaturated molecule is converted to a saturated molecule.

Another important reaction of alkenes is the addition of water. This reaction, called *hydration*, also requires a catalyst—in this case a strong acid, usually sulfuric acid.

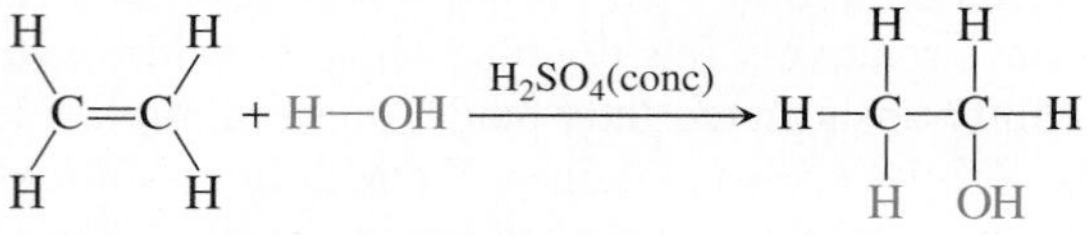

Ethene Ethyl alcohol

Vast quantities of ethyl alcohol for use as an industrial solvent are made in this way. Although the alcohol is identical to that used in alcoholic beverages, federal law requires that all beverage alcohol be produced by the natural process of fermentation.

Polymers

Perhaps the most important commercial reaction of alkenes is **polymerization**. In this reaction, one alkene molecule adds to another; the product is joined by a third alkene molecule; and so on. The reaction is started by an *initiator*, usually a free radical (see page 338). The free radical, symbolized as R·, pairs with one electron of the double bond. The other electron of that pair is now unpaired, giving rise to a new free radical. Thus, the initial step in the polymerization of ethylene is

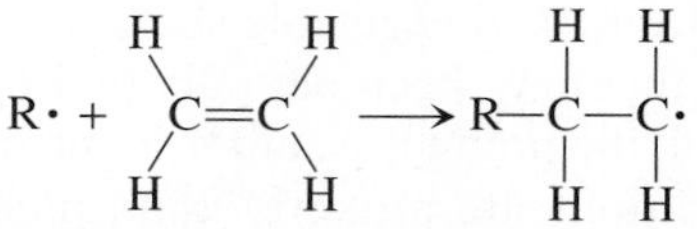

The new radical combines with another molecule of ethylene to form a larger free radical.

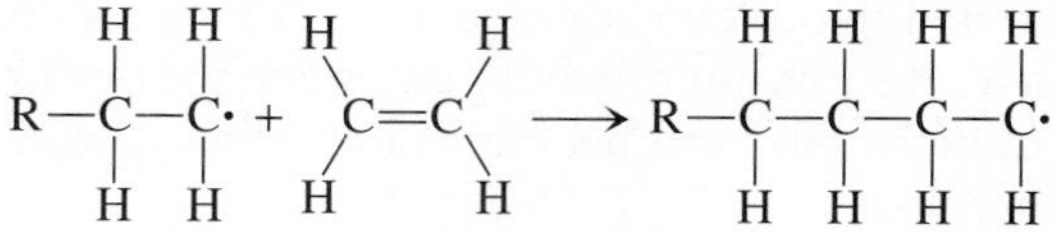

The process continues through many, many steps called *propagation* steps. Each step results in a new free radical, larger than the previous one by the unit CH_2CH_2. The process finally ends in a *termination* step, in which the product molecule no longer has an unpaired electron. One possible termination step is the combination of two radicals through an electron pair bond.

$$R(CH_2)_nCH_2\cdot + \cdot CH_2(CH_2)_nR' \longrightarrow R(CH_2)_nCH_2CH_2(CH_2)_nR'$$

The net result of this polymerization is a combination of hundreds or even thousands of ethylene molecules to form *polyethylene* (Figure 9.13).

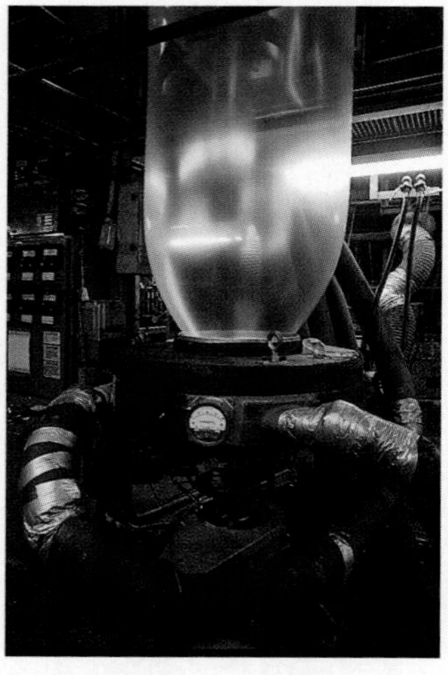

A giant bubble of tough, transparent plastic film emerges from the die of an extruding machine. The film is used in packaging, consumer products, and food service operations.

Figure 9.13 The formation of polyethylene. In the synthesis of polyethylene, many monomer units ($CH_2{=}CH_2$) join together to form huge polymer molecules. In this computer-generated representation, the yellow dots indicate a new bond forming as an ethylene molecule (upper left) is added to the growing polymer chain.

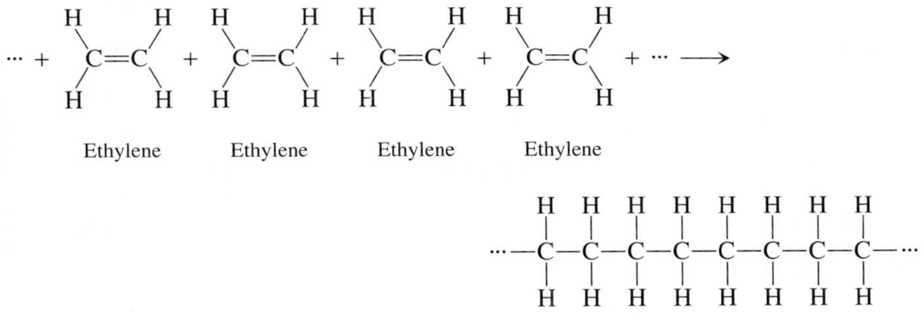

Polyethylene (polyethene)

Many small alkene units, called **monomers**, combine to produce a very large product molecule, a **polymer**. Polyethylene means many ethylene molecules combined. The name of the product is that of the starting material preceded by "poly." Even though it is called polyethylene, the product is not an alkene. This is an addition reaction, and in addition reactions the double bonds of the alkene starting material are converted to single bonds. Several of the many different possible alkene monomers and the polymers made from them are listed in Table 9.2.

Alkynes

In alkenes, carbon atoms in the double bond share two pairs of electrons. Carbon atoms can also share three pairs of electrons, forming triple bonds. Hydrocarbons whose molecules contain carbon-to-carbon triple bonds are called **alkynes**. The Lewis and structural formulas of the simplest alkyne are

$$H{:}C{:::}C{:}H \quad \text{or} \quad H{-}C{\equiv}C{-}H$$

The common name of this compound is *acetylene* (Figure 9.14). The IUPAC nomenclature parallels that for alkenes, except that the family ending is *-yne* rather than *-ene*. The official name for acetylene is *ethyne*.

The common name, acetylene, sounds very much like ethylene or propylene. Remember, however, that acetylene is an alkyne, whereas ethylene and propylene are alkenes.

TABLE 9.2 A Selection of Addition Polymers

Monomer	Polymer	Polymer Name	Some Uses
$H_2C{=}CH_2$	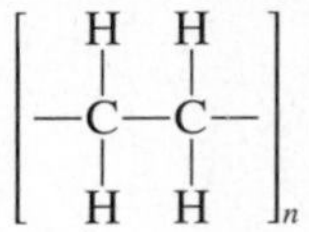	Polyethylene	Plastic bags, bottles, toys, electrical insulation
$H_2C{=}CH{-}CH_3$	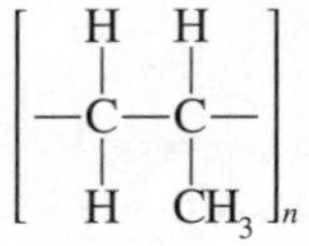	Polypropylene	Indoor–outdoor carpeting, bottles
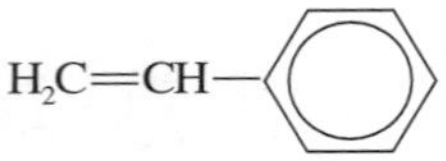 [a]	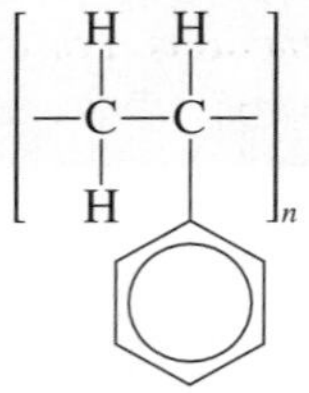	Polystyrene	Simulated wood furniture, styrofoam insulation and packing materials
$H_2C{=}CH{-}Cl$	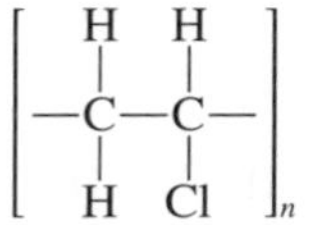	Poly(vinyl chloride), PVC	Plastic wrap, simulated leather (Naugahyde), phonograph records, garden hoses
$H_2C{=}CCl_2$	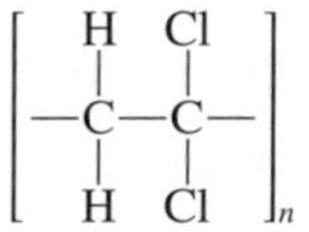	Poly(vinylidene chloride), Saran	Food wrap
$F_2C{=}CF_2$	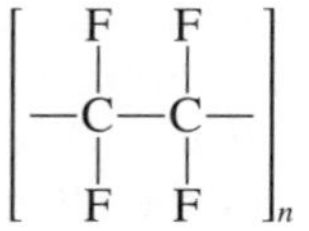	Polytetrafluoroethylene, Teflon	Nonstick coating for cooking utensils, electrical insulation

[a] The significance of the symbol  is discussed on page 350.

Figure 9.14 Molecular model of ethyne (acetylene).

Alkynes are similar to alkenes in most physical and chemical properties. Alkynes undergo addition reactions, but because either one or two pairs of electrons in the triple bond can participate, there is a possibility of adding twice as much of a reagent to an alkyne as to an alkene. For example, in hydrogenation, under carefully controlled conditions, acetylene can add one molecule of hydrogen to form ethene.

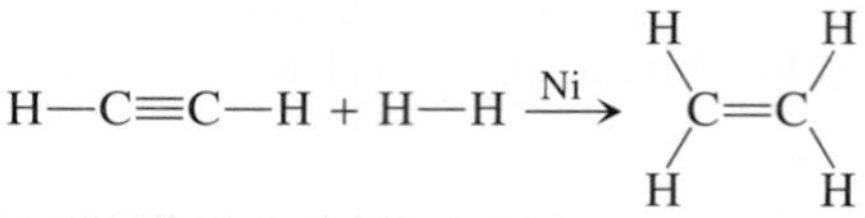

More likely, however, acetylene will add two molecules of hydrogen to form first ethene and then ethane. The overall reaction is

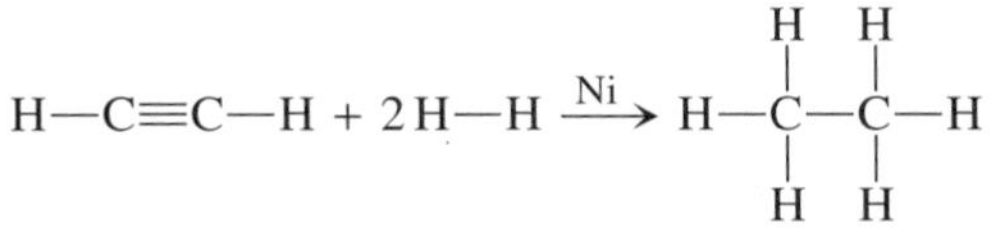

Benzene: Aromatic Compounds

Benzene, a liquid that smells like gasoline, boils at 80 °C and freezes at 5.5 °C. It was discovered by Michael Faraday in 1825, but the structure of the benzene molecule puzzled chemists for over a century. Its molecular formula suggests that it is an unsaturated hydrocarbon, but it does not undergo the addition reactions typical of the alkenes. Rather, like the alkanes, benzene undergoes substitution reactions. For example, with $FeCl_3$ as a catalyst, benzene reacts with chlorine to form chlorobenzene.

$$C_6H_6 + Cl_2 \xrightarrow{FeCl_3} C_6H_5Cl + HCl$$

In 1865, F. A. Kekulé proposed a cyclic structure for benzene that has six carbon atoms in a hexagonal ring with alternate single and double bonds between carbon atoms and a hydrogen atom bonded to each carbon atom. Two equivalent structures can be drawn that differ in the positions of the double bonds.

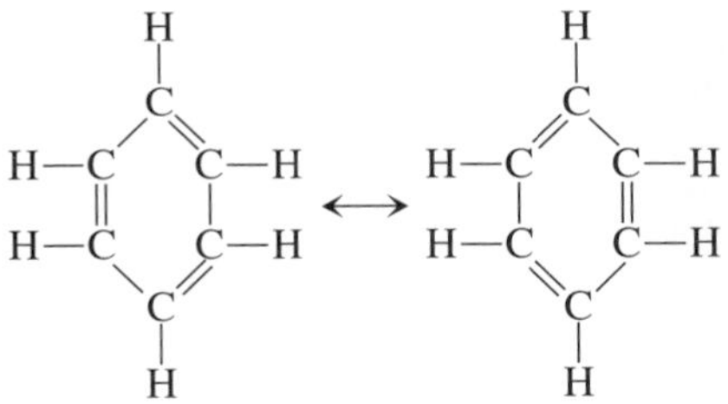

These Kekulé structures often are abbreviated as

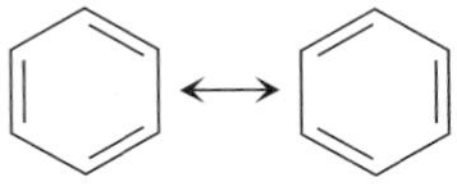

Kekulé structures still are inadequate because they describe an unsaturated hydrocarbon, and benzene doesn't act unsaturated.

We can use resonance theory to describe the true structure of benzene; it is a resonance hybrid that has the two Kekulé structures as its contributing structures. Instead of alternate double and single bonds, the bonds between carbon atoms in the real benzene molecule are all the same. They are intermediate in length and strength between double and single bonds. The six extra electrons that we ordinarily would place between carbon atoms to make three double bonds are spread out over all six carbon atoms; they are *delocalized*.

The modern symbol for benzene is simply a hexagon with a circle in it. Each corner stands for a carbon atom, with the sides representing bonds between carbon atoms. The circle symbolizes the six delocalized electrons shared by the ring as a whole. It is understood that there is a hydrogen atom on each carbon

Freidrich August Kekulé (1829–1896) claimed to have discovered the cyclic structure of benzene while dozing by a fire. (In some versions of the story, he was dozing on a bus.) He dreamed of atoms and molecules as snakes. Suddenly one of the snakes seized its own tail, forming a ring. Kekulé's other contributions included the idea that the carbon atom is tetravalent (1858) and an influential textbook in which he defined organic chemistry as the chemistry of carbon compounds.

atom, and these are not shown. If some other group were substituted for a hydrogen atom, its symbol(s) would be shown.

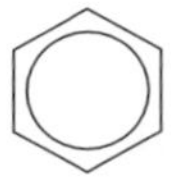

Many aromatic compounds, including benzene itself, are carcinogenic (cancer-causing). Modern workplace regulations severely restrict exposures to aromatic compounds.

Many of the first benzenelike compounds discovered had pleasant odors and hence acquired the name *aromatic*. In modern chemistry, the term **aromatic compounds** simply refers to compounds with structures and properties like those of benzene. Specifically, aromatic compounds have some delocalized bonding and tend to undergo substitution reactions. We will look again at bonding in the benzene molecule in Chapter 10 and refer to aromatic compounds from time to time in the remainder of the text.

SUMMARY

Lewis symbols of representative elements are related to their locations in the periodic table. The Lewis structure of an ionic compound combines the Lewis symbols of cations and anions. The net energy decrease in the formation of an ionic crystal from its gaseous ions is the lattice energy. The lattice energy and enthalpy of formation of an ionic compound, together with other atomic and molecular properties, can be related in a thermochemical cycle.

A covalent bond is created by the sharing of an electron pair between atoms. In a Lewis structure representing covalent bonds, electron pairs are either bonding pairs or lone pairs. Usually, each atom in a structure acquires a noble-gas electron configuration, which for many atoms is seen as a valence-shell octet of electrons.

In a covalent bond between atoms of different electronegativity (EN), electrons are displaced toward the atom with the higher EN. Electronegativity values are related to positions of the elements in the periodic table. In terms of electronegativity differences chemical bonds vary over the range: nonpolar to polar covalent to ionic.

In some cases of covalent bonding, one atom appears to provide both electrons in the bonding pair; the bond is coordinate covalent. In some cases, bonded atoms may share more than one pair of electrons between them, giving rise to *multiple* covalent bonding.

In the phenomenon of resonance, two or more Lewis structures have the same skeletal structure but different bonding arrangements. The best description of the actual structure (resonance hybrid) is obtained by combining plausible structures (contributing structures). The resonance hybrid must conform to experimental data, such as bond lengths and/or bond energies. Bond lengths can be estimated from atomic radii. Bond energies are usually tabulated as *average* values.

Exceptions to the octet rule are found in odd-electron molecules and molecular fragments called free radicals. A few structures appear to have too few electrons to complete all the octets. Some structures appear to have too many. In the latter case, a central atom may employ an "expanded" octet with five, six, or even seven electron pairs.

Unsaturated hydrocarbon molecules have one or more multiple bonds between carbon atoms. Alkenes have double bonds and alkynes, triple bonds. Alkenes differ from alkanes in chemical properties. Typical reactions of alkanes are *substitution* reactions, whereas those of alkenes and alkynes involve *addition*. In a special kind of addition reaction that involves free radicals, small alkene units (monomers) can combine into high molecular weight polymers.

Some unsaturated hydrocarbons, known as aromatic hydrocarbons, react by substitution rather than addition. Resonance is required in order to explain the structures of aromatic compounds by the Lewis theory, which indicates that some bonding electrons are *delocalized* throughout the molecule. The simplest aromatic hydrocarbon is benzene, C_6H_6.

KEY TERMS

addition reaction (9.10)
alkene (9.10)
alkyne (9.10)
aromatic compound (9.10)
average bond energy (9.7)
bond dissociation energy (*D*) (9.7)
bond length (9.7)
bond order (9.7)
bonding pair (9.4)
central atom (9.8)
chemical bond (Introduction)
contributing structure (9.8)
coordinate covalent bond (9.6)
covalent bond (9.4)

crystal (9.1)
double bond (9.6)
electron dot-symbol (9.2)
electronegativity (9.5)
expanded octet (9.9)
formal charge (9.8)
free radical (9.9)
ionic bond (9.1)
lattice energy (9.3)
Lewis structure (9.4)
Lewis symbol (9.2)
lone pair (9.4)
monomer (9.10)
multiple bond (9.6)
nonpolar (9.5)
octet rule (9.2)
polar (9.5)
polymer (9.10)
polymerization (9.10)
resonance (9.8)
resonance hybrid (9.8)
resonance structure (9.8)
single bond (9.6)
skeletal structure (9.8)
substitution reaction (9.10)
terminal atom (9.8)
triglyceride (9.10)
triple bond (9.6)
unsaturated hydrocarbon (9.10)

REVIEW QUESTIONS

1. Which group of elements in the periodic table is characterized by especially unreactive electron configurations?

2. What is the structural difference between a sodium ion and a neon atom? What is the similarity between them?

3. What are the structural differences among chlorine atoms, chlorine molecules, and chloride ions?

4. Write Lewis symbols for each of the following elements. You may use a periodic table.
a. sodium **b.** oxygen
c. fluorine **d.** aluminum

5. Write Lewis symbols for each of the following elements. You may use a periodic table.
a. silicon **b.** rubidium
c. calcium **d.** bromine
e. arsenic

6. Explain how Lewis symbols and *spdf* notation differ in their representation of electron spin.

7. Use *spdf* notation *and* Lewis symbols to represent the electron configuration of each of the following.
a. K^+ **b.** S^{2-}
c. F^- **d.** Al^{3+}

8. Use *spdf* notation and Lewis symbols to indicate the electron configuration of each of the following.
a. Mg^{2+} **b.** Cl^-
c. Li^+ **d.** N^{3-}

9. Use Lewis symbols to show the transfer of electrons (**a**) from calcium atoms to bromine atoms and (**b**) from magnesium atoms to sulfur atoms. In each case assume that the ions formed have noble gas electron configurations.

10. Use Lewis symbols to show the transfer of electrons (**a**) from aluminum atoms to sulfur atoms and (**b**) from magnesium atoms to phosphorus atoms. In each case assume that the ions formed have noble gas electron configurations.

11. What is meant by the lattice energy of an ionic compound?

12. Use Lewis symbols to show the sharing of electrons between two iodine atoms to form an iodine molecule. Label all electron pairs as bonding pairs or lone pairs.

13. Use Lewis symbols to show the sharing of electrons between a hydrogen atom and a fluorine atom. Label the ends of the molecule with symbols that indicate polarity.

14. Classify the following bonds as ionic or covalent. For those bonds that are covalent, indicate whether they are polar or nonpolar.
a. KF **b.** IBr **c.** MgS
d. NO **e.** CaO **f.** NaBr
g. Br_2 **h.** F_2 **i.** HCl

15. What is the electronegativity of an atom? How does it differ from electron affinity?

16. If atoms of the two elements in each set below are joined by a covalent bond, which atom will more strongly attract the electrons in the bond?
a. N and S **b.** B and Cl
c. As and F **d.** S and O

17. Using only the periodic table (inside front cover), indicate which element in each set is more electronegative.
a. Br or F **b.** Br or Se
c. Cl or As **d.** N or H

18. Why is it so difficult to find a proper place for hydrogen in the periodic table?

19. Why does neon tend not to form chemical bonds?

20. Draw an electron-charge cloud picture for the HF molecule. Use the symbols δ+ and δ– to indicate the polarity of the molecule.

21. How many single covalent bonds do each of the following atoms usually form in molecules that have only single covalent bonds? You may refer to the periodic table.
a. H **b.** C **c.** O
d. F **e.** N **f.** Br

22. What is a coordinate covalent bond? Can you tell which is the coordinate covalent bond in the ammonium ion? Explain.

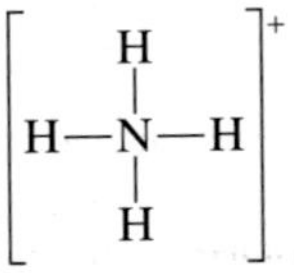

23. Use Lewis structures to show the formation of a coordinate covalent bond between BF_3 and NH_3.

24. Name four elements that readily form double bonds.

25. Name two elements that readily form triple bonds.

26. List the three main types of compounds that are exceptions to the octet rule. Give an example of each.

27. Where in the periodic table do you expect to find the elements whose atoms can accommodate expanded octets when forming covalent bonds?

28. Describe the phenomenon of resonance. What is the difference between a resonance structure and a resonance hybrid?

29. Are all C-to-H bonds in a molecule of the same strength? Is the same amount of energy required to break the first and second of these bonds in methane, CH_4? Explain.

30. What are unsaturated hydrocarbons?

31. Which of the following compounds are unsaturated hydrocarbons?
a. $HC{\equiv}CCH_3$
b. $CH_3CHCH{=}CH_2$
c. $HC{\equiv}CCH_2CH_3$
d. $CH_3CH_2CH_2CH_2CH_3$

32. Which of the compounds in Question 31 is (are) alkenes? Which is (are) alkynes?

33. Define and give an example of an addition reaction.

34. What is the major difference in properties between alkenes and alkanes? How is this difference related to the structures of the two types of compounds?

35. Define each of the following.
a. monomer **b.** polymer
c. polymerization **d.** free radical

36. What is the essential feature of the monomer molecules that undergo addition polymerization?

37. What is meant by a delocalized bond?

38. What are the Kekulé structures of benzene? What type of experimental evidence indicates that a single Kekulé structure is inadequate for describing the benzene molecule?

39. How is the resonance hybrid of benzene represented?

40. What is the chemical meaning of the term *aromatic compound*?

PROBLEMS

Ions and Ionic Bonding

41. Which of the following ions have noble gas electron configurations?
a. Cr^{3+} **b.** Sc^{3+} **c.** Zn^{2+}
d. Te^{2-} **e.** Zr^{4+} **f.** Cu^{+}

42. Use noble-gas-core-abbreviated *spdf* notation to give the electron configuration for the simple ion most likely to be formed by each of the following elements.
a. Ba **b.** K **c.** Se
d. I **e.** N **f.** Te

43. Use noble-gas-core-abbreviated *spdf* notation to write the electron configurations of Sn, Sn^{2+}, and Sn^{4+}.

44. Use noble-gas-core abbreviated *spdf* notation to write the electron configurations of Fe, Fe^{2+}, and Fe^{3+}. (*Hint:* 4*s* electrons are lost before 3*d*.)

45. Write Lewis structures for the following ionic compounds:
a. KI **b.** CaO **c.** $BaCl_2$ **d.** Rb_2S.

46. Write Lewis structures for the following ionic compounds:
a. potassium bromide **b.** sodium selenide
c. barium fluoride **d.** strontium nitride

Energetics of Ionic Bond Formation

47. The lattice energy of sodium fluoride is –914 kJ/mol NaF. Use this value together with other values found in Section 9.3 (including Example 9.3) to determine the enthalpy of formation of NaF(s). Compare your result with the value listed in Appendix D.

48. The lattice energy of potassium chloride is –701 kJ/mol KCl. The enthalpy of sublimation of K(s) is 89.24 kJ/mol, and the first ionization energy of K(g) is 419 kJ/mol. Use these values together with other values found in Section 9.3 to determine the enthalpy of formation of KCl(s). Compare your result with the value listed in Appendix D.

49. The enthalpy of formation of cesium chloride is

$$\text{Cs(s)} + \tfrac{1}{2}\text{Cl}_2\text{(g)} \longrightarrow \text{CsCl(s)} \qquad \Delta H^\circ = -442.8 \text{ kJ/mol}$$

The enthalpy of sublimation of cesium is

$$\text{Cs(s)} \longrightarrow \text{Cs(g)} \qquad \Delta H^\circ = 77.6 \text{ kJ/mol}$$

Use these data, with other data from Section 9.3, to calculate the lattice energy of CsCl(s).

50. The enthalpy of formation of sodium iodide is

$$\text{Na(s)} + \tfrac{1}{2}\text{I}_2\text{(s)} \longrightarrow \text{NaI(s)} \qquad \Delta H^\circ = -288 \text{ kJ/mol}$$

The enthalpy of sublimation of iodine is

$$\text{I}_2\text{(s)} \longrightarrow \text{I}_2\text{(g)} \qquad \Delta H^\circ = +62 \text{ kJ/mol}$$

Use these data, with other data from Section 9.3, to calculate the lattice energy of NaI(s).

Molecules: Single Covalent Bonds

51. Write Lewis structures for the simplest covalent molecules formed with the following atoms, assuming the octet rule (duet rule for hydrogen) is followed in each case.
a. P and H **b.** C and F

52. Write Lewis structures for the simplest covalent molecules formed with the following atoms, assuming the octet rule (duet rule for hydrogen) is followed in each case.
a. Si and H **b.** N and Cl

Electronegativity: Polar Covalent Bonds

53. Without referring to figures or tables in the text (other than the periodic table), arrange each of the following sets of atoms in their order of increasing electronegativity.
a. B, F, N **b.** As, Br, Ca **c.** C, O, Ga

54. Without referring to figures or tables in the text (other than the periodic table), arrange each of the following sets of atoms in their order of increasing electronegativity.
a. I, Rb, Sb **b.** Cs, Li, Na **c.** Cl, P, Sb

55. Use differences in electronegativity values to arrange each of the following sets of bonds in order of increasing polarity. Use the symbols δ+ and δ– to indicate partial charges, if any, in the bonds.
a. Cl—F, F—F, Br—F, H—F, I—F
b. H—Br, H—Cl, H—F, H—H, H—I

56. Use difference in electronegativity values to arrange each of the following sets of bonds in order of increasing polarity. Use the symbols δ+ and δ– to indicate partial charges, if any, in the bonds.
a. H—C, H—F, H—H, H—N, H—O
b. C—Br, C—C, C—Cl, C—F, C—I

57. Use appropriate data from Table 9.1 to estimate bond lengths for the following bonds.
a. I—Cl **b.** C—F

58. Use appropriate data from Table 9.1 to estimate the bond lengths for the following bonds.
a. O—Cl **b.** N—I

59. Use bond energies from Table 9.1 to estimate the enthalpy change (ΔH) for the reaction

$$\text{H}_2\text{(g)} + \text{F}_2\text{(g)} \longrightarrow 2\ \text{HF(g)}$$

60. Use bond energies from Table 9.1 to estimate the enthalpy change (ΔH) for the reaction

$$\text{CH}_4\text{(g)} + \text{Cl}_2\text{(g)} \longrightarrow \text{CH}_3\text{Cl(g)} + \text{HCl(g)}$$

Lewis Structures

61. Write Lewis structures that follow the octet rule for the following covalent molecules.
a. CH_3OH **b.** CH_2O **c.** NH_2OH
d. N_2H_4 **e.** COF_2 **f.** PCl_3

62. Write Lewis structures that follow the octet rule for the following covalent molecules:
a. NF_3 **b.** C_2H_2 **c.** C_2H_4
d. CH_3NH_2 **e.** H_2SiO_3 **f.** HCN

Formal Charges

63. Assign formal charges to each atom in each of the following structures. Based on these formal charges, which is the preferred Lewis structure for this ion, the cyanate ion?

$$[:\ddot{\text{C}}{=}\text{N}{=}\ddot{\text{O}}]^- \qquad [:\text{N}{\equiv}\text{C}{-}\ddot{\underset{..}{\text{O}}}:]^-$$

64. Assign formal charges to each atom in each of the following structures. Based on these formal charges, which is the preferred Lewis structure for this molecule, the dinitrogen monoxide molecule?

a. :N≡N—Ö: **b.** :N̈=N=Ö:

Resonance Structures

65. Write two equivalent resonance structures for the acetate ion, $CH_3CO_2^-$. The skeletal structure is

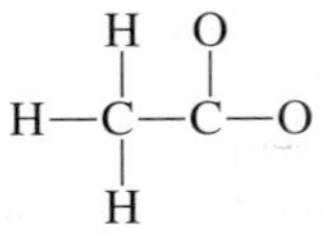

Describe the resulting resonance hybrid structure for the acetate ion.

66. Write two equivalent resonance structures for the bicarbonate ion, $HOCO_2^-$. Describe the resulting resonance hybrid structure for the bicarbonate ion. (*Hint:* All O atoms are bonded to the C atom.)

67. Write three equivalent resonance structures for the carbonate ion, CO_3^{2-}. Describe the resulting resonance hybrid structure for the carbonate ion.

68. Chlorobenzene (C_6H_5Cl) is derived from benzene by replacing one hydrogen atom with a chlorine atom. Draw the two Kekulé structures for chlorobenzene. Describe the resulting resonance hybrid structure for chlorobenzene.

More Lewis Structures

69. Write the simplest Lewis structure for each of the following molecules. Comment on any unusual features of the structures.

a. NO **b.** BCl_3
c. ClO_2 **d.** $Be(CH_3)_2$

70. Write a Lewis structure for each of the following molecules.

a. ClF_3 **b.** SeF_4 **c.** PF_5
d. IF_5 **e.** XeF_6

71. Write a Lewis structure for each of the following covalently bonded polyatomic anions.

a. I_3^- **b.** IF_4^- **c.** ICl_2^- **d.** SF_5^-

72. Write Lewis structures for the following molecules. Where appropriate, use the concepts of formal charge and resonance to choose the most likely structure(s).

a. SSF_2 **b.** HNO_3 **c.** SO_2Cl_2
d. H_2CO_3 **e.** XeO_4

73. Write Lewis structures for the following covalently bonded anions. Where appropriate, use the concepts of formal charge, expanded octets, and resonance to choose the most likely structure(s).

a. OH^- **b.** CN^- **c.** NO_2^-

74. Write Lewis structures for the following covalently bonded anions. Where appropriate, use the concepts of formal charge, expanded octets, and resonance to choose the most likely structure(s).

a. ClO_2^- **b.** BrO_3^- **c.** IO_4^-

Alkenes and Alkynes

75. Give the condensed structural formula for each of the following.

a. ethene **b.** 1-butene
c. propyne **d.** 2-pentyne

76. Give the condensed structural formula for each of the following.

a. ethyne **b.** 1-butyne
c. propene **d.** 3-hexyne

77. Write equations to represent each of the following addition reactions.

a. hydrogen to ethene
b. bromine to propene
c. two molecules of hydrogen to ethyne.

78. Write equations to represent each of the following addition reactions.

a. chlorine to 1-butene
b. one molecule of hydrogen to ethyne
c. water to ethene.

Polymers

79. With the aid of Table 9.2, draw an eight-carbon-atom section of each of the following.

a. polyethylene **b.** poly(vinyl chloride)
c. polypropylene

80. Draw four-monomer-unit sections of the polymers formed from the following monomers.

a. tetrafluoroethylene, $F_2C{=}CF_2$
b. vinylidene chloride, $H_2C{=}CCl_2$
c. acrylonitrile, $H_2C{=}CH{-}CN$

ADDITIONAL PROBLEMS

81. Fill in this table assuming that elements W, X, Y, and Z are all main-group elements. Use the first column (W) as an example.

Element	W	X	Y	Z
Group number	7A	1A	____	____
Lewis symbol	:Ẇ̤:	____	·Y·	____
Charge on ion	1–	____	____	2–

82. Borazine ($B_3N_3H_6$) has a structure analogous to that of benzene. How might you represent its Lewis structure?

83. Two different molecules have the formula C_2H_6O. Write Lewis structures for the two molecules.

84. Draw four resonance structures for the oxalate ion, $C_2O_4^{2-}$. (*Hint*: The oxygen atoms are terminal atoms.)

85. Draw the structural formulas of five hydrocarbons that have the formula C_4H_6. (*Hint:* Use multiple bonds, cyclic structures, and combinations of these.)

86. Draw plausible Lewis structures for the following molecules, anions, or free radicals. Which would you expect to be diamagnetic and which paramagnetic?

a. OH^- **b.** OH **c.** NO_3
d. SO_3 **e.** SO_3^{2-} **f.** HO_2

87. Draw acceptable Lewis structures for *two* isomers with the formula S_2F_2.

88. A compound is found to have the following mass percent composition: 24.3% C, 71.6% Cl, and 4.1% H.

a. What is the *empirical* formula of this compound?
b. Draw a Lewis structure based on the empirical formula and comment on its deficiencies.
c. Propose a *molecular* formula for the compound that results in a more plausible Lewis structure. Draw the Lewis structure.

89. The reaction for forming one mole of gaseous ion pairs Na^+Cl^- from gaseous *atoms* is

$$Na(g) + Cl(g) \longrightarrow Na^+Cl^-(g) \qquad \Delta H = -304 \text{ kJ}$$

What is the enthalpy of formation of one mole of gaseous ion pairs from gaseous *ions*?

$$Na^+(g) + Cl^-(g) \longrightarrow Na^+Cl^-(g) \qquad \Delta H = ?$$

What additional amount of energy is associated with the clustering of one mole of gaseous ion pairs into one mole of the solid ionic crystal, NaCl(s)?

$$Na^+Cl^-(g) \longrightarrow NaCl(s) \qquad \Delta H = ?$$

(*Hint:* Obtain additional data from pages 317–318.)

90. Calculate ΔH_f° for the hypothetical compound MgCl(s) by using the following data.

Enthalpy of sublimation of Mg:	+150 kJ/mol
First ionization energy of Mg(g):	+738 kJ/mol
Enthalpy of dissociation of Cl_2(g):	+243 kJ/mol
Electron affinity of Cl(g):	−349 kJ/mol
Lattice energy of MgCl(s):	−676 kJ/mol

91. Use the data of Problem 90, together with the second ionization energy of Mg(g) (+1451 kJ/mol) and the lattice energy of $MgCl_2$(s) (−2500 kJ/mol) to calculate ΔH_f° for $MgCl_2$(s). Explain why you would expect $MgCl_2$(s) to be much more stable than MgCl(s).

92. Estimate the N-to-F and N-to-O bond lengths in the molecule nitryl fluoride, NO_2F. To do so, draw plausible resonance structures for this molecule and use data from Table 9.1

93. A 225-mL sample of propylene gas (C_3H_6) at 22 °C and 735 mmHg pressure is polymerized. If it were possible to produce polypropylene molecules that all had the formula $\text{+}CH_2CH(CH_3)\text{+}_n$, where $n = 875$, how many polymer molecules would form?

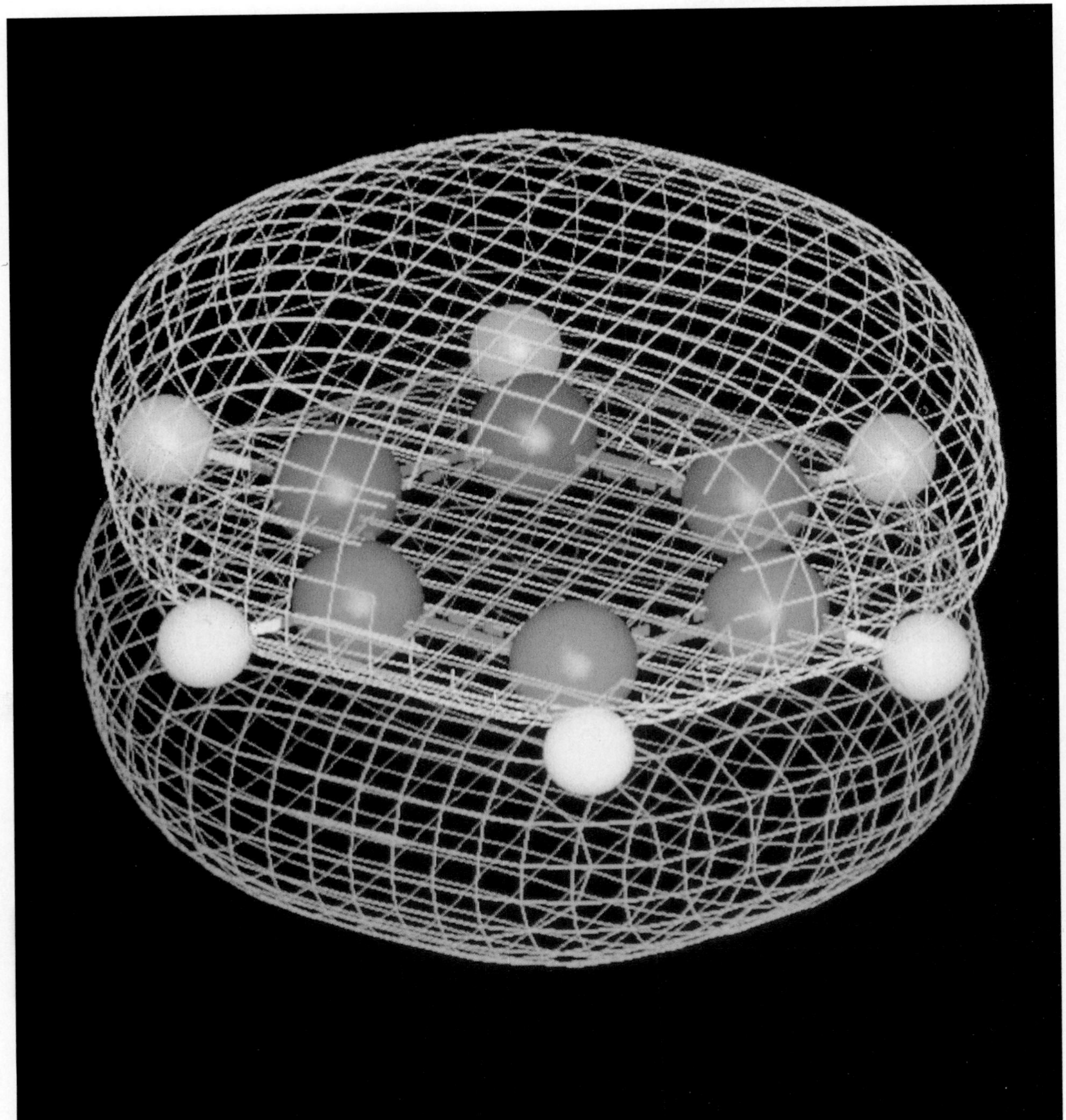

Shown here is a computer-generated representation of bonding in the benzene (C_6H_6) molecule. Part of the bonding is described by one of the theories discussed in this chapter, and part by another of the theories considered here.

10

Bonding Theory and Molecular Structure

CONTENTS

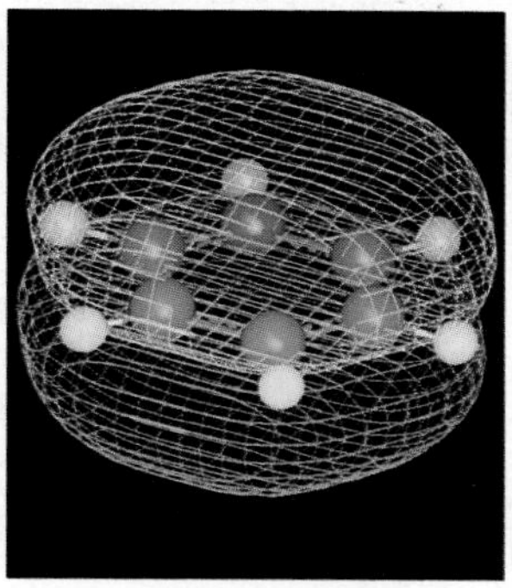

Our treatment of chemical bonding through Lewis structures in the previous chapter was not framed in quantum mechanical terms; Lewis's theory predated modern quantum mechanics. In this chapter we will find that quantum mechanics can advance our understanding of chemical bonding, just as it advanced our understanding of atomic structure in Chapter 6.

In one quantum-mechanical approach, called the *valence bond method*, we continue to think about the atoms in a molecule in terms of atomic orbitals but focus on the particular orbitals involved in covalent bond formation. In a second approach, called the *molecular orbital theory*, we start "from scratch." We fix the atomic nuclei at certain positions in a molecule and then assign electrons to orbitals that belong to the molecule as a whole—molecular orbitals.

However, before turning to quantum mechanics, we will learn how to extract additional information from Lewis structures to make predictions about the three-dimensional structures of molecules. Then we will begin to see how the properties of a substance depend on the structure of its molecules, a theme that recurs in later chapters.

10.1 Molecular Shapes

Let us first clarify what we mean by the shape of a molecule. If we could magnify oxygen, nitrogen, or carbon dioxide molecules to a size that we can see, we might describe these molecules as having shapes like jelly beans. But "jelly bean" is not a term that chemists would use. Chemists refer to these molecules as being *linear*. Although all molecules occupy three-dimensional space, chemists specify the structure of a molecule in terms of the geometric figure that best describes the positions of the nuclei of the atoms that are bonded together.

Diatomic molecules like O_2 and N_2 have only two nuclei; the molecules are *linear* because two points determine a straight line. If the three nuclei of a *triatomic* molecule happen to fall along the same straight line, as they do in CO_2, the molecule is also described as linear. If the three nuclei are not in a straight line, the molecule is *angular* or *bent*. Examples are H_2O and SO_2. In general, the shapes of polyatomic molecules are more complex than linear or angular. Shortly we will describe some common shapes.

Actual molecular shapes can only be determined by experiment, but in many cases it is possible to make fairly good predictions. Writing plausible Lewis structures is an important first step in a method that is quite successful in making these predictions.

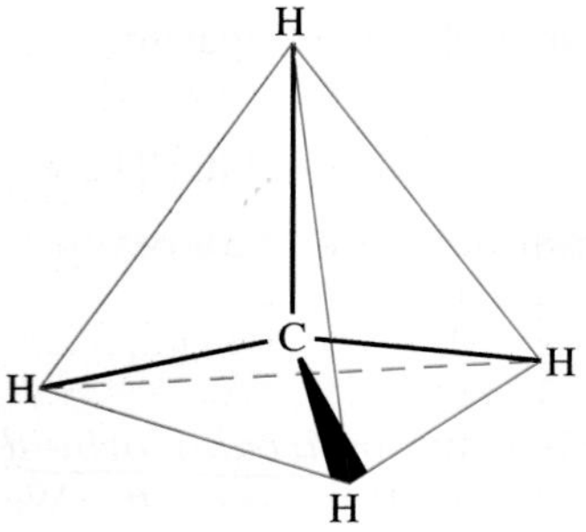

VSEPR notation: AX_4

Figure 10.1 Molecular shape of methane. The chemical bonds in CH_4 are represented by the black lines, and the shape of the molecule—a regular tetrahedron—is outlined by the red lines. All H—C—H bond angles are 109.5°.

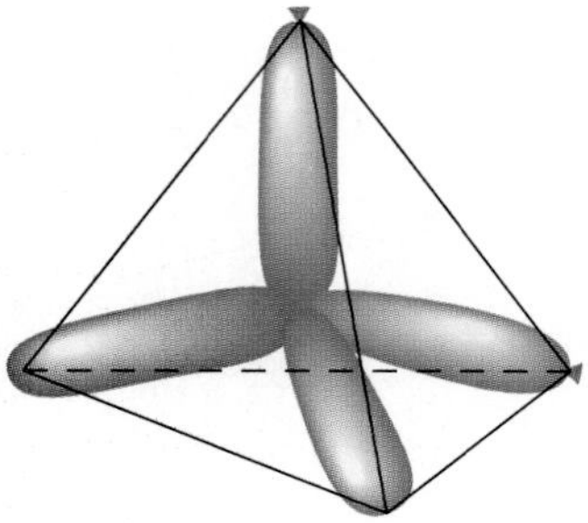

Figure 10.2 Valence-shell electron-pair repulsion pictured. The lobes of the balloons represent electron pairs. Like electron pairs, the lobes are directed to the corners of a tetrahedron. A tetrahedron has four faces, each an equilateral triangle.

Valence-Shell Electron-Pair Repulsion (VSEPR) Theory

The basis of the *v*alence-*s*hell *e*lectron-*p*air *r*epulsion or VSEPR (pronounced "vesper") theory is the fact that pairs of valence electrons in bonded atoms repel one another. This mutual repulsion pushes electron pairs as far away from each other as possible and causes terminal atoms to assume certain preferred orientations about the central atom to which they are bonded. The resulting arrangement of the atoms in a molecule gives the molecule a distinctive geometric shape.

Consider the Lewis structure of the methane molecule, CH_4. The central C atom acquires a valence-shell octet consisting of four bonding pairs of electrons.

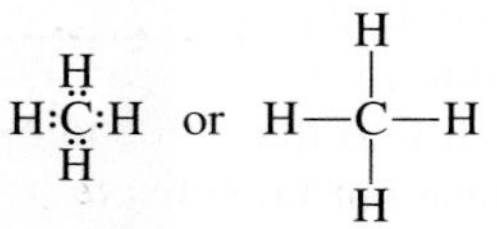

The energy of the molecule is minimized when the four bonding pairs of electrons assume an orientation that puts them as far from each other as possible. The arrangement that does this has the C atom at the center of a regular tetrahedron and the electron pairs directed to the four corners, where the H atoms are located (Figure 10.1). Both the predicted and experimentally observed H—C—H bond angles are 109.5°. In VSEPR notation, AX_4 signifies that four atoms (X) are bonded to a central atom (A) through electron-pair covalent bonds.

A useful analogy to electron-pair repulsion can be seen by twisting together two elongated air-filled balloons. The balloons separate into four lobes, corresponding to four electron pairs. Each lobe is directed to a corner of a regular tetrahedron (Figure 10.2). We can push any two lobes closer together, but this takes work (energy). Similarly, we can spread any two lobes farther apart, but this also takes work. The tetrahedral arrangement is the lowest energy state.

In NH_3 and H_2O the central atom also acquires a valence-shell octet—four pairs of electrons—but the structures of these molecules are different from that of methane. VSEPR theory predicts the orientation of the valence-shell electron pairs about a central atom. The **molecular geometry** or the shape of the molecule, on the other hand, is described by the geometric figure formed when the appropriate atomic nuclei are joined by straight lines. In CH_4 the electron-pair geometry and the molecular geometry are the same; they are both tetrahedral.

In the NH_3 molecule, three of the electron pairs in the valence shell of the central N atom are bonding pairs and the fourth is a lone pair (Figure 10.3). In VSEPR notation the molecule is referred to as AX_3E, where E represents the lone pair of electrons. When the nucleus of the N atom is joined to those of the H atoms, the geometric figure outlined is that of a pyramid with an N atom at the apex and with the three H atoms at the vertices of the triangular base. The molecular geometry is described as *trigonal pyramidal*. In this structure the H—N—H bond angles are predicted to be the regular tetrahedral bond angle (109.5°), but the measured bond angle—about 107°—is slightly less. We can rationalize the smaller bond angle by saying that the hydrogen atoms help to localize the bonding electron pairs, whereas the charge cloud of the lone-pair electrons spreads out. This spreading out of the lone-pair electronic charge pushes the bonding pairs into closer proximity, reducing the bond angle. In the balloon analogy of Figure 10.2, this corresponds to one of the four lobes being larger than the others and pushing the remaining three closer together.

In the H_2O molecule, two of the four electron pairs are bonding pairs and two are lone pairs, represented in VESPR notation as AX_2E_2. The geometric figure obtained by joining the two H nuclei with the nucleus of the O atom is an angle made up of two segments with a common endpoint (the O nucleus). The molecular shape is described as *angular* or *bent* (Figure 10.4). The H—O—H bond angle in H_2O is 104.5° rather than the regular tetrahedral angle of 109.5° we might have expected. We can think of the effect of lone-pair electrons on reducing the bond angle as being even greater in H_2O than in NH_3 because there are two lone pairs in H_2O and only one lone pair in NH_3.

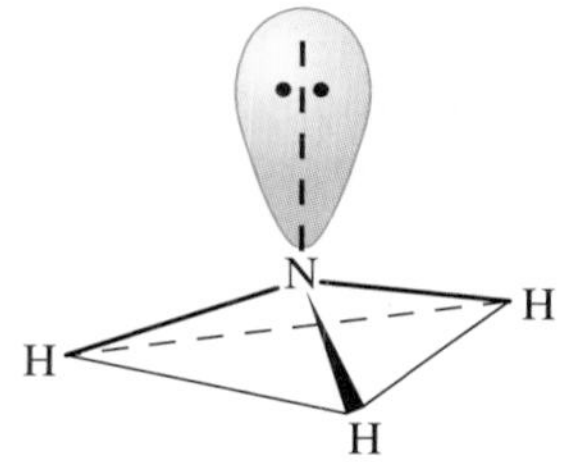

VSEPR notation: AX_3E

Figure 10.3 Molecular shape of ammonia. The bonds in NH_3 are represented by the heavy black lines. The shape of the molecule—a trigonal pyramid—is outlined by the solid lines and broken lines. The lone pair of electrons, which is directed to a "missing" corner of a tetrahedron, is shown in blue.

Possibilities for Electron-Group Geometries

So far we have described three molecules in some detail: CH_4, NH_3, and H_2O. All three have a molecular geometry based on the *tetrahedral* distribution of four electron pairs. To extend our discussion to a greater variety of structures, we need to introduce a few additional ideas. First is the idea that repulsions occur not just between pairs of electrons but between groups of electrons. An **electron group** is any collection of valence electrons localized in a region around a central atom that exerts repulsions on other groups of valence electrons. Particularly common among electron groups, of course, are lone pairs of electrons and bonding pairs in single covalent bonds. In a double covalent bond *two* pairs of electrons constitute a single electron group, and in a triple covalent bond, *three* pairs. A lone unpaired electron also constitutes an electron group. The most common numbers of electron groups distributed about a central atom are 2, 3, 4, 5, and 6. The geometric orientations assumed by these electron groups, called the **electron-group geometry**, are

You may find it interesting to compare the molecular shapes outlined in Figures 10.1, 10.3, and 10.4 with the shapes suggested through the space-filling molecular models in Figure 9.8.

- **2 electron groups:** linear
- **3 electron groups:** trigonal planar
- **4 electron groups:** tetrahedral
- **5 electron groups:** trigonal bipyramidal
- **6 electron groups:** octahedral

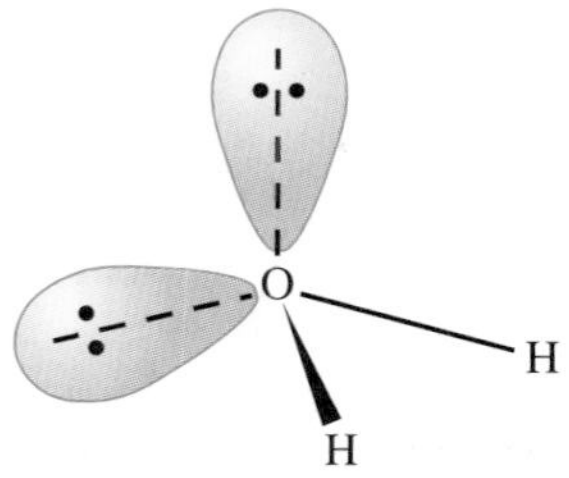

VSEPR notation: AX_2E_2

Figure 10.4 Molecular shape of water. The bonds in H_2O are represented by solid black lines and outline the shape of the molecule—angular or bent. The two lone pairs of electrons are directed to "missing" corners of a tetrahedron and are shown in blue.

The balloon analogy for electron-group geometries is extended to these additional cases in Figure 10.5.

When we compared CH_4, NH_3, and H_2O, we saw that the molecular geometry and the electron-group geometry were different if there were lone pairs of electrons. Only species that have the VSEPR notation AX_n (that is, AX_2, AX_3, AX_4, AX_5, and AX_6) have molecular geometries and electron-group geometries that are the same. Table 10.1 (pages 362–363) shows sketches of all the common molecular geometries. Those for the AX_n molecules are illustrated with ball-and-stick models.

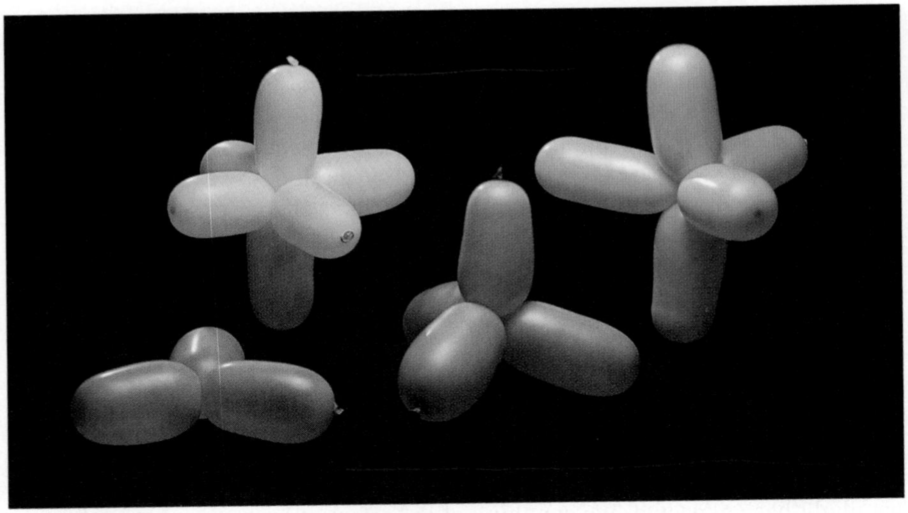

Figure 10.5 Electron-group geometries for 3, 4, 5, and 6 electron groups. The electron-group geometries are trigonal planar (orange), tetrahedral (green), trigonal bipyramidal (lavender), and octahedral (yellow). Molecular geometries based on these electron-group geometries are summarized and illustrated in Table 10.1

Example 10.1

Silicon tetrachloride is used in the manufacture of ultrapure silicon for transistors. What is the geometric shape of the silicon tetrachloride molecule?

Solution

From its name, we deduce that the formula of silicon tetrachloride is $SiCl_4$. Then we draw a Lewis structure for this molecule using the general method developed in Chapter 9.

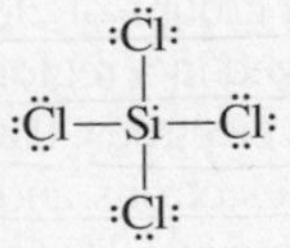

From this Lewis structure we see that the VSEPR notation of the $SiCl_4$ molecule is AX_4. Because molecules with the VSEPR notation AX_n have the same electron-group and molecular geometries, this molecule is *tetrahedral.*

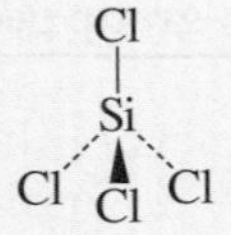

Exercise 10.1

Predict the shape of a molecule of antimony pentachloride.

Predictions of the type illustrated in Example 10.1 cover only a few cases, those of the type AX_n. As indicated in Table 10.1, if one or more of the electron groups are lone pairs, that is, for the types AX_mE_n, molecular geometries differ from electron-group geometries. Moreover, to understand some of the examples in Table 10.1 we need two additional ideas.

- *The closer together two groups of electrons are forced, the stronger is the repulsion between them.* Thus, the force of repulsion between two electron groups is stronger at 120° than at 180°, and much stronger still at 90°.
- *Lone-pair electrons spread out more than do bond-pair electrons.* The repulsion of one lone pair of electrons for another lone pair is therefore greater than, say, the repulsion between two bonding pairs. In fact, the strength of repulsive forces, from strongest to weakest, is

lone pair–lone pair (LP–LP) > lone pair–bonding pair (LP–BP)
> bonding pair–bonding pair (BP–BP)

Some of these ideas are illustrated in Example 10.2.

Example 10.2

Use VSEPR theory to rationalize the linear structure of the XeF_2 molecule shown on page 363 (Table 10.1).

Solution

The Lewis structure of XeF_2 suggested in Table 10.1 features an expanded octet for the central Xe atom. We can write three alternatives for the molecular structure of XeF_2.

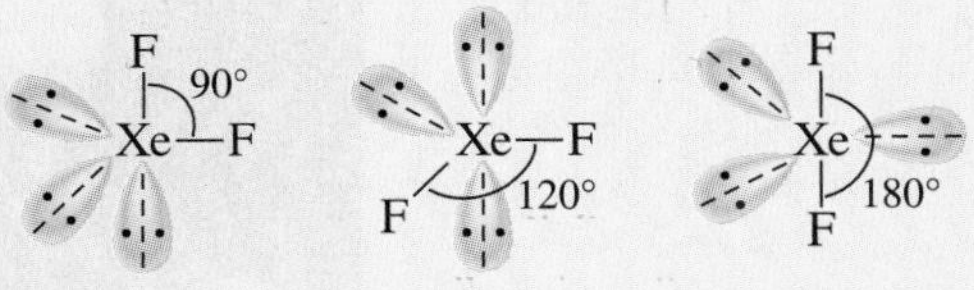

(I) (II) (III)

Although the electron-group geometry is trigonal bipyramidal, this cannot be the molecular geometry. The VSEPR notation for XeF_2 is AX_2E_3 (two bonding pairs, three lone pairs), *not* AX_5. We can choose among the three structures by focusing on the nature of the lone pair–lone pair repulsions because the LP–LP repulsions are the most severe.

Structure (I) has one LP–LP interaction of 120° and *two* that are 90°. In structure (II) one LP–LP interaction is at 180° and, again, *two* are at 90°. In structure (III), on the other hand, there are *no* LP–LP interactions of 90°; rather, all are at 120°. Because repulsions of electron pairs at 90° are much more severe than are those at 120° or 180°, structures (I) and (II) are at a higher energy and therefore less stable than structure (III). We should expect the Xe and two F atoms to lie on a straight line, a linear structure (structure III), and that is what we find experimentally for XeF_2.

Exercise 10.2

Use VSEPR theory to rationalize the "seesaw" structure of SF_4 shown below.

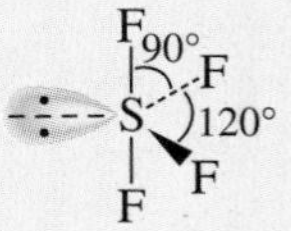

TABLE 10.1 **VSEPR Notations, Electron-Group Geometry, and Molecular Geometry**

Number of Electron Groups	Electron-Group Geometry	Number of Lone Pairs	VSEPR Notation	Molecular Geometry	Ideal Bond Angles	Example
2	linear	0	AX_2	X—A—X (linear)	180°	$BeCl_2$
3	trigonal planar	0	AX_3	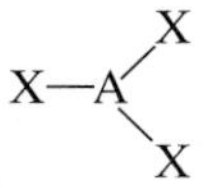 (trigonal planar)	120°	BF_3
	trigonal planar	1	AX_2E	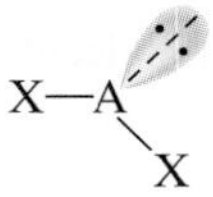 (angular)	120°	SO_2
4	tetrahedral	0	AX_4	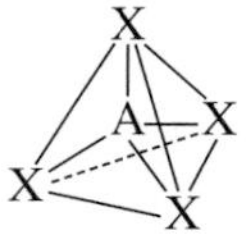 (tetrahedral)	109.5°	CH_4
	tetrahedral	1	AX_3E	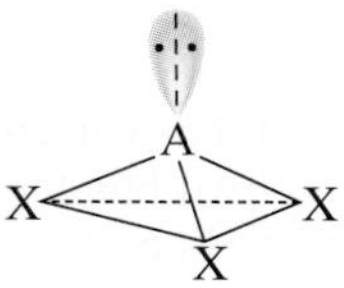 (trigonal pyramidal)	109.5°	NH_3
	tetrahedral	2	AX_2E_2	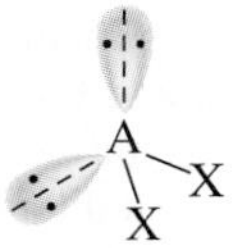 (angular)	109.5°	OH_2
5	trigonal bipyramidal	0	AX_5	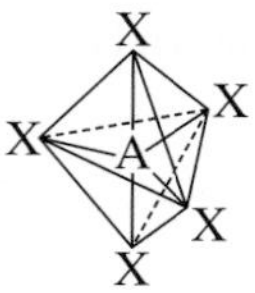 (trigonal bipyramidal)	90°, 120°	PCl_5

($BeCl_2$)

(BF_3)

(CH_4)

(PCl_5)

TABLE 10.1 (Continued)

Number of Electron Groups	Electron-Group Geometry	Number of Lone Pairs	VSEPR Notation	Molecular Geometry	Ideal Bond Angles	Example
	trigonal bipyramidal	1	AX_4E	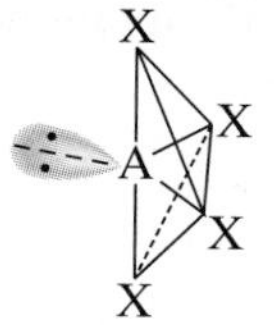 (seesaw)	90°, 120°	SF_4
	trigonal bipyramidal	2	AX_3E_2	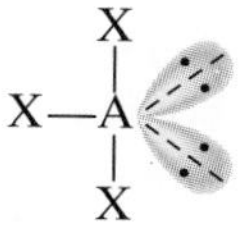 (T-shaped)	90°	ClF_3
	trigonal bipyramidal	3	AX_2E_3	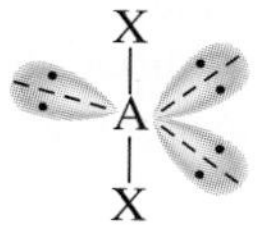 (linear)	180°	XeF_2
6	octahedral	0	AX_6	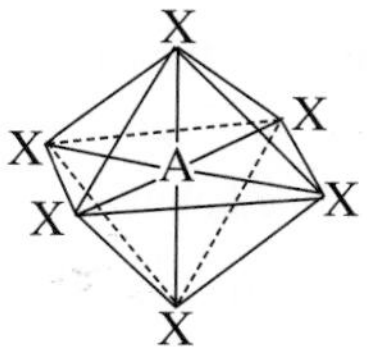 (octahedral)	90°	SF_6
	octahedral	1	AX_5E	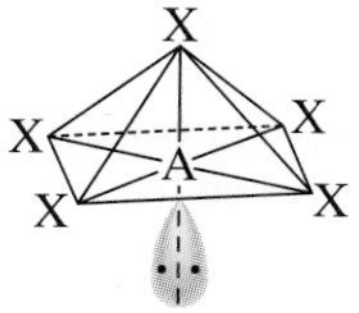 (square pyramidal)	90°	BrF_5
	octahedral	2	AX_4E_2	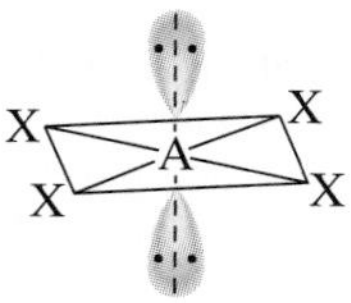 (square planar)	90°	XeF_4

(SF_6)

A Strategy for Applying VSEPR Theory

Now that we have learned several of its important features, let's consider a general strategy for applying the VSEPR theory. The following four-step procedure usually leads to a credible prediction of the geometric shape of a molecule or polyatomic ion. Remember, though, that the predicted structure may not match the true structure in all details, and that the actual structure can be determined only *by experiment*. First, let's apply the strategy to cases where all the electron groups are pairs of electrons.

1. Draw a Lewis structure of the molecule or polyatomic ion. The structure does *not* have to be the "best" possible structure but simply plausible. That is, the structure may have formal charges, or it may be just one of several contributing structures to a resonance hybrid.
2. Determine the number of electron pairs around the central atom, and identify each pair as either a bonding pair or a lone pair.
3. Establish whether the geometric orientation of the electron pairs around the central atom—the electron-group geometry—is linear, trigonal planar, tetrahedral, trigonal bipyramidal, or octahedral.
4. Describe the geometric shape based on the positions around the central atom that are occupied by other atoms (not lone-pair electrons). Use the information in Table 10.1 as a guide.

Example 10.3

Use the VSEPR theory to predict the shape of the chlorate ion, ClO_3^-.

Solution

Use the general strategy outlined above.

1. Write a plausible Lewis structure. Begin by determining the number of valence electrons.

From Cl +	From O +	Electron to establish charge of 1−	
(1×7) +	(3×6) +	1	= 26

The atom with the lower electronegativity, Cl, is the central atom. Use electron pairs to form bonds between the Cl and O atoms and to provide each terminal atom with an octet.

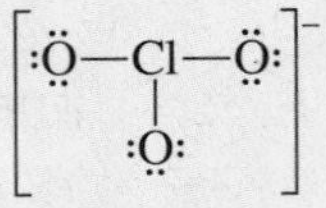

Then, add a lone pair of electrons to the Cl atom.

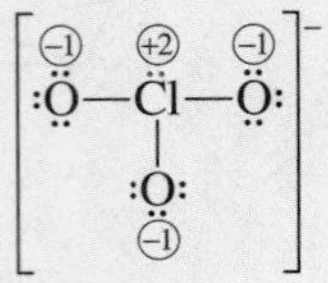

A structure with double bonds between the Cl and two of the O atoms has no formal charges on either the Cl or those two O atoms. This is the "best" structure that we could write according to the rules of Chapter 9, but to make a prediction with the VSEPR theory we don't need the "best" structure. The one shown here will do.

2. Count and identify the types of electron pairs around the central atom. Of the *four* electron pairs, *three* are *bonding* pairs and *one* is a *lone* pair (blue dots).
3. Describe the electron-group geometry. It is *tetrahedral.*
4. Describe the geometric shape of the ion. In VSEPR notation, the ClO_3^- ion is AX_3E. The ion has a *trigonal pyramidal* shape (see Figure 10.6).

Exercise 10.3

Use the VSEPR theory to predict the shape of the ICl_4^- ion.

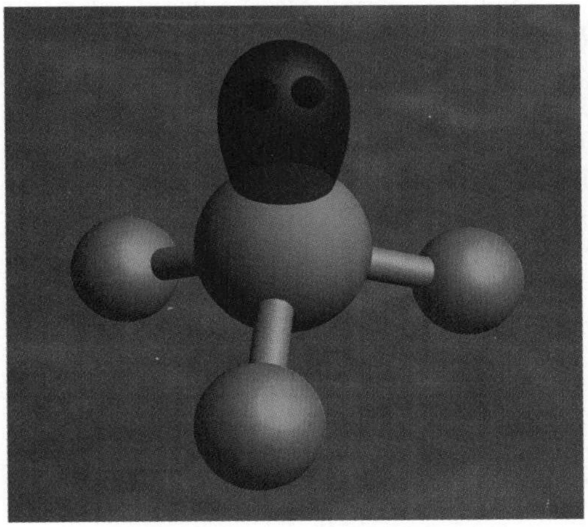

Figure 10.6 Structure of the chlorate anion, ClO_3^-. The measured bond angle in this trigonal pyramidal structure (~107°) is close to the 109.5° angle predicted by the VSEPR theory. It is totally consistent with VSEPR predictions when deviations due to the stronger repulsions exerted by lone-pair electrons are considered.

Multiple Covalent Bonds in VSEPR Theory

All the electrons in a double or triple covalent bond are confined to the region between the bonded atoms and constitute a single electron group, just as does a pair of electrons in a single covalent bond. In applying VSEPR theory to structures with multiple covalent bonds, we will stress the notion of electron groups.

Consider the ozone molecule, O_3. In the Lewis structure we find $3 \times 6 = 18$ valence electrons. Although there are two equivalent contributing structures to the resonance hybrid of O_3, we need write only one of these.

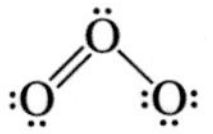

We treat the two pairs of electrons in the O-to-O double bond as a single electron group. The pair of electrons in the O-to-O single bond is a second electron group, and the lone pair of electrons on the central O atom is a third electron group. The electron-group geometry for three electron groups around the central O atom is *trigonal planar.* Because one of the electron groups is a lone pair, the VSEPR notation for the O_3 molecule is AX_2E. The actual shape of the molecule, the molecular geometry, is *angular* or *bent.* The predicted bond angle is 120° and the experimentally determined bond angle is 117°. This slight reduction in bond angle is presumably due to the interaction of the lone pair of electrons with the bond pairs.

Example 10.4

Use VSEPR theory to predict the geometric shape of the dinitrogen monoxide (nitrous oxide) molecule, N_2O.

Solution

As usual in applying the VSEPR theory, we start by writing a plausible Lewis structure. The two structures shown below should come to mind, but the true structure is a resonance hybrid of the two. Note that regardless of the structure we choose to work with, we will arrive at the same conclusion.

(I) (II)

In structure (I) the electrons in the two double bonds constitute electron groups. The electron-group geometry for two groups of electrons is linear. Structure (I) is *linear.* In structure (II) electrons in the triple bond are one electron group, and those in the N-to-O single bond are a second group. Structure (II) is also *linear.* Structures (I) and (II) are the most plausible Lewis structures, and we predict that the N_2O molecule is *linear.* Experimental evidence shows that it is.

Exercise 10.4

Draw a Lewis structure for the sulfuryl chloride molecule, SO_2Cl_2, and then predict the molecular geometry—its shape.

VSEPR Treatment of Molecules with More Than One Central Atom

VSEPR theory predicts the orientation of valence-shell electron groups around a central atom. So far our examples—CH_4, NH_3, H_2O, $SiCl_4$, XeF_2, and so on—have had just one central atom. But many of the molecules of interest to us have more than one central atom. We have already encountered, for example, the alkane hydrocarbons C_2H_6, C_3H_8, . . ., the alkenes C_2H_4, C_3H_6, . . ., and the alkynes C_2H_2, C_3H_4, ... We can now use VSEPR theory to predict their geometric shapes. We do it as follows: We simply work out the orientation of other atoms or groups of atoms around each central atom and combine our results into an overall description of the molecular shape.

Example 10.5

Describe the geometric shape of the nitric acid molecule, HNO_3, to the extent possible by the VSEPR theory.

Solution

The first challenge is to identify the central atom(s). We can eliminate H; it is always a terminal atom. The N atom has a lower electronegativity than the O atoms, so we expect it to be a central atom. However, a Lewis structure having N as the only central atom would violate the octet rule. A plausible Lewis structure requires that both the N and one of the O atoms be central atoms. A contributing structure to the resonance hybrid of HNO_3 is shown below.

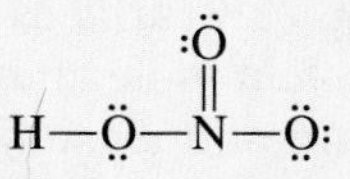

By treating the N-to-O double bond as a single electron group, we count *three* electron groups around the N atom. The orientation of the three O atoms around the N atom (AX_3) is the same as the electron-group geometry, *trigonal planar*. The O—N—O bond angles should be about 120°. The O atom having a terminal H atom has a valence-shell octet. Two pairs of these electrons are bonding pairs and two are lone pairs (AX_2E_2). The electron-group geometry is tetrahedral, and so the H—O—N portion of the molecule is angular or bent, with an expected bond angle of about 109°.

Figure 10.7 is a model of the HNO_3 molecule predicted by VSEPR theory. One feature of the molecular geometry that the theory does not address is where the H atom is located with respect to the plane of the N and O atoms.

Exercise 10.5

Describe the geometric shape of the ethanol molecule, C_2H_5OH, to the extent that you can with the VSEPR theory.

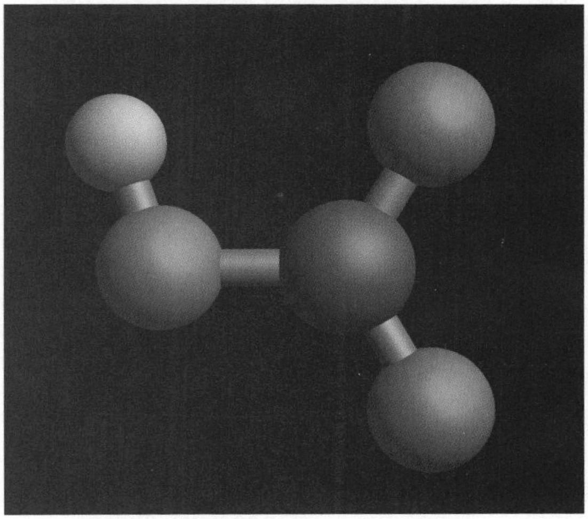

Figure 10.7 Predicted geometrical shape of the HNO_3 molecule by the VSEPR theory. The prediction is that the N and O atoms lie in the same plane. The O—N—O bond angles of about 120° are determined by trigonal planar electron-group geometry around the central N atom. The N—O—H bond angle of about 109° is determined by tetrahedral electron-group geometry around the O atom. (Experimental measurements place the H atom in about the same plane as the other atoms.)

Polar Molecules and Dipole Moments

We learned in Section 9.5 that in most covalent bonds the electron-attracting powers of the bonded atoms differ. If the opposite ends of the bond acquire slight charges, indicated in a Lewis structure by δ+ and δ–, the bond is *polar covalent*. For a *diatomic* molecule having a polar covalent bond, such as HCl,

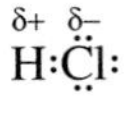

Molecular Shape and Drug Action

Many drugs target specific organs or tissues. For example, morphine, the principal narcotic found in opium poppies, acts on the brain. Most drugs act on protein molecules on cell surfaces or inside cells, but they don't act just anywhere on the protein molecules. They fit specific sites, called *receptor sites,* that have three-dimensional shapes designed to receive *endogenous* molecules. Endogenous substances like epinephrine (adrenaline) are substances that our bodies produce naturally. [Greek: *endon* (within) and *genes* (born, produced)]. An effective drug molecule, by virtue of its size, shape, and bond polarities, fits a receptor site. It acts either by mimicking the action of natural molecules on the receptor site or by blocking the receptor site from these natural molecules.

Substances produced in the body during strenuous physical activity may account for the phenomenon of "runner's high."

Recall that the symbol δ suggests a small magnitude of partial charge, less than the charge on an electron.

we can describe a quantity called the *dipole moment*, which is a measure of the separation between the centers of positive and negative charge. The **dipole moment (μ)** is defined as the product of the magnitude of the charge (δ) at either end of the dipole multiplied by the distance (*d*) that separates the charges.

$$\mu = \delta \cdot d$$

A molecule having a dipole moment is said to be **polar**.

Dipole moments have the units of electric charge times distance: coulomb × meter (C · m). A quantity called 1 **debye (D)** is equal to 3.34 ×

Drug molecules that fit a receptor site and cause their own characteristic response are called *agonists*. Those that bind to a receptor site and block the action of another substance are called *antagonists*. The interaction of a drug molecule and receptor site is something like that of a lock and key. The agonist is the key that fits and opens (activates) the lock (receptor). An antagonist is like a key that breaks off in a lock, blocking the insertion of other keys (agonist molecules). The lock-and-key model is too restrictive, however. Receptor sites are not fixed in shape like the inner mechanism of a lock. Rather, they are flexible; they are more like a glove that can stretch to fit a hand.

If both agonist and antagonist are present, there is competition for the receptor sites. In general, an antagonist binds more tightly than an agonist, so that a small quantity of antagonist can block the action of a larger quantity of agonist. For example, a 1-mg dose of naloxone, a morphine antagonist, when given intravenously, can block the action of 25 mg of heroin, a morphine-related agonist.

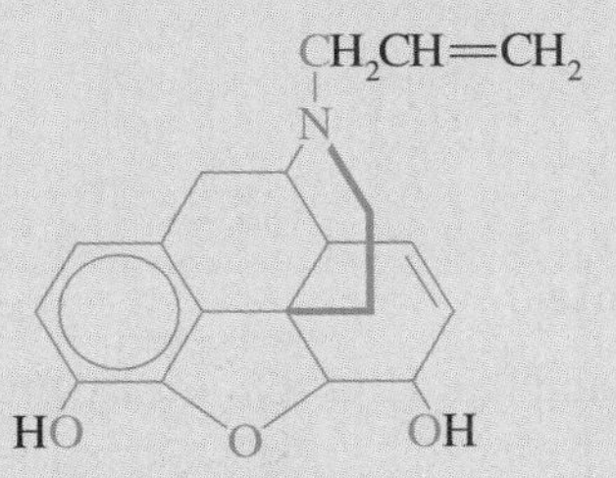

Naloxone

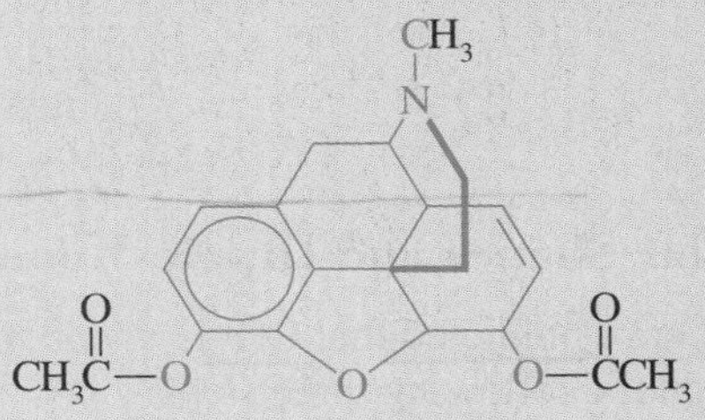

Heroin

Why should the human brain have receptors for a plant-derived drug like morphine? The answer seems to lie in the existence of several substances called *endorphins* ("endogenous morphines") formed naturally by the human body. For example, the production of these compounds during strenuous athletic activity may account for the fact that athletes can sometimes continue to compete after an injury. They don't feel the pain until the event is over. The so-called "runner's high" experienced by some marathon runners and joggers may also be due to endorphin production.

10^{-30} C · m, and dipole moments are generally expressed in debyes. For example, the measured dipole moment of HCl is $\mu = 1.07$ D.

Bond Dipoles and Molecular Dipoles

Because of electronegativity differences between C and O atoms, the C-to-O bonds in carbon dioxide are *polar* covalent bonds, but the molecule as a whole is found to be *nonpolar*. To understand this fact we need to distinguish between **bond moments**, which describe the magnitude of charge separation in individual bonds, and molecular or **resultant dipole moments**, which describe charge separation in the molecule as a whole by taking every bond into account. In heteronuclear *diatomic* molecules, such as HCl, the bond moment and the resultant dipole moment are the same, and the reason is clear: there is only one bond.

The situation with CO_2 can be represented as follows.

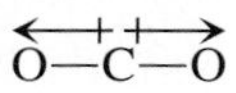

$\mu = 0$

The cross-based arrows show the directions of the bond moments; electronic charge is displaced toward the O atoms. Because the molecule is *linear*, these two bond moments, which are equal in magnitude and opposite in direction, cancel each other and the *resultant* dipole moment is zero.

Starting with the experimental fact that water is a polar molecule with $\mu = 1.84$ D, we can immediately conclude that H_2O cannot be a linear molecule. The O—H bond moments must combine in such a way as not to cancel but to yield a resultant dipole moment of 1.84 D. The following structure is consistent with the experimental evidence.

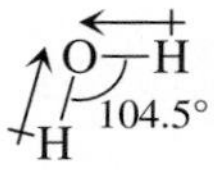

Molecular Shapes and Dipole Moments

We could have predicted that CO_2 is *nonpolar* and that H_2O is *polar* by proceeding through the following three steps:

1. Use electronegativity data to predict the existence of individual bond moments.
2. Use VSEPR theory to predict the molecular shape.
3. From the molecular shape, determine whether bond moments cancel to give a nonpolar molecule or whether they combine to produce a resultant dipole moment.

The guiding principle is that any structure with the maximum symmetry possible for that type is nonpolar.

Example 10.6

Explain whether you expect the following molecules to be polar or nonpolar.
a. CCl_4 **b.** $CHCl_3$.

Solution

The Lewis structures of these two molecules are

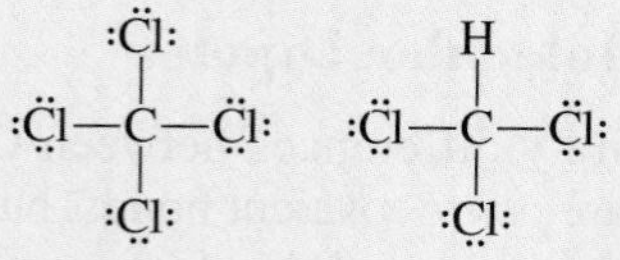

(a) (b)

In both molecules, the electron-group geometry and the molecular geometry are *tetrahedral* (VSEPR notation: AX_4). The symmetrical distribution of bond moments in CCl_4 is shown in Figure 10.8. The CCl_4 molecule itself has no resultant dipole moment and is *nonpolar*.

When an H atom is substituted for one of the Cl atoms in CCl_4, one of the bond moments is greatly reduced (that is, the C—H bond moment is much smaller than the C—Cl bond moment). The portion of the molecule with the three C—Cl bonds acquires a slight negative charge, and the portion with the C—H bond, a slight positive charge. There is a net resultant dipole moment, and the $CHCl_3$ molecule is *polar*. Its dipole moment is found to be 1.01 D (Figure 10.8).

Exercise 10.6

Explain whether you expect the following molecules to be polar or nonpolar:
a. BF_3 **b.** SO_2 **c.** BrCl **d.** SO_3

CONCEPTUAL EXAMPLE 10.1

Of the two compounds, nitrosyl fluoride, NOF, and nitryl fluoride, NO_2F, one has a resultant dipole moment, $\mu = 1.81$ D, and the other $\mu = 0.47$ D. Which dipole moment do you think is associated with each compound? Explain.

Solution

Our basic task is to identify bond moments and determine how they combine to produce a resultant dipole moment. This in turn requires us to have a sense of the geometric structures of the two molecules, which we can get from the VSEPR theory. Let's begin with Lewis structures. In each structure we expect N as the central atom; it has the lowest electronegativity of the three types of atoms.

	NOF	NO_2F
Valence electrons	18	24
Lewis structures	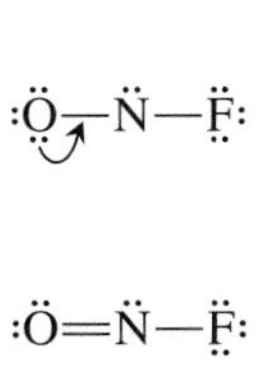	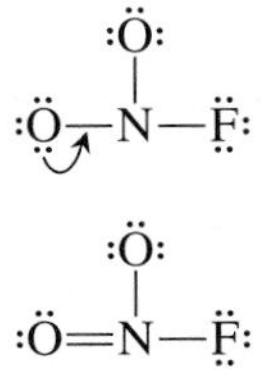
Electron groups around central atom	3	3
Electron-group geometry	trigonal planar	trigonal planar
VSEPR notation	AX_2E	AX_3
Molecular geometry	angular	trigonal planar
Bond moments (electronegativities F > O > N)	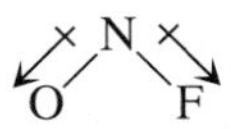	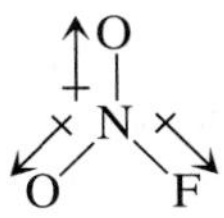

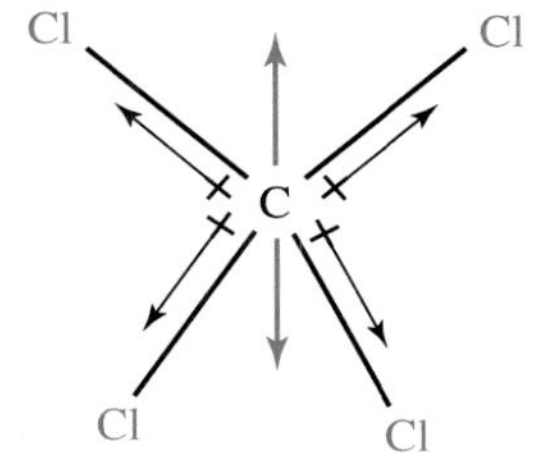

(a) CCl_4: a nonpolar molecule

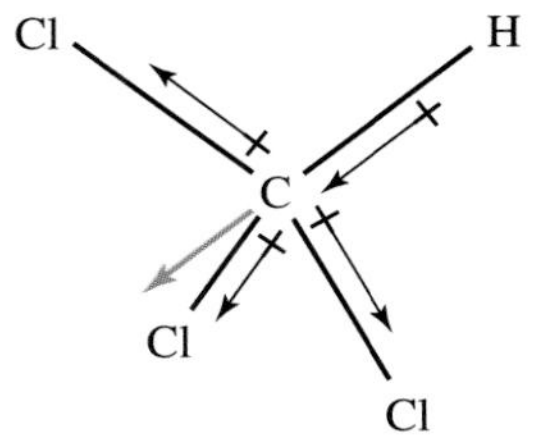

(b) $CHCl_3$: a polar molecule

Figure 10.8 **Molecular shapes and dipole moments.** The individual bond moments are represented by the cross-based arrows (+⟶). (a) The red arrow is the resultant of the bond moments to the two Cl atoms shown in red, and the blue arrow is the resultant of the bond moments to the two Cl atoms shown in blue. The red and blue arrows are of equal magnitude and point in opposite directions. The resultant dipole moment for the entire molecule is zero. (b) The individual bond moments do not cancel; they combine to produce a resultant dipole moment of 1.01 D for $CHCl_3$ (green arrow).

We can readily see that in NOF two bond moments point "down" and lead to a net "downward" displacement of electron charge density. In NO_2F, one of the bond moments acts in opposition to the other two. We should expect a smaller net displacement of electron charge density. Our prediction is NOF, $\mu = 1.81$ D; NO_2F, $\mu = 0.47$ D.

Conceptual Exercise 10.1

Of the two molecules NOF and NO_2F, one has a measured F—N—O bond angle of 110° and the other, 118°. Which bond angle do you think is associated with each molecule? Explain.

10.2 Valence Bond Theory

Lewis theory provides a simple, qualitative way to describe covalent bonding, and VSEPR theory allows us to carry the description further. It enables us to predict probable molecular shapes. Let us now consider a quantum mechanical approach to covalent bonding that confirms much of what we described through VSEPR theory, but which allows us to carry that description further still.

Imagine two hydrogen atoms approaching one another. As they get close, their electron charge clouds begin to merge. We describe this intermingling as the *overlap* of the 1*s* orbitals of the two atoms. The overlap results in an increased electron charge density in the region between the atomic nuclei. The increased density of negative charge serves to hold the positively charged atomic nuclei together. This new approach, called the **valence bond (VB) method**, views a covalent bond in this way:

A covalent bond is a region of high electron charge density that results from the overlap of atomic orbitals between two atoms.

In general, the more extensive the overlap between two orbitals, the stronger is the bond between two atoms. As the two atoms are brought more closely together, however, the repulsion of the atomic nuclei becomes more important than the electron-nucleus attractions and the bond becomes unstable. For each bond, then, there is a condition of optimal orbital overlap that leads to a maximum bond strength (bond energy) at a particular internuclear distance (bond length). The valence bond method attempts to find the best approximation to this condition for all the bonds in a molecule.

The bonding of two hydrogen atoms into a hydrogen molecule through the overlap of their 1*s* orbitals is pictured in Figure 10.9. When we apply the valence bond method to the H_2S molecule (Figure 10.10), we see more clearly these aspects of the method.

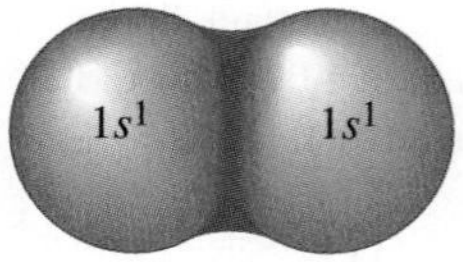

Figure 10.9 Atomic orbital overlap and bonding in H_2. Each 1*s* atomic orbital contains one electron. As a result of the overlap of the two orbitals, the electrons become paired and produce a region of high electron charge density (and also high electron probability) between the atomic nuclei—a covalent bond.

- Most of the electrons in a molecule remain in the same orbital locations as they did in the separated atoms.
- Bonding electrons are *localized* (fixed) in the region of atomic orbital overlap.

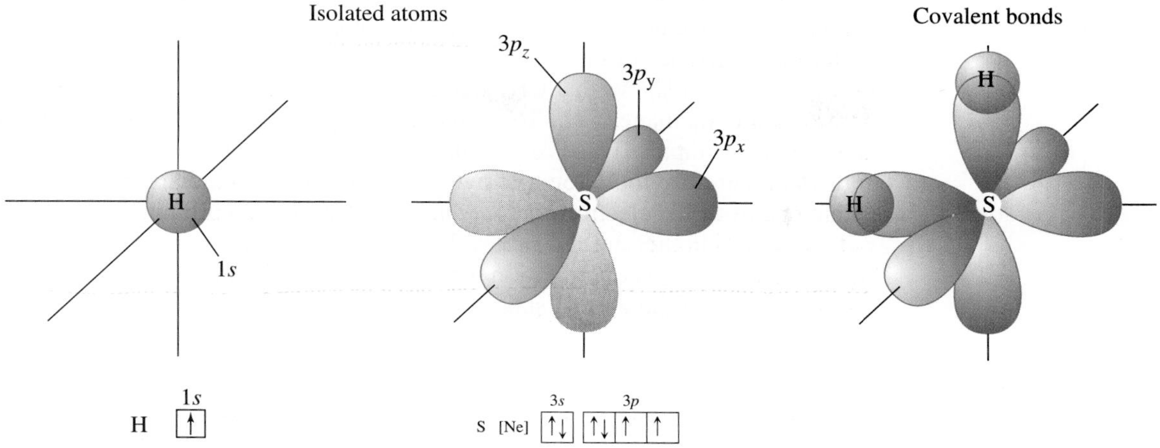

Figure 10.10 Atomic orbital overlap and bonding in H_2S. Orbitals with a single electron are shown in gray, and those with an electron pair, in color. For S, only $3p$ orbitals are shown, and in representations that have been somewhat simplified. The $1s$ orbitals of the two H atoms overlap with the $3p_x$ and $3p_z$ orbitals of the S atom, producing an H_2S molecule with a predicted bond angle of 90°.

- For orbitals with directional lobes (all except *s* orbitals), maximum overlap occurs when atomic orbitals overlap end-to-end.
- The molecular geometry depends on the geometric relationships among the atomic orbitals of the central atom that participate in bonding. The two $3p$ orbitals of the S atom that overlap with $1s$ orbitals of the H atoms are perpendicular to one another. The predicted H—S—H bond angle in H_2S is 90°.

As a first approximation, VSEPR theory predicts a tetrahedral bond angle (109.5°) in H_2S. However, if we take into account the strong repulsions between the lone pairs of electrons on the S atom and the bond pairs, we expect the bonds in H_2S to be forced into a smaller angle. The measured bond angle in the H_2S molecule is 92.1°, suggesting that the valence bond method describes the covalent bonding in H_2S better than the VSEPR theory does. There are not very many cases, however, where the unmodified VB method produces superior results. We will be most successful in describing molecular structures by using a combination of VSEPR theory and a modification of the VB method considered in the next section.

10.3 Hybridization of Atomic Orbitals

In applying the valence bond method to the ground-state electron configuration of carbon, we can assume that the filled $1s$ orbital will not be involved in the bonding and focus our attention on the valence-shell orbitals.

	$1s$	$2s$	$2p$		
Ground-state C:	↑↓	↑↓	↑	↑	

We observe two unpaired electrons in the 2*p* subshell, and *predict* the simplest hydrocarbon molecule to be CH_2 with a bond angle of 90°. However, we might question the existence of this molecule because it does not follow the octet rule. In fact, experiment shows that CH_2 is not a stable molecule. The simplest *stable* hydrocarbon is methane, CH_4. To account for this, we need an orbital diagram that shows *four* unpaired electrons in the valence shell of carbon, requiring four bonds (and therefore four atoms of hydrogen). To get such a diagram we imagine that one of the 2*s* electrons is *promoted* to the empty 2*p* orbital. To boost the 2*s* electron to a higher energy sublevel requires that energy must be absorbed. The resulting electron configuration is that of an *excited state* that has an energy greater than that of the ground state.

		2*s*	2*p*		
Excited-state C:	[He]	↑	↑	↑	↑

The three mutually perpendicular 2*p* orbitals of this excited-state configuration lead us to predict a molecule with three C—H bonds at angles of 90°. The fourth C—H bond, based on the spherical 2*s* orbital, would be oriented in a direction that interfered least with the other three C—H bonds. By experiment, however, we observe that all four H—C—H bond angles are the same; they are the tetrahedral angles of 109.5° (see again Figure 10.1). This is the structure predicted by VSEPR theory, as shown in Table 10.1. The excited-state electron configuration of carbon allows for the correct number of C-to-H bonds, but not for the correct bond angles.

Our inaccurate predictions for CH_4 are based on an untested assumption. We have assumed that *bonded* atoms have exactly the same kinds of orbitals (*s*, *p*, . . .) as do isolated *nonbonded* atoms. This assumption works well in a few cases (for example, H_2S), but in most cases it does not.

*sp*³ Hybridization

Consider once more the excited-state electron configuration of the carbon atom in CH_4. Suppose we could blend together the one 2*s* and three 2*p* orbitals of that atom to produce four new orbitals that are equivalent to each other in energy and in shape, but pointing in different directions. This blending is called **hybridization**, it is a *hypothetical* process (not an observed one) that can be carried out as a quantum mathematical calculation. Figure 10.11 illustrates the hybridization of an *s* orbital and three *p* orbitals into the new set of four **hybrid orbitals**. The new orbitals are called $\boldsymbol{sp^3}$ hybrid orbitals.

In a hybridization scheme *the number of new hybrid orbitals is equal to the total number of atomic orbitals that are combined*. The symbols used for hybrid orbitals identify the kinds and numbers of atomic orbitals used to form the hybrids. Thus, sp^3 signifies the hybridization of *one s* and *three p* orbitals. We can also use an orbital diagram of the valence shell to represent hybridization.

	sp^3			
sp^3 *hybridization* in C:	↑	↑	↑	↑

Figure 10.12 suggests how sp^3 hybrid orbitals are involved in bond formation in methane.

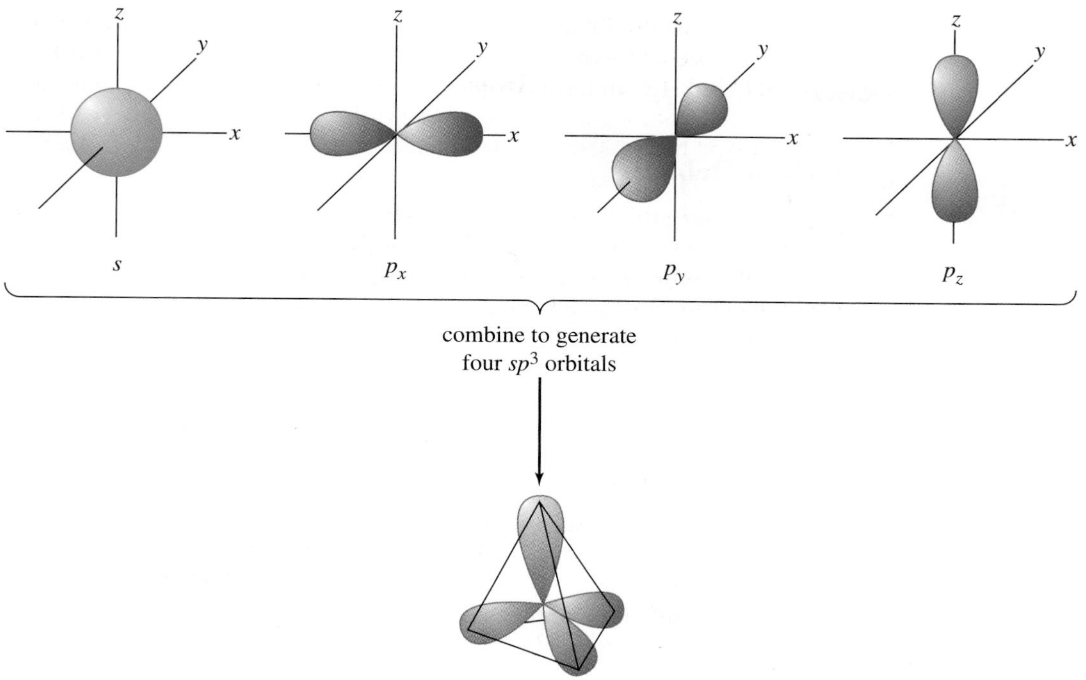

Figure 10.11 The sp^3 hybridization scheme illustrated.

Hybridization is not an actual physical phenomenon that can be observed. It is a way of rationalizing a particular molecular structure that has been determined by experiment. It is possible, however, to assess the energy changes in the hypothetical process of promoting an electron from the ground state to an excited state, hybridizing orbitals in the excited state, and using the hybrid orbitals to form bonds. A hybridization scheme is chosen that minimizes the energy of the molecular structure, while satisfactorily accounting for the observed geometry.

We might also expect to use sp^3 hybridization not only for structures of the type AX_4 (as in CH_4), but also for AX_3E (as in NH_3) and AX_2E_2 (as in H_2O). For example, sp^3 hybridization of the central N atom in NH_3

$$sp^3 \text{ hybridization in N: } \quad \overset{sp^3}{\boxed{\uparrow\downarrow}\boxed{\uparrow}\boxed{\uparrow}\boxed{\uparrow}}$$

accounts for the formation of three N—H bonds and a lone pair of electrons on the N atom (Figure 10.13). The predicted H—N—H bond angles of 109.5° are close to the experimentally observed angles of 107°. A similar scheme for H_2O

$$sp^3 \text{ hybridization in O: } \quad \overset{sp^3}{\boxed{\uparrow\downarrow}\boxed{\uparrow\downarrow}\boxed{\uparrow}\boxed{\uparrow}}$$

accounts for the formation of two O—H bonds and two lone pairs of electrons on the O atom. The predicted H—O—H bond angle of 109.5° is also

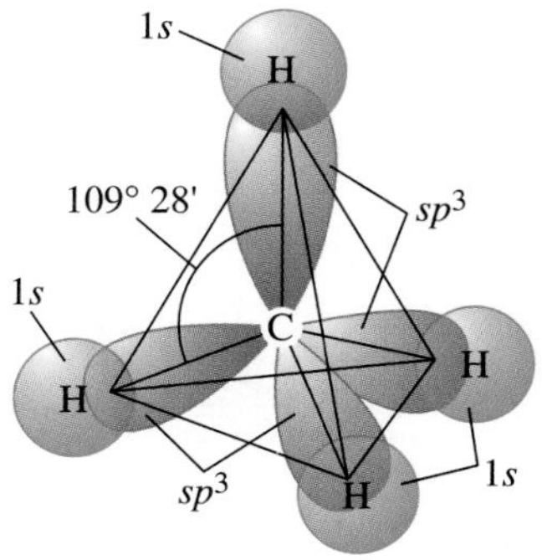

Figure 10.12 sp^3 hybrid orbitals and bonding in CH_4. The four carbon orbitals are sp^3 hybrid orbitals (purple). The hydrogen orbitals are 1*s* (red). The molecular structure is tetrahedral; the H—C—H bond angles are 109.5°.

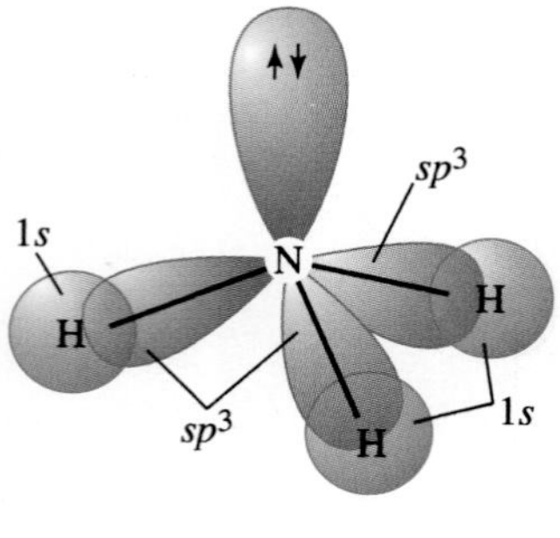

Figure 10.13 ***sp*³ hybrid orbitals and bonding in NH_3.** The three N—H bonds are formed by the overlap of three *sp*³ hybrid orbitals of the N atom and 1*s* orbitals of the H atoms. The molecular geometry of NH_3 is trigonal pyramidal. The lone pair electrons on the N atom reside in the fourth *sp*³ orbital.

reasonably close to the observed 104.5° angle. As in our discussion of the VSEPR theory, we can explain the somewhat smaller-than-tetrahedral bond angles in NH_3 and H_2O in terms of repulsions involving lone-pair electrons.

*sp*² Hybrid Orbitals

Now let's turn our attention to boron, a Group 3A element. The boron atom has four orbitals but only three electrons in its valence shell. In most boron compounds the hybridization scheme combines one 2*s* and two 2*p* orbitals into three ***sp*²** orbitals. Using orbital diagrams to represent this hybridization, we have

	2*s*	2*p*			
Ground-state B:	↑↓	↑			
	2*s*	2*p*			
Excited-state B:	↑	↑	↑		
		*sp*²			2*p*
*sp*² *hybridization* in B:		↑	↑	↑	

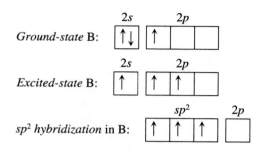

As shown in Figure 10.14, the *sp*² hybrid orbitals are distributed geometrically within a plane at 120° angles. The valence bond method predicts that BF_3 is a trigonal planar molecule with 120° F—B—F bond angles. This indeed is what is observed experimentally. By far the most common examples of *sp*² hybridization are found in organic molecules with double bonds, as we shall see in Section 10.4.

Figure 10.14 The *sp*² hybridization scheme illustrated.

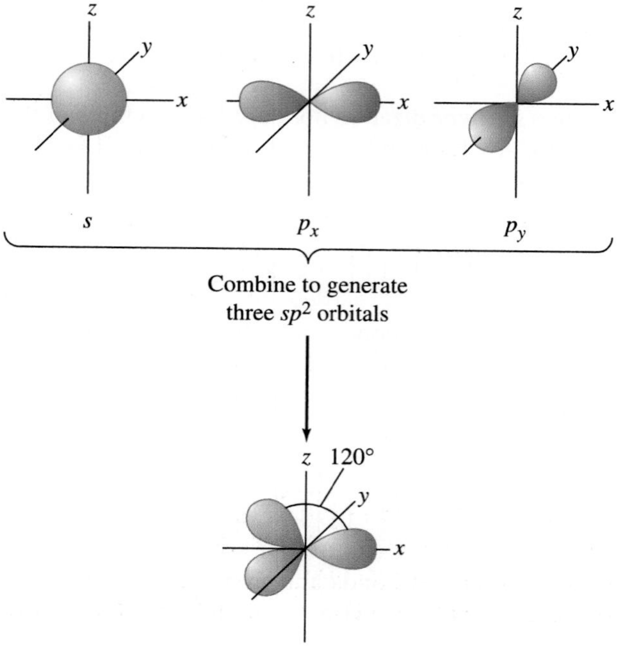

sp Hybrid Orbitals

Next, let's consider beryllium, a Group 2A element. The Be atom has *two* electrons and *four* orbitals in its valence shell. In the triatomic molecule $BeCl_2$, which is present in gaseous $BeCl_2$ at high temperatures, the 2*s* and one of the 2*p* orbitals of the Be atom are hybridized into ***sp*** hybrid orbitals. The remaining two 2*p* orbitals are left unhybridized and unoccupied in the orbital diagram.

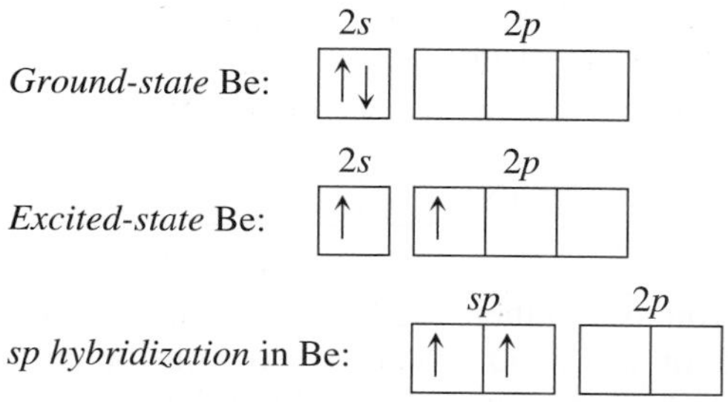

As shown in Figure 10.15, the two *sp* hybrid orbitals are directed along a straight line, 180° apart. We predict that the $BeCl_2$ molecule should be linear, and this prediction is confirmed by experimental evidence. As with sp^2 hybridization, most examples of *sp* hybridization are found in organic molecules, especially those with triple covalent bonds.

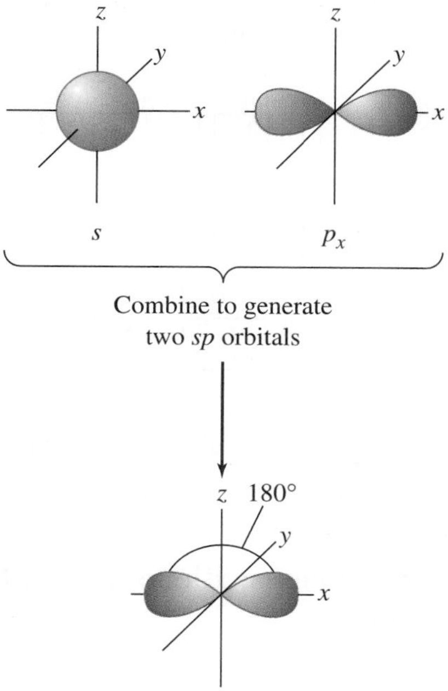

Figure 10.15 The *sp* hybridization scheme illustrated.

Hybrid Orbitals Involving *d* Subshells

A maximum of *eight* valence electrons can be accommodated by any hybridization scheme involving only *s* and *p* orbitals. Hybridization schemes for structures involving *expanded* octets must include additional orbitals, and these extra orbitals can come from a *d* subshell.

sp^3d orbitals

Figure 10.16 The sp^3d hybrid orbitals. The sp^3d hybrid orbitals, in a trigonal bipyramidal arrangement, are deployed by a phosphorus atom in the molecule PCl_5 (see page 362).

For example, we need *five* hybrid orbitals to describe bonding in PCl_5. We get these by combining *one s*, *three p*, and *one d* orbital, as suggested by the following orbital diagrams.

		3*s*	3*p*			3*d*				
Ground-state P:	[Ne]	↑↓	↑	↑	↑					
Excited-state P:	[Ne]	↑	↑	↑	↑	↑				

		sp^3d					3*d*			
sp^3d *hybridization* in P:	[Ne]	↑	↑	↑	↑	↑				

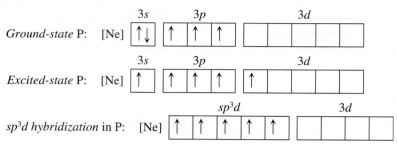

The $\boldsymbol{sp^3d}$ hybrid orbitals and their trigonal bipyramidal orientation are shown in Figure 10.16. Like other hybrid orbitals we've considered, the five sp^3d hybrid orbitals are arranged in a symmetrical fashion, but unlike the other cases, the five sp^3d orbitals are not all equivalent. Three of the orbitals are directed in the plane of the central atom at 120° angles with one another; the remaining two are perpendicular to the plane of the other three. The three positions in the central plane are called *equatorial* positions and the two positions perpendicular to the plane are called *axial* positions.

Another structure featuring an expanded octet is SF_6. Here, *six* hybrid orbitals are required in order to describe bonding. These are obtained through the hybridization scheme sp^3d^2, represented by the orbital diagrams

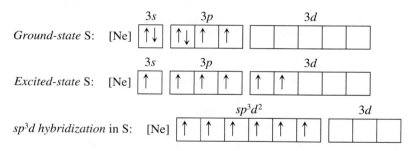

		3*s*	3*p*			3*d*				
Ground-state S:	[Ne]	↑↓	↑↓	↑	↑					
Excited-state S:	[Ne]	↑	↑	↑	↑	↑	↑			

		sp^3d^2						3*d*		
sp^3d *hybridization* in S:	[Ne]	↑	↑	↑	↑	↑	↑			

sp^3d^2 orbitals

Figure 10.17 The sp^3d^2 hybrid orbitals. The sp^3d^2 hybrid orbitals, in an octahedral arrangement, are deployed by a sulfur atom in the molecule SF_6 (see page 363).

The $\boldsymbol{sp^3d^2}$ hybrid orbitals and their octahedral orientation are shown in Figure 10.17.

Predicting Hybridization Schemes: Blending VSEPR Theory and the Valence Bond Method

In hybridization schemes, one hybrid orbital is produced for every simple atomic orbital involved. In a molecule, each of the hybrid orbitals of the central atom acquires an electron pair, either a bond pair or a lone pair. And the hybrid orbitals have the same orientation as the electron-group geometry predicted by VSEPR theory.

When we describe the probable hybridization scheme for a structure, we must choose a scheme that conforms to *experimental* evidence. We can do this by blending the VSEPR and valence bond approaches. The procedure is outlined in the following four steps and illustrated in Example 10.7.

Of course, we don't always have experimental evidence, so in many cases we have to make an educated guess—a prediction.

1. Write a plausible Lewis structure for the species.
2. Use VSEPR theory to predict the electron-group geometry of the central atom.
3. Select the hybridization scheme that corresponds to the VSEPR prediction.
4. Describe the orbital overlap.

Example 10.7

Iodine pentafluoride, IF_5, is used commercially as a fluorinating agent—a substance that produces other fluorine-containing compounds in its chemical reactions. Describe a hybridization scheme for the central atom and sketch the molecular geometry of the IF_5 molecule.

Solution

1. The key step in writing the Lewis structure of this molecule is to obtain the correct number of valence-shell electrons: $6 \times 7 = 42$. The I atom is the central atom in a structure with an expanded octet. First, we indicate the five I—F bonds and then complete the octets of the F atoms.

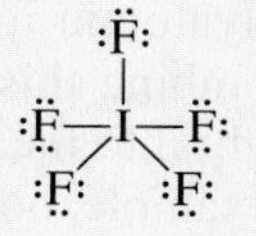

This accounts for 40 of the 42 valence electrons. The remaining electron pair (shown in blue) must go on the central I atom.

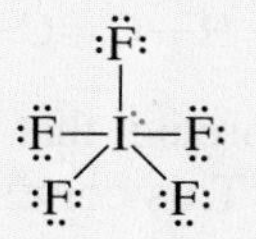

2. VSEPR theory predicts an octahedral electron-group geometry for six electron pairs.
3. The hybridization scheme corresponding to octahedral electron-group geometry is sp^3d^2.
4. The six sp^3d^2 hybrid orbitals are directed to the corners of an octahedron, but one of the orbitals is occupied by a lone pair of electrons (shown in blue below).

		sp^3d^2						$5d$	
sp^3d^2 hybridization of I:	[Kr]$4d^{10}$	↑↓	↑	↑	↑	↑	↑		

The resulting molecular geometry is that of a *square pyramid* with bond angles of approximately 90° (Figure 10.18).

Note that because of the similarity in formulas we might expect IF_5 to have the same structure as PCl_5, trigonal bipyramidal. It doesn't, however, because PCl_5 is of the VSEPR type AX_5, whereas IF_5 is AX_5E.

Exercise 10.7

Describe a hybridization scheme for the central atom and the molecular geometry of the triiodide ion, I_3^-.

Figure 10.18 Bonding scheme for iodine pentafluoride, IF_5. The central I atom is hybridized sp^3d^2. One of the hybrid orbitals is occupied by lone-pair electrons (blue) The other orbitals are the bonding orbitals. Each bond involves the overlap of an sp^3d^2 orbital with a $2p$ orbital of a terminal F atom. Because of repulsions between the lone-pair electrons and the I—F bond-pair electrons, the plane of the four F atoms is raised slightly above the I atom.

10.4 Hybrid Orbitals and Multiple Covalent Bonds

We have already learned how to use VSEPR theory to predict the geometric structures of molecules and polyatomic ions with double and triple covalent bonds (page 365). When we combine this knowledge with the valence bond method, we gain additional insight into the some of the essential characteristics of multiple covalent bonds (for example, bond energies).

In Chapter 9 we found the Lewis structure of ethylene, C_2H_4, to be

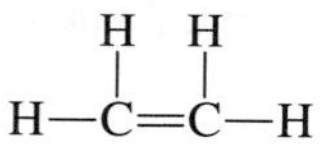

With VSEPR theory we would predict that the electron-group geometry around each C atom is *trigonal planar*. This corresponds to H—C—H bond angles of 120°. To account for these bond angles in the valence bond method, we assume that the $2s$ and two of the $2p$ orbitals of the valence shell combine to produce sp^2 hybridization of the two carbon atoms.

sp^2 *hybridization* in first C: sp^2 [↑][↑][↑] $2p$ [↑]

sp^2 *hybridization* in second C: sp^2 [↑][↑][↑] $2p$ [↑]

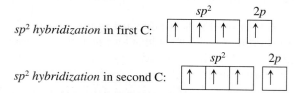

All the C—H bonds in C_2H_4 are formed by the overlap of sp^2 hybrid orbitals of the C atoms with $1s$ orbitals of the H atoms, with these orbitals directed along hypothetical lines joining the nuclei of the bonded atoms. Covalent bonds formed by this "end-to-end" overlap of atomic orbitals, regardless of their type, are called **sigma (σ) bonds**. All single covalent bonds are σ bonds. The double bond between the two C atoms has two components: One of the bonds involves sp^2 orbitals overlapping along the line joining the two carbon nuclei; it, like the C—H bonds, is a σ bond. The other bond between the two C atoms results from the overlap of the half-filled unhybridized $2p$ orbitals that extend above and below the plane of the C and H atoms. These orbitals overlap in a "parallel" or "side-by-side" fashion, not along a line joining the carbon nuclei. A bond formed by this type of orbital overlap is called a **pi (π) bond**. A double covalent bond consists of *one* σ and *one* π bond. The features of a double bond described here are illustrated in Figure 10.19.

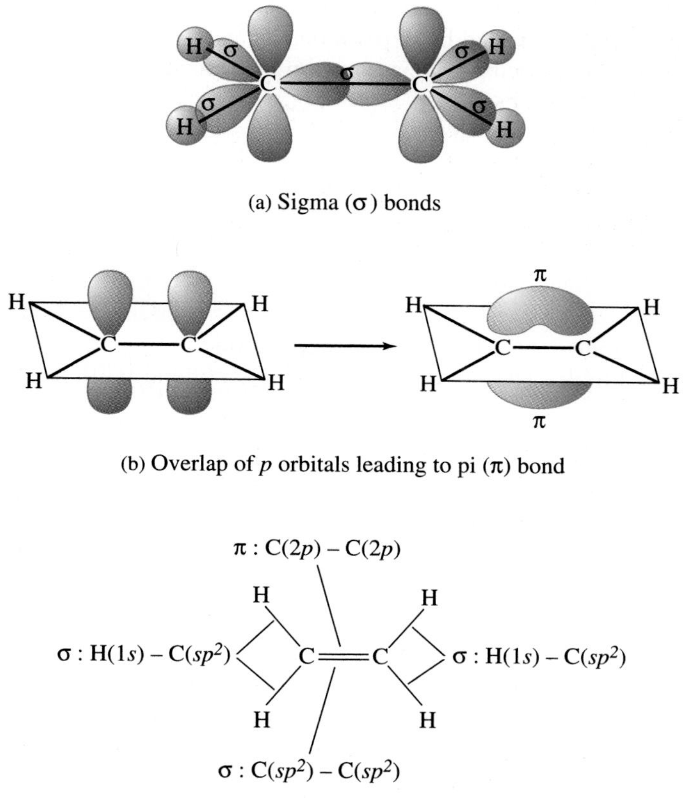

Figure 10.19 Bonding in ethylene, C_2H_4.

Note particularly the method of representing hybridization schemes and orbital overlaps introduced in Figure 10.19c.

VSEPR theory describes $H_2C{=}CH_2$ as consisting of two planar CH_2 groups, each with a 120° H—C—H bond angle. But it does not tell us how the two groups are oriented with respect to one another. Are they both in the same plane? Is one perpendicular to the other? The valence bond description of a double bond gives us an answer. The maximum sidewise overlap between the unhybridized $2p$ orbitals to form a π bond occurs when both CH_2 groups lie in the same plane. Ethylene is a *planar* molecule. If one CH_2 group is twisted out of the plane of the other, the extent of overlap of the $2p$ orbitals is reduced, the π bond is weakened, and the molecule becomes less stable.

Aside from the restriction in rotation about a double bond just described, electron pairs in π bonds do not fix the positions of bonded atoms to one another. These positions are established by the σ bonds, the so-called σ-bond framework, and there is only one σ-bond in a multiple covalent bond. VSEPR theory predicts an electron-group geometry equivalent to the σ-bond framework because all the electrons in a multiple covalent bond constitute a single electron group.

We can describe a triple covalent bond in a manner similar to that just used for a double bond. Consider the acetylene molecule, C_2H_2. Its Lewis structure is

$$H—C≡C—H$$

The molecule is *linear*, with 180° H—C—C bond angles, as predicted by VSEPR theory and confirmed by experiment. To account for these bond angles with the valence bond method, we assume *sp* hybridization of the valence shell orbitals of the two C atoms.

sp hybridization in first C atom: *sp* [↑ | ↑] 2*p* [↑ | ↑]

sp hybridization in second C atom: *sp* [↑ | ↑] 2*p* [↑ | ↑]

In the C≡C bond in C_2H_2, as in all triple bonds, *one* bond is a σ bond and *two* are π bonds. The bonding scheme in acetylene is illustrated in Figure 10.20.

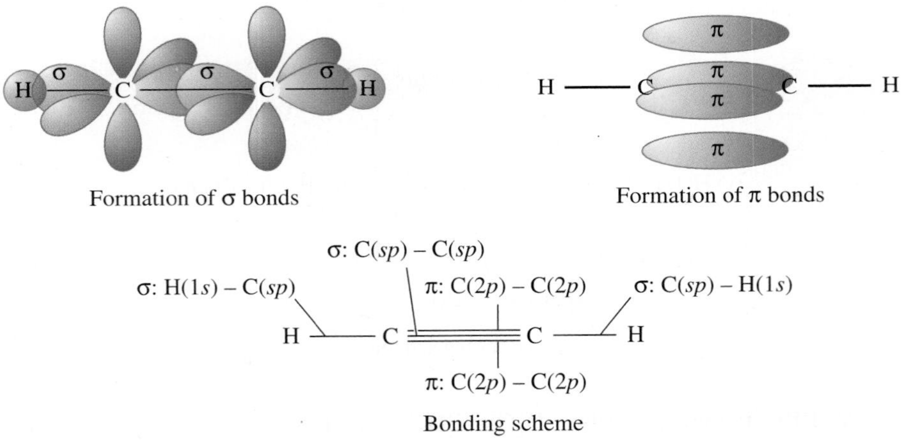

Figure 10.20 **Bonding in acetylene, C_2H_2.** The σ-bond framework joins the atoms in a linear structure through the overlap of 1*s* orbitals of the H atoms and *sp* orbitals of the C atoms. Each π bond can be thought of as two parallel cigar-shaped segments. In fact, however, when two π bonds are present, the segments merge into a hollow and symmetric cylindrical shell with the C-to-C σ bond as its axis.

Geometric Isomerism

We introduced alkenes in Section 9.10 and focused on the first two simple alkenes, ethylene, $H_2C{=}CH_2$, and propylene, $H_2C{=}CHCH_3$. There are four C_4H_8 alkene isomers. 1-Butene is related to butane and isobutylene is related to isobutane.

$H_2C{=}CHCH_2CH_3$ (1-Butene) $\qquad$ $H_2C{=}C(CH_3){-}CH_3$ (Isobutylene)

A third structure comes to mind, also related to butane, in which the double bond connects the second and third carbon atoms.

$$CH_3CH{=}CHCH_3$$

"2-Butene"

What is not immediately apparent is that there are *two* isomers with this structural formula. There are two different 2-butenes. We have not encountered this type of isomerism before. It occurs because of the restricted rotation about the double bond in the molecule.

The two arrangements of 2-butene are shown in Figure 10.21. At first, it appears that we should be able to convert one form to the other just by breaking the double bond, rotating one end of the molecule, and reforming the double bond. However, this does not readily happen because breaking the double bond requires too much energy. The two isomers of 2-butene are two distinctly different compounds, and to distinguish between them we call one *cis*-2-butene and the other *trans*-2-butene.

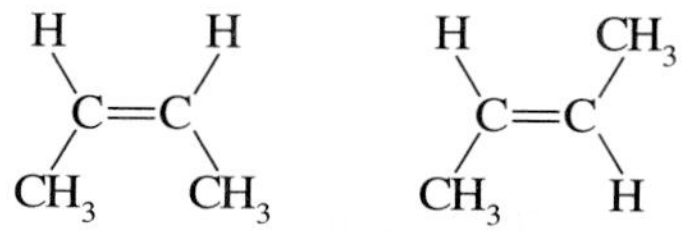

cis-2-Butene *trans*-2-Butene

In Latin, cis means "on this side of" and trans means "across." With these prefixes it's not difficult to figure out the meanings of the adjectives cisatlantic and transatlantic.

The **cis** isomer is the one with both methyl groups (CH_3) on the same side of the molecule. In the **trans** isomer, the methyl groups are on opposite sides or across the double bond from one another. (Draw a straight line passing through the two

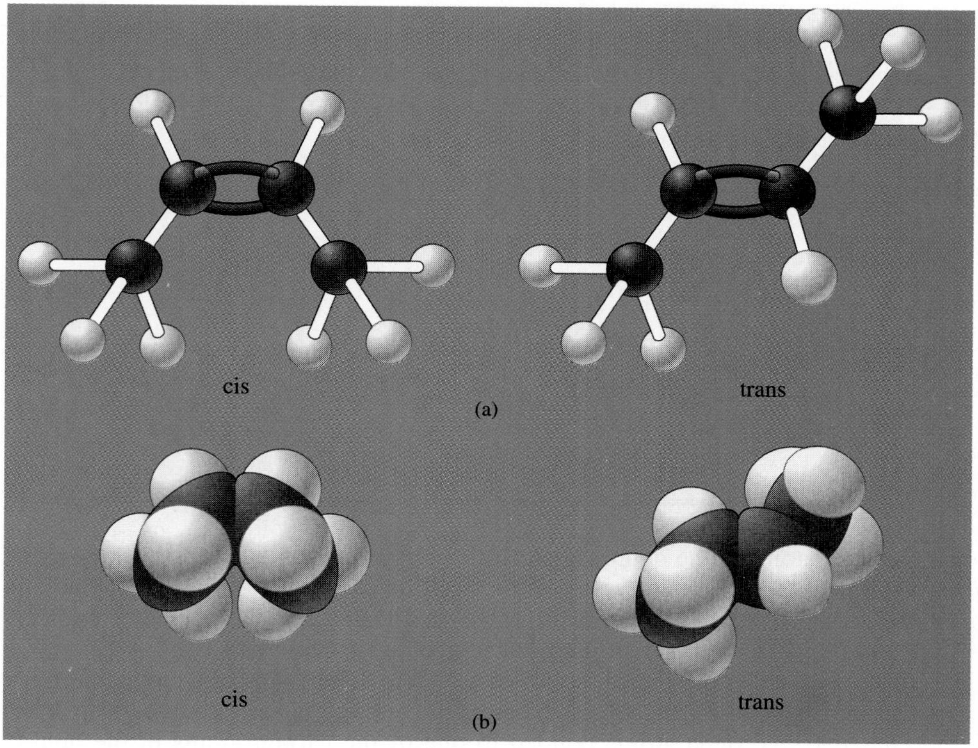

Figure 10.21 Geometric (cis-trans) isomerism in 2-butene. (a) Ball-and-stick models. (b) Space-filling models.

Geometric Isomerism and Vision

Our vision depends on the conversion by light of one geometric isomer to another. The light-sensitive pigment found in the receptor cells of the retina of the eye is rhodopsin, a complex of a compound called 11-*cis*-retinal and a protein called opsin. When light strikes rhodopsin, a reaction called *isomerization* occurs. The cis isomer is converted to the trans isomer.

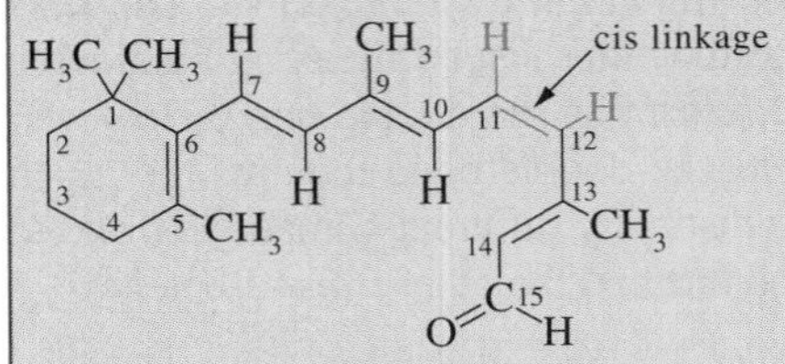

11-*cis*-Retinal

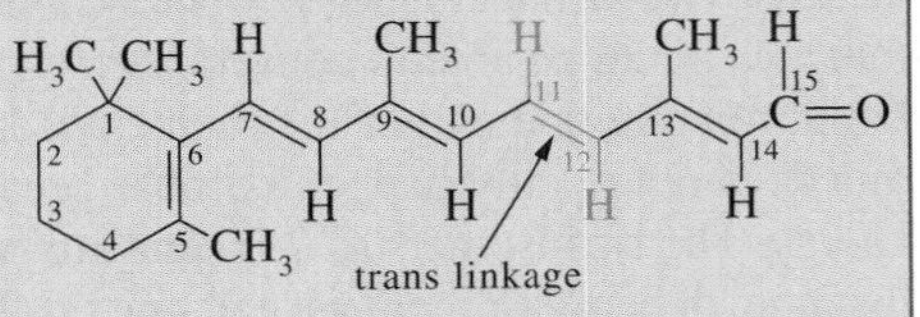

11-*trans*-Retinal

Cis to trans isomerization of 11-*cis*-retinal is accompanied by an electrical impulse that is transmitted to the brain through the optic nerve. The brain translates the nerve impulse into an image—we see. Following transmission of the impulse, the rhodopsin complex splits into opsin and free retinal, and an enzyme converts 11-*trans*-retinal back to 11-*cis*-retinal. The 11-*cis*-retinal again complexes with opsin and is primed for the next pulse of light.

Some retinal is lost during the regeneration of opsin from rhodopsin and must be replaced by vitamin A from the bloodstream, making vitamin A an essential vitamin for proper vision. Vitamin A differs from 11-*trans*-retinal only in that it has a terminal *alcohol* group, $—CH_2OH$, instead of the *aldehyde* group:

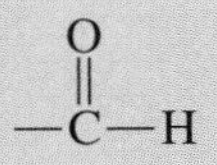

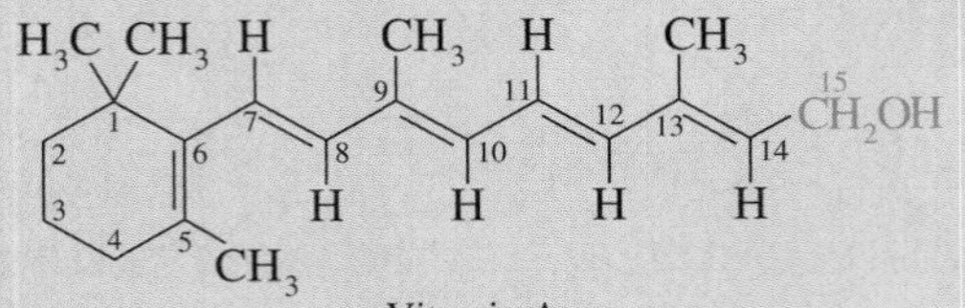

Vitamin A

Chapter 23 includes a discussion of cis and trans isomers of *inorganic* substances.

carbon atoms in the double bond. If the methyl groups fall on the same side of the line, the compound is *cis*. If they fall on opposite sides, the compound is *trans*.) Cis and trans isomers differ only in the geometrical arrangement of certain substituent groups; they are called **geometric isomers**.

If either of the carbon atoms in a double bond has two identical atoms or groups bonded to it, cis-trans isomerism is not possible. Propene (propylene) has two hydrogen atoms bonded to one of the carbon atoms in the double bond; it does not exhibit cis-trans isomerism. Although we can draw two structural formulas,

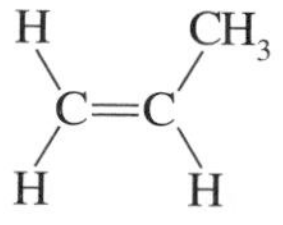

(I)

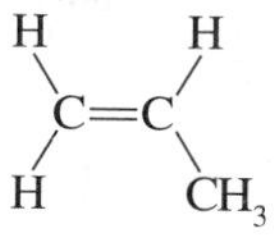

(II)

structure (II) is really no different from structure (I). You need only to imagine picking structure (II) up from the page and flipping it over to see that the two structures are identical.

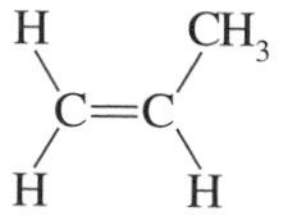

(I)

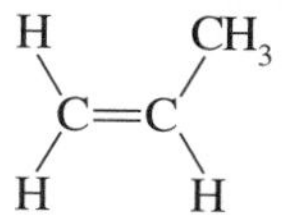

(II flipped over)

CONCEPTUAL EXAMPLE 10.2

Formic acid, HCOOH, is the irritant released in an ant bite (Latin *formica*, ant).

a. Predict a plausible molecular geometry for this molecule.

b. Propose a hybridization scheme for the central atoms that is consistent with the predicted geometry.

c. Sketch a bonding scheme for the molecule.

Solution

a. We begin by writing a plausible Lewis structure. This is based on $(2 \times 1) + (2 \times 6) + 4 = 18$ valence electrons, which are deployed in single bond pairs and as lone pairs on the O atoms. To complete the octet for C, we need to form a C-to-O double bond.

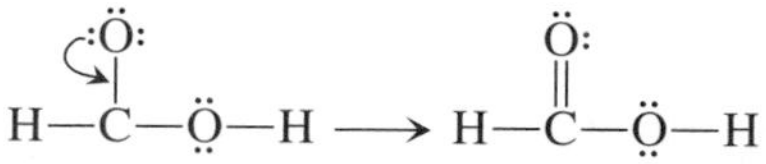

Then, with VSEPR theory, we describe the electron-group geometry about the central atoms, C and O. The orientation of three electron groups about the C atom—two single bonds and a double bond—is trigonal planar. The orientation of four electron groups about the central O atom—two bonding pairs and two lone pairs—is tetrahedral.

Because the three electron groups in the valence shell of the C atom are in chemical bonds, the molecular geometry around the C atom is the same as the electron-group geometry—*trigonal planar*. The H—C—O and the O—C—O bond angles are expected to be about 120°.

The molecular geometry around the central O atom is *angular* or *bent*, with an expected bond angle of about 104.5° (based on the VSEPR notation AX_2E_2 and the observed bond angle in H_2O).

b. The hybridization schemes consistent with the electron-group geometries described in (a) are sp^2 for the central C atom and sp^3 for the central O atom.

c. One of the bonds in the C=O double bond is a π bond. All the other bonds in the molecule are σ bonds. These bonds and the orbital overlaps producing them are suggested by the schematic diagram

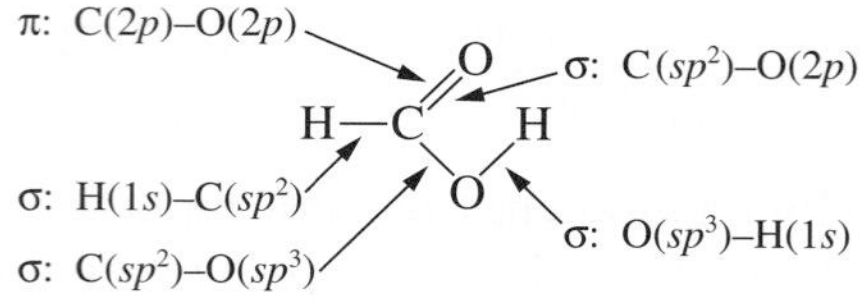

Conceptual Exercise 10.2

Methanol, CH_3OH, is the simplest alcohol; it shows promise as a future motor fuel.

a. Predict a plausible molecular geometry for this molecule.

b. Propose a hybridization scheme for the central atoms that is consistent with the predicted geometry.

c. Sketch a bonding scheme for the molecule.

10.5 A Brief Survey of Molecular Orbital Theory

The quantum mechanical treatment of electrons in *atoms* as matter waves yields *atomic* orbitals. A similar treatment applied to electrons in *molecules* yields **molecular orbitals (MO)**, which are mathematical descriptions of regions of high electron charge density in a molecule. By identifying these regions we can learn about the structure of a molecule.

It is easy to understand **molecular orbital theory** in principle: We seek an arrangement of appropriately placed atomic nuclei and electrons to produce an energetically favorable, stable molecule. The difficulty comes in practice, in trying to construct a wave equation for a system of several particles. The usual approach, which we will not attempt to pursue, is to write approximate wave equations by relating them to atomic orbitals.

Typical results are like those shown for the H_2 molecule in Figure 10.22. In place of the atomic orbitals of the separated atoms we obtain molecular orbitals for the united atoms, and these are of two types. One type, a **bonding molecular orbital**, places a high electron charge density between atoms. The other, an **antibonding molecular orbital**, places a high electron charge density away from the region between atoms. Electrons in bonding orbitals contribute to and electrons in antibonding orbitals detract from bond formation. (In some cases a third type of molecular orbital arises, a *nonbonding* orbital which neither contributes nor detracts from bonding.)

The **bond order** is simply one-half the difference between the number of electrons in bonding MOs and the number in antibonding MOs. The H_2 molecule has its two electrons in a bonding molecular orbital and a bond order of $\frac{1}{2} \times 2 = 1$; the H—H bond is a single covalent bond.

Three basic ideas that we used in writing the electron configurations of atoms carry over into the assignment of electrons to molecular orbitals.

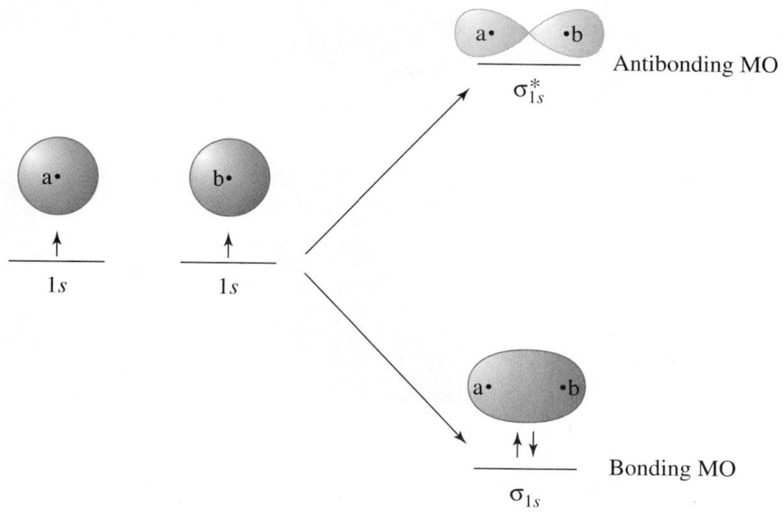

Figure 10.22 Molecular orbitals and bonding in the H_2 molecule. The short horizontal line segments represent the relative energy levels of the atomic and molecular orbitals. The molecular orbital, labeled σ_{1s} is a bonding orbital and σ^*_{1s} is an antibonding orbital. Distributions of electron charge density are suggested by the sketches above the energy levels. The small arrows represent electrons in the H atoms and in the H_2 molecule.

1. Electrons normally seek the lowest energy MOs available to them.
2. The maximum number of electrons that can be present in an MO is *two* (Pauli exclusion principle).
3. Electrons enter MOs of identical energies singly before they pair up (Hund's rule).

With these rules and the energy level diagram of Figure 10.22, we can see that the one-electron ion H_2^+ should have one electron in a bonding molecular orbital and a bond order of $\frac{1}{2}$. The three-electron ion, He_2^+, has two electrons in a bonding molecular orbital, one electron in an antibonding orbital, and a bond order of $\frac{1}{2}(2 - 1) = \frac{1}{2}$. Both of these ions exist as stable ions; they have bond energies that are about one-half that of the H_2 molecule. Molecular orbital theory more easily explains bonding in odd-electron species than does the valence bond method.

Bonding in the O_2 Molecule

The Lewis structure

$$:\ddot{O}=\ddot{O}:$$

which indicates an O-to-O double bond, is consistent with the observed bond length and bond energy listed in Table 9.1. However, it is not consistent with the experimentally determined fact that oxygen is *paramagnetic*—O_2 molecules have unpaired electrons.

Molecular orbital theory accounts for *both* the double bond character of O_2 and its paramagnetism. The eight molecular orbitals available to the valence electrons of the O atoms in O_2 are represented in the energy-level diagram of Figure 10.23. The 12 valence-shell electrons of the two O atoms enter the molecular orbitals in the order in which they are numbered in the figure. The last two electrons enter two degenerate antibonding molecular orbitals, singly. These are the two unpaired electrons responsible for the paramagnetism of the

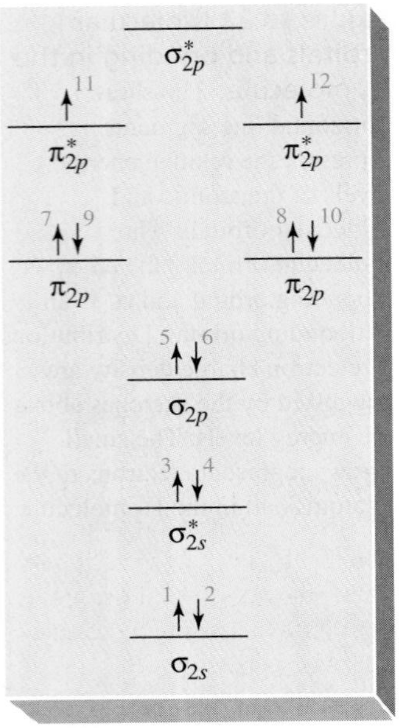

Figure 10.23 **Molecular-orbital energy diagram for O_2.** The eight valence-shell atomic orbitals associated with the two O atoms in O_2 are replaced by eight molecular orbitals. The molecular orbitals are arranged from lowest energy (bottom) to highest energy (top). The symbols under the energy-level lines are orbital designations used in MO theory. Those marked with an asterisk (*) are antibonding orbitals; the others are bonding orbitals. Twelve valence electrons are assigned to the orbitals in the order in which the arrows are numbered.

The paramagnetism of liquid oxygen is seen through the attaction of the liquid into the magnetic field of a large magnet.

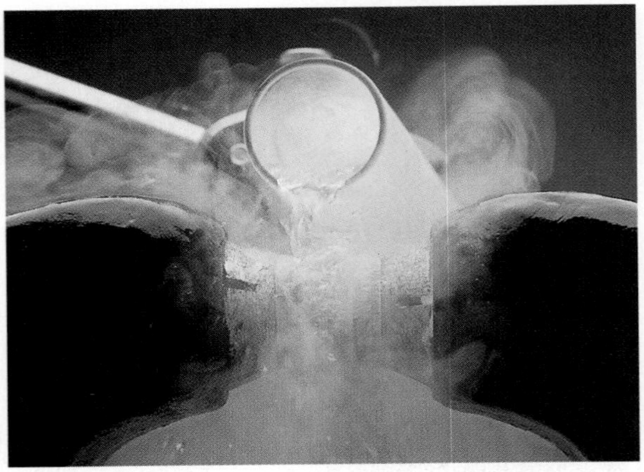

O_2 molecule. The total number of electrons in bonding orbitals is eight, and in antibonding orbitals, four. The bond order is $(8 - 4)/2 = 2$, corresponding to an O-to-O double covalent bond.

Bonding in the Benzene Molecule

In Section 9.10 we described the benzene molecule, C_6H_6, as the resonance hybrid of two principal contributing structures—the Kekulé structures.

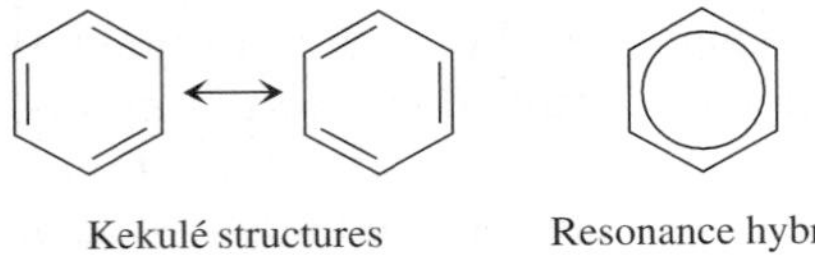

Representing the resonance hybrid as a hexagon with an inscribed circle conveys the idea that six of the bonding electrons are *delocalized.* That is, these six electrons are shared by carbon atoms throughout the ring rather than being localized between any particular pairs of atoms. Molecular orbital theory gives us another way of looking at delocalized electrons.

Consider that each of the six carbon atoms in C_6H_6 has three sp^2 hybridized orbitals, just like the two carbon atoms in ethylene, C_2H_4. This corresponds to the following valence-shell orbital diagram.

sp^2 hybridization in 6 C atoms:

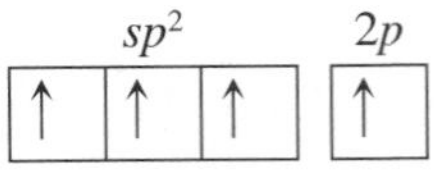

In addition, consider that each C atom is bonded to two other C atoms and an H atom through overlaps involving those *three* sp^2 hybrid orbitals. This creates a σ-bond framework in which all the atoms lie in the same plane and all the bond angles are 120°. The *six* half-filled 2*p* atomic orbitals presumably combine to produce *six* molecular orbitals, *three* of which are bonding and *three,* antibonding. Finally, as shown in Figure 10.24, the six 2*p* electrons occupy the bonding molecular orbitals and leave the antibonding orbitals empty. The bonding attributed to these electrons in delocalized molecular orbitals is referred to as π-bonding.

All the carbon-to-carbon bonds in the σ-bond framework are single bonds. The bond order associated with the six electrons in the -bonding molecular orbitals is (6−0)/2 = 3. If we apportion these three bonds among the six carbon atoms, this amounts to an additional $\frac{1}{2}$ bond between each pair of C atoms, suggesting that the C-to-C bond order overall is $1\frac{1}{2}$. This is the same conclusion that we reach by "averaging" the two Kekulé structures. Figure 10.25 is a computer-generated model of the benzene molecule emphasizing the delocalized electrons.

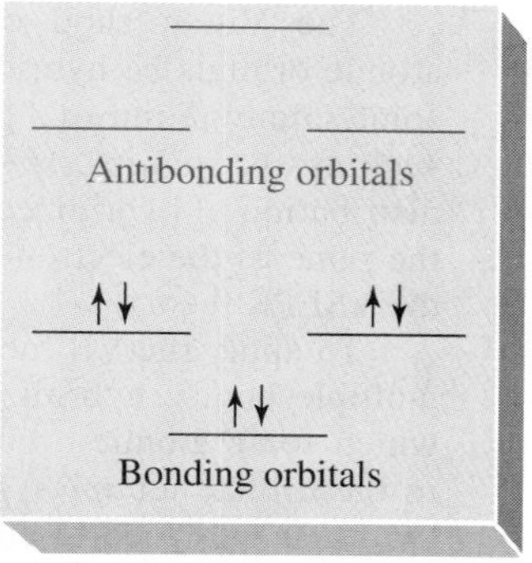

Figure 10.24 **π Molecular-orbital diagram for C_6H_6.** The three lower energy bonding molecular orbitals are filled with electron pairs. The three higher energy antibonding molecular orbitals remain empty.

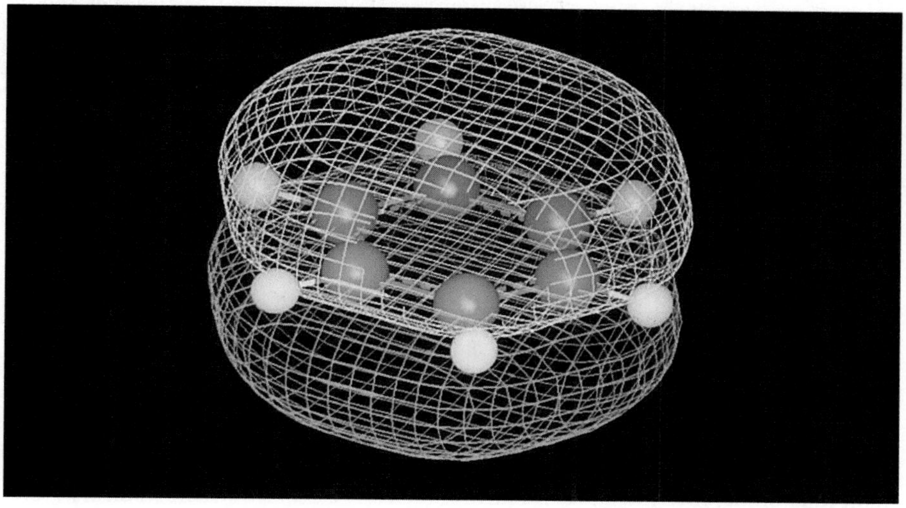

Figure 10.25 **Representation of bonding in C_6H_6.** The σ-bond framework of 12 coplanar atoms (six C and six H) is represented with a ball-and-stick model. The π molecular orbitals (doughnut-shaped regions) above and below the plane of C and H atoms are highlighted.

SUMMARY

VSEPR theory predicts the geometrical structures of molecules and polyatomic ions based on mutual repulsions among valence-shell electron groups.. The theory requires determining the number of valence-shell electron groups for each central atom, and assessing the geometrical distribution of these electron groups. If all electron groups consist of bonding electrons, the molecular (actual) geometry is the same as the electron-group geometry. If some of the electron groups are lone pairs, the molecular and electron-group geometry are related, but not identical. Multiple bonds, whether they consist of four electrons (double bond) or six electrons (triple bond), are treated as single electron groups.

The separation of the centers of positive and negative charge in a polar covalent bond creates a bond moment. Whether a molecule as a whole is polar, that is, whether there is a resultant dipole moment, is established by bond moments *and* molecular geometry. A symmetrical distribution of identical bond moments about a central atom can result in a cancellation of all bond moments, with the result that the molecule is nonpolar, as is the case with CCl_4.

In the valence bond (VB) method a covalent bond is viewed as the overlap of atomic orbitals of the bonded atoms in a region between the atomic nuclei. Molecular geometry is determined by the spatial orientations of the atomic orbitals involved in bonding.

The VB method often requires that bonding atomic orbitals be hybridized. A hybridized orbital is some combination of "pure" *s*, *p*, *d*, and *f* orbitals, such as *sp*, sp^2, sp^3, sp^3d, and sp^3d^2. The geometric distribution of hybridized orbitals in the VB method is the same as the electron-group geometry proposed by the VSEPR theory.

To apply the VB method to structures containing multiple bonds, hybridization schemes are needed in which some atomic orbitals are left unhybridized, as in the orbital set $sp^2 + p$. Hybrid orbitals overlap in the usual way (end-to-end) and form σ bonds. Unhybridized *p* orbitals overlap in a side-by-side manner and give rise to π bonds. A double bond consists of one σ bond and one π bond; a triple bond, one σ bond and two π bonds.

Acceptable solutions to wave equations written for the electrons in a molecule are called molecular orbitals (MO). The two main types of MOs are *bonding* molecular orbitals, which concentrate electron charge density between atoms, and *antibonding* molecular orbitals, which concentrate electron charge densities away from the internuclear bonding region. Electrons can be assigned to molecular orbitals by a scheme similar to the aufbau process. MO theory provides more satisfactory descriptions for certain structures than does the VB method, for example, some odd-electron species and the O_2 molecule.

KEY TERMS

antibonding molecular orbital (10.5)
bond order (10.5)
bond moment (10.1)
bonding molecular orbital (10.5)
cis (10.4)
debye (D) (10.1)
dipole moment (μ) (10.1)
electron group (10.1)
electron-group geometry (10.1)
geometric isomers (10.4)
hybridization (10.3)
hybrid orbital (10.3)
molecular geometry (10.1)
molecular orbital (MO) (10.5)
molecular orbital theory (10.5)
pi (π) bond (10.4)
resultant dipole moment (10.1)
sigma (σ) bond (10.4)
sp (10.3)
sp^2 (10.3)
sp^3 (10.3)
sp^3d (10.3)
sp^3d^2 (10.3)
trans (10.4)
valence bond (VB) method (10.2)
valence-shell electron-pair repulsion (VSEPR) theory (10.1)

REVIEW QUESTIONS

1. Explain why all diatomic molecules are said to be linear.

2. Is it possible for a molecule consisting of three or more atoms to be linear? Explain.

3. In VSEPR theory, what is the distinction between the terms *electron-group geometry* and *molecular geometry*?

4. In VSEPR theory electron pairs are designated as bonding pairs (BP) or lone pairs (LP). Which of the electron-pair repulsions are the strongest? Explain.

5. In VSEPR theory, what is the meaning of the notation AX_2E_2?

6. What is the VSEPR notation that should be written for the central atom in each of the following?
a. PH_3 **b.** OCl^- **c.** NO_2^-

7. What approximate bond angle is expected in triatomic molecules having the following electron-group geometries about the central atom?
a. linear
b. trigonal planar
c. tetrahedral

8. Is it possible to have a *linear* molecule in which the electron-group geometry is *tetrahedral*? If so, give an example.

9. Explain why it is not necessary to find the "best" Lewis structure when applying the VSEPR theory.

10. Explain how VSEPR theory can be applied to a molecule or ion containing more than one central atom.

11. Must every chemical bond have a bond dipole moment? Explain.

12. Must every molecule have a resultant dipole moment? Explain.

13. Explain why SO_2 is a polar molecule whereas SO_3 is not.

14. Water has a resultant dipole moment of 1.84 D. Explain why this fact proves that the H_2O molebule must have a bent shape.

15. Explain how the valence bond method is able to predict a 90° bond angle in H_2S, whereas the VSEPR prediction is 109.5°.

16. Why is it necessary to introduce the idea of *hybridized* atomic orbitals in the valence bond approach to chemical bonding?

17. In valence bond theory, what is the meaning of the designation sp^3?

18. Which hybridization scheme in valence bond theory accounts for trigonal planar electron-group geometry? For octahedral electron-group geometry?

19. What hybridization scheme for the central atom will yield the expected molecular geometry of each of the following?

a. CCl_4 **b.** CO_2

20. Nitrogen and phosphorus are both in Group 5A of the periodic table. Apparently, a phosphorus atom can employ sp^3d hybridization in forming bonds to other atoms. Can a nitrogen atom do likewise? Explain.

21. Explain the difference between a σ and a π bond.

22. Why does the valence bond method account for σ and π bonds whereas Lewis structures do not?

23. Is the following a valid statement? "All bonds in an alkane hydrocarbon are σ bonds." Explain.

24. Is the following a valid statement? "All bonds in an alkene hydrocarbon are π bonds." Explain.

25. Describe the difference between an *atomic* orbital and a *molecular* orbital.

26. How do *antibonding* molecular orbitals differ from *bonding* molecular orbitals?

27. Can electrons occupy antibonding molecular orbitals? If so, under what conditions?

28. How does molecular orbital theory define the term *bond order*? Under what conditions can there be a fractional bond order (e.g., $\frac{1}{2}$ or $\frac{3}{2}$)?

PROBLEMS

VSEPR Theory

29. Predict whether each of the following species is probable. For those that seem improbable, tell why they are.

a. a linear H_2O molecule
b. a planar SO_3 molecule
c. a planar PH_3 molecule

30. Predict whether each of the following species is probable. For those that seem improbable, tell why they are.

a. a tetrahedral $GeCl_4$ molecule
b. a trigonal bipyramidal NCl_5 molecule
c. a bent HCN molecule

31. Explain why the BF_3 molecule is trigonal planar, whereas a molecule with a similar formula, ClF_3, is T-shaped.

32. Explain why the ICl_4^- ion is square planar, whereas an ion with a similar formula, BF_4^- is tetrahedral.

33. Predict the molecular geometry of each of the following.

a. PCl_3 **b.** ClO_4^-
c. XeF_4 **d.** OCN^-

34. Predict the molecular geometry of each of the following.

a. PH_4^+ **b.** NI_3
c. Cl_2CO **d.** NSF

35. Describe the geometric shape of each of the following.

a. H_2O_2 **b.** HOCN **c.** HNO_2

36. Describe the geometric shape of each of the following.

a. $HClO_3$ **b.** N_2O_4 **c.** CH_3CN

Valence Bond Method

37. Indicate the hybridization scheme expected for the central atom in each of the following.

a. OF_2 **b.** NH_4^+ **c.** CO_2 **d.** $COCl_2$

38. Indicate the hybridization scheme expected for the central atom in each of the following.

a. BF_4^- **b.** SO_3 **c.** NO_2^- **d.** XeF_4

39. Indicate the hybridization schemes expected for the central atoms in each of the following
a. C_2N_2 **b.** HNCO
c. NH_2OH **d.** CH_3COOH

40. Indicate the hybridization schemes expected for the central atoms in each of the following
a. CH_3CN **b.** CH_3NH_2
c. $CH_3C{\equiv}CCH_3$ **d.** CH_3NCO

41. Propose a simple Lewis structure (or resonance structures), the molecular geometry, and a bonding scheme for each of the following
a. $ClNO_2$ **b.** OF_2 **c.** CO_3^{2-}

42. Propose a simple Lewis structure (or resonance structures), the molecular geometry, and a bonding scheme for each of the following.
a. HNO_3
b. AsF_6^-
c. CH_3CCH

43. In both of the ions, $IC1_2^+$ and $IC1_2^-$, an iodine atom is bonded to two C1 atoms. Do you expect that the same hybridization scheme for the central I atom applies in each case? Explain.

44. Both in the molecule OSF_4 and the ion SF_5^- a sulfur atom is bonded to five other atoms. Do you expect that the same hybridization scheme for the central S atom applies in each case? Explain.

45. The structure of oxalate ion, $C_2O_4^{2-}$, is represented below.

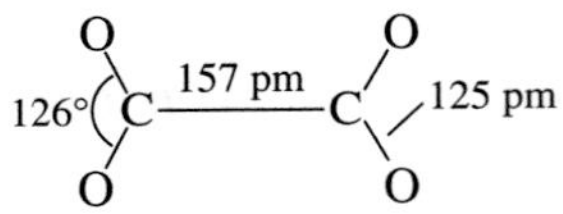

Propose hybridization and bonding schemes consistent with this structure. (*Hint:* Use data from Table 9.1.)

46. The structure of hydrazoic acid, HN_3, is indicated below.

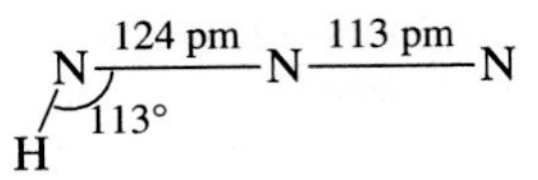

Propose hybridization and bonding schemes consistent with this structure. (*Hint*: See Table 9.1.)

Geometric Isomerism

47. Which of the following compounds can exist as cis-trans isomers?
a. $CH_3CH_2CH{=}CHCH_2CH_3$
b. $CH_2{=}CHCH_2CH_2CH_3$
c. $CH_3CH_2CH{=}CHCH_3$

48. Which of the following compounds can exist as cis-trans isomers?

a. $CH_3C{=}CHCH_2CH_3$
  |
  CH_3

b. $CH_3CH{=}CHCH_2CH_2CH_3$

c. $CH_3CHCH{=}CHCH_3$
  |
  CH_3

Molecular Orbital Theory

49. Assuming the same molecular orbitals as found in H_2, explain why you would not expect the molecule He_2 to exist.

50. The oxygen molecule ion, O_2^+, has a *stronger* bond than does the O_2 molecule itself. Explain how this can be.

ADDITIONAL PROBLEMS

51. Do you think the following is a valid statement? "The greater the electronegativity difference between the atoms in a molecule, the greater is the resultant dipole moment of that molecule." Explain.

52. The NO_2F molecule depicted in Conceptual Example 10.1 (p. 371) is a symmetrical molecule with bond angles of about 120°. Why doesn't it have a dipole moment of 0?

53. Apply VSEPR theory to predict the geometric shape of the BrF_4^+ ion.

54. Apply VSEPR theory to predict the geometric shape of the XeF_5^+ ion.

55. Draw a Lewis structure for the N_2 molecule and then predict a hybridization scheme consistent with this structure. Could bonding in N_2 be described by the use of pure (unhybridized) atomic orbitals? Explain.

56. Phosphorus pentachloride is molecular in the gas phase, but in the solid phase it is an ionic compound consisting of the ions $[PCl_4]^+$ and $[PCl_6]^-$. Propose hybridization and bonding schemes to represent these molecular and ionic forms of phosphorus pentachloride.

57. Carbon suboxide, C_3O_2, a foul smelling gas, is a less stable oxide than either CO or CO_2. The C—O bond lengths in the molecule are 116 pm and the C—C bond lengths are 128 pm. Draw a plausible Lewis structure for this molecule, propose hybridization and bonding schemes, and predict its geometric shape. (*Hint*: Use data from Table 9.1.)

58. The structure of the molecule allene, CH_2CCH_2, is indicated at the right. Propose hybridization schemes for the C atoms in this molecule.

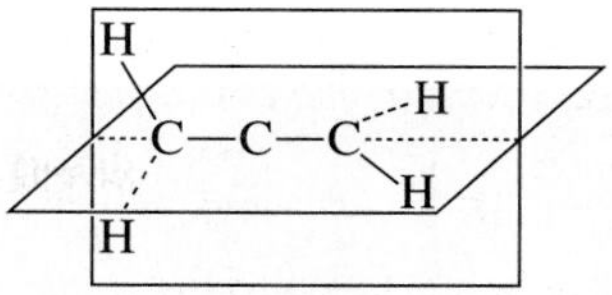

59. In Problem 49 it is stated that the molecule He_2 does not exist. Could this molecule exist in an electronically excited state? Explain.

60. Under appropriate conditions potassium forms the normal oxide K_2O, but more often it forms the superoxide, KO_2. Represent these ionic compounds through Lewis structures, and comment on the difficulty presented by KO_2. Show how the molecular orbital theory can help to resolve this difficulty.

Water exists in three physical states in this view of Paradise Bay in Antarctica. Water vapor is invisible, but condensed into liquid droplets, water is visible in clouds. The liquid water in the bay and the solid water in the snow and ice on the mountains and in the glacier represent the two condensed states that are the focus of this chapter.

11

Liquids, Solids, and Intermolecular Forces

CONTENTS

Liquids and solids are more complex than gases, and many of the techniques we used with gases will not work with liquids and solids. For example, we cannot write an equation that is comparable to the ideal gas equation for either liquids or solids. Instead of doing quantitative calculations, our emphasis in this chapter will be on acquiring a *qualitative* understanding of

natural phenomena such as

- why ice floats on liquid water;
- why dry ice (solid CO_2) does not normally melt;
- why it takes longer to boil an egg in the mountains than at the sea shore;
- why diamond, one form of carbon, is so hard that it can scratch glass, but graphite, another form of carbon, is soft enough to use in pencils.

11.1 Intermolecular Forces and the States of Matter

In the previous two chapters our main concern was the nature of the forces that bind atoms to one another *within* molecules. We called these forces chemical bonds, but we could also have called them *intra*molecular forces (see Figure 11.1). We learned that the geometric shapes and resulting chemical properties of molecules are determined by these intramolecular forces.

intra, Latin for "within"; *inter*, Latin for "between."

Attractive forces that exist *between* molecules—*inter*molecular forces—are equally important (see Figure 11.1). Intermolecular forces determine the *physical* properties of liquids and solids. In fact, if there were no intermolecular forces, there would be no liquids or solids—everything would be in a gaseous state.

In our physical model of gases, we visualize speedy, energetic molecules undergoing frequent collisions and never coming to rest or clumping together. Intermolecular forces are relatively unimportant in gases; in fact, in ideal gases we assume that they do not even exist. If intermolecular forces are sufficiently strong, however, molecules cluster together into the liquid or solid states. In our discussion of liquids and solids, we need to say something about the different kinds of intermolecular forces and their relative strengths. However, before we discuss this matter, in Section 11.5, let's first consider some fundamental phenomena involving liquids and solids.

We can acquire a pretty good initial understanding of liquids and solids with this simple observation:

> *Molecules have a tendency to remain apart from each other. Intermolecular forces of attraction are most likely to overcome this tendency at low temperatures, where thermal energies are diminished, and at high pressures, where molecules are forced close together.*

We can summarize this observation with the diagram in Figure 11.2, which is a plot of the pressure versus the temperature under which a substance is maintained. A substance exists as a gas at points that correspond to high temperatures and low pressures; here, molecules are far apart. The solid state exists at points that represent low temperatures and moderate to high pressures; here, molecules are most closely packed. The liquid state—a sort of intermediate condition—exists at points that correspond to intermediate temperatures and moderate to high pressures.

The adjectives "high" and "low" are relative terms. A temperature of 1000 K is far above the temperatures at which oxygen exists as a liquid. On the other hand, 1000 K is a low temperature in relation to the melting point of iron.

The collections of points just described define *areas* in the pressure–temperature graph, one area for the gaseous state of a substance, one for its liquid state, and one for its solid state. The different areas are separated by *curves*. Points on these curves represent conditions under which a substance exists in two different states of matter at the same time: liquid and gas, liquid and solid, and solid and gas.

In its entirety, the pressure–temperature graph in Figure 11.2 is called a phase diagram. A **phase diagram** indicates the conditions of temperature and pressure under which a substance exists as a solid, a liquid, a gas, or some combination of these in equilibrium. Although the terms "phase" and "state" of matter are often used interchangeably, they have a slightly different meaning. A *phase* is a physically distinct portion of a sample of matter having a unique set of properties. A mixture of diamond and graphite exists in *two* phases (all of the diamond is one phase and all of the graphite is the other) in a *single* state of matter: solid. We will discuss some specific phase diagrams in Section 11.4, but first, in the next two sections, let's consider some of the features of the general phase diagram of Figure 11.2.

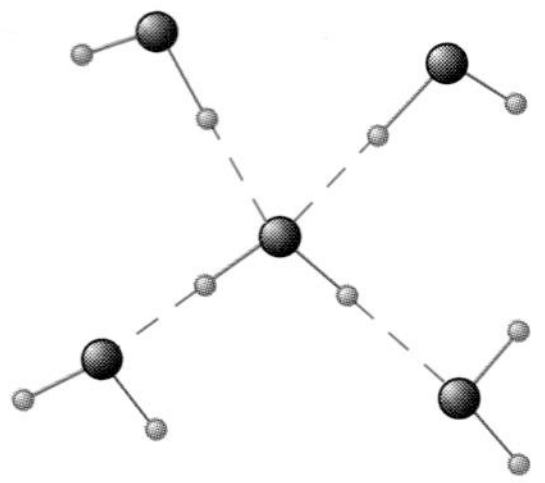

Figure 11.1 *Inter*molecular and *intra*molecular forces compared. In the hypothetical situation described here *intra*molecular forces—chemical bonds—are represented by the green lines. Forces between molecules—*inter*molecular forces—are suggested by the broken red lines.

11.2 Vaporization and Vapor Pressure

To clean up a water spill on a tile floor, we use a mop to spread the water into a thin film. The floor dries as the water then evaporates. Evaporation, or **vaporization**, is the conversion of a liquid to a gas (vapor).

Let's try to picture the process of vaporization at the molecular level. The molecules of a liquid have a range of kinetic energies, with the *average* kinetic energy determined by the temperature. Molecules at the surface of a liquid that have considerably more kinetic energy than the average are able to overcome the intermolecular attractive forces of neighboring molecules in and below the surface. They fly off into the gaseous or vapor state—they vaporize.

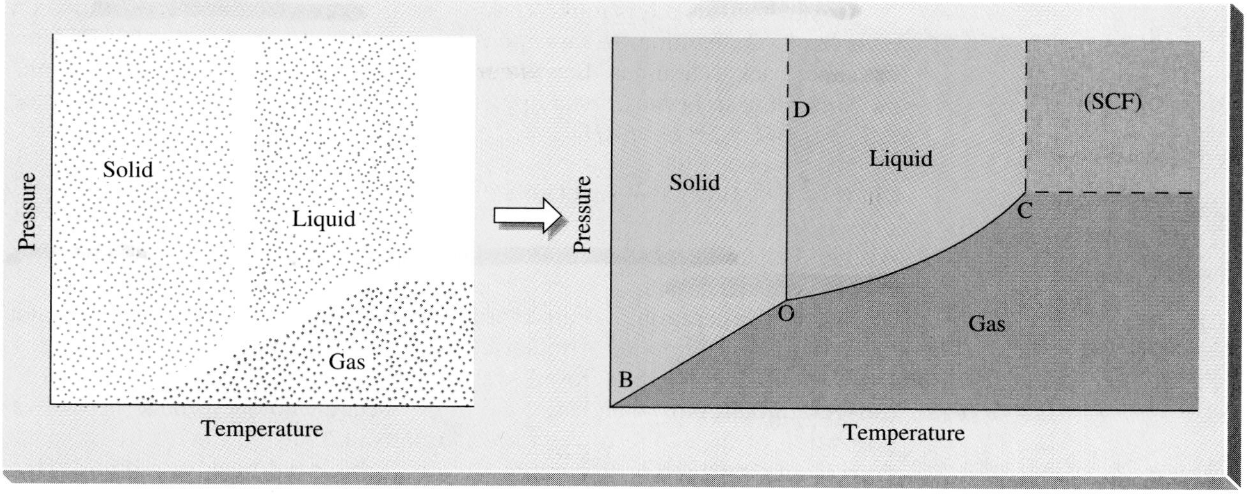

Figure 11.2 **A phase diagram: representing temperatures, pressures, and states of matter.** The collection of temperature–pressure points suggested at the left leads to the diagram at the right—a phase diagram. Details of the diagram are discussed in the several sections that follow.

As more energetic molecules leave the surface of a liquid through vaporization, the average kinetic energy of the molecules remaining in the liquid decreases. The temperature of the liquid *decreases*, as in the familiar cooling sensation on the skin when a volatile liquid such as rubbing alcohol (isopropyl alcohol) evaporates.

Enthalpy (Heat) of Vaporization

To maintain a liquid at a *constant* temperature while it vaporizes, we must heat it to replace the excess kinetic energy carried away by the vaporizing molecules. The **enthalpy (heat) of vaporization** is the quantity of heat that must be absorbed to vaporize a given amount of liquid at a constant temperature. Usually this is expressed in kilojoules per mol (kJ/mol), often in conjunction with an equation, such as

$$H_2O(l) \longrightarrow H_2O(g) \qquad \Delta H = 44.0 \text{ kJ/mol (at 298 K)}$$

We can also signify that an enthalpy change is for a vaporization process by attaching the subscript "vapn" to the ΔH symbol. Several ΔH_{vapn} values are listed in Table 11.1.

TABLE 11.1 **Some Enthalpies (Heats) of Vaporization at 298 K[a]**

Liquid	ΔH_{vapn}, kJ/mol
Carbon disulfide, CS_2	27.4
Carbon tetrachloride, CCl_4	37.0
Methanol, CH_3OH	38.0
Octane, C_8H_{18}	41.5
Ethanol, CH_3CH_2OH	43.3
Water, H_2O	44.0
Aniline, $C_6H_5NH_2$	52.3

[a] ΔH_{vapn} values are somewhat temperature dependent.

Condensation refers to the conversion of a gas (vapor) to a liquid. It is the reverse of vaporization. If we vaporize a quantity of liquid and then condense the vapor back to liquid at the same temperature, the total enthalpy change must be *zero*. This is because enthalpy is a function of state. As a consequence, $\Delta H_{\text{vapn}} + \Delta H_{\text{condn}} = 0$, or $\Delta H_{\text{condn}} = -\Delta H_{\text{vapn}}$. Thus,

$$H_2O(g) \longrightarrow H_2O(l) \qquad \Delta H = -44.0 \text{ kJ/mol (at 298 K)}$$

Also note that vaporization is always an *endothermic* process and condensation is an *exothermic* one.

Most refrigeration and air-conditioning systems are based on a repeated cycle of vaporization and condensation. The refrigerant, a volatile liquid, is allowed to vaporize in a closed system. The surroundings—the interior of a refrigerator, a room, or an entire building—become colder as heat (heat of vaporization of the liquid refrigerant) is removed. Outside the enclosed space, the gaseous refrigerant is compressed and condensed back to a liquid. Heat (heat of condensation) is expelled. We can see, then, that through repeated cycles of vaporization and condensation, a refrigerator cools its contents and heats the room where it is kept.

Example 11.1

How much heat, in kilojoules, is required to vaporize 175 g methanol, CH_3OH, at 25 °C?

Solution

The ΔH_{vapn} value listed in Table 11.1 is for one *mole* of methanol at 298 K (25 °C). We must convert the quantity of methanol to moles and then multiply by the ΔH_{vapn} value.

$$? \text{ kJ} = 175 \text{ g } CH_3OH \times \frac{1 \text{ mol } CH_3OH}{32.04 \text{ g } CH_3OH} \times \frac{38.0 \text{ kJ}}{1 \text{ mol } CH_3OH} = 208 \text{ kJ}$$

Exercise 11.1

To vaporize a 1.50-g sample of liquid benzene, C_6H_6, requires 652 J of heat. Calculate ΔH_{vapn} of benzene in kJ/mol.

Estimation Exercise 11.1

Which liquid in Table 11.1 requires the greatest quantity of heat to vaporize 1.00 kg of the liquid at 25 °C? Which requires the smallest quantity of heat?

Vapor Pressure

When a liquid is placed in an *open* container, eventually all the liquid molecules escape into the vapor state and disperse throughout the atmosphere. Something different happens with a liquid in a *closed* container. At first the volume of liquid decreases as some of the liquid is converted to vapor, but then evaporation seems to stop. The volume of liquid remains constant, and it appears that nothing further is happening.

Now let's see what's going on at the molecular level (Figure 11.3). From the very start, as soon as molecules appear in the vapor state, some of the vapor molecules strike the liquid surface, are captured, and return to the liquid state. Condensation occurs simultaneously with vaporization. We can represent these two opposing simultaneous processes by arrows pointing in opposite directions:

$$\text{Liquid} \underset{\text{condensation}}{\overset{\text{vaporization}}{\rightleftharpoons}} \text{Vapor}$$

At first, many more molecules pass from the liquid to the vapor than in the reverse direction. The rate of vaporization is greater than the rate of condensation. As the number of vapor molecules increases, however, so does the rate of condensation. Eventually the rate of condensation becomes equal to the rate of vaporization. When this happens the number of molecules per unit volume in the vapor state remains constant with time.

The **vapor pressure** of a liquid is the pressure exerted by the vapor in dynamic equilibrium with a liquid at a constant temperature. **Dynamic equilibrium** occurs whenever two opposing processes take place at exactly the same rate. Although nothing further seems to happen in an equilibrium condition, things keep going on at the molecular level. The condition is *dynamic,* not static. The time required to reach liquid–vapor equilibrium at a particular temperature depends on several factors, including the volume of the container and the surface area of the liquid. The equilibrium vapor pressure, on the other hand, depends only on the identity of the liquid and the equilibrium temperature.

Liquid–vapor equilibrium is a *physical* equilibrium; it involves opposing physical processes. In Chapter 16 we will consider *chemical* equilibrium, where the opposing processes are chemical reactions.

As the temperature of a liquid is raised, more molecules have enough energy to escape from the liquid state. The rate of vaporization increases. At equilibrium, the rates of vaporization and condensation are again equal, but the pressure exerted by the vapor is higher at the higher temperature. The vapor pressures of liquids increase with temperature.

Table 11.2 lists the vapor pressure of water at various temperatures. Perhaps you recall that we have already used data from this table when applying Dalton's

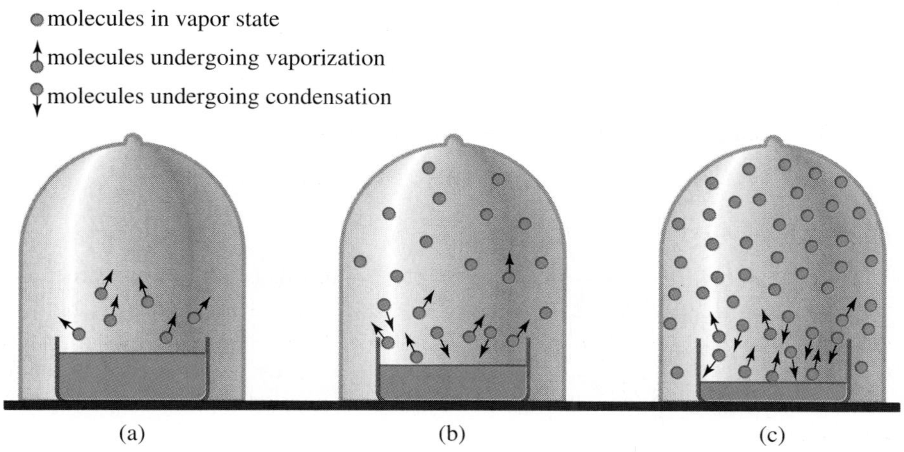

Figure 11.3 Liquid–vapor equilibrium and vapor pressure. (a) Vaporization of a liquid begins. (b) Condensation begins as soon as the first vapor molecules appear, though in this illustration the rate of condensation is still less than the rate of vaporization. (c) The rates of vaporization and condensation have become equal. The maximum number of molecules that can be accommodated in the vapor state has been reached. The partial pressure exerted by these molecules—the vapor pressure of the liquid—remains constant.

law of partial pressures to the collection of gases over water (page 138). Example 11.2 illustrates how vapor pressure may also enter into an ideal gas law calculation.

TABLE 11.2 Vapor Pressure of Water at Various Temperatures

Temperature, °C	Pressure, mmHg	Temperature, °C	Pressure, mmHg	Temperature, °C	Pressure, mmHg
0.0	4.6	29.0	30.0	93.0	588.6
10.0	9.2	30.0	31.8	94.0	610.9
20.0	17.5	40.0	55.3	95.0	633.9
21.0	18.7	50.0	92.5	96.0	657.6
22.0	19.8	60.0	149.4	97.0	682.1
23.0	21.1	70.0	233.7	98.0	707.3
24.0	22.4	80.0	355.1	99.0	733.2
25.0	23.8	90.0	525.8	100.0	760.0
26.0	25.2	91.0	546.0	110.0	1074.6
27.0	26.7	92.0	567.0	120.0	1489.1
28.0	28.3				

Example 11.2

Suppose that the equilibrium illustrated in Figure 11.3 is between liquid hexane, C_6H_{14}, and its vapor at 298.15 K. A sample of the equilibrium vapor is found to have a density of 0.701 g/L. What is the vapor pressure of hexane at 298.15 K, expressed in mmHg?

Solution

If we assume that hexane vapor behaves as an ideal gas, the pressure that we calculate with the ideal gas equation is the quantity we are seeking, the vapor pressure of hexane. Let's base the calculation on 1.00 L of the equilibrium vapor. The mass of this volume of vapor is 0.701 g. The number of moles of gas in 1.00 L of vapor is

$$n = 0.701 \text{ g } C_6H_{14} \times \frac{1 \text{ mol } C_6H_{14}}{86.18 \text{ g } C_6H_{14}}$$

The other quantities that we need are

$$V = 1.00 \text{ L}$$
$$R = 0.0821 \text{ L} \cdot \text{atm} \cdot \text{mol}^{-1} \cdot \text{K}^{-1}$$
$$T = 298.15 \text{ K}$$

We can rearrange the ideal gas equation and substitute the known data as follows:

$$P = \frac{nRT}{V}$$

$$= \frac{0.701 \text{ g } C_6H_{14} \times \dfrac{1 \text{ mol } C_6H_{14}}{86.18 \text{ g } C_6H_{14}} \times 0.0821 \text{ L} \cdot \text{atm} \cdot \text{mol}^{-1} \cdot \text{K}^{-1} \times 298.15 \text{ K}}{1.00 \text{ L}}$$

$$= 0.199 \text{ atm}$$

$$P = 0.199 \text{ atm} \times \frac{760 \text{ mmHg}}{1 \text{ atm}} = 151 \text{ mmHg}$$

Exercise 11.2

Suppose that the equilibrium illustrated in Figure 11.3 is between liquid water and its vapor at 22 °C. What mass of water vapor would be present in a vapor volume of 275 mL at 22 °C? Use data from Table 11.2.

Vapor pressure data in handbooks are rarely tabulated, as in Table 11.2, or graphed, as in Figure 11.4. They are usually presented through mathematical equations based on experimentally determined vapor pressures (see Problem 75).

A graph of vapor pressure as a function of temperature is called a **vapor pressure curve** (see Figure 11.4). The vapor pressure curve is the boundary between the liquid and gas (vapor) areas in a phase diagram (the curve *OC* in Figure 11.2). The two phases coexist in equilibrium at all points lying on the curve. Specifically, each point on the curve represents the vapor pressure of the liquid at a given temperature.

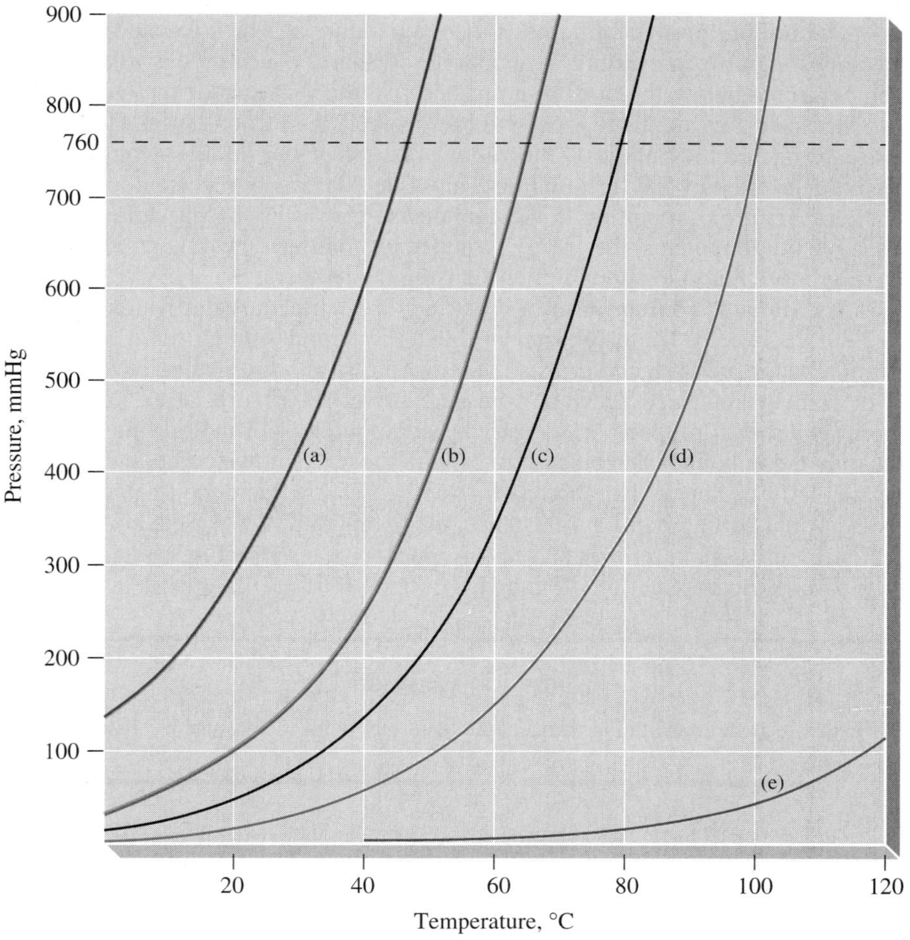

Figure 11.4 Vapor pressure curves of several liquids. (a) carbon disulfide, CS_2; (b) methanol, CH_3OH; (c) ethanol, CH_3CH_2OH; (d) water, H_2O; (e) aniline, $C_6H_5NH_2$. The temperature of the intersection of the line P = 760 mmHg with a vapor pressure curve is the normal boiling point.

If intermolecular forces within a liquid are weak, many molecules escape from the liquid surface before equilibrium is established. The vapor pressure is high, and we say that the liquid is *volatile*. If the intermolecular forces are strong, the equilibrium vapor pressure is low and we say that the liquid is *nonvolatile*. Diethyl ether, with a vapor pressure of 534 mmHg at 25 °C, is highly volatile. At 25 °C, water has a vapor pressure of 23.8 mmHg; it is moderately volatile. Compared to diethyl ether and water, mercury would be considered *nonvolatile*. Nevertheless, mercury vapor in equilibrium with liquid mercury is present at a high enough partial pressure in air to be toxic. Gasoline is a mixture of hydrocarbons, some of which are quite volatile and others less so. Gasoline for use in cold weather is formulated to have a higher proportion of volatile components, so that the cold gasoline can be ignited more readily.

Boiling Point

We observe a special vaporization phenomenon when we heat a liquid in an open container. When a particular temperature is reached, vaporization occurs, not only at the surface, but *throughout the liquid*. Vapor produced in the interior of the liquid forms bubbles that rise to the surface and escape. The liquid boils.

The **boiling point** of a liquid is the temperature at which its vapor pressure becomes equal to prevailing atmospheric pressure. Boiling does not occur at temperatures below the boiling point because the pressure of the atmosphere would cause any incipient vapor bubbles in the liquid to collapse. That is, the pressure inside the bubbles—the vapor pressure of the liquid—would be less than that exerted by the surrounding atmosphere. It is also important to note that the temperature of a boiling liquid cannot rise above the boiling point. As long as some liquid remains, the energy provided by continued heating goes into converting more liquid to vapor, not into raising the temperature.

The **normal boiling point** of a liquid is the temperature at which its vapor pressure is exactly 1 atm. We can establish the normal boiling point of a liquid from a vapor pressure curve. It is the temperature at which a line at P = 1 atm (760 mmHg) intersects the vapor pressure curve (see Figure 11.4). The boiling point is a useful property in identifying compounds. Normal boiling points of unknown compounds can be compared with those of known compounds readily available in chemistry handbooks. In a few cases, the handbook entry may not be a normal boiling point. For instance, the listing of 319^{741} for the boiling point of antipyrine (a pain reliever and fever reducer) means a boiling point of 319 °C at a barometric pressure of 741 mmHg.

Reducing the air pressure in the bell jar can cause water to boil at room temperature.

Estimation Exercise 11.2

Estimate the normal boiling points of the first four liquids represented in Figure 11.4.

Figure 11.4 helps to explain the common observation that the boiling points of water and other liquids vary with altitude and weather conditions. Shift the line at P = 760 mmHg to higher or lower pressures and the boiling point changes, increasing as the barometric pressure rises and decreasing as it falls. Barometric pressures below 1 atm are commonly encountered at high altitudes. In the mile-high city of Denver, Colorado (1609 m), the average atmospheric pressure is about 630 mmHg. The boiling point of water at this pressure is 95 °C (203 °F).

The chemical reactions that occur when food is cooked are quite temperature dependent. The rates of these reactions, and hence the rate of cooking, increase sharply even with small temperature increases and decrease sharply with small decreases in temperature. When water boils at 100 °C an egg can be soft-boiled to perfection in 3 min. In Denver, it might take 5 to 6 min to boil the egg to the same state of excellence. On top of Mount Everest (8848 m) it would take much longer still.

The increase in boiling point with increased external pressure is the principle used in pressure cookers and autoclaves. The chemical reactions involved in the cooking of a tough piece of meat proceed more rapidly at the higher-than-normal boiling temperatures attainable in a pressure cooker, where the pressure of the water vapor (steam) can rise to much more than 1 atm. In an autoclave, bacteria (even resistant spores) are killed more rapidly than in boiling water, not directly by the increased pressure but by the higher temperatures attained.

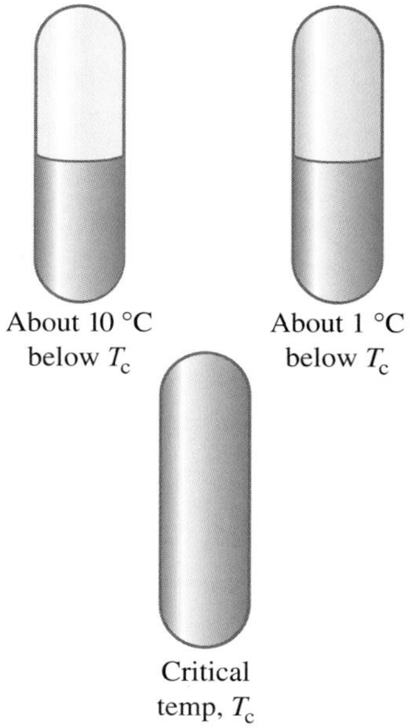

Figure 11.5 The critical point. The meniscus separating a liquid (bottom) from its vapor (top) disappears at the critical point. The liquid state does not exist above the critical temperature, regardless of the pressure that might be applied to it.

The Critical Point

If an appropriate quantity of a liquid is heated in a *closed* container, boiling does not occur. Instead, as we raise the temperature we observe (Figure 11.5) that the density of the liquid decreases, the density of the vapor increases, and the boundary (meniscus) between the liquid and vapor becomes blurred and disappears. Finally, only the gaseous state remains. The **critical temperature, T_c**, is the highest temperature at which a liquid can coexist in equilibrium with its vapor. The vapor pressure at this temperature is called the **critical pressure, P_c**. The point corresponding to a temperature of T_c and a pressure of P_c, called the **critical point**, is the final point on the vapor pressure curve. It is point *C* in the phase diagram of Figure 11.2.

Another way to describe the critical temperature is that it is the highest temperature at which a gas can be condensed into a liquid solely by increasing its pressure. Thus, if a gaseous substance has its T_c *above* room temperature we can liquefy it at room temperature just by applying sufficient pressure. If the T_c is *below* room temperature, however, we must both apply pressure *and* lower the temperature to a value below T_c. Sometimes we use the term *vapor* to refer to the gaseous state of a substance below its T_c and *gas* to the gaseous state above T_c. Put in these terms, a vapor can be condensed to a liquid simply by the application of pressure and a gas cannot. Table 11.3 lists critical temperatures and pressures for several substances.

TABLE 11.3 Critical Constants of Various Substances

Substance	T_c, K	P_c, atm
H_2	33.0	12.8
N_2	126.3	33.5
O_2	154.8	50.1
CH_4	190.6	45.4
CO_2	304.2	72.9
C_2H_6	305.4	48.2
HCl	324.6	81.5
C_3H_8	369.8	41.9
NH_3	405.6	111.3
SO_2	430.6	77.9
H_2O	647.3	218.3

CONCEPTUAL EXAMPLE 11.1

To keep track of how much gas remains in a cylinder, we can weigh the cylinder when it is empty, when it is filled, and after each usage. In some cases, though, we can equip the cylinder with a pressure gauge and simply relate the amount of gas to the measured gas pressure. Which method should we use to keep track of the bottled propane, C_3H_8, in a gas barbecue?

Solution

The propane is at a temperature below its T_c (369.8 K) and is present in the cylinder as a mixture of a liquid and vapor. The volume of liquid in the cylinder decreases and that of the vapor increases as the fuel is consumed. However, the vapor pressure of the liquid propane does not depend on the amounts of liquid and vapor present. The pressure will remain constant (assuming a constant temperature) as long as some liquid remains. Only after the last of the liquid has vaporized will the pressure drop. Measuring the pressure doesn't tell us how much propane is left in the cylinder until it's just about all gone. We would have to monitor the contents of the cylinder by weighing.

Conceptual Exercise 11.1

Which of the two methods described above would you use if the fuel in the cylinder were methane, CH_4? Explain.

Supercritical Fluids

What lies beyond the critical point? This is the portion of the generalized phase diagram in Figure 11.2 in a gray color. It really is a "gray" area, in that it's hard to know what to call it. Because liquids and gases both flow readily—they are *fluids*—and because the liquid and gaseous states become indistinguishable at the critical point and remain so somewhat beyond this point, the term *supercritical fluid* (SCF) is commonly used. Although an in-depth study of supercritical fluids has been made only in recent years, SCFs are already being used in many practical applications.

11.3 Phase Changes Involving Solids

In this section we consider phase changes that involve the solid state. We especially emphasize the energy changes that accompany these phase changes.

Melting, Melting Point, and Heat of Fusion

The structural units (atoms, ions, or molecules) of a crystalline solid exhibit little motion other than vibrations about fixed points. As the temperature is raised, these vibrations become more vigorous, finally reaching a point where the vibrations are so forceful that the crystalline structure collapses. The solid becomes a liquid; the process is called **melting** or **fusion**. The temperature at which a solid melts is called its **melting point**. The reverse of melting—the conversion of a liquid to a solid—is called **freezing**; the temperature at which it occurs is the **freezing point**. Whether we think of it as a melting solid or a freezing liquid, solid and liquid coexist at equilibrium, and the freezing point and melting point are identical. The freezing point of water and the melting point of ice are the same temperature, 0 °C.

The data in tables of melting points are usually the *normal* melting points—that is, for melting that occurs under a pressure of 1 atm. Melting points do vary slightly with pressure, however. In a phase diagram, the variation of melting point with pressure is called the *fusion curve*. In the generalized phase diagram of Figure 11.2, the curve *OD* is a fusion curve. The broken-line portion of the curve signifies that the fusion curve has no known ending. That is, there is no point on the fusion curve corresponding to the critical point that terminates the vapor pressure curve.

Supercritical Fluids in the Food Processing Industry

Supercritical fluids are versatile solvents because their properties can be varied significantly through changes in temperature and pressure. In this way an SCF can be used to dissolve one component of a mixture while leaving others unaffected. Supercritical fluid extraction is now widely used in laboratories and in industries. It is used to measure the fat content of foods, to remove environmental contaminants such as diesel fuel and polychlorinated biphenyls (PCBs) from soil, and to prepare other samples (particularly nonvolatile ones such as polymers) for analysis.

Supercritical carbon dioxide is particularly attractive as a solvent in the food-processing industry. For many years decaffeinated coffee was made by extracting the caffeine with solvents such as methylene chloride (dichloromethane, CH_2Cl_2). At high concentrations, CH_2Cl_2 has been shown to cause several health problems. Even though no CH_2Cl_2 has been found in the beverage brewed from coffee beans, some companies have shifted to supercritical CO_2 to dissolve and carry off the caffeine. In one process, green coffee beans are exposed to CO_2 at about 90 °C and 160 to 220 atm. The caffeine content, normally about 1 to 3%, is reduced to around 0.02%. When the pressure on the system is reduced, the separated CO_2 is converted to the gaseous state and the caffeine precipitates; no toxic residue is left in the coffee. The CO_2 is recycled.

Supercritical fluids also are used to extract cholesterol from eggs, butter, lard, and other fatty foods, making them more suitable for low-cholesterol diets. Potato chips made by cooking in oils can be treated with supercritical CO_2 to reduce their normally high fat content, improving both the nutritional value and the shelf life. SCFs can also be used to extract the chemical compounds that impart flavor and fragrance to such products as lemons, black pepper, almonds, and nutmeg. These extracts can then be used to flavor other foods or household products.

Supercritical fluid extraction is used in the production of decaffeinated coffee and several other food products

The quantity of heat required to melt a given amount of solid is called the **enthalpy (heat) of fusion**. As shown below for the melting of ice, melting is an *endothermic* process:

$$H_2O(s) \longrightarrow H_2O(l) \qquad \Delta H = +6.01 \text{ kJ/mol}$$

Freezing, the reverse of melting, is an *exothermic* process. The enthalpy change for freezing is the negative of the enthalpy of fusion:

$$H_2O(l) \longrightarrow H_2O(s) \qquad \Delta H = -6.01 \text{ kJ/mol}$$

Table 11.4 lists some typical enthalpies of fusion.

TABLE 11.4 **Some Enthalpies (Heats) of Fusion**

Substance	Melting Point, °C	ΔH_{fusion}, kJ/mol
Mercury, Hg	−38.9	2.30
Ethanol, CH_3CH_2OH	−114	5.01
Water, H_2O	0.0	6.01
Benzene, C_6H_6	5.5	9.87
Silver, Ag	960.2	11.95
Iron, Fe	1537	15.19

Estimation Exercise 11.3

Which of the substances in Table 11.4 requires the greatest quantity of heat to melt a 1.00-g sample? Which requires the smallest quantity?

We can determine the freezing point of a liquid by measuring its temperature as it slowly cools. The temperature falls regularly with time, but when freezing begins, the temperature remains constant until all the liquid has frozen. Then the temperature falls regularly as the solid cools. A temperature-versus-time graph plotted with data obtained in this way is called a **cooling curve**. A typical cooling curve is shown in Figure 11.6. Because the freezing point of a liquid corresponds to the straight-line portion of a cooling curve, we can generally get a good indication of the freezing point of a pure liquid by observing this temperature; we don't have to plot the cooling curve.

To determine a melting point, we can work backwards: start with the pure solid and add heat. The temperature rises until melting begins, remains constant until melting is completed, and then rises again. This plot of temperature versus time is called a **heating curve**. As expected, the freezing point obtained from a cooling curve and the melting point from a heating curve are identical.

Sometimes the temperature of a cooling liquid may drop below the freezing point without a solid appearing. This condition is called **supercooling**. It may occur if a liquid is quickly cooled below the freezing point before the liquid molecules can gather into solid crystals. Crystal formation requires centers or sites on which crystals can grow, and supercooling is more likely to occur if a liquid is kept free of such centers, for example, dust particles. A supercooled liquid is unstable; once freezing begins, the temperature rises to the freezing point and remains there until freezing is completed.

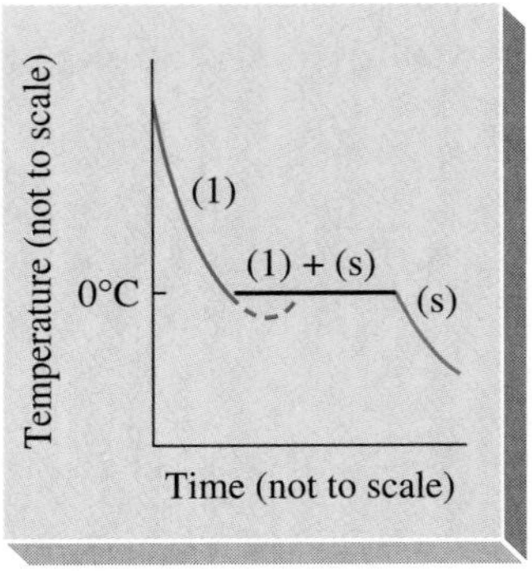

Figure 11.6 Cooling curve for water. The broken line portion represents the phenomenon of supercooling described in the text. (l) = liquid; (s) = solid.

Sublimation

Although few solids are as volatile as some familiar liquids, solids do vaporize, even if to a limited extent. The passage of molecules directly from the solid to the vapor state (Figure 11.7) is called **sublimation**. The reverse process, the condensation of a vapor to a solid, is generally called *deposition*. A dynamic equilibrium is reached when the rates of sublimation and deposition become equal, and the solid exerts a characteristic vapor pressure, sometimes called the *sublimation pressure*.

A plot of the vapor pressure of a solid versus temperature is called a **sublimation curve**. This is the curve (*OB*) that separates the areas representing solid and gas in a phase diagram (Figure 11.2). The sublimation curve ends at the origin of the pressure–temperature graph. That is, at the absolute zero of temperature, 0 K, the sublimation pressure of a solid is *zero*.* At this point, no molecules are found in the vapor state.

Although no liquid phase is involved when a solid sublimes, it is useful in one respect to think of sublimation as equivalent to a two-step process: melting followed by vaporization. This helps us to understand that the **enthalpy (heat) of sublimation** is simply the sum of the enthalpies of fusion and vaporization:

$$\Delta H_{\text{subln}} = \Delta H_{\text{fusion}} + \Delta H_{\text{vapn}}$$

Figure 11.7 Sublimation of iodine. Even at 70 °C, well below the melting point of iodine (114 °C), solid iodine has a high vapor pressure; it sublimes. The purple iodine vapor deposits as a solid on the colder walls of the flask.

ESTIMATION EXAMPLE 11.1

Estimate the enthalpy of sublimation of ice at 0 °C using data from tables in this chapter.

Solution

The enthalpy of vaporization is found in Table 11.1 and the enthalpy of fusion in Table 11.4.

$$\Delta H_{\text{subln}} = \Delta H_{\text{fusion}} + \Delta H_{\text{vapn}} = 6.01 + 44.0 = 50.0 \text{ kJ/mol}$$

This result is only an estimate because we have used the enthalpy of vaporization of water at 25 °C, and the enthalpy of vaporization is somewhat temperature dependent. (Using ΔH_{vapn} at 0 °C, 44.92 kJ/mol, would give a more precise value: $\Delta H_{\text{subln}} = 50.93$ kJ/mol.)

Estimation Exercise 11.4

The enthalpy of sublimation of benzene at its melting point is 44.2 kJ/mol. Estimate the enthalpy of vaporization of benzene at 25 °C

People living in cold climates are especially familiar with the phenomenon of sublimation. Snow disappears from the ground and ice from the windshield of an automobile even though the temperature stays below 0 °C. This occurs through sublimation, not melting; there is no liquid water at any point in the process. The vapor pressure of ice at 0 °C is 4.58 mmHg.

The Triple Point

We have now discussed all the features of the generalized phase diagram of Figure 11.2 except for one eminently important *point*: the point *O*. This point,

* The only known exception to this statement is helium. Liquid helium persists at all temperatures including, presumably, 0 K. However, at pressures greater than 25 atm, solid helium can be obtained at temperatures below about 1 K.

Figure 11.8 Polymorphism of mercury(II) iodide, $HgI_2(s)$. The red solid is the stable form at room temperature. At 127 °C the red solid changes to the yellow form. Mercury(II) iodide also sublimes, and a deposit of a mixture of the red and yellow solids can be seen on the colder walls of the flask.

called the **triple point**, defines the *only* temperature and pressure at which the *three* phases—solid, liquid, and gas—can coexist. It is the only point common to the vapor pressure, sublimation, and fusion curves. The triple point for water is at 0.0098 °C and 4.58 mmHg.

You may wonder why the triple point temperature is not exactly 0 °C. After all, can't solid water (ice), liquid water, and water vapor exist together in the open atmosphere at 0 °C? They can, but here the system has two different pressures. The vapor exists at a pressure of 4.58 mmHg—its partial pressure—while the pressure on the ice and liquid water is atmospheric pressure (760 mmHg). To establish a true triple point, a system must consist of a pure substance existing only under the pressure of its own vapor. There can be no extraneous substances present (such as the gases in air).

11.4 Phase Diagrams of Some Common Substances

In this section we will extend the discussion of the generalized phase diagram of Figure 11.2 to three specific cases. Each case will introduce an important idea that we have not previously considered.

Mercury(II) Iodide, HgI_2

Our interest in mercury(II) iodide is that it exists in two different solid forms (Figure 11.8). *Polymorphism*, the existence of two or more forms of a solid, is a common phenomenon—so much so that the phase diagram in Figure 11.9 is more typical than the generalized diagram of Figure 11.2.

Each solid form of mercury(II) iodide has its own pressure–temperature region in Figure 11.9. Red $HgI_2(s)$, which is the form we would see in a storeroom bottle, is stable up to 127 °C. At this temperature red $HgI_2(s)$ converts to yellow $HgI_2(s)$. Yellow $HgI_2(s)$ is the exclusive form of the solid up to the melting point of 259 °C.

Figure 11.9 Phase diagram of mercury(II) iodide, HgI_2. The two polymorphic forms of solid HgI_2 and the liquid and gaseous states are all represented. Also noted are the triple point for the equilibrium of red and yellow $HgI_2(s)$ and gaseous HgI_2 at 127 °C, the normal melting point of yellow $HgI_2(s)$ at 259 °C, and the boiling point of $HgI_2(l)$ at 354 °C.

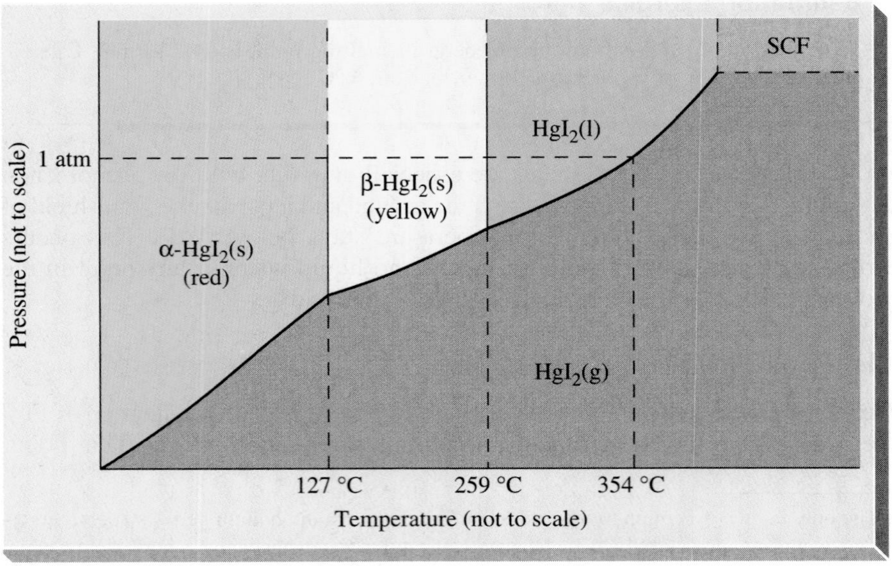

Note that of the two solid forms only the pressure–temperature field of the yellow HgI_2(s) is adjacent to that of liquid HgI_2. Red HgI_2(s) does not have a melting point.

Carbon Dioxide, CO_2

The phase diagram for carbon dioxide in Figure 11.10 also introduces some new ideas. First, the diagram suggests that if a solid–liquid fusion curve, *OD*, is followed to sufficiently high pressures (several thousand atmospheres in the case of CO_2), the curve extends beyond the critical temperature. At first, this observation may seem unlikely. However, if we think about how closely together molecules are forced at extremely high pressures, the solid state is the one we should expect. Furthermore, although the critical temperature may seem like a high temperature, for carbon dioxide T_c is 304.2 K, about room temperature.

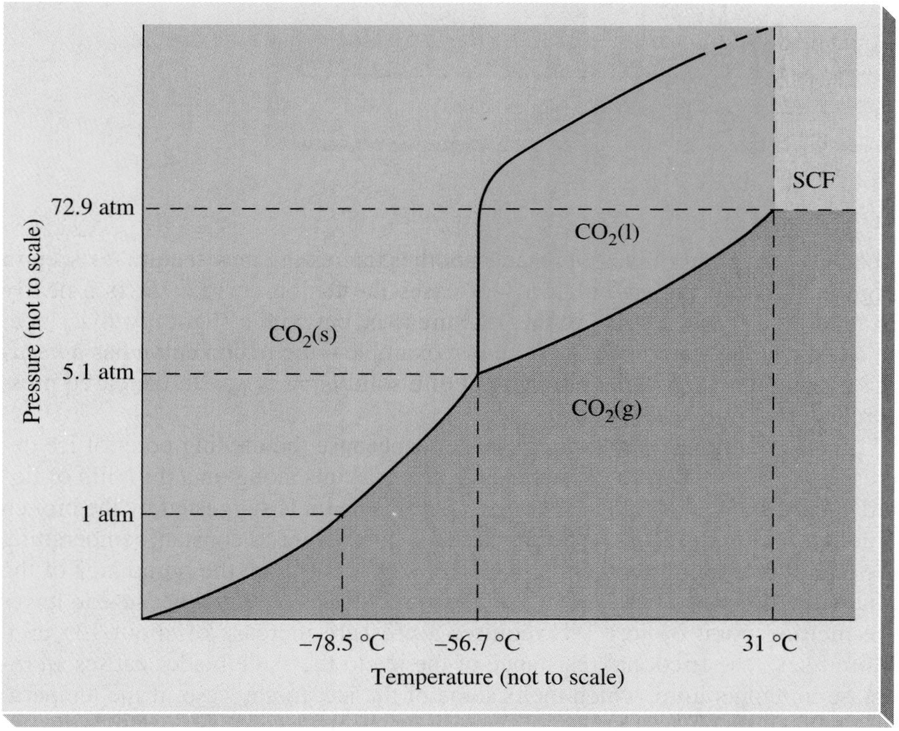

Figure 11.10 **Phase diagram of carbon dioxide, CO_2.** Noted in the diagram are the normal sublimation temperature of CO_2(s), –78.5 °C; the triple point, –56.7 °C, 5.1 atm; and the critical point, 31 °C, 72.9 atm. The fusion curve slopes away from the pressure axis and at very high pressures reaches temperatures above the critical temperature.

Another interesting feature of the CO_2 phase diagram is that the triple point pressure is 5.1 atm. Liquid CO_2 can be maintained only at pressures greater than 5.1 atm, and certainly not at normal atmospheric pressure. Liquid CO_2 does not have a normal boiling point. Neither does solid CO_2 have a normal melting point. When solid CO_2 is heated at 1 atm pressure, it sublimes at –78.5 °C. Solid CO_2, called "dry ice," is a useful coolant for two reasons: Because no melting occurs, the materials being cooled do not come in contact with a liquid and remain dry. And the temperature of the remaining dry ice stays at –78.5 °C, no matter how much of it has already sublimed. Thus, the cooling effect of the dry ice can be maintained for a relatively long time.

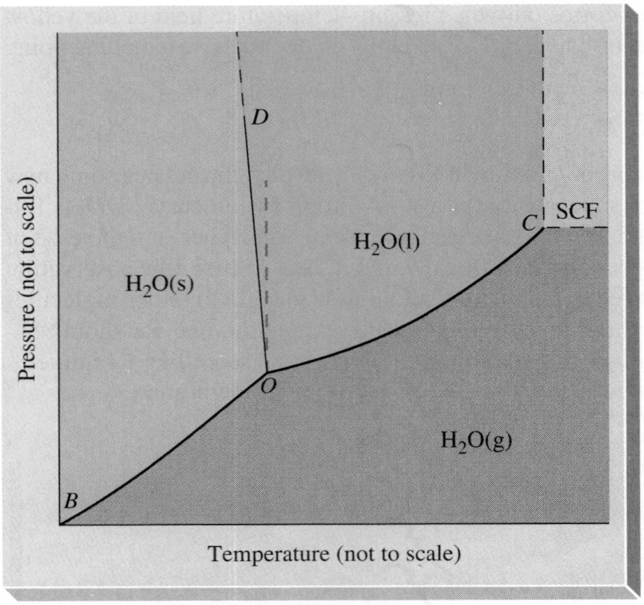

Figure 11.11 Phase diagram of water, H_2O. The triple point, *O*, is at +0.0098 °C and 4.58 mmHg. The critical point, *C*, is at 374.1 °C and 218.2 atm. The negative slope of the fusion curve *OD* is greatly exaggerated. Not shown are the several polymorphic forms of ice that exist at high pressures (in excess of 2045 atm). The significance of the broken blue line is described in the text.

Water, H_2O

The phase diagram of water presents another interesting new feature, as seen in Figure 11.11. In the vast majority of cases the fusion curve, *OD*, is a nearly straight line, nearly parallel to the pressure axis, but with a slight *positive* slope. In a few cases—water is the best known example—the fusion curve has a *negative slope*. Generally, the melting point of a solid *increases* with increased pressure, but that for ice *decreases*.

Some claim that ice skating is possible because the melting point of ice decreases with increased pressure, and the skater skims along on a thin film of liquid water from the melted ice. The basis of this claim is suggested by the broken blue line in Figure 11.11. With an increase in pressure at constant temperature, the equilibrium point on the fusion curve, *OD*, moves into the liquid area of the phase diagram. However, this explanation probably is incorrect because to lower the melting point by just 1°C requires a pressure increase of about 125 atm. More likely, the frictional resistance of the ice to the skate blades causes an increase in temperature, which melts some of the ice. In any case, if the temperature is below −22 °C (−8 °F), pressure-induced melting cannot occur. At lower temperatures water can exist only in one or more polymorphic forms of ice.

CONCEPTUAL EXAMPLE 11.2

In Figure 11.12, 50.0 mol $H_2O(g)$ (steam) at 100.0 °C and 1.00 atm is added to an insulated cylinder with 5.00 mol $H_2O(s)$ (ice) at 0 °C. Use the following data to determine which of the four possible final conditions is achieved.

$$\Delta H_{\text{fusion}} = 6.01 \text{ kJ/mol}$$

$$\Delta H_{\text{vapn}} = 40.6 \text{ kJ/mol (at 100 °C)}$$

$$\text{molar heat capacity of } H_2O(l) = 76 \text{ J} \cdot \text{mol}^{-1} \cdot \text{°C}^{-1}$$

Solution

Two changes that must occur are the melting of ice and the condensation of steam. Let's calculate the quantities of heat involved if each of these changes were to go to completion. Note that the heat of condensation of the steam is simply the *negative* of the heat of vaporization of $H_2O(l)$.

$$\text{Melting of ice:} \quad 5.00\ \text{mol} \times \frac{+6.01\ \text{kJ}}{1\ \text{mol}} = +30.1\ \text{kJ}$$

$$\text{Condensation of steam:} \quad 50.0\ \text{mol} \times \frac{-40.6\ \text{kJ}}{1\ \text{mol}} = -2030\ \text{kJ}$$

Because the heat evolved in the condensation of steam is far and away greater than required to melt the ice, no ice can remain. Possibility (a) is eliminated.

Suppose that the melted ice (now liquid water) is heated to 50.0 °C, the temperature for condition (b). The amount of heat required is

$$?\ \text{kJ} = 5.00\ \text{mol} \times \frac{76\ \text{J}}{\text{mol} \cdot {}^\circ\text{C}} \times (50.0 - 0.0)\ {}^\circ\text{C} \times \frac{1\ \text{kJ}}{1000\ \text{J}} = 19\ \text{kJ}$$

More than enough heat is available from the condensation of steam to melt the ice *and* raise its temperature to 50.0 °C. Possibility (b) is eliminated.

Because the condensation of steam produces $H_2O(l)$, liquid water must be present. Possibility (d) is eliminated. The final condition, then, must be one of liquid *and* gaseous water at 100 °C; condition (c). (See also Problem 76.)

Conceptual Exercise 11.2

A 1.05-mol sample of $H_2O(g)$ is introduced into a 2.61-L flask at 30.0 °C. Use a point on the phase diagram of Figure 11.11 to indicate the final condition reached.

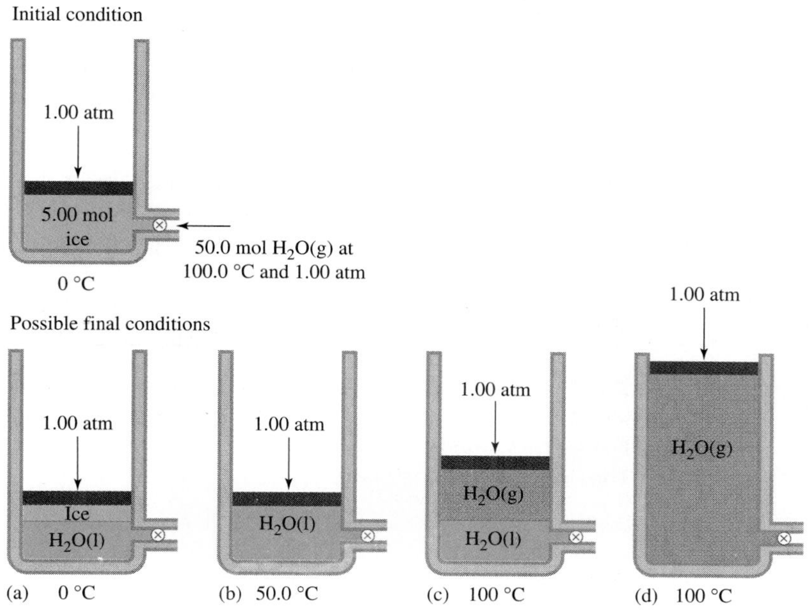

Figure 11.12 Visualizing Conceptual Example 11.2.

Liquid Crystals

As first observed in 1888, a cholesterol derivative, cholesteryl benzoate, melts sharply at 145.5 °C to produce a milky fluid and then abruptly changes to a clear liquid at 178.5 °C. Between 145.5 and 178.5 °C, cholesteryl benzoate has the fluid properties of a liquid, the optical properties of a crystalline solid, and some unique properties of its own. It is in a form commonly called *liquid crystal*. Long regarded only as laboratory curiosities, liquid crystals are now known to every user of LCD (liquid crystal display) devices such as digital watches, calculators, thermometers, and computer screens.

Liquid crystal formation is relatively common among organic compounds that have molar masses in the range of a few hundred grams per mole and that consist of rodlike molecules. Three possibilities for the orientation of these rodlike molecules are suggested in Figure 11.13.

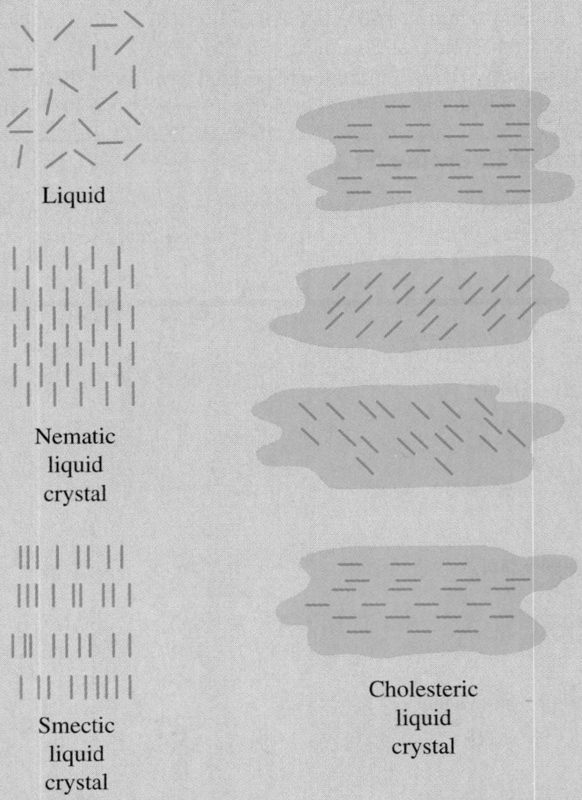

Figure 11.13 The liquid crystalline state.

In *nematic* (threadlike) liquid crystals, molecules are arranged parallel to one another. The molecules can move in any direction, and they can rotate on their long axes in the same way that a single pencil can be moved in a loosely packed box of pencils. In *smectic* (greaselike) liquid crystals, molecules are arranged in layers, with the long axes of the molecules perpendicular to the planes of the layers. The molecules can rotate on their long axes and move within the layers.

Cholesteric liquid crystals are related to the smectic form, but the orientation of the molecules in each layer is different from that in the layer above and below. A collection of layers forms a repeating sequence of orientations. The distance between two layers with the same orientation of molecules is a distinctive property of a cholesteric liquid crystal. When a beam of white light strikes a film of cholesteric liquid crystal, the color of the reflected light depends on this characteristic distance. Because this distance changes with temperature, so does the color of the reflected light. Some liquid-crystal temperature-sensing devices exhibit color changes for temperature changes as small as 0.01 °C.

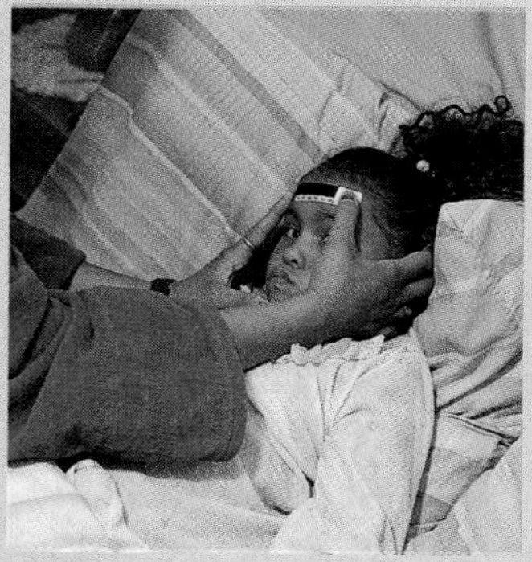

A liquid crystal strip being used to determine a young patient's temperature. Color and color changes produced by changes in temperature or other variables are characteristic of liquid crystals.

Liquid-crystal displays operate on the principle that the orientation of molecules in a thin film of a nematic liquid crystal changes in the presence of an electric field. This change in orientation produces changes in optical properties of the film. In a digital watch or calculator display, electrodes coated with films of liquid crystals are arranged in patterns in the shape of numbers. When an electric field is imposed through the electrodes, the patterns of the electrodes (the numbers) become visible.

11.5 Intermolecular Forces of the van der Waals Type

We now return to the topic with which we opened the chapter—intermolecular forces. We will survey the main types, and we will see how these forces explain some of the properties and phenomena described in the preceding three sections, such as why some liquids are volatile and others are not, and why some solids sublime and others do not. The two types of intermolecular forces that we explore in this section manifest themselves in various ways. For example, they explain deviations from ideal gas behavior. Because these intermolecular forces are well accounted for in the van der Waals equation (page 146), it is customary to refer to them collectively as *van der Waals forces*.

Dispersion Forces

In earlier chapters, we have seen that helium atoms have a high ionization energy and no affinity for an additional electron. In short, they do not form bonds with other atoms. Because helium atoms form no chemical bonds, we might also expect them to "keep their distance" from one another at all temperatures. If they did, helium would remain a gas to 0 K. However, helium does condense to a liquid at temperatures below about 5 K. This tells us that there must be some kind of intermolecular force among the He atoms to cause them to come close enough together to enter the liquid state. What is the source of this force?

To answer that question, we first must realize that the electron charge density pictures we have used since Chapter 6 represent *average* situations only. On average, the electron charge density associated with helium's two 1*s* electrons is evenly distributed about the nucleus. However, at any given instant, purely by chance, the electron charge density may become uneven. The normally nonpolar atom becomes momentarily polar; an *instantaneous* dipole is formed. This transitory dipole, in turn, can displace electrons in a neighboring helium atom, also producing a dipole. One dipole *induces* the other, and the newly formed dipole is therefore called an *induced* dipole. Taken together, these two events lead to an intermolecular force of attraction (Figure 11.14). The force of attraction between the instantaneous dipole and the induced dipole is known as a **dispersion force** (also called a *London* force, named for Fritz London, a professor of chemical physics at Duke University who offered a theoretical explanation of these forces in 1928.)

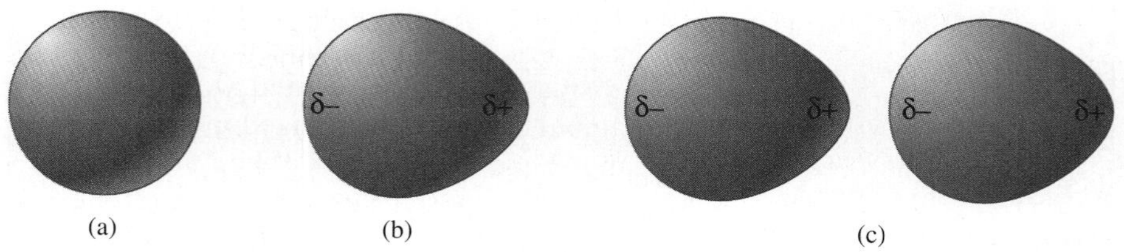

Figure 11.14 Dispersion forces. (a) *Unpolarized molecule.* The electron charge distribution is symmetrical. (b) *Instantaneous dipole.* A displacement of electron charge density (to the left) produces an instantaneous dipole. (c) *Induced dipole.* The instantaneous dipole on the left induces charge separation in the molecule on the right, making it also a dipole. The attraction between the two dipoles constitutes an intermolecular force.

Polarizability is a measure of the ease with which electron charge density is distorted by an external electric field—that is, the ease with which a dipole can be induced in an atom or molecule. Large atoms have more electrons and larger electron clouds than do small ones. In large atoms, the outer electrons are more loosely bound; they can shift toward another atom more readily than can the more tightly bound electrons in small atoms. Large atoms and molecules are therefore more polarizable than small ones. Atomic and molecular sizes are also closely related to atomic and molecular weights, which means that polarizability increases with increased molecular weight. And the greater the polarizability of molecules, the stronger are the intermolecular forces between them.

This trend is easily seen in the physical properties of the Group 7A elements, the halogens. All are nonpolar. The first member, fluorine (F_2), is a *gas* at room temperature (bp −188 °C). The second member, chlorine (Cl_2), is also a *gas* (bp −34 °C), but it is more easily liquefied than is fluorine. Bromine (Br_2), is a *liquid* (bp 58.8 °C), and iodine (I_2) is a *solid* (mp 184 °C). Because large molecules are easily polarizable, intermolecular forces between them are strong enough to form liquids or even solids.

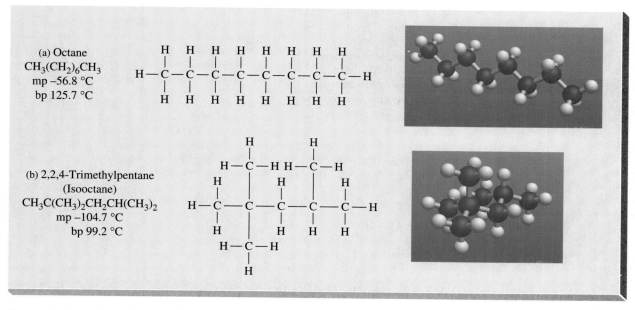

(a) Octane
$CH_3(CH_2)_6CH_3$
mp –56.8 °C
bp 125.7 °C

(b) 2,2,4-Trimethylpentane
(Isooctane)
$CH_3C(CH_3)_2CH_2CH(CH_3)_2$
mp –104.7 °C
bp 99.2 °C

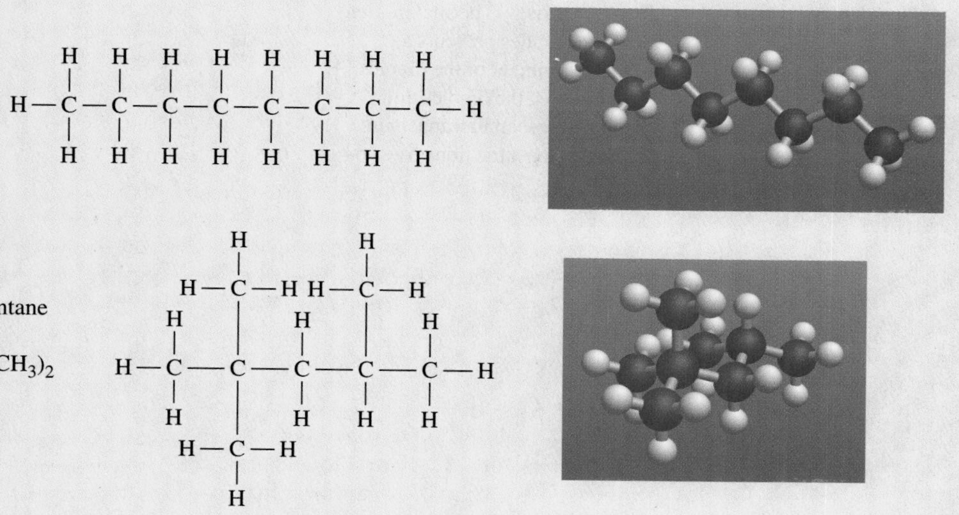

Figure 11.15 Molecular shapes and polarizability. The elongated octane molecules are more easily polarized than are the more compact isooctane molecules, and intermolecular forces are therefore stronger than in isooctane. Octane has a higher melting point and boiling point than does isooctane.

Another factor that affects polarizability is molecular shape. Electrons in elongated molecules are more easily displaced than are those in compact molecules. Figure 11.15 describes two isomers of octane found in gasoline. They have identical molecular weights but different molecular shapes and different physical properties.

Dipole–Dipole Forces

We have just shown that instantaneous and induced dipoles can form in a nonpolar substance. A *polar* substance, because of its molecular shape and differences in electronegativities between bonded atoms, has *permanent* dipoles. Permanent dipoles attempt to align themselves with the positive end of one dipole directed toward the negative ends of neighboring dipoles and vice versa (Figure 11.16). Even if attractive forces between permanent dipoles are fairly weak, when these are considered *in addition to* the dispersion forces found in all molecular substances, the result is a total intermolecular force that is greater in a polar substance than in a nonpolar substance of about the same molecular weight. We see this effect in the series: nitrogen, nitrogen monoxide, and oxygen.

	N_2	NO	O_2
MW	28.0 u	30.0 u	32.0 u
μ	0	0.15 D	0
bp	−196 °C	−152 °C	−183 °C

The more polar a molecule—that is, the greater its dipole moment—the more pronounced is the effect of dipole–dipole forces on physical properties.

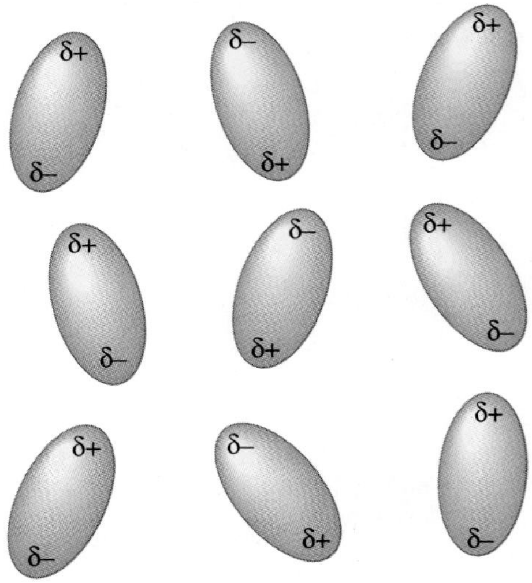

Figure 11.16 Dipole–dipole interactions. Thermal motion of the molecules prevents a perfect alignment of the dipoles. Nevertheless, the dipoles do maintain a general arrangement leading to the attractions δ+ ··· δ–.

We can see this by comparing two substances of nearly identical molecular weights: *propane*, $CH_3CH_2CH_3$ (MW 44.10 u), the gas used in gas-fired barbecue grills, and *acetaldehyde*,

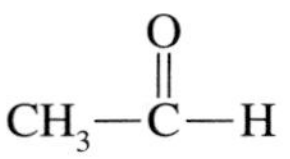

(MW 44.05 u), used in the manufacture of artificial flavors. The electronegativity difference between C and H atoms is so small that propane is a nonpolar substance. In acetaldehyde the electronegativity difference between C and O is large. This produces a large bond moment, not offset by other bond moments, and a large resultant dipole moment (μ = 2.69 D). We expect acetaldehyde to have a considerably higher boiling point than propane. The boiling point of propane is –42.1 °C and that of acetaldehyde is 20.2 °C.

Predicting Physical Properties of Molecular Substances

We can make predictions about such properties as melting points, boiling points, and enthalpies (heats) of vaporization by assessing the impact of both types of van der Waals forces—dispersion forces and dipole–dipole forces. The following summary should help.

- Dispersion forces become stronger with increasing molecular weight and elongation of molecules. *In comparing nonpolar substances, molecular weight and molecular shape are the essential factors to consider.*
- Dipole–dipole forces refer to those between permanent dipoles and are associated with polar substances. *In comparing polar substances to nonpolar substances of comparable molecular weights, intermolecular forces are expected to be stronger in the polar substances, and the more polar the substance, the greater should be the intermolecular force.*
- Because dispersion forces occur in all molecular substances, they must always be considered.

Example 11.3

Arrange the following substances in the expected order of increasing boiling point: carbon tetrabromide, CBr_4; butane, $CH_3CH_2CH_2CH_3$; fluorine, F_2; acetaldehyde,

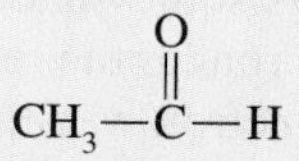

Solution

The first three substances are *nonpolar*: F_2 because the atoms in the molecules are identical; CBr_4 because of the symmetrical tetrahedral molecular structure ($\mu = 0$); butane because the electronegativities of the C and H atoms are so nearly alike. We expect the boiling points of the three to increase with increasing molecular weight:

$$F_2 \text{ (MW 38.00 u)} < CH_3CH_2CH_2CH_3 \text{ (MW 58.12 u)} < CBr_4 \text{ (MW 331.6 u)}$$

As we noted on page 416, the C-to-O bond moment in acetaldehyde is not offset by any other bond moments. Even though the molecular weight of acetaldehyde (44.05 u) is somewhat smaller than that of butane, because acetaldehyde is polar ($\mu = 2.69$ D) we should expect it to have a higher boiling point than butane.

The comparison of acetaldehyde and CBr_4 is more difficult. As in the comparison with butane, the polar character of acetaldehyde argues for it to have the higher boiling point. But because the molecular weight of CBr_4 is so much greater than that of acetaldehyde (by over 250 u), we should expect CBr_4 to have the higher boiling point. Our prediction of the order of increasing boiling points is

$$F_2 < CH_3CH_2CH_2CH_3 < CH_3CHO < CBr_4$$

(The observed boiling points are −188.1, −0.50, 20.2, and 189.5 °C, respectively.)

Exercise 11.3

Of the two substances BrCl and IBr, one is a solid at room temperature and the other is a gas. Which do you think is which? Explain.

11.6 Hydrogen Bonds

Suppose we were asked to predict the relative boiling points of water and acetaldehyde. If we reason as in Example 11.3, we would conclude that acetaldehyde, CH_3CHO ($\mu = 2.69$ D), has a higher boiling point than water H_2O ($\mu = 1.84$ D). Both are polar molecules, but acetaldehyde has a larger dipole moment and also a higher molecular weight—44.05 u compared to 18.02 u. However, our prediction would be wrong. Water has the higher boiling point, 100 °C compared to 20.2 °C. Our incorrect prediction suggests that there is an *additional* kind of intermolecular force in water that is not found in acetaldehyde. There is indeed such a force, and it is a kind known as a *hydrogen bond.*

A **hydrogen bond** between molecules is an intermolecular force in which a hydrogen atom covalently bonded to a nonmetal atom in one molecule is *simultaneously* attracted to a nonmetal atom of a neighboring molecule. Although in rare cases the nonmetal atoms may include chlorine and sulfur, the strongest hydrogen bonds are formed if the nonmetal atoms are *small* and *highly electronegative*. Only nitrogen, oxygen, and fluorine routinely fill this bill.

Think of a hydrogen bond in this way: In a covalent bond an electron cloud joins a hydrogen atom to another atom—oxygen, for example. The electron cloud is much denser (the electron charge density is greater) at the oxygen end of the bond. The bond is polar, with δ– on the O atom and δ+ on the H atom. This leaves the hydrogen nucleus somewhat exposed, and an oxygen atom of a neighboring molecule can approach the exposed hydrogen nucleus rather closely and share some of its "electron wealth" with the hydrogen atom. Hydrogen bonding in water is suggested in Figure 11.17, where we follow the customary convention of representing hydrogen bonds by dotted lines.

Example 11.4

For each of the following substances, comment on whether hydrogen bonding is an important intermolecular force: N_2, HI, HF, $(CH_3)_2O$, CH_3OH.

Solution

N_2: *No.* N is a highly electronegative atom, but we can't have hydrogen bonds without H atoms.

HI: *No.* H atoms are present, but they must be bonded to small, highly electronegative nonmetal atoms. Iodine atoms do not fit this requirement.

HF: *Yes.* Both of the requirements stated on page 417 are met: H is bonded to a highly electronegative nonmetal.

$(CH_3)_2O$: *No.* Both H and a highly electronegative nonmetal (oxygen) are present, but the H atoms are bonded to *carbon*, not oxygen.

CH_3OH: *Yes.* Both H and highly electronegative O are present, and one of the H atoms is bonded to O.

Exercise 11.4

For each of the following substances, comment on whether hydrogen bonding is an important intermolecular force.

$$NH_3,\ CH_4,\ C_6H_5OH,\ CH_3\overset{\overset{\displaystyle O}{\|}}{C}OH,\ H_2S,\ H_2O_2$$

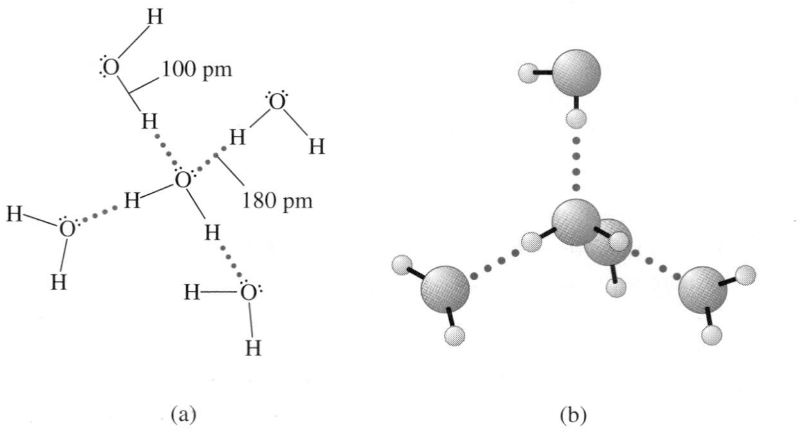

Figure 11.17 Hydrogen bonds in water. As suggested through (a) Lewis structures and (b) ball-and-stick models, each water molecule is linked to four others through hydrogen bonds. Each H atom lies along a line that joins two O atoms. The shorter distances (100 pm) correspond to O—H covalent bonds, and the longer distances (180 pm) to the hydrogen bonds.

Water: Some Unusual Properties

Figure 11.18 shows a model for ice. There we see how hydrogen bonds hold water molecules in a rigid but open structure. As ice melts, some of the hydrogen bonds are broken. Water molecules move into some of the "holes" that were in the ice structure. The molecules are therefore closer together in liquid water than in ice. This means that liquid water at 0 °C is *more dense* than ice. Water is most unusual in this regard. For the vast majority of substances the liquid state is *less dense* than the solid.

If we continue to heat liquid water just above the melting point, more hydrogen bonds break. The molecules become still more closely packed, and the density of liquid water increases to a maximum density at 3.98 °C. Above 3.98 °C the density of water decreases with temperature, as is normally expected for a liquid. These density phenomena explain why a freshwater lake freezes from the top down. In winter, when the water temperature falls below 4 °C, the more dense water sinks to the bottom of the lake. The colder surface water freezes first. The ice at the top of the lake then insulates the water below the ice from further heat loss. Except for relatively shallow lakes in extremely cold climates, lakes generally do not freeze solid in the winter time.

The high boiling point of water—100 °C—is also unusual. In fact, a number of substances with considerably higher molecular weights than water are *gases* at room temperature, among them CO, CO_2, SO_2, and SO_3. Methanol, CH_3OH, is another substance with a low molecular weight (MW 32.04 u) that is a liquid at room temperature because of strong hydrogen bonding between molecules.

(a)

(b)

Figure 11.18 Hydrogen bonds in ice. (a) Oxygen atoms are arranged in layers of distorted hexagonal rings. Hydrogen atoms lie between pairs of O atoms, closer to one (covalent bond) than to the other (hydrogen bond). (b) At the macroscopic level, this structural pattern is revealed in the hexagonal shapes of snowflakes.

Hydrogen Bonding and Life Processes

The hydrogen bond may seem merely an interesting piece of chemical theory, but its importance to life and health cannot be overstated. The structure of proteins, substances essential to life, is determined partly by hydrogen bonding. The action of enzymes, the protein molecules that catalyze the reactions that sustain life, depends in part on the forming and breaking of hydrogen bonds. The heredity that one generation passes on to the next is carried in nucleic acids joined in an elegant arrangement through hydrogen bonds. Certain bonds in the DNA and protein structure must be easy to break and reform. Of all the kinds of intermolecular forces, only hydrogen bonds have just the right amount of energy for this—from about 15 to 40 kJ/mol. By contrast, covalent chemical bonds have energies that range from about 150 to several hundred kJ/mol, and van der Waals forces have energies in the range of only 2 to 20 kJ/mol.

Hydrogen Bonding in Organic Substances

Hydrocarbons form no hydrogen bonds. Many other organic compounds contain oxygen or nitrogen, however, so hydrogen bonding is commonly encountered in organic chemistry.

Acetic acid, CH_3COOH, meets the requirements that we have set forth for hydrogen bonding. However, it has a much lower heat of vaporization than we

Hydrogen Bonding in Proteins

Proteins are giant polymeric molecules with molecular weights ranging from several thousand to several million. The proteins of all living species, from bacteria to humans, are constructed from a basic set of 20 amino acids that serve as monomers. Almost all amino acids are characterized by an amino group ($-NH_2$) bonded to the carbon atom adjacent to a carboxylic acid group, the so-called alpha (α) carbon atom, $RCH(NH_2)COOH$. The individual amino acids have different R groups.

Two amino acids join by forming a peptide bond.

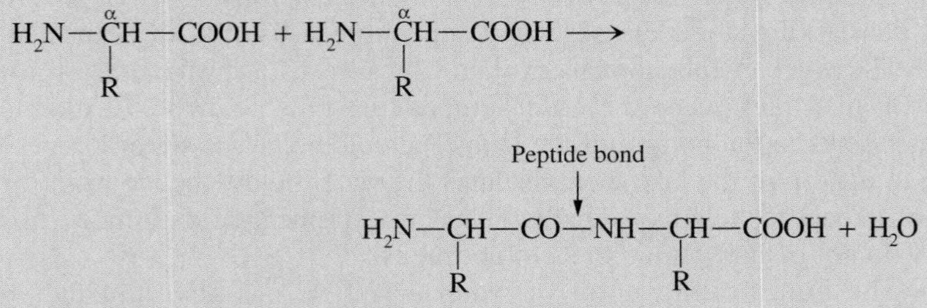

The groups at the ends of this dipeptide can each join with another amino acid molecule to produce a chain segment with four amino acid monomers and three peptide bonds. That chain segment can add two more amino acid molecules, and so on, until a long chain forms.

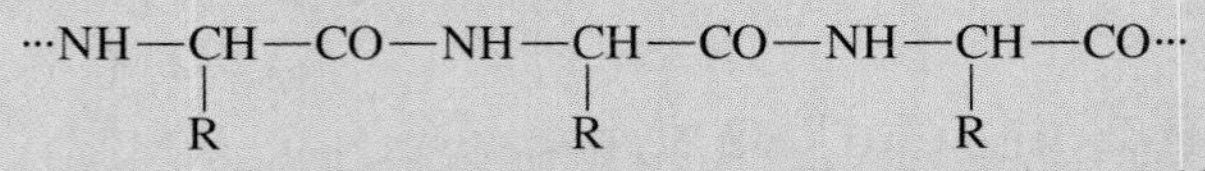

The *sequence* of amino acids in the protein chain is specified by the genetic code and is called the *primary structure* of the protein. The chains, in turn, can twist and fold to give a variety of shapes to protein molecules. These shapes are called the *secondary structure* of the protein. As suggested by Figure 11.19, hydrogen bonding is of great importance in determining the two principal kinds of secondary structures found in all proteins, the α helix and the β sheet.

The many proteins differ in the *sequence* of the amino acids along the protein chain, the lengths of the chains, and in the number of chains. In some proteins, the chains are twisted about one another into larger cables and fibers, which are used for connections, support, and structure. These are the *fibrous proteins*; they are found, for example, in hair, skin, and muscle, and in insect fibers such as silk. In other proteins, the chains are folded back on themselves to make compact *globular proteins*. Hemoglobin, enzymes, and the gamma globulins that serve as protective antibodies are globular proteins. Enzymes, the proteins that catalyze nearly all the reactions in living cells, are considered further in Chapter 15.

A protein molecule functions only when it is in the proper configuration. Heat, ultraviolet light, and some chemical substances *denature* proteins—that is, change them in ways that destroy their function. They do so by disrupting the hydrogen bonds that hold the molecules in their proper arrangement.

Most proteins are denatured when heated above 50 °C. Heat, ultraviolet radiation, and organic compounds such as ethyl alcohol and isopropyl alcohol are used to disinfect things by denaturing the proteins in bacteria and thus killing them. We cook most of our protein-containing food because denatured proteins are usually easier to chew and easier for digestive enzymes to break down.

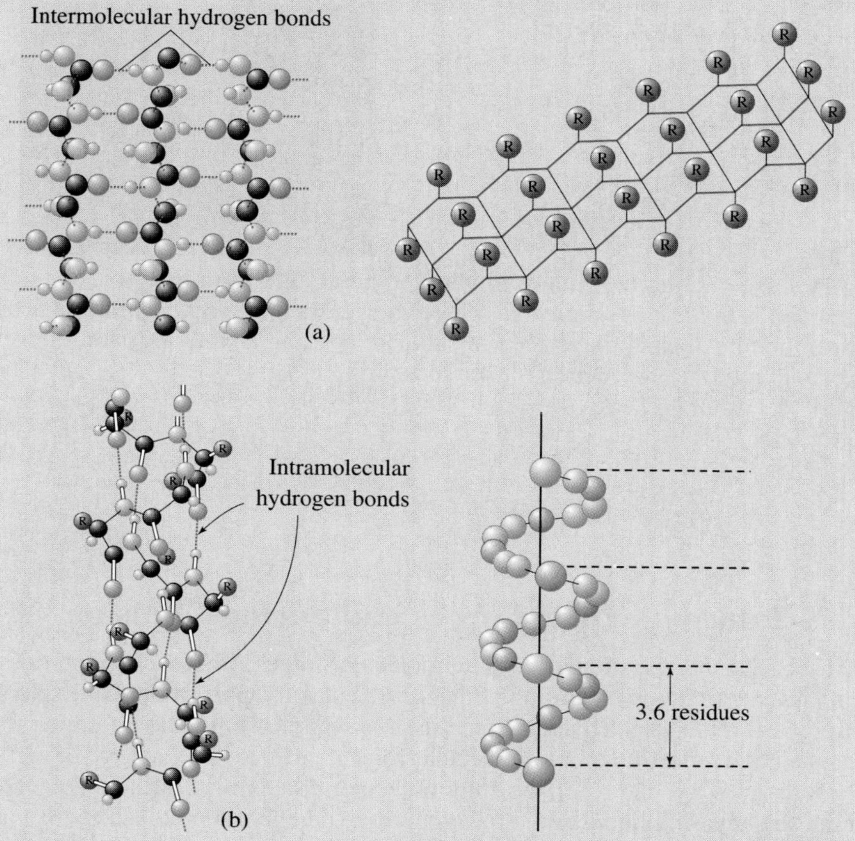

Figure 11.19 The two principal secondary structures of proteins. (a) In the pleated sheet arrangement, two protein chains—or two parts of the same chain folded back on itself—can run parallel to one another in opposite directions. The antiparallel chains are held together by hydrogen bonds between them, like rungs in a ladder. Each hydrogen bond joins the —NH— group on one chain to a —CO— group on another. The ball-and-stick model (left) shows the hydrogen bonds. The schematic drawing (right) emphasizes the pleats. (b) In the alpha-helical arrangement, the protein chain is coiled into a right-handed helix. Each —NH— group is hydrogen bonded to a —CO— group one helical turn away in the same chain. There are 3.6 amino acids per turn of the helix. This gives a fairly rigid cylindrical structure, with side chains on the outside. The ball-and-stick model (left) shows the hydrogen bonds. The skeletal drawing (right) best shows the helix.

would expect for a substance with strong intermolecular forces. Why should this be? What we find is that hydrogen bonding does occur in acetic acid, and the hydrogen bonds are strong enough for *dimers* (double molecules) to form. When acetic acid vaporizes, many of the dimer molecules hold together. Not all the hydrogen bonds are broken during vaporization. It takes less energy to transfer a given quantity of liquid to vapor than expected, and the heat of vaporization is abnormally low. Figure 11.20 shows the presumed structure of a dimer of acetic acid.

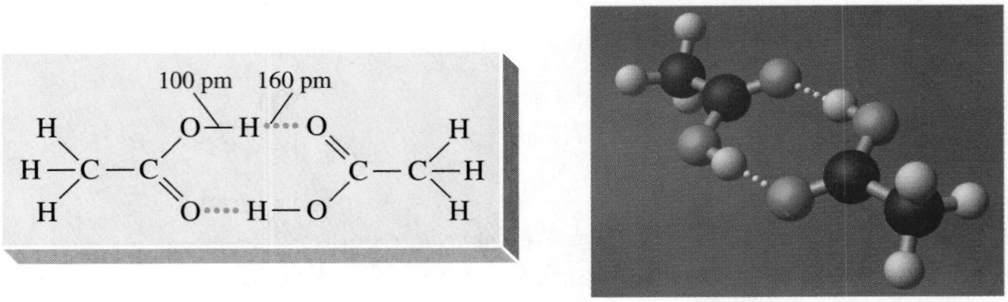

Figure 11.20 Hydrogen bonding in acetic acid. The two acetic acid molecules are joined through two hydrogen bonds (dotted lines) into a "double" molecule—a dimer. Note that the hydrogen-bond lengths (H···O) are longer than the H—O covalent bond lengths.

In some organic molecules a hydrogen bond may form between two nonmetal atoms *in the same molecule*. These molecules have an *intra*molecular hydrogen bond. An example is seen in salicylic acid, an analgesic (pain reliever) and antipyretic (fever reducer) related to aspirin.

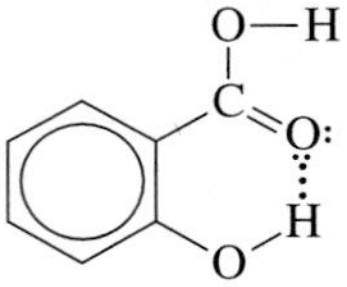

Salicylic acid

11.7 Intermolecular Forces and Surface Tension

We have seen that the strengths of intermolecular forces affect such properties as densities, melting points, boiling points, and heats of vaporization. Other physical properties are affected as well, and in this section we look at a familiar one that is a good reflection of intermolecular forces.

Whatever the nature of the intermolecular forces in a liquid, molecules within the bulk of a liquid are attracted to more neighboring molecules than are surface molecules (Figure 11.21). Because they experience a greater net intermolecular attractive force, molecules in the bulk of a liquid are in a lower energy state than are surface molecules. Molecules crowd into the interior of a liquid to

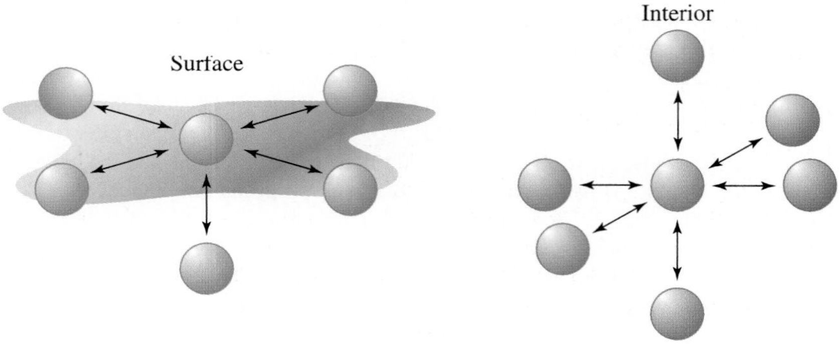

Figure 11.21 Intermolecular forces in a liquid. Molecules at the surface of a liquid are attracted only by other molecules at the surface and by molecules below the surface. Molecules in the interior of a liquid experience forces from neighboring molecules in all directions.

the greatest extent possible, leaving a minimum of surface area. A sphere has a smaller ratio of surface area to volume than any other three-dimensional figure. This explains the tendency of free-falling liquids to form spherical drops.

It takes energy to increase the surface area of a liquid because doing so requires that some molecules pass from the lower energy bulk liquid to the higher energy surface state. **Surface tension** is the amount of work required to extend a liquid surface. We usually express surface tension (γ) in the unit joule per square meter, J/m^2—that is, the work required to extend the surface area of a liquid by one square meter. Some typical values at 20 °C are

hexane, $CH_3(CH_2)_4CH_3$:	$\gamma = 1.84 \times 10^{-2}\ J/m^2$
water, H_2O:	$\gamma = 7.28 \times 10^{-2}\ J/m^2$
mercury, Hg:	$\gamma = 47.2 \times 10^{-2}\ J/m^2$

We can account for the fact that the surface tension of water is significantly greater than that of hexane by comparing intermolecular forces. Strong intermolecular hydrogen bonding makes it more difficult to extend the liquid surface of water than of hexane, where the intermolecular forces are dispersion forces. The unusually high surface tension of mercury can be attributed to the special type of bonding, called metallic bonding, that exists between mercury atoms.

A liquid's reluctance to expand its surface makes it seem as if the liquid is covered with a tight skin. Water striders and other insects are able to walk on water (Figure 11.22) because the energy required to spread the surface (that is, to break through the surface) is greater than can be offset by the force of gravity on the insects.

Figure 11.22 Surface tension illustrated. The surface tension of the water keeps the water strider from breaking through the surface.

When a drop of liquid spreads across a surface, we say that the liquid "wets" the surface. The energy required to spread the drop comes from the collapse of the drop under the force of gravity. Another factor, though, involves the comparative strengths of adhesive and cohesive forces. **Adhesive forces** are intermolecular forces between unlike molecules, and **cohesive forces** are those between like molecules. If the cohesive forces in a liquid are sufficiently strong compared to adhesive forces between the liquid and a surface, the liquid will not wet the surface.

In order to clean a surface with water, the water must wet the surface. This happens readily, for example, with glass and certain fabric surfaces. If a glass surface is covered with a film of grease or oil, though, water won't wet the surface. That is, water does not wet the oily film; instead, water droplets bead on the greasy glass. The function of a dishwashing detergent is to disperse the grease in water (page 302). As shown in Figure 11.23, we can tell at a glance whether a piece of glassware is clean. To waterproof a fabric (a canvas tent, for example), we purposely coat the material with an oil that prevents water from wetting the fabric.

Figure 11.23 Adhesive and cohesive forces. Adhesive forces between water and glass cause a thin film of water to spread over the surface of the glass (left). Cohesive forces between water molecules in the drops of water are stronger than the adhesive forces between water and a film of grease coating the glass. Drops of water are left standing on the glass (right).

If a liquid wets the surface of its container, the liquid is drawn up the walls of the container to a slight extent. The interface between the liquid and the air above it, called a **meniscus**, has a *concave* or saucerlike shape (◡). On the other hand, if a liquid does not wet the walls of its container, the liquid pulls away from the wall and exhibits a *convex* meniscus, like an inverted saucer (◠). Water in a glass container produces a concave meniscus, and mercury forms a convex one (Figure 11.24). Meniscus formation is greatly exaggerated in tubes of very small diameter, called *capillary* tubes. This phenomenon is referred to

Figure 11.24 Meniscus formation. Because it wets glass, water forms a *concave* meniscus in a glass container (left). Mercury does not wet glass (right) and forms a *convex* meniscus.

as *capillary* action. As shown in Figure 11.25, water is drawn up several centimeters into a small glass capillary. The rise of water in a sponge also occurs through capillary action.

11.8 Network Covalent Solids

In most covalently bonded substances, bonds between atoms in molecules, *intra*molecular forces, are quite strong. In these same substances, attractions between molecules, *inter*molecular forces, are much weaker, perhaps only a few percent as strong. Many molecular substances exist as gases at room temperature. The rest are mostly liquids or solids with low to moderate melting points. In a few covalently bonded substances, called **network covalent solids,** covalent bonds extend throughout a crystalline solid. In these cases *intra-* and *inter-*molecular forces are indistinguishable, and the crystalline solid is held together by exceptionally strong forces. Two of the best examples are the two principal naturally occurring forms of carbon: diamond and graphite.

Diamond

With the knowledge that the carbon-to-carbon bonds in diamond are *single* covalent, we can attempt to write a Lewis structure.

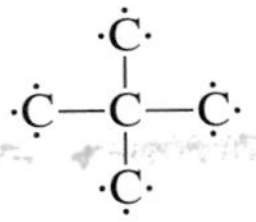

We can continue to add C atoms to this structure, but no matter how many more we add, we can never turn it into a satisfactory Lewis structure unless we accept the possibility that all the atoms in a diamond crystal belong to one molecule. The C atoms of a diamond must form a network covalent solid.

In the extremely small portion of a diamond crystal suggested by Figure 11.26, each carbon atom is bonded to four other carbon atoms in a tetrahedral arrangement. This is the bonding arrangement that we associate with sp^3 hybridization of the carbon atoms. All the electrons in the diamond structure are *localized*. They cannot be set in motion easily by an electric field, meaning that diamond is a *nonconductor* of electricity.

To scratch or break a diamond crystal, we have to break many covalent bonds. This is quite hard to do, and as a result diamond is the hardest natural substance known. Nothing can scratch a diamond, but a diamond can scratch other solids such as glass. Diamond is the best abrasive available. To melt a diamond, we must also break covalent bonds; this accounts for its exceptionally high melting point of more than 3500 °C.

Because silicon is in the same group as carbon (Group 4A) we might expect Si atoms to be able to substitute for some of the C atoms in the diamond structure. Think of this as occurring in the compound silicon carbide, SiC. Silicon carbide is best known as the abrasive Carborundum, extensively used to make grindstones.

Figure 11.25 Capillary action. The spread of a thin film of water up the capillary walls produces a slight drop in pressure below the meniscus. Atmospheric pressure pushes a column of water up the capillary to offset the pressure difference.

Graphite

Another bonding scheme that accommodates all of the valence electrons has each C atom bonded to three other C atoms in the same plane. This model has three of the four valence electrons of each C atom *localized* in sp^2 hybrid orbitals. The fourth valence electron is in a $2p$ orbital perpendicular to the plane of the sp^2 orbitals. These electrons are located between planar layers of C atoms, and they are *delocalized* in molecular orbitals formed when the unhybridized $2p$ orbitals combine. The crystal structure of this modification of carbon is suggested by Figure 11.27.

Graphite has some interesting properties that are consistent with the bonding scheme just described:

1. C—C bond lengths within layers are much shorter (142 pm) than between layers (335 pm), and they are comparable to those found in benzene (139 pm).
2. The weak bonding between layers allows the layers to glide over one another rather easily. This makes graphite a good lubricant* and is also what makes graphite pencils work.
3. Graphite is a good electrical conductor because the p electrons between layers of carbon atoms are delocalized and can be set in motion by an external electric field. Graphite is extensively used as electrodes in batteries and in electrolysis reactions.

The solids diamond and graphite are *polymorphic* forms of carbon, but they are more than that. Two or more forms of an *element* that differ in their basic *molecular* structure are called **allotropes**. Diamond and graphite are allotropes of carbon.

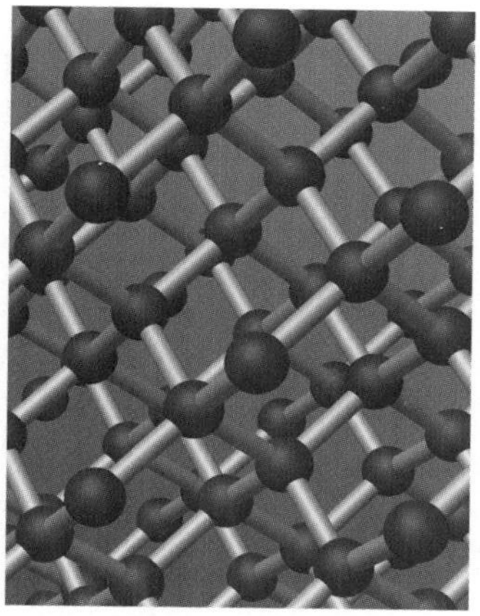

Figure 11.26 The crystal structure of diamond. Each carbon atom is bonded to four others in a tetrahedral fashion.

Fullerenes, which comprise a third allotropic form of carbon, were discovered in 1985. Fullerenes consist of molecules having formulas such as C_{60}, C_{70}, C_{74}, and C_{82} and shapes that are roughly spherical.

11.9 Ionic Bonds as "Intermolecular" Forces

As we noted in introducing ionic compounds in Chapter 2, and again in describing ionic bonding in Chapter 9, there is no such thing as a "molecule" of a solid ionic compound. So, there are really no intermolecular forces. There are simply interionic attractions in which each ion is simultaneously attracted to several ions of the opposite charge. These interionic attractions—ionic bonds—extend throughout an ionic crystal.

We have already identified a property that measures the strength of interionic attractions, the *lattice energy* (page 318). To make qualitative comparisons, we do not need to use actual values of lattice energies. Instead, the following generalization, illustrated in Figure 11.28, works pretty well.

The attractive force between a pair of oppositely charged ions increases as the charges on the ions increase and as the ionic radii decrease. Lattice energies increase accordingly.

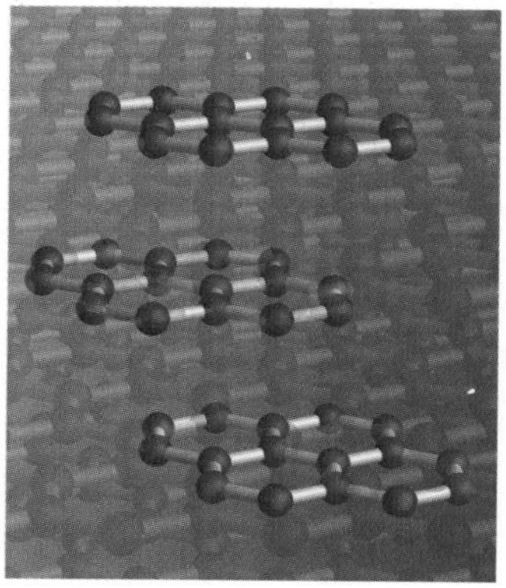

Figure 11.27 The crystal structure of graphite.

*After being strongly heated in a vacuum, graphite is a much poorer lubricant. Additional factors are probably needed to explain its lubricating properties.

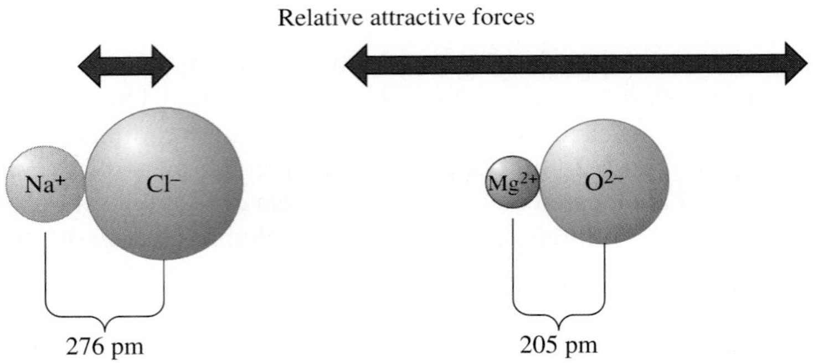

Figure 11.28 Interionic forces of attraction. Because of the higher charges on the ions and the reduced interionic distance, the attractive force between Mg^{2+} and O^{2-} is about seven times as great as between Na^+ and Cl^-. The interionic distances are the sums of the ionic radii listed in Figure 7.11.

Most lattice energies are high enough that ionic solids do not readily sublime. We can melt ionic solids if we supply enough thermal energy to break down the crystalline lattice. In general, the greater its lattice energy, the higher is the melting point of an ionic solid.

Two requirements are necessary for electrical conductivity: (1) charged particles must be present, and (2) the particles must be free to move in an electric field. In a metal, the charged particles are electrons, and their movement is made possible by the nature of metallic bonding. Solid ionic compounds meet the first requirement; the charged particles are ions. But ionic solids do not meet the second requirement. The ions are fixed in a crystalline lattice. Consequently, ionic solids do not conduct electricity. However, when we melt an ionic solid or dissolve it in a suitable solvent, such as water, the ions are free to move. Solutions of ionic compounds are good electrical conductors. We'll have more to say about the electrical conductivities of aqueous solutions in the next chapter.

Example 11.5

Arrange the following three ionic solids in the expected order of increasing melting point: MgO, NaCl, and NaBr.

Solution

Mg^{2+} and O^{2-} ions have higher charges than do Na^+, Cl^-, and Br^-. Also, Mg^{2+} is a smaller ion than Na^+, and O^{2-} is smaller than both Cl^- and Br^-. As a result, by the generalization stated on page 425, we expect the lattice energy of MgO to be much greater than that of either NaCl or NaBr. With its highest lattice energy, MgO should have the *highest* melting point.

Comparing NaCl and NaBr, we see that all cations carry a charge of 1+ and all anions, 1−. Also, the cation radii are the same in the two compounds. The only difference is in the anion radii. The Br^- ion is a larger ion than the Cl^- ion. Consequently, we expect the interionic attractions in NaBr to be somewhat weaker than in NaCl. We expect NaBr to have a lower melting point than NaCl. The expected order is

$$\text{NaBr} < \text{NaCl} < \text{MgO}$$

The observed melting points are 747 °C for NaBr, 801 °C for NaCl, and 2832 °C for MgO.

Exercise 11.5

Arrange the following ionic solids in the expected order of increasing melting point: KCl, MgF_2, KI, and CsBr.

11.10 The Structure of Crystals

In a scientific sense, a **crystal** is a piece of a solid substance that has plane surfaces, sharp edges, and a regular geometric shape. The fundamental units—atoms, ions, or molecules—are assembled in a regular, repeating manner extending in three dimensions throughout the crystal. An essential feature of a crystal is that we must be able to figure out its entire structure from just a tiny portion of it. Some solids, like glass, lack this long-range order and are said to be *amorphous*.

Crystal Lattices

Dealing with repeating patterns is something most of us have done, one time or another, whether stringing beads, cutting out fabric, or laying floor tiles. Usually, though, our need has been limited to two dimensions (length and width). To describe crystals we need to work with *three*-dimensional patterns. The framework on which we outline the pattern is called a *lattice*. We would need 14 different lattices to describe all crystalline solids, but we will confine our discussion to the *cubic* type.

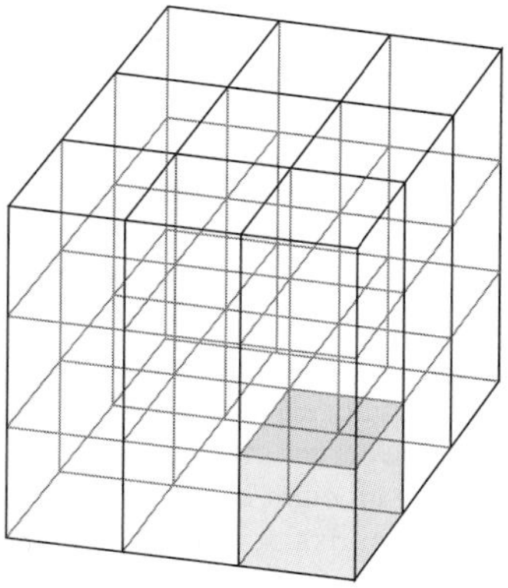

Figure 11.29 The cubic lattice. The entire lattice can be generated by straight-line displacements (left and right, front and back, up and down) of the unit cell shaded in color.

The lattice shown in Figure 11.29 consists of three sets of equidistant, mutually perpendicular planes. A geometric figure called a *parallelepiped*, is outlined in color. It has *six* faces formed by the intersection of three pairs of parallel planes. This parallelepiped is a special kind, called a *cube*, because the distance between each pair of parallel planes is the same and the intersecting planes form 90° angles. A single parallelepiped that can be used to generate the entire lattice by simple, straight-line displacements is called a **unit cell.** The cube in color is a unit cell of the lattice in Figure 11.29.

The simplest unit cell has structural particles (atoms, ions, or molecules) only at its corners, but sometimes we can better describe a crystal structure in terms of a unit cell having more structural particles. The **body-centered cubic (bcc)** structure has an additional structural particle at the center of the cube. The **face-centered cubic (fcc)** structure has an additional particle at the center of each face. These three unit cells are shown in Figure 11.30. The common metals

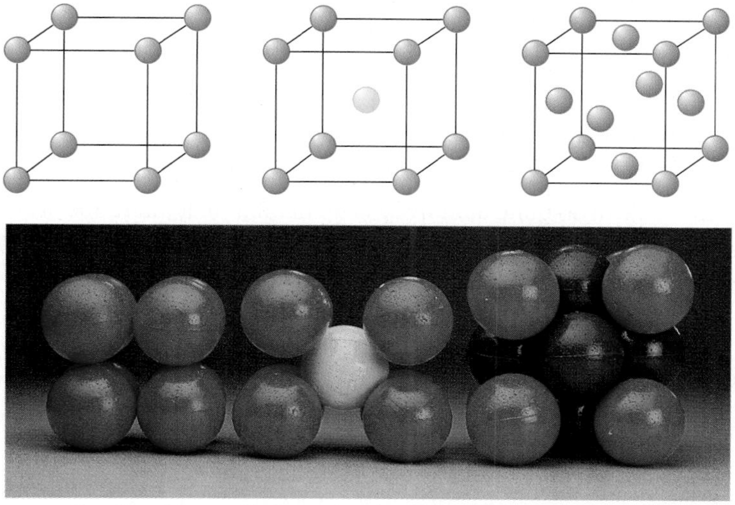

Simple cubic Body-centered cubic Face-centered cubic

Figure 11.30 Unit cells in cubic crystal structures. Only the centers of spheres (atoms) are shown in the top row. The space-filling models in the bottom row show that certain of the spheres are in direct contact.

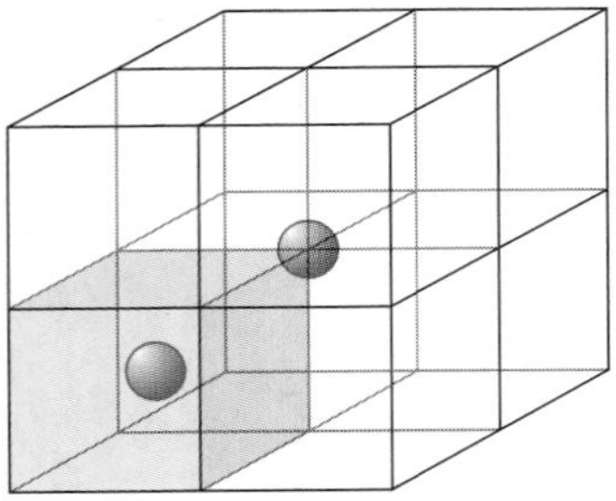

Figure 11.31 Counting atoms in a unit cell. The eight unit cells outlined here are body-centered cubic. Two atoms are shown. One is in the exact center of the unit cell in color. The other, a corner atom, is shared among all eight unit cells. As shown in Example 11.6, the effective number of atoms in a bcc unit cell is *two*.

Fe, K, Na, and W have a bcc crystal structure, and Al, Cu, Pb, and Ag have an fcc structure.

Just as with marbles in a box, in a crystal each structural particle (assume they are atoms) is in contact with a number of others. In the bcc unit cell in Figure 11.30, the center atom is in contact with each of the eight corner atoms. We call the number of atoms with which a given atom is in contact in a crystal the *coordination number.* For the simple cubic cell this number is 6, for the bcc structure it is 8, and for the fcc structure, 12.

We need nine atoms to construct a bcc unit cell, but with help from Figure 11.31, Example 11.6 illustrates why we should say that the unit cell contains *two* atoms, not nine.

Example 11.6

Show that the number of atoms in the bcc unit cell is 2.

Solution

Only the atom at the very center belongs entirely to the unit cell shown in color in Figure 11.31. Each of the corner atoms is shared among *eight* unit cells. In effect, only one-eighth of the corner atom shown belongs to the unit cell in color. There are eight corner atoms so, collectively, they contribute $8 \times \frac{1}{8} = 1$ atom to the unit cell. The total number of atoms in the unit cell, then, is 1 (center atom) + $8 \times \frac{1}{8}$ (corner atoms) = 2 atoms.

Exercise 11.6

Show that the number of atoms in the fcc unit cell is 4. (*Hint*: Here the two types of positions you must consider are the centers of the faces and the corners. You may find it helpful to sketch a grouping of several unit cells, similar to Figure 11.31.)

Ionic Crystal Structures

Ionic crystal structures are somewhat more complicated than those of metals for two reasons.

1. There are two different types of structural particles, cations and anions, rather than atoms of a single kind.
2. The sizes of the cations are different from those of the anions.

Still, we can do quite a bit if we keep in mind that interionic attractions bring oppositely charged ions into contact, whereas interionic repulsions keep like-charged ions from making contact.

To define a unit cell of an ionic crystal we must choose a portion of the crystal that

- generates the entire crystal by straight-line displacements in three dimensions;
- indicates the coordination numbers of the ions;
- is consistent with the formula of the compound.

A unit cell of CsCl is pictured in Figure 11.32 and one of NaCl in Figure 11.33.

Example 11.7

What is the coordination number of each type of ion in CsCl? Show that the unit cell in Figure 11.32 meets the requirements set forth on page 428.

Solution

We see from the figure that the Cs^+ ion in the center of the cell is in contact with *eight* Cl^- ions. By extending the crystal we could easily show that each Cl^- is also at the center of a unit cell with *eight* Cs^+ ions at the corners. (This Cl^--centered unit cell would look different from the one in Figure 11.32 but would generate the same lattice.) The coordination number of each ion is *eight.*

Now we count ions in the unit cell in the same way as we counted atoms in Example 11.6. We conclude that there is *one* Cs^+ at the center of the unit cell and $8 \times \frac{1}{8}$ $Cl^- = 1$ Cl^- ion in the corners. Because a unit cell of cesium chloride has the equivalent of one Cs^+ and one Cl^- ion, this is consistent with the formula CsCl.

Exercise 11.7

Show that the coordination number of each type of ion in NaCl is *six*. Show that the unit cell in Figure 11.33 contains the equivalent of *four* formula units of sodium chloride and is thus consistent with the formula NaCl. (*Hint*: Note that there are now *four* types of positions in the unit cell: center of the cell, center of a face, center of an edge, and corner of the cell. Na^+ ions occupy two types of positions and Cl^- ions the other two.)

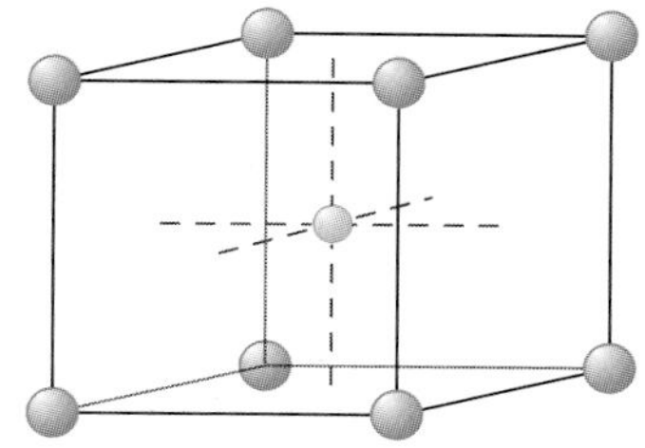

Figure 11.32 A unit cell of cesium chloride. In the unit cell shown here, a Cs^+ ion is at the very center of the cell and Cl^- ions in each of the eight corners. The Cs^+ and the Cl^- ions are in actual contact along the four diagonals of the cube.

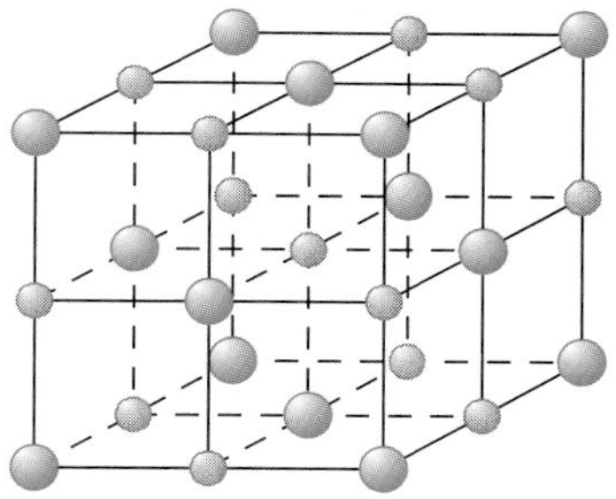

Figure 11.33 A unit cell of sodium chloride. Only the centers of the ions are shown, though oppositely charged ions are actually in contact along edges of the unit cell. In the unit cell shown here, Cl^- ions are at the corners and the centers of the faces. Na^+ ions are at the centers of the edges and at the very center of the cell.

Experimental Determination of Crystal Structures

Crystal structure determinations are essential in many areas of science, including physics, chemistry, biochemistry, geology, metallurgy, and ceramics. Let us briefly survey the basic experimental approach that is used.

We cannot see the patterns of atoms, ions, and molecules in crystalline solids with our eyes, even with the help of a microscope, because these particles are much too small. To figure out the patterns we need to use radiation with wavelengths comparable to the dimensions of unit cells. X-rays are ideal for this purpose.

Figure 11.34 suggests the interaction of X-rays with a crystal and a geometric analysis proposed by the father–son team, W. H. Bragg and W. L. Bragg, in 1912. In the illustration, wave *a* is reflected by one plane of atoms or ions and wave *b* by the next plane below. Wave *b* travels farther than does wave *a*. To achieve a maximum intensity of the reflected radiation, waves *a* and *b* must reinforce each other—that is, their crests and troughs must line up. For this to happen, the additional distance traveled by wave *b* must be a multiple, *n*, of the wavelength, λ, of the X-rays:

$$n\lambda = 2d \sin \theta$$

By measuring the angle θ at which scattered X-rays of known wavelength have their greatest intensity, we can calculate the spacing *d* between atomic planes. If we repeat the process for different orientations of the crystal, we eventually obtain the complete crystal structure.

Types of Crystalline Solids: A Summary

Throughout this section we concentrated on atoms and ions as structural particles, but the structural particles can also be molecules. For example, solid

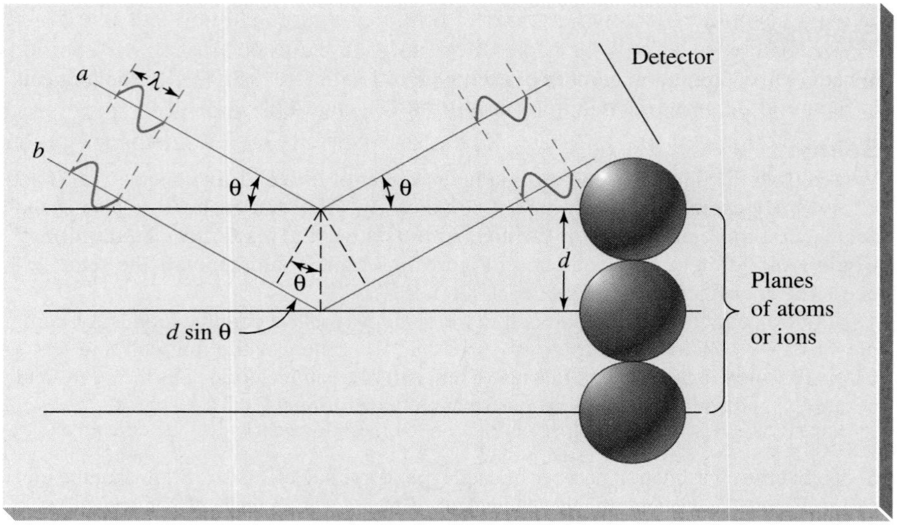

Figure 11.34 X-Ray determination of crystal structure. The hypotenuse of each triangle is equal to the interatomic distance, *d*. The side opposite the angle θ has a length of $d \sin \theta$. Wave *b* travels farther than *a* by the distance $2d \sin \theta$, and this distance is an integral multiple *n* of the X-ray wavelength λ. Thus, $n\lambda = 2d \sin \theta$ where $n = 1, 2, 3, \ldots$.

TABLE 11.5 Some Characteristics of Crystalline Solids

Structural Particles	Principal Forces Between Particles	Typical Properties	Examples
Ionic			
Cations and anions	electrostatic attractions	hard; brittle; moderate to very high melting points; nonconductors as solids, but good electrical conductors as liquids; many are soluble in polar solvents such as water	$NaCl$, CaF_2 K_2S, MgO
Network Covalent			
Atoms	covalent bonds	most are very hard; sublime or melt at very high temperatures; most are nonconductors of electricity	C (diamond), C (graphite), SiC, AlN, SiO_2
Metallic			
Cations and delocalized electrons	metallic bonds	hardness varies from soft to very hard; melting points vary from low to very high; lustrous; ductile; malleable; good to excellent conductors of heat and electricity	Na, Mg, Al, Fe, Cu, Zn, Mo, Ag, Cd, W, Pt, Hg, Pb
Molecular			
Nonpolar			
Atoms or nonpolar molecules	dispersion forces	from extremely low to moderate melting points depending on molecular weight; sublime readily in some cases; soft; soluble in nonpolar solvents	Ar, H_2, I_2, CH_4, CCl_4, CO_2
Polar			
Polar molecules	dispersion forces and dipole–dipole attractions	low to moderate melting points; soluble in other polar and some nonpolar solvents	HCl, H_2S, $CHCl_3$, $(CH_3)_2O$, $(CH_3)_2CO$
Hydrogen-bonded			
Molecules with H bonded to N, O, or F	hydrogen bonds	low to moderate melting points; soluble in other hydrogen-bonded liquids and some polar liquids	H_2O, HF, NH_3, CH_3OH, CH_3COOH

methane has a face-centered cubic crystalline structure. Think of the tetrahedral CH_4 molecules as being spherelike in overall shape and occupying the positions shown for the face-centered cubic structure in Figure 11.30. As we would expect for a low-molecular-weight, nonpolar substance, the intermolecular forces within a methane crystal are weak: The vapor pressure of solid methane is about 90 mmHg at its melting point of −182.5 °C; its boiling point is −161.5 °C.

Table 11.5 summarizes the main types of crystalline solids, indicates the nature of the intermolecular forces in the crystals, and gives some of their characteristic properties.

SUMMARY

Vaporization is the passage of molecules from the liquid to the gaseous (vapor) state; condensation is the reverse process. Dynamic equilibrium is established when vaporization and condensation occur at the same rate, and the pressure exerted by the vapor is called the vapor pressure. A graph of vapor pressure as a function of temperature is called a vapor pressure curve. The normal boiling point of a liquid is the temperature at which its vapor pressure is 1 atm. The critical point is the highest pressure–temperature point on a vapor pressure curve, and the critical temperature is the highest temperature at which a vapor can be liquefied by applying pressure alone.

A phase diagram is a pressure–temperature plot that shows the conditions under which a substance exists as a solid, a liquid, a gas, or some combination of these phases. Two-phase equilibria are represented by curves in the diagram: the vaporization (vapor pressure) curve, the sublimation curve, and the fusion curve. The triple point is the unique temperature and pressure at which the solid, liquid, and vapor phases of a substance coexist.

Fluctuations in electron charge density in a molecule produce an instantaneous dipole, which in turn induces dipoles in neighboring molecules. Attractions between instantaneous and induced dipoles, called dispersion forces, are the intermolecular forces found in nonpolar substances. Polar substances consist of permanent dipoles and have additional intermolecular forces of the dipole–dipole type. Lumped together and called van der Waals forces, dispersion forces and dipole–dipole forces determine physical properties such as melting points and boiling points.

A strong intermolecular force—a hydrogen bond—forms when an H atom attached to an O, N, or F atom of one molecule is simultaneously attracted to an O, N, or F atom of another molecule. Hydrogen bonding accounts for some unusual properties of water (for example, an unusually high boiling point, and a greater density as a liquid than as a solid). It also governs aspects of the behavior of biologically active molecules such as proteins and the nucleic acids.

Differences in the intermolecular forces between molecules in the bulk of a liquid and at the surface give rise to the property of surface tension. Several familiar natural phenomena are related to surface tension.

In some solids, covalent bonds extend throughout a crystal. Such solids generally are harder, have much higher melting points, and are less volatile than other molecular solids. In ionic solids, interionic attractions bind all the ions together in a crystal. The strengths of these attractions depend primarily on ionic charges and sizes and partly determine physical properties like melting points.

The structure of a crystal is described in terms of a three-dimensional pattern called a lattice. In the cubic lattice there are three possibilities: simple cubic, body-centered cubic, and face-centered cubic. An important concept relating to crystal lattices is that of the unit cell. Crystalline solids can also be described in relation to their structural particles, principal intermolecular forces, and characteristic properties (Table 11.5).

KEY TERMS

adhesive force (11.7)
allotrope (11.8)
body-centered cubic (bcc) (11.10)
boiling point (11.2)
cohesive force (11.7)
condensation (11.2)
cooling curve (11.3)
critical point (11.2)
critical pressure (11.2)
critical temperature (11.2)
crystal (11.10)
dispersion force (11.5)
dynamic equilibrium (11.2)
enthalpy (heat) of fusion (11.3)
enthalpy (heat) of sublimation (11.3)
enthalpy (heat) of vaporization (11.2)

face-centered cubic (fcc) (11.10)
freezing (11.3)
freezing point (11.3)
heating curve (11.3)
hydrogen bond (11.6)
melting (fusion) (11.3)
melting point (11.3)
meniscus (11.7)
network covalent solid (11.8)
normal boiling point (11.2)
phase diagram (11.1)
polarizability (11.5)
sublimation (11.3)
sublimation curve (11.3)
supercooling (11.3)
surface tension (11.7)
triple point (11.3)
unit cell (11.10)
vaporization (11.2)
vapor pressure (11.2)
vapor pressure curve (11.2)

REVIEW QUESTIONS

1. What do we mean when we say that a system is in *dynamic equilibrium*?

2. What do we mean when we say that a liquid is *volatile*? *nonvolatile*?

3. What is the distinction intended in the terms *vaporization* and *boiling*?

4. What is the distinction between the terms: boiling point of a liquid and *normal* boiling point of a liquid?

5. What is meant by the term *critical point*?

6. What is a *phase diagram*?

7. In a phase diagram for a pure substance, how many different phases are present in portions of the diagram represented by areas, curves, and points?

8. What is a *triple point*? Is it the same as a *normal* melting point? Explain.

9. Do all substances have a critical point? a normal boiling point? Explain.

10. How can the following be represented in phase diagrams?
a. polymorphic solids **b.** supercritical fluids

11. Describe several features found in the phase diagrams of all substances and one or two features found only in some phase diagrams.

12. Which of the following variables do you expect to affect the vapor pressure of a liquid?
a. temperature
b. volume of liquid in the liquid–vapor equilibrium
c. volume of vapor in the liquid–vapor equilibrium
d. area of contact between liquid and vapor
Explain.

13. Although the terms gas and vapor are often used interchangeably, sometimes a distinction is made between the two terms. What is this distinction?

14. What is the difference between an *intra*molecular force and an *inter*molecular force. Can they ever be the same? Explain.

15. What is the difference between an *instantaneous* and an *induced* dipole?

16. What is a dispersion force?

17. What is the difference in meaning of the terms *polar molecule* and *polarizability* of a molecule?

18. Why does a polar liquid generally have a higher normal boiling point than a nonpolar liquid of the same molecular weight?

19. State the principal reasons why CH_4 is a gas at room temperature whereas H_2O is a liquid.

20. Methane and ethane are both gases at room temperature and 1 atm pressure, yet the corresponding alcohols, methanol and ethanol, are liquids. Explain why this is an expected observation.

21. State *several* reasons why you would expect the boiling point of hexanoic acid, $CH_3(CH_2)_4COOH$, to be *higher* than that of 2-methylbutane,

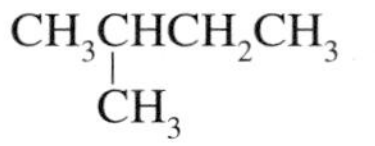

22. Only one of these substances is a solid at STP: C_6H_5COOH, $CH_3(CH_2)_8CH_3$, CH_3OH, $(CH_3CH_2)_2O$. Which do you think it is and why?

23. Only one of these substances is a gas at STP:

NI_3, BF_3, PCl_3, 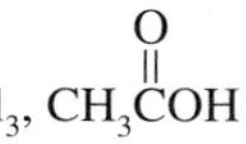

Which do you think it is and why?

24. Diamond and graphite are both network covalent solids. Explain why their properties are so dissimilar.

25. What exactly is meant by the statement that water "wets" glass? Does water "wet" all materials? Explain.

26. Explain why, over a certain small temperature range, it is possible for liquid water to have the same density at *two* different temperatures. (*Hint*: Approximately what is this temperature range?)

27. What is a meniscus, and what determines its shape (that is, whether convex or concave)?

28. Describe the general way in which ionic charges and sizes affect lattice energy, and the general way that lattice energy affects the melting point of an ionic solid.

29. What is meant by the term *crystal lattice*?

30. What is the difference between a formula unit and a unit cell of an ionic compound?

PROBLEMS

Heat of Vaporization

31. Explain why burns produced by steam are often more severe than those produced by boiling water, even when both are at 100 °C.

32. How many kilojoules are required to convert 25.0 g H_2O from liquid at 18.0 °C to vapor at 25.0 °C? (*Hint*: What is the specific heat of water, and what is its heat of vaporization?)

33. How many kilojoules are released when 1.25 mol $CH_3OH(g)$ at 25.0 °C is condensed and cooled to 15.5 °C? The specific heat of $CH_3OH(l)$ is 2.53 $J \cdot g^{-1} \cdot {}^\circ C^{-1}$.

Vaporization, Vapor Pressure

34. Conventional wisdom tells us to spread spilled water into a film so that the floor will dry faster. A classmate tells you this is because the vapor pressure of water increases with increased surface area. Do you agree?

35. By referring to Figure 11.4, estimate the approximate value of (**a**) the vapor pressure of carbon disulfide at 30 °C; (**b**) the boiling point of ethanol when the barometric pressure is 720 mmHg.

36. Refer to Figure 11.4 and estimate the approximate value of (**a**) the vapor pressure of aniline at 100 °C; (**b**) the boiling point of methanol when the barometric pressure is 680 mmHg.

37. What is the *lowest* temperature at which liquid water can be made to boil? Explain how this can be done.

38. How many silver atoms are present in a vapor volume of 486 mL when Ag(l) and Ag(g) are in equilibrium at 1360 °C? (Vapor pressure of silver at 1360 °C = 1.00 mmHg.)

39. Even though the vapor pressure of liquid mercury is very low, enough atoms are present in the vapor to present a significant health hazard when inhaled over time. Determine the number of Hg atoms in a tightly sealed room of 27.5 m^3 volume when equilibrium is established between Hg(l) and Hg(g) at 22.0 °C. (Vapor pressure of mercury at 22.0 °C = 0.00143 mmHg.)

40. Equilibrium is established between a small quantity of $CCl_4(l)$ and its vapor at 40.0 °C in a flask having a volume of 285 mL. The total mass of vapor present is 0.480 g. What is the vapor pressure of CCl_4, in mmHg, at 40.0 °C?

41. The density of acetone vapor in equilibrium with liquid acetone, $(CH_3)_2CO$, at 32 °C is 0.876 g/L. What is the vapor pressure of acetone, in mmHg, at 32 °C?

42. A 1.82-g sample of H_2O is injected into a 2.55-L flask at 30.0 °C. In what form(s) will the H_2O be present, that is, solid and/or liquid and/or gas? Explain.

43. A 0.625-g sample of H_2O is injected into a 178.5-L flask at 0 °C. In what form(s) will the H_2O be present, that is, solid and/or liquid and/or gas? Explain.

Phase Changes

44. Following are several characteristic enthalpy changes of a substance: ΔH_{vapn}; ΔH_{fusion}; ΔH_{condn}; ΔH_{subln}. Arrange them in order of increasing magnitude (that is, increasing numerical value and without regard for signs).

45. How much heat, in kilojoules, is required to melt a 3.55-kg block of ice? (Refer to Table 11.4.)

46. How many kilojoules are required to melt an ice "cube" having the dimensions 3.5 cm × 2.6 cm × 2.4 cm? The density of ice is 0.92 g/cm^3. (Refer to Table 11.4.)

47. Approximately how much heat, in kilojoules, is required to sublime 1.00 kg of ice at 0 °C? (*Hint*: Refer to Tables 11.1 and 11.4.)

48. A 0.506-kg chunk of ice at 0.0 °C is added to an insulated container holding 315 mL of water at 20.2 °C. Will any ice remain after thermal equilibrium is established?

49. An ice cube weighing 25.5 g at a temperature of 0.0 °C is added to 125 mL of water at 26.5 °C in an insulated container. What will be the final temperature after the ice has melted?

50. It is often observed that a liquid can be cooled to a temperature below its freezing point without freezing. What makes this possible? Do you think that a solid can be similarly heated above its melting point without melting? Explain.

51. If $CO_2(s)$ is heated under a pressure greater than 5.1 atm, it melts rather than sublimes. If the pressure is then reduced below 5.1 atm, the liquid boils to produce $CO_2(g)$. How does the enthalpy change for this two-step process compare to the enthalpy change for the direct sublimation of CO_2?

Phase Diagrams

52. Sketch a phase diagram for iodine, including the labeling of significant points, given that the triple point is at 114 °C and 91 mmHg, the normal boiling point is at 184°C, and the critical point is at 512 °C and 116 atm.

53. Sketch a phase diagram for hydrazine, including the labeling of significant points, given that the triple point is at 2.0 °C and 3.4 mmHg, the normal boiling point is 113.5 °C, and the critical point is at 380 °C and 145 atm.

54. Indicate where each of the following points should fall on the phase diagram for water, that is, whether in a one-phase region (solid, liquid, or gas) or on a two-phase curve (vaporization, sublimation, or fusion). Use information supplied with Figure 11.11 and elsewhere in the chapter.
a. 88.15 °C and 0.954 atm pressure
b. 25.0 °C and 0.0313 atm pressure
c. 0 °C and 2.50 atm pressure
d. −10 °C and 0.100 atm pressure

55. Describe the phase changes that occur when a sample of solid CO_2 is slowly heated under a constant pressure of 10 atm from −100 °C to 100 °C. (*Hint*: Consider that the heating is done in a device like that pictured in Figure 11.12, and use the phase diagram in Figure 11.10.)

Intermolecular Forces

56. Which of the following would you expect to have the *lower* boiling point: carbon disulfide or carbon tetrachloride? Why?

57. Which of the following would you expect to have the *lower* boiling point: phosphorus triiodide or phosphorus trifluoride? Why?

58. Which of the following would you expect to have the *higher* boiling point: hexane, C_6H_{14}, or 2,2-dimethylbutane, $CH_3CH_2C(CH_3)_3$? Why?

59. Which of the following would you expect to have the *higher* boiling point: hexane, C_6H_{14}, or 2,2-dimethyl-1-butanol, $CH_3CH_2C(CH_3)_2CH_2OH$? Why?

60. Which of the following would you expect to have the *higher* melting point: pentane, $CH_3CH_2CH_2CH_2CH_3$, or diethyl ether, $(CH_3CH_2)_2O$? Why?

61. Which of the following would you expect to have the *higher* melting point: 1-pentanol, $CH_3CH_2CH_2CH_2CH_2OH$, or 3,3-dimethylpentane, $CH_3CH_2C(CH_3)_2CH_2CH_3$? Why?

62. Arrange the following substances in the expected order of *increasing* boiling point: C_4H_9OH, NO, C_5H_{12}, N_2, $(CH_3)_2O$. Give the reasons for your ranking.

63. Arrange the following substances in the expected order of *increasing* boiling point: H_2O, NH_3, CH_4, CH_3CH_3. Give the reasons for your ranking.

64. Arrange the following substances in the expected order of *increasing* melting point: Cl_2, CsCl, CCl_4, $MgCl_2$. Give the reasons for your ranking.

65. Arrange the following substances in the expected order of *increasing* melting point: NaOH, CH_3OH, LiOH, C_6H_5OH. Give the reasons for your ranking.

Surface Tension

66. Which of the following liquids would you expect to have the higher surface tension: octane, C_8H_{18}, or 1-octanol, $C_8H_{17}OH$? Explain.

67. Which of the following liquids would you expect to have the higher surface tension: isopropyl alcohol, $(CH_3)_2CH_2OH$, or ethylene glycol, CH_2OHCH_2OH? Explain.

Crystal Structures

68. Use appropriate data from Chapter 7 to determine the length of the unit cell of NaCl pictured in Figure 11.33. (*Hint*: The ions along an edge of the cube are in direct contact.)

69. Magnesium oxide has a crystal structure similar to that of sodium chloride. Sketch a unit cell of magnesium oxide. Use the appropriate data from Chapter 7 to determine the length of the unit cell. (*Hint*: See Problem 68.)

ADDITIONAL PROBLEMS

70. Which has the greater heat of sublimation, ΔH_{subln}, $CO_2(s)$ (that is, dry ice) or MgO(s)? Explain.

71. The purpose of a "wetting agent" is to make water better able to wet a surface. What property of water must be affected by the wetting agent, and in what way?

72. Explain the phenomenon shown in the photograph of water boiling in a paper cup. That is, why doesn't the paper cup burn? Why do you suppose the temperature is 99.9 °C (rather than 100.0 °C)?

73. A 150.0-mL sample of $N_2(g)$ at 25.0 °C and 750.0 mmHg is passed through benzene, $C_6H_6(l)$, until the gas

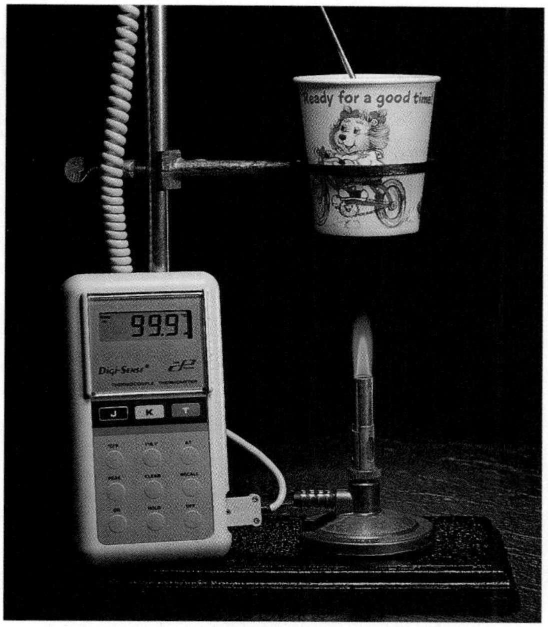

becomes saturated with $C_6H_6(g)$. The new volume of the gas is 172 mL at a total pressure of 750.0 mmHg. What is the vapor pressure of benzene at 25.0 °C?

74. The following data are given for CCl_4: normal melting point, −23 °C; normal boiling point, 77 °C; density of liquid, 1.59 g/mL; heat of fusion, 3.28 kJ/mol; vapor pressure at 25.0 °C, 110 mmHg.

a. How much heat must be absorbed to convert 10.0 g of solid CCl_4 to liquid at −23 °C?

b. What is the volume occupied by 1.00 mol of the saturated vapor at 77 °C?

c. What phases—solid, liquid, and/or vapor—are present if 3.5 g CCl_4 is kept in an 8.21-L volume at 25.0 °C?

75. A handbook gives this equation for the vapor pressure of hexane, C_6H_{14}, as a function of the *Celsius* temperature, *t*:

$$\log P\ (\text{mmHg}) = 6.876 - \frac{1171}{t + 224.4}$$

a. What is the vapor pressure of hexane at 25 °C?

b. Determine the normal boiling point of hexane.

76. In the final condition reached in Figure 11.12, that is, condition (c), what mass of water is present as liquid and what mass as vapor?

77. A mixture of 1.00 g H_2 and 10.00 g O_2 is ignited in a 3.15-L flask. What is the pressure in the flask when it is cooled to 25 °C? (*Hint*: What is the reaction that occurs? What are the contents of the flask?)

78. Sketch a phase diagram for tin, including the labeling of significant points in the diagram, given that there are three polymorphic forms of the solid: α, gray tin below 19 °C; β, white tin between 19 and 161 °C; and γ, brittle tin from 161 °C to the melting point at 232 °C. The normal boiling point of tin is 2623 °C. Sketch the cooling curve you would expect to obtain if a sample of liquid tin at 250 °C is slowly cooled to 0 °C.

79. A 1.00-g sample of $H_2O(g)$ is injected into a 40.0-L flask at a temperature of 35.0 °C and cooled. To the nearest degree Celsius, determine the temperature at which the first liquid water will condense.

80. Intramolecular bonding in salicylic acid was illustrated on page 422. Salicylic acid actually forms both intramolecular and intermolecular hydrogen bonds. Draw the structure of a dimer of salicylic acid that shows four hydrogen bonds altogether.

81. The quantity of CO_2 present in a CO_2 fire extinguisher is determined by weighing the charged cylinder and subtracting off the mass of the empty cylinder (tare weight). In this way the mass of CO_2 in a 11.5-L cylinder is found to be 1510 g. In what state(s) of matter is the CO_2 in this cylinder at 25 °C?

82. In both NaCl and CsCl the coordination number of the cation is the same as that of the anion. Would this also be true for CaF_2? for MgO? Explain.

83. Aluminum has an atomic radius of 143 pm and crystallizes in a face-centered cubic structure. What is the volume of a unit cell of aluminum? (*Hint*: Use Figure 11.30 as an aid in establishing the length of a diagonal of a face and the length of an edge of a unit cell.)

84. Use the result of Problem 83 and other facts in the text to predict the density of aluminum. (*Hint*: How many Al atoms are present in the unit cell? What is their mass?)

85. Atoms in a crystal structure tend to pack as tightly as possible, that is, to minimize the amount of empty space (called voids) around them. Show that the percent voids in the simple cubic structure is much greater than in the face-centered cubic structure (Figure 11.30). (*Hint*: What is the number of atoms of radius *r* per unit cell in each structure?)

The reaction of copper with nitric acid is an oxidation-reduction reaction, one of the three types of chemical reactions considered in this chapter.

12

Chemical Reactions in Aqueous Solutions

CONTENTS

Three quarters of Earth's surface is covered with water, but little of this natural water is pure. Mostly it is in the form of *aqueous* solutions, that is, as water that contains dissolved substances. Solutions are formed by the contact of liquid water with gases in Earth's atmosphere, with solid materials in Earth's solid crust, and with materials released to the environment by human activities. Water in living organisms is also in the form of aqueous solutions.

Water is the medium most often used for carrying out chemical reactions in the laboratory. Reactions in aqueous solutions fall into a few basic categories, all of which have important applications. We will explore these types of reactions in this chapter and return to each of them later in the text.

12.1 Some Characteristics of Aqueous Solutions

Water is such a good solvent for so many solutes that it has been called the universal solvent. It dissolves most simple ionic substances and many molecular substances. A useful way of characterizing an aqueous solution is by whether the solute particles are ions, molecules, or a mixture of ions and molecules. Figure 12.1, which compares the electrical conductivities of three solutions of equal molarity, allows us to establish three categories based on the observations summarized below.

- *The lamp bulb lights up brightly*. The electrical conductivity is *high*, indicating that the solution contains a significant concentration of ions. The substance in solution, the solute, is called a **strong electrolyte**. (We observe this even for a dilute solution of a strong electrolyte.)
- *The bulb doesn't light*. There are few, if any, ions present. The solute is present in *molecular* form and is called a **nonelectrolyte**.
- *The bulb lights but glows only dimly*. The electrical conductivity is *low*, corresponding to a low concentration of ions. The solute is present only partly

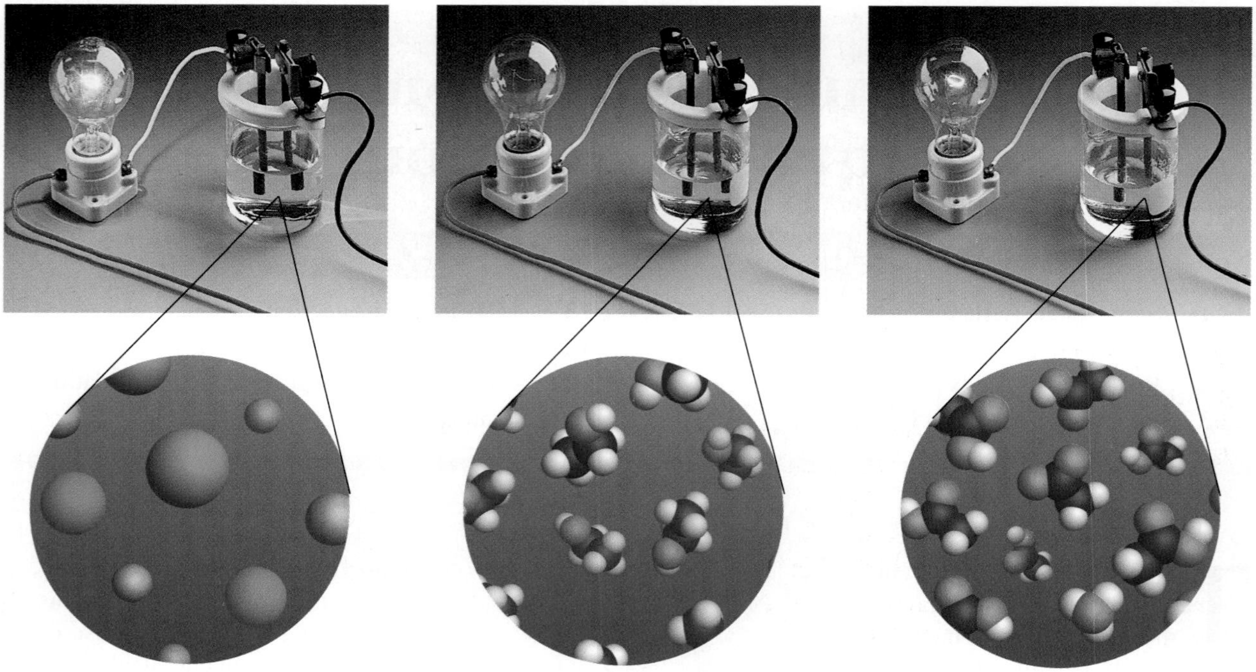

(a) 1 M NaCl(aq)
Strong electrolyte

(b) 1 M $CH_3CH_2OH(aq)$
Nonelectrolyte

(c) 1 M $CH_3COOH(aq)$
Weak electrolyte

Figure 12.1 **Electrolytic properties of aqueous solutions.** If electric current is to flow, cations and anions must be present and free to flow between the graphite electrodes.

in ionic form; it is called a **weak electrolyte**. (We observe this even for a concentrated solution of a weak electrolyte.)

If we were to test a large number of water-soluble substances by the method suggested in Figure 12.1, we would arrive at the following generalization.

- Water-soluble ionic compounds and a few molecular compounds (a few acids) are *strong electrolytes*.
- Most molecular compounds are either *nonelectrolytes* or *weak electrolytes*.

Among water-soluble organic compounds, most are nonelectrolytes (for example, alcohols and sugars), but carboxylic acids (Section 2.11) and amines (a family of organic bases related to ammonia) are weak electrolytes.

These generalizations can help us decide how best to represent the solute in an aqueous solution. For a solution of a *nonelectrolyte*, we write the molecular formula. Thus, we represent an aqueous solution of ethanol as $CH_3CH_2OH(aq)$.

When we dissolve a crystal of a water-soluble ionic compound—a *strong electrolyte*—the cations and anions dissociate from one another and appear in solution as independent solute particles. Thus, for NaCl, we can write

$$NaCl(s) \xrightarrow{H_2O} Na^+(aq) + Cl^-(aq)$$

For most purposes, we will continue to represent aqueous sodium chloride as NaCl(aq), but we will write $Na^+(aq) + Cl^-(aq)$ when it is more appropriate to do so.

Example 12.1 illustrates how we can calculate the concentrations of cations and anions in an aqueous solution of an ionic compound. In this example we introduce a new symbol. The "bracket" symbol, [ ], signifies the molar concentration of the ion or molecule whose formula is enclosed within the brackets.

Example 12.1

Calculate the molarities of the ions present in an aqueous solution that is both 0.00384 M Na_2SO_4 and 0.00202 M $Al_2(SO_4)_3$.

Solution

We must first recognize that this is an aqueous solution of two ionic compounds and that each compound is dissociated into its ions.

$$Na_2SO_4(s) \xrightarrow{H_2O} 2\,Na^+(aq) + SO_4^{2-}(aq)$$

$$Al_2(SO_4)_3(s) \xrightarrow{H_2O} 2\,Al^{3+}(aq) + 3\,SO_4^{2-}(aq)$$

Next, we can establish the molarities of the ions derived from each solute with the help of the conversion factors shown in color.

$$[Na^+] = \frac{0.00384 \text{ mol } Na_2SO_4}{1 \text{ L}} \times \frac{2 \text{ mol } Na^+}{1 \text{ mol } Na_2SO_4} = \frac{0.00768 \text{ mol } Na^+}{1 \text{ L}}$$

$$= 0.00768 \text{ M}$$

$$[SO_4^{2-}] = \frac{0.00384 \text{ mol } Na_2SO_4}{1 \text{ L}} \times \frac{1 \text{ mol } SO_4^{2-}}{1 \text{ mol } Na_2SO_4} = \frac{0.00384 \text{ mol } SO_4^{2-}}{1 \text{ L}}$$

$$= 0.00384 \text{ M}$$

$$[Al^{3+}] = \frac{0.00202 \text{ mol } Al_2(SO_4)_3}{1 \text{ L}} \times \frac{2 \text{ mol } Al^{3+}}{1 \text{ mol } Al_2(SO_4)_3} = \frac{0.00404 \text{ mol } Al^{3+}}{1 \text{ L}}$$

$$= 0.00404 \text{ M}$$

$$[SO_4^{2-}] = \frac{0.00202 \text{ mol } Al_2(SO_4)_3}{1 \text{ L}} \times \frac{3 \text{ mol } SO_4^{2-}}{1 \text{ mol } Al_2(SO_4)_3} = \frac{0.00606 \text{ mol } SO_4^{2-}}{1 \text{ L}}$$

$$= 0.00606 \text{ M}$$

The SO_4^{2-} ions derived from the $Al_2(SO_4)_3$ are no different from those derived from Na_2SO_4. To obtain the sulfate ion concentration in solution, we must add together the contributions from these two sources. Each of the cations Na^+ and Al^{3+} has only one source.

$$Na^+ = 0.00768 \text{ M}$$

$$Al^{3+} = 0.00404 \text{ M}$$

$$SO_4^{2-} = 0.00384 \text{ M} + 0.00606 \text{ M} = 0.00990 \text{ M}$$

Exercise 12.1

Each year, oral rehydration therapy (ORT)—feeding of an electrolyte solution—saves the lives of a million children worldwide who become dehydrated from diarrhea. The solution contains 3.5 g NaCl, 1.5 g KCl, 2.9 g sodium citrate ($Na_3C_6H_5O_7$), and 20.0 g glucose ($C_6H_{12}O_6$) per liter. Calculate the molarity of each of the species present in the solution. (*Hint*: Sodium citrate is a strong electrolyte and glucose is a nonelectrolyte. Notice that there is more than one source of Na^+ and Cl^-.)

12.2 Reactions of Acids and Bases

We learned about Arrhenius's concept of acids and bases in Section 2.9, and we saw how acid–base reactions are used to perform chemical analyses by the method of titration in Section 3.11. In this section we learn more about this important type of reaction.

Strong and Weak Acids

We will adopt a different acid–base theory in Chapter 17, but for now we will continue to use the Arrhenius view that an acid is a substance that produces hydrogen ion (H^+)* in aqueous solution. However, we can now extend this definition to reflect the electrolytic properties of aqueous acid solutions.

Acids that are completely ionized in water solution are *strong* electrolytes and are called **strong acids**. To represent the ionization of a strong acid, such as hydrochloric acid, we can write the equation

$$HCl(g) \xrightarrow{H_2O} H^+(aq) + Cl^-(aq)$$

We can therefore represent an aqueous solution of aqueous hydrochloric acid either as HCl(aq) or $H^+(aq) + Cl^-(aq)$. We can calculate the concentrations of cations and anions in an aqueous solution of a strong acid in the same way as we did for ionic compounds in Example 12.1. For example, in 0.0010 M HCl,

$$[H^+] = 0.0010\ M; \qquad [Cl^-] = 0.0010\ M; \qquad [HCl] = 0$$

Most carboxylic acids with from one to four carbon atoms are water soluble.

The vast majority of acids, because they are only partially ionized in aqueous solution, are *weak* electrolytes and are called **weak acids**. This category includes, for example, essentially all water-soluble carboxylic acids. One important difference between weak acids and strong acids is in how we represent them in solution. The ionization of a weak acid in aqueous solution is a *reversible* reaction, represented in an equation with a double arrow. For the most common weak acid, acetic acid, we can imagine these two steps

$$CH_3COOH(l) \xrightarrow{H_2O} CH_3COOH(aq)$$

* The simple hydrogen ion, H^+, does not exist in aqueous solutions. Instead it is associated with several H_2O molecules; that is, it is present as $H(H_2O)_n^+$. Usually, n is taken to be 1; $H(H_2O)^+ = H_3O^+$ is called the *hydronium* ion. We will use the abbreviation H^+ in the current discussion and switch to H_3O^+ in Chapter 17.

followed by

$$\underset{\text{Acetic acid}}{CH_3COOH(aq)} \rightleftharpoons H^+(aq) + \underset{\text{Acetate ion}}{CH_3COO^-(aq)}$$

Acetic acid, CH_3COOH

When this reversible reaction reaches equilibrium, most of the acetic acid molecules remain nonionized; in 1 M $CH_3COOH(aq)$, fewer than 1% of the molecules ionize. Because the predominant solute species are acetic acid molecules, an aqueous solution of acetic acid is best represented as $CH_3COOH(aq)$, *not* as $H^+(aq) + CH_3COO^-(aq)$.

To calculate concentrations in an aqueous solution of a weak electrolyte—for example, $[H^+]$, $[CH_3COO^-]$, and $[CH_3COOH]$ in $CH_3COOH(aq)$—requires a kind of calculation different from that of Example 12.1. We will defer such calculations to Chapter 17.

Some acids, called *polyprotic* acids, can produce two or more hydrogen ions per molecule of the acid. Sulfuric acid, H_2SO_4, and phosphoric acid, H_3PO_4, are two common examples. Sulfuric acid is interesting in that it is a *strong* acid in its first ionization and a *weak* acid in its second.

$$H_2SO_4(aq) \longrightarrow H^+(aq) + HSO_4^-(aq)$$

$$HSO_4^-(aq) \rightleftharpoons H^+(aq) + SO_4^{2-}(aq)$$

Phosphoric acid, on the other hand, is a *weak* acid in each of its three ionization steps.

$$H_3PO_4(aq) \rightleftharpoons H^+(aq) + H_2PO_4^-(aq)$$

$$H_2PO_4^-(aq) \rightleftharpoons H^+(aq) + HPO_4^{2-}(aq)$$

$$HPO_4^{2-}(aq) \rightleftharpoons H^+(aq) + PO_4^{3-}(aq)$$

We will say more about polyprotic acids in Chapter 17.

So, you may wonder, how can you tell if a substance is an acid and, if so, whether it's a strong or weak acid? Chemists generally identify *ionizable* hydrogen atoms by writing formulas in one of two ways.

1. *A molecular formula with ionizable H atoms has the H atoms written first.* HNO_3, H_2SO_4, and H_3PO_4 are acids with one, two, and three ionizable H atoms, respectively. Methane, CH_4, has four H atoms, but they are *not* ionizable; CH_4 is *not* an acid. From its name alone we know that acetic acid is an acid. When its formula is written as $HC_2H_3O_2$, we see that it has four H atoms, but only one that is ionizable—the H atom that is shown first.
2. *A condensed structural formula shows H atoms where they are bonded.* For acetic acid, a carboxylic acid, the formula CH_3COOH identifies the carboxylic acid functional group,—COOH. The formula shows that the three nonionizable H atoms are bonded to a C atom. The fourth H atom is bonded to an O atom, and it is this one that is ionizable.

The simplest way to tell whether an acid is strong or weak is to note that the *common* strong acids are limited to those in Table 12.1. Unless you are given information to the contrary, assume that any other acid is a weak acid.

TABLE 12.1 **Common Strong Acids and Strong Bases**

Acids	Bases
Hydrogen halides	*Alkali metal hydroxides*
HCl	LiOH
HBr	NaOH
HI	KOH
	RbOH
	CsOH
Oxoacids	*Alkaline earth hydroxides*
HNO_3	$Ca(OH)_2$
H_2SO_4[a]	$Sr(OH)_2$
$HClO_4$	$Ba(OH)_2$

[a] H_2SO_4 is a strong acid in its first ionization step but weak in its second ionization step.

Strong and Weak Bases

By the Arrhenius definition we gave in Chapter 2, a base is substance that produces hydroxide ions, OH^-, in aqueous solution. Soluble ionic hydroxides such as NaOH are obviously bases. Moreover, because soluble ionic compounds are *strong* electrolytes, those that are hydroxides are **strong bases**.

$$NaOH(s) \xrightarrow{H_2O} Na^+(aq) + OH^-(aq)$$

Some compounds produce OH^- ions by *reacting* with water, not just by dissolving in it. Such substances are also bases. The equations below suggest that gaseous ammonia dissolves in water, and then a reaction between ammonia and water produces an equilibrium mixture of ions and molecules.

$$NH_3(g) \xrightarrow{H_2O} NH_3(aq)$$

$$NH_3(aq) + H_2O(l) \rightleftharpoons NH_4^+(aq) + OH^-(aq)$$

As in the case of acetic acid, most of the ammonia molecules in $NH_3(aq)$ remain nonionized. Ammonia is a *weak* electrolyte and therefore a **weak base**.* Most molecular substances that act as bases are *weak* bases.

Methylamine, CH_3NH_2

Organic compounds with OH groups are either acidic (the carboxylic acids) or neutral (the alcohols).

The family of organic compounds called **amines** have molecules in which one or more of the H atoms of NH_3 is replaced by a hydrocarbon group. Some examples are *methylamine*, CH_3NH_2; *dimethylamine*, $(CH_3)_2NH$; *trimethylamine,* $(CH_3)_3N$; and *ethylamine*, $CH_3CH_2NH_2$. The amines with one to four carbon atoms are water soluble, and, like ammonia, they are *weak* bases.

$$CH_3NH_2(aq) + H_2O(l) \rightleftharpoons CH_3NH_3^+(aq) + OH^-(aq)$$

How can we recognize a base and how do we know whether the base is weak or strong? If an ionic compound contains OH^- ions, expect it to be a base and, if water soluble, to be strong base. NaOH and KOH are strong bases. On the other hand, methanol, CH_3OH is *not* a base. The OH group is not present as OH^-; it is covalently bonded to the C atom. Similarly, acetic acid, CH_3COOH, is *not* a base. It does not contain OH^-, nor does it produce $OH^-(aq)$ in water; rather it produces $H^+(aq)$ and is therefore an *acid.* To identify a weak base, you usually need a chemical equation for the ionization reaction. However, you can identify many by using these facts:

> *There are only a few common strong acids and strong bases; these are listed in Table 12.1. The most common weak bases are ammonia and the amines.*

Acid–Base Reactions: Neutralization

In the reaction of an acid and a base, called **neutralization**, the identifying characteristics of the acid and base cancel out or neutralize each other. The acid and base are converted to an aqueous solution of an ionic compound, called a **salt**.

---

* In Arrhenius's time, chemists generally believed that a substance must *contain* OH groups to be a base; $NH_3(aq)$ was thought to be NH_4OH (ammonium hydroxide). Even though there is no compelling evidence for the existence of NH_4OH, this formula is still often seen as a representation of $NH_3(aq)$.

If we use conventional formulas for the acid and base, we can write what we might call a "complete formula" equation* for a neutralization reaction as follows.

$$HCl(aq) + NaOH(aq) \longrightarrow NaCl(aq) + H_2O(l)$$

But this "complete formula" equation is not the best way to show what happens in the neutralization. We can better do that by indicating the actual ions and molecules present in solution; we write the equation in *ionic* form.

$$\underset{\text{(Acid)}}{H^+(aq) + Cl^-(aq)} + \underset{\text{(Base)}}{Na^+(aq) + OH^-(aq)} \longrightarrow \underset{\text{(Salt)}}{Na^+(aq) + Cl^-(aq)} + \underset{\text{(Water)}}{H_2O(l)}$$

When we eliminate "spectator" ions—those that just "look on" and that appear unchanged in the ionic equation—we find that the equation reduces to the even more informative **net ionic equation**.

$$H^+(aq) + OH^-(aq) \longrightarrow H_2O(l)$$

The essence of a neutralization reaction, then, is that H^+ ions from an acid and OH^- ions from a base combine to form water. If the spectator ions form a soluble salt, they remain in solution and are able to conduct electricity. If the water is evaporated, the soluble salt is left as a solid.

Example 12.2 provides additional examples of the different types of equations described here.

Example 12.2

Barium nitrate is used to produce a green color in fireworks. It can be made by the reaction of nitric acid with barium hydroxide. Write (**a**) "complete formula," (**b**) ionic, and (**c**) net ionic equations for this neutralization reaction.

Solution

a. Write chemical formulas for the substances involved in the reaction and balance the equation.

$$HNO_3(aq) + Ba(OH)_2(aq) \longrightarrow Ba(NO_3)_2(aq) + H_2O(l) \quad (\textit{not balanced})$$

$$2\,HNO_3(aq) + Ba(OH)_2(aq) \longrightarrow Ba(NO_3)_2(aq) + 2\,H_2O(l) \quad (\textit{balanced})$$

b. Now, represent the strong electrolytes with the formulas of their ions and the nonelectrolyte water with its molecular formula.

$$2\,H^+(aq) + 2\,NO_3^-(aq) + Ba^{2+}(aq) + 2\,OH^-(aq) \longrightarrow$$
$$Ba^{2+}(aq) + 2\,NO_3^-(aq) + 2\,H_2O(l)$$

c. Cancel the spectator ions (Ba^{2+} and NO_3^-) in the above equation to give

$$2\,H^+(aq) + 2\,OH^-(aq) \longrightarrow 2\,H_2O(l)$$

or, more simply,

$$H^+(aq) + OH^-(aq) \longrightarrow H_2O(l)$$

Exercise 12.2

Calcium hydroxide is used to neutralize a waste stream of hydrochloric acid. Write (**a**) "complete formula," (**b**) ionic, and (**c**) net ionic equations for this neutralization reaction.

---

* The "complete formula" equation is commonly called a "molecular" equation, but this term is misleading. Many of the formulas written in such an equation—for example, NaCl(aq)—represent formula units, not actual molecules.

In discussing the stoichiometry of acid–base titrations in Section 3.11, we used only "complete formula" equations for neutralization reactions. In Example 12.3 we show that the titration calculation can also be done by using a net ionic equation.

Example 12.3

A 10.00-mL sample of an aqueous solution saturated in $Ca(OH)_2$ requires 23.30 mL of 0.02000 M $HNO_3(aq)$ for its neutralization. What is the molarity of the saturated $Ca(OH)_2(aq)$?

Solution

We can calculate the number of moles of $Ca(OH)_2$ in the saturated solution in a single setup based on the following numbered factors. From the volume (1) and molarity (2) of $HNO_3(aq)$, and knowing that HNO_3 is a strong acid that produces *one* H^+ ion for every molecule of HNO_3 (3), we can calculate the number of moles of H^+ used in the titration. According to the net ionic equation

$$H^+(aq) + OH^-(aq) \longrightarrow H_2O(l)$$

one mole of OH^- is neutralized for every mole of H^+ (4). The number of moles of OH^- in the saturated solution is related to the number of moles of $Ca(OH)_2$: *two* moles of OH^- for every *one* mole of $Ca(OH)_2$ (5). Thus,

$$? \text{ mol } Ca(OH)_2 = 23.30 \text{ mL} \times \underset{(1)}{\frac{1 \text{ L}}{1000 \text{ mL}}} \times \underset{(2)}{\frac{0.02000 \text{ mol } HNO_3}{1 \text{ L}}} \times \underset{(3)}{\frac{1 \text{ mol } H^+}{1 \text{ mol } HNO_3}}$$

$$\times \underset{(4)}{\frac{1 \text{ mol } OH^-}{1 \text{ mol } H^+}} \times \underset{(5)}{\frac{1 \text{ mol } Ca(OH)_2}{2 \text{ mol } OH^-}} = 2.330 \times 10^{-4} \text{ mol } Ca(OH)_2$$

The amount of $Ca(OH)_2$ just calculated is found in a 10.00-mL (0.01000 L) sample. We can then use the definition of molarity to write

$$\text{Molarity} = \frac{2.330 \times 10^{-4} \text{ mol } Ca(OH)_2}{0.01000 \text{ L}} = 0.02330 \text{ M } Ca(OH)_2$$

Exercise 12.3

What volume of 0.550 M NaOH(aq) is required to titrate a 10.00-mL sample of vinegar that is 4.12% by mass of acetic acid, $HC_2H_3O_2$? Assume that the vinegar has a density of 1.01 g/mL. [*Hint*: The net ionic equation is $HC_2H_3O_2(aq) + OH^-(aq) \longrightarrow H_2O(l) + C_2H_3O_2^-(aq)$.]

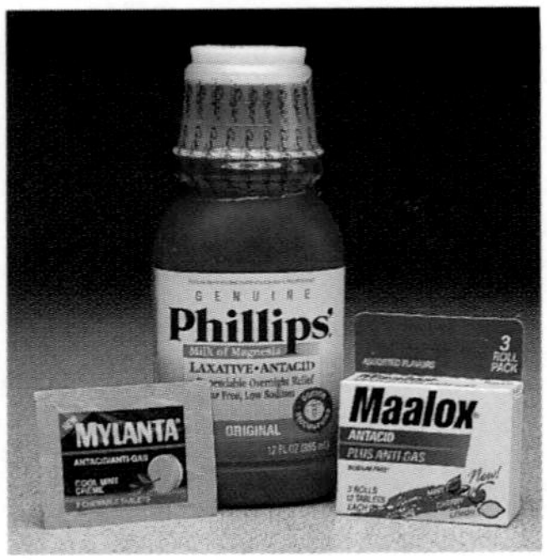

Water-insoluble hydroxides, such as $Mg(OH)_2$ (milk of magnesia) and $Al(OH)_3$, are used as antacids. One should *never* use a water-soluble ionic hydroxide. In high concentrations $OH^-(aq)$ is strongly basic, and it causes severe burning and scarring of tissue.

Acid–Base Reactions: More Net Ionic Equations

Magnesium hydroxide is an ionic compound, but it is only slightly soluble in water. In a net ionic equation showing how a water suspension of $Mg(OH)_2(s)$ neutralizes excess stomach acid, we should represent the magnesium hydroxide through its complete formula. Nevertheless, OH^- from $Mg(OH)_2$, a base, combines with H^+ of the acid to form water.

$$Mg(OH)_2(s) + 2\,H^+(aq) \longrightarrow Mg^{2+}(aq) + 2\,H_2O(l)$$

The net equation for the neutralization of HCl(aq) with NH_3(aq) looks rather different from that of a strong acid and strong base for two reasons: NH_3 is a weak base and so we should write its complete formula, and the NH_3 contains no OH^-. Instead of the net equation $H^+(aq) + OH^-(aq) \longrightarrow H_2O(l)$, we have

$$H^+(aq) + NH_3(aq) \longrightarrow NH_4^+(aq)$$

That is, think of H^+ from the acid as combining directly with NH_3 molecules in solution. This occurs through the formation of a coordinate covalent bond between H^+ and NH_3, using the lone pair electrons on the N atom (page 325).

CONCEPTUAL EXAMPLE 12.1

With the help of a net ionic equation, explain the observations made in Figure 12.2 on page 446.

Solution

The dimly lit bulb in Figure 12.2(a) indicates that acetic acid is a weak acid. As we have previously written,

$$CH_3COOH(aq) \rightleftharpoons H^+(aq) + CH_3COO^-(aq)$$

The dimly lit bulb in Figure 12.2(b) indicates that ammonia is a weak base.

$$NH_3(aq) + H_2O(l) \rightleftharpoons NH_4^+(aq) + OH^-(aq)$$

When the two solutions are mixed, we can think of three processes occurring simultaneously: these two ionizations and the combination of H^+(aq) from the acid and OH^-(aq) from the base to form H_2O(l). Because the neutralization reaction goes to completion, it forces the ionization reactions to do the same. The net equation for the three reactions is

$$\begin{array}{rl} CH_3COOH(aq) & \rightleftharpoons H^+(aq) + CH_3COO^-(aq) \\ NH_3(aq) + H_2O(l) & \rightleftharpoons NH_4^+(aq) + OH^-(aq) \\ H^+(aq) + OH^-(aq) & \longrightarrow H_2O(l) \\ \hline \textit{Net:}\ CH_3COOH(aq) + NH_3(aq) & \longrightarrow NH_4^+(aq) + CH_3COO^-(aq) \end{array}$$

The two weak electrolyte solutions are replaced by an aqueous solution of an ionic compound, a strong electrolyte. This accounts for the brightly lit bulb in Figure 12.2(c). We can also describe the reaction as the direct transfer of H^+ from CH_3COOH to NH_3, as we did for the hydrochloric acid–ammonia reaction.

Conceptual Exercise 12.1

In a situation similar to that in Figure 12.2, describe the observations you would make if the two original solutions were CH_3NH_2(aq) and HNO_3(aq).

In the same way that it can combine with OH^- to form H_2O, H^+ can combine with certain other anions to produce weak electrolytes or nonelectrolytes. In a broad sense these reactions are also acid–base reactions.

In baking, carbon dioxide is released from baking powders, causing dough to rise. The basis of the process is the action of an acid component on baking soda, $NaHCO_3$. Specifically, H^+ from a weak acid (let's call it HA) reacts with

(a) (b) (c)

Figure 12.2 **Change in electrical conductivity in a chemical reaction.** The change in electrical conductivity on mixing 1 M $CH_3COOH(aq)$ (a) and 1 M $NH_3(aq)$ (b) is described in Conceptual Example 12.1.

hydrogen carbonate (bicarbonate) ion, HCO_3^-, to form the weak acid carbonic acid, H_2CO_3. Carbonic acid is unstable and decomposes to $H_2O(l)$ and $CO_2(g)$.

$$\begin{array}{rcl} HA(aq) + HCO_3^-(aq) & \longrightarrow & H_2CO_3(aq) + A^-(aq) \\ H_2CO_3(aq) & \longrightarrow & H_2O(l) + CO_2(g) \\ \hline HA(aq) + HCO_3^-(aq) & \longrightarrow & A^-(aq) + H_2O(l) + CO_2(g) \end{array}$$

The $CO_2(g)$ provides the leavening action when baking powder is used in baking (see Figure 12.3).

Figure 12.3 **The leavening action of baking soda.** When acidified, here with citric acid from a lemon, baking soda ($NaHCO_3$) reacts to produce carbonic acid (H_2CO_3), which decomposes to carbon dioxide and water. The carbon dioxide gas produces a "lift" in the dough being baked.

Looking Ahead

In Chapter 17 we will discuss

- theories that expand the definitions of acids and bases beyond just H^+ and OH^-;
- how the acidity of a solution is described through its pH;
- how to calculate the concentrations of ions and molecules in solutions of weak acids and weak bases;
- how acid strength is related to molecular structure.

12.3 Reactions That Form Precipitates

Certain cations and anions form ionic compounds that are insoluble in water.* When these ions are brought together in an aqueous solution, the insoluble ionic compound separates or precipitates from solution. A chemical reaction between ions that produces a precipitate is called a **precipitation reaction**. Table 12.2 lists some precipitation reactions that are of industrial importance, and Figure 12.4 pictures one of these.

TABLE 12.2 Some Precipitation Reactions of Practical Importance

Reaction in Aqueous Solution	Application
$Al^{3+} + 3\,OH^- \longrightarrow Al(OH)_3(s)$	Water purification. (The gelatinous precipitate carries down suspended matter.)
$Al^{3+} + PO_4^{3-} \longrightarrow AlPO_4(s)$	Removal of phosphates from wastewater in sewage treatment.
$Mg^{2+} + 2\,OH^- \longrightarrow Mg(OH)_2(s)$	Precipitation of magnesium from seawater. (First step in Dow process for extracting magnesium from seawater.)
$Ag^+ + Br^- \longrightarrow AgBr(s)$	Preparation of AgBr for use in photographic film.
$Zn^{2+} + SO_4^{2-} + Ba^{2+} + S^{2-} \longrightarrow ZnS(s) + BaSO_4(s)$	Production of lithopone, a mixture used as a white pigment.
$Cd^{2+} + 2\,OH^- \longrightarrow Cd(OH)_2(s)$ $Pb^{2+} + CO_3^{2-} \longrightarrow PbCO_3(s)$ $Hg^{2+} + S^{2-} \longrightarrow HgS(s)$ $Ba^{2+} + CrO_4^{2-} \longrightarrow BaCrO_4(s)$	Removal of metal ions from industrial waste waters.

Figure 12.4 An important industrial precipitation reaction. The precipitation of $Mg(OH)_2(s)$ from seawater is carried out in huge vats. This is the first step in the Dow process for extracting magnesium from seawater (see page 293).

* In principle, all ionic compounds dissolve in water to some extent. However, some of them yield so few ions in solution that they can be ignored; these compounds are said to be *insoluble*. If the amount of solid that dissolves is less than about 0.01 mol/L, we consider it to be insoluble.

When you are asked to *predict* a chemical reaction, you are given the left side of an equation—for example,

$$AgNO_3(aq) + KI(aq) \longrightarrow ?$$

and are asked to complete the right side, either by writing formulas for the expected products or by writing "no reaction" if none can occur. Perhaps the task is easier to visualize if we rewrite the expression in *ionic* form.

$$Ag^+(aq) + NO_3^-(aq) + K^+(aq) + I^-(aq) \longrightarrow ?$$

We can get products that differ from the starting reactants only by forming KNO_3 and AgI. This reaction will occur *only* if one or both of these products is insoluble; that is, if one or both settle out of solution (precipitate). If a *precipitate* is formed, a cation–anion combination is removed from solution that is different from either of the cation–anion combinations originally present.

In short, to make predictions we need to know which ionic compounds are water soluble and which are not. This requires that we either look up solubility data in handbooks or memorize solute solubilities. Memorizing solute solubilities is not difficult because we can boil down a lot of data into just a few **solubility rules** (Table 12.3).

TABLE 12.3 **General Rules for the Water Solubilities of Common Ionic Compounds**[a]

Compounds that are mostly *soluble:*
All nitrates.
Alkali metal (Group 1A) and ammonium compounds.
Chlorides, bromides, and iodides, *except* for those of Pb^{2+}, Ag^+, Hg_2^{2+}.
Sulfates, *except* for those of Sr^{2+}, Ba^{2+}, Pb^{2+}, and Hg_2^{2+}. ($CaSO_4$ is slightly soluble.)
Compounds that are mostly *insoluble:*
Carbonates, hydroxides, and sulfides, *except* for ammonium compounds and those of the Group 1A metals. (The hydroxides and sulfides of Ca^{2+}, Sr^{2+}, and Ba^{2+} are slightly to moderately soluble.)

[a] In general, if the concentration of a saturated solution is greater than about 0.10 M, we say that the solute is *soluble;* if less than 0.01 M, the solute is *insoluble.* For concentrations between these limits, we might say that the solute is slightly soluble.

From Table 12.3 we quickly conclude that KNO_3 is soluble (all nitrates are soluble) and that AgI is not (all chlorides, bromides, and iodides are soluble *except* those of Pb^{2+}, Ag^+, and Hg_2^{2+}). We complete the equation by showing AgI as a solid and KNO_3 as separate ions.

$$Ag^+(aq) + NO_3^-(aq) + K^+(aq) + I^-(aq) \longrightarrow AgI(s) + K^+(aq) + NO_3^-(aq)$$

Usually our interest is only in the net ionic equation, so we can eliminate the spectator ions and write

$$Ag^+(aq) + I^-(aq) \longrightarrow AgI(s)$$

Whatever their sources, if Ag^+ and I^- are found in the same solution, they should form a precipitate of AgI(s). Our prediction is confirmed by the experimental result pictured in Figure 12.5.

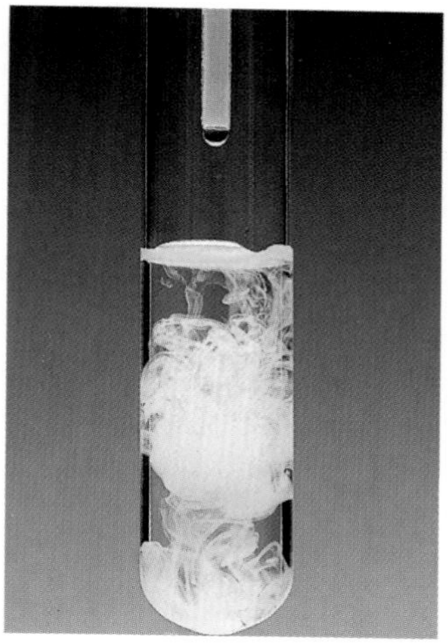

Figure 12.5 The precipitation of silver iodide, AgI(s). When a clear, colorless, aqueous solution of silver nitrate is added to one of potassium iodide, a yellow precipitate of silver iodide is formed.

$$Ag^+(aq) + I^-(aq) \longrightarrow AgI(s)$$

Example 12.4

Predict whether a reaction will occur in each of the following cases. If so, write a net ionic equation for the reaction.

a. $Na_2SO_4(aq) + MgCl_2(aq) \longrightarrow ?$
b. $BaS(aq) + Cu(NO_3)_2(aq) \longrightarrow ?$
c. $(NH_4)_2CO_3(aq) + ZnCl_2(aq) \longrightarrow ?$

Solution

a. The initial reactants are in aqueous solution; they are water soluble. The possible products, NaCl and $MgSO_4$, are also both water soluble. All common sodium compounds are, as are most sulfates, including that of magnesium. There is *no reaction*.

b. Here the possible products are $Ba(NO_3)_2$ and CuS. All nitrates are water soluble, but most sulfides are not, including CuS. The reaction that occurs is

$$Ba^{2+}(aq) + S^{2-}(aq) + Cu^{2+}(aq) + 2\ NO_3^-(aq) \longrightarrow CuS(s) + Ba^{2+}(aq) + 2\ NO_3^-(aq)$$

Ba^{2+} and NO_3^- are spectator ions and can be canceled out to give the net ionic equation

$$S^{2-}(aq) + Cu^{2+}(aq) \longrightarrow CuS(s)$$

c. All common ammonium compounds are soluble, and among carbonates, only those of the Group 1A metals are soluble. We should expect a precipitate of $ZnCO_3(s)$ and an aqueous solution containing NH_4^+ and Cl^- ions.

$$2\ NH_4^+(aq) + CO_3^{2-}(aq) + Zn^{2+}(aq) + 2\ Cl^-(aq) \longrightarrow ZnCO_3(s) + 2\ NH_4^+(aq) + 2\ Cl^-(aq)$$

The net ionic equation is

$$CO_3^{2-}(aq) + Zn^{2+}(aq) \longrightarrow ZnCO_3(s)$$

Exercise 12.4

Predict whether a reaction will occur in each of the following cases. If so, write a net ionic equation for the reaction.

a. $MgSO_4(aq) + KOH(aq) \longrightarrow ?$
b. $FeCl_3(aq) + Na_2S(aq) \longrightarrow ?$
c. $CaCO_3(s) + NaCl(aq) \longrightarrow ?$

CONCEPTUAL EXAMPLE 12.2

Figure 12.6 shows that the dropwise addition of $NH_3(aq)$ to $FeCl_3(aq)$ produces a precipitate. What is the precipitate?

Solution

The initial reactants, NH_3 and $FeCl_3$, both are water soluble. All common ammonium compounds are water soluble, so the precipitate is not likely to contain

Figure 12.6 Predicting the product of a precipitation reaction. Addition of $NH_3(aq)$ to $FeCl_3(aq)$ produces a precipitate.

NH_4^+; it must contain Fe^{3+}. But what is the anion? We need to recall that NH_3 is a weak *base*; it produces OH^- in aqueous solution.

$$NH_3(aq) + H_2O(l) \rightleftharpoons NH_4^+(aq) + OH^-(aq)$$

The $OH^-(aq)$ can then combine with $Fe^{3+}(aq)$ to form a precipitate of $Fe(OH)_3(s)$.

$$Fe^{3+}(aq) + 3\,OH^-(aq) \longrightarrow Fe(OH)_3(s)$$

To get the net ionic equation for the precipitation reaction, we need to multiply the first equation by *three*. That is,

$$3\,NH_3(aq) + 3\,H_2O(l) \rightleftharpoons 3\,NH_4^+(aq) + 3\,OH^-(aq)$$
$$Fe^{3+}(aq) + 3\,OH^-(aq) \longrightarrow Fe(OH)_3(s)$$

$$\textit{Net: } Fe^{3+}(aq) + 3\,NH_3(aq) + 3\,H_2O(l) \longrightarrow 3\,NH_4^+(aq) + Fe(OH)_3(s)$$

In conclusion, the hydroxide ion found in $Fe(OH)_3(s)$ does not appear on the left side of the equation. It is formed by the ionization of $NH_3(aq)$ as an intermediate step in the net precipitation reaction.

Conceptual Exercise 12.2

Suppose that a large quantity of HCl(aq) is added to the beaker in Figure 12.6 after all the $Fe(OH)_3(s)$ has precipitated. Describe what you would expect to see, and write a net ionic equation for this change.

Looking Ahead

In Chapter 18 we will learn how to

- calculate the actual solubilities of "insoluble" and slightly soluble ionic compounds;
- alter the solubility of a precipitate through the presence of other substances in solution;
- use precipitation reactions in analytical chemistry.

12.4 Oxidation–Reduction

Chemists often refer to oxidation–reduction reactions by turning the words around and abbreviating them to "redox" reactions; redox is a little easier to say.

The third category of chemical reactions, known as *oxidation–reduction* reactions, is perhaps the largest of all. It includes all combustion processes, most metabolic reactions in living organisms, the extraction of metals from their ores, the manufacture of countless chemicals, and many of the reactions occurring in our natural environment.

The term "oxidation" was originally used to describe reactions in which a substance combines with oxygen. The opposite process, the removal of oxygen, was described by the term "reduction." To encompass the wide range of reactions described above, however, we need a much broader definition of oxidation and reduction. First, we introduce a concept that assists us in formulating this broader definition.

Oxidation State

In a general way, **oxidation state** or **oxidation number** refers to the number of electrons transferred, shared, or otherwise involved in the formation of the chemical bonds in a substance. For example, in the formation of sodium chloride Na atoms lose electrons and Cl atoms gain them, forming the ions Na^+ and Cl^-. Na is in the oxidation state +1 and Cl is in the oxidation state −1.

In the compound $CaCl_2$, chlorine is also in the oxidation state −1, existing as Cl^- ions. The oxidation state of calcium, however, is +2; it is present as Ca^{2+} ions. The total of the oxidation states of the atoms (ions) in a formula unit of $CaCl_2$ is $+2 - 1 - 1 = 0$.

In the formation of a molecule no electrons are actually transferred; they are shared. We can, however, *arbitrarily* assign oxidation states. In a binary molecule we should assign a negative oxidation state to the more electronegative element. This means that in H_2O the oxidation state of H should be positive and that of O should be negative. We assign H the oxidation state +1. We also require that the total of the oxidation states of the three atoms in H_2O be *zero*. This means that the oxidation state of O must be −2; that is, $+1 + 1 - 2 = 0$.

In the H_2 molecule the H atoms are identical and we should assign them the same oxidation state. If we require the sum of these oxidation states to be *zero*, then the oxidation state of each H atom must also be 0.

From these examples you can see that we need some rules in assigning oxidation states. We can deal with the great majority of compounds with the following rules. Important exceptions to the rules are listed in the marginal notes. The rules are listed below *by priority*, with the highest priority given first. If two rules contradict one another, use the rule with the higher priority, and this should generally take care of exceptions. Some examples are listed for each rule and all the rules together are applied in Example 12.5.

1. *The total of the oxidation states* **(O.S.)** *of all the atoms in a neutral species (an isolated atom, a molecule, or a formula unit) is* 0.
   [Examples: The O.S. of an isolated Fe atom is 0. The sum of the O.S. of all the atoms in each of the molecules Cl_2, S_8, and $C_6H_{12}O_6$ is 0. The sum of the O.S. of the ions in $MgBr_2$ is 0.]
   The total of the O.S. of all the atoms in an ion is equal to the charge on the ion.
   [Examples: The O.S. of Cr in the Cr^{3+} ion is +3. The sum of the O.S. of all the atoms in PO_4^{3-} is −3; the sum in NH_4^+ is +1.]

Because all the atoms in a molecule of an element are alike, each Cl atom in Cl_2 and each S atom in S_8 has an O.S. of 0.

2. *In their compounds, the Group 1A metals all have an O.S. of* +1, *and the Group 2A metals have an O.S. of* +2.
   [Examples: The O.S. of Na in Na_2SO_4 is +1 and that of Ca in $Ca_3(PO_4)_2$ is +2.]
3. *In its compounds, the O.S. of fluorine is* −1.
   [Examples: The O.S. of F is −1 in HF, OF_2, ClF_3, SO_2F_2.]
4. *In its compounds, hydrogen has an O.S. of* +1.
   [Examples: The O.S. of H is +1 in HCl, H_2O, NH_3, and CH_4.]
5. *In its compounds, oxygen has an O.S. of* −2.
   [Examples: The O.S. of O is −2 in CO, CH_3OH, $C_6H_{12}O_6$, and ClO_4^-.]
6. *In their binary (two-element) compounds with metals, Group 7A elements have an O.S. of* −1, *Group 6A elements have an O.S. of* −2, *and Group 5A elements have an O.S. of* −3.

The principal exception to rule 4 is when H is bonded to an element that is less electronegative than itself, as in metal hydrides.

The principal exceptions to rule 5 are when oxygen is bonded to itself, as in peroxides (for example, H_2O_2) and superoxides (for example, KO_2).

[Examples: The O.S. of Br is −1 in $CaBr_2$, that of S is −2 in Na_2S, and that of N is −3 in Mg_3N_2.]

Example 12.5

What is the oxidation state (O.S.) of each element in the following substances?

a. $KClO_4$ **b.** $Cr_2O_7^{2-}$ **c.** CaH_2 **d.** Na_2O_2 **e.** Fe_3O_4

Solution

a. The O.S. of K is +1 (rule 2) and that of O is −2 (rule 5). The total for *four* O atoms is −8. For K and O together, the total is +1 + (−8) = −7. The O.S. of the Cl atom must be +7, so that the total for all atoms in the formula unit is 0 (rule 1). The oxidation states are +1 for K, +7 for Cl, and −2 for O.

b. The O.S. of O is −2 (rule 5), and the total for *seven* O atoms is −14. The total of the O.S. for all atoms in this *ion* must be −2 (rule 1). This means that the total O.S. of *two* Cr atoms is +12, and that of one Cr atom is +6. The oxidation states are +6 for Cr and −2 for O.

c. The O.S. of Ca is +2 (rule 2). The total for the formula unit must be 0 (rule 1). Even though the O.S. of H is usually +1 (rule 4), here it must be −1, so that the total for the *two* H atoms is −2. Rule 2 takes priority over rule 4. The oxidation states are +2 for Ca and −1 for H.

d. The O.S. of Na is +1 (rule 2), and for the *two* Na atoms, +2. The total for the formula unit must be 0 (rule 1). Even though the O.S. of O is usually −2 (rule 5), here it must be −1, so that the total for the *two* O atoms is −2. The oxidation states are +1 for Na and −1 for O.

e. The O.S. of O is −2 (rule 5). For *four* atoms the total is −8. The total for the formula unit must be 0 (rule 1). The total for *three* Fe atoms must be +8, and for each Fe atom, $+\frac{8}{3}$. The oxidation states are $+2\frac{2}{3}$ for Fe and −2 for O.

Usually fractional oxidation states signify an average. The compound Fe_3O_4 is actually $Fe_2O_3{\bullet}FeO$. Two of the Fe atoms are in the O.S. of +3 and one has the O.S. of +2. The average is $(3 + 3 + 2)/3 = +\frac{8}{3}$.

Exercise 12.5

What is the oxidation state (O.S.) of each element in the following substances?

a. Al_2O_3 **b.** P_4 **c.** $NaMnO_4$ **d.** ClO^- **e.** $HAsO_4^{2-}$
f. $HSbF_6$ **g.** CsO_2 **h.** CH_3F **i.** $CHCl_3$ **j.** CH_3COOH

(The assignment of oxidation states in CH_3F, $CHCl_3$, and CH_3COOH demonstrates the variability of the oxidation state of carbon in organic compounds.)

Figure 12.7 The thermite reaction:
$2\,Al(s) + Fe_2O_3(s) \longrightarrow 2\,Fe(l) + Al_2O_3(s)$
Oxidation produces an increase in oxidation state, and reduction produces a decrease in oxidation state. Oxidation and reduction always occur together.

Identifying Oxidation–Reduction Reactions

The spectacular reaction pictured in Figure 12.7, called the *thermite* reaction, is used to produce liquid iron for welding large iron objects.

$$2\,Al(s) + Fe_2O_3(s) \longrightarrow 2\,Fe(l) + Al_2O_3(s)$$

Even by the limited definitions we gave at the start of this section, we can call this an *oxidation–reduction* reaction. Al is *oxidized* to Al_2O_3; aluminum atoms take on or *gain* oxygen atoms. Fe_2O_3 is *reduced* to Fe; iron(III) oxide *loses* oxygen atoms.

Now let's assign oxidation states to the elements involved in the thermite reaction. These are the small numbers written above the chemical symbols in the equation.

$$2\,\overset{0}{Al}(s) + \overset{+3}{Fe_2}\overset{-2}{O_3}(s) \longrightarrow 2\,\overset{0}{Fe}(l) + \overset{+3}{Al_2}\overset{-2}{O_3}(s)$$

In the thermite reaction, the oxidation state of Al atoms *increases* from 0 to +3, and the oxidation state of Fe atoms *decreases* from +3 to 0.

> *In an* oxidation–reduction *reaction the oxidation state of one or more elements* increases—*an* **oxidation** *process. The oxidation state of one or more elements* decreases—*a* **reduction** *process. Oxidation and reduction must always occur together.*

The reaction pictured in Figure 12.8 is strikingly different in appearance from the thermite reaction, but the expanded definition identifies this also as an oxidation–reduction reaction. Oxidation states are noted in the equation below.

$$\overset{0}{Mg}(s) + \overset{+2}{Cu}^{2+}(aq) \longrightarrow \overset{+2}{Mg}^{2+}(aq) + \overset{0}{Cu}(s)$$

This equation also suggests that the reaction can be visualized as two **half-reactions** that occur simultaneously. In one, the oxidation half-reaction, Mg atoms *lose* two electrons and are *oxidized* to Mg^{2+} ions.

$$\textit{Oxidation:}\quad Mg(s) \longrightarrow Mg^{2+}(aq) + 2\,e^-$$

In the other half-reaction, the reduction half-reaction, Cu^{2+} ions *gain* two electrons and are *reduced* to Cu atoms.

$$\textit{Reduction:}\quad Cu^{2+}(aq) + 2\,e^- \longrightarrow Cu(s)$$

This interpretation of an oxidation–reduction reaction suggests still another useful definition.

> *An* oxidation–reduction *reaction consists of two* half-reactions: *an* oxidation *half-reaction in which electrons are "lost" and appear on the* right *side of the half-equation, and a* reduction *half-reaction in which electrons are "gained" and appear on the* left *side of the half-equation.*

Figure 12.8 An oxidation–reduction reaction:

$Mg(s) + Cu^{2+}(aq) \longrightarrow Mg^{2+}(aq) + Cu(s)$

In the top photograph, a coil of magnesium ribbon is added to a solution of $CuSO_4(aq)$. After several hours, all of the $Cu^{2+}(aq)$ has been displaced from solution, leaving a deposit of red-brown copper metal, some unreacted magnesium, and clear, colorless $MgSO_4(aq)$.

Example 12.6

Is the chemical reaction represented by the following equation a plausible one?

$$MnO_2(s) + O_2(g) + H^+(aq) \longrightarrow Mn^{2+}(aq) + H_2O \quad \textit{(not balanced)}$$

Solution

By designating oxidation states in the equation

$$\overset{+4}{Mn}\overset{-2}{O_2}(s) + \overset{0}{O_2}(g) + \overset{+1}{H}^+(aq) \longrightarrow \overset{+2}{Mn}^{2+}(aq) + \overset{+1}{H_2}\overset{-2}{O} \quad \textit{(not balanced)}$$

we see that there are *two* reduction processes and *no* oxidation. That is, MnO_2 is reduced to Mn^{2+} (the O.S. of Mn decreases from +4 to +2), and O_2 is reduced to H_2O (the O.S. of O decreases from 0 to −2). A reduction cannot occur without an accompanying oxidation. The reaction is not plausible.

We could reach the same conclusion by separating the equation into half-equations; both would be reduction half-equations.

Exercise 12.6

Is the chemical reaction represented by the following equation a plausible one?

$$I_2(s) + Cl_2(g) + H_2O \longrightarrow IO_3^-(aq) + H^+(aq) + Cl^-(aq) \quad \textit{(not balanced)}$$

Balancing Oxidation–Reduction Equations

Some equations for oxidation–reduction reactions can be easily balanced by inspection. In fact, we have been balancing such equations since Chapter 3. However, we can use special methods for oxidation–reduction equations that are often easier than balancing by inspection.

One particularly effective method builds from the half-reaction description of oxidation–reduction reactions. It is based on the principle that *all the electrons "lost" in an oxidation half-reaction must be "gained" in a reduction half-reaction.*

What we must do is

- Separate an oxidation–reduction equation into half-equations for oxidation and reduction.
- Balance each half-equation both atomically and for electric charge.
- Adjust coefficients in the half-equations so that the same number of electrons appears in each half-equation.
- Add together the two adjusted half-equations to obtain a *net* oxidation–reduction equation.

A specific six-step procedure for accomplishing this is presented in detail in Example 12.7.

Example 12.7

Microorganisms convert wine to vinegar by oxidizing ethanol to acetic acid. We can produce the same oxidation chemically, as in the reaction

$$CH_3CH_2OH(aq) + Cr_2O_7^{2-}(aq) + H^+(aq) \longrightarrow CH_3COOH(aq) + Cr^{3+}(aq) + H_2O(l)$$

Balance this equation by the *half-reaction* method.

Solution

Step 1. *Identify the species undergoing oxidation and reduction and write "skeleton" half-equations.* A surefire way to do this is to assign oxidation states to the various elements and look for changes in O.S. The average O.S. of the carbon atoms *increases* from −2 in CH_3CH_2OH to 0 in CH_3COOH; this is an oxidation process. The O.S. of chromium *decreases* from +6 in $Cr_2O_7^{2-}$ to +3 in Cr^{3+}; this is a reduction process. An easier way is just to note that one of the half-equations must involve the carbon-containing compounds and the other, the chromium-containing species.

$$CH_3CH_2OH \longrightarrow CH_3COOH \quad (\textit{not balanced})$$

$$Cr_2O_7^{2-} \longrightarrow Cr^{3+} \quad (\textit{not balanced})$$

Step 2. *Balance the numbers of atoms in each half-equation in the order*

- atoms other than H and O;
- O atoms by adding molecules of H_2O, as needed;
- H atoms by adding H^+ ions, as needed.

The atoms other than H and O are already balanced in the ethanol–acetic acid half-equation.

$$CH_3CH_2OH \longrightarrow CH_3COOH \quad \text{(2 C atoms on each side)}$$

For the other half-equation, we write

$$Cr_2O_7^{2-} \longrightarrow 2\,Cr^{3+} \quad \text{(2 Cr atoms on each side)}$$

To balance O atoms we now add H_2O.

$$CH_3CH_2OH + H_2O \longrightarrow CH_3COOH \quad \text{(2 O atoms on each side)}$$

$$Cr_2O_7^{2-} \longrightarrow 2\,Cr^{3+} + 7\,H_2O \quad \text{(7 O atoms on each side)}$$

To balance H atoms we add H^+.

$$CH_3CH_2OH + H_2O \longrightarrow CH_3COOH + 4\,H^+ \quad \text{(8 H atoms on each side)}$$

$$Cr_2O_7^{2-} + 14\,H^+ \longrightarrow 2\,Cr^{3+} + 7\,H_2O \quad \text{(14 H atoms on each side)}$$

Step 3. *Balance each half-equation for electric charge.* The half-equations in step 2 violate an important principle—electric charge cannot be created or destroyed in a chemical reaction. To achieve a balance add the correct number of electrons (e^-) to the appropriate sides of the half-equations.

Oxidation: $CH_3CH_2OH + H_2O \longrightarrow CH_3COOH + 4\,H^+ + 4\,e^-$ (each side, 0 net charge)

Reduction: $Cr_2O_7^{2-} + 14\,H^+ + 6\,e^- \longrightarrow 2\,Cr^{3+} + 7\,H_2O$ (each side, +6 net charge)

It is also helpful to label the half-equations as oxidation or reduction at this point.

Step 4. *Combine half-equations into a net equation after multiplying by the appropriate coefficients so that the number of electrons in the oxidation half-equation equals the number in the reduction half-equation.* Multiply the oxidation half-equation by *three* and the reduction half-equation by *two*.

Oxidation: $3\,CH_3CH_2OH + 3\,H_2O \longrightarrow 3\,CH_3COOH + 12\,H^+ + \cancel{12\,e^-}$

Reduction: $2\,Cr_2O_7^{2-} + 28\,H^+ + \cancel{12\,e^-} \longrightarrow 4\,Cr^{3+} + 14\,H_2O$

Net: $3\,CH_3CH_2OH + 2\,Cr_2O_7^{2-} + 3\,H_2O + 28\,H^+ \longrightarrow 3\,CH_3COOH + 4\,Cr^{3+} + 12\,H^+ + 14\,H_2O$

Step 5. *Simplify*. The net equation should not contain the same species on both sides. Subtract *three* H_2O molecules and *twelve* H^+ ions from each side of the equation, leaving 11 H_2O on the right and 16 H^+ on the left.

$$3\,CH_3CH_2OH + 2\,Cr_2O_7^{2-} + 16\,H^+ \longrightarrow 3\,CH_3COOH + 4\,Cr^{3+} + 11\,H_2O \quad \textit{(balanced)}$$

Step 6. *Verify*. Check the net equation to ensure that both atoms and electric charge are balanced. This verification is summarized below.

	C	Cr	O	H	Electric charge
Left side:	6	4	17	34	$(-2 \times 2) + 16 = +12$
Right side:	6	4	17	34	$(+3 \times 4) = +12$

Exercise 12.7

Use the half-reaction method to balance the equation for the following redox reaction in acidic solution.

$$MnO_4^- + C_2O_4^{2-} \longrightarrow Mn^{2+} + CO_2(g) \quad \textit{(incomplete and not balanced)}$$

(a) (b) (c)

The reaction of ethanol and dichromate ion in an acidic solution. (a) Orange $K_2Cr_2O_7(aq)$ about to be added to aqueous ethanol acidified with H_2SO_4. (b) The ethanol solution acquires a pale orange color due to the presence of $Cr_2O_7^{2-}$. (c) After the reaction, the solution acquires a pale violet color, signifying the absence of $Cr_2O_7^{2-}$ and the presence of $Cr^{3+}(aq)$.

Two common variations in balancing oxidation–reduction equations involve reactions that

1. occur in *basic* rather than acidic solution;
2. have the same substance undergoing both oxidation and reduction (**disproportionation reactions**).

Hydroxide ion, OH^-, and not H^+ should appear in the net equation for an oxidation–reduction reaction in basic solution. We can achieve this result by adding H_2O and OH^- rather than H_2O and H^+ when balancing the half-equations. However, because H_2O and OH^- both contain O and H atoms, it is sometimes hard to tell at first glance to which side of a half-equation to add each one.

The method in Example 12.8 admittedly is somewhat artificial, but it does allow us to follow essentially the same method as shown in Example 12.7 and to arrive at a correct net equation. We first balance the equation *as if* the reaction occurs in acidic solution. Then, to *each* side of the *net* equation we add a number of OH^- ions equal to the number of H^+ ions appearing in the equation. As a result, one side of the equation will have H^+ and OH^- ions present in equal number; they can be combined and replaced with H_2O molecules. The other side will have OH^- ions. Example 12.8 illustrates a *disproportionation* reaction in basic solution.

Example 12.8

In *basic* solution, Br_2 disproportionates to bromate and bromide ions. Use the half-reaction method to balance this equation.

$$Br_2(l) \longrightarrow Br^-(aq) + BrO_3^-(aq) \quad \textit{(incomplete and not balanced)}$$

Solution

Step 1. *Identify the species undergoing oxidation and reduction.* This is a *disproportionation* reaction; that is, some of the Br_2 is oxidized and some is reduced.

$$Br_2 \longrightarrow Br^-$$
$$Br_2 \longrightarrow BrO_3^-$$

Step 2. *Balance the numbers of atoms in each half-equation.* First we balance the Br.

$$Br_2 \longrightarrow 2\ Br^-$$
$$Br_2 \longrightarrow 2\ BrO_3^-$$

Then we balance the O,

$$Br_2 \longrightarrow 2\ Br^-$$

$$Br_2 + 6\ H_2O \longrightarrow 2\ BrO_3^-$$

Finally balance the H.

$$Br_2 \longrightarrow 2\ Br^-$$

$$Br_2 + 6\ H_2O \longrightarrow 2\ BrO_3^- + 12\ H^+$$

Step 3. *Balance each half-equation for electric charge.* We do this by adding electrons (e^-). At this point we can also label each half-equation as either an oxidation or a reduction.

Reduction: $Br_2 + 2\ e^- \longrightarrow 2\ Br^-$

Oxidation: $Br_2 + 6\ H_2O \longrightarrow 2\ BrO_3^- + 12\ H^+ + 10\ e^-$

Step 4. *Combine half-equations into a net equation after multiplying by the appropriate coefficients so that the same number of electrons appears in each half-equation.* Multiply the reduction half-equation by *five*; leave the oxidation half-equation as is.

Reduction: $5\,Br_2 + \cancel{10\,e^-} \longrightarrow 10\,Br^-$

Oxidation: $Br_2 + 6\,H_2O \longrightarrow 2\,BrO_3^- + 12\,H^+ + \cancel{10\,e^-}$

Net: $6\,Br_2 + 6\,H_2O \longrightarrow 10\,Br^- + 2\,BrO_3^- + 12\,H^+$

Step 5. *Simplify.* All the coefficients in the net equation of step 4 can be divided by *two*:

$$3\ Br_2 + 3\ H_2O \longrightarrow 5\ Br^- + BrO_3^- + 6\ H^+$$

Step 6. *Convert from acidic to basic solution.* Add *six* OH^- ions to each side of the equation of step 5.

$$3\ Br_2 + 3\ H_2O + 6\ OH^- \longrightarrow 5\ Br^- + BrO_3^- + 6\ H^+ + 6\ OH^-$$

On the right side, combine H^+ and OH^- into H_2O.

$$3\ Br_2 + 3\ H_2O + 6\ OH^- \longrightarrow 5\ Br^- + BrO_3^- + 6\ H_2O$$

Simplify by subtracting *three* H_2O molecules from each side.

$$3\ Br_2 + 6\ OH^- \longrightarrow 5\ Br^- + BrO_3^- + 3\ H_2O$$

Step 7. *Verify.* The balanced condition has 6 atoms each of Br, H, and O on the left and on the right. The net charge on each side is –6.

Exercise 12.8

Cyanide ion in waste solutions from gold mining operations can be destroyed by treatment with hypochlorite ion in basic solution. Write a balanced oxidation–reduction equation for this reaction.

$$CN^-(aq) + OCl^-(aq) + OH^-(aq) \longrightarrow CO_3^{2-}(aq) + N_2(g) + Cl^-(aq) + H_2O(l)$$

(*Hint*: Notice that in the oxidation half-reaction CN^- yields two products.)

Because the half-reaction method depends on the use of H_2O and H^+ to balance half-equations, you might think that it is limited to reactions in aqueous solution, where these species occur. However, we can generally treat chemical reactions *as if* they occur in aqueous solution and continue to use the half-reaction method. In these cases H^+ ions and any extraneous H_2O molecules will cancel out.

For example, let's balance the equation for the following reaction that converts the noxious air pollutant NO to harmless N_2.

$$NH_3(g) + NO(g) \longrightarrow N_2(g) + H_2O(g)$$

Treating this reaction *as if* it occurs in aqueous solution, we obtain the following balanced half-equations and net equation.

$$\begin{array}{lrcl} \textit{Oxidation:} & 4\,NH_3 & \longrightarrow & 2\,N_2 + 12\,H^+ + \cancel{12\,e^-} \\ \textit{Reduction:} & 6\,NO + 12\,H^+ + \cancel{12\,e^-} & \longrightarrow & 3\,N_2 + 6\,H_2O \\ \hline \textit{Net:} & 4\,NH_3 + 6\,NO & \longrightarrow & 5\,N_2 + 6\,H_2O \end{array}$$

Note that H^+ cancels out, as it should, but that H_2O, a product of the reaction, remains.

In Example 12.9 we consider a different method for balancing the equation for this reaction. Because it is based on assessing changes in oxidation states, it is called the *change in oxidation state* method.

Example 12.9

Balance the following redox equation by the change in oxidation state method.

$$NH_3(g) + NO(g) \longrightarrow N_2(g) + H_2O(g)$$

Solution

Step 1. *Identify the elements whose oxidation states change in the reaction.* In this case only nitrogen is involved. Its O.S. increases in the change $NH_3 \longrightarrow N_2$ (oxidation) and decreases in the change $NO \longrightarrow N_2$ (reduction).

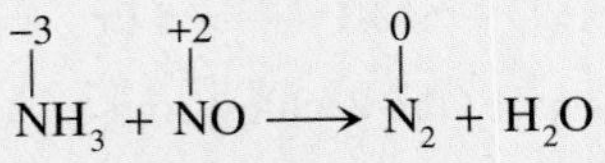

Step 2. *Determine the number of electrons corresponding to each change in oxidation state, per atom.* For each N atom in NH_3 that is oxidized, the O.S. increases by *three* units (from −3 to 0). This is equivalent to a *loss* of 3 electrons per N atom in NH_3. For each N atom in NO that is reduced, the O.S. decreases by *two* units (from +2 to 0). This is equivalent to a *gain* of 2 electrons per N atom in NO. These changes are summarized in the scheme below.

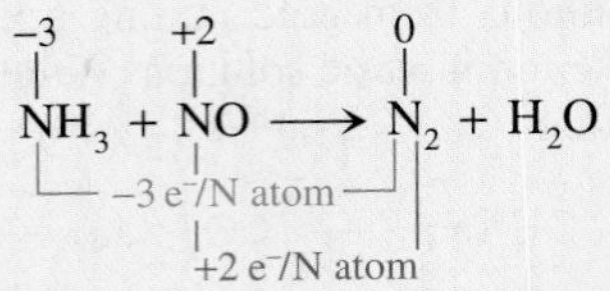

Step 3. *Adjust coefficients so that the total electron loss corresponding to the increase in O.S. is equal to the total electron gain corresponding to the decrease in O.S.* This

condition is achieved by using the coefficient *two* for NH_3 and *three* for NO. That is, $2 \times 3 = 3 \times 2 = 6$.

$$\overset{-3}{2\,NH_3} + \overset{+2}{3\,NO} \longrightarrow \overset{0}{N_2} + H_2O$$

loss of 6 e⁻ (from NH_3 to N_2); gain of 6 e⁻ (from NO to N_2)

Step 4. *Adjust the remaining coefficients by inspection.* The coefficients of "2" and "3" on the left fix the number of N atoms at *five*. There must be five N atoms on the right, and this requires the fractional coefficient "$\frac{5}{2}$" for N_2. The six H and three O atoms on the left require a coefficient of *three* for H_2O.

$$2\,NH_3 + 3\,NO \longrightarrow \tfrac{5}{2}\,N_2 + 3\,H_2O$$

Step 5. *Simplify.* Multiply all coefficients by *two* so as to eliminate the fractional coefficient.

$$4\,NH_3(g) + 6\,NO(g) \longrightarrow 5\,N_2(g) + 6\,H_2O(g)$$

Step 6. *Verify.* The numbers of atoms involved on each side of the equation are 10 N, 12 H, and 6 O. In some cases an equation may need to be checked for net charge as well.

Exercise 12.9

Balance the following redox equation by the change in oxidation state method.

$$Fe(OH)_3(s) + OCl^- + OH^- \longrightarrow FeO_4^{2-} + Cl^- + H_2O$$

(*Hint*: Use the requirement of a balance in electric charge to help balance H and O atoms.)

Example 12.9 and Exercise 12.9 demonstrate that the change in oxidation state method works for balancing a wide range of redox equations, just as does the half-reaction method. However, because we will have a particular need later in the text (Chapter 20) to describe redox reactions through half-reactions, we have stressed the half-reaction method of balancing redox equations.

12.5 Oxidizing and Reducing Agents

An inorganic chemistry treatise lists dinitrogen tetroxide, N_2O_4, as a "fairly strong oxidizing agent" and hydrazine, N_2H_4, as a "powerful reducing agent." Such terms are common in chemistry, and chemists generally readily understand their meanings. The terms describe the way substances participate in oxidation–reduction reactions.

In an oxidation–reduction reaction, the substance that is *oxidized*—because it causes some other substance to be reduced—is called a **reducing agent**. Similarly, the substance that is *reduced*—because it causes another substance to be oxidized—is called an **oxidizing agent**. By analogy, a travel agent does not take a trip nor does a real estate agent move into a new house. These agents make it possible for others to do the traveling or the moving. Similarly, an oxidizing agent is *not* oxidized and a reducing agent is *not* reduced; rather, they cause other substances to undergo these changes.

We might well predict that nitrogen tetroxide, a "fairly strong oxidizing agent," and hydrazine, a "powerful reducing agent," should react with one another, and they do. The following reaction, which is accompanied by the release of large quantities of heat, is the basis of a rocket propulsion system.

$$N_2O_4(l) + 2\ N_2H_4(l) \longrightarrow 3\ N_2(g) + 4\ H_2O(g)$$

Even though the changes in O.S. occur in N atoms, we do not refer to the *atoms* as the oxidizing or reducing agents. Rather, the *substances* in which these atoms are found, that is, N_2O_4 and N_2H_4, respectively, are given these labels.

In the reaction, N_2O_4 is *reduced* to N_2 (O.S. of N decreases from +4 to 0); N_2O_4 is the *oxidizing agent*. N_2H_4 is *oxidized* to N_2 (O.S. of N increases from −2 to 0); N_2H_4 is the *reducing agent*.

Oxidation States of Nonmetals

Some compounds and ions that contain the nonmetallic elements nitrogen, sulfur, and chlorine are listed in Figure 12.9. They are arranged in order of decreasing oxidation state (O.S.) and in columns that correspond to the periodic table. Let's use this figure to illustrate some further ideas about oxidizing and reducing agents.

- The *maximum* oxidation state exhibited by a nonmetal is equal to the number of the group in the periodic table in which it is found: +5 for Group 5A, +6 for Group 6A, and +7 for Group 7A. (There are some exceptions to this statement, notably oxygen and fluorine.) The *minimum* oxidation state is the group number minus eight.

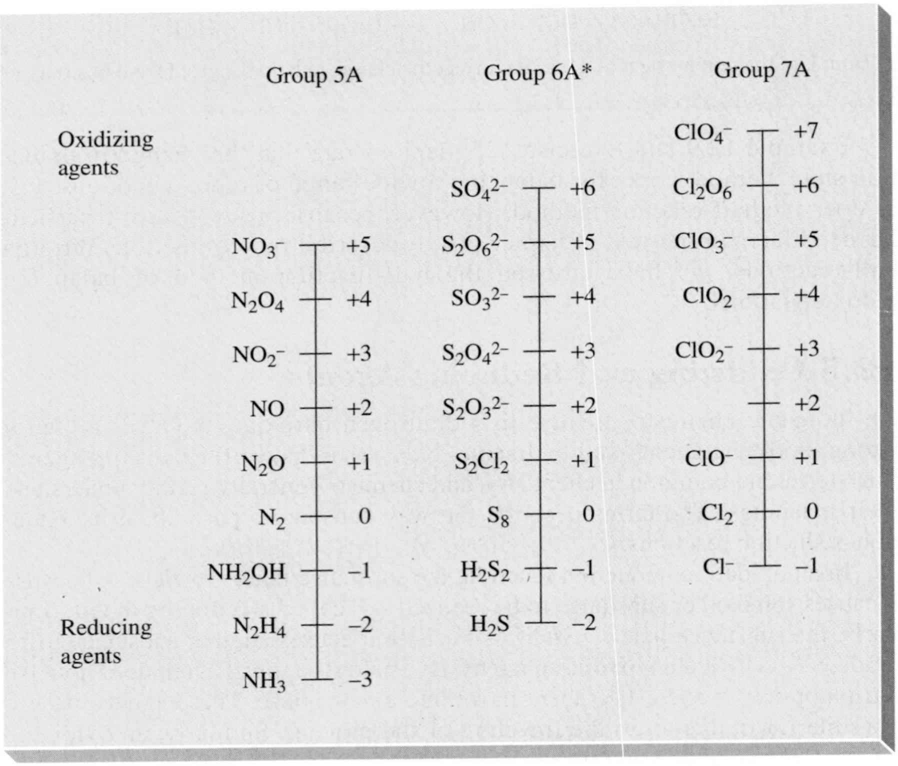

Figure 12.9 Oxidation states of some nitrogen-, sulfur-, and chlorine-containing species.

* The O.S. of −1 in H_2S_2 is an average of −2 for one S atom and 0 for the other. In $S_2O_3^{2-}$, the O.S. of S is an average of +6 and −2. By the oxidation state rules on page 451 we might assign an O.S. of +7 to S in peroxodisulfate ion, $S_2O_8^{2-}$, but the O.S. is actually +6 (see Problem 77).

- Species having the nonmetal atom in the maximum oxidation state are invariably oxidizing agents. The only possible oxidation-state change in their redox reactions is for the oxidation state of the nonmetal atom to decrease. Species having the nonmetal atom in its minimum oxidation state are reducing agents. Thus, in a redox reaction NO_3^- can only be an oxidizing agent and H_2S can only be a reducing agent.
- In principle, a species having the nonmetal atom in an intermediate oxidation state can be either an oxidizing or a reducing agent, depending on the particular reaction. In practice, one role or the other is generally more commonly encountered. Thus, N_2O_4 is almost always an oxidizing agent and N_2H_4 is almost always a reducing agent. Even though it is high on the oxidation-state scale for sulfur, sulfite ion, SO_3^{2-}, usually acts as a reducing agent; $Cl_2(g)$, though low on the oxidation-state scale for chlorine, is generally an oxidizing agent. In other words, knowing the oxidation state of an element in a compound does not necessarily tell us whether the compound usually functions as an oxidizing or a reducing agent.

Metals as Reducing Agents

In all common metallic compounds, the metal atom is in a positive oxidation state. Metals in the free state are in their lowest oxidation state (O.S. 0). Metals are *reducing* agents. Their strengths as reducing agents vary widely, however. The atoms of some metals, such as those of Groups 1A and 2A, lose electrons easily. They are readily oxidized, and as a result are powerful reducing agents. Other metals, such as silver and gold, are oxidized with great difficulty. They are exceptionally poor reducing agents. Figure 12.10 is a listing of some common metals in a form called the **activity series of the metals**.

A metal will displace from solution the ions of any metal that lies *below it* in the series. With the activity series we can describe the reaction pictured in Figure 12.8 (page 453) much more clearly.

$$Mg(s) + Cu^{2+}(aq) \longrightarrow Mg^{2+}(aq) + Cu(s)$$

Mg(s), a good reducing agent, reduces Cu^{2+} to Cu(s). In the reaction, $Cu^{2+}(aq)$ is the oxidizing agent. With the activity series we can also confidently predict that if we substitute Ag(s) for Mg(s),

$$Ag(s) + Cu^{2+}(aq) \longrightarrow \text{no reaction}$$

Ag(s) is a poor reducing agent; it is unable to reduce $Cu^{2+}(aq)$ to Cu(s).

Including H_2 in the activity series of the metals greatly increases the usefulness of the series. This permits us to say that any metal *above* hydrogen in the series will react with an acid to produce $H_2(g)$. For example,

$$Mg(s) + 2\,H^+(aq) \longrightarrow Mg^{2+}(aq) + H_2(g)$$

And any metal *below* hydrogen will not react with an acid to produce $H_2(g)$. For example,

$$Ag(s) + H^+(aq) \longrightarrow \text{no reaction}$$

Strength as a reducing agent	
	K
"Powerful"	Na
	Ca
"Strong"	Mg
	Al
	Cr
"Good"	Zn
	Fe
	Cd
"Fair"	Ni
	Sn
	Pb
	H_2
"Poor"	Cu
	Ag
	Hg
"Very poor"	Au

Figure 12.10 Activity series of some metals.

Thus, $Mg(s)$ is a sufficiently good reducing agent to reduce $H^+(aq)$ to $H_2(g)$, but $Ag(s)$ is not. Or, to put the matter in another way, $H^+(aq)$ is a good enough oxidizing agent to oxidize $Mg(s)$ to $Mg^{2+}(aq)$ but not strong enough to oxidize $Ag(s)$ to $Ag^+(aq)$.

CONCEPTUAL EXAMPLE 12.3

Explain the difference in the behavior of a copper penny toward hydrochloric and nitric acids pictured in Figure 12.11. Write a net ionic equation for any probable reaction(s) that occurs.

Solution

Copper lies below hydrogen in the activity series of the metals. $Cu(s)$ is unable to reduce $H^+(aq)$ to $H_2(g)$ and itself be oxidized to $Cu^{2+}(aq)$. Or, put differently, H^+ is not a good enough oxidizing agent to oxidize $Cu(s)$ to $Cu^{2+}(aq)$ and itself be reduced to $H_2(g)$. Chloride ion in $HCl(aq)$ can only be a *reducing* agent. So, we expect no reaction between $Cu(s)$ and $HCl(aq)$.

In the case of nitric acid, there are *two* potential oxidizing agents: $H^+(aq)$ *and* $NO_3^-(aq)$. From Figure 12.9 we see that NO_3^- is an oxidizing agent because the N atom is present in its highest possible oxidation state. Figure 12.9 also suggests that NO_3^- might be reduced to any one of several products. In the half-reaction below we assume that it is reduced to $NO_2(g)$, a red-brown gas. To combine the half-equations into a net equation, we must multiply the reduction half-equation by the factor *two*.

Nitrate ion is a strong oxidizing agent only in acidic solution. In neutral or basic solution, $NO_3^-(aq)$ will not oxidize $Cu(s)$ to $Cu^{2+}(aq)$.

Oxidation: $$Cu(s) \longrightarrow Cu^{2+}(aq) + 2\,e^-$$

Reduction: $$2\{NO_3^-(aq) + 2\,H^+(aq) + e^- \longrightarrow NO_2(g) + H_2O\}$$

Net: $$Cu(s) + 4\,H^+(aq) + 2\,NO_3^-(aq) \longrightarrow Cu^{2+}(aq) + 2\,NO_2(g) + 2\,H_2O$$

Conceptual Exercise 12.3

Potassium dichromate, $K_2Cr_2O_7$, also exhibits a difference in behavior when heated with hydrochloric and nitric acids. With one acid, a gas is evolved and the solution color changes from red-orange to green. With the other, no reaction occurs. Explain this difference in behavior.

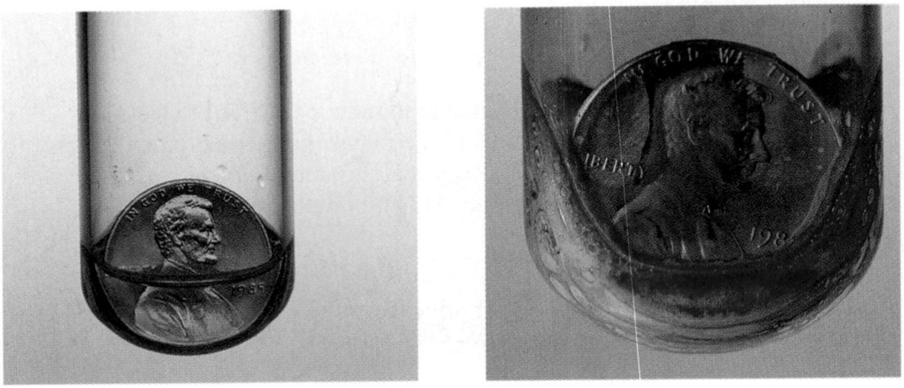

Figure 12.11 The action of $HCl(aq)$ and $HNO_3(aq)$ on copper. As seen in these photographs, a copper penny does not react with hydrochloric acid (left), but it does react with concentrated nitric acid, producing red-brown fumes and a green-blue solution (right).

Looking Ahead

In Chapter 20 we will reconsider oxidation–reduction, including how to

- rank oxidizing and reducing agents on a *quantitative* scale (rather than by terms such as "powerful," "good," and "poor") and explain the activity series of the metals;
- determine whether a given combination of oxidation and reduction half-reactions will indeed result in an oxidation–reduction reaction;
- design batteries that use oxidation–reduction reactions to produce electricity;
- use electrolysis to bring about oxidation–reduction reactions that would not otherwise occur;
- explain corrosion in terms of oxidation and reduction.

12.6 Some Practical Applications of Oxidation and Reduction

Now that we have established a few key ideas about oxidation and reduction, we are better able to explore some applications. We do so through a few of the settings in which oxidation–reduction reactions are encountered.

In Analytical Chemistry

Permanganate ion, usually from $KMnO_4$, is one of the most commonly used oxidizing agents in the chemical laboratory. For example, it can be used to oxidize Fe^{2+} to Fe^{3+} in a titration for determining the percent iron in an iron ore.

$$5\ Fe^{2+}(aq) + MnO_4^-(aq) + 8\ H^+(aq) \longrightarrow 5\ Fe^{3+}(aq) + Mn^{2+}(aq) + 4\ H_2O$$

The titration is illustrated in Figure 12.12 and some sample data are presented in Example 12.10.

(a) (b) (c)

Figure 12.12 A redox titration using permanganate ion as an oxidizing agent. (a) The solution in the flask contains an unknown amount of Fe^{2+}, and the buret contains $KMnO_4(aq)$ of a known concentration. (b) The solution is immediately decolorized as MnO_4^- reacts with Fe^{2+}. (c) When all the Fe^{2+} has been oxidized to Fe^{3+}, the next drop of $KMnO_4(aq)$ produces a lasting pink coloration of the solution.

Example 12.10

An 0.2865-g sample of an iron ore is dissolved in acid and the iron obtained as $Fe^{2+}(aq)$. To titrate the solution, 26.45 mL of 0.02250 M $KMnO_4(aq)$ is required. What is the mass percent of iron in the ore?

Solution

As indicated through the factors numbered (1) through (6), we can determine the number of moles of $KMnO_4$ consumed in the titration as the product of the volume, in liters (1), and the molarity (2). With the next factor (3) we convert from mol $KMnO_4$ to mol MnO_4^-, the actual oxidizing agent. We then use a stoichiometric factor from the balanced redox equation (4) to convert from mol MnO_4^- to mol Fe^{2+}. We need to recognize that every mole of Fe^{2+} ion in solution is produced from one mole of Fe (5). And our final factor (6) is the molar mass of iron.

$$? \text{ g Fe} = 26.45 \text{ mL} \times \overset{(1)}{\frac{1 \text{ L}}{1000 \text{ mL}}} \times \overset{(2)}{\frac{0.02250 \text{ mol KMnO}_4}{1 \text{ L}}} \times \overset{(3)}{\frac{1 \text{ mol MnO}_4^-}{1 \text{ mol KMnO}_4}}$$

$$\times \overset{(4)}{\frac{5 \text{ mol Fe}^{2+}}{1 \text{ mol MnO}_4^-}} \times \overset{(5)}{\frac{1 \text{ mol Fe}}{1 \text{ mol Fe}^{2+}}} \times \overset{(6)}{\frac{55.847 \text{ g Fe}}{1 \text{ mol Fe}}}$$

$$= 0.1662 \text{ g Fe}$$

Finally, the mass percent of iron is

$$\% \text{ Fe} = \frac{0.1662 \text{ g Fe}}{0.2865 \text{ g iron ore}} \times 100\% = 58.01\%$$

Exercise 12.10

Suppose the titration described in Example 12.10 were carried out with 0.02250 M $K_2Cr_2O_7(aq)$ rather than 0.02250 M $KMnO_4(aq)$. What volume of the $K_2Cr_2O_7(aq)$ would be required?

$$6 \text{ Fe}^{2+}(aq) + \text{Cr}_2\text{O}_7^{2-}(aq) + 14 \text{ H}^+ \longrightarrow 6 \text{ Fe}^{3+}(aq) + 2 \text{ Cr}^{3+}(aq) + 7 \text{ H}_2\text{O}$$

(*Hint*: The required volume will be a fraction of the 26.45 mL of Example 12.10.)

An oxidation–reduction titration widely used in analytical chemistry is that of iodine, I_2, by sodium thiosulfate, $Na_2S_2O_3(aq)$, with a starch indicator.

$$\text{I}_2 + 2 \text{ S}_2\text{O}_3^{2-}(aq) \longrightarrow 2 \text{ I}^-(aq) + \text{S}_4\text{O}_6^{2-}(aq)$$

Iodine forms a deep blue complex in a dilute starch solution. The end point of the titration is marked by the disappearance of the blue color of the iodine–starch complex (see Figure 12.13).

Suppose, for example, that we add an excess of $I^-(aq)$ to a hydrogen peroxide solution. Hydrogen peroxide oxidizes iodide ion to free iodine, and is itself reduced to water.

$$\text{H}_2\text{O}_2(aq) + 2 \text{ I}^-(aq) + 2 \text{ H}^+(aq) \longrightarrow \text{I}_2 + 2 \text{ H}_2\text{O}$$

The volume of a standard sodium thiosulfate solution required to titrate the liberated iodine is directly related to the amount of H_2O_2 in the solution being analyzed.

(a) (b) (c)

Figure 12.13 The iodine–thiosulfate ion titration using starch indicator:

$$I_2 + 2\,S_2O_3^{2-} \longrightarrow 2\,I^- + S_4O_6^{2-}$$

(a) The solution in the flask contains I_2 and starch indicator. The buret contains $Na_2S_2O_3(aq)$. (b) The solution retains the deep blue color of the starch–iodine complex as long as any iodine remains. (c) When just enough $Na_2S_2O_3(aq)$ has been added to reduce the last of the iodine to I^-, the starch–iodine complex is destroyed and the solution becomes colorless.

What makes the thiosulfate–iodine titration so versatile is that many oxidizing agents are able to oxidize I^- to I_2. Thus, the titration can be used to determine manganese as MnO_4^-; chromium as $Cr_2O_7^{2-}$; ozone (O_3) in air-pollution studies; and so on.

In Organic Chemistry

We pointed out a use of dichromate ion, $Cr_2O_7^{2-}$, as an oxidizing agent in analytical chemistry in Exercise 12.10. Another use is to oxidize alcohols. The product of the reaction depends on the location of the OH group in the alcohol molecule, the relative proportions of alcohol and dichromate ion, and reaction conditions such as temperature. If the OH group is on a terminal carbon atom and the product is distilled off continuously, an **aldehyde** is formed. The characteristic functional group of an aldehyde is

$$-\overset{\overset{\displaystyle O}{\|}}{C}-H$$

$$\underset{\substack{\text{Ethanol}\\ \text{bp 78 °C}}}{3\,CH_3CH_2OH} + Cr_2O_7^{2-} + 8\,H^+ \longrightarrow \underset{\substack{\text{Acetaldehyde}\\ \text{bp 21 °C}}}{3\,CH_3\overset{\overset{\displaystyle O}{\|}}{C}H} + 2\,Cr^{3+} + 7\,H_2O$$

Hydrogen bonding accounts for the higher boiling point of ethanol compared to acetaldehyde. Thus, acetaldehyde can be boiled off from the reaction mixture while the ethanol is retained. If the acetaldehyde is not removed as it is formed, it is further oxidized to acetic acid. The oxidation of ethanol to acetic acid was the subject of Example 12.7.

$$\underset{\text{Ethanol}}{3\,CH_3CH_2OH} + 2\,Cr_2O_7^{2-} + 16\,H^+ \longrightarrow \underset{\text{Acetic acid}}{3\,CH_3\overset{\overset{\displaystyle O}{\|}}{C}-OH} + 4\,Cr^{3+} + 11\,H_2O$$

When the OH group of an alcohol is bonded to an interior carbon atom, the product of its oxidation is called a **ketone** and has a functional group known as the *carbonyl* group.

$$-\overset{\overset{\displaystyle O}{\|}}{C}-$$

The simplest ketone is derived from 2-propanol (isopropyl alcohol). It is the common solvent *acetone*, $(CH_3)_2CO$, used to dissolve substances as varied as varnishes and lacquers, rubber cement, and fingernail polish.

$$\underset{\substack{\text{2-Propanol}\\ \text{(Isopropyl alcohol)}}}{3\,CH_3\overset{\overset{\displaystyle OH}{|}}{C}HCH_3} + Cr_2O_7^{2-} + 8\,H^+ \longrightarrow \underset{\substack{\text{2-Propanone}\\ \text{(Acetone)}}}{3\,CH_3\overset{\overset{\displaystyle O}{\|}}{C}CH_3} + 2\,Cr^{3+} + 7\,H_2O$$

In Industry

The most widely used oxidizing agent in industrial processes, and certainly the cheapest and least objectionable environmentally, is oxygen itself. In the conversion of iron to steel, high-pressure oxygen gas is blown over molten impure iron (pig iron). This process burns off carbon and sulfur as gaseous oxides. The oxygen also converts the elements Si, P, and Mn to oxides. Lime (CaO) is added to convert these oxides to a *slag*, which is drawn off as a liquid.

Oxygen is used to oxidize hydrogen or acetylene in welding and cutting metals. In the oxyacetylene torch, for example, the reaction is

$$2\,C_2H_2(g) + 5\,O_2(g) \longrightarrow 4\,CO_2(g) + 2\,H_2O(g) \qquad \Delta H = -2.5 \times 10^3\text{ kJ}$$

The heat released in the reaction provides the high temperatures needed to melt metals.

Another group of important industrial oxidizing agents consists of chlorine gas and chlorine compounds in positive oxidation states. Chlorine gas and solutions containing hypochlorite ion, OCl^-, are used in water treatment plants to kill pathogenic (disease-causing) microorganisms and in the paper and textile industries for bleaching.

The principal industrial reducing agents are carbon and hydrogen. Carbon is extensively used in large-scale metallurgical operations, usually in a form known as *coke*. Coke is produced by driving off the volatile matter from coal. In the reduction of iron ore to iron in a blast furnace, the principal reducing agent is CO(g) produced from coke.

$$C(s) + O_2(g) \longrightarrow CO_2(g)$$

$$C(s) + CO_2(g) \longrightarrow 2\,CO(g)$$

$$Fe_2O_3(s) + 3\,CO(g) \longrightarrow 2\,Fe(l) + 3\,CO_2(g)$$

Hydrogen is used as a reducing agent in smaller scale metallurgical processes and in cases where a metal might react with carbon to form an objectionable metal carbide. As an example, tungsten metal is produced by passing a stream of $H_2(g)$ over the oxide WO_3 at 1200 °C.

$$WO_3(s) + 3\ H_2(g) \longrightarrow W(s) + 3\ H_2O(g)$$

In Everyday Life

Oxygen is undoubtedly the most important oxidizing agent in all aspects of life. We use it to oxidize fuels in heating our homes and in propelling automobiles. Oxygen corrodes metals—that is, oxidizes them back to positive oxidation states, as in the rusting of iron. It is also involved in the rotting of wood and the weathering of rocks. Oxygen even "burns" the foods we eat, to release the energy needed for all physical and mental activity.

A common household oxidizing agent is hydrogen peroxide, H_2O_2, usually in the form of an aqueous solution with 3% H_2O_2. An advantage of hydrogen peroxide over other oxidizing agents is that in most reactions it is converted to water, an innocuous product. The 3% solution is used in medicine as an *antiseptic*. (Antiseptics are substances that are applied to living tissue to kill microorganisms or prevent their growth.) It can be used to treat minor cuts and abrasions. An enzyme in blood catalyzes its decomposition, a *disproportionation* reaction.

$$2\ H_2O_2(aq) \longrightarrow 2\ H_2O + O_2(g)$$

The escaping oxygen gas bubbles help to carry dirt and germs out of the wound.

Benzoyl peroxide, $(C_6H_5COO)_2$, a powerful oxidizing agent, has long been used at 5% and 10% concentrations for treating acne. In addition to its antibacterial action, benzoyl peroxide acts as a skin irritant, causing old skin to slough off and be replaced with newer, fresher-looking skin. When used on areas exposed to sunlight, however, benzoyl peroxide is thought to promote skin cancer; such use is therefore discouraged.

Chlorine and its compounds are commonly encountered in daily life. Cl_2 is used to kill microorganisms, both in drinking water and in wastewater treatment. Swimming pools are usually disinfected by "chlorination." In very large pools the chlorine is generally introduced from steel cylinders in which it is stored as a liquid. The chlorine reacts with water to form hypochlorous acid (the actual sanitizing agent).

$$Cl_2(g) + H_2O \longrightarrow H^+(aq) + Cl^-(aq) + HOCl(aq)$$

The pool water soon becomes too acidic because of the presence of HCl(aq) and must be neutralized by adding a base (typically sodium carbonate).

Small swimming pools are often chlorinated with sodium hypochlorite, NaOCl, dissolved in NaOH(aq). These pools become too alkaline because of the added NaOH(aq), and the excess basicity must be neutralized by adding hydrochloric acid. Calcium hypochlorite, $Ca(OCl)_2$, is used to disinfect clothing and bedding in hospitals and nursing homes.

Nearly any oxidizing agent can be used as a bleach to remove unwanted color from fabrics, hair, or other materials. However, some are too expensive, some harm fabrics, some produce undesirable products, and some are simply unsafe.

Pool "chlorine" used to treat small home swimming pools is actually an alkaline solution of sodium hypochlorite, formed by the reaction of $Cl_2(g)$ with NaOH(aq).

Oxidation–Reduction in Bleaching and Stain Removal

A colored material absorbs certain wavelength components of white light to boost loosely bound electrons to higher energy levels. The color it displays is that of the reflected light. Bleaching agents do their work by removing or tying down these loosely bound electrons. In aqueous solution, for example, hypochlorite ions take up electrons and are reduced.

$$OCl^-(aq) + H_2O + 2e^- \longrightarrow Cl^-(aq) + 2\,OH^-(aq)$$

The colored component of the material being bleached is oxidized. Hypochlorite bleaches are safe and effective for cotton and linen because they do not oxidize the carbohydrates that make up the fabric itself. Hypochlorites do oxidize protein and proteinlike molecules, however, and should not be used for wool, silk, or nylon. Chlorine dioxide, ClO_2, and sodium chlorite, $NaClO_2$, are equally good oxidizing agents, and do less damage to materials than do chlorine and hypochlorites.

Other bleaching agents include hydrogen peroxide, sodium perborate (often represented as $NaBO_2 \cdot H_2O_2$ to indicate a loose association of $NaBO_2$ and H_2O_2), and a variety of chlorine-containing organic compounds that release Cl_2 in water.

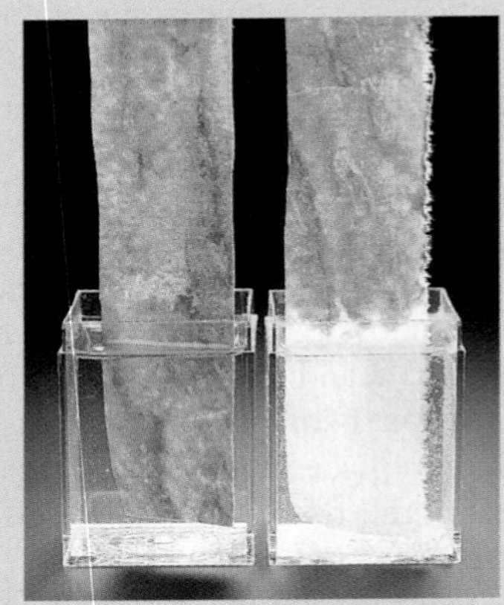

Pure water (left) has little ability to remove a dried tomato sauce stain. Sodium hypochlorite, NaOCl (aq) (right), easily bleaches (oxidizes) the colored components of the sauce to colorless products.

Stain removal is not nearly so simple a process as bleaching. A few stain removers are oxidizing agents or reducing agents; others have quite different chemical natures. Nearly all stains require rather specific stain removers.

Hydrogen peroxide in cold water removes bloodstains from cotton and linen fabrics. Potassium permanganate can be used to remove most stains from white fabrics (except rayon). The purple permanganate stain then can be removed in an oxidation–reduction reaction with oxalic acid.

$$5\,H_2C_2O_4(aq) + 2\,MnO_4^- + 6\,H^+(aq) \longrightarrow 2\,Mn^{2+}(aq) + 8\,H_2O + 10\,CO_2(g)$$

Sodium thiosulfate readily removes iodine stains by an oxidation–reduction reaction that we have already seen.

$$I_2 + 2\,S_2O_3^{2-}(aq) \longrightarrow 2\,I^-(aq) + S_4O_6^{2-}(aq)$$

When vitamin C ($C_6H_8O_6$) acts as an antioxidant, it is converted to dehydroascorbic acid ($C_6H_6O_6$). The half-reaction is

$$C_6H_8O_6 \longrightarrow C_6H_6O_6 + 2\,H^+ + 2\,e^-$$

Vitamin E reacts as a free-radical scavenger; it combines with free radicals to stop the chain reactions by which fats are oxidized and become rancid.

In food chemistry, the substances known as *antioxidants* are reducing agents. Vitamin C, which is water soluble, is thought to retard potentially damaging oxidation of living cells. Tocopherol (vitamin E) is a fat-soluble antioxidant.

Green plants carry out the redox reaction that makes possible all life on Earth. They do this through photosynthesis—the process by which carbon dioxide and water are converted to glucose, a simple sugar. The process requires the catalyst chlorophyll and radiant energy ($h\nu$) from the sun.

$$6\,CO_2 + 6\,H_2O + h\nu \xrightarrow{\text{chlorophyll}} C_6H_{12}O_6 + 6\,O_2$$

In this reaction CO_2 is *reduced* to glucose and H_2O is *oxidized* to oxygen gas. Other reactions convert the simple sugar to more complex carbohydrates and to plant proteins and oils. Animals feeding on plants are additional sources of fats and proteins.

In a way, crop farming is a process of reduction, and the energy captured in cultivated plants is the basis of human life. Equally important to human life are the processes by which this energy is released, and these are oxidation processes. We can think of them as oxidation half-reactions.

Carbohydrates (for example, sucrose: a disaccharide, a sugar):

$$\underset{\text{Sucrose}}{C_{12}H_{22}O_{11}} + 13\,H_2O \longrightarrow 12\,CO_2 + 48\,H^+ + 48\,e^-$$

Proteins (for example, a polypeptide of the amino acid glycine):

$$\underset{\text{Polypeptide}}{C_{12}H_{20}N_6O_7} + 14\,H_2O \longrightarrow 9\,CO_2 + \underset{\text{Urea}}{3\,CO(NH_2)_2} + 36\,H^+ + 36\,e^-$$

Fats (for example, the fatty acid lauric acid):

$$\underset{\text{Lauric acid}}{C_{11}H_{23}COOH} + 22\,H_2O \longrightarrow 12\,CO_2 + 68\,H^+ + 68\,e^-$$

These representative half-reactions for 12-carbon molecules suggest a reason that the energy content of fats exceeds that of carbohydrates and proteins (page 186). The fatty acid is in a "more reduced" state than is the sugar or polypeptide. More electrons must be lost in its oxidation than in the oxidation of the sugar or polypeptide, and this means that more energy is liberated.

SUMMARY

Soluble ionic compounds and a few molecular compounds are strong electrolytes. They are completely dissociated into ions in aqueous solution. Most water-soluble molecular compounds are either nonionized (nonelectrolytes) or only partially ionized (weak electrolytes) in aqueous solution.

Strong acids are strong electrolytes, but the majority of acids are weak acids; they are only partially ionized in aqueous solution. Polyprotic acids have more than one ionizable H atom per molecule. The strong bases are water soluble ionic hydroxides. Weak bases are molecular compounds that partially ionize in aqueous solution. Amines are a family of weak bases in which other groups replace one or more of the H atoms in NH_3. Neutralization reactions between acids and bases are conveniently represented by ionic and net ionic equations.

Another important type of reaction in solution is one in which ions combine to form an insoluble solid—a precipitate. Many precipitation reactions can be predicted with a set of solubility rules.

The oxidation state (O.S.) of an atom is related to the number of electrons that the atom transfers or shares in forming bonds. Oxidation entails an increase in O.S. and reduction, a decrease. Oxidation and reduction always occur simultaneously in an oxidation–reduction (redox) reaction. The primary method of balancing oxidation–reduction equations in this chapter is the half-reaction method. A second method is the change in oxidation state method.

In oxidation–reduction, the reactant that undergoes reduction is the oxidizing agent. The reactant that undergoes oxidation is the reducing agent. Among the strong oxidizing agents are a few of the free nonmetals and polyatomic anions or compounds with an element in a high oxidation state. Among the strong reducing agents are active metals and certain polyatomic ions or compounds with an element in a low oxidation state. The activity series is a relative ranking of metals by their strengths as reducing agents. In a disproportionation reaction, the same substance acts both as the oxidizing and the reducing agent.

Permanganate ion and dichromate ion are important oxidizing agents in the laboratory. For example, dichromate ion in acidic solution oxidizes certain alcohols to aldehydes and others to ketones. Also, it oxidizes aldehydes to carboxylic acids. Some commonly used oxidizing agents in industry are oxygen, chlorine, and hypochlorite ion. Carbon (coke) and hydrogen are common industrial reducing agents.

KEY TERMS

activity series of the metals (12.5)
aldehyde (12.6)
amine (12.2)
disproportionation reaction (12.4)
half-reaction (12.4)
ketone (12.6)
net ionic equation (12.2)
neutralization (12.2)
nonelectrolyte (12.1)
oxidation (12.4)
oxidation state (oxidation number) (12.4)
oxidizing agent (12.5)
precipitation reaction (12.3)
reducing agent (12.5)
reduction (12.4)
salt (12.2)
solubility rules (12.3)
strong acid (12.2)
strong base (12.2)
strong electrolyte (12.1)
weak acid (12.2)
weak base (12.2)
weak electrolyte (12.1)

REVIEW QUESTIONS

1. In aqueous solution, which of the following substances are *strong* electrolytes, which are *weak* electrolytes, and which are *non*electrolytes?
a. CH_3OH **b.** KCl
c. HI **d.** HCOOH
e. NaOH **f.** HNO_2
g. HBr **h.** $CH_2OHCHOHCH_2OH$

2. Identify each of the following substances using the terms: strong acid, weak acid, strong base, weak base, or salt.
a. Na_2SO_4 **b.** KOH
c. $CaCl_2$ **d.** CH_3CH_2COOH
e. HBr **f.** NH_3
g. NH_4I **h.** $Ca(OH)_2$

3. Which of the following solutions has the *highest* and which has the *lowest* concentration of NO_3^-?
(a) 0.10 M KNO_3 (b) 0.040 M $Ca(NO_3)_2$
(c) 0.040 M $Al(NO_3)_3$ (d) 0.050 M $Mg(NO_3)_2$

4. Which of the following solutions have $[Al^{3+}]$ greater than that in 0.0030 M $AlPO_4$?
(a) 0.0036 M $Al(NO_3)_3$ (b) 0.0025 M $AlCl_3$
(c) 0.0020 M $Al_2(SO_4)_3$ (d) 0.0028 M $AlNa(SO_4)_2$

5. Which of the following solutions has the highest *total* concentration of ions?
(a) 0.012 M $Al_2(SO_4)_3$ (b) 0.030 M KCl
(c) 0.022 M $CaCl_2$ (d) 0.025 M K_2SO_4

6. Which of the following aqueous solutions is the *best* electrical conductor: 0.10 M NaCl, 0.10 M CH_3CH_2OH, 0.10 M CH_3COOH, 0.10 M $C_6H_{12}O_6$? Explain.

7. Which of the following aqueous solutions has the highest concentration of H^+ ion: 0.10 M HCl, 0.10 M H_2SO_4, 0.10 M CH_3COOH, 0.10 M NH_3?

8. According to the Arrhenius theory, are all hydrogen-containing compounds acids? Explain.

9. According to the Arrhenius theory, are all compounds containing OH groups bases? Explain.

10. Slaked lime $[Ca(OH)_2(s)]$ can be used to reduce excess acidity in natural waters such as lakes. Write a net ionic equation for the reaction that occurs.

11. With continued use, automatic coffee makers often develop a mineral deposit ($CaCO_3$). The manufacturer's instructions generally call for removing the deposit by treatment with vinegar. Write a net ionic equation for the reaction that occurs. (*Hint*: Recall that vinegar contains acetic acid, CH_3COOH.)

12. A paste of sodium hydrogen carbonate (sodium bicarbonate) and water can be used to relieve the pain of an ant bite. The irritant in the ant bite is formic acid (HCOOH). Write a net ionic equation for the reaction that occurs.

13. Addition of HCl(aq) to a solution containing several different cations produces a white precipitate. What conclusion can you draw from this single observation?

14. Only one of the following compounds is *insoluble* in water: $Ba(NO_3)_2$, $ZnCl_2$, $CuSO_4$, $PbCrO_4$. Which one must it be? Explain.

15. Which of the following compounds reacts to precipitate Mg^{2+} from an aqueous solution of $MgCl_2$: Na_2S, NaI, Na_2CO_3, $NaNO_3$?

16. What simple test can you perform to determine whether a particular magnesium compound is $MgSO_4(s)$ or $Mg(OH)_2(s)$?

17. What simple test can you perform to determine whether a particular barium compound is $BaSO_4(s)$ or $BaCO_3(s)$?

18. What simple test can you perform to determine whether a particular aqueous solution is $Na_2SO_4(aq)$ or $MgSO_4(aq)$?

19. When aqueous solutions of copper(II) nitrate and potassium hydroxide are mixed, a precipitate forms. Write the net ionic equation for this reaction.

20. When aqueous solutions of iron(III) chloride and sodium sulfide are mixed, a precipitate forms. Write the net ionic equation for this reaction.

21. What is the usual oxidation state of hydrogen in its compounds? What is that of oxygen in its compounds? What are some exceptions?

22. What happens to the oxidation state of one of its elements when a compound is oxidized and when it is reduced?

23. Is it possible for the same compound to be both an oxidizing agent and a reducing agent? Can this occur in the same reaction? Explain.

24. Are there any circumstances under which an oxidation half-reaction can occur unaccompanied by a reduction half-reaction? Explain.

25. Indicate the oxidation state of the underlined element in each of the following.

a. $\underline{Cr}$ **b.** $\underline{Cl}O_2^-$ **c.** $K_2\underline{Se}$ **d.** $\underline{Te}F_6$
e. $\underline{P}H_4^+$ **f.** $Ca\underline{Ru}O_3$ **g.** $Sr\underline{Ti}O_3$ **h.** $\underline{P}_2O_7^{4-}$
i. $\underline{S}_4O_6^{2-}$ **j.** $\underline{N}H_2OH$

26. Indicate the average oxidation state of the carbon atoms in the following organic compounds.
a. C_2H_6
b. HCOOH
c. CH_3CHO
d. $(CH_3)_2O$
e. $CH_3(CH_2)_3CH_2OH$

27. In the reaction

$$Cu(s) + 2H_2SO_4(aq) \longrightarrow CuSO_4(aq) + 2\,H_2O + SO_2(g)$$

is the $H_2SO_4(aq)$ oxidized or reduced or neither? Explain.

28. Which of the following metals would you expect to react with HCl(aq) to liberate $H_2(g)$ gas: zinc, silver, iron, aluminum, copper, tin? Which of the metals react with nitric acid? Explain.

PROBLEMS

Ion Molarities

29. Determine the molarity of
a. Li^+ in 0.0385 M $LiNO_3$
b. Cl^- in 0.035 M $CaCl_2$
c. Al^{3+} in 0.0112 M $Al_2(SO_4)_3$
d. Na^+ in 0.12 M Na_2HPO_4

30. Determine the molarity of
a. I^- in 0.0185 M KI
b. CH_3COO^- in 1.04 M $(CH_3COO)_2Ca$
c. Li^+ in 0.053 M Li_2CO_3
d. NO_3^- in 0.0112 M $Al(NO_3)_3$

31. A solution is prepared by dissolving 0.112 g $Mg(NO_3)_2 \cdot 6H_2O$ in 125 mL of water solution. What is $[NO_3^-]$ in this solution?

32. A solution has 25.0 mg K_2SO_4/mL. What is $[K^+]$ in this solution?

33. A solution is 0.0554 M NaCl and 0.0145 M Na_2SO_4. What are $[Na^+]$, $[Cl^-]$, and $[SO_4^-]$ in this solution?

34. A solution is 0.015 M each in LiCl, MgI_2, Li_2SO_4, and $AlCl_3$. What is the molarity of each ion in this solution?

35. What volume of 0.0250 M $MgCl_2$ should be diluted to 250.0 mL to obtain a solution with $[Cl^-] = 0.0135$ M?

36. A solution is prepared by mixing 100.0 mL 0.438 M NaCl, 100.0 mL 0.0512 M $MgCl_2$, and 250.0 mL of water. What are the concentrations of Na^+, Mg^{2+}, and Cl^- in the resulting solution?

Acid–Base Reactions

37. Write equations to show the ionization of the following acids.
a. HI(aq) **b.** CH_3CH_2COOH(aq)
c. HNO_2(aq) **d.** $H_2PO_4^-$(aq)

38. Write equations to show the ionization of the following acids and bases.
a. HNO_3(aq) **b.** KOH(aq)
c. HCOOH(aq) **d.** CH_3NH_2(aq)
e. $HC_2O_4^-$(aq) **f.** $Ba(OH)_2$(aq)
g. $HClO_2$(aq) **h.** $C_6H_5NH_2$(aq)

39. Calculate the volume, in milliliters, of 0.0195 M HCl required to titrate (**a**) 25.00 mL 0.0365 M KOH(aq); (**b**) 10.00 mL 0.0116 M $Ca(OH)_2$(aq); (**c**) 20.00 mL 0.0225 M NH_3(aq).

40. Calculate the volume, in milliliters, of 0.0108 M $Ba(OH)_2$(aq) required to titrate (**a**) 20.00 mL 0.0265 M H_2SO_4(aq); (**b**) 25.00 mL 0.0213 M HCl(aq); (**c**) 10.00 mL 0.0868 M CH_3COOH(aq).

41. Vinegar is an aqueous solution of acetic acid, CH_3COOH. A 10.00-mL sample of a particular vinegar requires 31.45 mL of 0.2560 M KOH for its titration. What is the molarity of acetic acid in the vinegar? (*Hint*: Write a net ionic equation for the titration reaction.)

42. Most window cleaners are aqueous solutions of ammonia. A 10.00-mL sample of a particular window cleaner requires 39.95 mL of 1.008 M HCl for its titration. What is the molarity of ammonia in the window cleaner? (*Hint*: Write a net ionic equation for the titration reaction.)

43. What volume of CO_2(g), measured at 748 mmHg and 22 °C, is produced by the reaction of excess H_2SO_4(aq) with 148 kg Na_2CO_3(s)?

$$Na_2CO_3(s) + H_2SO_4(aq) \longrightarrow Na_2SO_4(aq) + H_2O + CO_2(g)$$

44. What volume of SO_2(g), measured at 764 mmHg and 26 °C, is produced by the reaction of excess HCl(aq) with 212 kg Na_2SO_3(s)?

$$Na_2SO_3(s) + 2\,HCl(aq) \longrightarrow 2\,NaCl(aq) + H_2O + SO_2(g)$$

Ionic Equations

45. Rubidium chloride has been used in medical studies as an antidepressant. It can be made by the reaction of an aqueous solution of rubidium hydroxide and hydrochloric acid. Write the (**a**) complete formula, (**b**) ionic, and (**c**) net ionic equations for this reaction.

46. Strontium iodide can be made by the reaction of solid strontium carbonate with hydroiodic acid. Write the (**a**) complete formula, (**b**) ionic, and (**c**) net ionic equations for this reaction.

47. Complete each of the following as a *net ionic equation*. If no reaction occurs, write "no reaction."
a. $K^+ + I^- + Pb^{2+} + 2\,NO_3^- \longrightarrow$
b. $Mg^{2+} + 2\,Br^- + Zn^{2+} + SO_4^{2-} \longrightarrow$
c. $Cr^{3+} + 3\,Cl^- + Li^+ + OH^- \longrightarrow$
d. $H^+ + Cl^- + CH_3COOH \longrightarrow$
e. $Ba^{2+} + 2\,OH^- + H^+ + I^- \longrightarrow$
f. $K^+ + HSO_4^- + Na^+ + OH^- \longrightarrow$

48. Complete each of the following as a *net ionic equation*. If no reaction occurs, write "no reaction."
a. $Ba^{2+} + 2\,Cl^- + 2\,Na^+ + CO_3^{2-} \longrightarrow$
b. $Pb^{2+} + 2\,NO_3^- + Mg^{2+} + SO_4^{2-} \longrightarrow$
c. $2\,Na^+ + SO_4^{2-} + Cu^{2+} + 2\,Cl^- \longrightarrow$
d. $K^+ + OH^- + Na^+ + HSO_4^- \longrightarrow$
e. $Na^+ + OH^- + NH_4^+ + Cl^- \longrightarrow$
f. $CH_3CH_2COOH + Ba^{2+} + 2\,OH^- \longrightarrow$

49. Predict whether a reaction is likely to occur in each of the following cases. If so, write a *net ionic equation* for the reaction.
a. MgO(s) + HI(aq) $\longrightarrow$
b. HCOOH(aq) + NH_3(aq) $\longrightarrow$
c. CH_3COOH(aq) + H_2SO_4(aq) $\longrightarrow$
d. $CuSO_4$(aq) + Na_2CO_3(aq) $\longrightarrow$
e. KBr(aq) + $Zn(NO_3)_2$(aq) $\longrightarrow$

50. Predict whether a reaction is likely to occur in each of the following cases. If so, write a *net ionic equation* for the reaction.
a. BaS(aq) + $CuSO_4$(aq) $\longrightarrow$
b. $Cr(OH)_3$(s) + HBr(aq) $\longrightarrow$

c. $NH_3(aq) + H_2SO_4(aq) \longrightarrow$
d. $MgBr_2(aq) + ZnSO_4(aq) \longrightarrow$
e. $NaOH(aq) + NH_4NO_3(aq) \longrightarrow$

Oxidation–Reduction Reactions

51. Complete and balance the following half-equations and indicate whether oxidation or reduction is involved.
a. $ClO_2(g) \longrightarrow ClO_3^-(aq)$ (acidic solution)
b. $MnO_4^-(aq) \longrightarrow MnO_2(s)$ (acidic solution)
c. $BrO^-(aq) \longrightarrow Br_2(l)$ (basic solution)
d. $SbH_3(g) \longrightarrow Sb(s)$ (basic solution)

52. Complete and balance the following half-equations and indicate whether oxidation or reduction is involved.
a. $VO_2^+(aq) \longrightarrow VO^{2+}(aq)$ (acidic solution)
b. $P_4(s) \longrightarrow H_3PO_4(aq)$ (acidic solution)
c. $MnO_2(s) \longrightarrow MnO_4^-(aq)$ (basic solution)
d. $CH_3CH_2OH(aq) \longrightarrow CO_2(g)$ (basic solution)

53. Balance the following redox equations, except in any case where this is not possible.
a. $Ag(s) + H^+ + NO_3^- \longrightarrow Ag^+ + H_2O + NO(g)$
b. $H_2O_2(aq) + MnO_4^- + H^+ \longrightarrow Mn^{2+} + H_2O + O_2(g)$
c. $PbO(s) + V^{3+} + H^+ \longrightarrow PbO_2(s) + VO^{2+} + H_2O$

54. Balance the following redox equations, except in any case where this is not possible.
a. $Mn^{2+} + ClO_3^- + H_2O \longrightarrow MnO_2(s) + Cl^- + H^+$
b. $O_2(g) + NO_3^- + H^+ \longrightarrow H_2O_2 + H_2O + NO(g)$
c. $S_8(s) + O_2(g) + H_2O \longrightarrow H_2SO_4 + H^+$

55. Balance the following redox equations for reactions occurring in basic solutions.
a. $CN^- + BrO_3^- + OH^- \longrightarrow OCN^- + Br^- + H_2O$
b. $S_8(s) + OH^- \longrightarrow S_2O_3^{2-} + S^{2-} + H_2O$

56. Balance the following redox equations for reactions occurring in basic solutions.
a. $CrO_4^{2-} + AsH_3(g) + H_2O \longrightarrow Cr(OH)_3(s) + As(s) + OH^-$
b. $CH_3OH + MnO_4^- + OH^- \longrightarrow HCOOH + Mn^{2+} + H_2O$

57. Balance the following redox equations by either the half-reaction or the change in oxidation state method.
a. $NO(g) + H_2(g) \longrightarrow NH_3(g) + H_2O(g)$
b. $Fe_2S_3(s) + O_2(g) + H_2O \longrightarrow Fe(OH)_3(s) + S_8(s)$

58. Balance the following redox equations by either the half-reaction or the change in oxidation state method.
a. $(NH_4)_2Cr_2O_7(s) \longrightarrow Cr_2O_3(s) + N_2(g) + H_2O(g)$
b. $CrI_3(s) + H_2O_2(aq) + OH^- \longrightarrow CrO_4^{2-} + IO_4^- + H_2O$

59. Write balanced net redox equations for the following reactions.
a. The action of potassium permanganate on oxalic acid (HOOCCOOH) in acidic solution. The products are $Mn^{2+}(aq)$ and carbon dioxide gas.
b. The action of permanganate ion on acetaldehyde (CH_3CHO) in basic solution. The products are manganese(IV) oxide and acetic acid (CH_3COOH).

60. Write balanced net redox equations for the following reactions.
a. The action of concentrated nitric acid on zinc metal. The products are zinc(II) nitrate and nitrogen dioxide gas.
b. The action of nitrate ion on zinc metal in basic solution. The products are $Zn^{2+}(aq)$ and $NH_3(g)$.

61. Identify the oxidizing and reducing agents in the redox reactions in Problems 53 and 54.

62. Identify the oxidizing and reducing agents in the redox reactions in Problems 55 and 56.

63. $Mn^{2+}(aq)$ can be determined by titration with $MnO_4^-(aq)$ in basic solution.

$$Mn^{2+} + MnO_4^- + OH^- \longrightarrow MnO_2(s) + H_2O \quad (\textit{not balanced})$$

A 25.00-mL sample of $Mn^{2+}(aq)$ requires 34.77 mL of 0.05876 M $KMnO_4(aq)$ for its titration. What is the molarity of the $Mn^{2+}(aq)$?

64. A $KMnO_4(aq)$ solution is to be standardized by titration against $As_2O_3(s)$. A 0.1156-g sample of $As_2O_3(s)$ requires 27.08 mL of the $KMnO_4(aq)$ for its titration. What is the molarity of the $KMnO_4(aq)$?

$$As_2O_3 + MnO_4^- + H_2O + H^+ \longrightarrow H_3AsO_4 + Mn^{2+} \quad (\textit{not balanced})$$

65. To titrate a 5.00-mL sample of a saturated aqueous solution of sodium oxalate, $Na_2C_2O_4$, requires 25.82 mL of 0.02140 M $KMnO_4(aq)$. How many grams of $Na_2C_2O_4$ would be present in 250.0 mL of the saturated solution?

$$C_2O_4^{2-}(aq) + MnO_4^-(aq) + H^+(aq) \longrightarrow Mn^{2+}(aq) + H_2O(l) + CO_2(g) \quad (\textit{not balanced})$$

66. A 10.00-mL sample of an aqueous solution of H_2O_2 is treated with an excess of KI(aq). The liberated I_2 requires 28.91 mL of 0.1522 M $Na_2S_2O_3$ for its titration. Is the H_2O_2(aq) up to full strength (3% H_2O_2 by mass) as an antiseptic solution? Assume that the density of the H_2O_2(aq) is 1.00 g/mL.

$$H_2O_2(aq) + H^+(aq) + I^-(aq) \longrightarrow H_2O(l) + I_2 \quad (\textit{not balanced})$$

$$I_2 + S_2O_3^{2-}(aq) \longrightarrow S_4O_6^{2-}(aq) + I^-(aq) \quad (\textit{not balanced})$$

ADDITIONAL PROBLEMS

67. Which of the following 0.010 M solutions has the highest [H^+]: CH_3CH_2COOH(aq), HI(aq), NH_3(aq), H_2SO_4(aq), $Ba(OH)_2$(aq)? Explain.

68. Which of the following solids, when added to an aqueous solution of $(NH_4)_2SO_4$, greatly reduces the electrical conductivity of the solution: PbS(s), $Ba(OH)_2$(s), HOOCCOOH (oxalic acid)? Explain.

69. Which of the following solutions has the higher molarity of K^+: 10.5 L of a solution containing 2.46 mg KNO_3 or a solution with 45.5 ppm K^+? (Assume a density of 1.00 g/mL.)

70. A sample of ordinary table salt is 98.8% NaCl and 1.2% $MgCl_2$ by mass. What is [Cl^-] if 6.85 g of this mixture is dissolved in 500.0 mL of an aqueous solution?

71. A solution is 0.0240 M KI and 0.0146 M MgI_2. What volume of water should be added to 100.0 mL of this solution to produce a solution with [I^-] = 0.0500 M?

72. A white solid is known to be either $MgCl_2$, $MgSO_4$, or $MgCO_3$. A water solution prepared from the solid yields a white precipitate when treated with $Ba(NO_3)_2$. What must the solid be?

73. What reagent solution (including pure water) would you use to separate the cations in the following pairs—that is, so that one cation will appear in solution and the other in a precipitate?
a. $BaCl_2$(s) and NaCl(s)
b. $MgCO_3$(s) and Na_2CO_3(s)
c. $AgNO_3$(s) and KNO_3(s)
d. $PbSO_4$(s) and $CuCO_3$(s)
e. $Mg(OH)_2$(s) and $BaSO_4$(s)

74. A railroad tank car carrying 1.5×10^3 kg of concentrated sulfuric acid derails and spills its load. The acid is 93.2% H_2SO_4 and has a density of 1.84 g/mL. What mass, in kg, of sodium carbonate (soda ash) is needed to neutralize the acid? (*Hint*: What is the neutralization reaction?)

75. To 125 mL of 1.05 M Na_2CO_3(g) is added 75 mL of 4.5 M HCl(aq). Then the solution is evaporated to dryness. What mass of NaCl(s) is obtained?

76. A home swimming pool is disinfected by the daily addition of 0.50 gal of a "chlorine" solution—NaOCl in NaOH(aq). To maintain the proper acidity in the pool, the NaOH(aq) in the "chlorine" solution must be neutralized. By experiment it is found that about 220 mL of an HCl(aq) solution that is 31.4% HCl by mass (d = 1.16 g/mL) is required to neutralize 0.50 gal of the "chlorine" solution. What is the [OH^-] of the "chlorine" solution? (1 gal = 3.785 L)

77. Show that the oxidation state assigned to S in the peroxodisulfate ion, $S_2O_8^{2-}$, by the general rules on page 451 is +7. Then draw a plausible Lewis structure for the ion and show that the O.S. of S is really +6. (*Hint*: The term "peroxo" signifies a peroxide-type linkage in the ion: —O—O—.)

78. Draw a Lewis structure of O_2^- and use it to justify the oxidation state of $-\frac{1}{2}$ assigned to oxygen in CsO_2.

79. When phosphorus, P_4, is heated with water it *disproportionates* to phosphine, PH_3, and phosphoric acid, H_3PO_4. Write a balanced equation for this reaction.

80. Cyanide wastes can be detoxified by the addition of chlorine to a basic solution of the wastes.

$$NaCN(aq) + NaOH(aq) + Cl_2(g) \longrightarrow NaOCN(aq) + NaCl(aq) + H_2O$$

Following the addition of some acid so that the solution is not quite so basic, a further reaction occurs.

$$NaOCN(aq) + NaOH(aq) + Cl_2(g) \longrightarrow NaCl(aq) + H_2O + NaHCO_3(aq) + N_2(g)$$

Balance the two redox equations.

81. Balance the following redox equations.
a. $I_2 + H_5IO_6 \longrightarrow IO_3^- + H_2O + H^+$
b. $SCl_2 + NH_3 \longrightarrow S_4N_4 + S_8 + NH_4Cl$
c. $XeF_6 + OH^- \longrightarrow XeO_6^{4-} + Xe + O_2 + F^- + H_2O$
d. $S_4N_4 + OH^- + H_2O \longrightarrow S_2O_3^{2-} + SO_3^{2-} + NH_3$

82. The following reactions are associated with wood-eating termites: (**a**) Certain bacteria that live in the hindgut of the termites convert glucose

($C_6H_{12}O_6$) to acetic acid (CH_3COOH), carbon dioxide, and hydrogen. (**b**) Other bacteria convert part of the carbon dioxide and hydrogen to more acetic acid. (**c**) The termite then meets its respiratory energy requirements through the reaction of acetic acid and oxygen to produce carbon dioxide and water. Write balanced equations for these three reactions and identify the oxidizing and reducing agents.

83. Incineration of a chlorine-containing toxic waste such as a polychlorinated biphenyl (PCB) produces CO_2 and HCl.

$$C_{12}H_4Cl_6 + O_2 + \longrightarrow CO_2 + HCl$$

Balance the equation for this combustion reaction. Comment on the advantages and disadvantages of incineration as a method of disposal of such wastes.

84. Methanol, CH_3OH, can reduce chlorate ion to chlorine dioxide in an acidic solution. The methanol is oxidized to carbon dioxide. How many milliliters of methanol (d = 0.791 g/mL) are needed to produce 125 L ClO_2(g), at 25 °C and 748 mmHg? (*Hint*: Write a balanced equation for the reaction.)

85. The exact concentration of an aqueous solution of oxalic acid, HOOCCOOH (that is, $H_2C_2O_4$), is determined by an acid–base titration. Then the oxalic acid solution is used to determine the concentration of $KMnO_4$(aq) by a redox titration in acidic solution. The titration of 25.00-mL samples of the oxalic acid solution requires 32.15 mL of 0.1050 M NaOH and 28.12 mL of the $KMnO_4$(aq). What is the molarity of the $KMnO_4$(aq)? (*Hint*: Write equations for the two titration reactions.)

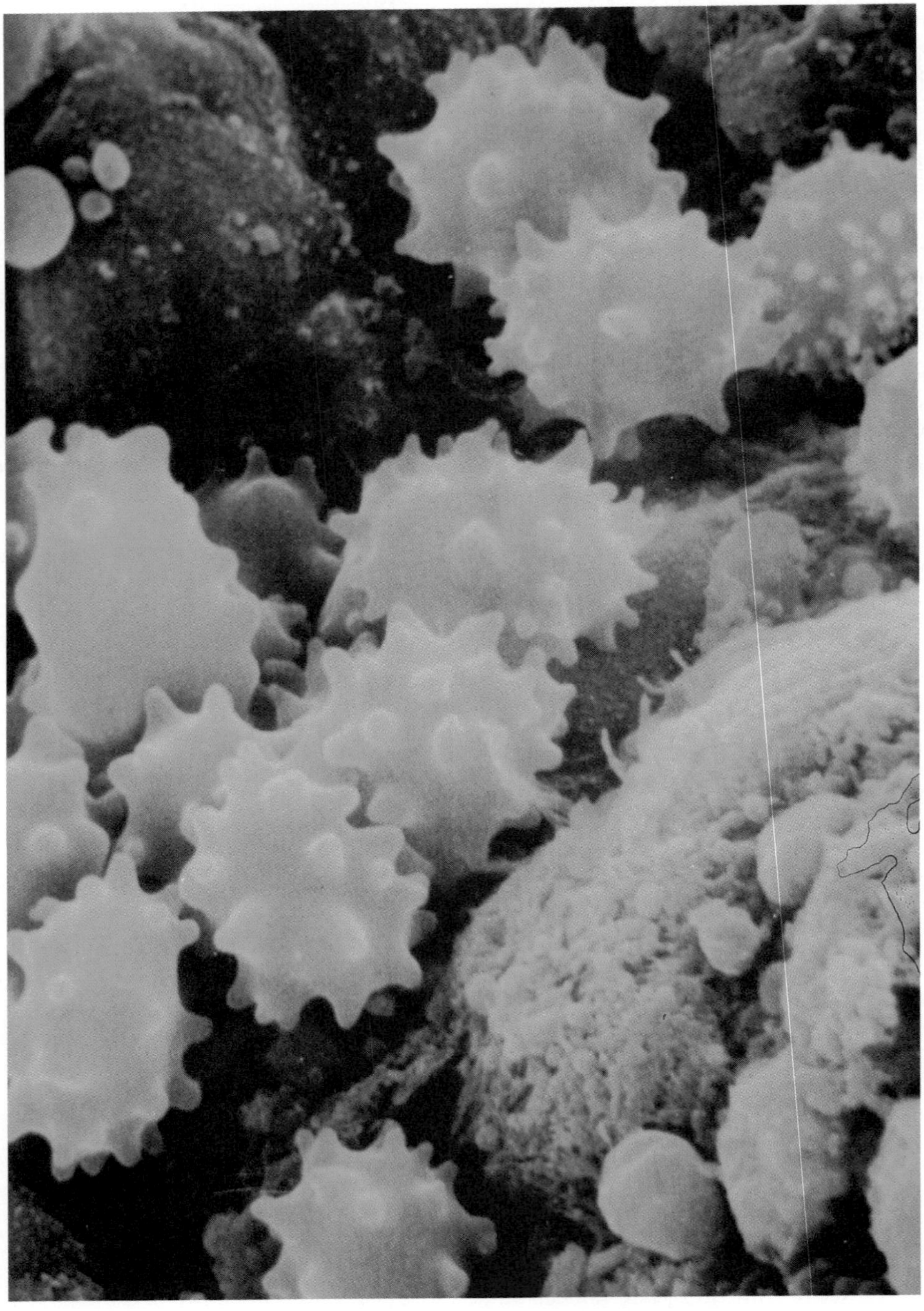

These red blood cells are shrunken in shape because they are suspended in a solution that has a greater osmotic pressure than normal red blood cells. Osmotic pressure is one of the solution properties presented in this chapter.

13

Solutions: Some Physical Properties

CONTENTS

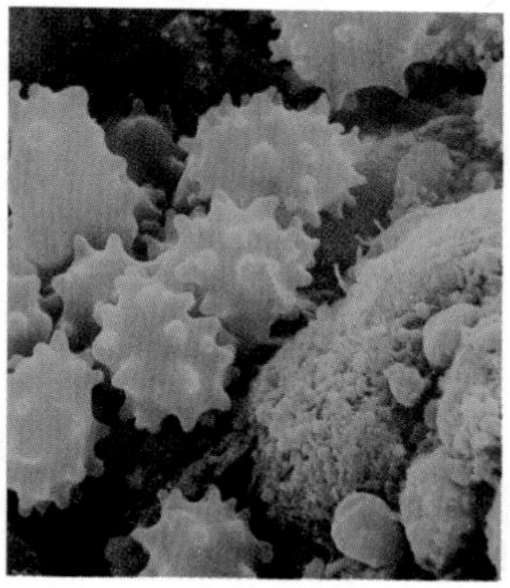

Chapter 12 dealt with the *chemical* properties of aqueous solutions. In this chapter we examine some *physical* properties of solutions and phenomena related to them.

Consider the coolant in an automobile engine. Water alone could be used to cool the engine, but we usually prefer the qualities of a solution of ethylene glycol (CH_2OHCH_2OH) in water. This solution has both a higher *boiling point* and a lower *freezing point* than does water. The solution is less likely to boil away, and it is at the same time a coolant and an antifreeze. In the manufacture of gasoline, we separate the hydrocarbons in petroleum (a solution) by fractional distillation, a process related to the *vapor pressures* of the solution components. In the intravenous injections of fluids, pure water cannot be used; it would enter blood cells, causing them to swell and burst. Instead, aqueous solutions with the proper value of a property known as *osmotic pressure* are required.

Boiling points, freezing points, vapor pressures, and osmotic pressures are a set of interrelated physical properties of solutions that we explore in this chapter. We also consider some fundamental questions about solution formation: which substances dissolve in one another and why; what conditions favor the formation of ideal solutions; and how the solubility of a substance varies with temperature.

13.1 Solutions: Some Terms and Definitions

We have introduced a number of key terms pertaining to solutions in earlier chapters: homogeneous mixture (solution), solute, solvent, concentrated, dilute. If these terms are not fresh in your memory, you should review them.

Keep in mind that the solvent—and hence the solution itself—can exist in any of the three states of matter, although solutions in the liquid state are the most familiar. Table 13.1 gives some examples of a variety of types of solutions.

TABLE 13.1 Some Common Types of Solutions

Solute	Solvent	Solution	Examples
Gas	gas	gas	air (O_2, Ar, CO_2, …, in N_2); natural gas (C_2H_6, C_3H_8, …, in CH_4)
Gas	liquid	liquid	club soda (CO_2 in H_2O); blood substitute (O_2 in perfluorodecalin)
Liquid	liquid	liquid	vodka (CH_3CH_2OH in H_2O); vinegar (CH_3COOH in H_2O)
Solid	liquid	liquid	saline solution (NaCl in H_2O); racing fuel (naphthalene in gasoline)
Gas	solid	solid	hydrogen (H_2) in palladium (Pd)
Solid	solid	solid	14-karat gold (Ag in Au); yellow brass (Zn in Cu)

To study the physical properties of solutions, we need several different concentration units. In the next section, we review some of the concentration units that we have seen before and introduce some new ones.

13.2 Solution Concentration

From the definition of molarity,

$$\text{Molarity (M)} = \frac{\text{amount of solute (mol)}}{\text{volume of solution (L)}}$$

we see that if we know the volume and molarity of a solution, we can calculate the number of moles of a solute. We use this kind of information in titrations, for example, which explains why we introduced the concept of molarity early in the text and have used it so often. However, molarity is not generally useful for some of the applications in this chapter.

Percent by Mass, Percent by Volume, and Mass/Volume Percent

For many practical applications, we often express solution compositions in percentage composition. Then, if we require a precise quantity of solution, we simply measure out a mass or volume. For example, commercial sulfuric acid is supplied as a solution that is 35.7% H_2SO_4 for use in storage batteries, 77.7% H_2SO_4 for the manufacture of phosphate fertilizers, and 93.2% H_2SO_4 for pickling steel. Each of these figures is a *percent by mass*: 35.7 g H_2SO_4 per 100 g sulfuric acid solution, and so on.

Example 13.1

How would you prepare 750 g of an aqueous solution that is 2.5% NaOH by mass?

Solution

We know that the mass of solution is to be 750 g. We can determine the required mass of NaOH by using the definition of percent composition by mass. The required mass of water will be 750 g minus this mass of NaOH.

$$\% \text{ NaOH} = \frac{\text{mass NaOH}}{\text{total mass}} \times 100\% = \frac{\text{mass NaOH}}{750 \text{ g}} \times 100\% = 2.5\% \text{ NaOH}$$

$$\text{mass NaOH} = \frac{2.5 \times 750 \text{ g}}{100} = 19 \text{ g}$$

$$\text{mass } H_2O = 750 \text{ g solution} - 19 \text{ g NaOH} = 731 \text{ g } H_2O$$

To make 750 g of solution, weigh out 19 g NaOH and add it to 731 g of water.

Exercise 13.1

What is the percent by mass of a solution made by dissolving 163 g of glucose in 755 g of water?

We don't need the formula of glucose. The percent by mass composition of a solution of 163 g of solute dissolved in 755 g water is the same, regardless of the formula of the solute.

Estimation Exercise 13.1

Which of the following solutions has the greatest mass percent of solute?
a. 15.5 g NaCl in 100.0 g solution **b.** 15.5 g Na_2SO_4 in 100.0 mL water
c. 155 g Na_3PO_4 per kg water?

If both the solute and solvent are liquids, *percent by volume* is often used, because liquid volumes are so easily measured. Ethanol used for medicinal purposes is generally of a grade referred to as USP (an abbreviation of United States Pharmacopeia, the official publication of standards for pharmaceutical products). USP ethanol is 95% C_2H_5OH, by volume. That is, it consists of 95 mL C_2H_5OH(l) per 100 mL of aqueous solution.

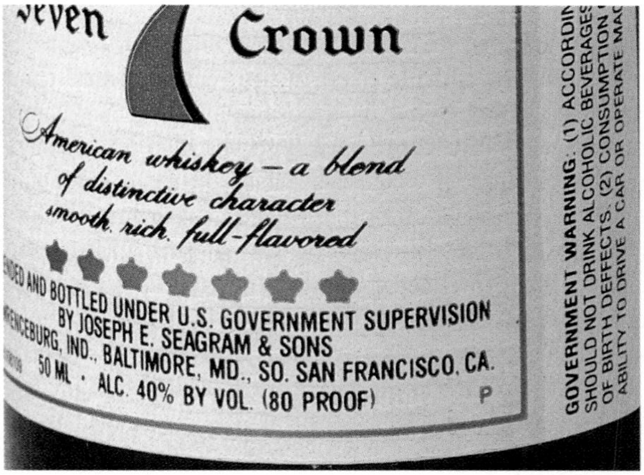

The alcoholic content of a beverage is often given by its "proof" number, which is twice the volume percent of alcohol. The 80-proof whiskey shown here contains 40% C_2H_5OH by volume. The term "proof" originated in seventeenth-century England. A sample of whiskey was poured on gunpowder and ignited. If the gunpowder ignited after the whiskey burned off, this was "proof" that the whiskey had not been watered down. The minimum content to give a positive test was 50% ethanol. Thus, the 50% solution became known as "100 proof."

Example 13.2

USP ethanol is an aqueous solution containing 95.0% ethanol by volume. At 20 °C, pure ethanol has a density of 0.789 g/mL, and the USP ethanol, 0.813 g/mL. What is the mass percent ethanol in USP ethanol?

Solution

We can think of USP ethanol as containing 95.0 mL of pure ethanol in 100.0 mL of the solution. Then we can determine the mass of ethanol and of the solution.

$$\text{Mass of ethanol} = 95.0 \text{ mL} \times \frac{0.789 \text{ g}}{1 \text{ mL}} = 75.0 \text{ g}$$

$$\text{Mass of solution} = 100 \text{ mL} \times \frac{0.813 \text{ g}}{1 \text{ mL}} = 81.3 \text{ g}$$

If there is 75.0 g of ethanol in 81.3 g of solution, the mass percent of ethanol in the solution is

$$\frac{75.0 \text{ g ethanol}}{81.3 \text{ g solution}} \times 100\% = 92.3\% \text{ ethanol, by mass}$$

Exercise 13.2

What volumes, in milliliters, of ethanol and water would you mix to obtain 200 mL of a 40.0 percent-by-volume solution of ethanol in water? What assumption is required in your calculation?

Estimation Exercise 13.2

Which aqueous solution has the higher mass percent of solute: (**a**) 25.0% ethanol, *by volume*, or (**b**) 25.0% methanol, *by mass*? The densities of the pure solutes are 0.789 g/mL for ethanol and 0.791 g/mL for methanol. The density of solution (a) is 0.968 g/mL and that of solution (b) is 0.959 g/mL. (*Hint:* Some of the density data are not needed, and for those that are needed, a simple comparison should suffice. In short, you don't need to do a detailed calculation.)

Another percentage unit widely used in medicine is that of *mass/volume percent*. For example, a solution of sodium chloride used in intravenous injections has the composition 0.89% (mass/vol) NaCl; that is, it contains 0.89 g NaCl per 100 mL solution. A volume of 100 mL—one-tenth of a liter—is also a *deciliter*. If the mass of solute is expressed in *milligrams*, then the mass/volume concentration unit is *milligrams per deciliter* (mg/dL). A blood cholesterol reading of 187, for example, means 187 mg cholesterol/dL blood. The use of milligrams per deciliter avoids the sometimes cumbersome use of decimal numbers. For example, in mass/volume percent the cholesterol reading just cited is 0.187%.

Mass/volume percent and mass/mass percent are nearly the same for *dilute aqueous* solutions because their densities are approximately 1.00 g/mL. That is, 100 mL of such a solution weighs 100 g.

Parts per Million, Parts per Billion, and Parts per Trillion

For solutions that are extremely dilute, concentrations are often expressed in parts per million (**ppm**), parts per billion (**ppb**), or even parts per trillion (**ppt**). For example, in fluoridated drinking water, fluoride ion is maintained at about 1 ppm. A typical level of the contaminant chloroform in a municipal drinking water taken from the lower Mississippi River is 8 ppb.

Example 13.3

The maximum allowable level of nitrates in drinking water set by the State of California is 45 mg NO_3^-/L. What is this level expressed in parts per million (ppm)?

Solution

The density of water, even if it contains traces of dissolved substances, is essentially 1.00 g/mL. One liter of water has a mass of 1000 g. So, the allowable nitrate level is

$$\frac{45 \text{ mg } NO_3^-}{1000 \text{ g water}}$$

To obtain this level in ppm, we need to do two things: (1) express the numerator and the denominator in the same units, and (2) write the denominator as one million times the unit chosen. Let's choose the unit, mg, and proceed as follows.

$$\frac{45 \text{ mg } NO_3^-}{1000 \text{ g water} \times \dfrac{1000 \text{ mg water}}{1 \text{ g water}}} = \frac{45 \text{ mg } NO_3^-}{1{,}000{,}000 \text{ mg water}} = 45 \text{ ppm } NO_3^-$$

Exercise 13.3

What is the concentration in **(a)** ppb and **(b)** ppt corresponding to a maximum allowable level in water of 0.1 μg/L of the pesticide chlordane?

Consider these comparisons: One cent in \$10,000 is one ppm, and one cent in \$10,000,000 is one ppb. Five dollars is about one ppt of the current United States national debt. A single individual in the city of San Diego, CA, represents about one ppm, and a single individual in the People's Republic of China, about one ppb.

Molality

Figure 13.1 illustrates an important feature of the *molarity* of a solution: It varies with temperature, decreasing as the temperature increases. This variation occurs because a solution's volume varies with temperature. If we need a concentration unit that is *independent* of temperature, then it must be based only on mass, not volume. Molality is such a unit. The **molality** of a solution is the number of moles of solute per *kilogram of solvent* (not of solution).

Percent by mass is independent of temperature; percent by volume and mass/volume percent are not.

$$\text{molality } (m) = \frac{\text{amount of solute (mol)}}{\text{mass of solvent (kg)}}$$

(a) (b)

Figure 13.1 The effect of temperature on molarity. (a) A 0.1000 M HCl solution prepared at 20.0 °C. (b) The solution expands on being warmed to 25.0 °C. At this temperature, because the same amount of solute is present in a larger volume of solution than at 20.0 °C, the molarity is *less than* 0.1000 M HCl.

Setting Environmental Standards

Current concern over clean air, the purity of drinking water, and soil contamination centers on trace amounts of potentially dangerous compounds. For example, benzene has been shown to produce leukemia symptoms in laboratory animals and humans. The U.S. Supreme Court has dealt with the question of whether the concentrations of benzene in the air breathed by workers should be limited to 10 ppm or 1 ppm. The Court decided that industries could not be required to lower the concentration from 10 to 1 ppm unless the higher concentration was *proven* to be dangerous.

As technology becomes more sophisticated, our ability to detect minute quantities of materials increases. This increase in the sensitivity of analytical techniques raises important questions for which there are no set answers: When a substance is first detected in the ppb or ppt range, is it a new contaminant in the environment or has it been there all along at levels that were previously undetectable? What is the relationship between the level at which a substance can be detected and the level at which it is injurious to the health of individuals or the environment?

For example, a solution of 0.1 mol of ethanol in 100 g (0.1 kg) of water is 1 *m* ethanol. The molality of a solution of 0.2 mol ethanol in 500 g of water is

$$\frac{0.2 \text{ mol ethanol}}{0.5 \text{ kg water}} = 0.4\ m \text{ ethanol}$$

Note that the mixture described in Example 13.4 is a solid at room temperature and a liquid solution only at temperatures above 53 °C, the melting point of the solvent. On a molarity basis we can prepare a solution only at a temperature at which the solvent is a liquid.

Example 13.4

What mass of benzoic acid (C_6H_5COOH) must be dissolved in 50.0 g of *para*-dichlorobenzene ($C_6H_4Cl_2$) to produce a solution that is 0.100 *m* benzoic acid in *para*-dichlorobenzene?

Solution

In the defining equation for molality, let's replace the number of moles of solute by (mass of solute/molar mass of solute). Then we can solve for the mass of solute.

$$\text{Molality} = \frac{\text{moles of solute}}{\text{kg solvent}} = \frac{\dfrac{\text{mass of solute}}{\text{molar mass of solute}}}{\text{kg solvent}}$$

$$\text{Mass of solute} = \text{molality} \times \text{kg solvent} \times \text{molar mass of solute}$$

In the substitution of known data that follows, note that the molality unit *m* is (mole solute/kg solvent).

$$? \text{ g } C_6H_5COOH = \frac{0.100 \text{ mol } C_6H_5COOH}{1 \text{ kg solv.}} \times 0.0500 \text{ kg solv.} \times \frac{122.1 \text{ g } C_6H_5COOH}{1 \text{ mol } C_6H_5COOH}$$

$$= 0.611 \text{ g } C_6H_5COOH$$

Exercise 13.4

What is the molality of a solution prepared by dissolving 5.00 mL C_2H_5OH (d = 0.789 g/mL) in 75.0 mL C_6H_6 (d = 0.877 g/mL)?

Mole Fraction and Mole Percent

Some solution properties depend on the relative amounts of *all* the solution components, expressed on a *mole* basis. The **mole fraction** of a solution component (χ_i) is the fraction of all the molecules in the solution contributed by that component, a quantity expressed as

$$\chi_i = \frac{\text{amount of component } i \text{ (mol)}}{\text{total amount of solution components (mol)}}$$

The **mole percent** of a solution component is the *percent* of all the molecules in a solution contributed by that component. To obtain a mole percent, we simply multiply a mole fraction by 100%.

Recall that we introduced both mole fraction and mole percent for mixtures of gases in Chapter 4.

Quite often, a solution composition is known in one unit, but we need to express this composition in other units, as illustrated in Example 13.5.

Example 13.5

An aqueous solution of ethylene glycol used as an automobile coolant is 40.0% CH_2OHCH_2OH, by mass, and has a density of 1.05 g/mL. What are the (**a**) molarity, (**b**) molality, and (**c**) mole fraction of CH_2OHCH_2OH in this solution?

Solution

For calculations of this type it is generally easiest to start with a fixed quantity of solution, say 1.00 kg or 1.00 L, and determine information about the solution components in the necessary units. Let's choose 1.00 L.

a. The mass of 1.00 L of the solution is

$$? \text{ g} = 1.00 \text{ L} \times \frac{1000 \text{ mL}}{1 \text{ L}} \times \frac{1.05 \text{ g}}{1 \text{ mL}} = 1050 \text{ g}$$

The mass of ethylene glycol in the solution is

$$? \text{ g } CH_2OHCH_2OH = 1050 \text{ g soln} \times \frac{40.0 \text{ g } CH_2OHCH_2OH}{100 \text{ g soln}}$$

$$= 420 \text{ g } CH_2OHCH_2OH$$

The number of moles of ethylene glycol is

$$? \text{ mol } CH_2OHCH_2OH = 420 \text{ g } CH_2OHCH_2OH \times \frac{1 \text{ mol } CH_2OHCH_2OH}{62.07 \text{ g } CH_2OHCH_2OH}$$

$$= 6.77 \text{ mol } CH_2OHCH_2OH$$

The molarity of the solution is

$$\frac{6.77 \text{ mol } CH_2OHCH_2OH}{1 \text{ L}} = 6.77 \text{ M } CH_2OHCH_2OH$$

b. The mass of solvent (H_2O) in the solution is

$$\text{Mass of solution} - \text{mass of solute} = 1050 \text{ g} - 420 \text{ g} = 630 \text{ g}$$

The mass of solvent, in *kilograms*, is

$$630 \text{ g} \times \frac{1 \text{ kg}}{1000 \text{ g}} = 0.630 \text{ kg}$$

The number of moles of solute determined in part (**a**) is 6.77 mol CH_2OHCH_2OH. The molality of the solution, then, is

$$\frac{6.77 \text{ mol } CH_2OHCH_2OH}{0.630 \text{ kg}} = 10.7\ m\ CH_2OHCH_2OH$$

c. The amount of CH_2OHCH_2OH is 6.77 mol. The number of moles of H_2O is

$$? \text{ mol } H_2O = 630 \text{ g } H_2O \times \frac{1 \text{ mol } H_2O}{18.02 \text{ g } H_2O} = 35.0 \text{ mol } H_2O$$

The mole fraction of CH_2OHCH_2OH is

$$\chi_{CH_2OHCH_2OH} = \frac{6.77 \text{ mol } CH_2OHCH_2OH}{6.77 \text{ mol } CH_2OHCH_2OH + 35.0 \text{ mol } H_2O} = 0.162$$

Exercise 13.5

A 2.90 *m* solution of methanol (CH_3OH) in water has a density of 0.984 g/mL. What are the (**a**) mass percent, (**b**) molarity, and (**c**) mole percent of methanol in this solution?

Estimation Exercise 13.3

Which of the following aqueous solutions has the greatest mole fraction of CH_3OH?
(a) 1.0 *m* CH_3OH (b) 10.0% CH_3OH by mass (c) $\chi_{CH_3OH} = 0.10$

13.3 Intermolecular Forces and Solution Formation

Why do some substances dissolve in a given solvent, but others do not? Why does ethanol dissolve in water, but not the natural gas component ethane? Why does sodium carbonate dissolve in water, but not magnesium carbonate? Why does grease dissolve in kerosene, but not in water? And so on. To answer these questions let's look at intermolecular forces in mixtures.

Ideal Solutions

If the intermolecular forces between solvent and solute molecules in a mixture, between solvent molecules in the pure solvent, and between solute molecules in the pure solute are all of about the same strength, we should expect a random intermingling of molecules when solute and solvent are mixed. We should expect a *homogeneous* mixture or *solution* to form. Moreover, it is reasonable to expect properties of the solution to be related to properties of its components. For example, we expect the volume of such a solution to be the sum of the volumes of the solution components. And we expect no energy change in the formation of the solution. That is, ΔH should be zero, with the solution process being neither endothermic nor exothermic. A solution that meets these requirements is called an **ideal solution.**

Probably no real solution meets the requirements of an ideal solution exactly, but several come close enough that we can consider them ideal. One example is the mixture of hydrocarbons that make up gasoline. Another example is a solution of toluene in benzene. We expect this solution to be ideal because of the similarities in molecular structure of benzene and toluene (Figure 13.2). Ideal solutions form in accordance with the rule of thumb, "like dissolves like," where "like" refers to similarities in structures and in polar/nonpolar characteristics of molecules.

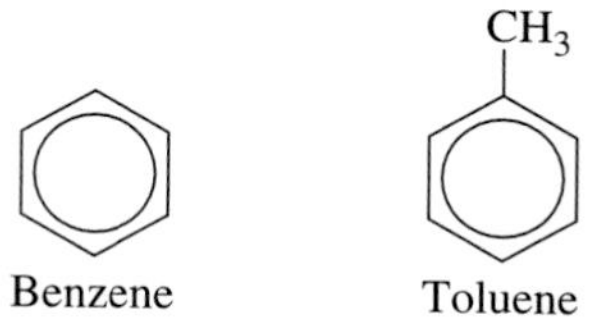

Figure 13.2 Two molecules with similar structures. The structure of benzene is a six-carbon ring with an H atom attached to each C atom. Toluene differs only in that a $—CH_3$ group substitutes for one of the H atoms.

Nonideal Solutions

If intermolecular forces between solvent and solute molecules are particularly strong compared to those within the individual components, we also expect a solution to form, but we should expect it to be *nonideal.* When ethanol is dissolved in water, hydrogen bonding between ethanol and water molecules is stronger than it is in pure ethanol. The volume of solution is *less than* the combined volumes of the water and ethanol because hydrogen bonds bring the solute and solvent molecules closer together than would be the case for an ideal solution (see Figure 13.3). Heat is evolved ($\Delta H < 0$); the solution process is exothermic. Ethanol–water solutions are nonideal.

Figure 13.3 A nonideal solution: ethanol and water. Identical volumetric flasks are filled to the 50.0-mL mark with ethanol and with water. When the two liquids are mixed, the volume is seen to be less than the expected 100.0 mL; it is only about 95 mL.

Even when intermolecular forces between solvent and solute molecules are not particularly strong, the molecules may still be able to form a solution, again a *nonideal* one. Such is the case, for example, in solutions of carbon disulfide, CS_2, a *nonpolar* liquid, in acetone, $(CH_3)_2CO$, a *polar* one.

This oil slick produced in the Exxon "Valdez" oil spill in Alaska in 1989 is a reminder that hydrocarbons and water do not mix.

Of course, if intermolecular forces between solvent and solute molecules are especially weak, the molecules will not intermingle at all. They remain clustered together into separate phases and form a *heterogeneous* mixture. Gasoline does not dissolve in water and vice versa. Gasoline and water form heterogeneous mixtures. Gasoline consists of nonpolar hydrocarbon molecules in which the intermolecular forces are dispersion forces. Intermolecular forces between polar water molecules are chiefly hydrogen bonds, but hydrogen bonds cannot form between hydrocarbon and water molecules. Gasoline and water conform to another familiar homily: "oil and water don't mix."

Aqueous Solutions of Ionic Compounds

Now consider ionic compounds dissolving in water. The forces within an ionic solid are interionic attractions. The intermolecular forces causing the solid to dissolve are *ion–dipole* forces—the attraction of water dipoles for cations and anions (Figure 13.4). Ion–dipole forces cause all ions in an aqueous solution to have a certain number of water molecules associated with them. The ions are said to be *hydrated.*

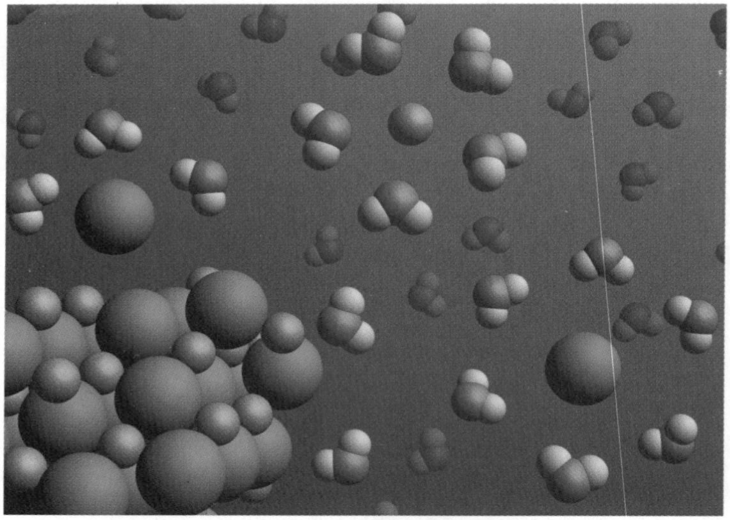

Figure 13.4 Ion-dipole forces in the dissolving of an ionic crystal. Water dipoles attract ions in the crystal lattice, causing them to enter the solution. The ion-dipole forces persist within the solution as well; the ions are hydrated.

The important determinant of the extent to which an ionic solid dissolves in water is how the energy released in the interactions of ions and water dipoles (hydration energy) compares to the energy required to destroy the crystalline lattice (lattice energy). If the ion-dipole interactions predominate, the solids tend to be water soluble. This is the case for the solubility rule (Table 12.3, page 448) that common compounds of the Group 1A metals and ammonium compounds are water soluble. If interionic attractions are the dominant factor, the solids tend to be insoluble. This corresponds to the solubility rule that carbonates, hydroxides, and sulfides are mostly insoluble, with the exception of the Group 1A and ammonium compounds.

It is often difficult to predict the water solubility of a specific compound. For example, we would expect the interionic attractions to be strong between the highly charged Al^{3+} and SO_4^{2-} ions, leading to a high lattice energy and perhaps a limited water solubility. Yet, $Al_2(SO_4)_3$ is quite soluble in water. Ion–dipole attractions must be especially strong in the solution, probably because the negative charge centers of water dipoles can closely approach the small, highly charged Al^{3+} ions. Notice, however, that $Al_2(SO_4)_3$ conforms to the solubility rule that most sulfates are soluble. In general, our best approach to predicting the solubilities of ionic compounds will be with the solubility rules.

Example 13.6

Predict whether each of the following mixtures is likely to be homogeneous—a solution—or heterogeneous.

a. methanol, CH_3OH, and water, HOH.
b. pentane, $CH_3(CH_2)_3CH_3$, and hexane, $CH_3(CH_2)_4CH_3$.
c. sodium chloride, NaCl, and carbon tetrachloride, CCl_4.
d. 1-decanol, $CH_3(CH_2)_8CH_2OH$, and water, HOH.

Solution

a. The most important structural feature of the molecules is the —OH group. This group is attached to an H atom in water and a CH_3— group in methanol. The molecules are similar in this regard: hydrogen bonding is the major intermolecular force in both pure liquids and in their mixtures. We expect the mixtures to be *homogeneous* (solutions).

b. The hydrocarbons pentane and hexane are quite similar; they differ only by one CH_2 group. Intermolecular forces are all of the dispersion type and of similar magnitude. We expect the mixtures to be *homogeneous* ("like dissolves like"). Moreover, the solutions are likely to be *ideal*, or nearly so.

c. Sodium chloride and carbon tetrachloride are both chlorides, but the similarity ends here. NaCl is an ionic compound and CCl_4 is molecular. Moreover, the tetrahedral CCl_4 molecule is *nonpolar*, so there is no opportunity for ion–dipole forces to separate the ions in NaCl(s). We expect the mixture to be *heterogeneous* (dissolving does not occur).

d. Although this case may seem similar to (a)—a hydrocarbon chain substituted for an H atom in HOH—the hydrocarbon chain is ten carbon atoms long. This long hydrocarbon chain is the principal structural feature of 1-decanol. Dispersion forces are the dominant intermolecular forces in 1-decanol. We expect 1-decanol-water mixtures to be *heterogeneous* (no significant dissolving occurs). Think of this as a case of "oil and water don't mix."

Exercise 13.6

Following are four organic compounds. Rank them in order of increasing solubility in water. That is, list the one that dissolves to the smallest extent first, and so on.

a. acetic acid, CH_3COOH **b.** heptane, $CH_3(CH_2)_5CH_3$
c. 1-octanol, $CH_3(CH_2)_6CH_2OH$ **d.** pentanoic acid, $CH_3(CH_2)_3COOH$

Methanol, ethanol, 1-propanol, and 2-propanol are soluble in water in all proportions. The solubility of 1-butanol is limited (a maximum of 9 g per 100 g of water at room temperature). The solubilities of alcohols fall off rapidly as the hydrocarbon chain length increases beyond four.

13.4 Equilibrium in Solution Formation

Some substances can dissolve in each other in all proportions, as is the case with ethanol and water, pentane and hexane, and benzene and toluene. In most cases, though, as in solutions of ionic compounds in water, only a limited quantity of solute can be dissolved in a given quantity of solvent. This limit varies with the nature of the solute and of the solvent, and with temperature.

Suppose that we place 40 g NaCl in 100 g of water at 20 °C. What happens? Initially, many of the Na^+ and Cl^- ions are plucked from the surface of crystals by ion–dipole forces and they wander about at random through the solution; *dissolving occurs*. In their wanderings, some of the ions pass near a crystal surface and are attracted back to the surface by ions of the opposite charge, becoming once more a part of the crystal lattice; *crystallization occurs*. As more and more NaCl dissolves, the number of "wanderers" that return to be trapped once again in the solid crystals increases. Eventually (when 36 g NaCl has dissolved), the number of ions returning to the surfaces of undissolved crystals equals the number leaving the crystal surfaces. The rate of crystallization becomes equal to the rate of dissolving. A condition of *dynamic equilibrium* is established. The net quantity of NaCl in solution remains the same, despite the fact that there is a lot of activity as ions come and go from the surface of the crystals. The net quantity of undissolved NaCl crystals also remains constant (in this example, 4 g), even though individual crystals may change in size and shape. Some small crystals may even disappear as others grow larger. This process is illustrated in Figure 13.5.

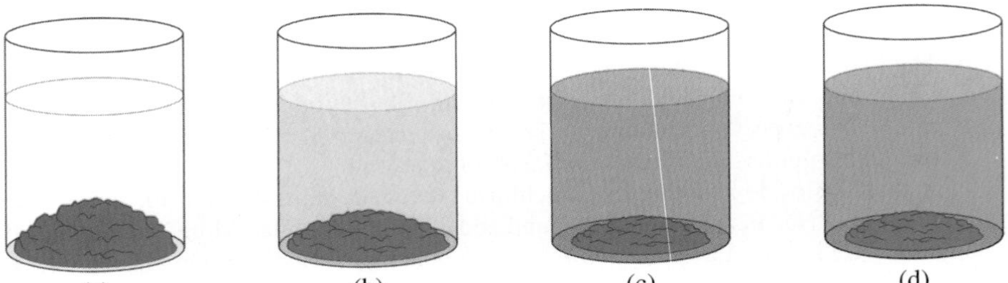

Figure 13.5 Formation of a saturated solution. (**a**) A solid solute is added to a fixed quantity of water. (**b**) After a brief period of time, the solution has acquired a color due to the dissolved solute. The quantity of undissolved solute is visibly less than in (a). (**c**) After a longer period of time, the solution color has deepened, and the quantity of undissolved solute has decreased still further from that in (b). The solution in (b) must be unsaturated. (**d**) After a much longer period of time, the solution color and the quantity of undissolved solute appear to be the same as in (c). Dynamic equilibrium must have been attained in (c) and persists in (d). In both (c) and (d), the solution is saturated.

When dynamic equilibrium is established between undissolved solute and a solution, we say that the solution is **saturated**. We refer to the concentration of the saturated solution as the **solubility** of the solute. The water solubility of NaCl at 20 °C is 36 g NaCl/100 g H_2O. Any solution containing less solute than it could hold at equilibrium is **unsaturated**. At 20 °C, a solution with 24 g NaCl/100 g H_2O is unsaturated; so is one with 32 g NaCl/100 g H_2O.

It is important to understand that the terms unsaturated and saturated are not related to whether a solution is dilute or concentrated. For example, a saturated aqueous solution of $Ca(OH)_2$ at 20 °C is 0.023 M, quite a dilute solution. On the other hand, at 20 °C a 10 M NaOH solution, though quite concentrated, is still unsaturated.

Solubility as a Function of Temperature

Solubility varies with temperature, so when we cite solubility data, we must always indicate the temperature. Later in the text we'll show that the effect of temperature on the solubility of a substance is related to the heat of solution. For the present, though, let's just use this generalization regarding the solubilities of ionic compounds in water.

The majority—*about 95% of them—have aqueous solubilities that increase significantly with temperature. Most of the remainder have solubilities that change little with temperature; a few, for example, some sulfates and selenates, even have solubilities that decrease with temperature.*

Solubility usually is listed in a handbook in grams of solute per 100 g of *solvent*. A graph of solubility as a function of temperature is called a **solubility curve**. Figure 13.6 shows the solubility curves of several ionic compounds.

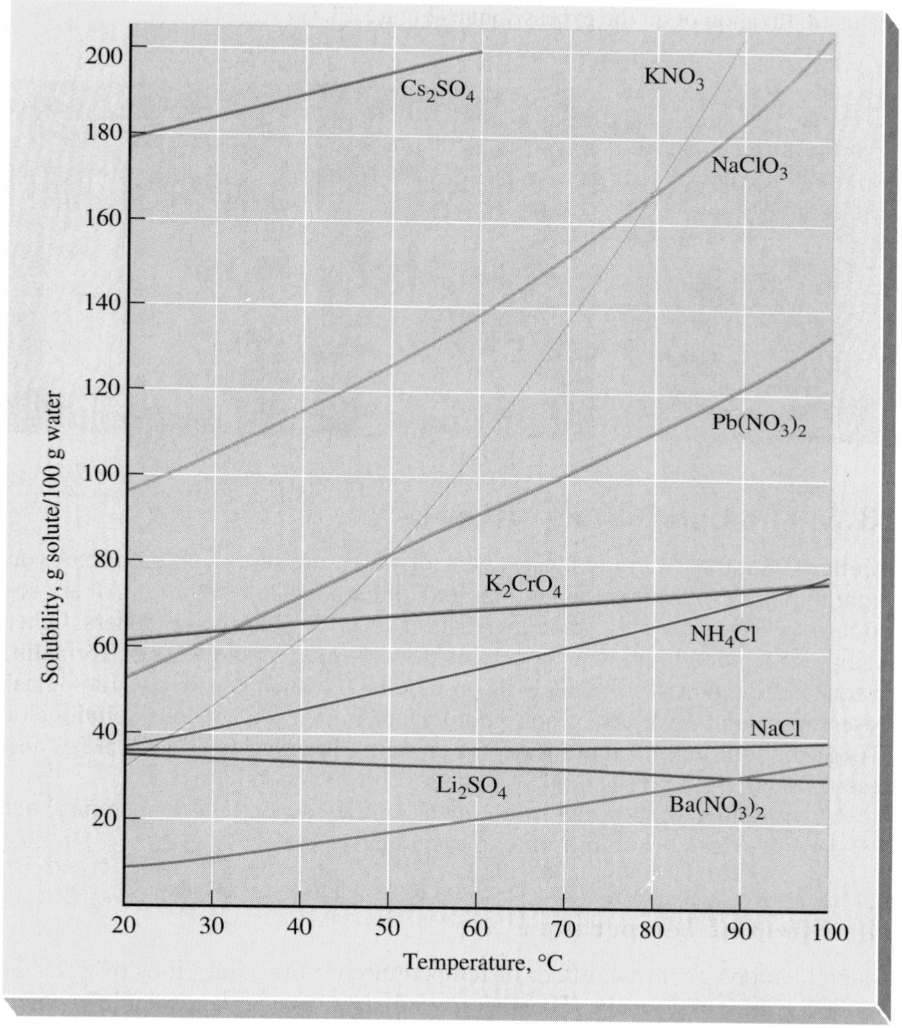

Figure 13.6 The solubility curves for several salts in water.

If we cool a saturated solution of lead nitrate in contact with excess $Pb(NO_3)_2(s)$, more solute crystallizes until solubility equilibrium is once again established at the lower temperature. For example, consider a saturated solution of lead nitrate at 90 °C. For each 100 g water the solution contains 122 g of dissolved $Pb(NO_3)_2$. When the solution is cooled to 20 °C, the solution can contain only 54 g $Pb(NO_3)_2$. The excess of 68 g crystallizes out, increasing the quantity of undissolved solute.

Now consider what would happen if we started to cool a saturated solution of lead nitrate with no excess solute present. Would lead nitrate crystallize? It might. Then again, it might not. There is no dynamic equilibrium—no crystals to capture wandering ions. We might be able to cool the solution to 20 °C without crystallization occurring. Such a solution, containing more solute than it would normally contain at equilibrium, is said to be **supersaturated**. Solute may crystallize when a supersaturated solution is stirred or if the inside of the container is scratched with a glass rod (producing microscopic fragments of glass on which precipitation can occur). Addition of a "seed" crystal of solute will nearly always result in the sudden crystallization of all the excess solute (Figure 13.7).

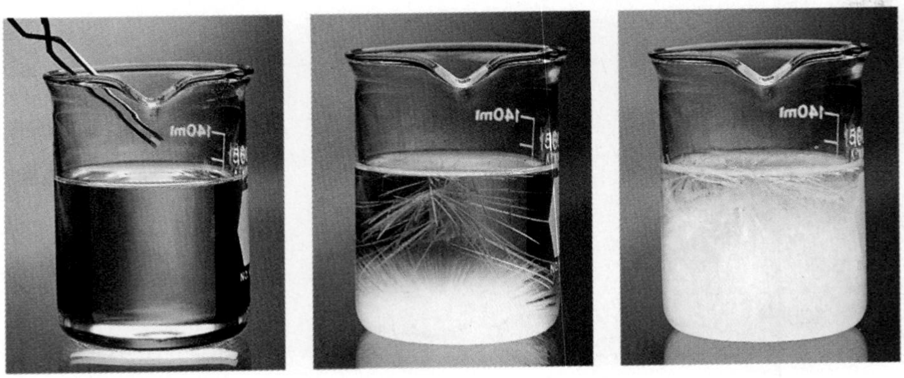

Figure 13.7 Seeding a supersaturated solution. A tiny seed crystal is added to a supersaturated solution of sodium acetate (left). Crystallization begins (center) and continues until all the excess solute has crystallized from the solution (right).

13.5 The Solubilities of Gases

Solutions of gases in liquids, especially in water, are more common than you might think. First, there is the familiar case of carbonated beverages. All are solutions of $CO_2(g)$ in water, sometimes with added flavors and sweeteners. Other examples include blood, which contains dissolved $O_2(g)$ and $CO_2(g)$; formalin, an aqueous solution of formaldehyde gas (HCHO), which is used as a biological preservative; and a variety of household cleaners that are aqueous solutions of $NH_3(g)$. In addition, *all* natural waters contain dissolved $O_2(g)$ and $N_2(g)$ and traces of other gases.

As with other solutes, the solubilities of gases depend on temperature, but they depend even more significantly on pressure.

The Effect of Temperature

Generalizations about the effect of temperature on the solubilities of gases in liquids are difficult to make. The majority of gases become *less* soluble in water as the temperature increases. However, the situation is often the reverse, increased solubility with increased temperature, for gases in organic solvents. The solubilities of the noble gases in water pass through a minimum, for example, at 35 °C for helium at 1 atm pressure.

The water solubility of air at standard atmospheric pressure at different temperatures is represented through the graph in Figure 13.8. Although oxygen makes up only about 23% of air, by mass, it constitutes about 35% of the air dissolved in water; $O_2(g)$ is more water soluble than is $N_2(g)$.

Some Supersaturated Solutions in Nature

Supersaturated solutions are not just laboratory curiosities; they occur naturally. Honey is one example; the principal solute is the sugar glucose. If honey is left to stand, the glucose crystallizes. We say, not very scientifically, that the honey has "turned to sugar." Supersaturated sucrose (cane sugar) solutions are fairly common in cooking. Jellies are one example. Sucrose often crystallizes from jelly that has been standing for a long time.

Some wines have high concentrations of potassium hydrogen tartrate, $KHC_4H_4O_6$. When chilled, the solution may become supersaturated. After a time, crystals may form and settle out if the wine is stored in the consumer's refrigerator. Modern wineries solve this problem—and render the wine less acidic—by a process known as cold stabilization. The wine is chilled to near 0 °C, a temperature below that commonly found in refrigerators. Tiny seed crystals of $KHC_4H_4O_6$ are added to the supersaturated wine. Crystallization is complete after a period of time, and the excess crystals are filtered off. At one time, wine making was the principal source of potassium hydrogen tartrate, the "cream of tartar" used in baking.

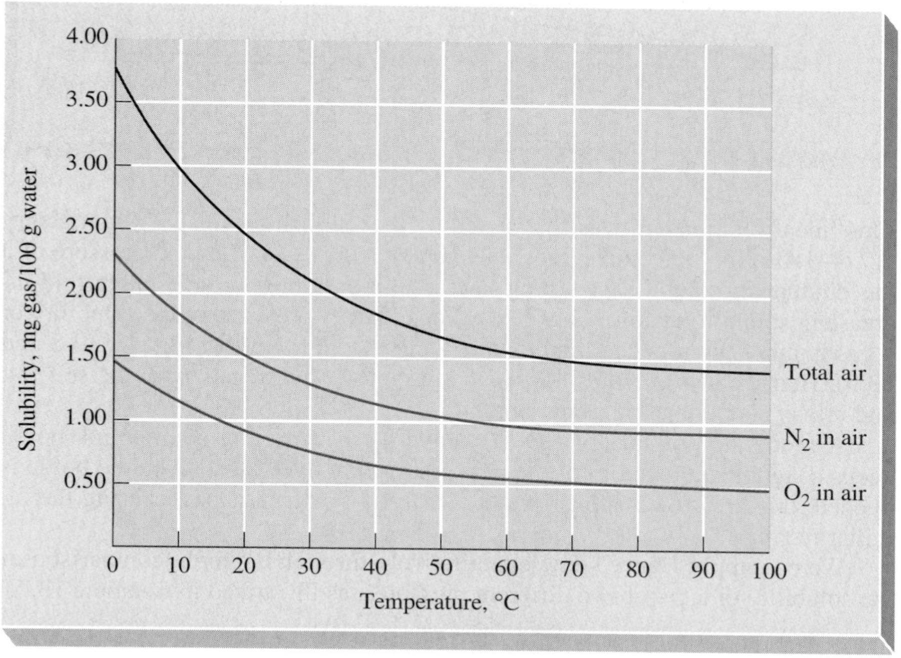

Figure 13.8 Solubility of air in water as a function of temperature at 1 atm pressure. The solubility curve for air (black) is essentially the sum of the curves for N_2 in air (red) and O_2 in air (blue). (N_2 and O_2 together make up about 99% of air.) Although more N_2 than O_2 is present in the dissolved air, O_2 is actually the more soluble of the two gases when a comparison is made at equal partial pressures of the gases. In air, the partial pressure of N_2 is almost four times as great as that of O_2.

The limited solubility of air is essential to aquatic life. Fish depend on dissolved air for O_2(g). Moreover, the fact that the solubility *decreases* with temperature explains why many fish (trout, for example) cannot survive in warm water—they don't get enough oxygen. At 30 °C the amount of dissolved O_2(g) in water is only about one-half of what is found at 0 °C.

The world's major ocean fisheries are located in *cold* regions such as the Bering Sea and the Grand Banks off Newfoundland.

The Effect of Pressure: Henry's Law

As shown in Figure 13.9, at constant temperature the solubility of a gas in water (S) is directly proportional to the pressure of the gas in equilibrium with the aqueous solution (P_{gas}). That is,

$$S \propto P_{gas} \quad \text{or} \quad S = k \cdot P_{gas}$$

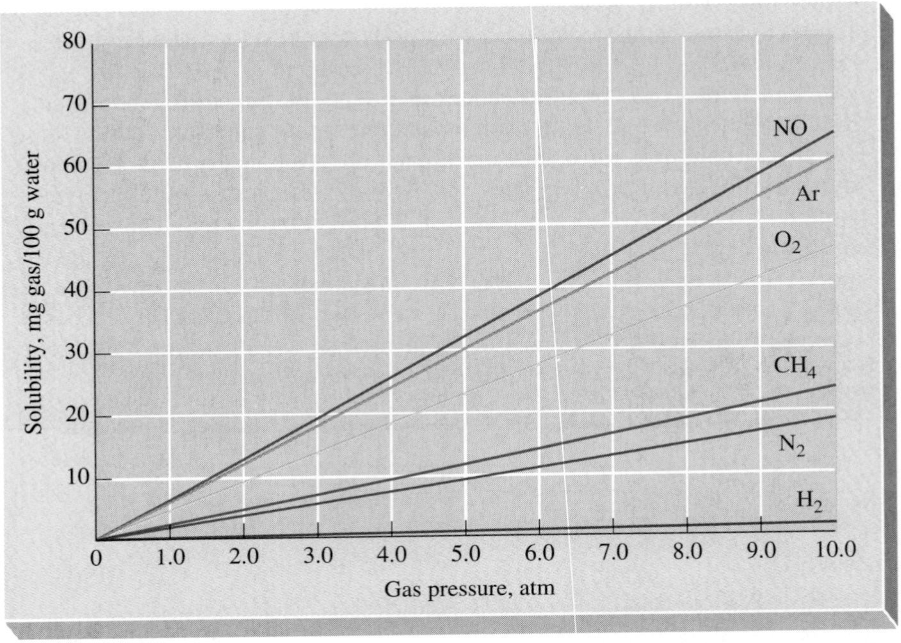

Figure 13.9 The effect of gas pressure on aqueous solubilities of gases at 20 °C.

This linear relationship was first stated by the English chemist William Henry (1774–1836) in 1803 and is known as **Henry's law**. Doubling the pressure of a gas doubles its solubility; a tenfold increase in pressure produces a ten-fold increase in solubility; and so on. Clearly, the effect of pressure on the solubility of a gas is more pronounced than that of temperature (compare Figures 13.8 and 13.9). Henry's law may not hold if the gaseous solute reacts with the solvent, and it may also fail at high pressures.

A moderate pressure of $CO_2(g)$ above the beverage in a soft-drink bottle keeps a significant quantity of the gas dissolved in the water. When the bottle is opened, this pressure is released and dissolved $CO_2(g)$ escapes, causing the familiar fizzing.

We can apply Henry's law quantitatively through a simple comparison of the solubility of a gas at two different pressures, as illustrated in Example 13.7.

The shooting cork and foam show that the carbon dioxide gas in champagne is maintained under pressure.

Example 13.7

A 225-g sample of pure water is shaken in contact with air under a pressure of 1 atm and a temperature of 20 °C. What mass of Ar(g) will be present in the water when solubility equilibrium is reached? Use (1) data from Figure 13.9, (2) the fact that the mole percent of Ar in air is 0.934%, and (3) the method of Example 4.19.

Solution

Air is a gaseous mixture, and Ar is one of its minor components. The method of Example 4.19 relates the partial pressures and mole fractions of gaseous mixtures

$$\frac{P_{Ar}}{P_{total}} = \frac{n_{Ar}}{n_{total}} = \chi_{Ar}$$

Because a percentage represents "parts per hundred," the mole fraction of Ar(g) is 0.934/100 = 0.00934. The total pressure is 1 atm. The partial pressure of Ar in air is

$$P_{gas} = 0.00934 \times 1\ \text{atm} = 0.00934\ \text{atm}$$

We now have the value of P_{gas} that we need in the Henry's law expression: $S = k \cdot P_{gas}$, but we need a value of the constant k before we can calculate S. A way to get a value of k is to select a point from the argon graph in Figure 13.9 and solve for k. Suppose we choose the point S = 60 mg Ar/100 g water and P_{gas} = 10 atm. Then,

$$k = \frac{S}{P_{gas}} = \frac{60\ \text{mg Ar/100 g water}}{10\ \text{atm}}$$

Now, we can use this value of k (1), the partial pressure of Ar in air (2), and the mass of water that is to be saturated with Ar (3) to calculate the mass of Ar present.

$$? \text{ mg Ar} = \overset{\textbf{(1)}}{\frac{60\ \text{mg Ar/100 g water}}{10\ \text{atm}}} \times \overset{\textbf{(2)}}{0.00934\ \text{atm}} \times \overset{\textbf{(3)}}{225\ \text{g water}}$$

$$= 0.13\ \text{mg Ar}$$

Exercise 13.7

At 25 °C and 1 atm gas pressure, the solubility of CO_2(g) is 149 mg/100 g water. When air at 25 °C and 1 atm is in equilibrium with water, what is the quantity of dissolved CO_2, in mg/100 g water? Air contains 0.036 mole % CO_2(g).

It is not difficult to rationalize Henry's law. Solubility equilibrium is established when the rate at which dissolved gas molecules escape becomes equal to the rate at which they enter a solution. When we increase the pressure of a gas, we produce an increase in the number of molecules per unit volume in the gas above the liquid. This causes the rate at which gas molecules enter the solution to increase. In order for the rate of escape of dissolved gas molecules to once again become equal to their rate of entry, more of them must be present in the solution—the solubility of the gas must increase.

13.6 Vapor Pressure of Solutions

In the next few sections, we will consider several physical properties of solutions, all of which can be related to the property of vapor pressure. The first systematic studies of the vapor pressures of solutions were reported by the French chemist F. M. Raoult in 1887. Raoult's finding, now known as **Raoult's law**, was that the addition of a solute lowers the vapor pressure of the solvent, and that the fractional lowering of the vapor pressure is equal to the mole fraction of the solute.

Divers' Bends: A Painful Application of Henry's Law

Divers who dive deeply into the sea must carry their own supply of air. We normally breath about 800 L of air per hour. To have enough air for an hour or so of underwater exploring, a diver must carry air that is compressed to a much smaller volume. As we should expect from Henry's law, compressed air is much more soluble in blood and other body fluids than is air at normal pressures. One of the dissolved gases, nitrogen, can cause problems. Divers must be careful to return to the surface slowly or spend considerable time in a decompression chamber where the pressure is gradually lowered. If they don't, excess $N_2(g)$ quickly escapes from solution (that is, blood) as tiny bubbles that block blood flow in capillaries. The divers experience decompression sickness, commonly called the *bends*, a condition characterized by severe pains in joints and muscles. The person may faint or suffer deafness, paralysis, or even death. People who work in deep mines or tunnels, where compressed air is used to keep water from infiltrating, have similar problems.

If the return to normal pressures is slow enough, the excess gases leave the blood gradually. Excess O_2 is used in metabolism, and excess N_2 is removed to the lungs and expelled by normal breathing. About 20 min of slow decompression is needed for each atmosphere of pressure above normal that a person experiences.

Bends can also be avoided by using a compressed helium–oxygen mixture as a substitute for compressed air. Helium is considerably less soluble than is nitrogen, so its absorption into the bloodstream is quite limited. Excess O_2 presents no problem because it is consumed in metabolism.

Underwater divers who surface too quickly may experience the painful and dangerous condition known as the "bends."

That is, if we represent the vapor pressure of solvent about a solution as P_A, the vapor pressure of the pure solvent as P_A°, and the mole fraction of the solute as χ_B,

$$\frac{P_A^\circ - P_A}{P_A^\circ} = \chi_B$$

The usual form in which Raoult's law is expressed involves a slight rearrangement of this expression.

$$\frac{P^\circ_A}{P^\circ_A} - \frac{P_A}{P^\circ_A} = \chi_B$$

$$1 - \frac{P_A}{P^\circ_A} = \chi_B$$

$$\frac{P_A}{P^\circ_A} = (1 - \chi_B)$$

Now, let's use the fact that the sum of the mole fractions of the components in a solution is 1. For a solution of solute B in solvent A, $\chi_A + \chi_B = 1$, or $\chi_A = 1 - \chi_B$. This allows us to write the final form of Raoult's law, as

$$P_A = \chi_A \cdot P^\circ_A$$

The most general statement of Raoult's law, then, is that the vapor pressure of a volatile component above a solution is equal to the vapor pressure of the pure component times the mole fraction of that component in solution. Because χ_A must always be less than one, the vapor pressure of a component of a solution is less than that of the pure component.

All the volatile components in an *ideal* solution obey Raoult's law, but the law may fail for *nonideal* solutions. However, even in nonideal solutions the solvent usually obeys Raoult's law reasonably well if the solutions are *dilute*.

In Examples 13.8 and 13.9 we focus on a *binary* (two-component) *ideal* solution in which both the solute and solvent are volatile. First we calculate the partial pressures of the two components above the solution, and then the composition of the vapor in equilibrium with the solution.

Example 13.8

The vapor pressures of pure benzene (C_6H_6) and toluene (C_7H_8) at 25 °C are 95.1 and 28.4 mmHg, respectively. A solution is prepared having equal mole fractions of C_6H_6 and C_7H_8. Determine the vapor pressures of C_6H_6 and C_7H_8 and the total vapor pressure above this solution. Consider the solution to be ideal.

Solution

Let's use the subscripts "b" and "t" for benzene and toluene, and note that the solution has $\chi_b = \chi_t = 0.500$. The vapor pressures are

$$P_b = \chi \cdot P^\circ_b = 0.500 \times 95.1 \text{ mmHg} = 47.6 \text{ mmHg}$$

$$P_t = \chi_t \cdot P^\circ_t = 0.500 \times 28.4 \text{ mmHg} = 14.2 \text{ mmHg}$$

$$P_{total} = P_b + P_t = 47.6 \text{ mmHg} + 14.2 \text{ mmHg} = 61.8 \text{ mmHg}$$

Exercise 13.8

The vapor pressure of pure water at 20.0 °C is 17.5 mmHg. What is the vapor pressure of water above a solution that is 1.00 *m* $C_{12}H_{22}O_{11}$ (sucrose)? (*Hint*: Start with the definition of molality. How many moles of H_2O are present in 1.00 kg H_2O? What is the mole fraction of H_2O?)

Example 13.9

What is the composition of the vapor in equilibrium with the benzene–toluene solution of Example 13.8, expressed on a mole fraction basis?

Solution

As indicated in Example 13.7, the expression that relates the mole fraction (χ_i) of a component (i) of a gaseous mixture to its partial pressure (P_i) and the total pressure (P_{total}) is

$$\chi_i = \frac{n_i}{n_{total}} = \frac{P_i}{P_{total}}$$

Applied to the data on benzene and toluene from Example 13.8,

$$\chi_b = \frac{47.6 \text{ mmHg}}{61.8 \text{ mmHg}} = 0.770$$

$$\chi_t = \frac{14.2 \text{ mmHg}}{61.8 \text{ mmHg}} = 0.230$$

Note that, as expected, $\chi_b + \chi_t = 0.770 + 0.230 = 1.000$.

Exercise 13.9

Above which of the following solutions does the vapor have the greater mole fraction of benzene at 25 °C: a solution with equal masses of benzene and toluene, or one having equal numbers of moles of benzene and toluene? (*Hint*: You will need information from Examples 13.8 and 13.9, but you do not need to do precise calculations of any sort.)

The results of Examples 13.8 and 13.9 illustrate an important idea: The vapor phase in equilibrium with an ideal solution of two volatile components is always richer in the *more volatile* component than is the liquid phase. Thus, if a vapor with $\chi_b = 0.770$ is condensed and the liquid revaporized, the new vapor will again be richer in benzene than the liquid. *Fractional distillation* is a method of separating the components of a mixture based on the changes in composition that occur in the vaporization of a mixture of liquids of differing volatilities. (The fractional distillation of petroleum was discussed on page 58.)

To prove this statement to yourself, redo Examples 13.8 and 13.9, starting with a solution in which $\chi_b = 0.770$. You will find that the equilibrium vapor has $\chi_b = 0.918$.

CONCEPTUAL EXAMPLE 13.1

Figure 13.10 shows two different aqueous solutions placed in the same enclosure. After a time, the solution level has risen in container A and dropped in container B. Explain how and why this happens.

Solution

The only way to transport water from one container to the other is through the vapor. There must be a net vaporization from container B and a net condensation into container A. And this requires that the vapor pressure of H_2O above solution

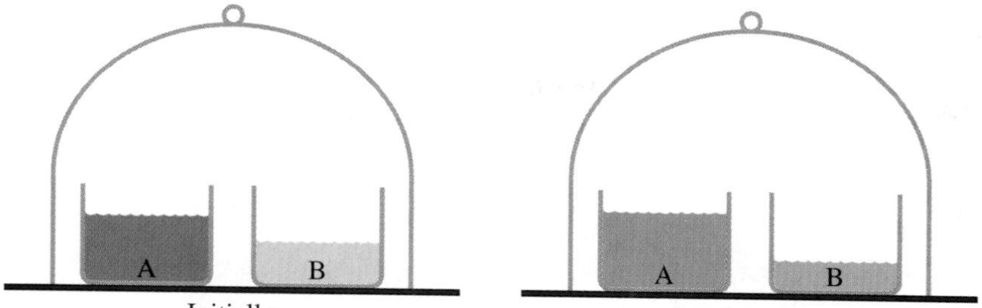

Figure 13.10 A phenomenon related to the vapor pressures of solutions. The changes that occur here are described in this example.

B be greater than that above solution A. According to Raoult's law, the vapor pressure of water above both solutions is given by the expression:

$$P_{H_2O} = \chi_{H_2O} \cdot P^\circ_{H_2O}$$

Thus, the mole fraction of water must be greater in the solution in container B than in container A. The solution in B must be more dilute. Water passes, through the vapor, from the more dilute to the more concentrated solution.

Conceptual Exercise 13.1

Would the process depicted in Figure 13.10 continue until the solution in container B evaporated to dryness? Explain.

13.7 Colligative Properties: Freezing Point Depression and Boiling Point Elevation

We don't often measure the lowering of the vapor pressure of a solvent by a dissolved solute. We do, however, measure properties related to vapor pressure lowering: freezing point depression, boiling point elevation, and osmotic pressure. These three properties, together with vapor pressure lowering, constitute a group known as **colligative properties.** The key feature of colligative properties is that their values depend on the concentration of solute particles in a solution but not on the identity of the solute. Thus, the lowering of the vapor pressure of water in an 0.1 M solution of a nonelectrolyte is the same, whether the solute is glucose, sucrose, or urea. All have the same concentration of particles (molecules): 0.1 mol/L.

In this section we focus on binary solutions in which the solute is (a) *nonvolatile*; (b) a *nonelectrolyte*; and (c) soluble in the liquid solvent but not in the solid solvent. In solutions with these characteristics, the vapor pressure is simply that of the solvent, and at all temperatures this vapor pressure is lower than that of the pure solvent. This statement is illustrated in Figure 13.11, in which a partial phase diagram for the solution (in red) is superimposed onto the phase diagram of the pure solvent (in blue).

In the partial phase diagram for the solution, a line at $P = 1$ atm intersects the fusion curve at a *lower* temperature and the vapor pressure curve at a *higher* temperature than is the case for the pure solvent. In this way we see that the freezing point of the solvent is *depressed* and the boiling point is *elevated*. Only

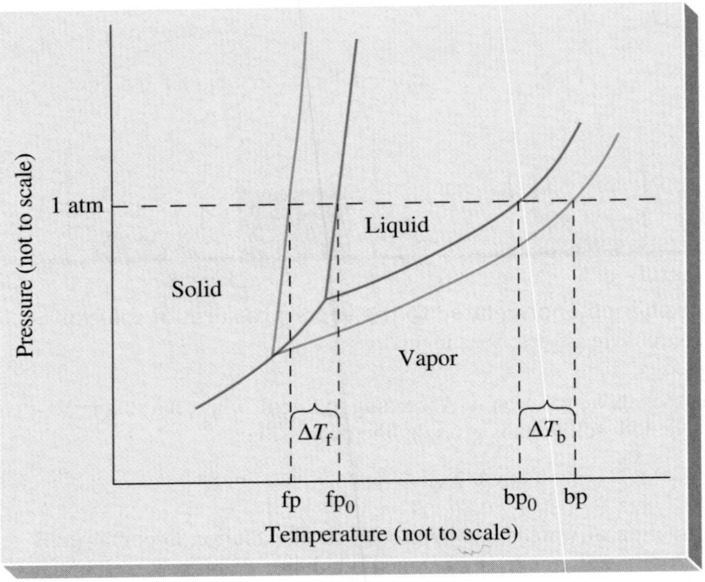

Figure 13.11 Vapor pressure lowering by a nonvolatile solute. In a solution, the vapor pressure of the solvent is lowered and the fusion curve is displaced to lower temperatures (red curves). As a consequence the freezing point of the solvent is lowered by the amount ΔT_f and the boiling point is raised by ΔT_b.

the solvent freezes from solution, and only the solvent escapes as vapor during boiling. The solute(s) remain in solution.

The extent of the lowering of the vapor pressure depends on the *mole fraction* of solute present. So too do the freezing point depression and boiling point elevation. However, because in *dilute* solutions the molality of solute is proportional to its mole fraction, we can write

In *dilute aqueous* solutions, molarity and molality are nearly equal, and both are proportional to the mole fraction of solute.

$$\Delta T_f = -K_f \times m$$

$$\Delta T_b = K_b \times m$$

In these equations ΔT_f and ΔT_b are the freezing point depression and boiling point elevation, respectively, and m is the molality of solute. K_f and K_b are proportionality constants: K_f is related to the melting point, enthalpy of fusion, and molar mass of the solvent; K_b is related to the boiling point, enthalpy of vaporization, and molar mass. The freezing point depression equation has a minus sign because ΔT_f is a negative quantity. Table 13.2 lists some typical values of K_f and K_b.

TABLE 13.2 **Freezing Point Depression and Boiling Point Elevation Constants**

Solvent	Normal Freezing Point, °C	K_f, °C m^{-1}	Normal Boiling Point, °C	K_b, °C m^{-1}
Acetic acid	16.63	3.90	117.90	3.07
Benzene	5.53	5.12	80.10	2.53
Cyclohexane	6.55	20.0	80.74	2.79
Nitrobenzene ($C_6H_5NO_2$)	5.8	8.1	210.8	5.24
Water	0	1.86	100	0.512

Example 13.10

What is the freezing point of an aqueous sucrose solution that has 25.0 g $C_{12}H_{22}O_{11}$ per 100.0 g H_2O?

Solution

From the data given we can calculate the molality of the solution.

$$\frac{\dfrac{\text{mass of solute}}{\text{molar mass of solute}}}{\text{kg solvent}} = \frac{\dfrac{25.0\text{ g }C_{12}H_{22}O_{11}}{342.3\text{ g }C_{12}H_{22}O_{11}/\text{mol }C_{12}H_{22}O_{11}}}{0.1000\text{ kg }H_2O} = 0.730\ m$$

Then we can use this molality in the equation

$$\Delta T_f = -K_f \times m = -1.86\text{ °C } m^{-1} \times 0.730\ m = -1.36\text{ °C}$$

The depression of the freezing point $\Delta T_f = -1.36$ °C. Because the freezing point of pure water is 0.00 °C, the freezing point of the sucrose solution is –1.36 °C.

Exercise 13.10

What mass of sucrose, $C_{12}H_{22}O_{11}$, should be added to 75.0 g H_2O to raise the boiling point to 100.35 °C? (*Hint*: What must be the molality of the solution? What mass of sucrose is required to produce a solution of this molality? Use data from Table 13.2.)

Example 13.11

Sorbitol is a sweet substance found in fruits and berries and sometimes used as a sugar substitute. An aqueous solution containing 1.00 g of sorbitol in 100.0 g of water is found to have a freezing point of –0.102 °C.

a. What is the molar mass of sorbitol?

b. Elemental analysis indicates that sorbitol consists of 39.56% C, 7.75% H, and 52.70% O by mass. What is the molecular formula of sorbitol?

Solution

a. First, we can determine the molality of the sorbitol solution by using the freezing point data:

$$\Delta T_f = -0.102\text{ °C} - 0.000\text{ °C} = -0.102\text{ °C} \qquad K_f = 1.86\text{ °C} \cdot m^{-1}$$

$$\Delta T_f = -K_f \times \text{molality}$$

$$\text{Molality} = \frac{\Delta T_f}{-K_f} = \frac{-0.102\text{ °C}}{-1.86\text{ °C} \cdot m^{-1}} = 0.0548\ m$$

Now, we can use this value, together with the definition of molality—mol solute/kg solvent—to establish the molar mass of sorbitol.

$$\text{Molality} = \frac{1.00\text{ g sorbitol} \times \dfrac{1}{\text{molar mass of sorbitol}}}{0.1000\text{ kg water}} = 0.0548\ m$$

$$\text{Molar mass} = \frac{1.00\text{ g of sorbitol}}{0.1000\text{ kg water} \times 0.0548\ \dfrac{\text{mol sorbitol}}{\text{kg water}}} = 182\text{ g/mol sorbitol}$$

b. We can establish the number of moles of C, H, and O in one mole of sorbitol from the percent composition data and the molar mass.

$$? \text{ mol C} = 182 \text{ g cpd} \times \frac{39.56 \text{ g C}}{100 \text{ g cpd}} \times \frac{1 \text{ mol C}}{12.01 \text{ g C}} = 5.99 \text{ mol C}$$

$$? \text{ mol H} = 182 \text{ g cpd} \times \frac{7.75 \text{ g H}}{100 \text{ g cpd}} \times \frac{1 \text{ mol H}}{1.008 \text{ g H}} = 13.99 \text{ mol H}$$

$$? \text{ mol O} = 182 \text{ g cpd} \times \frac{52.70 \text{ g O}}{100 \text{ g cpd}} \times \frac{1 \text{ mol O}}{16.00 \text{ g O}} = 5.99 \text{ mol O}$$

The molecular formula of sorbitol is $C_6H_{14}O_6$.

We can also determine the molecular formula by using the methods of Examples 3.9 and 3.11.

Exercise 13.11

A 1.065-g sample of an unknown substance is dissolved in 30.00 g of benzene; the freezing point of the solution is 4.25 °C. The substance is found to consist of 50.69% C, 4.23% H, and 45.08% O by mass. What is the molecular formula of the substance? (*Hint*: Use data from Table 13.2.)

Alcohol whose molecules contain multiple hydroxy groups (OH) are commonly used for de-icing aircraft.

Freezing point depression and boiling point elevation have practical applications. Both are involved, for example, in the action of an antifreeze like ethylene glycol ($HOCH_2CH_2OH$) in an automobile cooling system. If water alone were used as the engine coolant, it would boil away in the heat of summer and freeze in the depths of northern winters. Addition of antifreeze raises the boiling point of the coolant and also prevents it from freezing when the temperature drops below 0 °C. With the proper ratio of ethylene glycol and water, it is possible to protect an automotive cooling system to temperatures as low as –48 °C.

Salts such as NaCl and $CaCl_2$ are scattered on icy sidewalks and streets to melt ice. The ice melts as long as the outdoor temperature is above the lowest freezing point of a salt–water mixture. Below this temperature the ice will not melt. NaCl can melt ice at temperatures as low as –21 °C (–6 °F) but no lower. $CaCl_2$ is effective at temperatures as low as –55 °C (–67 °F).

13.8 Colligative Properties: Osmotic Pressure

Everyday experience tells us that we can separate coffee grounds from brewed coffee by passing the mixture through filter paper. However, we cannot separate the caffeine from the brewed coffee with filter paper. Paper is *permeable* to water, to other solvents, and to solutes in solution. We also know that some materials are *impermeable* to liquids and solutions as well as to solids. Water and solutions do not pass through the metal walls of cans nor through the glass walls of jars and bottles. Are there, perhaps, materials with intermediate properties? Materials that will pass solvent molecules but not solute molecules? Materials that are permeable to some solutes but not others? The answer is yes. **Semipermeable membranes** are sheets or films of a material containing a network of submicroscopic holes or pores through which small solvent molecules can pass but which severely restrict the flow of solute particles. These may be natural materials of animal or vegetable origin, such as pig's bladder and parchment, or synthetic materials such as cellophane. Cell membranes, the lining of the digestive tract, and the walls of blood vessels are all semipermeable; they allow certain substances to go through while holding others back.

The action of a semipermeable membrane is illustrated in Figure 13.12. Here, a semipermeable membrane, such as a film of cellophane, separates pure water from an aqueous solution in a U-tube. The film is permeable to water molecules from either direction, but it is impervious to solute molecules. After a time, the liquid level rises in the arm of the tube containing the solution. There is a net flow of water molecules *from* pure water *into* the solution. We saw a similar situation in Conceptual Example 13.1 and Figure 13.10, where a net flow of H_2O molecules occurred, through the vapor phase, from a more dilute to a more concentrated solution.

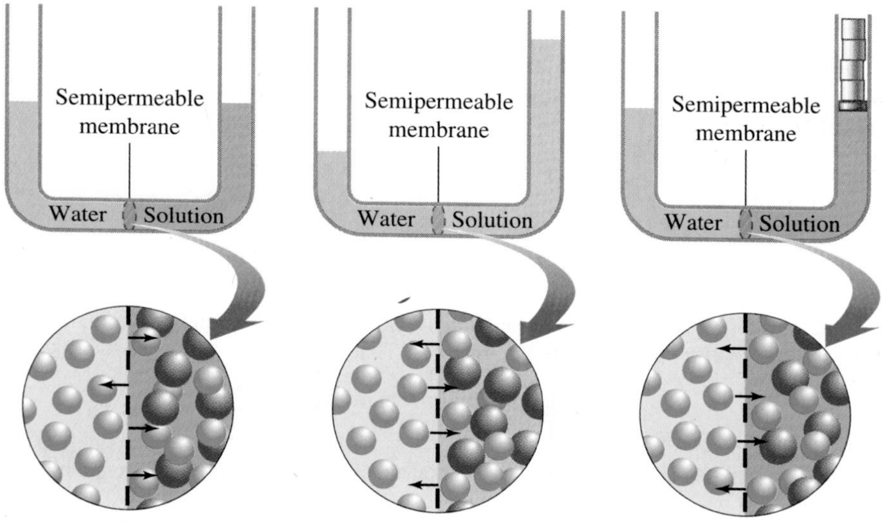

Figure 13.12 Osmosis and osmotic pressure illustrated. (a) An aqueous solution (colored green) is separated from pure water (colored blue) by a semipermeable membrane. Initially the heights of liquid in each arm of the U-tube are equal. (b) After a period of time, the height of the solution is greater than that of the water and the solution has become diluted. Although water molecules can pass through the membrane from either direction, there is a net flow from the pure water into the solution. The osmotic pressure of the solution corresponds to the maximum height that the solution reaches before the rates of transfer of water through the membrane in both directions become equal and there is no longer a net flow of water. (c) Alternatively, the osmotic pressure is the pressure that would have to be applied to the solution to increase the flow of H_2O molecules from the solution into the pure water to the point where there is no net flow of water through the semipermeable membrane.

The net flow of water molecules from pure water through a semipermeable membrane into a solution is called **osmosis.** For a 20% sucrose solution, osmotic flow could lift a column of solution to a height of 150 m. Osmosis can also occur between two solutions. A net flow of solvent molecules occurs from the more dilute to the more concentrated solution. We can think of the tendency of solvent molecules to equalize their concentrations between two solutions in contact as the driving force behind osmosis. The concentration of *solvent* molecules is greater in a dilute solution than in a concentrated one.

We can reduce the net flow of solvent molecules through a semipermeable membrane into a solution by increasing the flow of solvent molecules *out* of the solution. And this we can do by applying pressure to the solution. At a sufficiently high pressure there is no net flow of solvent through the semipermeable membrane. The pressure required to stop osmosis is called the **osmotic pressure** of the solution. For a 20% sucrose solution the osmotic pressure is about 15 atm.

Like other colligative properties, osmotic pressure depends only on the number of solute particles per unit volume of solution (and on temperature). An expression that works quite well for calculating osmotic pressures of *dilute* solutions of nonelectrolytes bears a striking resemblance to the ideal gas equation:

$$\pi = \frac{n}{V} \cdot RT$$

In this equation, π is the osmotic pressure in atm, n is the number of moles of solute in a solution volume of V (liters), R is the universal gas constant, and T is the Kelvin temperature. Because the ratio of number of moles of solute to a solution volume (L) is the *molarity* concentration (c), we can simply write

$$\pi = c \cdot RT$$

Osmotic pressure measurements and freezing point data complement each other nicely for determining molar masses. For most molecular substances, osmotic pressures are so high that they are difficult to measure. Freezing point data are preferred. On the other hand, for *macromolecular* substances (polymers) freezing point depressions are too small to measure. Osmotic pressure measurements are appropriate, as indicated in Example 13.12.

Low osmotic pressures are often measured and expressed in mmH_2O. By the method of Example 4.2, we could show that 0.00342 atm = 2.60 mmHg = 35.3 mmH_2O.

Example 13.12

An aqueous solution is prepared by dissolving 1.50 g of hemocyanin, a protein obtained from crabs, in 0.250 L of water. The solution has an osmotic pressure of 0.00342 atm at 277 K. What is the molar mass of hemocyanin?

Solution

The key to this problem is to use the fundamental expression for osmotic pressure in a form in which we substitute the mass of solute (m) in grams divided by its molar mass in g/mol for the amount of solute (n) in moles. That is,

$$\pi = \frac{n}{V} \cdot RT = \frac{(m/\text{molar mass})RT}{V}$$

Now, we can solve this equation for the molar mass and substitute the known data.

$$\text{Molar mass} = \frac{mRT}{\pi V} = \frac{1.50\text{ g} \times 0.08206\text{ L} \cdot \text{atm} \cdot \text{mol}^{-1} \cdot \text{K}^{-1} \times 277\text{ K}}{0.00342\text{ atm} \times 0.250\text{ L}}$$

$$= 3.99 \times 10^4\text{ g/mol}$$

Exercise 13.12

An aqueous solution is prepared by dissolving 1.08 g of human serum albumin, a protein obtained from blood plasma, in 50.0 mL of water. The solution has an osmotic pressure of 5.85 mmHg at 298 K. What is the molar mass of the albumin?

Practical Applications of Osmosis

Examples of osmosis are found in living organisms everywhere: Cells are much like semipermeable bags filled with solutions of ions, small and large molecules, and still larger cell components. If we place red blood cells in pure water, a net osmotic flow of water into the cells causes them to burst. On the other hand, if we place the cells in a solution that is 0.92% NaCl (mass/vol) there is no net flow of water through the cell membranes and the cells are stable. The fluids inside the cells have the same osmotic pressure as the sodium chloride solution. A solution having the same osmotic pressure as body fluids is said to be **isotonic**. If the concentration of NaCl in a saline solution is greater than 0.92%, a net flow of water *out* of the cells causes them to shrink. The saline solution is said to be **hypertonic**; it has a higher osmotic pressure than red blood cells. If the concentration

Medical Applications of Osmosis

The rupture of a cell by a *hypotonic* solution is called *plasmolysis*. If the cell is a red blood cell, the more specific term is *hemolysis*. The shrinkage of a cell in a *hypertonic* solution, called *crenation*, can lead to the death of a cell.

In replacing body fluids intravenously, it is important that the replacement fluid be isotonic. Otherwise, hemolysis or crenation results and the patient's well-being is jeopardized. As we have already described, an 0.92% NaCl (mass/vol) solution, called physiological saline, is isotonic with the fluid in red blood cells. The "D5W" so often referred to by television's doctors and paramedics is a 5.5% solution of glucose (also called dextrose, D) in water (W). It also is isotonic with the fluid in red blood cells. The 0.92% NaCl (mass/vol) is about 0.16 M, and the 5.5% glucose solution is approximately 0.31 M.

There are limits to intravenous feeding. There is a limit to how much water a patient can handle in a day—about 3 L. If an isotonic solution of 5.5% glucose is used, 3.0 L of this solution supplies only about 160 g of glucose, yielding an energy value of about 640 kcal (640 food Calories) per day. This is woefully inadequate. Even a resting patient requires about 1400 kcal/day. And for a person suffering from serious burns, for example, requirements may be as high as 10,000 kcal/day. With carefully formulated solutions containing other vital nutrients as well as glucose, the feeding of a patient can be increased to about 1200 kcal/day. This still falls short of the requirements of many seriously ill people.

One answer to the problem is to use solutions that are about six times as concentrated as isotonic solutions. But instead of being administered through a vein in an arm or a leg, this solution is infused directly through a tube into the superior vena cava, a large blood vessel leading to the heart. The large volume of blood flowing through this vein quickly dilutes the solution to levels that do not damage the blood. With this technique patients have been given 5000 kcal/day and have even gained weight.

In Section 13.9, we'll see why these two isotonic solutions have different molarities.

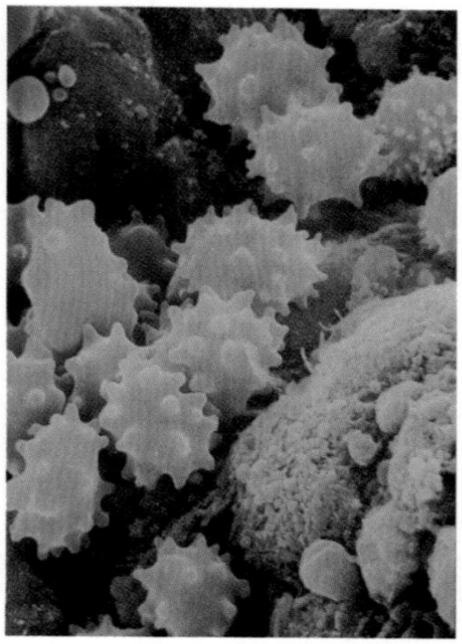

Red blood cells in a hypertonic solution, as seen through an electron microscope.

of NaCl is less than 0.92%, water flows *into* the cells. The solution is **hypotonic**; it has a lower osmotic pressure than red blood cells.

One modern application of osmosis, called **reverse osmosis**, is based on *reversing* the normal net flow of water molecules through a semipermeable membrane. That is, by applying to a solution a pressure *exceeding* the osmotic pressure, water can be driven from a solution into pure water (recall Figure 13.12). In this way pure water can be extracted from brackish water, seawater, or industrial wastewater. The success of reverse osmosis requires using membranes that can withstand high pressures. Reverse osmosis is widely used in ships at sea and in water-poor nations of the Middle East.

13.9 Colligative Properties: Solutions of Electrolytes

So far our discussion has centered on solutions of nonelectrolytes. What about electrolyte solutions? Do they have colligative properties? We've already implied that they do when we mentioned the osmotic properties of isotonic saline

Jacobus Hendricus van't Hoff (1852–1911) received the first Nobel Prize in chemistry in 1901 for his work on the physical properties of solutions. However, he is also known for proposing (simultaneously with the French chemist Joseph Le Bel) the tetrahedral geometries of carbon compounds (for example, CH_4). Van't Hoff was 22 years old at the time.

solutions and the use of NaCl and $CaCl_2$ in deicing roads. The major difference in treating solutions of electrolytes is in how we assess solute concentrations. Nonelectrolyte solutions have *molecules* as their solute particles. Strong electrolyte solutions have *ions*, not molecules, as solute particles. Weak electrolyte solutions have *both* molecules and ions as solute particles.

To assess the colligative properties of a solution we need the expressions

$$\Delta T_f = -i \cdot K_f \cdot m$$

$$\Delta T_b = i \cdot K_b \cdot m$$

$$\pi = i \cdot c \cdot RT$$

where the factor i is called the **van't Hoff factor.**

For nonelectrolyte solutions, $i = 1$. For a solution of the strong electrolyte NaCl, $i = 2$; *two* moles of ions are present for each mole of solute dissolved. For Na_2SO_4 and $CaCl_2$ we would expect $i = 3$. For a solution of a weak electrolyte, i is somewhat greater than 1 but less than 2. That is, a weak electrolyte is only partially ionized. The degree of ionization of a weak electrolyte depends on the concentration of the solution, and so the value of i is a function of concentration.* A solution that is 0.10 m CH_3COOH is about 1% ionized ($i \approx 1.01$), whereas 0.010 m CH_3COOH is about 4% ionized ($i \approx 1.04$).

ESTIMATION EXAMPLE 13.4

Estimate the freezing point of an aqueous solution of 0.15 m K_2SO_4.

Solution

Each formula unit of K_2SO_4 produces two K^+ and one SO_4^{2-} ion in the solution; $i = 3$. Now calculate the freezing point *depression*, ΔT_f.

$$\Delta T_f = -i \cdot K_f \cdot m = -3 \times 1.86\ °C\ m^{-1} \times 0.15\ m = -0.84\ °C$$

The estimated freezing point of the solution is 0.84 °C below that of water, that is, −0.84 °C.

Estimation Exercise 13.4

Estimate the molality of an aqueous solution of magnesium chloride if it is found to have a freezing point of −0.12 °C.

* Even in strong electrolyte solutions the value of i depends on the concentration. Only if the solution is quite dilute will i be an integer (that is, 2, 3, . . .). Otherwise it will be a noninteger. For example, values of i are 1.81 in 1.0 m NaCl, 1.87 in 0.10 m NaCl, 1.94 in 0.010 m NaCl, and 1.97 in 0.0010 m NaCl. In concentrated solutions, attractive forces between cations and anions cause the solution to behave as if dissociation of the solute were incomplete. In this text we will assume solutions are sufficiently dilute that strong electrolytes have an integral value of i.

CONCEPTUAL EXAMPLE 13.2

Place the following in order of increasing osmotic pressure.

a. 0.01 M $CO(NH_2)_2(aq)$ at 25 °C

b. 0.01 M $C_6H_{12}O_6(aq)$ at 37 °C

c. 0.01 *m* $KNO_3(aq)$ at 25 °C

d. a solution of 1.00 g polystyrene (molecular weight: 3.5×10^5 u) in 100 mL benzene at 25 °C?

Solution

Let's begin by writing the most general equation for the osmotic pressure of a solution.

$$\pi = i \cdot \frac{n}{V} \cdot RT$$

a. Let's designate the osmotic pressure of 0.01 M $CO(NH_2)_2(aq)$, a *nonelectrolyte* solution ($i = 1$), as $\pi_{(a)}$. Then we'll compare the other osmotic pressures to $\pi_{(a)}$.

b. This is also an 0.01 M solution of a *nonelectrolyte* ($i = 1$). The same number of particles are present per unit volume as in (a). However, because the temperature is *slightly higher*, the osmotic pressure of this solution should be slightly greater than that in (a), by a factor of 310/298: $\pi_{(b)} > \pi_{(a)}$.

c. The solution is 0.01 *m* rather than 0.01 M. However, in dilute aqueous solutions, because $d \approx 1.0$ g/mL, molality and molarity are essentially equal. This is a solution of an *electrolyte* ($i = 2$). The osmotic pressure is about twice as great as that of solution (a). It is also greater than that of solution (b) because a factor of 2 is greater than one of 310/298.

$$\pi_{(c)} > \pi_{(b)} > \pi_{(a)}$$

d. Each of the other three solutions has at least 0.001 mol of solute particles in a 100-mL sample. The mass of 0.001 mol of the polystyrene is $0.001 \times 3.5 \times 10^5 = 350$ g, but the mass of polystyrene in the solution is only 1.00 g. The solution is much more dilute than 0.01 M. The fact that the solute is a polymer and the solvent is an aromatic hydrocarbon has no bearing on the situation. Solution (d) has the lowest osmotic pressure by far; solution (c) has the highest:

$$\pi_{(c)} > \pi_{(b)} > \pi_{(a)} > \pi_{(d)}$$

Conceptual Exercise 13.2

Which of the following aqueous solutions has the *lowest* and which has the *highest* freezing point: 0.010 *m* $CO(NH_2)_2(aq)$, 0.0080 M $HCl(aq)$, 0.0050 *m* $MgCl_2(aq)$, 0.0030 M $Al_2(SO_4)_3(aq)$?

13.10 Colloids

The particles in a solution—atoms, ions, or molecules—are of submicroscopic size. Once the solute and solvent are thoroughly mixed, the *solute does not settle out*; molecular motion keeps the particles randomly distributed. The mixture is *homogeneous*. Thus, the sugar in a bottle of soda pop does not settle to the bottom, and the last drop is just as sweet as, but no sweeter than, the first.

If we try to dissolve sand (silica, SiO_2) in water, the two substances may momentarily appear to be mixed, but the sand rapidly settles to the bottom of the container. The temporary dispersion of sand in water is called a *suspension*. We can separate the sand and water by pouring the suspension into a funnel fitted with filter paper; water passes through the paper and sand remains behind. The mixture of water and sand is obviously *heterogeneous*, for part of it is clearly sand with one set of properties and part of it is water with another set of properties.

Is there no halfway point between true solutions, with particles of atomic, ionic, or molecular size, and suspensions, with gross chunks of insoluble matter? Yes, there is something else: *colloids*.

Even though silica (SiO_2) is insoluble in water, it's possible to prepare a dispersion of silica in water with up to 40% SiO_2 by mass that is stable for years. Such a dispersion does not involve ordinary grains of silica, nor is the suspended silica of ionic or molecular size. The dispersion is called a colloidal mixture.

Recall that 1 nm = 1×10^{-9} m. Simple optical microscopes are not able to resolve particles smaller than about 1000 nm. Colloidal particles generally cannot be seen under a microscope.

A material is called a colloid, not because of the kind of matter in a dispersion, but because of the extent of its subdivision. True solutions have solute and solvent particles less than about 1 nm in diameter. Ordinary suspensions have particle dimensions of about 1000 nm or more. A dispersion in which the dispersed matter has one or more dimensions (length, width, or thickness) in the range from about 1 nm to 1000 nm is said to be a **colloid** (see Figure 13.13).

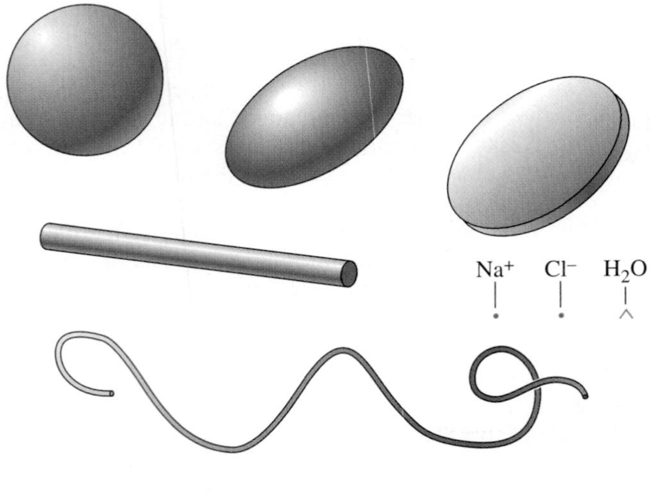

Figure 13.13 Colloidal dimensions. The shapes of dispersed matter in colloids can vary widely, but in every case at least one dimension is in the range from about 1 to 1000 nm. The particles in a colloidal dispersion of silica in water are spherical. Tobacco mosaic viruses are rod shaped, whereas gamma globulin in blood plasma forms disk-shaped particles. Cellulose and other natural fibers form long, randomly coiled filaments. Even a thin film of oil on water (not shown here) is colloidal. The relative sizes of Na^+ and Cl^- ions and of H_2O molecules are included for comparative purposes.

There are eight different kinds of colloids, based on the physical state of the particles themselves (the dispersed phase) and of the "solvent" (the dispersion medium). These are listed, together with examples, in Table 13.3. Much of living matter consists of colloidal particles, mainly in the form of sols and

TABLE 13.3 Some Common Types of Colloids

Dispersed Phase	Dispersion Medium	Type	Examples
Solid	liquid	sol	starch solution, clay sols,[a] jellies
Liquid	liquid	emulsion	milk, mayonnaise
Gas	liquid	foam	whipped cream, meringues
Solid	gas	aerosol[b]	fine dust or soot in air
Liquid	gas	aerosol[b]	fog, hair sprays
Solid	solid	solid sol	ruby glass
Liquid	solid	solid	pearl, opal, butter emulsion
Gas	solid	solid foam	floating soap (Ivory)

[a] In water purification it is sometimes necessary to precipitate clay particles or other suspended colloidal materials. This is often done by treating the water with an appropriate electrolyte. Clay sols are also suspected of adsorbing organic substances, such as pesticides, and distributing them in the environment.

[b] Smogs are complex materials that are at least partly colloidal. The suspended particles are both solid (smoke) and liquid (fog): smoke + fog = smog. Other constituents are molecular, such as SO_2, CO, NO, and O_3.

emulsions. Substances with high molecular weights, such as starches and proteins, generally form colloidal dispersions rather than true solutions in water.

As you might expect, the properties of colloidal dispersions are different from those of true solutions and suspensions (see Table 13.4). Colloids often appear milky or cloudy. Even those that appear clear reveal their colloidal nature by scattering a beam of light passed through them. This phenomenon, first studied by John Tyndall in 1869, is known as the **Tyndall effect** (see Figure 13.14). The spectacular sunsets often seen in the desert are caused, at least in part, by the preferential scattering of blue light as sunlight passes through dust-laden air (an aerosol). The transmitted light is deficient in the color blue and thus appears a reddish color.

TABLE 13.4 Properties of Solutions, Colloids, and Suspensions

Property	Solution	Colloid	Suspension
Particle size	<1 nm	1–1000 nm	>1000 nm
Settles on standing?	no	no	yes
Filter with paper?	no	no	yes
Separate by dialysis?	no	yes	yes
Homogeneous?	yes	borderline	no

A cube of solid that is 1 cm on an edge has a surface area of 6 cm^2. If this cube is divided into 1000 cubes, each with an edge of 0.1 cm, the total surface area is 60 cm^2. Imagine how large the surface area becomes if the solid is subdivided into cubes that are only about 10 nm on edge.

We have already noted that the properties of surfaces are in some ways different from those of bulk matter (recall our discussion of surface tension on page 422). A special property of surfaces is the ability to attach or *adsorb* ions to

Figure 13.14 The Tyndall effect. The flashlight beam is not visible as it passes through a true solution (left), but it is readily visible as it passes through the colloidal dispersion of iron(III) oxide (right).

themselves from a solution. This effect is not significant in suspensions, where the particles are relatively large, but it can be crucial in colloidal dispersions.

Colloidal particles generally selectively adsorb certain cations or anions. The SiO_2 particles in colloidal silica selectively adsorb OH^-. As a result, the particles carry a net *negative* charge (even though the mixture as a whole is electrically neutral). Because all the silica particles are negatively charged, they repel one another. These repulsions are strong enough to overcome the force due to gravity, and the particles do not settle. On the other hand, a high concentration of an electrolyte can *coagulate* or *precipitate* a colloid (see Figure 13.15). It does so by neutralizing the charges on colloidal particles.

Figure 13.15 Coagulation of a colloid. The red colloidal hydrous iron(III) oxide, $Fe_2O_3 \cdot xH_2O$, on the left was obtained by adding concentrated $FeCl_3(aq)$ to boiling water. When a few drops of $Al_2(SO_4)_3(aq)$ are added, the suspended particles rapidly coagulate into a precipitate of $Fe_2O_3 \cdot xH_2O$ (right).

Coagulation can be prevented and a colloidal dispersion made more stable by removing excess ions. This can be done through *dialysis*, a process similar to osmosis. In dialysis, ions pass from the colloidal dispersion through a semipermeable membrane into pure water, but the much larger colloidal particles do not. In *electrodialysis* the process is facilitated by the attractions of ions to an electrode having the opposite charge. A human kidney acts as a dialyzer; it removes excess electrolytes from blood, a colloidal mixture. In some diseases the kidney loses its dialyzing ability, but fortunately this function can be served by a dialysis machine.

SUMMARY

Several different concentrations units are used in describing solutions. In molarity concentration, the amount of solute is expressed in moles and the volume of solution in liters. In molality concentration, the amount of solute is in moles and the quantity of solvent in kilograms. A solution can also be described in terms of what fraction of all the molecules each component contributes; these concentration units are called mole fractions. Various percent concentration units are of practical importance.

A mixture in which all intermolecular forces are of essentially equal magnitude is an ideal solution. Where the magnitudes differ appreciably, solute(s) and solvent remain segregated into a heterogeneous mixture, or, if dissolving does occur, the solution is nonideal. Ion-dipole forces predominate in aqueous solutions of ionic compounds.

The solubility of a solute refers to the concentration of a saturated solution of the solute. In the majority of cases the solubility of a solid solute in water increases with temperature, but a few solutes become less soluble with an increase in temperature. The solubilities of gases in water generally decrease with an increase in temperature. They increase with an increase in the pressure of the gas above the solution.

The vapor pressure of a solvent above a solution is lowered by the presence of a solute(s). The vapor pressure lowering follows Raoult's law for ideal solutions and for dilute solutions even if nonideal. The basis of fractional distillation is that the equilibrium vapor above a solution is richer in the more volatile component than is the solution.

Vapor pressure lowering, freezing point depression, and boiling point elevation (together with osmotic pressure) are colligative properties. They depend on the particular solvent and the concentration of solute particles but not on the identity of the solute.

Osmotic pressure is the pressure that must be applied to a solution to prevent the flow of solvent molecules through a semipermeable membrane from pure solvent into the solution. Osmotic pressure measurements can be used to establish molecular weights, especially of macromolecular materials (polymers).

Equations relating to colligative properties must be modified for solutes that are strong or weak electrolytes. This is done through the van't Hoff factor, *i*. The value of *i* is determined by the extent to which a solute ionizes and the ions interact in solution.

Colloids are dispersions in which the dispersed matter has one or more dimensions in the range from 1 to 1000 nm. Colloids are intermediate between true solutions and heterogeneous mixtures in much of their behavior. One of their distinctive features is the ability to scatter light.

KEY TERMS

colligative property (13.7)
colloid (13.10)
Henry's law (13.5)
hypertonic solution (13.8)
hypotonic solution (13.8)
ideal solution (13.3)
isotonic solution (13.8)
molality (m) (13.2)
mole fraction (χ) (13.2)
mole percent (13.2)
osmosis (13.8)
osmotic pressure (13.8)
parts per billion (ppb) (13.2)
parts per million (ppm) (13.2)
parts per trillion (ppt) (13.2)
Raoult's law (13.6)
reverse osmosis (13.8)
saturated solution (13.4)
semipermeable membrane (13.8)
solubility (13.4)
solubility curve (13.4)
supersaturated solution (13.4)
Tyndall effect (13.10)
unsaturated solution (13.4)
van't Hoff factor (i) (13.9)

REVIEW QUESTIONS

1. Define or explain and, where possible, illustrate these terms:
   a. percent by mass **b.** percent by volume
   c. molality **d.** mole fraction.

2. Define or explain and, where possible, illustrate these terms:
   a. unsaturated solution
   b. saturated solution
   c. supersaturated solution

3. Define or explain and, where possible, illustrate these terms:
   a. colligative property
   b. semipermeable membrane
   c. osmosis
   d. osmotic pressure
   e. isotonic solution
   f. hypotonic solution
   g. hypertonic solution

4. Define or explain and, where possible, illustrate these terms:
   a. suspension **b.** colloid
   c. Tyndall effect **d.** dialysis

5. In a dynamic equilibrium involving a saturated solution, describe the two processes for which the rates are equal.

6. Precipitation is induced in a supersaturated solution by the addition of a seed crystal. When no more solid crystallizes, is the solution saturated, unsaturated, or supersaturated? Explain.

7. List some of the characteristics of an *ideal* solution. Would you expect a solution of methanol (CH_3OH) in water to be ideal? Explain.

8. Arrange the following in order of increasing concentration: 1% by mass, 1 mg/dL, 1 ppb, 1 ppm, 1 ppt.

9. Explain why molarity is temperature dependent whereas molality is not. Is mole fraction temperature dependent? Is volume percent?

10. Would you expect NaCl to dissolve in benzene, C_6H_6? Explain.

11. Would you expect motor oil to dissolve in water? In benzene (C_6H_6)? Explain.

12. Indicate which of the following substances is highly soluble in water, which is slightly soluble, and which is insoluble:
    a. chloroform, $CHCl_3$
    b. benzoic acid, C_6H_5COOH
    c. propylene glycol, $CH_3CHOHCH_2OH$

13. One of the following substances is soluble *both* in water and in benzene, C_6H_6. Which would you expect it to be? Explain.
    a. chlorobenzene, C_6H_5Cl
    b. ethylene glycol, $HOCH_2CH_2OH$
    c. phenol, C_6H_5OH

14. When the cap is removed from a bottle of pop, carbon dioxide gas escapes. Explain why this is so.

15. Would it be a good idea to sterilize the water to be used in a fishbowl by boiling? Explain.

16. The vapor in equilibrium with a pure liquid has the same composition as the liquid. Is this statement also true for solutions? Explain.

17. Contrast a true solution and a colloid in terms of (**a**) the size of the solute particles; (**b**) the nature of the distribution of solute and solvent particles; (**c**) the color and clarity of the solution; (**d**) the Tyndall effect.

18. Compare and contrast a colloidal dispersion and a suspension.

19. Is it possible to have a colloidal dispersion of one gas in another? Explain.

20. Compare and contrast osmosis and dialysis with respect to the kinds of particles that pass through the semipermeable membrane.

21. If two containers of a gas at different pressures are connected, the net diffusion of gas is from the *higher* to the lower pressure. The net diffusion in osmosis is from the solution of *lower* to that of higher concentration. How can you explain this apparent difference?

22. Can a nonvolatile solute and solvent in a solution be separated by filtration? Describe how this separation can be achieved by distillation and by freezing.

23. What is meant by the term *reverse osmosis*? Give an example of its use.

24. The text describes an isotonic solution of NaCl as being about 0.16 M, whereas an isotonic solution of glucose is about 0.31 M. Explain why the concentrations of these isotonic solutions are not the same.

25. What would be the effect on red blood cells if they were placed in (**a**) 5.5% NaCl(aq); (**b**) 0.92% glucose(aq)?

26. For each pair of solutions at 25 °C, indicate which member has the higher osmotic pressure.
a. 0.1 M $NaHCO_3$, 0.05 M $NaHCO_3$
b. 1 M NaCl, 1 M glucose
c. 1 M NaCl, 1 M $CaCl_2$
d. 1 M NaCl, 3 M glucose

27. Arrange the following 0.0010 *m* aqueous solutions in order of *decreasing* freezing point: $Al_2(SO_4)_3$, CH_3COOH, CH_3OH, $MgBr_2$, NaCl.

PROBLEMS

Percent Concentration

28. Describe how you would prepare 5.00 kg of an aqueous solution that is 10.0% NaCl *by mass*.

29. Describe how you would prepare exactly 2.00 L of an aqueous solution that is 2.00% acetic acid *by volume*.

30. Describe how you would prepare 250.0 mL of an aqueous solution of $MgSO_4$ that is 1.5% (mass/vol).

31. What is the mass percent concentration of each of the following solutions?
a. 4.12 g NaOH in 100.0 g water
b. 5.00 mL ethanol (d = 0.789 g/mL) in 50.0 g water
c. 1.50 mL glycerol (d = 1.324 g/mL) in 22.25 mL H_2O (d = 0.998 g/mL)

32. What is the mass percent concentration of each of the following solutions?
a. 175 mg NaCl/g solution
b. 0.275 L methanol (d = 0.791 g/mL)/kg water
c. 4.5 L ethylene glycol (d = 1.114 g/mL) in 6.5 L of propylene glycol (d = 1.036 g/mL)

33. What is the volume percent concentration of each of the following solutions?
a. 35.0 mL of water in 725 mL of ethanol
b. 10.00 g acetone of (d = 0.789 g/mL) in 1.55 L of water
c. 1.05% by mass 1-butanol (d = 0.810 g/mL) in ethanol (d = 0.789 g/mL)

34. What is the volume percent concentration of each of the following solutions?
a. 58.0 mL of water in 625 mL of an ethanol–water solution
b. 10.00 g of methanol (d = 0.791 g/mL) in 75.00 g of ethanol (d = 0.789 g/mL)
c. 24.0% by mass ethanol (d = 0.789 g/mL) in a water solution with d = 0.963 g/mL

35. On average, glucose makes up about 0.10% of human blood, by mass. What is the approximate concentration in mg/dL?

36. A vinegar sample has a density of 1.01 g/mL and contains 5.88% acetic acid by mass. What mass of acetic acid is contained in a one pint bottle of the vinegar? (8 pt = 1 gal; 1 gal = 3.785 L.)

Parts per Million, Parts per Billion, Parts per Trillion

37. Express the following aqueous concentrations in the unit indicated.
a. 1 μg benzene/L water, as ppb benzene
b. 0.0035% NaCl, by mass, as ppm of NaCl
c. 2.4 ppm F^-, as molarity of fluoride ion, $[F^-]$

38. Express the following aqueous concentrations in the unit indicated:
a. 5 μg trichloroethylene/L water, as ppb trichloroethylene
b. 0.0025 g KI/L water, as ppm of KI
c. 45 ppm NO_3^-, as molarity of nitrate ion, $[NO_3^-]$

Molarity and Molality

39. Calculate the *molality* of a solution prepared by dissolving 18.0 g of glucose, $C_6H_{12}O_6$, in 80.0 g of water.

40. Calculate the *molality* of a solution prepared by dissolving 125 mL of pure methanol, CH_3OH, (d = 0.791 g/mL) in 275 g of ethanol.

41. A typical commercial grade phosphoric acid is 75% H_3PO_4, by mass, and has d = 1.57 g/mL. What are the molarity and the molality of H_3PO_4 in this acid?

42. An aqueous solution is prepared in which 3.30 mL of acetone, CH_3COCH_3, (d = 0.789 g/mL) is diluted with water to a final volume of 75.0 mL to produce a solution with a density of 0.993 g/mL. What are the molarity and molality of acetone in this solution?

43. A typical dilute sulfuric acid used in the chemical laboratory is 3.0 M H_2SO_4 and has d = 1.18 g/mL. What is the mass percent of H_2SO_4 in this acid?

44. What mass of ethylene glycol, $HOCH_2CH_2OH$, is present in 2.30 L of 6.27 *m* $HOCH_2CH_2OH$ (d = 1.035 g/mL)?

45. A solution is prepared by mixing 25.00 mL of methanol, CH_3OH (d = 0.791 g/mL), and 25.00 mL of water (d = 0.998 g/mL). The solution can be described as (a) the solute methanol in water or (b) the solute water in methanol. Explain why the molality is different depending on which of these two descriptions is used. In which case, (a) or (b), is the molality greater? (*Hint*: An actual calculation of molalities is not required.)

Mole Fraction and Mole Percent

46. Explain why the mole fraction of solute in a 1.00 *m* *aqueous* solution of a *nonelectrolyte* is independent of the solute. Is the same statement true for a 1.00 M solution? Explain.

47. What is the mole fraction of naphthalene, $C_{10}H_8$, in a solution prepared by dissolving 23.5 g $C_{10}H_8(s)$ in 315 g $C_6H_6(l)$?

48. What is the mole fraction of naphthalene, $C_{10}H_8$, in an 0.250 *m* $C_{10}H_8$ solution in benzene, C_6H_6? (*Hint*: Recall the definition of molality.)

49. What mass of sucrose, $C_{12}H_{22}O_{11}$, must be dissolved per liter of water (d = 0.998 g/mL) to obtain a solution with 2.50 mole percent $C_{12}H_{22}O_{11}$? (*Hint*: Think of the required number of moles of sucrose as an unknown, x.)

50. Which of the following solutions has the greatest mole fraction of solute: (**a**) 1.00 *m* CH_3OH (in aqueous solution); (**b**) 5.0% CH_3CH_2OH, by mass, in water; (**c**) 10.0% sucrose ($C_{12}H_{22}O_{11}$), by mass, in water? Explain.

51. Which of the following aqueous solutions has the greater mass percent of urea, $CO(NH_2)_2$: 0.10 *m* urea or χ_{urea} = 0.010? (*Hint*: Computation can be kept to a minimum.)

Saturated, Unsaturated, and Supersaturated Solutions

52. Use data from Figure 13.6 to determine (**a**) whether a solution containing 48 g NH_4Cl per 100 g water at 60 °C is saturated, unsaturated, or supersaturated; (**b**) the additional mass of water that should be added to a mixture of 35 g K_2CrO_4 and 35 g H_2O at 25 °C to dissolve all the solute; (**c**) the approximate temperature to which a mixture of 50.0 g KNO_3 and 75.0 g water must be heated so that complete dissolving of the KNO_3 occurs.

53. Use data from Figure 13.6 to determine (**a**) the mass percent $NaClO_3$ in a saturated aqueous solution at 50 °C; (**b**) the molality of saturated $Li_2SO_4(aq)$ at 50 °C; (**c**) the additional mass of $Pb(NO_3)_2$ that can be dissolved in an aqueous solution containing 325 mg $Pb(NO_3)_2$/g solution at 30 °C.

54. Can you think of any way(s) in which solute can be crystallized from an unsaturated solution without changing the solution temperature? Explain.

55. Are there any exceptions to the general rule that a supersaturated solution can be made to deposit excess solute by cooling? Explain.

Solubilities of Gases

56. The solubility of $O_2(g)$ in water is 4.43 mg O_2/100 g H_2O at 20 °C when the gas pressure is maintained at 1 atm.

a. What is the molarity of the saturated solution?

b. What pressure of $O_2(g)$ would be required to produce a saturated solution that is 0.010 M O_2?

57. When 250 g of water saturated with air is heated from 20 °C to 80 °C, estimate the quantity of air released in terms of its (**a**) mass, in mg; (**b**) volume, in mL (at STP). Use data from Figure 13.8, and assume 28.96 g/mol as the molar mass of air.

Liquid–Vapor Equilibrium

58. A solution has a 1:4 mole ratio of pentane to hexane. The vapor pressures of the pure hydrocarbons at 20 °C are 441 mmHg for pentane and 121 mmHg for hexane.

a. What are the partial pressures of the two hydrocarbons above the solution?

b. What is the mole fraction composition of the vapor?

59. A solution has a 2:3 mass ratio of toluene (C_7H_8) to benzene (C_6H_6). The vapor pressures of toluene and benzene at 25 °C are 28.4 mmHg and 95.1 mmHg, respectively.

a. What are the partial pressures of the two hydrocarbons above the solution?

b. What is the mole fraction composition of the vapor?

60. What is the vapor pressure of water above an 0.10 *m* glucose ($C_6H_{12}O_6$) solution at 20 °C? Use vapor pressure data from Table 11.2.

61. In many cases the boiling point of a solution is greater than that of the pure solvent. Yet, there are

also many cases where the boiling point of a solution may be less than that of the solvent. Explain this situation.

Freezing Point Depression and Boiling Point Elevation

62. Use data from Table 13.2 to determine the freezing point of each of the following solutions.
a. 0.25 *m* urea in water
b. 0.050 *m* toluene in benzene
c. 10.0% by mass of hexane (C_6H_{14}) in cyclohexane (C_6H_{12})

63. Use data from Table 13.2 to determine (**a**) the freezing point of a solution that is 0.02 M $C_6H_{12}O_6$(aq); (**b**) the molality of a toluene in benzene solution that has a freezing point of 3.85 °C; (**c**) the molality of a naphthalene in benzene solution that has the same freezing point as 0.55 *m* sucrose in water.

64. Which of the following aqueous solutions has the *lower* freezing point, 0.10 *m* glucose or 1.00% 1-butanol ($CH_3CH_2CH_2CH_2OH$) by mass? Explain your reasoning.

65. Which of the following aqueous solutions has the *higher* boiling point, 0.080 M urea [$CO(NH_2)_2$] or $\chi_{urea} = 0.0010$? Explain your reasoning.

66. A 2.11-g sample of naphthalene ($C_{10}H_8$) dissolved in 35.00 g of *para*-xylene has a freezing point of 11.25 °C. The pure solvent has a freezing point of 13.26 °C. What is the K_f of *para*-xylene?

67. A 1.45-g sample of an unknown compound is dissolved in 25.00 mL of benzene (d = 0.874 g/mL). The solution freezes at 4.25 °C. What is the molar mass of the unknown?

68. An unknown compound consists of 33.81% C, 1.42% H, 45.05% O, and 19.72% N by mass. A 1.505-g sample of the compound, when dissolved in 50.00 mL of benzene (d = 0.874 g/mL), lowers the freezing point of the benzene to 4.70 °C. What is the molecular formula of the compound?

Osmotic Pressure

69. Pickles are made by soaking cucumbers in a salt solution. Which has the higher osmotic pressure, the salt solution or the liquid in the cucumber? Explain.

70. Two aqueous solutions of sucrose ($C_{12}H_{22}O_{11}$) are separated by a semipermeable membrane. Solution A has 3.00% sucrose (mass/volume) and solution B is 0.10 M sucrose. In which direction will a net flow of water occur, from solution A to B or from solution B to A?

71. What is the osmotic pressure at 37 °C of an aqueous solution that is 1.80% CH_3CH_2OH (mass/vol). Is this solution isotonic, hypotonic, or hypertonic? [*Hint*: What is the osmotic pressure of a 5.5% (mass/vol) solution of glucose, $C_6H_{12}O_6$?]

72. At 25 °C an 0.325-g sample of polystyrene (a polymer used in plastics materials, e.g., Styrofoam) in 50.00 mL of benzene has an osmotic pressure capable of supporting a column of the solution (d = 0.88 g/mL) 5.3 mm in height. What is the molecular weight of the polystyrene? [*Hint*: What is the height of mercury (d = 13.6 g/mL) equivalent to that of the polymer solution?]

Colligative Properties of Strong Electrolytes, Weak Electrolytes, and Nonelectrolytes

73. Hydrogen chloride is soluble both in water and in benzene. The freezing point *depression* of 0.01 *m* HCl(aq) is about 0.04 °C, and for 0.01 *m* HCl (in benzene) it is about 0.05 °C. Are these the results you would expect for the two solutions? Explain.

74. Predict approximate freezing points of the following aqueous solutions: (**a**) 0.10 *m* glucose ($C_6H_{12}O_6$); (**b**) 0.10 *m* $CaCl_2$; (**c**) 0.10 *m* CH_3COOH; (**d**) 0.10 *m* KI. Which of these predictions is probably the most precise? Explain.

75. The boiling point of water, when the barometric pressure is 744 mmHg, is 99.4 °C. What approximate mass of NaCl would you add to 3.50 kg of the boiling water to raise the boiling point to 100.0 °C?

76. The text lists –21 °C as the lowest temperature at which NaCl can melt ice. What is the approximate mass percent of NaCl in an aqueous solution having this freezing point. Why is the calculation only approximate?

77. Arrange the following aqueous solutions in order of *decreasing* freezing point (from the highest to the lowest) and state your reasons:
(a) 0.15 *m* CH_3COOH
(b) 0.15 *m* $CO(NH_2)_2$ (urea)
(c) 0.10 *m* H_2SO_4
(d) 0.10 *m* $Mg(NO_3)_2$
(e) 0.10 *m* NaBr

78. Arrange the following five solutions in the order of *increasing* boiling elevation, ΔT_b and state your reasons:
(a) 0.25 *m* sucrose(aq)
(b) 0.15 *m* KNO_3(aq)

(c) 0.048 m $C_{10}H_8$ (naphthalene) in benzene
(d) 0.15 m $CH_3COOH(aq)$
(e) 0.15 m H_2SO_4.

Colloids

79. Complete the exercise suggested by the marginal note on page 507 and determine the total surface area associated with all of the cubes, 10 nm on edge, that can be obtained by subdividing a cube of material that is 1 cm on edge.

80. Aluminum sulfate is commonly used to coagulate or precipitate colloidal suspensions of clay particles in municipal water treatment. Why do you suppose this electrolyte is more effective than sodium chloride?

ADDITIONAL PROBLEMS

81. An aqueous solution with density 0.980 g/mL at 20 °C is prepared by dissolving 11.3 mL CH_3OH (d = 0.793 g/mL) in enough water to produce 75.0 mL of solution. What is the percent CH_3OH, expressed as (**a**) volume percent; (**b**) mass percent; (**c**) mass/volume percent; (**d**) mole percent?

82. The liquids water and triethylamine, ($(CH_3CH_2)_3N$), are only partially miscible at 20 °C—that is, triethylamine dissolves in water to some extent and water dissolves in triethylamine to some extent. In a mixture of 50.0 g of water and 50.0 g of triethylamine, 40.0 g of a phase consisting of 84.5% water and 15.5% triethylamine is obtained—a saturated solution of triethylamine in water. What is the percent by mass of water in the second phase—a saturated solution of water in triethylamine?

83. Carbon dioxide, $CO_2(g)$, is much more soluble in NaOH(aq) than it is in water. Can you explain why?

84. Example 13.2 shows that the mass percent ethanol in an ethanol–water solution is less than the volume percent. Would you expect this to be the case for all ethanol–water solutions? for all solutes in water solution? Explain.

85. Use data from Figure 13.8 to establish the solubility of air in water at 20 °C in the units mL air (STP)/100 g water. This is the unit commonly used in listing gas solubilities in handbooks. (*Hint*: Use 28.96 g/mol as the apparent molar mass of air.)

86. You plan to do some molecular weight determinations by freezing point depression. You plan to use 50.00 mL of benzene (d = 0.874 g/mL) as the solvent, and you want the freezing point depression to be between 2 and 3 °C. What mass of the unknown would you use if its anticipated molecular weight is (**a**) 85 u; (**b**) 125 u?

87. Some vitamins are water soluble and some are fat soluble. Indicate which of the following should be a water-soluble vitamin and which should be fat soluble.

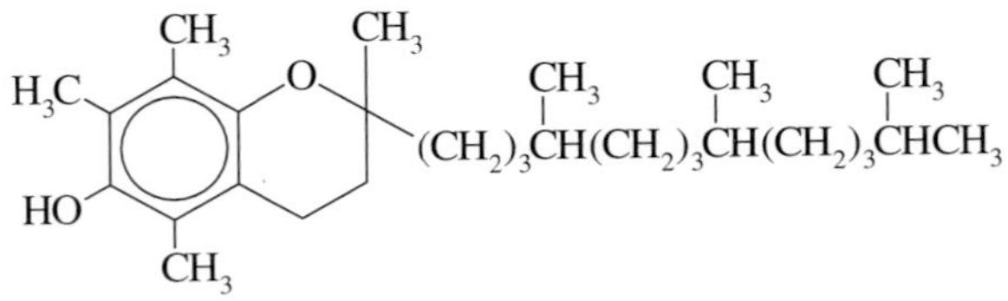

Vitamin E (α-Tocopherol)

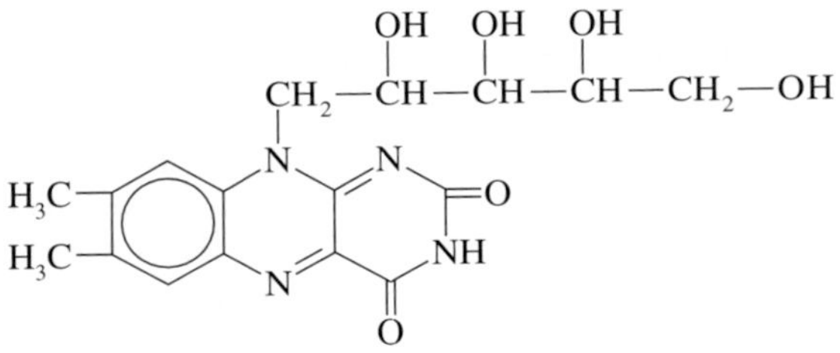

Vitamin B_2 (Riboflavin)

88. An aqueous solution that is 0.205 m in urea [$CO(NH_2)_2$] is observed to boil at 100.025 °C. Is the prevailing barometric pressure above or below standard atmospheric pressure? Explain.

89. In the derivation of equations for freezing point depression and boiling point elevation, these quantities are found to be proportional to the *mole fraction* of solute. In this text we have limited our treatment to solutions that are quite dilute and have written expressions based on the *molality* of solute. Show that in sufficiently dilute solutions the molality of a solute is proportional to its mole fraction.

90. An aqueous, isotonic solution can be described in terms of freezing point depression as well as osmotic pressure. It has a freezing point of –0.52 °C. Show that this definition describes isotonic

NaCl(aq) reasonably well [that is, 0.92% NaCl (mass/vol)]. Describe factors that could produce greater agreement.

91. In Chapter 12 we described a solution that is used in oral rehydration therapy (ORT) (Exercise 12.1, page 440). Show that this solution is essentially isotonic. Use the definition of an isotonic solution given in Problem 90.

92. The solubility of CO_2(g) is 149 mg CO_2/100 g H_2O at 20 °C when the CO_2(g) pressure is maintained at 1 atm. What is the concentration of CO_2 in water that is saturated with air at 20 °C? Express this concentration as mL CO_2 (STP)/100 g water. Use the fact that the mole percent of CO_2 in air is 0.036%.

93. A 1.684-g sample of an unknown oxygen derivative of a hydrocarbon yields 3.364 g CO_2 and 1.377 g H_2O upon complete combustion. A 0.605-g sample of the same compound dissolved in 34.89 g of water lowers the freezing point of the water to −0.244 °C. What is the molecular formula of the compound?

94. Suppose that, initially, solution A in Figure 13.10 is 200.0 g of an aqueous solution with a mole fraction of urea, $CO(NH_2)_2$, of 0.100 and solution B is 100.0 g of an aqueous solution with a mole fraction of urea of 0.0500. What will be the mass and mole fraction concentration of each solution when equilibrium is established, that is, when there is no longer a net transfer of water between the two solutions?

Weather phenomena, such as this typhoon over the Pacific Ocean southeast of New Zealand, occur in the troposphere, the region of the atmosphere closest to the Earth's surface.

14

Chemicals in Earth's Atmosphere

CONTENTS

Astronauts have walked on the surface of Earth's barren, airless moon. Photographs of Mercury, the planet nearest the sun, from spacecraft show a dusty desolation much like that of the moon. Robotic probes have dropped through clouds of sulfuric acid and a thick blanket of carbon dioxide to land on the hot, inhospitable surface of Venus. Other space probes have descended through the thin, dusty atmosphere of Mars in a vain search for life on its surface, and they have examined the crushing, turbulent atmospheres of Jupiter, Saturn, Uranus, and Neptune. Pluto has also been studied, but only from vast distances. Simple life forms exist in many seemingly inhospitable places on Earth, from hot springs to parched deserts to Antarctic ice, and similar organisms could possibly exist elsewhere in the solar system; however, Earth alone has an atmosphere hospitable to higher life forms as we know them.

In this chapter we resume a discussion of descriptive chemistry that we began in Chapter 8. Here we examine the chemistry of some elements obtained from the atmosphere, some of their compounds, and some substances introduced into the atmosphere by human activities.

14.1 The Atmosphere

Without food, we can live about a month; without water, only a few days. But without air we would die within minutes. Air is vital to humans because it contains free oxygen (O_2), an element essential to the basic processes of respiration and metabolism. But we need more than the oxygen in air; life as we know it could not exist in an atmosphere of pure oxygen. The oxygen in air is diluted with nitrogen; this lessens the tendency for everything in contact with air to become oxidized. Carbon dioxide and water vapor are but minor components in air, yet they are the primary raw materials of the plant kingdom, and plants produce the food on which we and all other animal life depend. And even ozone, a gas present only in trace quantities, plays vital roles in shielding Earth's surface from harmful ultraviolet radiation and in maintaining a proper energy balance in the atmosphere.

Air, the material that makes up the lower atmosphere, is a mixture of gases. On a mole percent basis, dry air consists of about 78% N_2, 21% O_2, and 1% Ar. Among the minor constituents of dry air, the most abundant is carbon dioxide. The concentration of CO_2 in air has increased from about 275 ppm in 1880 to its present value of more than 359 ppm. The concentration most likely will continue to rise as more and more fossil fuels (coal, oil, and natural gas) are burned. The composition of dry air is summarized in Table 14.1, and some phenomena associated with water vapor in moist air are presented in Section 14.2.

Applied to gaseous mixtures, ppm, ppb, and ppt are on a number basis. That is, 359 ppm CO_2 signifies that of every million molecules of air, 359 are CO_2 molecules. In liquid solutions, on the other hand, ppm, ppb, and ppt are generally on a mass basis.

TABLE 14.1 **Composition of Dry Air (Near Sea Level)**

Component	Mole Percent[a]
Nitrogen (N_2)	78.084
Oxygen (O_2)	20.946
Argon (Ar)	0.934
Carbon dioxide (CO_2)	0.0359
Neon (Ne)	0.001818
Helium (He)	0.000524
Methane (CH_4)	0.0002
Krypton (Kr)	0.000114
Hydrogen (H_2)	0.00005
Dinitrogen monoxide(N_2O)	0.00005
Xenon (Xe)	0.000009

Plus traces of:
- Ozone (O_3)
- Sulfur dioxide (SO_2)
- Nitrogen dioxide (NO_2)
- Ammonia (NH_3)
- Carbon monoxide (CO)
- Iodine (I_2)

[a] The compositions of gaseous mixtures are often expressed in percentages by volume. Volume percent and mole percent compositions are the same.

Altogether, the thin blanket of gases that make up the atmosphere are spread over a surface area of 5.0×10^8 km^2 and have a mass of about 5.2×10^{15} metric tons (1 metric ton = 1000 kg). That is about 10 million tons of air over each square kilometer of surface, 10 tons over each square meter, or about 1 ton (nearly the mass of one small automobile) over each square foot.

How deep is the atmosphere? It is hard to say, for the atmosphere does not end abruptly but gradually fades away as the distance from Earth's surface increases. We know, though, that 99% of the mass of the atmosphere lies within 30 km of Earth's surface—a thin layer of air indeed, akin to the peel of an apple, only relatively thinner.

Rather arbitrarily, we divide the atmosphere into the layers shown in Figure 14.1. The layer nearest Earth's surface, the *troposphere*, extends for about 12 km above the surface and contains about 90% of the mass of the atmosphere. Weather occurs in the troposphere, which also harbors almost all living things

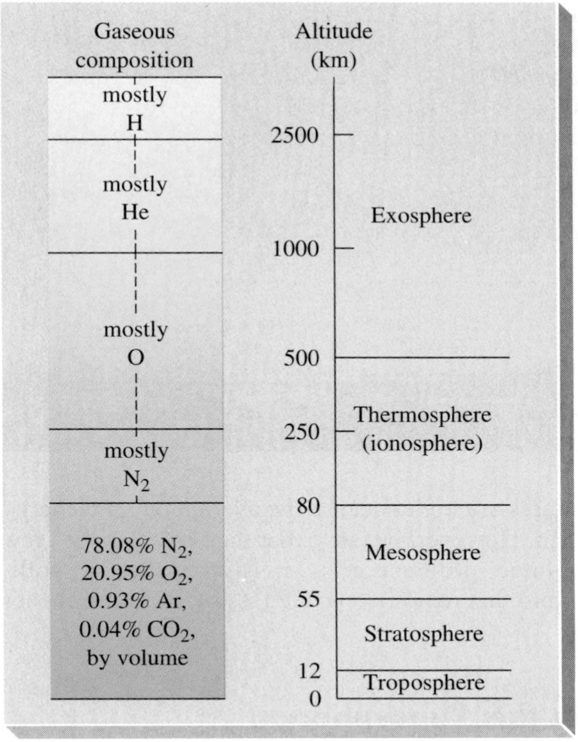

Figure 14.1 Layers of the atmosphere. The altitudes of the several layers of the atmosphere are only approximate. For example, the height of the troposphere varies from about 8 km at the poles to 16 km at the equator. The approximate temperatures in these layers are cited in the text.

and nearly all human activity. Temperatures generally decline with altitude in the troposphere, ranging from a maximum of about 320 K at Earth's surface to a minimum of about 220 K in the upper troposphere.

The next layer, the *stratosphere,* extends from about 12 to 55 km above Earth's surface. The ozone layer that shields living creatures from deadly ultraviolet radiation lies within the stratosphere, from about 15 to 30 km above Earth's surface. Supersonic aircraft fly in the lower part of the stratosphere. Temperatures in the stratosphere are fairly constant at about 220 K from 12 to 25 km, then rise with increasing altitude to about 280 K at 50 km.

The density of air is about 1.3 *grams* per liter at sea level, but air gets "thinner" (less dense) with altitude. Beyond the stratosphere the density drops into the range of *micrograms* and even *nanograms* per liter. The next layer above the stratosphere, ranging from about 55 to 80 km, is called the *mesosphere*. Here the temperature falls continuously to about 180 K. The layer above the mesosphere is known as the *thermosphere* or *ionosphere*. In this region molecules can absorb such highly energetic electromagnetic radiation from the sun that they dissociate into atoms; some of the atoms are further broken down into positive and negative ions and free electrons.

Temperatures in the ionosphere (thermosphere) rise to about 1500 K,* but high temperatures in this region don't have the same significance as on

Mt. Cayambe in Ecuador is only a few kilometers from the equator, but its peak is covered with snow, illustrating the decrease in temperature with increased altitude in the troposphere. At an altitude of 5.79 km, the top of the mountain extends about halfway through the troposphere.

*The temperature in the thermosphere depends upon the amount of solar radiation and varies between night and day and with sunspot activity.

This meteor seen against a starry sky at dusk is an extraterrestrial chunk of matter that has entered Earth's atmosphere. It emits light because it is heated to a high temperature. This heating does not occur in the high-temperature thermosphere, however. It occurs through the frictional resistance the particle encounters as it passes through gases lower in the atmosphere (about 80 to 100 km).

Earth's surface. The temperatures are high because the average kinetic energy of the gaseous particles is high. However, because there are so relatively few of these particles per unit volume, little energy is transferred through collisions. A cold object brought into this region does not get hot; it never reaches temperature equilibrium.

14.2 Water Vapor in the Atmosphere

Unless it has been specially dried, air invariably contains water vapor. Atmospheric water vapor is one of the key players in the **hydrologic (water) cycle**—the series of natural processes by which water is recycled through the environment (see Figure 14.2).

Humidity

The proportion of water vapor in air is quite variable. It ranges from trace amounts to about 4% by volume. The general term describing the water vapor content of air is *humidity*. The *absolute* humidity is the actual quantity of water vapor present in an air sample, usually expressed in g H_2O/m^3 air. *Relative* humidity expresses water vapor content as a percentage of the maximum possible. Relative humidity compares the actual partial pressure of water vapor in an air sample to the maximum partial pressure that could exist at the given temperature—the vapor pressure of water. That is,

$$\text{Relative humidity (R.H.)} = \frac{\text{partial pressure of water vapor}}{\text{vapor pressure of water}} \times 100\%$$

A number of different experimental methods exist for determining the relative humidity of air. A crude but colorful method is illustrated in Figure 14.3.

Example 14.1

The partial pressure of water vapor in an air sample at 20.0 °C is found to be 12.8 mmHg. What is the relative humidity of this air?

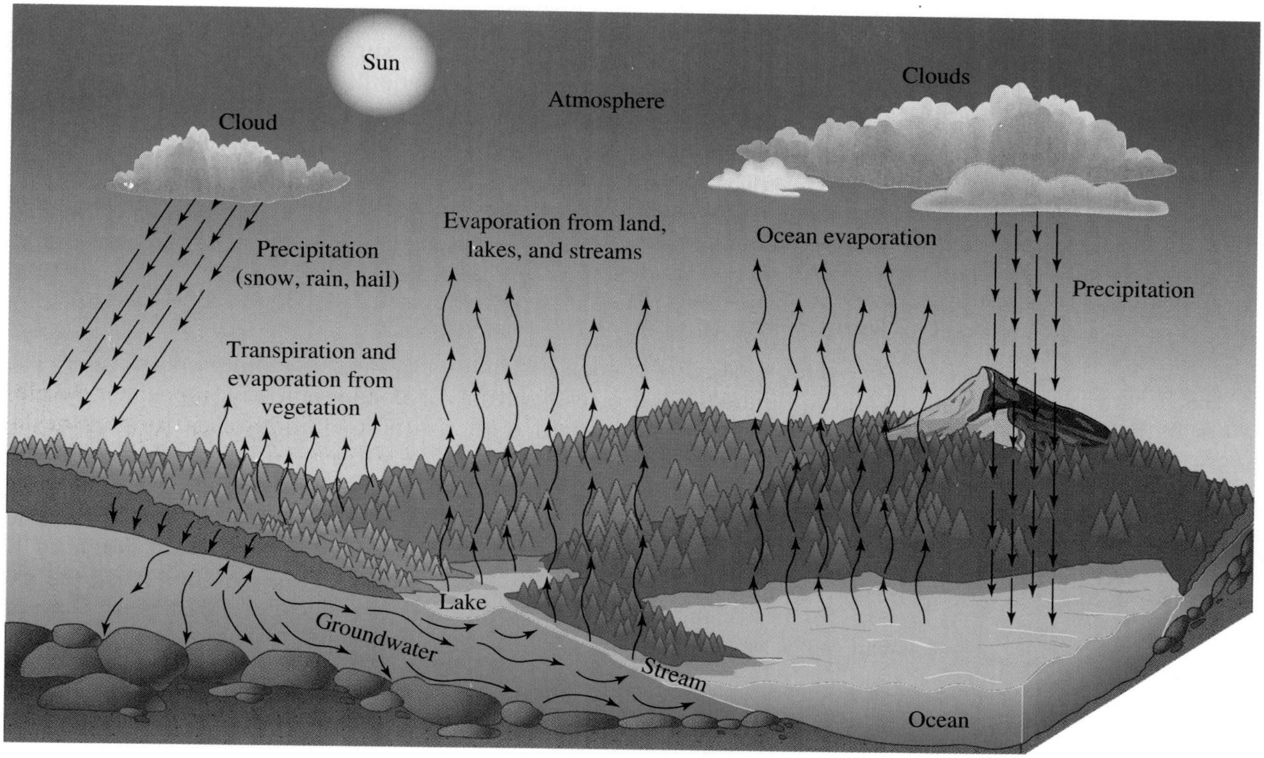

Figure 14.2 The hydrologic (water) cycle. The oceans on Earth are vast reservoirs of water. Water evaporates from the oceans, producing moist air masses that move over the land. When the moist air is cooled, the water vapor forms clouds and the clouds produce rain. Rainwater replenishes groundwater and forms lakes and streams. The water eventually returns to the ocean. Water is also returned to the atmosphere by evaporation throughout the cycle.

Solution

We need two quantities to establish the relative humidity: the partial pressure of water vapor in an air sample (here, 12.8 mmHg) and the vapor pressure of water at the given temperature (here, 20.0 °C). We can find the vapor pressure of water in a tabulation like Table 11.2.

$$\text{R.H.} = \frac{12.8 \text{ mmHg}}{17.5 \text{ mmHg}} \times 100\% = 73.1\%$$

Exercise 14.1

If an air sample at 20.0 °C has a relative humidity of 38.5%, what is the partial pressure of water vapor in this sample?

Figure 14.3 A measure of relative humidity. The strips of filter paper were impregnated with an aqueous solution of cobalt(II) chloride and allowed to dry. In dry air the strip is blue, the color of *anhydrous* $CoCl_2$. In more humid air, the strip acquires the red color of the *hexahydrate*, $CoCl_2 \cdot 6H_2O$.

If we warm the air sample described in Example 14.1, the relative humidity *decreases*. This is because the vapor pressure of water increases with temperature, whereas the partial pressure of water vapor in the air sample would remain approximately constant.

If, on the other hand, we cool the air sample in Example 14.1, the relative humidity *increases*. If the vapor pressure of the water remains at 12.8 mmHg, the relative humidity at 15.0 °C is 100%. We can say that the air is now satu-

Dill covered with morning dew. When the temperature of air in contact with a surface falls to the point that the absolute humidity of the air is greater than the vapor pressure of water, water vapor condenses in a characteristic and familiar form—dew.

rated with water vapor. At temperatures below 15.0 °C, the relative humidity would exceed 100%, and the air would be supersaturated with water vapor. This is an unstable situation that is not at equilibrium; it cannot persist. Some of the water vapor would condense as droplets of liquid water that we call *dew*. The highest temperature at which condensation of water vapor from an air sample can occur is known as the *dew point*. When the dew point is below the freezing point of water (0 °C), the water condenses as *frost* without going through the liquid state.

Deliquescence

The water vapor that condenses from air at the dew point is pure water. The liquid in Figure 14.4 also results from the condensation of atmospheric water vapor, but it is not pure water. Rather, it is a solution of calcium chloride, formed in this way: If the partial pressure of water vapor in the air exceeds that of a saturated solution of $CaCl_2 \cdot 6H_2O$, water vapor condenses on the solid $CaCl_2 \cdot 6H_2O$ and dissolves some of it, producing some saturated solution. This condensation of water vapor on a solid followed by solution formation is called **deliquescence**; it continues until all the solid has dissolved and the vapor pressure of the solution (now unsaturated) equals the partial pressure of water vapor in the air.

14.3 Nitrogen

Figure 14.4 Deliquescence of calcium chloride. Water vapor from the air condenses onto the solid $CaCl_2{\cdot}6H_2O$ and produces a solution of $CaCl_2(aq)$. Here, the solution is saturated, but eventually all the solid would dissolve and the solution would become unsaturated. The *deliquescence* of $CaCl_2{\cdot}6H_2O$ occurs only when the relative humidity exceeds 32%. Other water soluble solids deliquesce under other conditions of relative humidity.

Nitrogen is one of the few elements (the noble gases are the others) found in greater abundance in the atmosphere than anywhere else. The mass percent of nitrogen in Earth's atmosphere is 76.8%, but that in Earth's solid crust is only 0.002%. There are only two important mineral sources of nitrogen—KNO_3 (niter or saltpeter) and $NaNO_3$ (sodaniter or Chile saltpeter). Because both of these compounds are water soluble, they are found on Earth's surface mainly in a few desert regions. Nitrogen compounds occur in all living matter, and some organic nitrogen compounds can be extracted from plant or animal sources or from the fossilized remains of ancient plant life, such as coal.

For some uses, as in the synthesis of ammonia, pure nitrogen is not required; air (78.1 mol % N_2) can be used instead. In other uses, such as providing an inert blanketing atmosphere for metallurgical operations, pure nitrogen is essential. The only important commercial method of producing nitrogen is the fractional distillation of liquid air (see Figure 14.5).

Oxidation States of Nitrogen: Nine Possibilities

Nitrogen atoms have the electron configuration $1s^22s^22p^3$. In combining with other atoms, nitrogen atoms acquire a noble gas electron configuration (that of neon; $1s^22s^22p^6$) by gaining or (more likely) sharing *three* electrons. It is also possible for a nitrogen atom to use all *five* of its valence electrons in bond formation. As a result, nitrogen can display all oxidation states from −3 to +5. Examples of nitrogen atoms in each of these oxidation states were given in Figure 12.9 (page 460). We again encounter a few of these oxidation states in the nitrogen compounds discussed in this section.

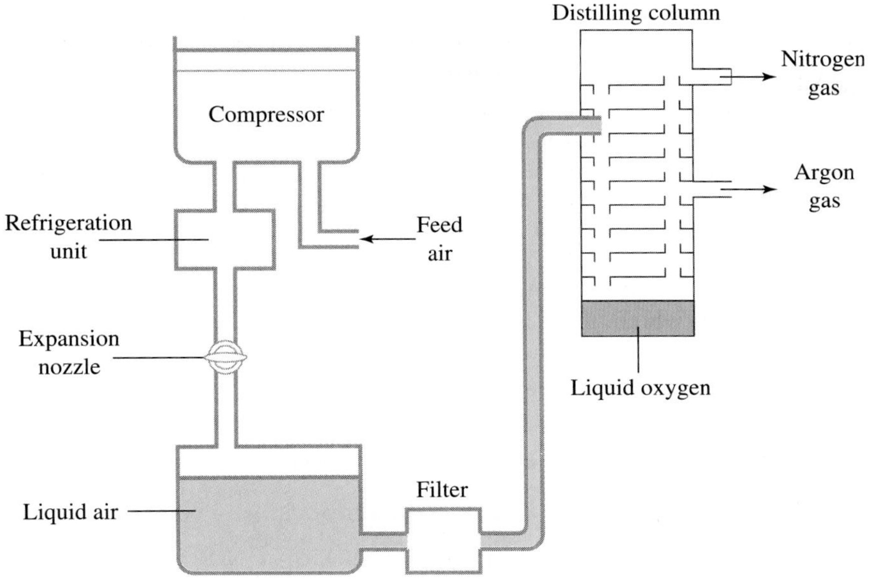

Figure 14.5 The fractional distillation of liquid air. Clean air is compressed and then cooled by refrigeration. Upon expanding, the air further cools and liquefies. The liquid air is filtered to remove CO_2(s) and then distilled. Nitrogen is the most volatile component, with a normal boiling point of 77.4 K (−195.8 °C); it passes off as a gas. Gaseous argon, which boils at 87.5 K (−185.7 °C) is removed from the middle of the column, and liquid oxygen, the least volatile component with a normal boiling point of 90.2 K (−183.0 °C), collects at the bottom of the column.

Bonding in the N_2 Molecule

When two nitrogen atoms combine to form a molecule of N_2, they acquire valence-shell octets by forming a *triple* covalent bond between them. The nitrogen-to-nitrogen triple bond is one of the strongest chemical bonds known. The rupture of one mole of these bonds requires the absorption of 945.4 kJ of energy.

$$N{\equiv}N(g) \longrightarrow 2\,N(g) \qquad \Delta H = +945.4 \text{ kJ}$$

Chemical reactions in which strong bonds are replaced by weaker bonds are often *endothermic*. This is because the energy required to break bonds is not offset by the energy released in forming new bonds. As a result, many nitrogen compounds have positive enthalpies of formation. For example,

$$N_2(g) + O_2(g) \longrightarrow NO(g) \qquad \Delta H_f^\circ = +90.25 \text{ kJ}$$

Generally, highly endothermic reactions do not occur to any appreciable extent, and this is why N_2(g) and O_2(g) can coexist so peacefully in the atmosphere. If the formation of NO(g) from N_2(g) and O_2(g) occurred to any appreciable extent at room temperature, which we would expect if the formation reaction were exothermic, the amount of this highly noxious gas in the atmosphere would be significant. At the same time, the amount of life-sustaining O_2(g) would be reduced. There would be no life as we know it on Earth.

Nitrogen Fixation: The Nitrogen Cycle

Nitrogen is an essential element to living organisms; it occurs in all proteins and in many other biologically important molecules. Although it exists in large quantities in the atmosphere, N_2(g) cannot be used directly by higher plants or animals. The N_2 molecules first must be "fixed," that is, converted to compounds that are more readily usable by living organisms. This conversion of atmospheric nitrogen into nitrogen compounds is called **nitrogen fixation**.

Nitrogen fixation occurring during electrical storms is an important part of the natural nitrogen cycle.

Certain bacteria, found in water (in blue-green algae, for example) or soil or attached to root nodules of leguminous plants (clover, soybeans, peas, . . .), are able to convert atmospheric nitrogen to ammonia. Other plants take up nitrogen atoms in the form of nitrate ion (NO_3^-) or ammonium ion (NH_4^+). Nitrogen atoms in plants, combined with carbon compounds from photosynthesis, form amino acids, the building blocks of proteins (page 420). The food chain for animals originates with plant life, and the decay of plant and animal life returns nitrogen to the environment as nitrates and ammonia. Eventually nitrogen is returned to the atmosphere as N_2 by the action of *denitrifying* bacteria on nitrates.

Lightning also "fixes" some atmospheric nitrogen by creating a high-energy environment in which nitrogen and oxygen can combine. Nitrogen monoxide and nitrogen dioxide are formed in this way.

$$N_2(g) + O_2(g) \xrightarrow{\text{lightning}} 2\,NO(g)$$

$$2\,NO(g) + O_2(g) \longrightarrow 2\,NO_2(g)$$

Nitrogen dioxide reacts with water to form nitric acid.

$$3\,NO_2(g) + H_2O(l) \longrightarrow 2\,HNO_3(aq) + NO(g)$$

The nitric acid falls in rainwater, adding to the available nitrates in sea and soil. The net effect of all these natural activities is a constant recycling of nitrogen atoms in the environment in a **nitrogen cycle** (see Figure 14.6).

Scientists and engineers have intervened in the nitrogen cycle by industrial fixation, mainly by the manufacture of nitrogen fertilizers. This intervention has greatly increased the world's food supply because the availability of fixed nitrogen is often the limiting factor in the production of food. Not all the consequences of this intervention have been favorable, however. Excessive runoff of dissolved nitrogen fertilizers has led to serious water pollution problems in some areas, but modern methods of high-yield farming demand "chemical" fertilizers.

Nitrogen Fixation: The Synthesis of Ammonia

For centuries, farmers depended on manure as a source of fixed nitrogen. Later, the discovery of deposits of sodium nitrate (Chile saltpeter) in the deserts of northern Chile led to use of this substance as a supplemental source of fixed nitrogen. But these sources were still not sufficient; a rapid rise in population in the late nineteenth and early twentieth centuries placed increasing pressure on the available food supply. The atmosphere was seen as an almost unlimited source of nitrogen, if only this nitrogen could be converted to a useful compound form.

The crucial breakthrough in nitrogen fixation came through the efforts of Fritz Haber in Germany in 1908, on the eve of World War I. Haber worked out the reaction conditions for the combination of nitrogen and hydrogen to make ammonia. Carl Bosch, also in Germany, did the necessary engineering work required to transform Haber's laboratory studies into a commercial manufacturing process.

Choosing reaction conditions for the ammonia synthesis reaction requires balancing the need to speed up a chemical reaction (Chapter 15) and the need to obtain the maximum yield of a product in a reversible reaction (Chapter 16). We will consider this matter again later in the text.

The basic problem in ammonia synthesis is that under most conditions the reaction does not go to completion. It is a *reversible* reaction.

$$N_2(g) + 3\,H_2(g) \rightleftharpoons 2\,NH_3(g)$$

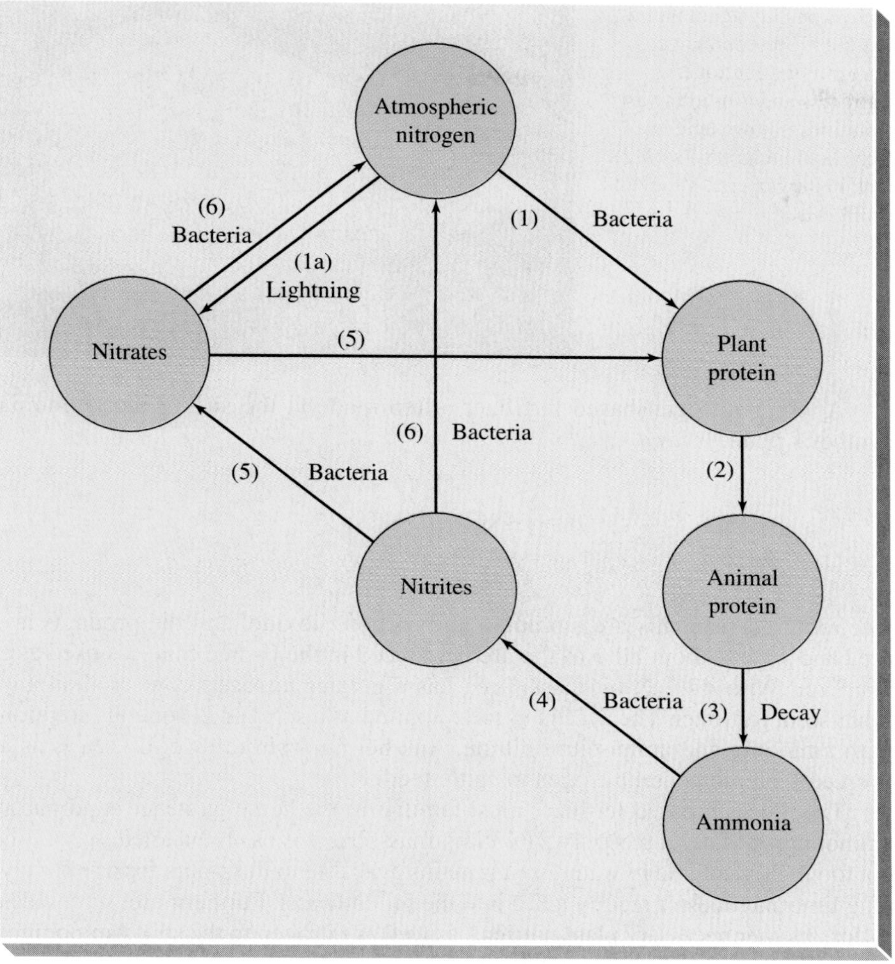

Figure 14.6 The nitrogen cycle. Bacteria fix atmospheric nitrogen and convert it through chemical reactions into plant proteins (1). Plants also convert nitrates to proteins (5). Animals feed on plants and other animals (2). The decay of plant and animal proteins produces ammonia (3), which through a series of bacterial actions is converted to nitrites and nitrates (4 and 5). Denitrifying bacteria decompose nitrites and nitrates, returning N_2O and N_2 to the atmosphere (6). Some atmospheric nitrogen is converted to nitrates during electric storms (1a). A significant amount of nitrogen fixation and denitrification occurs in oceans, and modern manufacturing and agricultural processes also play important roles in the cycle.

To get a good yield of ammonia, a high temperature (400 °C), a high pressure (about 200 atm), and an appropriate catalyst (Fe_3O_4 with small amounts of Al_2O_3, MgO, CaO, and K_2O) are required. Also, by (1) liquefying and thus removing the ammonia to prevent reaction reversal and (2) recycling unreacted $N_2(g)$ and $H_2(g)$, it is possible to get essentially a 100% yield. Air is the source of N_2 for the process. In some processes, pure N_2 is obtained by fractional distillation of liquid air, but in most cases air is used directly. The oxygen gas is removed by the partial combustion of a fuel, leaving mostly N_2. The required $H_2(g)$ is usually made from natural gas (page 276).

Fertilizers: Ammonia and Compounds Derived from It

Ammonia usually ranks fifth or sixth, by mass, among chemicals produced in the United States. Worldwide, annual production is about 140 million tons. About 85% of this production goes into the manufacture of fertilizers. Pure liquid ammonia, in a form called *anhydrous ammonia*, is the most used of nitrogen-based fertilizers because of its exceptionally high (82%) mass percent of nitrogen.

Anhydrous liquid ammonia has the highest percent of nitrogen of all nitrogen fertilizers and is used in vast quantities. Shown here is a large liquid ammonia storage tank in the Imperial Valley of California.

Another nitrogen-based fertilizer, often made at the site of the ammonia synthesis plant, is *urea*.

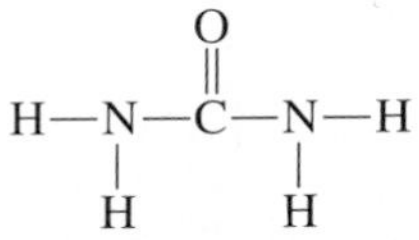

The required reactants are ammonia and carbon dioxide, and the products are urea and water. About 80% of the urea produced in the United States is used as a fertilizer. With 47% N by mass, urea has a greater nitrogen content than any other solid fertilizer. The urea is usually applied as a solid or in aqueous solution with ammonia and ammonium nitrate. Another agricultural use of urea is as a source of supplemental nitrogen in cattle feed.

The nitrogen-based fertilizer most familiar to the home gardener is probably ammonium sulfate. It has only 21% N by mass, but it is easily handled. Also, it is not too highly soluble in water, so it remains available to the plants for a relatively long time and doesn't readily leach into the groundwater. Furthermore, it provides sulfur, another necessary plant nutrient, as well as nitrogen to the soil. Ammonium sulfate can be made by the reaction of ammonia and sulfuric acid, but mostly it is obtained as a byproduct of other chemical manufacturing processes.

Another group of dual-purpose fertilizers are the ammonium phosphates, produced by neutralizing phosphoric acid with aqueous ammonia. These fertilizers supply the two vital plant nutrients N and P. They represent one of the fastest growing segments of the fertilizer industry. Since 1960, U.S. production has increased about 25-fold. Currently ammonium phosphates account for about two-thirds of the total production of phosphate fertilizers in the United States. An example of the ammonium phosphates is ammonium hydrogen phosphate, $(NH_4)_2HPO_4$, referred to commercially as *diammonium phosphate (DAP)*.

Estimation Exercise 14.1

Without the use of a calculator, show that the mass percents N in ammonium sulfate and diammonium phosphate (DAP) are nearly identical. Which of the two has the slightly higher percent N?

By 1913, after the development of the Haber process for ammonia, the Germans were able to make ammonium nitrate by neutralizing $NH_3(aq)$ with

Fritz Haber

Fritz Haber (1868–1934) was awarded the 1919 Nobel Prize in chemistry for establishing the foundation for the synthesis of ammonia. The award, strongly denounced by many scientists, was ironic in many ways. Alfred Nobel, a Swedish chemist and the inventor of dynamite, endowed the Nobel Prizes (including the Nobel Peace Prize) with a fortune derived from his own work with explosives. During his lifetime, Nobel was bitterly disappointed that the explosives he developed for excavation and mining were put to such disastrous use in wars.

Haber's motivation in developing the ammonia synthesis reaction was partly to make Germany independent of nitrate supplies from Chile. This independence made possible the almost unlimited production of explosives and prolonged World War I by at least two years. Haber contributed to the German war effort in other direct ways as well. In 1915 he supervised the release of chlorine gas in the first poison gas attack on Allied troops, and two years later, the much more lethal mustard gas.

After the war Haber tried to use his chemical knowledge to benefit his country by attempting, unsuccessfully, to develop a process for extracting gold from seawater. This would have provided Germany with the means to pay war reparations. Despite his efforts in behalf of his homeland, Haber, a Jew, was forced out of his position as Director of the Kaiser Wilhelm Institute of Physical Chemistry by Nazi racial laws and driven into exile in 1933. Haber's life ended with a heart attack less than a year later.

HNO_3(aq). They were mainly interested in ammonium nitrate as a high explosive, but it turned out to be a valuable nitrogen fertilizer as well. The same discovery that enabled Germany to prolong World War I probably helped postpone predicted world famine for several decades thereafter.

It is largely through nitrogen fertilizers that we are able to feed the large human population that inhabits Earth. Unfortunately, our current ways of fixing nitrogen require great quantities of energy. The petroleum shortages and higher energy prices of the 1970s led to scarcities and an inflation in fertilizer prices. Many scientists are currently looking for ways to fix nitrogen gas at lower temperatures, the way microorganisms do. Others seek to modify the bacteria that fix nitrogen in legumes so that they can do the same thing in corn and other plants.

Example 14.2

Information about chemical reactions is frequently given only in words. You may have to supply equations. Write chemical equations for the commercial preparation of two of the fertilizers described in this section: (**a**) diammonium phosphate (DAP) and (**b**) urea.

Solution

Refer to the paragraphs in which these substances are described. The information given about the first reaction indicates that aqueous solutions are involved. Reaction conditions for the second reaction are not stated.

a. $2\ NH_3(aq) + H_3PO_4(aq) \longrightarrow (NH_4)_2HPO_4(aq)$

b. $2\ NH_3 + CO_2 \longrightarrow NH_2CONH_2 + H_2O$

Friedrich Wilhelm Ostwald (1853–1932), a Russian-German chemist, was one of the of the principal founders of modern physical chemistry. He championed Arrhenius's views on solutions of electrolytes and acid–base theory, and he translated into German the scientific papers of J. Willard Gibbs, which established thermodynamics as a central organizing theme of physical chemistry. He was awarded the Nobel Prize in 1909 for his work on catalysis.

Exercise 14.2
Write plausible equations for the preparation of (**a**) ammonium nitrate and (**b**) ammonium sulfate.

Nitric Acid and Nitrates

Another process in nitrogen chemistry, one that is perhaps second in importance only to the ammonia synthesis reaction, was developed by Friedrich Wilhelm Ostwald. The Ostwald process is the catalyzed oxidation of $NH_3(g)$ to $NO(g)$.

$$4\,NH_3(g) + 5\,O_2(g) \xrightarrow{\text{Pt/Rh catalyst}} 4\,NO(g) + 6\,H_2O(g)$$

This is the first step in the commercial preparation of nitric acid. The additional steps involve the oxidation of $NO(g)$ to $NO_2(g)$ and the reaction of $NO_2(g)$ with water.

$$2\,NO(g) + O_2(g) \longrightarrow 2\,NO_2(g)$$
$$3\,NO_2(g) + H_2O(l) \longrightarrow 2\,HNO_3(aq) + NO(g)$$

The $NO(g)$ formed in the second reaction is recycled in the first reaction.

Some of the industrial uses of nitric acid, such as the manufacture of ammonium nitrate, are based on its acidic properties, namely its ability to neutralize bases.

$$NH_3(aq) + HNO_3(aq) \longrightarrow NH_4NO_3(aq)$$

Most of the nitric acid produced in the United States goes into the manufacture of ammonium nitrate and other fertilizers.

Nitric acid is also used as an *oxidizing agent*. The actual agent is nitrate ion, NO_3^-, in an acidic solution. For example, when metals react with nitric acid, the metal is oxidized to aqueous metallic ions, but hydrogen gas is rarely obtained as a reduction product. The product has nitrogen atoms in an oxidation state lower than +5: usually $NO(g)$ or $NO_2(g)$, but even $N_2O(g)$ or NH_4^+ in some instances. The reaction of copper with nitric acid was the subject of Conceptual Example 12.3 (page 462).

$$Cu(s) + 4\,H^+(aq) + 2\,NO_3^-(aq) \longrightarrow Cu^{2+}(aq) + 2\,NO_2(g) + 2\,H_2O$$

Ammonium nitrate has nitrogen in both the oxidation states −3 (in NH_4^+) and +5 (in NO_3^-). This means that it can participate in oxidation–reduction reactions with no other agent required to react with it. Nitrate ion is an oxidizing agent, and ammonium ion is a reducing agent. When heated gently at about 200 °C, ammonium nitrate forms dinitrogen monoxide and water.

$$NH_4NO_3(l) \xrightarrow{\Delta} N_2O(g) + 2\,H_2O(g)$$

Stronger heating or mechanical shock can induce a much more vigorous reaction that is the basis of ammonium nitrate's use as an explosive.

$$2\,NH_4NO_3(l) \xrightarrow{\Delta} 2\,N_2(g) + 4\,H_2O(g) + O_2(g)$$

The explosive power of ammonium nitrate is increased severalfold by mixing it with fuel oil.

The explosive potential of ammonium nitrate in contact with organic materials was dramatically revealed in an enormous explosion of a cargo ship in Texas City, Texas, in 1947. Nearly 600 people were killed and several thousand injured. Explosive mixtures of ammonium nitrate fertilizer and fuel oil were used in the terrorist attacks on the World Trade Center in New York in 1993 and on the federal building in Oklahoma City in 1995.

Oxides of Nitrogen and Air Pollution

The common oxides of nitrogen include examples of nitrogen in every oxidation state from +1 to +5. Their names, formulas, and some uses are listed in Table 14.2. Three of these oxides—N_2O, NO, NO_2—were known before Dalton's time. In fact, Dalton cited them in establishing his law of multiple proportions (page 33).

TABLE 14.2 **Some Common Oxides of Nitrogen**

O.S. of N	Formula	Name	Uses
+1	N_2O	dinitrogen monoxide (nitrous oxide)	anesthetic in dentistry and surgery; propellant in aerosol cans; production of sodium azide for use in air bags in automobiles.
+2	NO	nitrogen monoxide (nitric oxide)	intermediate in the production of nitric acid; bleaching rayon.
+3	N_2O_3	dinitrogen trioxide	oxidant in specialty fuels.
+4	NO_2	nitrogen dioxide	intermediate in the production of nitric acid; nitrating agent; catalyst.
+4	N_2O_4	dinitrogen tetroxide	oxidant in rocket fuels for space exploration.
+5	N_2O_5	dinitrogen pentoxide	(no common uses.)

Earlier in this section we considered methods of producing three of the oxides of nitrogen: $N_2O(g)$ by the mild decomposition of ammonium nitrate, and NO(g) and $NO_2(g)$ by the reduction of nitrate ion in acidic solution (such as in the reactions of metals with nitric acid). NO(g) is produced commercially by the oxidation of ammonia in the Ostwald process.

The roles of NO(g) in air pollution and as an intermediate in the production of nitric acid have been well known for decades. In 1992, to the surprise of everyone, NO was shown to be an important messenger molecule in the human body. It is the first gas to have been identified as a messenger in mammals. NO is essential to maintaining blood pressure. It aids the immune response in killing foreign invaders. It mediates the relaxation phase of intestinal contractions in the movement of food through the digestive tract. NO is essential to the establishment of long-term memory, and it dilates the blood vessels that allow blood flow into the penis to cause an erection. Much remains to be learned about the role of this fascinating molecule in our bodies.

These and many other findings led *Science* magazine to name NO the 1992 molecule of the year.

Photochemical Smog

One *unwelcome* source of NO(g) is high-temperature combustion in air.

$$N_2(g) + O_2(g) \longrightarrow 2\ NO(g)$$

This reaction takes place in power plants that burn fossil fuels and in incinerators. The greatest source of such NO(g), though, is in the exhaust fumes of high-compression, high-temperature, internal-combustion automobile engines.

An **air pollutant** is a substance found in air in greater abundance than normally occurs naturally and that has some harmful effect(s) on the environment. NO(g) in sufficiently high concentrations can react with hemoglobin in blood and rob it of its oxygen-carrying ability, just as CO(g) does—but these concentrations are rarely reached in polluted air. NO(g) is considered an air pollutant primarily because it participates in various reactions that yield several other objectionable pollutants.

Nitrogen dioxide is an irritant to the eyes and respiratory system. Tests with laboratory animals indicate that chronic exposure to levels of NO_2 in the range 10 to 25 ppm might lead to emphysema or other degenerative lung diseases. Red-brown $NO_2(g)$ is often seen in polluted air in major urban centers, but like NO(g), $NO_2(g)$ is objectionable mostly for its chemical reactions. In the presence of sunlight ($h\nu$) NO_2 decomposes.

$$NO_2(g) + h\nu \longrightarrow NO(g) + O(g)$$

Oxygen *atoms* produced by the photochemical (light-induced) decomposition of NO_2 are highly reactive. The O atoms can react with many substances that are generally available in polluted air. For example, they may react with O_2 molecules to form ozone.

$$O(g) + O_2(g) \longrightarrow O_3(g)$$

A considerably higher than normal concentration of ozone (O_3) in air is one of the hallmarks of photochemical smog. **Photochemical smog** is air that is polluted with oxides of nitrogen and unburned hydrocarbons, together with ozone and several other components produced by the action of sunlight. The smog component known as PAN is the organic compound *peroxyacetyl nitrate* formed by the combination of two free radicals.

NO_2 is a free radical in the sense that it is an odd-electron molecule, that is, a molecule with an unpaired electron. We introduced free radicals in Chapter 9 (page 338).

$$CH_3\overset{\overset{\displaystyle O}{\|}}{C}—O—O\cdot + \cdot NO_2 \longrightarrow CH_3\overset{\overset{\displaystyle O}{\|}}{C}—O—ONO_2$$

PAN

Ozone in the stratosphere protects us from harmful ultraviolet radiation, but ground-level ozone is the main cause of the breathing difficulties that some people experience during smog episodes. Another effect of ozone is that it causes rubber to crack and deteriorate. PAN is a powerful *lacrimator*: it makes the eyes form tears. In addition to their role in smog formation, NO(g) and $NO_2(g)$ contribute to the fading and discoloration of fabrics. By forming nitric acid, they contribute to the acidity of rainwater, which accelerates the corrosion of metals and building materials. They also produce crop damage, although specific effects of these gases are difficult to separate from those of other pollutants. The components of photochemical smog also reduce visibility (Figure 14.7).

A normal radish plant (left) and one damaged by air pollution (right).

The reactions involved in photochemical smog formation are exceedingly complex; they are still not completely understood. We cannot pursue this reaction chemistry in any depth, but we can point out a few of its features.

We have already mentioned two reactions: the photochemical decomposition of NO_2, followed by the production of ozone. If ozone formation is to

Figure 14.7 Photochemical smog. Photochemical smog is characterized by an amber haze, like that seen in this view across the city of Buenos Aires, Argentina.

continue, there must be a continuous source of NO_2. One reaction that we've previously seen for the formation of NO_2 is

$$2\,NO(g) + O_2(g) \longrightarrow 2\,NO_2(g)$$

However, at the low concentrations of NO(g) in a smoggy atmosphere and at normal air temperatures this reaction proceeds too slowly to produce much NO_2. The NO(g) appears to be converted to $NO_2(g)$ in another—much faster—way, and this involves hydrocarbons, mostly from automotive exhaust.

For example, a hydrocarbon molecule, RH, can react with an oxygen atom to produce two free radicals.

$$RH + O \longrightarrow R\cdot + \cdot OH$$

The OH radical can react with another hydrocarbon molecule.

$$RH + \cdot OH \longrightarrow R\cdot + H_2O$$

The hydrocarbon radicals, R·, can react with O_2 to produce new radicals, called peroxyl radicals.

$$R\cdot + O_2 \longrightarrow RO_2\cdot$$

And then the peroxyl radicals can react with NO to form NO_2.

$$RO_2\cdot + NO \longrightarrow RO\cdot + NO_2$$

We have stressed the contribution of automotive exhaust to the production of photochemical smog, but geographic factors are also important. Smog is especially likely to occur in a region like the Los Angeles basin, which is surrounded by mountains: in the absence of strong winds, the only direction in which mixing and dilution of air pollutants can occur is vertically into the atmosphere. At times the region may also experience a *temperature inversion*—a mass of warm air overlays a body of colder air below. This inversion layer acts like a lid on a pressure cooker. Through the action of sunlight, smog components are formed and then trapped in the cooler stagnant layer of air. The most severe smog episodes occur during periods of strong temperature inversions.

Control of Photochemical Smog

Measures to reduce the levels of photochemical smog tend to focus on automobiles, but all potential sources of smog precursors are included, ranging from power plants to charcoal lighter fluid. In many parts of the world, automobiles are now equipped with *catalytic converters*. The first function of these converters is to oxidize carbon monoxide and unburned hydrocarbons to CO_2 and H_2O in the presence of a catalyst such as palladium (Pd) or platinum (Pt). NO can be removed from automotive exhaust by reducing it to N_2. This requires a *reduction* catalyst, which is different from an oxidation catalyst. Some automobiles use a dual-catalyst system. Another method of lowering the NO(g) content of automotive exhaust is to use a fuel-rich mixture of fuel and air. This results in some unburned hydrocarbons and CO(g), which can reduce NO to N_2, as in the reaction

A cutaway view of a dual-bed automobile catalytic converter.

$$2\,CO(g) + 2\,NO(g) \longrightarrow 2\,CO_2(g) + N_2(g)$$

The excess unburned hydrocarbons and CO(g) are then oxidized to CO_2 and H_2O in the catalytic converter.

We can reduce air pollution by driving less—walking or using a bicycle—and by driving smaller cars that get better gas mileage. More radical control measures, which may be required in the future—at least in the areas that have the worst air pollution problems—are the use of alternative fuels (such as natural gas, methanol, or hydrogen) that burn more cleanly than gasoline, and the development of electric-powered automobiles. Reductions in smog-forming substances can also be achieved by improving the efficiency of gasoline engines, ride-sharing, and restrictions on the use of gasoline-powered automobiles.

14.4 Oxygen and Oxides

Nitrogen is far more abundant than oxygen in the atmosphere, but the situation is reversed in Earth's solid crust. Nitrogen compounds occur only in trace amounts in the crust, but those of oxygen are so numerous that oxygen is the most abundant element, making up 45.5% by mass of Earth's crust. The main reason for this difference is that nitrogen gas, with its triple bond, is rather inert, and oxygen gas is quite reactive.

We encounter oxygen compounds throughout the text because oxygen forms compounds with all the elements except the lighter noble gases (He, Ne, and Ar). Oxygen is therefore central to a study of chemistry. Moreover, a study of the properties of oxygen and its compounds helps us to understand many chemical principles.

We explore the relationship of oxygen to the other Group 6A members in Chapter 21.

Oxygen atoms have the electron configuration $1s^22s^22p^4$. The principal oxidation states of oxygen are −2 and 0. In the atmosphere oxygen occurs mainly as diatomic molecules O_2. In normal oxides, oxygen is in the −2 oxidation state

(O.S.) and is present as covalently bonded atoms (as in H_2O) or as O^{2-} ions (as in Li_2O). In peroxides, such as H_2O_2 and Na_2O_2, the O.S of oxygen is −1. In a superoxide ion, O_2^- (in KO_2, for example), one O atom has O.S. −1, and the other has O.S. 0, giving an *average* O.S. of $-\frac{1}{2}$.

The chief reactions of elemental, atmospheric oxygen are oxidation processes: combustion (rapid burning), rusting and other forms of corrosion, and respiration. Oxygen reacts rapidly with active metals. Magnesium, for example, burns with a brilliant white flame when ignited in air.

$$2\,Mg(s) + O_2(g) \longrightarrow 2\,MgO(s) + \text{heat} + \text{light}$$

This same reaction occurs in a more subdued manner on the surface of freshly prepared magnesium at room temperature. Similar reactions occur with aluminum and titanium. The oxides formed in these reactions are thin, transparent coatings that are impervious to air, preventing further oxidation of the metal. It is only because of these protective oxide coatings that magnesium, aluminum, and titanium have such a wide range of uses, from building materials to aircraft.

Iron also forms an oxide coating when exposed to atmospheric oxygen. This coating also shields the metal from air and adequately protects it from further reaction but only as long as the surface is dry. Iron cookware won't rust as long as it is kept dry or covered with a protective coating of grease. If iron is left wet or the grease film is destroyed, and particularly if electrolytes such as NaCl are present, the iron is converted quickly to iron(III) oxide (rust). This oxide is not tightly bound to the surface of the metal. It flakes off, and corrosion continues.

Preparation and Uses of Oxygen

Oxygen gas, mostly obtained by the fractional distillation of liquid air (recall Figure 14.5), is an important commercial chemical. Typically, it ranks third in quantity produced in the United States, following sulfuric acid and nitrogen. Some important uses of oxygen are listed in Table 14.3.

Occasionally, small quantities of oxygen are prepared by the decomposition of compounds containing oxoanions, usually in the presence of a catalyst. For example, the decomposition of potassium chlorate is catalyzed by MnO_2.

$$2\,KClO_3(s) \xrightarrow{MnO_2(s)} 2\,KCl(s) + 3\,O_2(g)$$

The decomposition of certain oxides also yields oxygen gas. Joseph Priestley discovered oxygen in 1774 by heating mercury(II) oxide.

$$2\,HgO(s) \longrightarrow 2\,Hg(l) + O_2(g)$$

Lavoisier, who gave oxygen its name, used this decomposition reaction as the source of oxygen in his early studies of combustion.

Oxygen can also be produced by the decomposition of aqueous solutions of hydrogen peroxide.

$$2\,H_2O_2(aq) \longrightarrow 2\,H_2O(l) + O_2(g)$$

The reaction is slow but can be speeded up greatly by using a catalyst, such as $Br^-(aq)$ or $I^-(aq)$. Another oxygen-producing reaction uses potassium superoxide.

$$4\,KO_2(s) + 2\,CO_2(g) \longrightarrow 2\,K_2CO_3(s) + 3\,O_2(g)$$

TABLE 14.3
Common Uses of Oxygen

Manufacture of iron, steel and other metals
Fabrication of metals (cutting and welding)
Manufacture of chemicals
Water treatment
Oxidizer of rocket fuels
Respiration therapy and other medical uses
Petroleum refining

This reaction, because it consumes CO_2 while producing O_2, is used in emergency breathing apparatus and in recycling the air in submarines and spacecraft.

Oxygen is also produced, together with hydrogen, during the electrolysis of water (Section 8.1).

Ozone

By now you are no doubt familiar with the form of oxygen known as *ozone*, O_3. We described its molecular structure (a resonance hybrid) in Section 9.8 and its role in the formation of photochemical smog in the preceding section. Forms of an element that differ in their bonding and molecular structure are called *allotropes*. So O_2 and O_3 are allotropes of oxygen, just as graphite and diamond are allotropes of carbon (Section 11.8).

Allotropy, the existence of allotropes, is limited to a small number of nonmetals and metalloids—for example, carbon, tin, phosphorus, antimony, oxygen, and sulfur.

The concentration of ozone in the troposphere is quite low, about 0.04 ppm near ground level. In polluted air its concentration can rise to as high as about 0.5 ppm. Levels above 0.12 ppm are considered unhealthy.

The production of $O_3(g)$ directly from $O_2(g)$ is a highly endothermic reaction and occurs only rarely in the troposphere, chiefly during electrical storms.

$$3\,O_2(g) \longrightarrow 2\,O_3(g) \qquad \Delta H^\circ = +285\text{ kJ}$$

Ozone, which can be identified by its pungent odor, is occasionally formed around heavy-duty electrical equipment and xerographic office copiers.

To produce ozone commercially, $O_2(g)$ is held in a high-energy environment by passing either an electric discharge or ultraviolet radiation through it. Because it is unstable and decomposes back to $O_2(g)$, ozone is usually generated at the point where it is to be used (see Figure 14.8).

Figure 14.8 An ozone generator used in water purification. Ozone is an excellent oxidizing agent, second only to $F_2(g)$ among common substances. The most important use of ozone is as a substitute for chlorine in purifying drinking water. Its advantages are that ozone does not impart a taste to the water and it does not form the potentially carcinogenic products that chlorination can. Its main disadvantage is that ozone is unstable and quickly disappears after the water has been treated. Thus, the water is not as well protected after it leaves the waterworks as is water that has been chlorinated.

The Ozone Layer

A band of the stratosphere about 20 km thick, with its center at about 25 to 30 km altitude, has a much higher ozone concentration than the rest of the atmosphere. This band is known as the **ozone layer**. It protects life on Earth through ozone's ability to absorb ultraviolet (UV) radiation.

UV radiation, invisible to human eyes, has profound effects on living matter. Proteins, nucleic acids, and other cell components are broken down by UV radiation with wavelengths shorter than about 290 nm. Of this radiation, O_2 and other atmospheric components effectively filter out wavelengths below 230 nm. Ozone alone absorbs UV radiation in the wavelength range from 230 to 290 nm. Radiation with wavelengths in the range 290 to 320 nm, so-called UV-B radiation, produces sunburns and can also cause eye damage and skin cancer. UV-B radiation is only partially absorbed by ozone, and the amount reaching Earth's surface is critically dependent on the concentration of $O_3(g)$ in the ozone layer. Society faces a crucial challenge to maintain the appropriate concentration of ozone in the ozone layer.

When two opposing processes occur, one producing and the other consuming a substance, the result will be a "steady-state" concentration. Let's consider the processes that lead to a natural steady-state concentration of about 8 ppm of ozone in the ozone layer. Ozone is *produced* in the upper atmosphere in a sequence of two reactions. First, an O_2 molecule absorbs UV radiation and dissociates into two O atoms.

$$O_2 + h\nu \longrightarrow O + O$$

Then atomic and molecular oxygen react to form ozone. (A "third body," M, such as an N_2 molecule, carries off excess energy. Otherwise the energetic O_3 would simply decompose back to O_2 and O.)

$$O_2 + O + M \longrightarrow O_3 + M$$

Ozone is *decomposed* when it absorbs UV radiation.

$$O_3 + h\nu \longrightarrow O_2 + O$$

The atomic oxygen then reacts with ozone, forming O_2 molecules and releasing heat to the environment.

$$O + O_3 \longrightarrow 2\,O_2 \qquad \Delta H = -391.9\text{ kJ}$$

The heat released in this reaction accounts for the characteristic temperature increase with altitude in the stratosphere described on page 519.

Other naturally occurring species also contribute to the decomposition of ozone. Atmospheric NO, produced mainly from N_2O released by soil bacteria, decomposes ozone through this sequence of reactions.

$$\begin{array}{l} (1)\ NO + O_3 \longrightarrow NO_2 + O_2 \\ (2)\ NO_2 + O \longrightarrow NO + O_2 \\ \hline \text{Net: } O_3 + O \longrightarrow 2\,O_2 \end{array}$$

The interesting feature of this sequence of events is that NO consumed in the first reaction is regenerated in the second. A little NO goes a long way. Moreover, we should expect that the injection of any additional NO into the stratosphere would increase the destruction of ozone and reduce its steady-state

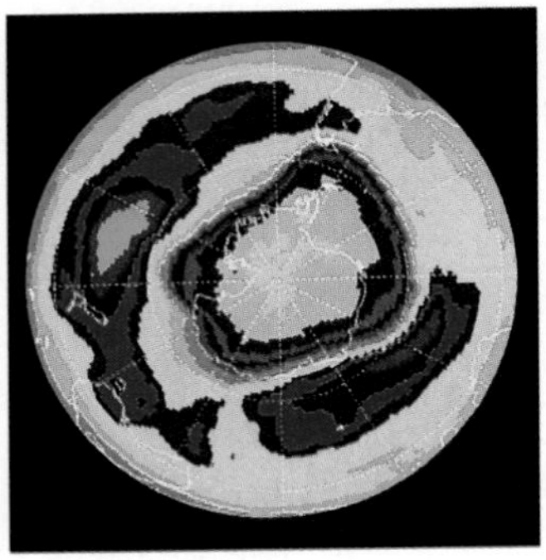

Figure 14.9 The ozone hole over Antarctica. In this measurement, recorded on October 2, 1994, the different colors represent different ozone concentrations. The grey, pink, and purple areas at the center constitute the ozone hole, with the ozone concentration being lowest in the grey region. Higher concentrations of ozone are found in the yellow, green, and brown areas.

The societal importance of the ozone depletion problem is reflected in the award of the 1995 Nobel Prize in chemistry to Paul Crutzen, F. Sherwood Rowland, and Mario Molina for working out the mechanism of ozone depletion, including the role of CFCs.

Worldwide production of CFCs reached a peak of about 1.3×10^6 tons in 1988. Current production levels are far below this peak.

concentration. This additional NO can come, for example, from combustion processes in supersonic jets operating in the stratosphere.

A more significant cause of ozone destruction through human activities is thought to be the use of chlorofluorocarbons (CFCs). These substances have a long lifetime in the atmosphere. Some of the molecules eventually rise and appear in low concentrations in the stratosphere. Here, by absorbing UV radiation, they decompose to produce atomic and molecular fragments (radicals).

$$CCl_2F_2 + h\nu \longrightarrow \cdot CClF_2 + Cl\cdot$$

The ozone-destroying cycles set up beyond this point are known to be quite complex, but a simplified representation is

$$\begin{array}{ll} (1) & Cl\cdot + O_3 \longrightarrow ClO\cdot + O_2 \\ (2) & ClO\cdot + O \longrightarrow Cl\cdot + O_2 \\ \hline \text{Net:} & O_3 + O \longrightarrow 2\,O_2 \end{array}$$

The chlorine atom that reacts in step (1) is regenerated in step (2); one chlorine atom can therefore destroy thousands of ozone molecules. (As in polymerization, the reaction proceeds until ended by the combination of two radicals.)

Perhaps the best evidence for the depletion of stratospheric ozone is that obtained from studies in Antarctica (Figure 14.9). The current strategy for dealing with the ozone depletion problem is to replace the offensive CFCs and other chlorine- or bromine-containing compounds that might diffuse into the stratosphere with more benign substances. Fluorocarbons, which have no Cl or Br to form radicals, are one kind of replacement. Another is to use hydrochlorofluorocarbons (HCFCs); these molecules break down more readily in the troposphere, and fewer ozone-destroying molecules reach the stratosphere (see Table 14.4).

TABLE 14.4 **Some CFCs, Halons, and Replacement Substances**

CFC or Halon[a]	Possible Replacement	Used in
CFC-12 (CCl_2F_2)	HFC-134a (CH_2FCF_3)	compressors in home refrigerators and car air conditioners
CFC-11 (CCl_3F)	HFC-141b (CH_3CCl_2F)	foamed plastics (blowing agent)
Halon-1301 (CF_3Br)	CF_3I	fire-extinguisher systems

[a] The numbers that follow the letter designations are industry codes.

14.5 Oxides of Carbon

At a concentration of 359 ppm CO_2 or 0.0359 mol %, carbon dioxide is a minor component of Earth's atmosphere. However, even this relatively small quantity of CO_2 has a profound effect on Earth's climate, and small increases in the concentration of CO_2 may have dramatic effects on its future climate.

In the atmosphere as a whole, carbon monoxide occurs to a much lesser extent than does CO_2, but high local concentrations of CO can cause problems. We examine some of the atmospheric chemistry of these two oxides in this section. Some sources and commercial uses of CO and CO_2 are listed in Table 14.5.

TABLE 14.5 **Some Sources and Uses of Oxides of Carbon**

Source	Uses
Carbon monoxide, CO	
Steam reforming of natural gas: $CH_4(g) + H_2O(g) \longrightarrow CO(g) + 3\,H_2(g)$	manufacture of methanol and other organic compounds from synthesis gas (CO/H_2 mixture)
	metallurgical reducing agent, as in blast furnace reaction: $Fe_2O_3(s) + 3\,CO(g) \longrightarrow 2\,Fe(l) + 3\,CO_2(g)$
Carbon dioxide, CO_2	
Decomposition (calcination) of limestone at about 900 °C: $CaCO_3(s) \longrightarrow CaO(s) + CO_2(g)$	refrigerant in freezing, storage, and transport of foods
	production of carbonated beverages
Recovery from stack gases in the combustion of fuels	petroleum recovery in oil fields
	fire extinguisher systems
Fermentation byproduct in the production of ethanol	

The Carbon Cycle

Natural processes cycle carbon atoms throughout Earth's solid crust, oceans, and atmosphere. In the process of *photosynthesis*, atmospheric CO_2 is converted to carbohydrates, the chief structural material of plants. For example, the photosynthesis of glucose, one of the simplest carbohydrates, occurs through dozens of sequential steps leading to the net change

$$6\,CO_2(g) + 6\,H_2O(l) \xrightarrow[\text{sunlight}]{\text{chlorophyll}} C_6H_{12}O_6(s) + 6\,O_2(g) \qquad \Delta H = +2.8 \times 10^3\ \text{kJ}$$

Animals acquire carbon compounds as they consume plants, and they return CO_2 to the atmosphere when they breathe (respiration). The decay of plant and animal matter also returns CO_2 to the air. A great deal of photosynthesis occurs in the oceans, where algae and related plants convert CO_2 into organic compounds. Some of Earth's carbon is locked up in fossilized forms—in coal, petroleum, and natural gas from decaying organic matter and in limestone from the shells of decayed mollusks in ancient seas. Figure 14.10 shows a simplified cycle.

As suggested by Figure 14.10, we humans today play a key role in the carbon cycle. We do this not so much through respiration and decay but by releasing carbon atoms as CO and CO_2 when we burn wood and fossil fuels. Without this human intervention, the cycle would exist in a nearly steady state for thousands of years.

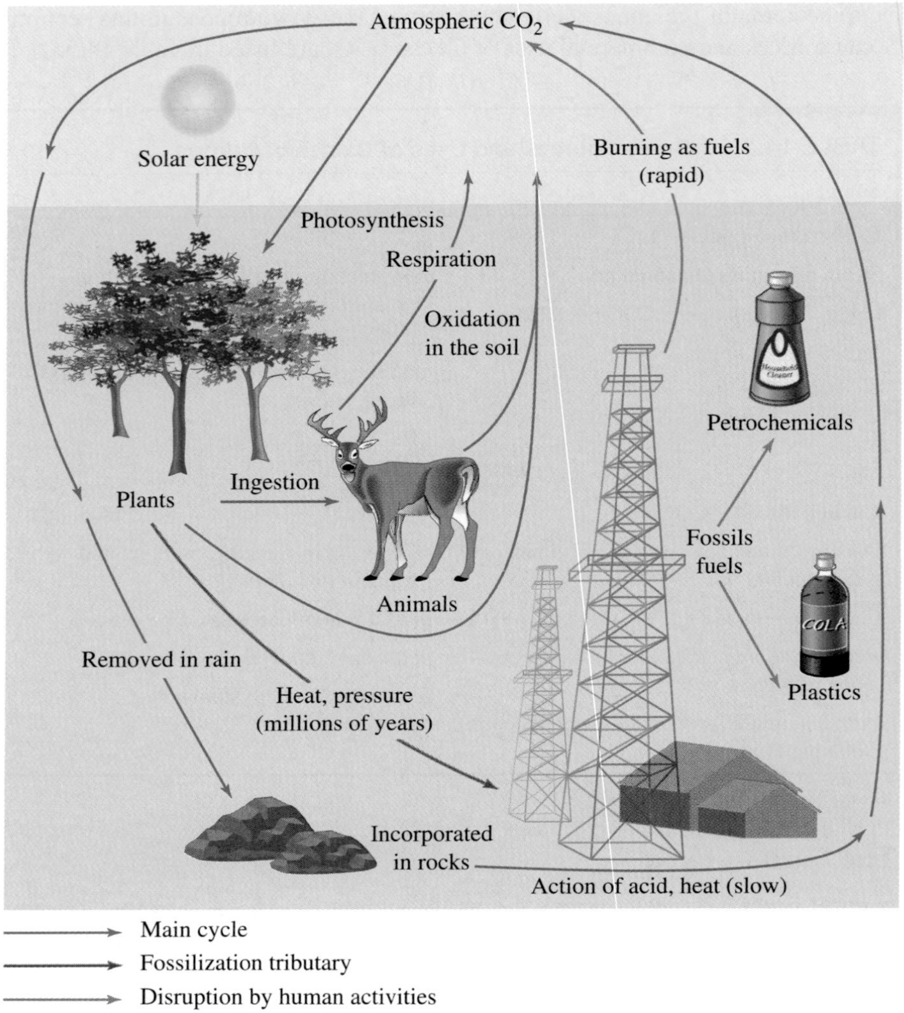

Figure 14.10 The carbon cycle. The natural main cycle is indicated by blue arrows. Some carbon atoms are locked up by the formation of fossil fuels and limestone deposits, so-called fossilization tributaries (brown arrows). Disruption of the cycle by human activities is becoming increasingly important (purple arrows).

Combustion of Carbon and Hydrocarbons

Carbon monoxide and carbon dioxide are formed in varying quantities when fossil fuels are burned. Coal is mostly carbon; natural gas and petroleum are mainly hydrocarbons. The combustion of methane, the major component of natural gas, yields carbon monoxide and carbon dioxide by the reactions

$$2\ CH_4(g) + 3\ O_2(g) \longrightarrow 2\ CO(g) + 4\ H_2O(l)$$

$$CH_4(g) + 2\ O_2(g) \longrightarrow CO_2(g) + 2\ H_2O(l)$$

With an *excess* of O_2, as is the case when air is readily available, the combustion products are almost exclusively $CO_2(g)$ and $H_2O(l)$. If the quantity of air is more limited, as in a dust- or dirt-clogged wall heater, CO(g) is also formed.

The chief source of CO(g) in polluted air is the incomplete combustion of gasoline hydrocarbons in automobile engines. Millions of metric tons (1 metric ton = 1000 kg) of this invisible but deadly gas are poured into the atmosphere each year, about 75% of it from automobile exhausts. The United States government has set danger levels at 9 ppm CO averaged over an 8-hour period and 35 ppm averaged over a 1-hour period. Even in off-street urban areas, levels often average 7 to 8 ppm. On streets and in parking garages, danger levels are exceeded much of the time. Such levels do not cause immediate death, but exposure over a long period can cause physical and mental impairment.

Because CO is an invisible, odorless, tasteless gas, we can't tell when it is around without using test reagents or instruments. Drowsiness is usually the only symptom of carbon monoxide poisoning, and drowsiness is not always unpleasant. How many auto accidents are caused by drowsiness or sleep induced by CO(g) escaping into the car from a faulty exhaust system? No one knows for sure.

Carbon monoxide exerts its insidious effect because CO molecules displace O_2 molecules normally bonded to Fe atoms in hemoglobin in blood (see Figure 14.11). The symptoms of carbon monoxide poisoning are those of oxygen deprivation. All except the most severe cases of the poisoning are reversible. The best antidote is the administration of pure oxygen. Artificial respiration may help if a tank of oxygen is not available. Because carbon monoxide poisoning impairs the blood's ability to transport oxygen, the heart has to work harder to supply oxygen to tissues. Chronic exposure, even to low levels of CO, as through cigarette smoking, puts an added strain on the heart and increases the chances of a heart attack.

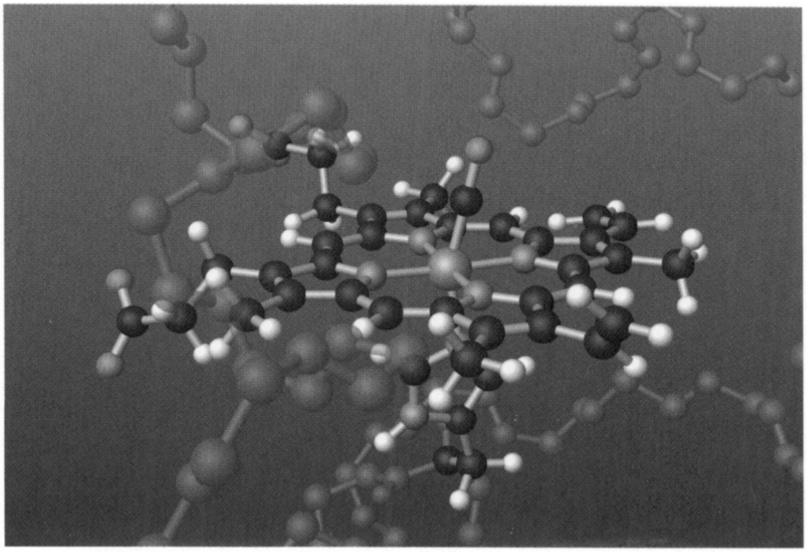

Figure 14.11 A molecular view of carbon monoxide poisoning. The hemoglobin molecule consists of thousands of atoms, but key portions of the molecule are four heme groups, one of which is shown here. Each heme has an iron atom (gray) in the center of a square formed by four nitrogen atoms. The heme group is able to bind one small molecule to the iron atom. The heme group has a much higher affinity for a CO molecule (shown here pointing up from the iron atom) than for O_2. As a result, even in the presence of low concentrations of CO(g), O_2 molecules are easily displaced by CO.

Carbon monoxide is a local pollution problem: It is a severe threat in urban areas with heavy traffic, but it does not appear to be a global threat. In laboratory tests, CO molecules survive about 3 years in contact with air. But nature is able to prevent the buildup of CO in the atmosphere, despite the large amounts entering the environment. Bacteria in the soil presumably convert CO to CO_2. In fact, it is estimated that on a global basis up to 80% of the CO in the atmosphere comes from natural sources. Except for highly localized situations that can be quite severe, nature seems to have carbon monoxide under control.

Carbon Dioxide, the Greenhouse Effect, and Global Warming

We all exhale carbon dioxide with every breath; it is a product of respiration. Even though it is also produced by the combustion of fossil fuels, we generally do not think of CO_2 as an air pollutant because low levels are not toxic. Increased levels of CO_2 in the atmosphere, however, could have a profound effect on the environment by producing a significant increase in the average global temperature.

When electromagnetic radiation from the sun reaches Earth, some is reflected back into space, some is absorbed by substances in the atmosphere, and some reaches Earth's surface, where it is absorbed. The surface gets rid of some of the energy by emitting infrared radiation toward outer space. Certain atmospheric gases, principally $CO_2(g)$ and $H_2O(g)$, absorb some of this infrared radiation. This radiant energy is retained in the atmosphere and warms it. The process is summarized in Figure 14.12. Because the process in a way resembles the retention of heat in a greenhouse, it has become known as the **greenhouse effect**. The effect is a natural one that is crucial to maintaining the proper temperature for life on Earth. Without the greenhouse effect, Earth would be an icehouse, permanently covered with snow and ice.

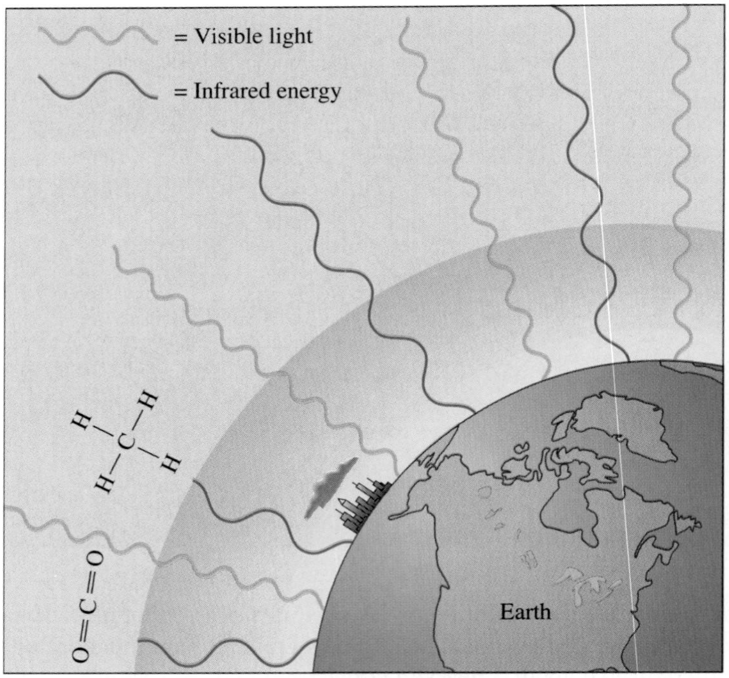

Figure 14.12 The "greenhouse" effect. Sunlight passing through the atmosphere is absorbed, warming Earth's surface. The warm surface emits infrared radiation. Some of this radiation is absorbed by CO_2, H_2O, and other gases and retained in the atmosphere as thermal energy.

From 1880 to 1990 the average CO_2 content of the atmosphere increased from 275 to 354 ppm. The 1995 level was about 359 ppm, and it currently increases about 1 ppm each year. These increases come from combustion of carbon-containing fuels—wood, coal, natural gas, gasoline—and the cutting of trees that would otherwise consume atmospheric CO_2 through photosynthesis.

Trees consume water vapor and carbon dioxide gas in the process of photosynthesis. The deforestation of tropical rainforests, as in this "slash and burn" clearing in a forest in Guyana, is a contributing factor in the increase of $CO_2(g)$ levels in the atmosphere.

Most computer models of the atmosphere indicate that a CO_2 buildup will cause **global warming** (that is, an increase in Earth's average temperature). There are many uncertainties, however. It is impossible to identify all the factors that should be included in computer models and to know how heavily to weight each factor. For example, warming of the atmosphere could lead to the increased evaporation of water and an accompanying increase in cloud cover. Because clouds reflect some incoming radiation into space, this could cause global *cooling* rather than global warming. Still, most models predict warming, with some saying that a doubling of the CO_2 content in air from its preindustrial values will cause a global temperature increase of from 1.5 to 4.5 °C. Major effects could be experienced as early as the middle of the twenty-first century.

Two probable effects of global warming are

- Local climate changes: The Great Plains states might become much drier with hotter summers.
- A rise in mean sea level caused by increased melting of polar ice caps and thermal expansion of sea water. The sea level could rise as much as several meters, flooding coastal cities and causing increased erosion of beaches.

We are not limited to computer modeling to assess the likelihood of global warming. Direct experimental evidence also exists. For example, the ice in the Greenland and Antarctic ice caps is laid down in layers much like the annual growth rings of trees. Analyses of tiny air bubbles trapped in these layers show a strong correlation between atmospheric CO_2 content and estimated global temperatures over the past 160,000 years—lower CO_2 levels correlate with lower temperatures and higher levels with higher temperatures.

The main strategy for countering a possible global warming is to curtail the use of fossil fuels, but this may not be enough. For example, several gases—methane, ozone, nitrous oxide (N_2O), and CFCs—are even better absorbers of infrared radiation than is carbon dioxide. More research is clearly needed. Whether we need to take immediate action and how drastic our efforts should be are subjects of much debate. The debate is likely to continue for years.

The number of huge icebergs that break off the continental ice shelf in Antarctica may increase as a result of global warming.

CONCEPTUAL EXAMPLE 14.1

For complete combustion of gasoline in an internal-combustion automotive engine, the mass ratio of air to fuel should be about 14.5 to 1. Use ideas from this chapter and elsewhere in the text to show that this is about what you would predict.

Solution

We are not given much data, so we must rely on several ideas that we have previously learned about the composition of gasoline, its combustion, and the behavior of gases. The following numbered items deal with these ideas.

1. Gasoline is a mixture of many hydrocarbons having from five to twelve carbon atoms in their molecules (page 58). On several occasions we have used octane, C_8H_{18}, as an "average" gasoline molecule.
2. The maximum quantity of heat is liberated in the complete combustion of the gasoline, that is, when CO_2 and H_2O are the products. A representative equation is

$$C_8H_{18}(l) + 12.5\ O_2(g) \longrightarrow 8\ CO_2(g) + 9\ H_2O(l)$$

3. From the balanced equation we can obtain the stoichiometric factor

$$\frac{12.5 \text{ mol } O_2}{1 \text{ mol } C_8H_{18}}$$

This is the mole ratio of O_2 to C_8H_{18} for complete combustion.

4. We need to replace O_2 by air in this ratio of reactants. To do this we can formulate a conversion factor (shown in red) from data in Table 14.1.

$$\frac{12.5 \text{ mol } O_2}{1 \text{ mol } C_8H_{18}} \times \frac{100 \text{ mol air}}{20.95 \text{ mol } O_2} = \frac{12.5 \times 100 \text{ mol air}}{20.95 \text{ mol } C_8H_{18}}$$

5. All that remains is to convert this mole ratio to a mass ratio. The molar mass of C_8H_{18} is 114 g/mol. Air is a mixture of gases, and its molar mass is a weighted average. To two significant figures, the molar masses and mole fractions of the three principal gases N_2, O_2, and Ar are

28 g/mol N_2 (mol fraction, 0.78)
32 g/mol O_2 (mol fraction, 0.21)
40 g/mol Ar (mol fraction, 0.0093)

$$\text{Molar mass of air} = (0.78 \times 28) + (0.21 \times 32) + (0.0093 \times 40) = 29 \text{ g/mol}$$

$$\frac{12.5 \times 100 \text{ mol air}}{20.95 \text{ mol } C_8H_{18}} \times \frac{1 \text{ mol } C_8H_{18}}{114 \text{ g } C_8H_{18}} \times \frac{29 \text{ g air}}{1 \text{ mol air}} = 15\ \frac{\text{g air}}{\text{g } C_8H_{18}}$$

Conceptual Exercise 14.1

A small gasoline-powered engine with a 1-L storage tank is inadvertently left running overnight in a large warehouse, 95 m × 38 m × 16 m. When workers arrive in the morning, are they likely to enter an environment in which the level of CO exceeds the danger level of 35 ppm? (*Hint*: Refer to Conceptual Example 14.1 for the composition of gasoline, and make other reasonable assumptions.)

14.6 Oxides of Sulfur

Like carbon, sulfur forms two principal oxides. Sulfur dioxide, SO_2, forms when sulfur burns or sulfide ores are heated in air. Sulfur trioxide, SO_3, forms by the further oxidation of sulfur dioxide. Both SO_2 and SO_3 are rare in unpolluted air.

Sulfur dioxide is sometimes a major component of volcanic gases, and the cone of a volcano may contain lethal levels. The concentration of SO_2 quickly diminishes, however, as distance from the source increases. The SO_2 is diluted by the great volume of clean air, and it is also quickly removed by chemical reactions.

This copper smelter emitted 900 tons of SO_2 daily before it ceased operation in January of 1987.

Industrial Smog

Polluted air associated with industrial activities is often called **industrial smog**. The presence of SO_2 and SO_3, collectively referred to as SO_x, is an important characteristic of this type of smog. A major source of SO_x in some areas is smelters, where sulfide ores are roasted as the first step in the production of metals such as copper, lead, and zinc. For example,

$$2\,ZnS(s) + 3\,O_2(g) \longrightarrow 2\,ZnO(s) + 2\,SO_2(g)$$

Smog is a term coined to describe a combination of *smoke* and *fog* that was once common in London. We now refer to the London-type smog as *industrial* smog.

Coal, especially soft coal from the eastern United States, has a relatively high sulfur content. When this coal is burned, sulfur compounds in it also burn, forming sulfur dioxide, a choking, acrid gas. Sulfur dioxide is readily absorbed in the respiratory system. It is a powerful irritant and is known to aggravate the symptoms of people who suffer from asthma, bronchitis, emphysema, and other lung diseases.

And things get worse. Although the actual reactions are more complicated than those we show here, the net results are the same. Some of the sulfur dioxide reacts further with oxygen in air to form sulfur trioxide.

$$2\,SO_2(g) + O_2(g) \longrightarrow 2\,SO_3(g)$$

Sulfur trioxide then reacts with water to form sulfuric acid.

$$SO_3(g) + H_2O(l) \longrightarrow H_2SO_4(aq)$$

Fine droplets of this acid form an aerosol mist that is even more irritating to the respiratory tract than sulfur dioxide.

Usually, industrial smog is also characterized by high levels of **particulate matter**, solid and liquid particles of greater than molecular size (Figure 14.13). The largest particles often are visible in air as dust and smoke. Smaller particles 1 μm or less in diameter are called *aerosols* and are often invisible to the naked eye.

Particulate matter consists in part of *soot* (unburned carbon). A larger portion is made up of the mineral matter that occurs in coal. These minerals do not burn. In the roaring fire of a huge boiler in a factory or power plant some of this solid mineral matter is left behind as *bottom ash*, but much is carried aloft in the tremendous draft created by the fire. This *fly ash* settles over the surrounding area, covering everything with dust. It is also inhaled, contributing to respiratory problems in animals and humans.

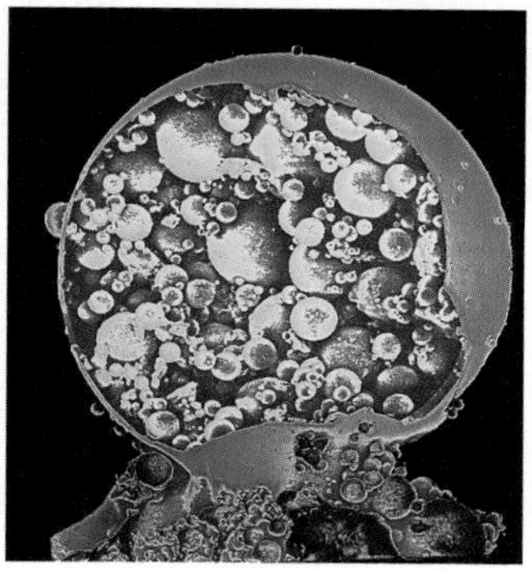

Figure 14.13 False–color scanning electron micrograph of fly ash from a coal-burning power plant.

Estimation Exercise 14.2

Assume that a typical urban air contains 100 μg of suspended particles per m^3 of air. Assume that the average particle is spherical in shape, with a diameter of 1 μm and a density of 1 g/cm^3. Estimate the *number* of particles per cm^3 of the air.

Health Effects of Industrial Smog

Perhaps the most insidious form of particulate matter is the sulfates. Some of the sulfuric acid in smog reacts with ammonia to form solid ammonium sulfate.

$$2\,NH_3(g) + H_2SO_4(aq) \longrightarrow (NH_4)_2SO_4(s)$$

The solid ammonium sulfate and minute liquid droplets of sulfuric acid are trapped in the lungs.

The harmful effects of sulfur dioxide and particulate matter may be considerably magnified by their interaction. A certain level of sulfur dioxide, without the presence of particulate matter, might be reasonably safe. A certain level of particulate matter, without sulfur dioxide around, might be fairly harmless. But take these same levels of the two together, and the effect might well be deadly. *Synergistic effects* such as this are quite common whenever certain chemicals are brought together. For example, some forms of asbestos are carcinogenic, and about 35 or 40 of the chemicals in cigarette smoke are carcinogens. Asbestos workers who smoke develop cancer at a much greater rate than do people who are exposed to one carcinogen but not the other.

When the pollutants in industrial smog come into contact with the alveoli of the lungs, the cells are broken down. The alveoli lose their resilience; it becomes difficult for them to expel carbon dioxide. Such lung damage leads to—or at least contributes to—pulmonary emphysema, a condition characterized by an increasing shortness of breath. Emphysema is one of the fastest growing causes of death in the United States. The principal factor in the rise of emphysema is cigarette smoking. However, air pollution is known to be a factor too. For instance, the incidence of the disease among smokers is three times as great in St. Louis, where air pollution is rather heavy, as in Winnipeg, Manitoba, where air pollution is rather mild.

The oxides of sulfur and the aerosol mists of sulfuric acid are damaging to plants. Leaves become bleached and splotchy when exposed to sulfur oxides. The yield and quality of farm crops can be severely affected. These compounds are also major ingredients in the production of acid rain.

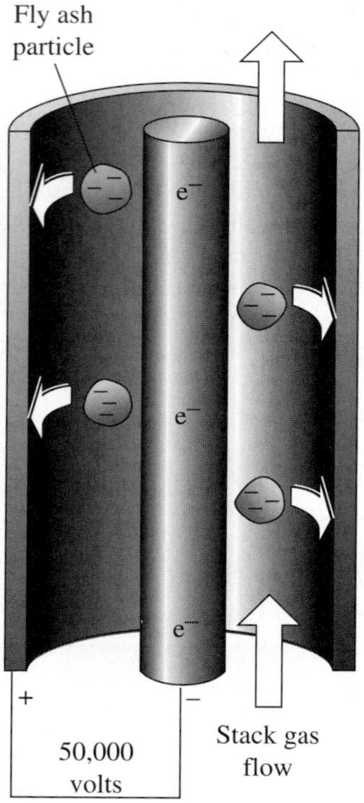

Figure 14.14 An electrostatic precipitator. Electrons emitted from the negatively charged electrode in the center attach themselves to the particles of fly ash, giving them a negative charge. The negatively charged particles are attracted to the positively charged cylindrical collector plate and are deposited there.

Controlling Industrial Smog

Much research has gone into reducing and preventing industrial smog. Soot and fly ash can be removed from smokestack gases in several ways. One method uses *electrostatic precipitators* (Figure 14.14). These devices induce electrical charges on the particles, which are then attracted to the oppositely charged plates and deposited on them. Another method uses *bag filtration*, a system that works much like the bag in a vacuum cleaner. Filters are placed in a bag house and arranged in a way that allows them to be shaken (Figure 14.15). The filters are cleaned by periodically blowing air through them, in the direction opposite to that taken by the stack gases. A third device, called a *cyclone separator*, is arranged so that the stack gases spiral upward with a circular motion. Suspended particles hit the walls of the cone-shaped device and settle out at the bottom. *Wet scrubbers* are also used. In these devices stack gases are passed through water, usually sprayed in a fine mist. Particles settle out with the water. This wastewater must be treated to remove the particulates, adding to the cost of this method.

The exact device used in a plant operation depends on the type of coal being burned, the size of the plant, and other factors. All require energy—electrostatic precipitators use 10% of a power plant's output—and the collected ash has to be put somewhere. Ash production in the United States is about 70 million metric tons (t) per year. About 16 million t of this is used, some to replace a part of the clay used in making cement. Some is melted and blown in air to make mineral wool for insulation. The rest has to be stored; 70% of it goes into ponds and the rest into landfills.

It is harder to remove SO_x than particulates. Sulfur can be removed from coal before burning. One way is to separate out pyrite (FeS_2), a mineral containing much of the coal's sulfur, by *flotation*. This method makes use of the different densities and wettabilities of pyrite and coal. Another method of desulfurizing coal is to gasify or liquefy the coal by reaction with $H_2(g)$. Mineral forms of sulfur, such as pyrite, are left behind. Both methods are expensive.

Another way to eliminate sulfur in coal-burning operations is through reactions that remove SO_2 from stack gases after the coal has been burned. For example, if finely powdered coal and limestone are burned together, the limestone decomposes and SO_2 reacts with the lime (CaO).

$$CaCO_3(s) \longrightarrow CaO(s) + CO_2(g)$$

$$CaO(s) + SO_2(g) \longrightarrow CaSO_3(s)$$

Sulfur dioxide can also be allowed to react with hydrogen sulfide, creating elemental sulfur that is easily recovered.

$$2\ H_2S(g) + SO_2(g) \longrightarrow 3\ S(s) + 2\ H_2O(l)$$

Sale of the sulfur can partially offset the cost of this process. Great technical problems have to be overcome, though, before this method can be put into widespread use. Moreover, if all the unwanted sulfur in stack gases could be converted to useful chemicals such as sulfuric acid, the quantity produced would greatly exceed current demand.

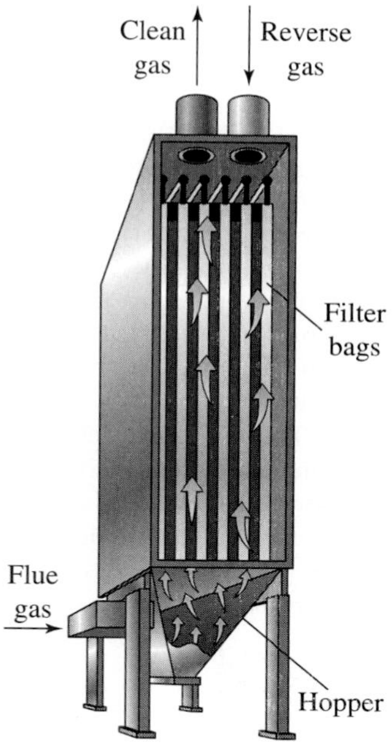

Figure 14.15 **Bag filtration.** Flue gases entering at the bottom of the bag collector contain particulate matter. Particles are held back by the walls of the bag. When the bags are vibrated, the particles are shaken loose and collect in the dust hopper at the bottom. Additional particles are detached from the bags when air is forced through the bags in the reverse direction.

Acid Rain: Air Pollution $\longrightarrow$ Water Pollution

We have seen how sulfur oxides are converted to sulfuric acid (page 543) and nitrogen oxides to nitric acid (page 524) in the atmosphere. These acids fall to Earth as acid rain or acid snow or are deposited from acid fog or adsorbed on (attached to) particulate matter. When rainfall is more acidic than it would be just by dissolving atmospheric $CO_2(g)$, it is called **acid rain**. Some rainfall has been reported that is even more acidic than vinegar or lemon juice. The best current evidence indicates that acid rain comes mainly from sulfur oxides emitted from power plants and smelters and from nitrogen oxides from automobiles. These acids may be carried for hundreds of kilometers before falling as rain or snow (Figure 14.16).

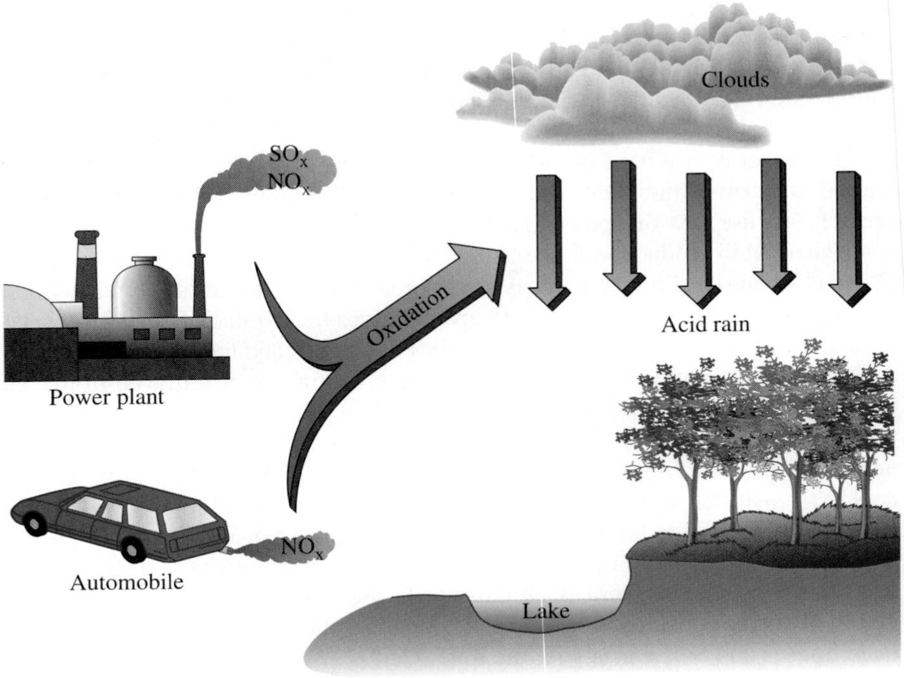

Figure 14.16 Acid rain. The two principal sources of acid rain are oxides of sulfur, SO_x, from power plants, and oxides of nitrogen, NO_x, from power plants and automobiles. These oxides are converted to SO_3 and NO_2, which then react with water to form H_2SO_4 and HNO_3. These acids may fall in rain hundreds of kilometers from their sources. They cross oceans and international boundaries, causing political as well as environmental problems.

14.7 The Noble Gases

In 1894, the eminent physicist John Rayleigh made this puzzling observation: The measured density of $N_2(g)$ extracted from air was always slightly greater than that of $N_2(g)$ prepared from nitrogen compounds. He suspected that impurities were present in the N_2 from one of these sources but not the other. The solution to this puzzle came when Rayleigh enlisted the aid of the chemist William Ramsay.

Ramsay recalled a long-neglected experiment performed more than a century earlier by Henry Cavendish, the discoverer of hydrogen. Cavendish, in trying to achieve the complete combination of nitrogen and oxygen by electric discharges through air, was always left with a small bubble of unreacted gas. Ramsay tried a similar experiment. He allowed the nitrogen obtained from a sample of air to react with magnesium, forming magnesium nitride.

$$3\,Mg(s) + N_2(g) \longrightarrow Mg_3N_2(s)$$

Again, a bubble of gas remained. Ramsay and Rayleigh did a spectroscopic analysis of this residual gas and found it to be a new element. Because of its chemical inertness, they called the gas argon, from the Greek *argos*, meaning "lazy one."

Because argon had no appropriate place in the periodic table in use at the time, Ramsay placed it in a new group and proposed that there should be other elements in this same group. Over the next six years he isolated the remaining

Natural Pollutants

Even before there were people there was air pollution. Volcanoes erupted, spewing ash and poisonous gases into the atmosphere. They still do. The eruption of Mount St. Helens in 1980 ejected ash that blanketed areas over several states. It ejected 1000 metric tons (t) of SO_2 per day during the eruption, and it continued to emit 160 t per day for several years following. El Chichon in Mexico in 1982 released 3.3 million t of SO_2 into the stratosphere, where the SO_2 was converted to sulfuric acid mist. Kilauea in Hawaii emits 200 to 300 t SO_2 per day. Acid rain downwind from this volcano has created a barren region called the Kau desert.

The June 1991 eruption of Mount Pinatubo in the Philippines injected such vast quantities of particulates into the stratosphere that the reflection of incoming solar radiation by these particles apparently produced a temporary reversal of global warming. Earth's average surface temperature fell from 15.47 °C in 1990 to 15.13 °C in 1992.

Dust storms, especially in arid regions, add massive amounts of particulate matter to the atmosphere. Dust from the Sahara Desert reaches the Caribbean area and South America. Swamps and marshes emit noxious gases such as hydrogen sulfide, a toxic gas with the odor of rotten eggs.

The greatest contributor of particulates to the atmosphere is the sea. Wave action causes seawater droplets to be suspended in the air, and when these droplets evaporate, salt particles are left behind. Tropical thunderstorms even carry chloride ions into the stratosphere where they can be converted to chlorine atoms and perhaps contribute to the destruction of ozone. The greatest component (by mass) of all the particulate matter in the atmosphere is ordinary salt (NaCl), and it comes from a perfectly natural source. Nature isn't always benign.

An eruption at Kilauea volcano on the island of Hawaii.

five members of what we now call Group 8A. Until 1962, these elements were called the inert gases; no chemical reactions were known for any of them. In that year, the first xenon compounds were prepared, and now the number of xenon compounds known is quite large. For example, the fluorides XeF_2, XeF_4, and XeF_6 can all be formed by the direct reaction of Xe(g) and F_2(g) under the appropriate conditions (Figure 14.17). A few compounds of krypton are known and evidence exists of radon compounds as well, but no compounds have yet been made of helium, neon, and argon.

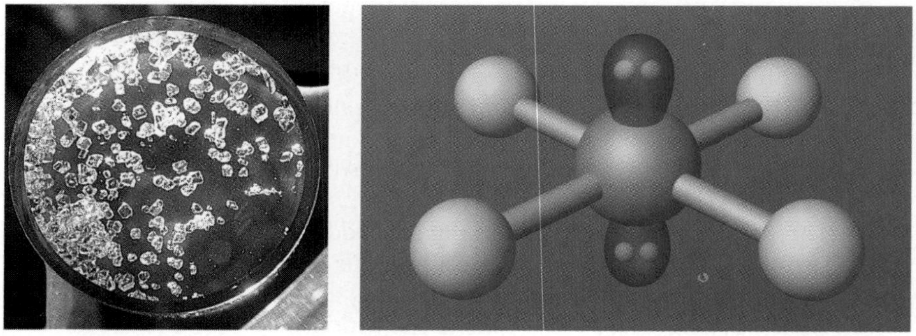

Figure 14.17 Crystals of xenon tetrafluoride, and the molecular structure of XeF_4.

Occurrence

The mole percentages of He, Ne, and Ar in air are 0.000524%, 0.001818%, and 0.934%, respectively. The proportions of Kr and Xe are much smaller: about 1 ppm and 0.05 ppm, respectively. Except for helium and radon, the noble gases are found only in the atmosphere. Helium is found in some natural gas deposits, particularly those underlying the Great Plains of the United States. Presumably, this helium is formed by the α-particle decay of naturally occurring radioactive isotopes. Alpha particles ($^4He^{2+}$) become helium atoms by acquiring two electrons. Radon, a radioactive element, is formed by the progressive decay of high-atomic-number naturally occurring radioactive isotopes, for example, beginning with ^{238}U.

The emission of α and β particles and γ rays was discussed in Section 6.2.

Although some helium is continually being formed in Earth's interior, the process is exceedingly slow. Today's helium has accumulated over eons—it is an essentially nonrenewable resource. Our future supplies are threatened because it is not economical to collect all the helium that is liberated in commercial operations. Much of it is simply released into the atmosphere.

Most of the noble gases except argon have escaped from the atmosphere since Earth was formed. The fact that argon is much more abundant than the other noble gases can be explained rather easily. It is a product of the decay of potassium-40, a fairly abundant naturally occurring radioactive isotope. We can show this through a *nuclear equation* in which $^{0}_{-1}e$ represents a β particle.

$$^{40}_{19}K \longrightarrow ^{40}_{20}Ar + ^{0}_{-1}e$$

Even though helium is also constantly formed through α-particle emissions by radioactive isotopes, it escapes from the atmosphere into outer space at a higher rate because the molar mass of helium is only one-tenth that of argon.

The abundance of helium on Earth is quite low, but in the universe as a whole it is second only to hydrogen.

Properties and Uses

Helium is used to fill balloons and dirigibles. Its lifting power is nearly as good as that of hydrogen, the least dense of all gases, but it has the important advantage of being nonflammable. Helium is also used to provide an inert atmosphere for the welding of metals that otherwise might be attacked by oxygen in air.

Liquid helium is used to achieve extremely low temperatures. It boils at −268.9 °C (only 4.2 K), and it exists as a liquid to temperatures approaching 0 K. All other substances freeze to solids at temperatures well above 0 K; even hydrogen has a freezing point of 14 K. Large quantities of liquid helium are used in cryogenics, the study of materials at extremely low temperatures. Metals become *superconductors* at liquid helium temperatures (that is, they essentially lose their resistance to the flow of electrical current). Powerful magnets can be created by immersing the metal-wire coils of electromagnets in liquid helium. These magnets are the key components in nuclear magnetic resonance (NMR) instruments used in research laboratories and magnetic resonance imaging (MRI) devices found in most hospitals.

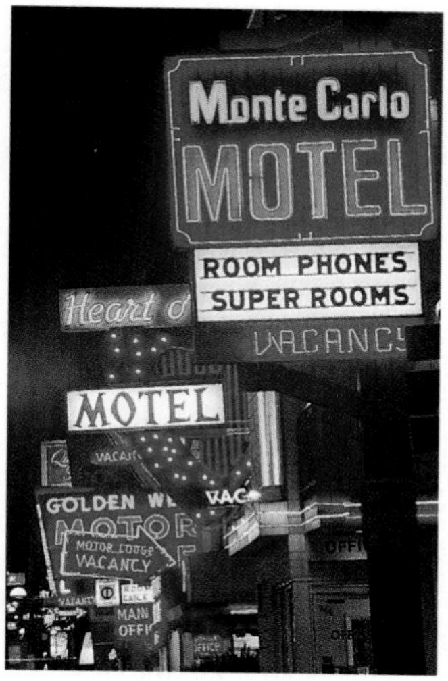

Display signs containing neon emit a characteristic red light when an electric discharge is passed through them. Other colors can be obtained by colored coatings on the discharge tubes.

Another interesting use of helium is in helium–oxygen breathing mixtures for deep-sea divers (page 494). Similar mixtures are also used in the treatment of asthma, emphysema, and other conditions involving respiratory obstruction. The same low molar mass that gives helium its lifting power also permits it to diffuse into partially obstructed areas of the lungs more rapidly than does nitrogen. The helium–oxygen mixture puts less strain on the muscles involved in breathing than does air.

Neon is used in lighted advertising signs. A tube with imbedded electrodes is shaped into letters or symbols and filled with neon at low pressure. An electric current passed through the gas causes the neon atoms to emit light of characteristic wavelengths. With a spectroscope we would see individual spectral lines, but with the unaided eye we merely see the familiar orange-red glow. Other colors can be obtained with mixtures of neon and argon or mercury vapor.

Argon, the most plentiful of the noble gases, can be separated from air rather inexpensively (recall Figure 14.5). Like helium, argon gas provides an inert atmosphere. In the laboratory, an argon atmosphere is an ideal medium for reactions in which one or more of the reactants or products is sensitive to air oxidation or to reaction with nitrogen. In industry, argon is used to blanket materials that need to be protected from nitrogen and oxygen, such as in certain types of welding and in the preparation of ultrapure semiconductor materials like silicon and germanium. A mixture of argon and nitrogen is used to fill incandescent light bulbs to increase their efficiency and life. Unlike oxygen, argon does not react with the tungsten filament. It also decreases the tendency of the filament to vaporize, thus extending the filament's life. Fluorescent light bulbs are filled with a mixture of argon and mercury vapor.

Krypton and xenon are too expensive to have many important commercial applications, although krypton has found some use in light bulbs. Both krypton and xenon are used in lasers and in flash lamps in photography. Radon, though exceedingly rare in the atmosphere, can be collected from the radioactive decay of radium. (In the following nuclear equation, ${}^{4}_{2}He$ represents an α-particle.)

$$ {}^{226}_{88}Ra \longrightarrow {}^{222}_{86}Rn + {}^{4}_{2}He $$

The radon can be sealed in small vials and used for radiation therapy of certain malignancies.

Recent studies indicate that perhaps 10% of all cases of lung cancer are caused by exposure to radon. Nearly all the rest are caused by cigarette smoking.

Radon in the Environment

One of the principal naturally occurring radioactive isotopes is uranium-238. This isotope—through a series of emissions of α- and β-particles—gives rise to radon-222. Radon-222 yields the isotope polonium-218.

$$^{222}_{86}\text{Rn} \longrightarrow {}^{218}_{84}\text{Po} + {}^{4}_{2}\text{He}$$

This step-by-step decay process continues until the nonradioactive isotope lead-206 is reached. What makes radon-222 unique in this decay process is that it is a *gas*, and because of this it can escape from Earth's surface and become trapped in the air in buildings.

The possible harmful effect of exposure to radon seems to be primarily an increased risk of lung cancer, but the issue is still unclear. Many scientists think the threat of radon has been exaggerated, and research and debate continue. Because radon is a gas, it readily moves through air passages in the body and is breathed in and out. It may be that any health hazards associated with radon are actually due to its decay products, such as polonium-218, which may become attached to dust particles in the air and then breathed into the lungs.

Some of the highest radon exposures are found near uranium mining waste facilities and residues from phosphate fertilizer production. In general, though, radon levels depend on the geological characteristics of an area and are known to vary widely. Fortunately, indoor radon is easily detected by its radioactivity, and measures can be taken to reduce its concentration. These measures include improving the ventilation systems of buildings and venting subsoil radon to keep it from entering buildings in the first place. In areas where radon is prevalent, measures to minimize indoor radon are now a conscious part of building construction.

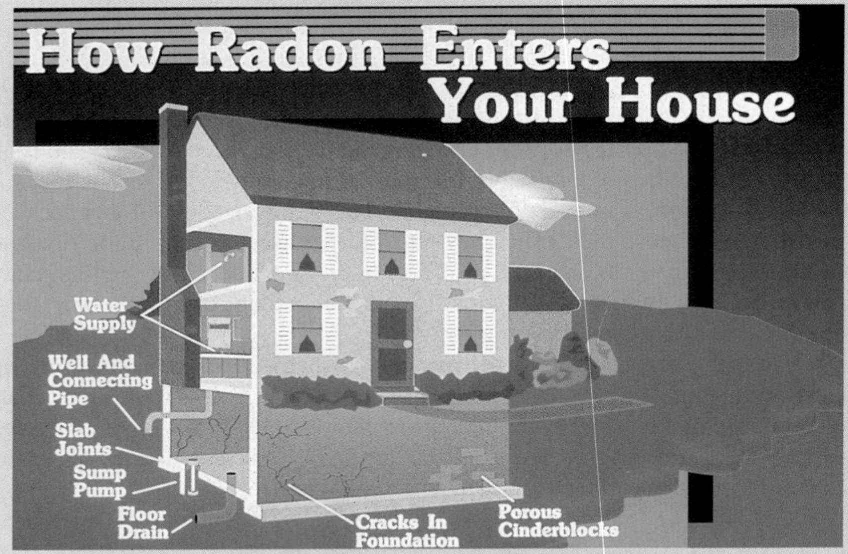

A chart produced by the Environmental Protection Agency (EPA) to show ways in which radon may enter a home.

SUMMARY

Starting at Earth's surface, the primary regions of the atmosphere are the troposphere, stratosphere, mesosphere, and thermosphere (ionosphere). The density of air and air pressure decrease sharply with increased altitude.

The chief components of dry air are N_2, O_2, and Ar. The water vapor content of air is expressed as humidity. Phenomena based on relative humidity include dew and frost formation and deliquescence. Water vapor is an important participant in the hydrologic (water) cycle.

Nitrogen must be "fixed" for use by plants, animals, and humans—naturally in the nitrogen cycle and artificially in the synthesis of ammonia. Ammonia is used as a fertilizer and in the manufacture of urea, ammonium sulfate, ammonium phosphates, nitric acid, and ammonium nitrate.

Oxides of nitrogen, in the presence of unburned hydrocarbons and sunlight, lead to photochemical smog. Temperature inversions also contribute to smog conditions. Smog control measures focus on catalytic converters for automobiles and the control of combustion processes to reduce emissions.

Oxygen forms compounds with all elements except the lighter noble gases. Most oxygen is prepared by the fractional distillation of liquid air; small quantities are made by the decomposition of certain oxides and oxoanions.

Ozone, O_3, forms in a high-energy environment containing O_2. Ozone in the stratosphere protects living organisms by absorbing ultraviolet radiation, but the integrity of the ozone layer is currently threatened by human activities, principally those that release chlorofluorocarbons (CFCs) into the atmosphere.

Atmospheric CO_2 is the carbon source for carbohydrate synthesis in the carbon cycle. Some carbon is locked out of the cycle in fossil fuels (coal, natural gas, and petroleum), but the combustion of these fuels returns CO and CO_2 to the cycle. CO is an air pollutant, and the constant buildup of CO_2 in the atmosphere is creating the possibility of future global warming.

As atmospheric pollutants, SO_2 and SO_3 contribute to the problems of industrial smog and acid rain. SO_2 is oxidized to SO_3, which reacts with water to form H_2SO_4. Particulate matter is also present in industrial smog. Control measures for industrial smog are designed to reduce the emissions of both SO_2 and particles. SO_2 and particulate matter appear to have serious health effects.

Of the noble gases, argon is found in significant amounts in air, helium is present in certain natural gas deposits, and radon forms by a radioactive decay process that begins with uranium. Naturally occurring radon is a potential environmental hazard. The noble gases have uses that encompass low-temperature applications (He), the production of inert atmospheres (He, Ar), applications in lasers and photography (Kr, Xe), and radiation treatment (Rn).

KEY TERMS

acid rain (14.6)
air pollutant (14.3)
carbon cycle (14.5)
deliquescence (14.2)
global warming (14.5)
greenhouse effect (14.5)
humidity (14.2)
hydrologic (water) cycle (14.2)
industrial smog (14.6)
nitrogen cycle (14.3)
nitrogen fixation (14.3)
ozone layer (14.4)
particulate matter (14.6)
photochemical smog (14.3)

REVIEW QUESTIONS

1. Which layer of the atmosphere (**a**) lies nearest the surface of Earth? (**b**) contains the ozone layer?
2. List the three major components of dry air and give the approximate (nearest whole number) mole percent of each.
3. Explain the difference between the absolute and the relative humidity of an air sample.
4. Describe the formation of dew and frost from air.
5. What is nitrogen fixation? Why is it important?
6. What is the nitrogen cycle? How has industrial fixation of nitrogen to make fertilizers affected the nitrogen cycle?
7. Briefly describe each of the following terms dealing with a natural phenomenon.
   a. deliquescence **b.** allotropy
   c. photosynthesis **d.** the greenhouse effect

8. Briefly describe the importance of each of these commercial products.
a. anhydrous ammonia **b.** urea
c. diammonium phosphate **d.** fossil fuels

9. What specific materials are implied by these terms for atmospheric pollutant(s)?
a. PAN **b.** SO_x **c.** fly ash

10. By name and/or formula, indicate (**a**) the fertilizer with the highest nitrogen content; (**b**) two gases able to displace O_2 in blood hemoglobin; (**c**) two "greenhouse" gases, in addition to CO_2 and H_2O; (**d**) a radioactive noble gas; (**e**) a constituent of acid rain; (**f**) two gases commonly used for inert atmospheres; (**g**) a gaseous mixture used in respiration therapy.

11. Supply a name or formula for each of the following. (**a**) Mg_3N_2; (**b**) dinitrogen tetroxide; (**c**) K_2O_2; (**d**) potassium superoxide; (**e**) urea.

12. List two uses of chlorofluorocarbons (CFCs). How are CFCs implicated in the depletion of the ozone layer?

13. How does a neon sign work? What function does argon serve in an electric light bulb?

14. What is photochemical smog? What is the role of sunlight in its formation?

15. What is industrial smog? How is it formed?

16. What conditions favor the formation of carbon monoxide during the combustion of gasoline in an automobile engine?

17. How does carbon monoxide exert its poisonous effect?

18. What is synergism? Indicate one specific example of a synergistic effect concerning air pollution.

19. What are the health effects associated with ozone in the stratosphere and at ground level? Why are they not the same?

20. Trace the steps by which the burning of high-sulfur coal leads to acid rain.

21. How is each of the following used to reduce air pollution?
a. electrostatic precipitator
b. catalytic converter **c.** wet scrubber

22. Which of the following are important contributors to the formation of photochemical smog and which are not? Explain.
a. NO **b.** CO
c. hydrocarbon vapors **d.** SO_2.

23. What is a temperature inversion? How does a temperature inversion contribute to air pollution problems?

24. How are the following terms related to one another regarding air pollution: aerosol, fly ash, and particulate matter?

25. Describe measures that can be used to control the emission of nitrogen oxides in automotive exhaust and explain why these are not the same measures used to control emissions of hydrocarbons and carbon monoxide.

26. Is all air pollution the result of human activity? Explain.

27. What is the difference in meaning conveyed by the terms "noble gases" and "inert gases"?

PROBLEMS

The Atmosphere

28. The text states that 99% of the mass of the atmosphere lies within 30 km of the surface of Earth. Which of the following is a reasonable estimate of air pressure at an altitude of 30 km: (a) 0.1 mmHg, (b) 1 mmHg, (c) 10 mmHg, (d) 100 mmHg? Explain your reasoning. (*Hint*: Recall the basic ideas relating to pressure from Section 4.3.)

29. When present in a very small proportion in air, it is customary to indicate the concentration of a gas in parts per million (ppm) rather than in mole percent or volume percent. Use data in Table 14.1 to determine the parts per million in air of the noble gases that are listed there.

Water Vapor in the Atmosphere

30. What are the mole percent and ppm of H_2O in an air sample at STP in which the partial pressure of water vapor is 2.00 mmHg?

31. What is the relative humidity of a sample of air at 25 °C in which the partial pressure of water vapor is 10.5 mmHg? (*Hint*: Use data from Table 11.2.)

32. What is the partial pressure of water vapor in a sample of air having a relative humidity of 75.5% at 20 °C? (*Hint*: Use data from Table 11.2.)

33. A parcel of air has an *absolute* humidity, expressed as a partial pressure of water vapor, of 18.0 mmHg. At which of the following temperatures does the air have the greatest relative humidity: 25 °C, 30 °C, or 40 °C? Explain.

34. What is the dew point of the parcel of air described in Problem 33? (*Hint:* Use data from Table 11.2.)

35. Why is it that condensed water vapor (steam) can be seen above a kettle of boiling water even in a hot kitchen, whereas you can see your breath (steam) only on a cold day?

Nitrogen

36. Place the following in order of increasing mass percent of nitrogen. Explain the basis of your arrangement. (a) seawater; (b) the atmosphere; (c) natural gas; (d) the fertilizer ammonium nitrate.

37. Without doing detailed calculations, place the following in order of increasing mass percent nitrogen. (a) nitrous oxide (dinitrogen monoxide); (b) ammonia; (c) nitric oxide (nitrogen monoxide); (d) ammonium chloride.

38. Write an equation to represent (**a**) the production of nitrogen monoxide (nitric oxide) in an automobile engine; (**b**) the action of sunlight on nitrogen dioxide; (**c**) the reduction of nitrogen monoxide to nitrogen gas by carbon monoxide.

39. Write an equation to represent (**a**) the *complete* combustion of octane, C_8H_{18}; (**b**) the *incomplete* combustion of methane; (**c**) the oxidation of carbon monoxide in a catalytic converter.

40. Write a series of equations to represent the natural fixation of atmospheric nitrogen in an electrical storm.

41. Write a series of equations to represent the artificial fixation of nitrogen as ammonium nitrate via the Haber–Bosch (page 524) and Ostwald processes (page 528).

42. The text mentions the highly explosive nature of mixtures of ammonium nitrate and fuel oil. What are the gaseous products that you would expect for this explosive reaction? Explain.

43. Electric storms typically fix about 4 kg of nitrogen per year per acre of land. What mass of ammonium sulfate fertilizer, per year per acre, is required to produce the same quantity of nitrogen?

Oxygen

44. How do oxygen atoms, oxygen molecules, and ozone molecules differ in structure and properties? Is it appropriate to refer to all three as *allotropes*? Explain.

45. Write an equation to represent (**a**) the action of atmospheric oxygen on a freshly prepared aluminum surface; (**b**) the formation of oxygen by the decomposition of potassium chlorate; (**c**) the formation of oxygen by the action of water on sodium peroxide (sodium hydroxide is the other product).

46. Write an equation to represent (**a**) the formation of ozone by the passage of an electric discharge through air; (**b**) the formation of oxygen by the thermal decomposition of mercury(II) oxide; (**c**) the formation of potassium superoxide by the action of oxygen gas on potassium metal.

47. An excess of KO_2(s) is placed in a closed container of CO_2(g). After the reaction is completed, will the gas pressure be the same, greater, or less than the initial value? Explain. (*Hint*: What is the reaction that occurs?)

48. An electric arc was passed through a volume of oxygen gas, converting a part of it to ozone. After the temperature and pressure were readjusted to their initial values, the volume was found to have decreased by 525 mL. What volume of ozone, in milliliters, was formed?

Oxides of Carbon

49. The combustion of a hydrocarbon, especially if the quantity of oxygen is limited, produces a mixture of carbon dioxide and carbon monoxide. The decomposition of a metal carbonate by an acid produces only carbon dioxide, even if the quantity of acid is limited. Explain this difference in behavior.

50. Indicate a natural process or processes by which carbon atoms are (**a**) removed from the atmosphere; (**b**) returned to the atmosphere; (**c**) effectively withdrawn from the carbon cycle.

51. Write an equation that represents the *complete* combustion of the hydrocarbon hexane, C_6H_{14}(l). Explain why it is not possible to write a unique equation to represent its *incomplete* combustion.

52. Carbon monoxide is a poisonous gas, even in low concentrations, whereas carbon dioxide is not. Yet, except in some local situations, there is less environmental concern over carbon monoxide than over carbon dioxide. Explain why this is so.

53. The United States leads the world in per capita emissions of CO_2(g) with 19.8 metric tons (t) per person per year (1 metric ton = 1000 kg). What mass, in metric tons, of each of the following fuels would yield this quantity of CO_2?
a. CH_4 **b.** C_8H_{18}
c. coal that is 94.1% C by mass

54. Tabulations on carbon dioxide emissions often list cement manufacture as one of the sources. Describe *two* ways in which the manufacture of cement injects carbon dioxide into the atmosphere. (*Hint*: Review page 296.)

Oxides of Sulfur

55. Write equations for the following reactions.
   a. Sulfur burns in air forming sulfur dioxide.
   b. Zinc sulfide, heated in air, yields zinc oxide and sulfur dioxide.
   c. Sulfur dioxide reacts with oxygen, forming sulfur trioxide.
   d. Sulfur trioxide reacts with water, forming sulfuric acid.
   e. Sulfuric acid is completely neutralized by aqueous ammonia.

56. The workplace standard for $SO_2(g)$ in air is 5 ppm. Approximately what mass of sulfur could be burned in an enclosed workplace, 10.5 m × 5.4 m × 3.6 m, before this limit is exceeded?

57. *Per ton of material consumed*, which of the following would you expect to produce the greatest quantity of $SO_2(g)$: (a) smelting zinc sulfide, (b) smelting lead sulfide, (c) burning coal, or (d) burning natural gas? Explain.

Particulate Matter

58. Describe how the following particulate matter may be produced.
   a. sodium chloride from seawater
   b. sulfate particles in an industrial smog.

Noble Gases

59. It is feasible to extract helium from natural gas containing as little as 0.3% He by volume. How much more abundant is helium in this natural gas than it is in air?

60. Why is helium preferred to hydrogen for filling dirigibles, even though hydrogen has greater lifting power? What advantage does helium have over nitrogen in breathing mixtures for deep-sea divers? For someone with emphysema?

61. Why is it that helium is formed in so many natural radioactive decay processes whereas argon is formed in only a single natural radioactive decay process?

ADDITIONAL PROBLEMS

62. The text states that 5.2×10^{15} metric tons of atmospheric gases are spread over a surface area of 5.0×10^8 km^2. Use these facts, together with data from Appendix B, to estimate a value of standard atmospheric pressure.

63. At 20 °C the vapor pressure of a saturated solution of $CaCl_2 \cdot 6H_2O$ is 5.67 mmHg. If a quantity of this solution is placed in a large sealed container at 20 °C and the solution kept saturated by the presence of excess solid, what relative humidity will be maintained in the air in the container? How effective is $CaCl_2 \cdot 6H_2O$ in dehumidifying air?

64. A 12.012-L sample of air is saturated with water vapor at 25.0 °C. The air is then cooled to 20.0 °C. What mass of water (dew) will deposit on the walls of the container?

65. The discovery of the noble gases followed from the observation that the density of nitrogen derived from the atmosphere (actually, a nitrogen–argon mixture) was greater than that of nitrogen derived from nitrogen-containing compounds. Determine the percent difference between these two densities. (*Hint*: Use data from Table 14.1, and work with a convenient volume of a nitrogen–argon mixture, for example, 22.414 L at STP, to establish an apparent molar mass for the nitrogen–argon mixture derived from air.)

66. There are different ways to assess how much the combustion of various fuels contributes to the buildup of CO_2 in the atmosphere. One relates the mass of CO_2 formed to the mass of fuel burned; another relates the mass of CO_2 to the quantity of heat evolved in the combustion. Which of the three fuels C(graphite), $CH_4(g)$, or $C_4H_{10}(g)$ produces the *smallest* mass of CO_2 (**a**) per gram of fuel; (**b**) per kJ of heat evolved? (*Hint*: Use data from Appendix D, and assume that all the products of each combustion are gases.)

67. A large coal-fired electric plant burns 2500 tons of coal per day. The coal that is burned contains 0.65% S by mass. Assume that all of the sulfur is converted to SO_2 and, because of a thermal inversion, remains trapped in a parcel of air that is 45 km × 60 km × 0.40 km. Will the level of SO_2 in this air exceed the primary national air quality standard of 365 μg SO_2/m^3 air?

68. Use the following and other data from the text to show that if all the sulfur in coal used in electric power plants were converted to sulfuric acid, the quantity of acid produced would exceed current demand. (1) Annual U.S. coal consumption by electric power plants: approx. 8.7×10^8 ton. (2) Average SO_2 formation in the combustion of coal: 2 mg SO_2/kJ heat evolved. (3) Typical annual U.S. production of sulfuric acid: approx. 80×10^9 lb. [*Hint*: You need to estimate the heat of combustion of coal. To do this, assume that coal is 100% C (graphite) and use data from Appendix D.]

69. The world's termite population is estimated to be 2.4×10^{17}. Annually, these termites produce an estimated 4.6×10^{16} g CO_2. The atmosphere contains 5.2×10^{15} metric tons of air (1 metric ton = 1000 kg). On a number basis, the current CO_2 level in the atmosphere is 359 ppm. What percent increase in this CO_2 level would the termites cause if none of the CO_2 they produce were removed by natural processes? (*Hint:* Recall how we obtained a molar mass for air in Conceptual Example 14.1.)

70. An important variable in the combustion of gasoline in an internal-combustion engine is the air/fuel ratio. The accompanying figure shows how the emission of pollutants is related to the air/fuel ratio. Provide a plausible interpretation of this figure. (*Hint*: We verified the stoichiometric ratio in Conceptual Example 14.1. Recall also that RH represents hydrocarbons.)

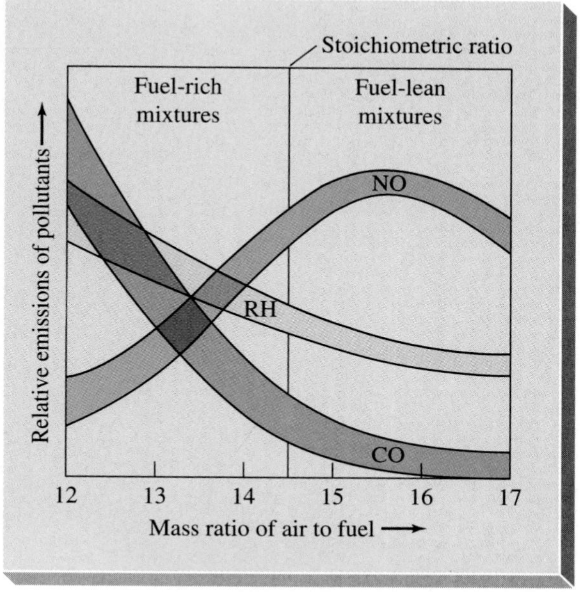

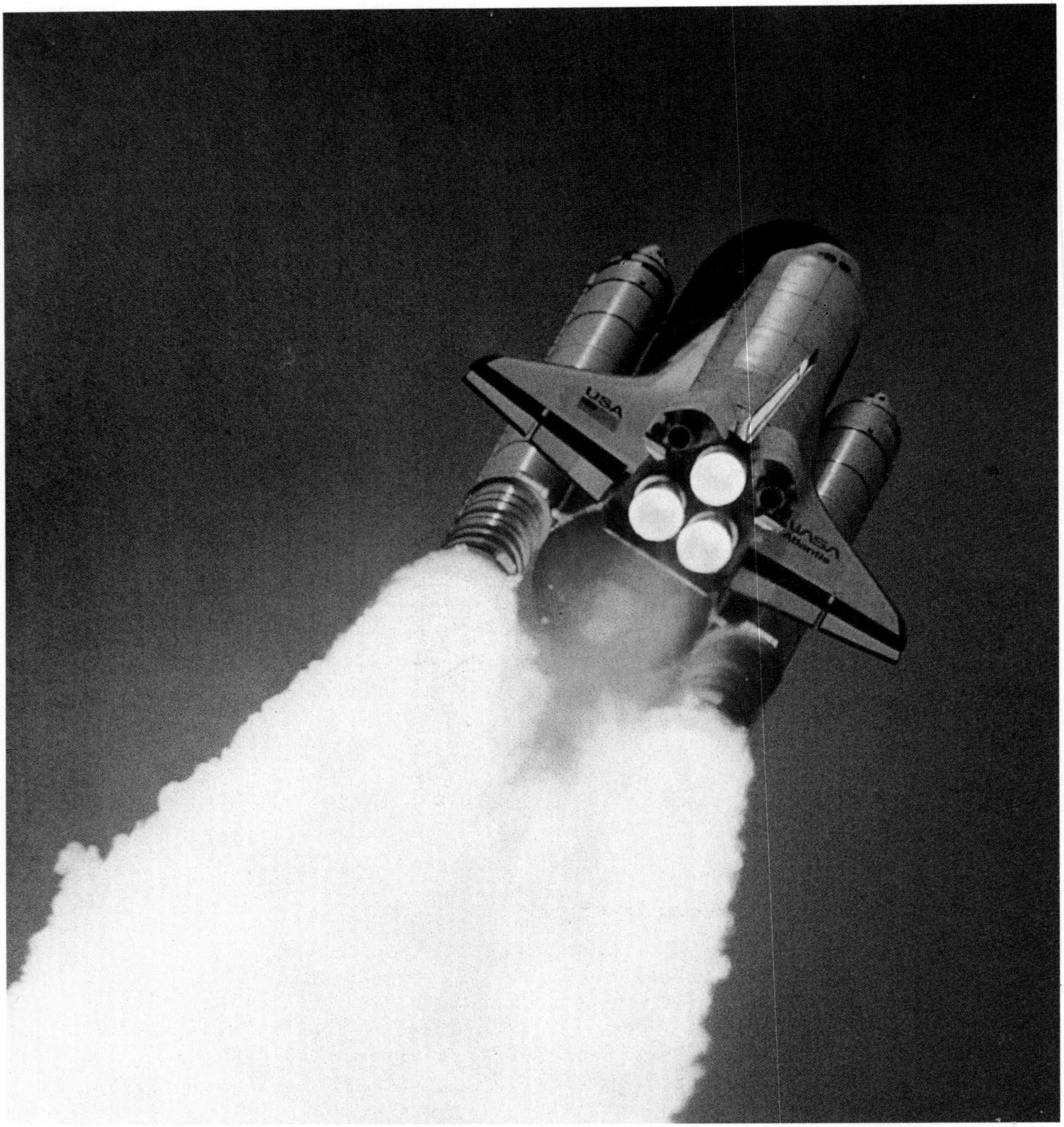

Chemical reactions proceed at vastly different rates. A rocket fuel, as shown here, must burn very quickly to provide thrust to the rocket engines. Hydrogen peroxide decomposes in the presence of a catalyst, but at a moderate rate (page 561), and iron rusts rather slowly in moist air.

15

Chemical Kinetics: Rates and Mechanisms of Chemical Reactions

CONTENTS

We have learned how to write equations for chemical reactions and have used these equations to relate quantities of reactants and products (Chapter 3). We have examined energy changes in chemical reactions (Chapter 5) and have paid particular attention to three important types of reactions: acid–base, precipitation, and oxidation–reduction (Chapter 12).

What more do we need to know about chemical reactions? Actually, quite a bit. Two other kinds of information are especially useful: How fast does a reaction go? And, in reversible reactions, what proportions of reactants are converted to products? We will deal with the first of these two fundamental questions in this chapter and with the second in several chapters following this one.

15.1 Chemical Kinetics—An Overview

Reactions proceed at different rates. Some are exceedingly slow. The disintegration of an aluminum can by atmospheric oxidation or of a plastic bottle by the action of sunlight can take years, decades, or even centuries. Some reactions are over in the blink of an eye. The decomposition of nitroglycerin, even of a large sample, occurs so fast it creates an explosion. Moreover, some reactions can take place at vastly different rates depending on conditions: Iron corrodes rather rapidly in a humid environment, but in a desert region iron rusts so slowly that objects discarded half a century ago may show only slight signs of rust. When hydrogen and fluorine are mixed at room temperature, they form hydrogen fluoride in a highly exothermic and very rapid reaction. The reaction of hydrogen and oxygen to form water is also highly exothermic, but the reaction does not occur to any measurable extent at room temperature. When the hydrogen–oxygen mixture is ignited at a high temperature, however, the reaction occurs with explosive violence (Figure 15.1).

Figure 15.1 The reaction of hydrogen and oxygen to form water. The hydrogen-filled soap bubbles, in contact with oxygen in the air, explode when they are ignited.

Both CO and NO are produced in large quantities in automobile engines (Section 14.3). Although they can react to form carbon dioxide and nitrogen,

$$2\,CO(g) + 2\,NO(g) \longrightarrow 2\,CO_2(g) + N_2(g)$$

there is little reaction before the exhaust gases escape into the environment. Yet, in the presence of a proper catalyst, the reaction can be used to reduce the concentrations of both pollutants.

Two of the goals of chemical kinetics are to *measure* and to *predict* the rates of chemical reactions. Measured rates of reaction can also contribute to another goal, that of working out a probable, step-by-step process by which a reaction occurs—a *reaction mechanism*. For example, plausible reaction mechanisms for the destruction of stratospheric ozone have implicated chlorofluorocarbons (CFCs) as important participants. To be able to predict the rates of chemical reactions means to be able to control them. Reactions can be speeded up or slowed down by adjusting certain variables. The principal variables, all of which are considered in this chapter, are

- *Concentration of reactants.* Reaction rates generally increase with the concentrations of reactants.
- *Temperature.* Reaction rates generally increase quite rapidly as the temperature increases.
- *Surface area.* Some reactions occur on a surface rather than within a homogeneous solution. For surface reactions, the rate of reaction increases with increased surface area.

The combustion of flour dust has been responsible for many flour mill explosions. The combustion reaction occurs so rapidly because of the large surface area of the dust.

- *Catalysis.* Some substances, called *catalysts*, speed up a reaction while themselves remaining unchanged.

15.2 The Meaning of the Rate of a Reaction

Rate or *speed* refers to how much something changes in a unit of time. Consider a person running at a speed of 16 km/h. The "something" that changes is the person's position, through a distance of 16 km. The unit of time is one hour. In a chemical reaction the "something" that changes is the concentration of a reactant or product, expressed as moles per liter ($mol \cdot L^{-1}$) or molarity (M). The unit of time is generally chosen to be 1 s, though it could be 1 min, 1 h, and so on. The rate of reaction, then, has units such as moles per liter per second: $mol \cdot L^{-1} \cdot s^{-1}$ or $M \cdot s^{-1}$.

Consider the steam reforming of methane, one of the chief industrial sources of hydrogen.

$$CH_4(g) + H_2O(g) \longrightarrow CO(g) + 3\ H_2(g)$$

One way to describe the rate of this reaction is in terms of the rate at which methane reacts. And the rate at which methane reacts is reflected in how the concentration of the remaining methane changes with time. The more rapid this change in concentration, the faster the reaction goes. In the expression shown below, the bracket symbol, [ ], refers to molar concentration and the delta symbol, Δ, signifies "change." Thus, $\Delta[CH_4]$ signifies a change in the molarity of CH_4. This change in molarity occurs between a time, t_1, and a later time, t_2, a time interval that we represent as Δt. For the rate of change of $[CH_4]$ we can write

We introduced the bracket symbol, [ ], for the molarity of a substance in a homogeneous mixture on page 439.

$$\text{Rate of change of } [CH_4] = \frac{[CH_4]_2 - [CH_4]_1}{t_2 - t_1} = \frac{\Delta[CH_4]}{\Delta t}$$

Because CH_4 is *consumed* in the reaction, we know that $[CH_4]_2$, the concentration of CH_4 at time t_2, is *smaller* than $[CH_4]_1$, the concentration at time t_1. $\Delta[CH_4]$ and the rate of change of $[CH_4]$ are both *negative* quantities. In order to make the rate of reaction a *positive* quantity, we introduce a *negative* sign in the expression

$$\text{Rate of reaction of } CH_4 = -(\text{rate of change of } [CH_4])$$

The rate of the steam-reforming reaction in terms of the rate at which H_2O reacts is

$$\text{Rate of reaction of } H_2O = -(\text{rate of change of } [H_2O])$$

Because CH_4 and H_2O are consumed on a 1:1 mole basis, we can also write that

$$\text{Rate of reaction of } CH_4 = \text{rate of reaction of } H_2O$$

We can also describe the rate of the reaction in terms of the rate of formation of a product, but with this important difference: we do *not* use a negative sign. Because the concentration of a product *increases* with time, the rate of change of concentration of a product is a *positive* quantity. Thus,

$$\text{Rate of formation of CO} = \text{rate of change of [CO]}$$

Finally, based on the coefficients in the balanced equation, we can write additional expressions such as

$$\text{Rate of formation of CO} = \text{rate of reaction of } CH_4$$

$$\text{Rate of formation of } H_2 = 3 \times \text{rate of reaction of } CH_4$$

Generally, there should be no difficulty in deciphering what is meant by a rate of reaction, but where ambiguity exists, we need to be specific about which participant we are referring to.

As we will see in Section 15.4, the rate of a reaction generally depends on the concentrations of the reactants. Thus, a reaction like the steam reforming of methane may start off at a high rate, but the rate drops continuously as more and more of the reactants are consumed. For this reason, when we use an equation that relates a rate of reaction to the ratio of change in concentration to change in time, the rate we calculate is an *average* rate for the time interval Δt. At the beginning of the interval, the rate of the reaction is faster than this average rate, and at the end of the interval it is slower. The situation is rather like taking your foot off the accelerator when driving an automobile at 50 mph and coming to a stop at a red light. Your average speed in this time interval may be 25 mph, but the actual speed at various times in the interval would range from 50 mph to 0 mph.

We calculate an average rate of reaction in Example 15.1 and illustrate different ways of expressing the rate of a reaction in Exercise 15.1. We'll determine more exact rates of reaction in the next section.

Example 15.1

Suppose that at some particular point in the hypothetical reaction: $A + 2\,B \longrightarrow 3\,C + 2\,D$, we find that $A = 0.4658$ M. Suppose that 125 s later $A = 0.4282$ M. What is the average rate of the reaction of A during this time period, expressed in $M \cdot s^{-1}$?

Solution

The average rate of change in the concentration of A is given by the *change* in its molarity, that is, $\Delta[A]$, divided by the time interval, Δt, over which the change occurs.

$$\Delta[A] = 0.4282\ M - 0.4658\ M = -0.0376\ M;\quad \Delta t = 125\ s$$

$$\text{Average rate of reaction of A} = -\{\text{rate of change of [A]}\}$$

$$\text{Average rate of reaction of A} = -\frac{\Delta[A]}{\Delta t} = \frac{-(-0.0376\ M)}{125\ s}$$

$$= 3.01 \times 10^{-4}\ M \cdot s^{-1}$$

Exercise 15.1

In the reaction $2\ A + B \longrightarrow 3\ C + D$, $-\Delta[A]/\Delta t$ is found to be $2.10 \times 10^{-5}\ M \cdot s^{-1}$. **(a)** What is the average rate of reaction of B? **(b)** What is the rate of formation of C?

15.3 Measuring Reaction Rates

The 3% aqueous solution of hydrogen peroxide many of us keep in our medicine cabinets loses its effectiveness as a mild antiseptic over a period of time. The H_2O_2 decomposes to oxygen gas and water, as suggested by Figure 15.2.*

$$2\ H_2O_2(aq) \longrightarrow 2\ H_2O(l) + O_2(g)$$

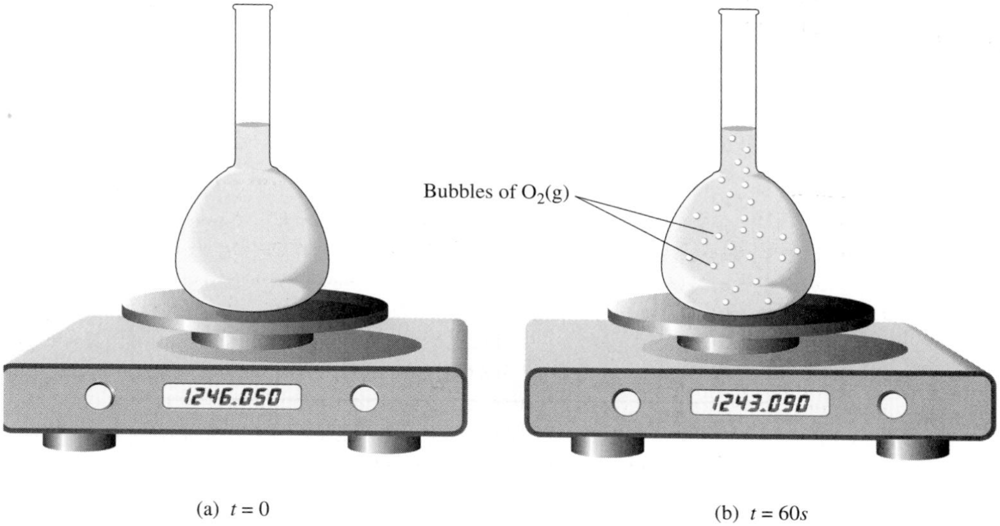

Figure 15.2 An experimental setup for determining the rate of decomposition of H_2O_2: $2\ H_2O_2(aq) \longrightarrow 2\ H_2O(l) + O_2(g)$. The mass of the reaction mixture decreases continuously as $O_2(g)$ escapes. The difference between the initial mass and the mass at a later time, t, is the mass of $O_2(g)$ that has been produced in the time interval. This sketch shows (a) the initial mass and (b) the mass 60 s later.

*Because this reaction occurs quite slowly, it is customary in laboratory studies to employ a catalyst such as $I^-(aq)$ to speed up the reaction. The data presented here are for the catalyzed decomposition of $H_2O_2(aq)$.

Figure 15.2 illustrates a simple method of determining the rate of this decomposition reaction: The $O_2(g)$ produced is allowed to escape from the reaction mixture, and the mass of the mixture is determined at various times. Some typical experimental data are tabulated in Table 15.1. The relationship between the mass of O_2 produced and the molarity of H_2O_2 remaining is described in a footnote to the table. In Figure 15.3 the molarity of H_2O_2 is plotted as a function of time. Here are some significant points brought out by Table 15.1 and Figure 15.3.

- In general, the greater the concentration of a reactant, the faster the reaction goes. In Table 15.1, note that in the first 60 s $[H_2O_2]$ drops by 0.185 M (from 0.882 M to 0.697 M). In the last 60 s interval listed, $[H_2O_2]$ drops by only 0.026 M (from 0.120 M to 0.094 M).
- The *negative* of the slope of the dashed purple line in Figure 15.3 gives the *average* rate of reaction during the interval from $t = 0$ to $t = 600$ s. This average rate can also be obtained by the method of Example 15.1. Note that the average rate is *not* the same as the rate of the reaction at 300 s (the midpoint of time in Figure 15.3).
- To get a rate of reaction at a particular time we must choose a very short time interval in the expression $-\Delta[H_2O_2]/\Delta t$, that is, a value of Δt approaching zero: $\Delta t \longrightarrow 0$. This rate is the *negative of the slope of a tangent line*, such

TABLE 15.1 Decomposition of H_2O_2[a]

Time, s	Accumulated mass O_2, g	$[H_2O_2]$, M[b]
0	0	0.882
60	2.960	0.697
120	5.056	0.566
180	6.784	0.458
240	8.160	0.372
300	9.344	0.298
360	10.336	0.236
420	11.104	0.188
480	11.680	0.152
540	12.192	0.120
600	12.608	0.094

[a] The decomposition of 1.00 L of 0.882 M, with I^- as the catalyst.
[b] Values in this column calculated as shown below for $t = 60$ s.

$$\text{no. mol } O_2 \text{ produced} = 2.960 \text{ g } O_2 \times \frac{1 \text{ mol } O_2}{32.00 \text{ g } O_2} = 0.0925 \text{ mol } O_2$$

$$\text{no. mol } H_2O_2 \text{ consumed} = 0.0925 \text{ mol } O_2 \times \frac{2 \text{ mol } H_2O_2}{1 \text{ mol } O_2} = 0.185 \text{ mol } H_2O_2$$

$$\text{no. mol } H_2O_2 \text{ remaining} = \left(1 \text{ L} \times \frac{0.882 \text{ mol } H_2O_2}{\text{L}}\right) - 0.185 \text{ mol } H_2O_2 = 0.697 \text{ mol } H_2O_2$$

$$[H_2O_2] = \frac{0.697 \text{ mol } H_2O_2}{1 \text{ L}} = 0.697 \text{ M}$$

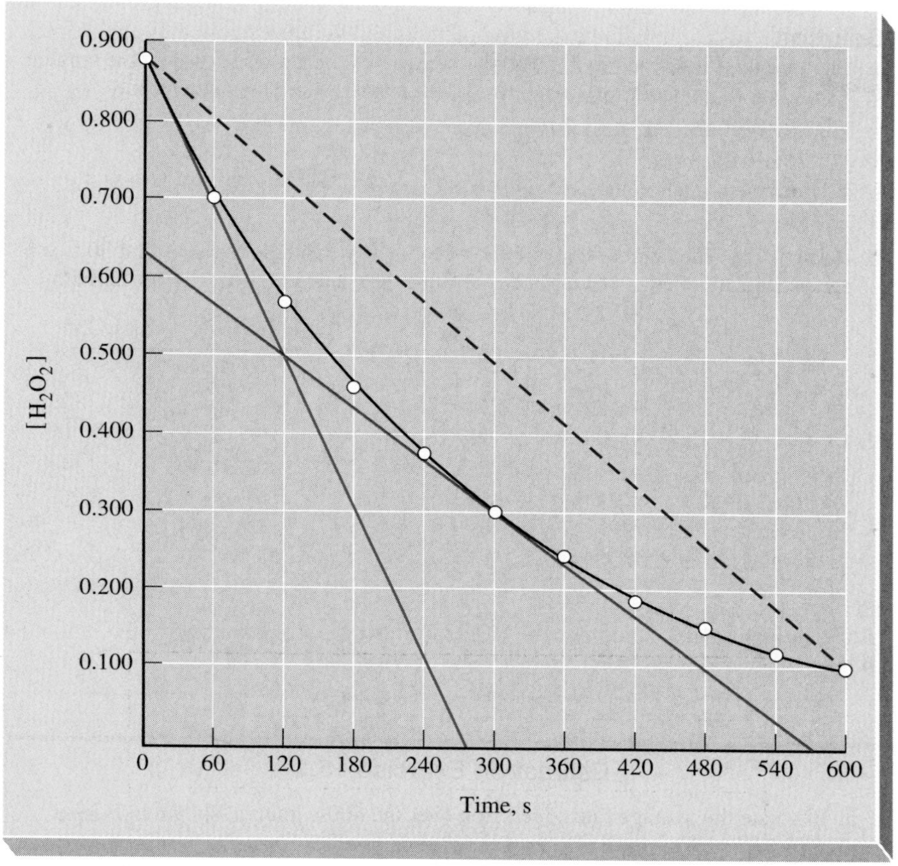

Figure 15.3 Kinetic data for the reaction $2\ H_2O_2(aq) \longrightarrow 2\ H_2O(l) + O_2(g)$. The significance of the dashed, red, and blue lines is described in the text.

as the red tangent line at t = 300 s. A rate of reaction derived from a tangent line to the concentration–time graph is called an **instantaneous rate of reaction**.

- The blue tangent line originating at the first point on the curve (0.882 M, t = 0) gives the **initial rate of reaction**. Because the tangent line and the curve coincide in the earliest stages of a reaction (about 30 s in Figure 15.3), the average and instantaneous rates of reaction are practically equal at the very start of the reaction. Thus, we can obtain an initial rate of reaction by calculation, without plotting a concentration–time graph.

Let's explore these points a bit further through Example 15.2 and Exercise 15.2.

Example 15.2

a. From Figure 15.3, determine the initial rate of reaction.
b. Calculate $[H_2O_2]$ at t = 30 s.

Solution

a. The initial rate of reaction is given by the negative of the slope of the blue tangent line. Let's evaluate the slope in terms of the points where the tangent line intersects the two axes.

$$\text{Initial rate} = -(\text{slope tangent line}) = \frac{-(0 - 0.882)\ \text{M}}{(275 - 0)\ \text{s}} = 3.21 \times 10^{-3}\ \text{M} \cdot \text{s}^{-1}$$

b. Because the blue tangent line and the curve nearly coincide over the first 30 s, we can use the initial rate from (**a**) and determine the change in concentration, $\Delta[H_2O_2]$.

$$\text{Rate of reaction} = -\{\text{rate of change of } [H_2O_2]\} = \frac{-\Delta[H_2O_2]}{\Delta t} = \frac{-\Delta[H_2O_2]}{30\ \text{s}}$$

$$= 3.21 \times 10^{-3}\ \text{M} \cdot \text{s}^{-1}$$

$$\Delta[H_2O_2] = 30\ \text{s} \times 3.21 \times 10^{-3}\ \text{M} \cdot \text{s}^{-1} = -0.096\ \text{M}$$

$$[H_2O_2]_{30\ \text{s}} = [H_2O_2]_{0\ \text{s}} + \Delta[H_2O_2]$$

$$= 0.882\ \text{M} - 0.096\ \text{M} = 0.786\ \text{M}$$

Exercise 15.2

a. From Figure 15.3, determine the instantaneous rate of reaction at $t = 300$ s.
b. Use the result of (**a**) to calculate a value of $[H_2O_2]$ at $t = 310$ s.

Estimation Exercise 15.1

a. Estimate the average rate of reaction over the 600-s interval shown in Figure 15.3. (*Hint*: What is the slope of the dashed line?)
b. Estimate the particular time at which the instantaneous rate of reaction is the same as the average rate found in (**a**). (*Hint*: Draw a tangent line that is parallel to the dashed line. Is more than one tangent line possible?)

15.4 The Rate Law of a Chemical Reaction

We have noted that one of the goals of chemical kinetics is to be able to *predict* the rates of reactions. This is done through mathematical equations known as **rate laws** or **rate equations.** For the general reaction

$$a\text{A} + b\text{B} + \cdots \longrightarrow g\text{G} + h\text{H} + \cdots$$

where A, B, ⋯, represent the reactants; G, H, ⋯, represent the products; and *a, b, g, h,* ⋯, represent the coefficients in the balanced equation. The rate law or rate equation can generally be written as*

$$\text{Rate of reaction} = k[\text{A}]^m[\text{B}]^n \cdots$$

*The reaction is assumed to go to completion. For a reversible reaction, this type of rate equation only holds for the earliest stages, when the reverse reaction is relatively unimportant.

The terms [A], [B], . . . , are molarities. The exponents m, n, . . . , which must be determined by experiment, are generally small whole numbers, *not* necessarily related to the coefficients a, b, ... The proportionality constant, k, is called the **rate constant**; it relates the rate of reaction to the concentrations of the reactants. The units of k depend on the particular rate law. Its numerical value depends on the particular reaction, the temperature, and the presence of a catalyst (if any). The larger the value of k, the faster the reaction goes.

We have already mentioned that greater concentrations of reactants tend to speed up the reaction rate—but by how much? Finding the *order of a reaction* is one way to predict just how strongly the concentrations of the reactants affect the rate of a given reaction. The **order of a reaction** is determined by the values of the exponents in the rate equation. If $m = 1$, we say the reaction is *first order* in A. If $n = 3$, the reaction is *third order* in B, and so on. The *overall* order of the reaction is the sum of the exponents in the rate equation: $m + n + \cdots$.

We will consider several ways of establishing the order of a reaction, with one of the simplest being the **method of initial rates**. Here is how we might apply the method to the general reaction

$$a\,\mathrm{A} + b\,\mathrm{B} + \cdots \longrightarrow g\,\mathrm{G} + h\,\mathrm{H} + \cdots$$

for which the rate law (or rate equation) is

$$\text{Rate of reaction} = k[\mathrm{A}]^m[\mathrm{B}]^n \cdots$$

Suppose we choose a set of reaction conditions and measure the *initial* rate of the reaction. Let's call the measured rate initial rate(1). Now let's repeat the reaction with all variables held constant, except for [A], which we will *double*: $[\mathrm{A}]_2 = 2 \times [\mathrm{A}]_1$. Again, we measure the initial rate of reaction and call it initial rate(2). Because the concentrations of all other reactants are held constant, let's combine these constants with the rate constant, k, and call this combined factor k'. That is,

$$\text{Initial rate(1)} = k'[\mathrm{A}]_1^m$$
$$\text{Initial rate(2)} = k'[\mathrm{A}]_2^m = k'(2 \times [\mathrm{A}]_1)^m = k' \times 2^m \times [\mathrm{A}]_1^m$$

Now let's write the *ratio* of initial rate(2) to initial rate(1).

$$\frac{\text{Initial rate(2)}}{\text{Initial rate(1)}} = \frac{k' \times 2^m \times [\mathrm{A}]_1^m}{k'[\mathrm{A}]_1^m} = 2^m$$

Finally, let's consider some possible results.

- If the initial reaction rate stays the same,

$$\frac{\text{Initial rate(2)}}{\text{Initial rate(1)}} = 1 \qquad \text{and} \qquad m = 0 \quad (\text{that is, } 2^m = 2^0 = 1)$$

  The reaction is *zero order* in A. If a reaction is zero order with respect to a reactant, the reaction proceeds at a rate that is *independent* of the concentration of that reactant.

- If the initial reaction rate doubles,

$$\frac{\text{Initial rate(2)}}{\text{Initial rate(1)}} = 2 \qquad \text{and} \qquad m = 1 \quad (\text{that is, } 2^m = 2^1 = 2)$$

  The reaction is *first order* in A. If a reaction is first order with respect to a reactant, the rate of reaction is directly proportional to the concentration of that reactant.

- If the initial reaction rate increases by a factor of four,

$$\frac{\text{Initial rate(2)}}{\text{Initial rate(1)}} = 4 \qquad \text{and} \qquad m = 2 \quad (\text{that is, } 2^m = 2^2 = 4)$$

  The reaction is *second order* in A. If a reaction is second order with respect to a reactant, the rate of reaction is directly proportional to the *square* of the concentration of that reactant.

- If the initial reaction rate increases by a factor of eight,

$$\frac{\text{Initial rate(2)}}{\text{Initial rate(1)}} = 8 \qquad \text{and} \qquad m = 3 \quad (\text{that is, } 2^m = 2^3 = 8)$$

  The reaction is *third order* in A. If a reaction is third order with respect to a reactant, the rate of reaction is directly proportional to the *cube* of the concentration of that reactant. Third order reactions are quite rare compared to those of lower reaction orders.

- If the initial reaction rate changes by a factor other than a power of two (1, 2, 4, 8, . . .), as occasionally happens, the reaction order is not an integer. For example, if doubling the concentration of a reactant causes the rate to increase by the factor 1.41, the order is 0.50—that is, $2^{0.5} = 1.41$.

To establish the order of the reaction with respect to another reactant, say B, all variables are held constant except [B], which is doubled between the first and second experiment. And so on for other reactants.

When the method of initial rates is applied to the decomposition of H_2O_2(aq),

$$2\,H_2O_2(aq) \longrightarrow 2\,H_2O(l) + O_2(g)$$

we find, on doubling the initial concentration of H_2O_2(aq), that initial rate(2) = 2 × initial rate(1). That is, $m = 1$. The reaction is first order with respect to H_2O_2. Because there are no other reactants, the overall reaction order is first order. Note again that there is no relationship between the order of a reaction and the coefficients in the balanced equation.

The method of initial rates is further illustrated in Example 15.3.

Example 15.3

Three experiments were done for the reaction

$$2\,NO(g) + O_2(g) \longrightarrow 2\,NO_2(g)$$

The following data were obtained for the rate of reaction of NO:

Experiment	Initial [NO]	Initial $[O_2]$	Initial rate of reaction
1	0.0125 M	0.0255 M	$0.0281\ M \cdot s^{-1}$
2	0.0125 M	0.0510 M	$0.0561\ M \cdot s^{-1}$
3	0.0250 M	0.0255 M	$0.112\ M \cdot s^{-1}$

What is the rate law and what is the value of the rate constant?

Solution

The rate law has the form

$$\text{Rate of reaction of NO} = k[NO]^m[O_2]^n$$

First, compare Experiments 1 and 2. The initial reaction rate doubles when [NO] is held constant and $[O_2]$ is doubled. The reaction is first order in O_2. Now compare the first and third experiments. The initial reaction rate increases fourfold when [NO] is doubled and $[O_2]$ is held constant. The reaction is second order in NO. The reaction is third order overall ($m + n$), and the rate law is

$$\text{Rate of reaction of NO} = k[NO]^2[O_2]$$

(The agreement between the exponents in the rate law and the coefficients in the balanced equation is just coincidental.)

We can substitute data from the three experiments into the rate law to calculate the rate constant, k. For Experiment 1,

$$\text{Rate of reaction of NO} = 2.81 \times 10^{-2}\ M \cdot s^{-1} = k(0.0125\ M)^2(0.0255\ M)$$

$$k = \frac{2.81 \times 10^{-2}\ M \cdot s^{-1}}{(0.0125\ M)^2(0.0255\ M)} = 7.05 \times 10^3\ M^{-2} \cdot s^{-1}$$

The values of k based on Experiments 2 and 3 are $7.04 \times 10^3\ M^{-2} \cdot s^{-1}$ and $7.03 \times 10^3\ M^{-2} \cdot s^{-1}$, respectively. Notice how in this third order reaction the concentration terms, collectively, produce the unit M^3. As a consequence, the units of k must be $M^{-2} \cdot s^{-1}$, so that the units for the rate of reaction are the required $M \cdot s^{-1}$. That is, $M^3 \times M^{-2} \cdot s^{-1} = M \cdot s^{-1}$.

The units of k depend on the overall reaction order. In every case the units must be such that the rate of reaction from the rate law has the units, $M \cdot (\text{time})^{-1}$.

Exercise 15.3

Consider now a possible fourth experiment in the kinetic study of Example 15.3. In this case the initial [NO] = 0.200 M and that of $[O_2]$ = 0.400 M. Predict the initial rate of reaction. (*Hint*: Use the value of k found in Example 15.3.)

In the next two sections we will focus on first-order reactions, a common reaction order with many important applications. Following that, we will briefly describe some characteristics of zero-order and second-order reactions.

15.5 First-Order Reactions

In this section, we focus our discussion on first-order reactions in which a single reactant yields products—that is, reactions of the type

$$A \longrightarrow \text{products}$$

The decomposition of H_2O_2 is an example. The rate law for such a reaction is written

$$\text{Rate of reaction of A} = k[\text{A}]$$

Concentration as a Function of Time—The Integrated Rate Equation

In Example 15.3, we saw a way to establish the order of a reaction and a value of k. In Exercise 15.3 we used the value of k to calculate a rate of reaction. As we have seen, however, the rate of a reaction generally depends on the concentrations of the reactants, and so it keeps changing as a reaction proceeds. Usually, what we really want to know is this: What will be the concentration of a reactant at a later time if we know how much is present initially? To perform this calculation we need an equation derived from the rate law and called the **integrated rate equation.** The integrated rate equation for a first-order reaction is

A brief derivation of this integrated rate equation is given in Appendix A.

$$\ln \frac{[\text{A}]_t}{[\text{A}]_0} = -kt \qquad \text{or} \qquad \ln\, [\text{A}]_t = -kt + \ln\, [\text{A}]_0$$

In this equation, $[\text{A}]_t$ is the concentration of A at some time t and $[\text{A}]_0$ is its concentration initially—that is, at $t = 0$. Notice, however, that it is not the concentrations themselves that appear in the equation but their natural logarithms, denoted by the symbol **ln**. The constant k is the rate constant for the reaction, the same constant that appears in the rate law.

The method of initial rates gave us one way to use experimental data to determine if a reaction is first order. The integrated rate equation gives us two others. If a plot of the logarithm (ln) of the concentration of reactant as a function of time yields a *straight line*, the reaction is first order.

Equation of straight line:	y	=	mx	+	b
Integrated rate equation for first-order reaction:	$\ln\, [\text{A}]_t$	=	$(-k)t$	+	$\ln\, [\text{A}]_0$

The rate constant k is the *negative* of the slope of the straight line. If the reaction is not first order, the graph of ln [A] versus t will not be a straight line.

Data for the first-order decomposition of H_2O_2 from Table 15.1 are presented again in Table 15.2, this time also listing $\ln\, [H_2O_2]$. The data are plotted in Figure 15.4. The value of the rate constant, derived from the slope of the straight-line plot, is

$$k = -(\text{slope}) = -(-3.66 \times 10^{-3}\ \text{s}^{-1}) = 3.66 \times 10^{-3}\ \text{s}^{-1}$$

Note that the unit of k for a first-order reaction, s^{-1}, is consistent with the rate law for the reaction. That is, the product $k[H_2O_2]$ has the units $\text{M} \cdot \text{s}^{-1}$, the required units for a rate of reaction.

The same data that might be used to plot ln [A] versus t can be used to establish whether a reaction is first order by direct calculation. Substitute concentration–time data into the integrated rate equation to calculate values of k.

$$k = \frac{-\ln\, [\text{A}]_t/[\text{A}]_0}{t}$$

TABLE 15.2 Decomposition of H_2O_2: Data Required to Test for a First-Order Reaction

Time, s	$[H_2O_2]$, M	ln $[H_2O_2]$
0	0.882	−0.126
60	0.697	−0.361
120	0.566	−0.569
180	0.458	−0.781
240	0.372	−0.989
300	0.298	−1.21
360	0.236	−1.44
420	0.188	−1.67
480	0.152	−1.88
540	0.120	−2.12
600	0.094	−2.36

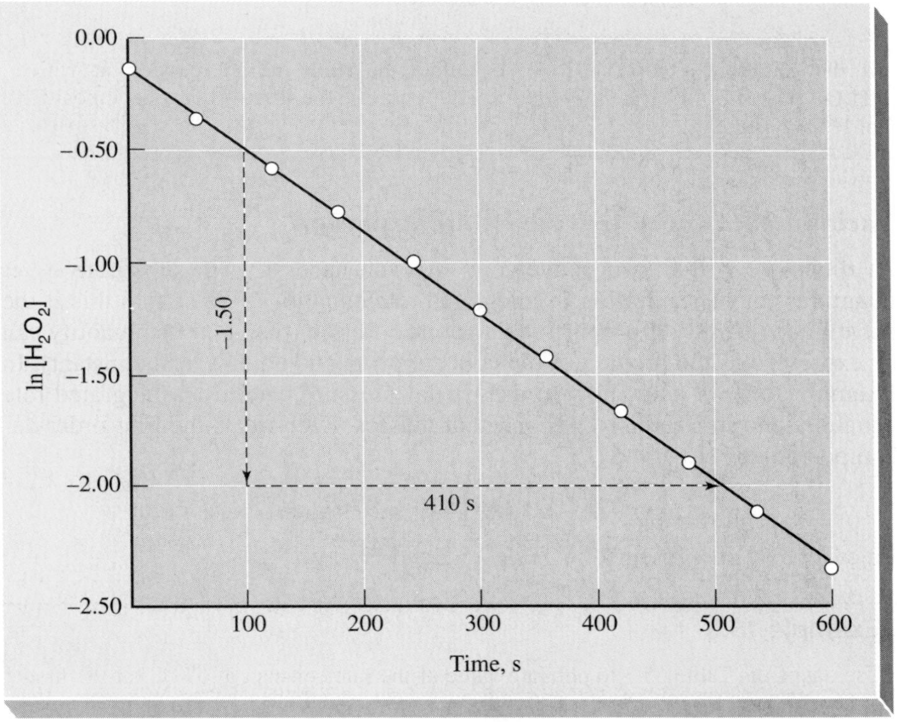

Figure 15.4 Test for a first-order reaction decomposition of $H_2O_2(aq)$. The data plotted are from Table 15.2. The straight-line plot of ln $[H_2O_2]$ vs. time proves the reaction to be first order. The slope of the line is

$$m = \frac{-1.50}{410\text{ s}} = -3.66 \times 10^{-3}\text{ s}^{-1}$$

The rate constant,

$$k = -m = 3.66 \times 10^{-3}\text{ s}^{-1}$$

If all the concentration–time data give essentially the same value of k, the reaction is first order. If the k values vary significantly, the reaction is not first order.

Example 15.4 shows how to *calculate* the concentration of a reactant at some given time in a reaction. Exercise 15.4 illustrates the inverse: calculating the time in a reaction at which a particular concentration of reactant remains.

Example 15.4

Starting with an initial $[H_2O_2] = 0.882$ M, calculate $[H_2O_2]$ after 225 s. Use the value $k = 3.66 \times 10^{-3}\ s^{-1}$ for the first-order decomposition of $H_2O_2(aq)$.

Solution

We have three of the four terms in the equation

$$\ln \frac{[A]_t}{[A]_0} = -kt$$

We can rearrange the equation to the form

$$\ln [A]_t - \ln [A]_0 = -kt$$

$$\ln [A]_t = -kt + \ln [A]_0$$

substitute the known data, and solve for $[A]_t$.

$$\ln [A]_t = -\{(3.66 \times 10^{-3}\ s^{-1}) \times 225\ s\} + \ln 0.882$$

$$= -0.824 - 0.126 = -0.950$$

$$[A]_t = e^{-0.950} = 0.387\ M$$

As shown in the discussion of logarithms in Appendix A, the relationship between a number, N, and its logarithm, $\ln N$, is that $N = e^{\ln N}$. Thus, $[A]_t = e^{\ln [A]_t} = e^{-0.950} = 0.387$ M.

Exercise 15.4

If the initial $[H_2O_2] = 0.882$ M, calculate the time in the reaction at which $[H_2O_2] = 0.500$ M. Use $k = 3.66 \times 10^{-3}\ s^{-1}$ for the first-order decomposition of $H_2O_2\,(aq)$.

Variations of the Integrated Rate Equation

At times we will find it convenient, or even necessary, to substitute other quantities for concentration in integrated rate equations. We can do this if the substituted quantity is proportional to concentration (that is, if the quantity can be expressed as the product of the concentration and a numerical constant). In Example 15.5 we show how to use partial pressure data in an integrated rate equation, and in Exercise 15.5, reactant masses. Each treats the first-order decomposition of N_2O_5 at 67 °C.

$$2\ N_2O_5(g) \longrightarrow 4\ NO_2(g) + O_2(g)$$

Relevant data are presented in Table 15.3 and Figure 15.5.

TABLE 15.3
Decomposition of N_2O_5 at 67 °C

Time, s	$P_{N_2O_5}$, mmHg
0	800
60	564
120	398
180	279
240	197
300	138

Example 15.5

Use data from Table 15.3 to obtain a value of the rate constant at 67 °C for the first-order reaction

$$2\ N_2O_5(g) \longrightarrow 4\ NO_2(g) + O_2(g)$$

Solution

By our previous approach, we would substitute known data into the integrated rate equation

$$\ln \frac{[A]_t}{[A]_0} = -kt$$

and solve for k. However, we are not given concentrations of N_2O_5, nor can we calculate them. We don't know the initial amount of N_2O_5 or the reaction volume.

For an ideal gas, $P = nRT/V$. As long as T and V are constant, P is directly proportional to n, the number of moles of gas. And n/V is simply the molar concentration of N_2O_5. We can therefore substitute partial pressures of N_2O_5 for molarities. The data in the following expression are the initial partial pressure $P_0 = 800$ mmHg at $t = 0$ and an arbitrarily chosen second point, $P_t = 279$ mmHg at $t = 180$ s.

$$\ln \frac{P_t}{P_0} = -kt$$

$$\ln \frac{279 \text{ mmHg}}{800 \text{ mmHg}} = -k(180 \text{ s})$$

$$-1.05 = -k(180 \text{ s})$$

$$k = \frac{1.05}{180 \text{ s}} = 5.83 \times 10^{-3} \text{ s}^{-1}$$

Exercise 15.5

A 45.0-g sample of N_2O_5 is allowed to decompose at 67 °C. What mass of N_2O_5 remains *undecomposed* after 5.00 min? Use the rate constant established in Example 15.5. (*Hint*: Can you show that the mass of a substance is proportional to the number of moles of substance and, in a constant reaction volume, to molarity?)

Half-life of a Reaction

We often describe a first-order reaction in terms of the time required to reduce a reactant concentration (or mass or partial pressure) to a *fraction* of its initial value. Thus, in the decomposition of N_2O_5, we can speak of the point in a reaction where one-half ($\frac{1}{2}$) of the N_2O_5 remains, or one-quarter ($\frac{1}{4}$), or 0.35, or 38%, and so on. We are especially interested in the time in which one-half of a reactant has been consumed, a time called the **half-life** of the reaction. At this time, which we will denote as $t = t_{1/2}$, $[A]_t = \frac{1}{2}[A]_0$, which means that

$$\ln \frac{[A]_t}{[A]_0} = \ln \frac{\frac{1}{2}[A]_0}{[A]_0} = \ln \tfrac{1}{2} = -kt_{1/2}$$

$$t_{1/2} = \frac{-\ln \frac{1}{2}}{k} = \frac{0.693}{k}$$

Thus, *for a first-order reaction the half-life, $t_{1/2}$, is constant and depends only on the value of k. If either $t_{1/2}$ or k is known, the other can be readily calculated.* These ideas are illustrated in Figure 15.5 and Example 15.6. In Exercise 15.6 we see that the idea of the half-life can be extended to other periods of time that are powers of $\frac{1}{2}$. That is, the time for the concentration of reactant to be reduced to $\frac{1}{4}$ is $2 \times t_{1/2}$; reduced to $\frac{1}{8}$, $3 \times t_{1/2}$; and so on. Estimation Exercise 15.2 shows that estimates about reaction rates can be made without detailed calculations.

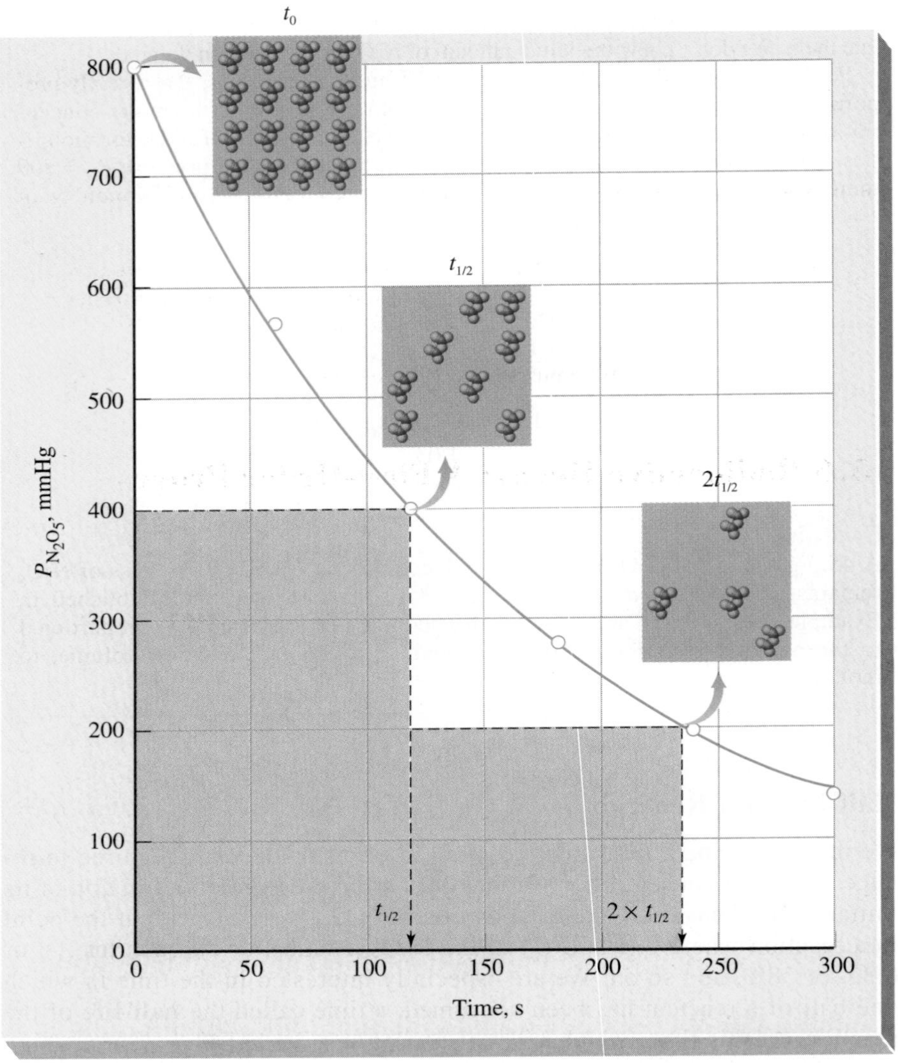

Figure 15.5 Decomposition of N_2O_5 at 67 °C. The significance of the time periods $t_{1/2}$ and $2 \times t_{1/2}$ is discussed in the text and illustrated in Example 15.6.

Example 15.6

a. Use Figure 15.5 to evaluate the half-life of the decomposition of N_2O_5 at 67 °C.
b. What is the rate constant at 67 °C?

Solution

a. In the first half-life period, the partial pressure of N_2O_5 falls from 800 mmHg to 400 mmHg, and this occurs in just under 120 s, perhaps about 118 s. The conclusion of the second half-life period, in which the partial pressure of N_2O_5 falls from 400 mmHg to 200 mmHg, appears at about 235 s. This confirms that $t_{1/2}$ is just about 118 s.
b. The rate constant, k, is calculated using $t_{1/2} = 118$ s.

$$k = \frac{0.693}{t_{1/2}} = \frac{0.693}{118 \text{ s}} = 5.87 \times 10^{-3} \text{ s}^{-1}$$

This agrees well with our earlier calculation, $k = 5.83 \times 10^{-3}\ \text{s}^{-1}$, from Example 15.5.

Compare the partial pressures of N_2O_5 at the times 0, 120, and 240 s in Table 15.3. Can you see that $t_{1/2}$ should be slightly less than 120 s?

Exercise 15.6

Use the result of Example 15.6 to determine the time required to reduce the quantity of N_2O_5 to $\frac{1}{32}$ of its initial value. (*Hint*: How many half-life periods are involved?)

Estimation Exercise 15.2

Which seems like a probable time required to consume 90% of a sample of N_2O_5 decomposing at 67 °C: (a) 200 s, (b) 300 s, (c) 400 s, (d) 500 s? (*Hint*: What is the fraction of N_2O_5 remaining? How is this fraction related to powers of $\frac{1}{2}$?)

15.6 Radioactive Decay: A First-Order Process

One of the most familiar and important examples of a first-order process is radioactive decay, a process governed by the **radioactive decay law**:

> *The rate of disintegration of a radioactive isotope, called the decay rate or activity, A, is directly proportional to the number of atoms present.*

$$\text{Rate of decay} = A = \lambda N$$

Consider a 5,500,000-atom sample disintegrating at the rate of 85 atoms per second. $N = 5.5 \times 10^6$ atom and $A = 85$ atom s^{-1}. The *decay constant*, λ, is

$$\lambda = \frac{A}{N} = \frac{85 \text{ atom s}^{-1}}{5.5 \times 10^6 \text{ atom}} = 1.5 \times 10^{-5} \text{ s}^{-1}$$

Note that λ is analogous to k for a first-order reaction. Both have the unit $(\text{time})^{-1}$.

By analogy to other first-order processes, we can write the equations

$$\ln \frac{N_t}{N_0} = -\lambda t \qquad \text{or} \qquad \ln N_t = -\lambda t + \ln N_0 \qquad \text{and} \qquad t_{1/2} = \frac{0.693}{\lambda}$$

In these equations N_0 represents the initial number of atoms at $t = 0$. N_t is the number of atoms at time t; λ is the decay constant; and $t_{1/2}$ is the half-life of the decay process. Applied to a radioactive isotope, the **half-life** is the length of time required for one-half of the atoms of the isotope to undergo decay. At the end of a half-life period there are only half as many radioactive atoms present as at the beginning.

Iodine-131, a radioactive isotope used in studies of the thyroid gland, has a half-life of 8 days. Thus, half the atoms in a sample undergo decay in an 8-day period; the activity, A, falls to half its initial value in 8 days. The activity is down to one-fourth of its initial value in 16 days, to one-eighth in 24 days, and so on.

The shorter the half-life, the larger the value of λ and the faster the decay proceeds. Half-lives range from microseconds—radium-218 has a half-life of 14 μs—to billions of years (like uranium-238). Table 15.4 lists several radioisotopes, their half-lives, and some uses.

TABLE 15.4 **Half-Lives of Representative Isotopes**

Isotope	Half-Life[a]	Typical Use
Hydrogen-3	12.26 y	biochemical tracer
Carbon-11	20.39 min	PET scans
Carbon-14	5730 y	dating of artifacts
Sodium-24	14.659 h	tracer, cardiovascular system
Phosphorus-32	14.3 d	biochemical tracer
Potassuim-40	1.25×10^9 y	dating of rocks
Iron-59	44.496 d	tracer, red blood cell lifetime
Cobalt-57	271.77 d	tracer, uptake of vitamin B_{12}
Cobalt-60	5.271 y	radiation treatment of cancer
Strontium-90	28.5 y	no present use (component of radioactive fallout)
Iodine-131	8.040 d	tracer, thyroid studies
Gadolinium-153	241.6 d	determining bone density
Radium-226	1.60×10^3 y	radiation therapy for cancer
Uranium-238	4.51×10^9 y	dating of rocks and Earth's crust

[a] s = second, min = minute, h = hour, d = day, y = year.

Example 15.7

The isotope sodium-24 (Table 15.4) is used to detect constrictions and obstructions in the circulatory system. It emits β^- particles. **(a)** What is the decay constant, in s^{-1}, for sodium-24? **(b)** What is the activity of a 1.00-mg (1.00×10^{-3} g) sample of sodium-24? **(c)** What will be the approximate rate of decay of the 1.00-mg sample after 73 h? **(d)** What will be the rate of decay of the 1.00-mg sample after 1 week (168 h)?

Solution

a. We can relate λ to the value of $t_{1/2}$ in Table 15.4. We also must convert from h^{-1} to s^{-1}.

$$\lambda = \frac{0.693}{t_{1/2}} = \frac{0.693}{14.659\ \text{h}} \times \frac{1\ \text{h}}{60\ \text{min}} \times \frac{1\ \text{min}}{60\ \text{s}} = 1.31 \times 10^{-5}\ \text{s}^{-1}$$

b. First let's find the number of atoms (N) in 1.00 mg of sodium-24. Then we can multiply this number by the decay constant (λ) to get the decay rate.

$$N = 1.00 \times 10^{-3}\ \text{g}\ {}^{24}\text{Na} \times \frac{1\ \text{mol}\ {}^{24}\text{Na}}{24.0\ \text{g}\ {}^{24}\text{Na}} \times \frac{6.022 \times 10^{23}\ \text{atom}\ {}^{24}\text{Na}}{1\ \text{mol}\ {}^{24}\text{Na}}$$

$$= 2.51 \times 10^{19}\ \text{atoms of}\ {}^{24}\text{Na}$$

$$A = \lambda N = 1.31 \times 10^{-5}\ \text{s}^{-1} \times 2.51 \times 10^{19}\ \text{atom}\ {}^{24}\text{Na}$$

$$= 3.29 \times 10^{14}\ \text{atom} \cdot \text{s}^{-1}$$

c. The simplest answer to this question requires us to see that 73 h is almost exactly *five* half-life periods, that is, 73 h/14.659 h ≈ 5. In five half-life periods the activity, A, falls to $(\frac{1}{2})^5 = \frac{1}{32}$ of its initial value.

$$A = \frac{1}{32} \times 3.29 \times 10^{14}\ \text{atom} \cdot \text{s}^{-1} \approx 1.03 \times 10^{13}\ \text{atom} \cdot \text{s}^{-1}$$

d. The method of part (**c**) will not work here. The number of half-life periods would be 168 h/14.569 h = 11.5, a nonintegral number. Instead let's combine the radioactive decay law, $A = \lambda N$ or $N = A/\lambda$, and the integrated rate equation, $\ln N_t/N_0 = -\lambda t$

$$\ln \frac{N_t}{N_0} = \ln \frac{A_t/\lambda}{A_0/\lambda} = \ln \frac{A_t}{A_0} = -\lambda t$$

and

$$\ln A_t = -\lambda t + \ln A_0$$

We have a value of λ from part (**a**), and A_0 from part (**b**). We must convert time from hours to seconds and solve the following expression for A_t.

$$\ln A_t = -(1.31 \times 10^{-5}\ \text{s}^{-1} \times 168\ \text{h} \times 3600\ \text{s/h}) + \ln (3.29 \times 10^{14}) = 25.50$$

$$A_t = e^{25.50} = 1.19 \times 10^{11}\ \text{atom} \cdot \text{s}^{-1}$$

Exercise 15.7

The half-life of plutonium-239 is 2.411×10^4 y. How long would it take for a sample of plutonium-239 to decay to 1.00% of its present activity? (*Hint*: Use the method of Example 15.7 (**d**). What is the ratio A_t/A_0, in this case?)

Estimation Exercise 15.3

Use the method of Example 15.7(**c**) to show that the answer obtained in Example 15.7 (**d**) is a reasonable one. (*Hint*: Between what two values should the answer in Example 15.7 (**d**) fall?)

Radiocarbon Dating

We can use the decay rate of naturally occurring carbon-14, a β^- emitter with a half-life of 5730 y, to determine the age of organic matter. The method works fairly well for objects from a few hundred to about 50,000 years old.

Carbon-14 appears to be formed at a constant rate in the upper atmosphere by the bombardment of nitrogen-14 with neutrons from cosmic radiation. When a neutron is absorbed by the nucleus of a nitrogen-14 atom, a proton is ejected. The result is that the nitrogen-14 atom is converted to a carbon-14 atom. We can show this through a *nuclear* equation in which the neutron is represented as ${}^1_0n$ and the proton as ${}^1_1\text{H}$.

$${}^{14}_{7}\text{N} + {}^{1}_{0}\text{n} \longrightarrow {}^{14}_{6}\text{C} + {}^{1}_{1}\text{H}$$

The ^{14}C isotope is eventually incorporated into atmospheric carbon dioxide. At the same time that ^{14}C is being formed, ^{14}C in the environment is constantly disintegrating, resulting in an essentially constant concentration of about one atom of ^{14}C for every 10^{12} atoms of ^{12}C. A living plant consumes CO_2; animals consume plants; and the plants and animals are in equilibrium with ^{14}C in the atmosphere. That is, they replace ^{14}C atoms that have undergone radioactive decay with "fresh" ^{14}C atoms from their environment.

Carbon-14 in living matter has an activity of about 15 disintegrations per minute (dis min^{-1}) per gram of carbon. But consider what happens when a tree is cut. The ^{14}C that decays is not replaced, and as the concentration of ^{14}C falls, the disintegration rate falls as well. From the reduced disintegration rate at some later time, we can estimate the age of an object incorporating wood from the tree.

Example 15.8

A wooden object from an Egyptian tomb is subjected to radiocarbon dating. The decay rate observed for its carbon-14 content is 7.2 dis min^{-1} per g C. What is the age of the wood in the object (and, presumably, of the object itself)? The half-life of ^{14}C is 5730 y.

Solution

First, we must determine the decay constant.

$$\lambda = \frac{0.693}{t_{1/2}} = \frac{0.693}{5730\ \text{y}} = 1.21 \times 10^{-4}\ \text{y}^{-1}$$

Now, we need to use the integrated radioactive decay equation. Here, as in Example 15.7(**d**), we can use the radioactive decay law to write

$$N_0 = A_0/\lambda \qquad \text{and} \qquad N_t = A_t/\lambda$$

Then we can solve the following equation for time, t.

$$\ln\frac{N_t}{N_0} = \ln\frac{7.2/\cancel{\lambda}}{15/\cancel{\lambda}} = \ln\frac{7.2}{15} = -\lambda t = -(1.21 \times 10^{-4}\ \text{y}^{-1}) \times t$$

$$t = \frac{-\ln 7.2/15}{1.21 \times 10^{-4}\ \text{y}^{-1}} = 6.1 \times 10^{-3}\ \text{y}$$

Exercise 15.8

Tritium (^3H), a β^--emitting hydrogen isotope, can be used to determine the age of items up to about 100 years. A sample of brandy, stated to be 25 years old and offered for sale at a premium price, has tritium with half the activity of that found in new brandy. Is the claimed age of the beverage authentic? Use data from Table 15.4.

15.7 Reactions of Other Orders

We have focused most of our attention on first-order reactions, but there are reactions of other orders. We will briefly consider some of those.

Zero-Order Reactions

The rate of a **zero-order** reaction is independent of the concentration of reactant(s); in the rate equation, the sum of the exponents: $m + n + \cdots = 0$. Some factor other than concentration determines the rate of reaction—for example, light intensity for a photochemical reaction or surface area in a surface-catalyzed reaction (page 592).

The decomposition of ammonia on a tungsten surface is a zero-order reaction.

$$2\,NH_3(g) \xrightarrow{W} N_2(g) + 3\,H_2(g)$$

$$\text{Rate of reaction} = k \times [NH_3]^0 = k$$

Concentration–time data for two experiments at 1100 °C are plotted in Figure 15.6, illustrating that

- The graph of concentration versus time is a *straight line* with a *negative* slope. The slope is $-3.40 \times 10^{-6}\ M \cdot s^{-1}$.
- The rate of the reaction, which remains *constant* throughout the reaction, is the *negative* of the slope.

$$\text{Rate of reaction} = -(-3.40 \times 10^{-6}\ M \cdot s^{-1}) = 3.40 \times 10^{-6}\ M \cdot s^{-1}.$$

The units of k for a zero-order reaction are the same as the units of the rate of reaction—$M \cdot s^{-1}$.

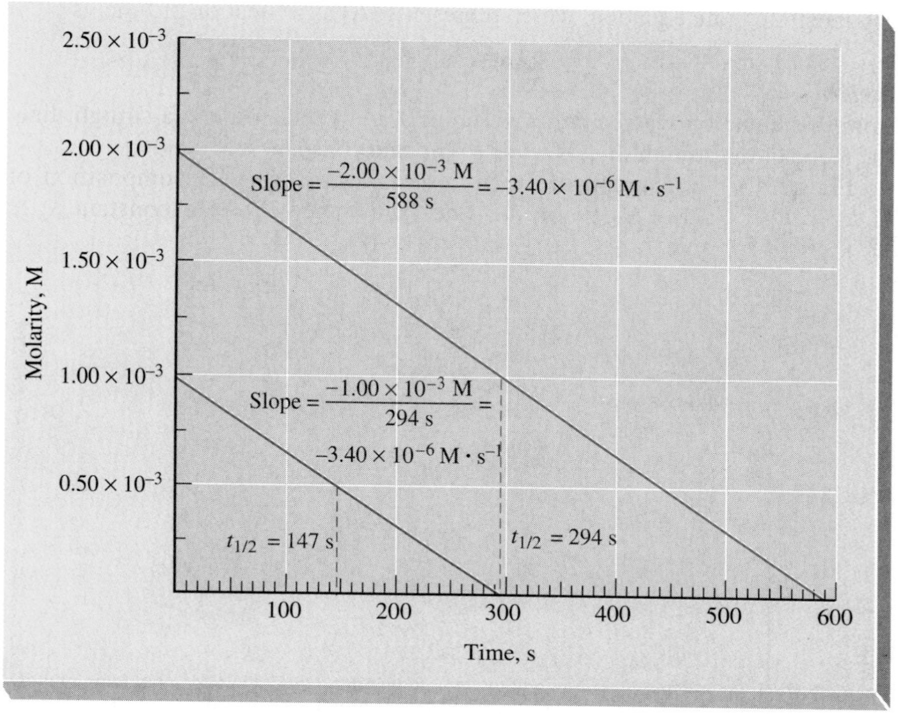

Figure 15.6 **The decomposition of ammonia on a tungsten surface at 1100 °C: a zero-order reaction.** The concentration of $NH_3(g)$ is plotted as a function of time for two different initial concentrations. The parallel, straight-line graphs show this to be a zero-order reaction. The slopes of the lines are indicated, as are the half-lives for the two experiments.

- The two straight lines are *parallel*. They have the same slope. This indicates that the rate of reaction is independent of the initial concentration of reactant.
- The half-life is *different* for the two experiments. The time required for half of a reactant to be consumed increases as the initial concentration increases.

We'll encounter zero-order reactions again in the discussion of enzyme-catalyzed reactions in Section 15.12.

Second-Order Reactions

A **second-order** reaction has a rate equation in which the sum of the exponents, $m + n + \cdots = 2$. One example is the reaction of NO(g) and $O_3(g)$.

$$NO(g) + O_3(g) \longrightarrow NO_2(g) + O_2(g)$$

This reaction is first order in NO and in O_3 and second order overall. The rate of reaction $= k[NO][O_3]$.

The decomposition of HI(g) is also second order.

$$2\,HI(g) \longrightarrow H_2(g) + I_2(g)$$

This reaction is of the type A $\longrightarrow$ products—that is, a single reactant yields products. The rate equation for this type of second-order reaction is

$$\text{Rate of reaction of A} = k[A]^2$$

A brief derivation of this integrated rate equation is given in Appendix A.

The integrated rate equation, which represents [A] as a function of time, is

$$\frac{1}{[A]_t} = kt + \frac{1}{[A]_0}$$

From this equation we see that a graph of 1/[A] versus time is a straight line. The slope of the straight line is equal to the rate constant, k.

Figure 15.7 is a graph of 1/[HI] versus time for the decomposition of HI at 321.4 °C. The slope of the line, and hence the rate constant k, is $2.3 \times 10^{-4}\ M^{-1} \cdot min^{-1}$.

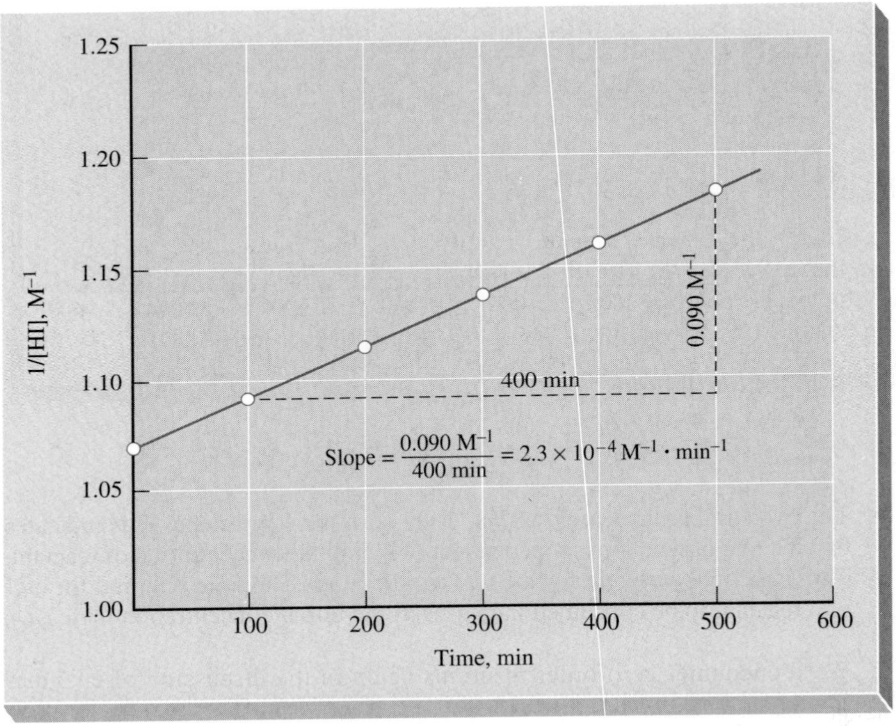

Figure 15.7 The decomposition of hydrogen iodide at 321.4 °C: A second-order reaction. The *reciprocal* of the concentration of HI(g), 1/[HI], is plotted as a function of time. The rate constant for this second-order reaction is equal to the slope of the line: $k = 2.3 \times 10^{-4}\ M^{-1} \cdot min^{-1}$.

To establish the half-life of a second-order reaction of the type, A $\longrightarrow$ products, we substitute the quantities $[A]_t = \frac{1}{2}[A]_0$ and $t = t_{1/2}$ into the integrated rate equation.

$$\frac{1}{\frac{1}{2}[A]_0} = kt_{1/2} + \frac{1}{[A]_0}$$

$$\frac{2}{[A]_0} - \frac{1}{[A]_0} = kt_{1/2}$$

$$\frac{1}{[A]_0} = kt_{1/2}$$

$$t_{1/2} = \frac{1}{k[A]_0}$$

As with a zero-order reaction, the half-life of second-order reaction depends on the initial concentration as well as on the rate constant k.

Example 15.9

What is the half-life of the decomposition reaction represented in Figure 15.7?

Solution

The value of $1/[A]_0$—that is, the initial point of the graph—appears to be at 1.065 M^{-1}, and the rate constant is $k = 2.3 \times 10^{-4}\ M^{-1}\cdot min^{-1}$.

$$t_{1/2} = \frac{1}{k}\left(\frac{1}{[A]_0}\right) = \frac{1}{2.3 \times 10^{-4}\ M^{-1} \cdot min^{-1}} \times 1.065\ M^{-1} = 4.6 \times 10^3\ min$$

Exercise 15.9

Show that the next half-life period for the reaction in Figure 15.7 is *twice* as long as the first half-life period, that is, 9.2×10^3 min. (*Hint*: What is the value of $1/[A]_0$ at the start of the second half-life period?)

Pseudo-First-Order Reactions

To simplify their kinetic treatment, at times, we treat higher order reactions in a way that makes them resemble reactions of a lower order. If a higher order reaction is made to behave like a first-order reaction, it is called *pseudo* first order. For example, the conversion of sucrose into the simpler sugars glucose and fructose (a reaction known as the inversion of sucrose) is first order in each reactant and second order overall.

$$\underset{\text{Sucrose}}{C_{12}H_{22}O_{11}(aq)} + H_2O(l) \longrightarrow \underset{\text{Glucose}}{C_6H_{12}O_6(aq)} + \underset{\text{Fructose}}{C_6H_{12}O_6(aq)}$$

$$\text{Rate of reaction} = k[C_{12}H_{22}O_{11}][H_2O]$$

In dilute aqueous solutions the molarity of H_2O far exceeds that of the sucrose. Therefore, when all the sucrose has reacted, the amount of H_2O consumed is very small compared to its initial amount; $[H_2O]$ remains essentially unchanged in the reaction. If our interest is in how the concentration of sucrose changes during the reaction, we can treat $[H_2O]$ as a constant and combine its value with that of k to obtain a new rate constant, k'.

$$\text{Rate of reaction} = k[H_2O] \times [C_{12}H_{22}O_{11}] = k'[C_{12}H_{22}O_{11}]$$

As a result, the reaction displays all the qualities of a first-order reaction.

CONCEPTUAL EXAMPLE 15.1

Figure 15.8 is a graph of [A] versus time for two different experiments dealing with the reaction

$$A \longrightarrow \text{products}$$

What is the order of this reaction?

Solution

We are seeking the exponent, m, in the rate equation

$$\text{Rate of reaction of A} = k[\text{A}]^m$$

Test for Zero-Order Reaction. The rate law of a zero-order reaction is rate = $k[\text{A}]_0 = k$. If the reaction were zero order, the rate of reaction would be constant, meaning that the graph of [A] versus time would be a straight line. Because the graphs are not linear, the reaction cannot be zero order.

Test for First-Order Reaction. One test for a first-order reaction is that the half-life is constant: $t_{1/2} = 0.693/k$. In Experiment 1, [A] falls from 1.00 M to one-half this value, 0.50 M, in 1000 s. In Experiment 2, [A] falls from 2.00 M to 1.00 M in 500 s. The half-life is *not* constant; the reaction is *not* first order.

Test for Second-Order Reaction. We can show that the reaction is second order by

1. *The method of initial rates:* In the initial stages of a reaction, where the tangent line and the concentration–time curve coincide, we can approximate the reaction rate by the expression

$$\text{Rate of reaction of A} = -\Delta[\text{A}]/\Delta t$$

In the present case an appropriate Δt is 50 s.
For Experiment 1,

$$\text{Initial rate} = -(0.95 - 1.00)\ \text{M}/50\ \text{s} = 1 \times 10^{-3}\ \text{M} \cdot \text{s}^{-1}$$

For Experiment 2,

$$\text{Initial rate} = -(1.82 - 2.00)\ \text{M}/50\ \text{s} = 3.6 \times 10^{-3}\ \text{M} \cdot \text{s}^{-1}$$

Doubling the initial concentration causes essentially a quadrupling of the initial rate of a reaction. The reaction must be *second* order.

2. *A half-life method:* If the reaction is second order, the half-life can be expressed through the equation

$$t_{1/2} = \frac{1}{k[\text{A}]_0}$$

but we can also write

$$k = \frac{1}{t_{1/2}[\text{A}]_0}$$

As we noted above, when $[\text{A}]_0 = 2.00$ M, $t_{1/2} = 500$ s, and when $[\text{A}]_0 = 1.00$ M, $t_{1/2} = 1000$ s. Substituted into the above equation, these data yield the same value of k, suggesting that the reaction is second order.

$$k = \frac{1}{500\ \text{s} \times 2.00\ \text{M}} = 1.00 \times 10^{-3}\ \text{M}^{-1} \cdot \text{s}^{-1}$$

$$k = \frac{1}{1000\ \text{s} \times 1.00\ \text{M}} = 1.00 \times 10^{-3}\ \text{M}^{-1} \cdot \text{s}^{-1}$$

Conceptual Exercise 15.1

Use the data in Figure 15.8 to show, through an appropriate *straight line* graph, that the reaction is second order.

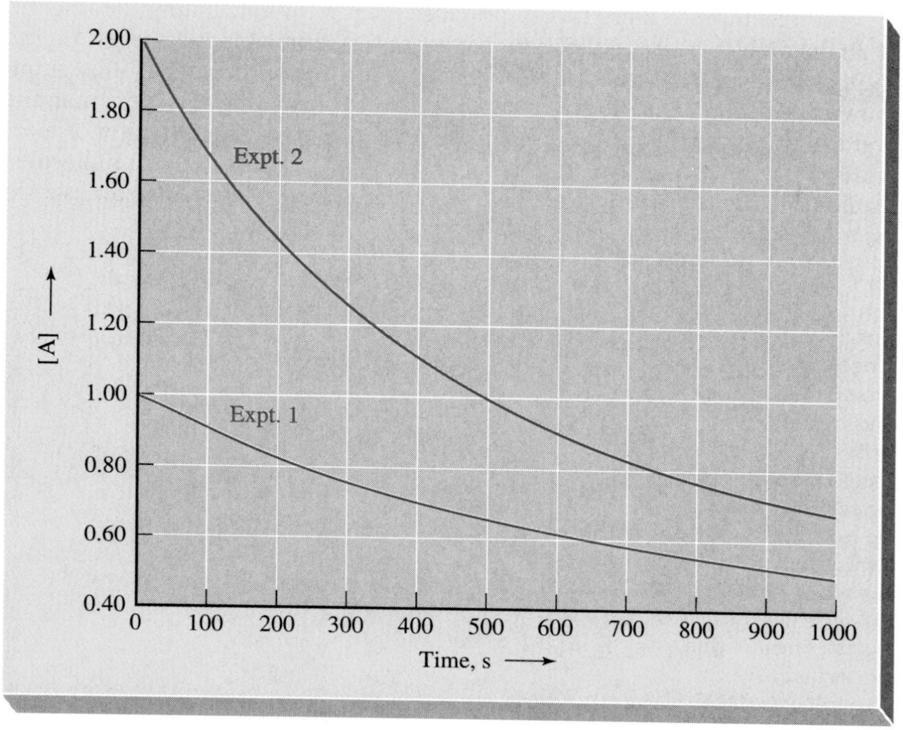

Figure 15.8 Conceptual Example 15.1 illustrated.

15.8 Theories of Chemical Kinetics

So far, we have explored a number of practical matters, and none of these has required us to look at what happens at the molecular level. Now, to answer some basic questions: Why are some reactions first order, others second order, or other orders? Why do some reactions go so fast and others so slowly? How do catalysts work? We do need to look at the molecular level, and we will do this in much of the remainder of the chapter.

Collision Theory

Before atoms, molecules, or ions can react, they must first get together—they must *collide*. However, if all collisions between molecules produced chemical reaction, the rates of chemical reactions would be much greater than what we generally observe. Some of the collisions must be ineffective. Thus, the rate of a chemical reaction is proportional to the product of the frequency of molecular collisions *and* the fraction of these collisions that are effective.

An effective collision between two molecules puts enough energy into certain key bonds to break them. Such a collision may occur between two fast-moving molecules, or perhaps when an especially fast one collides with a slower one. A collision between two slow-moving molecules will most likely be ineffective. The minimum total kinetic energy that two colliding molecules must bring to their collision for a reaction to occur is called the **activation energy** of the reaction.

The kinetic-molecular theory provides a basis for calculating the fraction of all the molecules in a collection that possess a certain kinetic energy. Figure 15.9 represents the situation at two different temperatures. As the graph shows, the average kinetic energy at T_1, the lower temperature, is less than the average kinetic energy at T_2. Seen another way, T_1 has a *large* fraction of molecules at *low* kinetic energies, compared with T_2. If we assume that molecules require kinetic energies in excess of the value marked by the red arrow, we are led to two conclusions:

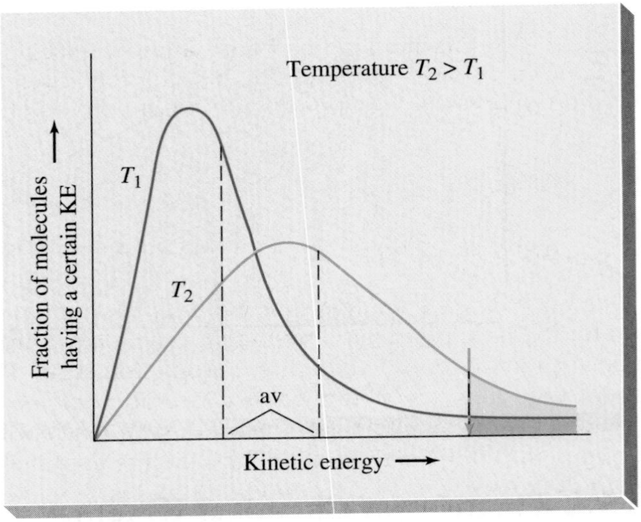

Figure 15.9 Distribution of kinetic energies of molecules. The fraction of the molecules having energies in excess of the value marked by the red arrow is small compared to the total number of molecules. However, this fraction increases rapidly with temperature.

- Only a small fraction of the molecules at either temperature are energetic enough to react, but
- this fraction becomes larger as the temperature increases.

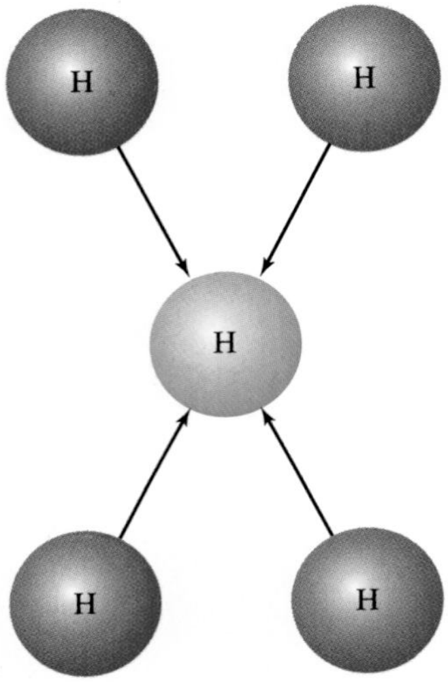

Figure 15.10 A reaction in which the orientation of colliding molecules is unimportant. No matter from which direction an oncoming hydrogen atom (blue) approaches the "target" hydrogen atom (red), its "view" of the impending collision is the same. There is no preferred direction for one hydrogen atom to approach another when they react to form a hydrogen molecule.

The rates of many reactions cannot be accounted for just in terms of the collision frequency and the fraction of activated molecules. Another factor is involved—the orientation of the colliding molecules. As shown in Figure 15.10, because of the symmetrical distribution of the electron cloud of a hydrogen atom, every approach of one H atom to another—front, back, top, bottom, . . . —is the same. The orientation of the colliding molecules is not a factor in the reaction

$$H\cdot + \cdot H \longrightarrow H_2$$

In most cases, however, the orientation of the colliding molecules is a factor. Consider the reaction of iodide ion with methyl bromide to produce methyl iodide and bromide ion.

$$I^- + CH_3Br \longrightarrow CH_3I + Br^-$$

As shown in Figure 15.11, a collision of an I^- ion with the C atom of CH_3Br can be effective in leading to a reaction. A collision of an I^- ion with the Br atom of CH_3Br is not.

We shall see another example of the importance of a proper orientation of reacting molecules in the action of enzymes as catalysts (Section 15.12).

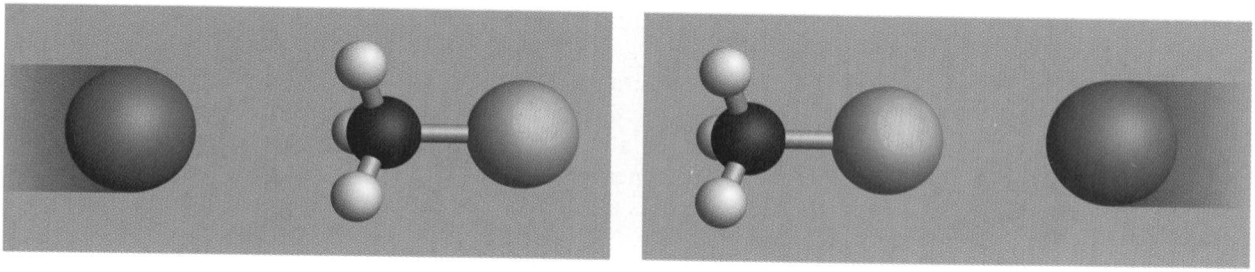

(a) (b)

Figure 15.11 The importance of orientation of colliding molecules. The I^- ion in (a) collides with the C atom of CH_3Br, a favorable collision for reaction to occur. The I^- ion in (b) collides with the Br atom of CH_3Br, an unfavorable collision for chemical reaction.

Transition State Theory

In addition to recognizing that reactions result from molecular collisions, it is useful to think in terms of an intermediate state that lies between the reactants and products; we call this the **transition state**. The aggregate of atoms in the transition state is formed by a favorable collision, and is called the **activated complex**. The total kinetic energy that colliding molecules must bring to the collision to form the high-energy activated complex is called the *activation energy*.

For the reaction in which iodide ion displaces bromide ion from methyl bromide, we can represent the activated complex in this way:

$$\underset{\text{Reactants}}{I^- + CH_3{-}Br} \longrightarrow \underset{\text{Activated complex}}{\overset{\delta-}{I}\cdots\cdot CH_3\cdots\cdot \overset{\delta-}{Br}} \longrightarrow \underset{\text{Products}}{I{-}CH_3 + Br^-}$$

This representation suggests that there is no bond between iodine and carbon in the reactants, a partial bond in the activated complex (indicated by the dotted line to show that a bond is forming), and a full covalent bond in the product, methyl iodide. The situation between bromine and carbon is the reverse: a full bond (reactant) $\longrightarrow$ a partial bond (activated complex) $\longrightarrow$ no bond (product). Another view of the situation is that there is a 1– charge on an iodine atom as a reactant, a 1– charge on a bromine atom as a product, and a partial negative charge (indicated as $\delta-$) on each of these atoms in the activated complex.

Another way of thinking about the activation energy of a reaction is illustrated in Figure 15.12. This representation, called a **reaction profile**, shows energy plotted as a function of a parameter called the progress of the reaction. Think of the progress of the reaction as representing how far the reaction has gone. That is, the reaction begins with reactants on the left, passes through a transition state, and ends with products on the right.

Figure 15.12 is a reaction profile of the reaction

$$CO(g) + NO_2(g) \longrightarrow CO_2(g) + NO(g) \qquad \Delta H = -226 \text{ kJ}$$

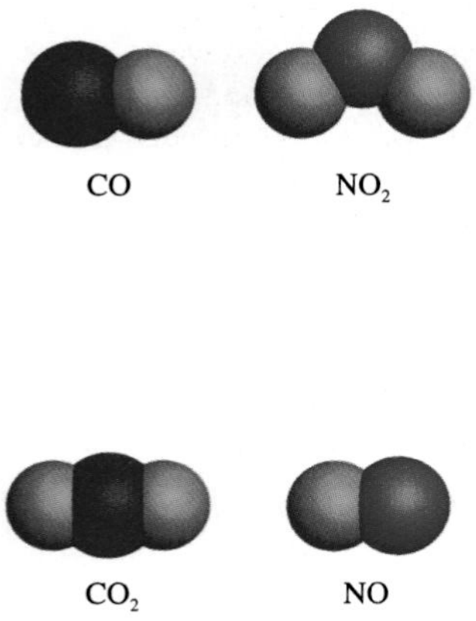

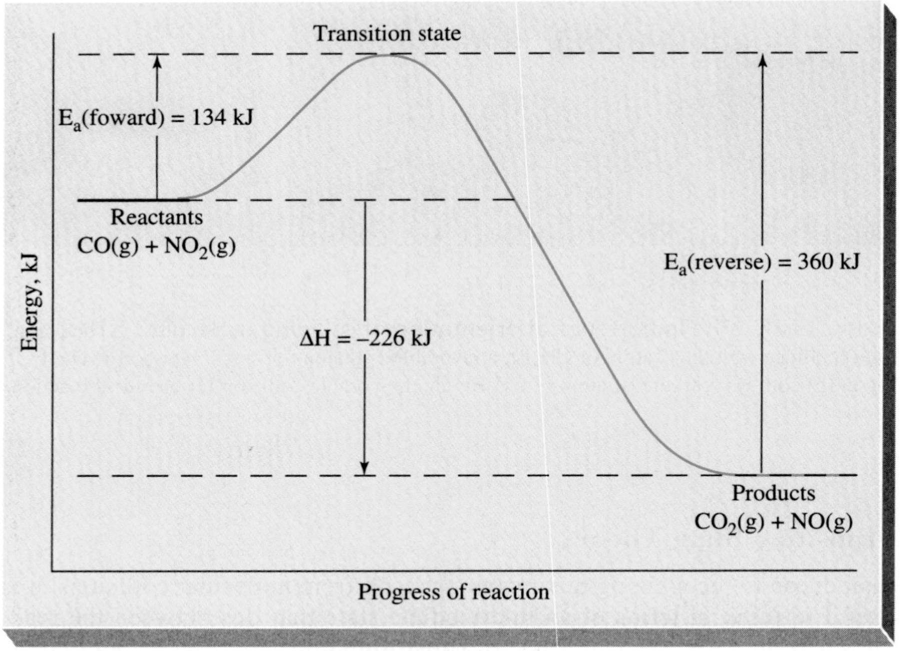

Figure 15.12 A reaction profile for the reaction CO(g) + NO_2(g) ⟶ CO_2(g) + NO(g). This reaction profile follows energy changes during the course of a reaction. From left to right, we proceed from reactants through the transition state to products.

at temperatures above 600 K. The difference in energies between the products and reactants is ΔH for the reaction*; the reaction is exothermic. An energy barrier separates the products from the reactants, and reactant molecules must have enough energy to surmount this barrier if a reaction is to occur. The difference in energy between the top of the barrier—the transition state—and that of the reactants is the activation energy of the forward reaction, E_a(forward). Its value signifies that, collectively, one mole of CO molecules and one mole of NO_2 molecules must bring 134 kJ of energy into their collisions to form one mole of activated complex. The activated complex then dissociates into product molecules.

Figure 15.12 also describes the reverse reaction, the reaction of CO_2 and NO to form CO and NO_2. The activation energy of the reverse reaction, E_a(reverse) = 360 kJ, is greater than that of the forward reaction.

Two important ideas brought out in Figure 15.12 are that (1) ΔH of a reaction is equal to the difference in activation energies of the forward and reverse reactions,

$$\Delta H = E_a(\text{forward}) - E_a(\text{reverse})$$

and (2) for an *endothermic* reaction the activation energy must always be equal to or greater than the enthalpy of reaction. Figure 15.13 provides an analogy to a reaction profile and activation energy.

---

* The difference in potential energy between reactants and products is ΔE of a reaction. For this reaction, $\Delta H = \Delta E$, but as we learned in Section 5.4, even for reactions where they are not equal, the difference between ΔH and ΔE is usually quite small.

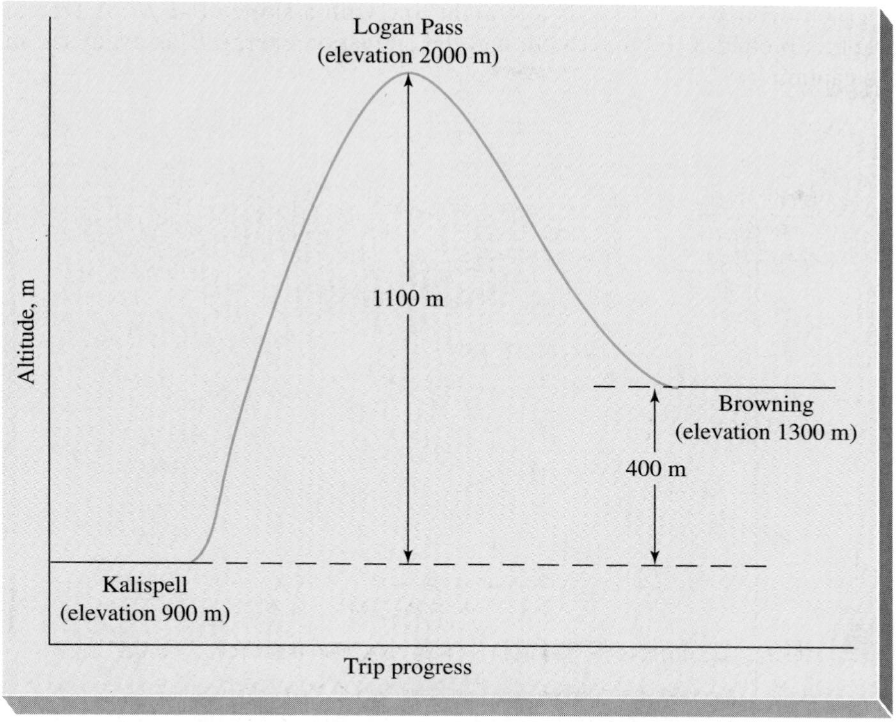

Figure 15.13 An analogy to the profile of a reaction and activation energy. If you were in Kalispell, Montana (the reactants), and wished to travel to Browning (the products), you could choose to drive the scenic Going-to-the-Sun Highway (the reaction profile) through Glacier National Park. To do this you would have to cross the continental divide at Logan Pass (the transition state). First, you would have to climb 1100 m (the activation energy), but then it would be downhill the rest of the way.

15.9 Effect of Temperature on the Rates of Reactions

Charcoal (carbon) reacts so slowly with oxygen at room temperature that we see no change in it at all. When we raise the temperature of the charcoal with flames from lighter fluid, the charcoal begins to burn more rapidly. As it does so, the heat of combustion raises the temperature even more, and the charcoal burns faster still. To slow down reactions, we generally lower the temperature. We refrigerate butter to slow down the reactions that cause it to go rancid. In the laboratory, H_2O_2(aq) is usually stored in the refrigerator to slow down its decomposition.

It seems reasonable that raising the temperature should speed up a reaction. The average kinetic energies of molecules increase, leading to more frequent collisions. But increased collision frequency is not the most important factor. As we saw in Figure 15.9, higher temperatures mean that more molecules are especially energetic (energetic enough, that is, to create a reaction). So not only are there more collisions, but the percentage of *effective* collisions is also greater, and the rate of a chemical reaction increases.

The activation energy usually does not change with temperature, so the energy requirement of the reactant molecules also does not change.

Quantitatively, the effect of temperature on the rate of a chemical reaction is expressed through its effect on the rate constant, k. Svante Arrhenius was the first to propose this relationship, in 1889. The *Arrhenius equation* is

$$\ln k = \frac{-E_a}{RT} + \ln A$$

A graph of ln k versus $1/T$ is a straight line with a slope of $-E_a/R$. A typical graph is plotted in Figure 15.14, and the activation energy E_a is evaluated in the caption.

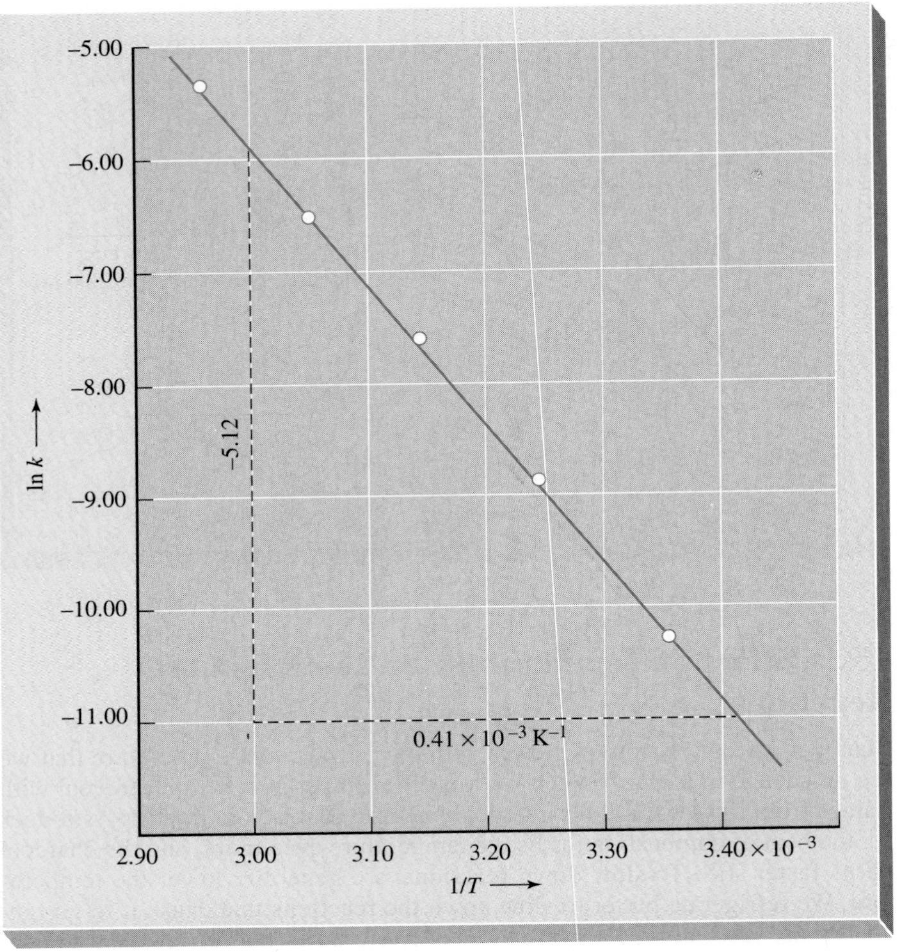

Figure 15.14 A plot of ln k vs. $1/T$ for the reaction $2\ N_2O_5(g) \longrightarrow 4\ NO_2(g) + O_2(g)$

Evaluation of E_a:

$$\text{slope} = \frac{-5.12}{0.41 \times 10^{-3}\ K^{-1}} = -1.2 \times 10^4\ K = \frac{-E_a}{R}$$

$$E_a = 1.2 \times 10^4\ K \times 8.3145\ J \cdot mol^{-1} \cdot K^{-1} = 1.0 \times 10^5\ J/mol = 1.0 \times 10^2\ kJ/mol$$

An alternative that requires no graphing is to measure the rate constant at two different temperatures and to apply the Arrhenius equation twice. This eliminates the constant term ln A, yielding the equation

$$\ln \frac{k_2}{k_1} = \left(\frac{E_a}{R}\right)\left(\frac{1}{T_1} - \frac{1}{T_2}\right)$$

where k_1 and k_2 are the rate constants at the Kelvin temperatures T_1 and T_2, and E_a is the activation energy. The equation can be used to calculate any one of the five quantities from known values of the other four. The required form of the gas constant R is $8.3145\ J \cdot mol^{-1} \cdot K^{-1}$.

The derivation of this form of the Arrhenius equation from that shown on page 585 is given in Appendix A.

Example 15.10

Calculate a value of k at 375 K for the reaction

$$2\ N_2O_5(g) \longrightarrow 4\ NO_2(g) + O_2(g)$$

given that $E_a = 1.0 \times 10^2$ kJ/mol and that $k = 2.5 \times 10^{-3}\ s^{-1}$ at $T = 332$ K.

Solution

We have the necessary data to substitute into the equation

$$\ln \frac{k_2}{k_1} = \left(\frac{E_a}{R}\right)\left(\frac{1}{T_1} - \frac{1}{T_2}\right)$$

Let's label the unknown k as k_2. The corresponding temperature is $T_2 = 375$ K.

$$\ln \frac{k_2}{2.5 \times 10^{-3}} = \frac{1.0 \times 10^2\ \text{kJ/mol} \times 1000\ \text{J/kJ}}{8.3145\ \text{J} \cdot \text{mol}^{-1} \cdot \text{K}^{-1}}\left(\frac{1}{332\ \text{K}} - \frac{1}{375\ \text{K}}\right)$$

$$= 1.2 \times 10^4\ \text{K} \times (0.00301 - 0.00267)\ \text{K}^{-1}$$

$$= 4.1$$

$$\frac{k_2}{2.5 \times 10^{-3}} = e^{4.1} = 60$$

$$k_2 = 1.5 \times 10^{-1}\ \text{s}^{-1}$$

Exercise 15.10

Use data from Example 15.10 to determine the temperature at which the rate constant $k = 1.0 \times 10^{-5}\ s^{-1}$ for the reaction

$$2\ N_2O_5(g) \longrightarrow 4\ NO_2(g) + O_2(g)$$

(*Hint*: In the setup shown above, use $k_2 = 1.0 \times 10^{-5}\ s^{-1}$ and solve for T_2.)

15.10 Reaction Mechanisms

A few simple reactions consist only of the single step suggested by the balanced equation for the reaction. The reaction between methyl bromide and iodide ion (Section 15.8) is an example. That is, the reaction occurs as the direct result of one effective collision: I^- and CH_3Br.

$$I^- + CH_3Br \longrightarrow CH_3I + Br^-$$

Now consider the reaction between nitrogen monoxide and oxygen, a key reaction in the formation of photochemical smog (Section 14.3).

$$2\ NO(g) + O_2(g) \longrightarrow 2\ NO_2(g)$$

It is highly unlikely that two NO molecules and one O_2 molecule collide *simultaneously*, just as it is unlikely that *three* basketballs might simultaneously collide in midair as basketball players take warm-up shots at the basket. Instead, the overall reaction is accomplished in steps.

A **reaction mechanism** portrays a chemical reaction as a series of simple steps, called elementary steps, that ultimately lead from the initial reactants to the final products. An **elementary step** represents, at the molecular level, a single stage in the progress of the overall reaction. The changes that occur in an elementary step involve an altering of the energy or geometry of starting molecules or the formation of a new molecule(s). Two requirements must be met by a plausible reaction mechanism. The mechanism must

- account for the *experimentally determined* rate law, and
- be consistent with the stoichiometry of the overall or net reaction.

Let's explore the nature of elementary steps a bit further.

Elementary Processes

When proposing a reaction mechanism, we rely on these important characteristics of elementary steps:

1. The exponents of the concentration terms in the rate equation *for an elementary step* are the same as the coefficients in the balanced equation for the step. (This is often *not* the case with the rate equation for the overall reaction.)
2. An elementary step in which a single molecule dissociates is a **unimolecular** step; one in which two molecules collide effectively is a **bimolecular** step. Both unimolecular and bimolecular steps are much more probable than a step requiring the simultaneous collision of three molecules (a **termolecular** step).
3. Elementary steps are reversible, and some may reach a state of equilibrium in which the rates of the forward and reverse steps are equal.
4. Species formed in one step of a reaction mechanism and consumed in another are called **intermediates**. Intermediates must not appear in the net chemical equation nor in the overall rate law.
5. One elementary step may be much slower than all the others. In many cases this is the **rate-determining step**, the crucial step in establishing the rate of the overall reaction.

A Mechanism with a Slow Step Followed by a Fast Step

Let's return to the decomposition of hydrogen peroxide. We learned earlier in this chapter that this is a first-order reaction with the rate law: rate of reaction = $k[H_2O_2]$. The reaction is speeded up in the presence of $I^-(aq)$, a fact that we represent through the equation

$$2\,H_2O_2(aq) \xrightarrow{I^-} 2\,H_2O + O_2(g)$$

In the presence of iodide ion, the rate of reaction is found experimentally to be first order in both H_2O_2 and I^-. Any mechanism we propose for the reaction must account for this fact. It must also account for the fact that I^- does not appear in the overall or net equation.

Here's a plausible reaction mechanism:

$$\begin{array}{ll} \textit{Slow:} & H_2O_2 + I^- \longrightarrow H_2O + OI^- \\ \textit{Fast:} & H_2O_2 + OI^- \longrightarrow H_2O + O_2 + I^- \\ \hline \textit{Net:} & 2\,H_2O_2 \longrightarrow 2\,H_2O + O_2 \end{array}$$

We say that a reaction mechanism is "plausible" if it explains the *experimental* facts. We don't say that it is "the" reaction mechanism because there might be some other equally or more plausible mechanism.

The two steps, added together, give the *net* or overall reaction. The *slow* step is the *rate-determining* step. The OI^- ion is an *intermediate* in the reaction; as soon as OI^- is formed in the first step, it is consumed in the second step. The rate of the reaction of H_2O_2 is determined solely by the rate of the first *elementary step.*

$$\text{Rate of reaction of } H_2O_2 = \text{rate of slow step} = k[H_2O_2][I^-]$$

What is the role of I^- in the mechanism for the decomposition of H_2O_2? It is consumed in the first step but produced once more in the second step. I^- does not appear in the net equation, yet it is *not* an intermediate. It is there at the beginning of the reaction and remains at the end. It merely serves to speed up the reaction; I^- is a *catalyst.* Because $[I^-]$ remains constant throughout the reaction, we can incorporate its value with that of the rate constant k. That is, the product of two constants is also a constant, so that $k[I^-] = k'$. This is what makes it possible for us to treat the decomposition of H_2O_2 as a simple first-order reaction, which we did earlier in the chapter.

$$\text{Rate of reaction of } H_2O_2 = k'[H_2O_2]$$

The value of k', of course, depends on the initial concentration of I^-. The greater the concentration of I^- that is present, the larger the rate constant k', and the faster the reaction goes.

As an analogy to a slow first step followed by a fast second step, consider driving to a market a few hundred yards away but having to cross a nearby bridge under construction. If traffic on the bridge is limited to one lane controlled by a road worker and with an average wait of 15 min, the trip to the market, on average, will take just over 15 min. The time to complete the trip is almost entirely determined by the rate of the first slow step: crossing the bridge.

Reaction mechanisms in which there is no slow step, or in which the reversibility of an elementary step is important, are more complicated than the one we have just described. We will not consider such cases in this text.

Example 15.11

At certain temperatures, the reaction of CO and NO_2 is thought to follow the mechanism

$$\begin{array}{ll} \textit{Step 1}: & 2NO_2(g) \longrightarrow NO_3(g) + NO(g) \\ \textit{Step 2}: & NO_3(g) + CO(g) \longrightarrow NO_2(g) + CO_2(g) \end{array}$$

The experimentally determined rate law is

$$\text{Rate of reaction of } NO_2 = k[NO_2]^2$$

(**a**) Write the equation for the net reaction. (**b**) Which species is an intermediate? (**c**) Which step is likely to be the rate-determining step?

Solution

a. The net equation is the sum of the equations for the two elementary steps:

$$NO_2(g) + CO(g) \longrightarrow NO(g) + CO_2(g)$$

b. $NO_3(g)$ is an intermediate. It is formed in the first step, consumed in the second step, and not present in the net equation.

c. The first is probably the rate-determining step. The rate equation for this elementary step is the same as the experimentally determined overall rate law. The second step cannot be rate determining. Its rate equation, rate $= k[NO_3][CO]$, contains a term, $[NO_3]$, that does not occur in the experimentally determined rate law.

At temperatures above 600 K, the reaction appears to proceed by a one-step mechanism, $CO(g) + NO_2(g) \longrightarrow CO_2(g) + NO(g)$ with the rate law rate of reaction $= k[CO][NO_2]$. The reaction profile in Figure 15.12 is for this one-step mechanism.

Exercise 15.11

Consider the reaction between nitrogen monoxide and chlorine gases to form NOCl(g). The experimentally determined rate law is

$$\text{Rate of reaction} = k[NO][Cl_2]$$

A proposed mechanism is

Step 1: $NO(g) + Cl_2(g) \longrightarrow NOCl_2(g)$

Step 2: $NOCl_2(g) + NO(g) \longrightarrow 2\,NOCl(g)$

(**a**) Write the equation for the net reaction. (**b**) Which species is an intermediate? (**c**) Which step is likely to be the rate-determining step?

15.11 Catalysis

We have seen that raising the temperature generally speeds up a reaction, sometimes dramatically. But when applied to reactions in a living organism, raising the temperature can have a disastrous effect: it can kill the organism. Living organisms rely on a different means of speeding up reactions—catalysis. Many industrial and laboratory processes require catalysis as well.

Recall the decomposition of potassium chlorate (Section 14.4), a reaction formerly used to produce small quantities of oxygen in the laboratory.

$$2\,KClO_3(s) \longrightarrow 2\,KCl(s) + 3\,O_2(g)$$

Without a catalyst, the $KClO_3(s)$ must be heated to over 400 °C to produce $O_2(g)$ at a useful rate. However, if we add a small quantity of $MnO_2(s)$, we can get the same rate of oxygen evolution at just 250 °C. Further, after the reaction is complete, we can recover all the manganese dioxide, unchanged. A **catalyst** is a substance that, like $MnO_2(s)$ in the decomposition of $KClO_3(s)$, increases the rate of a reaction without itself being changed by the reaction. In general, a catalyst works by changing the mechanism of a chemical reaction. The pathway of a catalyzed reaction has a lower activation energy than the pathway of an uncatalyzed reaction. If the activation energy is lowered, then more molecules have sufficient energies to engage in effective collisions. Figure 15.15 suggests an analogy to the reaction profiles of a catalyzed and an uncatalyzed reaction.

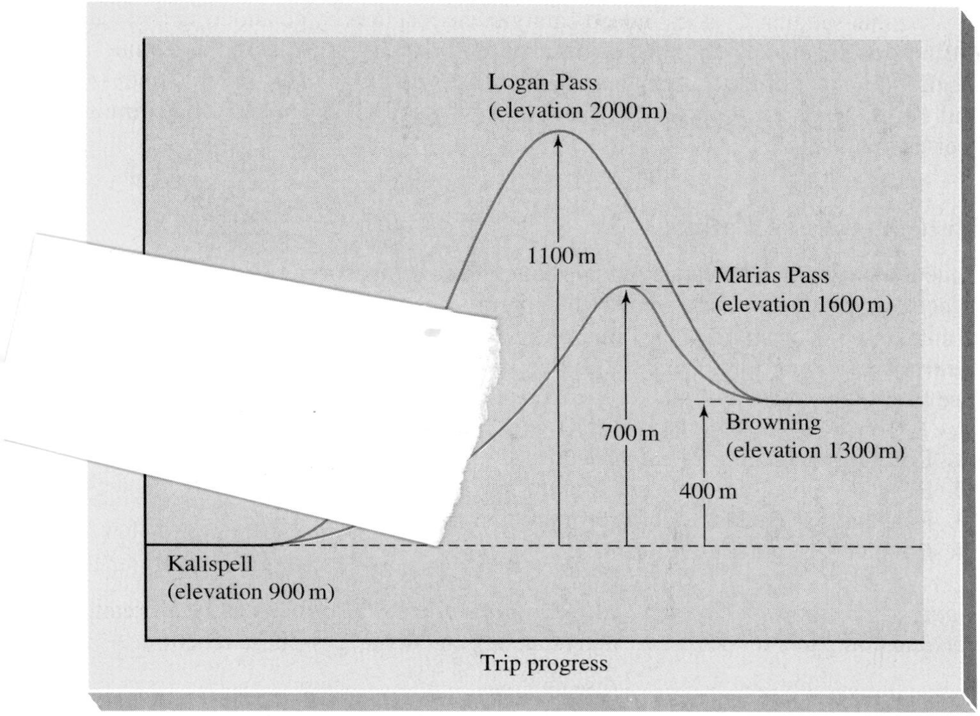

Figure 15.15 An analogy to the reaction profile and activation energy in a catalyzed reaction. To return to our analogy of the trip from Kalispell to Browning (Figure 15.13), it is possible to take an alternate route. U.S. Highway 2 crosses the continental divide through Marias Pass (catalyzed reaction). This route involves a climb of only 700 m (lower activation energy), compared with 1100 m via Logan Pass (uncatalyzed reaction). This alternate route is analogous to the different pathway provided by a catalyst in a chemical reaction.

Let's now reconsider two or three reactions discussed earlier in the text in terms of catalysis.

Homogeneous Catalysis

If the reactants, products, and catalyst all are present within the same *homogeneous* mixture—gaseous or liquid—the reaction involves *homogeneous* catalysis. One simple mechanism for catalysis in a reaction in which two reactants, A and B, yield two products, C and D, is

$$\begin{array}{ll} & \text{Reactant A + catalyst} \longrightarrow \text{intermediate + product C} \\ & \text{Reactant B + intermediate} \longrightarrow \text{product D + catalyst} \\ \hline \textit{Net:} & \text{Reactant A + reactant B} \longrightarrow \text{product C + product D} \end{array}$$

As expected, neither the intermediate nor the catalyst appears in the overall or net equation.

We can compare this general mechanism with the mechanism presented in Section 14.4 for ozone destruction. Here, the catalyst is Cl and the intermediate is ClO. Reactant A is O_3; reactant B is O. Product C and product D are both O_2.

$$\begin{array}{ll} & O_3 + Cl \longrightarrow ClO + O_2 \\ & O + ClO \longrightarrow O_2 + Cl \\ \hline \textit{Net:} & O_3 + O \longrightarrow 2\,O_2 \end{array}$$

We can see that Cl does indeed catalyze the reaction by considering that (1) the activation energy for the reaction of O_3 with Cl—0.4 kJ/mol—is much smaller than that of the direct, uncatalyzed reaction of O_3 and O—17.1 kJ/mol—and (2) the rate constant for the reaction of O_3 and Cl is about 50,000 times greater than that of O_3 and O.

Heterogeneous Catalysis

Reactions often can be catalyzed by the surfaces of appropriate solids. In order to act as a catalyst, a surface must be able to *adsorb* (bind) reactant molecules from a gaseous or liquid phase. Because a surface-catalyzed reaction occurs in a *heterogeneous* mixture, the catalytic action is called *heterogeneous* catalysis. The basic steps involved in heterogeneous catalysis are these:

1. Reactant molecules are *adsorbed.*
2. Reactant molecules diffuse along the surface.
3. Reactant molecules react to form product molecules.
4. Product molecules are *desorbed.*

Figure 15.16 shows a hypothetical reaction profile for a surface-catalyzed reaction and compares it with an uncatalyzed, homogeneous gas-phase reaction.

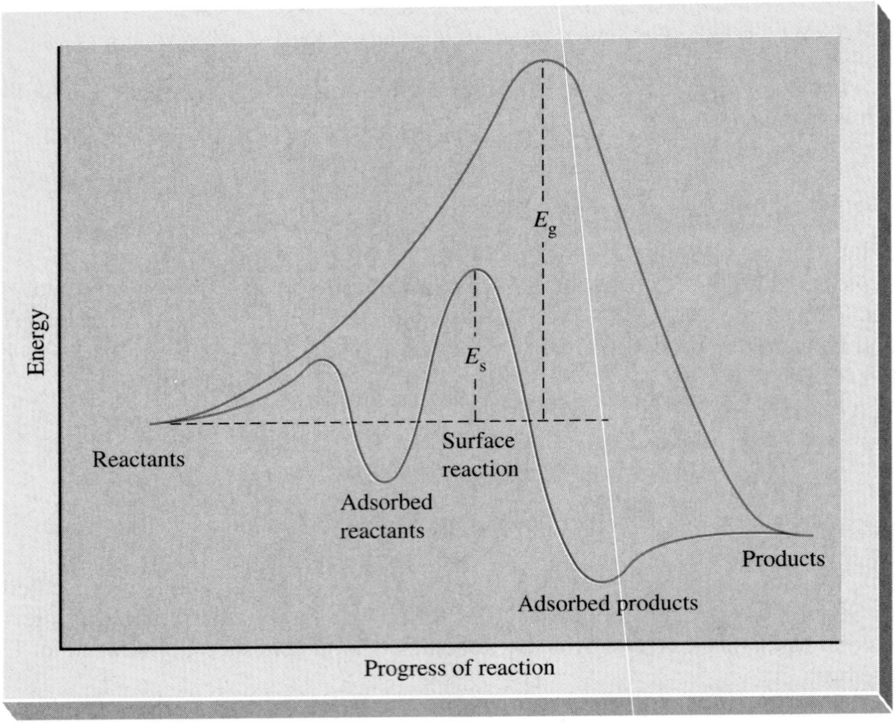

Figure 15.16 Reaction profile for a surface-catalyzed reaction. In the reaction profile (red) for the surface-catalyzed reaction, the activation energy for the reaction step, E_s, is considerably less than in the reaction profile (blue) for the uncatalyzed gas-phase reaction, E_g.

In Section 8.1 we described the hydrogenation of food oils to produce solid or semisolid fats. A simplified mechanism of an analogous surface-catalyzed reaction, the conversion of oleic to stearic acid, is suggested in Figure 15.17.

The Manufacture of Sulfuric Acid

Sulfuric acid has long been number one in production (by mass) among manufactured chemicals. It is so widely used that it is considered the "work horse" of the chemical industry. In the United States, the bulk of the sulfuric acid produced (70%) is used in the manufacture of fertilizers. Other uses include metals processing (5%), oil refining (5%), ores processing (5%), manufacture of the white pigment titanium dioxide (2%), and a host of minor uses, such as in automobile storage batteries.

The crucial step in the manufacture of sulfuric acid is the oxidation of SO_2 to SO_3.

$$SO_2(g) + \tfrac{1}{2}O_2(g) \longrightarrow SO_3(g)$$

In the gaseous phase, this reaction occurs very slowly, but the reaction can be greatly speeded up on the surface of a catalyst having vanadium pentoxide, V_2O_5, as its key component.

$$SO_2(g) + \tfrac{1}{2}O_2(g) \xrightarrow{V_2O_5} SO_3(g)$$

Following this catalytic conversion, the $SO_3(g)$ is dissolved in 98 or 99% H_2SO_4, forming a product known commercially as *oleum*. The oleum is then diluted with water to the desired strength. If we use the formula $H_2S_2O_7$ (pyrosulfuric acid) for oleum, the reactions are

$$SO_3(g) + H_2SO_4(l) \longrightarrow H_2S_2O_7(l)$$

$$H_2S_2O_7(l) + H_2O \longrightarrow 2\,H_2SO_4(l)$$

$$H_2SO_4(l) + \text{water} \longrightarrow H_2SO_4(aq)$$

Sulfuric acid was first made commerically about the middle of the eighteenth century in a process in which a mixture of S and KNO_3 was burned. The gaseous products were dissolved in water in lead-lined chambers, yielding $H_2SO_4(aq)$. Later it was discovered that oxides of nitrogen, formed by heating KNO_3, catalyzed the conversion of $SO_2(g)$ to $SO_3(g)$. A simplified description of this homogeneous catalysis is

$$\begin{array}{l} NO(g) + \tfrac{1}{2}O_2(g) \longrightarrow NO_2(g) \\ NO_2(g) + SO_2(g) \longrightarrow SO_3(g) + NO(g) \\ \hline SO_2(g) + \tfrac{1}{2}O_2(g) \longrightarrow SO_3(g) \end{array}$$

The lead-chamber process is now obsolete. However, the use of oxides of nitrogen to convert $SO_2(g)$ to $SO_3(g)$, with a subsequent conversion of $SO_3(g)$ to $H_2SO_4(aq)$, is being revived as an environmental control measure, for example, to remove $SO_2(g)$ emissions from ore smelters.

The decomposition of hydrogen peroxide, $H_2O_2(aq)$, to $H_2O(l)$ and $O_2(g)$ is a highly exothermic reaction that is strongly catalyzed on the surface of platinum metal.

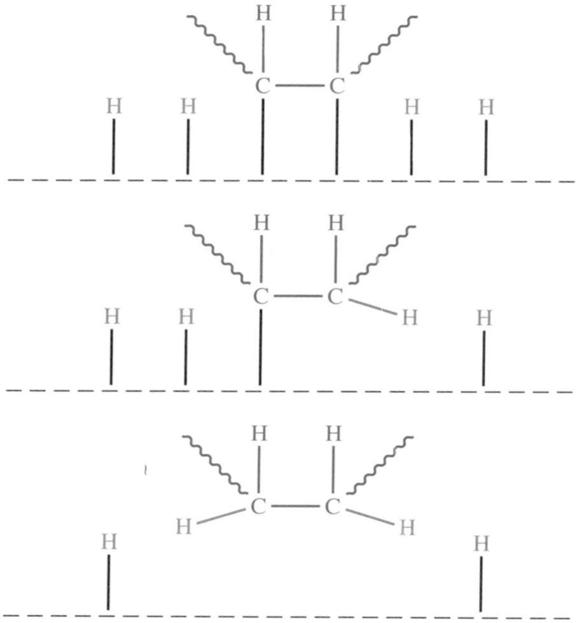

Figure 15.17 Heterogeneous catalysis. In the reaction

$$\underset{\text{Oleic acid}}{CH_3(CH_2)_7CH{=}CH(CH_2)_7COOH} + H_2 \xrightarrow{Ni} \underset{\text{Stearic acid}}{CH_3(CH_2)_7CH_2CH_2(CH_2)_7COOH}$$

a molecule of oleic acid attaches itself to the surface through the C atoms at the double bond site, and the double bond converts to a carbon-to-carbon single bond. The ends of the molecule are unattached and wave freely above the surface. Hydrogen molecules dissociate into hydrogen atoms when adsorbed on the surface of the nickel catalyst. First one, and then a second, H atom attaches to a surface-bonded C atom. The resulting molecule—stearic acid—desorbs (detaches itself) from the surface.

15.12 Enzyme Catalysis

Unlike platinum and nickel, which are able to catalyze a variety of reactions, the substances that catalyze chemical reactions in living organisms are quite specific in their catalytic action. Many of these catalysts, high molecular weight proteins called **enzymes**, catalyze one specific reaction and no others. In living organisms, even a simple reaction like the conversion of carbon dioxide to carbonic acid is catalyzed by an enzyme, called *carbonic anhydrase*. Each enzyme molecule can convert 100,000 CO_2 molecules per second.

$$CO_2(g) + H_2O \xrightarrow{\text{carbonic anhydrase}} H_2CO_3(aq)$$

Biochemists often use a simple model pictured in Figure 15.18 to explain enzyme action; it is called the *lock-and-key* model. The reacting substance, called the **substrate** (S), attaches itself to an area on the enzyme (E), called the **active site**, to form an enzyme–substrate complex (ES). The complex decomposes to form products (P) and the enzyme is regenerated.

$$E + S \longrightarrow ES \longrightarrow E + P$$

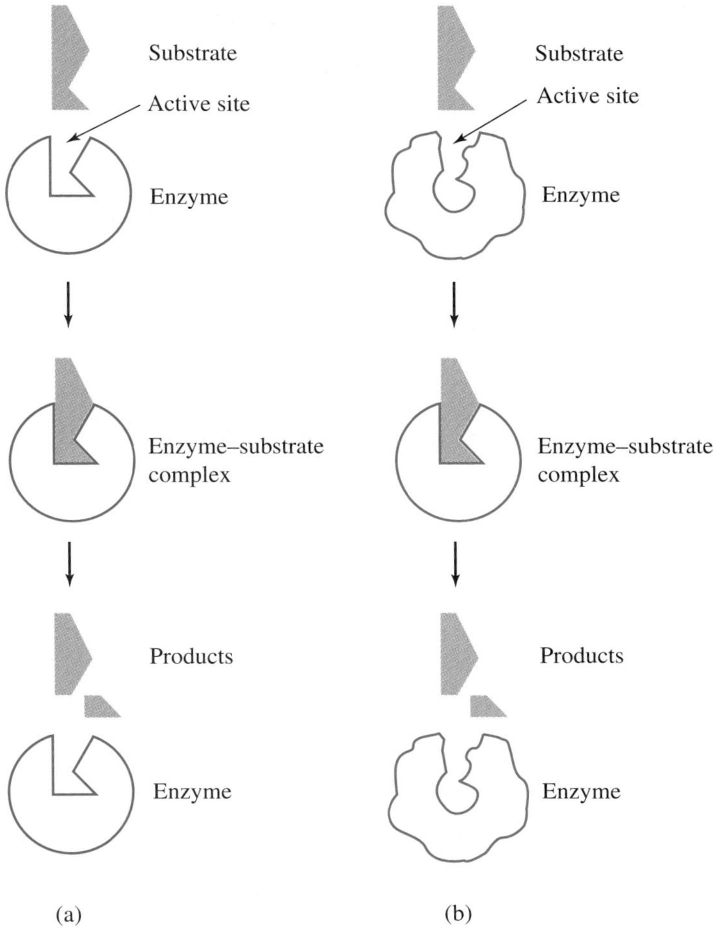

Figure 15.18 Models of enzyme action. (a) The lock-and-key model. The model assumes a perfect fit between the substrate and the active site on the enzyme molecule, much like the perfect match between a key and the key chamber in the lock that the key opens. (b) The induced-fit model. The shapes of the substrate and the active site are not perfectly complementary, but the active site adapts to fit the substrate, much like a glove molds to fit the hand that is inserted into it.

Factors Influencing Enzyme Activity

The rates of reactions involving enzymes are influenced by several factors, including concentration of substrate, concentration of enzyme, acidity of the medium, and temperature.

Enzyme and substrate must come together for a reaction to occur. The chance of this happening depends on the concentration of each substance. Figure 15.19 shows the reaction rate as the concentration of substrate increases while the enzyme concentration is held constant. The reaction rate increases as the concentration of substrate, S, is increased, but only up to a point. Eventually, a concentration is reached that saturates all the active sites on the enzyme molecules. The rate levels off, and a further increase in substrate concentration leaves the rate unchanged.

At low substrate concentrations, the rate of the reaction is first order in S, because the rate of ES formation is proportional to [S].

$$\text{Rate of reaction} = k[\text{S}]$$

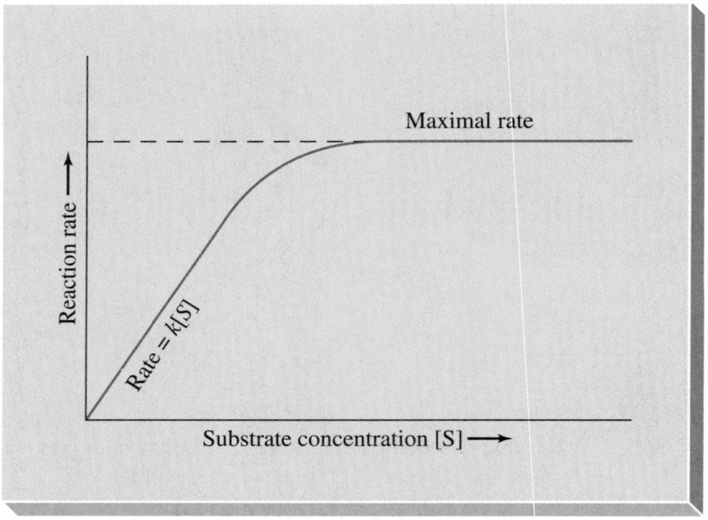

Figure 15.19 The effect of substrate concentration [S] on reaction rate with enzyme concentration held constant. When [S] is low, the reaction rate increases with [S]. At high values of [S], the reaction rate is independent of [S], and a fixed maximal rate is observed.

At high substrate concentrations, adding more substrate cannot accelerate the reaction. The rate of the reaction is independent of substrate concentration, and the reaction is zero order.

$$\text{Rate of reaction} = k'[S]^0 = k'$$

It's as if 10 taxis (enzyme molecules) were waiting at a taxi stand to take people (substrate) on a 10-min trip to a concert hall, one passenger at a time. If only 5 people are present at the stand, the rate of their arrival at the concert hall is 5 people in 10 min. If the number of people at the stand is increased to 10, the rate increases to 10 arrivals in 10 min. With 20 people at the stand, the rate would still be 10 arrivals in 10 min. The taxis have been "saturated." If the taxis could carry 2 or 3 passengers each, the same principle would apply. The rate would simply be higher (20 or 30 people in 10 min) before it leveled off.

If the concentration of substrate is held constant and *in excess* (if we always have more people than taxis), the rate of a reaction is proportional to the concentration of enzyme. The more enzyme there is, the faster the reaction goes (the more taxis, the more people transferred). This relationship holds over a wide range of enzyme concentrations (Figure 15.20).

Enzymes are proteins with acidic groups (such as —COOH) and basic groups (such as $—NH_2$) , and enzyme activity depends on the concentration of H^+ ions present in the medium surrounding the enzyme. Each enzyme has its own *optimum acidity*. Those that operate in the stomach work best in highly acidic media. Those that catalyze reactions in the small intestine, on the other hand, show optimum activity in slightly basic media.

As an example, consider *lysozyme*, an enzyme that breaks certain bonds in polysaccharides found in the cell walls of bacteria. Lysozyme is active only when two conditions are met simultaneously:

1. Aspartic acid, one of the amino acid components of lysozyme, must be ionized; that is, the —COOH group must be converted to $—COO^-$.
2. Glutamic acid, another amino acid component of lysozyme, must be nonionized; that is, the —COOH group of glutamic acid must be intact.

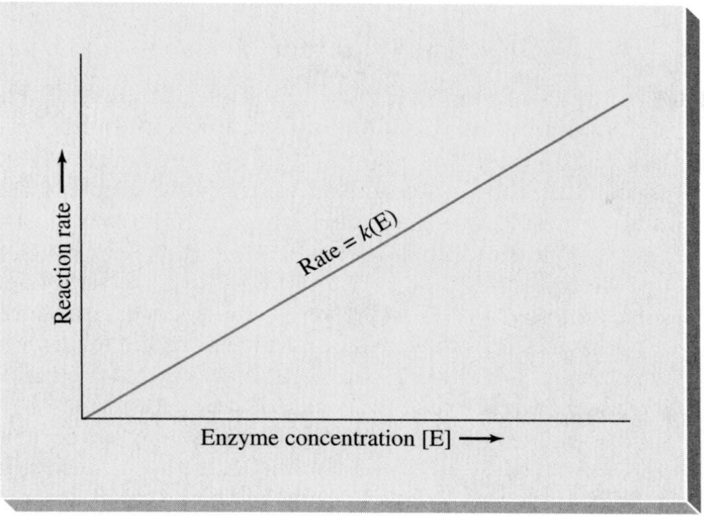

Figure 15.20 The effect of enzyme concentration on reaction rate, with the concentration of substrate held constant and in excess.

When the medium is slightly acidic, both conditions are met. In highly acid solutions, however, both acids are present in their —COOH form, and in basic solutions, both are present as $—COO^-$. In either highly acid or highly basic solutions the enzyme doesn't function.

Enzyme action is sensitive to temperature changes. For living cells there is often a rather narrow range of optimum temperatures. Both higher and lower temperatures can be disabling, if not deadly. A change in temperature changes the shape of the protein molecule; the active site no longer fits the substrate. The temperature range for the activity of enzymes generally is 10 to 50 °C, although some enzymes found in *thermophilic* (heat-loving) bacteria operate well near 100 °C. The *optimum temperature* for many enzymes in the human body is 37 °C—body temperature (Figure 15.21).

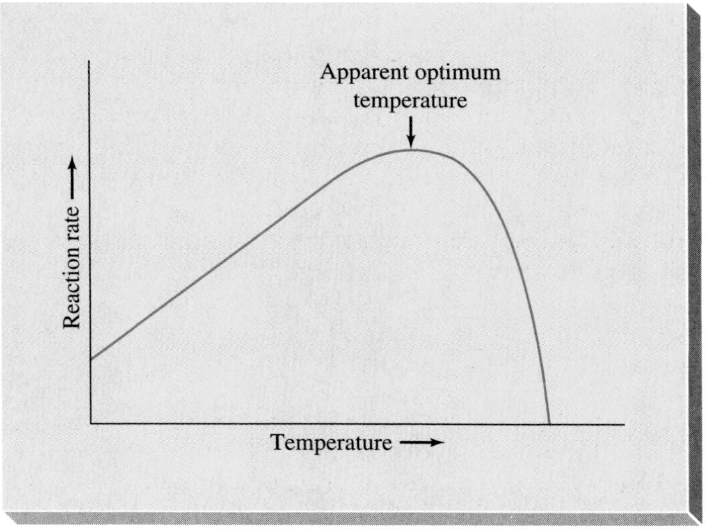

Figure 15.21 Enzyme activity as a function of temperature. If the hypothetical reaction whose rate is suggested here occurs in humans, the optimum temperature is likely to be 37 °C—normal body temperature.

Enzyme Inhibition: Poisons and Drugs

A substance making an enzyme less active or completely inactive is called an *inhibitor*. Some inhibitors bind at the active site of the enzyme and block out the substrate. Others bind elsewhere in the enzyme molecule but change the shape of the active site or otherwise hinder a substrate molecule's access to the site. Inhibition is an important, naturally occurring process that is essential in growth and metabolism, but some substances that we ingest—in food or otherwise—also act as inhibitors. Let's look at some inhibitors that kill—poisons—and others that sustain life—drugs.

Enzymes are proteins. Many proteins contain the amino acid *cysteine*, which has a sulfhydryl group (—SH) attached to each unit. Heavy metal ions such as Hg^{2+} and Pb^{2+} can deactivate enzymes by reacting with sulfhydryl groups to form sulfides. This poisoning can occur at a position removed from the active site, distorting or destroying the site (Figure 15.22).

Enzymes play a key role in the chemistry of the nervous system. When an electric signal from the brain reaches the end of a fiber (called an axon) that extends from a nerve cell, the substance *acetylcholine* is liberated and carries the signal across a tiny gap, called a *synapse*, to a receptor cell. Once acetylcholine has carried the impulse, it rapidly breaks down into acetic acid and choline in a reaction catalyzed by the enzyme *cholinesterase*.

$$\underset{\text{Acetylcholine}}{CH_3COOCH_2CH_2N^+(CH_3)_3} + H_2O \underset{\text{acetylase}}{\overset{\text{cholinesterase}}{\rightleftharpoons}} \underset{\text{Acetic acid}}{CH_3COOH} + \underset{\text{Choline}}{HOCH_2CH_2N^+(CH_3)_3}$$

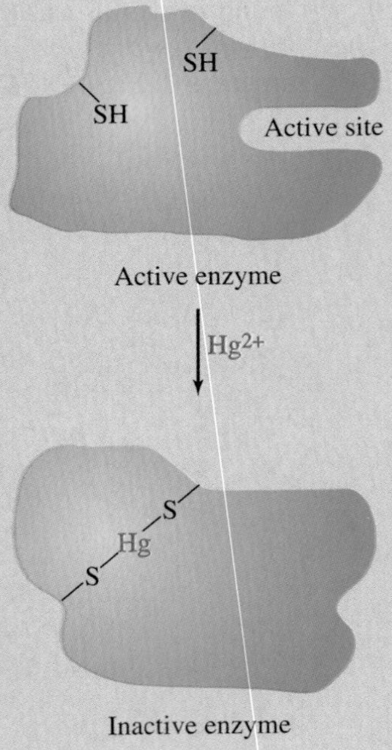

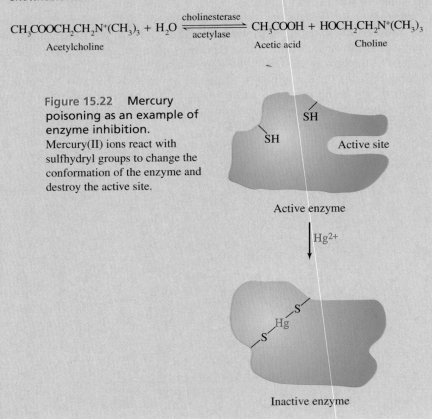

Figure 15.22 Mercury poisoning as an example of enzyme inhibition. Mercury(II) ions react with sulfhydryl groups to change the conformation of the enzyme and destroy the active site.

In effect, the breakdown of acetylcholine resets the receptor to the "off" position, making it ready to receive the next impulse. Other enzymes, such as acetylase, convert the acetic acid and choline back to acetylcholine, completing the cycle.

Anticholinesterase poisons block the action of cholinesterase. The insecticides *malathion* and *parathion* and the warfare agents *tabun* and *sarin* (Figure 15.23) are well-known nerve poisons. By displacing an atom or group (the F, as F^-, of sarin, for example), the cholinesterase binds the molecule of the poison tightly (Figure 15.24); this prevents the enzyme from performing its normal function. If the breakdown of acetylcholine is blocked, the concentration of this messenger compound builds up, causing the receptor nerve cell to "fire" repeatedly, that is, to be continuously "on." This overstimulates the muscles, glands, and organs. The heart beats wildly and irregularly. The victim goes into convulsions and dies quickly.

Sulfa drugs are an example of beneficial enzyme inhibitors. Here is how they work: Bacteria require folic acid as a coenzyme. (A coenzyme is a nonprotein part of an enzyme necessary for the enzyme to function.) Bacteria must synthesize folic acid from *p*-aminobenzoic acid. Sulfa drugs have structures similar to *p*-aminobenzoic acid, and a bacterial enzyme readily incorporates the sulfa drug into a false folic acid. This altered folic acid not only cannot function as a proper coenzyme but it inhibits the enzyme. Invading bacteria die because they are unable

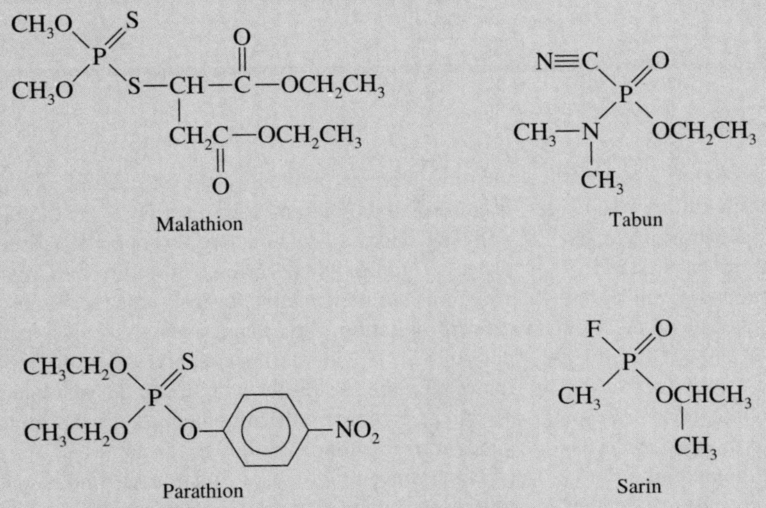

Figure 15.23 Some organic phosphorus compounds. Malathion and parathion are insecticides. Tabun and sarin are nerve gases for use in chemical warfare. Sarin was used in the 1995 terrorist attack on the Tokyo subway that killed 10 and sickened thousands.

to make vital compounds. The human patient lives because we obtain folic acid, a B vitamin, in our diet; we don't need to make folic acid.

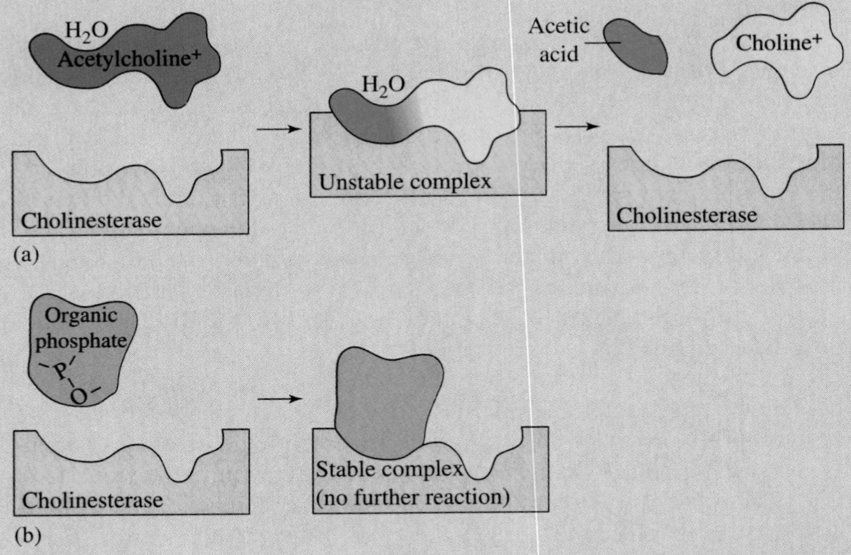

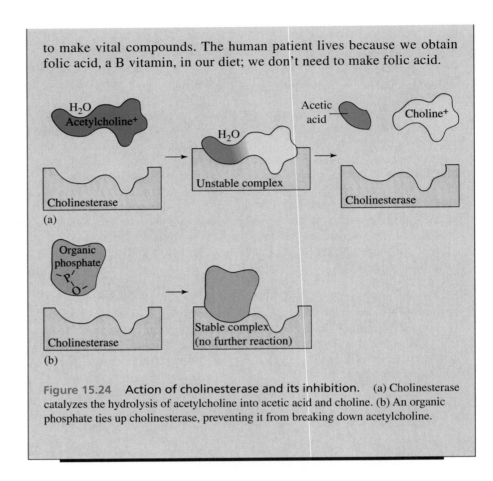

Figure 15.24 **Action of cholinesterase and its inhibition.** (a) Cholinesterase catalyzes the hydrolysis of acetylcholine into acetic acid and choline. (b) An organic phosphate ties up cholinesterase, preventing it from breaking down acetylcholine.

SUMMARY

The rate of a reaction refers to the rate of change of concentration of a reactant or of a product with time. To establish a rate of reaction, the mass, concentration, partial pressure, . . . of a reactant or product is measured at different times. An average rate of reaction is given by the expression $\Delta[A]/\Delta t$. An instantaneous rate of reaction can be obtained from the slope of a tangent line to a graph of concentration versus time.

A rate law or rate equation has the form: rate of reaction = $k[A]^m[B]^n$ . . . . The order of the reaction is established by the exponents, m, n, . . .; the overall order is $m + n + \cdots$. The proportionality constant, k, is called the rate constant.

An integrated rate equation relates concentration and time. For a first-order reaction, $\ln [A]_t/[A]_0 = -kt$. The rate constant, k, is the negative of the slope of the straight line graph: $\ln [A]_t$ versus t. The half-life of a reaction is the time in which one-half of the reactant initially present is consumed. For a first-order reaction, $t_{1/2} = 0.693/k$.

The decay of radioactive isotopes is a first-order process, generally described through half-lives and equations similar to those for first-order reactions. In radiocarbon dating, the carbon-14 activity of a sample is determined. By comparing this activity with that in living matter, the age of the sample can be established.

In a zero-order reaction, the rate of reaction is independent of the concentration of reactant, and a graph of reactant concentration as a function of time is a straight line. In a second-order reaction, the sum of the exponents in the rate equation is *two*. In the second-order reaction $A \longrightarrow$ products, a plot of $1/[A]$ versus time yields a straight line. For both zero-order and second-order reactions, the half-life depends on concentration as well as on the value of k.

The activated complex in a reaction is a transitory species formed through collisions between molecules having sufficient kinetic energies and the proper orientation. The excess energy of the activated complex over that of the reactants is the activation energy of the reaction. Energy changes in a reaction can be shown in a reaction profile. Reactions generally go faster at higher temperatures because of the large increase in the fraction of molecular collisions effective in producing the activated complex.

A plausible reaction mechanism generally consists of a series of elementary steps. The mechanism must yield the observed net equation, and the rate law deduced from the mechanism must be the same as that found experimentally.

A catalyst speeds up a reaction by changing the mechanism to one of lower activation energy; the catalyst is regenerated in the reaction. In enzyme catalysis, the reactants (substrate) bind to an active site on the enzyme to form a complex that dissociates into product molecules. Enzyme activity depends on the concentrations of substrate and enzyme, acidity of the medium, and temperature.

KEY TERMS

activation energy (15.8)
activated complex (15.8)
active site (15.12)
bimolecular (15.10)
catalyst (15.11)
elementary step (15.10)
enzyme (15.12)
first-order reaction (15.5)
half-life (15.5, 15.6)
initial rate of reaction (15.3)
instantaneous rate of reaction (15.3)
integrated rate equation (15.5)
intermediate (15.10)
method of initial rates (15.4)
order of a reaction (15.4)
radioactive decay law (15.6)
rate constant (15.4)
rate-determining step (15.10)
rate law (rate equation) (15.4)
reaction mechanism (15.10)
reaction profile (15.8)
second-order reaction (15.7)
substrate (15.12)
termolecular (15.10)
transition state (15.8)
unimolecular (15.10)
zero-order reaction (15.7)

REVIEW QUESTIONS

1. What are the quantities that must generally be measured to establish the rate of a chemical reaction.
2. Cite three or four factors that affect the rate of a chemical reaction.
3. For the reaction $2\ H_2O_2(aq) \longrightarrow 2\ H_2O + O_2(g)$, explain why the rate of formation of O_2 is not equal to the rate of disappearance of H_2O_2. How are they related?
4. At what point in a chemical reaction does the reaction usually go fastest? Explain.
5. What is the difference between the average rate and the instantaneous rate of a reaction? Can the average rate ever be the same as the instantaneous rate? Explain.
6. Explain the difference in meaning of these terms when applied to a chemical reaction: rate, rate constant, and rate law.
7. How is the rate constant, k, of a first-order reaction, $A \longrightarrow$ products, related to a graph of ln [A] versus time?
8. In a reaction, $[A] \longrightarrow$ products, we find that when [A] has fallen to half its initial value the reaction proceeds at only one-quarter of its initial rate. Is the reaction zero order, first order, or second order? Explain.
9. The rate equation for a reaction relates the rate of the reaction to the concentrations of reactants. What variables are related through the *integrated* rate equation? Write these two equations for the first-order reaction, $A \longrightarrow$ products.
10. What is meant by the half-life of a reaction? Does the half-life have a constant value for a zero-order reaction? a first-order reaction? a second-order reaction? Explain.
11. What is the radioactive decay law?
12. Which has the greater activity A, a radioactive isotope with a very short half-life or one with a very long half-life? Explain.
13. What is the relationship, if any, between the average kinetic energies of the molecules in a reaction mixture and the activation energy of the reaction?

14. With regard to the collision theory, what factor is most responsible for the fact that the rate of a chemical reaction generally increases sharply with a rise in temperature?

15. Why is the orientation of the colliding molecules an important factor in determining the rate of a reaction in some reactions but not in all reactions?

16. What is meant by the transition state of a chemical reaction? the activated complex?

17. Refer to the following reaction profile.

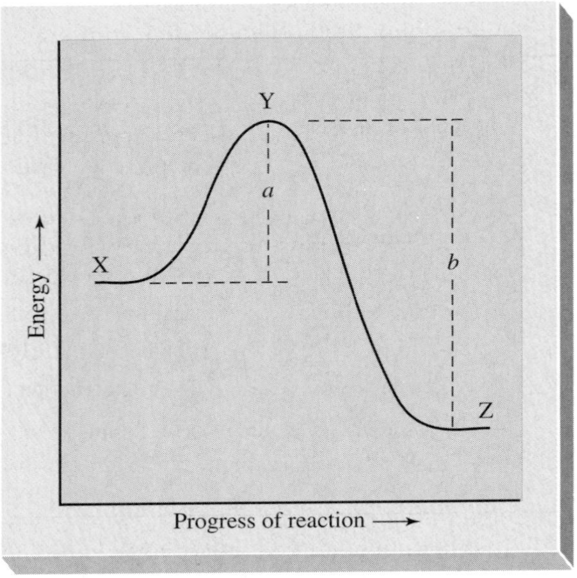

a. Which letter in the diagram refers to the products?
b. Which letter refers to the reactants?
c. What does the letter Y represent?
d. What does the broken line *a* represent?
e. What does the broken line *b* represent?

18. How is the activation energy of a reaction related to the heat of reaction for (**a**) an endothermic reaction, and (**b**) an exothermic reaction?

19. What plot of experimental data can be used to evaluate the activation energy, E_a, of a reaction? How is E_a related to this plot?

20. What are the chief requirements that must be met by a plausible reaction mechanism? Why do we say "plausible" mechanism rather than "correct" mechanism?

21. What is meant by an elementary step in a reaction mechanism?

22. What is the difference between a *unimolecular* and a *bimolecular* elementary step in a reaction mechanism?

23. What is the difference between a *transition state* and an *intermediate* in a reaction mechanism?

24. What is meant by the rate-determining step in a reaction mechanism? Which elementary step in a reaction mechanism is often the rate-determining step.

25. State *two* important requirements of a substance in order for it to be considered a catalyst in a chemical reaction.

26. What is the difference between *homogeneous* and *heterogeneous* catalysis?

27. To what class of macromolecular substances do enzymes belong?

28. What is the substrate in an enzyme-catalyzed reaction?

29. Describe the lock-and-key model of an enzyme-catalyzed reaction.

30. What happens to the rate of an enzyme-catalyzed reaction if the concentration of substrate X is doubled, (**a**) when the concentration of X is low, (**b**) when the concentration of X is very high?

31. Besides substrate concentration, what other factors affect the rate of an enzyme-catalyzed reaction?

PROBLEMS

Rate of Reaction and Reaction Order

32. In a reaction, A $\longrightarrow$ products, the initial concentration of A is 0.1108 M, and 44 s later, 0.1076 M. What is the initial rate of this reaction, expressed in (**a**) $M \cdot s^{-1}$; (**b**) $M \cdot min^{-1}$?

33. In a reaction, A $\longrightarrow$ products, $[A]_0 = 0.2546$ M and the initial rate of reaction is 4.0×10^{-5} $M \cdot s^{-1}$. What will [A] be at $t = 35$ s?

34. In the reaction $2\,A + 2\,B \longrightarrow C + 2\,D$, the rate of reaction of A is found to be 2.2×10^{-4} $M \cdot s^{-1}$. (**a**) What is the rate of reaction of B? (**b**) What is the rate of formation of C?

35. In the reaction $A + 2\,B \longrightarrow C + 3\,D$, the rate of reaction of B is found to be 6.2×10^{-4} $M \cdot s^{-1}$. (**a**) What is the rate of reaction of A? (**b**) What is the rate of formation of D?

36. For a reaction, A $\longrightarrow$ products, a graph of [A] versus time is found to be a straight line. What is the order of this reaction?

37. The rate of a reaction and the rate constant, *k*, have the same units only for reactions of a particular

overall order. Verify this statement and identify the reaction order.

38. The initial rate of a reaction, A + 2 B $\longrightarrow$ C + 2 D, is determined for four different initial conditions. The results obtained are tabulated below. (**a**) What is the order of the reaction with respect to A and to B? (**b**) What is the overall reaction order? (**c**) What is the value of the rate constant, k?

Expt.	[A]	[B]	Initial rate, $M \cdot s^{-1}$
1	0.22	0.16	6.1×10^{-4}
2	0.44	0.16	1.2×10^{-3}
3	0.44	0.32	2.4×10^{-3}
4	0.22	0.32	1.2×10^{-3}

39. The initial rate of a reaction, 2 A + B $\longrightarrow$ 2 C + D, is determined for four different initial conditions. The results obtained are tabulated below. (**a**) What is the order of the reaction with respect to A and to B? (**b**) What is the overall reaction order? (**c**) What is the value of the rate constant, k?

Expt.	[A]	[B]	Initial rate, $M \cdot s^{-1}$
1	0.38	0.24	2.8×10^{-3}
2	0.45	0.12	7.0×10^{-4}
3	0.19	0.48	1.1×10^{-2}
4	0.76	0.24	2.8×10^{-3}

40. Use a value of $k = 3.66 \times 10^{-3}\ s^{-1}$ to establish the instantaneous rate of decomposition of 2.05 M H_2O_2(aq). (*Hint*: Use the first-order rate law.)

41. What should be the instantaneous rate of decomposition of 3% H_2O_2(aq), by mass? Assume that the solution has a density of 1.00 g/mL and that $k = 3.66 \times 10^{-3}\ s^{-1}$. (*Hint*: Use the first-order rate law.)

First-Order Reactions

42. A first-order reaction, A $\longrightarrow$ products, has a rate of reaction of 0.00250 $M \cdot s^{-1}$ when [A] = 0.484 M. What is the rate constant, k, for this reaction?

43. A first-order reaction, A $\longrightarrow$ products, has a rate constant, k, of 0.0462 min^{-1}. What is [A] at the time when the reaction is proceeding at a rate of 0.0150 $M \cdot min^{-1}$?

44. In a first-order reaction, A $\longrightarrow$ products, [A] = 0.620 M initially and 0.520 M after 15.0 min. (**a**) What is the value of the rate constant, k? (**b**) What is the half-life of the reaction, $t_{1/2}$? (**c**) What will [A] be after 2.00 h?

45. In a first-order reaction, A $\longrightarrow$ products, [A] = 0.280 M initially, 0.264 M after 10.0 min, and 0.197 M after 1.00 h. (**a**) What is the value of the rate constant, k? (**b**) What is the half-life of the reaction, $t_{1/2}$? (**c**) At what time after the start of the reaction will [A] = 0.0350 M?

46. The smog constituent peroxyacetyl nitrate (PAN) dissociates into peroxyacetyl radicals and NO_2(g) in a first-order reaction with a half-life of 32 min.

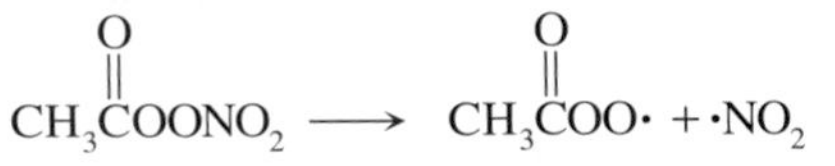

PAN   Peroxyacetyl radical

If the initial concentration of PAN molecules in an air sample is 5.0×10^{14} molecules/L, what will be the concentration 1.50 h later?

47. The decomposition of dimethyl ether at 504 °C is a first-order reaction with a half-life of 27 min.

$$(CH_3)_2O(g) \longrightarrow CH_4(g) + H_2(g) + CO(g)$$

a. What will be the partial pressure of $(CH_3)_2O(g)$ after 1.00 h, if its initial partial pressure was 626 mmHg?

b. What will be the total gas pressure after 1.00 h? [*Hint*: $(CH_3)_2O$ and its decomposition products are the only gases present in the reaction vessel.]

Radioactive Decay

48. Why is it that chemical reactions can be of different orders—zero, first, second, . . . —whereas radioactive decay is always a first-order process?

49. Many chemical reactions can be catalyzed. Do you think it is equally possible to catalyze a radioactive decay process? Explain.

50. According to Table 15.4, the isotope iodine-131 has a shorter half-life than does phosphorus-32. Yet, the activity of a sample of phosphorus-32 is greater than that of a sample of iodine-131 of the same mass. Explain why this is so.

51. The text states that radiocarbon dating is applicable to objects from a few hundred to about 50,000 years old. Why do you think that it is not applicable to objects that are only a few years old or much more than 50,000 years old?

52. A certain radioactive isotope is decaying at the rate of 225 disintegrations per second per one million atoms. What is the half-life, in minutes, of this isotope?

53. A 1.00-μg sample of plutonium-239 is found to decay at a rate of 2.27×10^3 disintegrations per second. What is the half-life of plutonium-239? (*Hint*: How many atoms are present in the sample?)

54. We follow the activity of a sample containing the radioactive isotope sulfur-35. At the start of the experiment we detect 138 disintegrations per minute, and after 20.0 d, the decay rate is 118 dis min^{-1}. What is the half-life of sulfur-35?

55. The half-life of iodine-131 is 8.04 d. We follow the activity of a sample of iodine-131 while it falls to 10% of its initial value. *Estimate* how long this will take.

56. Calculate a more exact value of the time required in Problem 55. (*Hint*: Does the sample size matter?)

Reactions of Other Orders

57. The decomposition of ammonia on a tungsten surface at 1100 °C was described by Figure 15.6. Refer to this figure and determine $[NH_3]$ at t = 335 s, if initially $[NH_3]$ = 0.0452 M.

58. Consider Problem 57: What is the half-life for the reaction in which $[NH_3]_0$ = 0.0452 M?

59. The rate constant for the second-order decomposition of hydrogen iodide at 324.1 °C is $k = 2.3 \times 10^{-4}$ $M^{-1} \cdot min^{-1}$. In a reaction in which, initially, [HI] = 0.56 M, what will [HI] be at t = 2.00 h?

60. Consider Problem 59: At what time will [HI] = 0.28 M for the reaction described?

61. The half-lives of zero-order and second-order reactions both depend on the initial concentration as well as on the rate constant k. In one case the half-life gets longer as the initial concentration increases, and in the other case it gets shorter. Which is which, and why isn't the situation the same for both?

Collision Theory; Activation Energy

62. Chemical reactions occur as a result of molecular collisions, and the frequency of molecular collisions can be calculated with the kinetic-molecular theory. Calculations of the rates of chemical reactions, however, are generally not very successful. Explain why this is so.

63. A temperature increase of 10 °C causes an increase in collision frequency of only a few percent, yet it can cause the rate of a chemical reaction to increase by a factor of two or more. How do you explain this apparent discrepancy?

64. A mixture of hydrogen and oxygen gases is indefinitely stable at room temperature; however, if struck by a spark, the mixture immediately explodes. What explanation can you offer for this observation?

65. For the reaction

$$H_2(g) + I_2(g) \longrightarrow 2\,HI(g)$$

the activation energy, E_a = 171 kJ/mol and the enthalpy of reaction, ΔH = −9.5 kJ/mol. Construct, roughly to scale, the reaction profile of this reaction.

66. Rate constants for the first-order decomposition of acetonedicarboxylic acid

$$\underset{\text{Acetonedicarboxylic acid}}{CO(CH_2COOH)_2(aq)} \longrightarrow \underset{\text{Acetone}}{CO(CH_3)_2(aq)} + 2\,CO_2(g)$$

are $k = 4.75 \times 10^{-4}$ at 293 K and $k = 1.63 \times 10^{-3}$ at 303 K. What is the activation energy, E_a, of this reaction?

67. The decomposition of di-*t*-butyl peroxide (DTBP) is a first-order reaction with a half-life of 320 min at 135 °C and 100 min at 145 °C. Calculate E_a for this reaction.

$$\underset{\text{DTBP}}{C_8H_{18}O_2(g)} \longrightarrow \underset{\text{Acetone}}{2\,(CH_3)_2CO(g)} + \underset{\text{Ethane}}{C_2H_6(g)}$$

68. The decomposition of ethylene oxide at 652 K

$$(CH_2)_2O(g) \longrightarrow CH_4(g) + CO(g)$$

is a first-order reaction with $k = 0.0120\ min^{-1}$. The activation energy of the reaction is 218 kJ/mol. Calculate the rate constant of the reaction at 525 K.

69. For the reaction in Problem 68, calculate the temperature at which the rate constant $k = 0.0100\ min^{-1}$.

Reaction Mechanisms

70. It is easy to see how a bimolecular elementary step in a reaction mechanism can occur as a result of a collision between two molecules. How do you suppose a *unimolecular* process is able to occur?

71. Why shouldn't we necessarily expect the rate equation of a net reaction to be the same as the rate equation of one of the elementary steps in a plausible reaction mechanism? Cite *two* situations, however, in which this may indeed be the case.

72. The following is proposed as a plausible reaction mechanism:

$$\text{Slow:}\quad A + B \longrightarrow I$$
$$\text{Fast:}\quad I + B \longrightarrow C + D$$

a. What is the net reaction described by this mechanism?

b. What is a plausible rate equation for the reaction?

73. The reaction $A + 2\,B \longrightarrow C + 2\,D$ is found to be first order in A and first order in B. A proposed mechanism for the reaction involves the following first step:

$$\text{Slow:}\quad A + B \longrightarrow I + D$$

a. Write a plausible second step in a two-step mechanism.

b. Is the second step slow or fast? Explain.

Catalysis

74. The decomposition of $H_2O_2(aq)$ is usually studied in the presence of $I^-(aq)$. The reaction is first order in both H_2O_2 and I^-. Why can we treat the reaction as if it were only first order overall rather than second order?

75. Describe in a general way how the reaction profile for the surface-catalyzed reaction of SO_2 and O_2 to form SO_3 differs from the reaction profile for the noncatalyzed, homogeneous gas-phase reaction.

76. Explain the difference between the lock-and-key and the induced-fit models of enzyme action.

77. A bacterial enzyme has an optimum temperature of 35 °C. Will the enzyme be more or less active at normal body temperature? Will it be more or less active if the patient has a fever of 40 °C?

78. The kinetics of some surface-catalyzed reactions are similar to enzyme-catalyzed reactions. Explain this connection.

79. Describe ways in which an enzyme inhibitor may function.

ADDITIONAL PROBLEMS

80. In the first-order reaction, A $\longrightarrow$ products, the following concentrations were found at the indicated times: $t = 0$ s, [A] = 0.88 M; 25 s, 0.74 M; 50 s, 0.62 M; 75 s, 0.52 M; 100 s, 0.44 M; 125 s, 0.37 M; 150 s, 0.31 M. Calculate the instantaneous rate of reaction at $t = 125$ s.

81. The decomposition of $H_2O_2(aq)$ can be followed by removing samples from the reaction mixture and titrating them with $MnO_4^-(aq)$.

$$2\ MnO_4^- + 5\ H_2O_2 + 6\ H^+ \longrightarrow 2\ Mn^{2+} + 8\ H_2O + 5\ O_2(g)$$

Assume that at each of the times listed in Table 15.1 a 5.00-mL sample of the remaining $H_2O_2(aq)$ is removed and titrated with 0.0500 M $KMnO_4$. Add a column to the table listing the volume of $KMnO_4$ required for the titration, initially, at $t = 60$ s, at $t = 120$ s, and so on.

82. Refer to Problem 81: Because the volumes of 0.0500 M $KMnO_4$ required for the titrations are proportional to the remaining $[H_2O_2]$, you can plot ln (mL $KMnO_4$) versus time and determine a value of k for the decomposition of $H_2O_2(aq)$. Show that the same value is obtained as in Figure 15.4.

83. The half-life of the first-order decomposition of nitramide

$$NH_2NO_2(aq) \longrightarrow H_2O + N_2O(g)$$

is 123 min at 15 °C. How long will it take for a sample of $NH_2NO_2(aq)$ to be 85% decomposed?

84. The decomposition of di-*t*-butyl peroxide (DTBP) is a first-order reaction with a half-life of 80.0 min at 147 °C.

$$\underset{\text{DTBP}}{C_8H_{18}O_2(g)} \longrightarrow \underset{\text{Acetone}}{2\ (CH_3)_2CO(g)} + \underset{\text{Ethane}}{C_2H_6(g)}$$

If a 4.50-g sample of DTBP is introduced into a 1.00-L flask at 147 °C, (**a**) what is the initial gas pressure in the flask? (**b**) the *total* gas pressure in the flask after 80.0 min? (**c**) the *total* gas pressure after 125 min?

85. Refer to Problem 46. The half-life for the decomposition of PAN at 25 °C is 32 min and the activation energy of the reaction is 113 kJ/mol. (**a**) What is the half-life at 35 °C? (**b**) What is the initial rate of decomposition of PAN if 6.0×10^{14} molecules of PAN are injected into an air sample at 35 °C?

86. For a zero-order reaction, A $\longrightarrow$ products, write equations involving the terms $[A]_t$, $[A]_0$, k, and t to represent (**a**) [A] as a function of time, t; (**b**) the length of time it takes for the reaction to go to completion; and (**c**) the half-life of the reaction.

87. What is the minimum activity, N_p, that can be detected by radiocarbon dating, if the activity of carbon-14 in living organic matter is 15 dis min^{-1} per gram of carbon? (*Hint:* Recall that the maximum age of objects that can be dated by the radiocarbon method is about 50,000 years.)

88. Assume that when the activity of a radioactive sample falls to about 1% of its initial value the material is essentially no longer radioactive. What is the maximum length of time that a source of cobalt-60 can be used in radiation therapy? (*Hint:* Use data from Table 15.4.)

89. Cobalt-60 and strontium-90 are both radioactive isotopes. What mass of strontium-90 should be present in a sample if it is to have the same activity (that is, the same number of disintegrations per unit time) as 1.00 g of cobalt-60. (*Hint:* Use data from Table 15.4.)

90. A "rule of thumb" in chemical kinetics states that for many reactions the rate of reaction approximately doubles for a temperature rise of 10 °C. What must be the activation energy of a reaction if the rate is indeed found to double between 25 °C and 35 °C?

91. For which type of reaction would you expect the rate to increase more rapidly with increasing temperature, one with high or one with low activation energy? Explain.

92. The enzyme acetylcholinesterase has —COOH, —OH, and $—NH_2$ groups on side chains of amino acids at the active site. The enzyme is inactive in acidic solution, but the activity increases as the solution becomes more basic. Explain.

From this ammonia production facility in Texas, liquid ammonia is transferred to railroad tank cars, to tanker trucks, and, through an 800-mile pipeline terminating in southern Minnesota, throughout the farm belt in the central United States. Principles of chemical equilibrum are crucial to the synthesis of ammonia (page 624).

16

Chemical Equilibrium

C O N T E N T S

From time to time we have spoken of reversible reactions, but our calculations have been limited to reactions that go to completion because we need more than the principles of stoichiometry to establish outcomes for *reversible* reactions.

In this chapter we will examine dynamic equilibrium in reversible reactions—the condition in which the rate of the reverse reaction equals the rate of the forward reaction. We will describe equilibrium through an equilibrium constant, and we will use equilibrium constants, here and in later chapters, to make both qualitative and quantitative predictions about reversible reactions. Reversible reactions are important in the laboratory, in chemical industry, and in natural phenomena.

16.1 The Dynamic Nature of Equilibrium

We have already encountered the idea that equilibrium involves opposing processes occurring at equal rates. In vapor pressure equilibrium, the rate of evaporation of a liquid equals the rate of condensation of its vapor. In solubility equilibrium between a solid and a solution, the rate of dissolution of the solid equals its rate of crystallization from solution.

We considered the formation of a saturated aqueous solution of NaCl in Section 13.4 (page 488) and stressed that the equilibrium between NaCl(s) and its solution is *dynamic*. Even after a solution has become saturated, NaCl(s) continues to dissolve and the solid continues to crystallize from solution. A particularly effective way to demonstrate this is by adding to saturated NaCl(aq) a small amount of NaCl(s) containing a trace of radioactive sodium-24 (Figure 16.1a). Immediately, radioactivity shows up in the saturated solution as well as in the undissolved solid (Figure 16.1b). This means that some dissolution of solid occurs. And, if dissolution occurs, so must crystallization, in order that the concentration of the saturated solution remain constant.

Figure 16.2 shows how concentrations vary with time in the decomposition of hydrogen iodide to hydrogen and iodine at 698 K. The curves are similar to some

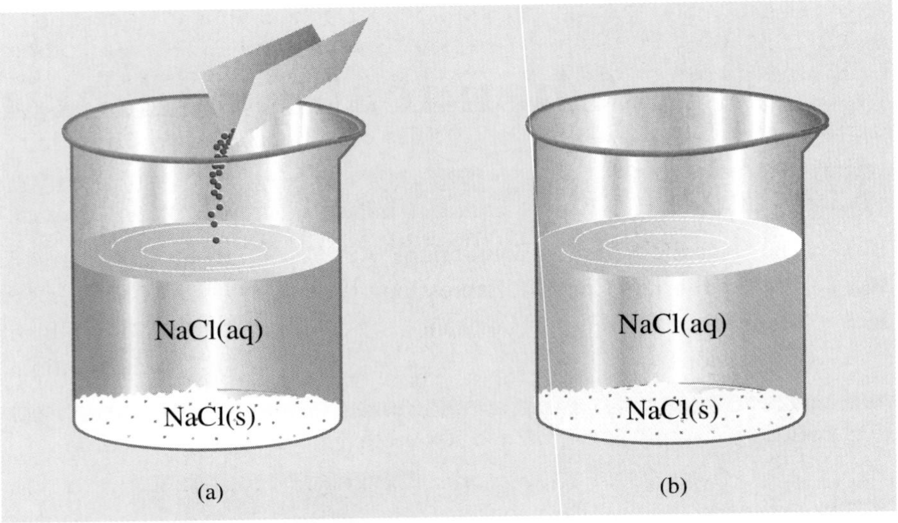

Figure 16.1 Dynamic equilibrium in saturated solution formation. If a trace of radioactive NaCl(s) (red) is added to saturated NaCl(aq) (a), radioactivity immediately appears in the solution phase (b). This proves that the dissolution process does not stop when a solution becomes saturated.

of the concentration–time graphs we saw in Chapter 15, but with this important difference: After the time marked t_e, the curves level off. The forward reaction does not go to completion. This type of behavior is a sign that the reaction is reversible:

$$2\ HI(g) \rightleftharpoons H_2(g) + I_2(g)$$

Beyond the time t_e, the system is at **equilibrium**: a forward and a reverse reaction proceed at equal rates and the concentrations of reactants and products remain constant. We could show that equilibrium in this reaction is dynamic by introducing into the equilibrium mixture a trace of radioactive iodine-131, that is, $^{131}I_2(g)$. The radioactivity would soon show up in the HI(g) as well as in the $I_2(g)$.

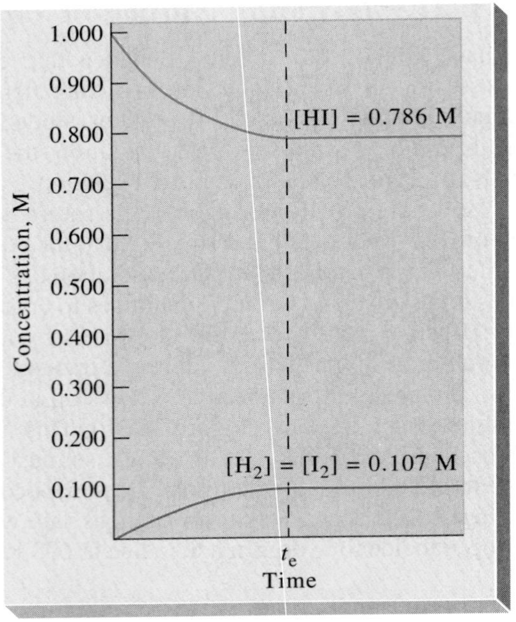

Figure 16.2 Concentration versus time graph for the reversible reaction $2\ HI(g) \rightleftharpoons H_2(g) + I_2(g)$ at 698 K. After the time t_e, the reaction is at equilibrium and the concentrations of reactant and products undergo no further change. The data shown here, together with those for two other experiments, are listed in Table 16.1.

In chemical kinetics the primary concern is with the portion of concentration versus time graphs preceding t_e. In a study of chemical equilibrium, concern is with what transpires after t_e.

As an analogy to dynamic equilibrium in a reaction, think of bailing out a leaking rowboat. Water leaking into the boat is analogous to a forward reaction and pouring buckets of water overboard is analogous to the reverse reaction. If water can be bailed out just as fast as it leaks into the boat, the pool of water at the bottom of the boat can be kept at a constant depth, analogous to the constant concentrations at equilibrium.

16.2 The Equilibrium Constant Expression

In the experiment described in Figure 16.2, when HI(g) is initially introduced into the reaction vessel, only the forward reaction occurs. However, as soon as some H_2(g) and I_2(g) form, the reverse reaction begins. As time passes, the forward reaction slows down because of the decreasing concentrations of HI(g). The reverse reaction speeds up as more H_2(g) and I_2(g) accumulate. Eventually, the forward and reverse reactions go at the same rate and the reaction mixture is at equilibrium.

In Section 15.7 we used the decomposition of HI(g) as an example of a second-order reaction: Rate of decomposition of HI(g) $= k[\text{HI}]^2$

TABLE 16.1 **Three Sets of Equilibrium Conditions for the Reaction $2\,\text{HI(g)} \rightleftharpoons \text{H}_2\text{(g)} + \text{I}_2\text{(g)}$ at 698 K**

Experiment	Initial Concn, M	Equilibrium Concns, M	$\frac{[H_2][I_2]}{[HI]}$	$\frac{[H_2][I_2]}{2 \times [HI]}$	$\frac{[H_2][I_2]}{[HI]^2}$
1	[HI] = 1.000	[HI] = 0.786 $[H_2]$ = 0.107 $[I_2]$ = 0.107	1.46×10^{-2}	7.28×10^{-3}	1.85×10^{-2}
2	[HI] = 0.800	[HI] = 0.629 $[H_2]$ = 0.0854 $[I_2]$ = 0.0854	1.16×10^{-2}	5.80×10^{-3}	1.84×10^{-2}
3	[HI] = 0.600	[HI] = 0.472 $[H_2]$ = 0.0640 $[I_2]$ = 0.0640	8.68×10^{-3}	4.34×10^{-3}	1.84×10^{-2}

Table 16.1 gives equilibrium conditions for the reaction described in Figure 16.2 and two similar experiments. At first glance, the equilibrium conditions seem to have nothing in common among the three experiments. The initial concentrations of HI are different, as are the equilibrium concentrations of HI, H_2, and I_2. The fourth column in the table lists the ratio of concentrations:

$$\frac{[H_2][I_2]}{[HI]}$$

and the fifth column, this one:

$$\frac{[H_2][I_2]}{2 \times [HI]}$$

Both ratios differ for each of the three experiments. But the ratio of concentrations in the sixth column, known as the **equilibrium constant expression**, has a constant value.

$$K_c = \frac{[H_2][I_2]}{[HI]^2} = 1.84 \times 10^{-2} \text{ (at 698 K)}$$

The symbol $\boldsymbol{K_c}$ is known as the *equilibrium constant.* Its value at 698 K is 1.84×10^{-2}. The subscript "c" signifies that concentrations (molarities) are used in the expression.*

Another example of a reversible reaction is the oxidation of NO(g) to NO_2(g).

$$2\,NO(g) + O_2(g) \rightleftharpoons 2\,NO_2(g)$$

Here the ratio of equilibrium concentrations having a constant value is

$$K_c = \frac{[NO_2]^2}{[NO]^2[O_2]}$$

Perhaps just from these two examples we can begin to see the general nature of an equilibrium constant expression:

- Concentrations of the products appear in the numerator, and concentrations of the reactants appear in the denominator.
- The exponents of the concentration terms are the same as the coefficients of the corresponding species in the balanced equation. Thus the general equilibrium constant expression, written for the hypothetical reaction

$$a\,A + b\,B + \cdots \rightleftharpoons g\,G + h\,H + \cdots$$

is

$$K_c = \frac{[G]^g[H]^h \cdots}{[A]^a[B]^b \cdots}$$

To get a sense of the significance of an equilibrium constant expression, consider a case where the decomposition of HI(g) produces an equilibrium concentration of I_2 of 0.0250 M. Because H_2 and I_2 are produced in equimolar amounts, $[H_2] = [I_2] = 0.0250$ M. To find the equilibrium concentration of HI,

$$K_c = \frac{[H_2][I_2]}{[HI]^2} = \frac{(0.0250)(0.0250)}{[HI]^2} = 1.84 \times 10^{-2}$$

$$[HI]^2 = \frac{(0.0250)(0.0250)}{1.84 \times 10^{-2}}$$

$$[HI] = \sqrt{\frac{(0.0250)(0.0250)}{1.84 \times 10^{-2}}} = 0.184\text{ M}$$

One method of determining $[I_2]$ in the equilibrium mixture is to extract a sample from the mixture, followed by the titration with $Na_2S_2O_3$(aq) described in Section 12.6 (page 466).

We will consider a variety of calculations based on equilibrium constant expressions later in the chapter.

*Because the unit "M" appears the same number of times in the numerator and in the denominator, the value of K_c for the decomposition of HI(g) has no units. In a more general case, the unit would be M^x, where x might be positive or negative and either an integer or a fraction. However, we will follow a common practice of not writing units for K_c. It is understood that all terms in the K_c expression are molarities.

Example 16.1

If, at 395 °C, the equilibrium concentrations of Cl_2 and $COCl_2$ are found to be the same, find the equilibrium concentration of CO in the reaction

$$CO(g) + Cl_2(g) \rightleftharpoons COCl_2(g) \qquad K_c = 1.2 \times 10^3 \text{ at } 395\ °C$$

Solution

$$K_c = \frac{[COCl_2]}{[CO][Cl_2]} = 1.2 \times 10^3$$

In this case, however, $[Cl_2] = [COCl_2]$, so we have

$$\frac{\cancel{[COCl_2]}}{[CO]\cancel{[Cl_2]}} = \frac{1}{[CO]} = 1.2 \times 10^3$$

As a result,

$$[CO] = \frac{1}{K_c} = \frac{1}{1.2 \times 10^3} = 8.3 \times 10^{-4}\ M$$

Thus, for the conditions stated here, although there are many possible values for $[Cl_2]$ and $[COCl_2]$, there is only *one* for [CO].

Exercise 16.1

If the equilibrium requirement in Example 16.1 were that $[CO] = [Cl_2]$, would there similarly be just one possible value of $[COCl_2]$? Explain.

16.3 More About Equilibrium Constant Expressions

Sometimes we'll want to modify an equilibrium constant expression to make it more useful in describing a particular equilibrium condition. We consider a few important modifications in this section.

Modifying the Balanced Chemical Equation

To describe the *formation* of $NO_2(g)$ at 298 K, we would likely write

$$2\ NO(g) + O_2(g) \rightleftharpoons 2\ NO_2(g) \qquad K_c = ?\ (\text{at } 298\ K)$$

Of course, to establish the numerical value of K_c we would have to obtain experimental data comparable to those in Table 16.1. From the appropriate experiments, we would find that

$$K_c = \frac{[NO_2]^2}{[NO]^2[O_2]} = 4.67 \times 10^{13} \quad (\text{at } 298\ K)$$

If our interest is in the *decomposition* of $NO_2(g)$ at 298 K, we would likely write the equation as

$$2\ NO_2(g) \rightleftharpoons 2\ NO(g) + O_2(g) \qquad K_c' = ?\ (\text{at } 298\ K)$$

We don't need to do another set of experiments to establish the value of K_c', if we consider that

$$K_c' = \frac{[NO]^2[O_2]}{[NO_2]^2} = \frac{1}{[NO_2]^2/[NO]^2[O_2]} = \frac{1}{K_c} = \frac{1}{4.67 \times 10^{13}} = 2.14 \times 10^{-14}$$

The general rule we have just established is that *when an equation is reversed, the value of K_c is inverted.*

Suppose we decide to describe the decomposition of $NO_2(g)$ based on *one* mole of reactant.

$$NO_2(g) \rightleftharpoons NO(g) + \tfrac{1}{2}O_2(g) \qquad K_c'' = ?\ \text{(at 298 K)}$$

Again, we don't need any additional experimental data if we see that

$$K_c'' = \frac{[NO][O_2]^{1/2}}{[NO_2]} = \left(\frac{[NO]^2[O_2]}{[NO_2]^2}\right)^{1/2} = \left(K_c'\right)^{1/2} = \left(\frac{1}{K_c}\right)^{1/2} = 1.46 \times 10^{-7}$$

The general rule here is that *if the coefficients in a balanced equation are multiplied by a common factor, the equilibrium constant is taken to the power corresponding to that factor*. Here the factor was $\frac{1}{2}$, so the power was $\frac{1}{2}$ (square root). In other cases the factor and power might be 2 (square), 3 (cube), and so on.

In summary, then, the form of a K_c expression depends on exactly how the equation for a reversible reaction is written.

Example 16.2

The equilibrium constant for the reaction

$$\tfrac{1}{2}H_2(g) + \tfrac{1}{2}I_2(g) \rightleftharpoons HI(g)$$

at 718 K is 7.07. What is the value of K_c at 718 K (**a**) for the reaction $HI(g) \rightleftharpoons \frac{1}{2}H_2(g) + \frac{1}{2}I_2(g)$? (**b**) What is the value of K_c at 718 K for the reaction $H_2(g) + I_2(g) \rightleftharpoons 2\ HI(g)$?

Solution

a. The reaction in question is the *reverse* of the one for which the equilibrium constant is 7.07. We need to invert this value.

$$K_c = \frac{1}{7.07} = 0.141$$

b. Here, the coefficients of the original equation have been doubled. We need to *square* the value of the original equilibrium constant.

$$K_c = (7.07)^2 = 50.0$$

Exercise 16.2

K_c for the reaction $SO_2(g) + \frac{1}{2}O_2(g) \rightleftharpoons SO_3(g)$ is 20.0 at 973 K. Calculate K_c at 973 K for the reaction $2\ SO_3(g) \rightleftharpoons 2\ SO_2(g) + O_2(g)$.

Equilibria Involving Gases

Just as we can use partial pressures of gases in place of molarities in equations relating to chemical kinetics (recall Example 15.5), we can also use them in equilibrium constant expressions for reactions involving gases. The partial pressure of gas A in a mixture of gases can be expressed as

$$P_A = \frac{n_A}{V} \cdot RT$$

and this equation rearranges to

$$[A] = \frac{n_A}{V} = \frac{P_A}{RT}$$

For the formation of $NO_2(g)$ from NO(g) at 298 K,

$$2\,NO(g) + O_2(g) \rightleftharpoons 2\,NO_2(g) \qquad K_c = 4.67 \times 10^{13}$$

we can write

$$K_c = \frac{[NO_2]^2}{[NO]^2[O_2]} = \frac{(P_{NO_2}/RT)^2}{(P_{NO}/RT)^2(P_{O_2}/RT)} = \frac{(P_{NO_2})^2}{(P_{NO})^2 P_{O_2}} \times RT$$

We give a special name to the ratio of partial pressures. It is called the partial pressure equilibrium constant, $\boldsymbol{K_p}$.

$$K_p = \frac{(P_{NO_2})^2}{(P_{NO})^2 P_{O_2}}$$

For this reaction, K_p and K_c are related as follows.

$$K_p = \frac{K_c}{RT} = K_c(RT)^{-1}$$

In a similar manner we could show that for the general reaction

$$K_p = K_c(RT)^{\Delta n_{gas}}$$

where Δn_{gas} is the sum of the coefficients of the gaseous products minus the sum of the coefficients of the gaseous reactants in the balanced equation. Thus, for the reaction

$$2\,NO(g) + O_2(g) \rightleftharpoons 2\,NO_2(g) \qquad K_c = 4.67 \times 10^{13} \text{ at } 298\text{ K}$$

$\Delta n_{gas} = 2 - (2 + 1) = -1$, leading to the equation $K_p = K_c(RT)^{-1}$, as we have already seen. The numerical value of K_p at 298 K is

$$K_p = K_c(RT)^{-1} = 4.67 \times 10^{13} \times (0.08206 \times 298)^{-1} = 1.91 \times 10^{12}$$

Unless otherwise indicated, K_p values are based on pressures expressed in atm. In most respects, we deal with K_p expressions just as we do with K_c expressions, as illustrated in Example 16.3.

Example 16.3

At 319 K, for the reaction

$$N_2O_4(g) \rightleftharpoons 2\,NO_2(g)$$

$K_p = 0.660$. (**a**) What is the value of K_c for this reaction? (**b**) What is the value of K_p for the reaction: $2\,NO_2(g) \rightleftharpoons N_2O_4(g)$? (**c**) If the equilibrium partial pressure of $NO_2(g)$ is found to be 0.332 atm, what is the equilibrium partial pressure of $N_2O_4(g)$?

Solution

a. We need to use the expression

$$K_p = K_c(RT)^{\Delta n_{gas}} \quad (\text{where } \Delta n_{gas} = 2 - 1 = 1)$$

However, we need to solve for K_c

$$K_c = \frac{K_p}{(RT)^1} = \frac{0.660}{0.08206 \times 298} = 0.0270$$

b. The equation in question is the reverse of the one given. For the reaction $2\,NO_2(g) \rightleftharpoons N_2O_4(g)$,

$$K_p = \frac{1}{0.660} = 1.52$$

c. For this calculation we can use either the original K_p expression or the one derived in part (**b**). Let's show that the result is the same either way. For the reaction $N_2O_4(g) \rightleftharpoons 2\,NO_2(g)$

$$K_p = \frac{(P_{NO_2})^2}{P_{N_2O_4}} = \frac{(0.332)^2}{P_{N_2O_4}} = 0.660$$

$$P_{N_2O_4} = \frac{(0.332)^2}{0.660} = 0.167 \text{ atm}$$

For the reaction $2\,NO_2(g) \rightleftharpoons N_2O_4(g)$

$$K_p = \frac{P_{N_2O_4}}{(P_{NO_2})^2} = \frac{P_{N_2O_4}}{(0.332)^2} = 1.52$$

$$P_{N_2O_4} = 1.52 \times (0.332)^2 = 0.168 \text{ atm}$$

The difference between the two values—0.167 and 0.168—results from rounding off.

Exercise 16.3

Given $K_c = 1.8 \times 10^{-6}$ for the reaction $2\,NO(g) + O_2(g) \rightleftharpoons 2\,NO_2(g)$ at 457 K, derive the value of K_p at 457 K for the reaction $NO_2(g) \rightleftharpoons NO(g) + \frac{1}{2}O_2(g)$.

Equilibria Involving Liquids and Solids

So far, we have only considered reactions involving gases—*homogeneous* reactions. In the next chapter or two we will concentrate on reactions in aqueous solution—again, homogeneous reactions. For *heterogeneous* reactions, where the reactants and products do not coexist in the same phase, we need to make some accommodations in equilibrium constant expressions. Fortunately, these adaptations generally take the form of simplifying the expressions.

By convention, the equilibrium constant does not include terms for pure solids and liquids. Consider the calcination of limestone, a reversible reaction that we discussed in Section 8.3.

$$CaCO_3(s) \rightleftharpoons CaO(s) + CO_2(g)$$

Although we can write an expression of the form

$$K_c' = \frac{[CaO][CO_2]}{[CaCO_3]}$$

we should write only

$$K_c = [CO_2] \qquad \text{and} \qquad K_p = P_{CO_2}$$

Think of the matter in this way: An equilibrium constant expression only includes terms for reactants and products whose concentrations and/or partial pressures can *change* during a chemical reaction. The concentrations of reactants and products that coexist in a homogeneous mixture (e.g., a liquid solution) change in the course of a chemical reaction. In the decomposition of $CaCO_3(s)$, CO_2 is a gas whose concentration (n/V) and partial pressure [$P_{CO_2} = (n/V) \times RT$] change from their initial values until a final equilibrium is reached. By contrast, however, although the amounts of $CaCO_3(s)$ and $CaO(s)$ change during the reaction, the *concentration* of the matter *within* each pure solid phase remains unchanged.

Similarly, for the equilibrium between pure liquid water and its vapor,

$$H_2O(l) \rightleftharpoons H_2O(g)$$

we can write

$$K_c = [H_2O(g)] \qquad \text{and} \qquad K_p = P_{H_2O}$$

In fact, the equilibrium constant K_p is simply the water vapor pressure, P_{H_2O}.

In summary, by convention, *equilibrium constant expressions do* not *include terms for substances found in pure solid or liquid phases.* We further illustrate this idea in Example 16.4.

Example 16.4

The reaction of steam and coke (carbon) produces a mixture of carbon monoxide and hydrogen called water gas. The water-gas reaction has long been used to make combustible gases from coal.

$$C(s) + H_2O(g) \rightleftharpoons CO(g) + H_2(g)$$

Write an equilibrium constant expression, K_c, for this reaction.

Solution

The products CO and H_2 and the reactant H_2O—all gases—are represented in the expression, but C(s), a solid, is not.

$$K_c = \frac{[CO][H_2]}{[H_2O]}$$

Exercise 16.4

Write the K_p expression for the water-gas reaction of Example 16.4.

Equilibrium Constants: When Do We Need Them and When Don't We?

In principle, every reaction has an equilibrium constant, but often we don't use it. How can this be? Let's answer this question by considering three specific cases.

For the reaction of hydrogen and oxygen gases at 298 K

$$2\,H_2(g) + O_2(g) \rightleftharpoons 2\,H_2O(l)$$

$$K_p = \frac{1}{(P_{H_2})^2(P_{O_2})} = 1.4 \times 10^{83}$$

Note that we do not include a term for liquid H_2O in the K_p expression.

Starting with a 2:1 mole ratio of hydrogen to oxygen, the equilibrium partial pressures of $H_2(g)$ and $O_2(g)$ must become extremely small—approaching zero—in order for the K_p value to be so large. The hydrogen and oxygen are effectively totally consumed in the reaction. We say that a reaction in which one or more reactants is totally consumed *goes to completion*, and we don't need to use equilibrium constant expressions for such reactions. We can calculate the outcomes of reactions that go to completion with just the principles of stoichiometry from Chapter 3.

A very large numerical value of K_c *or* K_p *signifies that a reaction goes to completion, or very nearly so.*

For the decomposition of limestone at 298 K

$$CaCO_3(s) \rightleftharpoons CaO(s) + CO_2(g) \qquad K_p = P_{CO_2} = 1.9 \times 10^{-23}$$

Intuitively, we know that limestone doesn't decompose to any great extent at normal temperatures. If it did, there would be much more $CO_2(g)$ in the atmosphere than there is now. The K_p value tells us that the partial pressure of $CO_2(g)$ in equilibrium with $CaCO_3(s)$ and $CaO(s)$ is exceedingly small—1.9×10^{-23} atm.

A very small numerical value of K_c or K_p signifies that the forward reaction, as written, does not occur to any significant extent.

In fact, very often in these cases we consider that the forward reaction doesn't take place at all. This is why we sometimes write expressions such as

$$CaCO_3(s) \xrightarrow{298\text{ K}} \text{“no reaction”}$$

On the other hand, the situation is totally different for the decomposition of limestone at about 1200 K.

$$CaCO_3(s) \rightleftharpoons CaO(s) + CO_2(g) \qquad K_p = P_{CO_2} \approx 1$$

Here the forward and reverse reactions are both significant and we do need to use the K_p expression to describe equilibrium situations.

Equilibrium constants are most likely to be significant when their values are neither very large nor very small.

As a *very rough* approximation, let's say this refers to values in the range from 10^{-15} to 10^{15}.

Finally, we must always keep in mind that an equilibrium constant expression applies only to a reversible reaction *at equilibrium*. Rates of reaction determine how long it takes to reach equilibrium, and thus, indirectly, when the equilibrium constant expression can be used. Although K_p for the reaction of $H_2(g)$ and $O_2(g)$ at 298 K is very large, the reaction proceeds at an immeasurably slow rate because of its high activation energy. The reaction never reaches equilibrium at 298 K. It is only when the mixture is heated that the reaction occurs at an explosive speed.

Example 16.5

Is the reaction $CaO(s) + CO_2(g) \rightleftharpoons CaCO_3(s)$ likely to occur to any appreciable extent at 298 K?

Solution

This reaction is the *reverse* of that describing the decomposition of limestone. Its K_p value is the *reciprocal* of that for the decomposition of limestone. $K_p = 1/(1.9 \times 10^{-23}) = 5.3 \times 10^{22}$. With this large a value of K_p we do expect the forward reaction to occur to a very significant extent. In fact, over time the reaction should go to completion.

Exercise 16.5

Refer to Example 16.1 and determine if we can assume that the reaction $CO(g) + Cl_2(g) \longrightarrow COCl_2(g)$ goes essentially to completion at 395 °C. Explain your reasoning.

16.4 Qualitative Treatment of Equilibrium: Le Châtelier's Principle

Sometimes, nonnumerical answers or simple "ballpark" estimates serve our needs just as well as do precise, numerical results. This is very much the case in working with the condition of equilibrium. A useful qualitative guide to equilibrium was framed by Henri Le Châtelier in 1888. Le Châtelier stated his principle in a rather detailed and lengthy manner. Here is a rough paraphrase:

> *When a change (for example, a change in concentration, temperature, pressure,. . .) is imposed on a system at equilibrium, the system responds by attaining a new equilibrium condition that* partially *offsets the impact of the change.*

You should acquire a better feel for Le Châtelier's principle as we now apply it to some specific situations.

The distinctive aroma and flavor of oranges are due in part to the ester octyl acetate.

Changing the Amounts of Reacting Species

Let's look at the reaction for the formation of octyl acetate from 1-octanol and acetic acid.

$$\underset{\text{1-Octanol}}{CH_3(CH_2)_6CH_2OH} + \underset{\text{Acetic acid}}{CH_3COOH} \overset{H^+}{\rightleftharpoons} \underset{\text{Octyl acetate}}{CH_3(CH_2)_6CH_2OOCCH_3} + H_2O$$

An equilibrium mixture of the four components is homogeneous; it exists in a single liquid phase. Let's start with the equilibrium mixture corresponding to the expression

$$K_c = \frac{[CH_3(CH_2)_6CH_2OOCCH_3][H_2O]}{[CH_3(CH_2)_6CH_2OH][CH_3COOH]}$$

Then let's disturb the equilibrium by adding more acetic acid. We indicate this addition by using larger type for acetic acid in the expression below.

$$K_c > \frac{[CH_3(CH_2)_6CH_2OOCCH_3][H_2O]}{[CH_3(CH_2)_6CH_2OH]\large{[CH_3COOH]}}$$

Because we have increased the denominator, the ratio of concentrations is now smaller than K_c. The concentrations must change in such a way as to make this ratio once again equal to K_c. This requires that the numerator also become larger; a net reaction must occur in the forward direction. But when this happens, the concentrations of the reactants decrease. The net result is suggested in the expression below, where again the sizes of the type suggest how the concentrations in the new equilibrium are related to those in the initial equilibrium.

$$K_c = \frac{\large{[CH_3(CH_2)_6CH_2OCOCH_3][H_2O]}}{\small{[CH_3(CH_2)_6CH_2OH]}\large{[CH_3COOH]}}$$

Le Châtelier's principle allows us to arrive at the same conclusion but without having to work through equilibrium constant expressions. To counter the ef-

The Significance of Chemical Equilibrium, in the Words of Henri Le Châtelier

Henri Le Châtelier (1850–1936) was one of the first chemists to appreciate the power of thermodynamics in dealing with chemical problems. In particular, he fully understood the difference between a reaction that goes to completion and one that can only reach an equilibrium condition. He stated this distinction rather nicely in a journal article in 1888, from which the following quotation is taken. In reading Le Châtelier's account, think of a limited reaction as a reversible reaction at equilibrium.

"It is known that in the blast furnace the reduction of iron oxide is produced by carbon monoxide, according to the reaction

$$Fe_2O_3(s) + 3\,CO(g) \rightleftharpoons 2\,Fe(s) + 3\,CO_2(g)$$

but the gas leaving the chimney contains a considerable proportion of carbon monoxide, ... Because this incomplete reaction was thought to be due to an insufficiently prolonged contact between carbon monoxide and the iron ore, the dimensions of the furnaces have been increased. In England they have been made as high as thirty meters. But the proportion of carbon monoxide escaping has not diminished, thus demonstrating, by an experiment costing several hundred thousand francs, that the reduction of iron oxide by carbon monoxide is a limited reaction. Acquaintance with the laws of chemical equilibrium would have permitted the same conclusion to be reached more rapidly and far more economically."

fect of increasing the concentration of one of the components, a net reaction must occur that consumes some of the added component, partially offsetting the increase in concentration. Acetic acid is consumed in the *forward* reaction. In the new equilibrium, the reaction has gone farther in the forward direction. We might say that the equilibrium *shifts to the right*, with the following results.

- *Acetic acid.* There will be *more* acetic acid in the new equilibrium than in the original equilibrium, but the amount present will be *less than* the original amount plus the added amount. (Recall that equilibrium shifts in a way that only *partially* offsets a change.)
- *1-Octanol.* There will be *less* 1-octanol than in the original equilibrium. Some of the 1-octanol present in the original equilibrium reacts with some of the added acetic acid.
- *Octyl acetate and water.* There will be more of each of these products in the new equilibrium. They are formed as the forward reaction proceeds.

Organic chemists often use an excess of acetic acid—as much as 10 mol to 1 mol of the more expensive 1-octanol—to drive the equilibrium toward the product side, giving a good yield of octyl acetate. Another method of improving the equilibrium yield of octyl acetate is to remove water, as illustrated in Example 16.6.

Example 16.6

Describe the effect on equilibrium in the reaction of 1-octanol and acetic acid if water is removed from the equilibrium mixture.

$$\underset{\text{1-Octanol}}{CH_3(CH_2)_6CH_2OH} + \underset{\text{Acetic acid}}{CH_3COOH} \xrightleftharpoons{H^+} \underset{\text{Octyl acetate}}{CH_3(CH_2)_6CH_2OOCCH_3} + H_2O$$

Solution

To partially offset the effect of the *removal* of water, a product of the reaction, the *forward* reaction is favored, thereby replacing some of the water. Not all of the water that was removed is replaced, however, so that in the new equilibrium the amount of water is somewhat *less* than in the original equilibrium. The amount of octyl acetate in the new equilibrium is *greater* than in the original equilibrium, and the amounts of both 1-octanol and acetic acid are *less*.

Exercise 16.6

What should be the effect of each of the following changes on a constant-volume equilibrium mixture of N_2, H_2, and NH_3?

$$N_2(g) + 3\,H_2(g) \rightleftharpoons 2\,NH_3(g)$$

(a) Adding $H_2(g)$; **(b)** removing $N_2(g)$; **(c)** removing $NH_3(g)$.

The actual effect of adding or removing a reacting species from a *homogeneous* equilibrium mixture is to change its concentration. If the concentration of one reactant changes, so too must all the others to reestablish the constant value of K_c. If the component added or removed is a pure solid or liquid phase in a *heterogeneous* equilibrium mixture, there is *no change* in the equilibrium condition. Liquids and solids do not appear in the equilibrium constant expression. Thus, the pressure of the $CO_2(g)$ in equilibrium with $CaO(s)$ and $CaCO_3(s)$ is unaffected by the amounts of the two solids present.

$$CaCO_3(s) \rightleftharpoons CaO(s) + CO_2(g) \qquad K_p = P_{CO_2}$$

Similarly, the vapor pressure of water is not affected by the addition or removal of liquid water in the equilibrium

$$H_2O(l) \rightleftharpoons H_2O(g) \qquad K_p = P_{H_2O}$$

Changing Pressure or Volume in Gaseous Equilibria

We can increase or decrease the partial pressure of an *individual* component in a constant-volume, gaseous equilibrium mixture by adding or removing some of that component. The equilibrium shifts to the left or to the right in the manner that we have already described (see Exercise 16.6).

We can increase or decrease the partial pressures of *all* components in a gaseous equilibrium mixture by reducing or increasing the reaction volume. To assess the effect that such changes have on the condition of equilibrium, we need to use both Le Châtelier's principle and ideas derived from Avogadro and Gay-Lussac.

Consider once more the formation of $NO_2(g)$ from NO(g) and $O_2(g)$ at 298 K.

$$2\,NO(g) + O_2(g) \rightleftharpoons 2\,NO_2(g)$$

$$K_p = \frac{(P_{NO_2})^2}{(P_{NO})^2 P_{O_2}} = 1.91 \times 10^{12}$$

In the forward reaction *three* moles of gas are converted to *two* moles of gas. First let's describe what is necessary to maintain a constant K_p.

If we reduce the volume of an equilibrium mixture—say, by increasing the external pressure—the partial pressures of the three gases all increase by the same factor. For example, if we reduce the volume to one-half of its initial value, all the partial pressures double. The numerator and denominator both increase. However, because the factor by which the partial pressures increase (two) appears three times in the denominator and only twice in the numerator, the ratio of partial pressures is now less than K_p. To restore equilibrium, the numerator has to become larger and the denominator smaller. A net reaction occurs *to the right*, producing a restored equilibrium that has more NO_2 and less NO and O_2 than the original equilibrium.

Now let's apply Le Châtelier's principle. If the volume of an equilibrium mixture is reduced, the molecules are "crowded" into a smaller volume. Because *two* moles of the product can be contained in a smaller volume than the *three* moles of reactant gases, the "crowding" is partially offset by a net reaction in the direction that reduces the total number of molecules—the forward direction (Figure 16.3). Equilibrium *shifts to the right*, just as we concluded by working with the K_p expression. By contrast, increasing the reaction volume "thins out" the gas molecules. Equilibrium shifts in the direction that produces a greater number of molecules, *to the left*. The basic idea, then, is this:

> *When the volume is* decreased, *the equilibrium condition shifts in the direction producing a* smaller number *of moles of gas. When the volume of an equilibrium mixture of gases is* increased, *equilibrium shifts in the direction producing a* larger number *of moles of gas.*

If an inert gas, such as helium, is added to an equilibrium mixture at *constant pressure*, the volume of the system must increase to accommodate the added gas. This causes the same type of changes as would transferring the original mixture to a container of larger volume. That is, the equilibrium shifts in the direction producing the greater number of molecules. In the reaction

$$2\,NO(g) + O_2(g) \rightleftharpoons 2\,NO_2(g)$$

this shift would be *to the left*. The inert gas molecules, of course, do not participate in the reaction in any way other than having caused the volume of the reaction mixture to increase.

On the other hand, if the inert gas is added to a *constant-volume* equilibrium mixture, the partial pressures and concentrations of the reacting gases are unchanged. The relevant ratio of partial pressures remains equal to K_p, and the relevant ratio of concentrations remains equal to K_c. There is no effect on the equilibrium.

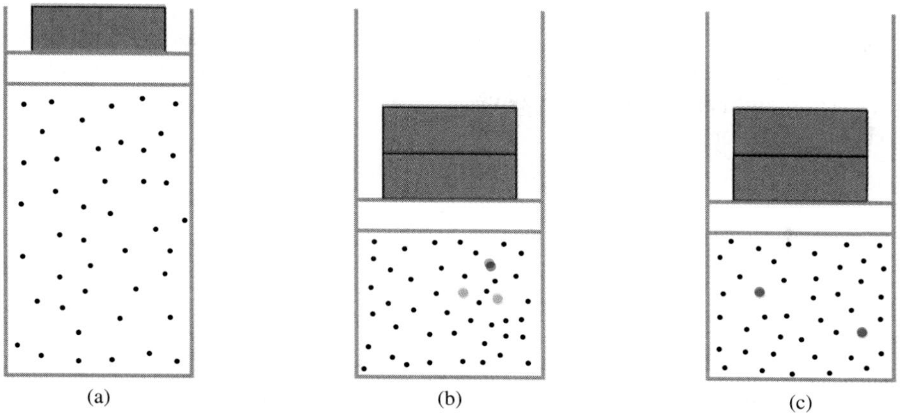

Figure 16.3 **Illustrating Le Châtelier's principle: Effect of volume in the reaction $2\,NO(g) + O_2(g) \rightleftharpoons 2\,NO_2(g)$.** (a) Equilibrium is established, and then the volume of the equilibrium mixture is (b) decreased by increasing the external pressure. Equilibrium is disturbed. Two representative NO molecules (tan) and one O_2 molecule (blue) are singled out in the equilibrium mixture. When these molecules react they are replaced by (c) two additional NO_2 molecules (red). Because *three* reactant molecules yield *two* product molecules, Le Châtelier's principle predicts that the forward reaction is favored by a decrease in volume (increase in pressure).

Example 16.7

An equilibrium mixture of $SO_2(g)$, $O_2(g)$, and $SO_3(g)$ is transferred from a 1.00-L flask to a 2.00-L flask. In what direction does a net reaction proceed to restore equilibrium?

$$2\,SO_3(g) \rightleftharpoons 2\,SO_2(g) + O_2(g)$$

Solution

Transferring the mixture to the larger flask reduces the partial pressure of each gas. From the standpoint either of an *increase* in volume or a *decrease* in pressure, we predict that the net reaction will be the one producing the larger number of moles of gas—the forward direction. The equilibrium shifts to the right. Some of the $SO_3(g)$ in the original equilibrium decomposes to $SO_2(g)$ and $O_2(g)$.

Exercise 16.7

How is the equilibrium amount of HI(g) in the reaction $H_2(g) + I_2(g) \rightleftharpoons 2\,HI(g)$ changed by changing the reaction volume? Explain.

If a certain change causes one reaction to be favored, the opposite change will cause the opposing reaction to be favored. Reducing the reaction volume favors the reverse reaction. Equilibrium shifts to the left.

Changing the Equilibrium Temperature

The changes we have described so far do not change the value of the equilibrium constant, but changing the temperature of an equilibrium mixture actually does change this value. If the equilibrium constant becomes larger, the forward reaction is favored, and equilibrium shifts to the right. If it becomes smaller, the reverse reaction is favored and equilibrium shifts to the left. In Chapter 19 we will learn how to calculate equilibrium constants as a function of temperature. For now, however, we can use Le Châtelier's principle to assess, qualitatively, the effect of temperature on equilibrium.

To change the temperature of a reaction mixture, we must add heat to raise the temperature or remove heat to lower the temperature. Adding heat to an equilibrium mixture will stimulate the reaction that can absorb some of the heat—the endothermic reaction. The removal of heat stimulates the exothermic reaction. In short,

> Raising *the temperature of an equilibrium mixture shifts equilibrium in the direction of the* endothermic reaction; lowering *the temperature shifts equilibrium in the direction of the* exothermic *reaction.*

Example 16.8

Will the amount of NO(g) formed from given amounts of $N_2(g)$ and $O_2(g)$ be greater at high or low temperatures?

$$N_2(g) + O_2(g) \rightleftharpoons 2\,NO(g) \qquad \Delta H° = +180.5\text{ kJ}$$

Solution

As written, the $\Delta H°$ value given is for the forward reaction. The forward reaction is endothermic, and the equilibrium condition is displaced to the right as the temperature is raised. The conversion of $N_2(g)$ and $O_2(g)$ to NO(g) is favored at high temperatures. This is one reason NO is found in the exhaust of high-compression automobile engines; they operate at high temperatures.

Exercise 16.8

Is the conversion of $SO_2(g)$ to $SO_3(g)$ more nearly complete at high or low temperatures?

$$2\,SO_2(g) + O_2(g) \rightleftharpoons 2\,SO_3(g) \qquad \Delta H° = -198\text{ kJ}$$

Perhaps you recall from Section 13.4 that the majority of solid solutes (95% or more) have aqueous solubilities that *increase* with temperature. In these cases solution formation is an *endothermic* process, and an endothermic process is favored at *higher* temperatures. This means that more of the solute dissolves before equilibrium is reached—that is, before the solution becomes saturated.

Adding a Catalyst

The reaction of $SO_2(g)$ with $O_2(g)$ to produce $SO_3(g)$ is greatly speeded up by a catalyst (such as platinum metal), but so too is the decomposition of $SO_3(g)$ to $SO_2(g)$ and $O_2(g)$—the reverse reaction.

$$2\,SO_2(g) + O_2(g) \xrightleftharpoons{Pt} 2\,SO_3(g) \qquad K_c = 2.8 \times 10^2 \text{ at } 1000\text{ K}$$

Because the rate of the forward and reverse reactions are increased to the same extent, the proportion of $SO_3(g)$ in the equilibrium mixture is the same as if no catalyst were present at all.

The function of a catalyst, as we learned in Chapter 15, is to change the reaction profile to a pathway with a lower activation energy, but this does not change either the initial state (reactants) or final state (products) in the chemical reaction. And, as we will learn in Chapter 19, the equilibrium constant is a function only of properties of the reactants and products.

CONCEPTUAL EXAMPLE 16.1

Flask A in Figure 16.4 contains an equilibrium mixture in the reaction

$$CO(g) + H_2O(g) \rightleftharpoons CO_2(g) + H_2(g) \qquad \Delta H = -41\ kJ \quad K_c = 9.03 \text{ at } 698\ K$$

The contents of flask B are described below. The valve between the two flasks is opened and equilibrium is reestablished. Describe, qualitatively, how the amount of each of the four gases in the final equilibrium mixture compares to its amount in the original equilibrium in flask A. If you are uncertain, explain why.

a. Flask B contains Ar(g) at 1 atm pressure.
b. Flask B contains 1.0 mol CO_2.
c. Flask B contains 1.0 mol CO, and the temperature of the A–B assembly is raised by 100 °C.

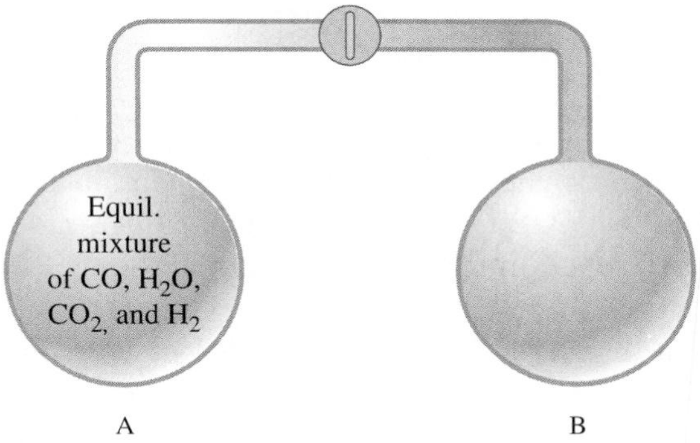

Figure 16.4 Conceptual Example 16.1 illustrated.

Solution

a. Ar(g) is an inert gas and has no effect on the reaction. Because the reaction involves the same number of reactant and product molecules, the equilibrium is unaffected by the change in volume. The amounts of CO, H_2O, CO_2, and H_2 are all unchanged.
b. Adding $CO_2(g)$ stimulates the *reverse* reaction. Again, increasing the volume has no effect. When equilibrium is reestablished, the amounts of CO and H_2O will be greater. H_2 is consumed in the reverse reaction, so its amount will be less. The effect of adding more CO_2 is only *partially* offset, and the amount of CO_2 will be greater in the final than in the original equilibrium.
c. Adding more CO(g) favors the *forward* reaction, but raising the temperature favors the endothermic or *reverse* reaction. Because these factors work in opposition, we cannot make a qualitative prediction.

Conceptual Exercise 16.1

Respond as directed in Conceptual Example 16.1 to these additional conditions.
a. Flask B contains 1.0 mol $H_2(g)$ at 1 atm pressure.
b. Flask B contains 1.0 mol $CO_2(g)$ and 1.0 mol $H_2O(g)$.
c. Flask B contains 1.0 mol $H_2O(g)$, and the temperature of the A–B assembly is lowered by 100 °C.

Chemical Equilibrium and the Synthesis of Ammonia

In Chapter 14 we described, in a general way, the production of ammonia (about sixth among industrially produced chemicals in the United States). Following are typical conditions used in its manufacture.

$$N_2(g) + 3\,H_2(g) \rightleftharpoons 2\,NH_3(g) \qquad \Delta H° = -92.22 \text{ kJ}$$

Reactants: 3:1 mol ratio of H_2 to N_2
Temperature: 400–600 °C
Pressure: 140–340 atm
Catalyst: Fe_3O_4 with small amounts of Al_2O_3, MgO, CaO, and K_2O, reduced to a metallic mixture before use.

Based on the chemical equation and Le Châtelier's principle, we would conclude that high yields of $NH_3(g)$ are favored by (1) *low* temperatures—the forward reaction is exothermic; (2) *high* pressures—the forward reaction is accompanied by a decrease in number of moles of gas; and (3) *continuous removal of* NH_3—removal of a product stimulates the forward reaction to form additional product (see Figure 16.5).

The synthesis reaction is indeed carried out at high pressures. The temperatures used, however, are moderately high, not low. The theoretical percent conversion of a 3 mol H_2:1 mol N_2 mixture to NH_3 is over 90% at high pressures and room temperature, but with these conditions it would take too long to reach equilibrium. To be commercially successful, a reaction must yield the desired product quickly. Thus, even though only 20% conversion of the reactants to NH_3 is achieved in an equilibrium mixture at 500 °C and 200 atm, with these conditions and a catalyst, equilibrium is reached in less than 1 min.

16.5 Some Illustrative Equilibrium Calculations

In the next two chapters we'll use equilibrium constant expressions in many problem-solving situations, and we will preview the subject in this section. For convenience, we will divide the problems into two basic types: those in which values of K_c or K_p are determined from experimental concentration or partial pressure data, and those in which values of K_c and K_p are used to calculate equilibrium concentrations or partial pressures.

Determining Values of Equilibrium Constants from Experimental Data

In Example 16.9, we seek to establish the value of an equilibrium constant, K_c. As suggested by Figure 16.6, we are given the *initial* amounts of the reactants

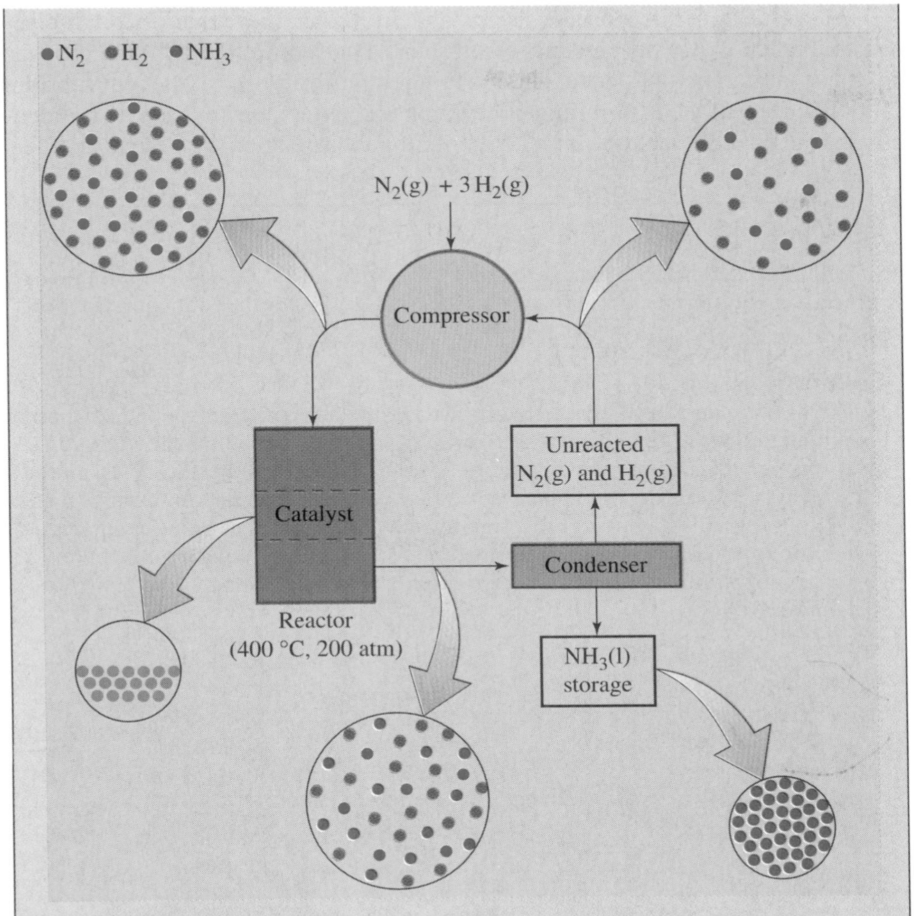

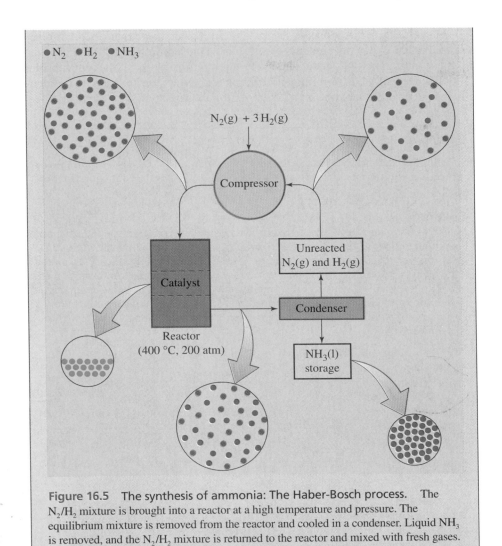

Figure 16.5 The synthesis of ammonia: The Haber-Bosch process. The N_2/H_2 mixture is brought into a reactor at a high temperature and pressure. The equilibrium mixture is removed from the reactor and cooled in a condenser. Liquid NH_3 is removed, and the N_2/H_2 mixture is returned to the reactor and mixed with fresh gases.

and the *equilibrium* amount of the product. From these data we must establish the equilibrium amounts, and then the equilibrium concentrations, of *all* the species involved in the equilibrium. Then, we can substitute equilibrium concentrations into the K_c expression.

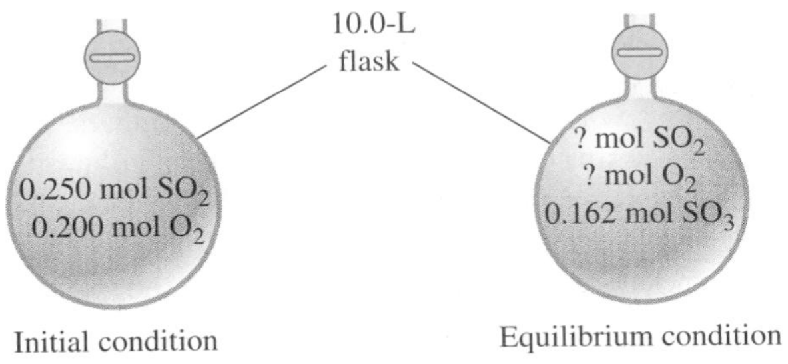

Figure 16.6 Example 16.9 illustrated. The key in Example 16.9 is to determine the amounts of SO_2 and O_2 *consumed* to reach equilibrium.

A useful general approach is to tabulate under the equation for the reversible reaction: (a) the amounts of substances present initially, (b) the changes in these amounts that occur in reaching equilibrium, (c) the equilibrium amounts, and (d) the equilibrium concentrations. Often, the key step is (b): identifying the changes that occur and determining their relationship to one another.

Example 16.9

At 1000 K, 0.250 mol SO_2 and 0.200 mol O_2 react in a 10.0-L vessel to form 0.162 mol SO_3 at equilibrium. Find the value of K_c, at 1000 K, for the following reaction.

$$2\,SO_2(g) + O_2(g) \rightleftharpoons 2\,SO_3(g)?$$

Solution

Let's set up the data in the suggested format. From the coefficients in the balanced equation we see the relationship: 2 mol SO_2:2 mol SO_3. The amount of SO_2 that must react, is the same as the amount of SO_3 at equilibrium: 0.162 mol. From the relationship: 2 mol SO_2:1 mol O_2, we see that only half as much O_2 reacts: 0.081 mol. In our set-up we use a *negative* sign for something that is consumed and a *positive* sign for something that is formed. We then obtain equilibrium amounts (moles), convert them to molar concentrations, place molarities in the K_c expression, and solve for K_c.

The reaction:	$2\,SO_2(g)$	+	$O_2(g)$	$\rightleftharpoons$	$2\,SO_3(g)$
Initial amounts:	0.250 mol		0.200 mol		0 mol
Changes:	−0.162 mol		−0.081 mol		+0.162 mol
Equil. amounts:	0.088 mol		0.119 mol		0.162 mol
Equil. concns, M:	$\frac{0.088}{10.0} = 0.0088$		$\frac{0.119}{10.0} = 0.0119$		$\frac{0.162}{10.0} = 0.0162$

$$K_c = \frac{[SO_3]^2}{[SO_2]^2[O_2]} = \frac{(0.0162)^2}{(0.0088)^2(0.0119)} = 2.8 \times 10^2$$

Exercise 16.9

A 1.00-kg sample of $Sb_2S_3(s)$ and 10.0 g $H_2(g)$ are allowed to react in a 25.0-L container at 713 K. At equilibrium, 72.6 g $H_2S(g)$ is present. Find the value of K_p at 713 K for the following reaction.

$$Sb_2S_3(s) + 3\,H_2(g) \rightleftharpoons 2\,Sb(s) + 3\,H_2S(g)$$

[*Hint*: You can use the data to calculate partial pressures of the gases and then K_p, or you can calculate K_c as in Example 16.9, and relate K_p to K_c. Do the amounts of $Sb_2S_3(s)$ and Sb(s) enter into the calculation in any way?]

Calculating Equilibrium Quantities from K_c and K_p Values

In one of the most common types of equilibrium calculations, illustrated in Figure 16.7 and Example 16.10, we start with initial reactants and no products, and we are given the value of the equilibrium constant. Our goal is to calculate the amounts of substances present at equilibrium. Typically, we use a symbol such as x to identify one of the changes that occurs in establishing equilibrium. Then, we relate all the other changes to x, substitute appropriate terms into the equilibrium constant expression, and solve the expression for x. The calculation in Example 16.10 is a relatively easy one because the algebra is simple.

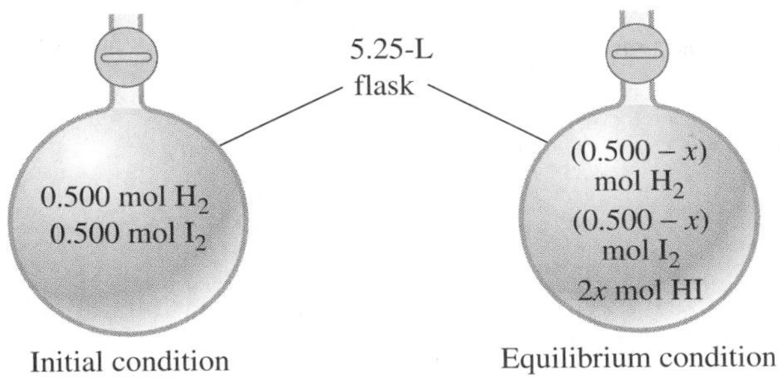

Figure 16.7 Example 16.10 illustrated. We can represent the equilibrium amounts in terms of the initial amounts and a quantity x. We solve for x by using the equilibrium constant expression K_c.

Example 16.10

Consider the reaction

$$H_2(g) + I_2(g) \rightleftharpoons 2\,HI(g) \qquad K_c = 54.3 \text{ at } 698\text{ K}$$

If we start with 0.500 mol $H_2(g)$ and 0.500 mol $I_2(g)$ in a 5.25-L vessel at 698 K, how many moles of each gas will be present at equilibrium?

Solution

If we let x = mol H_2 that react, the change in the amount of $H_2 = -x$. The change in amount of I_2 is also $-x$. The change in amount of HI is $+2x$.

The reaction:	$H_2(g)$	+	$I_2(g)$	$\rightleftharpoons$ 2 HI(g)
Initial amounts:	0.500 mol		0.500 mol	0 mol
Changes:	$-x$ mol		$-x$ mol	$+2x$ mol
Equil. amounts:	$(0.500 - x)$ mol		$(0.500 - x)$ mol	$2x$ mol
Equil. concns, M:	$\dfrac{(0.500 - x)}{5.25}$		$\dfrac{(0.500 - x)}{5.25}$	$\dfrac{2x}{5.25}$

Now, let's enter the equilibrium concentrations into the K_c expression. The first thing we notice is that the volume terms cancel. This will happen any time that the sums of the exponents in the numerator and denominator are equal (*two* in this case).

$$K_c = \frac{[HI]^2}{[H_2][I_2]} = \frac{(2x/5.25)^2}{[(0.500 - x)/5.25][(0.500 - x)/5.25]} = 54.3$$

$$= \frac{(2x)^2}{(0.500 - x)^2} = 54.3$$

Now, the simplest approach is to extract the square root of each side of the equation and solve for x.

$$\left(\frac{(2x)^2}{(0.500 - x)^2}\right)^{1/2} = \frac{2x}{(0.500 - x)} = (54.3)^{1/2}$$

$$2x = (54.3)^{1/2} \times (0.500 - x)$$

$$2x = 3.68 - 7.37x$$

$$9.37x = 3.68$$

$$x = 0.393$$

A useful check on this result is to substitute the calculated equilibrium amounts back into the K_c expression to see what value of K_c they yield:

$$K_c = \frac{[HI]^2}{[H_2][I_2]}$$

$$= \frac{(0.786/5.25)^2}{(0.107/5.25)(0.107/5.25)}$$

$$= 54.0$$

This is close enough to the given value of K_c (54.3) to confirm that our calculation is correct.

The equilibrium amounts are

H_2: $0.500 - x = 0.500 - 0.393 = 0.107$ mol H_2

I_2: $0.500 - x = 0.500 - 0.393 = 0.107$ mol I_2

HI: $2x = 2 \times 0.393 = 0.786$ mol HI

Exercise 16.10

What will be the mole fraction of NO(g) at equilibrium if an equimolar mixture of N_2(g) and O_2(g) is brought to equilibrium at 2500 K?

$$N_2(g) + O_2(g) \rightleftharpoons 2\ NO(g) \qquad K_c = 2.1 \times 10^{-3} \text{ at } 2500\ K$$

(*Hint*: Assume any initial amounts of N_2 and O_2, as long as they are equal. Does the volume of the system matter?)

In Example 16.11, the initial amounts of the reactants are unequal. The setup of the problem is identical to that in Example 16.10, but the algebraic solution requires use of the quadratic formula, a common requirement in equilibrium calculations.

Example 16.11

Suppose that in the reaction of Example 16.10 the initial amounts are 0.800 mol H_2 and 0.500 mol I_2. What will be the amounts of reactants and products when equilibrium is attained?

Solution

As in Example 16.10, we let x be the number of moles of H_2 that react. The change in the amount of $H_2 = -x$, and the change in amount of I_2 is also $-x$. The change in amount of HI is $+2x$.

The reaction:	$H_2(g)$	+	$I_2(g)$	$\rightleftharpoons$	2 HI(g)
Initial amounts:	0.800 mol		0.500 mol		0 mol
Changes:	$-x$ mol		$-x$ mol		$+2x$ mol
Equil. amounts:	$(0.800 - x)$ mol		$(0.500 - x)$ mol		$2x$ mol
Equil. concns, M:	$\frac{(0.800 - x)}{5.25}$		$\frac{(0.500 - x)}{5.25}$		$\frac{2x}{5.25}$

Now, let's enter the equilibrium concentrations into the K_c expression.

$$K_c = \frac{[HI]^2}{[H_2][I_2]} = \frac{(2x/5.25)^2}{[(0.800 - x)/5.25][(0.500 - x)/5.25]} = 54.3$$

$$= \frac{(2x)^2}{(0.800 - x)(0.500 - x)} = 54.3$$

$$4x^2 = 54.3(0.400 - 1.30x + x^2)$$

$$4x^2 = 21.7 - 70.6x + 54.3x^2$$

$$50.3x^2 - 70.6x + 21.7 = 0$$

This is a quadratic equation in the form $ax^2 + bx + c = 0$, for which the solutions are

$$x = \frac{-b \pm \sqrt{b^2 - 4ac}}{2a}$$

See also the discussion on quadratic equations in Appendix A.

Thus,

$$x = \frac{70.6 \pm \sqrt{(70.6)^2 - (4 \times 50.3 \times 21.7)}}{2 \times 50.3}$$

$$= \frac{70.6 \pm \sqrt{618}}{100.6} = \frac{70.6 \pm 24.9}{100.6}$$

Now we are faced with two choices. Either

$$x = \frac{70.6 + 24.9}{100.6} = 0.949$$

or

$$x = \frac{70.6 - 24.9}{100.6} = 0.454$$

Can you see that the correct answer must be $x = 0.454$ and not 0.949? The amount of I_2 consumed must be less than 0.500 mol, the initial amount. (There is no way that 0.949 mol I_2 could be consumed.)

The equilibrium amounts are

$$H_2\text{:}\quad 0.800 - x = 0.800 - 0.454 = 0.346 \text{ mol } H_2$$

$$I_2\text{:}\quad 0.500 - x = 0.500 - 0.454 = 0.046 \text{ mol } I_2$$

$$\text{HI:}\quad 2x = 2 \times 0.454 = 0.908 \text{ mol HI}$$

Exercise 16.11

How many moles of NO(g) will form at 2500 K when equilibrium is established in a mixture that initially has 0.78 mol N_2 and 0.21 mol O_2 (their proportions in one mole of air)?

$$N_2(g) + O_2(g) \rightleftharpoons 2\,NO(g) \qquad K_c = 2.1 \times 10^{-3} \text{ at } 2500 \text{ K}$$

(*Hint*: Does the volume of the system matter?)

An equilibrium calculation sometimes presents a special problem. If all reactants and products are present initially, in which direction does a net reaction occur to reach equilibrium? To answer this question, we can formulate a ratio of *initial* concentrations, called the **reaction quotient (Q)** that has the same form as the equilibrium constant expression. Then we can compare its value to K_c. For example, for the synthesis of HI(g) from H_2(g) and I_2(g) at 698 K,

$$H_2(g) + I_2(g) \rightleftharpoons 2\,HI(g) \qquad K_c = 54.3 \text{ at } 698 \text{ K}$$

$$Q = \frac{[HI]^2_{init.}}{[H_2]_{init.} \times [I_2]_{init.}}$$

If the value of Q is *less than* K_c ($Q < K_c$), the denominator is too large relative to the numerator for the mixture to be at equilibrium. A net reaction must occur in the *forward* direction. This increases the numerator and decreases the denominator to the point where the ratio becomes equal to K_c and the reaction reaches equilibrium. If the value of Q is *greater than* K_c ($Q > K_c$), a net reaction must occur in the *reverse direction* to establish equilibrium. Once we know the direction in which a net reaction occurs, we can proceed with an equilibrium calculation, as illustrated in Example 16.12.

Example 16.12

What will be the amounts of reactants and products when equilibrium is established in a gaseous mixture that *initially* has 0.0100 mol H_2, 0.0100 mol I_2, and 0.100 mol HI in a 5.25-L volume at 698 K?

$$H_2(g) + I_2(g) \rightleftharpoons 2\,HI(g) \qquad K_c = 54.3 \text{ at } 698\text{ K}$$

Solution

Since all reactants and products are present initially, we can evaluate Q to decide the direction in which a net reaction must occur.

$$Q = \frac{[HI]^2_{init.}}{[H_2]_{init.} \times [I_2]_{init.}}$$

$$Q = \frac{(0.100/5.25)^2}{(0.0100/5.25)(0.0100/5.25)} = 100$$

Because $Q > K_c$ (that is, 100 > 54.3), a net reaction must occur in the *reverse* direction. At equilibrium, the amounts of H_2 and I_2 will be greater than initially, and the amount of HI will be less. This guides us in labeling the changes in amounts as positive or negative.

The reaction:	$H_2(g)$	+	$I_2(g)$	$\rightleftharpoons$	$2\,HI(g)$
Initial amounts:	0.0100 mol		0.0100 mol		0.100 mol
Changes:	$+x$ mol		$+x$ mol		$-2x$ mol
Equil. amounts:	$(0.0100 + x)$ mol		$(0.0100 + x)$ mol		$(0.100 - 2x)$ mol
Equil. concns, M:	$\frac{(0.0100 + x)}{5.25}$		$\frac{(0.0100 + x)}{5.25}$		$\frac{(0.100 - 2x)}{5.25}$

$$K_c = \frac{[HI]^2}{[H_2][I_2]} = \frac{[(0.100 - 2x)/5.25]^2}{[(0.0100 + x)/5.25][(0.0100 + x)/5.25]} = 54.3$$

$$= \frac{(0.100 - 2x)^2}{(0.0100 + x)^2} = 54.3$$

$$\left(\frac{(0.100 - 2x)^2}{(0.0100 + x)^2}\right)^{1/2} = \frac{0.100 - 2x}{0.0100 + x} = (54.3)^{1/2}$$

$$0.100 - 2x = (54.3)^{1/2} \times (0.0100 + x)$$

$$0.100 - 2x = 0.0737 + 7.37x$$

$$9.37x = 0.0263$$

$$x = 0.00281$$

The equilibrium amounts are

$$H_2:\quad 0.0100 + x = 0.0100 + 0.00281 = 0.0128 \text{ mol } H_2$$

$$I_2:\quad 0.0100 + x = 0.0100 + 0.00281 = 0.0128 \text{ mol } I_2$$

$$HI:\quad 0.100 - 2x = 0.100 - (2 \times 0.00281) = 0.094 \text{ mol HI}$$

Exercise 16.12

What will be the amounts of reactants and products when equilibrium is established in a gaseous mixture that *initially* has 0.0100 mol H_2 and 0.100 mol HI in a 5.25-L volume at 698 K?

$$H_2(g) + I_2(g) \rightleftharpoons 2\,HI(g) \qquad K_c = 54.3 \text{ at } 698 \text{ K}$$

(*Hint*: Do you need to use Q to determine the direction of net change?)

As shown in Figure 16.8 and Example 16.13, our emphasis in a calculation based on K_p is on the partial pressures of the various gases and on the total gas pressure. Because the pressure of a gas in a constant-temperature, constant-volume mixture is proportional to the amount of gas, we can use changes in partial pressures (P in Example 16.13) for the changes in amounts, x, that we've used in previous examples.

In Exercise 16.13, gases are in equilibrium with a solid. The K_p expression, of course, does not include terms for the pure solid.

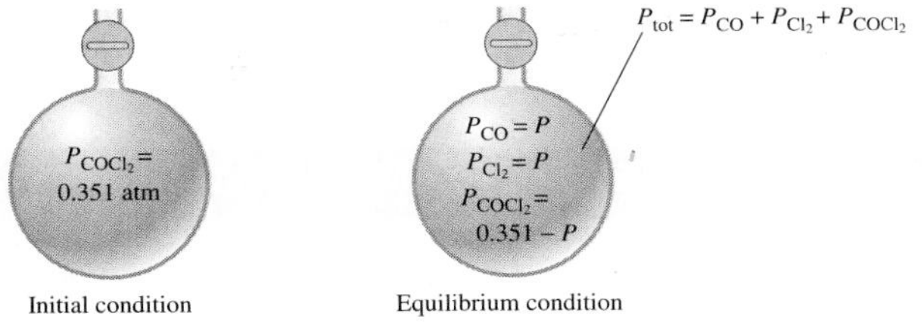

Figure 16.8 Example 16.13 illustrated. Because two moles of gaseous products are produced for every mole of reactant in the reaction $COCl_2(g) \rightleftharpoons CO(g) + Cl_2(g)$, we expect a higher total pressure at equilibrium than initially. Also, because $COCl_2$ must be consumed to produce CO and Cl_2, we expect P_{COCl_2} at equilibrium to be less than the initial P_{COCl_2}.

Example 16.13

A sample of phosgene, $COCl_2(g)$, is introduced into a constant-volume vessel at 395 °C and observed to exert an initial pressure of 0.351 atm. What will be the partial pressures of each gas and the total gas pressure when equilibrium is established?

$$CO(g) + Cl_2(g) \rightleftharpoons COCl_2(g) \qquad K_p = 22.5$$

Solution

Although we can work with the equation as given and describe a net reaction *to the left*, let's think of the partial dissociation of $COCl_2$ in terms of the *reverse* equation—that is, with a net reaction *to the right.* Of course, when we reverse the equation, the value of K_p is the reciprocal of the given value.

The reaction:	$COCl_2(g)$ $\rightleftharpoons$	$CO(g)$ +	$Cl_2(g)$	$K_p = 1/22.5 = 0.0444$
Initial pressures, atm:	0.351	0	0	
Changes, atm:	$-P$	$+P$	$+P$	
Equil. pressures, atm:	$(0.351 - P)$	P	P	

Now, substitute into the K_p expression and solve for P.

$$K_p = \frac{(P_{CO})(P_{Cl_2})}{P_{COCl_2}} = \frac{P \times P}{0.351 - P} = 0.0444$$

$$P^2 = 0.0156 - 0.0444P$$

$$P^2 + 0.0444P - 0.0156 = 0$$

$$P = \frac{-0.0444 \pm \sqrt{(0.0444)^2 + 4(0.0156)}}{2} = \frac{-0.0444 \pm 0.254}{2}$$

$$= 0.105 \text{ or } -0.149$$

The second root of this quadratic equation is –0.149, but we reject it because a partial pressure of –0.149 atm has no physical significance.

The equilibrium partial pressures are

$$P_{CO} = P_{Cl_2} = 0.105 \text{ atm}; \qquad P_{COCl_2} = 0.351 - 0.105 = 0.246 \text{ atm}$$

The total pressure is

$$P_{tot} = P_{CO} + P_{Cl_2} + P_{COCl_2} = 0.105 + 0.105 + 0.246 = 0.456 \text{ atm}$$

Exercise 16.13

Ammonium hydrogen sulfide dissociates readily, even at room temperature. What is the total pressure of the gases in equilibrium with $NH_4HS(s)$ at 25 °C?

$$NH_4HS(s) \rightleftharpoons NH_3(g) + H_2S(g) \qquad K_p = 0.108 \text{ at } 25\ °C$$

(*Hint*: What is the relationship between P_{NH_3} and P_{H_2S}?)

CONCEPTUAL EXAMPLE 16.2

A 0.100-mol sample of $SO_3(g)$ is introduced into a 15.0-L vessel at 900 K.

$$2\ SO_3(g) \rightleftharpoons 2\ SO_2(g) + O_2(g) \qquad K_p = 0.023$$

Which of these is a reasonable estimate of the total gas pressure at equilibrium: (a) 0.77 atm, (b) 0.73 atm, (c) 0.57 atm, (d) 0.80 atm, (e) 0.44 atm?

Solution

Because we'll be dealing with gas pressures, let's begin by calculating the initial pressure of $SO_3(g)$. We'll assume ideal gas behavior.

$$P = \frac{nRT}{V} = \frac{0.100 \text{ mol} \times 0.08206 \text{ L} \cdot \text{atm} \cdot \text{mol}^{-1} \cdot \text{K}^{-1} \times 900 \text{ K}}{15.0 \text{ L}} = 0.492 \text{ atm}$$

The key point in the tabulation below is that if the pressure of $O_2(g)$ becomes P, that of $SO_2(g)$ must be $2P$, because two moles of SO_2 are produced for every mole of O_2. The accompanying decrease in pressure of $SO_3(g)$ must also be $2P$.

The reaction:	$2\ SO_3$	$\rightleftharpoons$	$2\ SO_2(g)$	$+\ O_2(g)$
Initial pressures, atm:	0.492		0	0
Changes, atm:	$-2P$		$+2P$	$+P$
Equil. pressures, atm:	$(0.492 - 2P)$		$2P$	P

From these data, we can write two equations to describe the equilibrium condition:

$$P_{tot} = P_{SO_3} + P_{SO_2} + P_{O_2} = 0.492 - 2P + 2P + P = 0.492 + P$$

$$K_p = \frac{(P_{SO_2})^2(P_{O_2})}{(P_{SO_3})^2} = \frac{(2P)^2(P)}{(0.492 - 2P)^2} = \frac{4\,P^3}{(0.492 - 2P)^2} = 0.023$$

To obtain an *exact* value of P_{tot} we would need to solve the K_p expression for P. Solving a cubic equation is not easy, but to obtain an approximate answer, we don't have to.

Note that the *maximum* possible value of $P = 0.246$. This would occur if the reaction went to completion, for P_{SO_3} would then be zero: $(0.492 - 2P) = 0$. The corresponding $P_{tot} = 0.492 + P = 0.492 + 0.246 = 0.738$ atm. Because P_{tot} can't exceed this value, we can eliminate (a) and (d) as possible answers. Also, because P_{tot} would be 0.492 atm if no reaction occurred at all, we can eliminate any value of P_{tot} that is less than 0.492 atm; that rules out (e).

Of the other answers, (b), with $P_{tot} = 0.73$ atm, corresponds to an equilibrium point far to the right. But for this to be so, we should expect a large value of K_p, which is not the case. We are left with only (c), 0.57 atm, as a plausible answer.

Conceptual Exercise 16.2

Show that the estimated answer (c) above is in fact a reasonable one. (*Hint*: If the total pressure is 0.57 atm, what is P? When this value of P is substituted into the K_p expression, how closely does the result agree with the K_p that is given?)

SUMMARY

In a reversible reaction at equilibrium, the concentrations of all reactants and products remain constant with time. These concentrations of reactants and products must conform to the general expression

$$K_c = \frac{[G]^g[H]^h\cdots}{[A]^a[B]^b\cdots}$$

in which terms in the numerator are the products and those in the denominator are the reactants. The exponents correspond to the coefficients in the balanced equation.

If the balanced equation for a reversible reaction is modified, the equilibrium constant expression, K_c, must also be modified. If the equation is reversed, the K_c expression is inverted. If the coefficients of the equation are multiplied by a common factor, the K_c expression is raised to the corresponding power.

A K_p expression can be written for equilibria involving gases. It has the same form as K_c but uses partial pressures in place of concentrations. The relationship between the two is $K_p = K_c(RT)^{\Delta n_{gas}}$, where Δn_{gas} is the difference between the number of moles of gaseous products and gaseous reactants.

By convention, equilibrium constant expressions do not include terms for substances that exist in either a pure solid phase or a pure liquid phase.

In general, if K_c or K_p for a reaction is very large, the forward reaction goes to completion; if K_c or K_p is very small, the forward reaction occurs to a very limited extent. Calculations based on the equilibrium constant expression are necessary only when K_c or K_p values lie between these extremes.

Qualitative predictions about the effect of changes—changes in amounts of reactants or products, changes in volume or pressure, changes in temperature—can be based on Le Châtelier's principle: When a change is imposed on a system at equilibrium, the system responds by attaining a new equilibrium in which the impact of the change is partially offset. The presence of a catalyst has no effect on equilibrium.

The most common types of equilibrium calculations are (a) determining the value of an equilibrium constant from initial and equilibrium conditions, and (b) determining equilibrium amounts, concentrations, or partial pressures from initial conditions and equilibrium constant values. Some of these calculations require that algebraic equations be solved. Some calculations are facilitated by first evaluating the reaction quotient, Q. This is a ratio of concentrations (or partial pressures) that has the same form as an equilibrium constant expression but which uses *initial* values rather than equilibrium values.

KEY TERMS

equilibrium (16.1)
equilibrium constant expression (16.2)
K_c (16.2)
K_p (16.3)
Le Châtelier's principle (16.4)
reaction quotient (Q) (16.5)

REVIEW QUESTIONS

1. What is a reversible reaction?

2. What is the condition of equilibrium in a reversible reaction?

3. Why is the condition of equilibrium in a reversible reaction a *dynamic* one? Cite some experimental evidence to support this idea.

4. In everyday life we can encounter both *static* and *dynamic* equilibria. What is the difference between the two? Cite an example of each.

5. What is meant by the equilibrium constant expression and the equilibrium constant of a reversible chemical reaction?

6. What is the difference between K_c and K_p for a chemical reaction?

7. What is the difference between a *homogeneous* and a *heterogeneous* reaction?

8. Why can we say that the vapor pressure of a liquid at a given temperature can be expressed as an equilibrium constant?

9. What role does a catalyst play in a reversible chemical reaction?

10. In principle, all chemical reactions are reversible. Why is it that we are able to consider that some of them go to completion?

11. Why does the balanced chemical equation alone supply sufficient information for calculating the yield of products in a reaction that goes to completion, but it does not for a reversible reaction? What additional information is needed to calculate the yield of product(s) in a reversible reaction?

12. Write equilibrium constant expressions, K_c, for the following reactions.
a. $C(graphite) + CO_2(g) \rightleftharpoons 2\ CO(g)$
b. $H_2S(g) + I_2(s) \rightleftharpoons 2\ HI(g) + S(s)$
c. $4\ CuO(s) \rightleftharpoons 2\ Cu_2O(s) + O_2(g)$

13. Balance the following equations and write equilibrium constant expressions, K_c, for the reactions they represent.
a. $H_2S(g) \rightleftharpoons H_2(g) + S_2(g)$
b. $CS_2(g) + H_2(g) \rightleftharpoons CH_4(g) + H_2S(g)$
c. $CO(g) + H_2(g) \rightleftharpoons CH_4(g) + H_2O(g)$

14. Write equilibrium constant expressions, K_p, for the following reactions.
a. $CO(g) + H_2O(g) \rightleftharpoons CO_2(g) + H_2(g)$
b. $N_2(g) + 3\ H_2(g) \rightleftharpoons 2\ NH_3(g)$
c. $NH_4HS(s) \rightleftharpoons NH_3(g) + H_2S(g)$

15. Write an equilibrium constant expression, K_p, based on the formation of one mole of each of the following gaseous compounds from its elements.
a. $NO(g)$ **b.** $NH_3(g)$ **c.** $NOCl(g)$

16. Write an equilibrium constant expression, K_c, for each of the following reversible reactions.
   a. Carbon monoxide gas reduces nitrogen monoxide gas to gaseous nitrogen; carbon dioxide is the other product.
   b. Oxygen gas oxidizes gaseous ammonia to gaseous nitrogen monoxide; water vapor is the other product.
   c. Solid sodium hydrogen carbonate decomposes to form solid sodium carbonate, water vapor, and carbon dioxide gas.

17. For the following two reversible reactions, how are $K_c(a)$ and $K_c(b)$ related?
(a) $PCl_3(g) + Cl_2(g) \rightleftharpoons PCl_5(g)$ $K_c(a)$
(b) $PCl_5(g) \rightleftharpoons PCl_3(g) + Cl_2(g)$ $K_c(b)$

18. The reversible reaction $N_2O_4(g) \rightleftharpoons 2\ NO_2(g)$ has a value of $K_p = 0.113$ at 25 °C. Is the numerical value of K_p at 25 °C for the reaction $\frac{1}{2}\ N_2O_4(g) \rightleftharpoons NO_2(g)$ greater than, equal to, or less than 0.113? Explain.

19. Describe how you might be able to drive a reaction having a small value of K_c to completion.

20. Equilibrium is established in the reversible reaction

$$4\ HCl(g) + O_2(g) \rightleftharpoons 2\ H_2O(g) + 2\ Cl_2(g) \qquad \Delta H° = -114.4\ kJ$$

Describe four changes that can be made to this mixture to increase the amount of $Cl_2(g)$ at equilibrium.

21. In each of the following reactions, will the amount of product at equilibrium increase if the total gas pressure is raised from 1 atm to 10 atm? Explain.
   a. $SO_2(g) + Cl_2(g) \rightleftharpoons SO_2Cl_2(g)$
   b. $N_2(g) + O_2(g) \rightleftharpoons 2\ NO(g)$
   c. $SO_2(g) + \frac{1}{2}\ O_2(g) \rightleftharpoons SO_3(g)$

22. The extent of dissociation that occurs in one of the following reactions depends on the volume of the reaction vessel, and in the other it does not. Identify the situation for each reaction and explain why they are not the same.
   a. $2\ NO(g) \rightleftharpoons N_2 + O_2(g)$
   b. $2\ NOCl(g) \rightleftharpoons 2\ NO(g) + Cl_2(g)$

23. The extent of dissociation that occurs in one of the following reactions increases with an increase in temperature; with the other, it decreases with an increase in temperature. Identify the situation for each reaction and explain why they are not the same.
   a. $2\ NO(g) \rightleftharpoons N_2(g) + O_2(g)$ $\Delta H° = -180.5\ kJ$
   b. $2\ SO_3(g) \rightleftharpoons 2\ SO_2(g) + O_2(g)$ $\Delta H° = 197.8\ kJ$

PROBLEMS

Equilibrium Constant Relationships

24. Determine the values of K_p that correspond to the following values of K_c.
   a. $CO(g) + Cl_2(g) \rightleftharpoons COCl_2(g)$ $K_c = 1.2 \times 10^3$ at 668 K
   b. $2\ NO(g) + Br_2(g) \rightleftharpoons 2\ NOBr(g)$ $K_c = 1.32 \times 10^{-2}$ at 1000 K
   c. $2\ COF_2(g) \rightleftharpoons CO_2(g) + CF_4(g)$ $K_c = 2.00$ at 1000 °C

25. Determine the values of K_c that correspond to the following values of K_p.
   a. $SO_2Cl_2(g) \rightleftharpoons SO_2(g) + Cl_2(g)$ $K_p = 2.9 \times 10^{-2}$ at 303 K
   b. $2\ NO_2(g) \rightleftharpoons 2\ NO(g) + O_2(g)$ $K_p = 0.275$ at 700 K
   c. $CO(g) + Cl_2(g) \rightleftharpoons COCl_2(g)$ $K_p = 22.5$ at 395 °C

26. For the reaction $N_2(g) + O_2(g) \rightleftharpoons 2\ NO(g)$, $K_c = 4.08 \times 10^{-4}$ at 2000 K. What is the value of K_c at 2000 K for the reaction $NO(g) \rightleftharpoons \frac{1}{2}\ N_2(g) + \frac{1}{2}\ O_2(g)$?

27. For the reaction $SO_3(g) \rightleftharpoons SO_2(g) + \frac{1}{2}\ O_2(g)$, $K_c = 16.7$ at 1000 K. What is the value of K_c at 1000 K for the reaction $2\ SO_2(g) + O_2(g) \rightleftharpoons 2\ SO_3(g)$?

28. Is the equilibrium constant K_c for the reaction $A + B \rightleftharpoons \frac{1}{2}\ C + \frac{1}{2}\ D$ always smaller than the equilibrium constant K_c' for the reaction $2\ A + 2\ B \rightleftharpoons C + D$? That is, is $K_c < K_c'$ in all cases? Explain.

29. If the equilibrium concentrations found in the reaction $A(g) + 2\ B(g) \rightleftharpoons 2\ C(g)$ are [A] = 0.025 M, [B] = 0.15 M, and [C] = 0.55 M, calculate the value of K_c.

30. If the equilibrium concentrations found in the reaction $2\ A(g) + B(g) \rightleftharpoons C(g)$ are $[A] = 2.4 \times 10^{-2}$ M, $[B] = 4.6 \times 10^{-3}$ M, and $[C] = 6.2 \times 10^{-3}$ M, calculate the value of K_c.

31. In the reaction $CO(g) + Cl_2(g) \rightleftharpoons COCl_2(g)$, $K_c = 1.2 \times 10^3$ at 395 K. At equilibrium, these relationships among concentrations are found: $[CO] = 2[Cl_2] = \frac{1}{2}[COCl_2]$. What is the equilibrium value of $[COCl_2]$?

32. In the reaction $2\ H_2(g) + S_2(g) \rightleftharpoons 2\ H_2S(g)$, $K_c = 6.28 \times 10^3$ at 900 K. At equilibrium, this relationship among the concentrations is found: $[H_2S] = [S_2]^{1/2}$. What is the equilibrium value of $[H_2]$?

33. For the reaction $C(s) + CO_2(g) \rightleftharpoons 2\ CO(g)$, $K_p = 63$. What will be the *total* pressure above an equilibrium mixture if $P_{CO} = 10\ P_{CO_2}$?

34. For the reaction $H_2S(g) + I_2(s) \rightleftharpoons S(s) + 2\ HI(g)$, $K_p = 1.33 \times 10^{-5}$ at 333 K. What will be the *total* pressure above an equilibrium mixture if $P_{HI} = 0.010 \times P_{H_2S}$?

35. In the reaction $Sb_2S_3(s) + 3\ H_2(g) \rightleftharpoons 2\ Sb(s) + 3\ H_2S(g)$, equilibrium is established at a temperature at which $P_{H_2S} = P_{H_2}$. What must be the value of K_p at this temperature?

36. In the reaction $N_2O_4(g) \rightleftharpoons 2\ NO_2(g)$ equilibrium is established at a temperature at which $P_{NO_2} = 3(P_{N_2O_4})^{1/2}$. What must be the value of K_p at this temperature?

37. If a mixture is prepared at 395 °C in which $[CO] = 1.5$ M, $[Cl_2] = 2.0$ M, and $[COCl_2] = 0.10$ M, will the mixture be at equilibrium? If not, in which direction, left or right, will a net reaction occur to establish equilibrium? Explain.

$$CO(g) + Cl_2(g) \rightleftharpoons COCl_2(g) \qquad K_c = 1.2 \times 10^3 \text{ at } 395\ °C$$

38. If a mixture is prepared at 700 °C in which $P_{H_2S} = 0.10$ atm, $P_{CH_4} = 0.25$ atm, $P_{CS_2} = 0.65$ atm, and $P_{H_2} = 0.10$ atm, will the mixture be at equilibrium? If not, in which direction will the equilibrium be shifted, that is, to the left or to the right? Explain.

$$2\ H_2S(s) + CH_4(g) \rightleftharpoons CS_2(g) + 4\ H_2(g) \qquad K_p = 3.4 \times 10^{-4} \text{ at } 700\ °C$$

39. For the reaction $CO(g) + H_2O(g) \rightleftharpoons CO_2(g) + H_2(g)$, $K_p = 23.2$ at 600 K. Explain if it is possible to have an equilibrium mixture at 600 K in which

a. $P_{CO} = P_{H_2O} = P_{CO_2} = P_{H_2}$
b. $P_{H_2}/P_{H_2O} = P_{CO_2}/P_{CO}$
c. $(P_{CO_2})(P_{H_2}) = (P_{CO})(P_{H_2O})$
d. $P_{CO_2}/P_{H_2O} = P_{H_2}/P_{CO}$

Experimental Determination of Equilibrium Constants

40. Equilibrium is established in a sealed 1.75-L vessel at 250 °C in the reaction $PCl_5(g) \rightleftharpoons PCl_3(g) + Cl_2(g)$. The quantities found at equilibrium are 0.562 g PCl_5, 1.950 g PCl_3, and 1.007 g Cl_2. What is the value of K_c for this reaction? What is the value of K_p?

41. Equilibrium is established in a sealed 10.5-L vessel at 184 °C in the reaction $2\ NO_2(g) \rightleftharpoons 2\ NO(g) + O_2(g)$. The quantities found at equilibrium are 1.353 g NO_2, 0.0960 g NO, and 0.0512 g O_2. What is the value of K_c for this reaction? What is the value of K_p?

42. In the reaction $2\ HI(g) \rightleftharpoons H_2(g) + I_2(g)$ at 623 K, starting with 0.315 g HI in a 0.400-L bulb, the mass of $I_2(g)$ found at equilibrium is 0.0615 g. What is the value of K_c for this reaction? (*Hint*: How are $[H_2]$ and $[I_2]$ related at equilibrium?)

43. For the reaction $H_2(g) + I_2(g) \rightleftharpoons 2\ HI(g)$ at 623 K, starting with 25.4 g I_2 and 0.403 g H_2 in a 0.750-L bulb, the mass of I_2 found at equilibrium is 1.40 g. What is the value of K_c for this reaction?

44. In the dissociation of ammonium hydrogen sulfide at 25 °C, if we start with a sample of pure $NH_4HS(s)$, when equilibrium is established the *total* pressure of the gases is 0.658 atm. What is the value of K_p for the reaction $NH_4HS(s) \rightleftharpoons NH_3(g) + H_2S(g)$? (*Hint*: How are the partial pressures of the gases related to each other?)

45. In the dissociation of ammonium carbamate at 30 °C, if we start with a sample of pure $NH_2COONH_4(s)$, when equilibrium is established, the *total* pressure of the gases is 0.164 atm. What is the value of K_p for the reaction $NH_2COONH_4(s) \rightleftharpoons 2\ NH_3(g) + CO_2(g)$? (*Hint*: How are the partial pressures of the gases related to each other?)

Le Châtelier's Principle

46. $N_2O_4(g)$ is 12.5% dissociated into $NO_2(g)$ at 25 °C.

$$N_2O_4(g) \rightleftharpoons 2\ NO_2(g) \qquad \Delta H = 57.2\ \text{kJ}$$

Will the percent dissociation be greater or less than 12.5%, if **(a)** the reaction mixture is transferred to a vessel of twice the volume? **(b)** the temperature is raised to 50 °C? **(c)** a catalyst is added to the reaction vessel?

47. Explain why the extent of dissociation of diatomic molecules of an element into atoms of the element must increase with temperature, for example, $Cl_2(g) \rightleftharpoons 2\ Cl(g)$, $S_2(g) \rightleftharpoons 2\ S(g)$, $H_2(g) \rightleftharpoons 2\ H(g)$, and so on.

48. Is the statement made in Problem 47 equally valid for the dissociation of molecules of *compounds* into the constituent elements, for example, $2\ H_2O(g) \rightleftharpoons 2\ H_2(g) + O_2(g)$, $2\ NO(g) \rightleftharpoons N_2(g) + O_2(g)$, and so on? Explain. (*Hint*: Use data from Appendix D, if necessary.)

49. In the formation of the following gaseous compounds from their gaseous elements, which reactions occur to a greater extent at high pressures, and which reactions are unaffected by the total pressure: **(a)** $NO(g)$ from $N_2(g)$ and $O_2(g)$, **(b)** $NH_3(g)$ from $N_2(g)$ and $H_2(g)$, **(c)** $HI(g)$ from $H_2(g)$ and $I_2(g)$, **(d)** $H_2S(g)$ from $H_2(g)$ and $S_2(g)$?

50. Are there likely to be cases where the formation of a gaseous compound from its gaseous elements is favored at low pressures? Explain.

51. Use Le Châtelier's principle to develop an explanation of why the application of a high pressure causes ice to melt.

Calculations Based on K_c

52. The equilibrium constant for the isomerization of butane at 25 °C is $K_c = 7.94$.

$$CH_3CH_2CH_2CH_3 \rightleftharpoons CH_3CH(CH_3)CH_3$$

Butane — Isobutane

If 5.00 g butane is introduced into a 12.5-L flask at 25 °C, what mass of isobutane will be present when equilibrium is reached?

53. At 25 °C, the following equilibrium can be established in a homogeneous liquid solution.

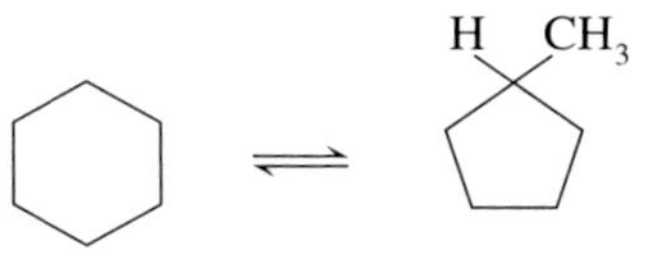

$K_c = 0.143$

C_6H_{12} Cyclohexane — $C_5H_9CH_3$ Methylcyclopentane

If equilibrium is established in a mixture initially consisting of 1.00×10^2 g cyclohexane, what mass of methylcyclopentane will be present? (*Hint*: Does the volume of solution matter?)

54. For the reaction

$$CO(g) + H_2O(g) \rightleftharpoons CO_2(g) + H_2(g) \qquad K_c = 23.2 \text{ at } 600\ \text{K}$$

if 0.250 mol each of CO and H_2O are introduced into a reaction vessel and equilibrium is established, how many moles each of CO_2 and H_2 will be present? (*Hint*: Does the volume of the reaction mixture matter?)

55. For the water-gas reaction

$$C(s) + H_2O(g) \rightleftharpoons CO(g) + H_2(g) \qquad K_c = 0.111 \text{ at about } 1100\ \text{K}$$

if 0.100 mol $H_2O(g)$ and 0.100 mol $H_2(g)$ are mixed with excess C(s) at this temperature and equilibrium is established in a 1.00-L container, how many moles of CO(g) will be present? No CO(g) is present initially.

56. For the synthesis of phosgene at 395 °C

$$CO(g) + Cl_2(g) \rightleftharpoons COCl_2(g) \qquad K_c = 1.2 \times 10^3$$

if 0.700 mol CO and 0.500 mol Cl_2 are placed in a 8.05-L reaction vessel at 395 °C and equilibrium is established, how many moles of $COCl_2$ will be present?

57. For the decomposition of carbonyl fluoride, COF_2,

$$2\ COF_2(g) \rightleftharpoons CO_2(g) + CF_4(g) \qquad K_c = 2.00 \text{ at } 1000\ °C$$

if 0.500 mol $COF_2(g)$ is placed in a 3.23-L reaction vessel at 1000 °C, how many moles of $COF_2(g)$ will remain undissociated when equilibrium is reached?

58. To establish equilibrium in the following reaction at 250 °C,

$PCl_3(g) + Cl_2(g) \rightleftharpoons PCl_5(g) \quad K_c = 26$ at 250 °C

0.100 mol each of PCl_3 and Cl_2 and 0.0100 mol PCl_5 are introduced into a 6.40-L reaction flask. How many moles of each of the gases should be present when equilibrium is established?

59. Refer to Problem 58. In a second experiment, the initial amounts are 0.100 mol of each gas. How many moles of each of the gases should be present in the reaction flask when equilibrium is established?

Calculations Based on K_p

60. In the reaction

$C(s) + S_2(g) \rightleftharpoons CS_2(g) \quad K_p = 5.60$ at 1009 °C

If, at equilibrium, $P_{CS_2} = 0.152$ atm, what must be **(a)** P_{S_2} and **(b)** the total gas pressure, P_{tot}?

61. In the reaction

$Sb_2S_3(s) + 3\,H_2(g) \rightleftharpoons 2\,Sb(s) + 3\,H_2S(g)$
$K_p = 0.429$ at 713 K

If, at equilibrium, $P_{H_2S} = 0.200$ atm, what must be **(a)** P_{H_2} and **(b)** the total gas pressure, P_{tot}?

62. For the reaction $C(s) + 2\,H_2(g) \rightleftharpoons CH_4(g)$, $K_p = 0.263$ at 1000 °C. Calculate the total pressure when 0.100 mol CH_4 and an excess of C(s) are brought to equilibrium at 1000 °C in a 4.16-L reaction vessel.

63. Solid molybdenum is kept in contact with $CH_4(g)$ and $H_2(g)$, each at a pressure of 1.00 atm, in a reaction vessel at 973 K. Calculate the total pressure when equilibrium is established.

$2\,Mo(s) + CH_4(g) \rightleftharpoons Mo_2C(s) + 2\,H_2(g)$
$K_p = 3.55$ at 973 K

ADDITIONAL PROBLEMS

64. For the reaction

$$CO(g) + Cl_2(g) \rightleftharpoons COCl_2(g)$$

which of the following ratios of concentrations would you expect to have a constant value and which would not (n represents an amount in moles)? Explain.

a. $\dfrac{[CO][Cl_2]}{[COCl_2]}$ **b.** $\dfrac{[COCl_2]}{[CO] + [Cl_2]}$ **c.** $\dfrac{n_{COCl_2}}{(n_{CO})(n_{Cl_2})}$

d. $\dfrac{[COCl_2]}{[CO][Cl_2]}$ **e.** $\dfrac{[COCl_2]^{1/2}}{[CO]^{1/2}[Cl_2]^{1/2}}$

65. Equilibrium is established in the homogeneous reaction of acetic acid and ethanol by starting with 1.51 mol CH_3COOH and 1.66 mol CH_3CH_2OH.

CH_3COOH (Acetic Acid) + CH_3CH_2OH (Ethanol) $\rightleftharpoons$ CH_3COOCH_2OH (Ethyl acetate) + H_2O

After equilibrium is reached, exactly one-hundredth of the equilibrium mixture is titrated with $Ba(OH)_2(aq)$ to determine the amount of acetic acid present.

$2\,CH_3COOH + Ba(OH)_2(aq) \longrightarrow (CH_3COO)_2Ba(aq) + 2\,H_2O$

The volume of 0.1025 M $Ba(OH)_2(aq)$ required for the titration is 22.44 mL. Use these data to calculate K_c for the formation of ethyl acetate.

66. For the isomerization of butane to isobutane, $K_c = 7.94$ (see Problem 52). Sketch a graph of concentration of butane and of isobutane as a function of time, in the manner of Figure 16.2. Start with any initial concentration of butane (and no isobutane) and show the equilibrium concentrations in their expected relative proportions.

67. A 0.0508-mol sample of N_2O_4 is added to an evacuated 2.85-L flask and equilibrated at 25 °C. What will be the total pressure in the flask when equilibrium is established?

$N_2O_4(g) \rightleftharpoons 2\,NO_2(g) \quad K_c = 4.61 \times 10^{-3}$ at 25 °C

68. At 500 °C, an equilibrium mixture in the reaction

$CO(g) + H_2O(g) \rightleftharpoons CO_2(g) + H_2(g)$

is found to contain 0.021 mol CO, 0.121 mol H_2O, 0.179 mol CO_2, and 0.079 mol H_2. If an additional 0.100 mol H_2 is added to the mixture, (**a**) in what direction must a net reaction occur to restore equilibrium, and (**b**) what will be the amounts of the four substances when equilibrium is reestablished?

69. An analysis of the gaseous phase [S_2(g) and CS_2(g)] present at equilibrium at 1009 °C in the reaction

$$C(s) + S_2(g) \rightleftharpoons CS_2(g)$$

shows it to be 13.71% C and 86.29% S, by mass. What is K_c for this reaction? [*Hint*: You can start with reactants on the left, and you can choose any initial amount of S_2(g) you wish. The volume of the reaction vessel is immaterial.]

70. For the dissociation of H_2S(g) at 750 °C,

$$2\ H_2S(g) \rightleftharpoons 2\ H_2(g) + S_2(g) \qquad K_c = 1.1 \times 10^{-6}$$

if 1.00 mol H_2S is placed in a 7.50-L reaction vessel and heated to 750 °C, how many moles of H_2 and S_2 will be present when equilibrium is reached? (*Hint*: If you assume that the amounts in question are small compared to 1.00, you can greatly simplify the algebraic solution.)

71. Estimate the total pressure at equilibrium in the reaction $2\ SO_3(g) \rightleftharpoons 2\ SO_2(g) + O_2(g)$, $K_p = 0.023$ at 900 K, starting initially with 0.255 mol SO_3(g) in a 18.5-L vessel at 900 K. (*Hint*: Use Conceptual Example 6.2 as a guide. Specifically, obtain an approximate solution of a cubic equation by "trial and error.")

72. For the synthesis of methanol at 500 K,

$$CO(g) + 2\ H_2(g) \rightleftharpoons CH_3OH(g) \qquad K_c = 14.5$$

starting with 1.00 mol each of CO and H_2 in a 10.0-L reaction vessel, estimate how many moles of CH_3OH will be present at equilibrium? (*Hint*: Obtain an approximate solution of a cubic equation by "trial and error.")

73. The decomposition of $CaSO_4$(s)

$$2\ CaSO_4(s) \rightleftharpoons 2\ CaO(s) + 2\ SO_2(g) + O_2(g)$$

has $K_p = 1.45 \times 10^{-5}$ at 1625 K. A sample of $CaSO_4$(s) is introduced into a reaction vessel filled with air at 1.0000 atm pressure at 1625 K. What will be the partial pressure of SO_2(g) when equilibrium is established? The mole percent O_2 in air is 20.95%.

74. If, at 25 °C, a large quantity of anhydrous (dry) Na_2HPO_4(s) is added to a 3.45-L flask containing 42 mg H_2O(g), will any of the hydrate $Na_2HPO_4 \cdot 2H_2O$(s) form?

$$Na_2HPO_4(s) + 2\ H_2O(g) \rightleftharpoons Na_2HPO_4 \cdot 2H_2O(s) \qquad K_p = 6.01 \times 10^3$$

If so, how many moles will form? (*Hint*: What is the equilibrium pressure of water vapor above a mixture of the hydrate and the anhydrous compound?)

We use acid–base chemistry in all aspects of our lives, including in the preparation and eating of foods. The acid–base reaction between lemon juice and fish is described on page 654.

17

Acids, Bases, and Acid–Base Equilibria

CONTENTS

The substances we call acids and bases are so important that we have referred to them early and often in this text. We introduced them and some of their common properties in Section 2.9, primarily so that we could thereafter refer to them by name and formula. In Section 3.11 we applied stoichiometric principles to acid–base titrations. In Section 12.2, we described the electrolytic properties of acids and bases (strong and weak) and discussed acid–base reactions as one of the three principal types that occur in aqueous solution.

In this chapter we will explore acid– base theory a little more deeply and examine factors that affect acid and base strength. We will describe what we mean by the pH of a solution—such as an acid rain having a pH of 4.2. We will learn how to calculate the pH of a solution, and how a constant pH can be maintained—such as a pH of 7.4 in blood. Also, we will discuss acid–base indicators and take another look at acid–base neutralization reactions carried out by titration. In particular, we will learn how to choose an appropriate indicator for a titration.

17.1 The Brønsted–Lowry Theory of Acids and Bases

The first successful theory of acids and bases was that of Arrhenius, which we introduced in Section 2.9 and have used from time to time since. In brief, Arrhenius proposed that an acid produces hydrogen ions, $H^+(aq)$, in aqueous solution, and a base produces hydroxide ion, $OH^-(aq)$. Also, he distinguished between *strong* acids and bases and *weak* acids and bases. A strong acid ionizes completely into $H^+(aq)$ and accompanying anions, and a strong base dissociates completely into $OH^-(aq)$ and accompanying cations. The ionizations of weak acids and bases are reversible and reach a state of equilibrium, in which only a small percent of the acid or base exists as ions. However, the Arrhenius theory has its limitations: it applies only to reactions in aqueous solution, and it does an inadequate job of explaining where the OH^- comes from in the ionization of weak bases such as ammonia, NH_3.

The Arrhenius theory seems to require that a base *contain* OH^- or the —OH group. As a consequence, aqueous ammonia came to be known as ammonium hydroxide, NH_4OH; it is still sometimes referred to by this name. However, it is doubtful that discrete molecules of NH_4OH exist in aqueous solution. We cannot, for example, write a satisfactory Lewis structure for NH_4OH.

The Brønsted–Lowry Theory

The shortcomings of the Arrhenius theory are largely overcome by a newer theory proposed independently, in 1923, by J. N. Brønsted in Denmark and T. M. Lowry in Great Britain. In their theory an acid is a **proton donor** and a base is a **proton acceptor**.

By a "proton" we mean an ionized hydrogen atom—that is, H^+.

The theory describes the behavior of ammonia as a base in this way:

$$\underset{\text{base(1)}}{NH_3} + \underset{\text{acid(2)}}{H_2O} \rightleftharpoons \underset{\text{acid(1)}}{NH_4^+} + \underset{\text{base(2)}}{OH^-}$$

Here are the main features of the Brønsted–Lowry notation, suggested also by Figure 17.1.

- The ionization reaction is reversible, denoted by the double arrow ($\rightleftharpoons$).
- The overall reaction consists of *two* combinations of acids and bases called *conjugate pairs*. NH_3 acts as a base, noted as base(1) above, by accepting a proton from H_2O. In the reverse reaction, NH_4^+, noted as acid(1) below, loses a proton to OH^-; NH_4^+ is the **conjugate acid** of NH_3. Similarly, OH^- is the **conjugate base** [base(2)] of the acid H_2O [acid(2)]; OH^- accepts a proton from NH_4^+ and H_2O donates a proton to NH_3.
- An acid–base conjugate pair differs in structure only by a proton (H^+): The conjugate acid of a species is that species *plus* a proton; the conjugate base of a species is that species *minus* a proton.

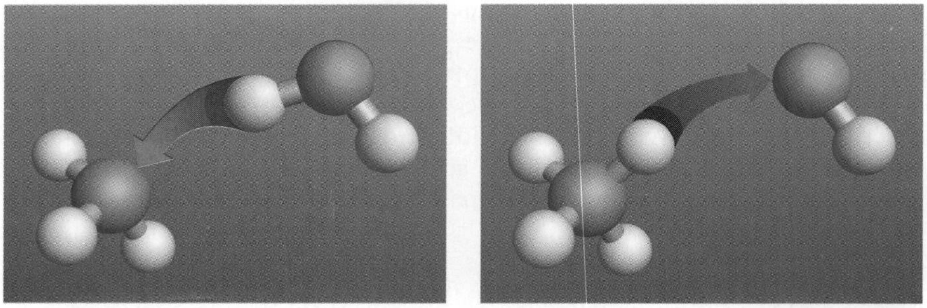

$$\underset{\text{base(1)}}{NH_3} + \underset{\text{acid(2)}}{H_2O} \rightleftharpoons \underset{\text{acid(1)}}{NH_4^+} + \underset{\text{base(2)}}{OH^-}$$

Figure 17.1 Ionization of NH_3 as a Brønsted–Lowry base. The red arrow represents proton transfer in the forward reaction and the blue arrow, in the reverse reaction. Because ammonium ion is a stronger acid than water, and because hydroxide ion is a stronger base than ammonia, the equilibrium condition lies quite far to the left. The relative strengths of some acids and bases are presented in Table 17.1 on page 645.

As with other reversible reactions, we can write an equilibrium constant expression.

$$NH_3 + H_2O \rightleftharpoons NH_4^+ + OH^-$$

$$K_c = \frac{[NH_4^+][OH^-]}{[NH_3][H_2O]}$$

In Chapter 16 we learned that by convention the concentrations of pure solids and liquids do not appear in equilibrium constant expressions. In the present case, water is the solvent in which the ionization occurs, but it is not a pure liquid phase; NH_3, NH_4^+, and OH^- are also present. In most dilute aqueous solutions, however, the quantity of solute is small compared to the quantity of water present. The solvent water is treated like a pure liquid,* and its concentration does not appear in the equilibrium constant expression. Therefore, for the ionization of the base ammonia, we use the expression

$$K_b = \frac{[NH_4^+][OH^-]}{[NH_3]} = 1.8 \times 10^{-5}$$

where the equilibrium constant is called the **base ionization constant** and is denoted by the symbol $\boldsymbol{K_b}$.

By the Brønsted–Lowry theory, the ionization of acetic acid is represented as

$$\underset{\text{acid(1)}}{CH_3COOH} + \underset{\text{base(2)}}{H_2O} \rightleftharpoons \underset{\text{acid(2)}}{H_3O^+} + \underset{\text{base(1)}}{CH_3COO^-}$$

In this reaction H_2O is a base, whereas it was an acid in the ionization of NH_3. A substance that can act either as an acid or a base is said to be **amphiprotic**. The conjugate acid of water is the **hydronium ion, $\boldsymbol{H_3O^+}$**. A free proton does not exist in water solution because H^+ has such a high positive charge density that it immediately seeks out centers of negative charge, such as the lone pairs of electrons on the O atoms of other H_2O molecules. The H^+ ion is probably associated with several H_2O molecules, for example, four H_2O molecules in the ion $H(H_2O)_4^+$ or $H_9O_4^+$. However, in applying the Brønsted–Lowry theory, we will assume the simplest hydrated hydrogen ion, H_3O^+. We can think of the bonding in this ion as involving a coordinate covalent bond.

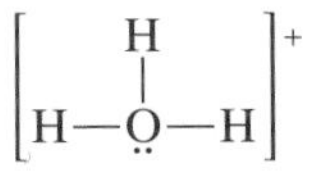

As in the case of K_b, we treat the water as a pure liquid, and its concentration does not appear in the equilibrium constant expression. Therefore, for the ionization of acetic acid, we use the expression

$$K_a = \frac{[H_3O^+][CH_3COO^-]}{[CH_3COOH]} = 1.8 \times 10^{-5}$$

It's just a coincidence that K_a of acetic acid has the same value as K_b of ammonia.

where the equilibrium constant is called the **acid ionization constant** and is denoted by the symbol $\boldsymbol{K_a}$.

*One liter of pure water (d = 1.00 g/mL) has a mass of 1000 g. The molar concentration of H_2O in water is

$$[H_2O] = \frac{1000 \text{ g } H_2O \times \dfrac{1 \text{ mol } H_2O}{18.02 \text{ g } H_2O}}{1.00 \text{ L}} = 55.5 \text{ M}$$

In a 1 M NH_3(aq) solution, there is less than 1 mol NH_3 (some has ionized) and only about 0.004 mol each of NH_4^+ and OH^-. These are indeed small concentrations compared to the $[H_2O]$, which remains at 55.5 M.

Example 17.1

Identify the Brønsted–Lowry acids and bases and their conjugates in each of the following ionizations.

a. $H_2S + NH_3 \rightleftharpoons NH_4^+ + HS^-$

b. $OH^- + H_2PO_4^- \rightleftharpoons H_2O + HPO_4^{2-}$

Solution

a. H_2S is converted to HS^- by losing a proton. H_2S is an acid; HS^- is its conjugate base. NH_3 accepts the proton lost by the H_2S. NH_3 is a base; NH_4^+ is its conjugate acid.

$$\underset{\text{acid(1)}}{H_2S} + \underset{\text{base(2)}}{NH_3} \rightleftharpoons \underset{\text{acid(2)}}{NH_4^+} + \underset{\text{base(1)}}{HS^-}$$

b. OH^- gains a proton from $H_2PO_4^-$. OH^- is a base; H_2O is its conjugate acid. $H_2PO_4^-$ donates a proton to OH^-; $H_2PO_4^-$ is an acid. HPO_4^{2-} is its conjugate base.

$$\underset{\text{base(1)}}{OH^-} + \underset{\text{acid(2)}}{H_2PO_4^-} \rightleftharpoons \underset{\text{acid(1)}}{H_2O} + \underset{\text{base(2)}}{HPO_4^{2-}}$$

Exercise 17.1

Identify the Brønsted–Lowry acids and bases and their conjugates in each of the following ionizations.

a. $HS^- + H_2O \rightleftharpoons H_2S + OH^-$

b. $HNO_3 + H_2PO_4^- \rightleftharpoons H_3PO_4 + NO_3^-$

To represent the ionization of hydrogen chloride, a *strong* acid, we note that the reaction goes essentially to completion by using a single arrow.

$$\underset{\text{acid(1)}}{HCl} + \underset{\text{base(2)}}{H_2O} \longrightarrow \underset{\text{acid(2)}}{H_3O^+} + \underset{\text{base(1)}}{Cl^-}$$

Because the ionization proceeds almost exclusively in the forward direction (left to right), we can conclude that Cl^- is a *weak* base. That is, the tendency for Cl^- ion to accept a proton from H_3O^+ is quite limited compared to the tendency for HCl to donate a proton to H_2O. As a generalization, we might say that

In a conjugate acid–base pair, reaction is favored in the direction from the stronger to the weaker member of the pair.

From this standpoint, we should expect the following reaction to occur almost exclusively in the *reverse* direction.

$$\underset{\text{acid(1)}}{CH_3COOH} + \underset{\text{base(2)}}{Br^-} \longleftarrow \underset{\text{acid(2)}}{HBr} + \underset{\text{base(1)}}{CH_3COO^-}$$

HBr, like HCl, is a strong acid, and Br^-, like Cl^-, is a weak base. Or, to look at the matter in another way, HBr, a strong acid, is more able to force protons onto CH_3COO^- ions than is CH_3COOH, a weak acid, able to force protons onto Br^- ions.

To apply these ideas more widely, we need a relative ranking of the strengths of acids and their conjugate bases, as is found in Table 17.1. As indi-

TABLE 17.1 **Relative Strengths of Some Brønsted–Lowry Acids and Their Conjugate Bases**

Increasing acid strength ↑

Increasing base strength ↓

K_a	Acid	Conjugate Base	K_b
10^9	HI (hydroiodic acid)	I^- (iodide ion)	10^{-23}
10^8	HBr (hydrobromic acid)	Br^- (bromide ion)	10^{-22}
1.3×10^6	HCl (hydrochloric acid)	Cl^- (chloride ion)	7.7×10^{-21}
1×10^3	H_2SO_4 (sulfuric acid)	HSO_4^- (hydrogen sulfate ion)	1×10^{-17}
3×10^1	HNO_3 (nitric acid)	NO_3^- (nitrate ion)	3×10^{-16}
1.1×10^{-2}	HSO_4^- (hydrogen sulfate ion)	SO_4^{2-} (sulfate ion)	9.1×10^{-13}
7.2×10^{-4}	HNO_2 (nitrous acid)	NO_2^- (nitrite ion)	1.4×10^{-11}
6.6×10^{-4}	HF (hydrofluoric acid)	F^- (fluoride ion)	1.5×10^{-11}
1.8×10^{-5}	CH_3COOH (acetic acid)	CH_3COO^- (acetate ion)	5.6×10^{-10}
4.4×10^{-7}	H_2CO_3 (carbonic acid)	HCO_3^- (hydrogen carbonate ion)	2.3×10^{-8}
5.6×10^{-10}	NH_4^+ (ammonium ion)	NH_3 (ammonia)	1.8×10^{-5}
4.7×10^{-11}	HCO_3^- (hydrogen carbonate ion)	CO_3^{2-} (carbonate ion)	2.1×10^{-4}
3.2×10^{-16}	CH_3OH (methanol)	CH_3O^- (methoxide ion)	3.1×10^1

cated by the K_a values, the strongest acids are at the top of the column on the left. The strongest bases are at the bottom of the column on the right, as shown by their K_b values. If you look closely at the K_a and K_b values for each conjugate pair, you will see that for an acid and its conjugate base, $K_a \times K_b = 1.0 \times 10^{14}$ (at 25° C). We will deal with this idea more explicitly later in the chapter.

17.2 Molecular Structure and Acid Strength

A central challenge in chemistry is to relate observable properties of a substance as a whole to its molecular structure and to atomic and molecular properties. Referring to Table 17.1, we would like to be able to explain, for example, why HF is a weak acid whereas HCl is strong, why HNO_3 is a stronger acid than HNO_2, and so on. In this section we'll consider several factors that affect acid strength, and this will help us answer such questions.

Strengths of Binary Acids

There are different ways to think about the strengths of binary acids. One way is to relate acid strength to the strength of the bond H—X. A strong H—X bond is difficult to break and is indicative of a *weak* acid. Conversely, a weaker bond signifies a stronger acid. Even though bond energies are for *gaseous* species and the acid ionizations occur in *aqueous solution*, this generalization still works rather well. Consider, for example, the order of decreasing acid strength of the binary acids of Group 7A, the hydrogen halides:

Bond energy (kJ/mol) increases:	297 < 368 < 431 < 569
	HI HBr HCl HF
Acid strength decreases:	$K_a = \sim 10^9 > \sim 10^8 > \sim 10^6 > 6.6 \times 10^{-4}$

We can also consider the ability of a halide ion to accept a proton. The halide ion most able to attract and hold a proton is the strongest base, and its conjugate acid is the weakest acid of the group. Conversely, the halide ion with the least attraction for a proton is the weakest base, and its conjugate acid is the strongest acid. Halide ions attract protons (H^+) because of their negative charge (X^-). As this negative charge becomes more concentrated (that is, as the negative ion becomes smaller), the attraction for a proton will be stronger; the base, X^-, will be stronger; and the acid, HX, will be weaker. The order of decreasing acid strength of the hydrogen halides is

Anion radius (pm) decreases:	216 > 195 > 181 > 136
	HI HBr HCl HF
Acid strength decreases:	$K_a = \sim 10^9 > \sim 10^8 > \sim 10^6 > 6.6 \times 10^{-4}$

The most important factor in establishing the relative acid strengths of binary halides in the same period of the periodic table appears to be the electronegativity difference (ΔEN) between the atoms H and X. A small difference denotes a covalent bond with essentially equal sharing of a pair of electrons, whereas a large difference indicates considerable charge separation. The loss of H^+ to a base occurs more readily where partial ionic charges already exist, and this corresponds to a stronger acid. The order of decreasing acid strength for the second period is

ΔEN *decreases:*	1.9 > 1.4 > 0.9 > 0.4
Acid strength decreases:	$HF > H_2O > NH_3 > CH_4$

HF is a weak acid with $K_a = 6.6 \times 10^{-4}$. The ability of H_2O to ionize as an acid is quite limited, as we shall see in Section 17.3. In aqueous solution, NH_3 ionizes as a weak base but not as an acid, nor does CH_4 ionize as an acid.

Strengths of Oxoacids

Oxoacids contain oxygen as well as hydrogen and a third nonmetallic element, E.

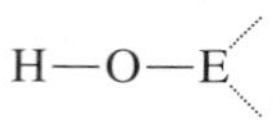

The broken lines suggest that in some cases other groups, such as O atoms or additional —OH groups, may also be bonded to the E atom. The strength of the O—H bond is affected by the attraction of the E atom for electrons in that bond. This electron-withdrawing effect increases with increased (a) electronegativity of E and (b) number of terminal O atoms attached to the E atom. As the electron-withdrawing effect of the E atom increases, the O—H bond becomes weaker and the acid strength increases. We can see the effect of the electronegativity of E in the three hypohalous acids.

Electronegativity increases:	2.5	<	2.8	<	3.0
Acid strength increases:	**HOI**		**HOBr**		**HOCl**
	$K_a = 2.3 \times 10^{-11}$	<	2.5×10^{-9}	<	2.9×10^{-8}

To test the effect of the second factor, let's look at the oxoacids of chlorine, whose structural formulas are shown below. We see that the acid strengths, measured by K_a values, are profoundly affected by the number of terminal O atoms. These O atoms act together with the E atom to withdraw electrons from the O—H bond.

Number of terminal O atoms increases:

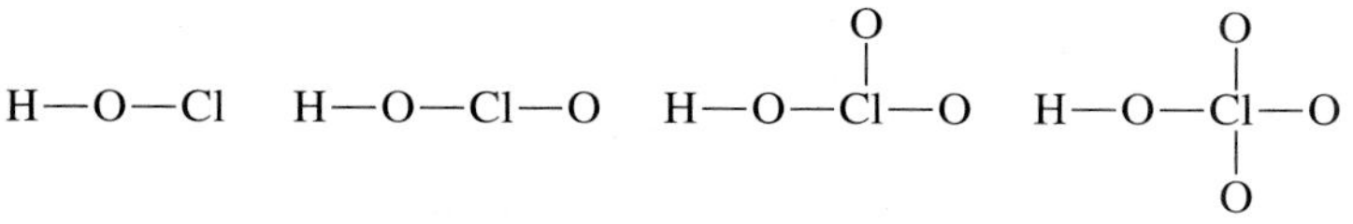

Acid strength increases: $K_a = 2.9 \times 10^{-8} < 1.1 \times 10^{-2} <$ ca. 1 $<$ ca. 10^8

Strengths of Carboxylic Acids

As with other acids, the strength of a carboxylic acid depends on the ease with which electrons can be withdrawn from an O—H bond so that a proton (H^+) can be liberated. Because all carboxylic acids share the —COOH group, we must look to differences in the R groups to explain variations in acid strength.

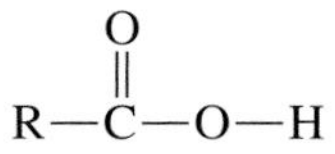

If the R group is simply a hydrocarbon chain, it has little effect on acid strength, as seen in the K_a values of the following two- and five-carbon acids.

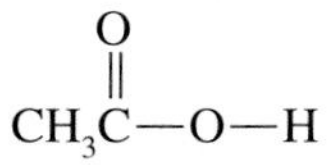

Ethanoic acid
(Acetic acid)
$K_a = 1.8 \times 10^{-5}$

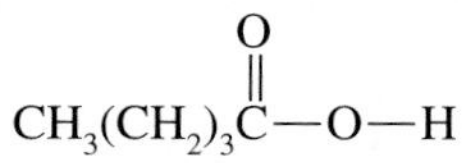

Pentanoic acid
(Valeric acid)
$K_a = 1.5 \times 10^{-5}$

On the other hand, if R = H, the effect is more significant.

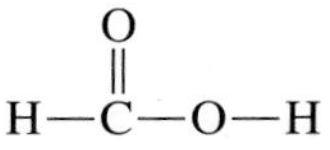

Methanoic acid
(Formic acid)
$K_a = 1.8 \times 10^{-4}$

If the R groups contain atoms of high electronegativity, these atoms can withdraw electrons from the O—H bond, thereby weakening the bond and increasing the acid strength. Two factors need to be considered: (1) how many atoms of high electronegativity are present and (2) how close they are to the —COOH group. These factors are illustrated by the following increasing order of acid strengths.

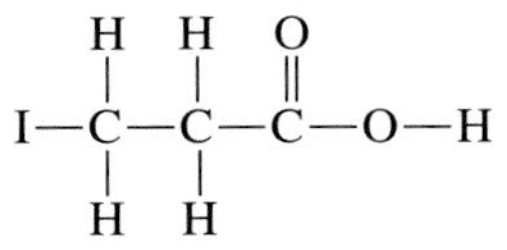

3-Iodopropanoic acid
$K_a = 8.3 \times 10^{-5}$

<

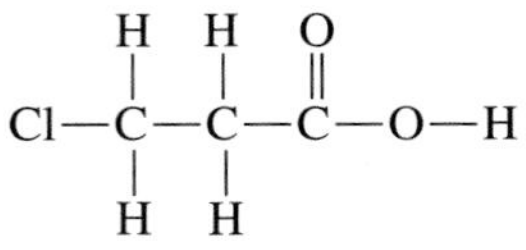

3-Chloropropanoic acid
$K_a = 1.0 \times 10^{-4}$

<

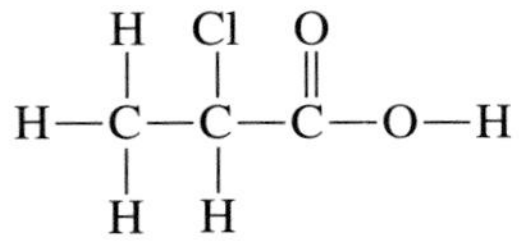

2-Chloropropanoic acid
$K_a = 1.4 \times 10^{-3}$

<

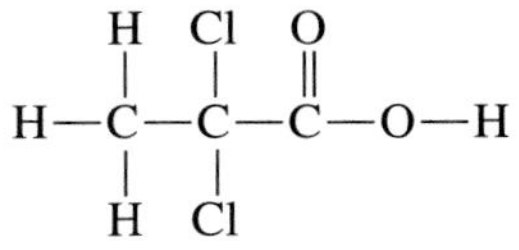

2,2-Dichloropropanoic acid
$K_a = 8.7 \times 10^{-3}$

Increasing acid strength ⟶

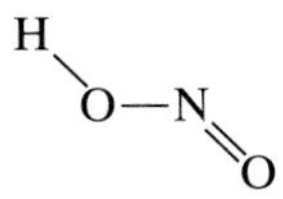

Nitrous acid

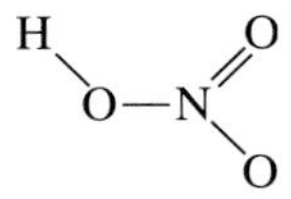

Nitric acid

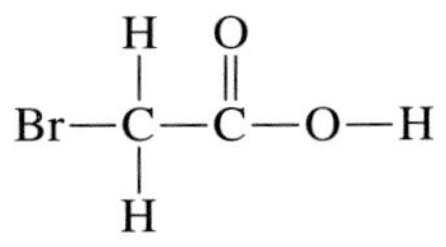

Bromoacetic acid

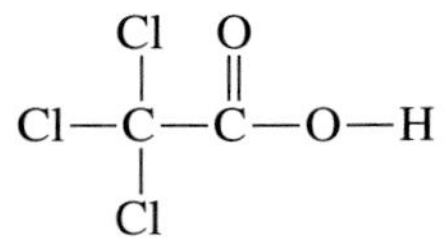

Trichloroacetic acid

Example 17.2

Select the stronger acid in each of the following pairs: **(a)** HNO_2 (or HONO) and HNO_3 (or $HONO_2$); **(b)** $BrCH_2COOH$ and CCl_3COOH.

Solution

a. The structures in the margin show that HNO_2 has one terminal O atom bonded to the central N atom and that HNO_3 has two terminal O atoms. We should expect HNO_3, because of its two terminal O atoms, to be the stronger of the two acids. (In fact, HNO_3 is a strong acid, whereas HNO_2 is weak.)

b. We should expect $BrCH_2COOH$ to be a somewhat stronger acid than CH_3COOH (acetic acid) because of the presence of the electronegative Br atom on the C atom adjacent to the —COOH group. However, CCl_3COOH, with *three* of the somewhat more electronegative Cl atoms adjacent to the C atom of the —COOH group, should be much stronger than both CH_3COOH and $BrCH_2COOH$. (The tabulated K_a values are 1.8×10^{-5}, for CH_3COOH; 1.3×10^{-3}, for $BrCH_2COOH$; and 3.0×10^{-1}, for CCl_3COOH.)

Exercise 17.2

Select the stronger acid in each of the following pairs: **(a)** H_2S or H_2Te; **(b)** $CH_3CH_2CH_2CH_2BrCOOH$ or $ClCH_2CH_2CH_2CH_2COOH$.

17.3 Self-ionization of Water—The pH Scale

Even the purest water conducts electricity, though it takes exceptionally sensitive measurements to detect this. Electrical conductivity requires the presence of ions, but where do they come from in pure water? The Brønsted–Lowry theory helps us to see how they form. Recall that water is *amphiprotic*. So, to a slight extent, water molecules can transfer protons among themselves. In the self-ionization of water, for every H_2O molecule that loses a proton, another H_2O molecule gains it. Hydronium and hydroxide ions are formed in equal numbers. Because H_3O^+ and OH^- are much stronger an acid and base, respectively, than H_2O molecules, the equilibrium point lies *far to the left*, as suggested below by the unequal arrow lengths.

$$\underset{\text{acid(1)}}{H_2O} + \underset{\text{base(2)}}{H_2O} \rightleftharpoons \underset{\text{acid(2)}}{H_3O^+} + \underset{\text{base(1)}}{OH^-}$$

The equilibrium constant expression for this self-ionization of water is

$$K_c = \frac{[H_3O^+][OH^-]}{[H_2O][H_2O]}$$

As in other cases, however, we do not include the concentration of water in the equilibrium constant expression. Rather we write,

$$K_w = K_c[H_2O]^2 = [H_3O^+][OH^-]$$

The experimentally determined equilibrium concentrations in pure water at 25 °C are

$$[H_3O^+] = [OH^-] = 1.0 \times 10^{-7}\ M$$

Thus, at 25 °C, the equilibrium constant for the self-ionization of water, called the **ion product of water** and represented as $\boldsymbol{K_w}$, is

Like other equilibrium constants, the value of K_w depends on temperature (see Problem 98).

$$K_w = [H_3O^+][OH^-] = (1.0 \times 10^{-7})(1.0 \times 10^{-7}) = 1.0 \times 10^{-14}$$

What makes this equilibrium constant so important is not just that it describes self-ionization in pure water. It applies as well to *all aqueous solutions*—that is, to solutions of acids, bases, salts, or nonelectrolytes. Consider, for example, a solution that is 0.00015 M HCl. Because HCl is a strong acid, its ionization is complete.

$$HCl + H_2O \longrightarrow H_3O^+ + Cl^-$$

and because $[H_3O^+]$ produced by the HCl is so much greater than that found in pure water, we can state with assurance that in this solution

$$[H_3O^+] = 0.00015\ M = 1.5 \times 10^{-4}\ M$$

With the K_w expression we can now calculate $[OH^-]$ in the solution.

$$[OH^-] = \frac{K_w}{[H_3O^+]} = \frac{1.0 \times 10^{-14}}{1.5 \times 10^{-4}} = 6.7 \times 10^{-11}\ M$$

pH and pOH

Although the exponential notation is a convenient way to express a small quantity like 1.5×10^{-4}, there's an even more convenient notation. In 1909 the Danish biochemist Søren Sørenson proposed a convention that is still used today. He let the term **pH** refer to "the potential of the hydrogen ion," and defined it as the *negative of the logarithm of [H+]*. Restated in terms of $[H_3O^+]$,*

Note that pH is defined in terms of the common logarithm, "log" (the base 10 logarithm), and *not* the natural logarithm, "ln" (the base *e* logarithm).

$$pH = -\log [H_3O^+]$$

*Because we can take logarithms only of dimensionless numbers, we use just the numerical value of the molarity of H_3O^+ and not the unit "M."

Refer to Appendix A for a discussion of why the number of significant figures in pH = 3.82 is said to be two (corresponding to the two in $[H_3O^+] = 1.5 \times 10^{-4}$) rather than three.

In a solution that is 0.00015 M HCl, in which

$$[H_3O^+] = 1.5 \times 10^{-4}\ M$$
$$pH = -\log [H_3O^+] = -\log (1.5 \times 10^{-4}) = -(-3.82) = 3.82$$

To determine the $[H_3O^+]$ corresponding to a given pH, we do an inverse calculation. In a solution with pH = 2.19,

$$-\log [H_3O^+] = pH$$
$$-\log [H_3O^+] = 2.19$$
$$\log [H_3O^+] = -2.19$$
$$[H_3O^+] = \text{antilog} (-2.19) = 10^{-2.19} = 6.5 \times 10^{-3}$$

We can also define **pOH** as

$$pOH = -\log [OH^-]$$

A 2.5×10^{-3} M NaOH solution has $[OH^-] = 2.5 \times 10^{-3}$ M and

$$pOH = -\log (2.5 \times 10^{-3}) = -(-2.60) = 2.60$$

As with pH and pOH, we can define **pK_w** as the negative logarithm of K_w.

$$pK_w = -\log K_w = -\log (1.0 \times 10^{-14}) = -(-14.00) = 14.00$$

Our purpose for doing this is to derive a simple relationship between the pH and pOH of a solution.

$$K_w = [H_3O^+][OH^-] = 1.0 \times 10^{-14}$$
$$-\log K_w = -\log ([H_3O^+][OH^-]) = -\log (1.0 \times 10^{-14})$$
$$pK_w = -\log [H_3O^+] - \log [OH^-] = 14.00$$
$$pK_w = pH + pOH = 14.00$$

Thus, the pH of 2.5×10^{-3} M NaOH is

$$pH = 14.00 - pOH = 14.00 - 2.60 = 11.40$$

In pure water, where $[H_3O^+] = [OH^-] = 1.0 \times 10^{-7}$ M, pH and pOH are both 7.00. Pure water and all aqueous solutions with pH = 7.00 are *neutral*. If the pH is less than 7.00 a solution is *acidic*; if the pH is above 7.00, a solution is *basic* or *alkaline*. As a solution becomes more acidic, $[H_3O^+]$ increases and pH decreases. As a solution becomes more basic, $[H_3O^+]$ decreases and pH increases.

Figure 17.2 gives the pH values of a number of familiar materials. You may want to use this as a frame of reference from time to time. Another point to keep in mind is that because pH is on a logarithmic scale, every unit change in pH represents a *tenfold* change in $[H_3O^+]$. Thus, lemon juice (pH ≈ 2.3) is somewhat more than ten times as acidic as orange juice (pH ≈ 3.5) and somewhat more than 100 times as acidic as tomato juice (pH ≈ 4.5).

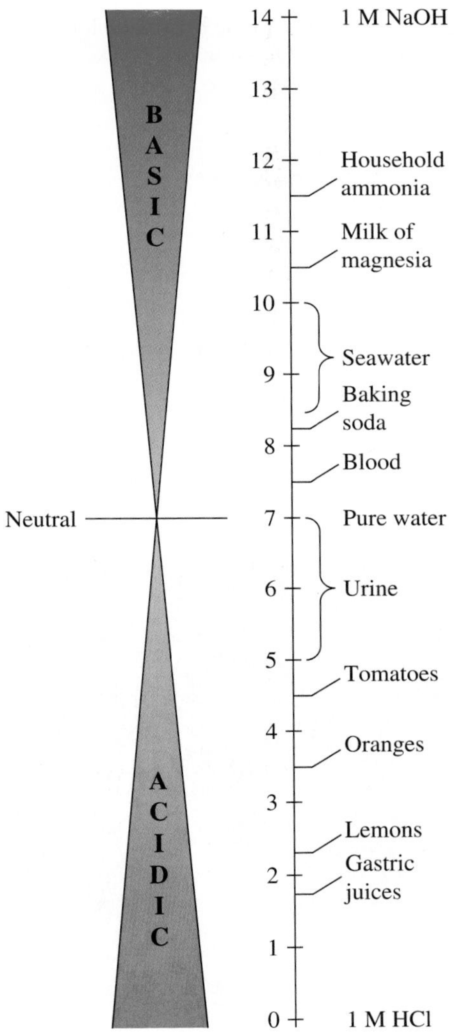

Figure 17.2 The pH scale.
The pH values of common substances range from about 0 to 14. Negative values are occasionally encountered ($[H_3O^+] = 10$ M corresponds to pH $= -1$), as are values somewhat greater than 14 ($[OH^-] = 10$ M corresponds to pH $= 15$).

Example 17.3

By the method suggested in Figure 17.3, a student determines the pH of "milk of magnesia"—a suspension of solid magnesium hydroxide in its saturated aqueous solution. A value of 10.52 is obtained. What is the molarity of $Mg(OH)_2$ in its saturated aqueous solution? The suspended, undissolved $Mg(OH)_2(s)$ does not affect the measurement.

Solution

We will assume that the dissolved magnesium hydroxide is completely dissociated into ions, so that we can write

$$Mg(OH)_2(aq) \longrightarrow Mg^{2+}(aq) + 2\,OH^-(aq)$$

This equation allows us to relate the molarity of $Mg(OH)_2$ to the hydroxide ion concentration. In turn, we can derive $[OH^-]$ from pOH, and pOH from pH. Thus,

$$\begin{aligned} pOH &= 14.00 - pH \\ &= 14.00 - 10.52 = 3.48 \\ \log [OH^-] &= -3.48 \\ [OH^-] &= 10^{-3.48} = 3.3 \times 10^{-4}\ M \end{aligned}$$

For the remainder of the calculation we have

$$\text{molarity} = \frac{3.3 \times 10^{-4}\ \text{mol OH}^-}{1\ \text{L}} \times \frac{1\ \text{mol Mg(OH)}_2}{2\ \text{mol OH}^-}$$

$$= 1.7 \times 10^{-4}\ \text{mol Mg(OH)}_2\text{/L} = 1.7 \times 10^{-4}\ \text{M Mg(OH)}_2$$

Exercise 17.3

What is the pH of a solution prepared by dissolving 0.0155 mol $Ba(OH)_2$ in water to give 735 mL of aqueous solution? Assume that the $Ba(OH)_2$ is completely dissociated.

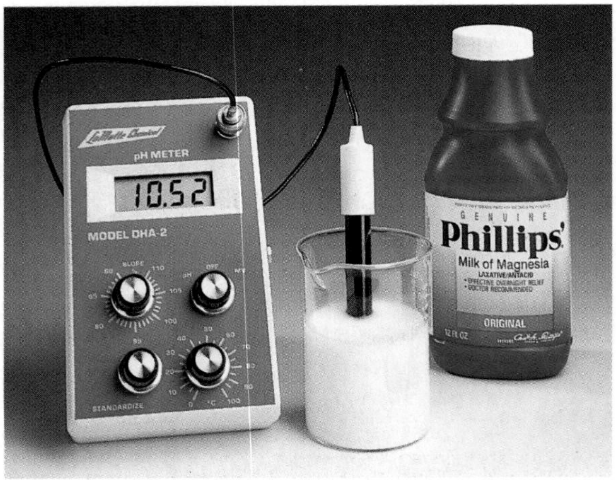

Figure 17.3
Measurement of pH with a pH meter—Example 17.3 illustrated. No modern laboratory is complete without an electrical measuring instrument known as a pH meter. We will consider the basic principles that underlie this method of pH measurement in Chapter 20.

CONCEPTUAL EXAMPLE 17.1

Is the solution 1.0×10^{-8} M HCl acidic, basic, or neutral?

Solution

Because the solution is one of a strong acid, our first thought is that the solution should be acidic. However, if we calculate the pH in the usual way,

$$HCl + H_2O \longrightarrow H_3O^+ + Cl^-$$

$$[H_3O^+] = 1.0 \times 10^{-8}\ M \quad \text{and} \quad pH = -\log[H_3O^+] = -\log(1.0 \times 10^{-8}) = 8.00$$

the solution appears to be basic.

The problem here is that the HCl(aq) is so dilute that the self-ionization of water actually produces more H_3O^+ than does ionization of the strong acid. As a result, the total $[H_3O^+]$ from the *two* sources is just slightly greater than 1.0×10^{-7} M. The pH of 1.0×10^{-8} M HCl is just slightly less than 7.00. The solution is *acidic*.

We can generally ignore the slight self-ionization of water if other ionization processes predominate. Usually, we cannot ignore it if the pH is within one unit or so of pH = 7.

An exact calculation of the pH is the subject of Problem 96.

Conceptual Exercise 17.1

Is a solution that is 1.0×10^{-8} M NaOH acidic, basic, or neutral? Explain.

17.4 Equilibrium in Solutions of Weak Acids and Weak Bases

Ordinary cider vinegar is an aqueous solution having acetic acid as its principal solute; it's what gives vinegar a tart, acidic flavor. A typical vinegar contains about 5 to 6% CH_3COOH by mass. This is equivalent to about 50 to 60 g CH_3COOH per liter, or about 1 M CH_3COOH. For 1 M HCl we expect a pH of about 0 (pH = –log 1 = 0), but we find the measured pH of vinegar to be much higher—about 2.4 (see Figure 17.4). CH_3COOH is a weak acid, whereas HCl is a strong one. With solutions of HCl, the conversion to H_3O^+ is essentially complete, but with solutions of acetic acid, the reaction does not go to completion. For solutions of the same solute molarity, there is a lot less H_3O^+ in $CH_3COOH(aq)$ than there is in HCl(aq). And remember, a lower $[H_3O^+]$ means a higher pH.

1 M HCl(aq) is essentially 100% ionized, whereas 1 M $CH_3COOH(aq)$ is only about 0.4% ionized. That is, only about 4 of every 1000 CH_3COOH molecules ionizes.

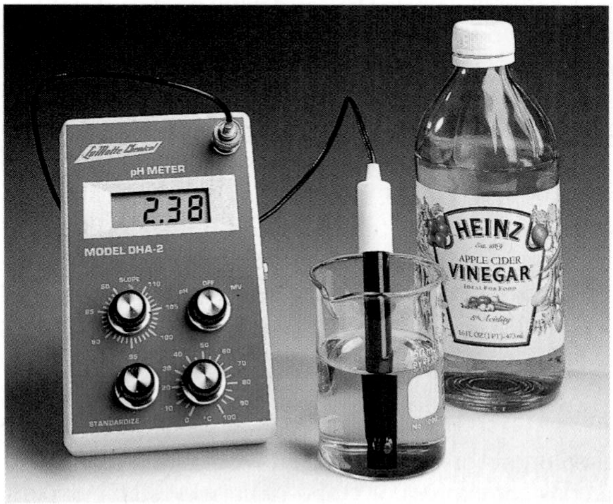

Figure 17.4 **Demonstrating that acetic acid is a weak acid: measuring the pH of vinegar.** The experimentally determined pH of vinegar—an aqueous solution of acetic acid—is considerably higher than that of a strong acid of the same molarity (about 1 M).

Organic Bases

As we noted in Section 12.2, organic compounds called amines are basic. Water-soluble amines can accept a proton from water. For example, trimethylamine dissolves in water and reacts to form a basic solution.

$$(CH_3)_3N{:}(aq) + H_2O(l) \rightleftharpoons [(CH_3)_3N{-}H]^+(aq) + OH^-(aq) \quad K_b = 6.3 \times 10^{-5}$$

Nearly all amines, including those that are not very soluble in water, will react with strong acids to form water-soluble salts. For example, aniline ($C_6H_5NH_2$), an aromatic amine used in the manufacture of dyes, drugs, resins, varnishes, and other commercial products, is only slightly soluble in water (3.5 g per 100 mL of water at 25 °C). Aniline ($K_b = 7.4 \times 10^{-10}$) reacts with hydrochloric acid to form a salt, commonly called "aniline hydrochloride," that is soluble in water to the extent of about 100 g per 100 mL of water.

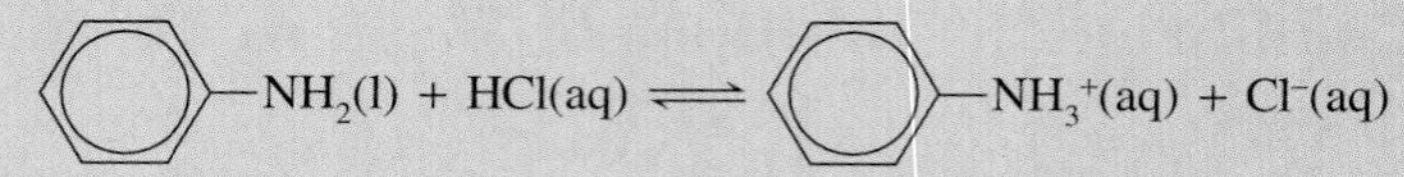

Aniline Aniline hydrochloride

Many medicines, both synthetic and naturally occurring, contain the amine functional group. *Alkaloids* are basic nitrogen-containing compounds produced by plants. Familiar alkaloids include caffeine, cocaine, morphine, nicotine, and strychnine (Figure 17.5). Many alkaloids have important physiological properties: they are medicines, poisons, and drugs of abuse.

Medicines that are amines are often converted to salts to enhance their water solubility. For instance, procaine (Figure 17.6) is soluble only to the extent of 0.5 g in 100 g of water. Its hydrochloride salt is soluble to the remarkable degree of 100 g in 100 g of water. Procaine hydrochloride, perhaps better known by the trade name Novocain, is a well known local anesthetic.

We also make use of the chemistry of amines when we put lemon juice on fish. The unpleasant fishy odor is due to amines. The citric acid in the juice converts the amines to nonvolatile salts, thus reducing the odor.

A solution of a weak base and its salt can be used as a buffer solution (Section 17.8) that will stabilize the pH of a solution in a basic range. An important example is tris(hydroxymethyl)aminomethane, often called simply "tris" (Figure 17.6). Tris and its salt, tris hydrochloride, buffer in the range of 7 to 9 pH units. These buffers find wide use in the cosmetic and textile industries, in cleaning compounds, and in biochemical research. Tris is also used in the treatment of metabolic acidosis (see the feature "Buffers in Blood" page 675).

To calculate pH values for solutions of weak acids and weak bases we need to use the appropriate ionization constant expressions, and to have available experimentally determined values of the acid and base ionization constants—K_a and K_b. You'll find a brief listing of these constants in Table 17.2 and a more extensive one

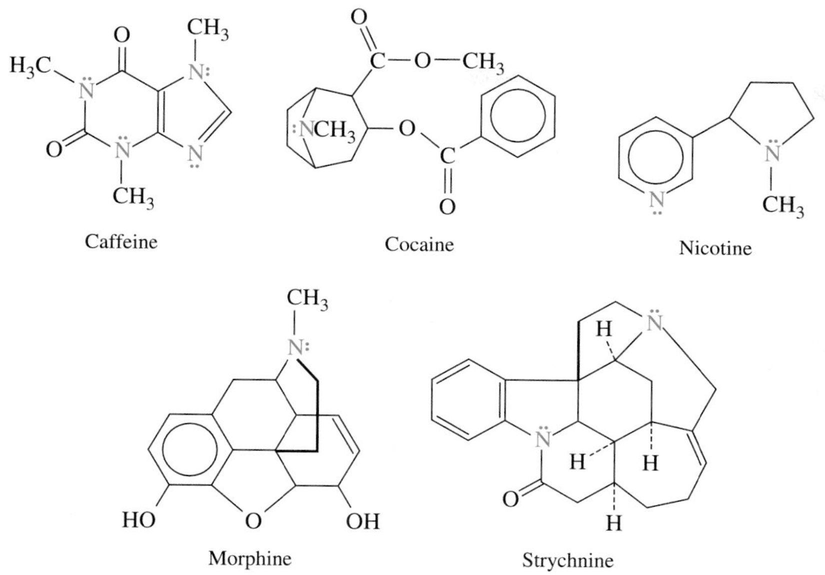

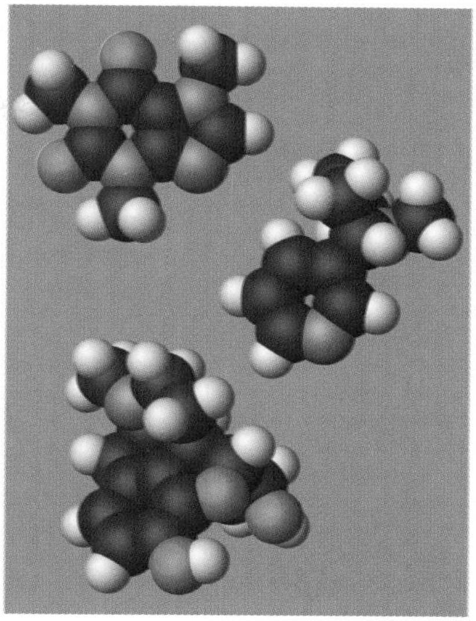

Figure 17.5 Some common alkaloids. Structural formulas are shown for five common alkaloids. The three molecular models, clockwise from the top, represent caffeine, nicotine, and morphine.

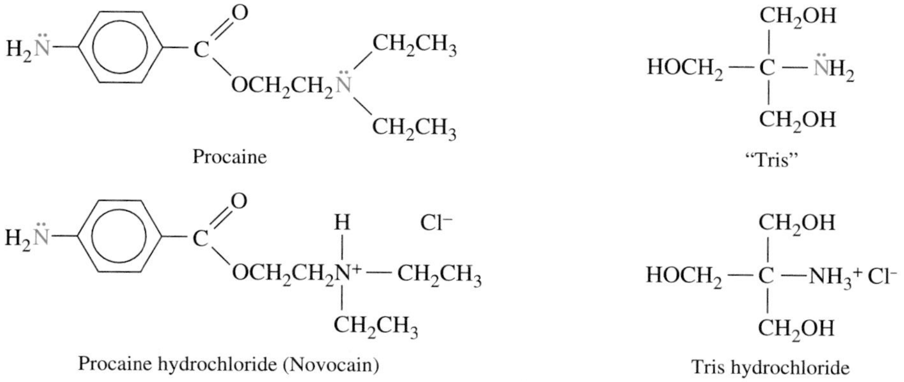

Figure 17.6 Some amines and their salts. Amine salts are more soluble in water than the "free bases" from which they are derived. The salts of amines with hydrochloric acid are often named as hydrochlorides.

in Appendix D. In these tabulations we also use a common shorthand designation: $\mathbf{p}\boldsymbol{K}_\mathbf{a} = -\log K_a$; $\mathbf{p}\boldsymbol{K}_\mathbf{b} = -\log K_b$. Small values of p$K_a$ and pK_b correspond to large values of K_a and K_b, just as a low value of pH corresponds to a high $[H_3O^+]$. In arranging the entries in Table 17.2 by increasing value of pK_a or pK_b, then, we are arranging them in the order of decreasing acid strength and base strength.

A Strategy for Acid–Base Equilibrium Calculations

The equilibrium calculations we do in this chapter are closely related to those of the preceding chapter. Because of this, much of what we said there applies here

TABLE 17.2 Ionization Constants of Some Weak Acids and Weak Bases in Water at 25 °C

	Ionization Equilibrium	Ionization Constant, K	pK
Inorganic Acids		K_a =	
Chlorous acid	$HClO_2 + H_2O \rightleftharpoons H_3O^+ + ClO_2^-$	1.1×10^{-2}	1.96
Nitrous acid	$HNO_2 + H_2O \rightleftharpoons H_3O^+ + NO_2^-$	7.2×10^{-4}	3.14
Hydrofluoric acid	$HF + H_2O \rightleftharpoons H_3O^+ + F^-$	6.6×10^{-4}	3.18
Hypochlorous acid	$HOCl + H_2O \rightleftharpoons H_3O^+ + OCl^-$	2.9×10^{-8}	7.54
Hypobromous acid	$HOBr + H_2O \rightleftharpoons H_3O^+ + OBr^-$	2.5×10^{-9}	8.60
Hydrocyanic acid	$HCN + H_2O \rightleftharpoons H_3O^+ + CN^-$	6.2×10^{-10}	9.21
Carboxylic Acids		K_a =	
Chloroacetic acid	$CH_2ClCOOH + H_2O \rightleftharpoons H_3O^+ + CH_2ClCOO^-$	1.4×10^{-3}	2.85
Formic acid	$HCOOH + H_2O \rightleftharpoons H_3O^+ + HCOO^-$	1.8×10^{-4}	3.74
Benzoic acid	$C_6H_5COOH + H_2O \rightleftharpoons H_3O^+ + C_6H_5COO^-$	6.3×10^{-5}	4.20
Acetic acid	$CH_3COOH + H_2O \rightleftharpoons H_3O^+ + CH_3COO^-$	1.8×10^{-5}	4.74
Inorganic Bases		K_b =	
Ammonia	$NH_3 + H_2O \rightleftharpoons NH_4^+ + OH^-$	1.8×10^{-5}	4.74
Hydrazine	$H_2NNH_2 + H_2O \rightleftharpoons H_2NNH_3^+ + OH^-$	8.5×10^{-7}	6.07
Amines		K_b =	
Diethylamine	$(CH_3)_2NH + H_2O \rightleftharpoons (CH_3)_2NH_2^+ + OH^-$	5.9×10^{-4}	3.23
Ethylamine	$CH_3CH_2NH_2 + H_2O \rightleftharpoons CH_3CH_2NH_3^+ + OH^-$	4.3×10^{-4}	3.37
Methylamine	$CH_3NH_2 + H_2O \rightleftharpoons CH_3NH_3^+ + OH^-$	4.2×10^{-4}	3.38
Hydroxylamine	$HONH_2 + H_2O \rightleftharpoons HONH_3^+ + OH^-$	9.1×10^{-9}	8.04
Pyridine	$C_5H_5N + H_2O \rightleftharpoons C_5H_5NH^+ + OH^-$	1.5×10^{-9}	8.82
Aniline	$C_6H_5NH_2 + H_2O \rightleftharpoons C_6H_5NH_3^+ + OH^-$	7.4×10^{-10}	9.13

as well—we write an equation for the reversible reaction, organize data under the equation, and so on. Some additional factors to consider are

- What are the molecular and ionic species present in solution, and are there some that can be neglected?
- Are there some equilibria that can be neglected, such as the self-ionization of water?
- Are there some assumptions that can be made to facilitate the solution of algebraic equations?
- What is a reasonable answer to the problem? For example, should the solution be acidic (pH < 7) or basic (pH > 7)?
- Once you've obtained an answer, can you justify your assumptions?

Illustrative Examples

To illustrate the general strategy just outlined, Example 17.4 will involve making a common assumption and justifying it. In Example 17.5 we will explore the limitations of this assumption.

Example 17.4

Calculate the pH of 1.00 M $CH_3COOH(aq)$.

Solution

First note that acetic acid is a much more significant source of H_3O^+ than is the self-ionization of water.

$$CH_3COOH + H_2O \rightleftharpoons H_3O^+ + CH_3COO^- \qquad K_a = 1.8 \times 10^{-5}$$
$$H_2O + H_2O \rightleftharpoons H_3O^+ + OH^- \qquad K_w = 1.0 \times 10^{-14}$$

We can neglect the self-ionization of water because K_w is so much smaller than K_a. Note also that the amount of solute and the solution volume are not given, and we don't need them. We can work directly with concentrations.

	CH_3COOH + H_2O $\rightleftharpoons$	H_3O^+ +	CH_3COO^-
Initial concns, M:	1.00	—	—
Changes, M:	$-x$	$+x$	$+x$
Equil. concns, M:	$(1.00 - x)$	x	x

$$K_a = \frac{[H_3O^+][CH_3COO^-]}{[CH_3COOH]} = \frac{x \cdot x}{1.00 - x} = 1.8 \times 10^{-5}$$

Just by inspecting this equation we see that it is a quadratic equation; the highest power is x^2. We can solve the equation with the quadratic formula, but we can simplify the solution by making an assumption (which we will examine below). Let us assume that x is *much* smaller than 1.00 ($x \ll 1.00$). If this is the case, $1.00 - x \approx 1.00$, and

The symbol $\ll$ means "much smaller than."

$$\frac{x^2}{1.00} = 1.8 \times 10^{-5}$$

$$x^2 = 1.8 \times 10^{-5}$$

$$x = [H_3O^+] = 4.2 \times 10^{-3}\,M$$

$$pH = -\log[H_3O^+] = -\log(4.2 \times 10^{-3}) = 2.38$$

To test the validity of our assumption, we see that $1.00 - x = 1.00 - 4.2 \times 10^{-3} = 1.00 - 0.0042 = 1.00$ (to two decimal places).

Exercise 17.4

Determine the pH of 0.250 M C_6H_5COOH (benzoic acid). Obtain the K_a value from Table 17.2.

The assumption we made in Example 17.4 amounts to saying that in a weak acid solution the proportion of the weak acid that ionizes (x) is negligible compared to the total concentration of acid (C_{acid}). That is, $C_{acid} - x \approx C_{acid}$. This assumption will generally be acceptable if x is less than about 5% of C_{acid}. We apply this idea again in Example 17.5, this time to the ionization of a weak base.

Example 17.5

What is the pH of 0.000100 M $NH_3(aq)$?

Solution

First, we assume that NH_3 is a much more significant source of OH^- than is the self-ionization of water.

$$NH_3 + H_2O \rightleftharpoons NH_4^+ + OH^- \qquad K_b = 1.8 \times 10^{-5}$$
$$H_2O + H_2O \rightleftharpoons H_3O^+ + OH^- \qquad K_w = 1.0 \times 10^{-14}$$

In the setup below we work directly with concentrations relating to the ionization equilibrium of ammonia.

	NH_3	$+ H_2O \rightleftharpoons$	NH_4^+	$+ OH^-$
Initial concns, M:	0.000100		—	—
Changes, M:	$-x$		$+x$	$+x$
Equil. concns, M:	$(0.000100 - x)$		x	x

$$K_b = \frac{[NH_4^+][OH^-]}{[NH_3]} = \frac{x \cdot x}{0.000100 - x} = 1.8 \times 10^{-5}$$

Let's begin by assuming that x is much smaller than 0.000100 ($x << 0.000100$). If this is the case, $0.000100 - x \approx 0.000100$, and

$$\frac{x^2}{0.000100} = 1.8 \times 10^{-5}$$

$$x^2 = 1.8 \times 10^{-9}$$

$$x = [OH^-] = 4.2 \times 10^{-5}\,M$$

$$pOH = -\log[OH^-] = -\log(4.2 \times 10^{-5}) = 4.38$$

$$pH = 14.00 - pOH = 14.00 - 4.38 = 9.62$$

To test the validity of our assumption, $0.000100 - x = 0.000100 - 4.2 \times 10^{-5} = 0.000100 - 0.000042 = 0.000058$. The assumption does *not* work. Here, x (0.000042 M) is 42% of C_{base} (0.000100 M).

We must go back to the expression

$$\frac{x^2}{0.000100 - x} = 1.8 \times 10^{-5}$$

and solve the quadratic equation

$$x^2 + 1.8 \times 10^{-5}x - 1.8 \times 10^{-9} = 0$$

$$x = \frac{-1.8 \times 10^{-5} \pm \sqrt{(1.8 \times 10^{-5})^2 + 7.2 \times 10^{-9}}}{2}$$

$$x = \frac{-1.8 \times 10^{-5} \pm 8.7 \times 10^{-5}}{2} = \frac{6.9 \times 10^{-5}}{2} = 3.5 \times 10^{-5}$$

$$[OH^-] = x = 3.5 \times 10^{-5}\,M$$

$$pOH = -\log[OH^-] = -\log(3.5 \times 10^{-5}) = 4.46$$

$$pH = 14.00 - pOH = 14.00 - 4.46 = 9.54$$

As with other equilibrium calculations, you can establish that $x = 3.5 \times 10^{-5}$ is an acceptable solution by evaluating the expression $x^2/(0.00010 - x)$ and comparing this value to $K_b = 1.8 \times 10^{-5}$. Note that the other root of the quadratic equation gives an impossible answer; $[OH^-]$ cannot have a negative value.

Exercise 17.5

Calculate the pH of 0.0010 M NH_3(aq). Show that even though the "5% rule" for the simplifying assumption fails, to *two* significant figures, the same $[OH^-]$ is obtained with and without the assumption.

It is generally not difficult to determine K_a of a weak acid or K_b of a weak base from experimental data. From a simple pH measurement we can readily obtain the equilibrium concentration of H_3O^+ or OH^-. In Example 17.6 we determine K_b for the weak base dimethylamine.

Example 17.6

The pH of an 0.164 M aqueous solution of dimethylamine is found to be 11.98. What must be the values of K_b and pK_b?

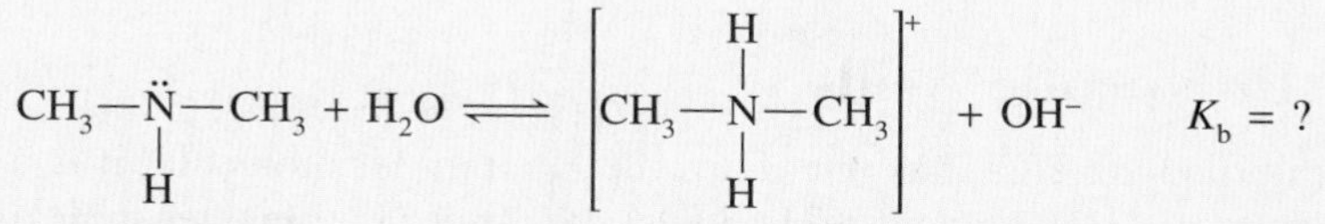

Dimethylamine   Dimethylammonium ion

Solution

From the measured pH we can first calculate pOH and then $[OH^-]$.

$$\begin{aligned} pOH &= 14.00 - pH = 14.00 - 11.98 = 2.02 \\ \log [OH^-] &= -2.02 \\ [OH^-] &= 10^{-2.02} = 9.5 \times 10^{-3}\ M \end{aligned}$$

Now let's use the format of Example 16.9 (page 626).

	$(CH_3)_2NH$	$+ H_2O \rightleftharpoons$	$(CH_3)_2NH_2^+$ +	OH^-
Initial concns, M:	0.164		—	—
Changes, M:	-9.5×10^{-3}		$+9.5 \times 10^{-3}$	$+9.5 \times 10^{-3}$
Equil. concns, M:	$(0.164 - 9.5 \times 10^{-3})$		9.5×10^{-3}	9.5×10^{-3}

$$K_b = \frac{[(CH_3)_2NH_2^+][OH^-]}{[(CH_3)_2NH]} = \frac{(9.5 \times 10^{-3})(9.5 \times 10^{-3})}{(0.164 - 0.0095)} = 5.8 \times 10^{-4}$$

$$pK_b = -\log K_b = -\log(5.8 \times 10^{-4}) = 3.24$$

Exercise 17.6

Suppose you discovered a new acid, HZ, and found that the pH of a 0.0100 M solution is 3.12. What are K_a and pK_a for HZ?

$$HZ + H_2O \rightleftharpoons H_3O^+ + Z^- \qquad K_a = ?$$

ESTIMATION EXAMPLE 17.1

Which solution has a greater $[H_3O^+]$, 0.030 M HCl or 0.050 M CH_3COOH?

Solution

HCl is a strong acid. In 0.030 M HCl, $[H_3O^+] = 0.030$ M.

CH_3COOH is a weak acid. The 0.050 M CH_3COOH would have to be more than 50% ionized to have $[H_3O^+]$ equal to that of 0.030 M HCl. This extensive an ionization would require K_a for acetic acid to be much larger than it is (from Table 17.2, $K_a = 1.8 \times 10^{-5}$). The 0.030 M HCl has the greater $[H_3O^+]$. (By the method of Example 17.4, we could show that $[H_3O^+]$ in 0.030 M HCl is over 30 times greater than in 0.050 M CH_3COOH.)

Estimation Exercise 17.1

Which solution has a greater $[OH^-]$, 0.025 M $NH_3(aq)$ or 0.030 M methylamine, $CH_3NH_2(aq)$? (*Hint*: Refer to Table 17.2.)

17.5 Polyprotic Acids

Hydrochloric acid, HCl, has one ionizable H atom per molecule; it is a strong *monoprotic* acid. Acetic acid, CH_3COOH, has four H atoms per molecule, but only one of them is ionizable; it is a weak *monoprotic* acid. Carbonic acid, H_2CO_3, has two H atoms, both ionizable; it is a weak *diprotic* acid. Phosphoric acid, H_3PO_4, has three H atoms, all ionizable; it is a *triprotic* acid. H_2CO_3 and H_3PO_4 fall in a general class of acids known as polyprotic acids. **Polyprotic acids** are acids that have more than one ionizable H atom per molecule. We will consider three important polyprotic acids in this section. Ionization constants of additional polyprotic acids are listed in Appendix D.

A typical cola drink has about 0.05% H_3PO_4 by mass (see Problem 53).

Phosphoric Acid

A key feature of all polyprotic acids is that ionization takes place in separate steps. This is illustrated below for phosphoric acid, but the situation would be similar for any polyprotic acid.

$$(1)\ H_3PO_4 + H_2O \rightleftharpoons H_3O^+ + H_2PO_4^- \qquad K_{a_1} = \frac{[H_3O^+][H_2PO_4^-]}{[H_3PO_4]} = 7.1 \times 10^{-3}$$

$$(2)\ H_2PO_4^- + H_2O \rightleftharpoons H_3O^+ + HPO_4^{2-} \qquad K_{a_2} = \frac{[H_3O^+][HPO_4^{2-}]}{[H_2PO_4^-]} = 6.3 \times 10^{-8}$$

$$(3)\ HPO_4^{2-} + H_2O \rightleftharpoons H_3O^+ + PO_4^{3-} \qquad K_{a_3} = \frac{[H_3O^+][PO_4^{3-}]}{[HPO_4^{2-}]} = 4.3 \times 10^{-13}$$

That the second ionization constant, K_{a_2}, is much smaller than the first, K_{a_1}, is not difficult to explain. In the first ionization, as a proton leaves an H_3PO_4 molecule, it must overcome the attraction of the $H_2PO_4^-$ ion left behind. In the second ionization, the departing proton is attracted to a negative ion with twice the charge: HPO_4^{2-}. Proton transfer occurs much less readily. In the final ionization, proton separation from the ion PO_4^{3-} is more difficult still, and this accounts for the fact that K_{a_3} is much smaller than K_{a_2}. This general observation about polyprotic acids leads to two conclusions.

The symbol $\gg$ means "much greater than."

1. Because generally $K_{a_1} \gg K_{a_2} \gg K_{a_3}$, in all but very dilute solutions we can usually assume that all of the H_3O^+ comes from the first ionization step alone.
2. Because K_{a_2} is so small, little of the anion produced in the first ionization step ionizes any further.

Applied to phosphoric acid, these statements mean that $[H_3O^+] \approx [H_2PO_4^-]$, and, *regardless of the solution molarity*,

$$[HPO_4^{2-}] \approx K_{a_2} = 6.3 \times 10^{-8}$$

That is,

$$K_{a_2} = \frac{\cancel{[H_3O^+]}[HPO_4^{2-}]}{\cancel{[H_2PO_4^-]}} = 6.3 \times 10^{-8}$$

Now, let's consider these ideas in some specific situations.

Example 17.7

An important industrial use of phosphoric acid is in removing deposits from boilers in power plants. Typically, the solution used is 5% H_3PO_4 by mass (d = 1.03 g/mL). What is the pH of this boiler-cleaning solution?

Solution

The key idea here is that essentially all the H_3O^+ comes from the first ionization step:

$$H_3PO_4 + H_2O \rightleftharpoons H_3O^+ + H_2PO_4^-$$

Our main concern is what to use for the molarity of H_3PO_4. The mass of a 1.00-L sample of the solution is

$$\text{mass} = 1.00 \text{ L} \times \frac{1000 \text{ mL}}{1 \text{ L}} = \frac{1.03 \text{ g}}{\text{mL}} = 1030 \text{ g}$$

The number of moles of H_3PO_4 in 1.00 L of the solution is

$$? \text{ mol } H_3PO_4 = 1030 \text{ g soln} \times \frac{5 \text{ g } H_3PO_4}{100 \text{ g soln}} \times \frac{1 \text{ mol } H_3PO_4}{98 \text{ g } H_3PO_4} = 0.5 \text{ mol } H_3PO_4$$

The boiler-cleaning acid is 0.5 M H_3PO_4.

Now, we can use a familiar format.

	H_3PO_4 + H_2O	$\rightleftharpoons$ H_3O^+	+ $H_2PO_4^-$
Initial concns, M:	0.5	—	—
Changes, M:	$-x$	$+x$	$+x$
Equil. concns, M:	$(0.5 - x)$	x	x

$$K_{a_1} = \frac{[H_3O^+][H_2PO_4^-]}{[H_3PO_4]} = \frac{x \cdot x}{0.5 - x} = 7.1 \times 10^{-3}$$

If we assume that $x \ll 0.5$, so that $(0.5 - x) \approx 0.5$, we can write

$$x^2 = 0.5 \times 7.1 \times 10^{-3} = 4 \times 10^{-3}$$
$$x = [H_3O^+] = 0.06 \text{ M}$$
$$\text{pH} = -\log [H_3O^+] = -\log (0.06) = 1.2$$

Our assumption appears not to work particularly well: x is more than 5% of $[H_3PO_4]$. That is, x = 0.06 is 12% of 0.5. On the other hand, our result can have only one significant figure, and the assumption is good enough.

A more precise calculation would yield $[H_3O^+] = 5.6 \times 10^{-2}$ M and pH = 1.25.

Exercise 17.7

What are the concentrations of dihydrogen phosphate ion, $H_2PO_4^-$, and hydrogen phosphate ion, HPO_4^{2-}, in the boiler-cleaning solution of Example 17.7?

Carbonic Acid

Carbon dioxide gas dissolves in water to form carbonic acid, a weak diprotic acid.

$$CO_2(g) + H_2O \rightleftharpoons H_2CO_3(aq)$$

The reaction is reversible. Carbonic acid is unstable and readily reverts to $CO_2(g)$ and H_2O. In an open vessel $CO_2(g)$ escapes and the reaction goes to completion *to the left*, just as we would expect from Le Châtelier's principle.

At one time or another we've probably all witnessed this fact when the taste of the soda pop in an opened can has gone flat.

The two ionization steps of H_2CO_3 and their K_a values are written below. Note how carbonic acid conforms to the generalization that we introduced in the discussion of phosphoric acid: $K_{a_1} \gg K_{a_2}$.

$$(1)\ H_2CO_3 + H_2O \rightleftharpoons H_3O^+ + HCO_3^- \qquad K_{a_1} = \frac{[H_3O^+][HCO_3^-]}{[H_2CO_3]} = 4.4 \times 10^{-7}$$

$$(2)\ HCO_3^- + H_2O \rightleftharpoons H_3O^+ + CO_3^{2-} \qquad K_{a_2} = \frac{[H_3O^+][CO_3^{2-}]}{[HCO_3^-]} = 4.7 \times 10^{-11}$$

In an indirect way we are already familiar with this two-step ionization. Neutralization of H_2CO_3 with 1 mol of OH^- in the first step produces salts with HCO_3^- ion, such as $NaHCO_3$, sodium hydrogen carbonate. Neutralization with 2 mol OH^- produces carbonate salts, such as Na_2CO_3.

Equilibria based on the ionization of carbonic acid are significant in several important natural phenomena, such as the formation of temporary hard water and limestone caves (Section 8.4). These equilibria are also essential in maintaining the proper pH of blood (page 675).

Sulfuric Acid

Sulfuric acid, another diprotic acid, is unusual in that its first ionization step goes essentially to completion, but its second step does not.

$$(1)\ H_2SO_4 + H_2O \longrightarrow H_3O^+ + HSO_4^- \qquad K_{a_1} \approx 10^3$$

$$(2)\ HSO_4^- + H_2O \rightleftharpoons H_3O^+ + SO_4^{2-} \qquad K_{a_2} = \frac{[H_3O^+][SO_4^{2-}]}{[HSO_4^-]} = 1.1 \times 10^{-2}$$

Following are three possible situations when considering sulfuric acid solutions:

1. *High Concentrations*: Because the first ionization step goes to completion, whereas the second is reversible, in moderately concentrated $H_2SO_4(aq)$, say greater than about 0.5 M, the H_3O^+ in solution is produced almost exclusively in the first step. Thus, we conclude that $[H_3O^+] \approx 1.00$ M in 1.00 M $H_2SO_4(aq)$. (A more precise calculation yields $[H_3O^+] = 1.01$ M.)

2. *Low Concentrations*: K_{a_2}, though quite small compared to K_{a_1}, is large enough that in sufficiently dilute solutions, say below about 0.0010 M H_2SO_4, the second ionization also goes nearly to completion. This statement leads us to conclude that in 0.0010 M $H_2SO_4(aq)$, $[H_3O^+] \approx 0.0020$ M. (A more precise calculation yields $[H_3O^+] = 0.0019$ M.)
3. *Intermediate Concentrations*: For intermediate concentrations, from about 0.0010 M to 0.50 M H_2SO_4, both ionizations must be considered when calculating ion concentrations in $H_2SO_4(aq)$.

Now, let's see how these ideas are used in some specific situations.

Example 17.8

What is the approximate pH of 0.71 M H_2SO_4?

Solution

This case fits the first of the three situations listed above. That is, essentially all the H_3O^+ comes from the first step in the ionization of H_2SO_4, in which ionization goes to completion.

$$H_2SO_4 + H_2O \longrightarrow H_3O^+ + HSO_4^-$$

Thus,

$$[H_3O^+] = 0.71 \text{ M; pH} = -\log [H_3O^+] = -\log 0.71 = 0.15$$

(Because a small additional amount of H_3O^+ is produced in the second ionization, the true pH is just slightly less than 0.15.)

Exercise 17.8

What is the approximate pH of 8.5×10^{-4} M H_2SO_4?

Estimation Exercise 17.2

Which of the following is likely to be closest to the measured $[H_3O^+]$ in 0.020 M H_2SO_4: (a) 0.020 M; (b) 0.025 M; (c) 0.040 M; (d) 0.045 M? Explain your reasoning.

17.6 Ions as Acids and Bases

A package of a common wall cleaner carries the warning, "Avoid contact with eyes and prolonged contact with skin." This is a common warning found on products that are either rather strongly acidic or strongly basic. The principal constituent of this particular cleaner is identified as sodium carbonate. On first sight, it might seem that Na_2CO_3 is neither an acid nor a base (we see no H atoms, nor OH groups, nor N atoms with lone-pair electrons). Yet, as shown in Figure 17.7, 1 M $Na_2CO_3(aq)$ is quite basic.

When the ionic compound $Na_2CO_3(s)$ dissolves, Na^+ and CO_3^{2-} ions enter the solution.

$$Na_2CO_3(s) \xrightarrow{H_2O} 2\,Na^+(aq) + CO_3^{2-}(aq)$$

Figure 17.7 **Hydrolysis of carbonate ion.** This sodium carbonate solution contains a few drops of thymolphthalein indicator. The blue color of the indicator shows that the pH of the solution is greater than 10.6. The solution is rather strongly basic as a result of the hydrolysis of CO_3^{2-} ion, that is, its ionization as a base.

Thymolphthalein indicator:	pH < 9.4	pH > 10.6
	colorless	blue

The Brønsted–Lowry theory shows how OH^- ions are produced.

$$\underset{\text{base(1)}}{CO_3^{2-}} + \underset{\text{acid(2)}}{H_2O} \rightleftharpoons \underset{\text{acid(1)}}{HCO_3^-} + \underset{\text{base(2)}}{OH^-}$$

This reaction raises $[OH^-]$ to a value much higher than 1.0×10^{-7} M. $[H_3O^+]$ decreases accordingly, and the pH rises well above 7.0.

Sodium ions do not react with water.

$$Na^+(aq) + H_2O \longrightarrow \text{no reaction}$$

The only possible source of H_3O^+ or OH^- from $Na^+(aq)$ would be from interactions between Na^+ and its associated water molecules in the hydrated cation, $Na^+(aq)$, but these interactions are not strong enough to cause ionization.

Although acid–base reactions of ions with water are fundamentally no different from other acid–base reactions, they are sometimes referred as hydrolysis reactions. In a general sense, **hydrolysis** is the reaction of a substance with water ("hydro") in which both the substance and water molecules are split apart ("lysis"). In $Na_2CO_3(aq)$, we say that CO_3^{2-} hydrolyzes and Na^+ does not. Other cations of Group 1A and Group 2A also do not hydrolyze. On the other hand, a number of metal cations do hydrolyze, particularly those with a small size and high charge. We will discuss the hydrolysis of certain metal cations in the next chapter, but for now let's consider some generalizations about hydrolysis. In the

We will discuss the ionization of certain hydrated cations as acids in Section 18.8.

accompanying photographs, the color of bromthymol blue indicator in solution depends on pH in this way.

pH < 7	pH = 7	pH > 7
yellow	green	blue

1. Salts of strong acids and strong bases form *neutral* solutions (pH = 7). Neither the cation nor the anion hydrolyzes in water. Examples: NaCl, KNO_3, BaI_2. Cl^-, NO_3^-, and I^- are conjugate bases of strong acids; they are all very weak bases. They do not hydrolyze, and neither do the Group 1A and Group 2A cations.
2. Salts of weak acids and strong bases form *basic* solutions (pH > 7). The anion ionizes as a base. Examples: Na_2CO_3, KNO_2, CH_3COONa. CO_3^{2-}, NO_2^-, and CH_3COO^- are the conjugate bases of weak acids; they are considerably stronger bases than are Cl^-, NO_3^-, and I^- (recall Table 17.1), and they do hydrolyze.

NaCl(aq)

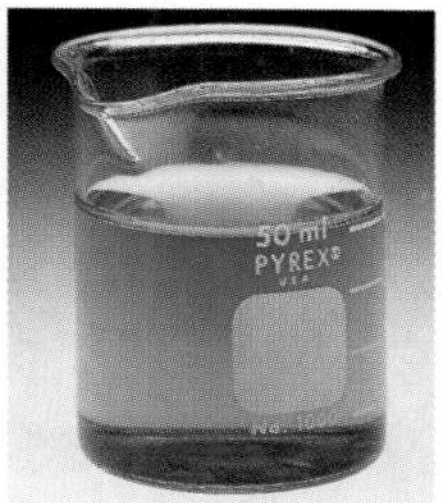

CH_3COONa(aq)

3. Salts of strong acids and weak bases form *acidic* solutions (pH < 7). The cation hydrolyzes as an acid. Examples: NH_4Cl, NH_4NO_3, and NH_4Br. NH_4^+ is the conjugate acid of the weak base NH_3.
4. Salts of weak acids and weak bases form solutions that are acidic in some cases, neutral or basic in others. The cations act as acids and the anions as bases, but the solution pH depends on the relative strengths of the weak acids and weak bases. Examples: NH_4CN, NH_4NO_2, and CH_3COONH_4.

NH_4Cl(aq)

CH_3COONH_4(aq)

Example 17.9

Indicate whether you expect each of the following solutions to be acidic, basic, or neutral: **(a)** NH_4I(aq), **(b)** CH_3COONH_4(aq).

Solution

a. NH_4I is the salt of a strong acid, HI, and weak base, NH_3. This corresponds to case 3 above. The cation, NH_4^+, hydrolyzes, yielding an *acidic* solution.

$$NH_4^+ + H_2O \rightleftharpoons NH_3 + H_3O^+$$

The anion, I^-, a very weak base, does not hydrolyze.

b. Ammonium acetate is the salt of a weak acid, CH_3COOH, and weak base, NH_3. It represents case 4 above; both ions hydrolyze.

$$NH_4^+ + H_2O \rightleftharpoons NH_3 + H_3O^+$$
$$CH_3COO^- + H_2O \rightleftharpoons CH_3COOH + OH^-$$

From the observation that aqueous ammonium acetate has pH ≈ 7 (photograph on page 665), we should expect the ionization constants for the two hydrolysis reactions to have about the same value. We will look next at how to obtain values of these ionization constants.

Exercise 17.9

Indicate whether you expect each of the following solutions to be acidic, basic, or neutral: **(a)** $NaNO_3(aq)$; **(b)** $CH_3CH_2CH_2COOK(aq)$.

To arrive at a definitive answer in Example 17.9(b), we need ionization constants for the two hydrolysis reactions. Their values are easily derived from two other, familiar ionization constants. Consider, for example, the hydrolysis

$$CH_3COO^- + H_2O \rightleftharpoons CH_3COOH + OH^- \qquad K_b = ?$$

$$K_b = \frac{[CH_3COOH][OH^-]}{[CH_3COO^-]} = ?$$

Two of the concentration terms in the K_b expression are the same as in the K_a expression for the ionization of acetic acid, the conjugate acid of CH_3COO^-. It seems, then, that K_b for CH_3COO^- and K_a for CH_3COOH should be related to each other. Suppose we multiply both the numerator and denominator of the K_b expression by $[H_3O^+]$.

$$K_b = \frac{[CH_3COOH][OH^-][H_3O^+]}{[CH_3COO^-][H_3O^+]} = \frac{K_w}{K_a} = \frac{1.0 \times 10^{-14}}{1.8 \times 10^{-5}} = 5.6 \times 10^{-10}$$

Note that the terms printed in blue are equivalent to K_w. Those in red represent the *inverse* of K_a for acetic acid, that is, $1/K_a$. A common way to write the result of what we've just done is

This is the relationship that we first encountered through the K_a and K_b values in Table 17.1.

$$K_a \times K_b = K_w$$

That is, the product of K_a and K_b of a conjugate acid–base pair equals the ion product of water, which at 25 °C is $K_w = 1.0 \times 10^{-14}$. Many tables of ionization constants list only pK_a values. To get a K_b value you need to calculate it from the expression $pK_a + pK_b = pK_w = 14.00$.

With this means of obtaining ionization constants, we can calculate the pH of solutions of salts that hydrolyze, as is illustrated in Example 17.10.

Example 17.10

Calculate the pH of a sodium acetate solution that is 0.25 M $CH_3COONa(aq)$.

Solution

We've already established that in this solution it is the acetate ion that ionizes as a base, and we just derived a numerical value of K_b for this hydrolysis. The data under the equation for the hydrolysis reaction are presented in the usual way.

$$CH_3COO^- + H_2O \rightleftharpoons CH_3COOH + OH^-$$

	CH_3COO^-	CH_3COOH	OH^-
Initial concns, M:	0.25	—	—
Changes, M:	$-x$	$+x$	$+x$
Equil. concns, M:	$(0.25 - x)$	x	x

$$K_b = \frac{[CH_3COOH][OH^-]}{[CH_3COO^-]} = \frac{K_w}{K_a} = \frac{1.0 \times 10^{-14}}{1.8 \times 10^{-5}} = 5.6 \times 10^{-10}$$

Let's assume that $x << 0.25$, so that $(0.25 - x) = 0.25$.

$$\frac{x^2}{0.25} = 5.6 \times 10^{-10}$$

$$x^2 = 1.4 \times 10^{-10}$$

$$x = [OH^-] = (1.4 \times 10^{-10})^{\frac{1}{2}} = 1.2 \times 10^{-5}$$

$$pOH = -\log[OH^-] = -\log(1.2 \times 10^{-5}) = 4.92$$

$$pH = 14.00 - pOH = 14.00 - 4.92 = 9.08$$

Exercise 17.10

Calculate the pH of a 0.052 M NH_4Cl solution.

Example 17.11

What molarity NH_4NO_3 solution has a pH = 5.20?

Solution

NH_4NO_3 is the salt of a strong acid (HNO_3) and a weak base (NH_3). In $NH_4NO_3(aq)$, NH_4^+ hydrolyzes and NO_3^- does not.

$$NH_4^+ + H_2O \rightleftharpoons H_3O^+ + NH_3 \qquad K_a = \frac{K_w}{K_b} = \frac{1.0 \times 10^{-14}}{1.8 \times 10^{-5}} = 5.6 \times 10^{-10}$$

Let's assume that essentially all the H_3O^+ in solution is produced in the hydrolysis reaction.

$$\log [H_3O^+] = -pH = -5.20$$
$$[H_3O^+] = 10^{-5.20} = 6.3 \times 10^{-6}\ M$$

In the setup below, the initial concentration of NH_4^+ is the unknown x.

$$NH_4^+ + H_2O \rightleftharpoons H_3O^+ + NH_3$$

	NH_4^+	H_3O^+	NH_3
Initial concns, M:	x	—	—
Changes, M:	-6.3×10^{-6}	$+6.3 \times 10^{-6}$	$+6.3 \times 10^{-6}$
Equil. concns, M:	$(x - 6.3 \times 10^{-6})$	6.3×10^{-6}	6.3×10^{-6}

We can now substitute equilibrium concentrations into the ionization constant expression for the hydrolysis reaction.

$$K_a = \frac{[H_3O^+][NH_3]}{[NH_4^+]} = \frac{(6.3 \times 10^{-6})(6.3 \times 10^{-6})}{(x - 6.3 \times 10^{-6})} = 5.6 \times 10^{-10}$$

Because K_a for the hydrolysis of NH_4^+ and K_b for the hydrolysis of CH_3COO^- are equal, we can see why a solution of CH_3COONH_4(aq) should be pH neutral (Example 17.9b).

Let's make the usual simplifying assumption, in this case that $6.3 \times 10^{-6} << x$, so that $x - 6.3 \times 10^{-6} \approx x$, and then solve for x.

$$\frac{(6.3 \times 10^{-6})^2}{x} = 5.6 \times 10^{-10}$$

$$x = \frac{(6.3 \times 10^{-6})^2}{5.6 \times 10^{-10}} = 0.071$$

The required solution is 0.071 M NH_4NO_3.

Exercise 17.11

What molarity CH_3COONa(aq) solution has a pH = 9.10?

Estimation Exercise 17.3

Which of the following 0.10 M solutions should have the higher pH: NH_4NO_2(aq) or NH_4CN(aq)? (*Hint*: What K_a and/or K_b values should you compare?)

17.7 The Common Ion Effect

The two solutions pictured in Figure 17.8 both contain acetic acid at the same molarity, but they have different pH values. The solution having both sodium acetate and acetic acid as solutes has a pH that is greater by about *two* units. $[H_3O^+]$ in this solution is only about one-hundredth of that in the solution with acetic acid as the only solute. There is no mystery here, however. This is a classic example of Le Châtelier's principle.

Suppose we produce an acetic acid–sodium acetate solution by starting with an acetic acid solution and then adding sodium acetate. If we increase the concentration of one of the products of a reversible reaction, in this case CH_3COO^-, the *reverse* reaction is stimulated and equilibrium is shifted to the *left*.

When a salt supplies CH_3COO^-, equilibrium shifts to the *left*.

$$\underset{\text{acid(1)}}{CH_3COOH} + \underset{\text{base(2)}}{H_2O} \rightleftharpoons \underset{\text{acid(2)}}{H_3O^+} + \underset{\text{base(1)}}{CH_3COO^-}$$

As CH_3COO^-, a base, is consumed in the reverse reaction, so too is H_3O^+, an acid. $[H_3O^+]$ decreases and the pH increases accordingly. Because it is found both in aqueous acetic acid and in sodium acetate, acetate ion is a *common ion*.

1.00 M CH_3COOH 1.00 M CH_3COOH–1.00M CH_3COONa

Figure 17.8 **The common ion effect.** Both solutions contain bromphenol blue indicator. The yellow color indicates that the pH < 3.0 in 1.00 M CH_3COOH, and the blue-violet color that pH > 4.6 in the solution that is 1.00 M in both CH_3COOH and CH_3COONa.

Bromphenol blue indicator: pH < 3.0 (yellow) pH > 4.6 (blue-violet)

The **common ion effect** is the suppression of the ionization of a weak acid or a weak base by the presence of a common ion from a strong electrolyte. This effect is the key to much of what we do in the remainder of the chapter.

In Example 17.12 we calculate the pH of the solution we have just described. In Exercise 17.12 we see the effect of NH_4^+ as a common ion on the ionization of $NH_3(aq)$.

Example 17.12

Calculate the pH of a solution that is both 1.00 M CH_3COOH and 1.00 M CH_3COONa.

Solution

We can make this calculation simpler by considering that first 1.00 M $CH_3COONa(aq)$ is prepared.

$$CH_3COONa(s) \xrightarrow{H_2O} Na^+(aq) + CH_3COO^-(aq)$$

Then, acetic acid is added to the solution until $[CH_3COOH] = 1.00$ M. Thus, in the usual solution format below, the initial concentrations of both CH_3COOH and $CH_3COO^- = 1.00$ M.

$$CH_3COOH + H_2O \rightleftharpoons H_3O^+ + CH_3COO^-$$

	CH_3COOH	H_2O	H_3O^+	CH_3COO^-
Initial concns, M:	1.00		—	1.00
Changes, M:	$-x$		$+x$	$+x$
Equil. concns, M:	$(1.00 - x)$		x	$(1.00 + x)$

$$K_a = \frac{[H_3O^+][CH_3COO^-]}{[CH_3COOH]} = \frac{x(1.00 + x)}{1.00 - x} = 1.8 \times 10^{-5}$$

If x is very small, $(1.00 - x) \approx (1.00 + x) \approx 1.00$.

$$\frac{x(\cancel{1.00})}{(\cancel{1.00})} = 1.8 \times 10^{-5}$$

$$x = [H_3O^+] = 1.8 \times 10^{-5}$$

$$pH = -\log[H_3O^+] = -\log(1.8 \times 10^{-5}) = 4.74$$

Compare this value of $[H_3O^+]$ with the $[H_3O^+]$ found in Example 17.4, and you will see just how effective CH_3COO^- is in suppressing the ionization of CH_3COOH.

Note that the assumption that x is very small works well:

$$(1.00 - 1.8 \times 10^{-5}) = 1.00 \quad \text{and} \quad (1.00 + 1.8 \times 10^{-5}) = 1.00.$$

Exercise 17.12

Calculate the pH of a solution that is 0.15 M NH_3 and 0.35 M NH_4NO_3.

$$NH_3 + H_2O \rightleftharpoons NH_4^+ + OH^- \qquad K_b = 1.8 \times 10^{-5}$$

17.8 Buffer Solutions

The acetic acid–sodium acetate solution of Example 17.12 acts as a buffer solution (also called a buffered solution). A **buffer solution** is one that changes pH *only slightly* when small amounts of a strong acid or a strong base are added.

Buffer solutions have many important applications in industry, in the laboratory, and in living organisms. Some chemical reactions consume acids, others produce acids, and many are catalyzed by H_3O^+. If we want to study the kinetics of these reactions, or simply to control their reaction rates, we often need to control the pH. We can do this by conducting the reactions in buffered solutions, keeping pH changes to a minimum. Enzyme-catalyzed reactions are particularly sensitive to pH changes. Studies that involve proteins usually are performed in buffered media because the magnitude and kind of electric charge carried by the protein molecules depend on the pH.

Figure 17.9 suggests just how effective the 1.00 M CH_3COOH–1.00 M CH_3COONa buffer is in resisting changes in pH. At the same time, it shows that pure water totally lacks buffering ability.

As we'll show in the next section, the special ingredients necessary for a buffer solution are

a weak acid and its salt (conjugate base)

or

a weak base and its salt (conjugate acid).

One of the buffer components, the acid, is able to neutralize small added amounts of OH^-, and the other, the base, is able to neutralize small added amounts of H_3O^+.

How a Buffer Solution Works

Let's look more closely at the acetic acid–sodium acetate buffer solution of Example 17.12. We can represent the buffer solution through the equation

$$CH_3COOH + H_2O \rightleftharpoons H_3O^+ + CH_3COO^-$$

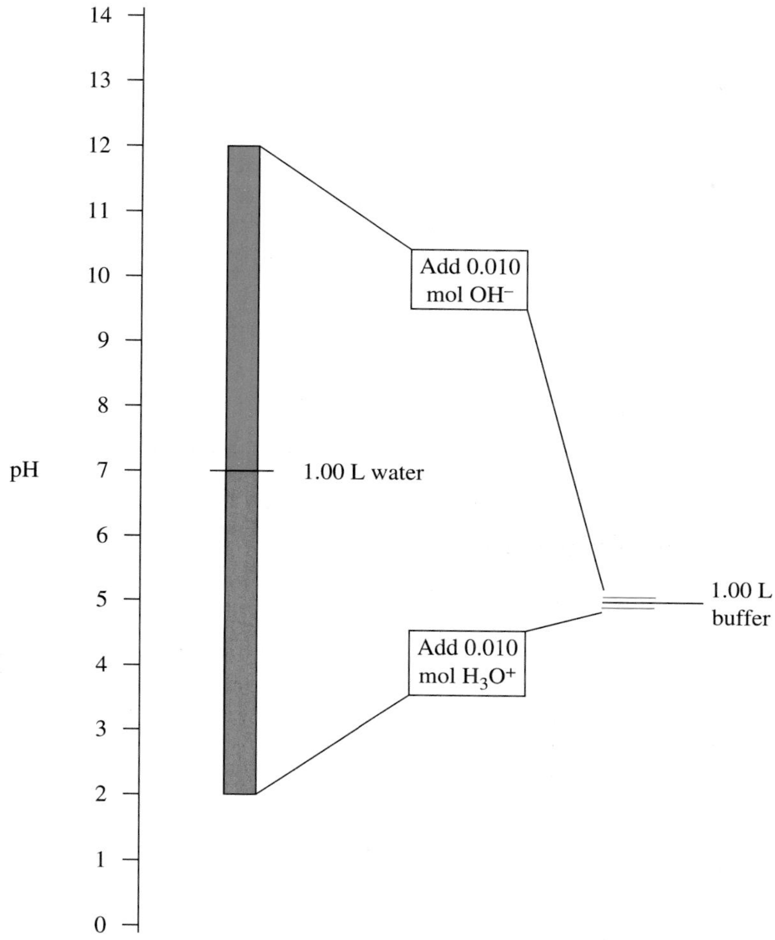

Figure 17.9 Representing buffer action. The addition of 0.010 mol of H_3O^+ or of OH^- to 1.00 L produces a huge change in pH of pure water and practically no change in the pH of a solution that is 1.00 M in both CH_3COOH and CH_3COONa. The acetic acid–sodium acetate solution is a buffer solution, and pure water has no buffering ability at all.

and equilibrium in the buffer solution by

$$K_a = \frac{[H_3O^+][CH_3COO^-]}{[CH_3COOH]} = 1.8 \times 10^{-5}$$

and $[H_3O^+]$ at equilibrium by

$$[H_3O^+] = K_a \times \frac{[CH_3COOH]}{[CH_3COO^-]} = 1.8 \times 10^{-5} \times \frac{[CH_3COOH]}{[CH_3COO^-]}$$

Thus, for the 1.00 M CH_3COOH–1.00 M CH_3COONa buffer solution

$$[H_3O^+] = K_a \times \frac{[CH_3COOH]}{[CH_3COO^-]} = 1.8 \times 10^{-5} \times \frac{1.00}{1.00} = 1.8 \times 10^{-5}$$

and

$$pH = -\log [H_3O^+] = -\log (1.8 \times 10^{-5}) = 4.74$$

Now, suppose we add enough of a strong base to the buffer solution to neutralize 2% of the acetic acid. This will reduce its concentration to 0.98 M and raise that of acetate ion to 1.02 M. A neutralization reaction in a buffer solution always converts some of one buffer component to the other.

	CH_3COOH +	OH^- $\rightleftharpoons$	H_2O + CH_3COO^-
Initial buffer:	1.00 M		1.00 M
Add:		0.02 M	
Changes:	−0.02 M	−0.02 M	+0.02 M
After neutralization:	0.98 M	≈0 M	1.02 M

After the added base is consumed,*

$$[H_3O^+] = K_a \times \frac{[CH_3COOH]}{[CH_3COO^-]} = 1.8 \times 10^{-5} \times \frac{0.98}{1.02} = 1.7 \times 10^{-5}$$

and

$$pH = -\log [H_3O^+] = -\log (1.7 \times 10^{-5}) = 4.77$$

The $[H_3O^+]$ and pH of the buffer solution have hardly been affected.

The situation for an added acid is quite similar. For the neutralization of 2% of the conjugate base (acetate ion) by added H_3O^+,

	CH_3COO^- +	H_3O^+ $\rightleftharpoons$	H_2O + CH_3COOH
Initial buffer:	1.00 M		1.00 M
Add:		0.02 M	
Changes:	−0.02 M	−0.02 M	+0.02 M
After neutralization:	0.98 M	≈0 M	1.02 M

$$[H_3O^+] = K_a \times \frac{[CH_3COOH]}{[CH_3COO^-]} = 1.8 \times 10^{-5} \times \frac{1.02}{0.98} = 1.9 \times 10^{-5}$$

$$pH = -\log [H_3O^+] = -\log (1.9 \times 10^{-5}) = 4.72$$

Again, the $[H_3O^+]$ and pH are hardly affected.

Because we often want to calculate the pH of a buffer solution, let's write an equation with which we can do so directly. First, let's take the *negative* logarithm of each side of the equation written above for $[H_3O^+]$.

$$-\log [H_3O^+] = -\log K_a - \log \frac{[CH_3COOH]}{[CH_3COO^-]}$$

Now let's substitute pH for $-\log [H_3O^+]$ and pK_a for $-\log K_a$.

$$pH = pK_a - (\log [CH_3COOH] - \log [CH_3COO^-])$$

Then we can rearrange the remaining log terms

$$pH = pK_a + (\log [CH_3COO^-] - \log [CH_3COOH])$$

* In a more detailed calculation, we would determine $[H_3O^+]$ by the method of Example 17.12. That is, the equilibrium concentrations would be $[H_3O^+] = x$, $[CH_3COOH] = 0.98 - x$, and $[CH_3COO^-] = 1.02 + x$. However, because x is so small, $0.98 - x = 0.98$ and $1.02 + x = 1.02$. The result is essentially the same either way.

and replace them by the logarithm of a ratio.

$$pH = pK_a + \log\frac{[CH_3COO^-]}{[CH_3COOH]}$$

This equation for the pH of an acetic acid–acetate ion buffer is a special case of a more general equation, known as the **Henderson–Hasselbalch** equation.

$$pH = pK_a + \log\frac{[\text{conjugate base}]}{[\text{weak acid}]}$$

In Example 17.13 we examine the action of a buffer solution in the alkaline pH region.

Example 17.13

A buffer solution is prepared that is 0.24 M NH_3 and 0.20 M NH_4Cl. (**a**) What is the pH of this buffer? (**b**) If 0.0050 mol NaOH is added to 0.500 L of this buffer, what will be the pH?

Solution

a. This calculation is quite similar to what we have just shown for an acetic acid–acetate ion buffer, except that the equilibrium involves a weak base and its salt. The relevant equations are

$$NH_3 + H_2O \rightleftharpoons NH_4^+ + OH^-$$

$$K_b = \frac{[NH_4^+][OH^-]}{[NH_3]} = 1.8 \times 10^{-5}$$

$$[OH^-] = K_b \times \frac{[NH_3]}{[NH_4^+]}$$

$$= 1.8 \times 10^{-5} \times \frac{0.24}{0.20} = 2.2 \times 10^{-5}$$

$$pOH = -\log[OH^-] = -\log(2.2 \times 10^{-5}) = 4.66$$

$$pH = 14.00 - pOH = 14.00 - 4.66 = 9.34$$

b. First, let's calculate the result of neutralizing the added NaOH. Adding 0.005 mol OH^- to 0.500 L of buffer produces an immediate $[OH^-]$ = 0.005 mol OH^-/0.500 L = 0.010 M. This $[OH^-]$ is reduced to near zero by the neutralization.

	NH_4^+	+ OH^-	$\rightleftharpoons$ H_2O +	NH_3
Initial buffer:	0.20 M			0.24 M
Add:		0.01 M		
Changes:	−0.01 M	−0.01 M		+0.01 M
After neutralization:	0.19 M	≈0 M		0.25 M

Now, let's apply the expression

$$[OH^-] = K_b \times \frac{[NH_4^+]}{[NH_3]} = 1.8 \times 10^{-5} \times \frac{0.25}{0.19} = 2.4 \times 10^{-5}\text{ M}$$

$$pOH = -\log[OH^-] = -\log(2.4 \times 10^{-5}) = 4.62$$

$$pH = 14.00 - pOH = 14.00 - 4.62 = 9.38$$

As an alternative, we can apply the Henderson–Hasselbalch equation:

$$pH = pK_a + \log\frac{[\text{conjugate base}]}{[\text{weak acid}]}$$

In this case, the weak acid is NH_4^+ and the conjugate base is NH_3. Recall that $pK_a + pK_b = 14.00$, and we have

$$pK_a = 14.00 - pK_b = 14.00 - [-\log(1.8 \times 10^{-5})]$$
$$= 14.00 - 4.74 = 9.26$$

Then,

$$pH = 9.26 + \log\frac{0.25}{0.19}$$
$$= 9.26 + 0.12 = 9.38$$

Exercise 17.13

What is the final pH if 0.03 mol HCl is added to 0.500 L of a 0.24 M NH_3–0.20 M NH_4Cl buffer solution? (*Hint*: This is the same buffer as in Example 17.13, but in the neutralization NH_3 is converted to NH_4^+.)

A common need in the laboratory is to prepare a buffer solution of a particular pH. As shown in Example 17.14, this calculation is easily performed with the Henderson–Hasselbalch equation.

Example 17.14

What must be the concentration of acetate ion in 0.500 M CH_3COOH to produce a buffer solution with pH = 5.00?

Solution

We have written the Henderson–Hasselbalch equation for an acetic acid–acetate ion buffer solution before. It is

$$pH = pK_a + \log\frac{[CH_3COO^-]}{[CH_3COOH]}$$

We can replace pH by 5.00, pK_a by $-\log K_a$, and $[CH_3COOH]$ by 0.500. Then we can solve for $[CH_3COO^-]$.

$$5.00 = -\log(1.8 \times 10^{-5}) + \log\frac{[CH_3COO^-]}{0.500}$$

$$5.00 = 4.74 + \log\frac{[CH_3COO^-]}{0.500}$$

$$\log\frac{[CH_3COO^-]}{0.500} = 5.00 - 4.74 = 0.26$$

$$\frac{[CH_3COO^-]}{0.500} = 10^{0.26} = 1.82$$

$$[CH_3COO^-] = 0.500 \times 1.82 = 0.910\ M$$

Exercise 17.14

What mass of NH_4Cl must be present in 0.250 L of 0.150 M NH_3 to produce a buffer solution with pH = 9.05? (*Hint*: What must $[NH_4^+]$ be?)

Buffers in Blood

Buffers are of utmost importance in our blood and other body fluids. Maintenance of proper pH is essential to the processes that occur in living organisms, primarily because enzyme function (Section 15.12) is sharply dependent on pH. The normal pH value of blood plasma is 7.4. Sustained variations of a few tenths of a pH unit can cause severe illness or death.

Acidosis, a condition in which there is a decrease in the pH of blood, can be brought on by heart failure, kidney failure, diabetes mellitus, persistent diarrhea, a long-term high-protein diet, or other factors. Prolonged, intense exercise can cause temporary acidosis.

Alkalosis, characterized by an increase in the pH of blood, may occur as a result of severe vomiting, hyperventilation (over breathing, sometimes caused by anxiety or hysteria), or exposure to high altitudes (altitude sickness). Arterial blood samples taken from climbers who reached the summit of Mount Everest (8848 m = 29,028 ft) without supplemental oxygen had pH values between 7.7 and 7.8. This alkalosis was caused by hyperventilation; to compensate for the very low partial pressures of O_2 (about 43 mmHg) at this altitude, the climber must breathe extremely rapidly.

A good measure of the buffering ability of human blood is that the addition of 0.01 mol HCl to one liter of blood lowers the pH only from 7.4 to 7.2, whereas the same amount of HCl added to a saline (NaCl) solution isotonic with blood lowers the pH from 7.0 to 2.0. The saline solution has no buffer capacity.

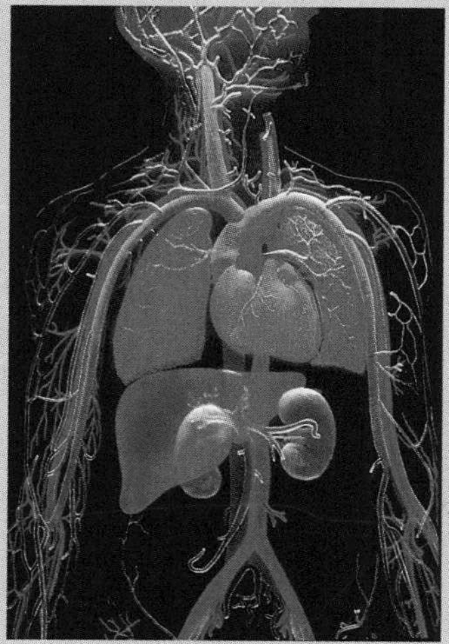

Blood carried by the human circulatory system is maintained at a pH of 7.4 in a highly buffered system.

Several buffers are involved in the control of the pH of blood. Perhaps the most important one is the HCO_3^- (bicarbonate ion) and H_2CO_3 (carbonic acid) system. [In this system we treat $CO_2(g)$ as if it were completely converted to H_2CO_3, and we consider only the first ionization of the diprotic acid H_2CO_3.]

(1) $$H_2CO_3 + H_2O \rightleftharpoons H_3O^+ + HCO_3^-$$

Carbon dioxide enters the blood from tissues as the byproduct of metabolic reactions. In the lungs $CO_2(g)$ is exchanged for $O_2(g)$, which is transported throughout the body by the blood.

The blood buffers must be able to neutralize excess acid, such as the lactic acid produced by exercise. A relatively high concentration of HCO_3^- helps in this regard; it reacts with excess acid to reverse the ionization reaction shown in equation (1).

Excess alkalinity is much less common than excess acidity. Should it occur, however, additional H_2CO_3 can be formed in the lungs by reabsorbing $CO_2(g)$ to build up the H_2CO_3 content of the blood.

$$CO_2(g) + H_2O \rightleftharpoons H_2CO_3$$

The H_2CO_3 then ionizes as needed to neutralize the excess alkali.

Other blood buffers include the dihydrogen phosphate ($H_2PO_4^-$)-monohydrogen phosphate (HPO_4^{2-}) system.

$$H_2PO_4^- + H_2O \rightleftharpoons H_3O^+ + HPO_4^{2-} \qquad (2)$$

The HPO_4^{2-} reacts with excess acid in the reverse of the reaction shown in equation (2). When excess alkali enters the blood, it is neutralized in part by reaction with $H_2PO_4^-$.

Some plasma proteins also act as buffers. The $—COO^-$ groups on the protein molecules can react with excess acid

$$—COO^- + H_3O^+ \rightleftharpoons —COOH + H_2O$$

and $—NH_3^+$ groups on the protein molecules can neutralize excess base.

$$—NH_3^+ + OH^- \rightleftharpoons —NH_2 + H_2O$$

The blood buffers do a remarkable job of maintaining the pH of blood at 7.4. They can be overwhelmed, however, if the body's metabolism goes badly amiss.

Buffer Capacity

If the buffer solution in Example 17.13 were required to eliminate 0.10 mol OH^-/L instead of 0.01 mol OH^-/L, the pH would rise from 9.34 to 9.79 rather than to 9.38. The resulting buffer solution would not be nearly as effective in resisting a change in pH. There is a limit to the capacity of a solution to function as a buffer, and obviously this limit comes before all of one of the buffer components is consumed. In the case of the 0.24 M NH_3–0.20 M NH_4Cl buffer, these outer limits are 0.20 mol OH^-/L (which would neutralize all the NH_4^+) and 0.24 mol H_3O^+/L (which would neutralize all the NH_3).

In general, the more concentrated a solution is in its buffer components, the more added acid or base it is capable of neutralizing. And, as a rule, a buffer is most effective against both added H_3O^+ and added OH^- if the buffer acid and its conjugate base are present in equal concentration. In an acetic acid–sodium acetate buffer, this corresponds to a pH = pK_a. That is,

$$\begin{aligned} pH &= pK_a + \log\frac{[CH_3COO^-]}{[CH_3COOH]} \\ &= pK_a + \log 1 \\ &= pK_a \end{aligned}$$

In general, the pH range over which a buffer solution is effective is about one pH unit on either side of pH = pK_a. In Example 17.14, we saw that pK_a = 4.74 for acetic acid, so the effective pH range for an acetic acid–acetate ion buffer is $3.74 < pH < 5.74$.

A similar situation holds for a weak base and its conjugate acid, for example, NH_3/NH_4^+, where pK_a is for the conjugate acid. In Example 17.13 we found that pK_a for NH_4^+ is 9.26. The effective pH range for an ammonia–ammonium ion buffer is $8.26 < pH < 10.26$.

17.9 pH Indicators

Figure 17.10 suggests an experiment with pH indicators (also called acid–base indicators) that you can do yourself using phenol red, the indicator commonly used to test the pH of home swimming pools. Add a few drops of phenol red to a small quantity of distilled water. The color of the water should appear *orange*. Now,

Figure 17.10 Phenol red—a pH indicator. Phenol red, when in an acidic solution, has a yellow color (left). In pure water, it is orange (center). In an alkaline solution, the indicator is red (right).

Phenol red indicator:	$pH < 6.6$	$pH \approx 7$	$pH > 8.2$
	yellow	orange	red

squeeze some lemon juice into the water and the color changes to *yellow*. Add some household ammonia to this yellow solution and the color changes to *red*.

To explain these color changes, let's call the yellow form of phenol red the indicator acid, HIn, and the red form, the indicator base, In^-. An equilibrium exists between the two, with the equilibrium concentrations of HIn and In^- depending on $[H_3O^+]$ and thus on pH.

$$\underset{\text{yellow}}{HIn} + H_2O \rightleftharpoons H_3O^+ + \underset{\text{red}}{In^-}$$

In an acidic solution, such as lemon juice, $[H_3O^+]$ is high. Because H_3O^+ is a common ion, it suppresses the ionization of the indicator acid. In accordance with Le Châtelier's principle, equilibrium shifts to the left, favoring the yellow form of the indicator. In a basic solution, such as $NH_3(aq)$, H_3O^+ from the indicator acid is neutralized. Equilibrium shifts to the right, favoring the red indicator anion. In a more nearly neutral solution the indicator acid and anion are present in about equal concentrations, and the observed color is orange—a mixture of red and yellow.

Phenol red indicator:	$pH < 6.6$	$6.6 < pH < 8.2$	$pH > 8.2$
	yellow	orange	red

One of the most commonly used pH indicators in the introductory chemistry laboratory is litmus, a material extracted from lichens. It is generally used in the form of paper strips that have been impregnated with a water solution of litmus and allowed to dry. The paper is moistened with the solution being tested. The color change of litmus occurs over a much broader pH range than of other pH indicators, so it can only be used to give a general indication of whether a solution is acidic or basic (see Figure 17.11).

Litmis indicator:	$pH < 4.5$	$pH > 8.3$
	red	blue

Figure 17.12 shows the pH ranges and colors for several other common pH indicators.

Acid–base indicators are most widely used in applications where a precise pH reading isn't necessary. One of the most familiar of these applications is in acid–base titrations, as we describe in the next section.

Figure 17.11 Litmus: A general purpose indicator. We often use litmus paper as a rough measure to determine whether a material is acidic or basic. The soil sample on the left turns red litmus blue and is basic. The soil sample on the right turns blue litmus red and is acidic.

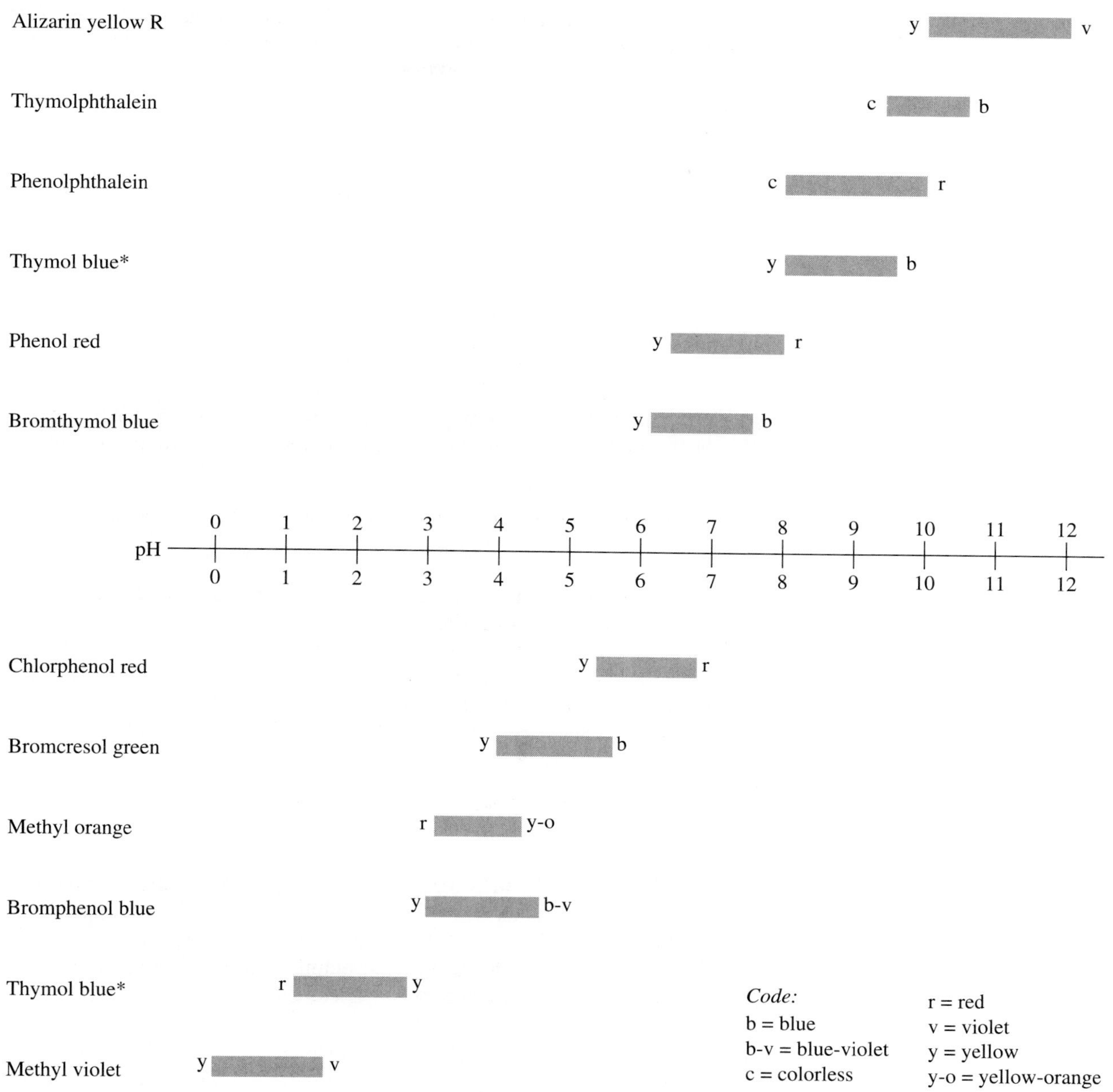

*Thymol blue is a diprotic acid, H_2In. In its ionization,

$$\underset{\text{red}}{H_2In} + H_2O \rightleftharpoons H_3O^+ + \underset{\text{yellow}}{HIn^-}$$

$$\underset{\text{yellow}}{HIn^-} + H_2O \rightleftharpoons H_3O^+ + \underset{\text{blue}}{In^{2-}}$$

Figure 17.12 pH ranges and colors of several common indicators.

CONCEPTUAL EXAMPLE 17.2

Explain the following series of color changes of *thymol blue* indicator and the reactions producing them (see Figure 17.13).

a. A few drops of thymol blue are added to HCl(aq). Solution color: red.

b. A quantity of sodium acetate is added to solution (a). Solution color: yellow.

c. A small quantity of sodium hydroxide is added to solution (b). Solution color: yellow.

d. An additional, larger quantity of sodium hydroxide is added to solution (c). Solution color: blue.

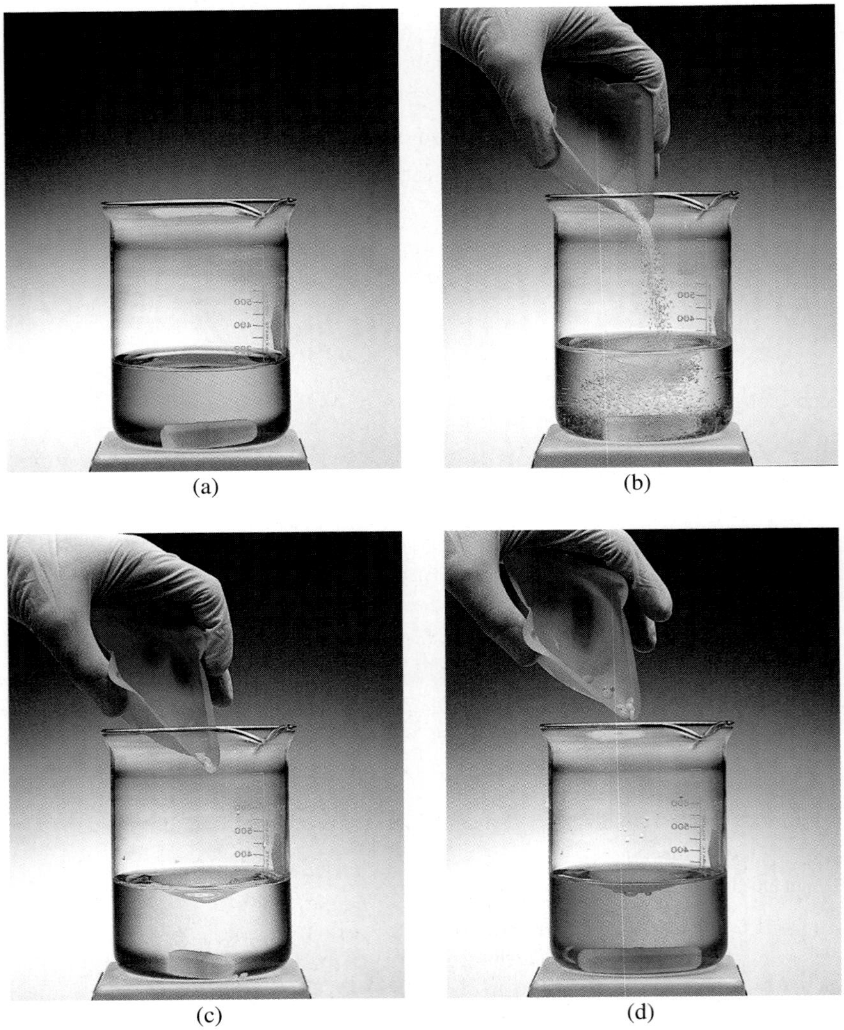

Figure 17.13 Conceptual Example 17.2 illustrated.

Solution

a. The red color of the thymol blue shows the solution to be rather strongly acidic: pH < 1.2 (see Figure 17.12). The HCl is completely ionized.

$$HCl + H_2O \longrightarrow H_3O^+ + Cl^-$$

b. The yellow color shows that the pH has risen to a value greater than 3.0. This corresponds to $[H_3O^+] < 1 \times 10^{-3}$. A reaction between excess acetate ion and H_3O^+ (limiting reactant) goes nearly to completion.

$$\underset{\text{(from HCl)}}{H_3O^+} + \underset{\text{(from }CH_3COONa)}{CH_3COO^-} \longrightarrow CH_3COOH + H_2O$$

The resulting solution is one of a weak acid, CH_3COOH, and its conjugate base, CH_3COO^-. It is a buffer solution with a pH somewhere around $pH = pK_a = -\log K_a = -\log (1.8 \times 10^{-5}) = 4.74$.

c. The small quantity of added NaOH is neutralized by the weak acid.

$$CH_3COOH + OH^- \longrightarrow H_2O + CH_3COO^-$$

This produces only a small change in the ratio $[CH_3COO^-]/[CH_3COOH]$ and correspondingly small changes in $[H_3O^+]$ and pH. The solution color remains yellow.

d. The quantity of NaOH added is sufficient to neutralize all the CH_3COOH in the buffer. The buffer capacity has been exceeded, and the buffer action is destroyed. Unreacted OH^- raises the pH to a value above 9.6, at which point thymol blue indicator has a blue color.

$$CH_3COOH + \underset{\text{(excess)}}{OH^-} \longrightarrow H_2O + CH_3COO^-$$

Conceptual Exercise 17.2

A solution is known to be one of the following four: (a) 1.00 M NH_4Cl; (b) 1.00 M NH_4Cl–1.00 M NH_3; (c) 1.00 M HCl–1.00 M HNO_3; (d) 1.00 M CH_3COOH–1.00 M CH_3COONa. A few drops of bromcresol green indicator are added to the solution, producing a green color (see Figure 17.12). Which solution(s) does this observation eliminate? What additional simple test involving a common laboratory acid and/or base could you perform to determine which of the four it is?

17.10 Neutralization Reactions and Titration Curves

In earlier discussions of acid–base reactions (Sections 3.11 and 12.2), we learned that *neutralization* is the reaction of an acid and a base, and that *titration* is a commonly used technique for conducting a neutralization. The critical point in a titration is the **equivalence point**, the point at which the acid and base have been brought together in stoichiometric proportions, with neither remaining in excess.

We can use the color change of an acid–base indicator to locate the equivalence point. The point in the titration at which the indicator changes color is called the **end point**, and the trick is to obtain a match between the indicator end point and the equivalence point of the neutralization. Specifically, we need an indicator whose color change occurs over a pH range that includes the pH at the equivalence point.

All of this suggests the value of having a **titration curve** for a neutralization reaction—a graph of pH versus volume of titrant (the solution added from a buret). We can measure the pH with a pH meter, and by connecting the pH meter to a recorder, we can automatically plot the titration curve. In this section we'll calculate the expected pH at some characteristic points on two types of titration curves. These calculations will also serve as a review of some of the acid–base equilibria presented earlier in this chapter.

In a typical titration the volume of titrant is less than 50 mL, and its molarity is generally less than 1 M. The typical amount of H_3O^+ or OH^- delivered from the buret is only a few thousandths of a mole—for example, 2.00×10^{-3} mol. We can avoid having to carry a lot of exponential terms in calculations if we work with the unit millimole (mmol) rather than mole (mol). We can best do this by defining molarity in this way.

$$\mathrm{M} = \frac{\mathrm{mol}}{\mathrm{L}} = \frac{\mathrm{mol}/1000}{\mathrm{L}/1000} = \frac{\mathrm{mmol}}{\mathrm{mL}}$$

Titration of a Strong Acid by a Strong Base

Imagine that we place 20.00 mL 0.500 M HCl in a small flask and then titrate this *strong* acid by adding to it 0.500 M NaOH (a *strong* base) from a buret. To establish data for a titration curve, we calculate the pH of the accumulated solution at different points in the titration. Then, we plot these pH values versus the volume of NaOH(aq) added. From the titration curve we can establish the pH at the equivalence point and identify appropriate indicators for the titration.

In Example 17.15 we calculate four representative points on the titration curve. In Exercise 17.15 we focus on the region near the equivalence point, illustrating the very rapid rise in pH with volume of titrant added.

Example 17.15

Calculate the pH at the following points in the titration of 20.00 mL of 0.500 M HCl by 0.500 M NaOH.

$$H_3O^+ + \cancel{Cl^-} + \cancel{Na^+} + OH^- \longrightarrow \cancel{Na^+} + \cancel{Cl^-} + 2\,H_2O$$

a. Before the addition of any NaOH (*the initial pH*).
b. After the addition of 10.00 mL of 0.500 M NaOH (*the half-neutralization point*). Half the original HCl is neutralized; half remains.
c. After the addition of 20.00 mL of 0.500 M NaOH (*the equivalence point*). Neither acid nor base is in excess.
d. After the addition of 21.00 mL of 0.500 M NaOH (*beyond the equivalence point*). Excess titrant is present.

Solution

a. Because HCl is a strong acid, it ionizes completely. Therefore, the initial solution has $[H_3O^+] = 0.500$ M, and

$$pH = -\log [H_3O^+] = -\log (0.500) = 0.301$$

b. The total amount of H_3O^+ to be titrated is

$$20.00 \text{ mL} \times 0.500 \text{ mmol } H_3O^+/\text{mL} = 10.0 \text{ mmol } H_3O^+$$

The amount of OH^- in 10.00 mL of 0.500 M NaOH is

$$10.00 \text{ mL} \times 0.500 \text{ mol } OH^-/\text{mL} = 5.00 \text{ mmol } OH^-$$

We can represent the progress of the neutralization reaction as follows.

	H_3O^+ +	OH^- $\longrightarrow$ $2H_2O$
Initial amounts, mmol:	10.0	
Add, mmol:		5.00
Changes, mmol:	−5.00	−5.00
After reaction, mmol:	5.0	≈0

The total volume is 20.00 mL + 10.00 mL = 30.00 mL, and

$$[H_3O^+] = \frac{5.0 \text{ mmol } H_3O^+}{30.00 \text{ mL}} = 0.17 \text{ M}$$

$$pH = -\log[H_3O^+] = -\log 0.17 = 0.77$$

c. The solution is simply NaCl(aq). Because neither Na^+ nor Cl^- hydrolyzes, the solution pH = 7.00.

d. The amount of OH^- in 21.00 mL of 0.500 M NaOH is

$$21.00 \text{ mL} \times 0.500 \text{ mmol } OH^-/\text{mL} = 10.5 \text{ mmol } OH^-$$

Again, we can represent the progress of the neutralization as follows.

	H_3O^+ +	OH^- $\longrightarrow$ $2H_2O$
Initial amounts, mmol:	10.0	
Add, mmol:		10.5
Changes, mmol:	−10.0	−10.0
After reaction, mmol:	≈0	0.5

The concentration of OH^-, the pOH, and the pH are

$$[OH^-] = \frac{0.5 \text{ mmol } OH^-}{(20.00 + 21.00) \text{ mL}} = 0.01 \text{ M}$$

$$pOH = -\log [OH^-] = -\log 0.01 = 2.0$$

$$pH = 14.00 - pOH = 14.00 - 2.0 = 12.0$$

Exercise 17.15

For the titration described in Example 17.15, determine the pH after the addition of the following volumes of 0.500 M NaOH: **(a)** 19.90 mL; **(b)** 19.99 mL; **(c)** 20.01 mL; **(d)** 20.10 mL.

Figure 17.14 illustrates these features of the titration curve for the titration of a *strong acid* by a *strong base*.

- The pH is *low* at the beginning of the titration.
- The pH changes slowly until just before the equivalence point.
- Just before the equivalence point, the pH rises sharply.
- At the equivalence point, the pH is 7.00.
- Just past the equivalence point, the pH continues its sharp rise.
- Slightly further beyond the equivalence point, the pH is high.
- Any indicator whose color changes in the pH range from about 4 to 10 can be used for this titration.

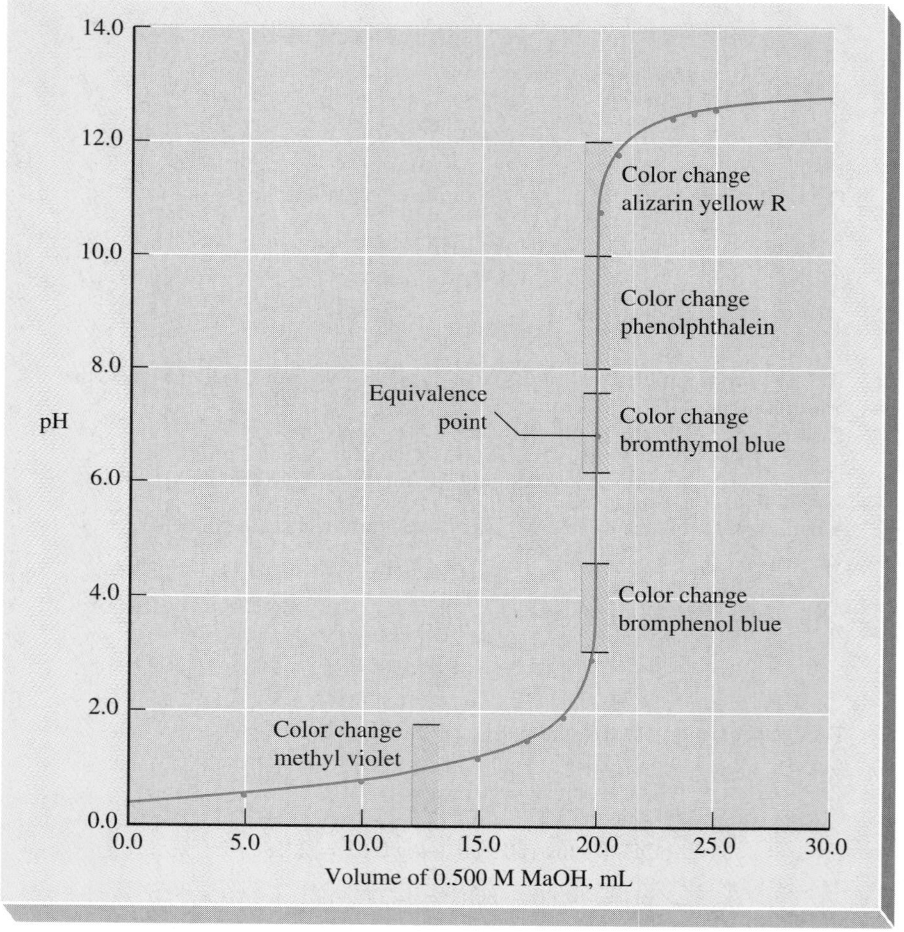

Figure 17.14 Titration curve for a strong acid by a strong base: 20.00 mL of 0.500 M HCl by 0.500 M NaOH. Any of the indicators that change color along the steep portion of the titration curve is suitable for the titration. Methyl violet changes color too soon and alizarin yellow R too late.

Titration of a Weak Acid by a Strong Base

If we use the same strong base as the titrant in the titration of two different solutions of equal molarity—one a strong acid and the other, a weak acid—

the titration curves we get have two features in common: (1) The volume of base required to reach the equivalence point is the same in both cases. (2) The portions of the curves after the equivalence points are substantially the same. (In either case the solution is simply one having excess OH^-.) However, there are several important differences between the two cases, as suggested by Figure 17.15.

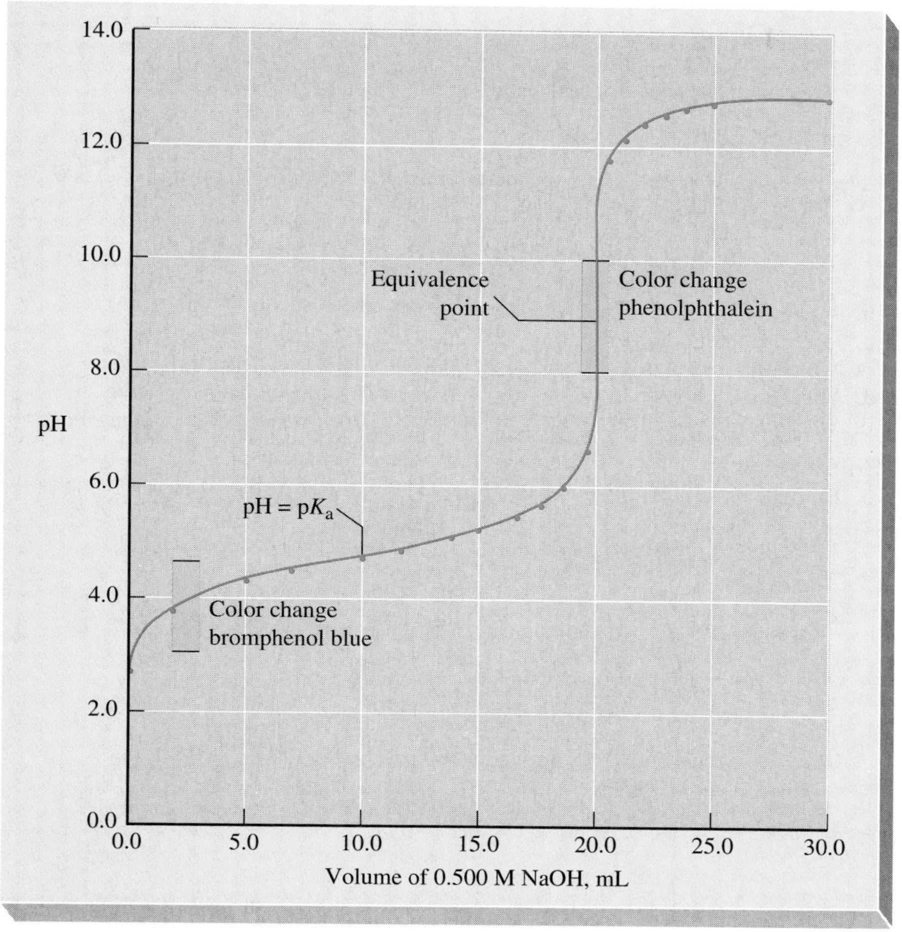

Figure 17.15 Titration curve for a weak acid by a strong base: 20.00 mL of 0.500 M CH_3COOH by 0.500 M NaOH. Phenolphthalein can be used as an indicator for this titration; bromphenol blue cannot. When exactly one-half the acid is neutralized, $[CH_3COOH] = [CH_3COO^-]$ and $pH = pK_a = 4.74$.

In contrast to a strong acid–strong base titration, in the titration of a weak acid by a strong base,

- The initial pH is higher. The weak acid is only partially ionized.
- At the half-neutralization point, $pH = pK_a$. The solution at this point is a buffer solution in which the concentrations of the weak acid and its conjugate base (the anion) are equal.
- The $pH > 7$ at the equivalence point because the anion of the weak acid hydrolyzes.

- The steep portion of the titration curve just prior to and just beyond the equivalence point is confined to a smaller pH range.
- The choice of indicators for the titration is more limited. Specifically, we can't use an indicator that changes color below pH = 7.

Some of these points should be made clearer by the calculations in Example 17.16 and Exercise 17.16.

Example 17.16

Calculate the pH at the following points in the titration of 20.00 mL of 0.500 M CH_3COOH by 0.500 M NaOH.

$$CH_3COOH + Na^+ + OH^- \longrightarrow Na^+ + CH_3COO^- + H_2O$$

a. Before the addition of any NaOH (*the initial pH*).
b. After the addition of 8.00 mL of 0.500 M NaOH (*the buffer region*).
c. After the addition of 10.00 mL of 0.500 M NaOH (*the half-neutralization point*).
d. After the addition of 20.00 mL of 0.100 M NaOH (*the equivalence point*).
e. After the addition of 21.00 mL of 0.100 M NaOH (*beyond the equivalence point*).

Solution

a. This calculation is similar to that of Example 17.4, except that the acid is 0.500 M.

	CH_3COOH	+ H_2O $\rightleftharpoons$	H_3O^+	+ CH_3COO^-
Initial concn., M:	0.500		—	—
Changes, M:	$-x$		$+x$	$+x$
Equil. concn., M:	$(0.500 - x)$		x	x

$$K_a = \frac{[H_3O^+][CH_3COO^-]}{[CH_3COOH]} = \frac{x \cdot x}{0.500 - x} = 1.8 \times 10^{-5}$$

If we make the usual assumption, that is, that $x << 0.500$,

$$K_a = \frac{x^2}{0.500} = 1.8 \times 10^{-5}$$

$$x^2 = 9.0 \times 10^{-6}$$

$$x = [H_3O^+] = 3.0 \times 10^{-3}$$

$$pH = -\log [H_3O^+] = -\log(3.0 \times 10^{-3}) = 2.52$$

b. The addition of 8.00 mL of 0.500 M NaOH represents the addition of 8.00 mL × 0.500 mmol OH^-/mL = 4.00 mmol OH^-. At this point in the titration,

	CH_3COOH	+ OH^- $\rightleftharpoons$ H_2O	+ CH_3COO^-
Initial amounts, mmol:	10.00		—
Add, mmol:		4.00	
Changes, mmol:	−4.00	−4.00	+4.00
After reaction, mmol:	6.00	≈0	4.00

The simplest approach is to use the Henderson–Hasselbalch equation, with $pK_a = -\log(1.8 \times 10^{-5})$ and the concentrations: $[CH_3COO^-] = 4.00$ mmol/28.00 mL and $[CH_3COOH] = 6.00$ mmol/28.00 mL.

$$pH = pK_a + \log\frac{[CH_3COO^-]}{[CH_3COOH]}$$

$$= 4.74 + \log\frac{(4.00/28.00)}{(6.00/28.00)}$$

$$= 4.74 + \log 0.667 = 4.74 - 0.18 = 4.56$$

c. At the half-neutralization point, the titration has progressed to this point:

$$CH_3COOH + OH^- \rightleftharpoons H_2O + CH_3COO^-$$

	CH_3COOH	OH^-	H_2O	CH_3COO^-
Initial amounts, mmol:	10.00			—
Add, mmol:		5.00		
Changes, mmol:	−5.00	−5.00		+5.00
After reaction, mmol:	5.00	≈0		5.00

Because the CH_3COOH and CH_3COO^- are present in equal amounts and in the same 30.00 mL of solution, their concentrations are equal. This means that

$$pH = pK_a + \log [CH_3COO^-]/[CH_3COOH] = pK_a + \log 1 = pK_a = 4.74$$

d. At the equivalence point, 10.0 mmol each of CH_3COOH and NaOH have reacted to produce 10.0 mmol of CH_3COONa in 40.00 mL of solution (20.00 mL acid + 20.00 mL base). The solution molarity is

$$\frac{10.0 \text{ mmol } CH_3COONa}{40.00 \text{ mL}} = 0.250 \text{ M } CH_3COONa$$

We calculated the pH of this solution in Example 17.10 in our discussion of hydrolysis and found pH = 9.08.

e. The addition of 21.00 mL of 0.500 M NaOH represents the addition of 21.00 mL × 0.500 mmol OH^-/mL = 10.50 mmol OH^-. This point is beyond the equivalence point, and

$$CH_3COOH + OH^- \rightleftharpoons H_2O + CH_3COO^-$$

	CH_3COOH	OH^-	H_2O	CH_3COO^-
Initial amounts, mmol:	10.00			—
Add, mmol:		10.50		
Changes, mmol:	−10.00	−10.00		+10.00
After reaction, mmol:	≈0	0.50		10.00

Acetate ion is a weak base compared to OH^-. The hydroxide concentration is simply

$$[OH^-] = \frac{0.50 \text{ mmol}}{41.00 \text{ mL}} = 0.012 \text{ M}$$

$$pOH = -\log [OH^-] = -\log(1.2 \times 10^{-2}) = 1.92$$

$$pH = 14.00 - pOH = 14.00 - 1.92 = 12.08$$

Exercise 17.16

For the titration described in Example 17.16, determine the pH after the addition of the following volumes of 0.500 M NaOH: **(a)** 12.50 mL; **(b)** 20.10 mL.

CONCEPTUAL EXAMPLE 17.3

The titration curve in Figure 17.16 involves 1.0 M solutions of an acid and a base. Identify the type of titration represented by this curve.

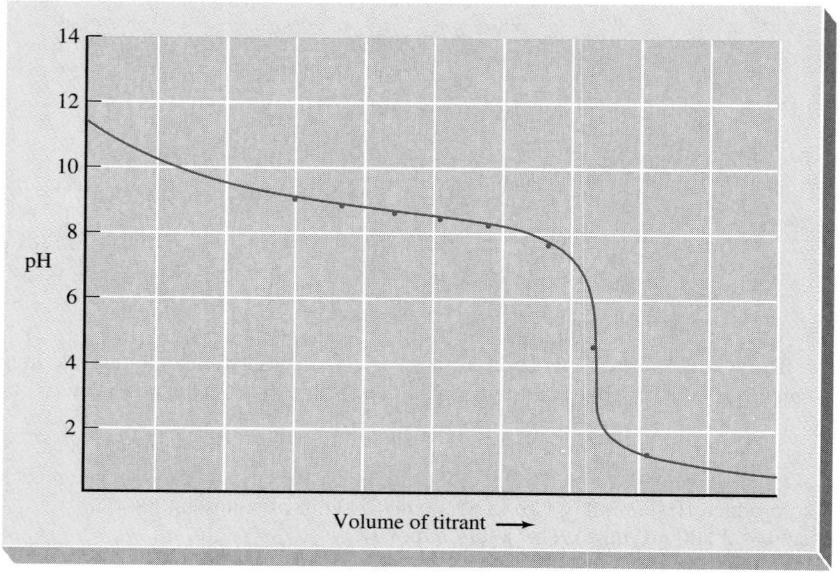

Figure 17.16 Conceptual Example 17.3 illustrated.

Solution

We can identify these features of the titration curve and draw certain conclusions from them.

- The pH starts high and decreases during the titration. The solution being titrated is a *base*; the titrant is an *acid*.
- The base must be a *weak base*. The initial pH is about 11.5, but a 1.0 M strong base would have $[OH^-] = 1.0$ M, $[H_3O^+] = 1.0 \times 10^{-14}$, and pH = 14.00.
- The pH at the equivalence point is *less than 7*. This is the pH expected for the hydrolysis of the salt of a *strong acid* and a *weak base*.
- The pH beyond the equivalence point drops to a low value (pH < 1), again suggesting that the titrant is a *strong acid*.

From these observations we conclude that the curve represents the titration of a weak base by a strong acid.

Conceptual Exercise 17.3

For the titration described in Conceptual Example 17.3, (**a**) estimate the value of K_b of the weak base, and (**b**) obtain a value of the pH at the equivalence point by calculation and by estimation from the titration curve.

SUMMARY

In Brønsted–Lowry theory an acid is a proton donor and a base is a proton acceptor. Every acid has a conjugate base, and every base has a conjugate acid. In an acid–base reaction, the forward reaction is between an acid and a base; the reverse reaction is between a conjugate base and a conjugate acid.

Brønsted–Lowry acids and bases can be ranked by relative strength, and the favored direction in an acid–base reaction is from stronger acid and base to weaker acid and base. Acid ionization requires bond breakage and release of a proton (H^+). The ease of bond breakage is affected by such factors as electronegativity, number of terminal O atoms in oxoacids, and the placement of substituent groups on carbon chains in carboxylic acids.

Water is amphiprotic; it can be either an acid or a base. The limited transfer of protons between H_2O molecules in pure water produces H_3O^+ and OH^-, in accordance with the ion product: $K_w = [H_3O^+][OH^-] = 1.0 \times 10^{-14}$ at 25 °C. In the notation of negative logarithms: $pH = -\log [H_3O^+]$; $pOH = -\log [OH^-]$; and $pH + pOH = pK_w = 14.00$. In pure water and in neutral aqueous solutions, pH = 7.00. In acidic solutions, pH < 7.00; in basic (alkaline) solutions, pH > 7.00.

The strategies for calculations involving weak acids and weak bases are similar to those used in Chapter 16, with a heavy reliance on simplifying assumptions.

Ions, as well as neutral molecules, can act as acids and bases. Reactions of ions with water molecules, called hydrolysis reactions, cause certain salt solutions to be either acidic or basic. In calculating the pH of a salt solution, it is often necessary to establish the ionization constant of an acid or a base from that of its conjugate: pK_a (acid) + pK_b (conjugate base) = 14.00.

A strong electrolyte that produces an ion common to the ionization equilibrium of a weak acid or a weak base suppresses the ionization of the weak electrolyte. One application of this common ion effect is in buffer solutions. A buffer is a mixture of a weak acid and its conjugate base or a weak base and its conjugate acid that maintains an essentially constant pH when small amounts of a strong acid or strong base are added. One buffer component neutralizes small added amounts of the acid, and the other component neutralizes small added amounts of the base.

Another application of the common ion effect is in the action of pH indicators. H_3O^+ from the solution being tested affects the concentrations of the indicator acid, HIn, and the indicator base, In^-. In turn, this determines the color exhibited by the pH indicator. An appropriate indicator for an acid–base titration undergoes a color change at the pH of the equivalence point. The pH at the equivalence point can be established from a titration curve—a graph of pH versus volume of titrant solution.

KEY TERMS

acid ionization constant (K_a) (17.1)
amphiprotic (17.1)
base ionization constant (K_b) (17.1)
buffer solution (17.8)
common ion effect (17.7)
conjugate acid (17.1)
conjugate base (17.1)
end point (17.10)
equivalence point (17.10)
Henderson–Hasselbalch equation (17.8)
hydrolysis (17.6)
hydronium ion (H_3O^+) (17.1)
ion product of water (K_w) (17.3)
pH (17.3)
pK_a (17.4)
pK_b (17.4)
pK_w (17.3)
pOH (17.3)
polyprotic acid (17.5)
proton acceptor (17.1)
proton donor (17.1)
titration curve (17.10)

REVIEW QUESTIONS

1. Write equations to represent the ionization of HI as an acid in both the Arrhenius and Brønsted–Lowry theories.
2. Can a substance be a Brønsted–Lowry acid if it does not contain H atoms? Are there any characteristic atoms that must be present in a Brønsted–Lowry base?
3. Must every Brønsted–Lowry acid have a conjugate base? every base have a conjugate acid? Explain.
4. Write equations to show how hydrogen phosphate ion, HPO_4^{2-}, can act either as a Brønsted–Lowry acid or as a Brønsted–Lowry base.

5. Write equations for the ionizations of each of the following as Brønsted–Lowry weak acids.
   a. HOClO b. CH_3CH_2COOH
   c. HCN d. C_6H_5OH

6. Write K_a expressions for each of the weak acids in Question 5.

7. For each of the following identify the conjugate acid–base pairs by using acid(1), base(1), and so on.
   a. $HOClO_2 + H_2O \rightleftharpoons H_3O^+ + OClO_2^-$
   b. $HSeO_4^- + NH_3 \rightleftharpoons NH_4^+ + SeO_4^{2-}$
   c. $HCO_3^- + OH^- \rightleftharpoons CO_3^{2-} + H_2O$
   d. $C_5H_5NH^+ + H_2O \rightleftharpoons C_5H_5N + H_3O^+$

8. Use Table 17.2 to place the following acids in order of increasing strength—that is, from weakest to strongest: (a) benzoic acid; (b) formic acid; (c) hydrofluoric acid; (d) nitrous acid.

9. Use Table 17.2 to place the conjugate bases of the acids in Question 8 in order of increasing strength.

10. Explain why nitric acid, HNO_3, is a strong acid, whereas nitrous acid, HNO_2, is weak.

11. Explain why perchloric acid, $HClO_4$, is a stronger acid than sulfuric acid, H_2SO_4.

12. Acetic acid has a $pK_a = 4.74$, whereas trifluoroacetic acid has a $pK_a = 0.50$. Explain why the difference in these pK_a values is so large.

13. What are the characteristic features of a *polyprotic* acid? Is CH_4 a polyprotic acid? Explain.

14. Explain why the K_{a_1} for a diprotic acid is larger than is K_{a_2}.

15. Describe the difference between the *ionization* of a weak acid and the *hydrolysis* of a weak acid anion.

16. What is the relationship between K_a for a Brønsted–Lowry acid and K_b for its conjugate base?

17. Explain what is meant by the *common ion effect.* How is this effect involved in the functioning of buffer solutions? of pH indicators?

18. Describe the difference between the *equivalence point* and the *end point* of an acid–base titration carried out with a pH indicator.

19. If a the titration of a *weak base* is carried out using a *strong acid* as the titrant, at what point on the titration curve will the pH be lowest and at what point will it be highest? At approximately what pH will the equivalence point be found?

20. Why are so many more acid–base indicators suitable for the titration of a strong acid with a strong base than are suitable for the titration of a weak acid with a strong base?

21. Why can a single pH indicator suffice to determine the equivalence point in a titration, whereas more than one indicator is often needed to determine the pH of a solution?

22. Without doing detailed calculations, determine which of the following solutions have a pH greater than 7.00 and which, lower than 7.00: (a) 0.0025 M HCl; (b) 0.037 M NH_3; (c) 0.225 M C_6H_5OH (phenol); (d) 0.42 M CH_3NH_2.

23. Which of the following 0.05 M aqueous solutions would you expect to have the *lowest* pH: (a) $KHSO_4$; (b) K_2SO_4; (c) K_2CO_3; (d) K_2HPO_4? Explain.

24. Arrange the following 0.10 M aqueous solutions in the order of *increasing* pH: (a) HCl, (b) CH_3COONa, (c) KCl, (d) H_3PO_4, (e) NH_3, (f) H_2SO_4, (g) NH_4NO_3, (h) KOH.

PROBLEMS

Brønsted–Lowry Acids and Bases

25. Make a sketch, similar to Figure 17.1, to illustrate the ionization of acetic acid in water. How would you modify the sketch to represent the ionization of hydrogen chloride?

26. Which of the following ions is amphiprotic; that is, which is able to lose or gain a proton in aqueous solution: (a) $H_2PO_4^-$; (b) SO_4^{2-}; (c) Cl^-; (d) NH_4^+?

27. Sodium bicarbonate, $NaHCO_3$, is a mildly alkaline material used as an antacid. How does the bicarbonate ion function as an antacid?

28. Liquid NH_3, like water, is an amphiprotic solvent. Write an equation for the self-ionization that occurs in liquid ammonia.

29. Use Table 17.1 to rank the following in order of increasing tendency for the reaction to go to completion (to the right).
   (a) $HSO_4^- + F^- \rightleftharpoons HF + SO_4^{2-}$
   (b) $NH_4^+ + Cl^- \rightleftharpoons NH_3 + HCl$
   (c) $HCl + CH_3COO^- \rightleftharpoons CH_3COOH + Cl^-$
   (d) $CH_3OH + Br^- \rightleftharpoons HBr + CH_3O^-$

30. With which of the following bases will the reaction of acetic acid, CH_3COOH, proceed furthest toward

completion (to the right): (a) CO_3^{2-}; (b) F^-; (c) Cl^-; (d) NO_3^-? Explain.

31. Aniline, $C_6H_5NH_2$, is a weak base in water, but it is a strong base in glacial acetic acid, $CH_3COOH(l)$. Explain this difference in behavior.

Molecular Structure and Acid Strength

32. H_2S and H_2SO_4 are both diprotic acids. Explain why H_2S is a weak acid in its first ionization step, but H_2SO_4 is a strong acid.

33. Refer to the ideas and data presented on page 648 and estimate a value of K_a for (**a**) 2,3-dichloropropanoic acid; (**b**) 4-chlorobutanoic acid.

34. The ionization of phenol, C_6H_5OH, as an acid is represented below.

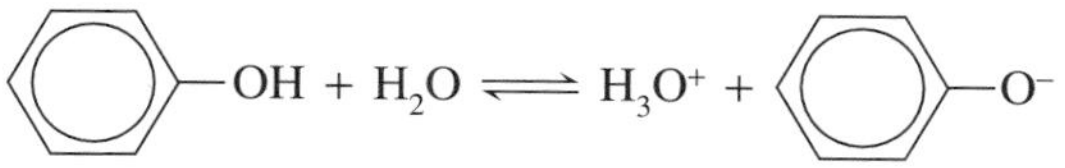

$$K_a = 1.0 \times 10^{-10}$$

Arrange the following substituted phenols in the expected order in which their K_a values *increase*. Where would you expect phenol itself to fit into this ranking?

(a) 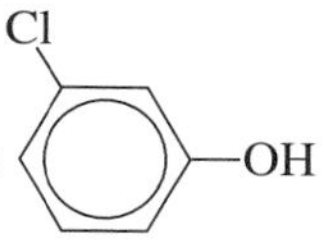

(b) 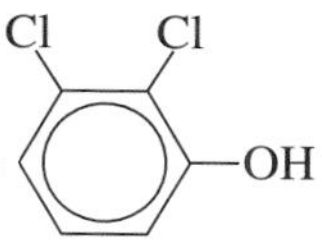

(c) 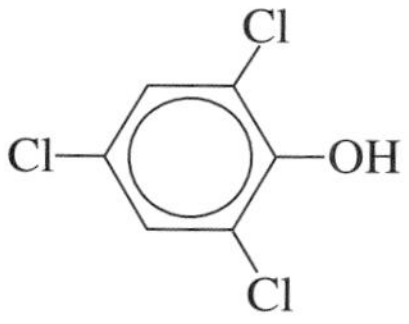

(d) 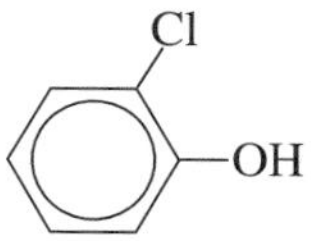

35. Given below are the structural formulas of several carboxylic acids. Arrange them in the expected order in which their K_a values *increase*. [*Hint*: Refer to Table 17.2. $K_a = 3.9 \times 10^{-4}$ for the acid in (c)].

(a) 

(b) 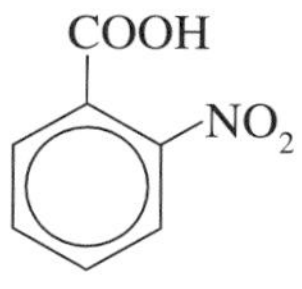

(c) 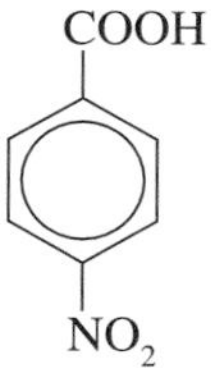

(d) 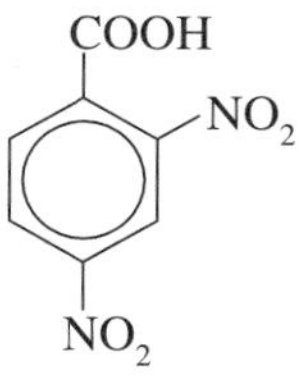

(e) $CH_3(CH_2)_6COOH$

The pH Scale

36. The pH of a cup of coffee at 25 °C is 4.32. What is the $[H_3O^+]$ in the coffee?

37. A detergent solution has a pH of 11.13 at 25 °C. What is the $[OH^-]$ in the solution?

38. What is the pH of each of the following solutions?
a. 0.0025 M HCl
b. 0.055 M NaOH
c. 0.015 M $Ba(OH)_2$
d. 1.6×10^{-3} M HBr

39. What is the pOH of each of the following solutions?
a. 2.5×10^{-3} M NaOH
b. 3.2×10^{-3} M HCl
c. 3.6×10^{-4} M $Ca(OH)_2$
d. 0.00022 M HNO_3

40. Which has the lower pH: 0.00048 M H_2SO_4 or a vinegar solution having pH = 2.42?

41. Which has the higher pH: 0.0062 M $Ba(OH)_2(aq)$ or an ammonia cleanser having a pH = 11.65?

42. What volume of 0.355 M KOH must be diluted with water to prepare 2.00 L of a solution having a pH of 12.50?

43. What mass of concentrated hydrochloric acid (37.2% HCl by mass) must be dissolved in water to prepare 0.250 L of a solution having pH = 1.75?

44. At body temperature, 37.0 °C, K_w for water is 2.4×10^{-14}. Calculate the pH of pure water at this temperature.

Equilibria in Solutions of Weak Acids and Weak Bases

45. Calculate the pH of 1.50 M formic acid, HCOOH.

$$HCOOH + H_2O \rightleftharpoons H_3O^+ + HCOO^- \qquad K_a = 1.8 \times 10^{-4}$$

46. Calculate the pH of a solution of pyridine that has 1.25 g in 125 mL of water solution.

$$C_5H_5N + H_2O \rightleftharpoons C_5H_5N^+ + OH^- \qquad K_b = 1.5 \times 10^{-9}$$

47. What molarity of hydrazoic acid, HN_3, is required to produce an aqueous solution with pH = 3.10?

$$HN_3 + H_2O \rightleftharpoons H_3O^+ + N_3^- \qquad K_a = 1.9 \times 10^{-5}$$

48. A water solution saturated with phenol, C_6H_5OH, is found to have pH = 4.90. What is the molarity of phenol in this solution?

$$C_6H_5OH + H_2O \rightleftharpoons H_3O^+ + C_6H_5O^- \quad K_a = 1.0 \times 10^{-10}$$

49. A 1.00-g sample of aspirin (acetylsalicylic acid) is dissolved in 0.300 L of water at 25 °C and its pH is found to be 2.62. What is the K_a of aspirin?

$$C_6H_4(OCOCH_3)COOH + H_2O \rightleftharpoons H_3O^+ + C_6H_4(OCOCH_3)COO^- \quad K_a = ?$$

50. Codeine, $C_{18}H_{21}NO_3$, a commonly prescribed painkiller, is a weak base. A saturated aqueous solution contains 1.00 g codeine in 120 mL of solution and has a pH of 9.8. What is the K_b of codeine?

$$C_{18}H_{21}NO_3 + H_2O \rightleftharpoons [C_{18}H_{21}NHO_3]^+ + OH^- \quad K_b = ?$$

Polyprotic Acids

51. Without doing detailed calculations, indicate which of the following polyprotic solutions will have the *lower pH*: 0.0045 M H_2SO_4 or 0.0045 M H_3PO_4.

52. Which of the following is the most likely pH for 0.010 M H_2SO_4: (a) 2.00, (b) 1.85, (c) 1.70, (d) 1.50?

53. Acids are added to cola drinks to lower the pH to about 2.5 to impart tartness. Show that a cola that is about 0.05% H_3PO_4 by mass has the requisite pH.

54. Oxalic acid, HOOCCOOH, is a weak dicarboxylic acid found in the cell sap of many plants (for example, rhubarb leaves) as its potassium or calcium salt. The solubility of oxalic acid at 25 °C is about 83 g/L. What are the (**a**) pH and (**b**) [$^-$OOCCOO$^-$] in the saturated solution?

$$HOOCCOOH + H_2O \rightleftharpoons H_3O^+ + HOOCCOO^- \quad K_a = 5.4 \times 10^{-2}$$

$$HOOCCOO^- + H_2O \rightleftharpoons H_3O^+ + {}^-OOCCOO^- \quad K_a = 5.3 \times 10^{-5}$$

Hydrolysis

55. Write an equation to represent the hydrolysis that occurs in the solution CH_3CH_2COOK(aq). Will the solution be acidic, basic, or neutral?

56. Write equations to represent the hydrolysis that occurs in the solution NH_4CN(aq). Will the solution be acidic, basic, or neutral?

57. Write an equation to represent the hydrolysis of carbonate ion in aqueous solution, CO_3^{2-}(aq). Determine the value of K_b for this hydrolysis reaction.

58. Write two equations, one to show the ionization of HPO_4^{2-} as an acid, and one to show its ionization (hydrolysis) as a base. Would you expect an aqueous solution of Na_2HPO_4 to be acidic or basic? Explain. (*Hint*: What are the relevant K_a and K_b values?)

59. Which of the following 0.100 M aqueous solutions has the *lowest* pH: (a) $NaNO_3$, (b) CH_3COOK, (c) NH_4I, (d) Na_3PO_4? Explain.

60. Which of the solutions in Problem 59 has the *highest* pH? Explain.

61. Calculate the pH of a solution that is 0.080 M NaOCl.

$$OCl^- + H_2O \rightleftharpoons HOCl + OH^- \quad K_b = 3.4 \times 10^{-7}$$

62. Calculate the pH of a solution that is 0.602 M NH_4Cl.

$$NH_4^+ + H_2O \rightleftharpoons H_3O^+ + NH_3 \quad K_a = 5.6 \times 10^{-10}$$

63. What is the molarity of a sodium acetate solution that is found to have a pH of 9.05? (*Hint*: What is the hydrolysis reaction and what is the K_b value for this reaction?)

64. What is the molarity of an ammonium bromide solution that is found to have a pH of 5.75? (*Hint*: What is the hydrolysis reaction and what is the K_a value for this reaction?)

Common Ion Effect

65. Which of the following will *suppress* the ionization of formic acid, HCOOH(aq): (a) NaCl; (b) KOH; (c) HNO_3; (d) $(HCOO)_2Ca$; (e) Na_2CO_3? Explain.

66. Which of the following will *increase* the ionization of formic acid, HCOOH(aq): (a) KI; (b) NaOH; (c) HNO_3; (d) HCOONa; (e) NaCN? Explain.

67. Calculate [NH_4^+] in a solution that is 0.15 M NH_3 *and* 0.015 M KOH.

$$NH_3 + H_2O \rightleftharpoons NH_4^+ + OH^- \quad K_b = 1.8 \times 10^{-5}$$

68. Calculate [$C_6H_5COO^-$] in a solution that is 0.035 M C_6H_5COOH and 0.051 M HCl.

$$C_6H_5COOH + H_2O \rightleftharpoons H_3O^+ + C_6H_5COO^- \quad K_a = 6.3 \times 10^{-5}$$

69. Calculate the pH of a solution that is 0.350 M CH_3CH_2COOH and 0.0786 M in CH_3CH_2COOK.

$$CH_3CH_2COOH + H_2O \rightleftharpoons H_3O^+ + CH_3CH_2COO^- \quad K_a = 1.3 \times 10^{-5}$$

70. Calculate the pH of a solution that is 0.132 M in diethylamine, $(CH_3)_2NH$, and 0.145 M in diethylammonium chloride, $(CH_3CH_2)_2NH_2Cl$.

$$(CH_3CH_2)_2NH + H_2O \rightleftharpoons (CH_3CH_2)_2NH_2^+ + OH^- \quad K_b = 6.9 \times 10^{-4}$$

Buffer Solutions

71. We have seen that a buffer solution contains an acid as one component and a base as the other. (**a**) Can the two components be HCl and NaOH? Explain. (**b**) Acetic acid contains both an acid, CH_3COOH, and its conjugate base, CH_3COO^-. Can an acetic acid solution alone be considered a buffer? Explain.

72. A solution is prepared that is 0.405 M HCOOH (formic acid) and 0.326 M HCOONa (sodium formate). What is the pH of this buffer solution?

$$HCOOH + H_2O \rightleftharpoons H_3O^+ + HCOO^- \quad K_a = 1.8 \times 10^{-4}$$

73. A solution is prepared that contains 12.5 mL of propylamine, $CH_3CH_2CH_2NH_2$ (d = 0.719 g/mL), and 15.00 g of propylamine hydrochloride, $CH_3CH_2CH_2NH_3Cl$. The volume of the aqueous solution is 725 mL. What is the pH of this buffer solution?

$$CH_3CH_2CH_2NH_2 + H_2O \rightleftharpoons CH_3CH_2CH_2NH_3^+ + OH^- \quad K_b = 3.5 \times 10^{-4}$$

74. If 1.00 mL of 0.250 M HCl is added to 50.0 mL of the buffer solution described in Problem 72, what will be the pH of the final solution?

75. If 2.00 mL of 0.0850 M NaOH is added to 75.0 mL of the buffer solution described in Problem 73, what will be the pH of the final solution?

76. What mass of sodium propionate, CH_3CH_2COONa, must be dissolved in 0.500 L of 0.525 M CH_3CH_2COOH to produce a buffer solution with pH = 4.75?

$$CH_3CH_2COOH + H_2O \rightleftharpoons H_3O^+ + CH_3CH_2COO^- \quad K_a = 1.3 \times 10^{-5}$$

77. What mass of CH_3NH_3Cl must be dissolved in 0.100 L of 0.350 M CH_3NH_2(aq) to produce a buffer solution with pH = 11.05?

$$CH_3NH_2 + H_2O \rightleftharpoons CH_3NH_3^+ + OH^- \quad K_b = 4.2 \times 10^{-4}$$

78. The salt "tris hydrochloride," $(HOCH_2)_3CNH_3^+Cl^-$, and the base "tris," $(HOCH_2)_3CNH_2$, are used to make buffers for biochemical reactions. Tris hydrochloride has pK_a = 8.40. What is K_b for tris?

79. Calculate the pH of a buffer that is 0.10 M tris and 0.15 M tris hydrochloride. Use data from Problem 78.

pH Indicators

80. Refer to Figure 17.12 and state the color you would expect for the following pH indicators in the given aqueous solution?
a. bromthymol blue in 0.10 M NH_4Cl
b. bromphenol blue in 0.10 M CH_3COOK
c. phenolphthalein in 0.10 M Na_2CO_3
d. methyl violet in 0.10 M CH_3COOH

81. An 0.100 M HCl(aq) solution contains thymol blue indicator and has a red color. An 0.100 M NaOH(aq) solution with phenolphthalein indicator also has a red color. What should be the color when equal volumes of the two solutions (with their indicators) are mixed?

82. When added to an unknown solution, bromcresol green indicator has a yellow color. When added to another sample of the same solution, thymol blue indicator also turns yellow. To within about one pH unit, what is the pH of the solution?

83. A particular buffer solution is believed to have a pH of about 9.8. Describe how you might use a pair of pH indicators to verify this. What two indicators would you use? Explain. (*Hint*: Use Figure 17.12.)

Neutralization Reactions and Titration Curves

84. Produce a listing comparable to that on page 688 to describe the titration of a strong base (for example, NaOH) with a strong acid (for example, HCl).

85. Produce a listing comparable to that on pages 685–686 to describe the titration of a weak base (e.g., NH_3) with a strong acid (e.g., HCl).

86. Can the same indicator be used in the titration of a strong base with a strong acid as in the titration of a strong acid with a strong base? Explain.

87. Can the same indicator be used in the titration of a weak base with a strong acid as in the titration of a weak acid with a strong base? Explain.

88. In the titration of 20.00 mL of 0.500 M CH_3COOH by 0.500 M NaOH, calculate the pH at the point in the titration where 7.45 mL of 0.500 M NaOH has been added. Compare your result with the value indicated in Figure 17.15.

89. In the titration of 20.00 mL of 0.500 M HCl by 0.500 M NaOH, calculate the volume of 0.500 M NaOH required to reach a pH of 2.0. Compare your result with the value indicated in Figure 17.14.

ADDITIONAL PROBLEMS

90. Ammonia, NH_3, has $K_b = 1.8 \times 10^{-5}$, whereas hydroxylamine, $HONH_2$, has $K_b = 9.1 \times 10^{-9}$. Explain why these K_b values should differ in this way.

91. A handbook lists for hydrazine, N_2H_4, that $pK_{b_1} = 6.07$ and $pK_{b_2} = 15.05$. Draw a structural formula for hydrazine and write equations to show how hydrazine can ionize as a base in two distinctive steps. Explain why pK_{b_2} is so much larger than pK_{b_1}.

92. The titration curve in the accompanying figure involves 1.00 M solutions of an acid and a base. In the manner of Conceptual Example 17.3 and Conceptual Exercise 17.3, (**a**) identify the type of acid–base titration represented by this curve; (**b**) estimate the p*K* value of the *titrant*; and (**c**) obtain a value of the pH at the equivalence point by calculation and by estimation from the titration curve.

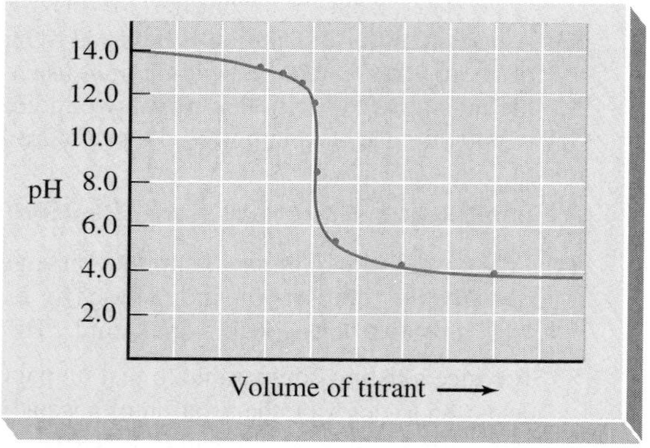

93. The term *percent ionization* refers to the percentage of the molecules of an acid or base that undergo ionization in solution.

a. What is the percent ionization of 1.00 M $CH_3COOH(aq)$?

b. What is the percent ionization of 0.100 M $CH_3COOH(aq)$?

c. What is the percent ionization of 1.0×10^{-4} M $CH_3COOH(aq)$?

d. The results of parts (**a**), (**b**), and (**c**) indicate that the percent ionization increases as a weak acid solution is made more dilute. Does the acidity of the solution, that is, $[H_3O^+]$, increase as well? Explain.

94. A batch of sewage sludge is dewatered to 28% solids; the remaining water has a pH of 6.0. The sludge is made quite alkaline to kill disease-causing microorganisms and make the sludge acceptable for spreading on farm land. How much quicklime (CaO) is needed to raise the pH of 1.0 ton of this sludge to 12.0? (*Hint*: See page 297).

95. What volume of concentrated nitric acid (70.4% HNO_3 by mass; $d = 1.42$ g/mL) must be dissolved in water to prepare 0.500 L of a solution having pH = 2.20?

96. Calculate the pH of a solution that is 1.0×10^{-8} M HCl. (*Hint*: What are the two sources of H_3O^+?)

97. Can a solution have $[H_3O^+] = 2 \times [OH^-]$? Can a solution have pH = 2 × pOH? If so, will the two solutions be the same?

98. For the neutralization reaction $H^+ + OH^- \longrightarrow H_2O$, $\Delta H° = -55.8$ kJ. Use this fact to establish whether the ion product of water, K_w, should increase or decrease in value with increasing temperature.

99. Normal rainfall is slightly acidic. This comes about from the dissolving of atmospheric CO_2 in the rainwater and ionization of the resulting carbonic acid, H_2CO_3. At a $CO_2(g)$ partial pressure of 1 atm pressure, a saturated solution is equivalent to 0.033 M H_2CO_3. Given that air contains 0.036% CO_2, by volume, estimate the pH of rainwater that is saturated with CO_2. [*Hint*: Recall Henry's law (page 492) and the ionization of carbonic acid (page 662).]

100. Citric acid is found in citrus fruit. Lemon juice is approximately 5% citric acid by mass. Citric acid is a triprotic acid having $pK_{a_1} = 3.13$, $pK_{a_2} = 4.76$, and $pK_{a_3} = 6.40$. Calculate the approximate pH of lemon juice.

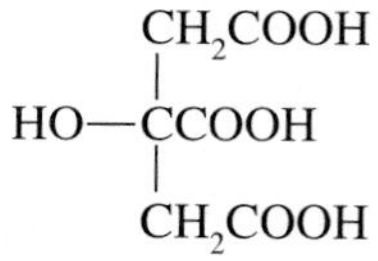

Citric acid

101. Although the titration of a strong acid with a strong base can be carried out easily and precisely, the titration of a weak acid with a weak base cannot be done precisely. Explain why this is the case.

102. The accompanying titration curve is that of 10.0 mL of 0.100 M H_3PO_4 titrated with 0.100 M NaOH.

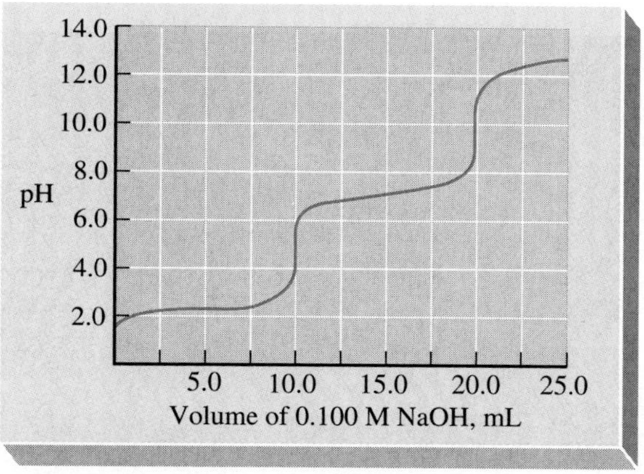

a. Write a net ionic equation for the neutralization reaction that occurs prior to the first equivalence point. What is the substance present in solution at the first equivalence point?

b. Write a net ionic equation for the neutralization reaction that occurs between the first and second equivalence points. What is the substance present in solution at the second equivalence point?

c. Select an appropriate indicator for titration to each of the equivalence points.

d. H_3PO_4 is a *tri*protic acid. Why do you suppose the titration curve is lacking a third equivalence point?

e. In what way would the titration curve be altered if the solution being titrated were 0.100 M in *both* H_3PO_4 and HCl?

103. A requirement of buffer solutions is that they contain components capable of reacting with added acid or base. The dihydrogen phosphate ion is capable of doing this—it is amphiprotic. Why isn't $NaH_2PO_4(aq)$ alone a particularly effective buffer? Is a solution containing NaH_2PO_4 and Na_2HPO_4 an effective buffer? Explain. (*Hint*: Use the titration curve accompanying Problem 102.)

104. Construct, by calculation, a titration curve for the titration of 40.00 mL of 0.200 M NH_3 with 0.500 M HCl. Specifically, what is the pH (**a**) initially, (**b**) after the addition of 5.00 mL of 0.500 M HCl, (**c**) at the half-neutralization point, (**d**) after the addition of 10.00 mL of 0.500 M HCl, (**e**) at the equivalence point, and (**f**) after the addition of 20.00 mL of 0.500 M HCl?

A precipitate of lead(II) iodide forms when an aqueous solution of potassium iodide is added to one of lead(II) nitrate. The conditions that lead to precipitate formation are considered on page 709.

18

Equilibria Involving Slightly Soluble Salts and Complex Ions

CONTENTS

In Chapter 16 we introduced general ideas about chemical equilibrium and applied them mostly to reactions involving gases. Much of Chapter 17 dealt with equilibria in aqueous solutions of weak acids and weak bases. In this chapter we will focus on equilibria between slightly soluble salts and their ions in solution, and on equilibria that involve complex ions. These equilibria,

in turn, are often interwoven with equilibria involving acids and bases.

Applications in this chapter are as close as our teeth—tooth decay is related to the solubility of slightly soluble substances—and the film in our cameras—the development of photographic film makes use of complex ion formation. We'll consider both of these applications and many others as well.

18.1 The Solubility Product Constant, K_{sp}

As you've probably concluded by now, much of the chemistry that we do is carried out in aqueous solutions. However, there are many important ionic compounds that are only slightly soluble in water—in fact, we often use the term "insoluble" to describe them. For example, the solubility of barium sulfate, $BaSO_4$, is only 0.000246 g per 100 g water at 25 °C. The comparative *insolubility* of $BaSO_4$ is one of its most important properties, particularly when it is used as a white paint pigment or as an opaque medium in X-ray photography of the gastrointestinal tract.

We can represent the equilibrium between undissolved $BaSO_4(s)$ and its ions in a saturated solution as

$$BaSO_4(s) \rightleftharpoons Ba^{2+}(aq) + SO_4^{2-}(aq)$$

The symbols (aq) indicate that water is the solvent in which this equilibrium occurs, but we do not represent water in any other way. Also, as we saw in Chapters 16 and 17, pure liquids and pure solids are not included in equilibrium constant expressions. As a result, the equilibrium constant expression consists simply of a product of ion concentration terms.

$$K_{sp} = [Ba^{2+}][SO_4^{2-}] = 1.1 \times 10^{-10} \text{ (at 25 °C)}$$

The **solubility product constant (K_{sp})** is the equilibrium constant for the equilibrium that exists between a solid ionic solute and its ions in a saturated aqueous solution. Like other equilibrium constants, K_{sp} depends on the particular reversible reaction and on the temperature. Table 18.1 lists some typical solubility product constants at about room temperature and the solubility equilibria to which they apply.

In applying the solubility product constant, we must be careful to match the K_{sp} expression to the appropriate equation representing the solubility equilibrium. We illustrate this idea in Example 18.1.

TABLE 18.1 **Some Solubility Product Constants at 25 °C**[a]

Solute	Solubility Equilibrium	K_{sp}
Aluminum hydroxide	$Al(OH)_3(s) \rightleftharpoons Al^{3+}(aq) + 3OH^-(aq)$	1.3×10^{-33}
Barium carbonate	$BaCO_3(s) \rightleftharpoons Ba^{2+}(aq) + CO_3^{2-}(aq)$	5.1×10^{-9}
Barium sulfate	$BaSO_4(s) \rightleftharpoons Ba^{2+}(aq) + SO_4^{2-}(aq)$	1.1×10^{-10}
Calcium carbonate	$CaCO_3(s) \rightleftharpoons Ca^{2+}(aq) + CO_3^{2-}(aq)$	2.8×10^{-9}
Calcium fluoride	$CaF_2(s) \rightleftharpoons Ca^{2+}(aq) + 2\,F^-(aq)$	5.3×10^{-9}
Calcium sulfate	$CaSO_4(s) \rightleftharpoons Ca^{2+}(aq) + SO_4^{2-}(aq)$	9.1×10^{-6}
Calcium oxalate	$CaC_2O_4(s) \rightleftharpoons Ca^{2+}(aq) + C_2O_4^{2-}(aq)$	4×10^{-9}
Chromium(III) hydroxide	$Cr(OH)_3(s) \rightleftharpoons Cr^{3+}(aq) + 3\,OH^-(aq)$	6.3×10^{-31}
Copper(II) sulfide	$CuS(s) \rightleftharpoons Cu^{2+}(aq) + S^{2-}(aq)$	8.7×10^{-36}
Iron(III) hydroxide	$Fe(OH)_3(s) \rightleftharpoons Fe^{3+}(aq) + 3\,OH^-(aq)$	4×10^{-38}
Lead(II) chloride	$PbCl_2(s) \rightleftharpoons Pb^{2+}(aq) + 2\,Cl^-(aq)$	1.6×10^{-5}
Lead(II) chromate	$PbCrO_4(s) \rightleftharpoons Pb^{2+}(aq) + CrO_4^{2-}(aq)$	2.8×10^{-13}
Lead II) iodide	$PbI_2(s) \rightleftharpoons Pb^{2+}(aq) + 2\,I^-(aq)$	7.1×10^{-9}
Magnesium carbonate	$MgCO_3(s) \rightleftharpoons Mg^{2+}(aq) + CO_3^{2-}(aq)$	3.5×10^{-8}
Magnesium fluoride	$MgF_2(s) \rightleftharpoons Mg^{2+}(aq) + 2\,F^-(aq)$	3.7×10^{-8}
Magnesium hydroxide	$Mg(OH)_2(s) \rightleftharpoons Mg^{2+}(aq) + 2\,OH^-(aq)$	1.8×10^{-11}
Magnesium phosphate	$Mg_3(PO_4)_2(s) \rightleftharpoons 3\,Mg^{2+}(aq) + 2\,PO_4^{3-}(aq)$	1×10^{-25}
Mercury(I) chloride	$Hg_2Cl_2(s) \rightleftharpoons Hg_2^{2+}(aq) + 2\,Cl^-(aq)$	1.3×10^{-18}
Mercury(II) sulfide	$HgS(s) \rightleftharpoons Hg^{2+}(aq) + S^{2-}(aq)$	2×10^{-53}
Silver acetate	$CH_3COOAg(s) \rightleftharpoons Ag^+(aq) + CH_3COO^-(aq)$	2.0×10^{-3}
Silver bromide	$AgBr(s) \rightleftharpoons Ag^+(aq) + Br^-(aq)$	5.0×10^{-13}
Silver chloride	$AgCl(s) \rightleftharpoons Ag^+(aq) + Cl^-(aq)$	1.8×10^{-10}
Silver iodide	$AgI(s) \rightleftharpoons Ag^+(aq) + I^-(aq)$	8.5×10^{-17}
Strontium carbonate	$SrCO_3(s) \rightleftharpoons Sr^{2+}(aq) + CO_3^{2-}(aq)$	1.1×10^{-10}
Strontium sulfate	$SrSO_4(s) \rightleftharpoons Sr^{2+}(aq) + SO_4^{2-}(aq)$	3.2×10^{-7}
Zinc sulfide	$ZnS(s) \rightleftharpoons Zn^{2+}(aq) + S^{2-}(aq)$	2×10^{-25}

[a] A more extensive listing of K_{sp} values is given in Appendix D.

Example 18.1

Write a solubility product constant expression for equilibrium in a saturated solution of each of the following slightly soluble salts.

a. iron(III) phosphate, $FePO_4$ **b.** chromium hydroxide, $Cr(OH)_3$

Solution

The K_{sp} expression must be based on a balanced equation. The coefficients of the ions are for the balanced equation representing the dissolution of 1 mol of the slightly soluble solute. That is, the coefficient on the left side of the equation is assumed to be 1. And, as with other equilibrium constant expressions, coefficients in the balanced equation appear as exponents in the K_{sp} expression.

a. $FePO_4(s) \rightleftharpoons Fe^{3+}(aq) + PO_4^{3-}(aq)$ $K_{sp} = [Fe^{3+}][PO_4^{3-}]$

b. $Cr(OH)_3(s) \rightleftharpoons Cr^{3+}(aq) + 3\ OH^-(aq)$ $K_{sp} = [Cr^{3+}][OH^-]^3$

Exercise 18.1

Write a K_{sp} expression for equilibrium in a saturated solution of each of the following slightly soluble solutes: (**a**) magnesium hydroxide, "milk of magnesia," and (**b**) copper(II) arsenate, $Cu_3(AsO_4)_2$, used as an insecticide.

18.2 Relationship Between Solubility and K_{sp}

It's not difficult to show that there's a close relationship between the solubility product constant of a solute, K_{sp}, and its molar solubility—the molarity of its saturated aqueous solution. This does *not* mean that they are equal or anywhere close to being equal, but that we can readily establish one from the other.

Although K_{sp} calculations are no more difficult than other equilibrium calculations—in fact in many cases they are easier—they are much more subject to error. This is because interionic attractions can make the "effective" concentration of an ion rather different from its *stoichiometric* concentration, which is based simply on the amount of solute dissolved. The "effective concentration," called *activity*, includes adjustments for interionic attractions. The differences between stoichiometric concentrations and activities can be considerable, especially when ion concentrations are high. This is one reason we don't use the solubility product concept for solutes that are moderately or highly soluble in water. However, let's not get bogged down by this distinction. We will continue to use concentrations as we have in other equilibrium calculations. Keep in mind, though, that at times K_{sp} calculations give only "ballpark estimates."

Like other equilibrium calculations, those dealing with solubility equilibrium fall into two broad categories: determining a value of K_{sp} from experimental data and calculating equilibrium concentrations when a value of K_{sp} is known. The first type of calculation is illustrated by Examples 18.2 and 18.3.

Example 18.2

Calcium oxalate, CaC_2O_4, is a substance found in the leaves of the rhubarb plant that makes that portion of the plant poisonous. A handbook lists the solubility of this salt as 0.00067 g/100 mL water at 18 °C. Calculate the K_{sp} for CaC_2O_4 at 18 °C.

$$CaC_2O_4(s) \rightleftharpoons Ca^{2+}(aq) + C_2O_4^{2-}(aq) \qquad K_{sp} = ?$$

Solution

Because concentrations in the K_{sp} expression must be in molarity, our first task is to convert the handbook data to molarity.

$$\frac{0.00067 \text{ g } CaC_2O_4}{0.100 \text{ L}} \times \frac{1 \text{ mol } CaC_2O_4}{128 \text{ g } CaC_2O_4} = 5.2 \times 10^{-5} \text{ M } CaC_2O_4$$

From the equation we see that 1 mol Ca^{2+} and 1 mol $C_2O_4^{2-}$ appear in solution for every mole of CaC_2O_4 that dissolves. The factors reflecting this fact are shown in blue below.

$$[Ca^{2+}] = \frac{5.2 \times 10^{-5} \text{ mol } CaC_2O_4}{1 \text{ L}} \times \frac{1 \text{ mol } Ca^{2+}}{1 \text{ mol } CaC_2O_4} = 5.2 \times 10^{-5} \text{ M}$$

$$[C_2O_4^{2-}] = \frac{5.2 \times 10^{-5} \text{ mol } CaC_2O_4}{1 \text{ L}} \times \frac{1 \text{ mol } C_2O_4^{2-}}{1 \text{ mol } CaC_2O_4} = 5.2 \times 10^{-5} \text{ M}$$

Now we can introduce these concentrations into the K_{sp} expression.

$$K_{sp} = [Ca^{2+}][C_2O_4^{2-}] = (5.2 \times 10^{-5})(5.2 \times 10^{-5}) = 2.7 \times 10^{-9}$$

Example 18.3

At 20 °C, a saturated aqueous solution of silver carbonate is found to have 25 mg Ag^+/L. Calculate the K_{sp} for Ag_2CO_3 at 20 °C.

$$Ag_2CO_3(s) \rightleftharpoons 2\,Ag^+(aq) + CO_3^{2-}(aq) \qquad K_{sp} = ?$$

Solution

Here, we are given information about the concentration of Ag^+ rather than the compound as a whole. We can convert this silver ion concentration to molarity, as follows.

$$[Ag^+] = \frac{25 \text{ mg } Ag^+}{\text{L}} \times \frac{1 \text{ g } Ag^+}{1000 \text{ mg } Ag^+} \times \frac{1 \text{ mol } Ag^+}{108 \text{ g } Ag^+} = 2.3 \times 10^{-4} \text{ M}$$

Then, we can relate $[CO_3^{2-}]$ to $[Ag^+]$ through the equation for the dissolution of $Ag_2CO_3(s)$. That is,

$$[CO_3^{2-}] = \frac{2.3 \times 10^{-4} \text{ mol } Ag^+}{\text{L}} \times \frac{1 \text{ mol } CO_3^{2-}}{2 \text{ mol } Ag^+} = 1.2 \times 10^{-4} \text{ M}$$

Now, we can write the K_{sp} expression for Ag_2CO_3 and substitute in the equilibrium concentrations of the ions.

$$K_{sp} = [Ag^+]^2[CO_3^{2-}] = (2.3 \times 10^{-4})^2(1.2 \times 10^{-4}) = 6.3 \times 10^{-12}$$

Exercise 18.2

The solubility of lead(II) fluoride listed in a handbook is 0.064 g PbF_2/100 mL H_2O at 20 °C. Calculate the K_{sp} for PbF_2 at 20 °C. (*Hint*: How are $[Pb^{2+}]$ and $[F^-]$ related to the molar solubility of PbF_2?)

Exercise 18.3

In Example 17.3 (page 651), we determined the molar solubility of magnesium hydroxide ("milk of magnesia") from the measured pH of its saturated solution. Use data from that example to determine the K_{sp} for $Mg(OH)_2$. (*Hint*: Is it necessary to start with the molar solubility? What is the relationship between $[Mg^{2+}]$ and $[OH^-]$?)

The second type of calculation involving solubility equilibrium is illustrated in Example 18.4, where we calculate the molar solubility of a solute from a tabulated value of its K_{sp}. This type of calculation has practical significance because many handbooks list only K_{sp} values and not molar solubilities.

Example 18.4

From its K_{sp} value, calculate for silver sulfate (**a**) its molar solubility and (**b**) the more typical handbook solubility in g Ag_2SO_4/100 mL of water.

$$Ag_2SO_4(s) \rightleftharpoons 2\,Ag^+(aq) + SO_4^{2-}(aq) \qquad K_{sp} = 1.4 \times 10^{-5} \text{ at } 25\text{ °C}$$

Solution

a. The equation shows that for every 1 mol $Ag_2SO_4(s)$ that dissolves, 1 mol SO_4^{2-} and 2 mol Ag^+ appear in solution. If we let s represent the number of moles of Ag_2SO_4 that dissolve per liter of solution, we see that

$$[SO_4^{2-}] = s \qquad \text{and} \qquad [Ag^+] = 2s$$

These concentrations must satisfy the K_{sp} expression.

$$K_{sp} = [Ag^+]^2[SO_4^{2-}] = (2s)^2(s) = 4s^3 = 1.4 \times 10^{-5}$$

$$s^3 = 3.5 \times 10^{-6}$$

$$s = (3.5 \times 10^{-6})^{1/3} = 1.5 \times 10^{-2}$$

The molar solubility of Ag_2SO_4 is

$$\frac{1.5 \times 10^{-2} \text{ mol } SO_4^{2-}}{\text{L}} \times \frac{1 \text{ mol } Ag_2SO_4}{1 \text{ mol } SO_4^{2-}} = 1.5 \times 10^{-2} \text{ M } Ag_2SO_4$$

b. We can readily convert this molarity to a solubility in g/mL.

$$\frac{1.5 \times 10^{-2} \text{ mol } Ag_2SO_4}{1 \text{ L}} \times \frac{312 \text{ g } Ag_2SO_4}{1 \text{ mol } Ag_2SO_4} \times \frac{1 \text{ L}}{1000 \text{ mL}} = 4.7 \times 10^{-3} \text{ g } Ag_2SO_4/\text{mL}$$

In 100 mL of solution, the mass of Ag_2SO_4 is 100 times as much as in 1 mL.

$$4.7 \times 10^{-3} \text{ g } Ag_2SO_4/\text{mL} \times 100 \text{ mL} = 0.47 \text{ g } Ag_2SO_4$$

In the handbook format, then, the solubility is 0.47 g Ag_2SO_4/100 mL.

Exercise 18.4

Calculate the molar solubility of silver arsenate, given that

$$Ag_3AsO_4(s) \rightleftharpoons 3\,Ag^+(aq) + AsO_4^{3-}(aq) \qquad K_{sp} = 1.0 \times 10^{-22}$$

At times, instead of calculating actual molar solubilities, we may need only to determine which of a group of similar solutes is the most soluble, which is the least soluble, and so on. We can do this by comparing K_{sp} values, as illustrated in Estimation Example 18.1.

ESTIMATION EXAMPLE 18.1

Use data from Table 18.1 to establish the order of *increasing* solubility of the silver halides: AgCl, AgBr, AgI.

Solution

The key to solving this problem is to recognize that all three solutes are of the same type—that is, the ratio of cation to anion in each is 1:1. As suggested by the following equation for the solubility equilibrium of AgCl, the ion concentrations are equal to each other and to the molar solubility, *s*.

$$AgCl(s) \rightleftharpoons Ag^+(aq) + Cl^-(aq)$$

$$s \text{ mol AgCl/L} \qquad s \text{ mol } Ag^+\text{/L} \qquad s \text{ mol } Cl^-\text{/L}$$

$$K_{sp} = [Ag^+][Cl^-] = (s)(s) = s^2 = 1.8 \times 10^{-10}$$

The solute solubility $s = \sqrt{K_{sp}}$

The order of increasing molar solubility is the same as the order of increasing K_{sp} values.

Molar solubilities: AgI < AgBr < AgCl

$$K_{sp}: \; 8.5 \times 10^{-17} < 5.0 \times 10^{-13} < 1.8 \times 10^{-10}$$

Estimation Exercise 18.1

Refer to Table 18.1 and arrange the following solutes in order of increasing molar solubility: MgF_2, CaF_2, $PbCl_2$, PbI_2.

We have just seen that we can easily compare the relative solubilities of solutes of the same type, that is, where all solutes have a 1:1 cation-to-anion ratio, where all solutes have a 1:2 cation-to-anion ratio, and so on. However, to compare molar solubilities of solutes of different types, for example, MX with MX_2 or with M_2X, we generally need to do some calculations.

Calculations of this type are required in Problems 27 and 28.

18.3 The Common Ion Effect in Solubility Equilibria

In Section 18.2 we stressed relationships between ion concentrations in a saturated solution. We saw that in some cases (for example, CaC_2O_4) cation and anion concentrations are equal, while in others they are related by a factor of two (for example, Ag_2SO_4), or three (for example, Ag_3AsO_4), and so on. These relationships among equilibrium concentrations apply only if there is a *single* source of the ions—the dissolved solute. In many situations, though, an *additional* source may exist for one of the ions. In these cases we must assess equilibrium ion concentrations in a different way.

Figure 18.1 illustrates our prediction of what should happen when concentrated $Na_2SO_4(aq)$ is added to a saturated solution of silver sulfate.

$$Ag_2SO_4(s) \rightleftharpoons 2\,Ag^+(aq) + SO_4^{2-}(aq)$$

Figure 18.1 The common ion effect in solubility equilibrium. A clear saturated solution of silver(I) sulfate from which excess undissolved solute has been removed by filtration is shown on the left. When concentrated $Na_2SO_4(aq)$ is added, the common ion, SO_4^{2-}, reduces the solubility of the slightly soluble Ag_2SO_4, causing some to precipitate (right). (Because the total mass of solute in the saturated solution is quite small, so too is the quantity of precipitate at the bottom of the beaker.)

Sulfate ion from the Na_2SO_4, a *common ion* in the solubility equilibrium, produces an increase in $[SO_4^{2-}]$, a stress that, according to Le Châtelier's principle, is relieved as equilibrium shifts in the *reverse* direction. To establish a new equilibrium,

- some additional $Ag_2SO_4(s)$ precipitates;
- $[Ag^+]$ is *reduced* from its original equilibrium value; and
- $[SO_4^{2-}]$ remains *greater* than in the original equilibrium.

A similar *common* ion effect can be achieved by adding $AgNO_3(aq)$ to saturated $Ag_2SO_4(aq)$, with the predicted effect being that

- some additional $Ag_2SO_4(s)$ precipitates;
- $[Ag^+]$ remains *greater* than in the original equilibrium; and
- $[SO_4^{2-}]$ is *reduced* from its original equilibrium value.

In diagrammatic fashion,

← When a salt supplies Ag^+ *or* SO_4^{2-}, equilibrium shifts to the *left*.

$$Ag_2SO_4(s) \rightleftharpoons 2\ Ag^+(aq) + SO_4^{2-}(aq)$$

To summarize the **common ion effect** in solubility equilibrium, *the solubility of a slightly soluble ionic compound is lowered in the presence of a second solute that furnishes a common ion.*

Qualitatively, the common ions Ag^+ and SO_4^{2-} have the same effect on the solubility of Ag_2SO_4; they reduce it. Quantitatively, however, the effect of Ag^+ is more pronounced than that of SO_4^{2-}, because $[Ag^+]$ appears to the *second* power in the K_{sp} expression for Ag_2SO_4, whereas $[SO_4^{2-}]$ appears to the first power. Example 18.5 shows how effective SO_4^{2-} is in reducing the solubility of Ag_2SO_4, and Exercise 18.5 shows that at the same concentration, Ag^+ is even more effective.

Example 18.5

Calculate the solubility of Ag_2SO_4 in 1.00 M $Na_2SO_4(aq)$.

Solution

Let's use 1.00 M $Na_2SO_4(aq)$ as the solvent, rather than pure water, to prepare a saturated solution of Ag_2SO_4. We assume that the Na_2SO_4 is completely dissociated and that Ag_2SO_4 has no effect on this. We will continue to use the symbol s to represent the number of moles of Ag_2SO_4 that dissolve per liter of saturated solution. From s mol Ag_2SO_4/L we get s mol SO_4^{2-}/L and $2s$ mol Ag^+/L. We tabulate this information, including the all-important 1.00 mol SO_4^{2-}/L, in the familiar format.

	$Ag_2SO_4(s) \rightleftharpoons$	$2\ Ag^+ +$	SO_4^{2-}	$K_{sp} = 1.4 \times 10^{-5}$
Initial concn, M:		—	1.00 (from the Na_2SO_4)	
From Ag_2SO_4 that dissolves, M:		$2s$	s	
Equil. concn, M:		$2s$	$(1.00 + s)$	

The assumption that $(1.00 + s) \approx 1.00$ amounts to saying that essentially all of the common ion comes from the added source, not from the slightly soluble solute. It works well here: $1.00 + (1.9 \times 10^{-3}) \approx 1.00$.

The usual K_{sp} relationship must be satisfied—that is,

$$K_{sp} = [Ag^+]^2[SO_4^{2-}] = (2s)^2(1.00 + s) = 1.4 \times 10^{-5}$$

To simplify the equation, let's assume that s is much smaller than 1.00 M, so that $(1.00 + s) \approx 1.00$.

$$(2s)^2(1.00) = 1.4 \times 10^{-5}$$

$$4s^2 = 1.4 \times 10^{-5}$$

$$s^2 = 3.5 \times 10^{-6}$$

$$s = \text{molar solubility} = (3.5 \times 10^{-6})^{1/2} = 1.9 \times 10^{-3} \text{ mol } Ag_2SO_4/L$$

Exercise 18.5

Calculate the molar solubility of Ag_2SO_4 in 1.00 M $AgNO_3$(aq).

The analytical chemistry laboratory is a place where we often encounter the common ion effect. Consider the following method of determining the percent calcium in limestone. Limestone is principally calcium carbonate, but with small amounts of other constituents.

First, calcium ion is obtained by dissolving the limestone sample in HCl(aq).

$$CaCO_3(\text{s, impure}) + 2\,H_3O^+ \longrightarrow Ca^{2+}(aq) + 3\,H_2O + CO_2(g)$$

After the excess, unreacted HCl(aq) is neutralized, an *excess* of ammonium oxalate solution, $(NH_4)_2C_2O_4$(aq), is added. By an excess we mean that we use more than enough $C_2O_4^{2-}$ ion to precipitate practically all the Ca^{2+} in the solution. The reaction is

$$Ca^{2+}(aq) + C_2O_4^{2-}(aq) \longrightarrow CaC_2O_4(s)$$

The presence of excess oxalate ion, a common ion, greatly reduces the solubility of CaC_2O_4(s). Without this common ion, the solubility of CaC_2O_4(s) is high enough that a small, but significant, percentage of Ca^{2+} would remain in solution. The precipitate is now removed by filtration and washed to free it of residual impurities. For this purpose, $(NH_4)_2C_2O_4$(aq) is used rather than pure water—again, so that the common ion, $C_2O_4^{2-}$, will keep the solubility very low.

Finally, the precipitate is heated to 500 °C, where the calcium oxalate decomposes to *pure* $CaCO_3$(s).

At 500 °C, any ammonium oxalate present decomposes to volatile products:

$$\underset{\text{acid(1)}}{2\,NH_4^+} + \underset{\text{base(2)}}{C_2O_4^{2-}} \longrightarrow \underset{\text{base(1)}}{2\,NH_3(g)} + \underset{\text{acid(2)}}{H_2C_2O_4}$$

$$H_2C_2O_4 \xrightarrow{\Delta} H_2O(g) + CO_2(g) + CO(g)$$

$$CaC_2O_4(s) \xrightarrow{\Delta} CaCO_3(\text{s, pure}) + CO(g)$$

By determining the mass of $CaCO_3$(s) obtained, it is a simple matter to calculate the percent Ca in the original limestone.

18.4 Will Precipitation Occur and Is It Complete?

Sometimes students mistakenly use municipal tap water instead of deionized water in preparing a solution. If the solute is NaCl, this may not cause a problem. However, if the solute is $AgNO_3$, the situation is different; a cloudy solution will probably result (see Figure 18.2). The cloudiness is due to a trace of a precipitate, principally AgCl. What concentration of Cl^- can be tolerated in water before a precipitate forms? We can easily establish a criterion to answer this and similar questions.

To Determine Whether Precipitation Occurs

Let's begin with the two expressions that describe equilibrium between AgCl(s) and its ions, Ag^+ and Cl^-.

$$AgCl(s) \rightleftharpoons Ag^+(aq) + Cl^-(aq) \qquad K_{sp} = [Ag^+][Cl^-] = 1.8 \times 10^{-10}$$

Suppose we use municipal tap water with $[Cl^-] = 1 \times 10^{-6}$ M as the solvent in which to prepare 0.100 M $AgNO_3$(aq). Now, let's formulate a reaction quotient in the form of the ion product Q_{ip} based on the *initial* concentrations.

$$Q_{ip} = [Ag^+]_{init} \times [Cl^-]_{init} = 0.100 \times (1 \times 10^{-6}) = 1 \times 10^{-7}$$

We see that Q_{ip} greatly exceeds the value of K_{sp}. As we established in Chapter 16, when a reaction quotient Q exceeds the value of the corresponding K, a net reaction should occur in the *reverse* direction. We might say that the solution is supersaturated, and that some Ag^+ and Cl^- should combine to form AgCl(s). That is, precipitation should occur, and it should continue until Q_{ip} falls to a value equal to K_{sp}. At this point the solid solute would be in equilibrium with its ions in a saturated solution.

Now, suppose we use a deionized water sample containing $[Cl^-] = 1 \times 10^{-10}$ M to prepare 0.100 M $AgNO_3$. In this case the value of Q_{ip} is

$$Q_{ip} = [Ag^+]_{init} \times [Cl^-]_{init} = 0.100 \times (1 \times 10^{-10}) = 1 \times 10^{-11}$$

Here, the reaction quotient, Q_{ip}, is smaller than K_{sp}. According to the criterion from Chapter 16, a net reaction should occur in the *forward* direction. However, this is not possible because there is no AgCl(s) present to dissolve, but certainly no AgCl will precipitate from the solution. We might say that the solution is unsaturated.

We can use these two examples to formulate a general rule describing what should happen when we mix solutions containing ions capable of precipitating as a slightly soluble ("insoluble") ionic solid.

- Precipitation *should occur* if $Q_{ip} > K_{sp}$
- Precipitation *cannot occur* if $Q_{ip} < K_{sp}$
- A solution is *just saturated* if $Q_{ip} = K_{sp}$.

Example 18.6 illustrates that in applying these criteria we can proceed in two steps: first determine the initial concentrations, and then evaluate Q_{ip}. In Exercise 18.7, where two solutions are mixed, we need to determine how the solutions dilute one another in order to establish the initial concentrations.

Figure 18.2 Precipitation of anions from municipal tap water. When a few drops of $AgNO_3$(aq) are added to municipal tap water, a white precipitate forms. The precipitate forms because certain ions present in the tap water (for example, Cl^-, CO_3^{2-}, and SO_4^{2-}) form slightly soluble silver compounds.

In the back cover photographs, $Q_{ip} > K_{sp}$ for lead(II) iodide before thorough mixing occurs (left), but $Q_{ip} < K_{sp}$ after mixing (right). In the front cover photograph, $Q_{ip} > K_{sp}$ at all times.

Example 18.6

If 1.00 mg of Na_2CrO_4 is added to 225 mL of 0.00015 M $AgNO_3$(aq), will a precipitate form?

$$Ag_2CrO_4(s) \rightleftharpoons 2\,Ag^+(aq) + CrO_4^{2-}(aq) \qquad K_{sp} = 2.4 \times 10^{-12}$$

Solution

Initial Concentrations: The initial concentration of Ag^+ is simply 1.5×10^{-4} M. To establish $[CrO_4^{2-}]$, we first need to determine the number of moles of CrO_4^{2-} placed in solution.

$$? \text{ mol } CrO_4^{2-} = 1.00 \times 10^{-3} \text{ g } Na_2CrO_4 \times \frac{1 \text{ mol } Na_2CrO_4}{162.0 \text{ g } Na_2CrO_4} \times \frac{1 \text{ mol } CrO_4^{2-}}{1 \text{ mol } Na_2CrO_4}$$

$$= 6.17 \times 10^{-6} \text{ mol } CrO_4^{2-}$$

$$[CrO_4^{2-}] = \frac{6.17 \times 10^{-6} \text{ mol } CrO_4^{2-}}{0.225 \text{ L}} = 2.74 \times 10^{-5} \text{ M}$$

Evaluation of Q_{ip}:

$$Q_{ip} = [Ag^+]^2[CrO_4^{2-}] = (1.5 \times 10^{-4})^2(2.74 \times 10^{-5}) = 6.2 \times 10^{-13}$$

Because $Q_{ip} < K_{sp}$, we conclude that *no precipitation occurs.*

Exercise 18.6

If 1.00 g $Pb(NO_3)_2$ and 1.00 g MgI_2 are both added to 1.50 L of H_2O, should a precipitate form?

$$PbI_2(s) \rightleftharpoons Pb^{2+}(aq) + 2\,I^-(aq) \qquad K_{sp} = 7.1 \times 10^{-9}$$

Example 18.7

If equal volumes of 0.0010 M $MgCl_2$(aq) and 0.020 M NaF(aq) are mixed, should a precipitate of MgF_2(s) form?

$$MgF_2(s) \rightleftharpoons Mg^{2+}(aq) + 2\,F^-(aq) \qquad K_{sp} = 3.7 \times 10^{-8}$$

Solution

To apply the precipitation criteria, we need to formulate a reaction quotient, Q_{ip}, from the initial concentrations of Mg^{2+} and F^-. These concentrations, of course, must be based on the *mixture* of the two solutions. We are not given actual volumes, but because the solution volumes are equal, each ion concentration is reduced by one-half as a result of dilution.

$$[Mg^{2+}] = \frac{1}{2} \times \frac{0.0010 \text{ mol } MgCl_2}{\text{L}} \times \frac{1 \text{ mol } Mg^{2+}}{1 \text{ mol } MgCl_2} = 5.0 \times 10^{-4} \text{ M}$$

$$[F^-] = \frac{1}{2} \times \frac{0.020 \text{ mol NaF}}{\text{L}} \times \frac{1 \text{ mol } F^-}{1 \text{ mol NaF}} = 1.0 \times 10^{-2} \text{ M}$$

Now we can compare Q_{ip} and K_{sp}.

$$Q_{ip} = (5.0 \times 10^{-4})(1.0 \times 10^{-2})^2 = 5.0 \times 10^{-8} > 3.7 \times 10^{-8}\ (K_{sp})$$

Because Q_{ip} exceeds K_{sp}, we conclude that a precipitate should form. However, because Q_{ip} exceeds K_{sp} only slightly, the solution could remain supersaturated in MgF_2. Also, because the quantity of precipitate that should form is quite small, it might not be discernible to the unaided eye.

Exercise 18.7

Exactly 100 mL of 0.020 M KI is mixed with 175 mL of 0.0025 M $Pb(NO_3)_2$. Will a precipitate of PbI_2(s) form?

$$PbI_2(s) \rightleftharpoons Pb^{2+}(aq) + 2\,I^-(aq) \qquad K_{sp} = 7.1 \times 10^{-9}$$

(*Hint*: What is the total solution volume from which precipitation might occur?)

To Determine Whether Precipitation Is Complete

The criterion for precipitation from solution—that $Q_{ip} > K_{sp}$—still leaves an unanswered question. How complete is the precipitation that does occur? That is, what percentage of the ion in question is precipitated and what percentage remains in solution? Although a slightly soluble solid can never be totally precipitated from solution, in general, if about 99.9% of the target ion is precipitated and only about 0.1% is left in solution, we consider precipitation to be essentially complete.

In Example 18.8 we return to the precipitation of calcium oxalate described at the end of Section 18.3 and consider a case where precipitation of CaC_2O_4 is not complete. Then, in Exercise 18.8, we see that precipitation is complete in the presence of a large excess of common ion. In general, conditions that favor completeness of precipitation are

- *a very small value of K_{sp}* (so that the concentration of the target ion in the saturated solution after precipitation will be quite small);
- *high initial ion concentrations* (so that the concentrations of the ions left in the saturated solution will be very small percentages of the initial concentrations);
- *a concentration of common ion that considerably exceeds* that of the target ion to be precipitated *(so that a large excess of the common ion remains after precipitation to reduce the solubility of the precipitate).*

Example 18.8

To a solution with $[Ca^{2+}] = 0.0050$ M we add sufficient solid ammonium oxalate to make the initial $[C_2O_4^{2-}] = 0.0051$ M. Will the precipitation of Ca^{2+} as $CaC_2O_4(s)$ be complete?

$$CaC_2O_4(s) \rightleftharpoons Ca^{2+}(aq) + C_2O_4^{2-}(aq) \qquad K_{sp} = 2.7 \times 10^{-9}$$

Solution

First, let's compare Q_{ip} with K_{sp}.

$$Q_{ip} = [Ca^{2+}][C_2O_4^{2-}] = (5.0 \times 10^{-3})(5.1 \times 10^{-3}) = 2.6 \times 10^{-5}$$

Because $Q_{ip} > K_{sp}$, precipitation should occur.

To simplify the algebraic solution of this problem let's approach it in this way: (1) Assume that all the Ca^{2+} precipitates as $CaC_2O_4(s)$ and then (2) calculate the solubility of $CaC_2O_4(s)$ in an aqueous solution containing the slight excess (0.0001 M) of unprecipitated $C_2O_4^{2-}$.

(1) *The Precipitation*:

	Ca^{2+} +	$C_2O_4^{2-}$	$\longrightarrow CaC_2O_4(s)$
Initial concn, M:	0.0050	0.0051	
Changes, M:	−0.0050	−0.0050	
Concn after pptn, M:	≈ 0	0.0001	

(2) *The Solubility Calculation*:

	$CaC_2O_4(s) \rightleftharpoons$	Ca^{2+} +	$C_2O_4^{2-}$
Initial concn, M:		—	0.0001
Changes, M:		$+s$	$+s$
Equil. concn, M:		s	$(0.0001 + s)$

Without the simplifying assumption, and by solving a quadratic equation, we would obtain $s = 2 \times 10^{-5}$. This does not alter our conclusion that precipitation is incomplete. The percent Ca^{2+} remaining in solution would be 0.4%.

As usual, let's assume that $s << 0.0001$ and that $(0.0001 + s) \approx 0.0001$. We then obtain

$$K_{sp} = [Ca^{2+}][C_2O_4^{2-}] = s(0.0001 + s) = s \times 1 \times 10^{-4} = 2.7 \times 10^{-9}$$

$$s = [Ca^{2+}] \approx 3 \times 10^{-5}\ M$$

The percentage of the Ca^{2+} remaining in solution, based on $[Ca^{2+}] = 3 \times 10^{-5}$ M, would be

$$\frac{3 \times 10^{-5}\ M}{5 \times 10^{-3}\ M} \times 100 = 0.6\%$$

Because complete precipitation requires that less than 0.1% of an ion should remain in solution, we conclude that *precipitation is incomplete.*

Exercise 18.8

To a solution with $[Ca^{2+}] = 0.0050$ M we add sufficient solid ammonium oxalate to make the $[C_2O_4^{2-}]$ also 0.0100 M. Will the precipitation of Ca^{2+} as $CaC_2O_4(s)$ be complete?

$$CaC_2O_4(s) \rightleftharpoons Ca^{2+}(aq) + C_2O_4^{2-}(aq) \qquad K_{sp} = 2.7 \times 10^{-9}$$

18.5 Effect of pH on Solubility

If we add HCl(aq) to a saturated solution of calcium oxalate in contact with the precipitated solid, the precipitate completely dissolves (see Figure 18.3). What happens, in part, is that H_3O^+, an acid, combines with $C_2O_4^{2-}$, a base, to form slightly ionized $HC_2O_4^-$. The $HC_2O_4^-$ ion is amphiprotic. In the presence of H_3O^+ it acts as a base and is converted to oxalic acid, $H_2C_2O_4$. The net result is

$$H_3O^+(aq) + C_2O_4^{2-}(aq) \longrightarrow \cancel{HC_2O_4^-(aq)} + H_2O$$
$$H_3O^+(aq) + \cancel{HC_2O_4^-(aq)} \longrightarrow H_2C_2O_4(aq) + H_2O$$

$$2\,H_3O^+(aq) + C_2O_4^{2-}(aq) \longrightarrow H_2C_2O_4(aq) + 2\,H_2O$$

Figure 18.3 The effect of pH on solubility. When an acid, HCl(aq), is added to solid calcium oxalate (left), the solid reacts and dissolves (right). The solubilities of calcium oxalate and many other solutes strongly depend on the pH of the solution.

According to Le Châtelier's principle, as $C_2O_4^{2-}(aq)$ is converted to the weak acid $H_2C_2O_4$, the dissolving of $CaC_2O_4(s)$ is stimulated.

$$CaC_2O_4(s) \rightleftharpoons Ca^{2+}(aq) + C_2O_4^{2-}(aq)$$

If excess HCl(aq) is used, the $CaC_2O_4(s)$ completely dissolves. The net reaction is

$$\begin{array}{l} CaC_2O_4(s) \rightleftharpoons Ca^{2+}(aq) + \cancel{C_2O_4^{2-}(aq)} \\ 2\,H_3O^+(aq) + \cancel{C_2O_4^{2-}(aq)} \longrightarrow H_2C_2O_4(aq) + 2\,H_2O \\ \hline CaC_2O_4(s) + 2\,H_3O^+(aq) \longrightarrow Ca^{2+}(aq) + H_2C_2O_4(aq) + 2\,H_2O \end{array}$$

Thus, the solubility of $CaC_2O_4(s)$ increases as the pH of a solution is lowered.

Calcium oxalate is the principal constituent of some kidney stones but, of course, we can't treat kidney stones with a strong acid like HCl(aq); the acid would be too corrosive to tissues. People prone to the formation of kidney stones can reduce stone formation by avoiding foods that contain oxalates—chocolates, spinach, rhubarb, and black tea.

In contrast to calcium oxalate, the solubility of silver chloride is *independent* of pH. For example, AgCl(s) does not dissolve in dilute $HNO_3(aq)$. Chloride ion, the conjugate base of the strong acid HCl, does not accept a proton from the added HNO_3.

$$H_3O^+(aq) + Cl^-(aq) \longrightarrow \text{no reaction}$$

Because of this, the precipitate does not dissolve.

AgCl(s) does dissolve in HCl(aq) with concentrations above about 0.3 M, but for an entirely different reason: the Ag^+ ion forms a complex ion with Cl^- (see Section 18.6).

Because carbonate ion is a moderately strong base, the solubilities of carbonates are strongly pH dependent. First, carbonate ions are converted to hydrogen carbonate ions, which react further to form carbonic acid, which then dissociates into $CO_2(g)$ and water.

$$CO_3^{2-}(aq) + H_3O^+(aq) \longrightarrow HCO_3^-(aq) + H_2O$$

$$HCO_3^-(aq) + H_3O^+(aq) \longrightarrow H_2CO_3(aq) + H_2O$$

$$H_2CO_3(aq) \longrightarrow H_2O(l) + CO_2(g)$$

Because of the escape of $CO_2(g)$, there is little chance of a reverse reaction; the carbonate dissolves completely in the presence of excess H_3O^+.

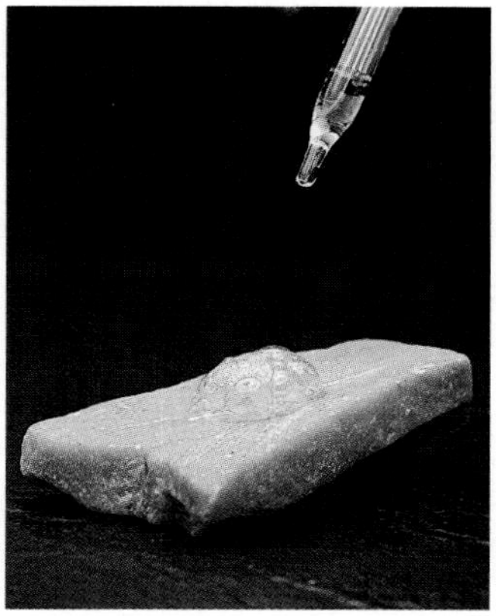

A simple geological field test for a carbonate mineral. When a drop of HCl(aq) is placed on a carbonate mineral, such as limestone, a characteristic "fizzing" occurs, signaling the formation of $CO_2(g)$.

Example 18.9

What is the solubility of $Mg(OH)_2(s)$ in a buffer solution having pH = 9.00?

$$Mg(OH)_2(s) \rightleftharpoons Mg^{2+}(aq) + 2\,OH^-(aq) \qquad K_{sp} = 1.8 \times 10^{-11}$$

Solution

Let's calculate pOH from the pH, and then $[OH^-]$ from pOH.

$$\text{pH} + \text{pOH} = 14.00$$

$$\text{pOH} = 14.00 - 9.00 = 5.00$$

$$\text{pOH} = -\log [OH^-]$$

$$-\text{pOH} = \log [OH^-] = -5.00$$

$$[OH^-] = \text{antilog}(-5.00) = 1.0 \times 10^{-5}\ \text{M}$$

As $Mg(OH)_2$ dissolves in the buffer solution, additional OH^- enters the solution. However, because the function of a buffer solution is to neutralize small added amounts of H_3O^+ or OH^- with substantially no change in pH, $[OH^-]$ remains constant at 1.0×10^{-5} M. This is the value that we substitute into the K_{sp} expression.

$$K_{sp} = [Mg^{2+}][OH^-]^2 = [Mg^{2+}] \times (1.0 \times 10^{-5})^2 = 1.8 \times 10^{-11}$$

$$[Mg^{2+}] = \frac{1.8 \times 10^{-11}}{1.0 \times 10^{-10}} = 0.18 \text{ M}$$

For every mole of $Mg^{2+}(aq)$ produced, one mole of $Mg(OH)_2(s)$ must have dissolved. The molar solubility of $Mg(OH)_2$ and the equilibrium concentration of Mg^{2+} are the same. The molar solubility of $Mg(OH)_2$ at pH 9.00 = 0.18 mol $Mg(OH)_2$/L.

Exercise 18.9

You slowly add NaOH(aq) to an aqueous solution of 0.10 M $FeSO_4(aq)$. At what pH will $Fe(OH)_2(s)$ begin to precipitate?

$$Fe(OH)_2(s) \rightleftharpoons Fe^{2+}(aq) + 2\,OH^-(aq) \qquad K_{sp} = 8.0 \times 10^{-16}$$

(*Hint*: Assume the NaOH(aq) is concentrated and that $[Fe^{2+}]$ remains constant at 0.10 M until precipitation occurs. What $[OH^-]$ is necessary for precipitation to begin?)

In Example 18.9 we found the molar solubility of $Mg(OH)_2$ in a buffer solution with pH = 9.00 to be 0.18 mol $Mg(OH)_2$/L. Previously, in Example 17.3, we saw that the measured pH of saturated $Mg(OH)_2(aq)$ is 10.52, corresponding to a molar solubility of 1.7×10^{-4} mol $Mg(OH)_2$/L. Although a solution with pH = 9.00 is basic, it is nevertheless more acidic than one with pH = 10.52. And the solubility is about a thousand times greater at the lower pH.

CONCEPTUAL EXAMPLE 18.1

In which of the following aqueous solutions will the molar solubility of $Mg(OH)_2$ be greatest: **(a)** 1.00 M NH_3, **(b)** 1.00 M NH_3–1.00 M NH_4Cl, **(c)** 1.00 M NH_4Cl?

$$Mg(OH)_2(s) \rightleftharpoons Mg^{2+}(aq) + 2\,OH^-(aq) \qquad K_{sp} = 1.8 \times 10^{-11}$$

Solution

At high pH values, OH^- in the solution will act as a common ion and *reduce* the solubility of the $Mg(OH)_2$. At lower pH values, OH^- produced by the dissolution of $Mg(OH)_2$ is neutralized, and this causes more $Mg(OH)_2$ to dissolve; thus *increasing* the solubility. We need to compare the pH values of the three solutions.

a. Ammonia is a weak base, and therefore the pH of 1.00 M NH_3 is considerably greater than 7.

b. Because $[NH_3] = [NH_4^+]$, the 1.00 M NH_3–1.00 M NH_4Cl buffer solution has a pH equal to the pK_a of NH_4^+.

$$\begin{aligned} pH = pK_a &= 14.00 - pK_b = 14.00 - (-\log K_b) \\ &= 14.00 + \log(1.8 \times 10^{-5}) \\ &= 14.00 - 4.74 = 9.26 \end{aligned}$$

c. In $NH_4Cl(aq)$, NH_4^+ hydrolyzes and Cl^- does not.

$$NH_4^+ + H_2O \rightleftharpoons H_3O^+ + NH_3 \qquad K_a = K_w/K_b$$

pH, Solubility, and Tooth Decay

Human teeth are composed principally of the mineral *hydroxyapatite*, $Ca_5(PO_4)_3OH$. Hydroxyapatite is insoluble in water but somewhat soluble in acidic solution, and therein lie the origins of tooth decay.

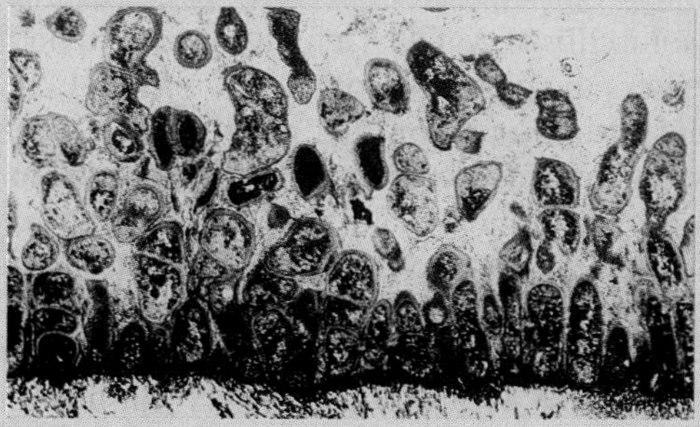

Plaque on tooth enamel, as seen with an electron microscope.

A carbohydrate–protein combination called *mucin* forms a film on teeth called *plaque*. If not removed by brushing and flossing, the buildup of plaque traps food particles. Bacteria ferment carbohydrates in these particles, producing lactic acid ($CH_3CHOHCOOH$). Saliva does not penetrate the plaque and thus cannot buffer against the buildup of acid. The pH drops to as low as 4.5. The H_3O^+ neutralizes the OH^- and converts the PO_4^{3-} of the hydroxyapatite to weakly ionized HPO_4^{2-}, whose calcium salt is water soluble. As a result, some of the hydroxyapatite dissolves.

$$Ca_5(PO_4)_3OH(s) + 4\,H_3O^+(aq) \longrightarrow 5\,Ca^{2+}(aq) + 3\,HPO_4^{2-}(aq) + 5\,H_2O$$

If unchecked, this dissolution of tooth enamel produces cavities. Tooth erosion is even more rapid in people with *bulimia*, an eating disorder characterized by binge eating followed by vomiting. Hydrochloric acid, vomited from the stomach, drops the pH in the mouth to as low as 1.5.

An effective strategy to fight tooth decay is to add fluorides to drinking water or to toothpaste; fluorides replace some of the hydroxyapatite in enamel with *fluorapatite*.

$$\underset{\text{Hydroxyapatite}}{Ca_5(PO_4)_3OH(s)} + F^-(aq) \rightleftharpoons \underset{\text{Fluorapatite}}{Ca_5(PO_4)_3F(s)} + OH^-(aq)$$

Fluorapatite is less soluble in acids than is hydroxyapatite because F^- is a weaker base than is OH^-. As a result, tooth decay is slowed.

The solution is acidic with a pH less than 7. Thus, $NH_4Cl(aq)$, the solution with the lowest pH, should dissolve the greatest amount of $Mg(OH)_2$.

Conceptual Exercise 18.1

Describe the dissolution of $Mg(OH)_2(s)$ in $NH_4Cl(aq)$ through a net acid–base reaction in which NH_4^+ is the acid.

18.6 Complex Ions and Coordination Compounds

On several occasions we've seen how the solubility of a slightly soluble solute is reduced by a common ion. At times, though, we must be careful how much common ion we use. Table 18.2 lists the solubility of AgCl(s) at progressively increasing concentrations of Cl^-. As expected, the solubility is less in the presence of excess Cl^- than it is in pure water, but only for low concentrations of Cl^-. With $[Cl^-]$ greater than about 0.3 M, the solubility of AgCl is actually *greater* than it is in pure water. A new phenomenon enters the picture—complex ion formation.

TABLE 18.2
Solubility of Silver Chloride in the Presence of Cl^-

$[Cl^-]$, M	mol AgCl/L × 10^5
0.000	1.1
0.00386	0.072
0.0176	0.131
0.0884	0.361
0.356	1.74
0.511	2.80
0.975	8.06

A **complex ion** is a polyatomic cation or anion that consists of a central cation to which are bonded other groups (molecules or ions) called **ligands.** Let's consider two examples.

The complex ion responsible for the increased solubility of AgCl(s) at high Cl^- concentrations consists of a central Ag^+ ion and two Cl^- ions as ligands. The complex ion is the entity represented below within brackets. Because it consists of a unipositive cation and two uninegative anions, the complex ion has a net charge of 1−; it is an anion.

$$Ag^+(aq) + 2\,Cl^-(aq) \longrightarrow [AgCl_2]^-(aq)$$

The complex ion formed between copper(II) ions and ammonia molecules is a complex cation—the central copper ion has a charge of 2+ and the ammonia molecules are neutral.

$$\underset{\text{blue}}{Cu^{2+}(aq)} + 4\,NH_3(aq) \longrightarrow \underset{\text{deep blue-violet}}{[Cu(NH_3)_4]^{2+}(aq)}$$

The formation of this complex ion is shown in Figure 18.4.

To obtain complex ions in a pure substance, of course, we need to have a combination of complex ions and counter ions of the opposite charge. For example, the complex ion $[Cu(NH_3)_4]^{2+}$ might be associated with simple sulfate ions, as in the ionic compound $[Cu(NH_3)_4]SO_4$. This is an example of a **coordination compound**—a compound containing complex ions that can be viewed as a combination of two simpler compounds: $CuSO_4$ and NH_3.

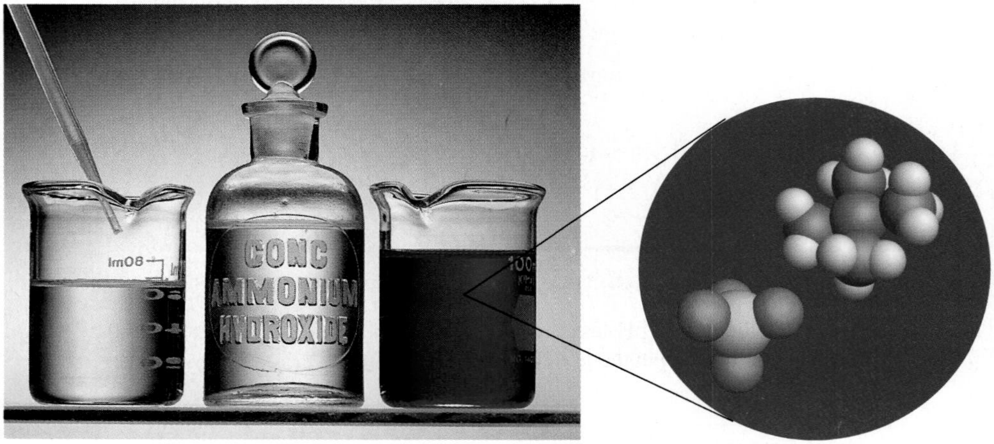

Figure 18.4 **Complex ion formation.** The solution on the left is $CuSO_4(aq)$. Its pale blue color, due to dilute $Cu^{2+}(aq)$, changes to the deep blue-violet color of $[Cu(NH_3)_4]^{2+}(aq)$ (right) upon addition of $NH_3(aq)$, here labeled "conc ammonium hydroxide." The molecular view shows the ions $[Cu(NH_3)_4]^{2+}$ and SO_4^{2-}.

18.7 Equilibria Involving Complex Ions

The equations we just wrote for the formation of complex ions are oversimplifications. Although many formation reactions go essentially to completion, the reactions are generally reversible. For example,*

$$Cu^{2+}(aq) + 4\ NH_3(aq) \rightleftharpoons [Cu(NH_3)_4]^{2+}(aq)$$

Our main concern regarding complex ions in this chapter is equilibrium in these reversible reactions. The **formation constant (K_f)** of a complex ion is the equilibrium constant for the reversible reaction by which a complex ion is formed. Thus, for the reaction

$$Ag^+(aq) + 2\ NH_3(aq) \rightleftharpoons [Ag(NH_3)_2]^+(aq)$$

$$K_f = \frac{[[Ag(NH_3)_2]^+]}{[Ag^+][NH_3]^2} = 1.6 \times 10^7$$

Table 18.3 lists a few representative formation constants.

TABLE 18.3 **Some Formation Constants for Complex Ions**[a]

Complex Ion	Equilibrium Reaction	K_f
$[Co(NH_3)_6]^{3+}$	$Co^{3+} + 6\ NH_3 \rightleftharpoons [Co(NH_3)_6]^{3+}$	4.5×10^{33}
$[Cu(NH_3)_4]^{2+}$	$Cu^{2+} + 4\ NH_3 \rightleftharpoons [Cu(NH_3)_4]^{2+}$	1.1×10^{13}
$[Fe(CN)_6]^{4-}$	$Fe^{2+} + 6\ CN^- \rightleftharpoons [Fe(CN)_6]^{4-}$	1×10^{37}
$[Fe(CN)_6]^{3-}$	$Fe^{3+} + 6\ CN^- \rightleftharpoons [Fe(CN)_6]^{3-}$	1×10^{42}
$[PbCl_3]^-$	$Pb^{2+} + 3\ Cl^- \rightleftharpoons [PbCl_3]^-$	2.4×10^1
$[Ag(NH_3)_2]^+$	$Ag^+ + 2\ NH_3 \rightleftharpoons [Ag(NH_3)_2]^+$	1.6×10^7
$[Ag(CN)_2]^-$	$Ag^+ + 2\ CN^- \rightleftharpoons [Ag(CN)_2]^-$	5.6×10^{18}
$[Ag(S_2O_3)_2]^{3-}$	$Ag^+ + 2\ S_2O_3^{2-} \rightleftharpoons [Ag(S_2O_3)_2]^{3-}$	1.7×10^{13}
$[Zn(NH_3)_4]^{2+}$	$Zn^{2+} + 4\ NH_3 \rightleftharpoons [Zn(NH_3)_4]^{2+}$	4.1×10^8
$[Zn(CN)_4]^{2-}$	$Zn^{2+} + 4\ CN^- \rightleftharpoons [Zn(CN)_4]^{2-}$	1×10^{18}
$[Zn(OH)_4]^{2-}$	$Zn^{2+} + 4\ OH^- \rightleftharpoons [Zn(OH)_4]^{2-}$	4.6×10^{17}

[a] A more extensive tabulation is given in Appendix D.

Consider the reactions that occur in Figure 18.5, the dissolution of silver chloride in aqueous ammonia. Ag^+ from AgCl(s) combines with NH_3 to form the complex ion $[Ag(NH_3)_2]^+$. The Cl^- from AgCl(s) goes into solution as $Cl^-(aq)$. The net reaction is

$$AgCl(s) + 2\ NH_3(aq) \longrightarrow [Ag(NH_3)_2]^+(aq) + Cl^-(aq)$$

but we can think of this as a combination of two equilibrium processes that occur simultaneously.

(1) $$AgCl(s) \rightleftharpoons Ag^+(aq) + Cl^-(aq) \qquad K_{sp} = 1.8 \times 10^{-10}$$

(2) $$Ag^+(aq) + 2\ NH_3(aq) \rightleftharpoons [Ag(NH_3)_2]^+(aq) \qquad K_f = 1.6 \times 10^7$$

---

* Even this equation is somewhat of an oversimplification. $Cu^{2+}(aq)$ itself is best represented as $[Cu(H_2O)_4]^{2+}(aq)$. The reaction is one in which NH_3, which forms stronger bonds to Cu^{2+} than H_2O, displaces H_2O molecules as ligands in the complex ion.

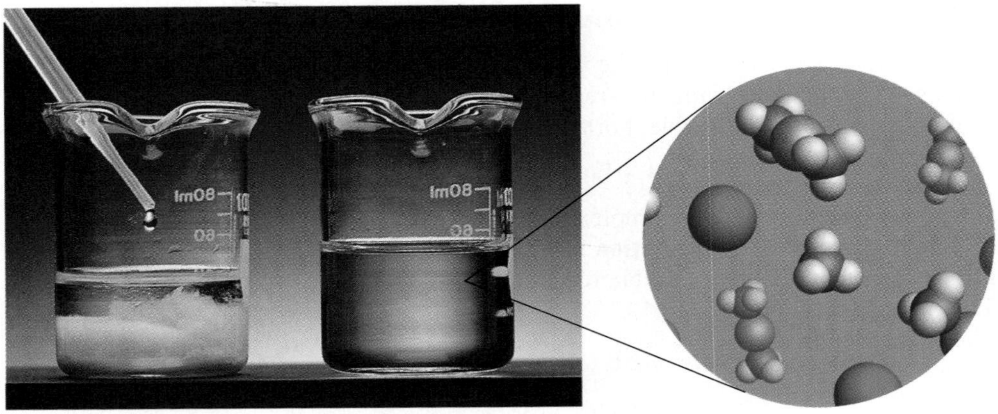

Figure 18.5 Complex ion formation and solute solubility. A precipitate of AgCl(s) (left) readily dissolves in a solution of NH_3 because of the formation of the complex ion $[Ag(NH_3)_2]^+$(aq) (right). The molecular view of the final solution shows $[Ag(NH_3)_2]^+$ and Cl^- ions and ammonia molecules.

Because $[Ag(NH_3)_2]^+$ is a stable complex ion with a large K_f, equilibrium in reaction (2) lies far to the right, and the equilibrium concentration of free Ag^+(aq) is kept low enough that Q_{ip}, the ion product $[Ag^+][Cl^-$, fails to exceed K_{sp} for AgCl. Reaction (1) goes to completion; AgCl(s) dissolves, replaced by soluble $[Ag(NH_3)_2]Cl$.

If we attempt to dissolve AgBr(s) in NH_3(aq), we find that little dissolves; with AgI(s), hardly any at all. We can write equations comparable to those on page 713 for AgCl(s).

$$\text{AgBr(s)} \rightleftharpoons \text{Ag}^+\text{(aq)} + \text{Br}^-\text{(aq)} \qquad K_{sp} = 5.0 \times 10^{-13}$$

$$\text{Ag}^+\text{(aq)} + 2\,\text{NH}_3\text{(aq)} \rightleftharpoons [\text{Ag(NH}_3)_2]^+\text{(aq)} \qquad K_f = 1.6 \times 10^{7}$$

and

$$\text{AgI(s)} \rightleftharpoons \text{Ag}^+\text{(aq)} + \text{I}^-\text{(aq)} \qquad K_{sp} = 8.5 \times 10^{-17}$$

$$\text{Ag}^+\text{(aq)} + 2\,\text{NH}_3\text{(aq)} \rightleftharpoons [\text{Ag(NH}_3)_2]^+\text{(aq)} \qquad K_f = 1.6 \times 10^{7}$$

Because of the very small values of K_{sp} for AgBr and AgI, the low Ag^+ ion concentrations that can be maintained in equilibrium with the complex ion $[Ag(NH_3)_2]^+$ may not be low enough to prevent the ion products $[Ag^+][Br^-]$ and $[Ag^+][I^-]$ from exceeding K_{sp}. In summary, we expect the solubilities of AgCl, AgBr, and AgI in NH_3(aq) to parallel their K_{sp} values:

Solubility in NH_3(aq) decreases:	AgCl	>	AgBr	>	AgI
K_{sp} decreases:	1.8×10^{-10}		5.0×10^{-13}		8.5×10^{-17}

Silver bromide, as an emulsion in gelatin, is used in photographic film. The photographic image is a deposit of metallic silver in the emulsion, produced by exposure of the film to light followed by treatment with a mild reducing agent (for example, hydroquinone, $C_6H_4(OH)_2$). After the film has been developed, it is necessary to remove the unexposed AgBr(s) so that the film will not darken on further exposure to light. For this purpose we can use sodium thiosulfate (also called sodium hyposulfite, or "hypo").

$$\text{AgBr(s)} + 2\,\text{S}_2\text{O}_3^{2-}\text{(aq)} \longrightarrow [\text{Ag(S}_2\text{O}_3)_2]^{3-}\text{(aq)} + \text{Br}^-\text{(aq)}$$

The dissolving of AgBr(s) in $Na_2S_2O_3$(aq), like that of AgCl(s) in NH_3(aq), is the net result of two simultaneous equilibrium processes.

$$\text{AgBr(s)} \rightleftharpoons \text{Ag}^+\text{(aq)} + \text{Br}^-\text{(aq)} \qquad K_{sp} = 5.0 \times 10^{-13}$$

$$\text{Ag}^+\text{(aq)} + 2\,\text{S}_2\text{O}_3^{2-}\text{(aq)} \rightleftharpoons [\text{Ag(S}_2\text{O}_3)_2]^{3-}\text{(aq)} \qquad K_f = 1.7 \times 10^{13}$$

These equations help us to see why AgBr(s) dissolves in $Na_2S_2O_3$(aq) but not appreciably so in NH_3(aq). Because of the much larger value of K_f for $[Ag(S_2O_3)_2]^{3-}$(aq) (1.7×10^{13}) compared to $[Ag(NH_3)_2]^+$(aq) (1.6×10^7), $[Ag^+]$ can be maintained at a low enough level in the presence of $S_2O_3^{2-}$(aq) so that the ion product $[Ag^+][Br^-]$ does not exceed K_{sp}.

In Example 18.10 we use a K_f value to calculate the concentration of the *free*, uncomplexed central cation in a solution containing a complex ion and an excess of the ligand. Then, in Example 18.11 we use this result in applying the criteria for precipitation of a slightly soluble solute.

Example 18.10

Calculate the concentration of free silver ion, $[Ag^+]$, in an aqueous solution that is initially 0.10 M $AgNO_3$ and also 3.0 M NH_3.

$$\text{Ag}^+\text{(aq)} + 2\,\text{NH}_3\text{(aq)} \rightleftharpoons [\text{Ag(NH}_3)_2]^+\text{(aq)} \qquad K_f = 1.6 \times 10^7$$

Solution

Because K_f is a large number, let's begin by assuming that the formation reaction goes almost to completion. The result of the reaction is

$$\text{Ag}^+ \rightleftharpoons 2\,\text{NH}_3 \longrightarrow [\text{Ag(NH}_3)_2]^+$$

	Ag^+	$2\,NH_3$	$[Ag(NH_3)_2]^+$
Initial concn, M:	0.10	3.0	—
Changes, M:	−0.10	−0.20	+0.10
Final concns, M:	≈0	2.8	0.10

Although we've indicated that $[Ag^+] \approx 0$, there is a very small concentration of Ag^+ in solution and it must be consistent with the K_f expression.

$$K_f = \frac{[[\text{Ag(NH}_3)_2]^+]}{[\text{Ag}^+][\text{NH}_3]^2} = \frac{0.10}{[\text{Ag}^+](2.8)^2} = 1.6 \times 10^7$$

$$[\text{Ag}^+] = \frac{0.10}{(2.8)^2(1.6 \times 10^7)} = 8.0 \times 10^{-10}\ \text{M}$$

Example 18.11

If 1.00 g KBr is added to 1.00 L of the solution described in Example 18.10, should any AgBr(s) precipitate from the solution?

$$\text{AgBr(s)} \rightleftharpoons \text{Ag}^+\text{(aq)} + \text{Br}^-\text{(aq)} \qquad K_{sp} = 5.0 \times 10^{-13}$$

Solution

The solution in Example 18.10 is initially 0.10 M in $AgNO_3$ and 3.0 M in NH_3. Our first task would be to determine $[Ag^+]$, but we did this in Example 18.10: $[Ag^+] = 8.0 \times 10^{-10}$ M. Next we need to determine the initial $[Br^-]$ obtained by dissolving 1.00 g KBr in 1.00 L of this solution.

$$[\text{Br}^-] = \frac{1.00\ \text{g KBr} \times \dfrac{1\ \text{mol KBr}}{119.0\ \text{g KBr}} \times \dfrac{1\ \text{mol Br}^-}{1\ \text{mol KBr}}}{1\ \text{L}} = 8.40 \times 10^{-3}\ \text{M}$$

Now we must compare Q_{ip} and K_{sp} for AgBr.

$$Q_{ip} = [Ag^+]_{init} \times [Br^-]_{init} = (8.0 \times 10^{-10})(8.40 \times 10^{-3})$$
$$= 67 \times 10^{-13} > 5.0 \times 10^{-13}\ (K_{sp})$$

Because $Q_{ip} > K_{sp}$, we conclude that some AgBr(s) should precipitate from the solution.

Exercise 18.10

Calculate the concentration of free silver ion, $[Ag^+]$, in an aqueous solution that is initially 0.10 M $AgNO_3$ and 1.0 M $Na_2S_2O_3$.

$$Ag^+(aq) + 2\ S_2O_3^{2-}(aq) \rightleftharpoons [Ag(S_2O_3)_2]^{3-}(aq) \qquad K_f = 1.7 \times 10^{13}$$

Exercise 18.11

If 1.00 g KI is added to 1.00 L of the solution described in Exercise 18.10, should any AgI(s) precipitate from the solution?

$$AgI(s) \rightleftharpoons Ag^+(aq) + I^-(aq) \quad K_{sp} = 8.5 \times 10^{-17}$$

Estimation Exercise 18.2

In which of the following 0.100 M solutions would you expect the solubility of AgI(s) to be greatest: NH_3(aq), $Na_2S_2O_3$(aq), NaCN(aq)? Explain.

CONCEPTUAL EXAMPLE 18.2

Figure 18.6 shows that a precipitate forms when HNO_3(aq) is added to the clear, colorless solution obtained in Figure 18.5. Write an equation(s) to show what happens.

Figure 18.6 **Destruction of a complex ion.** To the clear solution in Figure 18.5 (left) is added HNO_3(aq) (right). The reactions that occur are described in this example.

Solution

The principal solute species in the solution in Figure 18.5 are $[Ag(NH_3)_2]^+$(aq), Cl^-(aq), and uncomplexed, free NH_3(aq). Free Ag^+(aq) is a trace component. The added HNO_3(aq) is a source of H_3O^+, and a proton transfer occurs between H_3O^+, an acid, and uncomplexed, free NH_3, a base.

(a) $$H_3O^+(aq) + NH_3(aq) \longrightarrow NH_4^+(aq) + H_2O(l)$$

The stress imposed by this removal of NH_3(aq) stimulates the *decomposition* of $[Ag(NH_3)_2]^+$(aq), producing more NH_3(aq) and, simultaneously, more Ag^+(aq).

(b) $$[Ag(NH_3)_2]^+(aq) \longrightarrow Ag^+(aq) + 2\,NH_3(aq)$$

As $[Ag^+]$ increases, the ion product $[Ag^+][Cl^-]$ soon exceeds K_{sp} and AgCl(s) precipitates.

(c) $$Ag^+(aq) + Cl^-(aq) \longrightarrow AgCl(s)$$

The net reaction, then, is

(a): $2\,H_3O^+ + \cancel{2\,NH_3(aq)} \longrightarrow 2\,NH_4^+(aq) + 2\,H_2O(l)$
(b): $[Ag(NH_3)_2]^+(aq) \longrightarrow \cancel{Ag^+(aq)} + \cancel{2\,NH_3(aq)}$
(c): $\cancel{Ag^+(aq)} + Cl^-(aq) \longrightarrow AgCl(s)$

Net: $[Ag(NH_3)_2]^+(aq) + Cl^-(aq) + 2\,H_3O^+(aq) \longrightarrow AgCl(s) + 2\,NH_4^+(aq) + 2\,H_2O(l)$

Conceptual Exercise 18.2

Would you expect AgCl(s) to precipitate in Figure 18.6 if $NH_4NO_3(aq)$ were used instead of $HNO_3(aq)$? Explain.

18.8 Acid–Base Reactions of Complex Ions

Some complex ions containing water molecules as ligands—*aqua* complex ions—exhibit acidic properties. For example, in aqueous solutions of iron(III) compounds, the iron(III) exists mostly as the complex ion $[Fe(H_2O)_6]^{3+}$. The electron-withdrawing power of the small, highly charged Fe^{3+} central ion weakens an O—H bond in a ligand water molecule, causing it to lose a proton to a free water molecule in the solution. The result is the formation of a free H_3O^+ ion in solution and the conversion of an H_2O ligand to OH^-. This process is suggested by Figure 18.7 and by the equation

$$[Fe(H_2O)_6]^{3+}(aq) + H_2O(l) \rightleftharpoons [FeOH(H_2O)_5]^{2+}(aq) + H_3O^+(aq) \qquad K_a = 9 \times 10^{-4}$$

From the value of K_a, we see that aqueous solutions of iron(III) compounds have a comparable acidity to acetic acid solutions ($K_a = 1.8 \times 10^{-5}$).

In the complex ion $[Fe(H_2O)_6]^{2+}$, the electron-withdrawing effect of the larger, less highly charged Fe^{2+} central ion is weaker. As a consequence, $[Fe(H_2O)_6]^{2+}$ does not ionize as extensively as $[Fe(H_2O)_6]^{3+}$.

$$[Fe(H_2O)_6]^{2+}(aq) + H_2O(l) \rightleftharpoons [FeOH(H_2O)_5]^{+}(aq) + H_3O^+(aq) \qquad K_a = 1 \times 10^{-7}$$

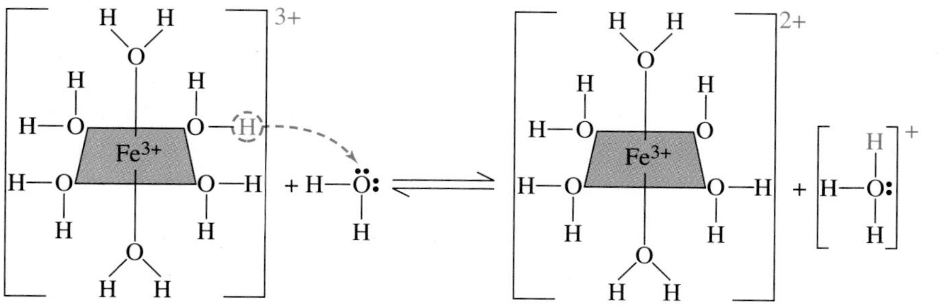

Figure 18.7 Ionization of $[Fe(H_2O)_6]^{3+}(aq)$ as an acid. The transfer of protons from ligand water molecules in $[Fe(H_2O)_6]^{3+}(aq)$ to solvent water molecules produces an increase in $[H_3O^+]$ in the solution.

Amphoterism

In some aqua complex ions, ligand water molecules appear to ionize in a stepwise fashion, with OH^- ions replacing H_2O molecules. In the hypothetical sequence of ionization reactions outlined below, OH^- ions replace H_2O molecules until eventually a complex *anion* is obtained.* In this sequence, complex cations are printed in black; complex anions, in blue; and neutral aluminum hydroxide, in red.

$$(1)\ [Al(H_2O)_6]^{3+} + H_2O \rightleftharpoons [AlOH(H_2O)_5]^{2+} + H_3O^+$$

$$(2)\ [AlOH(H_2O)_5]^{2+} + H_2O \rightleftharpoons [Al(OH)_2(H_2O)_4]^{+} + H_3O^+$$

$$(3)\ [Al(OH)_2(H_2O)_4]^{+} + H_2O \rightleftharpoons [Al(OH)_3(H_2O)_3](s) + H_3O^+$$

$$(4)\ [Al(OH)_3(H_2O)_3](s) + H_2O \rightleftharpoons [Al(OH)_4(H_2O)_2]^{-} + H_3O^+$$

We can use this series of equations to explain the fact that $Al(OH)_3(s)$, which is insoluble in water, reacts with and dissolves *both* in HCl(aq) and in NaOH(aq).

When $Al(OH)_3(s)$ is dissolved in HCl(aq) (Figure 18.8a), the high concentration of H_3O^+ from HCl(aq) strongly favors the *reverse* of reactions (3), (2), and (1), with the aluminum(III) ending up as $[Al(H_2O)_6]^{3+}(aq)$. We can represent this net reaction as

$$Al(OH)_3(s) + 3\ H_3O^+(l) \longrightarrow [Al(H_2O)_6]^{3+}(aq)$$

In the dissolution of $Al(OH)_3(s)$ in NaOH(aq) (Figure 18.8b), OH^- from NaOH(aq) neutralizes H_3O^+ and displaces equilibrium in reaction (4) *far to the right*. The actual form in which most of the aluminum(III) ends up is the complex anion $[Al(OH)_4]^-(aq)$. We can represent the net reaction simply as

$$Al(OH)_3(s) + OH^-(aq) + \longrightarrow [Al(OH)_4]^-(aq)$$

*The actual ionization sequence is more complicated than shown here. Most of the complex ions contain two or more Al^{3+} ions, with OH^- ions forming bridges between them.

(a) (b)

Figure 18.8 Amphoteric behavior of $Al(OH)_3(s)$. (a) Freshly precipitated $Al(OH)_3(s)$ reacts with HCl(aq). In the resulting colorless solution, the aluminum is present as the complex ion $[Al(H_2O)_6]^{3+}(aq)$. (b) Freshly precipitated $Al(OH)_3(s)$ reacts with NaOH(aq). Here the colorless aluminum-containing complex ion is $[Al(OH)_4]^-(aq)$.

Aluminum hydroxide is *amphoteric*. An **amphoteric** hydroxide is one that can react either with an acid or with a base. The hydroxides of zinc and chromium(III) are also amphoteric. The ions produced in acidic solution are $[Zn(H_2O)_4]^{2+}(aq)$ and $[Cr(H_2O)_6]^{3+}(aq)$, respectively. In basic solution they are $[Zn(OH)_4]^{2-}$ and $[Cr(OH)_4]^{-}$. The hydroxide of iron(III), on the other hand, is not amphoteric. $Fe(OH)_3$ reacts in acidic solutions to produce the cation $[Fe(H_2O)_6]^{3+}$ (aq); however, it does not react in basic solutions.

The oxides Al_2O_3, ZnO, and Cr_2O_3 behave in a similar way toward acids and bases as do their hydroxides. For example,

$$Al_2O_3(s) + 6\,H_3O^+(aq) + 3\,H_2O(l) \longrightarrow 2[Al(H_2O)_6]^{3+}(aq)$$
$$Al_2O_3(s) + 2\,OH^-(aq) + 3\,H_2O(l) \longrightarrow 2[Al(OH)_4]^-(aq)$$

Like hydroxides, oxides that can react with and dissolve in either an acid or a base are often said to be *amphoteric*.*

An Alternate View of Amphoterism

As we saw in Section 17.2, the strength of an acid is related to the ease with which an O—H bond can be broken in the linkage,

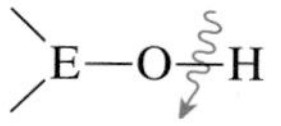

Protons are released and the substance acts as an *acid*. On the other hand, if the bond that breaks is the E—O bond,

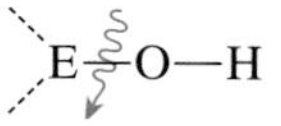

OH^- ions are released and the substance acts as a *base*.

In an *amphoteric* hydroxide, both of these possibilities exist. Thus, to explain the amphoterism of $Al(OH)_3(s)$, we can write

For $Al(OH)_3$, a base, reacting with the acid, H_3O^+:

$$Al(OH)_3(s) \rightleftharpoons Al^{3+}(aq) + \cancel{3\,OH^-(aq)}$$
$$3\,H_3O^+(aq) + \cancel{3\,OH^-(aq)} \longrightarrow 6\,H_2O(l)$$
$$\textit{Net:}\quad Al(OH)_3(s) + 3\,H_3O^+(aq) \longrightarrow Al^{3+}(aq) + 6\,H_2O(l)$$

For $Al(OH)_3$, an acid, reacting with the base, OH^-:

$$Al(OH)_3(s) + \cancel{H_2O(l)} \rightleftharpoons \cancel{H_3O^+(aq)} + \cancel{Al(OH)_2O^-}$$
$$\cancel{H_3O^+(aq)} + OH^-(aq) \longrightarrow \cancel{2\,H_2O(l)}$$
$$\cancel{Al(OH)_2O^-(aq)} + \cancel{H_2O(l)} \longrightarrow Al(OH)_4^-(aq)$$
$$\textit{Net:}\quad Al(OH)_3(s) + OH^-(aq) \longrightarrow Al(OH)_4^-(aq)$$

* The term *amphiprotic* (page 643) refers to hydrogen-containing ions or molecules that are able either to donate or accept a proton. It is best to restrict the term *amphoteric* to the ability of substances, particularly oxides and hydroxides, to react with both acids and bases. For example, although we might say that Al_2O_3 is amphoteric, we can't really call it amphiprotic because it contains no H atoms.

18.9 Qualitative Inorganic Analysis

Many concepts that we've studied in this text—particularly in this and the preceding chapter—have important applications in analytical chemistry. An area of analytical chemistry where acid–base chemistry, precipitation reactions, oxidation–reduction, and complex ion formation all come into sharp focus is *classical, qualitative inorganic analysis.*

The term qualitative signifies an interest in determining what is present (qualitative) but not how much (quantitative). The term "inorganic" signifies analyzing for inorganic constituents, typically as ions. By "classical" we mean analyses involving chemical reactions and by methods developed mostly in the last century. These days analytical chemists tend to use methods that rely on physical separations and the measurement of physical properties with sophisticated instrumentation. In most cases, modern methods are faster and more able to analyze for trace quantities of substances than are classical methods. Still, a study of classical analytical chemistry can be an important aid to learning chemical principles. And, we shouldn't overlook the powerful simplicity of some classical methods. The method of distinguishing between deionized water and municipal tap water illustrated in Figure 18.2 (page 705) is pretty hard to beat as an easy, rapid test.

Figure 18.9 outlines a scheme for determining the presence or absence of 25 common cations in samples that can be obtained in aqueous solution. The scheme first separates the cations into five groups. Further separations and testing for cations are done within each group. A separate scheme is used to test for anions. We'll consider the Group 1 cations in some detail and then briefly run through the remainder of the cation analysis scheme.

Cation Group 1

To an aqueous solution called the *unknown* and containing some combination of 25 possible cations, we add HCl(aq). Of these 25 ions, only Pb^{2+}, $Hg_2{}^{2+}$, and Ag^+ form *insoluble chlorides* (see Figure 18.10). If a precipitate forms at this point, we know that the unknown contains one or more of these ions. If there is no precipitate, these three ions are absent from the unknown. In either case, the solution (if there is no precipitate) or the filtrate (if there is a precipitate) is saved for separations and tests of the other four cation groups.

If we get a Group 1 precipitate, we must then separate the three insoluble chlorides from one another. Of the three chloride precipitates, $PbCl_2$(s) is the most water soluble. We can get enough of it to dissolve just by washing the precipitate with hot water. To establish the presence of Pb^{2+}, we precipitate from the hot water washings a lead compound that is less soluble than $PbCl_2$(s), for example, *yellow* lead chromate, $PbCrO_4$(s) (see Figure 18.10).

$$Pb^{2+}(aq) + CrO_4{}^{2-}(aq) \longrightarrow PbCrO_4(s) \qquad K_{sp} = 2.8 \times 10^{-13}$$

The K_{sp} values of the cation Group 1 precipitates are $PbCl_2$, 1.6×10^{-5}; Hg_2Cl_2, 1.3×10^{-18}; AgCl, 1.8×10^{-10}.

Next, we treat the undissolved portion of the chloride group precipitate with NH_3(aq). Any AgCl(s) present in this precipitate dissolves by the reaction we previously illustrated in Figure 18.5.

$$AgCl(s) + 2\,NH_3(aq) \longrightarrow [Ag(NH_3)_2]^+(aq) + Cl^-(aq)$$

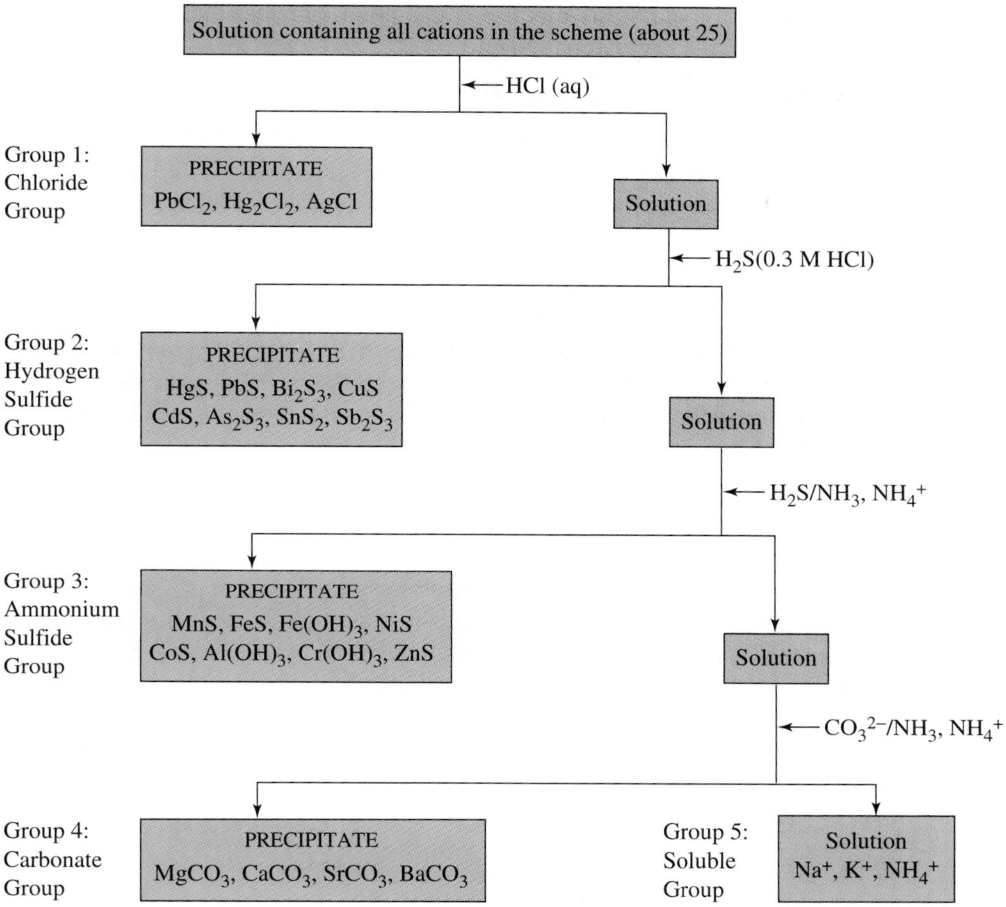

Figure 18.9 Outline of the qualitative analysis scheme for some common cations. Several aspects of this scheme are discussed in the text.

We can confirm that the AgCl(s) has dissolved by treating the $[Ag(NH_3)_2]^+$(aq) with HNO_3(aq), with the result previously illustrated in Figure 18.6.

$$[Ag(NH_3)_2]^+(aq) + Cl^-(aq) + 2\,H_3O^+(aq) \longrightarrow AgCl(s) + 2\,NH_4^+(aq) + 2\,H_2O(l)$$

At the same time that AgCl(s) dissolves in NH_3(aq), any Hg_2Cl_2(s) present undergoes an *oxidation–reduction* reaction. One of the products of the reaction is dark gray mixture of elemental mercury and $HgNH_2Cl$(s) (see Figure 18.10).

$$Hg_2Cl_2(s) + 2\,NH_3(aq) \longrightarrow \underbrace{Hg(l) + HgNH_2Cl(s)}_{\text{dark gray}} + NH_4^+(aq) + Cl^-(aq)$$

Figure 18.10 Cation Group 1 precipitates. Cation Group 1 precipitate: $PbCl_2$ (white); Hg_2Cl_2 (white); AgCl (white, left). Reaction product in test for Hg_2^{2+}: a mixture of Hg (black) and $HgNH_2Cl$ (white, middle). Reaction product in test for Pb^{2+}: $PbCrO_4$ (yellow, right) formed in the reaction of K_2CrO_4(aq) with saturated $PbCl_2$(aq).

The Hg_2Cl_2–NH_3 reaction is a *disproportionation* reaction. The same substance, Hg_2Cl_2, is both oxidized and reduced: mercury(I) is *oxidized* to mercury(II) (in $HgNH_2Cl$) and it is *reduced* to mercury (0) [in Hg(l)].

The presence of a dark gray mixture at this point indicates that mercury(I) ions were present in the original unknown.

Hydrogen Sulfide in the Qualitative Analysis Scheme

Once the cation Group 1 chlorides have been precipitated and separated, a key reagent for separating the remaining cations is hydrogen sulfide. H_2S is a *weak diprotic acid.*

$$H_2S(aq) + H_2O \rightleftharpoons H_3O^+(aq) + HS^-(aq) \qquad K_{a_1} = 1 \times 10^{-7}$$

$$HS^-(aq) + H_2O \rightleftharpoons H_3O^+(aq) + S^{2-}(aq) \qquad K_{a_2} = 1 \times 10^{-19}$$

From the extremely small value of K_{a_2} for hydrogen sulfide ion, HS^-, we expect very little ionization to occur. This suggests that HS^- is the precipitating agent, leading to the following net reaction for precipitation of a metal sulfide MS.

$$H_2S(aq) + H_2O \rightleftharpoons H_3O^+(aq) + HS^-(aq)$$

$$M^{2+}(aq) + HS^-(aq) + H_2O \rightleftharpoons MS(s) + H_3O^+(aq)$$

$$\textit{Net:}\quad M^{2+}(aq) + H_2S(aq) + 2\,H_2O \rightleftharpoons MS(s) + 2\,H_3O^+(aq)$$

Hydrogen sulfide gas has a familiar "rotten egg" odor that is especially noticeable in volcanic areas and near sulfur hot springs. Because the gas can produce headaches and nausea at levels of 10 ppm and can be lethal at about 100 ppm, it is generally produced only in small quantities and directly in the solution where it is to be used. For example, $H_2S(g)$ is slowly released when thioacetamide is heated in aqueous solution.

$$\underset{\text{Thioacetamide}}{CH_3\overset{\overset{\displaystyle S}{\|}}{C}NH_2(aq)} + H_2O(l) \longrightarrow \underset{\text{Acetamide}}{CH_3\overset{\overset{\displaystyle O}{\|}}{C}NH_2(aq)} + H_2S(aq)$$

Cation Groups 2, 3, 4, and 5

Applying Le Châtelier's principle to the net reaction for the precipitation of metal sulfides, we conclude that the reaction to the *left* is favored in *acidic* solutions and to the *right* in *basic* solutions. That is, metal sulfides are more soluble in acidic solutions and less soluble in basic solutions.

$$M^{2+}(aq) + H_2S(aq) + 2\,H_2O \underset{\text{acidic solution}}{\overset{\text{basic solution}}{\rightleftharpoons}} MS(s) + 2\,H_3O^+(aq)$$

Pb^{2+} is found in both Groups 1 and 2. $PbCl_2(s)$ is sufficiently soluble that enough $Pb^{2+}(aq)$ remains in the Group 1 filtrate to precipitate again in Group 2 as PbS(s), which is much less soluble than $PbCl_2(s)$.

This means that only the most insoluble sulfides can be precipitated in a strongly acidic solution of $H_2S(aq)$ where the concentration of HS^- is low. The sulfides in this category are the eight metal sulfides in Group 2. They are precipitated from $H_2S(aq)$ that is also 0.3 M in HCl. Of the eight cations in Group 3, five form sulfides that are soluble in acidic solution but insoluble in an alkaline NH_3/NH_4Cl buffer solution. The other three Group 3 cations form hydroxide precipitates in the alkaline solution.

The cations of Groups 4 and 5 form soluble sulfides, even in basic solution. Their hydroxides, except $Mg(OH)_2$, are also moderately or highly soluble. The Group 4 cations are precipitated as carbonates from a buffered alkaline solution. The cations of Group 5 remain soluble in the presence of all common reagents.

Solving Mysteries with Qualitative Analysis

Arsenic, usually as the oxide As_4O_6, has been widely popularized as a poison in mystery novels, stage plays, and motion pictures. A test for As^{3+} can be performed by the classical qualitative analysis scheme, where it precipitates as the sulfide, As_2S_3, in cation Group 2. However, a specific qualitative test for arsenic, known as the *Marsh test*, is easier to perform. This is the test most often cited in fictional works dealing with arsenic poisoning.

In the Marsh test, the suspected arsenic-containing material is treated with zinc in an acidic solution. Zinc reacts to produce hydrogen gas.

$$Zn(s) + 2\,H_3O^+(aq) \longrightarrow Zn^{2+}(aq) + 2\,H_2O(l) + H_2(g)$$

The hydrogen reduces arsenic in As_4O_6 to arsine, $AsH_3(g)$ (a gaseous compound analogous to ammonia).

$$As_4O_6(s) + 12\,H_2(g) \longrightarrow 4\,AsH_3(g) + 6\,H_2O(l)$$

When heated in a glass tube, the arsine decomposes.

$$2\,AsH_3(g) \longrightarrow 2\,As(s) + 3\,H_2(g)$$

Elemental arsenic deposits from the vapor as a thin film, which is detected as a mirror on the glass.

Today, forensic scientists use a wide variety of techniques, many of them employing chemical principles, to solve all kinds of mysteries, from identifying the source of a paint sample, to testing for drugs in urine or blood, to DNA "fingerprinting."

Arsenic poisoning through foods has been a recurrent theme in works of fiction, as in this scene from the 1944 film "Arsenic and Old Lace."

Within each of the qualitative analysis groups, further reactions are required to dissolve group precipitates and to separate and selectively precipitate individual cations for identification and confirmation. These reactions include oxidation–reduction, complex ion formation, and amphoteric behavior, as well as precipitation. Flame tests are a prominent feature for several of the ions in Groups 4 and 5.

Consider, for example, how we might detect the presence of Cr^{3+} in an unknown. When the Group 3 cations are precipitated, Groups 1 and 2 have already been disposed of, and cations of Groups 4 and 5 remain in the filtrate. The Group 3 precipitate is redissolved in a mixture of HCl(aq) and HNO_3(aq), and the ions possibly present are

$$Fe^{3+}, Mn^{2+}, Co^{2+}, Ni^{2+}, Al^{3+}, Zn^{2+}, \text{and } Cr^{3+}$$

Now, the solution is neutralized and made basic with NaOH(aq). Hydrogen peroxide, H_2O_2(aq), is also added. Although all the Group 3 hydroxides are insoluble in water, three of them are *amphoteric*. In addition to being basic, they have acidic properties that enable them to dissolve in the alkaline medium.

$$Al(OH)_3(s) + OH^-(aq) \longrightarrow Al(OH)_4^-(aq)$$

$$Zn(OH)_2(s) + 2\,OH^-(aq) \longrightarrow Zn(OH)_4^{2-}(aq)$$

$$Cr(OH)_3(s) + OH^-(aq) \longrightarrow Cr(OH)_4^-(aq)$$

The hydrogen peroxide, an *oxidizing* agent, brings about several oxidations, including that of $Cr(OH)_4^-$ to CrO_4^{2-}.

$$3\,H_2O_2(aq) + 2\,Cr(OH)_4^-(aq) + 2\,OH^-(aq) \longrightarrow \underset{\text{yellow}}{2\,CrO_4^{2-}(aq)} + 8\,H_2O(l)$$

Thus, Group 3 is separated into two subgroups: a precipitate and a filtrate. The filtrate possibly contains $Al(OH)_4^-$, $Zn(OH)_4^{2-}$, and CrO_4^{2-}. The appearance of a yellow color is a strong indication that CrO_4^{2-} is present; the other possible ions in the filtrate are colorless. The presence of CrO_4^{2-} can also be confirmed by other tests, such as the precipitation of yellow $BaCrO_4$(s).

SUMMARY

The solubility product constant, K_{sp}, represents equilibrium between a slightly soluble ionic compound and its ions in a saturated aqueous solution. The relationship between K_{sp} and molar solubility is such that either one can be established from a value of the other.

The solubility of a slightly soluble ionic compound is reduced, often significantly, in the presence of an excess of one of the ions involved in the solubility equilibrium. This *common ion effect* is particularly useful in procedures involving precipitates in quantitative analysis.

Criteria for determining what will happen when certain ions are brought together in solution depends on comparing the reaction quotient or ion product, Q_{ip}, with K_{sp}. A precipitate should form if $Q_{ip} > K_{sp}$. It will not form if $Q_{ip} < K_{sp}$, and a solution is just saturated if $Q_{ip} = K_{sp}$.

The usual criterion for completeness of precipitation is that no more than 0.1% of the ion in question remain in solution after precipitation has occurred. Completeness of precipitation is favored by a small value of K_{sp}, a high initial ion concentration, and the presence of a common ion.

The solubilities of some slightly soluble compounds depend strongly on pH. This is the case, for example, if anions of the solute are sufficiently basic to accept protons from H_3O^+. Hydroxides, carbonates, and oxalates all become more soluble as the solution pH is lowered.

Certain solutes become more soluble in the presence of species that can serve as ligands in complex ions. Thus, AgCl(s) is soluble in NH_3(aq) because Ag^+ can join with two NH_3 molecules to form the complex ion $[Ag(NH_3)_2]^+$. The combination of the complex cation and the Cl^- anion is the water soluble coordination compound $[Ag(NH_3)_2]Cl$. The extent to which a solute dissolves in the presence of complexing ligands depends on the values of K_{sp} and K_f. The formation constant of a complex ion, K_f, describes equilibrium between the complex ion, the free cation, and the free, uncomplexed ligands.

The ability of H_2O ligand molecules to donate protons accounts for the acidic character of some complex ions. And, the formation of complex ions with OH^- ions as ligands accounts for the amphoterism of some oxides and hydroxides [for example, those of aluminum, chromium(III), and zinc]. An amphoteric oxide or hydroxide can react with either an acid or a base.

Precipitation, acid–base, and oxidation–reduction reactions, together with complex ion formation and amphoteric behavior, are all used extensively in the classical scheme for the qualitative analysis of common cations.

KEY TERMS

amphoteric (18.8)
common ion effect (18.3)
complex ion (18.6)
coordination compound (18.6)
formation constant (K_f) (18.7)
ligand (18.6)
solubility product constant (K_{sp}) (18.1)

REVIEW QUESTIONS

(Use data from Tables 18.1 and 18.3 or Appendix D, as necessary.)

1. Write the solubility product constant expression for equilibrium in a saturated solution of iron(III) hydroxide.

$$Fe(OH)_3(s) \rightleftharpoons Fe^{3+}(aq) + 3\ OH^-(aq)$$

2. K_{sp} for zinc phosphate, $Zn_3(PO_4)_2$, is 9.0×10^{-33}. Write the equation for the reversible reaction to which this equilibrium constant applies.

3. Explain why the solubility product concept is limited to ionic compounds that are only slightly soluble in water. That is, why is the concept not useful in describing saturated solutions of NaCl or $NaNO_3$, for example?

4. What is the difference between the *solubility* and the *solubility product constant* of a slightly soluble ionic compound?

5. Which is the more soluble compound, $BaSO_4$ ($K_{sp} = 1.1 \times 10^{-10}$) or $PbSO_4$ ($K_{sp} = 1.6 \times 10^{-8}$)? Explain.

6. What is the reaction quotient, Q_{ip}, and how does it compare to K_{sp}? How is it used in establishing a criterion for precipitation?

7. Will a precipitate form if 1.0×10^{-6} mol each of $Pb(NO_3)_2$ and Na_2CrO_4 are added to 1.00 L of water?

8. What are the conditions that favor the precipitation of a slightly soluble solute going essentially to completion?

9. When does pH affect the solubility of a slightly soluble solute and when does it not? Give some examples.

10. In quantitative analysis, a precipitate is often washed with a dilute salt solution rather than with pure water. Explain why this is done. What type of salt solution should be used?

11. Which of the following reagents could be used to increase the aqueous solubility of $Fe(OH)_3$(s)? That is, which would produce a solution with a greater $[Fe^{3+}]$ than found in saturated $Fe(OH)_3$(aq): $FeCl_3$(aq), HCl(aq), NaOH(aq), CH_3COOH(aq), NH_3(aq)? Explain.

12. Which of the following reagent solutions will dissolve an appreciable quantity of $Cu(OH)_2$(s)? That is, which would produce a solution with a greater total concentration of copper(II) than found in saturated $Cu(OH)_2$(aq): H_2O(l), NH_3(aq), NaOH(aq), HCl(aq), $CuSO_4$(aq)?

13. If a concentrated aqueous solution of sodium acetate is mixed with an aqueous solution of silver nitrate, a precipitate of silver acetate forms. The precipitate dissolves readily when nitric acid is added. Write equations that explain what happens.

14. Describe the meaning of the following related terms: ligand, complex ion, and coordination compound.

15. Explain why $PbCl_2(s)$ is less soluble in 1 M $Pb(NO_3)_2$ than in pure water but somewhat more soluble in 1 M HCl(aq) than in pure water.

16. Explain why sodium thiosulfate, $Na_2S_2O_3(aq)$ ("hypo"), is used to remove excess AgBr from photographic film rather than $NH_3(aq)$, a cheaper reagent.

17. What are the properties that characterize an *amphoteric* oxide or hydroxide? Give an example of an amphoteric hydroxide.

18. What is the reagent used to separate the Group 1 cations from other cations in the qualitative analysis scheme? What reagents are used to separate the Group 2 cations from those of Groups 3, 4, and 5?

19. In qualitative analysis cation Group 1, what reagent is used to separate $PbCl_2(s)$ from the other chlorides? What reagent is used to separate AgCl(s) from $Hg_2Cl_2(s)$?

20. Explain how amphoterism is used to bring about the separation of the cations of qualitative analysis Group 3 into two subgroups.

PROBLEMS

(Use data from Tables 18.1 and 18.3 or Appendix D, as necessary.)

The Solubility Product Constant, K_{sp}

21. Write K_{sp} expressions for the following equilibria.
 a. $Fe(OH)_3(s) \rightleftharpoons Fe^{3+}(aq) + 3\ OH^-(aq)$
 b. $Hg_2Cl_2(s) \rightleftharpoons Hg_2^{2+}(aq) + 2\ Cl^-(aq)$
 c. $MgNH_4PO_4(s) \rightleftharpoons Mg^{2+}(aq) + NH_4^+(aq) + PO_4^{3-}(aq)$
 d. $Li_3PO_4(s) \rightleftharpoons 3\ Li^+(aq) + PO_4^{3-}(aq)$

22. Write equations for the solubility equilibria to which the following K_{sp} values apply: **(a)** $Hg_2(CN)_2$, 5×10^{-40}; **(b)** Ag_3AsO_4, 1.0×10^{-22}; **(c)** YF_3, 6.6×10^{-13}; **(d)** $Fe_4[Fe(CN)_6]_3$, 3.3×10^{-41}.

The Relationship Between Solubility and K_{sp}

23. Can the numerical values of the molar solubility and the solubility product constant of a slightly soluble ionic compound ever be the same? Which of the two is usually the larger quantity? Explain.

24. The solubility product constants, K_{sp}, of $CuCO_3$ and $ZnCO_3$ are 1.4×10^{-10} and 1.4×10^{-11}, respectively. Does this mean that $CuCO_3$ has ten times the solubility of $ZnCO_3$? Explain.

25. A handbook lists the water solubility of barium chromate as 0.0010 g $BaCrO_4$/100 mL. Calculate the K_{sp} of $BaCrO_4$.

26. A handbook lists the water solubility of cadmium iodate as 0.097 g $Cd(IO_3)_2$/100 mL. Calculate the K_{sp} of $Cd(IO_3)_2$.

27. Determine which of these slightly soluble solutes has the greater molar solubility: AgCl ($K_{sp} = 1.8 \times 10^{-10}$) or Ag_2CrO_4 ($K_{sp} = 2.4 \times 10^{-12}$).

28. Determine which of these slightly soluble solutes has the greater molar solubility: $Mg(OH)_2$ ($K_{sp} = 1.8 \times 10^{-11}$) or $MgCO_3$ ($K_{sp} = 3.5 \times 10^{-8}$).

29. Without doing a detailed calculation, indicate which of the following saturated aqueous solutions has the highest concentration of Ca^{2+} ion: $CaCO_3$, CaF_2, $CaSO_4$? Explain.

30. Without doing a detailed calculation, indicate which of the following saturated aqueous solutions has the highest concentration of PO_4^{3-} ion: $Ca_3(PO_4)_2$, $FePO_4$, $Mg_3(PO_4)_2$? Explain.

31. Calculate the concentration of Cu^{2+} in parts per billion (ppb) in a saturated solution of copper(II) arsenate, $Cu_3(AsO_4)_2(aq)$. (*Hint*: The term 1 ppb signifies 1 g Cu^{2+} per 10^9 g solution.)

32. Fluoridated drinking water contains about 1 part per million (ppm) of F^-. Is MgF_2 sufficiently soluble in water to be used as the source of fluoride ion for the fluoridation of drinking water? Explain. (*Hint*: The term 1 ppm signifies 1 g F^- per 10^6 g solution.)

The Common Ion Effect in Solubility Equilibria

33. In which of the following is the slightly soluble AgBr likely to be *most soluble*: pure water, 0.050 M KBr, or 1.25 M $AgNO_3$? Explain.

34. In which of the following is the slightly soluble Ag_2CrO_4 likely to be *least soluble*: pure water, 0.10 M K_2CrO_4, or 0.10 M $AgNO_3$? Explain.

35. Calculate the molar solubility of $Mg(OH)_2$ in 0.10 M $MgCl_2$.

36. Calculate the molar solubility of $Mg(OH)_2$ in 0.10 M $NH_3(aq)$. (*Hint*: What is $[OH^-]$ in 0.10 M $NH_3(aq)$?)

37. A 15.0-g sample of NaI is dissolved in water to make 0.250 L of solution. Then the solution is saturated with PbI_2. What will be $[Pb^{2+}]$ and $[I^-]$ in the saturated solution?

38. A solution is saturated with Ag_2SO_4. (**a**) Calculate $[Ag^+]$ in this saturated solution. (**b**) What mass of Na_2SO_4 must be added to 0.500 L of the solution to reduce $[Ag^+]$ in the saturated solution to 2.0×10^{-3} M?

Precipitation Criteria

39. What $[CrO_4^{2-}]$ must be present in 0.00105 M $AgNO_3(aq)$ to just cause $Ag_2CrO_4(s)$ to precipitate?

40. What must be the pH of a solution that is 0.050 M Fe^{3+} to just cause $Fe(OH)_3(s)$ to precipitate?

41. Will a precipitate form if 0.0010 mol $Hg_2(NO_3)_2$ and 0.0010 mol NaCl are added to 20.00 L of water?

42. Will a precipitate form if 0.48 mg $MgCl_2$ and 12.2 mg Na_2CO_3 are added to 225 mL of water?

43. Nineteen centuries ago, the Romans added calcium sulfate to wine. It clarified the wine and removed dissolved lead. (**a**) Calculate $[SO_4^{2-}]$ in wine saturated with $CaSO_4$. (**b**) What $[Pb^{2+}]$ remains in wine that is saturated with $CaSO_4$?

44. $[Ca^{2+}]$ in hard water is about 2.0×10^{-4} M. Water is fluoridated with 1.0 g F^- per 1.0×10^3 L of water. Will $CaF_2(s)$ precipitate from hard water upon fluoridation?

45. Is precipitation complete if $Mg(OH)_2(s)$ is precipitated from a pH = 9.20 buffer solution that is also 0.105 M $MgCl_2$? (*Hint*: What is $[Mg^{2+}]$ remaining in equilibrium with $Mg(OH)_2(s)$?)

46. What pH must be maintained by a buffer solution so that no more than 0.010% of the Mg^{2+} present in 0.360 M $MgCl_2(aq)$ remains in solution following the precipitation of $Mg(OH)_2(s)$? (*Hint*: What is the maximum $[Mg^{2+}]$ that can be permitted in the equilibrium between $Mg(OH)_2(s)$ and its saturated solution?)

Effect of pH on Solubility

47. Calculate the solubility of $Al(OH)_3(s)$ in a buffer solution with pH = 4.50.

48. Calculate the solubility of $Mg(OH)_2$ in a buffer solution that is 0.75 M NH_3 and 0.50 M NH_4Cl.

49. Which of the following would you add to a mixture of $CaCO_3(s)$ and its saturated solution to increase the molar solubility of $CaCO_3$: more water, Na_2CO_3, NaOH, $NaHSO_4$? Explain.

50. Which of the following solids are likely to be more soluble in acidic solution, and which in basic solution? Which are likely to have a solubility that is essentially independent of pH? Explain.
(a) C_6H_5COOH (b) $BaCO_3$ (c) $(CH_3COO)_2Cd$
(d) $LiNO_3$ (e) $CaCl_2$ (f) $Sr(OH)_2$

51. Boiler scale (mostly $CaCO_3$) is insoluble in water. Hydrochloric acid is sometimes used to remove the scale from boilers in commercial power plants. Write an equation to show what happens. Boiler scale is also produced in automatic coffee makers. The recommended method for its removal is to run vinegar (acetic acid) through the machine. Write an equation to show what happens.

Complex Ion Formation

52. A complex ion has Cr^{3+} as its central ion and four NH_3 molecules and two Cl^- ions as ligands. What is the formula of this complex ion?

53. Potassium ferricyanide is used in the manufacture of pigments and as a coating on blueprint paper. It is a coordination compound having the systematic name, potassium hexacyanoferrate(III). "Hexacyanoferrate(III)" signifies a complex ion with six cyanide ions, CN^-, as ligands and Fe^{3+} as the central ion. Write the chemical formula of potassium ferricyanide.

54. Which of the following complex ions would you expect to have the *lowest* concentration of *free* Ag^+ ion in a solution that is 0.10 M in the complex ion and 1.0 M in the free, uncomplexed ligand: $[Ag(NH_3)_2]^+$, $[Ag(CN)_2]^-$, $[Ag(S_2O_3)_2]^{3-}$?

55. Which of the following reagents will increase significantly the concentration of free Zn^{2+} ion when it is added to a solution of the complex ion $[Zn(NH_3)_4]^{2+}$?
(a) H_2O (b) $NaHSO_4^-$
(c) $NH_3(aq)$ (d) HCl(aq)

56. When a few drops of dilute aqueous NH_3 are added to $CuSO_4(aq)$, a pale blue precipitate is observed. When a larger quantity of $NH_3(aq)$ is added, the precipitate dissolves, forming a deep blue-violet solution. Write equations to represent the reactions that occur.

57. Figure 18.5 shows that $AgCl(s)$ can be dissolved in $NH_3(aq)$, and Figure 18.6 shows that $AgCl(s)$ can be reprecipitated by adding $HNO_3(aq)$ to an aqueous solution of $[Ag(NH_3)_2]Cl$. $AgCl(s)$ can also be dissolved in concentrated $HCl(aq)$, but it then cannot be reprecipitated by the addition of $HNO_3(aq)$. Explain this difference.

58. What is $[Zn^{2+}]$ in a solution that is 0.25 M in $[Zn(NH_3)_4]^{2+}$ and 1.50 M in free, uncomplexed NH_3?

59. What must be $[NH_3]$ in 0.15 M $[Ag(NH_3)_2]Cl$ if $[Ag^+]$ is to be kept at 1.0×10^{-8} M?

Amphoterism

60. Cooks know that they should not use aluminum cookware to cook highly acidic foods, such as tomato sauce. They also know that they should not use a strongly alkaline cleaner, such as oven cleaner, to clean aluminum. Explain these observations. (*Hint*: Aluminum metal normally is protected from corrosion in water by a thin, impervious, adherent film of Al_2O_3.)

61. Write plausible net ionic equations to represent what happens, if anything, in each of the following. If no reaction occurs, write N.R.
   a. $ZnCl_2(aq) + NaOH(concd\ aq) \longrightarrow$
   b. $Al_2O_3(s) + HCl(concd\ aq) \longrightarrow$
   c. $Fe(OH)_3(s) + NaOH(concd\ aq) \longrightarrow$

62. Write plausible net ionic equations to represent what happens, if anything, in each of the following. If no reaction occurs, write N.R..
   a. $Fe(OH)_3(s) + HCl(concd\ aq) \longrightarrow$
   b. $NaCl(aq) + NaOH(concd\ aq) \longrightarrow$
   c. $CrCl_3(aq) + NaOH(concd\ aq) \longrightarrow$

Qualitative Inorganic Cation Analysis

63. Both Pb^{2+} and Ag^+ form insoluble chlorides and sulfides. In the qualitative analysis scheme, Pb^{2+} appears both in cation Group 1 and Group 2 whereas Ag^+ appears only in Group 1. Give a plausible explanation for this observation.

64. Although metal sulfides are precipitated in both Groups 2 and 3 of the qualitative analysis scheme, explain why it is unlikely that the precipitating agent is $S^{2-}(aq)$. What is presumed to be the actual precipitating agent?

65. Based on the description in the text, write net ionic equations for the following qualitative analysis procedures.
   a. Precipitation of $PbCl_2(s)$ from a solution containing Pb^{2+}.
   b. The effect of concentrated $NH_3(aq)$ on a mixed precipitate of $AgCl(s)$ and $Hg_2Cl_2(s)$.
   c. The separation of Al^{3+} and Fe^{3+} when a solution containing both is treated with $NaOH(aq)$.

66. Based on the description in the text, write net ionic equations for the following qualitative analysis procedures.
   a. Separation of $PbCl_2(s)$ from $AgCl(s)$ and $Hg_2Cl_2(s)$ in the cation Group 1 precipitate.
   b. The precipitation of Pb^{2+} as $PbS(s)$ from an acidic aqueous solution of H_2S.
   c. The effect of $NaOH(aq)$ and $H_2O_2(aq)$ on an aqueous solution containing Al^{3+}.

67. A cation Group 1 unknown is treated with $HCl(aq)$ and yields a white precipitate. When $NH_3(aq)$ is added to the precipitate, it turns a gray color. According to these tests, indicate for each of the Group 1 cations whether it is definitely present, definitely absent, or whether its presence remains uncertain.

68. A cation Group 3 unknown is treated with $NaOH(aq)$ and H_2O_2. No precipitate forms, but the solution becomes bright yellow in color. Indicate for each of the Group 3 cations whether it is probably present, most likely absent, or whether its presence remains uncertain.

ADDITIONAL PROBLEMS

(Use data from Tables 18.1 and 18.3 or Appendix D, as necessary.)

69. $Ba^{2+}(aq)$ is poisonous when ingested. The lethal dosage in mice is about 12 mg Ba^{2+} per kg of body mass. Despite this fact $BaSO_4$ is used in medicine to obtain X-ray photographs of the gastrointestinal tract.
   a. Explain why $BaSO_4(s)$ is safe to take internally, even though $Ba^{2+}(aq)$ is poisonous.
   b. What is the concentration of Ba^{2+}, in mg/L, in saturated $BaSO_4(aq)$?
   c. $MgSO_4$ can be mixed with $BaSO_4(s)$ in this medical procedure. What function does the $MgSO_4$ serve?

70. Write equations to show why $Zn(OH)_2(s)$ is readily soluble in dilute $HCl(aq)$, $CH_3COOH(aq)$, $NH_3(aq)$, and $NaOH(aq)$.

71. Given the K_{sp} values for $PbCl_2$ of 1.6×10^{-5} at 25 °C and 3.3×10^{-3} at 80 °C. If 1.00 mL of saturated $PbCl_2(aq)$ at 80 °C is cooled to 25 °C, will a sufficient amount of $PbCl_2(s)$ precipitate to be visible? Assume that you can detect as little as 1 mg of the solid.

72. Should precipitation of AgCl(s) occur if 1.15 mg $AgNO_3$ is added to 455 mL of a water sample having 25.5 parts per million (ppm) of chloride ion?

73. The first step in the extraction of magnesium metal from seawater is the precipitation of Mg^{2+} as $Mg(OH)_2(s)$. $[Mg^{2+}]$ in seawater is about 0.059 M.
   a. If a seawater sample is treated so that its $[OH^-]$ is kept constant at 2.0×10^{-3} M, what will be $[Mg^{2+}]$ remaining in solution after precipitation has occurred?
   b. Is precipitation of $Mg(OH)_2(s)$ complete under these conditions?

74. The chief compound in marble is $CaCO_3$. Marble has been widely used for statues and ornamental work on buildings. Marble is readily attacked by acids. Determine the solubility of marble (that is, determine $[Ca^{2+}]$ in a saturated solution) in (**a**) normal rainwater with pH = 5.6 and (**b**) "acid" rainwater with pH = 4.22. Assume that the net reaction that can occur is

$$CaCO_3(s) + H_3O^+(aq) \rightleftharpoons Ca^{2+}(aq) + HCO_3^-(aq) + H_2O(l) \quad K_C = 6.0 \times 10^1$$

75. To a solution that is 0.010 M in CrO_4^{2-} and 0.010 M in SO_4^{2-}, a solution of 0.10 M $Pb(NO_3)_2(aq)$ is slowly added. Which solid, $PbCrO_4(s)$ or $PbSO_4(s)$, would you expect to precipitate first? What must be $[Pb^{2+}]$ at the point at which the first solid appears? What is $[Pb^{2+}]$ at the point at which the second solid begins to precipitate? What is the remaining concentration of the first anion to precipitate at the point where the second anion begins to precipitate? Can CrO_4^{2-} and SO_4^{2-} be effectively separated by this method of fractional precipitation?

76. Show that if an aqueous solution is 0.350 M $MgCl_2$ and also 0.750 M NH_3 a precipitate of $Mg(OH)_2(s)$ will form. The precipitate can be kept from forming if NH_4Cl is added to the solution. Explain why this is so. What minimum mass of NH_4Cl would have to be added to 0.500 L of the solution to prevent the precipitation of $Mg(OH)_2(s)$? (*Hint*: What is the source of OH^- in this solution?)

77. Calculate the minimum amount of NH_3, in moles, that must be added to prevent the precipitation of AgCl(s) from 0.250 L of 0.100 M $AgNO_3$ to which 0.100 mol NaCl is added.

78. A 0.100-L sample of saturated $SrC_2O_4(aq)$ is analyzed for its oxalate ion content by titration with permanganate ion in acidic solution. The volume of 0.00500 M $KMnO_4$ required is 3.22 mL. Use these data to determine K_{sp} for SrC_2O_4.

79. A *white* solid mixture consisting of *two* compounds, each containing a different cation, *partially* dissolves in water. The colorless solution obtained is treated with $NH_3(aq)$ and yields a *white* precipitate. The part of the original solid mixture that is insoluble in water dissolves in HCl(aq) with the evolution of a *gas*. The resulting solution is treated with $(NH_4)_2SO_4(aq)$ and produces a *white* precipitate. State which of the following cations *might* have been present and which *could not* have been present in the original solid mixture. Explain your reasoning.
   (a) Mg^{2+} (b) Cu^{2+} (c) Ba^{2+}
   (d) Na^+ (e) NH_4^+

80. The molar solubility of $CdCO_3(s)$ in 1.00 M KCl is found to be 5.7×10^{-5} mol/L. Use this information, together with $K_{sp} = 5.2 \times 10^{-12}$ for $CdCO_3$, to determine K_f for $[CdCl_4]^{2-}$. (*Hint*: Almost all of the Cd^{2+} is converted to the complex ion. What then are $[CO_3^{2-}]$ and $[Cd^{2+}]$ in the solution?)

This shipment of latex rubber examination gloves ignited through "spontaneous combustion," causing a fire at the Brooklyn Naval Yard on August 2, 1995. In common usage, we say that such a combustion is "spontaneous" because it happens suddenly and without warning. In this chapter we consider the scientific meaning of spontaneous: A spontaneous reaction is one that has a natural tendency to occur. However, to say that a reaction is spontaneous says nothing about whether the reaction will occur rapidly or immeasurably slowly.

19

Thermodynamics: Spontaneity, Entropy, and Free Energy

CONTENTS

The *thermo-* part of the word *thermodynamics* refers to heat; the *-dynamics* part pertains to motion. **Thermodynamics** deals with the relationship between heat and motion (work). More generally, it deals with transformations of energy from one form to another. What can this possibly have to do with chemistry? Recall that in Chapter 5 we discovered some important uses of *thermochemistry*, and thermochemistry is just one limited branch of thermodynamics.

In this chapter we will consider the powerful ideas of *spontaneity* (whether or not a process will go on its own), *entropy* (a measure of the degree of randomness among the atoms and molecules of a system), and *free energy* (a thermodynamic function that relates enthalpy and entropy to spontaneity). Most important, we will explore a key relationship between free energy change and the equilibrium constant.

19.1 Why Study Thermodynamics?

Having the right kind of knowledge and planning ahead can bring great rewards. Building a mountain home with a fire-resistant roof and keeping the area around the home free of brush may save it from being destroyed in a wildfire. With a knowledge of thermodynamics, and by making a few calculations before embarking on some new venture, scientists and engineers can save themselves and their employers a great deal of time, money, and frustration.

The practical value of thermodynamics was well recognized by G. N. Lewis and Merle Randall, who noted in the introduction to their landmark textbook on thermodynamics in 1923,*

*G. N. Lewis and M. Randall, *Thermodynamics and the Free Energy of Chemical Substances*, McGraw-Hill, New York, 1923.

> "To the manufacturing chemist thermodynamics gives information concerning the stability of his substances, the yield which he may hope to attain, the methods of avoiding undesirable substances, the optimum range of temperature and pressure, the proper choice of solvent ... ."

Although they admitted that ". . . the greatest development of applied thermodynamics is still to come," even Lewis and Randall may have been surprised at how widely thermodynamics is now applied. By the end of this chapter we will see that thermodynamics applies even to living systems, not just to reactions in beakers, bunsen burners, and blast furnaces.

19.2 Spontaneous Change

We know from experience that some processes occur by themselves without requiring us to do anything. The ice melts as a glass of iced tea stands at room temperature. In contact with an open flame, methane burns in air to form carbon dioxide and water vapor. An iron nail rusts when exposed to the oxygen and water vapor of moist air. These are all examples of spontaneous processes. From an appropriate initial set of conditions, a **spontaneous process** is one that can occur in a system left to itself; no action from outside the system is necessary to bring this about.

We know equally well that certain other processes don't occur by themselves. They are *nonspontaneous*. Water does *not* freeze to form ice at room temperature. Carbon dioxide and water vapor do *not* combine to form methane and oxygen. The iron oxide of a rusty nail does *not* revert to iron metal and oxygen gas. For a given set of initial conditions, a **nonspontaneous process** is one that will *not* take place in a system left to itself.

As we described earlier in the text, when sodium metal and chlorine gas come into contact a violent reaction occurs (see Figure 9.1, page 312).

$$2\ Na(s) + Cl_2(g) \longrightarrow 2\ NaCl(s)$$

If a process is spontaneous, the reverse process is nonspontaneous, and vice versa.

The reaction is *spontaneous*. Now consider the reverse reaction, the conversion of sodium chloride back to sodium metal and chlorine gas. This process is *nonspontaneous*.

Some nonspontaneous processes are impossible, such as freezing liquid water at a constant temperature of 25 °C. Others, like the conversion of sodium chloride to sodium metal and chlorine gas, are *not* impossible. To make a nonspontaneous process go, some action from outside the system is necessary. Solid sodium chloride can be melted, and the molten NaCl *can* be converted to sodium and chlorine, but only through the action of an electric current from outside the system (electrolysis). We'll consider a number of *nonspontaneous* electrolysis reactions in Chapter 20.

The term "spontaneous" signifies nothing about how fast a process occurs. The reaction between sodium and chlorine is very fast; the rusting of iron is much slower. By thermodynamic criteria, the reaction of H_2 and O_2 gases at room temperature is spontaneous. In fact, though, we never see evidence of this reaction at room temperature. Because of its very high activation energy, the reaction between the two gases at room temperature is so slow that it is immeasurable. We need to keep in mind that thermodynamics generally can tell us what processes are possible, but it is chemical kinetics (Chapter 15) that tells us how fast a process proceeds.

As illustrated in Example 19.1, even at this point in our discussion we can easily identify some spontaneous and nonspontaneous processes, but in other cases we may not be so sure.

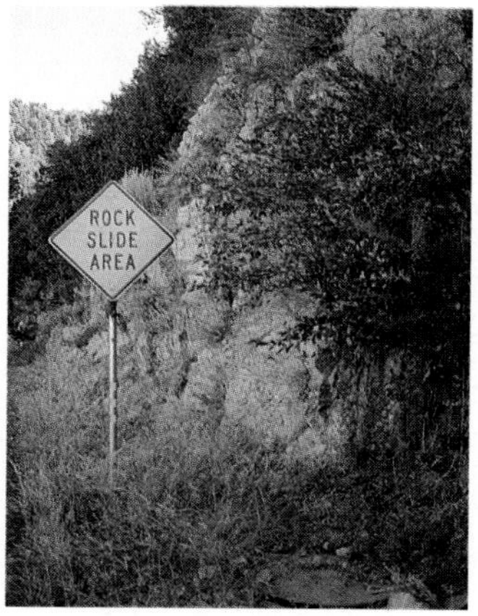

The sign "Rock Slide Area" warns of a spontaneous process. Moreover, the implication is that the barrier preventing a rock fall is weak (low activation energy). A heavy rainstorm or slight earth tremor may provide enough of a "push" to set some rocks in motion.

Example 19.1

Use your general knowledge to indicate whether each of the following is a spontaneous or a nonspontaneous process. Comment on cases where you cannot make a clear determination.

a. The action of toilet bowl cleaner [HCl(aq)] on "lime" deposits [$CaCO_3$(s)].
b. The boiling of water at normal atmospheric pressure and 65 °C.
c. The reaction of N_2(g) and O_2(g) to form NO(g) at room temperature.
d. The melting of an ice cube.

Solution

a. When we add the acid, the fizzing that occurs—the escape of a gas—indicates that the reaction occurs without any further action on our part. The net ionic equation is

$$CaCO_3(s) + 2\,H_3O^+(aq) \longrightarrow Ca^{2+}(aq) + 3\,H_2O(l) + CO_2(g)$$

The reaction is *spontaneous*.

b. We know that the normal boiling point of a liquid is the temperature at which the vapor pressure is equal to 1 atm. For water, this is 100 °C. Thus, the boiling of water at 65 °C and 1 atm pressure is *nonspontaneous*.

c. Nitrogen and oxygen gases occur mixed in air, and we see no evidence of their reaction at room temperature. This makes their reaction to form NO(g) appear to be *nonspontaneous*. On the other hand, this could be an example of a spontaneous reaction that occurs extremely slowly, like that of H_2 and O_2 to form H_2O. We'll have to develop some additional criteria before we can answer this question.

d. This is an interesting case. We know that ice melts spontaneously at temperatures above its normal melting point of 0 °C. Below this temperature it does not. So, the answer here is that whether the process is spontaneous depends on the temperature.

Exercise 19.1

Use your general knowledge to indicate whether each of the following processes is spontaneous or nonspontaneous. Comment on cases where you cannot make a clear determination.

a. The decay of a piece of lumber buried in soil.
b. The formation of sodium, Na(s), and chlorine, Cl_2(g), by vigorously stirring an aqueous solution of sodium chloride, NaCl(aq).
c. The formation of lime, CaO(s), and carbon dioxide, CO_2(g), at 1 atm pressure, from limestone, $CaCO_3$(s), at 600 °C.

What we can say at this point, then, is that

- If a process is *spontaneous* under given conditions, the reverse process is *nonspontaneous*.
- While both spontaneous and nonspontaneous processes are *possible*, only a spontaneous process will occur *naturally*, that is, in a system left to itself. Nonspontaneous processes require that we act upon the system in some way.

From Example 19.1 and Exercise 19.1, it is clear that we need a criterion for spontaneous change, one that tells us whether a spontaneous change will occur in the forward or the reverse direction, depending on the stated conditions. Common knowledge or intuition can help us a bit with this question. You know that if you leave your car in neutral and forget to set the parking brake, it will roll downhill. You also know that water in a stream always flows in only one direction—downhill. You probably also know why these things happen: In the absence of other forces, the force of gravity impels objects to move in the direction of *decreasing* potential energy—toward the center of Earth.

By analogous reasoning, early chemists proposed that spontaneous chemical reactions should occur in the direction in which energy decreases. In Chapter 5 we spoke of internal energy, E, as the total energy content of a system and ΔE as the change in internal energy accompanying a process such as a chemical reaction. But because most chemical reactions are carried out in vessels open to the atmosphere, we found it convenient to switch to an invented quantity, called *enthalpy*, H, whose change during a chemical reaction, ΔH, is equal to the measured heat of a reaction at constant temperature and pressure, q_p.

Figure 19.1 compares the energy change accompanying a mechanical process—the fall of water—to changes in enthalpy in chemical systems. By strict analogy, we should expect that only *exothermic* reactions occur spontaneously. Reactions that are highly exothermic (for example, combustion reactions) usually are spontaneous, and those that are very endothermic usually are not spontaneous. However, there are some exothermic reactions that are not spontaneous and some endothermic reactions that are. The vaporization of water represented in Figure 19.1 is a *spontaneous endothermic* process.

Let's not give up on common knowledge and intuition just yet, though. In seeking a criterion for spontaneous change, perhaps we need to look at some factor(s) *in addition to* the direction of decrease in energy.

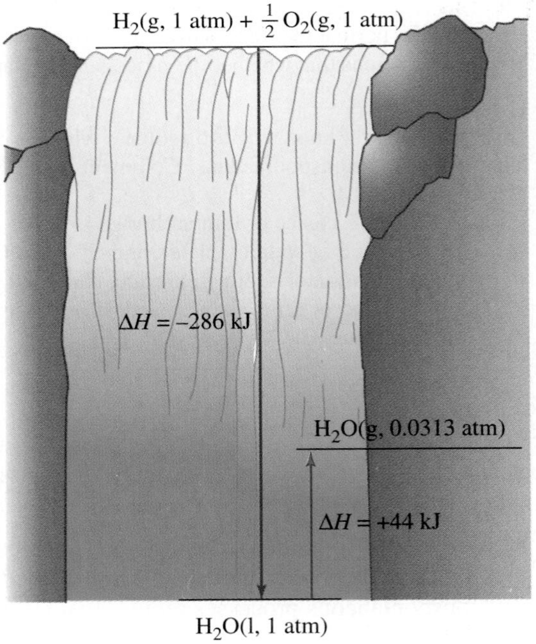

Figure 19.1 The direction of decrease in energy: a criterion for spontaneous change? The direction of the spontaneous flow of water is toward a lower elevation—a lower potential energy. There are exceptions to this idea in the realm of chemistry, however. The formation of $H_2O(l)$ from $H_2(g)$ and $O_2(g)$ at 25 °C and 1 atm is spontaneous, but so is the vaporization of liquid water at 25 °C to produce vapor at pressures up to 0.0313 atm. These two *spontaneous* processes, one *exothermic* and the other *endothermic*, are depicted here.

19.3 Entropy: Disorder and Spontaneity

Suppose you had a new deck of cards, arranged by suit and rank, and spread out on a card table. Now imagine that someone turns on a strong electric fan, and the cards are blown off the table. Which of the three arrangements of cards pictured in Figure 19.2 would you expect to see? You would certainly *not* expect arrangement (a). Some of the cards are still in midair, and the potential energy has *not* been minimized. Arrangements (b) and (c) are both in the minimum energy state available to them. Considering their energy states alone, we would say that the two arrangements are equally likely. Yet you know that arrangement (c) is closer to what you would observe. This scrambled or *disordered* arrangement is far, far more likely than the orderly arrangement in (b), in which all the cards are face up. You could spend a lifetime spilling decks of cards and never observe the arrangement in (b). This example suggests that we must answer *two* questions to assess whether or not a change will be spontaneous:

- Does the system gain or lose energy?
- Does the system become more or less disordered?

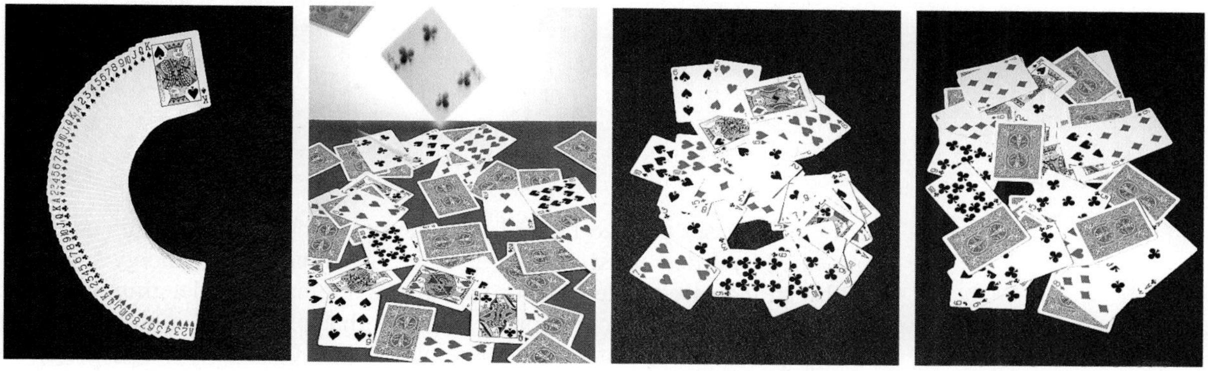

(a) (b) (c)

Figure 19.2 Entropy: the importance of randomness or disorder. A newly opened deck of cards is highly ordered, with the cards arranged according to suit and rank. When the cards are blown off the table by a blast of air, the arrangement in (c) is the one we expect.

When a process involves both a decrease in energy *and* an increase in disorder, as in Figure 19.2 (c), we can easily predict the direction of spontaneous change. In some cases, one of the two factors is negligible and the other takes precedence. An example is shown in Figure 19.3—the formation of an ideal solution of benzene and toluene. Because no chemical reaction occurs and also because intermolecular forces are essentially the same in the solution and in the pure liquids, $\Delta H \approx 0$. Energy change is not an important factor, but a significant change that we do see is in the degree of order. The condition immediately after mixing, with molecules of each type segregated from the other type, is highly organized and cannot persist. The driving force for the formation of the solution is the tendency of the molecules to become mixed up—to achieve the maximum state of *disorder* possible.

Figure 19.3 Formation of an ideal solution. The situation in which the benzene and toluene molecules remain segregated cannot persist; it is more ordered than the final solution with its randomly mixed molecules. Formation of the solution is driven by a natural tendency to achieve the maximum state of disorder possible.

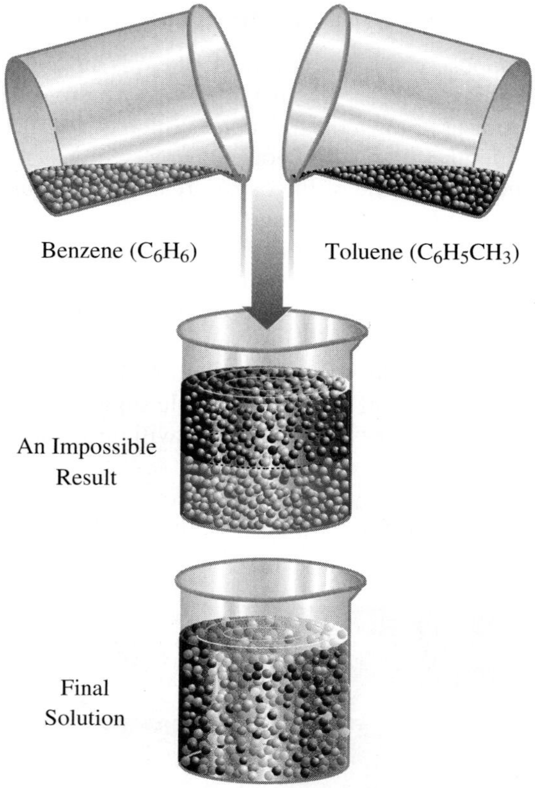

In many cases the two factors work in opposition: a decrease in energy accompanied by a decreased state of disorder, or an increase in energy accompanied by an increased state of disorder. Then it's a matter of determining which factor predominates. A common example is that of soluble ionic substances whose dissolution in water is endothermic ($\Delta H > 0$). Despite this energy increase, dissolution is driven by the increased disorder that is achieved.

The thermodynamic property related to the degree of disorder in a system is called **entropy** (***S***): *The greater the degree of disorder in a system, the greater is its entropy.*

Recall that in Section 5.4 we also defined enthalpy and enthalpy change in a way that made them functions of state.

Entropy is defined in such a way as to make it a function of state. That is, the entropy of a system has a unique value when the composition, temperature, and pressure of the system are specified, and the difference in entropy between two states, the **entropy change** (ΔS), also has a unique value.

In the formation of an ideal liquid solution, the greater disorder in the solution means that the entropy of the solution (S_{mixt}) is greater than the sum of the entropies of the separate liquids ($S_{A(l)} + S_{B(l)}$). The entropy change for the process is *positive* ($\Delta S > 0$).

$$\Delta S = S_{mixt} - (S_{A(l)} + S_{B(l)}) > 0$$

The increase in disorder and entropy that accompanies the spontaneous vaporization of water is suggested by Figure 19.4. You can imagine similar pictorial representations to illustrate these other general processes in which entropy *increases*.

- Solids melt to liquids.
- Solids or liquids vaporize to form gases.
- Solids or liquids mix with a solvent to form nonelectrolyte solutions.
- A chemical reaction produces an increase in number of molecules of *gases*.
- A substance is heated. (Increased temperature means increased atomic, ionic, or molecular motions. Increased motion means increased disorder.)

We apply these generalizations in Example 19.2.

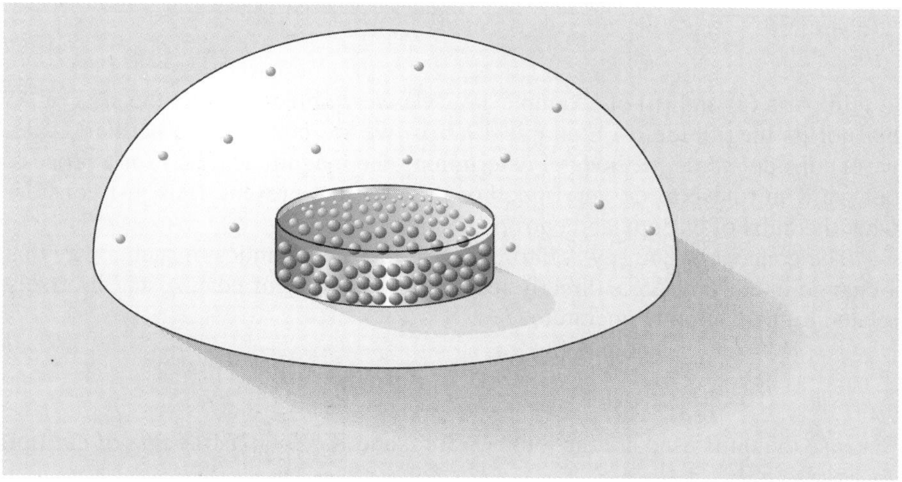

Figure 19.4 Increase of disorder and entropy in the vaporization of water. This illustration depicts the vaporization of water at both the microscopic (left) and macroscopic (right) levels.

Example 19.2

Predict whether each of the following leads to an increase or decrease in entropy. If a prediction is not possible, explain why.

a. The synthesis of ammonia.

$$N_2(g) + 3\,H_2(g) \longrightarrow 2\,NH_3(g)$$

b. Preparation of a sucrose solution.

$$C_{12}H_{22}O_{11}(s) \xrightarrow{H_2O} C_{12}H_{22}O_{11}(aq)$$

c. Evaporation to dryness of a solution of benzoic acid in benzene.

$$\underset{}{C_6H_5COOH[\text{in } C_6H_6(l)]} \longrightarrow \underset{\text{Benzoic acid}}{C_6H_5COOH(s)} + \underset{\text{Benzene}}{C_6H_6(g)}$$

Solution

a. Four moles of gaseous reactants produce two moles of gaseous product. Because of the *loss* of two moles of gas—a highly disorganized state of matter—we predict a more ordered final state and a *decrease* in entropy.

b. The sucrose molecules are highly ordered in the solid state whereas they are randomly distributed in water in an aqueous solution. We predict that entropy *increases*.

c. In the evaporation of benzene (the solvent), a liquid is converted to a gas, suggesting an *increase* in entropy. The benzoic acid, on the other hand, passes from its random distribution in the benzene solution to a highly ordered solid, suggesting a *decrease* in entropy. Overall, then, and without further information, we cannot say for sure whether entropy increases or decreases.

Exercise 19.2

Predict whether each of the following leads to an increase or decrease in entropy. If a prediction is not possible, explain why.

a. $NH_3(g) + HCl(g) \longrightarrow NH_4Cl(s)$
b. $2\ KClO_3(s) \longrightarrow 2\ KCl(s) + 3\ O_2(g)$
c. $CO(g) + H_2O(g) \longrightarrow CO_2(g) + H_2(g)$

In parts (a) and (b) of Example 19.2 we were able to predict the *sign* of ΔS but not its magnitude. In Example 19.2 (c) we were able to do neither. Later, we'll consider some methods of determining the magnitude of ΔS for a process, but we won't have to do very much of this. For the present, let's just consider what the units of entropy and entropy change might be.

The concept of entropy is introduced in thermodynamics in such a way that a change in entropy ΔS is directly related to a quantity of heat, q, and inversely related to the Kelvin temperature, T.

$$\Delta S \propto \frac{q}{T}$$

Because the units used are the joule (J) for q and K^{-1} for $1/T$, the unit of entropy change is $J \cdot K^{-1}$ or J/K.

Proof of the relationship among ΔS, q, and T is beyond the scope of this text, but we can rationalize its form by considering some facts about heating substances. Generally, when we heat a substance, no matter what state of matter it is in, the absorbed heat increases atomic and molecular motion. Greater randomness or disorder is introduced into the substance; its entropy *increases*. The greater the quantity of heat absorbed, the more disorder it can produce. It seems reasonable that ΔS should be directly related to the quantity of heat (q) absorbed.

Moreover, if a highly ordered system absorbs a quantity of heat, this produces more disordering than if the same quantity of heat is absorbed by a system that is already highly disordered. Figure 19.5 suggests an analogy to this proposition that may be easier to understand. We associate a high degree of order with substances at *low* temperatures (for example, for solid substances). Thus, the lower the temperature at which a quantity of heat is absorbed, the more disorder it should produce, in agreement with the fact that ΔS is inversely related to T.

The defining equation for entropy change looks simple, suggesting that it should be easy to use. Unfortunately, it isn't, and we won't be making direct use of the equation in calculations. The difficulty lies in the fact that S and ΔS are functions of state and q is not. The quantity of heat required to bring about a particular change depends on the path chosen. So, if we want ΔS to have a unique value, we must specify a particular path for the process. This path, called a *reversible process*, is suggested by Figure 19.6. For a reversible process, then

$$\Delta S = \frac{q_{rev}}{T}$$

Figure 19.5 An analogy to an entropy change, ΔS. A view of disorder produced by the January 17, 1994, Northridge, California, earthquake in a Los Angeles area market. An aftershock of comparable magnitude to the original 6.7-magnitude temblor would not produce much additional disorder to that shown here. The analogous situation with entropy is that a given input of heat to a rather ordered system at a low temperature produces more disorder than the same quantity of heat added to an already highly disordered system at a higher temperature.

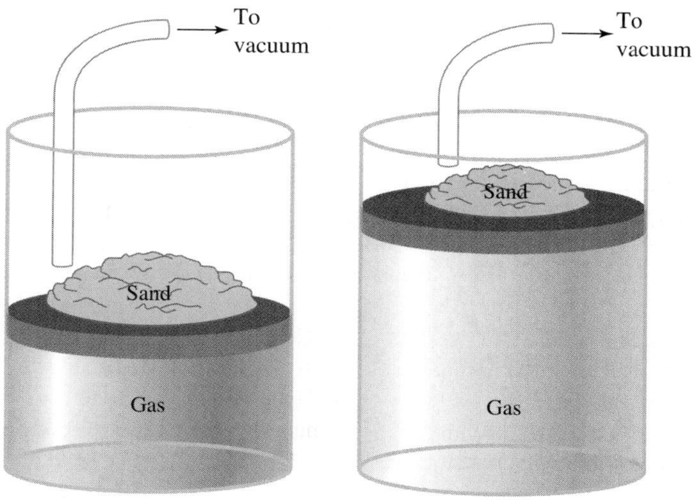

Figure 19.6 A nearly reversible process. A *reversible* process can be made to reverse direction with just an infinitesimal change in a system variable. Here, grains of sand are being removed one at a time and the gas very slowly expands. If at any time a grain of sand is *added* to the pile, the piston will reverse direction and compress the gas. The process is not quite reversible because grains of sand have more than an infinitesimal mass. If the expansion occurs at constant temperature, the amount of heat absorbed by the expanding gas is greater for a process that is reversible than for one that is not. Notice that the similar expansion of a gas pictured in Figure 5.5 was *not* reversible.

Absolute Entropies

The way we have described entropy, a system should have an entropy of *zero* under a condition of perfect order. The most perfect order we can think of is a pure, perfect crystal at the absolute zero of temperature. The **third law of thermodynamics** states that

The entropy of a pure, perfect crystal is zero at 0 K.

Starting with a pure crystalline substance at near 0 K, the entropy changes can be evaluated as the temperature and pressure of the substance are gradually changed to any given final point, such as 1 atm pressure and 25 °C. A small contribution must also be added for the entropy change in heating the substance from 0 K to the actual starting temperature. This can be done with a theoretical calculation. The result of all this effort is an *absolute* value of the entropy. Entropy is an *extensive* property (varying with the size of the sample), and tabulated values are usually on a molar basis. Moreover, if the entropy is for the substance in its standard state (1 atm), we can denote it as $S°$. Thermodynamic data are often tabulated for a substance in its standard state at 25 °C (298 K). We can denote such entropy values as $S°_{298}$.

There is an important distinction between entropy, S, and enthalpy, H, and internal energy, E: Only in the case of entropy can we obtain *absolute* values. With the other two properties we can only work with *differences*, ΔH and ΔE.

Appendix D contains a listing of standard molar entropies (absolute entropies) at 25 °C, and Example 19.3 illustrates how we can use such data to calculate the entropy change of a reaction.

Example 19.3

Use data from Appendix D to calculate the standard molar entropy change at 25 °C for the Deacon process, a high-temperature catalyzed reaction used to convert hydrogen chloride, a byproduct from organic chlorination reactions, into chlorine.

$$4\ HCl(g) + O_2(g) \longrightarrow 2\ Cl_2(g) + 2\ H_2O(g)$$

Solution

Because five moles of gaseous reactants yield four moles of gaseous products, we expect from the generalization on page 737 that entropy decreases in this reaction ($\Delta S < 0$). Let's see if that is the case.

Entropy is an extensive property, so we need to multiply each absolute molar entropy by the coefficient in the equation. That is,

$$\Delta S° = \Sigma \nu_p \times \Delta S°_{products} - \Sigma \nu_p \times \Delta S°_{reactants}$$

$$= 2 \times \Delta S°_{Cl_2(g)} + 2 \times \Delta S°_{H_2O(g)} - 4 \times \Delta S°_{HCl(g)} - \Delta S°_{O_2(g)}$$

$$= (2 \times 223.0) + (2 \times 188.7) - (4 \times 186.8) - 205.0$$

$$= -128.8\ J \cdot K^{-1}$$

$\Delta S°$ is indeed negative.

Exercise 19.3

Use data from Appendix D to calculate the standard molar entropy change at 25 °C for the following reaction.

$$CO(g) + H_2O(g) \longrightarrow CO_2(g) + H_2(g)$$

Factors Affecting Absolute Entropies

Several factors determine the magnitudes of standard molar entropies. We expect gases generally to have higher entropies than liquids, and liquids to have higher entropies than solids. Molecular disorder increases as we progress from solid to liquid to gas. We might wonder, though, why the molar entropy of $CO_2(g)$ [213.6 J • K^{-1} • mol^{-1}] is greater than that of CO(g) [197.6 J • K^{-1} • mol^{-1}].

When these gases are heated, some heat goes simply to raising the average kinetic energies of molecules—raising the temperature. But there are other ways for the heat energy to be used. One possibility, illustrated in Figure 19.7, is that the vibrational energies of the molecules can increase. In the *diatomic* molecule CO(g) only one type of vibration is possible, but in the *triatomic* molecule CO_2 there are three. With more ways of becoming disordered, $CO_2(g)$ has a higher entropy than CO(g) at any given temperature. This suggests that we add the following to the generalizations about entropy on page 737.

- In general, the more complex the molecule, the greater is its standard molar entropy.

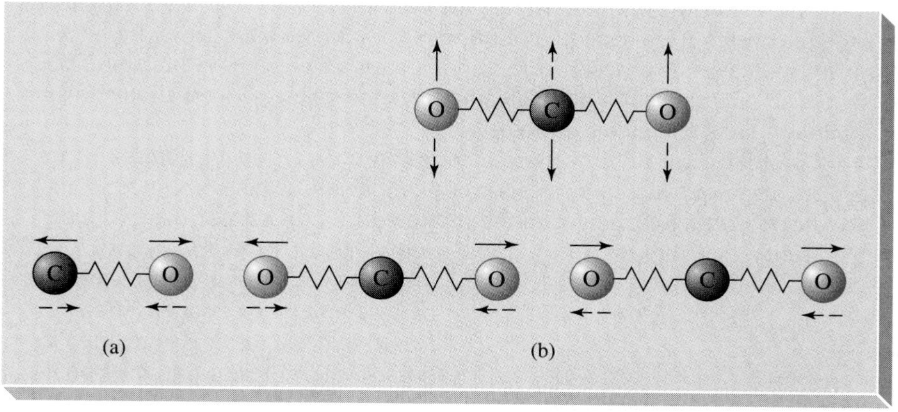

Figure 19.7 Vibrational energy and entropy. The movement of the atoms is suggested by the arrows. The CO molecule (a) has only one type of vibrational motion, whereas the CO_2 molecule (b) has three. The molar entropy of $CO_2(g)$ is greater than that of CO(g).

The Second Law of Thermodynamics

We have seen that an assessment of the energy change in a reaction (ΔH) cannot stand alone as a criterion for spontaneous change, because spontaneous endothermic reactions are possible ($\Delta H > 0$). Can the second factor dealing with spontaneous change, the tendency toward increasing disorder ($\Delta S > 0$), stand alone as a criterion?

In fact, it can, but only if we consider the entropy change of a system *and* of the *surroundings*. We refer to this total entropy change as the entropy change of the "universe."

$$\Delta S_{total} = \Delta S_{universe} = \Delta S_{system} + \Delta S_{surroundings}$$

The **second law of thermodynamics** establishes that *all spontaneous or natural processes increase the entropy of the universe*, which means that

$$\Delta S_{univ} = \Delta S_{syst} + \Delta S_{surr} > 0$$

If a process causes an entropy *increase* in *both* the system and the surroundings, then surely it is *spontaneous*. Just as surely, if both entropy changes are

Entropy and Pollution

Petroleum companies send people and machines to remote regions of Earth to find oil. The oil is pumped to the surface from porous underground rock, sent great distances through pipelines, and then carried across the seas on giant ships. In refineries it is distilled into fractions, and the fractions are processed into gasoline, fuel oil, and other useful products. The entropy of the oil is *decreased* as the oil is collected and separated into purer (and therefore more ordered) components. The production of petroleum products is a *nonspontaneous* process. Considerable outside intervention is necessary to make it happen, and entropy increases must be produced in the environment (surroundings) that more than offset the entropy decreases in the production of petroleum products.

On the other hand, when an oil tanker breaks apart on a shoal, vast quantities of oil spread out on the water and contaminate the shoreline. The entropy of the oil *increases* as the oil mingles with the water. No outside intervention is needed—pollution is a *spontaneous* process. Large entropy increases also occur in the combustion of petroleum fuels and the dispersal of combustion products into the atmosphere. Combustion and air pollution are spontaneous processes.

Another form of entropy-related pollution is *thermal pollution*. In a typical steam-powered electric power plant, heat is used to produce steam, the steam drives a turbine, and the turbine turns a generator that produces electricity (see Figure 19.8). Heat energy is lost in every step of this process, with the greatest amount being the heat of condensation released by steam as it is converted to liquid water for return to the boiler. This heat is captured by cooling water, usually taken from streams, lakes, or other large bodies of water. All of this absorbed heat produces temperature and entropy increases in the environment.

Over 600 oil-well fires like this one in Kuwait produced extensive air pollution at the close of the Persian Gulf war in 1991.

Thermal pollution affects fish and other aquatic life. Different types of algae grow at different temperatures. A higher temperature can cause an algae that is ideal food for fish to be displaced by one that is a poor food or even toxic. Also, because their metabolic rate goes up with temperature, fish need more food and oxygen as their water temperature rises—but the solubility of air (and thus the quantity of oxygen) in water decreases with

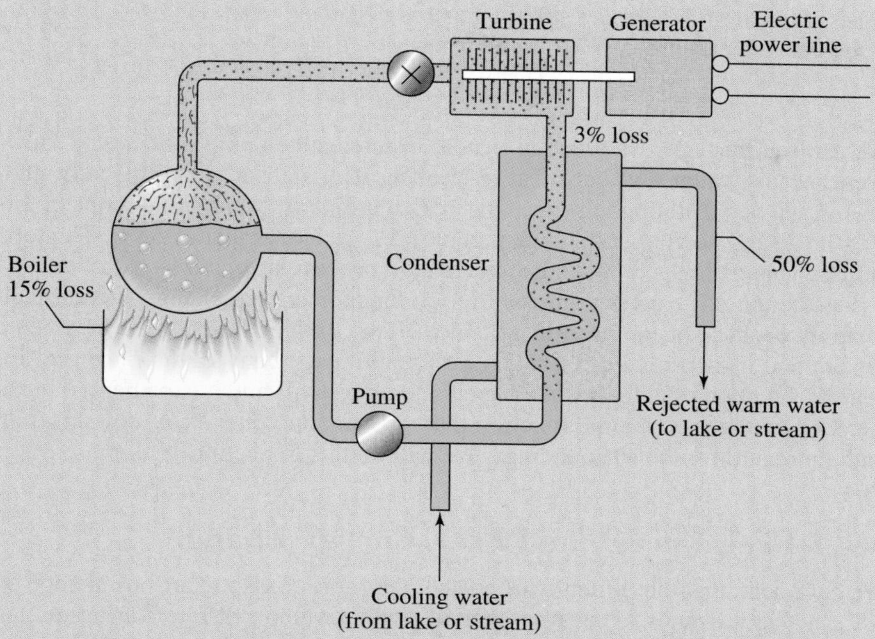

Figure 19.8 **The origin of thermal pollution in electric power generation.** Superheated steam produced in the boiler expands in the turbine where it does work by turning the rotor blades. The turbine drives the electric generator. If the steam were rejected to the surroundings, all the makeup water in the boiler would have to be cold water. This would require more fuel. The steam is condensed, and the condensed hot water is returned to the boiler. The cooling water used in the condensation carries away up to 50% of the heat released in the combustion of the fuel. This waste heat is an important source of thermal pollution.

increasing temperature. Temperature changes affect other physiological processes too. Some fish are killed just by the thermal shock of small temperature changes.

Some of these problems are avoidable. The cooling water from power plants can provide some of the hot water needs of industrial, commercial, and residential facilities. Electric power plants can be made more efficient, but only to a degree. The second law of thermodynamics places a strict limit on the efficiency of a heat engine, a device for converting heat to work (such as the steam turbine of Figure 19.8). This limitation arises because in the operation of the engine there must always be some unrecoverable waste heat. In a modern electric power plant, the conversion of heat to electricity typically cannot be more than about 40% efficient.

As we will see in the next chapter, electricity generated by chemical reactions does not require a heat-to-work conversion. Neither does electricity produced by solar energy devices or water power. In the future, society may need to look more to these sources as a way of reducing thermal pollution.

negative, the process is *nonspontaneous* and cannot occur. An interesting case arises if one term is positive and the other is negative. What will ΔS_{univ} be in this case? Consider the freezing of liquid water at −15 °C.

$$H_2O(l) \longrightarrow H_2O(s)$$

We can see that $\Delta S_{syst} < 0$: ice is a more ordered state than liquid water. However, because water gives off heat in freezing that is absorbed by the surroundings, $\Delta S_{surr} > 0$. Although the necessary calculations are beyond the scope of this text, it can be shown that the magnitude of ΔS_{surr} exceeds that of ΔS_{syst}. The total entropy change, $\Delta S_{univ} > 0$, and so the overall process is *spontaneous*.

Although ΔS_{univ} gives a single criterion for spontaneous change, it is difficult to apply because of the need to consider often complex interactions between a system and its surroundings. Just as we found it expedient to introduce the "invented" function of enthalpy in Chapter 5, here we will find it expedient to introduce a new thermodynamic function that we can apply *just to the system itself*, with no regard for the surroundings. We will do this in Section 19.4.

19.4 Free Energy and Free Energy Change

We have seen that the difficulty of basing a criterion for spontaneous change on ΔS_{univ} is that it cannot be easily evaluated. Let's develop a different criterion that will be easier to use.

To do so, imagine a process conducted at constant temperature and pressure and with work limited to pressure–volume work (page 163). For this process, $q_p = \Delta H_{syst}$. Because the process is at constant temperature, this heat must be exchanged with the surroundings, that is, $q_{surr} = -q_p = -\Delta H_{syst}$. With this information, we can evaluate ΔS_{surr}.*

$$\Delta S_{surr} = \frac{q_{surr}}{T} = \frac{-\Delta H_{syst}}{T}$$

Now, let's turn to the expression

$$\Delta S_{univ} = \Delta S_{syst} + \Delta S_{surr}$$

and substitute what we have just written for ΔS_{surr}.

$$\Delta S_{univ} = \Delta S_{syst} - \frac{\Delta H_{syst}}{T}$$

Next, we can multiply by T

$$T\Delta S_{univ} = T\Delta S_{syst} - \Delta H_{syst} = -(\Delta H_{syst} - T\Delta S_{syst})$$

and then by −1 (that is, change signs).

$$-T\Delta S_{univ} = \Delta H_{syst} - T\Delta S_{syst}$$

---

* To be valid, this equation requires that heat gained or lost by the surroundings must be by a reversible path, that is, $q_{surr} = q_{rev}$.

This equation is very significant because it gives us a way to calculate the elusive ΔS_{univ} in terms of two quantities, ΔH_{syst} and $-T\Delta S_{syst}$, each of which is based just on the system itself. In this way we don't have to consider the surroundings at all.

Now we can bring in an "invented" function. The term on the left of the equation is set equal to the change in a function called the **Gibbs free energy, *G*.**

$$\Delta G = -T\Delta S_{univ}$$

Then, for the **free energy change, Δ*G*,** for a process at constant temperature, the *Gibbs equation* is

$$\Delta G = \Delta H - T\Delta S$$

We can drop the subscripts "univ" and "syst," because in this equation all the quantities relate to the system itself and not the surroundings.

Because the criterion for spontaneous change is that $\Delta S_{univ} > 0$, and because $\Delta G = -T\Delta S_{univ}$, we can now state these criteria relating to spontaneous change in a process at constant T and P:

- If $\Delta G < 0$ (negative), a process is *spontaneous*.
- If $\Delta G > 0$ (positive), a process is *nonspontaneous*.
- If $\Delta G = 0$, a process is at *equilibrium*.

Using Δ*G* as a Criterion for Spontaneous Change

A little later in the chapter we'll use the Gibbs equation to do some quantitative calculations. However, let's look first at some qualitative applications in which we don't need numerical data. If ΔH is *negative* and ΔS is *positive*, then ΔG is *negative* and a reaction will be *spontaneous*. We can just as easily see that if ΔH is *positive* and ΔS is *negative*, then ΔG is *positive* and a reaction will be *nonspontaneous*.

The trickier situations are those in which ΔH and ΔS are *both positive* or *both negative*. Then, the question of whether a reaction is spontaneous or not—that is, whether ΔG has a negative or a positive value—depends on the temperature. You can usually expect, though, that the ΔH term will dominate at *low* temperature and the $T\Delta S$ term at *high* temperature, as suggested by Figure 19.9. Altogether, there are *four* possibilities for ΔG, depending on the signs of ΔH and ΔS. These four possibilities are outlined in Table 19.1 and illustrated in Example 19.4.

TABLE 19.1 Criterion for Spontaneous Change: $\Delta G = \Delta H - T\Delta S$

Case	ΔH	ΔS	ΔG	Result	Example
1	−	+	−	spontaneous at all T	$2\,O_3\,(g) \longrightarrow 3\,O_2\,(g)$
2	−	−	−	spontaneous toward low T	$N_2\,(g) + 3\,H_2\,(g) \longrightarrow 2\,NH_3\,(g)$
	−	−	+	nonspontaneous toward high T	
3	+	+	+	nonspontaneous toward low T	$2\,H_2O\,(g) \longrightarrow 2\,H_2 + O_2\,(g)$
	+	+	−	spontaneous toward high T	
4	+	−	+	nonspontaneous at all T	$2\,C\,(graphite) + 2\,H_2\,(g) \longrightarrow C_2H_4\,(g)$

Example 19.4

Predict which of the four cases in Table 19.1 you expect to apply to the following.

a. $C_6H_{12}O_6(s) + 6\,O_2(g) \longrightarrow 6\,CO_2(g) + 6\,H_2O(g)$ $\Delta H = -2540$ kJ

b. $Cl_2(g) \longrightarrow 2\,Cl(g)$

Solution

a. The reaction is exothermic; $\Delta H < 0$. Twelve moles of gaseous products replace six moles of gaseous reactant. We expect $\Delta S > 0$ for this reaction. With $\Delta H < 0$ and $\Delta S > 0$, the reaction is spontaneous at all temperatures (case 1 of Table 19.1).

b. Because one mole of gaseous reactant produces two moles of gaseous product, $\Delta S > 0$. No value is given for ΔH, but its *sign* must be *positive*: bonds are broken (an endothermic process), not offset by the formation of any new bonds (exothermic). With ΔH and ΔS both positive, we expect the reaction to be nonspontaneous at low temperatures and spontaneous at high temperatures (case 3 in Table 19.1).

This reaction is the first step in the mechanism for the chlorination of methane.

Exercise 19.4

Predict which of the four cases in Table 19.1 are likely to apply to each of the following reactions.

a. $N_2(g) + 2\,F_2(g) \longrightarrow N_2F_4(g)$ $\Delta H = -7.1$ kJ

b. $COCl_2(g) \longrightarrow CO(g) + Cl_2(g)$ $\Delta H = +110.4$ kJ

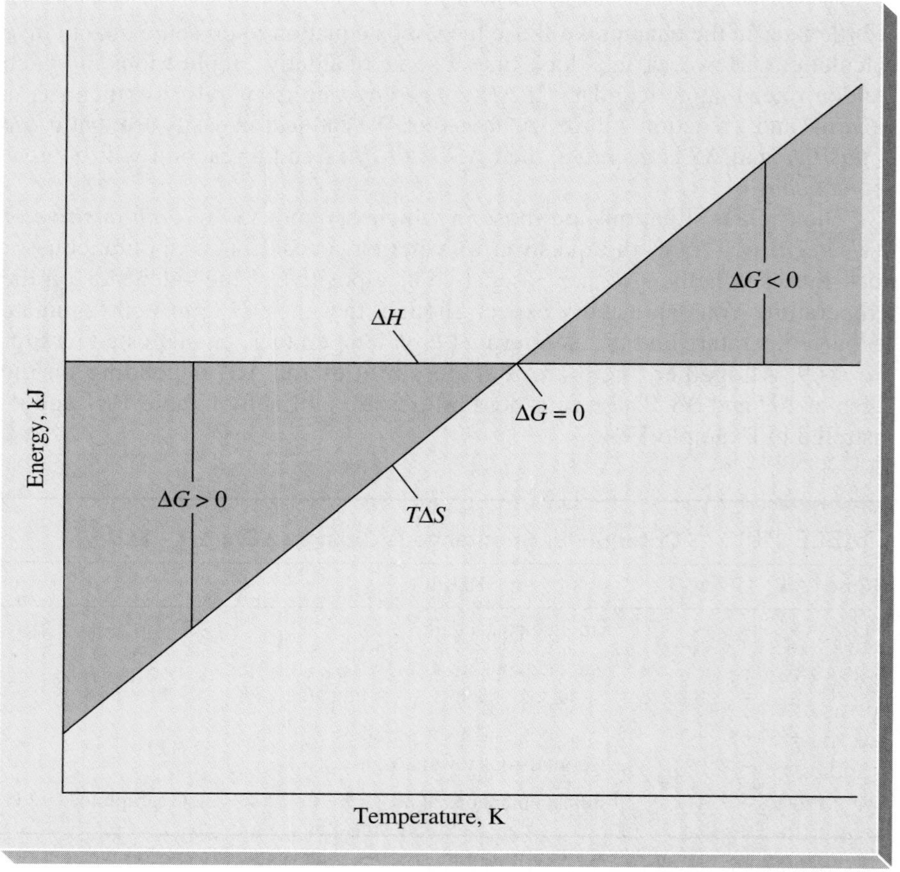

Figure 19.9 ΔG as a criterion for spontaneous change. ΔG, ΔH, and $T\Delta S$ all have the units of energy. The value of ΔG at any temperature is the value on the ΔH line *minus* the value on the $T\Delta S$ line. It is the distance between the two lines. At the temperature at which the lines intersect, this distance is *zero* ($\Delta G = 0$), and the system is at equilibrium. Below this temperature, $\Delta G > 0$ and the reaction is *nonspontaneous*. Above this temperature, $\Delta G < 0$, and the reaction is *spontaneous*. The situation described here is that of case 3 in Table 19.1.

CONCEPTUAL EXAMPLE 19.1

Molecules exist over only a relatively short range of possible temperatures—from near 0 K to a few thousand K. Explain why this is to be expected.

Solution

To cause a molecule to dissociate into its atoms, we need to supply the molecule with enough energy that its atoms can fly apart as a result of vigorous vibrations. This is an endothermic process ($\Delta H > 0$). On the other hand, a system of individual atoms is more disordered than the same atoms united into molecules ($\Delta S > 0$). The key factor is the temperature, T. At low temperatures, ΔH is the determining factor, and molecules are generally stable with respect to uncombined atoms. However, no matter how large the value of ΔH, eventually a temperature is reached where the magnitude of $T\Delta S$ exceeds that of ΔH. Then, ΔG is negative and the dissociation becomes spontaneous.

$$\Delta H - T\Delta S = \Delta G < 0$$

For all known molecules, this high temperature limit is no more than a few thousand K.

Conceptual Exercise 19.1

If the process described in Figure 19.9 is CO_2(s, 1 atm) $\longrightarrow$ CO_2(g, 1 atm), at what temperature will the two lines intersect? Explain. (*Hint*: Use data from Chapter 11.)

19.5 Standard Free Energy Change, ΔG°

In the next section we'll find a special use for the **standard free energy change, ΔG°**, of a reaction. This is the free energy change when reactants and products are in their standard states. The convention for the standard states of substances is the same as that used for ΔH° on page 177: The standard state of a solid or liquid is the pure element or compound at 1 atm pressure and at the temperature of interest. For a gas, the standard state is the pure gas behaving as an ideal gas at 1 atm pressure and the temperature of interest. One way to evaluate the standard free energy change of a reaction is to use standard enthalpy and entropy changes in the Gibbs equation.

$$\Delta G^\circ = \Delta H^\circ - T\Delta S^\circ$$

Another way is to use tabulated standard free energies of formation. The **standard free energy of formation**, ΔG_f°, is the free energy change that occurs in the formation of 1 mol of a substance in its standard state from the reference forms of its element in their standard states. The reference forms of the elements are their most stable forms at the given temperature and 1 atm pressure. Like standard enthalpies of formation, the standard free energies of formation of the elements in their most stable forms at 1 atm pressure have values of *zero*. Appendix D contains a listing of the standard free energies of formation of many substances at 25 °C (298 K), the temperature usually used for tabulations.

To obtain a standard free energy of reaction from standard free energies of formation, we use the same type of relationship that was presented for enthalpies on page 180.

$$\Delta G = \Sigma v_p \Delta G_f^\circ(\text{products}) - \Sigma v_r \Delta G_f^\circ(\text{reactants})$$

Both methods of calculating a standard free energy change are illustrated in Example 19.5.

Example 19.5

Calculate ΔG° at 298 K for the reaction

$$4\,HCl(g) + O_2(g) \longrightarrow 2\,Cl_2(g) + 2\,H_2O(g) \qquad \Delta H^\circ = -114.4\text{ kJ}$$

(**a**) with the Gibbs equation, and (**b**) from standard free energies of formation.

Solution

a. To use the Gibbs equation, we need values of both ΔH° and ΔS°. The value of ΔH° is given, and we can obtain ΔS° from standard molar entropies. We did this in Example 19.3 and obtained a value of $-128.8\text{ J}\cdot\text{K}^{-1}$. When we substitute into the Gibbs equation we must be sure to use the same energy unit for both ΔH° and ΔS°. For example, we can convert $-128.8\text{ J}\cdot\text{K}^{-1}$ to $-0.1288\text{ kJ}\cdot\text{K}^{-1}$.

$$\begin{aligned}\Delta G^\circ &= \Delta H^\circ - T\Delta S^\circ = -114.4\text{ kJ} - [298\text{ K} \times (-0.1288\text{ kJ}\cdot\text{K}^{-1})] \\ &= -114.4\text{ kJ} + 38.4\text{ kJ} = -76.0\text{ kJ}\end{aligned}$$

b. We must look up standard free energies of formation in Appendix D and use them in the expression

$$\begin{aligned}\Delta G^\circ &= 2\,\Delta G_f^\circ[Cl_2(g)] + 2\,\Delta G_f^\circ[H_2O(g)] - 4\,\Delta G_f^\circ[HCl(g)] - \Delta G_f^\circ[O_2(g)] \\ &= 2\,(0.00\text{ kJ}) + 2\,(-228.6\text{ kJ}) - 4\,(-95.30\text{ kJ}) - (0.00\text{ kJ}) \\ &= (-457.2 + 381.2)\text{ kJ} \\ &= -76.0\text{ kJ}\end{aligned}$$

Exercise 19.5

Determine the standard free energy change at 25 °C for these reactions:

a. $2\,NO(g) + O_2(g) \longrightarrow 2\,NO_2(g) \qquad \Delta H^\circ = -114.1\text{ kJ} \qquad \Delta S^\circ = -146.2\text{ J}\cdot\text{K}^{-1}$
using the Gibbs equation

b. $CS_2(l) + 2\,S_2Cl_2(g) \longrightarrow CCl_4(l) + 6\,S(s)$
using standard free energies of formation.

19.6 Free Energy Change and Equilibrium

We have said quite a bit about conditions where $\Delta G < 0$ (spontaneous process) and where $\Delta G > 0$ (nonspontaneous process). Now let's look at the condition where $\Delta G = 0$. For a system at equilibrium

$$\Delta G = \Delta H - T\Delta S = 0$$

and

$$\Delta H = T\Delta S$$

This latter expression allows us to determine ΔH from ΔS, or ΔS from ΔH, at the *equilibrium* temperature. For example, for a liquid *at its normal boiling point*, we can relate the *standard* enthalpy and *standard* entropy of vaporization to each other.

Consider the boiling of benzene at 80.10 °C.

At the normal boiling point, the liquid and its vapor are both in the standard state (1 atm).

$$C_6H_6(l, 1\ atm) \rightleftharpoons C_6H_6(g, 1\ atm) \qquad \Delta H° = 30.76\ kJ$$

$$\Delta S°_{vapn} = \frac{\Delta H°_{vapn}}{T_{bp}} = \frac{30.76\ kJ}{(80.10 + 273.15)\ K} = 0.08708\ kJ \cdot K^{-1}$$

Because this entropy change is for the conversion of one mole of benzene from the liquid to the vapor state, we can call it the standard molar entropy of vaporization of benzene, $\Delta S°_{vapn}$, at its normal boiling point.

Benzene conforms quite well to a rule known as **Trouton's rule.** For *nonpolar* liquids, the standard molar entropy of vaporization at the normal boiling point is approximately 87 $J \cdot K^{-1}$.

$$\Delta S°_{vapn} = \frac{\Delta H°_{vapn}}{T_{bp}} \approx 87\ J \cdot K^{-1}$$

The basis of the rule is that the amount of disorder created in transferring one mole of a nonpolar liquid to vapor at the normal boiling point is pretty much the same from one liquid to another (see Figure 19.10). If intermolecular forces of attraction in a liquid are particularly strong, a greater-than-usual disordering may occur in its vaporization. This will produce a standard molar entropy of vaporization larger than 87 $J \cdot K^{-1}$. With water, for example, hydrogen bonding is an important intermolecular force in the liquid and far less so in the vapor. The standard molar entropy of vaporization of water at the normal boiling point is especially large: 109 $J \cdot K^{-1}$.

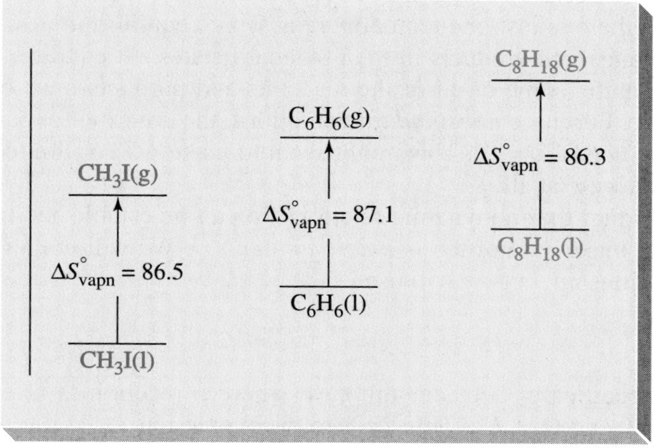

Figure 19.10 Illustrating Trouton's rule. Entropies of vaporization are given in $J \cdot K^{-1}$ per mole of liquid vaporized. The three liquids all have a standard molar entropy of vaporization of about 87 $J \cdot K^{-1}$. However, the following tabulation suggests that they have little else in common.

	MW	Absolute entropy, $S°_{298}$, $J \cdot K^{-1}$ (per mole)	BP, °C	$\Delta H°_{vapn}$, $kJ \cdot mol^{-1}$
CH_3I, iodomethane	142	163	42.4	27.3
C_6H_6, benzene	78	173	80.1	30.8
C_8H_{18}, octane	118	358	125.7	34.4

Estimation Exercise 19.1

The enthalpy of vaporization of pentadecane, $CH_3(CH_2)_{13}CH_3$, is 49.45 kJ/mol. What should be its approximate normal boiling point?

Relationship of $\Delta G°$ to the Equilibrium Constant, K_{eq}

Liquid water is in equilibrium with water vapor at 1 atm pressure and 100 °C. As we have seen, ΔG for this (or any) *equilibrium* process is *zero*. Moreover, because the liquid and vapor are both in their standard states, we can write

$$H_2O(l, 1\text{ atm}) \rightleftharpoons H_2O(g, 1\text{ atm}) \qquad \Delta G°_{373} = 0$$

We can write the same equation for the vaporization of water at 25 °C and determine $\Delta G°_{298}$ from tabulated standard free energies of formation.

$$H_2O(l, 1\text{ atm}) \rightleftharpoons H_2O(g, 1\text{ atm}) \qquad \Delta G°_{298} = +8.590\text{ kJ}$$

The *positive* value of $\Delta G°_{298}$ shows that the forward process is *nonspontaneous*. This does not mean that water will not vaporize at 25 °C, just that it will not produce a vapor *at 1 atm pressure*; equilibrium is displaced to the left. From Table 11.2 we see that the *equilibrium* vapor pressure of water at 25 °C is 23.8 mmHg (or 0.0313 atm).

$$H_2O(l, 0.0313\text{ atm}) \rightleftharpoons H_2O(g, 0.0313\text{ atm}) \qquad \Delta G_{298} = 0$$

To summarize, $\Delta G° = 0$ is a criterion for equilibrium at a single temperature. That is, there is just one temperature at which equilibrium can be attained with all reactants and products in their standard states. At equilibrium at every other temperature, some or all of the reactants and products must be in a *nonstandard* state. For these *nonstandard* conditions, the criterion for equilibrium is that $\Delta G = 0$ (*not* $\Delta G° = 0$). This might seem to make $\Delta G°$ of limited value, but such is not the case at all.

An important thermodynamic relationship can be established between ΔG and $\Delta G°$ that makes use of the *reaction quotient*, Q. We will not attempt to derive this relationship, but we will use it.

$$\Delta G = \Delta G° + RT \ln Q$$

Now, let's specifically consider the case where a reaction is at equilibrium, where $\Delta G = 0$ and $Q = K_{eq}$. This leads to the expression

$$0 = \Delta G° + RT \ln K_{eq}$$

and

$$\Delta G° = -RT \ln K_{eq}$$

Thus, if we have $\Delta G°$ for a reaction at a given temperature, we can *calculate* the equilibrium constant at that temperature. With the equilibrium constant we can then calculate equilibrium concentrations or partial pressures, just as we have done in the previous three chapters. In applying this equation, we should note that

- R, the gas constant, is expressed as 8.3145 J · mol^{-1} · K^{-1} or 0.0083145 kJ · mol^{-1} · K^{-1}.
- T, the temperature, is expressed in kelvins, K.
- K_{eq} must have the format described below.

The Equilibrium Constant, K_{eq}

We have chosen to write K_{eq} rather than K_c or K_p in the expression $\Delta G° = -RT \ln K_{eq}$, because K_{eq} may differ from both of them. The problem is essentially this: Because the equation relating the standard free energy change and the equilibrium constant includes the term "ln K_{eq}," K_{eq} must be a *dimensionless* number—that is, it can have no units—because we can't take the logarithm of a unit.

In Chapter 18 we noted that calculations of solute solubilities (page 699) would be more accurate if in solubility product constant expressions we used "effective" concentrations called *activities*, *a*, instead of ordinary molarities. However, we chose to avoid the added complexity involved in doing this. Another feature of activities is that they are defined to be *dimensionless* quantities. Thus, the substitution of activities into equilibrium constant expressions leads to *dimensionless* values of K_{eq}. However, we can achieve the goal of a dimensionless K_{eq} and continue to use solution molarities and gas pressures by adopting the following conventions:

- *For pure solid and liquid phases*: The activity $a = 1$.
- *For gases*: Assume ideal gas behavior and replace the activity by the gas pressure in atm.
- *For solutes in aqueous solution*: Assume that intermolecular or interionic attractions are negligible and replace solute activity by solute molarity.

We illustrate these ideas in Example 19.6.

Example 19.6

Write the equilibrium constant expression, K_{eq}, for the oxidation of chloride ion by manganese dioxide in an acidic solution.

$$MnO_2(s) + 4\,H^+(aq) + 2\,Cl^-(aq) \rightleftharpoons Mn^{2+}(aq) + Cl_2(g) + 2\,H_2O(l)$$

Solution

No term will appear for $MnO_2(s)$; it is a pure solid phase with $a = 1$. For Mn^{2+}, H^+, and Cl^-, ionic species in a dilute aqueous solution, we substitute molarities for activities. For the activity of chlorine gas, we substitute its partial pressure. Because H_2O is the preponderant species in the dilute aqeous solution, the activity of H_2O is essentially the same as in pure liquid water: $a = 1$.

$$K_{eq} = \frac{[Mn^{2+}](P_{Cl_2})}{[H^+]^4[Cl^-]^2}$$

Notice how this K_{eq} expression contains both molarities and a partial pressure. It is neither a K_c nor a K_p.

Exercise 19.6

Write an equation for the dissolution of magnesium hydroxide in an acidic solution, and then write the K_{eq} expression for this reaction.

Calculating Equilibrium Constants, K_{eq}

Now we're ready to combine several ideas into the calculation of an equilibrium constant, K_{eq}, and to consider its significance. Let's calculate K_{eq} for the vaporization of water at 25 °C, for which we have previously written

$$H_2O(l) \longrightarrow H_2O(g) \qquad \Delta G^\circ_{298} = +8.590 \text{ kJ}$$

We rearrange the equation

$$\Delta G^\circ = -RT \ln K_{eq}$$

to

$$\ln K_{eq} = \frac{-\Delta G^\circ}{RT}$$

and substitute values of ΔG°, R, and T. In this substitution, we must change the value of ΔG° from +8.590 kJ to 8.590×10^3 J. Also, to accomplish the cancellation of all units, we must add the unit "mol^{-1}" to the value of ΔG°, that is, $\Delta G^\circ = 8.590 \times 10^3 \text{ J} \cdot \text{mol}^{-1}$. Think of the additional "mol^{-1}" as simply signifying that quantities in the chemical equation are on a molar basis.

$$\ln K_{eq} = \frac{-8.590 \times 10^3 \text{ J} \cdot \text{mol}^{-1}}{8.3145 \text{ J} \cdot \text{mol}^{-1} \cdot \text{K}^{-1} \times 298.15 \text{ K}} = -3.465$$

$$K_{eq} = e^{-3.465} = 0.0313$$

According to the rules we established for writing K_{eq} expressions, the appropriate expression for this K_{eq} is

$$K_{eq} = P_{H_2O(g)} = 0.0313$$

That is, the *equilibrium* vapor pressure of water at 25 °C is 0.0313 atm (23.8 mmHg). This calculated vapor pressure agrees perfectly with the experimentally measured value listed in Table 11.2.

Example 19.7

Determine the value of K_{eq} at 25 °C for the reaction: $2\ NO_2(g) \rightleftharpoons N_2O_4(g)$.

Solution

Our first task is to obtain a value of ΔG°, which we can do from tabulated standard free energies of formation in Appendix D.

$$\Delta G^\circ = \Delta G^\circ_f[N_2O_4(g)] - 2 \times \Delta G^\circ_f[NO_2(g)]$$

$$\Delta G^\circ = (+97.82) - 2 \times (+51.30) = -4.78 \text{ kJ}$$

Next, after expressing ΔG° as $-4780 \text{ J} \cdot \text{mol}^{-1}$, we substitute into the expression

$$\ln K_{eq} = \frac{-\Delta G^\circ}{RT} = \frac{-(-4780 \text{ J} \cdot \text{mol}^{-1})}{8.3145 \text{ J} \cdot \text{mol}^{-1} \cdot \text{K}^{-1} \times 298.15 \text{ K}} = 1.93$$

$$K_{eq} = \text{antiln}\ (1.93) = e^{1.93} = 6.9$$

Exercise 19.7

Use data from Appendix D to determine the value of K_{eq} at 25 °C for the reaction $2\ HgO(s) \rightleftharpoons 2\ Hg(l) + O_2(g)$.

The Sign and Magnitude of $\Delta G°$: Their Significance

In Chapter 16 (page 615), we learned how to use the magnitude of K_{eq} to decide, roughly, whether to treat a reaction (a) as going essentially to completion, (b) occurring hardly at all in the forward direction, or (c) reaching an equilibrium condition that must be described through K_{eq}.

Now that we have an equation linking K_{eq} and $\Delta G°$,

$$\Delta G° = -RT \ln K_{eq}$$

we can use $\Delta G°$ alone to decide among these three possibilities. Figure 19.11 suggests how to do this.

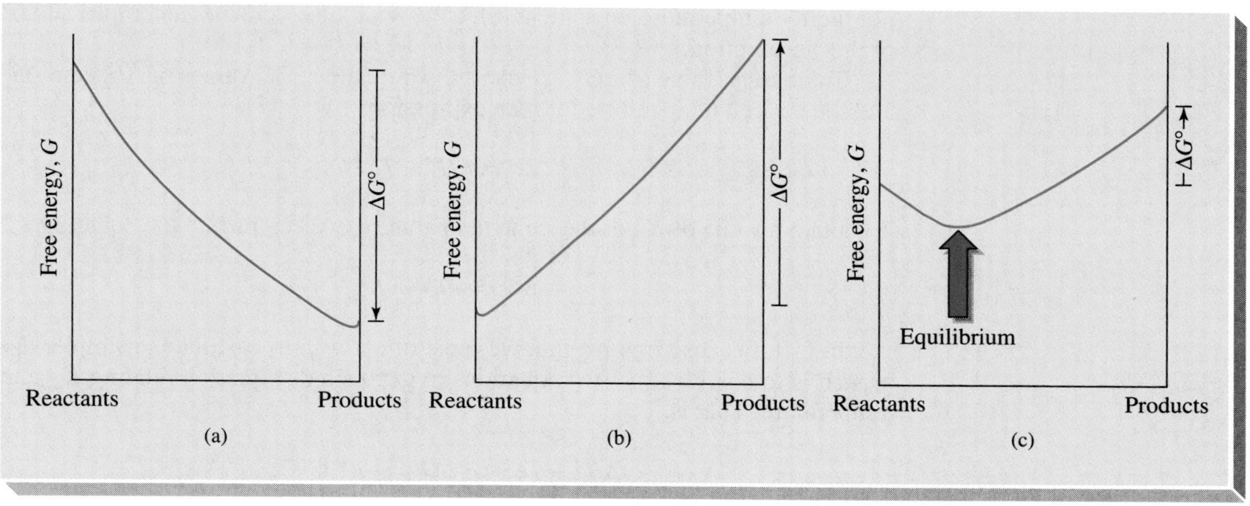

Figure 19.11 **$\Delta G°$ and the direction and extent of spontaneous change.** These reaction profiles show that the free energy decreases during a reaction, reaching a minimum value at the point of equilibrium. (a) *$\Delta G°$ is large and negative.* Equilibrium lies very near the right edge of the profile—reaction goes essentially to completion. For example, if $\Delta G° = -100$ kJ at 25 °C, $K_{eq} = 3 \times 10^{17}$. (b) *$\Delta G°$ is large and positive.* Equilibrium lies very near the left edge of the profile—reaction hardly occurs at all. For example, if $\Delta G° = +100$ kJ at 25 °C, $K_{eq} = 3 \times 10^{-18}$. (c) *$\Delta G°$ is neither very large nor very small and is either positive or negative.* Equilibrium lies well within the profile. K_{eq} must be used in calculations. For example, if $\Delta G° = +10.0$ kJ at 25 °C, $K_{eq} = 1.8 \times 10^{-2}$.

19.7 The Dependence of $\Delta G°$ and K_{eq} on Temperature

We have now established criteria for spontaneous change, and we have examined a number of qualitative and quantitative assessments that we can make with just these two equations:

$$\Delta G = \Delta H - T\Delta S$$

$$\Delta G° = -RT \ln K_{eq}$$

If you think about it, though, there appears to be one serious limitation: All our quantitative calculations have been at a single temperature of 25 °C. This is because tabulated thermodynamic data are given mostly for this single temperature. But we know that for practical purposes we must carry out reactions at a variety of temperatures, not just 25 °C. Our final need in this chapter is to obtain equilibrium constants at temperatures other than 25 °C.

Here's what we can do. We will assume that over reasonable temperature intervals, ΔH° and ΔS° remain constant. In particular, we will assume that values of ΔH° and ΔS° at 25 °C will apply at other temperatures as well. This assumption generally works because the enthalpies of formation and absolute molar entropies of the products and reactants all change in roughly comparable fashion with changes in temperature. As a result, ΔH° and ΔS°, which represent *differences* in enthalpies of formation and *differences* in absolute molar entropies, are not highly temperature dependent. Shortly, we'll also consider an empirical test of this assumption.

To obtain a value of ΔG°, we can substitute the 25 °C values of ΔH° and ΔS° and the desired temperature into the expression

$$\Delta G^\circ = \Delta H^\circ - T\Delta S^\circ$$

To obtain a value of K_{eq} at the same temperature, we can turn to the equation

$$\Delta G^\circ = -RT \ln K_{eq}$$

In fact, because it is K_{eq} that we most often wish to calculate, a simpler approach is to *combine* the two equations to get a single equation relating K_{eq} and temperature. That is,

$$\Delta H^\circ - T\Delta S^\circ = \Delta G^\circ = -RT \ln K_{eq}$$

and therefore

$$\ln K_{eq} = \frac{-\Delta H^\circ}{RT} + \frac{\Delta S^\circ}{R}$$

If we now introduce the assumption that ΔH° and ΔS° remain essentially constant over a range of temperatures, and replace the term $\Delta S^\circ/R$ by a constant, we can write

$$\ln K_{eq} = \frac{-\Delta H^\circ}{RT} + \text{constant}$$

This is the equation for a straight line, if we plot $\ln K_{eq}$ against $1/T$. The slope of the line is $-\Delta H^\circ/\mathrm{R}$. The best test of the assumption that ΔH° and ΔS° are independent of temperature is to plot some equilibrium data and see if they do yield a straight line. This is illustrated in Figure 19.12 for one of the reactions used to prepare $H_2(g)$ for use in the synthesis of ammonia.

By the method shown in Appendix A, we can replace the above equation with

$$\ln \frac{K_2}{K_1} = \frac{\Delta H^\circ}{R}\left(\frac{1}{T_1} - \frac{1}{T_2}\right)$$

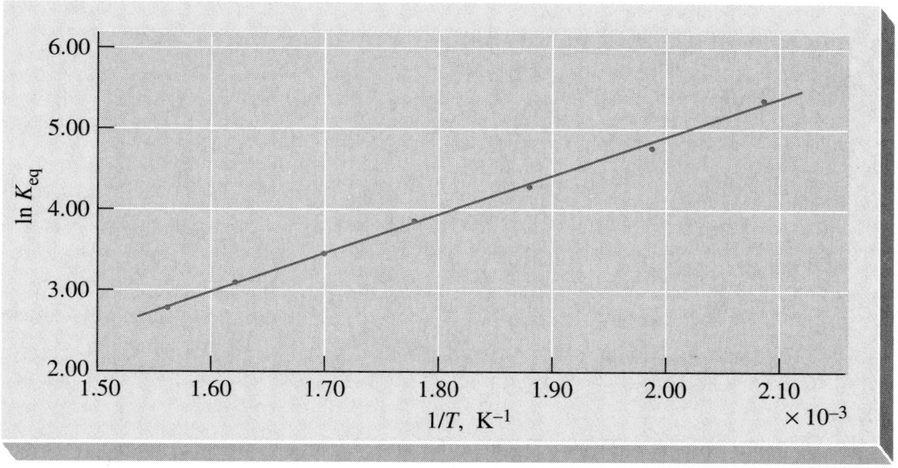

Figure 19.12 Temperature dependence of K_{eq} for the reaction $CO(g) + H_2O(g) \rightleftharpoons CO_2(g) + H_2(g)$. The value of $\Delta H°$ obtained from the slope of this graph is about –40 kJ, a value reasonably close to that of –41.2 kJ for $\Delta H°_{298}$, obtained from data in Appendix D. The data used in the graph are

T, K	$1/T$	K_{eq}	ln K_{eq}	T, K	$1/T$	K_{eq}	ln K_{eq}
478	2.09×10^{-3}	210	5.35	588	1.70×10^{-3}	31	3.43
503	1.99×10^{-3}	120	4.79	618	1.62×10^{-3}	22	3.09
533	1.88×10^{-3}	73	4.29	643	1.56×10^{-3}	16	2.77
563	1.78×10^{-3}	47	3.85				

This equation, called the *van't Hoff equation*, relates K_{eq} at two temperatures. The value of K_{eq} at temperature T_1 is K_1, and the value at temperature T_2 is K_2. $\Delta H°$ is the standard enthalpy change in the reaction, and R is the gas constant.

If the equilibrium constants are vapor pressures, we can substitute P for K and refer to the equation as the *Clausius–Clapeyron equation*.

Although it deals with different matters, the van't Hoff equation has the same form as the Arrhenius equation in Chapter 15 (page 586).

$$\ln \frac{P_2}{P_1} = \frac{\Delta H_{vapn}}{R}\left(\frac{1}{T_1} - \frac{1}{T_2}\right)$$

Example 19.8

At 25.0 °C, the enthalpy of vaporization of water (ΔH_{vapn}) is 44.0 kJ/mol and the vapor pressure is 23.8 mmHg. Calculate the vapor pressure of water at 40.0 °C and compare the result with data from Table 11.2.

Solution

The data we need to substitute into the Clausius–Clapeyron equation are

$P_1 = 23.8$ mmHg and $T_1 = (25.0 + 273.15) = 298.2$ K
$P_2 = ?$ and $T_2 = (40.0 + 273.15) = 313.2$ K
$\Delta H_{vapn} = 44.0 \times 10^3\ J \cdot mol^{-1}$ and $R = 8.3145\ J \cdot mol^{-1} \cdot K^{-1}$

It makes no difference which temperature we use as T_1 and which as T_2, so long as we match up the correct P value with each temperature. It is immaterial what units we use for pressure because the units cancel in the ratio, P_2/P_1.

$$\ln \frac{P_2}{P_1} = \frac{\Delta H_{vapn}}{R}\left(\frac{1}{T_1} - \frac{1}{T_2}\right)$$

$$\ln \frac{P_2}{23.8} = \frac{44.0 \times 10^3\ \text{J} \cdot \text{mol}^{-1}}{8.3145\ \text{J} \cdot \text{mol}^{-1} \cdot \text{K}^{-1}}\left(\frac{1}{298.2\ \text{K}} - \frac{1}{313.2\ \text{K}}\right)$$

$$\ln \frac{P_2}{23.8} = 5.29 \times 10^3(0.003353 - 0.003193)$$

$$\ln \frac{P_2}{23.8} = 5.29 \times 10^3 \times 0.000160 = 0.846$$

$$\frac{P_2}{23.8} = e^{0.846} = 2.33$$

$$P_2 = 2.33 \times 23.8 = 55.5\ \text{mmHg}$$

The experimentally determined vapor pressure at 40.0 °C listed in Table 11.2 is 55.3 mmHg. The calculated and experimental results are in good agreement.

Exercise 19.8

In Example 19.7, we determined the value of K_{eq} at 25 °C for the reaction $2\ NO_2(g) \rightleftharpoons N_2O_4(g)$. What is the value of K_{eq} for this reaction at 65 °C? (*Hint*: You will have to establish the value of $\Delta H°$ from tabulated data.)

CONCEPTUAL EXAMPLE 19.2

Estimate the value of $\Delta S°_{298}$ for the dissociation of calcium carbonate

$$CaCO_3(s) \rightleftharpoons CaO(s) + CO_2(g) \qquad \Delta H°_{298} = 178\ \text{kJ}$$

given that the equilibrium pressure of $CO_2(g)$ is 1.00 atm at 897 °C (1170 K).

Solution

Let's assume that $\Delta S°$ changes little with temperature, so that $\Delta S°_{298} \approx \Delta S°_{1170}$. We can calculate $\Delta S°_{1170}$ from the equation

$$\Delta G°_{1170} = \Delta H°_{1170} - T\Delta S°_{1170}$$

To do so we need $\Delta G°_{1170}$ and $\Delta H°_{1170}$. If we assume that $\Delta H°$ also changes little with temperature, then $\Delta H°_{1170} \approx \Delta H°_{298}$.

To obtain a value of $\Delta G°_{1170}$ we can use the expression

$$\Delta G°_{1170} = -RT \ln K_{eq}$$

The key, then, is to establish the value of K_{eq} at 1170 K.

Because the activities of the pure solid phases equal 1, for the reaction $CaCO_3(s) \rightleftharpoons CaO(s) + CO_2(g)$

$$K_{eq} = P_{CO_2}$$

The equilibrium pressure of $CO_2(g)$ at 1170 K is 1.00 atm. Therefore,

$$K_{eq} = 1.00 \qquad \text{and} \qquad \Delta G°_{1170} = -RT \ln (1.00) = 0$$

Now we see that

$$\Delta G°_{1170} = \Delta H°_{1170} - T\Delta S°_{1170}$$

becomes

$$0 = \Delta H°_{1170} - T\Delta S°_{1170}$$

Thus,

$$\Delta S°_{1170} = \frac{\Delta H°_{1170}}{1170 \text{ K}} = \frac{178 \text{ kJ}}{1170 \text{ K}} = 0.152 \text{ kJ} \cdot \text{K}^{-1}$$

Finally,

$$\Delta S°_{298} \approx \Delta S°_{1170} \approx 0.152 \text{ kJ} \cdot \text{K}^{-1}$$

Conceptual Exercise 19.2

As an alternative to the method used above, evaluate $\Delta S°_{298}$ for the reaction

$$CaCO_3(s) \longrightarrow CaO(s) + CO_2(g) \qquad \Delta H°_{298} = 178 \text{ kJ}$$

by using the van't Hoff equation to determine K_{eq} at 298 K, followed by

$$\Delta G°_{298} = \Delta H°_{298} - T\Delta S°_{298}$$

and show that the result is about the same as that obtained in Conceptual Example 19.2. Explain why you would expect it to be.

Thermodynamics and Living Organisms

In living organisms, potential energy—the energy stored in food—is changed into the kinetic energy associated with breathing, jogging, and twiddling thumbs. How do these energy changes come about? Do they obey the laws of thermodynamics?

A reaction that releases energy is *exergonic* ($\Delta G < 0$), and one that requires a net input of energy is *endergonic* ($\Delta G > 0$). Living organisms carry out exergonic reactions to maintain life processes, but there must also be energy-storing (endergonic) processes to provide for these exergonic reactions.

In living organisms complex molecules like proteins and DNA are produced from simpler ones like O_2, CO_2, and H_2O. Such processes usually involve a *decrease* in entropy ($\Delta S < 0$), an *increase* in enthalpy ($\Delta H > 0$), and consequently an *increase* in free energy ($\Delta G > 0$).

$$\Delta G = \Delta H - T\Delta S$$

We expect them to be *nonspontaneous*. How can nonspontaneous reactions essential to life proceed in directions that are not natural?

Consider this: If you drop a book it falls to the floor—a spontaneous process. The book won't leap back into your hands—a nonspontaneous process. Suppose, however, the book on the floor is tied by a rope to another, larger book that you hold. Now, you can grasp the rope and let the larger book fall. As it falls, the smaller book rises. What you have is a primitive pulley (see Figure 19.13). The spontaneous process (fall of the larger book) drives the nonspontaneous process (rise of the smaller book).

The reaction by which plants combine the simple sugars glucose and fructose into sucrose (common table sugar) has a positive ΔG; it is endergonic. The reaction must be coupled to one that is exergonic, such as the hydrolysis of ATP (adenosine triphosphate) to ADP (adenosine diphosphate). The sum of the two reactions has $\Delta G < 0$.

$$\text{Glucose} + \text{fructose} \longrightarrow \text{sucrose} + \mathrm{H_2O} \qquad \Delta G = +29.3 \text{ kJ}$$

$$\text{ATP} + \mathrm{H_2O} \longrightarrow \text{ADP} + \mathrm{P_i} \qquad \Delta G = -30.5 \text{ kJ}$$

$$\text{Glucose} + \text{fructose} + \text{ATP} \longrightarrow \text{sucrose} + \text{ADP} + \mathrm{P_i} \qquad \Delta G = -1.2 \text{ kJ}$$

P_i represents inorganic products such as HPO_4^{2-} and H^+.

The reactions are coupled through the *intermediate* glucose 1-phosphate. First, ATP reacts with glucose to form this intermediate. Some of the energy stored in the bonds of ATP is transferred to bonds in the intermediate.

$$\text{(a)} \quad \text{Glucose} + \text{ATP} \longrightarrow \text{glucose 1-phosphate} + \text{ADP}$$

When the intermediate reacts with fructose, the energy is transferred to the sucrose molecule.

$$\text{(b)} \quad \text{Glucose 1-phosphate} + \text{fructose} \longrightarrow \text{sucrose} + \mathrm{P_i}$$

Glucose 1-phosphate, an intermediate, drops out of the equation [(a) + (b)] for the net reaction

$$\text{Glucose} + \text{fructose} + \text{ATP} \longrightarrow \text{sucrose} + \text{ADP} + \mathrm{P_i} \qquad \Delta G = -1.2 \text{ kJ}$$

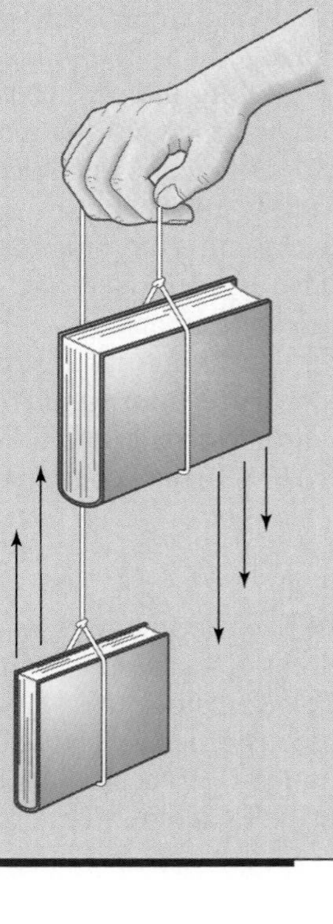

Figure 19.13 **An analogy to coupled reactions.** The small book will not rise on its own; that is a nonspontaneous process. When coupled by being attached by a rope to a larger book that is allowed to fall (a spontaneous process), the small book rises. The coupled process is spontaneous.

SUMMARY

A key objective of this chapter is to establish criteria for spontaneous change. A spontaneous change occurs by itself without outside intervention. The idea that spontaneous processes occur in the direction of decreasing enthalpy works in many cases. However, to account for exothermic and endothermic processes alike, it is necessary to use the concept of entropy, a measure of the degree of randomness or disorder in a system.

The direction of spontaneous change is that in which entropy (disorder) increases. The entropy change that must be assessed, however, is the *sum* of the entropy change of a system *and* of its surroundings. By the second law of thermodynamics, this so-called entropy of the universe, ΔS_{univ}, must always increase for a spontaneous process; that is, $\Delta S_{univ} > 0$.

The third law of thermodynamics states that the entropy of a pure, perfect crystal at 0 K is zero. This serves as a starting point for the experimental determination of absolute molar entropies. Absolute molar entropies can be used to calculate entropy changes in chemical reactions.

Free energy change, ΔG, which is equal to $-T\Delta S_{univ}$, applies *just to a system itself*, without regard for the surroundings. It is defined by the Gibbs equation.

$$\Delta G = \Delta H - T\Delta S$$

For a spontaneous process, ΔG must be negative; that is, $\Delta G < 0$. In many cases, the sign of ΔG can be predicted just by knowing the signs of ΔH and ΔS. The standard free energy change, $\Delta G°$, can be calculated (1) by substituting standard enthalpies and entropies of reaction and a Kelvin temperature into the Gibbs equation, or (2) by combining standard free energies of formation through the expression

$$\Delta G = [\Sigma \nu_p \Delta G^\circ_f(\text{products})] - [\Sigma \nu_r \Delta G^\circ_f(\text{reactants})]$$

The condition of equilibrium is one for which $\Delta G = 0$. The *standard* free energy change is a particularly useful property for describing equilibrium because of its relationship to the equilibrium constant, K_{eq}.

$$\Delta G^\circ = -RT \ln K_{eq}$$

One of the requirements of this equation is that the K_{eq} expression be based on molarities of reactants or products in solution and partial pressures (in atm) of gases.

Tabulated values of ΔG°_f, ΔH°_f, and $S°$ are generally for 25 °C. To use these data to obtain values of K_{eq} at other temperatures requires the assumption that $\Delta H°$ and $\Delta S°$ are independent of temperature. With this assumption, the *van't Hoff equation* can be derived to relate the equilibrium constant and temperature. The corresponding equation relating vapor pressure and temperature is known as the *Clausius–Clapeyron equation*.

KEY TERMS

entropy *(S)* (19.3)
entropy change *(ΔS)* (19.3)
equilibrium constant *(K_{eq})* (19.6)
free energy *(G)* (19.4)
free energy change *(ΔG)* (19.4)
nonspontaneous process (19.2)
second law of thermodynamics (19.3)
spontaneous process (19.2)
standard free energy change *(ΔG°)* (19.5)
standard free energy of formation *(ΔG°_f)* (19.5)
thermodynamics (introduction)
third law of thermodynamics (19.3)
Trouton's rule (19.6)

REVIEW QUESTIONS

1. What is meant by the term "a spontaneous process?"

2. Is it correct to say that a nonspontaneous process is an impossible process? Explain.

3. Use common knowledge only to determine which of the following are spontaneous changes: (**a**) the souring of milk; (**b**) obtaining copper metal from copper ore; (**c**) the rusting of a steel can in moist air. Explain.

4. Based on the relationship between entropy and disorder, indicate whether each of the following changes represents an increase or decrease in entropy of the system.

a. the freezing of acetic acid
b. the sublimation of the moth repellent *para*-dichlorobenzene ($C_6H_4Cl_2$)
c. the burning of gasoline

5. Make a sketch similar to Figure 19.3 to portray the change in entropy when a solid dissolves in a liquid to form a nonelectrolyte solution.

6. Why is the entropy change in a system not always a reliable predictor of whether the process producing the change is spontaneous?

7. What does the second law of thermodynamics tell us about the concept of entropy?

8. What is the basic idea underlying the third law of thermodynamics?

9. Which would you expect to have the higher absolute molar entropy at 25 °C, NOF_3(g) or NO_2F(g)? Explain.

10. What are the two tendencies that act as driving forces in establishing the direction of spontaneous change?

11. Under what conditions of temperature—low or high—would you expect a reaction to proceed farthest in the forward direction if the reaction has $\Delta H < 0$ and $\Delta S < 0$? Explain.

12. Would you expect the following reaction to occur spontaneously at room temperature? Explain.

$$H_2O(g, 1\ atm) \longrightarrow H_2O(l, 1\ atm)$$

13. Give an example of a phase change that is nonspontaneous at low temperatures and spontaneous at higher temperatures. What is the equilibrium temperature?

14. Give an example of a phase change that is spontaneous at low temperatures and nonspontaneous at higher temperatures. What is the equilibrium temperature?

15. What is the special significance of a reaction for which (**a**) $\Delta G = 0$; (**b**) $\Delta G° = 0$?

16. Would you expect each of the following quantities to be positive, negative, or zero for the mixing of ideal, inert gases: ΔH, ΔS, ΔG? Explain.

17. What is Trouton's rule? Why doesn't it work for all liquids?

18. Write the equilibrium constant expression K_{eq} for each of the following reactions.
a. $N_2(g) + 3\ Cl_2(g) \rightleftharpoons 2\ NCl_3(g)$
b. $PbI_2(s) \rightleftharpoons Pb^{2+}(aq) + 2\ I^-(aq)$
c. $SO_2(g) + 2\ H_2O(l) \rightleftharpoons H_3O^+(aq) + HSO_3^-(aq)$
d. $Ca(OH)_2(s) + CO_3^{2-}(aq) \rightleftharpoons CaCO_3(s) + 2OH^-(aq)$

19. How does thermodynamics make it possible to *calculate* the value of an equilibrium constant, K_{eq}? What kind of data are needed for this calculation?

20. If a reaction has $\Delta G° = 0$, what is K_{eq} for the reaction? Explain.

PROBLEMS

Disorder, Entropy, and Spontaneous Change

21. If you drop fifty pennies on the floor, explain why it is unlikely that all the pennies will land heads up. What is a much more likely result?

22. Predict whether the entropy change for each of the following processes is likely to be positive or negative.
a. $C_6H_6(l, 25\ °C) \longrightarrow C_6H_6(l, 55\ °C)$
b. $Fe(l, 1850\ K) \longrightarrow Fe(s, 1500\ K)$
c. $N_2(g, 745mmHg, 25\ °C) \longrightarrow N_2(g, 1\ atm, 25\ °C)$

23. Predict whether the entropy change for each of the following processes is likely to be positive or negative.
a. $Al(s, 125\ °C) \longrightarrow Al(s, 35\ °C)$
b. $O_2(g, STP) \longrightarrow O_2(g, 735\ mmHg, 5\ °C)$
c. $CO_2(g, 0.1\ atm, 25\ °C) \longrightarrow CO_2(s, 1\ atm, -78.5\ °C)$

24. For each of the following reactions, indicate whether you would expect the entropy of the system to increase or decrease. If you cannot tell just by inspecting the equation, explain why.
a. $CH_3OH(l) \longrightarrow CH_3OH(g)$
b. $N_2O_4(g) \longrightarrow 2\ NO_2(g)$
c. $CO(g) + H_2O(g) \longrightarrow CO_2(g) + H_2(g)$
d. $2\ KClO_3(s) \longrightarrow 2\ KCl(s) + 3\ O_2(g)$

25. For each of the following reactions, indicate whether you would expect the entropy of the system to increase or decrease. If you cannot tell just by inspecting the equation, explain why.
a. $CH_3COOH(l) \longrightarrow CH_3COOH(s)$
b. $N_2(g) + O_2(g) \longrightarrow 2\ NO(g)$
c. $N_2H_4(l) \longrightarrow N_2(g) + 2\ H_2(g)$
d. $2\ NH_3(g) + H_2SO_4(aq) \longrightarrow (NH_4)_2SO_4(aq)$

26. Compare the following processes and decide which of the two should have the *larger* value of ΔS. Explain your choice.
a. $H_2O(s, 1\ atm) \longrightarrow H_2O(l, 1\ atm)$
b. $H_2O(s, 1\ atm) \longrightarrow H_2O(g, 4.58\ mmHg)$

27. Compare the following processes and decide which of the two should have the *larger* value of ΔS. Explain your choice.

a. C(diamond, 1 atm) $\longrightarrow$ C(graphite, 1 atm)
b. CO_2(s, 1 atm) $\longrightarrow$ CO_2(g, 1 atm)

28. Which of the following statements must be true in order for a process to occur spontaneously? Explain.
a. The entropy of the system must increase.
b. The entropy of the surroundings must increase.
c. The entropy of the universe must increase.

29. Which of the following statements must be true in order for a process to occur spontaneously? Explain.
a. The entropy of the system and that of the surroundings must both increase.
b. The entropy of the surroundings must increase if that of the system decreases.
c. The entropy of the system and the surroundings must both decrease.

30. For the following reaction

$$2\,S(s,\ \text{rhombic}) + Cl_2(g) \longrightarrow S_2Cl_2(g)$$

Comment on the difficulty in deciding whether $\Delta S°$ is positive or negative when using the generalizations on page 737. Describe how the additional generalization on page 741 can help you decide the case. Calculate the actual value of $\Delta S°$ using data from Appendix D

31. The value of $\Delta S°_{298}$ obtained both in Conceptual Example 19.2 and in Conceptual Exercise 19.2 is only an estimate. What data can you use to obtain a more exact value? Do so, and compare the results.

32. Based on a consideration of entropy changes, why is it so difficult to eliminate environmental pollution?

33. An environmental problem posed by the use of aluminum cans is that their disintegration under environmental conditions appears to take an almost endless period of time. Can we say that the environmental disintegration of aluminum is a nonspontaneous process? Explain.

Free Energy and Spontaneous Change

34. Explain why free energy change is easier to use as a criterion for spontaneous change than is entropy change alone.

35. Explain why in a reversible reaction both the forward and the reverse reactions can be spontaneous, but not under the same conditions.

36. Make a sketch similar to Figure 19.9 that corresponds to case 2 of Table 19.1. Explain the significance of the plot in relation to the prediction in Table 19.1.

37. Make a sketch similar to Figure 19.9 that corresponds to case 1 of Table 19.1. Explain the significance of the plot in relation to the prediction in Table 19.1.

38. Would you expect each of the following reactions to be spontaneous at low temperatures, high temperatures, all temperatures, or not at all? Explain.
a. $PCl_3(g) + Cl_2(g) \longrightarrow PCl_5(g)$ $\Delta H° = -87.9$ kJ
b. $2\,NH_3(g) \longrightarrow N_2(g) + 3\,H_2(g)$ $\Delta H° = +92.2$ kJ
c. $2\,N_2O(g) \longrightarrow 2\,N_2(g) + O_2(g)$ $\Delta H° = -164.1$ kJ

39. Would you expect each of the following reactions to be spontaneous at low temperatures, high temperatures, all temperatures, or not at all? Explain.
a. $H_2O(g) + \frac{1}{2}O_2(g) \longrightarrow H_2O_2(g)$ $\Delta H° = +105.5$ kJ
b. $CH_4(g) + 2\,O_2(g) \longrightarrow CO_2(g) + 2\,H_2O(g)$ $\Delta H° = -802.3$ kJ
c. $2\,CO(g) + O_2(g) \longrightarrow 2\,CO_2(g)$ $\Delta H° = -566.0$ kJ

40. The following reaction is nonspontaneous at room temperature.

$$COCl_2(g) \longrightarrow CO(g) + Cl_2(g)$$

To make it a spontaneous reaction, would you raise or lower the temperature? Explain.

41. Would you expect the following reaction to be spontaneous?

$$3\,O_2(g) \longrightarrow 2\,O_3(g) \qquad \Delta H° = +285.4\ \text{kJ}$$

Will it become spontaneous by changing the temperature? Explain.

Standard Free Energy Change

42. Use data from Appendix D to determine $\Delta G°$ values for the following reactions at 25 °C.
a. $C_2H_4(g) + H_2(g) \longrightarrow C_2H_6(g)$
b. $SO_3(g) + CaO(s) \longrightarrow CaSO_4(s)$

43. Use data from Appendix D to determine $\Delta G°$ values for the following reactions at 25 °C.
a. $FeO(s) + H_2(g) \longrightarrow Fe(s) + H_2O(g)$
b. $CdO(s) + 2\,HCl(g) \longrightarrow CdCl_2(s) + H_2O(l)$

44. Use data from Appendix D to determine $\Delta H°$ and $\Delta S°$ at 298 K for the following reaction. Then determine $\Delta G°$ in two ways and compare the results.

$$C(\text{graphite}) + H_2O(g) \longrightarrow CO(g) + H_2(g)$$

45. Use data from Appendix D to determine $\Delta H°$ and $\Delta S°$ at 298 K for the following reaction. Then determine $\Delta G°$ in two ways and compare the results.

$$CS_2(l) + 3\,O_2(g) \longrightarrow CO_2(g) + 2\,SO_2(g)$$

Free Energy Change and Equilibrium

46. Estimate the normal boiling point of heptane, C_7H_{16}, given that at this temperature $\Delta H°_{\text{vapn}} = 31.69$ kJ/mol.

47. Estimate the normal boiling point of carbon tetrachloride, CCl_4, given that at this temperature $\Delta H^\circ_{vapn} = 30.0$ kJ/mol.

48. The normal boiling point of $Br_2(l)$ is 59.47 °C. Estimate ΔH°_{vapn} of bromine. Compare your result with a value based on data from Appendix D.

49. The normal boiling point of sulfuryl chloride $SO_2Cl_2(l)$ is 69.3 °C. Estimate ΔH°_{vapn} of sulfuryl chloride. Compare your result with a value based on data from Appendix D.

50. At 765 K, for the reaction $H_2(g) + I_2(g) \rightleftharpoons 2\ HI(g)$, $K_p = 46.0$.
   a. What is ΔG for a mixture at equilibrium at 765 K?
   b. What is ΔG° for this reaction at 765 K?

51. At 850 K, for the reaction $2\ SO_2(g) + O_2(g) \rightleftharpoons 2\ SO_3(g)$, $\Delta G^\circ = -36.3$ kJ.
   a. What is ΔG for a mixture at equilibrium at 850 K?
   b. What is K_p for this reaction at 850 K?

52. Write K_{eq} expressions for the following reactions. Do any of these expressions correspond to equilibrium constants that we have previously denoted as K_c, K_p, K_a, and so on?
   a. $2\ NO(g) + O_2(g) \rightleftharpoons 2\ NO_2(g)$
   b. $MgSO_3(s) \rightleftharpoons MgO(s) + SO_2(g)$
   c. $HCN(aq) + H_2O \rightleftharpoons H_3O^+(aq) + CN^-(aq)$

53. Write K_{eq} expressions for the following reactions. Do any of these expressions correspond to equilibrium constants that we have previously denoted as K_c, K_p, K_a, and so on?
   a. $2\ NaHSO_3(s) \rightleftharpoons Na_2SO_3(s) + H_2O(g) + SO_2(g)$
   b. $Mg(OH)_2(s) \rightleftharpoons Mg^{2+}(aq) + 2\ OH^-(aq)$
   c. $CH_3COO^-(aq) + H_2O \rightleftharpoons CH_3COOH(aq) + OH^-(aq)$

54. Use data from Appendix D to determine K_p at 298 K for the following reactions.
   a. $2\ N_2O(g) + O_2(g) \rightleftharpoons 4\ NO(g)$
   b. $2\ NH_3(g) + 2\ O_2(g) \rightleftharpoons N_2O(g) + 3\ H_2O(g)$

55. Use data from Appendix D to determine K_p at 298 K for the following reactions.
   a. $2\ SO_2(g) + O_2(g) \rightleftharpoons 2\ SO_3(g)$
   b. $CH_4(g) + 2\ H_2O(g) \rightleftharpoons CO_2(g) + 4\ H_2(g)$

56. The following are data for the vaporization of toluene at 298 K: $\Delta S^\circ = 99.7\ J \cdot K^{-1} \cdot mol^{-1}$; $\Delta H^\circ = 38.0\ kJ \cdot mol^{-1}$.

$$C_6H_5CH_3(l) \rightleftharpoons C_6H_5CH_3(g)$$

Calculate the equilibrium vapor pressure of toluene at 298 K.

57. The following data are given for the sublimation of naphthalene at 298 K: $\Delta S^\circ = 168.7\ J \cdot K^{-1} \cdot mol^{-1}$; $\Delta H^\circ = 73.6\ kJ \cdot mol^{-1}$. Calculate the partial pressure of naphthalene vapor in equilibrium with solid naphthalene at 298 K.

$$C_{10}H_8(s) \rightleftharpoons C_{10}H_8(g)$$

ΔG° and K_{eq} as Functions of Temperature

58. The vapor pressure of diethyl ether, $CH_3CH_2OCH_2CH_3$, at 10.0 °C is 291 mmHg and $\Delta H_{vapn} = 29.1$ kJ/mol. What is the vapor pressure of diethyl ether at 25.0 °C?

59. The vapor pressure of CCl_4 at 25 °C is 114 mmHg and $\Delta H_{vapn} = 32.5$ kJ/mol. What is the vapor pressure of CCl_4 at 15.0 °C?

60. The vapor pressure of cyclohexane, C_6H_{12}, is 97.6 mmHg at 25 °C, and $\Delta H_{vapn} = 23.1$ kJ/mol. What is the temperature at which the vapor pressure of cyclohexane is 175 mmHg?

61. The normal boiling point of carbon disulfide, CS_2, is 46.6 °C. Estimate the temperature at which the vapor pressure of CS_2 is 375 mmHg. (*Hint*: Estimate ΔH_{vapn} with Trouton's rule.)

62. For the reaction

$$CO(g) + H_2O(g) \rightleftharpoons CO_2(g) + H_2(g)$$

$\Delta H^\circ = -41.2$ kJ and $K_p \times 1.0 \times 10^5$ at 25 °C. Calculate a value of K_p at 315 °C.

63. Use data from Problem 62 to determine the temperature at which $K_p = 2.00$ for the reaction

$$CO(g) + H_2O(g) \rightleftharpoons CO_2(g) + H_2(g)$$

ADDITIONAL PROBLEMS

64. Which of the following substances would you expect to follow Trouton's rule most closely: (**a**) $CH_3(CH_2)_6CH_2OH$, (**b**) CH_3CH_2OH, (**c**) $CH_3(CH_2)_6CH_3$? Which should deviate most from the rule? Explain.

65. A handbook lists the following values for chloroform (trichloromethane) at 298 K:

$$\Delta H^\circ_f[CHCl_3(l)] = -132.2\ kJ/mol$$
$$\Delta H^\circ_f[CHCl_3(g)] = -102.9\ kJ/mol$$

Estimate the normal boiling point of chloroform.

66. A possible reaction for converting methanol to ethanol is

$$CO(g) + 2\ H_2(g) + CH_3OH(g) \rightleftharpoons CH_3CH_2OH(g) + H_2O(g)$$

Is the forward reaction favored at high or low temperatures? at high or low pressures?

67. Calculate a value of K_{eq} at 373 K for the reaction

$$CO(g) + Cl_2(g) \rightleftharpoons COCl_2(g)$$

(*Hint*: What other data do you need, and where can you find these data?)

68. For the decomposition of $NaHCO_3(s)$

$$2\ NaHCO_3(s) \rightleftharpoons Na_2CO_3(s) + H_2O(g) + CO_2(g)$$

a. Determine $\Delta H°$ and $\Delta S°$ at 298 K.
b. Estimate the temperature at which the total pressure of the gases above the solids is 1 atm.

69. The vapor pressure of hydrazine, N_2H_4, at 35.0 °C is 25.7 mmHg; at 50.0 °C, it is 57.0 mmHg. Estimate a value of the standard free energy change at 25 °C for the process

$$N_2H_4(l) \longrightarrow N_2H_4(g)$$

(*Hint*: How can you establish values of $\Delta H°_{vapn}$ and $\Delta S°_{vapn}$?) Determine a more exact value with data from Appendix D and compare the results.

70. When ethane is passed over a catalyst at 900 K and 1 atm, it is partly decomposed to ethylene and hydrogen.

$$CH_3CH_3(g) \rightleftharpoons CH_2{=}CH_2(g) + H_2(g) \qquad \Delta G°_{900\ K} = 22.4\ kJ/mol$$

Calculate the mole percent $H_2(g)$ at equilibrium.

71. Explain the meaning of the celebrated remark attributed to Rudolf Clausius in 1865. "Die Energie der Welt ist konstant; die Entropie der Welt strebt einem Maximum zu." ("The energy of the world is constant; the entropy of the world increases toward a maximum.")

72. Show that the effect of temperature on equilibrium that we predicted qualitatively in Chapter 16 is consistent with the effect of temperature on K_{eq} discussed in Section 19.7.

73. The free energy of solution formation is related to solute concentration by the expression: $\Delta G = \Delta G°_f + RT \ln c$ where c is the solute molarity. The concentration of creatine, an amino acid derivative formed during muscle contraction, is 2.0 mg/100 mL in blood and 75 mg/100 mL in urine. Calculate the free energy change for the transfer of creatine from blood to urine at 37 °C. (*Hint*: It is not necessary to convert concentrations from mg/100 mL to molarity. The proportionality constant between the two cancels out in the calculation.)

74. Use data from Appendix D to calculate the standard free energy change for burning one mole of solid glucose, $C_6H_{12}O_6$, to carbon dioxide gas and liquid water at 25 °C. Cite several factors that cause the free energy change for the metabolism of glucose in our bodies to differ from the calculated standard free energy change.

75. Following are values of K_p at different temperatures for the reaction $2\ SO_2(g) + O_2(g) \rightleftharpoons 2\ SO_3(g)$. At 800 K, $K_p = 9.1 \times 10^2$; at 900 K, $K_p = 4.2 \times 10^1$; at 1000 K, $K_p = 3.2$; at 1100 K, $K_p = 0.39$; and at 1170 K, $K_p = 0.12$. Plot $\ln K_p$ versus $1/T$ and determine $\Delta H°$ for this reaction.

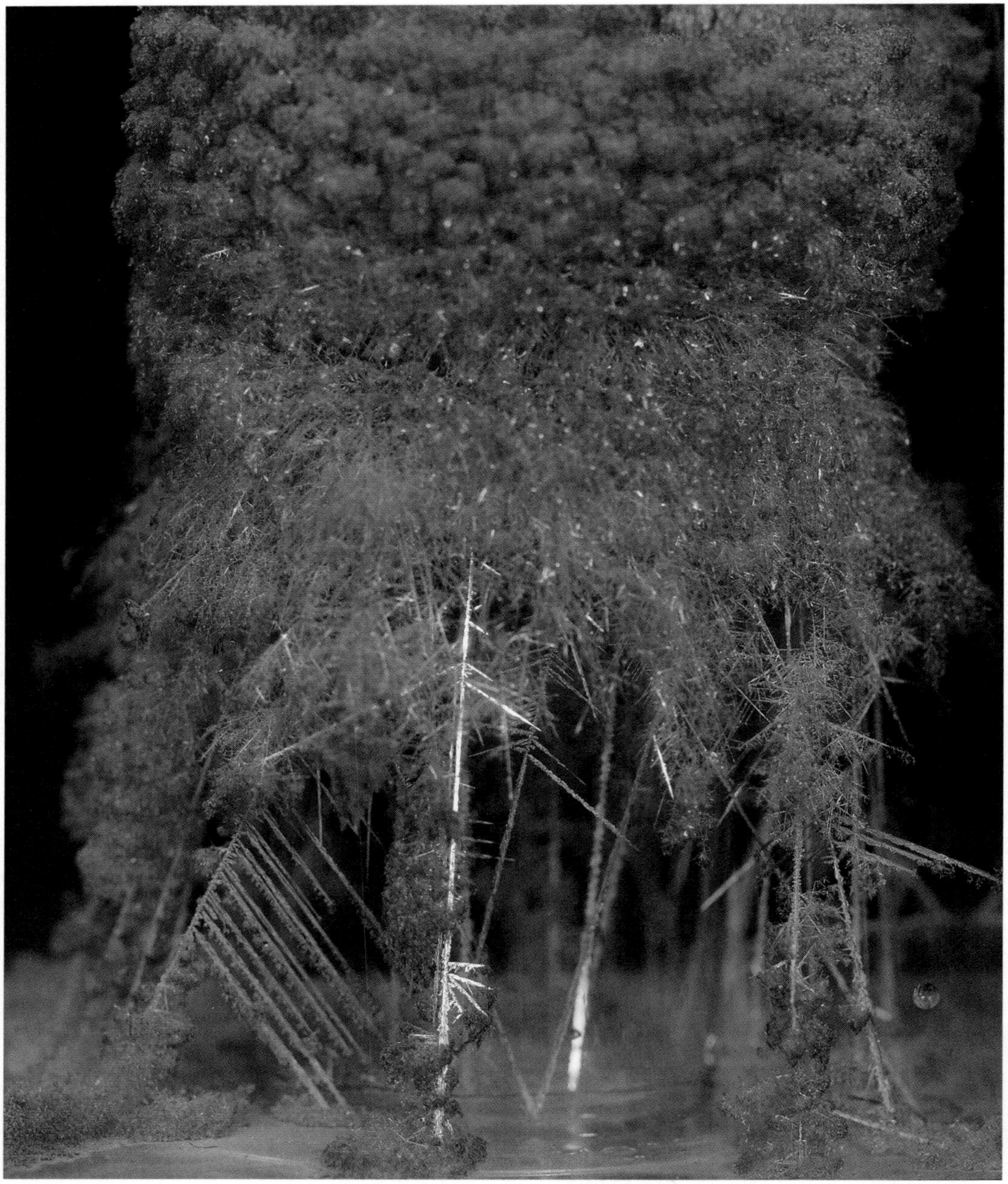

Solid copper metal displaces silver ions from solution. A shiny deposit of silver metal is formed. This photograph provides a close-up view of this reaction, which is also portrayed in Figure 20.6.

20

Electrochemistry

CONTENTS

Electricity is perhaps our most useful form of energy. It can be used to heat, cool, and run motors and other devices. We have seen that the most used method of generating electricity—burning fossil fuels—is inefficient (page 742). It requires the conversion of heat to mechanical work and then to electricity. The heat-to-work conversion is limited by thermodynamics to a maximum efficiency of about 40%. The conversion of heat to mechanical work in internal combustion engines in automobiles is even more restricted in its efficiency.

A much more efficient way to generate electricity is to avoid heat-to-work conversions and go directly from chemical energy (free energy) to electric energy. The links between chemical reactions and electricity are the subject matter of *electrochemistry*. Because electricity involves the flow of electrons, the types of chemical reactions associated with electricity are those in which electron transfers occur—*oxidation–reduction* reactions.

In this chapter we will see how *spontaneous* chemical reactions can produce electricity, and how electricity can be used to produce *nonspontaneous* reactions. We will also examine practical applications that range from batteries and fuel cells as electric power sources, to methods of controlling corrosion, to the manufacture of key chemicals and metals.

20.1 Electrochemical Cells

The strange device in Figure 20.1, strips of zinc and copper sticking into a lemon, produces electricity. We will consider how this happens later in the chapter, but first let's look at a similar but more ordinary case: one in which a zinc strip is dipped into zinc sulfate solution and a copper strip into copper sulfate solution.

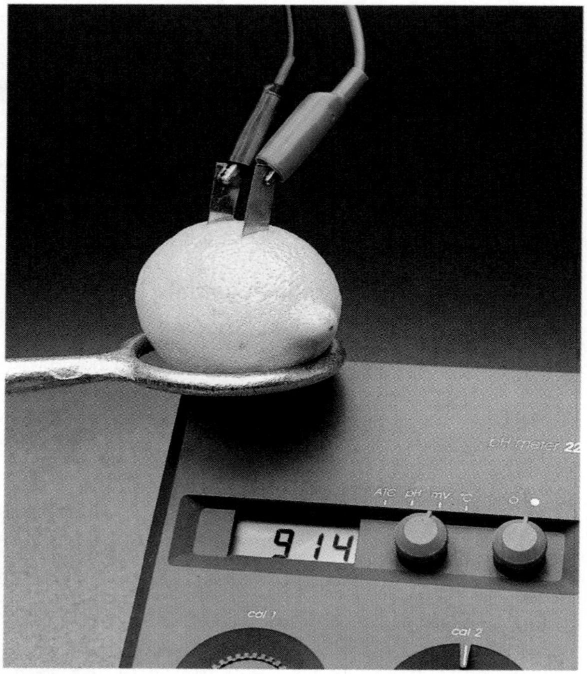

Figure 20.1 A simple voltaic cell. Two dissimilar metal strips (copper and zinc) and an electrolyte (lemon juice) are the key components of a device that converts chemical energy to electricity—a *voltaic cell.* The voltage of this cell is 0.914 V = 914 mV.

Electrode Equilibrium

When a zinc strip, called an electrode, is dipped into an aqueous solution of zinc sulfate, we can imagine that some of the zinc atoms are *oxidized* (Figure 20.2). Each Zn atom that is oxidized loses two electrons to the metal strip and enters the solution as a Zn^{2+} ion.

$$\textit{Oxidation}: \quad Zn(s) \longrightarrow Zn^{2+}(aq) + 2\,e^-$$

At the same time, electrons left behind on the metal strip attract Zn^{2+} ions from the solution. Some Zn^{2+} ions are *reduced* (Figure 20.2). Each Zn^{2+} ion that is reduced gains two electrons and deposits as a Zn atom.

$$\textit{Reduction}: \quad Zn^{2+}(aq) + 2\,e^- \longrightarrow Zn(s)$$

The opposing oxidation and reduction tendencies quickly come to an equilibrium, called an electrode equilibrium.

$$Zn(s) \underset{\text{reduction}}{\overset{\text{oxidation}}{\rightleftharpoons}} Zn^{2+}(aq) + 2\,e^-$$

A similar electrode equilibrium is established when a copper electrode is dipped into copper sulfate solution.

$$Cu(s) \underset{\text{reduction}}{\overset{\text{oxidation}}{\rightleftharpoons}} Cu^{2+}(aq) + 2\,e^-$$

In our discussion of oxidation–reduction in Chapter 12, we saw through the activity series of the metals (Figure 12.10) that zinc is a good reducing agent and is rather easily oxidized. By contrast, copper is a poor reducing agent; copper is not so readily oxidized. We should expect this difference to be reflected in the position of equilibrium at each electrode: the forward direction (oxidation)

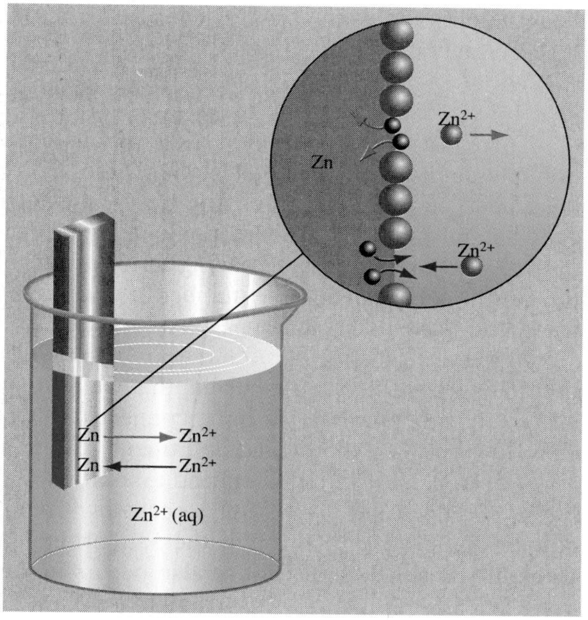

Figure 20.2 Electrode equilibrium. The zinc metal strip, called an electrode, is partially immersed in a solution containing Zn^{2+} ions. Oxidation and reduction occur at the electrode until a condition of equilibrium is reached. (The anions needed to produce an electrically neutral solution are not shown.)

should be favored for zinc over copper. Next, let's see how we can demonstrate this difference.

Voltaic (Galvanic) Cells: A Qualitative Description

In the device pictured in Figure 20.3, zinc and copper strips and their sulfate solutions are set up in separate compartments, and the metal strips are joined by a wire. Electrons are released at the zinc electrode, where oxidation—the loss of electrons—occurs.

$$\textit{Oxidation:}\quad Zn(s) \longrightarrow Zn^{2+}(aq) + 2\,e^-$$

Electrons flow through the wire to the copper electrode where the influx of electrons promotes the reverse process, reduction—the gain of electrons.

$$\textit{Reduction:}\quad Cu^{2+}(aq) + 2\,e^- \longrightarrow Cu(s)$$

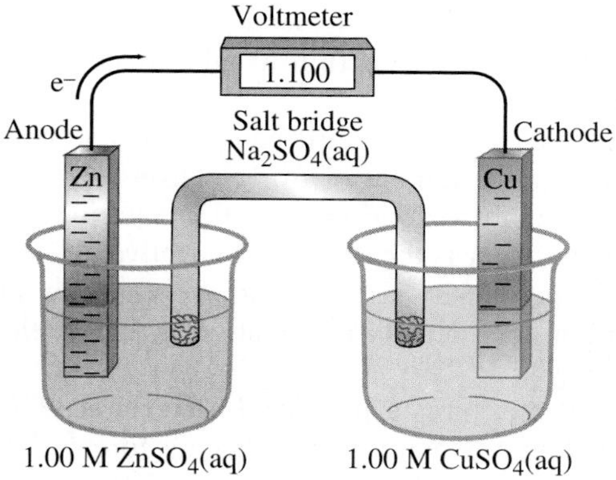

Figure 20.3 A Zn–Cu voltaic cell: the Daniell cell. Variants of this Zn–Cu cell, called Daniell cells, were used to power telegraph lines in the early days of the telegraph. The drawing suggests a greater negative charge density on the zinc electrode than on the copper. The flow of electrons (electricity) is propelled by the tendency of the electron density to even itself out. (The ends of the salt bridge are plugged with a material that allows ions to migrate but prevents bulk flow of the solution.)

The sum of the oxidation and reduction half-reactions is a *net* oxidation–reduction (redox) reaction.

$$Zn(s) + Cu^{2+}(aq) \longrightarrow Zn^{2+}(aq) + Cu(s)$$

There is a serious flaw in this description, however. The sulfate ions in solution, although not involved in the oxidation and reduction half-reaction, serve the crucial purpose of neutralizing the charges of metal ions to ensure electrically neutral solutions. As oxidation of the zinc occurs, the number of Zn^{2+} ions in the $ZnSO_4$(aq) increases, tending to make the solution positively charged. In the copper compartment, the number of Cu^{2+} ions in the $CuSO_4$(aq) decreases, tending to make that solution negatively charged.

In nature, creation of a charge imbalance occurs only at a considerable cost in energy. The electric current that we have just described doesn't flow at all. To fix this problem we must connect the two *solutions,* so that ions can also flow between the two compartments. We can make this connection with a *salt bridge,* an inverted U-shaped tube filled with a salt solution. For example, if the electrolyte is Na_2SO_4, a net flow of sulfate anions occurs from the copper compartment, through the salt bridge, into the zinc compartment. Generally, a net flow of cations also occurs in the opposite direction through a salt bridge.

To produce electricity with an oxidation–reduction reaction, then, we must *separate* the oxidation and reduction half-reactions and force electrons to flow from one part of the reaction system, through an external electric circuit, to another part of the system.

Some Useful Terminology

Chemists describing electrochemical devices generally use different terminology than we've been using. If we adopt some of their terminology, we will be able to continue our discussion somewhat more succinctly.

As we have already suggested, when used in electrochemical experiments, a strip of metal (M) is called an **electrode**. When the metal is partially immersed in an aqueous solution containing the metal ions (M^{n+}) the combination is called a **half-cell**. The combination of two half-cells, a salt bridge, and an electrical connection between the electrodes is an **electrochemical cell**. If an electrochemical cell produces electricity from an oxidation–reduction reaction, as in Figure 20.3, it is called a **voltaic** or **galvanic** cell. If electricity from an external source is passed through the cell (thus causing a nonspontaneous oxidation–reduction reaction to take place), it is called an **electrolytic** cell.

Credit for the discovery of current electricity is generally given to Luigi Galvani (1737–1798). However, Alessandro Volta (1745–1827) first constructed cells similar to the one in Figure 20.3 for the production of electricity.

Electrode potential is a property related to the electric charge density on an electrode. Think of the charge density as the number of free electrons per unit volume that accumulate on the electrode in the oxidation–reduction equilibrium. The greater the oxidation tendency, the greater is the density of negative electric charge. Even a slight *difference* in charge density is enough to set up a flow of electrons from the electrode with the higher density of negative charge to the one with a lower negative charge density.

This flow of electrons is analogous to the flow of water from a higher level (higher potential) to a lower level (lower potential), no matter how small the difference in levels.

A **voltmeter** is an instrument that measures a **potential difference**. The potential difference or difference in charge density is the "driving force" that pushes electrons from zinc to copper in the voltaic cell of Figure 20.3. Other names for the potential difference of a voltaic cell are **electromotive force (emf)** and **cell potential (E_{cell})**. The unit for measuring electrode potentials and potential differences is the **volt**. As a consequence, a cell potential is sometimes referred to as a *cell voltage*. For the voltaic cell in Figure 20.3, the cell potential or cell voltage is 1.100 V.

The focus of our discussion of voltaic cells will be on the driving force behind the flow of electrons, and thus on electrode potentials and potential differences or cell potentials. Later in the chapter, we will also be interested in the electron flow itself. The total electric *charge* that flows, expressed in **coulombs (C)**, is in effect a count of the number of electrons that pass through an electric circuit: the charge carried by one electron is -1.6022×10^{-19} C (page 199). Electric *current,* expressed in **amperes (A)**, is a measure of the *rate* at which electrons move through an electric circuit: one ampere corresponds to the passage of one coulomb of charge per second.

A more convenient way to represent an electrochemical cell than through drawings like Figure 20.3 is to construct a **cell diagram**, for which we use the following conventions:

- Place the **anode** (the electrode at which *oxidation* occurs) on the *left* side of the diagram.
- Place the **cathode** (the electrode at which *reduction* occurs) on the *right* side of the diagram.
- Use a *single* vertical line (|) to represent the boundary between different phases, such as between an electrode and a solution.
- Use a *double* vertical line (||) to represent a salt bridge separating two half-cells.

There are various mnemonic (memory) tricks for keeping track of these terms. Reading from left to right, the electrodes appear in alphabetical order: *a*node and *c*athode; so do the half-reactions that occur: *o*xidation and *r*eduction. Also, for the terms that are linked: *a*node and *o*xidation both begin with a vowel and *c*athode and *r*eduction both begin with a consonant.

The cell diagram for Figure 20.3 is

Anode ↘ Salt bridge ↓ Cathode ↙

$$Zn(s) \mid Zn^{2+}(aq) \parallel Cu^{2+}(aq) \mid Cu(s)$$

Half-cell (oxidation) Half-cell (reduction)

In many cases, an electrode does not itself participate in an oxidation–reduction equilibrium. The electrode is inert (unreactive) and simply furnishes the surface on which an electrode potential is established. For example, we cannot construct an electrode of $Cl_2(g)$ to measure the electrode potential for the oxidation–reduction equilibrium between chlorine gas and chloride ions. Instead, we can immerse a strip of platinum, an inert metal, into a solution of $Cl^-(aq)$ and bubble chlorine gas over its surface. The platinum electrode then acquires a potential that is characteristic of the oxidation–reduction equilibrium that takes place on its surface.

$$2\,Cl^-(aq) \xrightleftharpoons{\text{on Pt}} Cl_2(g) + 2\,e^-$$

The half-cell diagram for this electrode would be

As an anode (oxidation): $Pt, Cl_2(g) \mid Cl^-(aq)$

As a cathode (reduction): $Cl^-(aq) \mid Cl_2(g), Pt$

Example 20.1

Magnesium metal displaces silver ion from aqueous solution. (**a**) Write oxidation and reduction half-equations and a net ionic equation for this redox reaction. (**b**) Draw a cell diagram for a voltaic cell in which this redox reaction occurs.

Solution

a. Magnesium atoms go into solution as Mg^{2+}, and Ag^+ ions come out of solution as silver atoms. Mg is oxidized and Ag^+ is reduced. The half-equations are written, and coefficients are adjusted before combining them.

$$\begin{array}{ll} \textit{Oxidation:} & Mg(s) \longrightarrow Mg^{2+}(aq) + 2\,e^- \\ \textit{Reduction:} & 2\,\{Ag^+(aq) + e^- \longrightarrow Ag(s)\} \\ \hline \textit{Net:} & Mg(s) + 2\,Ag^+(aq) \longrightarrow Mg^{2+}(aq) + 2\,Ag(s) \end{array}$$

b. Oxidation occurs in the anode half-cell (left side) and reduction in the cathode half-cell (right side). The cell diagram is

$$Mg(s)\,|\,Mg^{2+}(aq)\,||\,Ag^+(aq)\,|\,Ag(s)$$

Example 20.2

Describe the half-reactions and the net reaction that occurs in the following voltaic cell.

$$Pt\,|\,Fe^{2+}(aq),\,Fe^{3+}(aq)\,||\,Cr_2O_7^{2-}(aq),\,Cr^{3+}(aq)\,|\,Pt$$

Solution

Each half-reaction occurs between ions in solution on the surface of an inert electrode (Pt). According to the conventions we have established, the electrode on the *left* is the *anode,* where *oxidation* occurs.

$$\textit{Oxidation:}\quad Fe^{2+}(aq) \longrightarrow Fe^{3+}(aq) + e^-$$

The electrode on the *right* is the *cathode,* where *reduction* occurs. The reduction involves the conversion of $Cr_2O_7^{2-}$, which has Cr in the oxidation state + 6, to Cr^{3+}, where the oxidation state is + 3. We can complete and balance the half-equation by the method we learned in Section 12.4.

$$\textit{Reduction:}\quad Cr_2O_7^{2-}(aq) + 14\,H^+(aq) + 6\,e^- \longrightarrow 2\,Cr^{3+}(aq) + 7\,H_2O(l)$$

To obtain an equation for the net reaction, we combine the oxidation and reduction half-reactions in the usual way.

$$\begin{array}{l} 6\,\{Fe^{2+}(aq) \longrightarrow Fe^{3+}(aq) + e^-\} \\ Cr_2O_7^{2-}(aq) + 14\,H^+(aq) + 6\,e^- \longrightarrow 2\,Cr^{3+}(aq) + 7\,H_2O(l) \\ \hline Cr_2O_7^{2-}(aq) + 6\,Fe^{2+}(aq) + 14\,H^+(aq) \longrightarrow 2\,Cr^{3+}(aq) + 6\,Fe^{3+}(aq) + 7\,H_2O(l) \end{array}$$

Exercise 20.1

Write the net equation for the redox reaction that occurs in the following voltaic cell.

$$Al(s)\,|\,Al^{3+}(aq)\,||\,Cu^{2+}(aq)\,|\,Cu(s)$$

Exercise 20.2

Suppose a voltaic cell is composed of an anode half-cell in which Zn(s) is oxidized to $Zn^{2+}(aq)$ and a cathode half-cell where $Cl_2(g)$ is reduced to $Cl^-(aq)$. Write an equation for the net cell reaction, and a cell diagram for the voltaic cell. (*Hint:* We have already seen diagrams for the two half-cells that make up the voltaic cell.)

20.2 Standard Electrode Potentials

Figure 20.3 shows how we can *measure* a cell voltage, but we can also *calculate* this voltage in an indirect way. We can do this by combining tabulated values of the electrode potentials for the two half-cells, but first we must find a way to assign potentials to individual electrodes.

It all begins with the half-cell pictured in Figure 20.4 and called the **standard hydrogen electrode (SHE).** Hydrogen gas at exactly 1 atm pressure is bubbled over an inert platinum electrode and into a hydrochloric acid solution. The concentration of the acid is adjusted so that the *activity* of H_3O^+ is exactly one ($a = 1$). This means that the hydrochloric acid solution is approximately 1 M HCl.* And, for simplicity, let's use the symbol H^+ in place of H_3O^+. Equilibrium between H_2 molecules and H^+ ions is established on the platinum surface; H_2(g) is oxidized to H^+(aq) and H^+(aq) is reduced to H_2(g). Finally, we will write the reversible half-equation with the *reduction* process as the forward direction (left to right) and the oxidation process as the reverse direction (right to left).

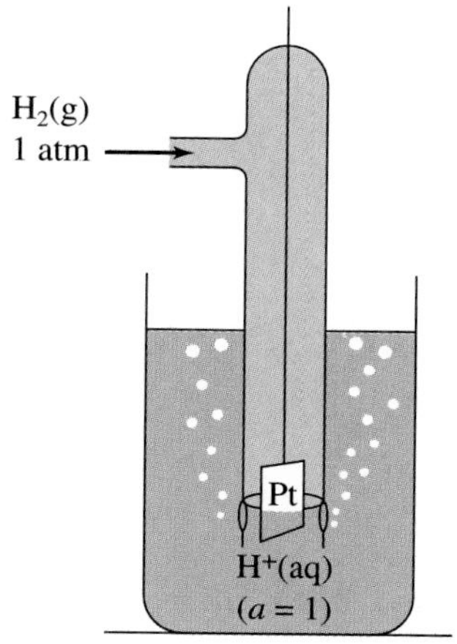

Figure 20.4 The standard hydrogen electrode (SHE). The inert platinum strip acquires a potential that is determined by the equilibrium

$$2\,H^+(a = 1) + 2\,e^- \rightleftharpoons H_2(g,\ 1\ atm)$$

The condition of $a_{H^+} = 1$ can be approximated by $[H^+] = 1$ M.

$$2\,H^+(a = 1) + 2\,e^- \underset{}{\overset{\text{on Pt}}{\rightleftharpoons}} H_2(g,\ 1\ atm) \qquad E^\circ = 0\ V\ \ (\textit{exactly})$$

By international agreement, a **standard electrode potential** (***E*°**) is based on the tendency for *reduction* to occur at the electrode. All solution species are present at unit activity ($a = 1$), which is about 1 M, and all gases are at 1 atm pressure. When no other metal is indicated as the electrode material, the potential is that for equilibrium established on an inert surface, such as platinum metal. The electrode potential for the standard hydrogen electrode (SHE) is arbitrarily set at *exactly* zero volts.

Now, using the SHE as a *reference* electrode in electrochemical cells like Figure 20.3, we can establish other standard electrode potentials. For example, if we construct an electrochemical cell with a SHE coupled with a standard copper electrode, we find that electrons flow *from* the hydrogen electrode (anode) *to* the copper electrode (cathode) at a measured voltage of 0.337 V. This indicates that Cu^{2+} ions are reduced to Cu(s) more easily than H^+ ions are reduced to H_2(g).

An electrode potential cannot be measured in isolation; it must be assessed through a measured potential *difference*. That is, the measurement requires that a flow of electrons occur between *two* half-cells.

$$\underset{\text{(anode)}}{\text{Pt, } H_2(g,\ 1\ atm)\,|\,H^+(1\ M)}\,||\,\underset{\text{(cathode)}}{Cu^{2+}(1\ M)\,|\,Cu(s)} \qquad E^\circ_{cell} = +0.337\ V$$

We can think of the cell voltage, called the **standard cell potential** (E°_{cell}) as the *sum* of two half-cell potentials. Let's see how we can add together two half-cell potentials at the same time that we combine half-equations to write a net equation for the cell reaction. In doing so, we will emphasize that the cell

*We have noted before that "effective" concentrations or activities should be used instead of stoichiometric concentrations for precise work dealing with solution properties, equilibrium constants, and the like. We will continue to use molarity in place of activity, but we will report results *as if* we had used activities. The results actually obtained when molarities are used are generally not as accurate as what we report.

reaction consists of an oxidation and a reduction half-reaction by using the symbols, E°_{ox} and E°_{red}. For example, we can write

$$\begin{array}{lll} \textit{Oxidation:} & H_2(g,\ 1\ atm) \longrightarrow 2\,H^+(1\ M) + 2\,e^- & E^\circ_{ox} = 0\ V \\ \textit{Reduction:} & Cu^{2+}(1\ M) + 2\,e^- \longrightarrow Cu(s) & E^\circ_{red} = ? \\ \hline \textit{Net:} & H_2(g,\ 1\ atm) + Cu^{2+}(1\ M) \longrightarrow 2\,H^+(1\ M) + Cu(s) & \end{array}$$

$$E^\circ_{cell} = E^\circ_{ox} + E^\circ_{red} = 0.337\ V$$
$$0\ V + E^\circ_{red} = 0.337\ V$$
$$E^\circ_{red} = 0.337\ V$$

The standard electrode potential, E°, is for the reduction of $Cu^{2+}(1\ M)$ to Cu(s). Thus,

$$E^\circ = E^\circ_{red} = 0.337\ V$$

Figure 20.5 is a schematic representation of standard electrode potentials in which we arbitrarily arrange them from lower (top) to higher (bottom) values. The standard electrode potential for the reduction of $Cu^{2+}(1\ M)$ to Cu(s) is seen to lie below that of the SHE by 0.337 V. The *positive* sign signifies that $Cu^{2+}(aq)$ is reduced more easily than is $H^+(aq)$.

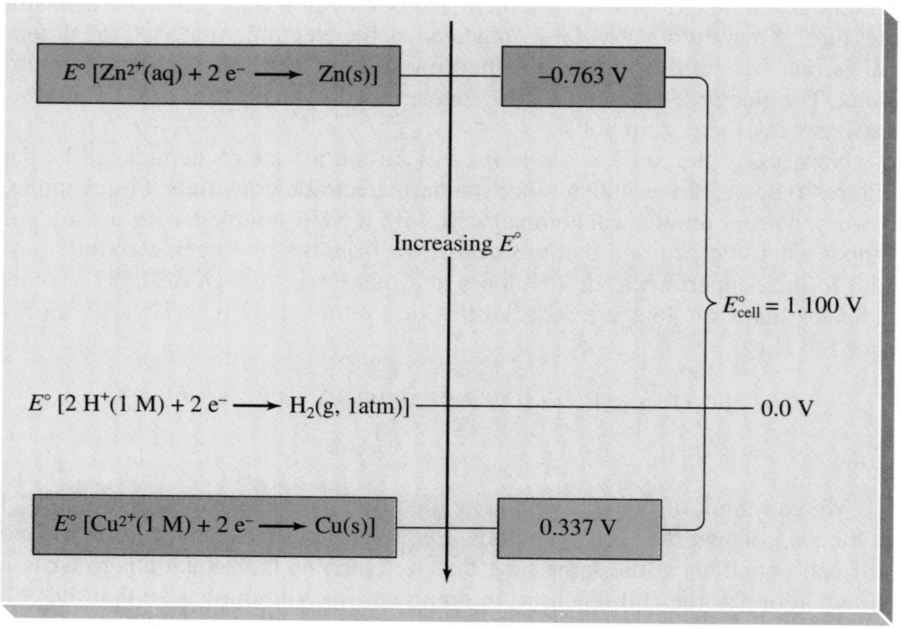

Figure 20.5 A representation of standard electrode potentials. The standard potentials for the copper and zinc electrodes are shown in relation to the standard hydrogen electrode potential. The potential *difference* between the copper and zinc electrodes—the cell voltage in Figure 20.3—is also shown.

If we couple a SHE with a standard zinc electrode, the zinc electrode is the anode and the SHE is the cathode. This signifies that H^+ ions are reduced to $H_2(g)$ more easily than Zn^{2+} ions are reduced to Zn(s).

$$Zn(s)\,|\,Zn^{2+}(1\ M)\ \|\ H^+(1\ M)\ |\ H_2(g,\ 1\ atm),\ Pt \qquad E^\circ_{cell} = 0.763\ V$$

The half-reactions and net cell reaction are

$$\begin{array}{lll} \textit{Oxidation:} & Zn(s) \longrightarrow Zn^{2+}(1\ M) + 2\,e^- & E^\circ_{ox} = ? \\ \textit{Reduction:} & 2\,H^+(1\ M) + 2\,e^- \longrightarrow H_2(g,\ 1\ atm) & E^\circ_{red} = 0\ V \\ \hline \textit{Net:} & Zn(s) + 2\,H^+(1\ M) \longrightarrow Zn^{2+}(1\ M) + H_2(g,\ 1\ atm) & \end{array}$$

$$E^\circ_{cell} = E^\circ_{ox} + E^\circ_{red} = 0.763\ V$$
$$E^\circ_{ox} + 0\ V = 0.763\ V$$
$$E^\circ_{ox} = 0.763\ V$$

Recall, however, that the standard electrode potential must be based on a *reduction* process: for the zinc electrode the reduction of Zn^{2+} ions to Zn(s). To obtain this value, we must *reverse* the oxidation half-reaction and *change the sign* of E°_{ox}. The standard electrode potential for the reduction

$$Zn^{2+}(1\ M) + 2\,e^- \longrightarrow Zn(s) \qquad E^\circ_{red} = ?$$

is

$$E^\circ = E^\circ_{red} = -E^\circ_{ox} = -0.763\ V$$

In the diagram of Figure 20.5, the electrode potential of zinc lies above that of the SHE by 0.763 V. The *negative* sign signifies that $Zn^{2+}(aq)$ is reduced with greater difficulty than is $H^+(aq)$. Figure 20.5 also suggests that we could have used the standard electrode potentials of the zinc and copper electrodes to *calculate* the voltage of the cell in Figure 20.3. Let's not forget, though, that experiments are required to establish standard electrode potentials in the first place.

Table 20.1 is a compilation of a number of standard electrode potentials and the reduction half-reactions that they describe. The arrangement is in order of increasing E° values. The most negative E° values are found at the top of the table; the most positive E° values are at the bottom.

Viewed from the *left* side of the half-equations in Table 20.1, the poorest oxidizing agents, those most difficult to reduce, appear high in the table (for example, Li^+, K^+, . . .). The best oxidizing agents, those most easily reduced, appear at the bottom of the table (for example, F_2, O_3, . . .).

Tables of standard electrode potentials are sometimes written in the reverse order, with the best oxidizing agents, those most easily reduced, at the top of the table.

Viewed from the *right* side of the half-equations, and thus from the standpoint of the reverse reactions, the best reducing agents, those most easily oxidized, are at the top of the table (for example., Li, K, . . .) and the poorest reducing agents, those oxidized with the greatest difficulty, are at the bottom of the table (for example, F^-, O_2 . . .).

We will use Table 20.1 a good deal. Our first application will be to combine electrode potentials—E° values—to obtain cell potentials, E°_{cell} values. We do this in a three-step procedure.

1. Write the reduction half-equation and its standard potential, E°_{red}. This will be an E° value taken from Table 20.1.
2. Write the oxidation half-equation and its standard potential, E°_{ox}, which is the *negative* of the value shown in Table 20.1. That is, $E^\circ_{ox} = -E^\circ_{red} = -E^\circ$.
3. Combine the half-equations into a net redox equation. *Add* the half-cell potentials to obtain E°_{cell}. That is, $E^\circ_{cell} = E^\circ_{red} + E^\circ_{ox}$. Adjusting the coefficients

TABLE 20.1 Some Selected Standard Electrode Potentials at 25 °C[a]

Reduction Half-reaction	$E°$, V
Acidic Solution	
$Li^+(aq) + e^- \longrightarrow Li(s)$	−3.040
$K^+(aq) + e^- \longrightarrow K(s)$	−2.924
$Ca^{2+}(aq) + 2\ e^- \longrightarrow Ca(s)$	−2.84
$Na^+(aq) + e^- \longrightarrow Na(s)$	−2.713
$Mg^{2+}(aq) + 2\ e^- \longrightarrow Mg(s)$	−2.356
$Al^{3+}(aq) + 3\ e^- \longrightarrow Al(s)$	−1.676
$Zn^{2+}(aq) + 2\ e^- \longrightarrow Zn(s)$	−0.763
$Fe^{2+}(aq) + 2\ e^- \longrightarrow Fe(s)$	−0.440
$Co^{2+}(aq) + 2\ e^- \longrightarrow Co(s)$	−0.277
$Sn^{2+}(aq) + 2\ e^- \longrightarrow Sn(s)$	−0.137
$Pb^{2+}(aq) + 2\ e^- \longrightarrow Pb(s)$	−0.125
$2\ H^+(aq) + 2\ e^- \longrightarrow H_2(g)$	0
$S(s) + 2\ H^+(aq) + 2\ e^- \longrightarrow H_2S(g)$	+0.14
$Sn^{4+}(aq) + 2\ e^- \longrightarrow Sn^{2+}(aq)$	+0.154
$SO_4^{2-}(aq) + 4\ H^+(aq) + 2\ e^- \longrightarrow 2\ H_2O + SO_2(g)$	+0.17
$Cu^{2+}(aq) + 2\ e^- \longrightarrow Cu(s)$	+0.337
$I_2(s) + 2\ e^- \longrightarrow 2\ I^-(aq)$	+0.535
$O_2(g) + 2\ H^+(aq) + 2\ e^- \longrightarrow H_2O_2(aq)$	+0.695
$Fe^{3+}(aq) + e^- \longrightarrow Fe^{2+}(aq)$	+0.771
$Ag^+(aq) + e^- \longrightarrow Ag(s)$	+0.800
$NO_3^-(aq) + 4\ H^+(aq) + 3\ e^- \longrightarrow NO(g) + 2\ H_2O$	+0.956
$Br_2(l) + 2\ e^- \longrightarrow 2\ Br^-(aq)$	+1.065
$2\ IO_3^-(aq) + 12\ H^+(aq) + 10\ e^- \longrightarrow I_2(s) + 6\ H_2O$	+1.20
$O_2(g) + 4\ H^+(aq) + 4\ e^- \longrightarrow 2\ H_2O$	+1.229
$MnO_2(s) + 4\ H^+(aq) + 2\ e^- \longrightarrow Mn^{2+}(aq) + 2\ H_2O$	+1.23
$Cr_2O_7^{2-}(aq) + 14\ H^+(aq) + 6\ e^- \longrightarrow 2\ Cr^{3+}(aq) + 7\ H_2O$	+1.33
$Cl_2(g) + 2\ e^- \longrightarrow 2\ Cl^-(aq)$	+1.358
$PbO_2(s) + 4\ H^+(aq) + 2\ e^- \longrightarrow Pb^{2+}(aq) + 2\ H_2O$	+1.455
$MnO_4^-(aq) + 8\ H^+(aq) + 5\ e^- \longrightarrow Mn^{2+}(aq) + 4\ H_2O$	+1.51
$H_2O_2(aq) + 2\ H^+(aq) + 2\ e^- \longrightarrow 2\ H_2O$	+1.763
$S_2O_8^{2-}(aq) + 2\ e^- \longrightarrow 2\ SO_4^{2-}(aq)$	+2.01
$O_3(g) + 2\ H^+(aq) + 2\ e^- \longrightarrow O_2(g) + H_2O$	+2.075
$F_2(g) + 2\ e^- \longrightarrow 2\ F^-(aq)$	+2.866
Basic Solution	
$2\ H_2O + 2\ e^- \longrightarrow H_2(g) + 2\ OH^-(aq)$	−0.828
$O_2(g) + 2\ H_2O + 4\ e^- \longrightarrow 4\ OH^-(aq)$	+0.401
$OCl^-(aq) + H_2O + 2\ e^- \longrightarrow Cl^-(aq) + 2\ OH^-(aq)$	+0.890
$O_3(g) + H_2O + 2\ e^- \longrightarrow O_2(g) + 2\ OH^-(aq)$	+1.246

[a] A more extensive listing of reduction half-reactions and their potentials is given in Appendix D.

of the half-equations does not affect values of $E°$ and $E°_{cell}$. They are *intensive* properties.

In applying this procedure, you will always be faced with this decision: Which is the reduction half-reaction and which is the oxidation half-reaction? The choice will always be the same: The reduction half-reaction will be the one with the larger value of $E°_{red}$, the one that is lower in Table 20.1. This choice will always make $E°_{cell} > 0$ for a voltaic cell.

That $E°_{cell}$ is an intensive property can be seen in the constant voltage (1.5 V) of dry cell batteries, whether they are large "D" batteries or smaller "AA" penlite batteries. The voltage does not depend on the amounts of substances involved in the cell reaction.

Example 20.3

A voltaic cell has a magnesium electrode in 1 M $Mg(NO_3)_2$(aq) and a silver electrode in 1 M $AgNO_3$(aq). What is $E°_{cell}$ of this voltaic cell?

Solution

Because $E°$ of the silver electrode has the larger value, *reduction* occurs at this electrode. Magnesium is oxidized at the other electrode.

Reduction: $2\,\{Ag^+(1\,M) + e^- \longrightarrow Ag(s)\}$ $\quad E°_{red} = 0.800\ V$

Oxidation: $Mg(s) \longrightarrow Mg^{2+}(1\,M) + 2\,e^-$ $\quad E°_{ox} = -E° = -(-2.356) = 2.356\ V$

Net: $Mg(s) + 2\,Ag^+(1\,M) \longrightarrow Mg^{2+}(1\,M) + 2\,Ag(s)$

$$E°_{cell} = E°_{red} + E°_{ox} = 0.800\ V + 2.356\ V = 3.156\ V$$

Exercise 20.3

Use the value of $E°$ from Table 20.1 for the reduction of Co^{2+}(aq) to Co(s), together with the following information

$$2\,Ce^{4+}(1\,M) + Co(s) \longrightarrow 2\,Ce^{3+}(1\,M) + Co^{2+}(1\,M) \qquad E°_{cell} = 1.887\ V$$

to determine $E°$ for the reduction half-reaction

$$Ce^{4+}(1\,M) + e^- \longrightarrow Ce^{3+}(1\,M)$$

[*Hint:* Use the same procedure as to establish the standard potential of the copper electrode when coupled with the standard hydrogen electrode (page 772).]

20.3 Electrode Potentials, Spontaneous Change, and Equilibrium

A reaction performs work when it takes place in a voltaic cell. We can think of this electric work as produced by moving electric charges. The total work done is the product of (1) the cell voltage, (2) the number of moles of electrons, n, transferred between electrodes, and (3) the electric charge per mole of electrons—a quantity called the **Faraday constant (*F*)** which equals 96,485 coulombs per mole.

$$w_{elec} = n \times F \times E_{cell}$$

The product (volt × coulomb) = joule. Thus, the unit of electric work is the joule (J).

The term "free energy" implies energy that is freely available as work, and $-\Delta G$ is the maximum amount of useful work that a system can do.

One of the most precise ways of measuring ΔG for a reaction is through E_{cell}. For example, for the magnesium–silver voltaic cell of Example 20.3,

$$\Delta G^\circ = -n \times F \times E^\circ_{cell} = -2 \text{ mol e}^- \times 96{,}485 \text{ C/mol e}^- \times 3.156 \text{ V} = -6.090 \times 10^5 \text{ V} \cdot \text{C} = -6.090 \times 10^5 \text{ J} = -609.0 \text{ kJ}$$

$$Mg(s) + 2\,Ag^+(1\ M) \longrightarrow Mg^{2+}(1\ M) + 2\,Ag(s) \qquad \Delta G^\circ = -609.0 \text{ kJ}$$

We won't attempt a derivation in this text, but it's not difficult to relate the quantity $-\Delta G$ to work that is done by a system. For the electric work associated with an electrochemical cell, $-\Delta G = w_{elec}$. Thus,

$$-\Delta G = n \times F \times E_{cell} \qquad \text{or} \qquad \Delta G = -n \times F \times E_{cell}$$

Notice that in this equation we have not written the superscript "°". The term E_{cell} signifies that the conditions at the electrodes are *not* standard conditions—solute concentrations are *not* 1 M and gas pressures are *not* 1 atm. Similarly, the term ΔG also implies *nonstandard* conditions. The equation is perfectly general, however, and we can apply it to a cell in which all substances are in the standard state. In that case we should write

$$\Delta G^\circ = -n \times F \times E^\circ_{cell}$$

Criteria for Spontaneous Change in Redox Reactions

We are not much interested in calculating quantities of electric work, but there are other important calculations that we can make with the equations presented above that we will consider later. For now, though, consider this fact: In order for a reaction to proceed spontaneously, $\Delta G < 0$, and if ΔG is *negative*, then E_{cell} must be *positive*. To our previous list of ideas for assessing the direction of spontaneous change, we can now add these:

- If E_{cell} is *positive*, a spontaneous reaction occurs in the *forward* direction (from left to right).
- If E_{cell} is *negative*, a spontaneous reaction occurs in the *reverse* direction (from right to left).
- If $E_{cell} = 0$, a reaction is at equilibrium.
- When a cell reaction is *reversed,* E_{cell} changes sign.

Our chief means of obtaining an E_{cell} value will be by combining standard electrode potentials from Table 20.1. The result will be a *standard* cell potential, E°_{cell}. This means that any predictions we make will be about reactions having reactants and products in their standard states. Usually, however, *qualitative* predictions based on standard-state conditions apply over a wide range of nonstandard conditions as well. We apply these new criteria for spontaneous change in Example 20.4 and Exercise 20.4.

Example 20.4

Will copper metal displace silver ion from aqueous solution? For example, does this reaction occur spontaneously from left to right?

$$Cu(s) + 2\,Ag^+(1\ M) \longrightarrow Cu^{2+}(1\ M) + 2\,Ag(s)$$

Solution

We must separate the net equation into its two half-equations, assign the appropriate E° values, and then recombine them to obtain E°_{cell}. If the resulting E°_{cell} is positive, then we know that the reaction is spontaneous.

Reduction:	$2\{Ag^+(1\ M) + e^- \longrightarrow Ag(s)\}$	$E^\circ_{red} = 0.800\ V$
Oxidation:	$Cu(s) \longrightarrow Cu^{2+}(1\ M) + 2\ e^-$	$E^\circ_{ox} = -E^\circ = -0.337\ V$
Net:	$Cu(s) + 2\ Ag^+(1\ M) \longrightarrow Cu^{2+}(1\ M) + 2\ Ag(s)$	

$$E^\circ_{cell} = E^\circ_{red} + E^\circ_{ox} = 0.800\ V - 0.337\ V = 0.463\ V$$

Because E°_{cell} is positive, the forward direction should be the direction of spontaneous change. Copper metal should displace silver ions from solution.

Exercise 20.4

Should the following reaction occur spontaneously in the forward direction?

$$Cu^{2+}(aq) + 2\ Fe^{2+}(aq) \longrightarrow 2\ Fe^{3+}(aq) + Cu(s)$$

Figure 20.6 confirms the prediction in Example 20.4, but it also shows that we can displace Ag^+ just by dropping copper metal into an aqueous solution containing Ag^+. *This is an important finding*: Although we use cell terminology to make predictions about the direction of spontaneous change in redox reactions, these predictions apply even if the reactions are not carried out in electrochemical cells.

Figure 20.6 The displacement of $Ag^+(aq)$ by Cu(s). Although we predict that Cu(s) will displace $Ag^+(aq)$ by using electrode potentials and a cell voltage, the reaction can be carried out as pictured here. (a) A coil of copper wire is immersed in $AgNO_3(aq)$. (b) The shiny deposit that forms on the copper wire is metallic silver. The blue color of the solution indicates the presence of $Cu^{2+}(aq)$.

The Activity Series of the Metals Revisited

We are now in a position to explain the activity series of the metals introduced in Section 12.5. According to the relationship between E_{cell} and the direction of spontaneous change, a metal will displace from a solution of its ions any metal lying *below* it in the table of standard electrode potentials (Table 20.1). And, treating the listing for hydrogen as just another entry in the table, a metal appearing *above* the SHE will react with a mineral acid (a solution having H^+ as the only oxidizing agent) to produce $H_2(g)$. A metal appearing *below* the SHE will not. These are the same conclusions we stated for the activity series of metals (Figure 12.10).

Estimation Exercise 20.1

a. Which of the following metals will displace Co^{2+} from aqueous solution: Pb, Zn, Ag, Al?

b. Which of the following metals will react with 1 M HCl(aq) to liberate $H_2(g)$: Co, Cu, Fe, Mg, Ag?

Figure 20.7 Copper–zinc assembly in HCl(aq).

CONCEPTUAL EXAMPLE 20.1

The photograph in Figure 20.7 shows a copper and zinc strip that are joined together and then dipped in HCl(aq). Explain what happens. That is, what are the gas bubbles on the zinc and on the copper, and how did they get there?

Solution

A metal should react with a mineral acid like HCl(aq), displacing $H_2(g)$ from solution, if the metal lies above $H_2(g)$ in the activity series of the metals (or in Table 20.1). Zinc fits this requirement, as we can establish by using $E°$ data.

Reduction: $2\,H^+(1\,M) + 2\,e^- \longrightarrow H_2(g, 1\,atm)$ $\quad E°_{red} = 0\,V$

Oxidation: $Zn(s) \longrightarrow Zn^{2+}(1\,M) + 2\,e^-$ $\quad E°_{ox} = -E° = -(-0.763) = 0.763\,V$

Net: $Zn(s) + 2\,H^+(1\,M) \longrightarrow Zn^{2+}(1\,M) + H_2(g, 1\,atm)$

$$E°_{cell} = E°_{red} + E°_{ox} = 0\,V + 0.7631\,V = 0.763\,V$$

The gas that forms on the zinc is $H_2(g)$, but what about the gas on the copper? Cu(s) lies *below* $H_2(g)$ in the activity series of the metals (and in Table 20.1). It should not displace $H_2(g)$ from HCl(aq).

Reduction: $2\,H^+(1\,M) + 2\,e^- \longrightarrow H_2(g, 1\,atm)$ $\quad E°_{red} = 0\,V$

Oxidation: $Cu(s) \longrightarrow Cu^{2+}(1\,M) + 2\,e^-$ $\quad E°_{ox} = -E° = -0.337\,V$

Net: $Cu(s) + 2\,H^+(1\,M) \longrightarrow Cu^{2+}(1\,M) + H_2(g, 1\,atm)$

$$E°_{cell} = E°_{red} + E°_{ox} = 0\,V - 0.337\,V = -0.337\,V$$

The reaction of Cu(s) with HCl(aq) is *nonspontaneous.*

Here is a plausible explanation. The gas on the Cu(s) *is* $H_2(g)$, but it is *not* formed by a reaction involving Cu(s). The electrons required for the reduction of H^+ to $H_2(g)$ on the copper come from the oxidation of the *zinc*. Think of the copper, a very good electric conductor, as being just an inert extension of the zinc electrode.

The device in Figure 20.7 is actually a voltaic cell. The portion of the reduction half-reaction that occurs on the copper is physically separated from the oxidation half-reaction that occurs only on the zinc. A flow of electrons occurs from the zinc to the copper.

Conceptual Exercise 20.1

Offer a plausible explanation of what occurs in the device in Figure 20.1 (page 766).

Equilibrium Constants for Redox Reactions

In the last chapter we discovered an important relationship between the standard free energy change and K_{eq} of a reaction. A similar one exists between K_{eq} and E°_{cell}.

$$\Delta G^\circ = -RT \ln K_{eq} = -n \times F \times E^\circ_{cell}$$

This allows us to write

$$E^\circ_{cell} = \frac{RT \ln K_{eq}}{nF}$$

In this equation, E°_{cell} is the standard cell potential, R is the gas constant expressed as 8.3145 $J \cdot mol^{-1} \cdot K^{-1}$, T is the Kelvin temperature, n is the number of moles of electrons involved in the reaction, and F is the Faraday constant. The traditional form of this equation uses common rather than natural logarithms (with the conversion, $\ln K_{eq} = 2.303 \log K_{eq}$).

$$E^\circ_{cell} = \frac{2.303RT \log K_{eq}}{nF}$$

E°_{cell} values are generally obtained from tables of E° at 25 °C, and the equation is almost always used at 25 °C. As a result, the quantity "$2.303RT/F$" has the same value from one calculation to another. It is easiest to replace it by its numerical equivalent "0.0592."

$$E^\circ_{cell} = \frac{2.303RT \log K_{eq}}{nF} = \frac{2.303 \times 8.3145 \times 298.15 \times \log K_{eq}}{n \times 96{,}485} = \frac{0.0592}{n} \log K_{eq}$$

The best way to establish n is from the half-cell reactions: It is the number of electrons in either the reduction or oxidation half-equation *after* the coefficients have been adjusted. That is, n is the number of electrons that cancel out on each side when the half-equations are combined to give the net equation.

Measuring E°_{cell} provides an easy and precise way of determining the equilibrium constant, K_{eq}, for a reaction. Once we obtain a value of K_{eq}, as in Example 20.5, we are then able to do all the types of equilibrium calculations of the past several chapters.

Example 20.5

Calculate the value of K_{eq} at 25 °C for the reaction

$$Cu(s) + 2\,Ag^+(1\text{ M}) \rightleftharpoons Cu^{2+}(1\text{ M}) + 2\,Ag(s) \qquad K_{eq} = ?$$

Solution

Look at the half-equations and the net equation written for this reaction in Example 20.4. The number of electrons that appeared in each half-equation before we combined them was 2, so in this case $n = 2$. We found E°_{cell} for the reaction to be +0.463 V.

$$E^\circ_{cell} = \frac{0.0592}{2}\log K_{eq} = 0.463$$

$$\log K_{eq} = \frac{2 \times 0.463}{0.0592} = 15.6$$

$$K_{eq} = 10^{15.6} = 4 \times 10^{15}$$

Based on the magnitude of K_{eq}, we can say that the displacement of $Ag^+(aq)$ by Cu(s) (page 777) goes essentially to completion.

Exercise 20.5

Calculate the value of K_{eq} at 25 °C for the reaction

$$Cu(s) + 2\,Fe^{3+}(1\text{ M}) \rightleftharpoons Cu^{2+}(1\text{ M}) + 2\,Fe^{2+}(1\text{ M})$$

20.4 Effect of Concentrations on Cell Voltage

We have now seen what we can do with *standard* cell voltages, E°_{cell}. However, most electrochemical cell measurements involve *nonstandard* conditions, as do many calculations. First, let's look *qualitatively* at the relationship between E°_{cell} and E_{cell}. To do this, let's return to the voltaic cell of Figure 20.3.

According to Le Châtelier's principle, the *forward* reaction should be favored by increasing $[Cu^{2+}]$ and/or decreasing $[Zn^{2+}]$ (represented below, in red). When the forward reaction is favored, $-\Delta G$ and E_{cell} *increase*. The *reverse* reaction should be favored by decreasing $[Cu^{2+}]$ and/or increasing $[Zn^{2+}]$ (represented below, in blue). When the reverse reaction is favored, $-\Delta G$ and E_{cell} *decrease*. Cell reactions and cell voltages for standard conditions and two nonstandard conditions are summarized below.

increase $[Cu^{2+}]$ $\longrightarrow$

$\longrightarrow$ decrease $[Zn^{2+}]$

Nonstandard:	$Zn(s) + Cu^{2+}(2.0\text{ M}) \longrightarrow Zn^{2+}(0.050\text{ M}) + Cu(s)$	$E_{cell} = 1.147$ V
Standard:	$Zn(s) + Cu^{2+}(1.0\text{ M}) \longrightarrow Zn^{2+}(1.0\text{ M}) + Cu(s)$	$E^\circ_{cell} = 1.100$ V
Nonstandard:	$Zn(s) + Cu^{2+}(0.050\text{ M}) \longrightarrow Zn^{2+}(2.0\text{ M}) + Cu(s)$	$E_{cell} = 1.053$ V

decrease $[Cu^{2+}]$ $\longleftarrow$

$\longleftarrow$ increase $[Zn^{2+}]$

Now let's look at the matter *quantitatively*. We can do this by combining several familiar expressions. We start with the equation introduced in Chapter 19 that relates ΔG, ΔG°, and Q.

$$\Delta G = \Delta G^\circ + RT\ln Q$$

Into this equation we can substitute the expressions

$$\Delta G = -nFE_{\text{cell}} \qquad \text{and} \qquad \Delta G^\circ = -nFE^\circ_{\text{cell}}$$

This leads to the equation

$$-nFE_{\text{cell}} = -nFE^\circ_{\text{cell}} + RT \ln Q$$

Next, let's change signs (multiply by −1),

$$nFE_{\text{cell}} = nFE^\circ_{\text{cell}} - RT \ln Q$$

and then solve for E_{cell}.

$$E_{\text{cell}} = E^\circ_{\text{cell}} - \frac{RT}{nF} \ln Q$$

Finally, let's switch from natural to common logarithms.

$$E_{\text{cell}} = E^\circ_{\text{cell}} - \frac{2.303 \times RT}{nF} \log Q$$

If we apply this equation at 25 °C, we can substitute "0.0592" for the combination of terms, $2.303 \times RT/F$, just as we did once before (page 779). In this way we obtain the equation known as the **Nernst equation**.

$$E_{\text{cell}} = E^\circ_{\text{cell}} - \frac{0.0592}{n} \log Q$$

In the Nernst equation, E_{cell} is based on the given *nonstandard* conditions, as is the reaction quotient, Q; n is the number of moles of electrons involved in the cell reaction.

Walther Nernst (1864–1941) is well known for several important achievements in chemistry. He formulated the Nernst equation in 1889, when he was only 25 years old. In the same year he suggested the solubility product concept. In 1906 he proposed a theorem that we now know as the third law of thermodynamics.

Example 20.6

Calculate the value of E_{cell} (the expected voltmeter reading) for the voltaic cell pictured in Figure 20.8.

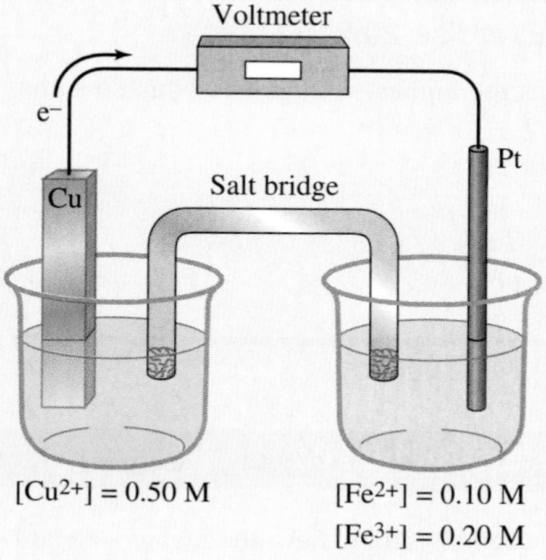

Figure 20.8 Example 20.6 illustrated. The object in this example is to determine the voltmeter reading expected for this voltaic cell.

Solution

Our first task is to determine the net cell reaction and a value of E°_{cell} with data from Table 20.1. The direction of electron flow shown in Figure 20.8 enables us to identify the anode and cathode.

Cathode (red.):	$2\,\{Fe^{3+}(1\ M) + e^- \longrightarrow Fe^{2+}(1\ M)\}$	$E^\circ_{red} = 0.771\ V$
Anode (ox.):	$Cu(s) \longrightarrow Cu^{2+}(1\ M) + 2e^-$	$E^\circ_{ox} = -E^\circ = -0.337\ V$
Net:	$Cu(s) + 2\,Fe^{3+}(1\ M) \longrightarrow Cu^{2+}(1\ M) + 2\,Fe^{2+}(1\ M)$	

$$E^\circ_{cell} = E^\circ_{red} + E^\circ_{ox} = 0.771\ V - 0.337\ V = 0.434\ V$$

We seek E_{cell} for a voltaic cell (Figure 20.8) in which the net reaction is

$$Cu(s) + 2\,Fe^{3+}(0.20\ M) \longrightarrow Cu^{2+}(0.50\ M) + 2\,Fe^{2+}(0.10\ M)$$

For this, we make the following substitutions into the Nernst equation: $E^\circ_{cell} = +0.434$ V; $n = 2$; $[Fe^{3+}] = 0.20$ M; $[Fe^{2+}] = 0.10$ M; $[Cu^{2+}] = 0.50$ M.

$$E_{cell} = E^\circ_{cell} - \frac{0.0592}{n}\log Q$$

$$E_{cell} = 0.434 - \frac{0.0592}{2}\log\left(\frac{[Cu^{2+}][Fe^{2+}]^2}{[Fe^{3+}]^2}\right)$$

$$E_{cell} = 0.434 - \frac{0.0592}{2}\log\left(\frac{0.50(0.10)^2}{(0.20)^2}\right)$$

$$E_{cell} = 0.434 - (0.0296 \times \log 0.125) = 0.434 - (-0.027) = +0.461\ V$$

Exercise 20.6

Use the Nernst equation to determine E_{cell} at 25 °C for the following voltaic cells.

a. $Zn(s)\,|\,Zn^{2+}(2.0\ M)\,||\,Cu^{2+}(0.050\ M)\,|\,Cu(s)$
b. $Zn(s)\,|\,Zn^{2+}(0.050\ M)\,||\,Cu^{2+}(2.0\ M)\,|\,Cu(s)$
c. $Cu(s)\,|\,Cu^{2+}(1.0\ M)\,||\,Cl^-(0.25\ M)\,|\,Cl_2(g,\ 0.50\ atm)$, Pt

Estimation Exercise 20.2

Which of the following voltaic cells has the highest voltage and which has the lowest?

(a) $Zn(s)\,|\,Zn^{2+}(2.0\ M)\,||\,Cu^{2+}(2.0\ M)\,|\,Cu(s)$
(b) $Zn(s)\,|\,Zn^{2+}(0.25\ M)\,||\,Cu^{2+}(2.0\ M)\,|\,Cu(s)$
(c) $Zn(s)\,|\,Zn^{2+}(0.20\ M)\,||\,Cu^{2+}(0.10\ M)\,|\,Cu(s)$
(d) $Zn(s)\,|\,Zn^{2+}(0.10\ M)\,||\,Cu^{2+}(1.0\ M)\,|\,Cu(s)$

CONCEPTUAL EXAMPLE 20.2

Which of the following reduction half-reactions should have the higher standard electrode potential—that is, the higher value of $E^\circ = E^\circ_{red}$?

(a) $AgCl(s) + e^- \longrightarrow Ag(s) + Cl^-(aq)$ $\quad E^\circ_{red} = ?$

(b) $AgI(s) + e^- \longrightarrow Ag(s) + I^-(aq)$ $\quad E^\circ_{red} = ?$

Solution

In both cases electrode equilibrium involves just Ag(s) and Ag^+(aq). However, the concentration of Ag^+ is controlled by another equilibrium, that of a slightly soluble solute and its ions in aqueous solution.

$$\text{(a) } Ag^+(aq) + e^- \underset{}{\overset{\text{on Pt}}{\rightleftharpoons}} Ag(s)$$

$$AgCl(s) \rightleftharpoons Ag^+(aq) + Cl^-(1\ M) \qquad K_{sp} = 1.8 \times 10^{-10}$$

$$\text{(b) } Ag^+(aq) + e^- \rightleftharpoons Ag(s)$$

$$AgI(s) \rightleftharpoons Ag^+(aq) + I^-(1\ M) \qquad K_{sp} = 8.5 \times 10^{-17}$$

The superscript ° in E°_{red} signifies that in (a) $[Cl^-] = 1$ M and in (b) $[I^-] = 1$ M. These anions are *common ions* that suppress the solubilities of the AgCl(s) and AgI(s), respectively. Their presence *reduces* the concentration of Ag^+ far below the $[Ag^+] = 1$ M associated with the standard silver electrode.

$$\text{From Table 20.1: } Ag^+(1\ M) + e^- \longrightarrow Ag(s) \qquad E^\circ_{red} = 0.800\ V$$

With a decreased $[Ag^+]$, the forward reaction is less favored, and $E^\circ_{red} < 0.800$ V. In comparing (a) and (b), we must compare $[Ag^+]$ in each case. We can calculate these concentrations from the K_{sp} expressions.

$$\text{(a) } [Ag^+] = K_{sp}/[Cl^-] = 1.8 \times 10^{-10}/1 = 1.8 \times 10^{-10}\ M$$

$$\text{(b) } [Ag^+] = K_{sp}/[I^-] = 8.5 \times 10^{-17}/1 = 8.5 \times 10^{-17}\ M$$

The silver ion concentration is much lower in half-reaction (b) than in (a), so we should expect $E^\circ_{red}(b) < E^\circ_{red}(a)$. That is,

$$E^\circ_{red}(b) < E^\circ_{red}(a) < 0.800\ V$$

(The actual tabulated values are $E^\circ_{red}(b) = -0.152$ V and $E^\circ_{red}(a) = 0.222$ V.)

Conceptual Exercise 20.2

A voltaic cell is constructed using the half-cells Ag(s), AgCl(s) | Cl^-(1 M), and Ag(s), AgI(s) | I^-(1 M). **(a)** Which electrode is the anode, and which is the cathode? **(b)** Write a net ionic equation for the spontaneous reaction that occurs. **(c)** What is the value of E°_{cell}? (*Hint*: Use information from Conceptual Example 20.2.)

Concentration Cells

The voltaic cell in Figure 20.9 looks different from others that we've seen before—the two electrodes are *identical*. However, the solutions have different concentrations, and because of this there is a potential difference between the electrodes. Let's determine the origin of this voltage and the nature of the cell reaction. First, for standard conditions,

Cathode (red.):	$Cu^{2+}(1\ M) + 2e^- \longrightarrow Cu(s)$	$E^\circ_{red} = 0.337$ V
Anode (ox.):	$Cu(s) \longrightarrow Cu^{2+}(1\ M) + 2e^-$	$E^\circ_{ox} = -E^\circ = -0.337$ V
Net:	$Cu^{2+}(1\ M) \longrightarrow Cu^{2+}(1\ M)$	

$$E^\circ_{cell} = E^\circ_{red} + E^\circ_{ox} = 0.337\ V - 0.337\ V = 0.000\ V$$

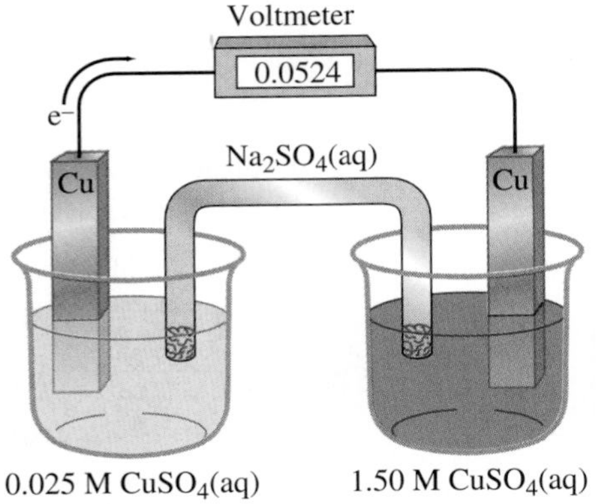

Figure 20.9 A concentration cell. The electrodes are identical, but the solution concentrations differ. The driving force for the cell reaction is the tendency for the solution concentrations to become equalized: $Cu^{2+}(1.50\ M) \longrightarrow Cu^{2+}(0.025\ M)$.

The *standard* cell voltage is *zero*, but this is what we should expect. One electrode reaction is simply the reverse of the other, and nothing happens. To calculate the cell voltage for the *nonstandard* conditions,

$$Cu^{2+}(1.50\ M) \longrightarrow Cu^{2+}(0.025\ M) \qquad E_{cell} = ?$$

we must use the Nernst equation.

$$E_{cell} = E^\circ_{cell} - \frac{0.0592}{n} \log Q$$

In this case, $E^\circ_{cell} = 0$, $n = 2$, and $Q = 0.025/1.50$.

$$E_{cell} = -\frac{0.0592}{2} \log \left(\frac{0.025}{1.50}\right) = -0.296 \times \log 0.017$$

$$= -0.0296 \times (-1.77) = 0.0524\ V$$

A cell whose potential is determined solely by a difference in concentration of solutes in equilibrium with identical electrodes is called a **concentration cell.** The net cell reaction is not a chemical reaction but equivalent to the migration of solute from a more concentrated to a more dilute solution. The more concentrated solution is diluted and the more dilute one becomes more concentrated. This is what would happen if we simply brought two solutions of different concentration in contact. Disorder and entropy increase, and the process is spontaneous. The unusual thing about a concentration cell is that it uses the natural tendency for solutions to mix as a basis of generating electricity.

pH Measurement

Suppose we construct a concentration cell consisting of two hydrogen electrodes. One is a standard hydrogen electrode (SHE), and the other is in a solution having an unknown pH—that is, with $[H^+] = x$. If $x < 1.0$ M—and this is generally the case—oxidation occurs at the nonstandard hydrogen electrode (the anode) and reduction at the SHE (the cathode).

Reduction:	$2\ H^+(1\ M) + 2\ e^- \longrightarrow H_2(g, 1\ atm)$
Oxidation:	$H_2(g, 1\ atm) \longrightarrow 2\ H^+(x\ M) + 2\ e^-$
Net:	$2\ H^+(1\ M) \longrightarrow 2\ H^+(x\ M)$

The voltage of this concentration cell is given by the Nernst equation, which takes the form

$$E_{\text{cell}} = -\frac{0.0592}{2} \log \frac{x^2}{1^2}$$

$$E_{\text{cell}} = -0.0296(\log x^2)$$

$$E_{\text{cell}} = -0.0296(2 \log x) = -0.0592 \log x$$

Because the acidity of a solution is often represented by the expression $\text{pH} = -\log [\text{H}^+] = -\log x$, we can also write for the concentration cell

$$E_{\text{cell}} = 0.0592 \text{ pH}$$

Thus, if we measure an E_{cell} value of 0.225 V in such a concentration cell at 25 °C, the unknown solution must have a pH of 0.225/0.0592 = 3.80.

The pH Meter

A SHE is a difficult electrode to construct and to maintain. In practice, to measure the pH of a solution we generally replace the SHE with some other reference electrode having a precisely known $E°$. We also replace the second hydrogen electrode by a special electrode known as a *glass electrode*. A glass electrode is a thin glass membrane containing HCl(aq) into which is immersed a silver wire coated with AgCl(s). When the electrode is dipped into a solution of unknown pH, H^+ ions are exchanged across the membrane. The potential developed on the silver wire depends on $[H^+]$ in the unknown solution. The potential difference between the glass electrode and the reference electrode similarly depends on $[H^+]$ in the unknown solution (see Figure 20.10).

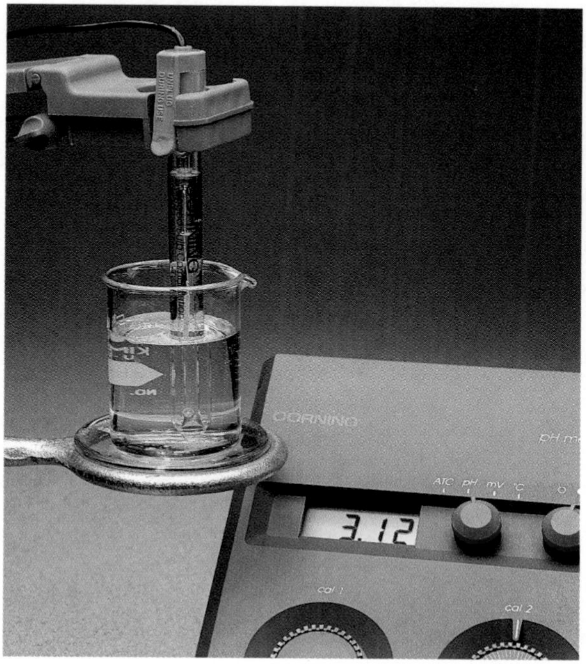

Figure 20.10 The glass electrode and the pH meter. A reference electrode and a glass electrode are both enclosed in the combination electrode shown. The potential difference between the two is converted to a pH value of the solution (3.12) and displayed by the meter.

The Electrochemistry of a Heartbeat

The heart is a pump made mostly of muscle, and like all muscles, it pulses to an electric signal. In muscle, the flow of electricity results from the movement of ions across cell membranes. A sodium ion–potassium ion pump moves these ions across the membranes of all cells. Sodium ion concentrations are high outside the cell and low inside; potassium ion concentrations are low outside the cell and high inside. These concentration gradients across cell membranes lead to a voltage difference, called the membrane potential, of about 90 millivolts (mV).

Unlike other muscles, the heart has its own stimulator and built-in circuitry—it keeps on beating even if outside nerve connections are cut. Cut

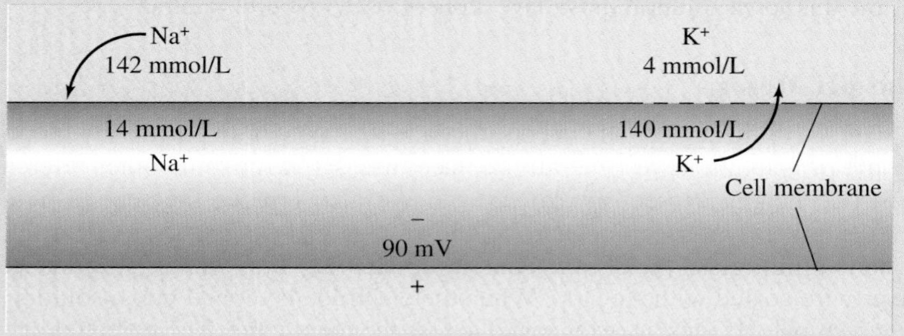

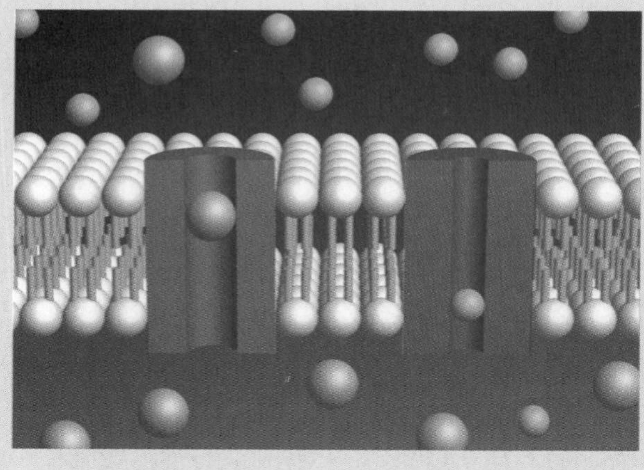

The diagram shows the concentration gradients of Na^+ and K^+ across the membrane of the axon of a nerve cell during the resting state. (An axon is a long extension of a nerve cell that connects to another nerve cell or to a muscle cell.) Ions tend to flow from areas of high concentration to areas of low concentration; thus there is a tendency for K^+ to flow out of the cell and Na^+ to flow into the cell. The membrane in the resting state is permeable to K^+ but almost impermeable to Na^+; some K^+ ions leak out, but few Na^+ ions leak into the cell. This charge imbalance leads to a potential difference across the membrane with the outside of the cell positive relative to the inside. The computer-generated rendering of a cell membrane shows a K^+ channel (blue) and an Na^+ channel (magenta). The area outside the cell (top) is rich in Na^+ and low in K^+. Inside the cell (bottom), the fluids are relatively rich in K^+ and low in Na^+.

the nerve connections to a pectoral muscle, on the other hand, and your arms are paralyzed for good.

The pacemaker cells of the heart exhibit a natural voltage that depends mainly on the concentrations of sodium and potassium ions. The cell membranes have a mechanism (like a tiny pump) to keep K^+ inside the cell and to expel Na^+. Some K^+ leaks out, however, and this gives the interior of the cell a negative potential with respect to the outside. When this potential difference reaches a critical value, channels open up in the pacemaker cell membranes and Na^+ ions rush in. This results in an electric discharge, a current flow that spreads almost instantaneously to other heart muscle cells. The cells act in concert, and the heart beats.

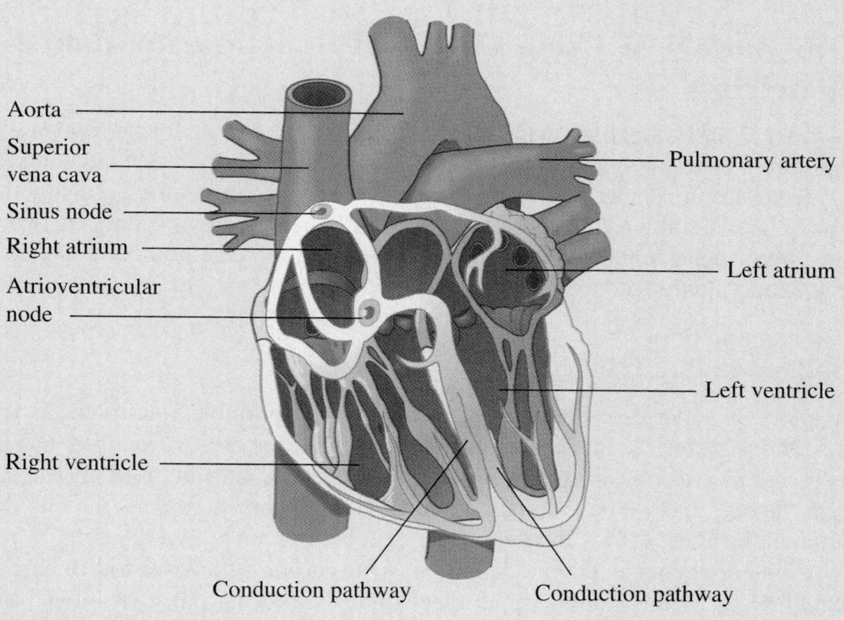

On the wall of the right atrium, a bundle of tissue called the sinus node ignites an impulse that triggers a contraction as it races down the wall to the atrioventricular (AV) node. The contraction sends blood into the ventricles. The AV node in turn flashes the electric impulse across the muscles of the ventricles, causing them to contract and pump the blood out of the heart and into the body. When the heart's own electrical system grows weak or faulty, an implanted battery-powered pacemaker can regulate the heartbeat by supplying electric impulses directly to the right ventricle.

Calcium ions also play a vital role in the heart beat. When the membrane is depolarized (discharged), Ca^{2+} ions flow into the cells and hold off repolarization for 0.15 s in atrial muscle and for 0.25 to 0.30 s in ventricular muscle. When the flow of Ca^{2+} ions into the cells ceases, K^+ ions again begin to leak out, repolarizing the membrane.

The proper level of K^+ is crucial. Normal blood levels are 3.5 to 5.0 mmol K^+/L blood. The normal ratio of $[K^+]$ between the inside and outside of the pacemaker cells is about 30 to 1. If $[K^+]$ in the blood is too *low*, the ionic imbalance goes beyond the critical stage, and the pacemaker cells are overcharged; they remain cocked like broken pistols, but do not fire. There is no heartbeat. If [K^+] in the blood is too *high*, the potential difference never gets large enough to fire off a proper signal. The heartbeat diminishes to nothingness. Both extremes—[K^+] too low or too high—lead to cardiac arrest.

The kidneys conserve Na^+ quite well, but there is no mechanism for retaining K^+. Dangerously low potassium levels are caused by a wasting of tissue from disease or starvation, by loss of fluids through prolonged diarrhea or excessive vomiting, and through the long-term use of diuretic drugs. Levels that are too high are caused by excessive use of dietary supplements or by kidney diseases in which the body does not excrete enough potassium ion.

20.5 Batteries: Using Chemical Reactions to Make Electricity

In everyday life, we call a device that stores chemical energy for later release of electricity a **battery**. A flashlight "battery" consists of a single voltaic cell with two electrodes in contact with one or more electrolytes. Sometimes, a distinction is made between the terms cell and battery. A battery is an assembly of two or more voltaic cells connected together. By this definition an automobile battery is a true battery. In this section we consider three different types of cells or batteries.

The Dry Cell

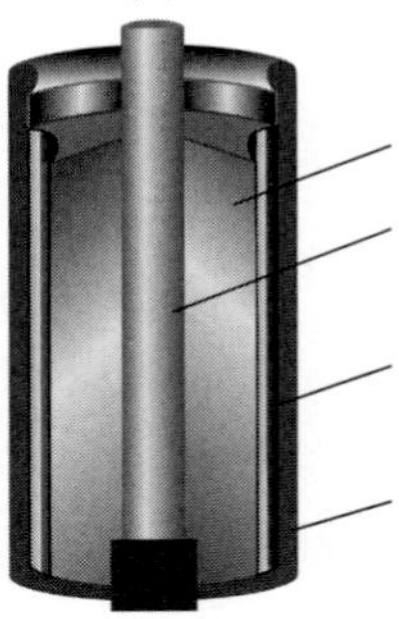

Figure 20.11 Cross-section of a Leclanché (dry) cell. The moist paste consists of $NH_4Cl(aq)$ and $ZnCl_2(aq)$; carbon black (a finely divided form of carbon) and $MnO_2(s)$ are also present.

The types of batteries used in most flashlights and portable electronic devices are *primary* batteries. In these batteries, cell reactions are irreversible. During use, reactants are converted to products; when the reactants are used up, the battery is "dead," and we have to buy a new one. A typical example is the one diagrammed in Figure 20.11, the *Leclanché cell*, better known as a dry cell.

A zinc container is the anode and an inert carbon (graphite) rod in contact with $MnO_2(s)$ is the cathode. The electrolyte is a moist paste of NH_4Cl and $ZnCl_2$. The cell is called a "dry" cell because there is no free-flowing liquid. The concentrated electrolyte solution is thickened into a gellike paste by an agent such as starch. The cell diagram for a Leclanché cell is

$$Zn(s) \mid NH_4Cl(aq), ZnCl_2(aq) \mid MnO_2(s), C(s)$$

The reactions that occur when electric current is drawn from a dry cell are quite complex and not completely understood. We will give only a simplified description. As suggested by the cell diagram, zinc metal is oxidized to Zn^{2+} at the anode.

$$\textit{Anode:}\quad Zn(s) \longrightarrow Zn^{2+}(aq) + 2\,e^-$$

The reduction reaction at the cathode appears to be

$$\textit{Cathode:}\quad MnO_2(s) + H^+(aq) + e^- \longrightarrow MnO(OH)(s)$$

followed by

$$2\,MnO(OH)(s) \longrightarrow Mn_2O_3(s) + H_2O(l)$$

The Leclanché cell is cheap to make and has a maximum cell voltage of 1.55 V, but it has two disadvantages: (1) The cell voltage drops rapidly during use due to a buildup of Zn^{2+} near the anode and a change of pH at the cathode, and (2) the zinc metal slowly reacts with the electrolyte, even when the cell electrodes are disconnected.

Alkaline cells cost more but last longer than ordinary dry cells. An alkaline cell has NaOH or KOH in place of NH_4Cl as the electrolyte. It uses essentially the same reduction half-reaction as the ordinary dry cell, but in an alkaline medium. Zinc ions produced in the oxidation half-reaction combine with $OH^-(aq)$ to form $Zn(OH)_2(s)$.

Reduction: $2\ MnO_2(s) + H_2O(l) + 2\ e^- \longrightarrow Mn_2O_3(s) + 2\ OH^-(aq)$ $\qquad E^\circ_{red} = 0.118\ V$

Oxidation: $Zn(s) + 2\ OH^-(aq) \longrightarrow Zn(OH)_2(s) + 2\ e^-$ $\qquad E^\circ_{ox} = 1.246\ V$

Net: $Zn(s) + 2\ MnO_2(s) + H_2O(l) \longrightarrow Zn(OH)_2(s) + Mn_2O_3(s)$

$$E^\circ_{cell} = E^\circ_{red} + E^\circ_{ox} = 0.118\ V + 1.246\ V = 1.364\ V$$

In practice, because the conditions in the cell are *nonstandard*, the cell potential is not equal to 1.364 V—it is actually somewhat higher.

Beside the fact that it has a longer shelf life and can be kept in service longer, an important advantage of the alkaline cell over the ordinary dry cell is that the voltage does not drop as rapidly when current is drawn. This is because the concentration of zinc ion does not build up at the anode and the pH remains more nearly constant at the cathode.

The Lead–Acid Storage Battery

The lead–acid storage battery used in automobiles is a *secondary* battery. The cell reaction can be reversed and the battery restored to its original condition. The battery can be used through repeated cycles of discharging and recharging. Figure 20.12 shows a portion of a cell in the lead–acid battery. Several anodes and several cathodes are connected together in each cell to increase its current-delivering capacity. Each cell has a voltage of 2 V, and six cells are connected together in "series" fashion, + to –, to form a 12-volt battery.

The anodes in the lead–acid storage cell are composed of a lead alloy, and the cathodes are made of a lead alloy impregnated with red lead dioxide. The electrolyte is dilute sulfuric acid. The half-reactions and net reaction on discharge are

Reduction: $PbO_2(s) + 4\ H^+(aq) + SO_4^{2-}(aq) + 2\ e^- \longrightarrow PbSO_4(s) + 2\ H_2O(l)$ $\qquad E^\circ_{red} = 1.69\ V$

Oxidation: $Pb(s) + SO_4^{2-}(aq) \longrightarrow PbSO_4(s) + 2\ e^-$ $\qquad E^\circ_{ox} = 0.36\ V$

Net: $Pb(s) + PbO_2(s) + 4\ H^+(aq) + 2\ SO_4^{2-}(aq) \longrightarrow 2\ PbSO_4(s) + 2\ H_2O(l)$

$$E^\circ_{cell} = E^\circ_{red} + E^\circ_{ox} = 1.69\ V + 0.36\ V = 2.05\ V$$

Even though the conditions in a lead–acid cell are nonstandard, the voltage delivered by a fully charged cell is just about the same as the E°_{cell} value.

As the cell reaction proceeds, $PbSO_4(s)$ precipitates and partially coats both electrodes; the water formed dilutes the $H_2SO_4(aq)$. The cell is *discharged*. By connecting the cell to an external electric energy source (of greater than 12 V)

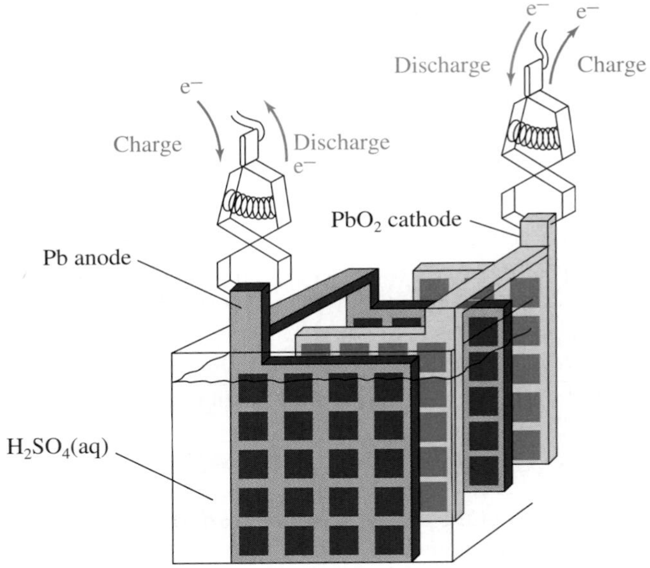

Figure 20.12 A lead–acid (storage) cell. The composition of the electrodes, the cell reaction, and the cell voltage are described in the text. Shown here are two anode plates and two cathode plates in "parallel" connections. This type of connection increases the surface area of the electrodes and the capacity of the cell to deliver current.

we can force electrons to flow in the opposite direction. The half-cell reactions and the net cell reaction are reversed and the battery is recharged.

Recharging reaction: $2\ PbSO_4(s) + 2\ H_2O(l) \longrightarrow Pb(s) + PbO_2(s) + 4\ H^+(aq) + 2\ SO_4^{2-}(aq)$

In an automobile, the battery is discharged when the engine is started. While running, the engine powers the alternator, which produces electric energy required to recharge the battery. The battery is constantly recharged as the automobile is driven. When a "dead" battery is unable to start the engine, the engine can often be "jump started" by connecting the weak battery to another fully charged battery. The engine gets its "start" from the good battery, and immediately the recharging of the weak battery begins. As a battery is discharged, the $H_2SO_4(aq)$ is diluted; it is reconcentrated during recharging. A simple method of determining the state of charge in a lead–acid battery is to measure the density of the acid. The more concentrated the acid, the greater is its density.

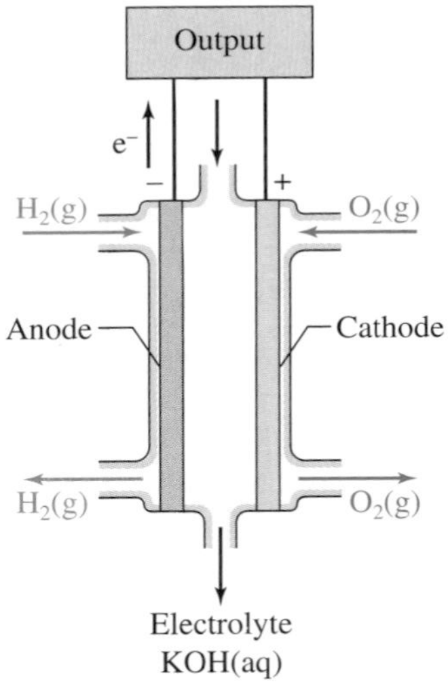

Figure 20.13 A hydrogen–oxygen fuel cell. The electrodes are porous to allow easy access of the gaseous reactants to the electrolyte. Also, the electrode material catalyzes the electrode reactions.

Fuel Cells

Modern civilization runs mainly on fossil fuels, but as we have noted previously, combustion of fuels and conversion of the evolved heat to electricity is quite limited in efficiency. A voltaic cell, on the other hand, is able to convert chemical energy to electricity with an efficiency of 90% or more. Because combustion reactions and voltaic cell reactions both involve oxidation and reduction, why not design a voltaic cell that has a combustion reaction as its cell reaction? It's already been done. Devices for doing this have been under active study for over 50 years. Such a device is called a *fuel cell*; its cell reaction is

$$\text{Fuel} + \text{oxygen} \longrightarrow \text{oxidation (combustion) products}$$

In the hydrogen–oxygen cell, hydrogen and oxygen gases flow over separate inert electrodes in contact with an electrolyte such as KOH(aq) (see Figure 20.13). The reactions are

Reduction:	$O_2(g) + 2H_2O + 4e^- \longrightarrow 4OH^-$	$E^\circ_{red} = 0.401$ V
Oxidation:	$2\{H_2(g) + 2OH^-(aq) \longrightarrow 2H_2O + 2e^-\}$	$E^\circ_{ox} = -E^\circ = -(-0.828) = 0.828$ V
Net:	$2H_2(g) + O_2(g) \longrightarrow 2H_2O(l)$	

$$E^\circ_{cell} = E^\circ_{red} + E^\circ_{ox} = 0.401\text{ V} + 0.828\text{ V} = 1.229\text{ V}$$

A fuel cell is like a battery with the contents outside the battery case rather than inside. It is a flow battery. In principle, a fuel cell will continue to operate as long as reactants are fed into it.

This E°_{cell} value is for 25 °C. Fuel cells of this type are generally operated at nonstandard conditions and at temperatures higher than 25 °C. Their operating voltages are about 1.0 to 1.1 V. Hydrogen–oxygen fuel cells have been widely used in space vehicles. In addition to the electricity, the water formed is also a valuable product.

In principle, a fuel cell can run directly on a hydrocarbon fuel, such as natural gas (mostly methane).

Reduction:	$2\{O_2(g) + 4H^+ + 4e^- \longrightarrow 2H_2O\}$
Oxidation:	$CH_4(g) + 2H_2O \longrightarrow CO_2(g) + 8H^+ + 8e^-$
Net:	$CH_4(g) + 2O_2(g) \longrightarrow CO_2(g) + 2H_2O(l)$

Fuel cells based on the direct conversion of a hydrocarbon fuel to oxidation products have been prepared on a research basis, but technical difficulties must be overcome before they can be commercialized. An indirect method that accomplishes the same objective involves steam reforming of the fuel to produce $H_2(g)$

$$CH_4(g) + H_2O(g) \longrightarrow CO(g) + 3H_2(g)$$

followed by use of $H_2(g)$ in a hydrogen–oxygen fuel cell. This approach also has drawbacks, however, because of the high temperature required in the steam reforming.

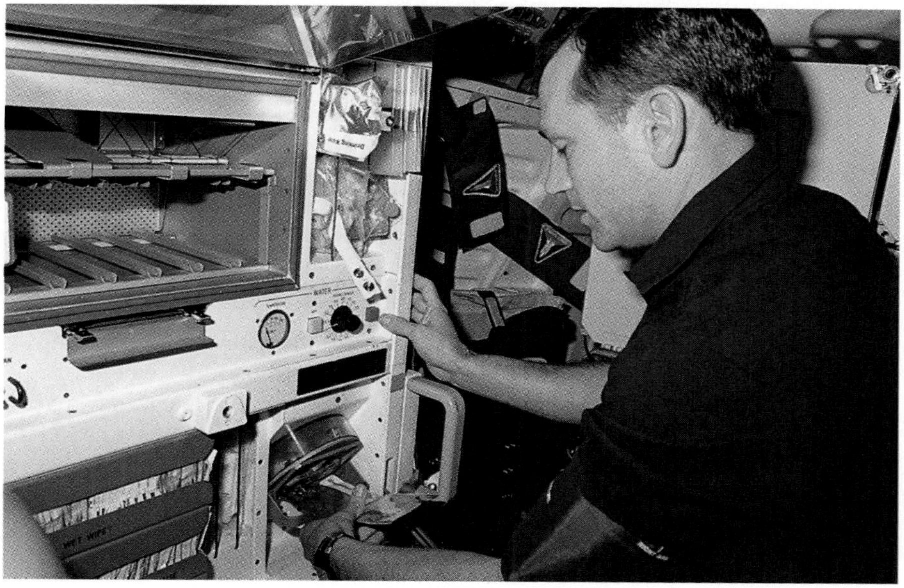

Astronauts can use water produced by a space shuttle's fuel cells to rehydrate packaged dried foods.

Continuing developments in fuel cell technology may one day lead to economically viable fuel cells for widespread use in automobiles. Those that use hydrogen–oxygen cells produce no undesirable products (beyond humid air) and are classified as zero-emission vehicles.

Air Batteries

An air battery is similar to a fuel cell in that $O_2(g)$ from air is used as the oxidizing agent. In an air battery, however, the reducing agent typically is a metal, such as zinc or aluminum.

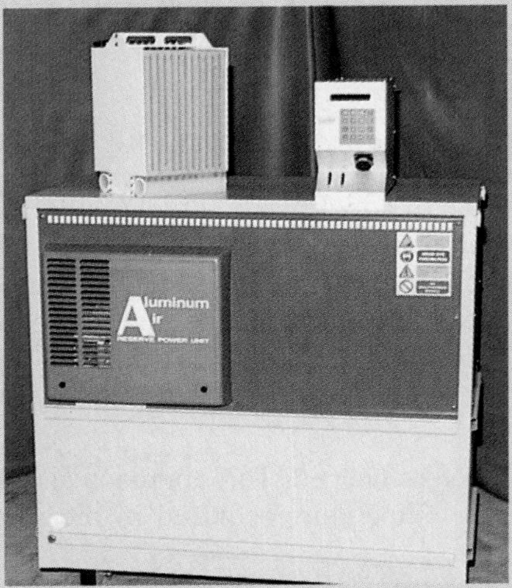

A reserve power unit for telecommunications using aluminum–air batteries. The unit is able to deliver electric power equivalent to two dozen lead–acid storage batteries.

In an aluminum–air battery, oxidation occurs at an aluminum anode and reduction at a carbon cathode. The electrolyte circulated through the battery is NaOH(aq). Because of the high concentration of OH^-, aluminum is oxidized to $[Al(OH)_4]^-$, not Al^{3+} (recall the discussion of amphoterism on pages 718–719).

Reduction:	$3\{O_2(g) + 2\,H_2O(l) + 4\,e^- \longrightarrow 4\,OH^-(aq)\}$	$E^\circ_{red} = 0.401\text{ V}$
Oxidation:	$4\{Al(s) + 4\,OH^-(aq) \longrightarrow [Al(OH)_4]^-(aq) + 3\,e^-\}$	$E^\circ_{ox} = 2.310\text{ V}$
Net:	$4\,Al(s) + 3\,O_2(g) + 6\,H_2O(l) + 4\,OH^-(aq) \longrightarrow 4\,[Al(OH)_4]^-(aq)$	

$$E^\circ_{cell} = E^\circ_{red} + E^\circ_{ox} = 0.401\text{ V} + 2.310\text{ V} = 2.711\text{ V}$$

During operation, the battery is fed chunks of aluminum metal and water. The electrolyte is circulated outside the battery, where the $[Al(OH)_4]^-(aq)$ is precipitated as $Al(OH)_3(s)$. Aluminum is later recovered from the $Al(OH)_3(s)$ in an aluminum manufacturing facility. An aluminum–air battery can power an automobile for several hundred miles before requiring maintenance.

20.6 Corrosion: Metal Loss Through Voltaic Cells

When carried out in voltaic cells, redox reactions are important sources of electricity. At the same time, the ability of redox reactions to produce electricity underlies the destruction caused by corrosion. This is a matter of great economic importance. It is estimated that, in the United States alone, corrosion costs $70 billion a year. (Perhaps 20% of all the iron and steel produced in the United States each year goes to replace corroded material.)

Let's look first at the corrosion of iron. In moist air, iron can be *oxidized* to Fe^{2+}, particularly at scratches, nicks, or dents. Regions where this occurs become pitted and are known as *anodic* areas. Electrons lost in the oxidation are conducted along the iron to other regions, called *cathodic* areas, where atmospheric $O_2(g)$ is reduced to OH^-.

$$\begin{array}{lll} \textit{Reduction:} & O_2(g) + 2\,H_2O(l) + 4\,e^- \longrightarrow 4\,OH^-(aq) & E^\circ_{red} = 0.401\text{ V} \\ \textit{Oxidation:} & 2\{Fe(s) \longrightarrow Fe^{2+}(aq) + 2\,e^-\} & E^\circ_{ox} = 0.440\text{ V} \\ \hline \textit{Net:} & 2\,Fe(s) + O_2(g) + 2\,H_2O(l) \longrightarrow 2\,Fe^{2+}(aq) + 4\,OH^-(aq) & \end{array}$$

$$E^\circ_{cell} = E^\circ_{red} + E^\circ_{ox} = 0.401\text{ V} + 0.440\text{ V} = 0.841\text{ V}$$

The corrosion of iron, then, looks every bit like a spontaneous redox reaction occurring in a voltaic cell, and it is. Oxidation and reduction can occur at separate points on the metal; electrons are conducted through the metal; and the circuit is completed by an electrolyte in aqueous solution. In the snow belt, this solution is often the slush from road salt and melting snow; in coastal areas it may be ocean spray.

Iron(II) ions migrate from anodic areas to cathodic areas, where they combine with OH^- ions to form insoluble iron(II) hydroxide.

$$Fe^{2+}(aq) + 2\,OH^-(aq) \longrightarrow Fe(OH)_2(s)$$

Usually, iron(II) hydroxide is further oxidized to iron(III) hydroxide.

$$4\,Fe(OH)_2(s) + O_2(g) + 2\,H_2O(l) \longrightarrow 4\,Fe(OH)_3(s)$$

Iron(III) hydroxide is generally represented as a hydrated iron(III) oxide of variable composition: $Fe_2O_3 \cdot xH_2O$. This material is the familiar *iron rust*. The overall process described here is illustrated in Figure 20.14.

In order to emphasize the resemblance to a voltaic cell, we have focused on a situation where the anodic and cathodic regions are some distance apart and where intermediate stages of the corrosion are identifiable. Often, though, all of the reactions occur simultaneously on the corroding portion of the iron. Then, all we really notice over time is the accumulation of rust.

Aluminum is more reactive than iron (more easily oxidized). We might expect aluminum to corrode more rapidly than iron, but the occasional litter of beer and soda pop cans that we see testifies to the fact that it doesn't. How can this be? Freshly prepared aluminum does form an oxide coating very rapidly. However, this thin, hard film of aluminum oxide is impervious to air and protects the underlying metal from further oxidation. Iron oxide (rust), on the other hand, flakes off an object and constantly exposes fresh surface to further corrosion. It is the difference in behavior of their corrosion products that explains why steel cans eventually disintegrate in the environment whereas aluminum cans seemingly last forever.

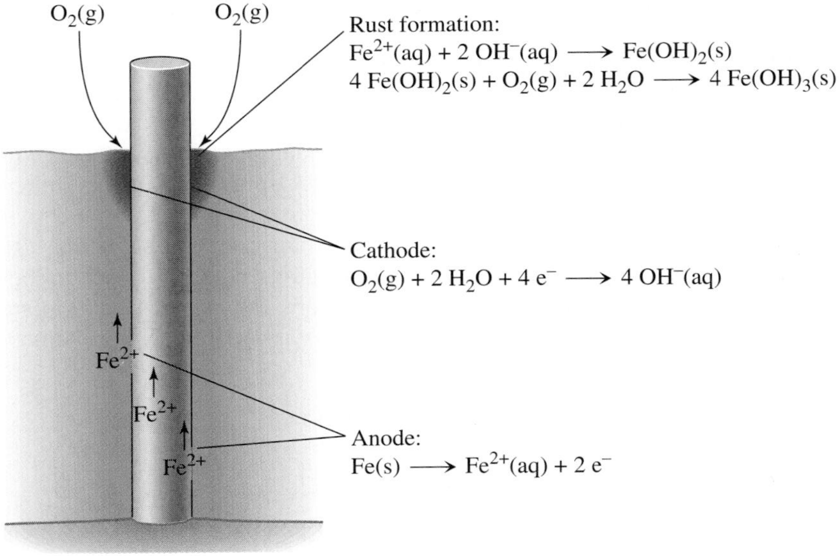

Figure 20.14 Corrosion of an iron piling: an electrochemical process. This schematic drawing illustrates the anodic and cathodic regions, their half-reactions, and the final formation of rust. The cathodic region is near the air–water interface where the availability of $O_2(g)$ is greatest. The anodic region is at greater depths below the water's surface. $Fe^{2+}(aq)$ from the anodic region migrates to the cathodic region where rust formation occurs.

The simplest line of defense against the corrosion of iron is to paint it. Another is to coat it with a thin layer of another, less active metal, as in coating a steel can with tin. Either way, however, the surface is protected only as long as the coating does not crack, chip, or peel. A tin can may begin to rust rapidly if it is dented, and a car body usually rusts first where the paint is scratched or chipped.

An entirely different approach is to protect iron with a *more* active metal, as in the zinc-clad iron known as *galvanized iron*. The oxide coating on zinc is like that on aluminum—thin, hard, and impervious to air. And it doesn't matter if a break occurs in the zinc plating. The zinc continues to corrode rather than the iron because zinc is more easily oxidized than is iron.

Another method that works on the same general principle as galvanized iron is known as *cathodic protection*. The iron object to be protected—a ship, a pipeline, a storage tank, a water heater, or the plumbing system in a swimming pool—is connected to a chunk of an active metal such as magnesium, aluminum, or zinc, either directly or with a wire. Oxidation occurs at the active metal and the metal slowly dissolves. The iron surface acquires electrons from the active metal and supports the *cathode* or reduction half-reaction. As long as some of the active metal remains, the iron is protected. The active metal is called a *sacrificial anode*. One of the more important uses of magnesium metal is in sacrificial anodes (about 12 million tons annually in the United States).

A long strip of an active metal (for example, zinc), when attached to a ship's hull, protects it from corrosion.

Another example of the electrochemical control of corrosion is seen in one method for the removal of tarnish from silver. The tarnish is generally black, insoluble silver sulfide formed when silver comes in contact with sulfur compounds, either in the atmosphere or in foods such as eggs. We can use a silver polish to remove the tarnish, but in doing so we also lose part of the silver.

The concentration of Ag^+ in an aqueous solution in contact with $Ag_2S(s)$ is extremely low. Yet aluminum is a good enough reducing agent to reduce this Ag^+ back to silver metal. A precious metal is conserved at the expense of a cheaper one.

$$3\ Ag^+(aq) + Al(s) \longrightarrow 3\ Ag(s) + Al^{3+}(aq)$$

Sodium hydrogen carbonate ($NaHCO_3$) works well as an electrolyte to sustain this desired voltaic cell reaction. The tarnished silver is placed in contact with aluminum foil, covered with $NaHCO_3(aq)$, and heated.

20.7 Electrolysis: Electricity as an Agent for Chemical Change

If we use a battery or other source of direct electric current to pass electricity through molten sodium chloride, here is what we observe: Yellow-green chlorine gas forms at one electrode and silvery, liquid, metallic sodium forms at the other electrode and floats on the molten salt (see Figure 20.15).

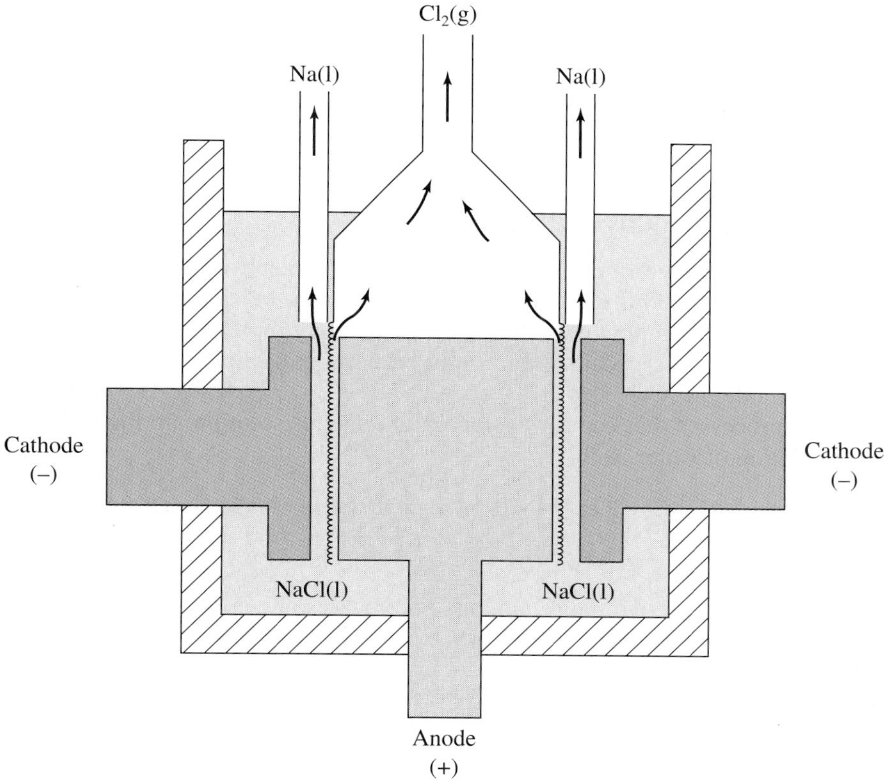

Figure 20.15 The electrolysis of molten sodium chloride. The electrolysis cell pictured here is called a *Down's cell.* The electrolyte is molten NaCl with a small amount of $CaCl_2$ added to lower its melting point. Liquid sodium forms at the steel cathode and gaseous chlorine at the graphite anode. The steel gauze diaphragm keeps the sodium and chlorine from recombining to form sodium chloride.

The decomposition of NaCl(l) into its elements is a *nonspontaneous* reaction, with a large positive value of ΔG. The function of the battery is to act as an "electron pump," forcing electrons onto the *cathode* of the electrolysis cell, where the *reduction* of sodium ions occurs.

$$Na^+ + e^- \longrightarrow Na \qquad E^\circ_{red} = -2.713\ V$$

The battery withdraws electrons from the *anode* of the electrolysis cell, where the *oxidation* of chloride ions occurs.

$$2\ Cl^- \longrightarrow Cl_2 + 2\ e^- \qquad E^\circ_{ox} = -E^\circ = -1.358\ V$$

Regardless of whether a cell is a voltaic or an electrolytic cell, the anode is the electrode at which oxidation occurs, and the cathode is the electrode at which reduction occurs.

The net reaction, E°_{cell}, and ΔG° are

$$2\ \text{NaCl(l)} \xrightarrow{\text{electrolysis}} 2\ \text{Na(l)} + \text{Cl}_2\text{(g)} \qquad E^\circ_{cell} = -4.071\ \text{V}$$

$$\Delta G^\circ = -nFE^\circ_{cell}(\text{s}) = -2\ \text{mol e}^- \times 96{,}485\ \text{C/mol e}^- \times (-4.071)\ \text{V}$$
$$= 7.86 \times 10^5\ \text{J} = 786\ \text{kJ}$$

The E° and E°_{cell} values are not quite right for the electrolysis, because they are based on a reaction in aqueous solution, whereas we are describing a reaction in molten NaCl. There are some other complicating factors as well. Still, though, these values give us a fair idea of the minimum voltage we must use for the electrolysis. The *negative* value of E°_{cell} and the large positive value of ΔG° confirm that the electrolysis is a *nonspontaneous* process. However, if we impose a voltage on the NaCl electrolytic cell of something in excess of 4 V (4.071 V according to the calculation), we can force electrons to flow in the opposite direction from the spontaneous reaction. **Electrolysis**, then, is a process of using electricity to produce a nonspontaneous chemical change.

Predicting Electrolysis Reactions

If we apply a voltage in excess of about 4 V to an electrolysis cell containing *aqueous* NaCl, we don't get the same products as in the electrolysis of molten NaCl. Electrolyses in aqueous solution are generally more complicated than in molten salts. This is because there may be competing reactions at the anode and/or cathode.

Two processes that can always occur in aqueous solution are the reduction and oxidation of water itself.

Reduction: $2\ \{2\ \text{H}_2\text{O} + 2\ \text{e}^- \longrightarrow \text{H}_2\text{(g)} + 2\ \text{OH}^-\text{(aq)}\}$ $\qquad E^\circ_{red} = -0.828\ \text{V}$

Oxidation: $2\ \text{H}_2\text{O} \longrightarrow 4\ \text{H}^+\text{(aq)} + \text{O}_2\text{(g)} + 4\ \text{e}^-$ $\qquad E^\circ_{ox} = -E^\circ = -1.229\ \text{V}$

When we add these two half-equations and make some combinations and cancellations, we get the net equation for the electrolysis of water.

$$6\,\text{H}_2\text{O} \longrightarrow 2\,\text{H}_2\text{(g)} + \text{O}_2\text{(g)} + 4\,\text{H}^+\text{(aq)} + 4\,\text{OH}^-\text{(aq)}$$
$$6\,\text{H}_2\text{O} \longrightarrow 2\,\text{H}_2\text{(g)} + \text{O}_2\text{(g)} + 4\,\text{H}_2\text{O}$$
$$\textit{Net:}\ 2\,\text{H}_2\text{O} \longrightarrow 2\,\text{H}_2\text{(g)} + \text{O}_2\text{(g)}$$

When the half-reactions cited above are the only ones feasible, the electrolysis of an aqueous solution has H_2(g) and O_2(g) as its products and the electrolysis of water as its net reaction. There are many cases, however, where only one, or perhaps neither, of those half-reactions occurs. As an example, let's return again to the electrolysis of NaCl(aq).

To obtain sodium metal by the reduction of Na^+(aq), we have to contend with an $E^\circ_{red} = -2.713$ V. This is more difficult to accomplish than the reduction of water, for which $E^\circ_{red} = -0.828$ V. Think of the matter in this way: As electrons are pumped into the cathode they are picked up so much more easily by H_2O molecules than by Na^+ ions that, in effect, all the electrons go to reducing

H_2O molecules and none to reducing Na^+ ions. In the electrolysis of NaCl(aq), we get H_2(g) and OH^- at the cathode, not sodium metal.

To obtain chlorine gas by the oxidation of Cl^-(aq), we have to contend with $E^\circ_{ox} = -1.358$ V. This is not much different from what is required for the oxidation of H_2O: $E^\circ_{ox} = -1.229$ V. Both oxidations can occur and that of Cl^- will predominate, especially in concentrated NaCl(aq). As a result, we can write

Reduction:	$2\,H_2O + 2\,e^- \longrightarrow H_2(g) + 2\,OH^-(aq)$	$E^\circ_{red} = -0.828$ V
Oxidation:	$2\,Cl^- \longrightarrow Cl_2(g) + 2\,e^-$	$E^\circ_{ox} = -E^\circ = -1.358$ V
Net:	$2\,Cl^-(aq) + 2\,H_2O \longrightarrow 2\,OH^-(aq) + H_2(g) + Cl_2(g)$	$E^\circ_{cell} = -2.186$ V

From this net ionic equation, we see that the Na^+ ions are mere spectators in the electrolysis. They cannot be reduced if water molecules are present.

Because E°_{ox} for the oxidation of Cl^-(aq) is more negative (−1.358 V) than is E°_{ox} for the oxidation of H_2O(l) (−1.229 V), why should Cl_2(g) predominate over O_2(g) in the electrolysis of NaCl(aq)? The reason lies in the phenomenon of overpotential or overvoltage. Because of various interactions at electrode surfaces, in some electrolysis half-reactions, particularly those involving gases, a higher-than-calculated voltage is required. The **overvoltage** of an electrode reaction is the excess voltage above that calculated from the E° value required to bring about the reaction. For example, the overvoltage for the formation of H_2(g) on a platinum surface is almost zero, but on a mercury surface it is about 1.5 V. The answer to the question posed above is that the overvoltage for the formation of O_2(g) is considerably higher than for the formation of Cl_2(g).

Example 20.7

Predict the net electrolysis reaction when $AgNO_3$(aq) is electrolyzed.

Solution

The principal species in the solution are Ag^+ and NO_3^- ions and H_2O molecules. Silver ion is easily reduced, as we can see from the half-reaction in Table 20.1

$$Ag^+(aq) + e^- \longrightarrow Ag(s) \qquad E^\circ_{red} = 0.800\text{ V}$$

This reduction should occur exclusively over that of water molecules.

$$2\,H_2O + 2\,e^- \longrightarrow H_2(g) + 2\,OH^-(aq) \qquad E^\circ_{red} = -0.828\text{ V}$$

Because NO_3^- has the N atom in its highest possible oxidation state (+5), it cannot be oxidized. The only possible oxidation that can occur is that of H_2O molecules.

$$2\,H_2O \longrightarrow O_2(g) + 4\,H^+(aq) + 4\,e^- \qquad E^\circ_{ox} = -E^\circ = -1.229\text{ V}$$

The net electrolysis reaction is

$$4\,Ag^+(aq) + 2\,H_2O \xrightarrow{\text{electrolysis}} 4\,Ag(s) + 4\,H^+(g) + O_2(g) \qquad E^\circ_{cell} = -0.429\text{ V}$$

Example 20.8

Predict the net electrolysis reaction when KBr(aq) is electrolyzed.

Solution

The principal species in the solution are K^+ and Br^- ions and H_2O molecules. We have already seen that H_2O molecules are reduced rather than Na^+ ions in NaCl(aq).

Potassium ion is even more difficult to reduce than Na^+. We expect the reduction half-reaction to be

$$2\,H_2O + 2\,e^- \longrightarrow H_2(g) + 2\,OH^-(aq) \qquad E^\circ_{red} = -0.828\text{ V}$$

The two possibilities for oxidation are that of water molecules to $O_2(g)$ and $Br^-(aq)$ to $Br_2(l)$. These oxidation half-reactions and their E°_{ox} values are

$$2\,Br^-(aq) \longrightarrow Br_2(l) + 2\,e^- \qquad E^\circ_{ox} = -E^\circ = -1.065\text{ V}$$

$$2\,H_2O \longrightarrow O_2(g) + 4\,H^+ + 4\,e^- \qquad E^\circ_{ox} = -E^\circ = -1.229\text{ V}$$

We should expect the oxidation of $Br^-(aq)$ to predominate and the net electrolysis reaction to be

$$2\,Br^-(aq) + 2\,H_2O \xrightarrow{\text{electrolysis}} Br_2(l) + H_2(g) + 2\,OH^-(aq) \qquad E^\circ_{cell} = -1.893\text{ V}$$

Exercise 20.7

Write plausible half-equations and a net equation for the electrolysis of KI(aq).

Exercise 20.8

Write plausible half-equations and a net equation for the electrolysis of $CuSO_4(aq)$.

To summarize some ideas illustrated by the preceding examples and exercises,

1. We expect the reduction at the cathode to be that of H_2O to $H_2(g)$ and $OH^-(aq)$, with $E^\circ_{red} = -0.828$ V, *unless* another reduction can occur that has $E^\circ_{red} > -0.828$ V.
2. We expect the oxidation at the anode to be that of H_2O to $O_2(g)$ and $H^+(aq)$, with $E^\circ_{ox} = -1.229$ V, *unless* another oxidation can occur that has $E^\circ_{ox} > -1.229$ V.
3. Oxidation of H_2O to $O_2(g)$ occurs exclusively in aqueous solutions of oxoanions in which the central atom is in its *highest* oxidation state (as was the case with NO_3^- in Example 20.7).

Of course, these first two expectations might not hold if overvoltage is an important factor [as in the electrolysis of NaCl(aq)], or if conditions are far from standard.

CONCEPTUAL EXAMPLE 20.3

Figure 20.16 shows two electrochemical cells joined together in a common circuit. Indicate whether a net flow of electrons occurs in the direction of the red arrows or the blue arrows.

Solution

Note that the two cells have the same electrodes in the same solutions, but that the solution concentrations differ. Also, the cells are joined in such a way that the cells oppose one another. We need to compare the force with which each zinc anode pushes out electrons toward the copper anode. The cell exerting the greater electromotive force (cell voltage) as a voltaic cell is the one that will establish the direction of electron flow. The two cells have the same E°_{cell} value.

Reduction (cathode): $Cu^{2+}(1\text{ M}) + 2\,e^- \longrightarrow Cu(s) \qquad E^\circ_{red} = 0.337\text{ V}$

Oxidation (anode): $Zn(s) \longrightarrow Zn^{2+}(1\text{ M}) + 2\,e^- \qquad E^\circ_{ox} = -E^\circ = 0.763\text{ V}$

Net: $Zn(s) + Cu^{2+}(1\text{ M}) \longrightarrow Zn^{2+}(1\text{ M}) + Cu(s) \qquad E^\circ_{cell} = 1.100\text{ V}$

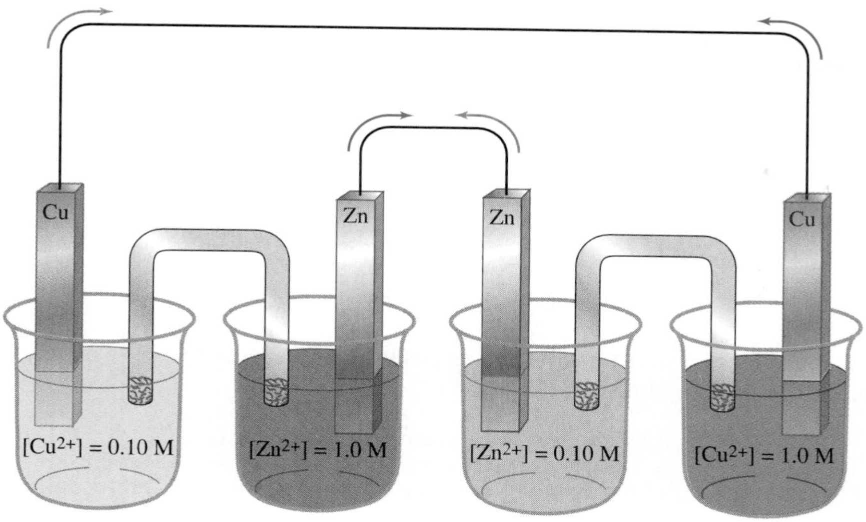

Figure 20.16 The joining of two electrochemical cells.

Now let's replace the standard state concentrations by the nonstandard concentrations given. We can substitute these directly into the net cell equation, and then state whether each E_{cell} is greater or less than E°_{cell}.

$$\textit{Cell A}: \quad Zn(s) + Cu^{2+}(0.10\ M) \longrightarrow Zn^{2+}(1.0\ M) + Cu(s)$$

Because of the lower concentration of Cu^{2+} compared to that in the standard-state cell, the forward reaction (left to right) is a little *less* favored than in the standard-state cell, and we should expect the cell voltage to be somewhat *less* than E°_{cell}. That is, $E_{cell} < 1.100$ V.

$$\textit{Cell B}: \quad Zn(s) + Cu^{2+}(1.0\ M) \longrightarrow Zn^{2+}(0.10\ M) + Cu(s)$$

Here, with $[Zn^{2+}]$ being lower than in the standard-state cell, the forward reaction (left to right) is a little *more* favored than in the standard-state cell, and we should expect the cell voltage to be somewhat *greater* than E°_{cell}. That is, $E_{cell} > 1.100$ V.

Our conclusion is that the direction of electron flow should be outward from the zinc anode in Cell B—the direction of the red arrows. Note also, that Cell B functions as a *voltaic* cell and Cell A becomes an *electrolytic* cell.

Conceptual Exercise 20.3

Write net ionic equations for the cell reactions in Cell A and Cell B when they are connected in the manner shown in Figure 20.16 and described in Conceptual Example 20.3. Explain why electric current cannot continue to flow indefinitely, and indicate at what point the current will stop flowing.

Michael Faraday (1791–1867) is known for his discoveries in electricity and magnetism as well as in electrochemistry. Faraday was unschooled in mathematics but had an unrivaled intuitive ability to visualize phenomena.

Quantitative Electrolysis

The quantitative basis of electrolysis was largely established by Michael Faraday. As we shall see, the quantity of a reactant consumed or product formed during electrolysis is related to (1) the molar mass of the substance, (2) the quantity of electric charge used, and (3) the number of electrons transferred in the electrode reaction.

We already know that quantities of electric charge are measured in coulombs (C), and that electric current, in amperes (A), describes the rate of flow of electric charge.

$$1 \text{ ampere (A)} = 1 \text{ coulomb (C)}/1 \text{ second (s)}$$

The total electric charge involved in an electrolysis reaction, then, is a product of the current and the time that it flows.

$$\text{Electric charge (C)} = \text{current (C/s)} \times \text{time (s)}$$

This expression and the Faraday constant—96,485 C/mol e^-—are usually required to calculate the outcomes of electrolysis reactions. Our general strategy will be to

1. Determine the amount of charge—the product of current and time.
2. Convert the amount of charge (C) to moles of electrons: 1 mol e^- = 96,485 C.
3. Relates moles of electrons to moles of a reactant or product through a half-equation. For example, the half-equation $Cu^{2}(aq) + 2\ e^- \longrightarrow Cu(s)$ yields the factor

$$\frac{1 \text{ mol Cu}}{2 \text{ mol } e^-}$$

4. Convert from moles of reactant or product to the final quality desired—mass, volume of gas, and so on.

You must be prepared to use variations of this general strategy. In Exercise 20.9, for example, each step of the strategy is inverted, and the steps are performed in the reverse order: Steps 4, 3, 2, and 1.

Example 20.9

We can use electrodeposition to determine the gold content of a sample. The sample is dissolved and all the gold converted to $Au^{3+}(aq)$. The reduction half-reaction at the cathode is $Au^{3+}(aq) + 3\ e^- \longrightarrow Au(s)$. What mass of gold will be deposited in 1.00 h by a current of 1.50 A?

Solution

Let's follow the four-step procedure.

1. The total electric charge involved is

$$\text{Charge} = \frac{1.50 \text{ C}}{\text{s}} \times 1.00 \text{ h} \times \frac{60 \text{ min}}{1 \text{ h}} \times \frac{60 \text{ s}}{1 \text{ min}} = 5.40 \times 10^3 \text{ C}$$

2. The number of moles of electrons is

$$? \text{ mol } e^- = 5.40 \times 10^3 \text{ C} \times \frac{1 \text{ mol } e^-}{96{,}485 \text{ C}} = 0.0560 \text{ mol } e^-$$

3. According to the reduction half-equation, 3 mol e^- are required for every mol Au deposited.

$$? \text{ mol Au} = 0.0560 \text{ mol e}^- \times \frac{1 \text{ mol Au}}{3 \text{ mol e}^-} = 0.0187 \text{ mol Au}$$

4. The mass of Au is simply the product of the amount in moles and the molar mass.

$$? \text{ g Au} = 0.0187 \text{ mol Au} \times \frac{197.0 \text{ g Au}}{1 \text{ mol Au}} = 3.68 \text{ g Au}$$

Exercise 20.9

For how many minutes must the electrolysis of a solution of $CuSO_4$(aq) be carried out with a current of 2.25 A to deposit 1.00 g of Cu(s) at the cathode? (*Hint*: What is the cathode half-reaction?)

Producing Chemicals by Electrolysis

We've already suggested several practical applications of electrolysis in this chapter: the production of sodium and chlorine from NaCl(l), the preparation of hydrogen and oxygen gases by the electrolysis of water, and the quantitative analysis of a metal by its electrodeposition. And there are many, many other applications. Electrolysis plays an important role in the manufacture or purification of sodium, calcium, magnesium, aluminum, copper, zinc, silver, hydrogen, chlorine, fluorine, hydrogen peroxide, sodium hydroxide, potassium dichromate, and potassium permanganate, among others.

Without doubt, one of the most commercially important electrolyses is that of NaCl(aq). We've already written a net ionic equation for this electrolysis (page 797), but let's rewrite it and add in the spectator ion, Na^+.

$$2\,Na^+(aq) + 2\,Cl^-(aq) + 2\,H_2O \longrightarrow \underbrace{2\,Na^+(aq) + 2\,OH^-(aq)}_{\text{Sodium hydroxide}} + \underbrace{H_2(g)}_{\text{Hydrogen}} + \underbrace{Cl_2(g)}_{\text{Chlorine}}$$

The most valuable products are the chlorine gas and sodium hydroxide solution. The various processes for the electrolysis of NaCl(aq)—and there are several—are called *chlor–alkali* processes.

A typical chlor–alkali electrolytic cell of the type called a diaphragm cell is pictured in Figure 20.17. The major challenge in a chlor–alkali process is to keep Cl_2(g) from coming into contact with the NaOH(aq). In an alkaline solution, Cl_2 disproportionates, producing ClO^- and Cl^-. The purpose of the diaphragm is to prevent this contact. Another requirement in a chlor–alkali cell is to prevent the mixing of Cl_2 and H_2. Mixtures of these gases can be explosive.

The voltage applied to a typical chlor–alkali cell is about 3.5 V and the current is about 1×10^5 A. Such a cell consumes about 2500 kilowatt hours (kWh) of electricity per ton of Cl_2 produced. (A typical household consumes about 600 kWh per month.) In all, the chlor–alkali industry in the United States consumes about 0.5% of all the electric power generated.

Electroplating

Electrolysis can be used to coat one metal onto another, a process called *electroplating*. Usually, the object to be electroplated, such as a spoon, is cast of

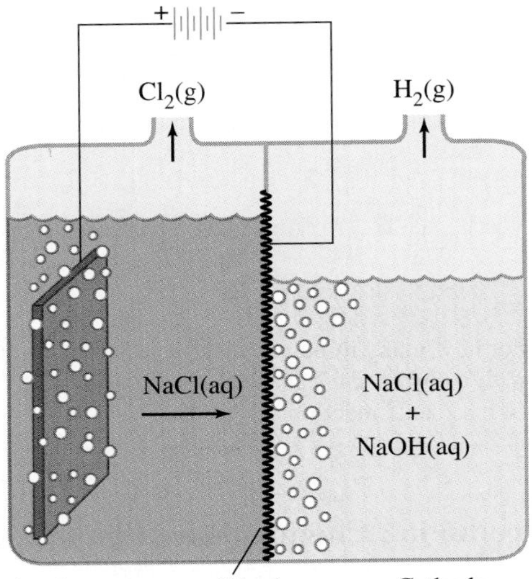

Figure 20.17 A diaphragm chlor–alkali cell. The anode is a specially treated titanium metal. The diaphragm and cathode are a composite unit consisting of an asbestos–polymer mixture deposited on a steel wire mesh.

a cheaper metal. It is then coated with a thin layer of a more attractive, more corrosion-resistant, and more expensive metal, such as silver or gold. The cost of the finished product is far less than if the object had been made entirely of the more expensive metal.

A cell for electroplating silver is shown in Figure 20.18. A piece of pure silver is the anode, and the spoon is the cathode. A solution of silver nitrate is used as the electrolyte. A battery or other external source of direct electric current removes electrons from the anode and forces them onto the cathode (the spoon). Silver ions from the solution are attracted to the cathode, where they gain electrons and are reduced to the free metal.

$$Ag^+(aq) + e^- \longrightarrow Ag(s, \text{cathode})$$

Silver atoms at the anode replace electrons withdrawn by the battery by undergoing oxidation and entering the solution as silver ions.

$$Ag(s, \text{anode}) \longrightarrow Ag^+(aq) + e^-$$

Thus, Ag^+ ions produced at the anode replace those that are removed from solution at the cathode. The sum of the reduction and oxidation half-equations is

$$Ag(s, \text{anode}) \longrightarrow Ag(s, \text{cathode})$$

The net process is simply the transfer of silver, as Ag^+, through $AgNO_3$(aq) from the bar to the spoon.

A similar, large-scale industrial use of electroplating is in refining copper to achieve the high purities required for applications that depend on copper's high electrical conductivity. A massive anode of impure copper and a thin cathode of pure copper are suspended in an electrolyte such as $CuSO_4$(aq). Just as in the case of silver plating, the net process is the transfer of copper atoms from the anode to the cathode.

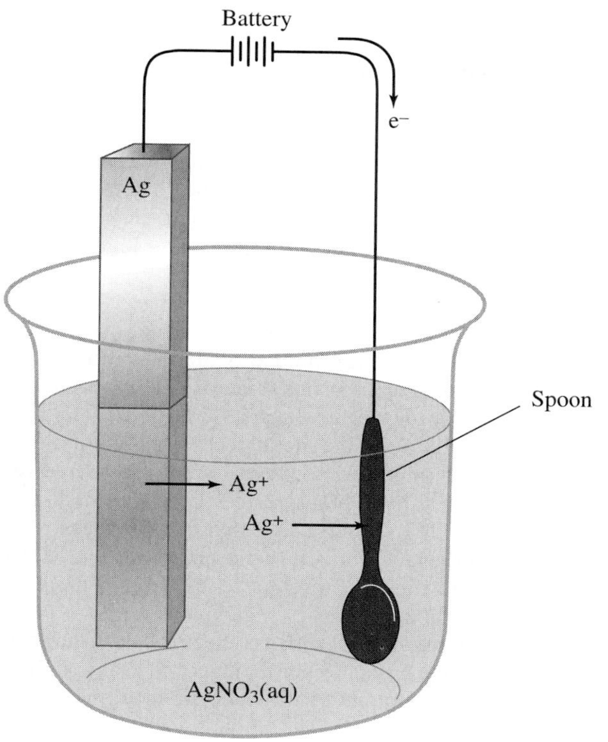

Figure 20.18 An electrochemical cell for silver plating. The anode is a silver bar and the cathode is an iron spoon.

Anode:	$Cu(s, impure) \longrightarrow Cu^{2+}(aq) + 2e^-$
Cathode:	$Cu^{2+}(aq) + 2e^- \longrightarrow Cu(s, pure)$
Net:	$Cu(s, impure) \longrightarrow Cu(s, pure)$

Electrolysis is continued until most of the anode has been consumed and the cathode has built up into a thick sheet.

Because copper is oxidized more readily than silver and gold, these metals, which may be present in the impure copper, drop to the bottom of the electrolysis cell below the anode. The value of the precious metals in this so-called *anode mud* is often sufficient to pay for the cost of the electrorefining.

Refining copper by electroplating.

SUMMARY

A half-cell consists of a metal strip (an electrode) partially immersed in an electrolyte solution. An electrochemical cell consists of two half-cells with their electrodes joined by a wire and their electrolyte solutions joined by a salt bridge. An oxidation half-reaction occurs in the anode half-cell and a reduction half-reaction in the cathode half-cell. In a voltaic (galvanic) cell, a spontaneous net cell reaction produces electricity.

A standard hydrogen electrode (SHE) has H^+ ion at unit activity (≈ 1 M) in equilibrium with H_2(g, 1 atm) on an inert platinum electrode. Standard electrode potentials, $E°$, are written for reduction half-reactions—that is, $E° = E°_{red}$. They are obtained by comparison to the SHE, which is assigned a potential of zero. $E°_{cell}$ for a reaction is the sum $E°_{red} + E°_{ox}$, where the standard potential for the oxidation half-reaction is $E°_{ox} = -E° = -E°_{red}$.

A redox reaction for which $E_{cell} > 0$ occurs spontaneously. If $E_{cell} < 0$, the reaction is nonspontaneous. Usually these criteria are applied for standard conditions, and $E°_{cell}$ values are obtained from tables of standard electrode potentials. Another important use of $E°_{cell}$ is in determining values of $\Delta G°$ and K_{eq}.

A cell voltage at nonstandard conditions is related to $E°_{cell}$ through the *Nernst equation*. At 25 °C, $E_{cell} = E°_{cell} - (0.0592/n) \log Q$. A pH meter uses a glass electrode, whose potential depends on [H^+], and a reference electrode; E_{cell} is related to pH. In concentration cells, the half-cells have identical electrodes and solutions of the same electrolyte but at different concentrations. The cell voltage depends on these concentrations.

Batteries are voltaic cells, either taken singly or joined together to yield higher voltages or higher currents. In primary batteries the electrode reactions are irreversible; in secondary batteries they are reversible. Thus, primary batteries cannot be recharged but secondary batteries can. In a fuel cell, a fuel and oxygen (from air) combine to produce oxidation products and the energy of the combustion is released as electricity.

A corroding metal consists of anodic areas, at which dissolution of the metal occurs, and cathodic areas, where atmospheric oxygen is reduced to hydroxide ion. A metal can be protected against corrosion by plating it with a second metal that corrodes less readily. In another method, cathodic protection, an active metal is sacrificed to protect a less active metal to which it is joined.

In an electrolytic cell, direct electric current from an external source produces nonspontaneous changes. Electrolysis is an important method of manufacturing chemicals and refining metals. The amount of chemical change produced in electrolysis depends on the amount of charge that is transferred at the electrodes. The amount of charge, in turn, depends on the magnitude of the electric current and the length of time it is used.

KEY TERMS

ampere (A) (20.1)
anode (20.1)
battery (20.5)
cathode (20.1)
cell diagram (20.1)
cell potential (E_{cell}) (20.1)
coulomb (C) (20.1)
concentration cell (20.4)
electrochemical cell (20.1)
electrode (20.1)
electrode potential (20.1)
electrolysis (20.7)
electrolytic cell (20.1)
electromotive force (emf) (20.1)
Faraday constant (F) (20.3)
galvanic cell (20.1)
half-cell (20.1)
Nernst equation (20.4)
overvoltage (20.7)
potential difference (20.1)
standard cell potential ($E°_{cell}$) (20.2)
standard electrode potential ($E°$) (20.2)
standard hydrogen electrode (SHE) (20.2)
volt (20.1)
voltaic cell (20.1)
voltmeter (20.1)

REVIEW QUESTIONS

1. What is meant by the term *electrode*?
2. What is a half-cell and what is an electrochemical cell?
3. What is meant by the terms *anode* and *cathode* in an electrochemical cell?
4. What is the function of a salt bridge in an electrochemical cell?
5. What are the two subgroups of electrochemical cells? In what ways are they similar and how do they differ?

6. Can a Brønsted–Lowry acid–base reaction be the basis of the net reaction in an electrochemical cell? Explain.
7. What is a cell diagram? What aspects of an electrochemical cell are incorporated in a cell diagram?
8. Which of the following pairs of terms have the same meaning and which have a different meaning? Explain.
   a. cell voltage and cell potential
   b. electrode potential and standard electrode potential
   c. half-cell and half-reaction
   d. voltaic cell and galvanic cell
9. Some standard electrode potentials have positive values and some have negative values. Why don't all electrode potentials have the same sign?
10. We have used the following symbols in working with electrode potentials: $E°$, $E°_{red}$, and $E°_{ox}$. What does each symbol signify, and how are they related to one another?
11. Explain the distinction between the symbols $E°$ and $E°_{cell}$.
12. Describe the standard hydrogen electrode (SHE). What is the potential assigned to it?
13. Explain the relationship between the activity series of the metals introduced in Chapter 12 and the table of standard electrode potentials introduced in this chapter.
14. What is the criterion for spontaneous chemical change based on cell potentials?
15. Which two of the following metals do *not* react with HCl(aq): Mg, Ag, Zn, Fe, Au?
16. Explain why predictions of redox reactions based on the sign of E_{cell} are not limited to reactions occurring within electrochemical cells.
17. If $E°_{cell} < 0$, does this mean that the cell reaction will not occur spontaneously, regardless of the conditions? Explain.
18. How is $E°_{cell}$ related to $\Delta G°$ and to K_{eq}?
19. What is the Faraday constant, F? What are its numerical value and units? Describe some situations where it is used.
20. What is the purpose served by the *Nernst equation* in electrochemistry?
21. The constant "0.0592" appears in several important equations of electrochemistry. Where does it come from and why is it useful?
22. What is a *concentration cell*? What is the value of $E°_{cell}$ for a concentration cell? What is the net change that occurs as electric current is drawn from the cell?
23. What is the difference between a *primary* and a *secondary* battery? Give an example of each.
24. What is a *fuel cell*? What is the basic principle involved in its operation?
25. Describe the electrochemical nature of *corrosion*. Use the corrosion of iron as an example.
26. Explain why iron corrodes much more readily than does aluminum, even though iron is the less active of the two metals.
27. What is cathodic protection? How does it provide protection against corrosion?
28. How do standard electrode potentials enter into determining the voltage required to carry out an electrolysis?
29. Should an object to be metal electroplated be made the anode or the cathode in an electrolytic cell? Explain.
30. What electrolysis is implied by the term *chlor–alkali* process? What are the products of the electrolysis?

PROBLEMS

Use data from Table 20.1 and Appendix D, as necessary.

Electrode Potentials and Voltaic Cells

31. For the reaction

$$H_2(g) + Hg_2^{2+}(aq) \longrightarrow 2\,H^+(aq) + 2\,Hg(l) \qquad E°_{cell} = +0.796\text{ V}$$

Determine the value of $E°_{red}$ for the reduction half-reaction

$$Hg_2^{2+}(aq) + 2\,e^- \longrightarrow 2\,Hg(l)$$

32. For the reaction

$$2\,Sc(s) + 6\,H^+(aq) \longrightarrow 2\,Sc^{3+}(aq) + 3\,H_2(g) \qquad E°_{cell} = 2.03\text{ V}$$

Determine the value of $E°_{red}$ for the half-reaction

$$Sc^{3+}(aq) + 3\,e^- \longrightarrow Sc(s)$$

33. For the reaction

$$2\,CuI(s) + Cd(s) \longrightarrow Cd^{2+}(aq) + 2\,I^-(aq) + 2\,Cu(s) \qquad E°_{cell} = +0.23\text{ V}$$

Given that for the cadmium electrode $E^\circ_{red} = -0.403$ V, determine the value of E°_{red} for the half-reaction

$$2\ CuI(s) + 2\ e^- \longrightarrow 2\ Cu(s) + 2\ I^-(aq)$$

34. For the reaction

$$3\ V(s) + 2\ SbO^+(aq) + 4\ H^+(aq) \longrightarrow 3\ V^{2+}(aq) + 2\ Sb(s) + 2\ H_2O \qquad E^\circ_{cell} = 1.387\ V$$

Given that for the vanadium electrode $E^\circ_{red} = -1.175$ V, determine the value of E°_{red} for the half-reaction

$$SbO^+(aq) + 2\ H^+(aq) + 3\ e^- \longrightarrow Sb(s) + H_2O$$

35. $E^\circ_{cell} = 1.47$ V for the voltaic cell

$$V(s)\,|\,V^{2+}(1\ M)\,||\,Cu^{2+}(1\ M)\,|\,Cu(s)$$

Determine the value of E°_{red} for the half-reaction

$$V^{2+}(aq) + 2\ e^- \longrightarrow V(s)$$

36. $E^\circ_{cell} = 3.73$ V for the voltaic cell

$$Y(s)\,|\,Y^{3+}(1\ M)\,||\,Cl^-(1\ M)\,|\,Cl_2(g,\ 1\ atm),\ Pt$$

Determine the value of E°_{red} for the half-reaction

$$Y^{3+}(aq) + 3e^- \longrightarrow Y(s)$$

37. Write equations for the half-reactions and net reaction and calculate E°_{cell} for each of the voltaic cells diagrammed below.

a. $Pb(s)\,|\,Pb^{2+}(aq)\,||\,H^+(aq)\,|\,H_2(g),Pt$

b. $Pt,I_2(s)\,|\,I^-(aq)\,||\,Cl^-(aq)\,|\,Cl_2(g),Pt$

38. Write equations for the half-reactions and net reaction and calculate E°_{cell} for each of the voltaic cells diagrammed below.

a. $Fe(s)\,|\,Fe^{2+}(aq)\,||\,Ag^+(aq)\,|\,Ag(s)$

b. $Pt\,|\,Fe^{2+}(aq),\ Fe^{3+}(aq)\,||\,Cr_2O_7^{2-}(aq),\ Cr^{3+}(aq)\,|\,Pt$

39. Each of the following reactions is made to take place in a voltaic cell. Write equations for the half-reactions and the net cell reaction. Write a cell diagram for the voltaic cell and calculate the value of E°_{cell}.

a. $Zn(s) + Ag^+(aq) \longrightarrow Ag(s) + Zn^{2+}(aq)$

b. $Fe^{2+}(aq) + O_2(g) + H^+(aq) \longrightarrow Fe^{3+}(aq) + H_2O(l)$

40. Each of the following reactions is made to take place in a voltaic cell. Write equations for the half-reactions and the net cell reaction. Write a cell diagram for the voltaic cell and calculate the value of E°_{cell}.

a. $Fe^{3+}(aq) + Sn^{2+}(aq) \longrightarrow Fe^{2+}(aq) + Sn^{4+}(aq)$

b. $Cu(s) + H^+(aq) + NO_3^-(aq) \longrightarrow Cu^{2+}(aq) + H_2O + NO(g)$

E°_{cell} *and the Spontaneity of Redox Reactions*

41. Predict whether a spontaneous reaction will occur in the forward direction in each of the following. Assume that all reactants and products are in their standard states.

a. $Sn(s) + Co^{2+}(aq) \longrightarrow Sn^{2+}(aq) + Co(s)$

b. $6\ Br^-(aq) + Cr_2O_7^{2-}(aq) + 14\ H^+(aq) \longrightarrow 2Cr^{3+}(aq) + 7\ H_2O + 3\ Br_2(l)$

42. Predict whether a spontaneous reaction will occur in the forward direction in each of the following. Assume that all reactants and products are in their standard states.

a. $Sn^{4+}(aq) + 2\ I^-(aq) \longrightarrow Sn^{2+}(aq) + I_2(s)$

b. $2\ MnO_2(s) + 3\ ClO^-(aq) + 2\ OH^-(aq) \longrightarrow 2\ MnO_4^-(aq) + 3\ Cl^-(aq) + H_2O$

43. Predict whether each of the following processes will proceed in the forward direction to any appreciable extent.

a. The displacement of $Cd^{2+}(aq)$ by Al(s).

b. The oxidation of $Cl^-(aq)$ to $Cl_2(g)$ by $Br_2(l)$.

c. The oxidation of $Cl^-(aq)$ to $ClO_3^-(aq)$ by $HO_2^-(aq)$ in basic solution.

44. Predict whether each of the following processes will proceed in the forward direction to any appreciable extent.

a. The reduction of $Sn^{4+}(aq)$ to $Sn^{2+}(aq)$ by Cu(s).

b. The oxidation of $I_2(s)$ to $IO_3^-(aq)$ by $O_3(g)$ in acidic solution.

c. The oxidation of $Cr(OH)_3(s)$ to $CrO_4^{2-}(aq)$ by $HO_2^-(aq)$ in basic solution.

45. Silver does not react with HCl(aq) but it does react with $HNO_3(aq)$. **(a)** Explain the difference in the behavior of silver toward these two acids. **(b)** Write a net ionic equation for the reaction of silver with $HNO_3(aq)$.

46. Can we use sodium metal to displace $Mg^{2+}(aq)$ from aqueous solution? If the reaction does occur, write the half-equations and the net equation. If the displacement reaction does *not* occur, write the equation for the reaction that does occur.

47. Rhodium is a rare metal used as a catalyst. The metal does not react with HCl(aq), but it does react with $HNO_3(aq)$, producing $Rh^{3+}(aq)$ and NO(g). Copper will displace Rh^{3+} from aqueous solution but silver will not. Estimate a value of E°_{red} for the half-reaction

$$Rh^{3+} + 3\ e^- \longrightarrow Rh(s)$$

48. Palladium is a rare metal used as a catalyst. Copper and silver will both displace Pd^{2+} from aqueous solution. The metal itself will react with $HNO_3(aq)$, producing $Pd^{2+}(aq)$ and NO(g). Estimate a value of E°_{red} for the half-reaction

$$Pd^{2+} + 2\ e^- \longrightarrow Pd(s)$$

E°_{cell}, ΔG°, K_{eq}

49. Determine the values of E°_{cell} and ΔG° for the following reactions.
a. $Al(s) + 3\,Ag^+(aq) \longrightarrow Al^{3+}(aq) + 3\,Ag(s)$
b. $4\,IO_3^-(aq) + 4\,H^+(aq) \longrightarrow 2\,I_2(s) + 2\,H_2O + 5\,O_2(g)$

50. Determine the values of E°_{cell} and ΔG° for the following reactions.
a. $O_2(g) + 4\,I^-(aq) + 4\,H^+(aq) \longrightarrow 2\,H_2O + 2\,I_2(s)$
b. $Cr_2O_7^{2-}(aq) + 3\,Cu(s) + 14\,H^+(aq) \longrightarrow 2\,Cr^{3+}(aq) + 3\,Cu^{2+}(aq) + 7\,H_2O$

51. Write the equilibrium constant expression for each of the following reactions and determine the numerical value of K_{eq} at 25 °C.
a. $Ag^+(aq) + Fe^{2+}(aq) \rightleftharpoons Fe^{3+}(aq) + Ag(s)$
b. $MnO_2(s) + 4\,H^+(aq) + 2\,Cl^-(aq) \rightleftharpoons Mn^{2+}(aq) + 2\,H_2O + Cl_2(g)$
c. $2\,OCl^-(aq) \rightleftharpoons 2\,Cl^-(aq) + O_2(g)$ (basic soln)

52. Write the equilibrium constant expression for each of the following reactions and determine the numerical value of K_{eq} at 25 °C.
a. $2\,Fe^{3+}(aq) + Cu(s) \rightleftharpoons 2\,Fe^{2+}(aq) + Cu^{2+}(aq)$
b. $PbO_2(s) + 4\,H^+(aq) + 2\,Cl^-(aq) \rightleftharpoons Pb^{2+}(aq) + 2\,H_2O + Cl_2(g)$
c. $3\,O_2(g) + 2\,Br^-(aq) \rightleftharpoons 2\,BrO_3^-(aq)$ (basic soln)

53. For the following displacement reaction, determine the ion concentrations at equilibrium if initially a strip of the metal is immersed in a 1.00 M solution of the ion being displaced. (*Hint*: What is the value of K_{eq}?)

$$Sn(s) + Pb^{2+}(aq) \rightleftharpoons Sn^{2+}(aq) + Pb(s)$$

54. For the following displacement reaction, determine the ion concentrations at equilibrium if initially a strip of the metal is immersed in a 1.00 M solution of the ion being displaced. (*Hint:* What is the value of K_{eq}?)

$$Cu(s) + 2\,Ag^+(aq) \rightleftharpoons Cu^{2+}(aq) + 2\,Ag(s)$$

Effect of Concentration on E_{cell}

55. Consider the redox reaction

$$Zn(s) + Pb^{2+}(1\text{ M}) \longrightarrow Zn^{2+}(1\text{ M}) + Pb(s) \qquad E^\circ_{cell} = +0.638\text{ V}$$

What is the value of E_{cell} for each of the following?
a. $Zn(s) + Pb^{2+}(1.55\text{ M}) \longrightarrow Zn^{2+}(0.025\text{ M}) + Pb(s)$
b. $Zn(s) + Pb^{2+}(0.50\text{ M}) \longrightarrow Zn^{2+}(0.50\text{ M}) + Pb(s)$
c. $Zn(s) + Pb^{2+}(3 \times 10^{-4}\text{ M}) \longrightarrow Zn^{2+}(0.082\text{ M}) + Pb(s)$

56. Consider the redox reaction

$$Co(s) + 2\,Ag^+(1\text{ M}) \longrightarrow Co^{2+}(1\text{ M}) + 2\,Ag(s) \qquad E^\circ_{cell} = 1.077\text{ V}$$

What is the value of E_{cell} for each of the following?
a. $Co(s) + 2\,Ag^+(0.20\text{ M}) \longrightarrow Co^{2+}(0.055\text{ M}) + 2\,Ag(s)$
b. $Co(s) + 2\,Ag^+(0.75\text{ M}) \longrightarrow Co^{2+}(0.75\text{ M}) + 2\,Ag(s)$
c. $Co(s) + 2\,Ag^+(2 \times 10^{-3}\text{ M}) \longrightarrow Co^{2+}(0.015\text{ M}) + 2\,Ag(s)$

57. The voltaic cell diagrammed below registers $E_{cell} = 0.108$ V.

Pt, H_2(g, 1 atm) | H^+(x M || H^+(1 M) | H_2(g, 1 atm), Pt

What is the pH of the unknown solution? (*Hint*: Write a net ionic equation for the net cell reaction, and then use the Nernst equation.)

58. A voltaic cell represented by the following cell diagram has $E_{cell} = +0.920$ V.

Fe(s) | Fe^{2+}(1.00 M) || Ag^+(x M) | Ag(s)

Calculate [Ag^+] in the cell. (*Hint*: Write a net ionic equation for the net cell reaction, and then use the Nernst equation.)

59. What is E_{cell} for the voltaic cell diagrammed below?

Pt, H_2(g, 1 atm) | 0.0025 M HCl || H^+(1 M) | H_2(g, 1 atm), Pt

(*Hint:* Write an equation for the net cell reaction. What is [H^+] in the HCl(aq)?)

60. What is E_{cell} for the voltaic cell diagrammed below?

Pt, H_2(g, 1 atm) | 0.0250 M KOH || 0.0675 M HCl | H_2(g, 1 atm), Pt

(*Hint*: Write an equation for the net cell reaction. What is [H^+] in each half-cell?)

Batteries

61. Describe plausible electrode reactions and the net cell reaction for a potential battery system having a Zn anode and an Cl_2 cathode. Determine its E°_{cell}.

62. Describe plausible electrode reactions and the net cell reaction for the potential battery system having a Mg anode and an O_2 cathode. Determine its E°_{cell}.

63. Silver–zinc cells—called button batteries—are tiny cells used in watches, electronic calculators, hearing aids, and cameras. The cell diagram is

Zn(s), ZnO(s) | KOH(sat'd aq) | Ag_2O(s), Ag(s)

Its storage capacity is about six times that of a lead–acid cell of the same size. Write equations for the half-reactions and the net reaction that occur when the cell is discharged. (*Hint*: Recall the discussion of writing and balancing half-equations in Section 12.4.)

64. A mercury battery, once widely used, is being phased out because of the problem of disposing of the toxic mercury. A simplified cell diagram for the battery is

$$Zn(s), ZnO(s) \,|\, KOH(sat'd\ aq) \,|\, HgO(s), Hg(l)$$

Write the equations for the half-reactions and the net reaction that occur when the cell is discharged. (*Hint*: Recall the discussion of writing and balancing half-equations in Section 12.4.)

Corrosion

65. Why are water, an electrolyte, and oxygen all required for the corrosion of iron?

66. Iron can be protected from corrosion by copper plating or by zinc plating. Both methods are effective so long as the plating remains intact. If a break occurs in the plating, however, zinc plating proves to be far superior to copper plating in providing protection to the underlying iron. Explain why this should be so.

67. Explain why the term "sacrificial anode" is an appropriate one in describing one method of protecting iron from corrosion.

68. Describe how silver tarnish can be removed without the loss of silver.

Electrolysis

69. Write a net ionic equation for the expected reaction when the electrolysis of $NiSO_4(aq)$ is conducted using **(a)** a nickel anode and an iron cathode; **(b)** a nickel anode and an inert platinum cathode; **(c)** an inert platinum anode and a nickel cathode.

70. Write a net ionic equation for the expected reaction when the electrolysis of $Cu(NO_3)_2(aq)$ is conducted using **(a)** a copper anode and a copper cathode; **(b)** an inert platinum anode and an iron cathode; **(c)** inert platinum for both electrodes.

71. Use electrode potential data to predict the probable products in the electrolysis with inert platinum electrodes of each of the following.
a. $BaCl_2(l)$ **b.** $HBr(aq)$ **c.** $NaNO_3(aq)$

72. Use electrode potential data to predict the probable products in the electrolysis with inert platinum electrodes of each of the following.
a. $ZnSO_4(aq)$ **b.** $MgBr_2(l)$ **c.** $NiCl_2(aq)$

73. How many grams of silver are deposited at a platinum cathode in the electrolysis of $AgNO_3(aq)$ by 1.73 A of electric current in 2.05 h? (*Hint*: What is the half-reaction at the cathode?)

74. How many mL of $H_2(g)$, measured at STP, are produced at a platinum cathode in the electrolysis of $H_2SO_4(aq)$ by 2.45 A of electric current in 5.00 min? (*Hint*: What is the half-reaction at the cathode?)

75. How many coulombs of electric charge are required to deposit 25.0 g Cu(s) at the cathode in the electrolysis of $CuSO_4(aq)$?

76. What is the electric current, in amperes, if 212 mg Ag(s) is deposited at the cathode in 1435 s in the electrolysis of $AgNO_3(aq)$?

77. Which of the following will yield the greatest mass of metal deposit on a platinum cathode when the solution is electrolyzed for exactly 1 h with a current of 1.00 A? Explain your choice.
(a) $Cu(NO_3)_2(aq)$ (b) $Zn(NO_3)_2(aq)$
(c) $AgNO_3(aq)$ (d) $NaNO_3(aq)$

78. Equal volumes of the following solutions are electrolyzed, using inert platinum electrodes and a current of 1.00 A, until the solution concentration falls to one-half of its initial value. Which solution will take the longest time? Explain your choice.
(a) 0.50 M $Cu(NO_3)_2$ (b) 0.75 M HNO_3
(c) 0.80 M $AgNO_3$ (d) 0.30 M $Zn(NO_3)_2$

ADDITIONAL PROBLEMS

79. Figure 20.2 describes what happens when a zinc electrode is partially immersed in $ZnSO_4(aq)$. Will a similar description apply to a sodium electrode partially immersed in NaCl(aq)? Explain.

80. Consider the silver plating of an iron spoon in Figure 20.18. Show that, in principle, Ag^+ should be reduced to Ag(s) simply by immersing the spoon in $AgNO_3(aq)$. Why do you suppose that electroplating is used rather than displacement of Ag^+ from solution?

81. What volume of $O_2(g)$, measured in milliliters at STP, should be liberated at a platinum anode at the same time that 1.02 g Ag(s) is deposited at a platinum cathode in the electrolysis of $AgNO_3(aq)$?

82. The electrolysis of 0.100 L of 0.785 M $AgNO_3(aq)$ using platinum electrodes is carried out with a current of 1.75 A for 25.0 min. What is the molarity of the $AgNO_3(aq)$ at this time?

83. Use standard electrode potential data to show that $MnO_2(s)$ should not react with HCl(aq) to liberate $Cl_2(g)$. Yet when $MnO_2(s)$ and concentrated HCl(aq) are heated together, $Cl_2(g)$ does form. It is a common laboratory method of generating small quantities of $Cl_2(g)$. Explain why the reaction occurs.

84. What minimum voltage is required to recharge a lead–acid storage battery? If a voltage much greater than this minimum is used, a danger exists that a potentially explosive mixture of gases could accumulate in the battery. Explain why this is so.

85. Calculate E_{cell} for the following voltaic cell.

Pt, H_2(g,1 atm) | CH_3COOH(0.45 M) || H^+(0.010 M) | H_2(g,1 atm), Pt

(*Hint*: What is the net cell reaction?)

86. Calculate E_{cell} for the following voltaic cell?

Pt, H_2(g,1 atm) | NH_3(0.45 M), NH_4^+(0.15 M) || H^+(0.010 M) | H_2(g,1 atm), Pt

(*Hint*: What is the net cell reaction?)

87. What is E_{cell} of the following voltaic cell?

Cu(s) | Cu^{2+}(0.10 M) || Ag_2CrO_4(sat'd aq) | Ag(s)

(*Hint*: What is $[Ag^+]$ in saturated $Ag_2CrO_4(aq)$?)

88. What $[Cl^-]$ should be maintained in the anode half-cell if the following voltaic cell is to have E_{cell} = 0.100 V?

Ag, AgCl(s) | Cl^-(x M) || Cu^{2+}(0.25 M) | Cu(s)

89. What happens to the voltage of the concentration cell in Figure 20.9 as it operates over a period of time? That is, does the voltage increase, decrease, or remain constant? If the cell operates continuously, does it stop producing electricity at some point? If so, what is the condition in each half-cell compartment at this point?

90. Consider that the reaction

$$Hg^{2+}(aq) + 2\,Fe^{2+}(aq) \longrightarrow 2\,Fe^{3+}(aq) + Hg(l)$$

occurs in a voltaic cell. If initially all the ion concentrations are 1.00 M, what will be the ion concentrations when the cell voltage drops to zero?

91. The electrolysis of $Na_2SO_4(aq)$ is conducted in two separate half-cells joined by a salt bridge, as suggested by the cell diagram

Pt | Na_2SO_4(aq) || Na_2SO_4(aq) | Pt

Write half-equations and a net ionic equation for the electrolysis. Suppose that after the electrolysis is stopped, the solutions in the two half-cells are mixed. Explain why the pH of the mixture is independent of the magnitude of the electrolysis current, the duration of the electrolysis, and the volumes and concentrations of the $Na_2SO_4(aq)$.

92. Use data for the Zn–Cu cell on page 780 to show that a graph of E_{cell} versus log $[Zn^{2+}]/[Cu^{2+}]$ is a straight line. Use the graph to determine the ratio: $[Zn^{2+}]/[Cu^{2+}]$ corresponding to E_{cell} = 1.000 V. Show that the ratio of $[Zn^{2+}]/[Cu^{2+}]$ determined graphically is the same as that calculated with the Nernst equation.

Shown here is the reaction between aluminum and liquid bromine. Aluminum, a typical metal, and bromine, a typical nonmetal, are both members of the *p*-block.

21

The *p*-Block Elements

CONTENTS

In this chapter we continue our discussion of descriptive chemistry that we began with the *s*-block elements in Chapter 8. Here, we look at the elements of periodic table Groups 3A through 8A—the *p*-block elements. The *p*-block includes the noble gases (except helium), all the nonmetals except hydrogen, all the metalloids, and even a few metals, such as aluminum, tin, and lead.

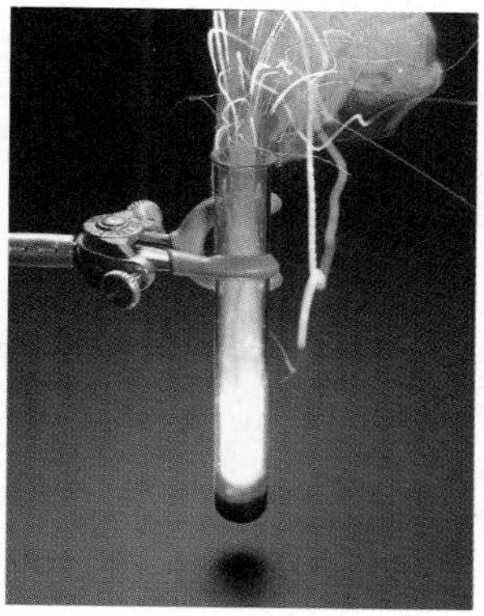

We will relate the properties of the elements and their compounds to their positions in the periodic table. We'll use principles of bonding theory, enthalpy and free-energy changes, electrode potentials, and equilibrium constants. This should serve a dual purpose: it will give greater meaning to the descriptive information, and it will reemphasize the power of fundamental principles to explain chemical phenomena.

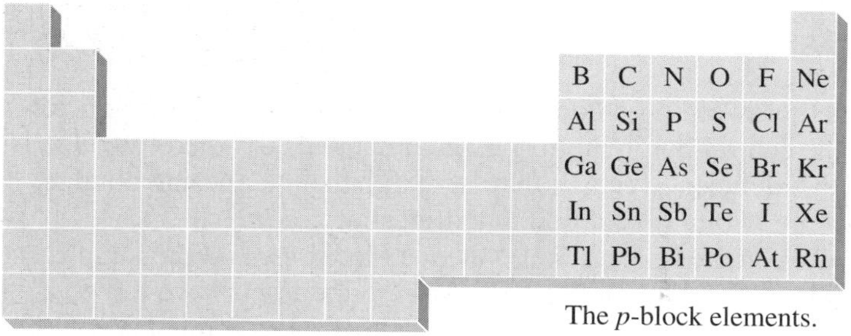

The *p*-block elements.

21.1 Group 3A: B, Al, Ga, In, and Tl

Group 3A elements all have the valence-shell electron configuration ns^2np^1. As we've seen before, the small atomic size of the first member of a family often causes it to differ in some respects from other members. For example, in Chapter 8, we identified Li and Be as the odd members of their families. We also identified *diagonal relationships*, in which one element shows marked similarities to another that appears diagonally adjacent on the periodic table (see again Figure 8.7, page 290). Boron is the odd member of Group 3A, and the diagonal relationship suggests that it might resemble silicon in Group 4A. Boron is a *non-metal*, whereas the other four members of Group 3A are metallic in appearance, in physical properties, and in many of their chemical properties. Several properties of the Group 3A elements are listed in Table 21.1.

TABLE 21.1 **Some Properties of the Group 3A Elements**

	B	Al	Ga	In	Tl
Atomic number	5	13	31	49	81
Atomic radius, pm[a]	88	143	122	163	170
Ionic radius, pm[b]	23	50	62	81	95
First ionization energy, kJ/mol	801	577	579	558	589
Electronegativity	2.0	1.5	1.6	1.7	1.8
Density at 20 °C, g/cm³	2.34	2.70	5.91	7.31	11.85
Melting point, °C	2300	660	30	157	304

[a] For B, the single covalent radius; for the others, the metallic radius.
[b] For the ion M^{3+}.

In this section we'll (1) describe a few aspects of the chemistry of boron, (2) take a more extensive look at aluminum, and (3) briefly consider gallium, indium, and thallium.

Boron: A Nonmetal

Boron atoms have the electron configuration $1s^2 2s^2 2p^1$. Two of the properties of boron, as listed in Table 21.1, are most suggestive of its nonmetallic character: (1) a high first ionization energy, suggesting that electrons are not easily lost by boron atoms, and (2) a high melting point, which can be attributed to network covalent bonding in the solid.

Much of the chemistry of boron compounds is based on the lack of an octet of electrons about the central boron atom. These compounds are therefore electron-deficient, and this deficiency causes them to exhibit some unusual bonding features that we explore in our look at boron hydrides.

Boron Hydrides Carbon atoms have four valence electrons, and the simplest hydrocarbon is methane, CH_4. By analogy, because boron atoms have three valence electrons, we might expect the simplest boron–hydrogen compound to be BH_3, *borane*. However, the boron atom in this compound lacks a valence-shell octet. It has only *six* electrons in its valence shell.

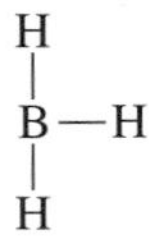

Borane does not exist as a stable compound.

Because the boron atom in the BH_3 grouping of atoms is one pair of electrons short of an octet, the BH_3 group can become part of a more extensive structure by forming a coordinate covalent bond with another atom that has a lone pair of electrons. In the following structure, this bond is between the B atom of BH_3 and the O atom of dimethyl ether. The compound, which results from the "addition" of one structure to another, is often called an **adduct**.

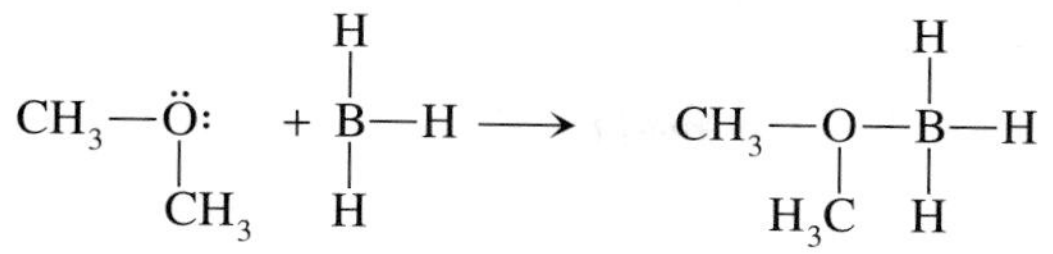

Dimethyl ether Borane Dimethyl ether–borane adduct

The simplest boron hydride that can be isolated is *diborane*, B_2H_6. But for many years following its isolation, diborane presented a bonding puzzle. What holds the two BH_3 units together?

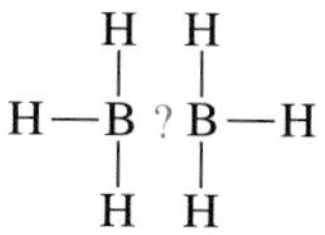

The problem is that the structure has a total of only *12* valence electrons, and the number required would seem to be *14* (as in ethane, C_2H_6).

The currently accepted structure of diborane is shown in Figure 21.1. The two B atoms and four of the H atoms lie in a plane. Think of each combination of one B atom and two H atoms as comprising a BH_2 unit. The B—H bonds in these units are regular two-electron, σ covalent bonds. The number of valence electrons in each BH_2 unit is five (three from B and one each from the two H atoms). Of these five, four are localized into the two B—H bonds. The two BH_2 units are joined together by the remaining two H atoms, which lie in a plane perpendicular to that of the BH_2 groups. One of these so-called "bridge" hydrogen atoms lies above the plane of the BH_2 units and one lies below. Altogether, there are *four* valence electrons to bond the BH_2 units together: one electron from each BH_2 unit and one electron from each bridge H atom. Thus, in each B–H–B bridge, *two* electrons must be shared among *three* atoms. For this reason, these bridges are called "three-center" bonds, a bond type that we have not seen before.

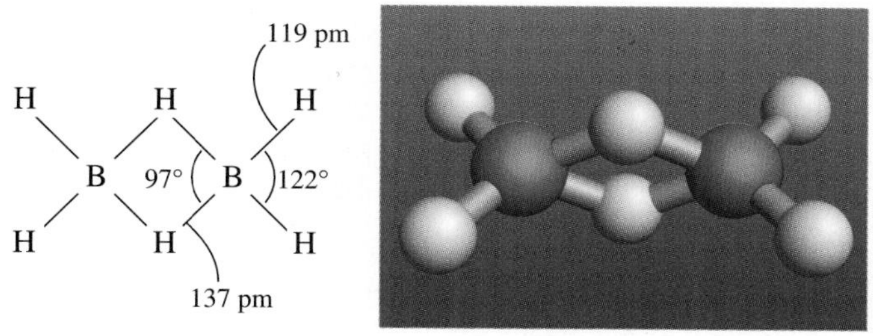

Figure 21.1 Structure of diborane, B_2H_6.

The valence bond theory does not satisfactorily explain the structure of B_2H_6. We would normally assume sp^3 hybridization to describe four bonds to a central atom, but the bond angles in the structure are not tetrahedral bond angles, nor are they all equal. Also, the B—H bond lengths in the B–H–B bridges are different from the lengths in the BH_2 units.

Only modern molecular orbital theory adequately explains bonding in B_2H_6. For example, it accounts for orbitals that are delocalized among several atoms (recall the case of benzene, page 388). In all, about two dozen different boron–hydrogen compounds (boranes) are now known, and their structures have been worked out with a combination of molecular orbital theory and experimental measurements. Boranes such as diborane and organic derivatives of borane are frequently used as reagents in the synthesis of organic compounds.

Borax, Boric Oxide, Boric Acid, and Borates Boron compounds are fairly widely distributed on Earth. Boron comprises about 9 ppm of Earth's crust and 4.8 ppm of seawater, and it is an essential element for some organisms. However, concentrated mineral deposits of boron are found only in a few locations, such as Turkey and the desert regions of California. *Borax*, $Na_2B_4O_7 \cdot 10\,H_2O$, a hydrated borate, is a typical ore.

Salt flats in Death Valley, California, an early source of the mineral borax. From here, the mineral was shipped by "twenty mule teams" to a railroad terminal in Mojave.

Borax is converted to boric acid, $B(OH)_3$, as the first step in the production of boron and boron compounds.

$$Na_2B_4O_7 \cdot 10\,H_2O + H_2SO_4 \longrightarrow 4\,B(OH)_3 + Na_2SO_4 + 5\,H_2O$$

Pyrolysis (decomposition by heating) dehydrates $B(OH)_3$ to boric oxide, B_2O_3.

$$2\,B(OH)_3(s) \xrightarrow{\Delta} B_2O_3(s) + 3\,H_2O(g)$$

Elemental boron and boron compounds are prepared from B_2O_3, as illustrated in Example 21.1.

Example 21.1

Write chemical equations to represent (**a**) the high-temperature reduction of B_2O_3 to elemental boron with magnesium as a reducing agent; (**b**) the preparation of boron trichloride by heating boric oxide with carbon and chlorine gas. (The carbon is oxidized to carbon monoxide.)

Solution

As we have seen before, the key to writing equations to represent chemical reactions is to have a complete description of the initial reactants, the final products, and the reaction conditions. Most of this information is provided in the statement of the problem.

a. $B_2O_3(s) + 3\,Mg(s) \xrightarrow{\Delta} 2\,B(s) + 3\,MgO(s)$

b. $B_2O_3(s) + 3\,C(s) + 3\,Cl_2(g) \xrightarrow{\Delta} 2\,BCl_3(s) + 3\,CO(g)$

Exercise 21.1

Write an equation to represent the preparation of pure boron by the reduction of $BCl_3(g)$ with hydrogen gas.

A nonmetal oxide whose reaction with water produces a ternary acid as the sole product is called an **acid anhydride**. B_2O_3 is the acid anhydride of $B(OH)_3$.

$$B_2O_3(s) + 3\ H_2O(l) \longrightarrow 2\ B(OH)_3(s)$$

Generally we write the formula of boric acid as $B(OH)_3$ rather than H_3BO_3. We do this because boric acid is an extremely weak *monoprotic* acid, not triprotic as the formula H_3BO_3 would indicate. Further, its acidity is displayed differently from that of most other weak acids. Instead of donating a proton, the acid accepts a hydroxide ion, forming the complex ion $[B(OH)_4]^-$. We can think of a hydroxide ion produced in the self-ionization of water as attaching itself to the B atom of $B(OH)_3$, an electron-deficient structure, through a coordinate covalent bond.

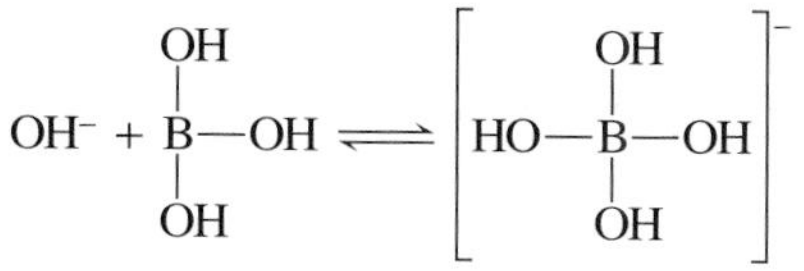

The net ionization reaction is

$$B(OH)_3(aq) + 2\ H_2O \rightleftharpoons H_3O^+(aq) + [B(OH)_4]^-(aq) \qquad K_a = 5.6 \times 10^{-10}$$

Boric acid is quite toxic if taken internally, but dilute solutions can be used externally as a mild antiseptic—for example, in eyewash solutions. Boric acid is also used as an insecticide against cockroaches and black carpet beetles.

Solutions of the salts of boric acid—borate solutions—are generally quite complex because they contain polymers of the borate anions as well as the simple anions $[B(OH)_4]^-$, BO_3^{3-}, and BO_4^{5-}.

Note that the oxidation state of the boron is +3 in all these various forms of the borate anion.

Borates hydrolyze in water to give basic solutions, and this is why borax is used in some cleaning agents. Sodium perborate, $NaBO_3 \cdot 4H_2O$, crystallizes from an aqueous solution of hydrogen peroxide and borax. It is used as a color-safe "oxygen" bleach for clothes that would be harmed by a "chlorine" bleach (sodium hypochlorite). Sodium perborate is also used in denture cleansers.

The empirical formula, $NaBO_3 \cdot 4H_2O$, that we have used for sodium perborate is deceptively simple. We might well calculate a seemingly improbable oxidation state of +5 for B. A more realistic—if somewhat more complicated—formula is $Na_2[B_2(O_2)_2(OH)_4] \cdot 6H_2O$, where the "$O_2$" is the peroxide ion, O_2^{2-}. These ions account for the bleaching action of sodium perborate.

Aluminum

The most important metal of Group 3A is aluminum. Over 5 million tons of the metal are produced per year in the United States; most of it is used in lightweight alloys. Let us consider the commercial production of the metal, its properties and uses, and some of its important compounds.

Production of Aluminum At 8.3% by mass of Earth's crust, aluminum is the third most abundant element and the most abundant metal. The metal was not isolated until 1825, when Hans Oersted produced some in impure form. For the next several decades it remained a semiprecious metal used mostly in jewelry and art work. It was still considered a semiprecious metal when an aluminum cap was placed atop the newly completed Washington Monument in 1884. Just a couple of years later, however, the situation changed completely. Charles Martin Hall in the United States and Paul Heroult in France discovered a fairly inexpensive way to make aluminum by electrolysis.

In much of the world, the name given to element number 13 is alumin*i*um (pronounced al-you-MIN-ee-um). In the United States, though, it is usually called aluminum.

Charles Martin Hall (1863–1914), was motivated by a professor at Oberlin College who remarked that anyone discovering a cheap method of producing aluminum would become rich and famous. Hall's discovery, in his home laboratory within eight months of his graduation, was the foundation of the aluminum industry in the United States.

The Hall–Heroult process involves two key features: First is the application of *amphoterism* to separate Fe_2O_3, the principal impurity, from Al_2O_3, the major constituent of *bauxite* ore. Al_2O_3 is amphoteric in the sense that it can accept protons, thus acting as a base, or—like boric acid—it can react with a hydroxide ion, thus acting as an acid. Fe_2O_3 has only basic properties. Thus, when the ore is treated with hot, concentrated NaOH(aq), only Al_2O_3 reacts.

$$Al_2O_3(s) + 2\,OH^-(aq) + 3\,H_2O \longrightarrow 2\,[Al(OH)_4]^-(aq)$$
$$Fe_2O_3(s) + OH^-(aq) \longrightarrow \text{no reaction}$$

The solution containing $[Al(OH)_4]^-$ is separated from the undissolved red mud (mostly Fe_2O_3, but also other impurities such as TiO_2, and SiO_2), diluted with water, and slightly acidified. $Al(OH)_3$ reprecipitates.

$$[Al(OH)_4]^-(aq) + H_3O^+(aq) \longrightarrow Al(OH)_3(s) + 2\,H_2O$$

The $Al(OH)_3(s)$ is heated to about 1200 °C and decomposes to pure $Al_2O_3(s)$.

$$2\,Al(OH)_3(s) \longrightarrow Al_2O_3(s) + 3\,H_2O(g)$$

The melting point of Al_2O_3 (2045 °C) is much too high to permit the electrolysis of molten Al_2O_3. Moreover, molten Al_2O_3 is not a particularly good conductor. Here is where Hall and Heroult made their crucial innovation: to use molten *cryolite*, Na_3AlF_6, as the electrolyte, with a few percent dissolved Al_2O_3. The mixture is a good electrical conductor and remains liquid even at about 950 °C.

The electrolysis cell pictured in Figure 21.2 uses a carbon-lined steel cell as the cathode and large chunks of carbon as the anodes. $Al_2O_3(s)$ is continually added to the bath as liquid aluminum is drawn off. The electrode reactions are complex, but the net electrolysis reaction is

$$\textit{Reduction:}\quad 4\,\{Al^{3+} + 3\,e^- \longrightarrow Al(l)\}$$
$$\textit{Oxidation:}\quad 3\,\{C(s) + 2\,O^{2-} \longrightarrow CO_2(g) + 4\,e^-\}$$
$$\textit{Net:}\quad 3\,C(s) + 4\,Al^{3+} + 6\,O^{2-} \longrightarrow 4\,Al(l) + 3\,CO_2(g)$$

The production of aluminum requires considerable energy, about 15,000 kWh per ton of Al. To produce a ton of steel requires only about one-fifth this much energy. On the other hand, because the density of aluminum (2.7 g/cm^3) is much less than that of steel (7.8 g/cm^3), less aluminum than steel is required for a particular application. Moreover, because it takes only about one-twentieth as much energy to recycle aluminum as it does to produce it from bauxite, much of the aluminum produced in the United States is recycled.

Paul Heroult (1863–1914), a student of Le Châtelier's, was, like Hall, 23 years old when he discovered the same method of producing aluminum. Heroult's discovery was the foundation of the aluminum industry in Europe.

Properties and Uses of Aluminum Because it is an active metal, aluminum is a good reducing agent. We see this, for example, in the electrode potential

$$Al^{3+}(aq) + 3\,e^- \longrightarrow Al(s) \qquad E^\circ_{red} = -1.676\text{ V}$$

The large negative value indicates that $Al^{3+}(aq)$ is reduced with difficulty; it is a poor oxidizing agent. On the other hand, Al(s) is a good reducing agent because it is easily oxidized.

$$Al(s) \longrightarrow Al^{3+}(aq) + 3\,e^- \qquad E^\circ_{ox} = 1.676\text{ V}$$

One interesting reaction based on the reducing power of aluminum is the highly exothermic *thermite* reaction, which can be used to produce molten iron metal that is used to weld large iron objects such as pipes and rails.

$$Fe_2O_3(s) + 2\,Al(s) \longrightarrow 2\,Fe(l) + Al_2O_3(s)$$

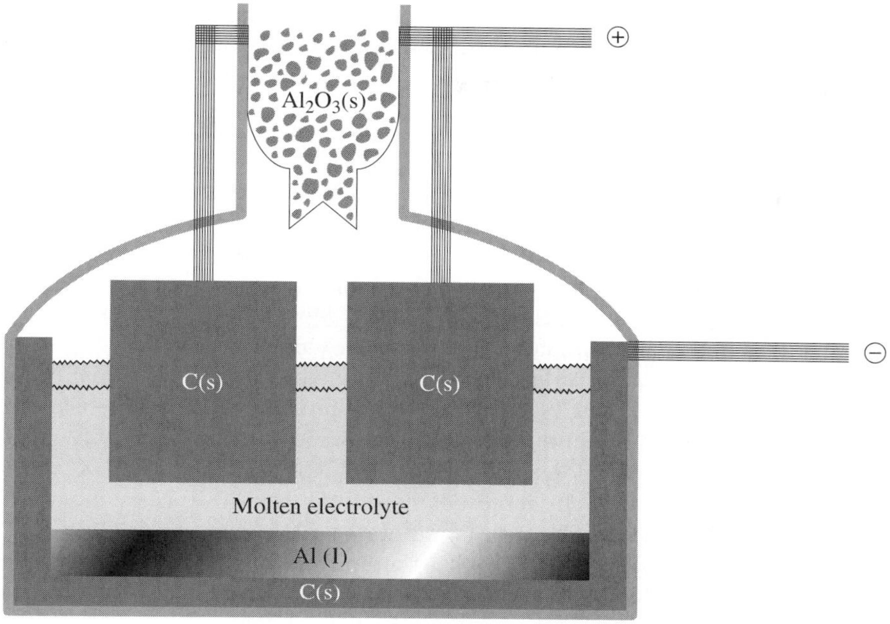

Figure 21.2 Electrolysis cell for aluminum production. The cathode is the carbon lining of a steel tank. The anodes are made of blocks of carbon. Liquid aluminum, which is more dense than the electrolyte, collects at the bottom of the tank and is removed. The electrolyte is alumina (Al_2O_3) dissolved in molten cryolite. Fresh alumina is continuously added from a hopper above the cell.

As we noted in Section 20.6, the oxide that readily forms on aluminum is a thin impervious film that protects the underlying metal from corrosion. The oxide film can be made thicker by making aluminum the *anode* in an electrolytic cell with dilute $H_2SO_4(aq)$ as the electrolyte. The anode (oxidation) half-reaction is

$$2\,Al(s) + 3\,H_2O \longrightarrow Al_2O_3(s) + 6\,H^+(aq) + 6\,e^-$$

This *anodized aluminum* can be colored by adding dyes to the electrolyte solution. Bronze and brown anodized aluminum are especially popular in modern buildings and as window frames in homes.

Anodized aluminum rivets.

As an active metal, aluminum readily reacts with acids to produce hydrogen gas.

$$2\,Al(s) + 6\,H^+(aq) \longrightarrow 2\,Al^{3+}(aq) + 3\,H_2(g)$$

However, aluminum also reacts with *basic* solutions. Here is a way of thinking about the reaction: First, the Al_2O_3 film on the metal reacts with $OH^-(aq)$. This is the same reaction that we described in the purification of bauxite.

$$Al_2O_3 + 3\,H_2O + 2\,OH^-(aq) \longrightarrow 2\,[Al(OH)_4]^-(aq)$$

Once the oxide film is removed from the aluminum, the metal is then free to display its true activity. It reacts with water in the alkaline solution, displacing hydrogen gas.

$$2\,Al(s) + 6\,H_2O + 2\,OH^-(aq) \longrightarrow 2\,[Al(OH)_4]^-(aq) + 3\,H_2(g) \qquad E^\circ_{cell} = 1.482\text{ V}$$

Some drain cleaners make use of this reaction. They consist of a mixture of solid sodium hydroxide and granules of aluminum metal. When added to water, the heat of solution of the NaOH(s) and the heat of reaction help to melt fat and grease. Hydrogen gas produced by the reaction helps to dislodge obstructions and unplug the drain. (Often, though, a plumber's snake is more effective than chemical drain cleaners.)

Because its combustion is a highly exothermic reaction, powdered aluminum is used as a component in rocket fuels, explosives, and fireworks.

$$2\,Al(s) + \tfrac{3}{2}\,O_2(g) \longrightarrow Al_2O_3(s) \qquad \Delta H^\circ = -1676\ kJ$$

Perhaps most familiar is aluminum's use in beverage cans, cookware, and as a foil for wrapping foods. The greatest quantity, though, is consumed in structural materials, usually alloyed with other metals to impart greater strength. Most modern aircraft use aluminum alloys, as do some automobile engines and the exterior trim of modern buildings. Because aluminum has good electrical conductivity (about 63.5% that of an equal volume of copper) and a low density, it finds important uses in the electrical industry. Most of the high-power transmission lines in the United States are now made of aluminum alloys.

Aluminum Compounds We've already mentioned some aspects of the chemistry of aluminum oxide and hydroxide. Of the aluminum halides, AlF_3 has some ionic properties: a high melting point and good electrical conductivity in the molten state. The other halides have covalent bonds and exist as Al_2X_6 molecules, although we still usually represent them simply as AlX_3. These molecules are *dimers* (double molecules), consisting of *two* AlX_3 units. In Al_2Cl_6, two Cl atoms are bonded exclusively to each Al atom. The remaining two Cl atoms form a bridge between the Al atoms. In the Lewis structure in Figure 21.3, we show two of the Al—Cl bonds as coordinate covalent bonds.

Figure 21.3 Bonding in Al_2Cl_6. Two Cl atoms form "bridges" between the $AlCl_3$ units, yielding the dimer Al_2Cl_6. Electrons donated by these Cl atoms to coordinate covalent bonds are represented by the colored arrows.

Aluminum chloride is especially important in organic chemistry. Like BH_3, $AlCl_3$ is an electron-deficient structure, and as such it forms complexes with most oxygen-containing organic compounds and other species having lone-pair electrons. For example, with 2-chloropropane the reaction is

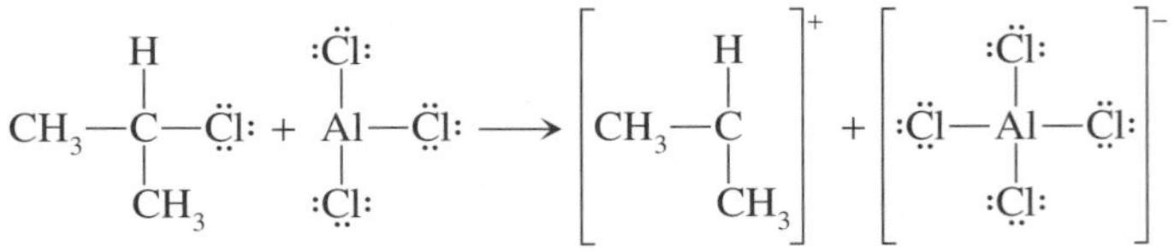

The hydrocarbon ion, called a **carbocation** because it has a positive charge on a carbon atom, can then attach itself to a benzene ring or to one of the double-bonded C atoms of an alkene. The net reaction, in which $AlCl_3$ acts as a catalyst, is one in which new carbon-to-carbon bonds are formed. For example,

Gemstones: Natural and Artificial

Natural gemstones are rare; they must contain just the right combination of substances, brought together under the appropriate conditions of temperature, pressure, concentrations, and so on, in order to form the proper crystals. One class of gemstones has aluminum oxide (corundum) as the principal constituent, together with small quantities of transition metal oxides as impurities. The colors are determined by the particular impurities present (see Table 21.2). Corundum gemstones with a pink, red, or dark red color are *rubies*; those of all other colors are *sapphires*.

TABLE 21.2 **Some Gemstones Based on Aluminum Oxide**

Gem	Impurity
White sapphire	none
Orange sapphire	Ni, Cr
Yellow sapphire	Ni
Green sapphire	Co, V, and/or Ni
Blue sapphire	Fe, Ti
Star sapphire	Ti
Ruby	Cr

Similar gemstones can be produced synthetically in furnaces of the type pictured in Figure 21.4. A mixed powder of Al_2O_3 and the appropriate transition metal oxide(s) is sprayed into the upper region of the furnace, melts in the hottest regions of the furnace, and deposits as a liquid layer that solidifies. The solidified material is gradually withdrawn from the furnace as layer after layer is added to it. Gemstones produced in this way have many industrial applications. The ruby—one of the rarest and most valuable of all gemstones—can now be manufactured for use in jewelry. Good quality natural rubies are still more valuable than synthetic ones. The natural gems are identified by their slight imperfections; the synthetic ones—as unbelievable as it seems—are too near to perfection.

Figure 21.4 Manufacture of artificial gemstones. In the furnace pictured here, a mixed powder of Al_2O_3 and the appropriate additive(s) is sprayed from above. The powder melts in the hottest regions of the furnace and deposits as a liquid layer that then solidifies. The solidified material is gradually withdrawn from the furnace as layer after layer is added to it.

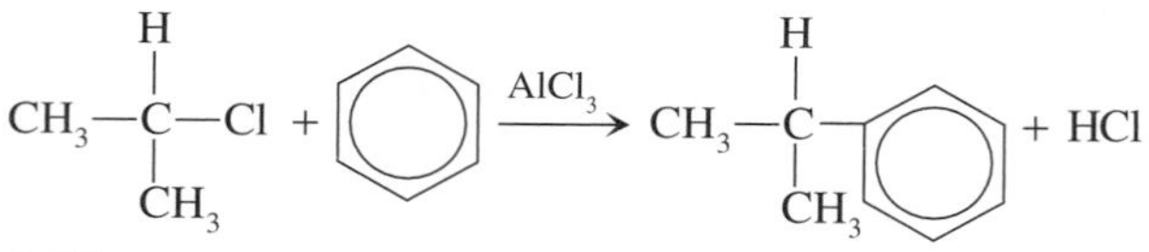

2-Chloropropane Benzene Isopropylbenzene

Reactions of this type are used in petroleum refining and in the manufacture of petrochemical products such as synthetic detergents.

Lithium aluminum hydride, $LiAlH_4$, is an important reducing agent in organic chemistry. If we think of hydride ion ($H:^-$) as a *pseudo*halide (some periodic tables place hydrogen at the top of Group 7A as well as at the top of Group 1A), then the AlH_4^- ion can be considered an adduct of AlH_3 and H^-. An example of its use in organic chemistry is the reduction of carboxylic acids, such as those from fats, to alcohols. For palmitic acid, the equation (not balanced*) is

$$\underset{\text{Palmitic acid (from palm oil)}}{CH_3(CH_2)_{14}COOH} + LiAlH_4 \longrightarrow \underset{\text{1-Hexadecanol}}{CH_3(CH_2)_{14}CH_2OH}$$

The fatty alcohols, such as 1-hexadecanol, are used to make specialty detergents such as those found in toothpaste and shampoos.

Aluminum sulfate, $Al_2(SO_4)_3$, is the most important industrial aluminum compound. It is prepared by the action of hot, concentrated $H_2SO_4(aq)$ on $Al_2O_3(s)$. The product that crystallizes from solution is $Al_2(SO_4)_3 \cdot 18H_2O$. About one-half of the 1 million tons of aluminum sulfate produced annually in the United States is used in water treatment. In this application the pH is raised to the point at which $Al(OH)_3(s)$ precipitates.

$$Al^{3+}(aq) + 3\ OH^-(aq) \longrightarrow Al(OH)_3(s)$$

As it settles, the gelatinous $Al(OH)_3(s)$ removes suspended solids from the water.

We've already mentioned some sources and uses of aluminum oxide. Additional important uses include refractory linings for high-temperature furnaces (Al_2O_3: mp, 2045 °C) and the manufacture of ceramic materials. The mineral *corundum* is a pure form of Al_2O_3, while *emery* is corundum contaminated with iron oxides (Fe_2O_3 and/or Fe_3O_4) and silica (SiO_2). Both corundum and emery have a hardness value of 9 on a ten-point scale with diamond as the hardest substance. They are both used in the manufacture of abrasive materials such as grinding wheels and sandpaper. Another important group of naturally occurring impure forms of aluminum oxide are many gemstones.

Corundum and emery are widely used in abrasive materials, as in the abrasive belt of this sander.

Gallium, Indium, and Thallium

Unlike aluminum, which forms only a 3+ ion, gallium, indium, and thallium also form 1+ ions. For example, gallium can give up its 4*p* electron, retain its two 4*s* electrons, and acquire the electron configuration $Ar3d^{10}4s^2$ for Ga^+. The heavier elements have an even greater tendency to form 1+ ions. In fact, Tl^+ is

* The mechanism of this reaction is quite complex. It proceeds through four steps, and the final mixture must be acidified to obtain the alcohol. In the course of the reaction, $LiAlH_4$ is decomposed to Li^+, Al^{3+}, and $H_2(g)$.

Gallium Arsenide

Rapid electron flow is an important property of modern semiconductor devices. For example, high-speed computers require a current to be switched on or off in less than a billionth of a second. There are two ways to speed up semiconductor devices: make them smaller or use a material through which electrons travel faster.

Silicon is the most common semiconductor material. Devices based on silicon are relatively easy to make and inexpensive. However, these devices have been pushed nearly to their limit in speed. Gallium arsenide (GaAs), isoelectronic with germanium and also a semiconductor, allows electrons to flow about 5 to 10 times faster than does silicon. This increased electron speed—and some desirable optical properties—make GaAs an important alternative material for use in supercomputers, radar devices, and communications satellites.

The method generally used to make GaAs semiconductors is quite different from most manufacturing processes. The gaseous reactants are trimethylgallium, $Ga(CH_3)_3$, and highly toxic arsine, AsH_3. The reaction is carried out at a high temperature and the GaAs is deposited as a thin film.

GaAs is more reactive than silicon, and GaAs wafers must be protected against air oxidation. As this and other technical difficulties are overcome, GaAs semiconductor materials should become an important part of the worldwide electronics market. The story of gallium arsenide illustrates how elements and compounds that were at one time just laboratory curiosities can become important players in the modern age of high technology.

Gallium bars and arsenic nuggets being fused into gallium arsenide.

more stable in aqueous solution than Tl^{3+}. This is reflected in the large *positive* potential for the reduction

$$Tl^{3+}(aq) + 2\,e^- \longrightarrow Tl^+(aq) \qquad E^\circ_{red} = +1.25\text{ V}$$

The valence-shell pair of electrons retained in the +1 ion (ns^2) is called an **inert pair**. This retention of the ns^2 pair is common in post-transition elements.

Gallium is similar to aluminum in some of its properties; for example, it forms an amphoteric hydroxide, $Ga(OH)_3$. Thallium resembles lead in its high density (11.85 g/cm^3), its softness, and the toxicity of some of its compounds. Thallium(I) compounds resemble alkali metal compounds in some respects. For example, because the Tl^+ ion (radius 140 pm) is about the same size as K^+ (radius 133 pm), the two ions can crystallize together in chlorate, perchlorate, sulfate, and phosphate salts. Like alkali metal hydroxides, TlOH is a strong base. In other respects Tl^+ resembles Ag^+: it forms a light-sensitive, insoluble chloride, TlCl, just as silver forms light-sensitive halide salts.

Gallium is a liquid from 30 °C to about 2400 °C, one of the longest liquid temperature ranges of any substance. This property makes gallium useful in some high-temperature thermometers. Ga, In, and Tl and their compounds are used commercially in alloys, transistors, photoconductors, and specialty glasses.

21.2 Group 4A: C, Si, Ge, Sn, and Pb

The members of Group 4A all have the valence electron configuration ns^2np^2, where n is the period number. Carbon, the first member, is a nonmetal. Carbon atoms use their four valence electrons to form four covalent bonds in nearly all their compounds. The next two members, silicon and germanium, also form covalent bonds for the most part. Both are metalloids and have interesting properties as semiconductors . Tin and lead are more metallic in their behavior. Both form 2+ and 4+ ions.

Carbon

The ground state electron configuration of carbon is $1s^22s^22p^2$. Hybridization of the valence-level electrons produces three different sets of orbitals: (1) four sp^3 orbitals, as in ethane, C_2H_6, (2) three sp^2 orbitals plus one p orbital, as in ethene (ethylene), C_2H_4, or (3) two sp orbitals plus two p orbitals, as in ethyne (acetylene), C_2H_2. These hybridization possibilities permit the formation of carbon chain and ring structures having multiple bonds between C atoms, as well as the more common single bonds. In these structures, C atoms also bond to H, O, N, S, halogens, and several other types of atoms. Taken together, these factors lead to a myriad of *organic* compounds. We've frequently referred to organic compounds since introducing them in Chapter 2.

An all-composite wing of a military jet aircraft.

Elemental carbon exists in nature mainly as the two allotropes diamond and graphite (Section 11.8). Graphite is perhaps most familiar as the "lead" in pencils. Because it conducts electric current, graphite is used for electrodes in batteries and in industrial electrolysis reactions. It can also withstand high temperatures, leading to its use in foundry molds, furnaces, and other high-temperature devices.

An important modern use of graphite is in the form of fibers. When carbon-based fibers such as rayon are heated to a high temperature, other elements are driven off as gaseous products and graphite fibers remain. These fibers are imbedded in plastic materials to make high-strength, lightweight composites. The composites are used in products as diverse as tennis rackets, golf clubs, canoes, and airplanes.

Diamonds are used in jewelry, but they also have industrial uses. Because they have a high thermal conductivity (they dissipate heat quickly) and are extremely hard, diamonds are used as abrasives and in drilling bits for cutting steel and other hard materials. In general, natural diamonds are used as gemstones and synthetic diamonds for industrial purposes.

Synthetic diamonds are made from graphite. Graphite is the more stable form of carbon at room temperatures and pressures and remains so up to temperatures of 3000 °C and pressures of 10^4 atm or more. The conversion of graphite to diamonds requires temperatures of 1000 to 2000 °C and pressures of 10^5 atm or more.

Carbon also exists in "amorphous" forms. The term *amorphous* implies a noncrystalline solid, but amorphous carbon is probably microcrystalline graphite. When coal is heated in the absence of air, volatile substances are driven off, leaving a high-carbon residue called *coke*. Coke is the principal metallurgical reducing agent. It is used in the reduction of iron oxide to iron in a blast furnace, for example. A similar *destructive distillation* of wood produces charcoal. As you've probably observed in an improperly adjusted bunsen burner, incomplete combustion of natural gas produces a smoky flame. This smoke can be deposited as a powdery soot called *carbon black*. Carbon black is used as a filler in rubber tires, as a pigment in printing inks, and as the transfer material in carbon paper, typewriter ribbons, and photocopying machines. The annual production of carbon black in the United States is about 3 billion pounds, which places it about 37th among industrial chemicals.

Activated carbon is formed by heating carbon black to 800–1000 °C in the presence of steam to expel all volatile matter. This form of carbon is highly porous, like sponges or honeycombs. Because of its high ratio of surface area to volume, it has a great capacity to adsorb substances from liquids and gases. Activated carbon is used in gas masks to adsorb poisonous gases from air, in water filters to remove organic contaminants from water, in the recrystallization of sugar solutions to remove colored impurities, in air conditioning systems to control odors, and in industrial plants for the control and recovery of vapors.

Carbon Compounds Carbon combines with most metals to form compounds called carbides. With active metals, the carbides are ionic. For example, calcium carbide is formed in the high-temperature reaction of lime and coke.

The two principal oxides of carbon, carbon monoxide and carbon dioxide, were discussed in Section 14.5.

$$CaO(s) + 3\,C(s) \xrightarrow{2000\ ^\circ C} CaC_2(s) + CO(g)$$

Calcium carbide, an ionic compound with a carbon-to-carbon triple bond in the anion$[Ca^{2+}(:C{\equiv}C:)^{2-}]$, reacts with water to produce acetylene, $H{-}C{\equiv}C{-}H$.

$$CaC_2(s) + 2\,H_2O \longrightarrow Ca(OH)_2(s) + \underset{\text{Acetylene}}{C_2H_2(g)}$$

Most acetylene for large-scale uses is prepared by the *flash pyrolysis* (decomposition by rapid heating) of methane.

$$2\,CH_4(g) \xrightarrow[0.01\text{–}0.1\ s]{2000\ ^\circ C} C_2H_2(g) + 3\,H_2(g)$$

Calcium carbide is a convenient reagent because we can transport the solid and then generate the gaseous fuel acetylene simply by adding water.

Two other binary compounds of carbon are carbon disulfide, CS_2, and carbon tetrachloride, CCl_4. Carbon disulfide is prepared by the reaction of methane with sulfur vapor in the presence of a catalyst.

$$CH_4(g) + 4\,S(g) \longrightarrow CS_2(l) + 2\,H_2S(g)$$

CS_2 is a flammable, volatile liquid that is a good solvent for sulfur, phosphorus, bromine, iodine, fats, and oils. Its toxicity somewhat limits this use, however. Carbon disulfide is an important intermediate in the manufacture of rayon and cellophane.

Carbon tetrachloride can be prepared by the direct chlorination of methane.

$$CH_4(g) + 4\,Cl_2(g) \longrightarrow CCl_4(l) + 4\,HCl(g)$$

Once extensively used as a solvent, dry-cleaning agent, and fire extinguisher, CCl_4 is declining in importance because it causes liver and kidney damage and is a suspected human carcinogen.

Certain groupings of atoms, called **pseudohalogens**, mimic the characteristics of a halogen atom. Several pseudohalogens contain C atoms: for example, —CN (cyanide), and —OCN (cyanate). The cyanide ion, CN^-, is similar to halide ions, X^-, in several ways: It forms an insoluble silver salt, AgCN, and an acid, HCN, though this acid (hydrocyanic acid) is quite weak. CN^- differs from halide ions in that it is very toxic. Despite its toxicity, HCN, a liquid that boils at about room temperature, is widely used in the manufacture of plastics. It is also used—carefully, by well-trained personnel—as a fumigant to kill rodents and insects on ships.

:N≡C—C≡N:

Cyanogen

Just as two Cl atoms combine to form the molecule Cl_2, two cyanide groups can combine to form the *cyanogen* molecule, $(CN)_2$, often represented by the formula C_2N_2. In basic solution, the (disproportionation) reaction of cyanogen

$$\overset{+3}{(CN)_2(g)} + 2\,OH^-(aq) \longrightarrow \overset{+2}{CN^-(aq)} + \overset{+4}{OCN^-(aq)} + H_2O$$

is similar to that of chlorine.

$$\overset{0}{Cl_2(g)} + 2\,OH^-(aq) \longrightarrow \overset{-1}{Cl^-(aq)} + \overset{+1}{OCl^-(aq)} + H_2O$$

Cyanogen is used as a reagent in organic synthesis, as a fumigant, and as a rocket propellant.

Silicon

Silicon is quite different from carbon, perhaps as different as the second and third period elements of any group in the periodic table. The Si—Si bond (bond energy = 226 kJ/mol) and Si—H bond (318 kJ/mol) are weaker than the corresponding C—C (347 kJ/mol) and C—H (414 kJ/mol) bonds, but this is probably not the primary reason why chains and rings of Si atoms lack the stability of those of C atoms. Instead, the activation energies of reactions of silicon chain and ring compounds are much lower than those of the corresponding carbon compounds; their rates of reaction are also correspondingly greater. Thus, disilane, Si_2H_6, spontaneously ignites on contact with oxygen (producing SiO_2 and H_2O), whereas the ignition of ethane, C_2H_6, requires a spark or open flame. In any event, strong Si—O bonds (464 kJ/mol) favor silicates as the predominant naturally occurring compounds of silicon.

A silicon atom, like a carbon atom, almost always forms four bonds. In elemental silicon, each atom uses its four valence electrons ($3s^23p^2$) in an sp^3 hybridization scheme. Silicon crystallizes in an *fcc* structure similar to diamond. There is no allotrope of silicon equivalent to graphite.

Silica In silica, SiO_2, each Si atom is bonded to four O atoms and each O atom, to two Si atoms. As suggested by Figure 21.5, the structure is that of a network covalent solid (recall Section 11.8). Quartz, a form of pure silica, is quite hard (a hardness of 7, compared to a diamond's hardness of 10), has a

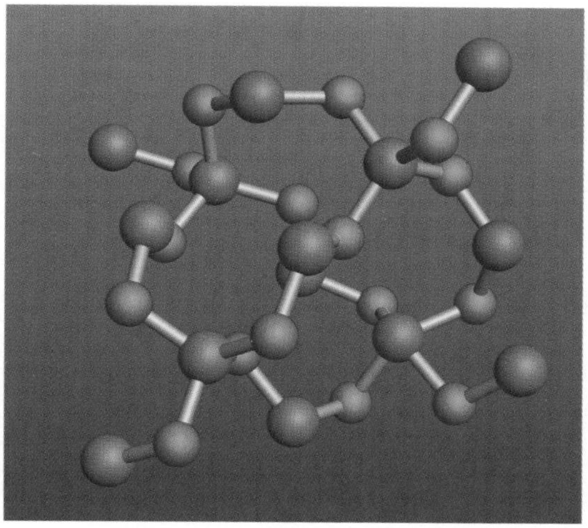

Figure 21.5 The structure of silica, SiO_2. Each Si atom forms bonds to four O atoms, and each O atom to two Si atoms, in this three-dimensional network covalent structure.

high melting point (about 1700 °C), and is a nonconductor of electricity. Silica is the basic raw material of the glass, ceramics, and refractory materials industries.

Ordinary soda–lime glass is a mixture of sodium and calcium silicates made by fusing sodium carbonate and calcium carbonate with sand (impure silica) at about 1500 °C. The liquid mixture becomes more viscous as it cools. Eventually it ceases to flow and becomes rigid; it becomes a *glass.* Glass is transparent to visible light. By varying the proportions of the three basic ingredients, and by adding other substances, we can alter the properties of glass. For example, incorporating boric oxide into soda–lime glass produces a borosilicate glass (one example is Pyrex). Because borosilicate glasses do not expand and contract as readily as soda–lime glass, they can be rapidly heated and cooled without breaking. Glass ovenware and most laboratory glassware are made of borosilicate glass.

Cobalt(II) oxide imparts a characteristic blue color to soda–lime glass.

Silicate Minerals Silicon is the second most abundant element (after oxygen) and makes up 27.2% by mass of Earth's solid crust. It is the key element of the mineral world. Silicate anions are often quite complex, but most have as a basic structural unit a tetrahedron with an Si atom at the center and O atoms at the four corners. Silicon is in the oxidation state +4 and oxygen, −2; the tetrahedral silicate anion has a charge of 4−, SiO_4^{4-} (see Figure 21.6). In silicate minerals, SiO_4 tetrahedra may be arranged in a variety of ways, leading to a host of mineral forms. Some of the possibilities are

- The anions may be simple SiO_4^{4-} tetrahedra. Typical minerals and their compositions are *thorite* ($ThSiO_4$) and *zircon* ($ZrSiO_4$).
- The anions may be combinations of two SiO_4 tetrahedra. In $Si_2O_7^{6-}$, the Si atoms of two tetrahedra share an O atom between them. A typical mineral is *thortveitite* ($Sc_2Si_2O_7$).
- SiO_4 tetrahedra may be joined together into long Si—O chains. Each Si atom shares two O atoms with Si atoms in adjacent tetrahedra. A typical mineral is *spodumene*, which has the empirical formula $LiAl(SiO_3)_2$; it is the principal natural source of lithium and its compounds.

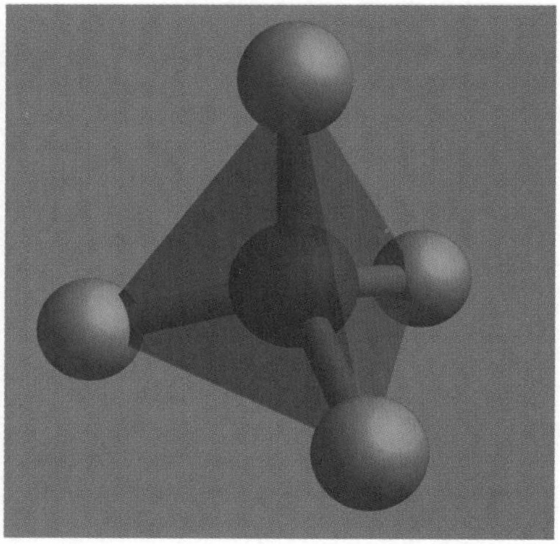

Figure 21.6 The silicate anion, SiO_{4-}. The Si atom is at the center of the tetrahedron and an O atom is at each of the four corners. The shaded figure (green) shows how the tetrahedron is typically represented in a mineral structure (see, for example, Figure 21.7).

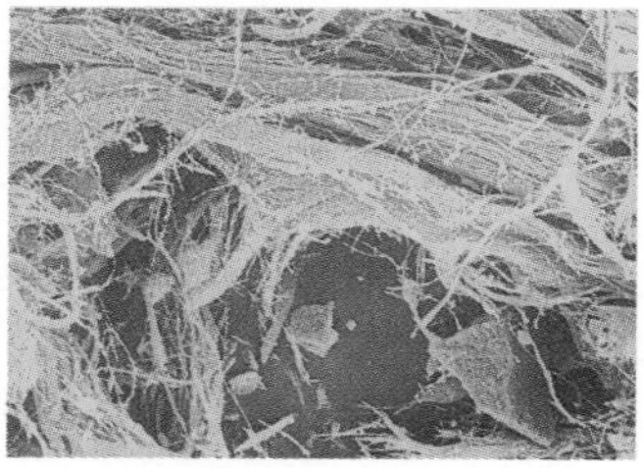

Chrysotile asbestos.

Muscovite mica.

- SiO_4 tetrahedra may be joined together into double Si—O chains. In *chrysotile asbestos*, the SiO_4 tetrahedra share three O atoms, and the double chains are bound together by cations, chiefly Mg^{2+}. The mineral has a fibrous appearance. The empirical formula is $Mg_3(Si_2O_5)(OH)_4$.
- SiO_4 tetrahedra may share three of their four O atoms with adjacent tetrahedra in two-dimensional sheets, as suggested by Figure 21.7. In *muscovite mica*, the counter ions to the silicate anions are mostly K^+ and Al^{3+}; its empirical formula is $KAl_2(AlSi_3O_{10})(OH)_2$. Because bonding within sheets is stronger than between sheets, mica flakes easily. Vermiculite, used as loose-fill insulation, filler, and packing material, has sheets of mica separated by double water layers.
- SiO_4 tetrahedra may share all four of their O atoms with adjacent tetrahedra in a three-dimensional array. This is the most common arrangement in silicate minerals in Earth's crust. It is the arrangement found in quartz itself, and in mineral forms such as *amethyst*, a purple quartz containing iron as an

Figure 21.7 A two-dimensional sheet in the structure of mica. Three of the four O atoms in each tetrahedron are shared with another tetrahedron in a two-dimensional array of tetrahedra. The cations found between the planes of silicate anions are mostly K^+ and Al^{3+}.

impurity and certain types of crystal defects; *agate*, an impure silica that crystallizes in colored bands; and in *petrified wood*, very old wood in which colored agate replaces organic matter but in which the microscopic structure of the wood is retained.

Petrified wood.

Organosilicon Compounds Although we have stressed the inorganic chemistry of silicon, there is also a chemistry that emulates the chemistry of carbon—an "organic" chemistry—although it is not nearly as rich as that of carbon. Silicon can form Si—Si bonds in chains of up to about a dozen Si atoms.

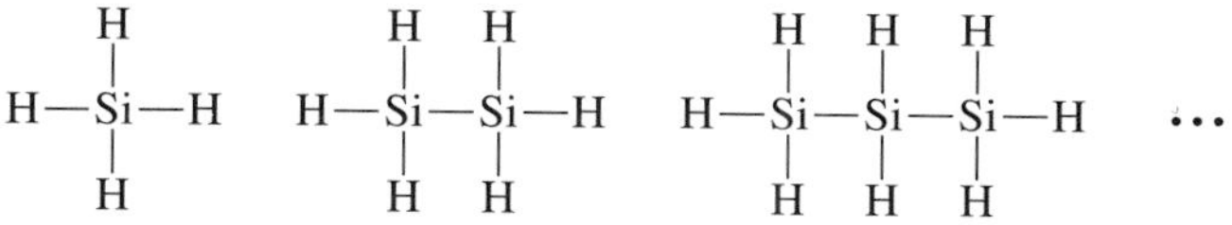

Silane Disilane Trisilane

The silanes are thermally unstable. When heated, the higher silanes decompose to lower silanes or to the elements. Like the hydrocarbons, the silanes are combustible; the combustion products are $SiO_2(s)$ and H_2O. As we have noted, however, unlike hydrocarbons, the silanes burst into flame in air. In this regard, silanes are like boranes (Section 21.1), a fact that illustrates the diagonal relationship of the two elements.

Other atoms can be rather easily substituted for H atoms in silanes. For example, if we substitute Cl for H atoms in SiH_4, we obtain successively SiH_3Cl, SiH_2Cl_2, $SiHCl_3$, and $SiCl_4$. The reaction of $(CH_3)_2SiCl_2$ with water produces dimethylsilanol, $(CH_3)_2Si(OH)_2$, a starting material for the production of a class of polymers called *silicones*.

$$(CH_3)_2SiCl_2 + 2\,H_2O \longrightarrow \underset{\text{Dimethylsilanol}}{(CH_3)_2Si(OH)_2} + 2\,HCl$$

Tin and Lead

Tin and lead are quite similar to each other. They are soft and malleable and melt at relatively low temperatures for metals. Their ionization energies are about equal, and their standard electrode potentials for the reduction $M^{2+}(aq) + 2\,e^- \longrightarrow M(s)$ are both slightly negative. Tin and lead are oxidized somewhat more easily than $H_2(g)$, which places them just above hydrogen in the activity series of metals (Figure 12.10).

There is only one solid form of lead, but tin exists in two allotropic forms. The α (gray) or nonmetallic form is stable below 13 °C, and the β (white) or metallic form is stable above 13 °C. When held below 13 °C for a long time, white tin changes to gray tin. The tin expands and crumbles to a powder. This transformation, called "tin disease," has led to the disintegration of organ pipes, buttons, medals, and other objects made of tin that are kept below 13 °C.

Nearly half of all the tin metal produced is used in tin plate, mostly in plating iron for use in cans to store foods. The next most important use (about 25%) is in the manufacture of solders—low-melting-point alloys used to join wires or pieces of metal. Other important alloys of tin are bronze (90% Cu, 10% Sn) and pewter (85% Sn, 7% Cu, 6% Bi, 2% Sb).

Lead Poisoning

The Latin name of lead, *plumbum*, reflects its long-time use in plumbing. Lead pipes have been used to carry water from the time of the Romans until well into the twentieth century. Some older homes still have lead plumbing systems, and many more have copper pipes fitted with lead solder. The U.S. Environmental Protection Agency now requires municipal water utilities to test for lead content. The maximum permitted level is 50 mg/L (50 ppb). In addition to contamination in drinking water, we are also exposed to lead through cooking and eating utensils and pottery glazes. In the past, the chief sources of lead contamination in the environment have been tetraethyllead, an antiknock additive for gasoline, and lead-based paints.

Mild forms of lead poisoning produce nervousness and mental depression. More severe cases can lead to permanent nerve, brain, and kidney damage, especially in children. Lead also interferes with the biochemical reactions that produce the iron-containing heme group in hemoglobin. Fortunately, average blood lead levels have dropped dramatically since leaded gasoline was phased out. However, lead contamination still persists in certain soils and in painted surfaces in some older homes.

Over half of the lead produced is used in lead–acid (storage) batteries. Other uses include the manufacture of solder and other alloys, lead shot, and radiation shields (to protect against X-rays and γ rays).

A significant difference between tin and lead is that the +4 oxidation state is much more stable in tin than in lead. For example, when tin is heated in air, the product is $SnO_2(s)$, whereas the air oxidation of lead produces PbO(s). Metallic tin reacts with $Cl_2(g)$ to form $SnCl_4$ (bp, 115 °C). (This reaction is used to recover tin from scrap tin plate.) Lead reacts with $Cl_2(g)$ to form $PbCl_2$ (mp, 501 °C). Lead forms only the sulfide, PbS. With tin, SnS can react further to produce SnS_2. A few uses of tin and lead compounds are listed in Table 21.3.

TABLE 21.3 Some Compounds of Tin and Lead

Compound	Use(s)
$SnCl_2$	reducing agent in the laboratory and in the manufacture of chemicals; tin galvanizing; catalyst
SnF_2	cavity preventative in toothpaste (stannous fluoride)
SnO_2	jewelry abrasive; ceramic glazes; preparation of perfume and cosmetics; catalyst
SnS_2	pigment; imitation gilding (tin bronze)
$Pb(NO_3)_2$	preparation of other lead compounds
PbO	used in glass, ceramic glazes, and cements
PbO_2	oxidizing agent; lead–acid battery electrodes; matches and explosives
$PbCrO_4$	a yellow pigment (chrome yellow) for industrial paints, plastics, and ceramics

Tin(II) chloride, stannous chloride, is a good reducing agent and is used in the laboratory to reduce Fe(III) to Fe(II), Hg(II) to Hg(I), and Cu(II) to Cu(I). Tin(II) fluoride, stannous fluoride, is used in toothpaste to decrease the incidence of cavities. SnO_2 is used as a jewelry abrasive, and SnS_2 (tin bronze) is used as a pigment and for imitation gilding.

21.3 Group 5A: N, P, As, Sb, and Bi

From top to bottom in Group 5A, the elements display properties that range from nonmetallic to metallic. All have the valence-shell electron configuration ns^2np^3, but their atoms can alter this electron configuration in different ways when forming compounds. In a few cases they can gain three electrons to produce a 3– ion with the noble gas electron configuration ns^2np^6. Far more often, though, they acquire this configuration by sharing three electrons, producing the –3 oxidation state. This is especially likely for the smaller atoms N and P. The larger atoms—As, Sb, and Bi—can give up the p^3 set of electrons. This leads to an electron configuration with a next-to-outermost shell of 18 and an outer shell of 2—the "inert pair" (page 822). This "18 + 2" configuration is found in several ionic species derived from metals following a transition series. In still other cases, all five valence electrons are involved in compound formation, leading to the +5 oxidation state.

Among the Group 5A elements we note the usual decrease of first ionization energy with increasing atomic number: 1402 kJ/mol for N and 1012, 947, 834, and 703 kJ/mol for P, As, Sb, and Bi, respectively. Electronegativities decrease from 3.0 for N to 2.1, 2.0, 1.9, and 1.9 for P, As, Sb, and Bi, respectively. These data suggest a high degree of nonmetallic character for nitrogen and progressively less of a nonmetallic character for the heavier members of Group 5A. None of the elements in Group 5A is highly metallic, however; that is, none has a metallic character comparable to the elements of Groups 1A and 2A.

Nitrogen

We presented a number of aspects of the chemistry of nitrogen in Chapter 14. Here we look at just a few additional ones.

Ammonia, as we know, is the principal commercial source of nitrogen compounds and the most common weak base. If we replace an H atom in NH_3 by the group $—NH_2$, the resulting molecule is NH_2NH_2, hydrazine. Because of its *two* N atoms, hydrazine is a weak base that can ionize in *two* steps.

$$NH_2NH_2(aq) + H_2O \rightleftharpoons NH_2NH_3^+(aq) + OH^-(aq) \qquad K_{b_1} = 8.5 \times 10^{-7}$$

$$NH_2NH_3^+(aq) + H_2O \rightleftharpoons NH_3NH_3^{2+}(aq) + OH^-(aq) \qquad K_{b_2} = 8.9 \times 10^{-16}$$

Some of the most important uses of hydrazine, however, are not as a base. It is an excellent reducing agent, meaning that it is easily oxidized to $N_2(g)$.

$$NH_2NH_3^+(aq) \longrightarrow N_2(g) + 5\,H^+(aq) + 4\,e^- \qquad E^\circ_{ox} = 0.23\text{ V}$$

One of its most important uses is the removal of dissolved O_2 from boiler water. The net reaction is

$$NH_2NH_3^+(aq) + O_2(aq) \longrightarrow 2\,H_2O + H^+(aq) + N_2(g) \qquad E^\circ_{cell} = +1.46\text{ V}$$

Recall that tabulated standard electrode potentials, E°, are for reduction half-reactions, $E^\circ = E^\circ_{red}$. That is,

$$N_2(g) + 5\,H^+(aq) + 4\,e^- \longrightarrow NH_2NH_3^+(aq) \qquad E^\circ_{red} = -0.23\text{ V}$$

The potential for an *oxidation* half-reaction is $E^\circ_{ox} = -E^\circ_{red}$.

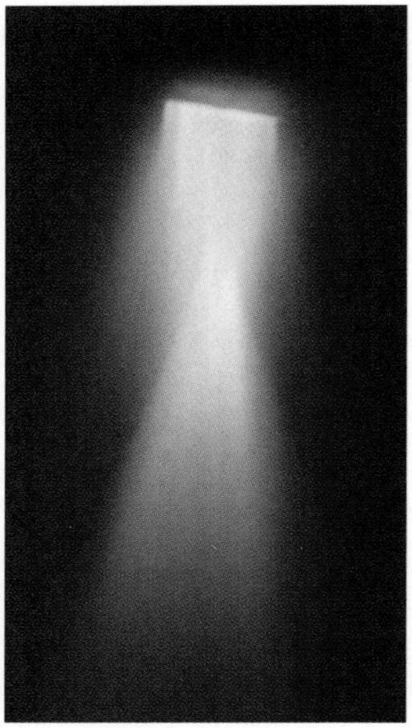

The upfiring thrusters of the Space Shuttle *Columbia,* shown here in an in-flight test, use methylhydrazine, CH_3NHNH_2, as a fuel.

The direct reaction of liquid hydrazine with oxygen gas can be used as the net reaction in a fuel cell and also for rocket propulsion.

$$N_2H_4(l) + O_2(g) \longrightarrow N_2(g) + 2\,H_2O(g) \qquad \Delta H° = -534.2\text{ kJ}$$

The oxidation of hydrazine in acidic solution by nitrite ion produces hydrogen azide, HN_3.

$$NH_2NH_3^+(aq) + NO_2^-(aq) \longrightarrow HN_3(aq) + 2\,H_2O$$

In aqueous solution, HN_3 is a weak acid ($K_a = 1.9 \times 10^{-5}$) called hydrazoic acid; its salts are called azides. Azides are extremely unstable; lead azide, for example, is used to make detonators. Sodium azide, NaN_3, also decomposes readily, releasing $N_2(g)$; but its decomposition can be sufficiently controlled for it to be the working substance in air-bag safety systems in automobiles.

Another unusual nitrogen-containing acid is NH_3 itself ($K_a \approx 10^{-34}$), under the appropriate conditions. In the presence of an exceptionally strong base such as hydride ion, H^-, NH_3 can donate a proton, yielding the amide ion, NH_2^-.

$$NaH + NH_3(l) \longrightarrow NaNH_2 + H_2(g)$$

Sodium amide (sodamide), $NaNH_2$, is used in a variety of organic syntheses where an especially strong base is required.

Phosphorus

Phosphorus is the eleventh most abundant element in Earth's crust. It occurs mainly in phosphate rock, produced in nature from the remains of marine animals deposited millions of years ago. Phosphate rock is a member of the class of minerals called *apatites* —fluorapatite, $3Ca_3(PO_4)_2 \cdot CaF_2$, is an example. Elemental phosphorus is prepared by heating phosphate rock, silica, and coke in an electric furnace. The net change is

$$2\,Ca_3(PO_4)_2(s) + 10\,C(s) + 6\,SiO_2(s) \xrightarrow{\Delta} 6\,CaSiO_3(l) + 10\,CO(g) + P_4(g)$$

The $P_4(g)$ is collected, condensed to a solid, and stored under water.

The solid phosphorus prepared in this way is a waxy, white, phosphorescent solid (a phosphorescent material glows in the dark). This solid, called *white* phosphorus, can be cut with a knife. It melts at 44.1 °C, is a nonconductor of electricity, and ignites spontaneously in air (that's why it is stored under water). White phosphorus is insoluble in water but soluble in nonpolar solvents like carbon disulfide (CS_2).

As shown in Figure 21.8, white phosphorus is made up of tetrahedral P_4 molecules, with a P atom at each corner of the tetrahedron. The P-to-P bonds in P_4 appear to involve the overlap of 3*p* orbitals almost exclusively. Normally such overlap produces 90° bond angles, but in P_4 the P—P—P bond angles are 60°. The bonds are said to be *strained*, and species with strained bonds, like P_4, are generally quite reactive.

When white phosphorus is heated to about 300 °C in the absence of air, it is transformed to *red* phosphorus. As suggested by Figure 21.8, what appears to happen at the molecular level is that one P—P bond per P_4 molecule breaks and the fragments join together into long chains. Red phosphorus and white phosphorus are allotropes; they have different fundamental structural units. The two allotropes differ appreciably in their properties; red phosphorus, for example, is less reactive than white phosphorus.

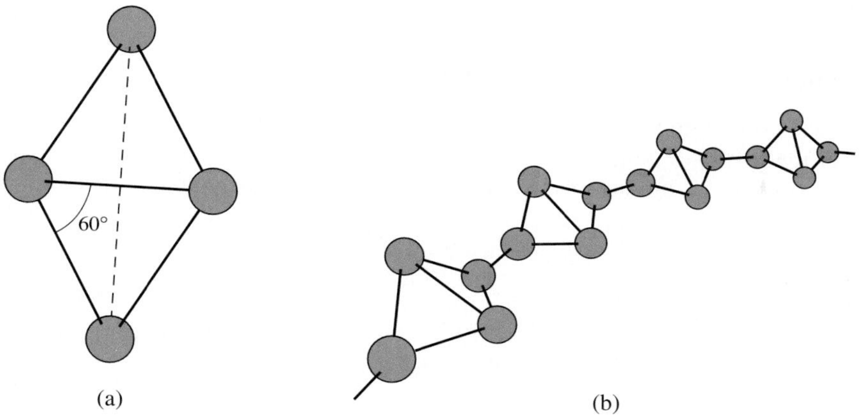

Figure 21.8 The molecular structures of white and red phosphorus. (a) White phosphorus is composed of P_4 molecules. (b) P_4 units join together in long chains in red phosphorus.

Phosphorus forms two important oxides. One, with P in the +3 oxidation state, has the empirical formula P_2O_3. The other, with P in the +5 oxidation state, has the empirical formula P_2O_5. Although these oxides have been for years called phosphorus *tri*oxide and phosphorus *pent*oxide, respectively, their true molecular formulas are double the empirical formulas, that is, tetraphosphorus hexoxide, P_4O_6, and tetraphosphorus decoxide, P_4O_{10}. The structures of these oxide molecules are shown in Figure 21.9. P_4O_6 forms in the reaction of P_4 with a limited quantity of $O_2(g)$, and P_4O_{10}, with excess $O_2(g)$. The oxides are the anhydrides of phosphorous acid and phosphoric acid. Thus, when the oxides are added to water the following reactions occur.

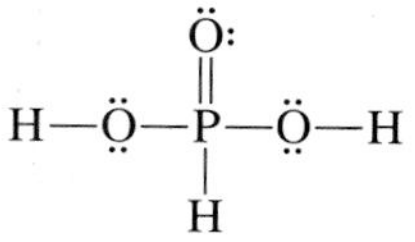

Phosphorous acid

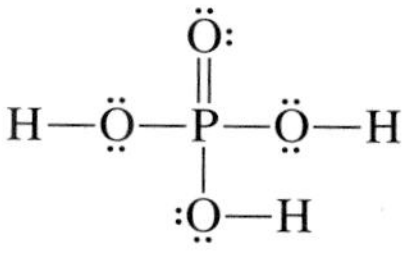

Phosphoric acid

$$P_4O_6 + 6\,H_2O \longrightarrow \underset{\text{Phosphorous acid}}{4\,H_3PO_3}$$

$$P_4O_{10} + 6\,H_2O \longrightarrow \underset{\text{Phosphoric acid}}{4\,H_3PO_4}$$

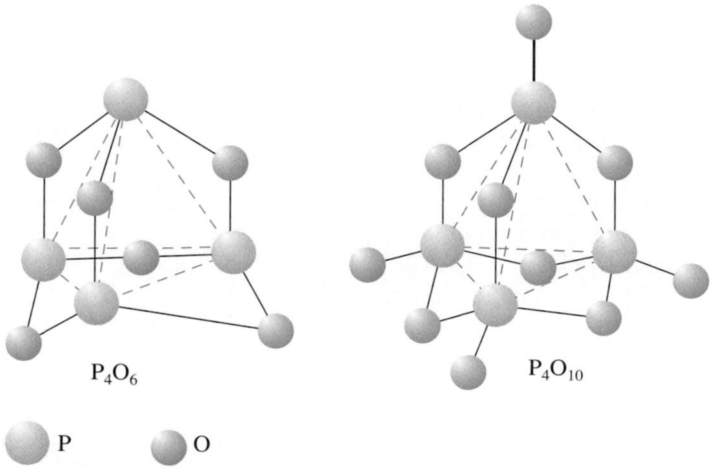

Figure 21.9 Molecular structures of P_4O_6 and P_4O_{10}. In P_4O_6 an O atom is inserted between each pair of P atoms of the basic P_4 structure, leading to a total of six O atoms in the molecule. In P_4O_{10} an additional O atom is bonded to each P atom, yielding a total of ten O atoms per molecule.

Phosphate Fertilizers

Although animal manure has been used since antiquity, artificial fertilizers have come into use only in more modern times. An early champion of the efficacy of inorganic chemicals in agriculture was the German chemist Justus von Liebig (1803–1873). Liebig was particularly aware of the importance of phosphorus, demonstrating, for example, the value of bones that had been treated with sulfuric acid as a plant nutrient. The early fertilizer industry then progressed to treating phosphate rock with sulfuric acid, yielding a mixture of calcium dihydrogen phosphate and insoluble calcium sulfate called *superphosphate*. The phosphorus content of superphosphate is only about 7 or 8% P by mass. The mass percent P can be nearly tripled—to 20 or 21% P—by the reaction of phosphoric acid with phosphate rock. This product is called *triple superphosphate*.

$$[3Ca_3(PO_4)_2 \cdot CaF_2] + 14\,H_3PO_4 + 10\,H_2O \longrightarrow \underset{\text{Triple superphosphate}}{10\,[Ca(H_2PO_4)_2 \cdot H_2O]} + 2\,HF$$

Other phosphate fertilizers can be formed by neutralizing phosphoric acid with aqueous ammonia. The two principal products are monoammonium phosphate (MAP), $NH_4H_2PO_4$, and diammonium phosphate (DAP), $(NH_4)_2HPO_4$. Together, these two products are the world's leading fertilizers. They have the advantage of providing two essential fertilizer elements (N and P) in a single substance.

Although phosphate fertilizers have helped immensely to increase the world's food supply, they present environmental problems, primarily due to runoff from fields and lawns treated with phosphate fertilizers. When this phosphate-enriched runoff enters freshwater bodies, it encourages algae growth. The algae die, and their decomposition depletes the oxygen content of the water, killing fish and promoting the growth of anaerobic bacteria—an overall process called *eutrophication*. Up to 98% of the phosphates in municipal and industrial wastewater can be removed by precipitation, but rainwater runoff from agricultural lands and lawns cannot be so easily treated. Part of the price we pay for increased food production and beautiful lawns is polluted waters.

An algae bloom leads to oxygen depletion in the water of this pond.

Phosphoric acid ranks about ninth among the chemicals produced in the United States, with an annual production of about 13 million tons. Most of it is used to make fertilizers, but it is also used to treat metals to make them more corrosion resistant. In the food industry it is used to impart tartness to soft drinks. Phosphate salts are used as dietary supplements, in baking powders, in dairy products, and as polishing agents in toothpaste.

Two other important phosphorus compounds are (1) phosphine, PH_3, the phosphorus analog of ammonia, used as a fumigant against rodents and insects and in the manufacture of flame retardants, and (2) phosphorus trichloride, PCl_3, used in the production of materials ranging from soaps and detergents to synthetic rubber, nylon, motor oils, insecticides, and herbicides.

Arsenic, Antimony, and Bismuth

In the remaining three members of Group 5A we find increasing evidence of metallic behavior. In fact, one of the main uses of these elements is in the manufacture of alloys. The addition of As and Sb to lead produces an alloy that has more desirable properties for use in electrodes in lead–acid batteries than does lead alone. Arsenic and antimony are also used in semiconductor materials (recall gallium arsenide, page 821). Because of their relatively low melting points, antimony (mp 631 °C) and bismuth (mp 271 °C) find some use in low-melting-point alloys for solders and fire-protection systems.

Elemental arsenic is obtained by heating sulfide minerals, such as FeAsS. The most important arsenic compound is the oxide, As_4O_6, known as *white arsenic*. It can be made by burning arsenic in air, and is the anhydride of arsenious acid, a weak acid.

$$As_4O_6(s) + 6\,H_2O \longrightarrow 4\,H_3AsO_3(aq) \qquad K_{a_1} = 6.6 \times 10^{-10}$$

Both inorganic and organic compounds derived from As_4O_6 have found extensive use as insecticides.

21.4 Group 6A: O, S, Se, Te, and Po

The elements of Group 6A have the valence-shell electron configuration ns^2np^4. Oxygen and sulfur are clearly nonmetallic, but the heavier elements, tellurium and polonium, exhibit some metallic properties and are classified as metalloids.

Oxygen

Oxygen is one of the most active nonmetals and one of the most important. It forms compounds with all the elements except the light noble gases. In general, its compounds with metals are ionic, and with other nonmetals, covalent. Oxygen ranks first in abundance in Earth's crust, about 45.5% by mass.

We've already discussed some of the chemistry of oxygen in Chapter 14, including its extraction from air. (Oxygen gas itself is one of the top chemicals—third in annual production in the United States at about 25 million tons.)

Actually, oxygen and its compounds are so central to the study of chemistry that we have been referring to them from the very beginning of the text. Oxygen and oxygen compounds are featured, for example, whenever we consider combustion reactions or the reactions of acids and bases in aqueous solutions. And oxygen is a key element in many of the functional groups of organic compounds: alcohols, ethers, aldehydes, ketones, carboxylic acids, esters. . . . In short, whenever we discuss the chemistry of an element, we invariably are led into a consideration of oxygen compounds.

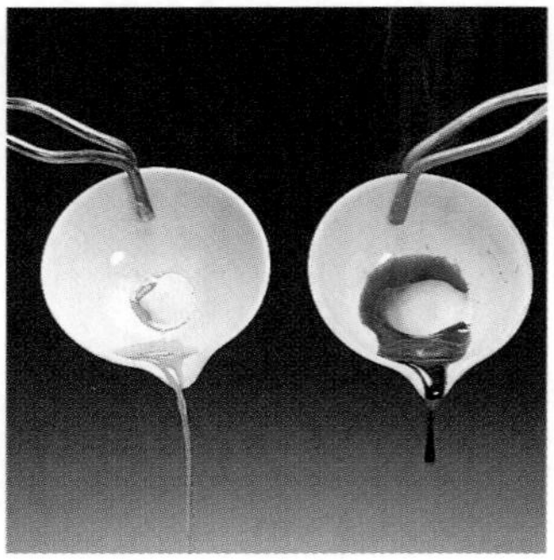

At its melting point of 119 °C, sulfur is a straw-colored, mobile liquid (left). At temperatures above 160 °C, the liquid becomes dark in color and very viscous (right).

Sulfur

Sulfur and oxygen are similar in several ways, as we expect from the electron configurations of their atoms. Both form ionic compounds with active metals, and both form some similar covalent compounds, such as H_2S and H_2O, CS_2 and CO_2, CH_3CH_2SH and CH_3CH_2OH. However, oxygen and sulfur compounds also differ in important ways. Let's consider two cases: (1) H_2O has an exceptionally high boiling point (100 °C) considering its low molecular weight (18 u). The boiling point of H_2S, on the other hand, is more normal (−61 °C) for its molecular weight (34 u). We can attribute this difference to the importance of hydrogen bonding (Section 11.6) as an intermolecular force in H_2O. Hydrogen bonding is not important when hydrogen atoms are bonded to S atoms. (2) S atoms are able to employ expanded octets (Section 9.9) in forming bonds with other atoms. Examples of expanded-octet structures are found in SF_4 and SF_6. Oxygen atoms cannot have expanded octets in Lewis structures.

Elemental sulfur exists as different molecular species ranging from monatomic S through S_2, S_6, and S_8, to polymeric forms S_n. Consequently, sulfur exists in more allotropic forms than any other element. *Rhombic* sulfur has sixteen S_8 rings in a unit cell. At 95.5 °C it converts to *monoclinic* sulfur, which is thought to have six S_8 rings in its unit cell. Monoclinic sulfur melts at 119 °C, forming a liquid that is also made up of S_8 molecules. This liquid is yellow, transparent, and mobile, but when the liquid is heated to 160 °C, a remarkable transformation occurs. The S_8 rings open up, and the S atoms join together into long spiral-chain molecules, resulting in a dark, thick, viscous liquid. At still higher temperatures, however, the chains break up and the viscosity decreases. Liquid sulfur boils at 445 °C, producing a vapor that consists of molecules ranging from S_2 to S_8, but predominantly S_8. At higher temperatures, S_2 predominates.

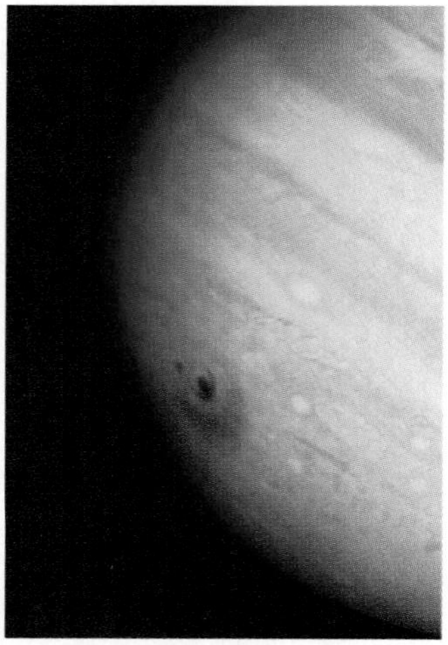

The 1994 impact of comet Shoemaker-Levy 9 with Jupiter (left center). The impact exhumed gases, including $S_2(g)$, from the depths of Jupiter's atmosphere. The molecules were identified by their characteristic spectra.

Sulfur is abundant in Earth's crust. It occurs as elemental sulfur, as mineral sulfides and sulfates, as $H_2S(g)$ in natural gas, and as organic sulfur compounds in oil and coal. Extensive deposits of elemental sulfur are found in Texas and Louisiana, some in offshore sites. This sulfur is mined in an unusual way known as the Frasch process (Figure 21.10). A mixture of superheated water and steam (at about 160 °C and 16 atm) is forced down the outermost of three concentric pipes into an underground bed of sulfur-containing rock. The sulfur melts and forms a liquid pool. Compressed air (at 20 to 25 atm) is pumped down the innermost pipe and forces liquid sulfur up the middle pipe. (The superheated water and steam descending in the outermost pipe

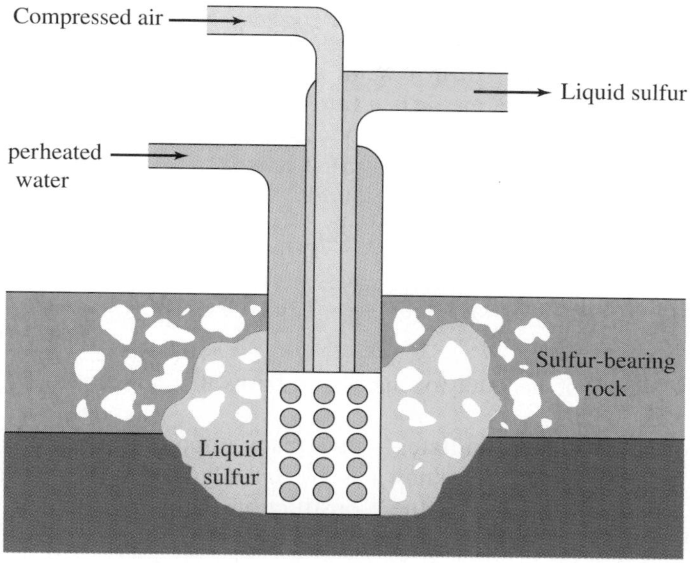

Figure 21.10 The Frasch process for mining sulfur. Sulfur is melted by superheated water forced down the outermost of three concentric pipes. Compressed air is blown down the innermost of the pipes, and the liquid sulfur is forced up the middle pipe.

help to keep the ascending liquid sulfur in the middle pipe from solidifying.) The Frasch process is no longer the most important way of producing sulfur in the United States. Because of the need to control emissions of oxides of sulfur, much of the sulfur now in use is recovered as a byproduct of industrial operations (page 545).

A small amount of elemental sulfur is used directly in vulcanizing rubber and as a pesticide (for example, in dusting grape vines). Most, however, is burned to $SO_2(g)$, and the greatest proportion of this is converted, by way of $SO_3(g)$, to H_2SO_4.

Sulfur dioxide and sulfur trioxide and their roles in air pollution were discussed in Section 14.6.

$SO_2(g)$ reacts with water to produce a solution of sulfurous acid, $H_2SO_3(aq)$, but pure H_2SO_3 is unstable; it has never been isolated. Sulfites—salts of sulfurous acid—are good reducing agents in reactions in which they are oxidized to sulfates. For example,

$$Cl_2(g) + SO_3^{2-}(aq) + H_2O \longrightarrow 2\ Cl^-(aq) + SO_4^{2-}(aq) + 2\ H^+(aq)$$

However, sulfites can also act as oxidizing agents, as in the reaction

$$2\ H_2S(g) + 2\ H^+(aq) + SO_3^{2-}(aq) \longrightarrow 3\ H_2O + 3\ S(s)$$

Sulfur dioxide and sulfites are widely used in the food industry as decolorizing agents and preservatives. The main use of sulfites, however, is in the pulp and paper industry. When wood is digested in an aqueous sulfite solution, chemical reactions occur that separate the cellulosic constituents of wood from unwanted materials such as lignin. The wood pulp thus formed is used to make paper and rayon.

Concentrated sulfuric acid is added to sucrose (cane sugar) on the left. The carbon produced in the reaction is seen on the right.

Sulfuric acid, H_2SO_4, is a typical strong acid. Aqueous solutions neutralize bases, react with metals to form $H_2(g)$, and react with carbonates to liberate $CO_2(g)$. Concentrated sulfuric acid, on the other hand, has some rather distinctive properties. For one, it is a fairly good oxidizing agent: It will react with copper metal, for example, producing $Cu^{2+}(aq)$ and $SO_2(g)$. It also is a strong dehydrating agent, so strong that it will extract H and O atoms in the form of H_2O from certain compounds. Concentrated sulfuric acid extracts water from a carbohydrate, leaving a carbon residue. The reaction with sucrose is

$$\underset{\text{Sucrose}}{C_{12}H_{22}O_{11}(s)} \xrightarrow{H_2SO_4(\text{concd})} 12\,C(s) + 11\,H_2O(l)$$

Reactions of this type account for the severe burns that are caused by the concentrated acid.

Salts of sulfuric acid—sulfates—have many important uses. The calcium sulfate hydrate $CaSO_4 \cdot 2H_2O$ is known as *gypsum*. Gypsum is the primary material used for wall board in the construction industry. As the hydrate $CaSO_4 \cdot \frac{1}{2}H_2O$, calcium sulfate is known as plaster of Paris, a substance used to make casts and molds. Aluminum sulfate is used in water treatment. $Al^{3+}(aq)$ from the aluminum sulfate is precipitated as gelatinous $Al(OH)_3(s)$, which clarifies the water by removing suspended matter. Copper(II) sulfate is used in electroplating, and it is also an effective fungicide and algaecide.

Thiosulfates are related to sulfates in that they have an S atom replacing one of the O atoms in SO_4^{2-} (see Figure 21.11). The average oxidation state of sulfur in $S_2O_3^{2-}$ is +2; but the two sulfur atoms are not equivalent. The central S atom is in the oxidation state +6, and the terminal S atom is in the oxidation state −2.

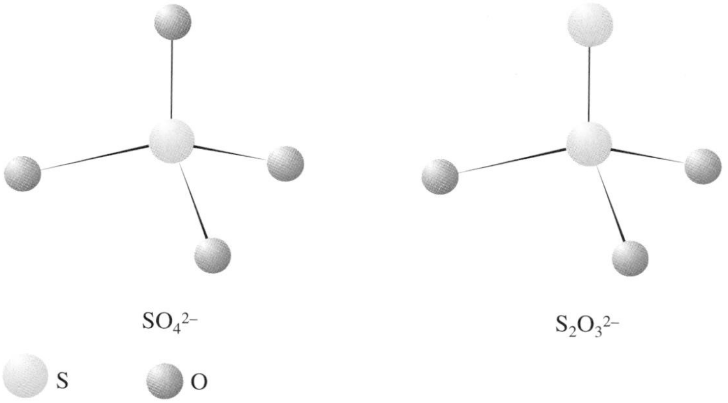

Figure 21.11 Structures of the sulfate and thiosulfate ions. (a) Sulfate ion. (b) Thiosulfate ion, formed by replacing one of the terminal O atoms of the sulfate ion with an S atom.

Thiosulfates can be prepared by boiling elemental sulfur in an alkaline solution of sodium sulfite. The sulfur is oxidized and the sulfite ion is reduced, both to thiosulfate ion.

$$SO_3^{2-}(aq) + S(s) \longrightarrow S_2O_3^{2-}(aq)$$

We have previously considered some important uses of sodium thiosulfate: as a fixer in photography (pages 714–715) and as a reducing agent in analytical chemistry (page 464).

Selenium, Tellurium, and Polonium

Selenium (a nonmetal) and tellurium (a metalloid) are similar to sulfur but more metallic. For example, sulfur is an electrical insulator, whereas selenium and tellurium are semiconductors. Polonium is an electrical conductor and has other metallic properties, such as the ability to form cations. Because it is highly radioactive, however, polonium has few uses. Polonium-218—an alpha particle emitter—has been implicated in the environmental problems associated with radon gas (page 550).

Se and Te occur as selenides and tellurides mixed in sulfide ores—for example, as Cu_2Se and Cu_2Te. The free elements are obtained mainly from the anode mud in the electrolytic refining of copper (pages 802–803). Both Se and Te find some use in alloys, and their compounds are used as additives to control the color of glass. Selenium is an important element in modern technology, primarily because it is a **photoconductor**; its electrical conductivity increases in the presence of light. Photoconductivity is the property that makes selenium useful in photocells for cameras and as the light-sensitive element in photocopying machines. Selenium is also used to make *rectifiers*, devices that convert alternating to direct electric current.

21.5 Group 7A: F, Cl, Br, I, and At

Atoms of the Group 7A elements—the halogens—have the valence-shell electron configuration ns^2np^5. They have high ionization energies, large negative electron affinities, and high to moderately high electronegativities. This means that they gain electrons readily and lose electrons only with difficulty—our expectation of a family of nonmetals.

Elemental halogens exist as X_2 molecules, where X is the generic symbol for a halogen atom. Much of the reaction chemistry of the halogens involves oxidation–reduction reactions in aqueous solution. Standard electrode potentials are the best guide to their reactivity.

$$F_2(g) + 2\,e^- \longrightarrow 2\,F^-(aq) \qquad E^\circ_{red} = 2.866\text{ V}$$

$$Cl_2(g) + 2\,e^- \longrightarrow 2\,Cl^-(aq) \qquad E^\circ_{red} = 1.358\text{ V}$$

$$Br_2(l) + 2\,e^- \longrightarrow 2\,Br^-(aq) \qquad E^\circ_{red} = 1.065\text{ V}$$

$$I_2(s) + 2\,e^- \longrightarrow 2\,I^-(aq) \qquad E^\circ_{red} = 0.535\text{ V}$$

By this measure, F_2 is clearly the strongest oxidizing agent of the halogens. In fact, of all the elements, F_2 has the greatest tendency to gain electrons; it is the most easily reduced and therefore the best oxidizing agent. Fluorine, chlorine, and bromine occur naturally only as the halide ions, X^-, in combination with appropriate cations. On the other hand, naturally occurring iodine, in addition to occurring as I^-, is often found in positive oxidation states, as in $NaIO_3(s)$. Although the tendency of $I_2(s)$ to be oxidized to IO_3^-

$$I_2(s) + 6\,H_2O \longrightarrow 2\,IO_3^-(aq) + 12\,H^+(aq) + 10\,e^- \qquad E^\circ_{ox} = -1.20\text{ V}$$

is not great, there are a number of oxidizing agents, including $O_2(g)$, strong enough to bring this about.

Astatine, at the bottom of Group 7A, is a radioactive element. All of its isotopes are short-lived, and only about 0.05 mg has ever been prepared at one time. These samples rapidly undergo radioactive decay, so that few of the properties of astatine have been determined with certainty. We will not consider the element further here.

Fluorine

The existence of fluorine has been known since early in the nineteenth century, but Henri Moissan was the first to succeed in preparing $F_2(g)$, in 1886. Moissan's method involves the electrolysis of HF dissolved in molten KHF_2.

$$2\,H^+ + 2\,F^- \xrightarrow{\text{electrolysis}} H_2(g) + F_2(g)$$

Fluorine is used in the production of $UF_6(g)$ which can then be separated into $^{235}U(g)$ and $^{238}U(g)$ by gaseous diffusion (page 145). Uranium enriched in the uranium-235 isotope is used in nuclear fuels and weapons.

Fluorine is also used to make $SF_6(g)$, which forms when sulfur is burned in fluorine gas. $SF_6(g)$ is a nonflammable, unreactive gas of low toxicity that is useful as an insulating gas in high-voltage electrical equipment. Interhalogen compounds of fluorine, such as ClF_3 and BrF_3, are used as substitutes for elemental fluorine in many reactions because they are easier to handle than the highly reactive F_2. An **interhalogen compound** is a compound of two different halogen elements (even three different halogens, in a few cases).

Like oxygen, fluorine forms compounds with all the other elements except the light noble gases. Hydrogen fluoride is the starting reagent for preparing most other fluorine compounds. An unusual property of HF is its ability to etch glass.

$$SiO_2(s) + 6\,HF(aq) \longrightarrow 2\,H_2O + 2\,H^+(aq) + SiF_6^{2-}(aq)$$

Ultimately, the glass is eroded away, and for this reason HF(aq) must be stored in special containers made of wax or Teflon.

Chlorine

Chlorine gas was first produced by Carl Wilhelm Scheele in 1774, using the reaction of $MnO_2(s)$ and HCl(aq).

$$MnO_2(s) + 4\,HCl(aq) \xrightarrow{\Delta} MnCl_2(aq) + 2\,H_2O + Cl_2(g)$$

This reaction is occasionally used in the laboratory, but the only significant commercial methods for preparing $Cl_2(g)$ today involve electrolysis. Some chlorine is made by the electrolysis of molten NaCl (page 795) or $MgCl_2$ (page 293), but most comes from the electrolysis of a concentrated aqueous solution of NaCl. Because the products of the electrolysis of NaCl(aq) are $H_2(g)$, $Cl_2(g)$ and NaOH(aq)—an alkali—the industry based on this electrolysis is called the *chlor–alkali* industry (page 801).

Elemental chlorine, with an annual United States production of about 12 million tons, generally ranks among the top ten chemicals produced. It has three main commercial uses: About 70% is used to produce chlorinated organic compounds, about 20% is used as a bleach in the paper and textile industries and for the treatment of municipal water, sewage, and swimming pools, and about 10% is used to produce chlorine-containing inorganic compounds.

Chlorine forms stable compounds with most of the elements except the noble gases. Especially important are chlorine-containing carbon compounds—that is, organic chlorine compounds. The reaction of chlorine with hydrocarbons produces chlorinated hydrocarbons. For example, in the direct reaction of $Cl_2(g)$ with $CH_4(g)$, chlorine atoms substitute for hydrogen atoms, and products ranging from CH_3Cl to CCl_4 are formed. HCl is also a product of the reactions.

$$CH_4(g) + Cl_2(g) \xrightarrow{\Delta} CH_3Cl(g) + HCl(g)$$

$$CH_3Cl(g) + Cl_2(g) \xrightarrow{\Delta} CH_2Cl_2(g) + HCl(g)$$

$$CH_2Cl_2(g) + Cl_2(g) \xrightarrow{\Delta} CHCl_3(g) + HCl(g)$$

$$CHCl_3(g) + Cl_2(g) \xrightarrow{\Delta} CCl_4(g) + HCl(g)$$

Similarly, the reaction of ethane, C_2H_6, with chlorine yields products with one to six Cl atoms substituted for H atoms. Several chlorinated hydrocarbons are used as solvents for nonpolar or slightly polar solutes.

The chlorofluorocarbons (CFCs) have F and Cl atoms bonded to carbon. The most widely used have been $CFCl_3$ and CF_2Cl_2. Because these substances are volatile liquids or easily condensable gases, they work well as refrigerants and as blowing agents in making foam plastics. We discussed the role of CFCs in the destruction of stratospheric ozone in Chapter 14 (page 536).

Bromine and Iodine

Commercially, bromine is extracted from subterranean brines, where it occurs primarily as Br^-. Brines from Arkansas contain 3800–5000 parts per million (ppm) of bromine, and waters in the Dead Sea, about 4500–5000 ppm. The water solution containing Br^- is concentrated by evaporation, and then treated with $Cl_2(g)$, which oxidizes Br^- to Br_2.

$$Cl_2(g) + 2\,Br^-(aq) \longrightarrow 2\,Cl^-(aq) + Br_2(l) \qquad E^\circ_{cell} = +0.293\ V$$

Brine is an aqueous solution of sodium chloride, usually containing other dissolved salts. Natural sources of brines are the oceans, desert lakes, and underground wells.

Iodine was once obtained in small quantities from dried seaweed. (Certain marine plants concentrate I^- selectively in the presence of Cl^- and Br^-.) Nowadays, iodine is also obtained from brines. Wells in Michigan, Oklahoma, and Japan yield brines with 100 ppm I^-. Elemental iodine is obtained from these brines by a process similar to that used for bromine.

Organic bromine compounds are used as pharmaceuticals, dyes, fumigants, and pesticides. Bromine compounds are also used in fire extinguishers and fire retardants. An important inorganic bromine compound is AgBr, the primary light-sensitive agent used in photographic film.

Iodine is of less commercial importance than chlorine and bromine, although iodine and its compounds do have applications as catalysts, in medicine, and in photographic emulsions (AgI). Iodine and its compounds also have important uses in analytical chemistry (page 464).

Iodide ion is used in the thyroid gland to synthesize thyroxin, a substance that helps to regulate metabolism. An iodide deficiency leads to an enlargement of the thyroid gland, a condition called goiter. Goiter is prevented by the use of iodized salt, a product that is mainly NaCl but includes a small quantity of added NaI or KI.

Hydrogen Halides

In aqueous solution, the hydrogen halides are called the hydrohalic acids. Except for HF, they are strong acids. Strong hydrogen bonding in HF accounts for the fact that HF(l) has the highest boiling point among the liquid hydrogen halides (Section 11.6), even though it has the lowest molecular weight. Hydrogen bonding may also play a role in making HF a weak acid rather than a strong acid.

The hydrogen halides can be prepared by the direct combination of the elements, but it generally is not convenient to do so.

$$H_2(g) + X_2(g) \longrightarrow 2\,HX(g)$$

The H_2–F_2 reaction occurs with explosive violence. Mixtures of $H_2(g)$ and $Cl_2(g)$ are stable in the dark, but react explosively in the presence of light. $Br_2(g)$ and $I_2(g)$ react more slowly with $H_2(g)$; a catalyst is required to get a reasonable reaction rate.

Another method that can be used to prepare the hydrogen halides is to heat a halide salt (for example, CaF_2 or NaCl) with a concentrated, *nonvolatile* acid (H_2SO_4 or H_3PO_4).

$$CaF_2(s) + H_2SO_4(\text{concd aq}) \longrightarrow CaSO_4(s) + 2\,HF(g)$$

$$NaCl(s) + H_2SO_4(\text{concd aq}) \longrightarrow NaHSO_4(s) + HCl(g)$$

To prepare HBr(g) or HI(g), H_2SO_4 can't be used because it's a sufficiently strong oxidizing agent to oxidize Br^- to Br_2 and I^- to I_2. To get around this difficulty we can use a *nonoxidizing*, nonvolatile acid, such as phosphoric acid. With NaBr, the reaction is

$$NaBr(s) + H_3PO_4(\text{concd aq}) \longrightarrow NaH_2PO_4(s) + HBr(g)$$

CONCEPTUAL EXAMPLE 21.1

We have just seen that HBr cannot be prepared by heating a bromide salt with concentrated sulfuric acid. Bromine gas is obtained instead.

a. Show that the reaction

$$2\,NaBr(s) + 2\,H_2SO_4(l) \longrightarrow Na_2SO_4(s) + 2\,H_2O(g) + SO_2(g) + Br_2(g)$$

should not be spontaneous at 25 °C.

b. Why does this reaction occur at higher temperatures?

Solution

a. The most direct approach is to evaluate the standard free energy change for the reaction. We can do this with standard free energies of formation from Appendix D and the expression

$$\Delta G^\circ = [\sum \nu_p \Delta G_f^\circ(\text{products})] - [\sum \nu_r \Delta G_f^\circ(\text{reactants})]$$

$$2\,NaBr(s) + 2\,H_2SO_4(l) \longrightarrow Na_2SO_4(s) + 2\,H_2O(g) + SO_2(g) + Br_2(g)$$

ΔG_f°, kJ/mol: −349.0 −690.1 −1270 −228.6 −300.2 3.14

$$\Delta G^\circ = [-1270 + 2(-228.6) + (-300.2) + 3.14] - [2(-349.0) + 2(-690.1)]$$

$$= -2024 - (-2078) = 54\text{ kJ}$$

The fact that $\Delta G^\circ = +54$ kJ tells us that the reaction will not occur spontaneously with reactants and products in their standard states at 25 °C. Because this value is large and positive, we wouldn't expect the reaction to be spontaneous at 25 °C even for most nonstandard conditions.

b. To show that the reaction is spontaneous at higher temperatures, the simplest approach is to use the Gibbs equation: $\Delta G = \Delta H - T\Delta S$. We don't need to do calculations, however. Just by inspecting the balanced equation we can see that ΔS is large and *positive* for this reaction: Two moles each of a solid and a liquid yield one mole of solid and *four* moles of gas. A reaction with a large, positive ΔS is favored at higher temperatures because this ensures a large, negative value of $-T\Delta S$, a value that at some temperature is certain to be large enough to overcome a positive value of ΔH. This situation, in which $\Delta H > 0$ and $\Delta S > 0$ corresponds to case 3 in Table 19.1. Le Châtelier's principle also indicates that the forward reaction should be even more favorable if the product gases are allowed to escape from the reaction mixture.

Conceptual Exercise 21.1

Use standard electrode potentials to determine whether this reaction should occur spontaneously as written.

$$2\,Br^-(aq) + 4\,H^+(aq) + SO_4^{2-}(aq) \longrightarrow 2\,H_2O(l) + SO_2(g) + Br_2(l)$$

Is your result consistent with the results obtained in parts (a) and (b) of Conceptual Example 21.1? Explain.

Oxoacids and Oxoanions of the Halogens

In its compounds, fluorine is always in the −1 oxidation state. The other halogens, however, can have positive oxidation states: +1, +3, +5, and +7. These oxidation states are found in the oxoacids of Cl, Br, and I listed in Table 21.4.

TABLE 21.4 **Some Oxoacids of the Halogens**[a]

Oxidation State of Halogen	Chlorine	Bromine	Iodine
+1	HOCl	HOBr	HOI
+3	HOClO	—	—
+5	$HOClO_2$	$HOBrO_2$	$HOIO_2$
+7	$HOClO_3$	$HOBrO_3$	$HOIO_3$

[a] In all these acids the H atom is bonded to an O atom, not to the central halogen atom. Nevertheless, the formulas of oxoacids are often written in order of increasing electronegativity—that is, in the form: HClO (for HOCl), $HClO_2$ (for HOClO), etc.

Chlorine forms a complete set of oxoacids, but bromine and iodine do not.

Hypochlorous acid, HOCl, is formed to some extent by the disproportionation of Cl_2 in water; chlorine with oxidation state 0 in Cl_2 goes to O.S. +1 in HOCl and O.S. −1 in Cl^-.

$$Cl_2(g) + H_2O \rightleftharpoons HOCl(aq) + H^+(aq) + Cl^-(aq) \qquad E^\circ_{cell} = -0.27\text{ V}$$

HOCl(aq) is an effective germicide, often used to disinfect water. HOCl exists only in solution; it cannot be isolated in the pure state.

If we dissolve Cl_2 in a basic solution such as NaOH(aq), equilibrium in the disproportionation reaction is displaced far to the right as HOCl is neutralized. An aqueous solution of a hypochlorite salt is formed.

$$Cl_2(g) + 2\,OH^-(aq) \longrightarrow OCl^-(aq) + Cl^-(aq) + H_2O \qquad E^\circ_{cell} = +0.84\text{ V}$$

Common household bleaches are aqueous alkaline solutions of NaOCl. Some drain cleaners are solutions of NaOCl and NaOH, as is a common solution used to chlorinate home swimming pools.

Chlorine dioxide, ClO_2, is a bleach for paper, fibers, and flour. When reduced by peroxide ion in aqueous solution, $ClO_2(g)$ is converted to chlorite ion.

$$2\,ClO_2(g) + O_2^{2-}(aq) \longrightarrow 2\,ClO_2^-(aq) + O_2(g)$$

Sodium chlorite is used as a bleaching agent for textiles.

When *hot* alkaline solutions are treated with $Cl_2(g)$, chlorate salts are formed.

$$3\,Cl_2(g) + 6\,OH^-(aq) \longrightarrow 5\,Cl^-(aq) + ClO_3^-(aq) + 3\,H_2O$$

Chlorates are good oxidizing agents. Solid chlorates are used in matches and fireworks. When heated, chlorates decompose to produce oxygen gas (Section 14.4), which supports the combustion of the other ingredients.

Electrolysis of chlorate salt solutions yields perchlorates; oxidation of ClO_3^- to ClO_4 occurs at a Pt anode. The half-reaction is

$$ClO_3^-(aq) + H_2O \longrightarrow ClO_4^-(aq) + 2\,H^+(aq) + 2\,e^- \qquad E^\circ_{ox} = -1.19\text{ V}$$

An important laboratory use of perchlorate salts is in solution studies; ClO_4^- has the least tendency of any anion to act as a ligand in complex ion formation. (Recall that a ligand is a group attached to the central ion of a complex ion.) In the presence of readily oxidizable substances, such as most organic compounds, perchlorates often react explosively. Mixtures of ammonium perchlorate and aluminum powder are used as solid propellants for rockets. Ammonium perchlorate is dangerous to handle: a reducing agent, NH_4^+, and an oxidizing agent, ClO_4^-, occur in the same compound.

A view of the widespread destruction produced by an explosion of ammonium perchlorate at a rocket fuel plant in Henderson, Nevada, in 1988.

21.6 Group 8A: He, Ne, Ar, Kr, Xe, and Rn

We conclude our survey of *p*-block elements with a final brief look at the unusual elements of Group 8A—the noble gases. For many years after their discovery, these elements were thought to be *inert*. This apparent inertness underlay the Lewis theory of bonding. Although we still use the Lewis theory, we no longer regard the noble gases as inert. Here is how the change in thinking came about.

In 1962 Neil Bartlett and D. H. Lohman identified a compound of O_2 and PtF_6 in a 1:1 mole ratio. Properties of the compound indicate that it is $(O_2)^+(PtF_6)^-$. It takes 1177 kJ/mol of energy to extract an electron from O_2 to form the ion O_2^+. Noting that this is almost identical to the first ionization energy of Xe (1170 kJ/mol) and that the size of the Xe atom is roughly the same as that of the O_2 molecule, Bartlett was able to substitute Xe for O_2 and obtain a yellow crystalline solid, apparently $XePtF_6$. [We now know the formula to be $Xe(PtF_6)_n$, where n is between 1 and 2.] Before long, chemists around the world had synthesized several additional noble gas compounds. For example, depending on the reaction conditions, XeF_2, XeF_4, and XeF_6 can be prepared by the direct reaction of Xe(g) and F_2(g). Some compounds of Kr and Rn exist, but no compounds of He, Ne, or Ar are known.

We can make qualitative predictions about the molecular structures of xenon compounds with the VSEPR theory. The number of valence-shell electrons that must be accommodated by the central Xe atom is 10 in XeF_2, and 12 in XeF_4. The VSEPR designations of these two molecules are AX_2E_3 and AX_4E_2, respectively. Their structures are shown in Figure 21.12.

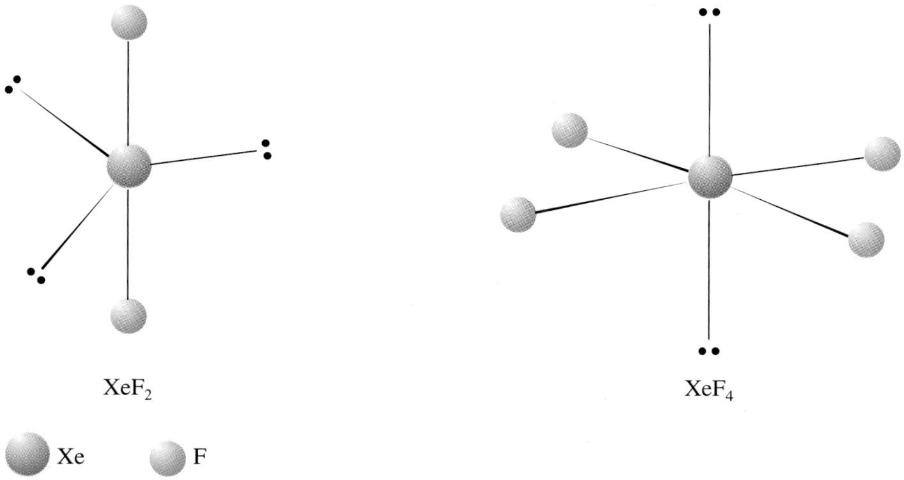

Figure 21.12 Geometric structures of XeF_2 and XeF_4. Lone-pair electrons are shown only for the central Xe atom.

It is more difficult to describe bonding in noble gas compounds in a way that is consistent with all the experimental facts, and we will not attempt to do this. In fact, the discovery of noble gas compounds has been a major stimulus to the development of modern bonding theories.

SUMMARY

Boron atoms, having only three valence-shell electrons, tend to form electron-deficient compounds, and this leads to some unusual bonding patterns. The chemistry of boron focuses on borax, boric oxide, boric acid, borates, and boron hydrides.

Aluminum, an active metal, is protected against corrosion by a film of $Al_2O_3(s)$. Both $Al_2O_3(s)$ and Al react with acids and strong bases. The amphoterism of $Al_2O_3(s)$ and its electrolysis in molten cryolite are the bases for the production of aluminum. The electron deficiency of aluminum chloride is a useful property in organic syntheses.

Carbon is the key element of organic chemistry, but the free element also has uses. Diamond is prized for hardness and thermal conductivity, graphite for electrical conductivity and refractory properties. Coke, carbon black, and activated carbon are amorphous forms of carbon.

Silicon forms some noncarbon "organic" compounds, but these do not occur naturally. Silicon is the key element of the mineral world, as silica, SiO_2, and as various minerals based on the silicate anion, SiO_4^{4-}.

Tin and lead are metals that are slightly more active than hydrogen. They can exist in both the +2 and +4 oxidation states, with Sn(II) being a good reducing agent and Pb(IV) a good oxidizing agent.

Much of the chemistry of nitrogen was considered in Chapter 14. Inorganic nitrogen compounds presented here are hydrazine, azides, and amides. The structure of phosphorus is based on the pyramidal molecule, P_4, both in the white and red modifications. The structures of the oxides P_4O_6 and P_4O_{10} are related to that of the P_4 molecule. The principal compounds of phosphorus are the phosphates. Two other important compounds are PH_3 and PCl_3. The chemistry of arsenic is mostly that of As_4O_6.

Oxygen chemistry was considered in Chapter 14 and elsewhere in the text. Sulfur differs from oxygen in important ways, such as in its variety of allotropic forms. Its important compounds are the oxides, oxoacids, sulfites, sulfates, and thiosulfates. Many of the reactions of these compounds are oxidation–reduction reactions.

The halogens—Group 7A—are nonmetals, with fluorine being the most nonmetallic of all elements. Fluorine and chlorine are prepared by electrolysis, and bromine and iodine by displacement reactions. Two different halogens can react to form an interhalogen compound. Halogen atoms can substitute for H atoms in hydrocarbons and other organic compounds. Hydrogen halides form by the direct combination of the elements or by the reaction of a halide salt with a nonvolatile acid. An important class of halogen compounds are the oxoacids and their salts. The chemical reactions of these compounds are mostly oxidation–reduction reactions.

Interest in the noble gases centers on their physical properties and inertness (Chapter 14). On the other hand, the ability of the heavier noble gases to form some chemical compounds provides important insights into bonding theory.

KEY TERMS

acid anhydride (21.1)
adduct (21.1)
carbocation (21.1)
inert pair (21.1)
interhalogen compound (21.5)
photoconductor (21.4)
pseudohalogen (21.2)

REVIEW QUESTIONS

1. What is the primary feature that characterizes a *p*-block element?
2. Which two *p*-block elements have some similar properties governed by the *diagonal relationship*?
3. Three of the *p*-block elements are the three most abundant elements in Earth's crust. Which are they?
4. Which is the most active metal and which is the most active nonmetal among the *p*-block elements?
5. What is a *bridge bond*? Describe the bridge bonds in B_2H_6.
6. Describe the bridge bonds in Al_2Cl_6. How do they differ from those in B_2H_6?

7. What is *anodized* aluminum? What purposes are served in anodizing aluminum?

8. What is the purpose served by each of these two common ingredients of common drain cleaners: pellets of NaOH(s) and granules of Al(s)?

9. A given copper wire is a better electrical conductor than an aluminum wire of the same diameter and length. Why is it, then, that most high-voltage transmission lines in the United States are made of aluminum rather than copper?

10. In 1866 aluminum metal cost $250 to $750 per kilogram. In 1924 it cost about $0.50/kg. Account for this dramatic difference in price.

11. What is *carbon black*? How is it prepared, and what are some of its uses?

12. What is *activated carbon*? How does it differ from carbon black, and what are some of its uses?

13. What is the basic composition of soda–lime glass, and how is this composition modified in Pyrex glass?

14. What is the basic structural unit of the silicate anion, and what are some of the ways in which these units are combined in silicate minerals?

15. Explain why H_2S is a gas at room temperature whereas H_2O, with only about one-half the molecular weight of H_2S, is a liquid.

16. Explain why copper does not react with HCl(aq) but does react with H_2SO_4(concd aq).

17. What is an *adduct*? Name two elements closely associated with adduct formation.

18. What is the *inert pair* effect? Name some elements in which this effect is seen.

19. What type of compound is suggested by the term interhalogen compound? Give an example.

20. What types of compounds are suggested by the terms chlorinated hydrocarbons and chlorofluorocarbons? Give two examples of each.

21. Although we have not specifically considered these compounds in the text, name them by using information from this chapter or principles discussed elsewhere in the text.
    a. AgN_3 **b.** KSCN
    c. At_2O **d.** H_2TeO_4

22. Although we have not specifically considered these compounds in the text, write plausible formulas for them by using information from this chapter or principles discussed elsewhere in the text.
    a. selenic acid **b.** hydrotelluric acid
    c. lead azide **d.** silver astatide

23. What are the formulas of the following compounds featured in this chapter: bauxite, boric oxide, corundum, cyanogen, hydrazine, silica?

24. Name key elements found in the following minerals featured in this chapter: apatite, borax, corundum, quartz.

PROBLEMS

Boron

25. Elemental boron can be prepared by heating magnesium metal and boric oxide. Write a plausible equation for this reaction.

26. Boron trifluoride reacts with lithium fluoride to form an ionic compound called lithium tetrafluoborate. Write a plausible equation for this reaction. (*Hint*: The tetrafluoborate ion is related to the borate ion but contains no O atoms.)

27. Although they have similar formulas, the compound BF_3 exists and the compound BH_3 does not. Offer a plausible explanation of this fact.

28. Although they have similar formulas, phosphoric acid, H_3PO_4, is a weak *triprotic* acid, whereas boric acid, H_3BO_3, is a weak, monoprotic acid. Explain this difference.

Aluminum

29. Write an equation to represent the production of aluminum sulfate from aluminum oxide and concentrated sulfuric acid described on page 820.

30. A typical baking powder contains baking soda ($NaHCO_3$) and alum [$NaAl(SO_4)_2$] as its active ingredients. During baking, the baking powder undergoes a reaction that yields $CO_2(g)$ and $Al(OH)_3(s)$ as two of its products. Write a plausible equation for this reaction.

31. What are the two key features that make possible the low-cost production of aluminum from bauxite?

32. What is the principle that underlies the fact that thallium forms both 1+ and 3+ ions, whereas aluminum forms only a 3+ ion?

33. Estimate the enthalpy change for the thermite reaction given on page 816. Why is this not an exact value?

34. Verify the value of E°_{cell} at 298 K for the reaction of aluminum with water in basic solution given on page 817.

35. Why cannot aluminum cookware be used in cooking strongly acidic foods? Should one scour an aluminum pot using a typical oven cleaner containing NaOH? Explain.

36. In the use of aluminum sulfate in water treatment (see page 820), the water to be treated is usually kept between pH 5 and pH 8. Why do you suppose this is the case?

Carbon

37. Sodium forms a carbide similar to calcium carbide—that is, containing the C_2^{2-} ion. Write the formula of sodium carbide and an equation for its reaction with water.

38. Unlike calcium carbide, which yields acetylene, $Al_4C_3(s)$ yields methane when it reacts with water. Write an equation for this reaction.

39. Write plausible Lewis structures for CS_2, CCl_4, and C_2N_2.

40. Write plausible Lewis structures for COS, CN^-, and OCN^-.

Silicon

41. Show the geometric structure and write a Lewis structure for the anion $Si_2O_7^{6-}$.

42. Write a formula and Lewis structure, and show the geometric structure, for an anion that consists of three SiO_4 units joined through O atoms.

43. Show that the empirical formula given on page 826 is consistent with the expected oxidation states of the elements in chrysotile asbestos.

44. Show that the empirical formula given on page 826 is consistent with the expected oxidation states of the elements in muskovite mica.

45. In many respects the naming of organosilicon compounds follows the same rules as organic carbon compounds. Write formulas for (**a**) tetramethylsilane, (**b**) dichlorodimethylsilane, and (**c**) triethylsilane.

46. What is the formula of tetrasilane? Write an equation for its combustion in air.

Tin and Lead

47. Use data from Appendix D to determine if $PbO_2(s)$ is a strong enough oxidizing agent to carry out the following oxidations in acidic solution.
a. $ClO_3^-(aq)$ to $ClO_4^-(aq)$
b. $H_2O(l)$ to $H_2O_2(aq)$
c. $Ag^+(aq)$ to $Ag^{2+}(aq)$
d. $Sn^{2+}(aq)$ to $Sn^{4+}(aq)$

48. Use data from Appendix D to determine if $Sn^{2+}(aq)$ is a strong enough reducing agent to carry out the following reductions.
a. $Cu^{2+}(aq)$ to $Cu^+(aq)$
b. $V^{3+}(aq)$ to $V^{2+}(aq)$
c. $Ag^+(aq)$ to $Ag(s)$
d. $Cr^{3+}(aq)$ to $Cr^{2+}(aq)$

49. Write plausible equations for the production of tin compounds described in a chemical dictionary in these ways.
a. Stannous chloride [tin(II) chloride]: "By dissolving tin in hydrochloric acid."
b. Stannic chloride [tin(IV) chloride]: "Treatment of stannous chloride [tin(II) chloride] with chlorine."
c. Stannic oxide [tin(IV) oxide]: "Precipitated from stannic chloride [tin(IV) chloride] solution by ammonium hydroxide [$NH_3(aq)$]."

50. Write plausible equations for the production of lead compounds described in a chemical dictionary in these ways.
a. Lead acetate: "By the action of acetic acid on litharge (PbO)."
b. Red lead oxide, Pb_3O_4: "By gently heating litharge (PbO) in a furnace in a current of air."

c. Lead dioxide: "By the action of an alkaline solution of calcium hypochlorite on lead(II) hydroxide."

Nitrogen

51. Write an equation to represent the complete combustion in oxygen of the rocket fuel dimethylhydrazine, $(CH_3)_2NNH_2$.

52. Write an equation to represent the ionization of hydrazoic acid in aqueous solution.

53. Complete and balance the following equations by proposing plausible reaction products. Some of these reactions are described in this chapter and others are based on ideas presented earlier in the text.
a. $NH_2NH_2(aq) + HCl(aq) \longrightarrow$
b. $Cu(s) + H^+(aq) + NO_3^-(aq) \longrightarrow$
c. $NO(g) + O_2(g) \longrightarrow$

54. Complete and balance the following equations by proposing plausible reaction products. Some of these reactions are described in this chapter and others are based on ideas presented earlier in the text.

a. $NO_2(g) + H_2O(l) \longrightarrow$

b. $NH_3(g) + O_2(g) \longrightarrow$

c. $NH_4NO_3(l) \xrightarrow{200\ °C}$

55. In acidic solution hydrazine reduces Fe^{3+} to Fe^{2+}. The hydrazine, which in acidic solution is present as $NH_2NH_3^+$, is oxidized to $N_2(g)$. Write half-equations and a net equation for this reaction. What is the value of $E°_{cell}$? (Hint: Use data from Appendix D and page 829.)

56. The text mentions that one use of hydrazine is to remove dissolved oxygen from boiler water. Write half-equations and a net equation for this reaction. What is the value of $E°_{cell}$? (Hint : Use data from Appendix D and page 829.)

Phosphorus

57. What are the two principal allotropic forms of phosphorus? Which is the more reactive? How do they differ in molecular structure?

58. What are the formulas of the oxides in which phosphorus is in the oxidation states of +3 and +5, respectively? What are their molecular structures?

59. The industrial preparation of phosphine, $PH_3(g)$, involves the reaction of white phosphorus with KOH(aq). The other reaction product is $KH_2PO_2(aq)$. Write a balanced equation for this redox reaction.

60. Phosphorus trichloride is used to produce many other phosphorus compounds. Write a balanced equation to show how phosphoryl chloride, $POCl_3$, can be made by the reaction of phosphorus trichloride, chlorine, and tetraphosphorus decoxide.

Sulfur

61. The equation given on page 836 for the formation of thiosulfate ion from sulfite ion and sulfur in an alkaline solution is the net equation for an oxidation–reduction reaction. Write equations for the half-reactions that occur. Derive the net equation from these half-equations.

62. When an aqueous solution of sodium thiosulfate is acidified, unstable thiosulfuric acid is formed. The acid decomposes immediately to sulfurous acid and sulfur. Write equations for the two reactions. (*Hint*: What is the formula for thiosulfuric acid?)

The Halogens

63. Which of the following would you add to an aqueous solution containing Br^- to oxidize $Br^-(aq)$ to $Br_2(l)$: I_2, I^-, Cl_2, Cl^-, F^-? Explain your choice or choices.

64. Can you think of a reagent that can be used to displace $F^-(aq)$ and produce $F_2(g)$? Explain.

65. If Br^- and I^- occur together in an aqueous solution, I^- can be oxidized to IO_3^- with $Cl_2(aq)$. Simultaneously, Br^- is oxidized to Br_2, which is extracted with $CS_2(l)$. Write chemical equations for the reactions that occur.

66. Write equations to illustrate the oxidation of $Fe^{2+}(aq)$ to $Fe^{3+}(aq)$ by the halogen, X_2. (*Hint*: How many different equations can you write?)

67. One of the following is a *pseudohalogen*, one is an *interhalogen*, and one fits neither of these categories: ICl, NaCl, $(CN)_2$. Identify each.

68. One of the following is a *halide*, one is a *halate*, and one fits neither of these categories: BrF_3, $MgBr_2$, $NaClO_3$. Identify each.

69. Write equations to represent the reaction of Cl_2 with each of the following.
a. $H_2O(l)$ **b.** cold NaOH(aq) **c.** hot NaOH(aq)

70. Write equations to represent the reaction of KI with each of the following.
a. H_2SO_4(concd aq) **b.** H_3PO_4(concd aq)

Noble Gases

71. Use VSEPR theory to predict the geometrical shapes of these molecules.
a. XeO_3 **b.** XeO_4

72. Use VSEPR theory to predict the geometrical shapes of these molecules.
a. XeF_4 **b.** $XeOF_4$

ADDITIONAL PROBLEMS

73. Predict the likely geometric structures of these interhalogen molecules.
a. BrF_3 **b.** IF_5

74. Predict the likely geometric structures of these polyatomic ions.
a. ICl_2^- **b.** I_3^-

75. The reaction of borax, calcium fluoride, and concentrated sulfuric acid yields sodium hydrogen sulfate, calcium sulfate, water, and boron trifluoride as products. Write a balanced equation for this reaction.

76. To prepare potassium aluminum alum, $KAl(SO_4)_2 \cdot 12H_2O$, aluminum foil is dissolved in KOH(aq). The solution obtained is treated with H_2SO_4(aq), and the alum is crystallized from the resulting solution. Write plausible equations for these reactions.

77. In 1986, fluorine was prepared by a *chemical* method (that is, not involving electrolysis). The reactions used were that of hexafluoromanganate(IV) ion, MnF_6^{2-}, with antimony pentafluoride to produce manganese(IV) fluoride and SbF_6^-, followed by the dissociation of manganese(IV) fluoride to manganese(II) fluoride and fluorine gas. Write equations for these two reactions.

78. Outline a scheme that you might use to make ethane gas (C_2H_6) if the only raw materials available are limestone, coal, and water.

79. Describe the physical properties and physical principles that make possible the Frasch process for mining sulfur.

80. Outline the phase changes and changes in molecular structure that occur when a sample of rhombic sulfur is heated from room temperature to about 450 °C.

81. In a reaction described on page 835, sulfite ion in acidic solution oxidizes $H_2S(g)$ to free sulfur. What volume of air, at 25 °C and 755 mmHg containing 1.5 mole percent H_2S, could be purged of H_2S by 2.50 L of 1.75 M Na_2SO_3?

82. What masses of C(s) and $Al_2O_3(s)$, in kilograms, are consumed to produce 1.00×10^3 kg Al by the electrolysis reaction of page 816? How many coulombs of electric charge are required for the electrode reactions?

83. To prevent air oxidation of $Sn^{2+}(aq)$ to $Sn^{4+}(aq)$, metallic tin can be kept in contact with $Sn^{2+}(aq)$. Use electrode potential data to explain the role of the metallic tin.

84. $E° = +2.32$ V for the reduction half-reaction

$$XeF_2(aq) + 2\,H^+(aq) + 2\,e^- \longrightarrow Xe(g) + 2\,HF(aq)$$

Show that XeF_2 decomposes in aqueous solution, producing $O_2(g)$. Write a net equation for this reaction and calculate $E°_{cell}$. Would you expect this decomposition to be favored in acidic or in basic solution? Explain.

85. Polonium is the only element known to crystallize in the simple cubic structure. In this structure, the interatomic distance is 335 pm. Use this description of the crystal structure to estimate the density of polonium. (*Hint*: How many atoms are in a unit cell? What are the mass and volume of this unit cell?)

86. The gemstone mineral beryl consists of SiO_4 tetrahedra arranged in six-membered rings with Be^{2+} and Al^{3+} in a 3:2 ratio as counter ions. Make a sketch of this structure and deduce the empirical formula of beryl.

Ammonium dichromate decomposes in a vigorous reaction. The $Cr_2O_7^{2-}$ ion is a strong oxidizing agent, and oxidation–reduction reactions are an important feature of the chemistry of the *d*-block elements.

22

The *d*-Block Elements

CONTENTS

In our exploration of the chemistry of the elements, we have previously studied the *s*-block (Chapter 8) and the *p*-block (Chapter 21). Our focus in this chapter is on the *d*-block, and we will also briefly consider the *f*-block. Taken together, of course, the *s*-, *p*-, *d*-, and *f*-blocks constitute the entire periodic table. The *d*- and *f*-blocks—the *transition* elements—account for about 60% of the known elements. In our discussion we will use most of the same principles of bonding and reactivity that we have used elsewhere, but with greater emphasis on physical properties, oxidation states, electrode potentials, and complex ion formation. Also, we will comment on the sources and abundances of the transition elements and uses of the elements and their compounds.

Some of the transition metals have many uses and are among the most important of all the elements. Others, because they are rare, have fewer uses. Yet, even among some of the rare ones there are certain applications for which only they can serve. And nine of the transition elements have biological functions that are essential to human life.

22.1 General Properties and Their Trends

The location of the *d*-block elements in the periodic table is indicated in Figure 22.1, and the elements emphasized in this chapter are shown. Before considering these elements in any detail, however, let's look at some general properties. We will consider trends in these properties and, to some extent, compare them to properties of the representative elements.

Some Properties of the *d*-Block Elements of the Fourth Period

First, let's consider the properties listed in Table 22.1 for the elements with atomic numbers from 21 to 30. Electron configurations correspond to elements in the fourth period—the valence electron shell is $n = 4$—and to *d*-block transi-

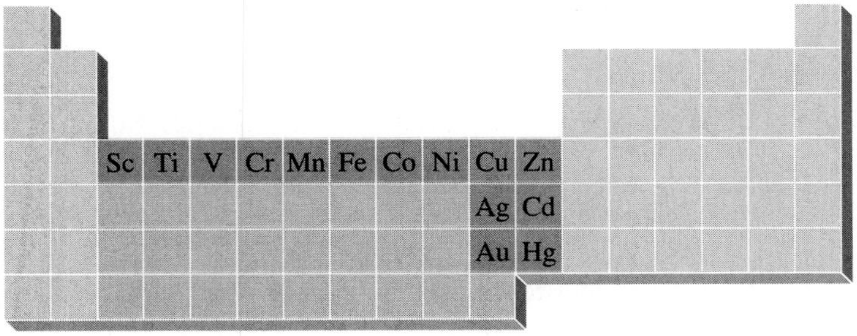

Figure 22.1 The *d*-block elements.

tion elements—the 3*d* subshell fills through this series of 10 elements. In fact, as we shall see in Section 22.4, both copper and zinc have filled 3*d* subshells.

One indication that the transition elements are metals is seen in their electronegativity values (EN). As we saw in Figure 9.4, nonmetals have EN greater than 2; metals, less than 2; and metalloids, about 2. The EN values of the first transition series range from 1.4 to 1.9, and the gradual increase in EN in moving from left to right across Table 22.1 conforms to one of the trends described in Section 9.5. However, according to Figure 9.6, we expect bonds between transition metal atoms and typical nonmetal atoms (for example, the halogens) to have somewhat less ionic character than the corresponding bonds involving the *s*-block metal atoms, which have electronegativities slightly below 1.

TABLE 22.1 Selected Properties of the *d*-Block Elements of the Fourth Period

	Sc	Ti	V	Cr	Mn	Fe	Co	Ni	Cu	Zn
Atomic number	21	22	23	24	25	26	27	28	29	30
Electron configuration[a]	$3d^14s^2$	$3d^24s^2$	$3d^34s^2$	$3d^54s^1$	$3d^54s^2$	$3d^64s^2$	$3d^74s^2$	$3d^84s^2$	$3d^{10}4s^1$	$3d^{10}4s^2$
Electronegativity	1.4	1.5	1.6	1.7	1.6	1.8	1.9	1.9	1.9	1.7
Common monoatomic cations	3+	2+, 3+	2+, 3+	2+, 3+	2+, 3+	2+, 3+	2+, 3+	2+	1+, 2+	2+
Common oxidation states[b]	3	2, 3, 4	2, 3, 4, 5	2, 3, 6	2, 3, 4, 6, 7	2, 3, 6	2, 3	2, 3	1, 2	2
Atomic radius, pm	161	145	132	125	124	124	125	125	128	133
$E°$, V[c]	−2.03	−1.63	−1.13	−0.90	−1.18	−0.440	−0.277	−0.257	+0.337	−0.763
Melting point, °C	1397	1672	1710	1900	1244	1530	1495	1455	1083	420
Density, g/cm³	3.00	4.50	6.11	7.14	7.43	7.87	8.90	8.91	8.95	7.14
Electrical conductivity[d]	3	4	6	12	1	16	25	23	95	27
Thermal conductivity[d]	4	5	7	22	2	19	23	21	93	27

[a]Each atom has an argon inner core configuration.
[b]The most important oxidation states are printed in red.
[c]For the reduction $M^{2+}(aq) + 2e^- \longrightarrow M(s)$ [except for Sc, where the ion is $Sc^{3+}(aq)$].
[d]Electrical and thermal conductivity are on an arbitrary scale relative to 100 for silver, the best metallic conductor.

Most of the common cations of the fourth-period transition series are produced by the loss of valence-shell electrons *and* one or more 3*d* electrons by the metal atoms. Only zinc atoms are unable to lose 3*d* electrons. The common oxidation states show that variability of oxidation state is typical for *d*-block elements. For the members of the fourth-period transition series up to and including manganese (Group 7B), the highest oxidation state corresponds to the periodic table group number: +3 for Sc (Group 3B), +4 for Ti (Group 4B), and so on. With the later members of the series (iron and beyond), the energy of the 3*d*-subshell is lowered to the point that it effectively becomes part of the electron core (Figure 22.2). Electrons in this subshell are less able to participate in chemical bonding and the range of oxidation states is more limited.

The electrode potentials, all negative except for that of Cu^{2+}, show that, except for Cu^{2+}, the M^{2+} cations of the first transition series are more difficult to reduce than is $H^+(aq)$.

$$M^{2+}(aq) + 2\,e^- \longrightarrow M(s) \qquad E^\circ_{red} < 0\,V$$

$$2\,H^+(aq) + 2\,e^- \longrightarrow H_2(g) \qquad E^\circ_{red} = 0\,V$$

Or, conversely, all the metals of the first transition series except Cu can displace $H_2(g)$ from an acidic solution.

$$2\,H^+(aq) + 2\,e^- \longrightarrow H_2(g) \qquad E^\circ_{red} = 0\,V$$

$$M(s) \longrightarrow M^{2+}(aq) + 2\,e^- \qquad E^\circ_{ox} > 0\,V$$

$$\textit{Net:}\quad M(s) + 2\,H^+(aq) \longrightarrow M^{2+}(aq) + H_2(g) \qquad E^\circ_{cell} > 0\,V \text{ and } \Delta G^\circ < 0$$

Unlike the *s*-block metals, most of which react with water to displace $H_2(g)$, among the fourth-period *d*-block elements only scandium is active enough as a metal to do so.

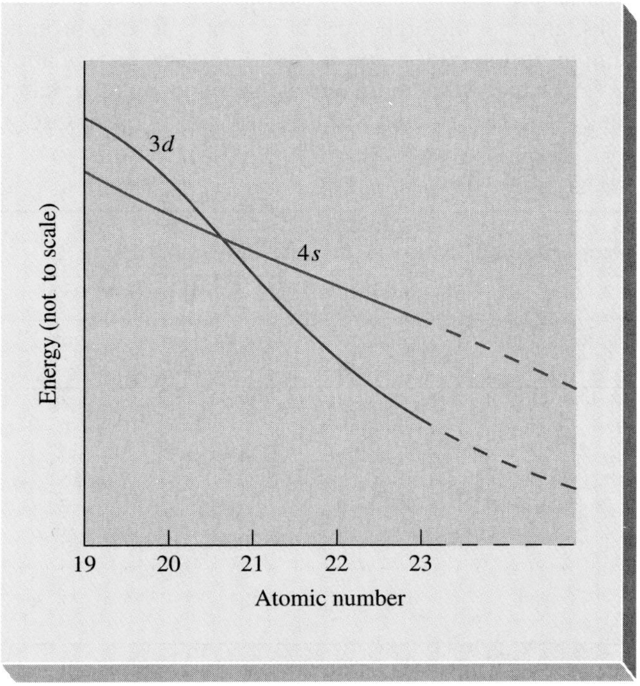

Figure 22.2 Some relative orbital energies in the fourth-period transition series. The energy level of the 3*d* subshell crosses and then drops below the energy level of the 4*s* subshell with increasing atomic number in the fourth period of the periodic table.

The physical properties of the first transition series are also consistent with those expected of most metals: moderate to high melting points and moderately high densities. (By comparison, sulfur, a typical nonmetal, has a melting point of 119 °C and a density at 25 °C of 2.07 g/cm^3.)

Perhaps the most distinctive physical properties of metals are the ability to conduct electricity and to conduct heat. Although only copper comes close to matching silver in these abilities, the values for the other metals are still significantly high. Compare them, for example, to values of the electrical and thermal conductivities for the nonmetal white phosphorus: 1.6×10^{-17} and 0.055, respectively (each relative to an arbitrary value of 100 for silver).

Other Properties and Comparisons

In comparing atomic (metallic) radii in groups of transition metals (Table 22.2) we see exceptions to an earlier generalization (page 256) that atoms become larger from the top to the bottom of a group. Between the fourth and fifth periods atomic radii do increase as expected, but between the fifth and sixth periods they remain nearly constant or even decrease slightly.

The sixth period is an especially long one of 32 members, because the 4*f* subshell must fill before 5*d*. The filling of the 4*f* subshell occurs in the lanthanide series. The geometric shapes of *f* orbitals are such that electrons in these orbitals are not very good at screening valence electrons from the nucleus. As a result, the increase in effective nuclear charge is greater than we would otherwise anticipate, as is the attraction of valence electrons to the nucleus. Atomic size does not increase between the fifth- and sixth-period members of a group of transition elements, a phenomenon known as the **lanthanide contraction**.

A comparison of common oxidation states within groups of transition elements (Table 22.2) shows that the heavier group members have a greater range of common oxidation states, and that the higher oxidation states tend to become more important. For example, in its oxides iron is found only in the oxidation states +2 (in FeO) and +3 (in Fe_2O_3). Ruthenium and osmium, on the other hand, occur in the oxidation states +4 (in RuO_2 and OsO_2) and +8 (in RuO_4 and OsO_4).

TABLE 22.2 Two Properties of Group 4B, 6B, and 8B Elements

Period	Group 4B		Group 6B		Group 8B	
	Atomic Radius, pm	Common O.S.[a]	Atomic Radius, pm	Common O.S.[a]	Atomic Radius, pm	Common O.S.[a]
4	Ti		Cr		Fe	
	145	3, 4	125	2, 3, 6	124	2, 3, 6
5	Zr		Mo		Ru	
	159	4	136	2, 3, 4, 5, 6	133	2, 3, 4, 6, 8
6	Hf		W		Os	
	156	4	137	2, 3, 4, 5, 6	134	2, 3, 4, 6, 8

[a]The most important oxidation states are printed in red.

Moreover, in compounds with the higher oxidation states there is a greater degree of covalency in the bonding. In line with this tendency, the melting points of FeO and Fe_2O_3 are 1337 and 1462 °C, respectively, whereas those of RuO_4 and OsO_4 are only 25 and 41 °C. RuO_4 and OsO_4 are molecular compounds; their melting points depend on the strengths of intermolecular forces rather than interionic attractions.

Finally, let's briefly note a major difference between transition (B-group) and representative (A-group) elements. In the representative elements, particularly those of the first and second period and those of Groups 1A and 2A, the nature of the chemical bonding is established by the *s* and *p* electrons of the valence shell. For *p*-block elements (3A through 8A) of the third period and beyond, *d* orbitals may also be involved in forming chemical bonds, but presumably only through half-filled hybrid orbitals in compounds such as SF_4, SF_6, and PF_5. With the transition elements, on the other hand, *d* orbitals are important in establishing physical and chemical properties. Differences in behavior between representative and transition elements regarding complex ion formation, color, magnetic properties, and catalytic activity can be traced to the differing roles played by *s*, *p*, and *d* orbitals.

22.2 The Elements Scandium Through Manganese

Now we shift our focus from a general overview of the transition elements to a closer look at specific elements. We begin with the first five elements of the fourth-period *d*-block elements, one element at a time.

Scandium

Scandium is not a particularly rare element. Its abundance in Earth's crust is about 25 ppm, which is greater than that of most of the other elements. Yet the history of scandium is much like that of an exceptionally rare element. Its existence was not discovered until its oxide was isolated in 1879. Several other compounds were prepared a few years later. The properties of these compounds corresponded closely with those expected of the element eka-boron, predicted by Mendeleev in his original periodic table. It was not until 1937 that the free element itself was isolated by the electrolysis of a mixture of molten chlorides including $ScCl_3$.

There are two reasons for the apparent rarity of scandium. (1) It is widely distributed in Earth's crust, but there is only one mineral, *thortveitite*, $Sc_2Si_2O_7$ or $Sc_2O_3 \cdot 2SiO_2$ (page 825), in which it is concentrated to any extent. (2) So far, scandium has no uses that cannot be met by cheaper alternatives, and so there has been little need to produce much of the metal or its compounds.

Nevertheless, because it is the first of the transition elements, the physical and chemical properties of scandium have been well characterized. Its chemistry is based mostly on the ion Sc^{3+}. This ion has the electron configuration of the noble gas argon; it is isoelectronic with the argon atom. In many ways scandium resembles not the transition metals but the representative metals, particularly aluminum. For example, scandium reacts with either acidic or basic solutions, and even with water itself, to liberate $H_2(g)$. Further similarities to aluminum include the fact that $Sc^{3+}(aq)$ is colorless and diamagnetic (all its electrons are paired), and that scandium hydroxide is amphoteric. $Sc(OH)_3(s)$ reacts with acidic solutions to produce Sc^{3+} and with alkaline solutions to form $[Sc(OH)_6]^{3-}$.

Titanium

Titanium is the ninth most abundant element, making up 0.63% of Earth's solid crust; it is the second most abundant transition metal (after iron). Its chief mineral sources are *rutile*, TiO_2, and *ilmenite*, $FeTiO_3$ ($FeO \cdot TiO_2$).

Most titanium is produced in a two-step process. First, $TiO_2(s)$ is heated with carbon and chlorine at about 800 °C to form gaseous $TiCl_4$. Then the $TiCl_4(g)$ is reduced with magnesium in an argon atmosphere at about 1000 °C.

Titanium is useful for many applications because of three desirable properties: (1) low density (4.5 g/cm^3), (2) high structural strength, even at high temperatures, and (3) corrosion resistance. The first two properties account for its use in the construction of aircraft, racing bicycles, and gas-turbine engines. The third property accounts for its use in the chemical industry, especially for reaction vessels containing mineral acids, alkaline solutions, chlorine, and a host of organic reagents.

The most important compound of titanium is the oxide, TiO_2, and its most important use is as a white pigment in paints, paper, and plastics. Naturally occurring rutile is impure TiO_2. It is usually colored because of the presence of impurities, and it requires extensive processing to yield the pure white powder required as a pigment.

White $TiO_2(s)$, mixed with other components to produce the desired color, is the leading pigment used in paints.

Example 22.1

From the description given in the text, write plausible equations for the two-step metallurgical process for producing titanium metal.

Solution

We can extract the necessary information for writing these equations from statements in the text. The first step is a high-temperature reaction (about 800 °C) having $TiO_2(s)$, $C(s)$, and $Cl_2(g)$ as reactants, and $TiCl_4(g)$ as a product. The products must also include an oxide of carbon. In actual practice, a mixture of $CO(g)$ and $CO_2(g)$ is probably obtained, but we can write a plausible equation using either oxide.

$$TiO_2(s) + 2\,C(s) + 2\,Cl_2(g) \xrightarrow{800\ °C} TiCl_4(g) + 2\,CO(g)$$

In the second step, $TiCl_4(g)$ reacts with magnesium at about 1000°C in an Ar(g) atmosphere to produce titanium metal. The other likely product is magnesium chloride. To indicate the probable physical forms of the reactants and products, we can find the melting point of Mg in Table 8.5 (649°C) and that of Ti in Table 22.1 (1672°C). To find the melting point of magnesium chloride (about 710°C), we would have to consult a handbook. A plausible equation is

$$TiCl_4(g) + 2\,Mg(l) \xrightarrow[Ar]{1000\ °C} Ti(s) + 2\,MgCl_2(l)$$

Exercise 22.1

Write plausible equations for the two-step "chloride" process for purifying rutile ore to obtain pure $TiO_2(s)$. In the first step, impure $TiO_2(s)$ is converted to $TiCl_4(g)$ at about 800 °C. In the second step, $TiCl_4(g)$ reacts at about 1200 °C with oxygen to produce $TiO_2(s)$ and chlorine. (*Hint*: Refer also to Example 22.1.)

Vanadium

Vanadium is one of the more abundant elements. At 136 ppm, it ranks just ahead of chlorine in its abundance in Earth's solid crust. Despite its abundance in the crust, vanadium is mostly obtained as a byproduct in the production of other metals, as in the extraction of uranium from the mineral *carnotite*, $K_2(UO_2)_2(VO_4)_2 \cdot 3H_2O$. Although vanadium metal can be prepared in high purity, mostly it is produced as an iron–vanadium alloy, *ferrovanadium*, with varying amounts of V.

The chief interests in vanadium are (1) as an alloying element in steel, (2) in the catalytic activity of some of its compounds, principally V_2O_5, and (3) in the variability of its oxidation states and the colors of its ions.

The inclusion of vanadium in steel imparts strength and toughness, making the steel ideal for use in springs and high-speed machine tools. The catalytic activity of vanadium pentoxide probably involves the reversible loss of O_2 that occurs when V_2O_5 is heated. V_2O_5 is the principal catalyst for the conversion of $SO_2(g)$ to $SO_3(g)$ in the manufacture of sulfuric acid (page 593). It is also useful in catalyzing the air oxidation of organic compounds, as in the conversion of naphthalene to phthalic anhydride, a compound used in the manufacture of resins.

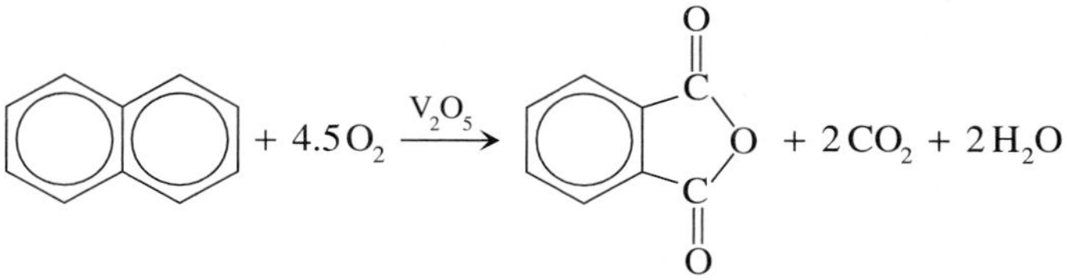

Figure 22.3 Some vanadium species in solution. The yellow solution has vanadium in the +5 oxidation state, as VO_2^+. In the blue solution the oxidation state is +4, in VO^{2+}. The green solution contains V^{3+}, and the violet solution contains V^{2+}.

Vanadium can be found in the +2, +3, +4, amd +5 oxidation states in solid compounds and in aqueous solutions. In its lower oxidation states, vanadium can exist as the monatomic cations V^{2+} and V^{3+}. In higher oxidation states, the vanadium is part of a polyatomic cation (VO^{2+} and VO_2^+) or a polyatomic anion (VO_4^{3-}). The distinctive colors of vanadium species are shown in Figure 22.3.

Chromium

The abundance of chromium in Earth's crust, 122 ppm, is comparable to that of vanadium; however, there is only one important source of the element: the mineral *chromite*, $FeCr_2O_4$ or $FeO \cdot Cr_2O_3$. An iron–chromium alloy, *ferrochrome*, is made by the reduction of chromite ore with carbon.

$$FeCr_2O_4 + 4\,C \xrightarrow{\Delta} \underbrace{Fe + 2\,Cr}_{\text{Ferrochrome}} + 4\,CO(g)$$

Ferrochrome and other alloying elements can be added directly to iron to produce stainless steel. When needed, pure chromium can be prepared by the reduction of Cr_2O_3 with aluminum (the thermite reaction described on page 816).

In addition to its use in alloys, chromium can also be plated onto other metals, generally by electrolysis from a solution containing CrO_3 in $H_2SO_4(aq)$. Because this electrolysis involves reducing Cr(VI) to Cr(0), it takes 6 mol of electrons to produce 1 mol of chrome plate: Chrome plating consumes more electric energy than do other types of plating. (Copper plating, for instance, requires 2 mol of electrons per mol Cu plated, and silver plating, 1 mol.) Chrome

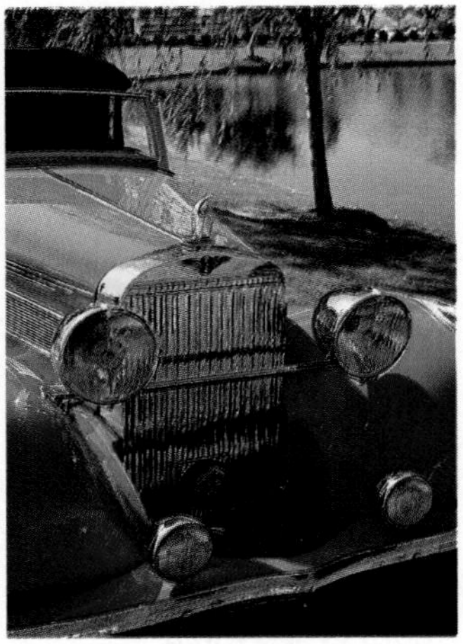

Chrome plating for decorative purposes is generally applied in an extremely thin layer—about 10 nm thick. The metal is first plated with a copper or nickel layer about 100 times thicker than the chrome plating. The function of the chromium is to provide an unusually bright surface.

plating is desirable because the metal is hard and bright. Moreover, the chromium is covered by an invisible oxide coating that gives it some protection from corrosion. To save money and better protect the underlying metal, chromium is often applied over nickel plating.

Most chromium compounds are prepared from sodium chromate, which in turn forms when chromite ore is heated in molten sodium hydroxide in the presence of air. [Oxygen in the air oxidizes Cr(III) in $FeCr_2O_4$ to Cr(VI) in Na_2CrO_4.]

Chromate ion in alkaline or neutral aqueous solutions is bright yellow (Figure 22.4). If the solution is made acidic, the color changes to a bright orange, due to the formation of dichromate ion, $Cr_2O_7^{2-}$. Chromate and dichromate ions participate in a reversible reaction.

$$2\ CrO_4^{2-}(aq) + 2\ H^+(aq) \rightleftharpoons Cr_2O_7^{2-}(aq) + H_2O$$

In solutions of low pH—acidic solutions—the forward reaction, the conversion of CrO_4^{2-} to $Cr_2O_7^{2-}$, is favored. In basic solutions, the removal of H^+ through the reaction $H^+ + OH^- \longrightarrow H_2O$ favors the reverse reaction, the conversion of $Cr_2O_7^{2-}$ to CrO_4^{2-}.

Dichromate ion is one of the primary oxidizing agents in organic chemistry. In an acidic solution, it is used ot oxidize alcohols to aldehydes, ketones, and carboxylic acids. It is also used extensively in the analytical chemistry laboratory; its standard electrode potential is large and *positive*.

$$Cr_2O_7^{2-}(aq) + 14\ H^+(aq) + 6\ e^- \longrightarrow 2\ Cr^{3+}(aq) + 7\ H_2O \qquad E^\circ_{red} = 1.33\ V$$

In analyzing for iron, the sample is (1) dissolved in a nonoxidizing mineral acid, (2) treated with a reducing agent to reduce any Fe^{3+} back to Fe^{2+}, and (3) titrated with a standard solution of $K_2Cr_2O_7(aq)$. The net ionic equation is

$$6\ Fe^{2+}(aq) + Cr_2O_7^{2-}(aq) + 14\ H^+(aq) \longrightarrow 6\ Fe^{3+}(aq) + 2\ Cr^{3+}(aq) + 7\ H_2O \qquad E^\circ_{cell} = 0.56\ V$$

By contrast, chromate ion in basic solution is not a good oxidizing agent; its standard electrode potential is slightly *negative*.

$$CrO_4^{2-}(aq) + 4\ H_2O + 3\ e^- \longrightarrow Cr(OH)_3(s) + 5\ OH^-(aq) \qquad E^\circ_{red} = -0.13\ V$$

Figure 22.4 Chromate and dichromate ions. Chromium(VI) exists as CrO_4^{2-} in basic solution (left) and as $Cr_2O_7^{2-}$ in acidic solution (right).

Instead, $CrO_4^{2-}(aq)$ is a precipitating agent. For example, in the qualitative analysis scheme (Section 18.9), lead chromate is precipitated in cation Group 1 and barium chromate in Group 4.

In industry, dichromates are used in chrome tanning. Hides are soaked in $Na_2Cr_2O_7(aq)$, followed by a reaction in which $SO_2(g)$ reduces Cr(VI) to Cr(III) in soluble $Cr(OH)SO_4$. Collagen, a protein in the hides, reacts to form an insoluble complex chromium compound, and the hides are converted to leather. Insoluble chromates, particularly $PbCrO_4$ and $ZnCrO_4$, find some use as paint pigments. These inorganic pigments are more stable to the action of sunlight and chemical agents than are organic dyes.

Chromium(IV) oxide, CrO_2, has electrical and magnetic properties that make it useful in magnetic recording tape.

The main oxides of chromium are Cr_2O_3 and CrO_3. Cr_2O_3 and the corresponding hydroxide, $Cr(OH)_3$, are amphoteric. The amphoterism of $Cr(OH)_3$ is illustrated in Figure 22.5. The ability of $Cr(OH)_3$ to dissolve in an alkaline solution, and the ease with which $[Cr(OH)_4]^-$ can be oxidized to CrO_4^{2-} play a role in the qualitative analysis scheme for cation Group 3 (page 724).

The oxide CrO_3, on the other hand, has only acidic properties. It dissolves in water to produce a strongly acidic solution.

Figure 22.5 Amphoterism of $Cr(OH)_3(s)$. Freshly precipitated $Cr(OH)_3(s)$ (center) is amphoteric. It reacts in acidic solution, here $HNO_3(aq)$, to produce violet $[Cr(H_2O)_6]^{3+}(aq)$ (left). $Cr(OH)_3$ also reacts in NaOH(aq) to form gray-green $[Cr(OH)_4]^-(aq)$ (right).

$$2\ CrO_3(s) + H_2O \longrightarrow 2\ H^+(aq) + Cr_2O_7^{2-}(aq)$$

$CrO_3(s)$ is produced by reversing this reaction; this happens when a saturated aqueous solution of a dichromate is treated with H_2SO_4(concd aq). Because of its strong oxidizing properties, an H_2SO_4–dichromate mixture has been used as a cleaning agent for laboratory glassware. It is particularly effective in oxidizing grease. However, because Cr(VI) is thought to be carcinogenic, this use has been largely discontinued.

The Greek word *chroma*, on which the name chromium is based, means color. The element is so named because of the many colors exhibited by its compounds (recall Figures 22.4 and 22.5). The variety of colors is accentuated by the fact that $Cr^{2+}(aq)$ and $Cr^{3+}(aq)$ can form so many different complex ions, a matter that we will explore in Chapter 23.

CONCEPTUAL EXAMPLE 22.1

Write a plausible equation to explain the photographs in Figure 22.6. Pure ammonium dichromate (left) is ignited with a match, producing pure chromium(III) oxide (right).

Solution

Ammonium dichromate, $(NH_4)_2Cr_2O_7(s)$, has Cr in the +6 oxidation state. Chromium(III) oxide, $Cr_2O_3(s)$, has Cr in the +3 oxidation state. Dichromate ion is an oxidizing agent and is *reduced*. If a reduction occurs, there must also be an oxidation.

In the ammonium ion, NH_4^+, nitrogen is in the –3 oxidation state, the lowest possible oxidation state for N. Ammonium ion is a reducing agent and can be *oxidized* to any of several higher oxidation states. Because the $Cr_2O_3(s)$ is pure, the other product(s) must be gaseous. One of these must be $H_2O(g)$, to account for the H and O atoms of the initial reactant. So far all the Cr and O atoms in the reactants are accounted for in the products.

$$(NH_4)_2Cr_2O_7(s) \longrightarrow Cr_2O_3(s) + 4\ H_2O \quad \textit{(not balanced)}$$

Figure 22.6 Decomposition of ammonium dichromate.

Another product must contain nitrogen. Although we might postulate it to be N_2, N_2O, NO, or NO_2, we see that no O atoms from $(NH_4)_2Cr_2O_7$ are available to form oxides of nitrogen, so the additional product must be N_2. (If any oxides were to form, the required oxygen would have to come from the air.)

$$(NH_4)_2Cr_2O_7(s) \longrightarrow Cr_2O_3(s) + 4\,H_2O(g) + N_2(g)$$

Conceptual Exercise 22.1

As noted on page 858, the principal source of chromium compounds is sodium chromate produced by heating chromite ore in molten NaOH in the presence of air. Write a plausible equation for this reaction. [*Hint*: Both Cr(III) and Fe(II) are oxidized.]

Manganese

Manganese is another abundant transition element (0.106% of Earth's crust by mass). Its principal mineral is *pyrolusite*, MnO_2. The bulk of the manganese produced is used in the manufacture of alloys, principally various types of wear-resistant and shock-resistant steel for use in railroad tracks, bulldozers, and road scrapers. Although relatively pure manganese can be produced by reducing MnO_2, common practice is to reduce a mixture of MnO_2 and Fe_2O_3 to obtain an iron–manganese alloy, *ferromanganese*.

$$MnO_2 + Fe_2O_3 + 5\,C \longrightarrow \underbrace{Mn + 2\,Fe}_{\text{Ferromanganese}} + 5\,CO(g)$$

We can find a clue to the behavior of manganese and its compounds in its electron configuration: $[Ar]3d^54s^2$.

	3d					4s
[Ar]	↑	↑	↑	↑	↑	↑↓

First, by giving up its two valence electrons, a manganese atom enters the +2 oxidation state. Then, by making use of its unpaired 3*d* electrons, the manganese atom can exhibit all additional oxidation states from +3 to +7.

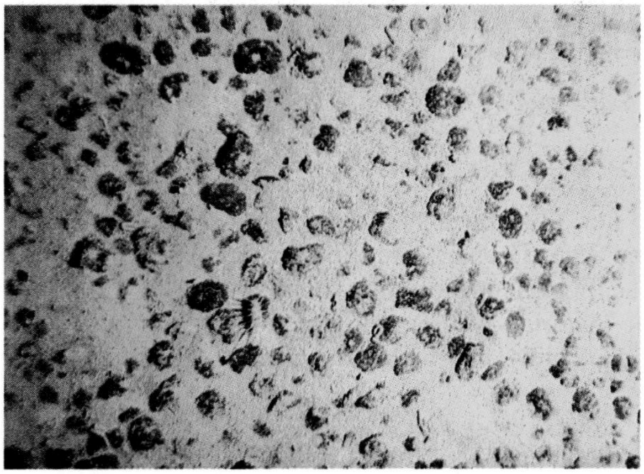

These manganese nodules are composed of layers of oxides of Mn and Fe, with small quantities of other metals such as Co, Cu, and Ni. The oxides are washed into the ocean and coalesce into rocklike objects that grow at a rate of only a few millimeters per million years. Marine organisms are believed to play a role in the formation of these nodules. The nodules are a potential source of several elements, but the technological problems involved are daunting.

Manganese dioxide is the starting point from which most other manganese compounds are prepared. To obtain the higher oxidation states of manganese, MnO_2 is first oxidized to a manganate salt, such as K_2MnO_4, which has Mn in the +6 oxidation state. This can be done by a strong oxidizing agent ($KClO_3$) in a basic medium (KOH).

$$3\,MnO_2(s) + ClO_3^-(aq) + 6\,OH^-(aq) \longrightarrow 3\,MnO_4^{2-}(aq) + Cl^-(aq) + 3\,H_2O$$

To obtain Mn(VII), K_2MnO_4 can be oxidized one step further to $KMnO_4$, with Cl_2 as the oxidizing agent, for instance.

Potassium permanganate is an important oxidizing agent both in the analytical and organic chemistry laboratory. Its oxidizing power also makes it useful as a medical disinfectant and as a substitute for Cl_2 in water purification.

The lower oxidation states of manganese can be obtained by reducing $MnO_2(s)$. For example, in the net reaction in a dry-cell battery (page 788), Zn is oxidized to Zn^{2+} and MnO_2 is reduced to Mn_2O_3. A starting point for the preparation of Mn(II) compounds is the reaction

$$MnO_2(s) + 4\,H^+(aq) + 2\,Cl^-(aq) \longrightarrow Mn^{2+}(aq) + 2\,H_2O + Cl_2(g)$$

Example 22.2

Show that permanganate ion in acidic solution is a more powerful oxidizing agent than either $O_2(g)$ or $Cr_2O_7^{2-}(aq)$.

Solution

An oxidizing agent undergoes *reduction* in a redox reaction. To compare the strengths of oxidizing agents, we can compare the standard electrode potentials for the half-reactions in which the oxidizing agents are reduced. For this purpose, let's turn to Appendix D, where we find these values:

$$MnO_4^-(aq) + 4\,H^+(aq) + 3\,e^- \longrightarrow MnO_2(s) + 2\,H_2O \qquad E^\circ_{red} = 1.70\ V$$

$$MnO_4^-(aq) + 8\,H^+(aq) + 5\,e^- \longrightarrow Mn^{2+}(aq) + 4\,H_2O \qquad E^\circ_{red} = 1.51\ V$$

$$O_2(g) + 4\,H^+(aq) + 4\,e^- \longrightarrow 2\,H_2O \qquad E^\circ_{red} = 1.229\ V$$

$$Cr_2O_7^{2-}(aq) + 14\,H^+(aq) + 6\,e^- \longrightarrow 2\,Cr^{3+}(aq) + 7\,H_2O \qquad E^\circ_{red} = 1.33\ V$$

The larger values of E°_{red} for the reduction half-reactions with $MnO_4^-(aq)$ suggest that in acidic solution $MnO_4^-(aq)$ is a stronger oxidizing agent than $O_2(g)$ and $Cr_2O_7^{2-}(aq)$.

Exercise 22.2

Show that $MnO_4^-(aq)$ is a good oxidizing agent in basic solution. (*Hint*: What are some oxidations that it is able to bring about?)

22.3 The Iron Triad: Fe, Co, and Ni

History is often divided into periods named for important materials that have shaped human culture: Stone Age, Bronze Age, and Iron Age. With annual worldwide production approaching one billion tons, iron is still today the most important metal. Although the materials of greatest current interest are the more exotic polymers, ceramics, composites, and bioengineered substances, the Iron Age is still with us.

Iron is widely distributed in Earth's crust; its abundance is 4.7% (or 47,000 ppm). Cobalt is much rarer; it makes up only 29 ppm of Earth's crust. However, Co is sufficiently concentrated in ore deposits such as *carrolite*, $CuCo_2S_4$ or $CuS \cdot Co_2S_3$, to support an annual production of millions of pounds. Cobalt is used primarily in alloys with other metals. Nickel's abundance in Earth's crust is 99 ppm and its ores include sulfides, oxides, silicates, and arsenides. Canada has particularly large deposits of a mixed iron and nickel sulfide. About 80% of the U.S. production of nickel goes into alloys and about 15% is used for electroplating. Smaller amounts of nickel are used as electrodes in batteries and fuel cells and as a catalyst. Another well-known use is in the manufacture of coins, such as the U.S. five-cent piece, the nickel.

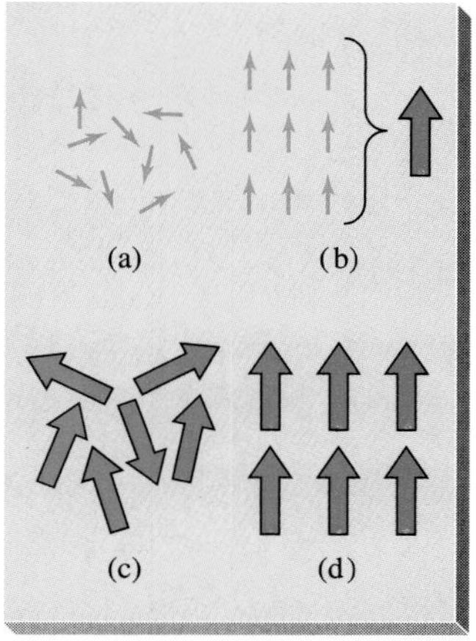

Figure 22.7 The phenomenon of ferromagnetism. (a) In ordinary paramagnetism the magnetic moments of the atoms or ions, represented by the thin red arrows, are randomly distributed. (b) In a ferromagnetic material the magnetic moments are aligned into domains, represented by the large blue arrows. (c) In an unmagnetized piece of ferromagnetic material the domains are randomly oriented. (d) In a magnetic field the domains are oriented in the direction of the field and the material becomes magnetized.

Ferromagnetism

Fe, Co, and Ni all have some unpaired *d* electrons, but so do most of the transition elements. This accounts for *paramagnetism* (page 252), but these three metals exhibit a much stronger magnetic effect known as **ferromagnetism**. With ferromagnetism, large numbers of atoms cluster into small regions called *domains*. The magnetic moments of the individual atoms in a domain are all oriented in the same direction, but ordinarily the domains themselves are mixed up every which way and their magnetic effects cancel. However, when a piece of Fe, Co, or Ni, is placed in a magnetic field, the domains line up and produce a strong resultant magnetic effect. The metal becomes magnetized and remains so even when the external magnetic field is removed. This description of ferromagnetism is illustrated schematically in Figure 22.7.

A key requirement for turning ordinary paramagnetism into ferromagnetism is to have just the right interatomic distances so that atoms can be arranged into domains. Iron, cobalt, and nickel meet the requirements, as do certain alloys of other metals, such as manganese in the combinations Al–Cu–Mn, Ag–Al–Mn, and Bi–Mn.

Metal Carbonyls

Another interesting property of the iron triad elements, though not unique to them, is the ability to form neutral complexes in which the ligand is carbon monoxide, CO. These complexes are called **metal carbonyls**. Table 22.3 lists three simple carbonyls and suggests these facts about them: (1) Each CO molecule contributes an electron pair to an empty orbital of the metal atom. (2) The metal carbonyls have no unpaired electrons; they are diamagnetic. (3) The central metal atom acquires the electron configuration of the noble gas Kr. Figure 22.8 suggests that we can use VSEPR theory to predict the structures of carbonyls by setting the number of valence-shell electron pairs equal to the number of CO ligands.

Metal carbonyls can be produced by direct combination of the metal atoms with CO(g). The reversible formation of nickel carbonyl has an important application in the metallurgy of nickel. In the *Mond process*, NiO contaminated with other oxides is treated with pure CO(g) at about 50 °C and 2 atm. The oxides are

TABLE 22.3 Three Metal Carbonyls

	Number of e^-		
	From Metal	From CO	Total
$Cr(CO)_6$	24	12	36
$Fe(CO)_5$	26	10	36
$Ni(CO)_4$	28	8	36

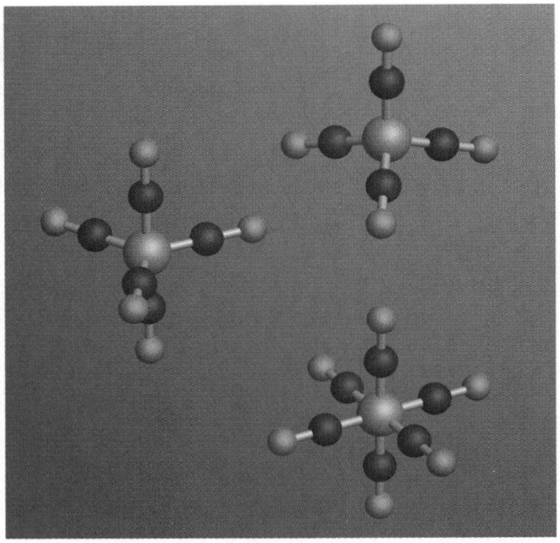

Figure 22.8 The structures of some simple carbonyls.

reduced to the metals, but nickel is carried off as $Ni(CO)_4(g)$. In a separate step, the $Ni(CO)_4$ is decomposed to pure nickel metal and CO(g) at 200°C.

$$Ni(s) + 4\,CO(g) \underset{200\ °C}{\overset{50\ °C}{\rightleftharpoons}} Ni(CO)_4(g)$$

Carbon monoxide poisoning results from a reaction similar to carbonyl formation. CO molecules coordinate with Fe^{2+} ions in hemoglobin, displacing the O_2 molecules normally carried by the hemoglobin. The metal carbonyls themselves are also very poisonous.

Oxidation States

All three iron triad elements form 2+ ions.

Fe^{2+} [Ar]$3d^6$ Co^{2+} [Ar]$3d^7$ Ni^{2+} [Ar] $3d^8$

For cobalt and nickel, the oxidation state +2 is the most stable, but for iron it is +3. When an iron atom loses its third electron, forming Fe^{3+}, the electron configuration is [Ar]$3d^5$. A half-filled $3d$ subshell and five unpaired electrons is a very stable configuration. As a result, iron(II) is readily oxidized to iron(III).

Fe^{2+}	$\longrightarrow$	$Fe^{3+} + e^-$
$3d$		$3d$
[Ar] ↑↓ ↑ ↑ ↑ ↑		[Ar] ↑ ↑ ↑ ↑ ↑

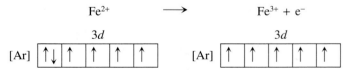

Furthermore, iron(III) can be oxidized further to iron FeO_4^{2-} only with great difficulty. The conversions of Co(II) and Ni(II) to Co(III) and Ni(III) are also difficult because the +3 oxidation states do not have a half-filled $3d$ subshell to stabilize them. The +3 oxidation state of cobalt can be stabilized in complex ions, such as $[Co(NH_3)_6]^{3+}$, but the +3 oxidation state is rare in nickel compounds.

Transition Elements in Living Matter

As suggested by Figure 22.9, only a fraction of the elements are essential to living matter. The elements that make up the bulk of living matter are those of low atomic number ($Z < 21$). Some elements are required only in trace amounts for most plant and animal life; most of these are transition elements, mainly those in the fourth period in the *d*-block. An adult human body that weighs 70 kg (154 lb), for example, contains only about 4 g of iron, but those 4 g are essential for good health.

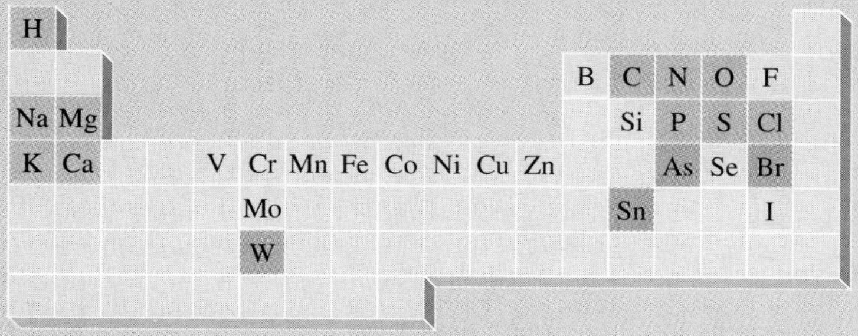

Figure 22.9 The elements in living matter.
elements that make up the bulk of living matter.
trace elements required for most plant and animal life.
trace elements possibly required by some life forms.
Most of these elements are relatively abundant in the natural environment suggesting that life forms develop around the elements available to them.

In biological matter the transition metals are often associated with large organic molecules. The protein hemoglobin, found in red blood cells and responsible for the color of blood, consists of iron-containing *heme* units attached to polypeptide molecules called *globins* (Figure 22.10). Hemoglobin transports O_2 molecules from the lungs, through the arteries, to all parts of the body for use in the metabolism of glucose. It then carries CO_2 produced in this metabolism, through the veins, back to the lungs.

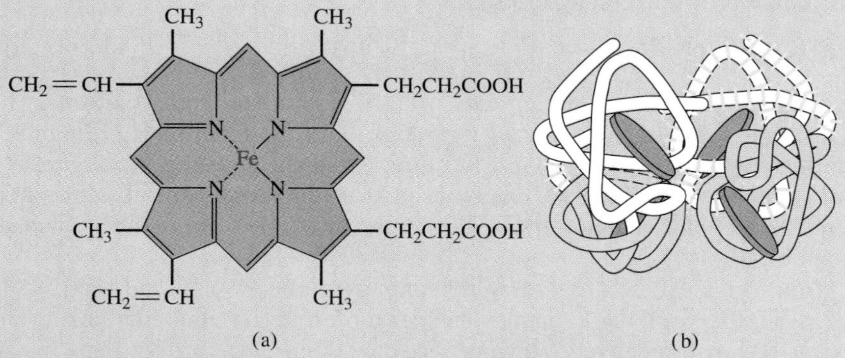

Figure 22.10 Heme and hemoglobin. (a) The structure of heme. (b) Four heme units and four coiled polypeptide chains are bonded together in a molecule of hemoglobin.

To perform its essential function, hemoglobin must have iron present as Fe^{2+}. If the iron is oxidized to Fe^{3+}, the resulting substance, called methemoglobin, is incapable of transporting oxygen. This oxidation is what happens in the disease called *methemoglobinemia*, which in infants is known as blue-baby syndrome. High concentrations of nitrate ion in ground water used for domestic purposes can cause this condition.

Another important biochemical process that involves both iron and molybdenum is the fixation of atmospheric nitrogen by certain bacteria through the action of the enzyme *nitrogenase*. Nitrogen molecules are apparently held to Mo atoms in the enzyme.

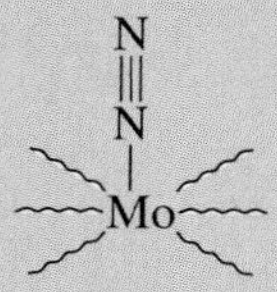

The function of the enzyme is to transfer six electrons to the N≡N molecule, so that the oxidation states of the N atoms are reduced from 0 to −3. The N atoms are then released as NH_3 molecules. The electron transfers appear to occur through changes in the oxidation states of the Fe and Mo atoms. Thus, the variability of oxidation states that we have noted throughout this chapter seems to play a key role in the biological activity of the transition elements.

Complex Ions

In aqueous solutions, ions of the iron triad metals typically are hydrated and colored. For example, $[Fe(H_2O)_6]^{2+}$ is pale green, and pure $[Fe(H_2O)_6]^{3+}$ is pale purple. Solutions of $Fe^{3+}(aq)$ rarely have a purple color, however, because other ligands may substitute for H_2O molecules. Thus, the ionization of ligand water molecules in $[Fe(H_2O)_6]^{3+}(aq)$ accounts for the fact that the solutions are acidic and yellow-brown in color. For example,

$$[Fe(H_2O)_6]^{3+}(aq) + H_2O \longrightarrow H_3O^+(aq) + [FeOH(H_2O)_5]^{2+}(aq) \qquad K_a = 8.9 \times 10^{-4}$$

Salts of the iron triad elements usually crystallize from solution as hydrates, and some of the water of hydration is in the form of H_2O ligands in complex ions. We have previously commented on how the hexahydrate of Co(II) chloride, $CoCl_2 \cdot 6H_2O$, can be used as a rough indicator of atmospheric humidity. Its color depends on the partial pressure of $H_2O(g)$ in the atmosphere (recall Figure 14.3).

Iron forms two common cyanide complexes, $[Fe(CN)_6]^{4-}$ and $[Fe(CN)_6]^{3-}$, that have the systematic names hexacyanoferrate(II) ion and hexacyanoferrate(III) ion, respectively. However, they are often called ferr*o*cyanide ion and ferr*i*cyanide ion. In aqueous solution iron(III) yields a dark blue precipitate, *Prussian blue*, when treated with potassium hexacyanoferrate(II).

$$4\,Fe^{3+}(aq) + 3\,[Fe(CN)_6]^{4-}(aq) \longrightarrow Fe_4[Fe(CN)_6]_3(s)$$

Because this precipitate is insoluble in water and organic solvents and unaffected by dilute acids, it finds use in paints, printing inks, blueprint paper, laundry bluing, and cosmetics (eye shadow).

$Fe^{3+}(aq)$ and $SCN^{-}(aq)$ react to form the blood-red complex ion $[FeSCN(H_2O)_5]^{2+}$. Formation of this complex ion is the basis of a sensitive test for iron; it is so intensely colored as to be detectable even at very low concentrations. We'll say more about this complex ion in Section 23.7.

22.4 Group 1B: Cu, Ag, and Au

The earliest coins were made of copper, silver, and gold. This use, which continues today, is based on the durability and corrosion-resistance of the metals. The data in Table 22.4 show that the metals have positive electrode potentials: The metal ions are easily reduced to the free metals, and, in turn, the metals are difficult to oxidize. This resistance to oxidation is what imparts "nobility" to the Group 1B metals.

Several ancient coins. The use of copper, silver, and gold in coins has earned the name "coinage metals" for the Group 1B metals.

Another insight from Table 22.4 concerns atomic radii and densities. That silver and gold have the same atomic radius is another example of the lanthanide contraction discussed earlier in the chapter (page 854). Although the two atoms are about the same size, gold atoms have nearly twice the mass of silver atoms. This largely accounts for the much higher density of gold compared to silver.

The Group 1B elements are exemplary metals, as is indicated by their physical properties. They are exceptionally malleable and ductile, and they have the highest electric and thermal conductivity of all metals. Like transition elements

TABLE 22.4 Some Properties of Copper, Silver, and Gold

	Cu	Ag	Au
Electron configuration	$[Ar]3d^{10}4s^1$	$[Kr]4d^{10}5s^1$	$[Xe]4f^{14}5d^{10}6s^1$
Atomic radius, pm	128	144	144
Electronegativity	1.9	1.9	2.4
Oxidation states[a]	1, 2	1, 2	1, 3
Electrode potential, V			
$M^{+}(aq) + e^{-} \longrightarrow M(s)$	+0.522	+0.800	+1.83
$M^{2+}(aq) + 2e^{-} \longrightarrow M(s)$	+0.337	+1.39	—
$M^{3+}(aq) + 3e^{-} \longrightarrow M(s)$	—	—	+1.52
Melting point, °C	1083	962	1064
Density, g/cm^3	8.96	10.5	19.3
Electrical conductivity[b]	95	100	68
Thermal conductivity[b]	93	100	74

[a]The most common oxidation state is printed in red.

[b]Electrical and thermal conductivity are on an arbitrary scale relative to 100 for silver, the best metallic conductor.

in general, the Group 1B metals display variable oxidation states; but the variability is limited, just as it is in the iron triad. Additional similarities to other transition elements include paramagnetism and color in some of their compounds and the ability to form complex ions.

In Mendeleev's original short form of the periodic table, the alkali metals (Group 1A) and Cu, Ag, and Au (Group 1B) were lumped together in Group I. The main similarity of the two subgroups is that atoms of elements in both groups have a single valence-shell electron. However, the Group 1B elements display multiple oxidation states, whereas Group 1A elements can exhibit only the oxidation state +1. All the Group 1A and 1B elements are metals; but Group 1A elements are the most active metals, whereas Group 1B elements are among the least active; they are unable, for example, to displace $H_2(g)$ from $H^+(aq)$. The long form of the periodic table does a better job than the short form in revealing the true nature of Group 1B.

Because of their durability, metallic luster, malleability, and ductility, the Group 1B metals are prized metals in the decorative arts and for making jewelry. Electrical wiring for residential and commercial purposes is almost always made of copper. Silver and gold are used in electronic components. Copper's corrosion-resistance is the property that mostly accounts for its use in plumbing, while good thermal conductivity is the basis of its use in cookware. Gold's primary use is a rather unusual one: as a monetary reserve for individuals and nations of the world.

As we have noted, the Group 1B metals do not displace $H^+(aq)$ from solutions; they do not react with HCl(aq). Copper and silver do react with H_2SO_4(concd aq) and $HNO_3(aq)$, producing $Cu^{2+}(aq)$ and $Ag^+(aq)$, along with $SO_2(g)$, and oxides of nitrogen, respectively, but no $H_2(g)$. Gold does not react with any single acid, but it will react with *aqua regia*, a mixture of 1 part HNO_3 and 3 parts of HCl. Two changes occur in the reaction: $HNO_3(aq)$ oxidizes gold to Au^{3+}, and HCl(aq) furnishes the Cl^- necessary to form the stable complex ion, $[AuCl_4]^-$.

$$Au(s) + 4\,H^+(aq) + NO_3^-(aq) + 4\,Cl^-(aq) \longrightarrow [AuCl_4]^-(aq) + 2\,H_2O + NO(g)$$

Gold is resistant to air oxidation, as is silver; but silver tarnishes by reacting with sulfur compounds to produce Ag_2S (page 794). Copper is corrosion resistant in dry air; in moist air it forms green basic copper carbonate, $Cu_2(OH)_2CO_3$.

Basic copper carbonate, $Cu_2(OH)_2CO_3$, is the compound that imparts a characteristic green color often seen in copper roofs and bronze statues. Unlike iron rust, the basic copper carbonate forms a tough, adherent film that protects the underlying metal.

The principal manufactured compound of copper is copper sulfate pentahydrate, $CuSO_4 \cdot 5H_2O$. It is used in electroplating, in batteries, and to prepare other copper salts. Cu^{2+} ion, though essential to life in trace amounts, is toxic at higher concentrations. As a result, copper sulfate is an important pesticide, used against bacteria, algae, and fungi.

Silver nitrate is the chief manufactured compound of silver. It too is used in electroplating, in batteries, and to prepare other silver compounds. Also, silver nitrate is an important laboratory reagent for performing chemical analyses. The silver compounds used in the greatest amount are the silver halides for photography. Gold compounds are used in electroplating, photography, medicinal chemistry, and in ruby glass and ceramics.

Superconductors

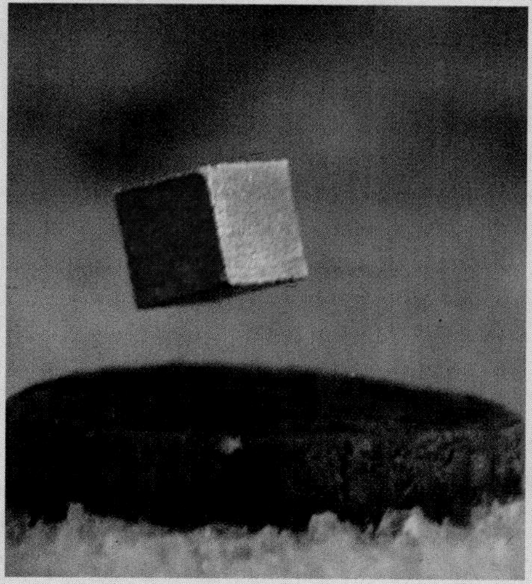

The small magnet induces an electric current in the superconductor. Associated with this current is another magnetic field that opposes the field of the small magnet, causing it to be repelled. The small magnet remains suspended as long as the superconducting current persists, which it does as long as the temperature of the superconductor is held at the boiling point of liquid nitrogen (77 K).

Atoms in a metal lattice vibrate about fixed positions, and these vibrations interfere with the flow of electrons in an electric current. Lattice vibrations increase with increased temperature. Consequently, a metal becomes a better electric conductor as its temperature is *lowered*. Theoretically, the electrical resistance of a metal should decrease to essentially zero at 0 K.

In 1911, Heike Onnes found that some metals become *superconductors*—they abruptly lose their electrical resistance—at liquid helium temperatures that are still well above 0 K. This was a big surprise to everyone. Below a certain critical temperature, an electric current in a superconducting material flows indefinitely without the loss of energy. Superconducting metals have been used mainly to construct powerful electromagnets for use in medical imaging, in particle accelerators, and in nuclear fusion research reactors. The current theory of superconductivity in metals is based on quantum theory.

Metallic superconductors function only at extremely low temperatures. The highest critical temperature among this group is 23 K. Many additional applications of superconductivity should be possible if it could be achieved at higher temperatures. We might even see widespread use of maglev trains—trains suspended and propelled in a magnetic field.

In 1986 a class of materials was discovered that become superconductors at higher temperatures than do metals. A number of these new superconductors have been characterized, including the three early examples listed in Table 22.5. The superconductor $YBa_2Cu_3O_7$ can be prepared by heating a mixture of yttrium oxide, barium carbonate, and copper oxide to 700 °C.

Although the new superconductors contain metal atoms, they are not metals. They are compounds with properties similar to those of ceramic

TABLE 22.5 **Some Ceramic Superconductors**

Critical Temperature for Superconductivity, K	Elements Present	Representative Formula
40	La, Sr, Cu, O	$La_{2-x}Sr_xCuO_4$[a]
90	Y, Ba, Cu, O	$YBa_2Cu_3O_7$
105	Th, Ca, Ba, Cu, O	$Th_2CaBa_2Cu_2O_8$

[a]$x < 0.3$

materials; they are called ceramic superconductors. Their structures are complex but seem to involve sheets of copper and oxygen atoms separated by the other constituents. Currently, several materials are known that are superconducting at liquid nitrogen temperatures (77 K), but the ultimate goal is to find one that functions at room temperature (about 300 K). Also, because ceramic superconductors are brittle, they are not easily made into wires of high current-carrying capacity. No practical ceramic superconductor device has yet been made, but superconductivity research is now one of the most active fields in materials science.

22.5 Group 2B: Zn, Cd, and Hg

Although they belong to the *d*-block, the Group 2B elements do not fit the usual definition of transition elements: the *d*-subshells of their next-to-outermost electronic shells are *filled*, both in the free atoms and in their ionic forms. They resemble the transition metals in some ways, such as in the ability to form complex ions, but they also differ. Compared to the Group 1B elements, for example, zinc and cadmium are more active metals, have lower melting points, and do not exhibit multiple oxidation states. Moreover, in some ways, zinc and cadmium resemble the Group 2A (alkaline earth) metals—Mendeleev placed Groups 2A and 2B together in Group II in the short form of the periodic table. Zinc and cadmium have a single principal oxidation state, +2; their ions are diamagnetic and colorless in solution; most of their compounds are colorless or white.

Mercury differs from zinc and cadmium in significant ways: (1) Mercury is a liquid at room temperature. (2) Mercury will not displace $H_2(g)$ from acidic solutions. (3) Mercury can exist in the oxidation state +1, in the diatomic ion Hg_2^{2+}, which features an Hg—Hg covalent bond. (4) Many mercury compounds are covalent (for example, $HgCl_2$ is only slightly ionized in aqueous solution). (5) Few mercury compounds are water soluble, and most of its compounds are not hydrated. (6) Mercury shows little tendency to oxidize, and its oxide, HgO, decomposes to the elements upon heating.

Large quantities of zinc are used to make alloys. As an example, brass is a copper alloy having from 20 to 45% zinc and small quantities of tin, lead, and iron. Brass has an attractive yellow metallic sheen, is corrosion resistant, and is a good electrical conductor. We have previously described zinc-coated iron, *galvanized iron*, and mentioned its corrosion resistance (page 794). Zinc is also an important metal for sacrificial anodes in cathodic protection against corrosion (page 794). Another use of zinc is in batteries such as dry cells (page 788).

Cadmium can substitute for zinc in coating metals for certain applications. Its primary uses are in alloys, such as low-melting solders, and batteries. Another specialized use, based on its neutron-absorbing capacity, is in fabricating control rods for use in nuclear reactors. The uses of cadmium are severely limited, however. Whereas zinc in trace amounts is an essential element for humans, cadmium, which closely resembles zinc, is quite toxic. Its effect may be to substitute for zinc in certain enzymes. Cadmium poisoning causes a painful skeletal disorder first observed among people who consumed rice from an area of Japan where effluents from a zinc mine had drained into rice fields. Cadmium poisoning can also cause liver damage, kidney failure, and pulmonary disease. Concern over cadmium poisoning is heightened by the fact that some cadmium is almost always found in zinc ores and thus in zinc metal itself and in zinc compounds.

In Japanese, the disorder is known as itai-itai kyo or ouch-ouch disease.)

The physical properties of mercury—its metallic and liquid qualities, and its high density—determine many of its uses. Mercury is found in thermometers, barometers, manometers, electrical relays and switches, and as electrodes for certain batteries and electrolysis cells. Mercury forms alloys called *amalgams* with most metals. Dental amalgam used to fill tooth cavities consists primarily of silver and tin with a small amount of mercury. The thermal properties of expansion and contraction of the amalgam closely match those of teeth. Currently, various metal powders, porcelain, and plastics are supplanting dental amalgam in many applications. Another important though limited use of mercury is as mercury vapor in fluorescent tubes and street lamps.

Liquid mercury was once treated as a laboratory plaything, but we now know that long-term exposure presents a significant health hazard. Mercury vapor is toxic, and levels exceeding 0.05 mg Hg/m^3 are unsafe. Even though liquid mercury has a low vapor pressure, the concentration of mercury in the saturated vapor at room temperature greatly exceeds the safe limit. Unsafe mercury vapor levels have been found in various locales in which the element is used—chlor–alkali plants, thermometer factories, smelters, and dental laboratories.

Mercury tends to accumulate in the body, and toxic levels affect the nervous system and cause brain damage. Mercury poisons the body's systems, in part, by interfering with sulfur-containing enzymes. Hg^{2+} reacts with —SH groups in the enzyme to change the shape of the enzyme and render it inactive. Some microorganisms can convert inorganic mercury to methylmercury (CH_3Hg^+) compounds, which can concentrate in the food chain and lead to unsafe mercury levels in fish. Environmental problems associated with cadmium and mercury notwithstanding, Group 2B compounds have some important uses, some of which are listed in Table 22.6.

Hatter's disease (which probably afflicted the Mad Hatter in *Alice's Adventures in Wonderland*) was a form of chronic mercury poisoning. Mercury compounds were used to convert fur to felt for felt hats.

The Lanthanide Elements (Rare Earths)

The elements cerium (Z = 58) through lutetium (Z = 71) comprise the first series of the *f*-block. They are called the *lanthanide* elements because they immediately follow lanthanum (La) in the periodic table. They are called *inner-transition* elements because the subshell that is being filled is *two* principal quantum levels below the valence shell. Filling of the 4*f* subshell precedes filling of the 5*d* in the sixth period. These elements are frequently called "rare earths," a name of historical origin but a clear misnomer because several are not rare at all. Cerium (Ce) and neodymium (Nd) are more abundant in Earth's crust than lead. All the lanthanides except Pm, which is radioactive, are more abundant than such well-known elements as Cd, Ag, Hg, and Au. The most striking feature of the lanthanide elements, which are often referred to by the general symbol Ln, is a similarity to each other and to La and the Group 3B metals.

In spite of their relative abundance, the term "rare earths" does seem appropriate for two reasons: (1) For many years after their discovery, there was only one known source of the elements, *monazite*, a mixed phosphate of La, Th, and the lanthanides. (2) The individual lanthanides are difficult to separate and isolate. (In the nineteenth century, their separation was a matter of *extreme* difficulty.)

When ions differ sufficiently we can often separate them in a one-step process. [When we add Cl^- to a solution of Ag^+ and Cu^{2+}, Ag^+ completely precipitates as AgCl(s) and Cu^{2+} remains in solution.] If we use differences in solubility to separate individual lanthanide cations (Ln^{3+}), we get only a slight separation. We have to go through many, many cycles of precipitation and redissolving to achieve total separation. Researchers took about 70 years, from 1839 to 1907, to isolate all of the lanthanide elements. (Promethium, found only as a product of nuclear fission, was not isolated until 1945.) In 1913, H. G. J. Moseley found that the frequency of X-rays is related to the atomic number of the target element in an X-ray tube. In just a matter of days, he was able to identify by atomic number all the lanthanide elements in a sample of monazite. Currently, complex ion formation and ion-exchange techniques provide a much easier separation of the lanthanides.

Recently, deposits of the mineral *bastnasite*, a mixed fluorocarbonate of lanthanum and the lanthanides found in the mountain pass region of the Mojave Desert in California, have come into production. These deposits are easier to work with than monazite because they do not contain radioactive thorium (a member of the actinide series).

Fortunately, many uses of the lanthanides do not require them to be separated. A mixture of the lanthanide metals with about 25% La is used in steel and magnesium-based alloys. Lanthanide–cobalt alloys are used to make permanent magnets. A number of lanthanide compounds also have commercial uses. Several of the oxides are used as phosphors in the screens of color television sets, as colorants for glass and ceramic glazes, and as catalysts. Cerium(IV) compounds are important oxidizing agents in analytical chemistry.

TABLE 22.6
Some Important Compounds of Zinc, Cadmium, and Mercury

Compound	Uses
$ZnCl_2$	textile processing; adhesives; dental cements and dentifrices; electroplating; antiseptic and deodorant solutions.
ZnO	pigment and reinforcing agent in rubber; paint pigment; dietary supplement; cosmetics; photoconductor in copying machines.
ZnS	pigment; luminous paints; fungicides; phosphors in X-ray and television screens.
$ZnSO_4$	manufacture of rayon; zinc plating; wood preservative; animal feeds; dietary supplement.
CdO	electroplating; batteries; catalyst; ceramic glazes; nematocide.
CdS	pigments and inks; fluorescent screens; photoconductor in copying machines; transistors and rectifiers; solar cells.
Hg_2Cl_2	fungicide; maggot control; pharmaceuticals; electrodes.
$HgCl_2$	manufacture of mercury compounds; disinfectant; fungicide; insecticide; wood preservative; catalyst; batteries.
HgO	pigment for paints and ceramics; pharmaceuticals; fungicide; batteries.

SUMMARY

The electron configurations of atoms and ions of the *d*-block elements feature partially filled *d* subshells; those of the *f*-block have partially filled *f* subshells. The *d*- and *f*-blocks make up the transition elements. All are metals. Unlike the representative metals, the transition metals exhibit a multiplicity of oxidation states, form many complex ions, and display a variety of colors in their ions. A comparison of the early (Sc, Ti, V, . . .) and later (Fe, Co, Ni, . . .) members of the fourth-period transition series shows that the early members tend to be more active as metals and to display a greater variety of oxidation states, with the highest oxidation state equal to the group number.

In comparing the lighter and heavier members of a group, we find that the fifth- and sixth-period transition metals have a greater tendency to exhibit their higher oxidation states than do those of the fourth period. Atomic radii increase in the expected manner between the fourth- and fifth-period members, but there is essentially no difference in radius between the fifth- and sixth-period members. This lack of increase in atomic radii, called the *lanthanide contraction*, is attributed to the poor shielding of valence-shell electrons by electrons in the *f* subshell.

Scandium resembles aluminum in its physical and chemical properties and has no important uses. Titanium is used for its high strength, low density, and corrosion resistance. The oxide TiO_2 also has important uses, for example, as a pigment. Vanadium is an alloying element in steel, and its compounds are of interest because of the variety they display in oxidation state and color. Chromium and manganese are also used in various steel alloys. Redox reactions are central to the chemistry of these two elements, with $Cr_2O_7^{2-}$and MnO_4^- being widely used oxidizing agents. $Cr_2O_7^{2-}$ participates in a pH-dependent equilibrium with CrO_4^{2-}, and CrO_4^{2-} is important as a precipitating agent.

Iron, cobalt, and nickel have similar properties and together are known as the iron triad. One unusual property of the three is ferromagnetism. Another property they share with several other *d*-block elements is the ability to form complexes with CO called metal carbonyls. The iron triad metals readily form complex ions.

Copper, silver, and gold—the coinage metals—differ from earlier members of their periods in that they are much less active metals. They are unable to displace $H_2(g)$ from acidic solutions. Most technological uses of copper, silver, and gold are based on their inertness toward air oxidation and their exceptional abilities to conduct heat and electricity.

Zinc, cadmium, and mercury, though in the *d*-block and resembling transition metals in some ways, are not transition elements in that their atoms and ions have filled *d*-subshells. Mercury differs from the lighter members of Group 2B in several important ways.

KEY TERMS

ferromagnetism (22.3)
lanthanide contraction (22.1)
metal carbonyls (22.3)

REVIEW QUESTIONS

1. Why are *d*-block elements first found in the fourth period of the periodic table and not sooner?
2. Why are *f*-block elements first found in the sixth period of the periodic table and not sooner?
3. What is the transition element having the electron configuration $[Ar]3d^7 4s^2$?
4. An atom has the electron configuration: $[Kr]4d^{10}5s^2 5p^1$. Is this a transition element? Explain.
5. Name at least three properties that distinguish transition metals from representative metals.
6. Which three of the transition metals are the best electric and thermal conductors of all the metals?
7. What is the lanthanide contraction and among which elements is it observed?
8. What is ferromagnetism and among which elements is it observed.
9. Name three *d*-block metals that are used to coat iron to provide it with corrosion protection.
10. Name three fourth-period, 3*d*-block metals that are commonly alloyed with iron to make steel.
11. Name one or more *amphoteric* transition metal hydroxides.
12. Name one or more transition metal hydroxides having only basic properties.
13. Indicate the chemical composition of the material known as (**a**) galvanized iron, (**b**) chromite ore, (**c**) basic copper carbonate.
14. Indicate a common name that is often used for the material that is (**a**) an iron–manganese alloy obtained in the metallurgy of manganese, (**b**) a copper–zinc alloy, (**c**) a mixture of nitric and hydrochloric acids.
15. The metallurgy of chromium mostly involves the reduction of chromite ore, with no further processing of the product. Why is this so?
16. Although rutile ore is a naturally occurring form of $TiO_2(s)$, extensive treatment of the ore is required before the $TiO_2(s)$ can be used commercially. Why is this so?
17. Which of the following elements is most often found in the +3 oxidation state in its compounds: Au, Ag, Ni, Zn?
18. Why are the most stable oxidation states of Co and Ni +2, whereas that of Fe is +3?
19. Write a name that adequately describes each of the following.
    a. $Sc(OH)_3$ **b.** Fe_2SiO_4
    c. Na_2MnO_4 **d.** $Os(CO)_5$
20. Write a formula that adequately describes each of the following.
    a. barium dichromate **b.** chromium trioxide
    c. mercury(I) bromide **d.** sodium vanadate
21. What happens when copper metal is added to HCl(aq)? to $HNO_3(aq)$? Why are the results different in the two cases?
22. What are the products when $Mg^{2+}(aq)$ and $Cr^{3+}(aq)$ each is treated with a limited amount of NaOH(aq)? With a large excess of NaOH(aq)? Why are the results different in the two cases?
23. Compounds and solutions containing Cr^{3+} do not all have the same color. Why is this so?
24. The color of fully hydrated $Fe^{3+}(aq)$ is pale purple, yet aqueous solutions with this ion often have a yellow to brown color. Why is this so?

PROBLEMS

Properties of the Transition Elements

25. Sc and Ca both have two valence-shell electrons ($4s^2$), yet Ca only exists in the oxidation state +2 in its compounds whereas Sc displays the oxidation state +3. Explain this difference.

26. In comparing some metal atoms and their ions, we find that Sc is paramagnetic and Sc^{3+} is diamagnetic, whereas both Ti and Ti^{2+} are paramagnetic and both Zn and Zn^{2+} are diamagnetic. Explain these differences.

27. Which of the following best expresses the approximate electronegativities of the transition elements? (a) EN < 1; (b) 1 < EN < 2; (c) EN = 2.5; (d) 3 < EN < 4. Explain.

28. Which of the following best expresses the approximate E°_{red} value for the reduction half-reaction

$$M^{2+}(aq) + 2\,e^- \longrightarrow M(s)$$

for the majority of the fourth-period members of the *d*-block? (a) $E^\circ_{\text{red}} < -2$ V; (b) $-2 \text{ V} < E^\circ_{\text{red}} < 0$ V; (c) $0 < E^\circ_{\text{red}} < 2$ V; (d) $E^\circ_{\text{red}} > 2$ V. Explain.

29. The two adjacent elements in the periodic table, K and Ca, have atomic (metallic) radii of 227 and 197 pm, respectively. Another two adjacent elements, Mn and Fe, each has an atomic radius of 124 pm. Explain why (**a**) the atomic radius of Ca is smaller than that of K; (**b**) why the atomic radius of Mn is smaller than that of Ca; and (**c**) why the atomic radii of Mn and Fe are the same.

30. The two consecutive elements in Group VB, Nb and Ta, each has an atomic (metallic) radius of 143 pm, whereas that of V, also a member of Group VB, is 131 pm. Explain why (**a**) all three atomic radii are not the same; and (**b**) if there is a reason why all three should not be the same, why are two of them still found to be the same.

Scandium Through Manganese

31. Use orbital diagrams to indicate the electron configurations of the following *d*-block atoms and ions.
a. Ti **b.** Ag **c.** Cr^{2+} **d.** Mn^{2+}

32. Use orbital diagrams to indicate the electron configurations of the following *d*-block atoms and ions.

a. Ni **b.** Zr **c.** V^{3+} **d.** Co^{3+}

33. What is the oxidation state of vanadium in each of the following ions: V^{2+}, VO^{2+}, VO_2^+, VO_4^{3-}?

34. What is the oxidation state of manganese in each of the following: MnO_2, MnO_4^{2-}, $Mn(OH)_2$, Mn_2O_3, MnO_4^-?

35. Write plausible equations for the following reactions.
a. the reaction of Sc(s) with HCl(aq)
b. the reaction of $Sc(OH)_3$(s) with HCl(aq)
c. the reaction of $Sc(OH)_3$(s) with excess NaOH(aq)
d. the electrolysis of $ScCl_3$(l)

36. Write plausible equations for the following reactions.
a. the reaction of Mn(s) with HCl(aq)
b. the reaction of $Mn(OH)_2$(s) with HCl(aq)
c. the reaction of $Cr(OH)_3$(s) with NaOH(aq)
d. the electrolysis of $Cr(NO_3)_3$(aq) using platinum electrodes

37. Suggest reactions with common chemicals by which the following syntheses can be accomplished.
a. $BaCrO_4$(s), starting with barium metal and K_2CrO_4(aq)
b. MnO_2(s), starting with $KMnO_4$(aq)

38. Suggest reactions with common chemicals by which the following syntheses can be accomplished.
a. $CrCl_3$(s), starting with $K_2Cr_2O_7$(aq)
b. $MnCO_3$(s), starting with $KMnO_4$(aq)

39. A qualitative analysis test for Mn^{2+} involves its oxidation to MnO_4^- in acidic solution by sodium bismuthate, $NaBiO_3$. The bismuthate ion is reduced to Bi^{3+}. Write a net ionic equation for the reaction.

40. The text suggests a two-step process for obtaining $KMnO_4$ from MnO_2(s) (page 862). Write a pair of balanced equations for the process.

41. When *colorless* $Pb(NO_3)_2$(aq) is added to *orange* $K_2Cr_2O_7$(aq), *yellow* $PbCrO_4$(s) precipitates. Explain how this comes about.

42. When *colorless* HCl(aq) is added to *orange* $K_2Cr_2O_7$(aq), a *green* solution is formed. Explain how this comes about.

The Iron Triad

43. Which of the following would you expect to be the best *reducing* agent: Fe(s), Fe^{2+}(aq), Co^{3+}(aq), Co(s)? Explain.

44. Which would you expect to be the better *oxidizing* agent, Fe^{2+}(aq) or Co^{3+}(aq)? Explain.

45. Although iron does not commonly occur in an oxidation state higher than +3, Fe_2O_3(s) can be oxidized to ferrate ion, FeO_4^{2-}, by Cl_2(g) in a strongly basic solution. Write a balanced redox equation for this reaction.

46. Show that an acidic solution of Co^{3+}(aq) is unstable and write a plausible balanced redox equation for the reaction that occurs. For the half-reaction $Co^{3+}(aq) + e^- \longrightarrow Co^{2+}(aq)$, $E^\circ_{red} = 1.92$ V.

Group 1B

47. Write plausible net ionic equations for the reactions described on page 868, that is,
a. $Cu(s) + H_2SO_4(\text{concd aq}) \longrightarrow$
b. $Cu(s) + HNO_3(aq) \longrightarrow$

48. Write plausible equations for the reactions described on page 868, that is,
a. $Ag(s) + H_2SO_4(\text{concd aq}) \longrightarrow$
b. $Ag(s) + HNO_3(aq) \longrightarrow$

49. Explain why it is that silver reacts with HNO_3(aq) and gold does not.

50. Explain how it is that gold, which reacts with neither HCl(aq) nor HNO_3(aq), does react with a mixture of the two acids. Write a net ionic equation.

Group 2B

51. Describe some important ways in which the Group 1B elements differ from those in Group 1A.

52. Describe some important ways in which mercury differs chemically from zinc and cadmium.

53. Write plausible net ionic equations to represent the following reactions.
a. Hg(l) with HNO_3(aq)
b. ZnO(s) with CH_3COOH(aq)

54. Write plausible net ionic equations to represent the following reactions.
a. Cd(s) with H_2SO_4(concd aq)
b. $ZnSO_4(aq) \xrightarrow{\text{electrolysis}}$

ADDITIONAL PROBLEMS

55. Give some reasons why the percent abundance in Earth's crust is not necessarily a good measure of the availability and importance of an element and its compounds.

56. Tin is considerably more toxic than zinc. (In small quantities Zn is an essential element.) Why, then, do you suppose that tin plate is used rather than galvanized iron in cans for food storage?

57. Manganese does not form the simple carbonyl $Mn(CO)_5$, but it does form $Mn_2(CO)_{10}$. Explain why this is reasonable to expect.

58. The text states that chromate ion is not a particularly good oxidizing agent. Yet, when Na_2CrO_4(s) is added to hot concentrated HCl(aq), chlorine gas is formed. Explain why, in fact, this is what you should expect to observe.

59. Write a plausible balanced equation to represent the air oxidation of copper to basic copper carbonate, $Cu_2(OH)_2CO_3$.

60. Manganate ion, MnO_4^{2-}, is unstable in acidic solution and disproportionates. Write a plausible net ionic equation for the disproportionation reaction. (*Hint*: What are the likely products?)

61. Table 22.4 indicates that silver can be found in the oxidation state +2. However, the only important compounds of Ag(II) are AgF_2 and AgO. Others, such as $AgCl_2$, $AgBr_2$, and AgS, are not stable. Suggest an explanation for this fact. (*Hint*: What are the probable products of the decomposition of the unstable compounds?)

62. Show that the oxidation of Fe^{2+}(aq) to Fe^{3+}(aq) by atmospheric oxygen in acidic solution is a spontaneous reaction when all reactants and products are in their standard states. Is the reaction also spontaneous if the partial pressure of $O_2(g) = 0.20$ atm, $[Fe^{3+}] = 1.0$ M, $[Fe^{2+}] = 0.10$ M, and pH $= 5$? (*Hint*: Either use the Nernst equation or determine K_{eq} for the reaction.)

63. You have available as reducing agents Sn^{2+}(aq), V^{2+}(aq), and Fe^{2+}(aq). For each one, determine whether it is capable of carrying out the following reductions in acidic solution.
a. MnO_2(s) to Mn^{2+}(aq) **b.** I_2(s) to I^-(aq)
c. H^+(aq) to H_2(g)

64. You have available as oxidizing agents: Fe^{3+}(aq), NO_3^-(aq), and V^{3+}(aq). For each one, determine whether it is capable of carrying out the following oxidations in acidic solution.
a. H_2(g) to H^+(aq) **b.** Sn^{2+}(aq) to Sn^{4+}(aq)
c. Hg(l) to Hg^{2+}(aq)

65. What mass of chromium can be electrodeposited in 3.25 h with a current of 2.57 A from a chrome-plating bath of the type described on page 857.

66. A procedure for determining iron by titration with $Cr_2O_7^{2-}$(aq) is outlined on page 858. A 1.2130-g sample of an iron ore is subjected to this analysis and requires 22.05 mL of 0.1050 M $K_2Cr_2O_7$(aq) for its titration. What is the mass percent of iron in the ore?

67. For the chromate–dichromate equilibrium,

$$2\,CrO_4^{2-}(aq) + 2\,H^+(aq) \rightleftharpoons Cr_2O_7^{2-}(aq) + H_2O \qquad K_{eq} = 3.2 \times 10^{14}$$

demonstrate that, above pH 9, $[Cr_2O_7^{2-}] \approx 0$ in a solution in which the total concentration of Cr(VI) is 1 M.

68. The vapor pressure of Hg as a function of temperature is given by the equation

$$\log P\,(\text{mmHg}) = \frac{-0.05223a}{T} + b$$

where a = 61,960; b = 8.118; T = Kelvin temperature. Show that the concentration of Hg(g) in equilibrium with Hg(l) at 25 °C greatly exceeds the maximum permissible level of 0.05 mg Hg/m³.

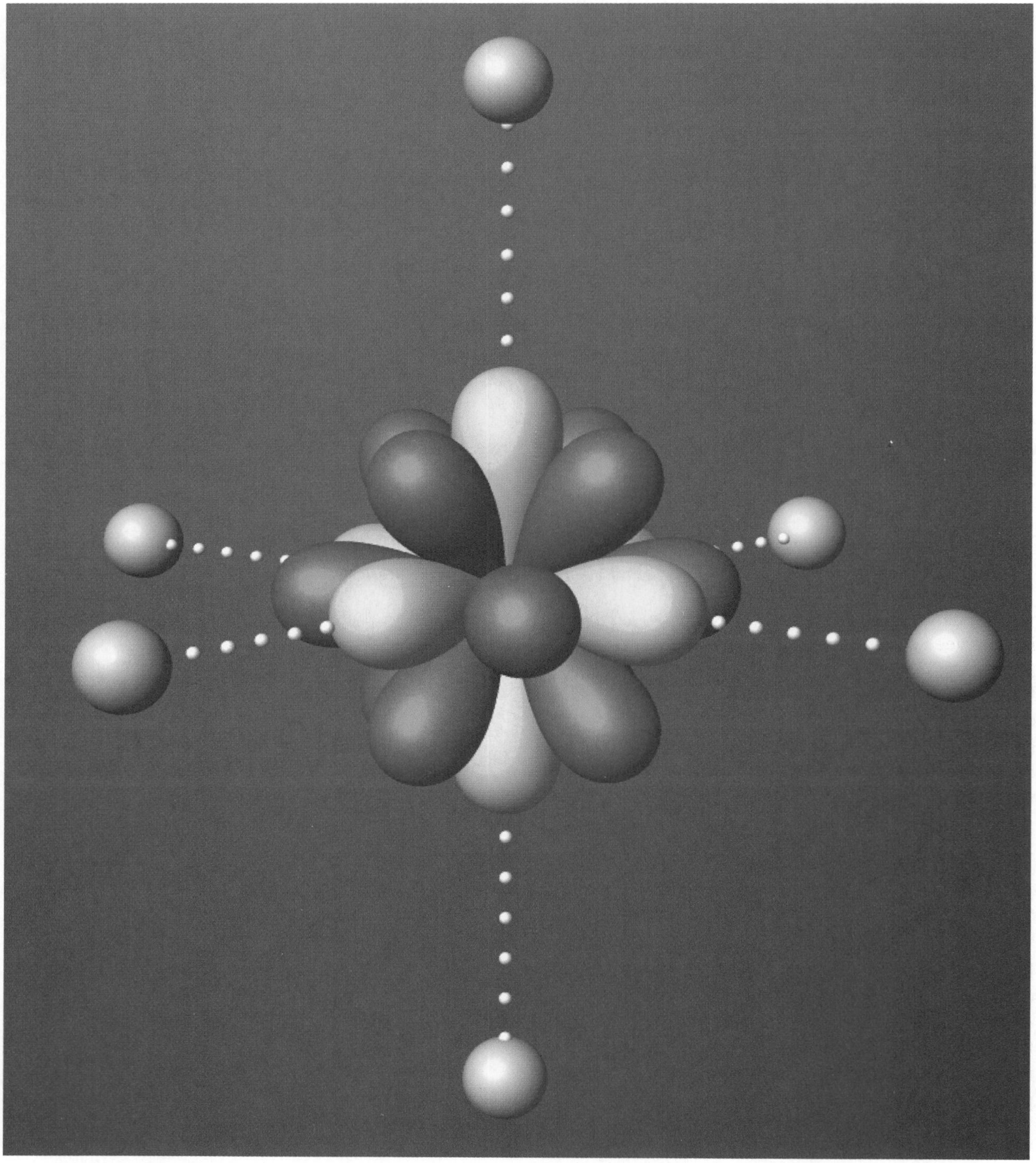

Visualizing the formation of a complex ion using the crystal field theory. See page 890.

23

Complex Ions and Coordination Compounds

CONTENTS

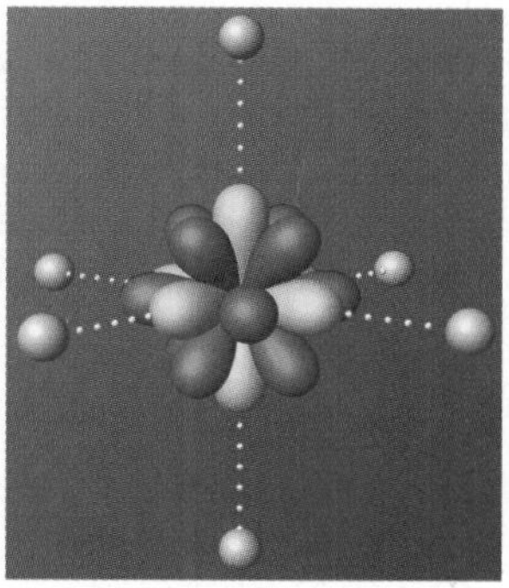

We first encountered complex ions and coordination compounds in Chapter 18, where we emphasized the equilibrium that exists between a complex ion and its component parts; a central metal ion and ligands. This equilibrium, we learned, can greatly influence other equilibria—for example, in enabling water-insoluble AgBr(s) to dissolve in $Na_2S_2O_3$(aq) in the fixing of photographic film. In Chapter 22 we saw that complex ion formation is an important property of the *d*-block elements. Complexes of iron, for example, range from Prussian blue pigment to hemoglobin in blood.

At this time we examine the nature of bonding in complex ions, their structures, and some properties related to their structures. We also briefly consider how complex ions and coordination compounds are named.

23.1 Werner's Theory of Coordination Chemistry

The two substances pictured in Figure 23.1 have quite similar formulas, but they look strikingly different. We can think of each one as consisting of two simpler compounds that are somehow joined together or coordinated; the substances are *coordination compounds*. The simpler compounds that make up the coordination compounds of Figure 23.1 are $CoCl_3$ and NH_3. One way to write formulas for coordination compounds is to separate the formulas of the simpler compounds by a dot.

$CoCl_3 \cdot 6NH_3$	$CoCl_3 \cdot 5NH_3$
Golden brown	Purple
(luteocobaltic chloride)	(purpureocobaltic chloride)

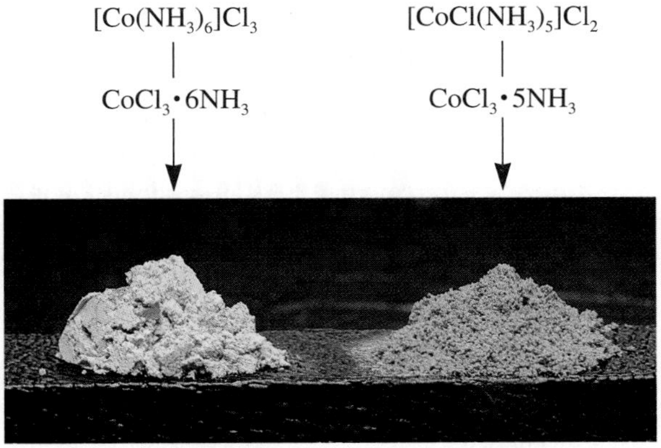

Figure 23.1 Two coordination compounds of cobalt(III). These compounds figured prominently in the development of Werner's coordination theory.

For a long time after their discovery about 200 years ago, these coordination compounds were known by the names in parentheses. Both were called cobaltic chloride because of the presence of Co(III) and chloride ions. The prefixes referred to the colors of the compounds (L. *luteus,* yellow; L. *purpureus*, purple). We'll consider a modern system of nomenclature in Section 23.3.

The two coordination compounds differ in more than color. Each substance has three moles of Cl^- per mole of compound, and, as we might expect, when one mole of the golden-brown compound is treated with $AgNO_3(aq)$, instantly *three* moles of AgCl(s) are obtained. However, only *two* moles of AgCl(s) are obtained from one mole of the purple compound.

In 1893 the Swiss chemist Alfred Werner explained the difference between the golden-brown and purple compounds by representing their dissociation in water in this way.

$$\underset{\text{Golden brown}}{[Co(NH_3)_6]Cl_3(s)} \xrightarrow{H_2O} [Co(NH_3)_6]^{3+}(aq) + 3\,Cl^-(aq)$$

$$\underset{\text{Purple}}{[CoCl(NH_3)_5]Cl_2(s)} \xrightarrow{H_2O} [CoCl(NH_3)_5]^{2+}(aq) + 2\,Cl^-(aq)$$

The purple compound provides only two-thirds as much free $Cl^-(aq)$ as the golden-brown one, and thus yields only two-thirds as much AgCl(s) when treated with an excess of $AgNO_3(aq)$.

The main proposition of Werner's coordination theory is that certain metal atoms have two types of valence or combining capacity:

Alfred Werner (1866–1919) claimed that the basic idea of coordination theory came to him, at age 25, during his sleep. He arose at 2 A.M. and had worked out the essentials by 5 A.M. Werner was awarded the 1913 Nobel Prize in chemistry for his work in coordination chemistry.

1. A *primary* valence displayed through the loss of electrons—for example, three electrons lost by a Co atom to form the Co^{3+} ion.
2. A *secondary* valence in the form of the attraction of a metal atom or ion for other groups, which are attached at specific positions around the central metal, leading to a distinctive geometric structure for the complex ion.

In a written formula the entity within brackets is properly called a *coordination entity*, but more commonly it is called a complex. A **complex** consists of a central atom, which is usually a metal ion, and attached groups called **ligands**. The ligands in a complex may be neutral molecules or anions (or, rarely, cations), and they need not all be of the same type. The electric charge on a complex may be positive, negative, or zero. If a complex carries a net electric charge, either positive or negative, it is generally called a **complex ion**. The **coordination number** is the total number of points around a central atom at whichbonding to ligands can occur. The most common coordination numbers are 2, 4, and 6, and the geometric structures associated with these coordination numbers are shown in Figure 23.2. A substance comprised of one or more complexes is a **coordination compound**. To summarize, for the coordination compound $[CoCl(NH_3)_5]Cl_2$,

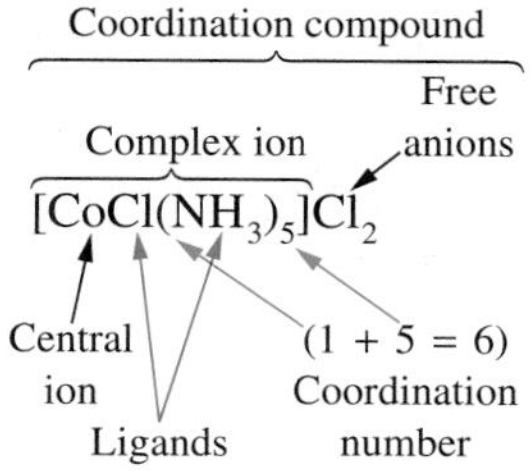

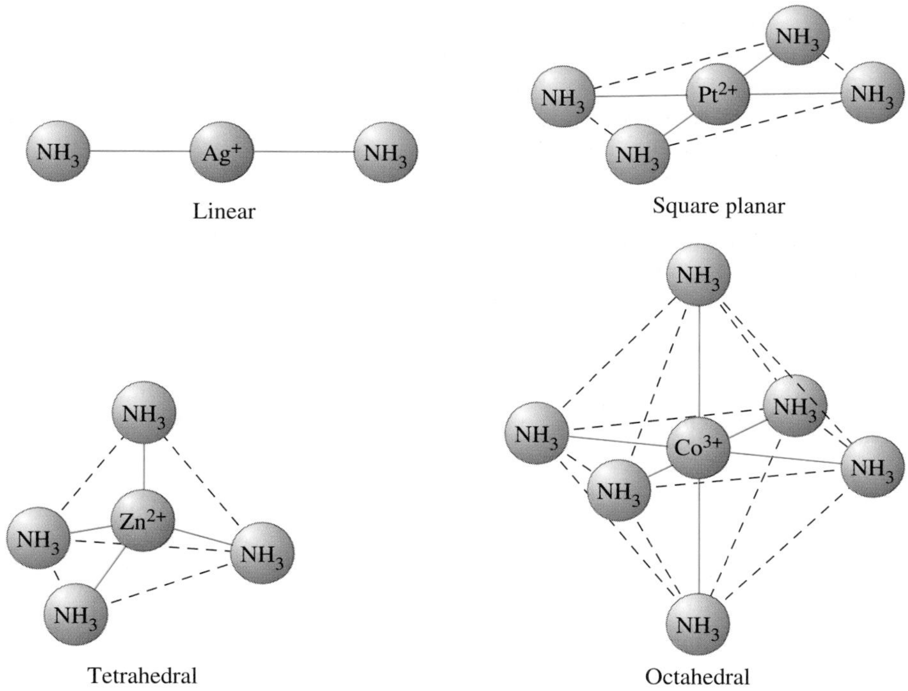

Figure 23.2 Four common structures of complex ions. In these complex ions the NH_3 molecules attach to the central metal ion through their lone-pair electrons.

Example 23.1

What are the coordination number and the oxidation state of the central atom in the complexes (**a**) $[CoCl_4(NH_3)_2]^-$ and (**b**) $Ni(CO)_4$?

Solution

a. Six ligand groups—four Cl^- ions and two NH_3 molecules—are attached to the central cobalt ion. The coordination number is *6*. This complex is a complex *anion* with a net charge of 1–. The NH_3 ligands carry no electric charge, and the charge associated with four Cl^- ions is 4–. The charge on the central cobalt ion, x, must be such that $x - 4 = -1$; $x = +3$. The central ion is Co^{3+}, which means the oxidation state of the cobalt is *+3*.

b. Four ligand groups—all CO molecules—are attached to a central Ni atom. The complex is electrically neutral, as are the CO molecules. The oxidation state of the nickel is *0*.

Exercise 23.1

What are the coordination number and the oxidation state of the central atom in the complexes **(a)** $[Co(SO_4)(NH_3)_5]^+$ and **(b)** $[Fe(CN)_6]^{4-}$?

23.2 More About Ligands: Lewis Acid–Base Theory

One description of the bonds between a central metal ion and its ligands is based on an acid–base theory proposed by G. N. Lewis in 1923. An atom, ion, or molecule is a **Lewis acid** if it can *accept* an electron pair, and a **Lewis base** if it can *donate* an electron pair. The reaction between a Lewis acid and a Lewis base results in the sharing of a pair of electrons between them. Because the electron pair is provided by just one member of a covalent bond (the Lewis base), the bond is a *coordinate* covalent bond.

Many metal atoms and ions, particularly those of the transition metals, have available empty orbitals into which electron pairs can be accommodated. The central metal of a complex is a Lewis acid. The characteristic to look for in ligands is lone-pair electrons. Ligands are Lewis bases.

The atom in a ligand that furnishes an electron pair is called a *donor* atom. Ligands with one donor atom have just one point of attachment to the central metal atom or ion. They are **monodentate** ("one-toothed") ligands. Chloride ions, water molecules, and ammonia molecules are **monodentate** ligands.

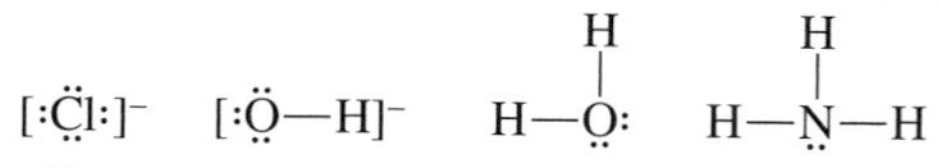

Ligand name: *chloro* *hydroxo* *aqua* *ammine*

Another common monodentate ligand is nitrite ion, NO_2^-, but here the donor atom can be either the N atom or one of the O atoms. The ligand is given a different name for each case.

The actual structure of NO_2^- is a resonance hybrid of two contributing structures, of which one is shown here. (What is the other?)

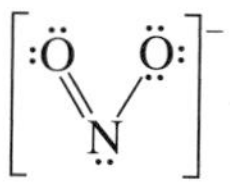

Ligand name: *nitro*, if N is donor atom
nitrito, if O is donor atom

Nitrite ion is monodentate because it can bond through only one of its two possible donor atoms. Ethylenediamine also has two donor atoms, but these are located far enough apart that the molecule can wrap itself around the central metal atom or ion and become attached to it through both donor atoms at the same time. It is a **bidentate** ("two-toothed") ligand.

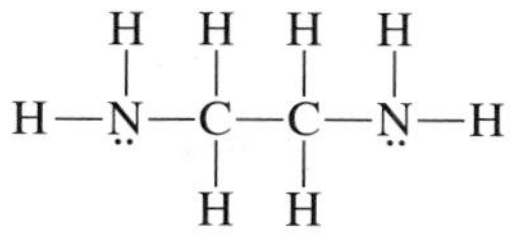

Ligand name: ethylenediamine (en)

The general name for ligands with multiple points of attachment is **polydentate** ("many-toothed"). Table 23.1 lists some common ligands.

TABLE 23.1 Some Common Ligands

MONODENTATE

Formula[a]	Name as Ligand[b]	Formula[a]	Name as Ligand[b]
	Neutral Molecules		
NH_3	ammine	NO	nitrosyl
H_2O	aqua	CH_3NH_2	methylamine
CO	carbonyl	C_5H_5N	pyridine
	Anions		
F^-	fluoro	NO_2^-	nitro
Cl^-	chloro	ONO^-	nitrito
Br^-	bromo	SCN^-	thiocyanato
I^-	iodo	NCS^-	isothiocyanato
OH^-	hydroxo	OSO_3^{2-}	sulfato
CN^-	cyano	SSO_3^{2-}	thiosulfato

POLYDENTATE

Name of Ligand[b]	Abbreviation	Formula[a]
ethylenediamine	en	$NH_2CH_2CH_2NH_2$
oxalato	ox	$[OOCCOO]^{2-}$
ethylenediaminetetraacetato	EDTA	$[(OOCCH_2)_2NCH_2CH_2N(CH_2COO)_2]^{4-}$

[a]Donor atoms are shown in red.
[b]Most neutral ligands carry the unmodified name. (Important exceptions are aqua, ammine, carbonyl, and nitrosyl.) Anion ligand names end in "o." This requires changing the terminal *-e* to *-o* (e.g., sulfate ⟶ sulfato). With many common anions, the entire *-ide* ending is changed to *-o* (e.g., cyanide ⟶ cyano).

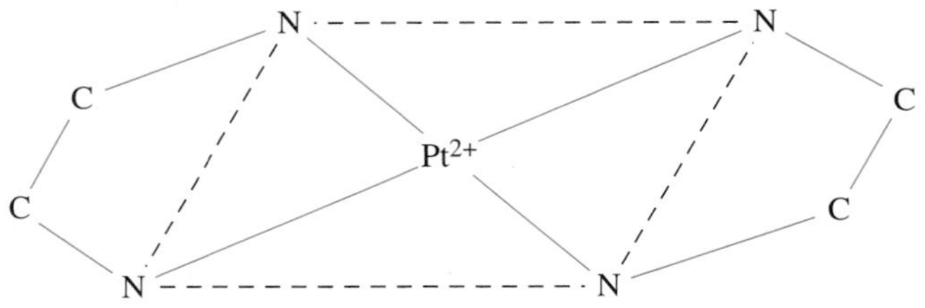

Figure 23.3 The chelate $[Pt(en)_2]^{2+}$. The N atoms of ethylenediamine (en) bond through lone-pair electrons. They attach at the corners along an edge of the square, but they *do not* bridge the square by attaching at opposite ends of a diagonal. For simplicity, H atoms are not shown.

One of the complex ions whose structure is shown in Figure 23.2 is $[Pt(NH_3)_4]^{2+}$. Suppose we replace each pair of NH_3 molecules along an edge of the square with the bidentate ligand ethylenediamine. The resulting complex ion, $[Pt(en)_2]^{2+}$, has the structure shown in Figure 23.3.

The interesting features in this structure are the two five-membered rings (pentagons) outlined in color. Each ring consists of one Pt, two N, and two C atoms. When a five- or six-membered ring is produced by the attachment of a polydentate ligand, we call the complex a **chelate**; the ligand is a *chelating agent*; and the process is called *chelation*. Chelate comes from the Greek word for claw. In the $[Pt(en)_2]^{2+}$ complex ion, it's as if the central Pt^{2+} ion is being pinched by two molecular claws. The ligand in Table 23.1 labeled EDTA is especially effective as a chelating agent. It has six donor atoms, and as noted in Figure 23.4, it forms complexes having five chelate rings, each with five members.

Chelate is pronounced "KEY-late."

23.3 Naming Complex Ions and Coordination Compounds

Some long-known complex ions and coordination compounds are still referred to by common or "trivial" names, though these names are no longer recommended. We encountered two examples in Chapter 22—potassium ferr*o*cyanide, $K_4[Fe(CN)_6]$, and potassium ferr*i*cyanide, $K_3[Fe(CN)_6]$. The "o" and the "i" indicate that the central ions are ferr*o*us (Fe^{2+}) and ferr*i*c (Fe^{3+}), respectively. With this information and the coordination number of 6, it's not hard to figure out the formulas of these two coordination compounds. What we need, though, is a

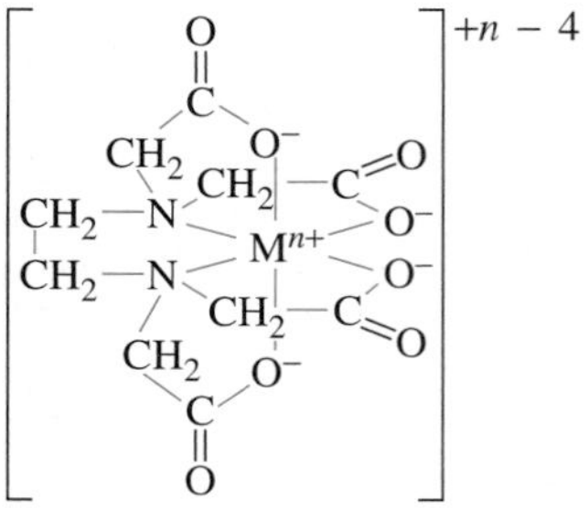

Figure 23.4 Structure of a metal–EDTA complex ion. The central metal ion M^{n+} (gray) can be any of many different cations: Ca^{2+}, Mg^{2+}, Fe^{2+}, Fe^{3+}, . . . The charge on the EDTA anion is 4–, and the net charge of the complex ion is that of the central ion (n+) plus that of the anion (4–), that is, $+n - 4$.

more general method for relating the names and formulas of complex ions. We can do this with data from Table 23.1 and a set of rules.

The list of rules needed to name all possible complexes is extensive and has undergone repeated revisions over the years. This means that a name for a given compound in the older chemical literature may differ from that in current publications, and that common practice followed by many chemists may differ from that of specialists in the field of coordination chemistry. We will just use a simplified list that generally achieves the goal of matching an acceptable name with a formula and an appropriate formula with a name.

1. When naming a complex, first name the ligands and then the central metal atom or ion.
2. List ligands in alphabetical order when naming a complex. Place anions before neutral molecules when writing the formula of a complex. (This rule is often not followed, however.)
3. Designate the number of ligands with a prefix: *mono* = 1, *di* = 2, *tri* = 3, *tetra* = 4, and so on. For ligands that have composite names—such as ethylene*di*amine, which itself has the prefix *di*—place parentheses around the ligand name and precede it with *bis* = 2, *tris* = 3, *tetrakis* = 4, and so on. (As in many other cases, the prefix mono is often not used.) The complex ion $[Pt(NH_3)_4]^{2+}$ in Figure 23.2 has the name tetraammineplatinum(II) ion; $[Pt(en)_2]^{2+}$ in Figure 23.3 has the name bis(ethylenediamine)platinum(II) ion.
4. Use the unmodified name for the central metal in a complex cation. Add the ending *ate* to the name of the central metal in a complex anion. Several common metals are given Latin-based names when they appear in complex anions (see Table 23.2). Whether in a complex cation or a complex anion, denote the oxidation state of the metal by a Roman numeral in parentheses. These rules are illustrated in Examples 23.2 and 23.3.

Example 23.2

Name the following: (**a**) $[CrCl_2(NH_3)_4]^+$; (**b**) $[Co(ox)_3]^{3-}$; (**c**) $K_4[Fe(CN)_6]$.

Solution

a. There are four (tetra) ammonia molecules (ammine) and two (di) chloride ions (chloro) as ligands. We name them in alphabetical order. *Two* units of negative charge are associated with the Cl^- ligands, and the net charge on the complex ion is *one plus*. The central ion must be Cr^{3+}. The name is tetraamminedichlorochromium(III) ion.

b. There are three *bi*dentate oxalate ions (ox) as ligands. Each oxalate ion carries a charge of 2–, and the central ion must be Co^{3+}. That is, the charge on the complex anion is (+3 – 6) = 3–. In naming this complex anion we must use the "ate" ending on cobalt: trioxalatocobaltate(III) ion.

c. This is a coordination compound made up of *four* K^+ cations for every complex anion, $[Fe(CN)_6]^{4-}$. The six (hexa) CN^- ligands are cyanide ions (cyano). Because the central Fe^{2+} ion is part of an anion, we must use the "ate" ending. Also, we use the Latin stem "ferr" for iron. The name of the complex anion is hexacyanoferrate(II), and the name of the coordination compound is potassium hexacyanoferrate(II).

TABLE 23.2
Names for Some Metals in Complex Anions

iron	*ferrate*
copper	*cuprate*
tin	*stannate*
silver	*argentate*
lead	*plumbate*
gold	*aurate*

Example 23.3

Write formulas for the following:

a. tetraammine copper(II) ion
b. triamminechlorodinitroplatinum(IV) ion
c. sodium hexanitrocobaltate(III)

Solution

a. This is a complex cation of Cu^{2+} having four (tetra) ammonia (ammine) molecules as ligands. Its formula is $[Cu(NH_3)_4]^{2+}$.

b. This is a complex cation of Pt^{4+} having three (tri) ammonia (ammine) molecules, one chloride (chloro) ion, and two (di) nitrite ions as ligands. The net charge on the complex ion is $+4 - 1 - 2 = 1+$. In writing the formula of the complex ion, the name "nitro" signifies that the donor atom of the nitrite ion is the N atom and that the ion should be represented as NO_2^-. Note also that we place the anions before the neutral ammonia molecules in the formula:

$$[PtCl(NO_2)_2(NH_3)_3]^+$$

c. This is a coordination compound made up of simple Na^+ ions and a complex anion. The anion has Co^{3+} as its central ion and six (hexa) nitrite (nitro) ions, NO_2^-, as ligands. The net charge on the anion is $(+3 - 6) = 3-$. There must be three Na^+ ions for every anion. The formula of the coordination compound is $Na_3[Co(NO_2)_6]$.

Exercise 23.2

Name the following: **(a)** $[Ag(NH_3)_2]^+$; **(b)** $[AuCl_4]^-$; **(c)** $[CoBr(NH_3)_5]Br_2$.

Exercise 23.3

Write formulas for the following:

a. (tris)ethylendiaminecobalt(III) ion
b. diamminetetrachlorochromate(III) ion
c. dichlorobis(ethylenediamine)platinum(IV) sulfate

23.4 Isomerism in Complex Ions and Coordination Compounds

On a number of occasions we have encountered isomerism—that is, situations where substances with the same formula have different structures and properties. Several types of isomerism are found among complex ions and coordination compounds.

One general type of isomers, called *structural* isomers, differ in the ligands that are attached to the central atom and/or the donor atoms through which they are bonded. For example, the following complex ions have the same ligands and in the same numbers. However, as seen on page 882, the donor atom in NO_2^- can be either the N atom or one of the O atoms. Formulas and names of the isomers of this complex ion are written below (with the donor atom shown in color).

$[Co(NO_2)(NH_3)_5]^{2+}$	$[Co(ONO)(NH_3)_5]^{2+}$
Pentaamminenitrocobalt(III) ion	Pentaamminenitritocobalt(III) ion

The two coordination compounds whose formulas and names are given on page 887 are also isomers. They have the same formula weight and the same percent composition by mass, but they differ in the ions present in their aqueous solutions. One compound has Cl^- as the free anions, and the other has SO_4^{2-}. The two compounds also differ in the ligands found in the complex cation.

$[Cr(SO_4)(NH_3)_5]Cl$ — Pentaamminesulfatochromium(III) chloride

$[CrCl(NH_3)_5]SO_4$ — Pentaamminechlorochromium(III) sulfate

Example 23.4

Indicate whether isomerism exists in each pair:

a. $[Co(NH_3)_6][Cr(CN)_6]$ and $[Cr(NH_3)_6][Co(CN)_6]$
b. $[Cr(H_2O)_5(NH_3)]^{3+}$ and $[Cr(NH_3)(H_2O)_5]^{3+}$
c. $[Co(H_2O)_5(NH_3)]^{3+}$ and $[Co(H_2O)(NH_3)_5]^{3+}$

Solution

a. The overall compositions of these two coordination compounds are the same, as are their formula weights. Also, they have the same numbers and types of ligands. What differs is the distribution of ligands between the complex cation and complex anion. For example, the cation in one compound is hexaamminecobalt(III) ion, whereas in the other it is hexaamminechromium(III) ion. The two compounds are isomers.

b. Each of these complex ions has one NH_3 and five H_2O molecules as ligands and Cr^{3+} as the central ion. The formulas are written differently, but the structures are identical. These two formulas do *not* represent isomers; they represent the same complex ion.

c. Each complex ion has Co^{3+} as the central ion and a total of six ligands. However, one ion has five H_2O and one NH_3 molecule, and the other has five NH_3 and one H_2O molecule. These two formulas represent *different* complex ions. This is *not* a case of isomerism.

Exercise 23.4

Indicate whether different isomers are possible of the following:
(a) $[Zn(H_2O)(NH_3)_3]^{2+}$; **(b)** $[Cu(NH_3)_4][PtCl_4]$.

In another type of isomerism, known as **stereoisomerism**, the number and types of ligands and their mode of attachment are all the same, but the manner in which the ligands occupy the space around the central atom differs. Let us next consider two important types of stereoisomerism.

Geometric Isomerism

Consider again the complex ion $[Pt(NH_3)_4]^{2+}$ pictured in Figure 23.2. Suppose we replace two of the NH_3 ligands by Cl^- ions. Figure 23.5 shows that there are *two* ways that we can do this: the two Cl^- ions can be either along the same edge of the square (cis) or on opposite corners (trans). This is similar to the cis-trans isomerism of alkenes that we encountered in Section 10.4.

The two isomers in Figure 23.5 are electrically neutral. They are complexes, but not complex ions.

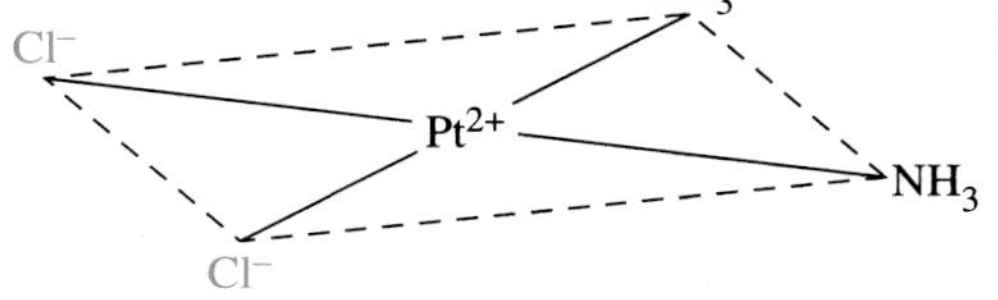

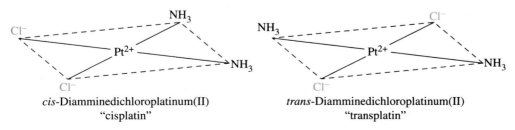

cis-Diamminedichloroplatinum(II) "cisplatin"

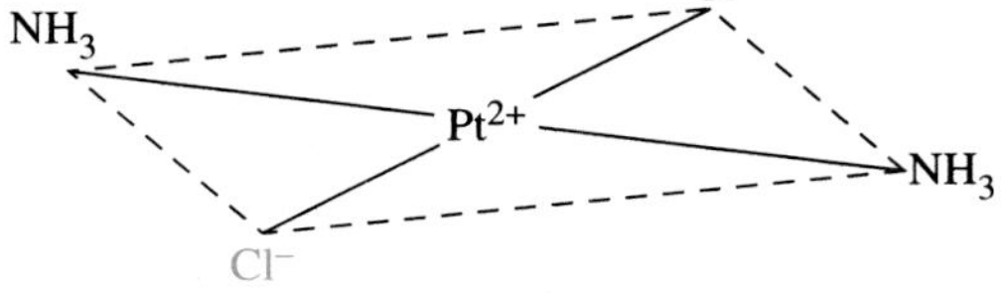

trans-Diamminedichloroplatinum(II) "transplatin"

Figure 23.5 Geometric isomerism in a square planar complex.

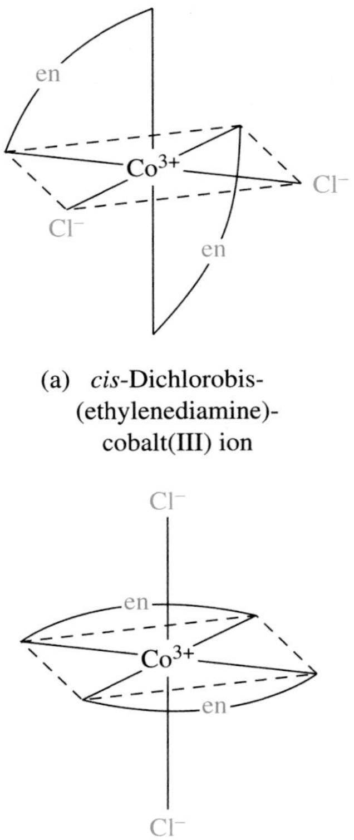

(a) *cis*-Dichlorobis-(ethylenediamine)-cobalt(III) ion

(b) *trans*-Dichlorobis-(ethylenediamine)-cobalt(III) ion

Figure 23.6 **Geometric isomerism in an octahedral complex ion.**

Some of the most important examples of optical isomers are found in organic chemistry, especially among biologically active molecules.

The two isomers of diamminedichloroplatinum(II) offer a striking example of the critical relationship between structure and properties. The cis isomer, known as "cisplatin," is a powerful antitumor drug used in chemotherapy. Apparently, the two Cl^- ligands are just the right distance apart so that the complex can latch onto the DNA in cancerous cells and inhibit further cell growth. In the trans isomer the Cl^- ligands are too far apart for the complex to be effective.

Figure 23.6 presents an example of cis and trans isomerism in an octahedral complex of Co^{3+}: $[CoCl_2(en)_2]^+$. The cis isomer has two Cl^- ions along the same edge, and the trans isomer has two Cl^- ions on opposite corners of the octahedron.

In Figure 23.5, if we replace the two NH_3 ligands of "cisplatin" by Cl^- or the two Cl^- by NH_3, there is no isomerism in the product ($[PtCl_4]^{2-}$ or $[Pt(NH_3)_4]^{2+}$). In Figure 23.6, if we replace the two Cl^- ligands of the cis isomer by a third ethylenediamine, on first thought, we again expect no isomerism. But, in fact, there are *two* different tris(ethylenediamine)cobalt(III) ions. How do we know?

Optical Isomers

The two ions are the same in almost all their properties, but they differ with respect to one property called *optical activity* (see Figure 23.7). The two ions are a type of optical isomer called **enantiomers**. The key to identifying enantiomers is to compare a structure with its mirror image. Every geometric structure has a mirror image, and in many cases the structure and its mirror image are identical. That is, following an appropriate rotation if necessary, the structure can be fitted over its mirror image. Enantiomers have *nonsuperimposable* mirror-image structures (see Figure 23.8) The structure of one of a pair of enantiomers cannot be fitted over that of the other (its mirror image), just as a right-handed glove does not fit a left hand. The only difference between a pair of enantiomers is in their ability to rotate the plane of polarized light (Figure 23.7). One enantiomer rotates the plane of polarized light to the right, and the other rotates it by exactly the same degree but to the left. Measurements of optical activity played an important role in the development of coordination chemistry.

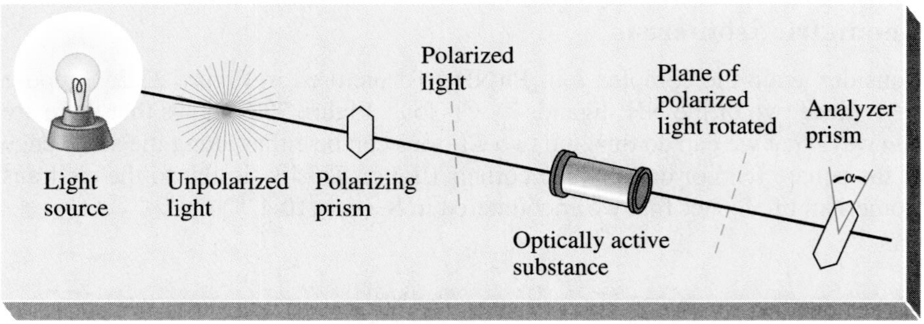

Figure 23.7 **Optical activity.** Ordinary light consists of electromagnetic waves that vibrate in all planes; it is unpolarized. Some substances are able to screen out all waves except those vibrating in a particular plane. Light having all vibrations in the same plane is *polarized*. Optically active substances are able to rotate the plane of polarized light. One optical isomer rotates the plane to the right (here, by the angle α), and the other optical isomer, to the left.

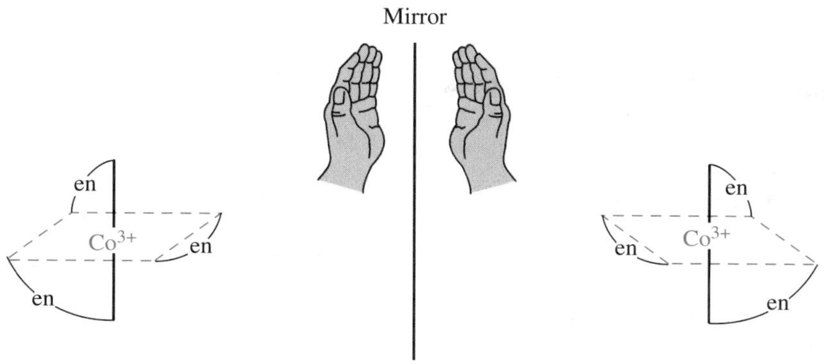

Figure 23.8 Optical isomers. The two structures are nonsuperimposable mirror images. They are like a right hand and a left hand. One cannot be superimposed onto the other.

CONCEPTUAL EXAMPLE 23.1

At one time the structure shown in Figure 23.9a was proposed as an alternative to the octahedral structure for coordination number 6. Assuming that the ethylenediamine (en) molecule can link only to adjacent coordination sites, determine the number of possible isomers of $[Co(en)_3]^{3+}$ based on this prism structure. Would any of these isomers exhibit optical activity?

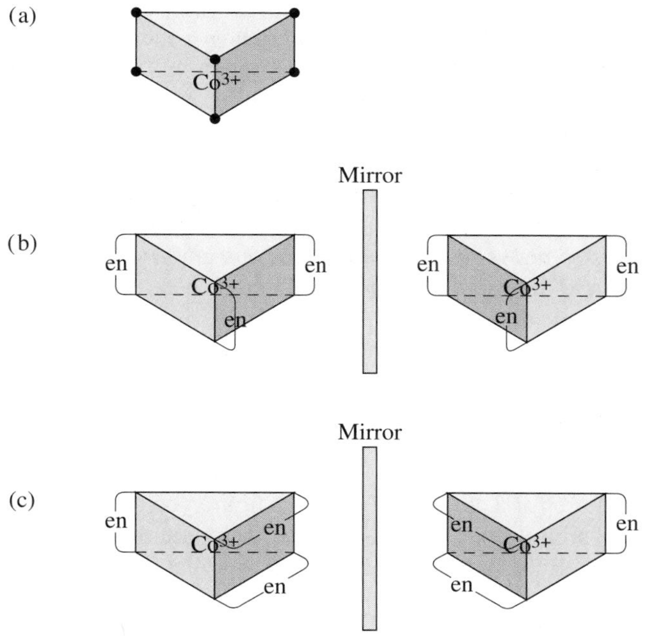

Figure 23.9 An alternate geometric structure at one time proposed for coordination number 6. The structures shown here are described in this example.

Solution

One possible structure, shown in Figure 23.9b, has (en) ligands along the three vertical edges of the prism. Another structure, shown in Figure 23.9c, has one (en) ligand along a vertical edge, and the remaining (en) ligands in the upper and lower triangular faces. These two structures are *geometric* isomers, and there are no other possible geometric isomers. (Can you see, for example, that there cannot be a structure with (en) ligands along just two of the vertical edges?)

To assess whether either of the geometric isomers also exhibits optical isomerism, we need to compare each structure with its mirror image. As we can see from Figure 23.9, each isomer yields a mirror image on which it is superimposable. There is no optical activity.

Conceptual Exercise 23.1

Another early suggestion for coordination number 6 was that of a planar hexagon.

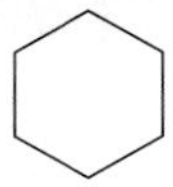

How many geometric and/or optical isomers would you predict for $[Co(en)_3]^{3+}$ based on the hexagonal structure?

23.5 Bonding in Complexes: Crystal Field Theory

The Lewis acid–base theory helped us explain how a metal ion and ligands may join to form a complex ion. But we need more than this to explain certain properties of complex ions, especially their characteristic colors and magnetic properties. A theory that works quite well is the *crystal field theory*.

Crystal field theory views the attractions between a central atom or ion and its ligands as being largely electrostatic: Lone-pair electrons on the ligands are attracted to the positively charged nucleus of the central metal. However, there are also repulsions between the ligand electrons and *d* electrons of the central metal. Crystal field theory focuses on the effect of these repulsions.

Recall from our discussion of atomic orbitals in Chapter 6 that *d* orbitals have a value of 2 for the angular momentum quantum number, l. When $l = 2$, the magnetic quantum number, m_l, may have the values −2, −1, 0, 1, and 2. Thus there are *five d* orbitals. We first showed these in Figure 6.27. We show the *d* orbitals again in Figure 23.10, this time as a group and with the approach of ligands along the x, y, and z axes. This is the direction of approach that produces an octahedral structure for a complex. From this figure you can see that electrons in the d_{z^2} and $d_{x^2-y^2}$ orbitals exert a maximum repulsion on the approaching ligands. This raises the energies of these *d* orbitals above what they would be in the free ion in the absence of ligands. The energies of the other *d* orbitals are not raised as much, and the differing interactions of ligands with *d* orbitals cause the *d* energy level of the central metal ion to split into two groups. We will designate the energy difference between the two groups by the symbol Δ.

Figure 23.11 is a schematic representation of (1) the energy levels of *d* orbitals in a hypothetical central ion free of ligands, (2) the average energy level to which the *d* orbitals are raised by the ligands, (3) the splitting of the energy levels that occurs in the case of (a) tetrahedral, (b) octahedral, and (c) square planar complex ions.

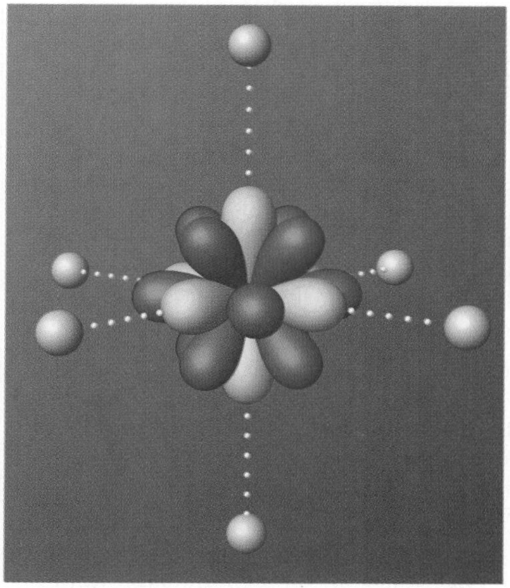

Figure 23.10 Ligand approach leading to formation of an octahedral complex. In the formation of an octahedral complex, ligands approach the central atom or ion along the x, y, and z axes. Maximum interference occurs with the d_{z^2} and $d_{x^2-y^2}$ orbitals (shown in yellow). The energies of these orbitals are raised with respect to those of the d_{xy}, d_{xz}, and d_{yz} orbitals (shown in red).

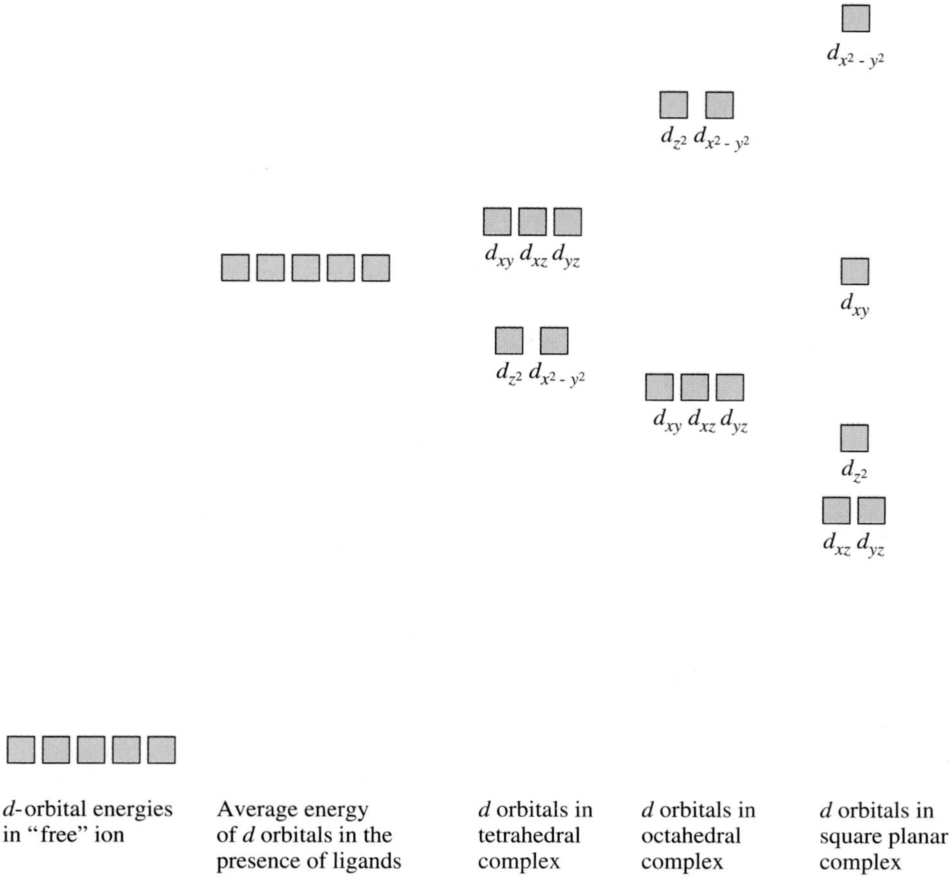

Figure 23.11 Schematic representation of d-level splitting.

As a reminder, iron has $Z = 26$. In Fe^{3+}, there are 23 electrons in the configuration $[Ar]3d^5$.

Of what use is all this, you may wonder. To see, let's consider how the five $3d$ electrons in Fe^{3+} are distributed among the $3d$ orbitals in two of its complex ions.

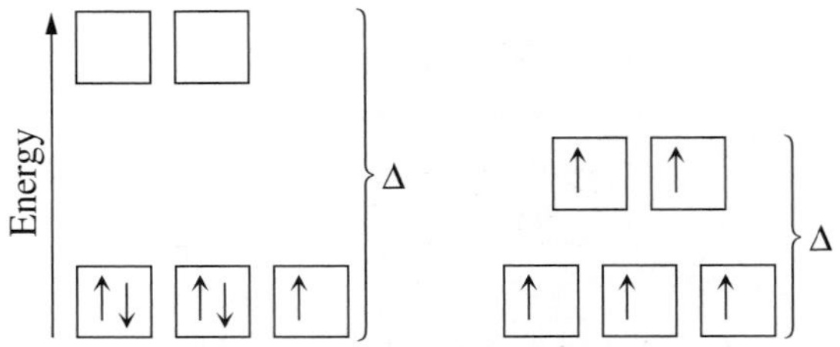

Fe^{3+} in $[Fe(CN)_6]^{3-}$, a "low spin" complex ion

Fe^{3+} in $[Fe(H_2O)_6]^{3+}$, a "high spin" complex ion

When we apply the "aufbau" principle to the five $3d$ electrons of Fe^{3+}, our tendency is to place these singly into $3d$ orbitals, all with parallel spins, as in $[Fe(H_2O)_6]^{3+}$. The aqua complex ion has *five* unpaired electrons and is known as a "high spin" complex ion. In $[Fe(CN)_6]^{3-}$, on the other hand, because the energy separation between the two groups of d orbitals is so large—larger than the energy benefit in having unpaired electrons—all the electrons are found in the lower-energy set of orbitals. Four of the electrons are paired and only *one* is unpaired; the complex ion is "low spin."

To predict the magnetic properties of complex ions, then, we need to know the structure of the complex ion (tetrahedral, octahedral, or square planar) and the ability of ligands to split the d-orbital energy level before we can determine the number of unpaired electrons. The following general arrangement, called the **spectrochemical series**, shows the relative abilities of some common ligands to split the d-orbital energy level. (Ligands can be referred to by terms such as strong field and weak field.)

Strong field *Intermediate field* *Weak field*

$$CN^- > NO_2^- > en > NH_3 > H_2O > OH^- > F^- > Cl^- > Br^- > I^-$$

The energy-level diagrams that we drew for the cyano and aqua complex ions of Fe^{3+} are consistent with the order suggested by the spectrochemical series.

Example 23.5

How many unpaired electrons would you expect for the octahedral complex ion $[CoF_6]^{3-}$?

Solution

We must do three things: (1) determine the number of $3d$ electrons in the central cobalt ion, (2) assess whether the energy difference between the two groups of d orbitals, Δ, is likely to be large or small, and (3) distribute electrons among the d orbitals in accordance with the aufbau principle.

1. The complex ion has six F^- ions as ligands and a net charge of 3–. The central ion must carry a charge of 3+, that is, Co^{3+}. The atomic number of cobalt is 27, and the number of electrons in Co^{3+} is 24, in the configuration $[Ar]3d^6$. The number of $3d$ electrons is six.
2. In the spectrochemical series, F^- is listed as a weak field ligand. We should expect the energy separation between the two sets of d orbitals, Δ, to be small.

3. Because of this small energy separation we should assign electrons to all five of the *d* orbitals singly before forming any pairs. There should be *four* unpaired electrons.

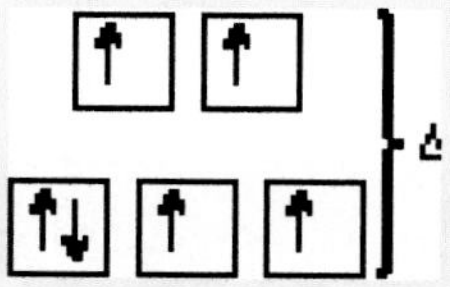

Exercise 23.5

How many unpaired electrons would you expect to find in each of the following complex ions: (**a**) the octahedral complex ion, $[Co(CN)_6]^{3-}$; (**b**) the tetrahedral complex ion, $[NiCl_4]^{2-}$? (*Hint*: Refer to Figure 23.11.)

23.6 Color in Complex Ions and Coordination Compounds

A substance appears colored if the energy of some component of visible light matches the energy required to excite an electron in the substance to a higher energy level. The substance absorbs that component and transmits others. That is, the substance absorbs one color and transmits another. This idea is illustrated in Figure 23.12, where the colors of five solutions encountered elsewhere in the text are pictured in relation to the color of light they absorb.

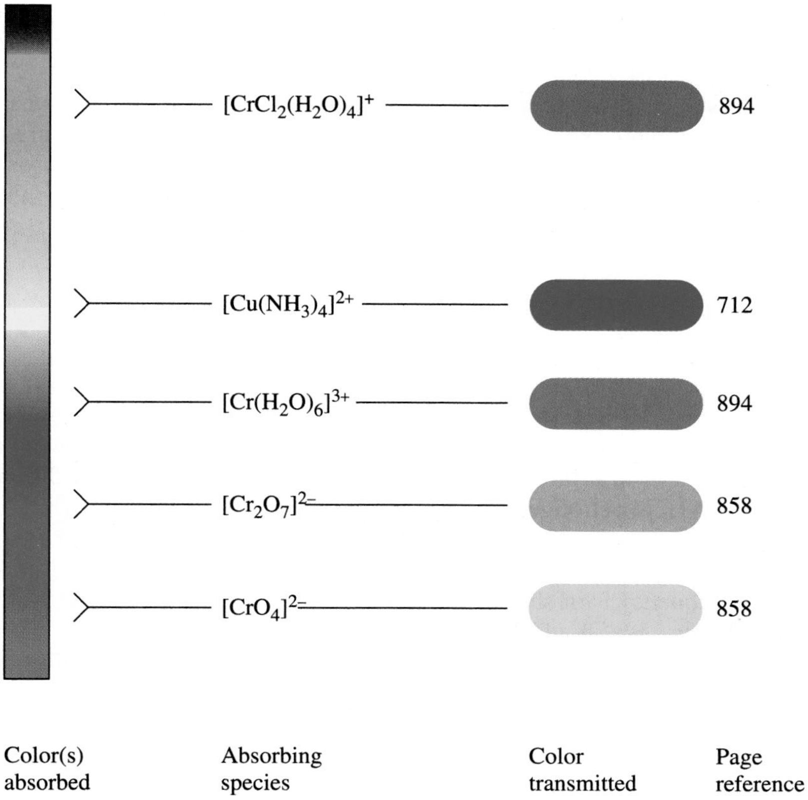

Figure 23.12 Visible spectrum and the origin of color.

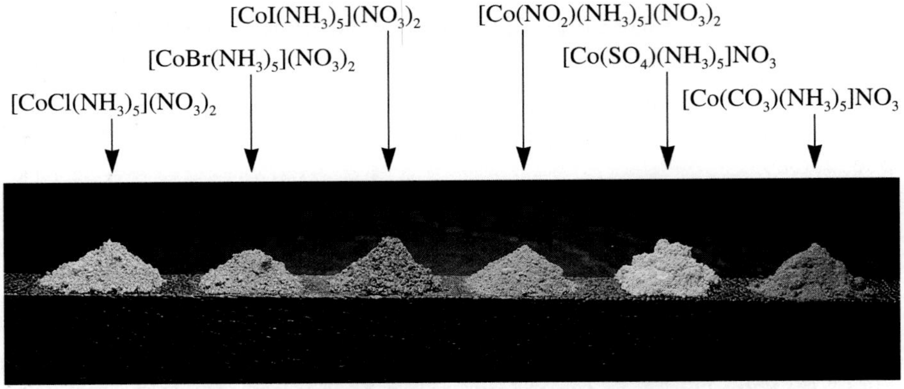

Figure 23.13 Effect of ligands on colors of coordination compounds.

Ions having a noble gas electron configuration (Na^+, for example, or Cl^-), an outer shell with 18 electrons (for example, Zn^{2+}), or an "18 + 2" configuration (for example, Sn^{2+}) have *no* electron transitions in the energy range of visible light. In aqueous solution these ions do not absorb visible light, and their solutions are colorless.

As shown in Figure 6.14, the wavelength range for visible light is from about 760 to 390 nm. This corresponds to a frequency range of about 3.9×10^{14} to 7.7×10^{14} Hz ($\nu = c/\lambda$) and an energy per photon of 2.6×10^{-19} to 5.1×10^{-19} J ($E = h\nu$).

Many complex ions are colored because the energy differences between *d* orbitals, Δ, match the energies of components of visible light. Substituting one ligand for another can produce subtle changes in the energy levels of the *d* orbitals and striking changes in the colors of complex ions, as seen in Figures 23.13 and 23.14.

Crystal field theory helps us to explain the colors of complex ions. For example, to explain why solutions of the complex ion $[Cr(H_2O)_6]^{3+}$ are *violet* in color whereas those of $[Cr(NH_3)_6]^{3+}$ are *yellow*, let's construct a *d*-orbital energy-level diagram for these octahedral complex ions. We begin by noting that the electron configuration of Cr is $[Ar]3d^54s^1$ and that of Cr^{3+} is $[Ar]3d^3$.

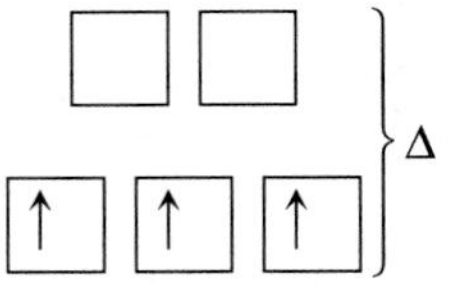

Figure 23.14 Colors of chromium(III) complex ions. When the green solid $CrCl_3 \cdot 6H_2O$ is dissolved in water, a green solution is obtained. The green color is due to $[CrCl_2(H_2O)_4]^+$ (left). A slow exchange of H_2O molecules for Cl^- ions as ligands leads to a reddish solution of $[Cr(H_2O)_6]^{3+}$ in one or two days (right).

The color of light absorbed in promoting an electron from a lower-energy to a higher-energy *d* orbital depends on the magnitude of the energy difference, Δ. Because NH_3 is a stronger field ligand than H_2O, Δ is larger for $[Cr(NH_3)_6]^{3+}$ than for $[Cr(H_2O)_6]^{3+}$. This means light of higher energy and thus greater frequency is absorbed by the ammine complex ion than by the aqua complex ion. Light of higher frequency is of shorter wavelength and lies more toward the violet end of the spectrum. We should expect $[Cr(NH_3)_6]^{3+}$ to absorb violet light and transmit yellow. Thus, the color of $[Cr(NH_3)_6]^{3+}(aq)$ should be yellow (similar to CrO_4^{2-} shown in Figure 23.12). The light absorbed by $[Cr(H_2O)_6]^{3+}$ is of lesser energy, lower frequency, and longer wavelength. Figure 23.12 shows that $[Cr(H_2O)_6]^{3+}$ absorbs in the green region of the spectrum and displays a violet color.

23.7 Some Practical Applications of Coordination Chemistry

There are wide-ranging applications of coordination chemistry in the laboratory and in a host of commercial products. We will briefly consider some of these applications.

Qualitative Analysis

We outlined a classical scheme of qualitative cation analysis in Chapter 18 (recall Figure 18.9) and mentioned the role of complex ions in two procedures: The formation of $[Ag(NH_3)_2]^+$ provides a means of separating $AgCl(s)$ from $Hg_2Cl_2(s)$ and $PbCl_2(s)$ in cation Group 1, and the formation of hydroxo complexes is the basis of separating Al^{3+}, Cr^{3+}, and Zn^{2+} from several other cations in Group 3.

Several of the confirmatory tests for cations involve the formation of colored complexes. The test for $Fe^{3+}(aq)$ involves the formation of the blood-red complex ion $[Fe(SCN)(H_2O)_5]^{2+}$ and the test for $Co^{2+}(aq)$, the blue complex ion $[Co(SCN)_4]^{2-}$. However, even trace amounts of $Fe^{3+}(aq)$ interfere with the test for cobalt(II) because the intense red color of $[Fe(SCN)(H_2O)_5]^{2+}$ completely masks the blue color of $[Co(SCN)_4]^{2-}$. Fortunately, this difficulty is readily overcome by adding NaF(aq), which destroys the $[Fe(SCN)(H_2O)_5]^{2+}$ and replaces it with the extremely stable, pale yellow $[FeF_6]^{3-}$.

$$[Fe(SCN)(H_2O)_5]^{2+}(aq) + 6\,F^-(aq) \longrightarrow [FeF_6]^{3-}(aq) + SCN^-(aq) + 5\,H_2O$$

The tests described here are shown in Figure 23.15.

Figure 23.15 Qualitative analysis tests for Co^{2+} and Fe^{3+}. $Co^{2+}(aq)$ combines with thiocyanate ions to form a blue complex ion, $[Co(SCN)_4]^{2-}(aq)$ (left). $Fe^{3+}(aq)$, even if present only in trace amounts, combines with thiocyanate ions, $SCN^-(aq)$, to produce a strongly colored blood-red complex ion, $[FeSCN(H_2O)_5]^{2+}(aq)$ (center). If a solution containing $Co^{3+}(aq)$ and a trace of $Fe^{3+}(aq)$ is treated with an excess of $F^-(aq)$, the $Fe^{3+}(aq)$ is tied up in the extremely stable, pale yellow complex ion, $[FeF_6]^{3-}(aq)$. The mixture of $[FeF_6]^{3-}(aq)$ and $[Co(SCN)_4]^{2-}(aq)$ is blue-green in color (right).

Sequestering Metal Ions

To sequester a metal ion means to tie it up in a form that effectively removes the ion from solution. One of the best ligands for sequestering metal ions is the ethylenediaminetetraacetate ion, $EDTA^{4-}$. This hexadentate ligand was shown in Table 23.1, and the structure of a typical metal chelate was given in Figure 23.4.

One application of EDTA chelation is in the treatment of boiler water. When they are the central ions in chelate structures, Ca^{2+}, Mg^{2+}, and Fe^{3+} are no longer able to form boiler scale (page 299). Another application of EDTA, in combination with preservatives, is to prevent the growth of certain bacteria in liquid soaps, shampoos, and other personal products. Chelation with EDTA removes

This iron chelate plant food is an iron–EDTA complex.

Ca^{2+} and Mg^{2+} ions, which are important constituents of the cell walls of these bacteria. The cell walls disintegrate, and the bacteria die.

In another use, a person suffering from lead poisoning can be fed a Ca–EDTA complex. Pb^{2+} ions displace Ca^{2+} from the complex, forming the still more stable Pb–EDTA complex, which is excreted.

$$Pb^{2+}(aq) + [CaEDTA]^{2-}(aq) \longrightarrow [PbEDTA]^{2-}(aq) + Ca^{2+}(aq)$$

$$K_f = 4 \times 10^{10} \qquad K_f = 2 \times 10^{18}$$

A similar treatment can be used to rid the body of radioactive isotopes, as in the treatment for plutonium poisoning.

Because of its hexadentate character and high negative charge, EDTA has the interesting effect of converting a simple cation into the central ion of an anion, for example, $Fe^{3+}(aq)$ to $[FeEDTA]^-$. For use as a plant food, iron can be effectively transported through soils as $[FeEDTA]^-(aq)$. Clay particles in soils have anions on their surfaces, and $Fe^{3+}(aq)$ cannot easily migrate through such a soil; rather it is immobilized by combining with surface anions. However, complex anions containing iron(III) are not held back by these surface anions, and the iron(III) is thus available to plants.

Porphyrins: One of Nature's Favorite Kinds of Ligand

Figure 23.16 shows the structure of a ligand commonly found in plant and animal matter. If the eight R groups are all H atoms, the molecule is called porphin. The central N atoms can give up their two H atoms, and a metal atom or ion can coordinate simultaneously with all four N atoms. The porphin is a tetradentate ligand or chelating agent for the central metal. The metal–ligand complex is called a *porphyrin*. Individual porphyrins differ in their central metal and in the R groups present on the porphin rings.

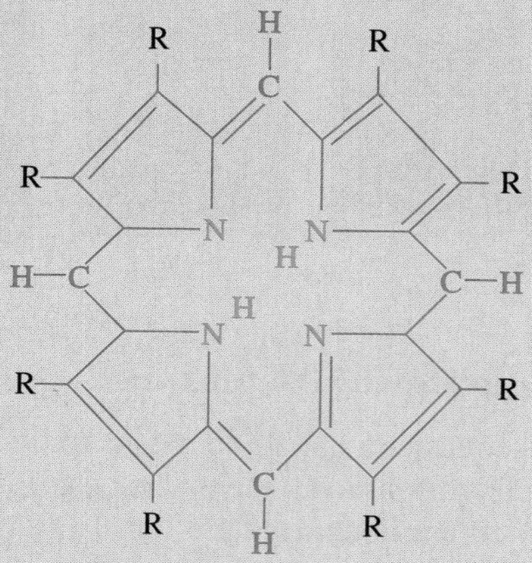

Figure 23.16 The structure of the ligand porphin.

The heme unit in hemoglobin, which we previously described in Figure 22.10, is a porphyrin that has Fe^{2+} in the center of its structure. Four of the coordination sites of the Fe^{2+} are occupied by N atoms of porphin, a fifth site perpendicular to the plane of the porphin rings is occupied by a nitrogen atom of the protein globin, and a sixth site, also perpendicular to the plane of the rings is the binding site for the O_2 molecule which the hemoglobin transports (Figure 23.17).

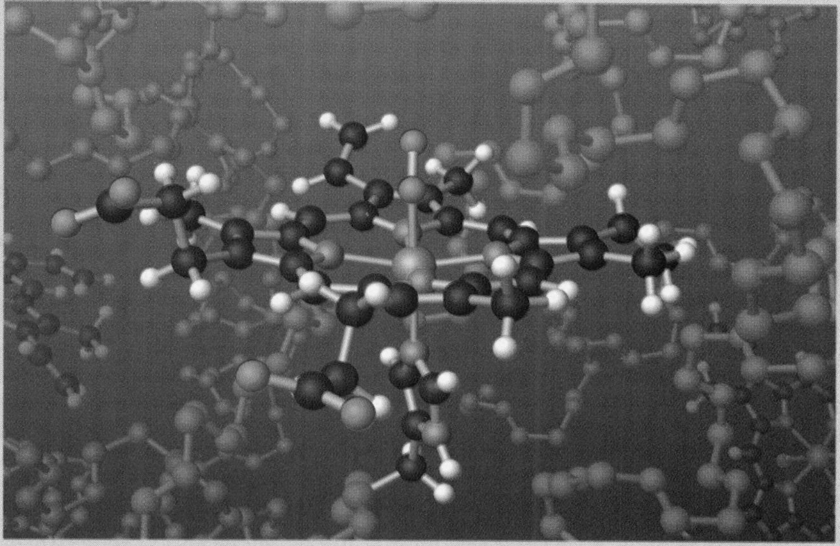

Figure 23.17 The heme unit and oxygen binding site in hemoglobin. The hemoglobin molecule consists of thousands of atoms, but the key portions of the molecule are four heme groups, one of which is shown here. Each heme has an iron atom (gray) in the center of a square formed by four nitrogen atoms. The heme group is able to bind one oxygen molecule (shown here pointing up) to the iron atom.

Other iron porphyrins, called cytochromes, differ from heme in that the fifth and sixth coordination positions are permanently coordinated to R groups of protein chains. The cytochromes cannot act as oxygen carriers. Instead, because the oxidation state of the central ion can change reversibly between Fe(II) and Fe(III), cytochromes are electron transfer agents.

In the various chlorophylls, which are catalytic agents in photosynthesis,

$$n\,CO_2 + n\,H_2O \xrightarrow[\text{chlorophyll}]{\text{sunlight}} \underset{\text{carbohydrate}}{(CH_2O)_n} + n\,O_2$$

the central atom position is occupied by Mg^{2+} (Figure 23.18).

Still another porphyrin is cyanocobalamine, vitamin B_{12}. Here the central metal is cobalt and one of the six coordination sites is occupied by cyanide ion. Vitamin B_{12} is found in liver, meat, eggs, and fish, and a deficiency of this vitamin causes pernicious anemia.

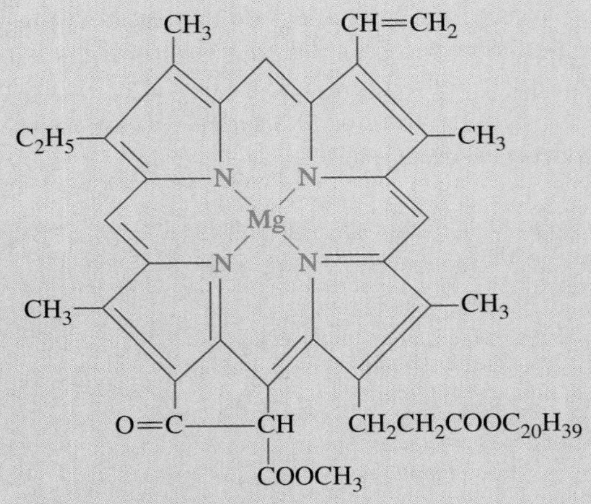

Figure 23.18 The structure of chlorophyll a.

SUMMARY

The central metal atom or ion of a metal complex is a Lewis acid. It forms coordinate covalent bonds by accepting lone-pair electrons from ligands, which are Lewis bases. A monodentate ligand attaches at a single coordination site of the central metal, and a polydentate ligand attaches at two or more sites. Chelates are complexes in which the attachment of polydentate ligands creates five or six-membered rings of atoms.

Some of the information that should be conveyed by the names and formulas of complexes are the identity and oxidation state of the central metal atom, the identities and numbers of ligands, and the net charge on the complex.

Isomerism among complexes is of two general types. Structural isomers differ in the ligands that are attached to the central atom and/or the donor atoms through which they are bonded. Stereoisomers differ in the manner in which ligands occupy the space around the central atom. Geometric isomerism (cis-trans) is one type of stereoisomerism and optical isomerism is another. Enantiomers, a type of optical isomers, bear the same relationship to one another as an object and its nonsuperimposable mirror image. The only physical property in which enantiomers differ is in their ability to rotate the plane of polarized light. One isomer rotates it to the right and the other to the left.

Interactions between ligand lone-pair electrons and electrons in the *d* orbitals of the central metal atom or ion produce a splitting of the *d* energy level. Electron transitions between *d* orbitals of different energy provide a way for a complex to absorb some wavelength components of visible light and transmit others, giving rise to color. Color is a common property of transition metal complexes, and it is strongly influenced by the particular ligands present. Explanations of the colors and magnetic properties of complexes are facilitated by a listing of common ligands in a form known as the spectrochemical series.

The chemistry of complexes is encountered in many of the subdisciplines of chemistry and plays an important role in biology as well. Practical applications of coordination chemistry are numerous and widespread, including many uses in consumer products.

KEY TERMS

bidentate (23.2)
chelate (23.2)
complex (23.1)
complex ion (23.1)
coordination compound (23.1)
coordination number (23.1)

crystal field theory (23.5)
enantiomers (23.4)
Lewis acid (23.2)
Lewis base (23.2)
ligand (23.1)
monodentate (23.2)
polydentate (23.2)
spectrochemical series (23.5)
stereoisomerism (23.4)

REVIEW QUESTIONS

1. What are the component parts of a complex ion?
2. What is the relationship between a complex ion and a coordination compound?
3. According to Werner's theory, what is the primary valence of the central metal atom in a complex ion? How is this related to the oxidation state of the central metal atom?
4. According to Werner's theory, what is the secondary valence of the central metal atom in a complex ion? How is this related to the coordination number of the central metal atom?
5. Explain how it is possible that a complex containing a cation may itself be an anion. Could the complex be an electrically neutral molecule?
6. Explain how it is possible that the coordination number and the number of ligands in a complex ion may not be the same.
7. What is the Lewis acid–base theory, and how is it used to describe complex ion formation?
8. Explain how it is that ammonia is both a Brønsted–Lowry base and a Lewis base.
9. Explain the meaning and give an example of a *monodentate* ligand and of a *polydentate* ligand.
10. What is a *chelate* and what conditions are required for its formation?
11. What are the ligands represented by the following names or symbols?
   a. aqua **b.** ammine **c.** en **d.** ox
12. What are the ligands represented by the following names?
   a. chloro **b.** carbonato
   c. nitrato **d.** hydrogensulfito
13. What do the terms *bis* and *tris* mean when they are used in the name of a complex ion?
14. What does the ending *ate* signify when referring to the metal atom in a complex ion, for example, cobaltate(III)?
15. What is the difference between a *nitro* and a *nitrito* ligand in a complex ion?
16. What is the difference between a *cis* and a *trans* isomer of a complex ion?
17. How does stereoisomerism differ from structural isomerism?
18. How does optical isomerism differ from geometrical isomerism?
19. What is meant by the terms "high spin" and "low spin" complex? "strong field" and "weak field" ligand?
20. What is the spectrochemical series?

PROBLEMS

Werner's Theory of Coordination Chemistry

21. What is the coordination number of the central metal atom in the following complexes?
   a. $[PtCl_3(NH_3)_3]^+$ **b.** $[Cu(en)_2]^{2+}$
   c. $Ni(CO)_4$ **d.** $[FeEDTA]^{2-}$
22. How many ligands of the following type would be found in an octahedral complex with Cr^{3+} as the central metal ion?
   a. CN^- **b.** $NH_2CH_2CH_2NH_2$
   c. $^-OOCCOO^-$ **d.** $EDTA^{4-}$
23. Indicate the oxidation state of the central metal atom in each of the following.
   a. $[Pt(en)_2]^{2+}$ **b.** $[CoCl_2(NH_3)_4]^+$
   c. $[Mn(CN)_6]^{3-}$ **d.** $[CrCl_4(H_2O)_2]^-$
24. Indicate the oxidation state of the central metal atom in each of the following.
   a. $[NiBr_4]^{2-}$ **b.** $[FeF_5(H_2O)]^{2-}$
   c. $[PtCl_2(NH_3)_4]^{2+}$ **d.** $[Fe(CO)_5]$.

Nomenclature

25. Name the following complex ions.
   a. $[Cu(NH_3)_4]^{2+}$ **b.** $[FeF_6]^{3-}$
   c. $[PtCl_2(NH_3)_4]^{2+}$ **d.** $[Cr(en)_3]^{3+}$
26. Name the following complex ions.
   a. $[Fe(OH)(H_2O)_5]^+$
   b. $[Co(en)_2(NO_2)_2]^+$
   c. $[Ag(S_2O_3)_2]^{3-}$
   d. $[Fe(ox)_3]^{3-}$

27. Write the formula of each of the following.
    a. hexaaquairon(II) ion
    b. tetraaminebromochlorochromium(III) ion
    c. trioxalatoaluminate(III) ion

28. Write the formula of each of the following.
    a. hexafluorocobaltate(III) ion
    b. tetraaquadichlorochromium(III) ion
    c. dibromobis(ethylenediamine)cobalt(III) ion

29. Name the following coordination compounds:
    a. $K_4[Cr(CN)_6]$ b. $K_3[Cr(ox)_3]$

30. Name the following coordination compounds:
    a. $K_2[PtCl_6]$ b. $NH_4[Cr(NCS)_4(NH_3)_2]$

31. Write formulas for the following coordination compounds:
    a. sodium tetrahydroxozincate(II)
    b. tris(ethylenediamine)chromium(III) sulfate
    c. dipotassium sodium hexanitrocobaltate(III)

32. Write formulas for the following coordination compounds:
    a. tetraamminecopper(II) tetrachloroplatinate(II)
    b. tris(ethylenediamine)chromium(III) hexacyanocobaltate(III)
    c. tetraammineplatinum(II) tetrachloroplatinate(II)

33. Each of the following names is in error. Point out the error and give the correct name.
    a. tetrahydroxozinc(II) ion
    b. iron(III)hexafluoride ion

34. Each of the following names is in error. Point out the error and give the correct name.
    a. tetraaquacuprate(II) ion
    b. tetraamminesulfatocobalt ion.

Isomerism

35. Which of the following pairs must be isomers? Explain your conclusion in each case.
    a. $[CrCl_2(NH_3)_2(H_2O)_2]Cl$ and $[CrCl_2(H_2O)_2(NH_3)_2]Cl$
    b. $[PtCl_2(NH_3)_4]Br_2$ and $[PtBr_2(NH_3)_4]Cl_2$
    c. $[Co(NH_3)_6]Cl_3$ and $[Co(NH_3)_6]Cl_2$
    d. $[Co(NO_2)(NH_3)_5]^{2+}$ and $[Co(ONO)(NH_3)_5]^{2+}$

36. Which of the following pairs must be isomers? Explain your conclusion in each case.
    a. $[Co(NCS)(NH_3)_5]^+$ and $[Co(SCN)(NH_3)_5]$
    b. $[PtCl(NH_3)_3]_2[PtCl_4]$ and $[Pt(NH_3)_4][PtCl_3(NH_3)]_2$
    c. $K_3[Fe(CN)_6]$ and $K_4[Fe(CN)_6]$
    d. $[Co(NO_3)(NH_3)_5]SO_4$ and $[Co(SO_4)(NH_3)_5]NO_3$

37. Draw appropriate sketches to show that the complex ion $[CrCl_4(en)]^-$ *does not* exhibit geometric isomerism.

38. Draw appropriate sketches to show that the complex $[CrCl_3(NH_3)_3]$ *does* exhibit geometric isomerism.

39. Would you expect the octahedral complex ion $[Cr(en)(ox)_2]^-$ to exist as optical isomers? Explain

40. Would you expect the tetrahedral complex ion $[ZnCl(NH_3)_3]^+$ to exist as optical isomers? Explain.

Magnetism and Color in Complexes

41. Show by a suitable diagram whether the following octahedral complex ions are diamagnetic or paramagnetic.
    a. $[Mn(H_2O)_6]^{2+}$ b. $[Fe(CN)_6]^{4-}$

42. Show by a suitable diagram whether the following tetrahedral complex ions are diamagnetic or paramagnetic.
    a. $[Zn(H_2O)_4]^{2+}$ b. $[CoCl_4]^-$

43. One of the following compounds has a yellow color and the other is violet: $[Cr(H_2O)_6]Cl_3$ and $[Cr(NH_3)_6]Cl_3$. Which do you think is the yellow compound and which is the violet? Explain.

44. One of the following compounds has a green color and the other has a yellow color: $Fe(NO_3)_2{\cdot}6H_2O$ and $K_4[Fe(CN)_6]{\cdot}3H_2O$. Which do you think is the green compound and which is the yellow? Explain.

ADDITIONAL PROBLEMS

45. Explain why one of the cis-trans isomers of $[CoCl_2(en)_2]$ exhibits optical isomerism and the other does not. Which is which?

46. Would you expect to find optical isomerism in either of the following hypothetical tetrahedral complexes: $[ZnA_2B_2]^{2+}$ and $[ZnABCD]^{2+}$ (where the ligands A, B, C, and D are neutral molecules)? Explain.

47. The complex $[PtCl_2(NH_3)_2]$ displays cis-trans isomerism, but $[ZnCl_2(NH_3)_2]$ does not. Why do you suppose these two cases are different?

48. Does the chelate $[M(EDTA)]^{+n-4}$ pictured in Figure 23.4 display optical isomerism? Explain.

49. The magnetic properties of the octahedral complex ion $[Cr(L)_6]^{3+}$ are independent of the identity of the ligands (L). How do you account for this fact?

50. In some of its complex ions, the paramagnetism of Cr^{2+} corresponds to *two* unpaired electrons and in others to *four*. How do you account for this difference?

51. A compound known before Werner's time was Magnus's green salt, having the empirical formula $PtCl_2 \cdot 2NH_3$. It is actually a coordination compound comprised of a dipositive complex cation and a dinegative complex anion. Propose a formula that is consistent with the information given here.

52. Zeise's salt has the empirical formula $PtCl_2 \cdot KCl \cdot C_2H_4$, but it is actually a coordination compound consisting of a unipositive cation and a complex anion. Propose a formula that is consistent with the information given here.

53. The complex ion $[Ni(CN)_4]^{2-}$ is *diamagnetic*. Use crystal field theory to determine whether the structure of this complex ion is octahedral, square planar, or tetrahedral.

54. Four structures are shown in the accompanying sketch. Indicate whether any of these structures are identical, whether any are geometric isomers, and whether any are optical isomers.

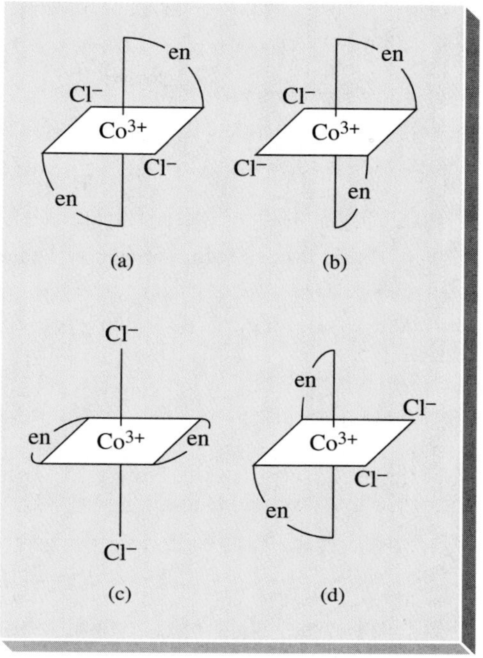

Appendix A
Some Mathematical Operations

A.1 Exponential Notation

A number is in *exponential form* when it is written as the product of a coefficient—usually with a value between 1 and 10—and a power of ten. For example,

$$4.18 \times 10^3 \quad \text{and} \quad 6.57 \times 10^{-4}$$

Expressing numbers in exponential form generally serves two purposes: (1) Very large or very small numbers can be written in a minimum of printed space and with a reduced chance of typographical error. (2) Explicit information is conveyed about the precision of measurements: The number of significant figures in a measured quantity is stated unambiguously.

In the expression 10^n, n is the exponent of 10. The number 10 is said to be raised to the *nth power*. If n is a *positive* quantity, 10^n has a value *greater than 1*. If n is a *negative* quantity, 10^n has a value *less than 1*. We are particularly interested in cases where n is an integer. For example,

Positive Powers of 10

$10^0 = 1$
$10^1 = 10$
$10^2 = 10 \times 10 = 100$
$10^3 = 10 \times 10 \times 10 = 1000$
and so on

The power of ten determines the number of zeros that follow the digit "1."

Negative Powers of 10

$10^0 = 1$
$10^{-1} = 1/10 = 0.1$
$10^{-2} = 1/(10 \times 10) = 0.01$
$10^{-3} = 1/(10 \times 10 \times 10) = 0.001$
and so on

The power of ten determines the number of places to the right of the decimal point where the digit "1" appears.

To express 612,000 in exponential form,

$$612{,}000 = 6.12 \times 100{,}000 = 6.12 \times 10^5$$

To express 0.0000505 in exponential form,

$$0.0000505 = 5.05 \times 0.00001 = 5.05 \times 10^{-5}$$

Here is a more direct approach to converting numbers to the exponential form.

- Count the number of places a decimal point must be moved to produce a coefficient having a value between 1 and 10.
- The number of places counted then becomes the power of 10.
- The power of 10 is *positive* if the decimal point is moved to the left.

$$612000 = 6.12 \times 10^5$$

5 4 3 2 1

- The power of 10 is *negative* if the decimal point is moved to the right.

$$0.0000505 = 5.05 \times 10^{-5}$$

1 2 3 4 5

To convert a number from exponential form to the conventional form, move the decimal point in the opposite direction. That is,

$$3.75 \times 10^6 = 3750000$$

1 2 3 4 5 6

$$7.91 \times 10^{-5} = 0.0000791$$

5 4 3 2 1

The key strokes required with your calculator may be different from those shown here. Check the specific instructions in the manual supplied with the calculator.

Most electronic calculators easily handle exponential numbers. A typical procedure is to enter the number, followed by the key EXP. The key strokes required for the number 2.85×10^7 are

| 2 | . | 8 | 5 | EXP | 7 |

and the result displayed is

| 2.85^{07} |

For the number 1.67×10^{-5}, the key strokes are

| 1 | . | 6 | 7 | EXP | 5 | ± |

and the result displayed is

| 1.67^{-05} |

Many calculators can be set to convert all numbers and calculated results to the exponential form, regardless of the form in which the numbers are entered. Generally, the calculator can also be set to display a fixed number of significant figures in results.

Addition and Subtraction

To add or subtract numbers in exponential notation, it is necessary to express each quantity as *the same power of ten*. In calculations, this treats the power of ten in the same way as a unit—it is simply "carried along." In the following, each quantity is expressed with the power 10^{-3}.

$$(3.22 \times 10^{-3}) + (7.3 \times 10^{-4}) + (4.8 \times 10^{-4})$$
$$= (3.22 \times 10^{-3}) + (0.73 \times 10^{-3}) + (0.48 \times 10^{-3})$$
$$= (3.22 + 0.73 + 0.48) \times 10^{-3}$$
$$= 4.43 \times 10^{-3}$$

Multiplication

To multiply numbers expressed in exponential form, *multiply* all coefficients to obtain the coefficient of the result, and *add* all exponents to obtain the power of ten in the result.

$$0.0803 \times 0.0077 \times 455 = (8.03 \times 10^{-2}) \times (7.7 \times 10^{-3}) \times (4.55 \times 10^{2})$$
$$= (8.03 \times 7.7 \times 4.55) \times 10^{(-2-3+2)}$$
$$= (2.8 \times 10^{2}) \times 10^{-3} = 2.8 \times 10^{-1}$$

Generally, an electronic calculator performs these operations automatically, and no intermediate results need be recorded.

Division

To divide two numbers in exponential form, *divide* the coefficients to obtain the coefficient of the result, and *subtract* the exponent in the denominator from the exponent in the numerator to obtain the power of ten. In the example below, multiplication and division are combined. First, the rule for multiplication is applied to the numerator and to the denominator, and then the rule for division is used.

$$\frac{0.015 \times 0.0088 \times 822}{0.092 \times 0.48} = \frac{(1.5 \times 10^{-2})(8.8 \times 10^{-3})(8.22 \times 10^{2})}{(9.2 \times 10^{-2})(4.8 \times 10^{-1})}$$
$$= \frac{1.1 \times 10^{-1}}{4.4 \times 10^{-2}} = 0.25 \times 10^{-1-(-2)} = 0.25 \times 10^{1}$$
$$= 2.5 \times 10^{-1} \times 10^{1} = 2.5 \times 10^{0} = 2.5$$

Raising a Number to a Power

To raise an exponential number to a given power, raise the coefficient to that power, and multiply the exponent by that power. For example, to raise the following number to the *third* power, that is, to *cube* the number,

$$(0.0066)^3 = (6.6 \times 10^{-3})^3 = (6.6)^3 \times 10^{3 \times (-3)}$$
$$= (2.9 \times 10^{2}) \times 10^{-9} = 2.9 \times 10^{-7}$$

Extracting the Root of an Exponential Number

To extract the root of an exponential number means to raise the number to a *fractional* power—one-half power for a square root, one-third power for a cube root, and so on. Most calculators have keys designed for extracting square roots and cube roots. Thus, to extract the square root of 1.57×10^{-5}, enter the number 1.57×10^{-5} into an electronic calculator, and use the "$\sqrt{\ }$" key.

$$\sqrt{1.57 \times 10^{-5}} = 3.96 \times 10^{-3}$$

To extract the cube root of 3.18×10^{10}, enter the number 3.18×10^{10} into an electronic calculator, and use the "$\sqrt[3]{\ }$" key

$$\sqrt[3]{3.18 \times 10^{10}} = 3.17 \times 10^{3}$$

Some calculators allow you to extract roots by keying the root in as a fractional exponent.

$$(2.18 \times 10^{7})^{1/5} = 2.94 \times 10^{1}$$

Another approach for extracting the roots of numbers is to use logarithms, which we will discuss next.

A.2 Logarithms

The common logarithm (log) of a number (N) is the exponent (x) to which the base 10 must be raised to yield the number.

$$\log N = x$$

$$\text{means that } N = 10^{x}$$

$$\text{or that } N = 10^{\log N}$$

In the expressions below, the numbers N are printed in blue and their logarithms ($\log N$) are printed in red.

$$\log 1 = \log 10^{0} = 0 \qquad \log 1 = \log 10^{0} = 0$$

$$\log 10 = \log 10^{1} = 1 \qquad \log 0.1 = \log 10^{-1} = -1$$

$$\log 100 = \log 10^{2} = 2 \qquad \log 0.01 = \log 10^{-2} = -2$$

$$\log 1000 = \log 10^{3} = 3 \qquad \log 0.001 = \log 10^{-3} = -3$$

The majority of numbers that we commonly encounter, of course, are not integral powers of ten, and their logarithms are not integral numbers. From the above pattern, though, we do have a general idea of what their logarithms might be. Consider, for example, the numbers 655 and 0.0078.

$$100 < 655 < 1000 \qquad 0.001 < 0.0078 < 0.01$$

$$2 < \log 655 < 3 \qquad -3 < \log 0.0078 < -2$$

Beyond this point, however, we must turn to a table of logarithms, or better still, the LOG key on an electronic calculator.

$$\log 655 = 2.816 \qquad \log 0.0078 = -2.11$$

In working with logarithmic relationships, an equally common requirement is to find the number that has a certain value for its logarithm. The number is sometimes called an *antilogarithm*, and it is easiest to think of it in these terms.

If $\log N = 3.076$,	If $\log N = -4.57$,
$N = 10^{3.076}$	$N = 10^{-4.57}$
$N = 1.19 \times 10^{3}$	$N = 2.7 \times 10^{-5}$

With an electronic calculator, the required operations are to enter the value of the logarithm (that is, 3.076 or –4.57) and then use the "10^x" key.

Significant Figures in Logarithms

At first sight, $\log N = 3.076$ appears to have four significant figures and $N = 1.19 \times 10^3$, only three. Actually, both values have only *three* significant figures. The idea here is that digits to the *left* of the decimal point in a logarithm simply correspond to the power of ten in the exponential form of a number. The only significant digits in a logarithm are those to the *right* of the decimal point, and this same number of digits should appear in the coefficient of the exponential form of the number. Thus, to express the logarithm of 2.5×10^{-12} to *two* significant figures, we would write: $\log 2.5 \times 10^{-12} = -11.60$.

Some Relationships Involving Logarithms

With the definition of logarithms we can write: $M = 10^{\log M}$ and $N = 10^{\log N}$. For the product $(M \times N)$ we can write either

$$(M \times N) = 10^{\log M} \times 10^{\log N} = 10^{(\log M + \log N)}$$

or

$$(M \times N) = 10^{\log (M \times N)}$$

This means that

$$(1) \quad \log (M \times N) = (\log M + \log N)$$

That is, the logarithm of the product of several terms is equal to the sum of the logarithms of the individual terms. Other relationships that can be established in a similar manner are

$$(2) \quad \log \frac{M}{N} = \log M - \log N$$

$$(3) \quad \log N^a = a \log \mathrm{N}$$

The only one of these relationships for which you may find explicit use in calculations is (3). It affords a simple method of extracting the roots of number. For example to determine $(2.75 \times 10^{-9})^{1/5}$, write

$$\log (2.75 \times 10^{-9})^{1/5} = 1/5 \times \log (2.75 \times 10^{-9})$$

$$= 1/5 \times (-8.561) = -1.712$$

$$(2.75 \times 10^{-9})^{1/5} = 10^{-1.712} = 0.0194$$

Natural Logarithms

Choosing the value of "10" as the base for common logarithms is an arbitrary choice. Another choice might have been made as well. For example, to the base "2," $\log_2 8 = 3$. This simply means that $2^3 = 8$. And $\log_2 10 = 3.322$ means that $2^{3.322} = 10$.

Several of the relationships in this text involve "natural" logarithms. The base for natural logarithms (ln) is the quantity e, which has the value, $e = 2.71828\cdots$. The "ln" function arises in circumstances where the rate at which a variable changes is proportional to the value of that variable at the time the rate is measured. These circumstances are common in physical science, including, for example, the rate of decay of a radioactive material.

If it is necessary to convert between natural and common logarithms, the required factor is $\log_e 10 = 2.303$. That is,

$$\ln N = 2.303 \log N$$

Generally one can work entirely within the natural logarithm system by using the calculator keys "ln" and "e^x" in place of LOG and "10^x."

A.3 Algebraic Operations

To solve an algebraic equation requires that we isolate one quantity—the unknown—on one side of the equation and the known quantities on the other side. This generally requires rearranging terms in the equation, and in these rearrangements the guiding principle is that *whatever is done to one side of the equation must be done to the other side as well.* Consider the equation

$$\frac{(5x^2 - 12)}{(x^2 + 4)} = 3$$

1. Multiply both sides of the equation by $(x^2 + 4)$.

$$\cancel{(x^2 + 4)} \times \frac{(5x^2 - 12)}{\cancel{(x^2 + 4)}} = 3 \times (x^2 + 4)$$

$$5x^2 - 12 = 3x^2 + 12$$

2. Subtract $3x^2$ from each side of the equation.

$$5x^2 - 3x^2 - 12 = \cancel{3x^2} - \cancel{3x^2} + 12$$

$$2x^2 - 12 = 12$$

3. Add 12 to each side of the equation.

$$2x^2 - \cancel{12} + \cancel{12} = 12 + 12 = 24$$

4. Divide each side of the equation by 2.

$$\frac{\cancel{2}x^2}{\cancel{2}} = \frac{24}{2} = 12$$

5. Extract the square root of each side of the equation.

$$\sqrt{x^2} = \pm\sqrt{12} = \pm\sqrt{4} \times \sqrt{3}$$

$$x = \pm 2\sqrt{3}$$

$$x = \pm 3.464$$

Quadratic Equations

A quadratic equation has "2" as the highest power of the unknown x. At times, quadratic equations are of the form

$$(x + n)^2 = m^2$$

To solve for x, extract the square root of each side.

$$(x + n) = \sqrt{m^2} = \pm m$$

and

$$x = m - n \qquad \text{or} \qquad x = -m - n$$

You will find an example of a quadratic equation of this type in Example 16.10 (page 627).

More often, however, the quadratic equation will be of the form

$$ax^2 + bx + c = 0$$

where a, b, and c are constants. To solve this equation for x, we can use the *quadratic formula*.

$$x = \frac{-b \pm \sqrt{b^2 - 4ac}}{2a}$$

Consider the solution of the quadratic equation

$$50.3x^2 - 70.6x + 21.7 = 0$$

$$x = \frac{70.6 \pm \sqrt{(70.6)^2 - (4 \times 50.3 \times 21.7)}}{2 \times 50.3}$$

$$x = \frac{70.6 \pm \sqrt{618}}{100.6} = \frac{70.6 \pm 24.9}{100.6}$$

$$x = \frac{70.6 + 24.9}{100.6} = 0.949 \qquad \text{and} \qquad x = \frac{70.6 - 24.9}{100.6} = 0.454$$

We encounter this particular quadratic equation in Example 16.11 (page 628), where we find that the only physically significant answer is $x = 0.454$.

Solving Equations by Approximation

If an equation is of higher degree than quadratic, the most direct solution is often one of successive approximations. We gather the terms involving the unknown on one side of the equation and a constant term on the other side of the equation. For example, consider the following equation obtained in Conceptual Example 16.2 (page 632).

$$\frac{4P^3}{(0.492 - 2P)^2} = 0.023$$

It is shown in the example that, based on the physical situation, P must be a positive quantity that cannot exceed 0.246. Suppose we "guess" at a possible value of P and then see, when we substitute this value of P into the equation, how close our result is to 0.023.

Try $P = 0.100$:

$$\frac{4P^3}{(0.492 - 2P)^2} = \frac{4 \times (0.100)^3}{[0.492 - (2 \times 0.100)]^2} = \frac{4 \times 10^{-3}}{[0.492 - 0.200]^2} = 0.0469$$

Because 0.0469 > 0.023, our "guess" is not good enough. Now let's make a second approximation.

Try P = 0.050:

$$\frac{4P^3}{(0.492 - 2P)^2} = \frac{4 \times (0.050)^3}{[0.492 - (2 \times 0.050)]^2} = \frac{5.0 \times 10^{-4}}{[0.492 - 0.100]^2} = 0.0033$$

Now our result 0.0033 < 0.023. We seem to be "on the other side" of the desired answer. We need to try a value 0.050 < P < 0.100. As a third approximation, let's

Try P = 0.080:

$$\frac{4P^3}{(0.492 - 2P)^2} = \frac{4 \times (0.080)^3}{[0.492 - (2 \times 0.080)]^2} = \frac{2.0 \times 10^{-3}}{[0.492 - 0.160]^2} = 0.018$$

In this third approximation we have 0.018 < 0.023. We are now much closer to an acceptable value of P. In some cases, this might be close enough, but suppose we try one more approximation, with 0.080 < P < 0.10.

Try P = 0.085:

$$\frac{4P^3}{(0.492 - 2P)^2} = \frac{4 \times (0.085)^3}{[0.492 - (2 \times 0.085)]^2} = \frac{2.5 \times 10^{-3}}{[0.492 - 0.170]^2} = 0.024$$

Now we are very close to the correct answer, since 0.024 ≈ 0.023. (A further approximation would show that 0.084 < P < 0.085.)

This method may seem laborious, but usually it is not. Once the format of the approximations is set up, the calculations can be quickly performed with an electronic calculator.

A.4 Graphs

Suppose we obtain the following data for the quantities x and y as a result of laboratory measurements:

$$x = 0,\ y = 2 \qquad x = 2,\ y = 6 \qquad x = 4,\ y = 10$$
$$x = 1,\ y = 4 \qquad x = 3,\ y = 8 \qquad \cdots$$

Just by inspecting these data, you can probably see that they fit the equation

$$y = 2x + 2$$

Sometimes an exact equation cannot be written from the experimental data, or the form of the equation may not be immediately obvious from the data themselves. In these cases it often proves helpful to graph the data. On page A-9, the data points listed above are plotted in a graph in which the x values are placed along the horizontal axis (abscissa) and the y values along the vertical axis (ordinate). For each point in the figure, the x and y values are listed in the order (x,y).

We see that the data indeed fall on a straight line. The equation for a straight line graph is

$$y = mx + b$$

To obtain a value of the *intercept*, b, we set $x = 0$; from the graph we see that $y = 2$. To obtain the slope of the line, we can work with two points on the graph:

$$y_2 = mx_2 + b \qquad \text{and} \qquad y_1 = mx_1 + b$$

The *difference* between these two equations is

$$y_2 - y_1 = m(x_2 - x_1) + \cancel{b} - \cancel{b}$$

and the value of m is

$$m = \frac{y_2 - y_1}{x_2 - x_1}$$

The slope is evaluated in the figure below; it is 2. Thus, the equation of the straight line is

$$y = mx + b = 2x + 2$$

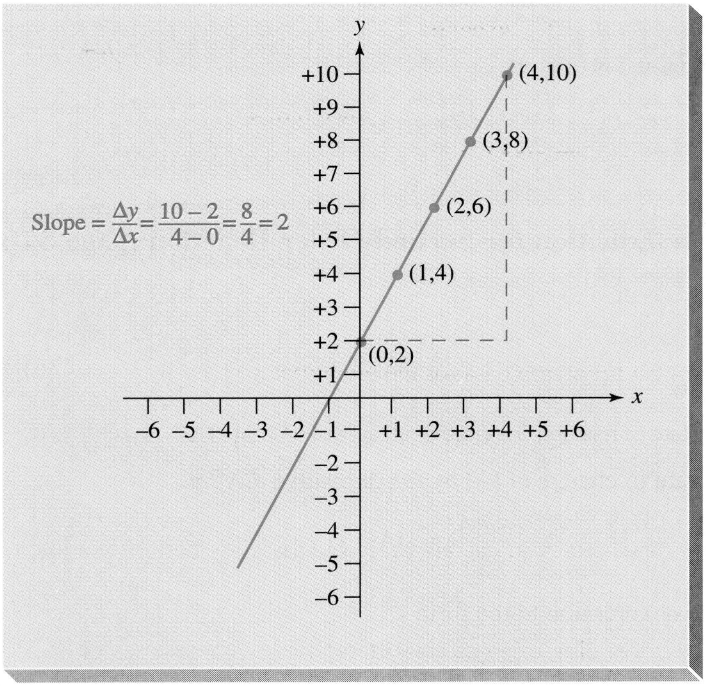

A straight-line graph: $y = mx + b$.

A.5 Some Key Equations

On several occasions in the text we refer to this appendix for details on the derivations of key equations or their manipulation into more useful forms. Abbreviated treatments follow. The first two require some prior knowledge of calculus.

Integrated Rate Equation for First-Order Reaction (page 568)

For the reaction

$$A \longrightarrow \text{products}$$

having the rate law:

$$\text{Rate of reaction of A} = -(\text{rate of change of [A]}) = k[\text{A}]$$

1. Replace the rate of change of [A] by the derivative $d[\text{A}]/dt$.

$$-\frac{d[\text{A}]}{dt} = k[\text{A}]$$

2. Rearrange this expression to the form

$$\frac{d[\text{A}]}{[\text{A}]} = -kdt$$

3. Integrate between the limits A_0 at time $t = 0$ and A_t at time t.

$$\int_{[\text{A}]_0}^{[\text{A}]_t} \frac{d[\text{A}]}{[\text{A}]} = -k\int_0^t dt$$

4. The result obtained is

$$\ln\frac{[\text{A}]_t}{[\text{A}]_0} = -kt$$

Integrated Rate Equation for Second-Order Reaction (page 578)

For the reaction

$$A \longrightarrow \text{products}$$

having the rate law:

$$\text{Rate of reaction} = -(\text{rate of change of [A]}) = k[\text{A}]^2$$

1. Replace the rate of change of [A] by the derivative $d[\text{A}]/dt$.

$$\frac{-d[\text{A}]}{dt} = k[\text{A}]^2$$

2. Rearrange this expression to the form

$$\frac{d[\text{A}]}{[\text{A}]^2} = -kdt$$

3. Integrate between the limits A_0 at time $t = 0$ and A_t at time t.

$$\int_{[\text{A}]_0}^{[\text{A}]_t} \frac{d[\text{A}]}{[\text{A}]^2} = -k\int_0^t dt$$

4. The result obtained is

$$-\frac{1}{[\text{A}]_t} + \frac{1}{[\text{A}]_0} = -kt \qquad \text{or} \qquad \frac{1}{[\text{A}]_t} = kt + \frac{1}{[\text{A}]_0}$$

The Arrhenius Equation (page 586)

Our goal is to convert the equation for the straight-line graph of Figure 15.14

$$\ln k = \frac{-E_a}{RT} + \ln A$$

into an equation that eliminates the constant term, ln A.

1. Write the equation for two different temperatures, T_1 and T_2, at which the rate constants are k_1 and k_2. (E_a and R are constants.)

$$\ln k_2 = \frac{-E_a}{RT_2} + \ln A \qquad \ln k_1 = \frac{-E_a}{RT_1} + \ln A$$

2. Subtract ln k_1 from ln k_2.

$$\ln k_2 - \ln k_1 = \frac{-E_a}{RT_2} + \ln A - \left(\frac{-E_a}{RT_1} + \ln A\right)$$

3. Use the relationship: ln M/N = ln M − ln N to write

$$\ln \frac{k_2}{k_1} = \frac{E_a}{RT_1} - \frac{E_a}{RT_2} + \cancel{\ln A} - \cancel{\ln A}$$

4. Rearrange the equation to the final form

$$\ln \frac{k_2}{k_1} = \frac{E_a}{R}\left(\frac{1}{T_1} - \frac{1}{T_2}\right)$$

The van't Hoff Equation (page 754)

Our goal is to convert the equation for the straight-line graph of Figure 19.12

$$\ln K_{eq} = \frac{-\Delta H^\circ}{RT} + \text{constant}$$

into an equation that eliminates the term "constant," represented below as A.

1. Write the equation for two different temperatures, T_1 and T_2, at which the equilibrium constants are K_1 and K_2. (ΔH° and R are constants.)

$$\ln K_2 = \frac{-\Delta H^\circ}{RT_2} + \ln A \qquad \ln K_1 = \frac{-\Delta H^\circ}{RT_1} + \ln A$$

2. Subtract ln K_1 from ln K_2.

$$\ln K_2 - \ln K_1 = \frac{-\Delta H^\circ}{RT_2} + \ln A - \left(\frac{-\Delta H^\circ}{RT_1} + \ln A\right)$$

3. Use the relationship: ln M/N = ln M − ln N to write

$$\ln \frac{K_2}{K_1} = \frac{\Delta H^\circ}{RT_1} - \frac{\Delta H^\circ}{RT_2} + \cancel{\ln A} - \cancel{\ln A}$$

4. Rearrange the equation to the final form

$$\ln \frac{K_2}{K_1} = \frac{\Delta H^\circ}{R}\left(\frac{1}{T_1} - \frac{1}{T_2}\right)$$

Appendix B
Some Basic Physical Concepts

B.1 Velocity and Acceleration

The speed of an object is the distance it travels per unit time. An automobile whose speedometer reads 105 km/h will, if it continues at this constant speed for exactly one hour, travel a distance of 105 km. For scientific work, *velocity* is a more appropriate term. Velocity has two components: a *magnitude* (speed) and a *direction* (up, down, east, southwest,. . .). The SI units of velocity are distance $\times$ time^{-1}, that is, $m \cdot s^{-1}$.

The velocity of an object changes if its speed changes, or if the direction of its motion changes. The rate of change of velocity is called *acceleration.* Acceleration has the units of velocity $\times$ time^{-1}, that is, $m \cdot s^{-1} \times s^{-1} = m \cdot s^{-2}$. For an object under a constant acceleration (a), its velocity (u) as a function of time (t) is

$$u = at \qquad \text{(B.1)}$$

and (by the methods of calculus) the distance (d) traveled is

$$d = \tfrac{1}{2}at^2 \qquad \text{(B.2)}$$

The constant *acceleration due to gravity* (g) experienced by a freely falling body is $9.81\ m \cdot s^{-2}$.

B.2 Force and Work

According to Newton's *first law* of motion, an object has a natural tendency—called *inertia*—to remain in motion at a constant velocity if it is moving or to remain at rest if it is not moving. A *force* is required to overcome the inertia of an object—that is, to give motion to an object at rest or to change the velocity of a moving object. Because a change in velocity is an acceleration, we can say that *a force is required to provide acceleration to an object.*

Newton's *second* law of motion describes the force (F) required to produce an acceleration (a) in an object of mass (m).

$$F = ma \qquad \text{(B.3)}$$

The SI unit of force is the *newton* (N). It is the force required to produce an acceleration of 1 $m \cdot s^{-2}$ in a 1 kg mass.

$$1\ \text{N} = 1\ \text{kg} \times 1\ \text{m} \cdot \text{s}^{-2} = 1\ \text{kg} \cdot \text{m} \cdot \text{s}^{-2} \qquad \text{(B.4)}$$

The weight W of an object is the force of gravity on the object. It is the mass of the object multiplied by the acceleration due to gravity.

$$W = F = mg$$

Work is done when a force acts through a distance.

$$\text{Work } (w) = \text{force } (F) \times \text{distance } (d)$$

A *joule* (J) is the work done when a force of one newton acts through a distance of one meter.

$$1\ \text{J} = 1\ \text{N} \times 1\ \text{m} = 1\ \text{N} \cdot \text{m}$$

From the definition of one newton given in (B.4), we see that the SI units of the joule are

$$1\ \text{J} = 1\ \text{kg} \times 1\ \text{m} \cdot \text{s}^{-2} \times 1\ \text{m} = 1\ \text{kg} \cdot \text{m}^2 \cdot \text{s}^{-2}$$

B.3 Energy

A moving object has a *kinetic energy* associated with its motion. Energy is the capacity to do work, and the work associated with the moving object is given by the previous expressions:

$$\text{Work} = F \times d = m \times a \times d$$

From equation (B.2) we can substitute $d = \frac{1}{2}at^2$.

$$\text{Work} = m \times a \times \tfrac{1}{2}at^2 = \tfrac{1}{2} \times m \times (at)^2$$

Now, substitute expression (B.1): $u = at$.

$$\text{Work} = \tfrac{1}{2} \times m \times u^2$$

Think of this work as that required to provide an object of mass m with a velocity u. This amount of work appears as the kinetic energy of the moving object.

$$\text{Kinetic energy (KE)} = \tfrac{1}{2}mu^2$$

In addition to kinetic energy associated with motion, an object may possess *potential energy*, which can be thought of as stored energy that can be released under appropriate circumstances. Think of it as energy that stems from the condition, position, or composition of an object. In principle, equations can be written for the various ways in which potential energy is stored in an object, but we do not specifically use such equations in the text.

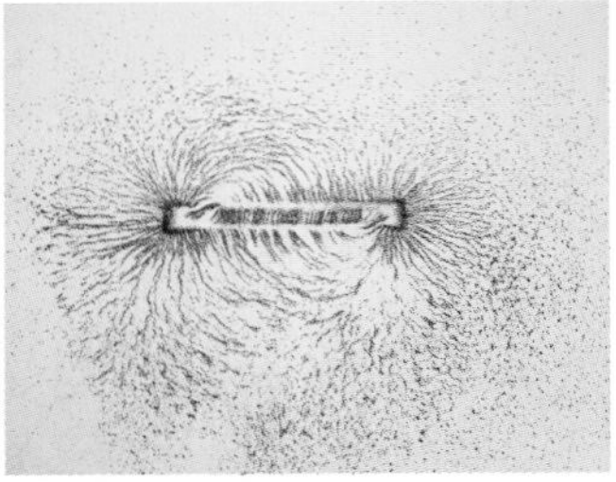

Visualizing a magnetic field. The sprinkling of iron filings outlines the magnetic field of a bar magnet. The density of the filings is heaviest near the two poles of the magnet, signifying that these are the regions where the magnetic field is strongest.

B.4 Magnetism

Attractive and repulsive forces associated with magnets are centered in regions of the magnets called *poles*. A magnet has a north pole and a south pole. If two magnets are arranged so that north pole of one magnet is brought near the south pole of another magnet, an attractive force develops between the two magnets. If like poles are brought into close proximity, either both north poles or both south poles, a repulsive force develops. *Unlike poles attract, and like poles repel.*

A *magnetic field* exists in the region surrounding a magnet in which the influence of the magnet can be felt. For example, a magnetic field can be detected through deflections of a compass needle, or the field can be visualized through the attractive forces that cause a characteristic alignment of iron filings.

B.5 Electricity

Electricity is a property closely related to magnetism. Ultimately, all bulk matter contains electrically charged particles—protons and electrons—but an object displays a net electric charge—positive or negative—only when the numbers of electrons and protons in the object are unequal. The basic expression dealing with stationary electrically charged particles—static electricity—is Coulomb's law: The magnitude of the force (F) between electrically charged objects is directly proportional to the magnitudes of the charges (Q) and inversely proportional to the *square* of the distance (r) between them.

$$F \propto \frac{Q_1 \times Q_2}{r^2}$$

Like charges repel. Whether both charges are positive or both are negative, their product is a positive quantity. A *positive* force is a *repulsive* force. *Unlike charges attract.* If one charge is positive and the other negative, their product is a negative quantity. A *negative* force is an *attractive* force.

An *electric field* exists in the region surrounding an electrically charged object in which the influence of the electric charge is felt. If an uncharged object is brought into the field of a charged object, an electric charge of the opposite sign may be *induced* in the previously uncharged object. This leads to a force of attraction between the two.

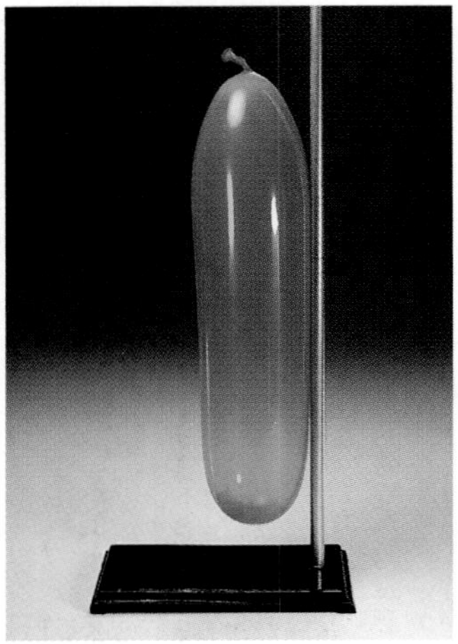

The phenomenon of induction. The attraction of a balloon to a surface is a common practical example of induction. The balloon develops a static electric charge when it is rubbed with a cloth. The charged ballon induces an electric charge of the opposite sign on the surface, and the balloon is then attracted to the surface.

Electric current is a flow of charged particles—electrons in metallic conductors and positive and negative ions in molten salts and in aqueous salt solutions. The unit of electric charge is the *coulomb* (C). The unit of electric current is the *ampere* (A). A current of one ampere is the flow of one coulomb of electric charge per second.

$$1\ \text{A} = 1\ \text{C}/1\ \text{s} = 1\ \text{C}\cdot\text{s}^{-1}$$

The magnitude of the electric current, I, in an electrical conductor is equal to the potential difference or voltage drop, E, expressed in volts (V), divided by the resistance, R, expressed in ohms (Ω). This relationship, known as Ohm's law, is

$$I = \frac{E}{R}$$

One joule of energy is associated with the passage of one coulomb of electric charge through a potential difference of one volt.

$$1\ \text{J} = 1\ \text{V} \times 1\ \text{C} = 1\ \text{V} \cdot \text{C}$$

Electric *power* refers to the rate of production (or consumption) of electric energy. The electric power unit, the *watt* (W), corresponds to the production (or consumption) of one joule of energy per second.

$$1\ \text{W} = 1\ \text{J} \cdot \text{s}^{-1} = 1\ \text{V} \cdot \text{C} \cdot \text{s}^{-1}$$

Because one coulomb per second ($\text{C} \cdot \text{s}^{-1}$) represents a current of one ampere (A), we can also write

$$1\ \text{W} = 1\ \text{V} \cdot \text{C} \cdot \text{s}^{-1} = 1\ \text{V} \times 1\ \text{A}$$

Thus, the electric power associated with the passage of 10.0 amp through a 110-volt electric circuit is

$$110\ \text{V} \times 10.0\ \text{A} = 1100\ \text{W}$$

B.6 Electromagnetism

A variety of relationships between electricity and magnetism, referred to collectively as electromagnetism, underlie some important practical applications: (1) Magnetic fields are associated with the flow of electrons, as in electromagnets (see the photograph below). (2) Forces are experienced by current-carrying conductors in a magnetic field, as in electric motors. (3) Electric currents are induced when electric conductors are moved through a magnetic field, as in electric generators. Several phenomena described in this text can be understood as electromagnetic effects.

An electromagnet. Electric current from the battery passes through the coil of wire wrapped around an iron bar. The electric current induces a magnetic field and causes the bar to act as a magnet, attracting small iron objects. When the electric current is cut off, the magnetic field dissipates and the bar loses its magnetism.

Appendix C
IUPAC Nomenclature for Selected Organic Compounds

In the early days of organic chemistry, compounds were given common names, often based on natural sources. For example, formic acid (HCOOH) was so named from the Latin *formica*, meaning ant, because it was first obtained from a species of red ants. As thousands of new organic compounds were discovered in the latter half of the nineteenth century, naming became utterly chaotic. To bring some order to the jumbled naming of the rapidly increasing number of newly discovered compounds, an international group of chemists, later known as the *International Union of Pure and Applied Chemistry* (IUPAC), held the first of many meetings on nomenclature (that is, a system for naming) in 1892. This conference established formal rules for naming compounds. Here we will examine some simple rules for naming five families of organic compounds. We start with the simplest of the hydrocarbons, the alkanes.

C.1 Alkanes

In IUPAC nomenclature, the ending of the name indicates the family of the compound. The following rules will enable us to name most alkanes.

1. Use the ending *-ane* to indicate that the compound is an alkane. (The stems of names of the continuous-chain alkanes having up to 10 carbon atoms are given in Table 2.6, page 53.)
2. The names of branched-chain alkanes are made up of two parts. Determine the longest continuous chain (LCC) of carbon atoms. This becomes the *parent chain* for naming the compound. For example, the compound

$$\begin{array}{l} CH_3\underset{\displaystyle CH_3}{\underset{|}{C}}HCH_2CH_2CH_3 \end{array}$$

is named as a derivative of pentane because there are five carbon atoms in the longest continuous chain. The *second* part of its name is *pentane.*

3. The *first* part of the name consists of prefixes that indicate the groups attached to the parent chain. If a group contains only carbon and hydrogen with no double or triple bonds, it is called an *alkyl* group. An alkyl group is derived by removing one H from, and is named after, the alkane with the same number of carbon atoms. For example, the group

$$—CH_3$$

is derived from methane (CH_4) and is called a *methyl* group. The alkyl group derived from ethane (CH_3CH_3) is

$$—CH_2CH_3$$

It is called an *ethyl* group. Two alkyl groups can be derived from propane.

$CH_3CH_2CH_2—$

Propyl group

Isopropyl group

There is only one alkane named propane, but a chain of three carbon atoms can be attached to a longer hydrocarbon chain in two different ways. One, the *propyl* group, is attached through an end carbon of the three-carbon chain; the other, the *isopropyl* group, through the middle carbon. There are other, larger alkyl groups, which need not concern us here.

4. Arabic numerals are used to indicate the position(s) at which the substituents (alkyl groups, in this case) are attached on the longest chain. Thus, to name the compound

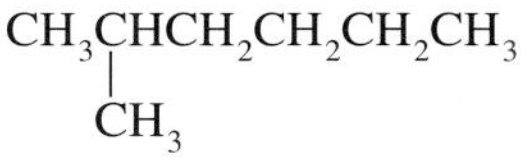

we first identify the longest continuous chain (red).

There are six carbon atoms in this chain. The compound is therefore a derivative of hexane. The group attached to the chain ($CH_3—$) is methyl. It is on the second carbon atom from the left end. Thus, the compound is 2-methylhexane.*

5. If two or more identical groups are attached to the main chain, a number is required to specify the location of each. Further, we must indicate whether there are two, three, or four identical groups attached to the parent chain, and we do this by using the prefix *di-* for two, the prefix *tri-* for three, and the prefix *tetra-* for four. Even if two identical groups are located at the same position, the number must be repeated for each group.

---

* We start numbering the main chain at the end that provides the lowest number for the position of the substituent. If we numbered the parent chain starting at the right, the substituent would be located at carbon number 5 of the parent chain. In order to obtain the lowest possible number for the substituent, then, we must count from *left* to right in this instance.

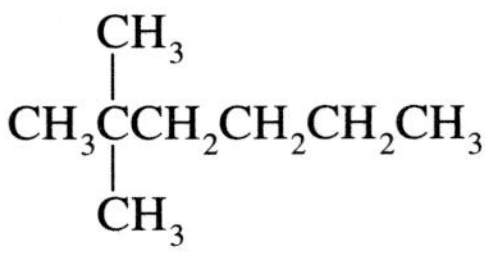

2,2-Dimethylhexane

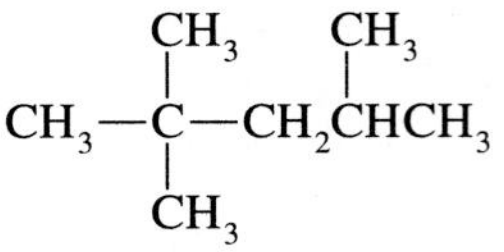

2,2,4-Trimethylpentane

(Notice from these examples that commas are used to separate numbers from each other and that hyphens are used to separate numbers from words.)

6. Groups are listed in alphabetical order. The proper name for the compound

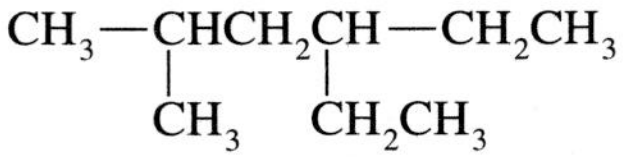

is 4-ethyl-2-methylhexane, not "2-methyl-4-ethylhexane."

The best way to learn nomenclature is by working out examples. Let's illustrate by naming the following compound.

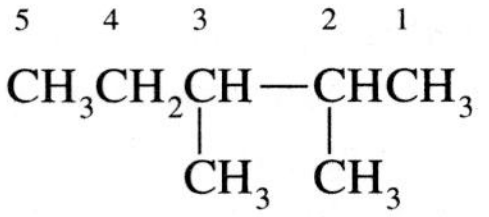

First, we note that the longest continuous chain has five carbon atoms; the compound is named as a derivative of pentane. Then we see that there are methyl groups attached to the second and third carbon atoms (not the third and fourth); use the lowest combination of numbers, counting from one end. The correct name is 2,3-dimethylpentane.

Now let's name the compound

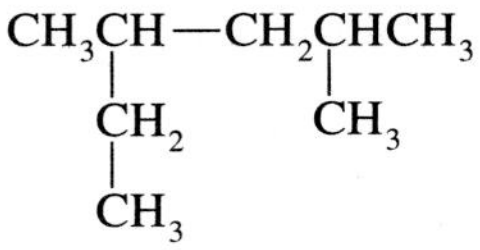

The correct name is 2,4-dimethylhexane, not 2-ethyl-4-methylpentane. This one is tricky. The parent compound is the longest continuous chain, not necessarily the chain drawn straight across the page. In this compound, the LCC (red) contains six, not five, carbon atoms.

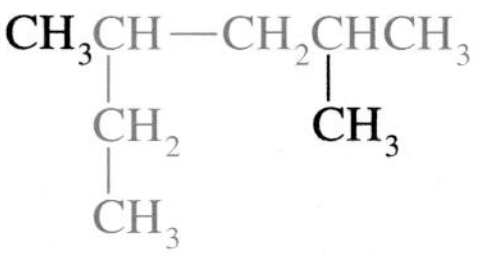

Now let's write a structure from a given name. Draw the structure of 4-isopropyl-2-methylheptane. To do so, we start with the parent chain, heptane in this case.

—C—C—C—C—C—C—C—

Then we add the substituent groups at the proper positions.

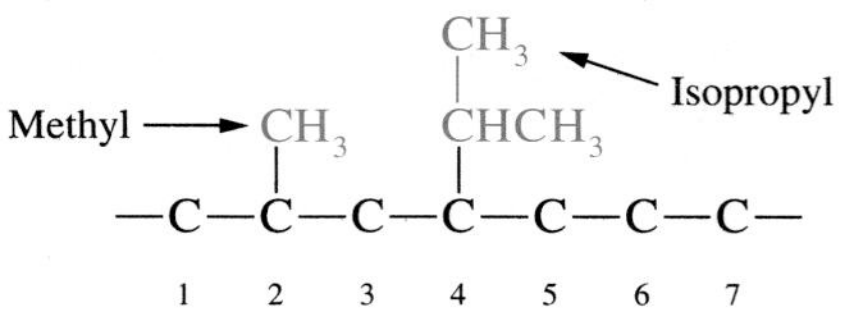

Finally, we add enough hydrogen atoms to give each carbon atom four bonds. The structure is

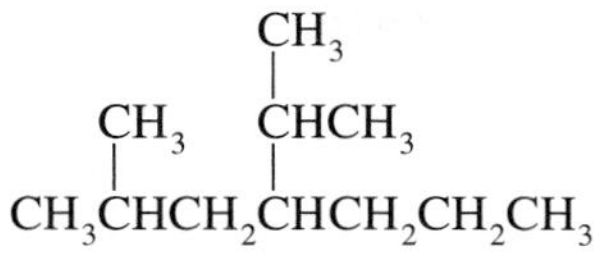

C.2 Alkenes

Alkenes are compounds with carbon-to-carbon double bonds. The simple alkenes are best known by common names (Section 9.10), but systematic names are needed for the many isomers that are possible with higher alkenes. Some of the IUPAC rules for alkenes are as follows.

1. All alkenes have names ending in *-ene.*
2. The longest chain of carbon atoms *containing the double bond* is the parent compound. The name has the same stem as the corresponding alkane (that is, the alkane with the same number of carbon atoms), but the ending is changed from *-ane* to *-ene.* The compound $CH_2{=}CH_2$ is ethene and the compound $CH_3CH{=}CH_2$, with three carbon atoms, is named *propene.*
3. When it is necessary to indicate the position of the double bond, the first carbon of the two that are doubly bonded is given the *lowest possible number* (that is, the carbon atoms are counted from the end of the chain nearer the first carbon atom of the double bond). For example, the compound $CH_3CH{=}CHCH_2CH_3$ has five carbon atoms with the double bond between the second and third. Its name is 2-pentene.
4. Substituent groups are named as usual. The position of each is indicated by a number. Thus,

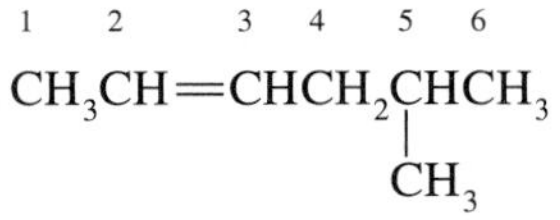

is 5-methyl-2-hexene. Note that we always number the parent chain in such a way as to give the double bond the lowest number, even if that forces a substituent to have a higher number. We say the double bond takes priority over substituent groups in numbering.

The rules are more easily learned through examples. Let's practice by naming the compound

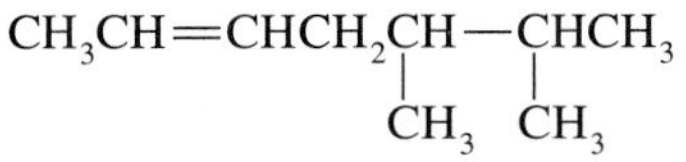

The longest continuous chain has seven carbon atoms. To give the first carbon atom of the double bond the lowest number, we start numbering from the left.

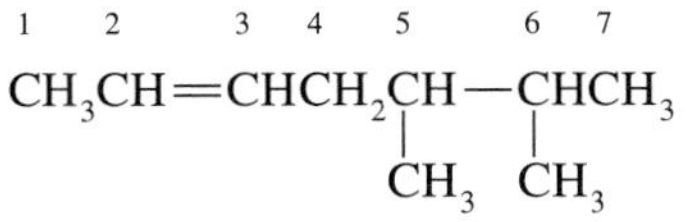

The name of the compound is 5,6-dimethyl-2-heptene.

Now let's name the compound

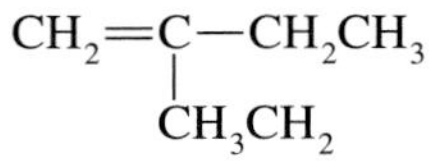

The name is 2-ethyl-1-butene. The longest continuous chain overall in the molecule contains five carbon atoms. However, the longest continuous chain *containing the double bond* incorporates only four carbon atoms, and this four-carbon chain serves as the parent compound.

C.3 Alkynes

The IUPAC nomenclature for alkynes, compounds with carbon-to-carbon triple bonds, parallels that of the alkenes, except that the family ending is *-yne* rather than *-ene.* The IUPAC name for acetylene, $HC{\equiv}CH$, is ethyne. Similarly, $CH_3C{\equiv}CH$ is propyne. The compounds $CH_3CH_2C{\equiv}CH$ and $CH_3C{\equiv}CCH_3$ are 1-butyne and 2-butyne, respectively.

C.4 Alcohols

In the IUPAC system, alcohol names are based on the alkane corresponding to the longest continuous chain of carbon atoms. The final *-e* of the alkane name is dropped and replaced by the ending *-ol.* If necessary, the position of the hydroxyl group is indicated by a number placed immediately in front of the name of the longest (parent) chain. The molecule is always numbered from the end that gives the hydroxyl group the lowest possible number.

The simplest alcohol (CH_3OH) is commonly called methyl alcohol. Its IUPAC name is derived from methane (CH_4). Merely drop the *-e* of methane, add *-ol*, and you have it: methanol. To name CH_3CH_2OH, commonly called ethyl alcohol, drop the final *-e* from the name of the corresponding alkane, ethane (CH_3CH_3), and add *-ol*. The IUPAC name is ethanol. There are two propyl alcohols. The IUPAC name of

$$\overset{3}{C}H_3\overset{2}{C}H_2\overset{1}{C}H_2OH$$

is 1-propanol (not 3-propanol) and that of

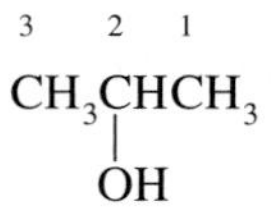

is 2-propanol.

The following alcohol is commonly called *tert*-butyl alcohol.

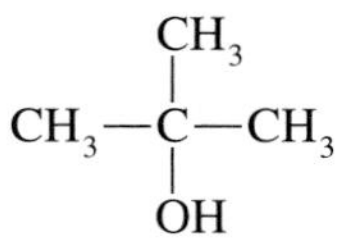

What is its IUPAC name? The longest continuous chain is three carbon atoms long, and the hydroxyl group is on the second carbon atom of this chain, yielding, for the moment, 2-propanol. There is also a methyl group attached to the second carbon atom of the parent chain, giving 2-methyl-2-propanol.

C.5 Carboxylic Acids

The simplest carboxylic acids are widely known by their common names. However, IUPAC names are readily derived from the names of alkanes with the same number of carbon atoms. Just drop the *-e* from the alkane name and add *-oic acid.* Thus, the IUPAC name for formic acid (HCOOH) is methanoic acid, and that for acetic acid (CH_3COOH) is ethanoic acid. The carboxylic acid with an LCC of eight carbon atoms is octanoic acid. For locating substituents, numbering begins with the carboxylic carbon atom as number 1. For example, 4-ethyl-6-methyloctanoic acid is

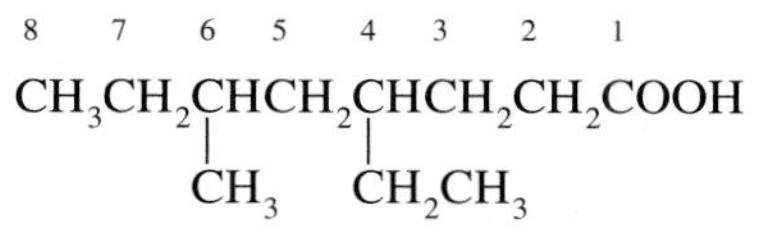

As a final example, let's derive the IUPAC name for

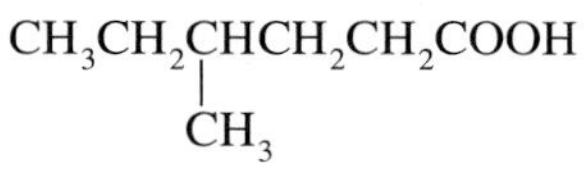

The longest continuous chain contains six carbon atoms; the compound is therefore named as a substituted hexanoic acid. The methyl substituent is at the fourth carbon atom; therefore the compound is 4-methylhexanoic acid.

C.6 Functional Groups

The idea of the organic functional group is introduced in Section 2.11, with an emphasis on the hydroxyl group (—OH) of the alcohols and the carboxyl group (—COOH) of the carboxylic acids. A sampling of some of the more important functional groups is provided in the accompanying table.

Some Classes of Organic Compounds and Their Functional Groups

Class	General Structural Formula[a]	Example	Name of Example
Alkane	R—H	$CH_3CH_2CH_2CH_2CH_2CH_3$	hexane
Alkene	$>C=C<$	$CH_2{=}CHCH_2CH_2CH_3$	1-pentene
Alkyne	—C≡C—	$CH_3C{\equiv}CCH_2CH_2CH_2CH_2CH_3$	2-octyne
Alcohol	R—OH	$CH_3CH_2CH_2CH_2OH$	1-butanol
Alkyl halide	R—X	$CH_3CH_2CH_2CH_2CH_2CH_2CH_2Br$	1-bromoheptane
Ether	R—O—R	$CH_3CH_2CH_2OCH_3$	1-methoxypropane (methyl propyl ether)[b]

Class	General Structural Formula[a]	Example	Name of Example
Aldehyde	$R-\overset{\overset{\displaystyle O}{\|}}{C}-H$	$CH_3CH_2CH_2\overset{\overset{\displaystyle O}{\|}}{C}-H$	butanal (butyraldehyde)[b]
Ketone	$R-\overset{\overset{\displaystyle O}{\|}}{C}-R$	$CH_3CH_2CH_2\overset{\overset{\displaystyle O}{\|}}{C}-CH_2CH_3$	3-hexanone (ethyl propyl ketone)[b]
Carboxylic acid	$R-\overset{\overset{\displaystyle O}{\|}}{C}-OH$	$CH_3CH_2CH_2\overset{\overset{\displaystyle O}{\|}}{C}-OH$	butanoic acid (butyric acid)[b]
Ester	$R-\overset{\overset{\displaystyle O}{\|}}{C}-OR$	$CH_3CH_2CH_2\overset{\overset{\displaystyle O}{\|}}{C}-OCH_3$	methyl butanoate (methyl butyrate)[b]
Amine	$R-NH_2$	$CH_3CH_2CH_2NH_2$	1-aminoproprane (propylamine)[b]
Amide	$R-\overset{\overset{\displaystyle O}{\|}}{C}-NH_2$	$CH_3CH_2CH_2\overset{\overset{\displaystyle O}{\|}}{C}-NH_2$	butanamide (butyramide)[b]
Arene	Ar—H[c]	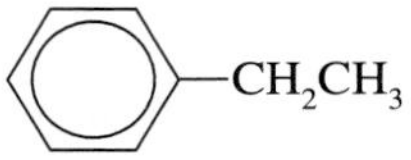	ethylbenzene
Phenol	Ar—OH	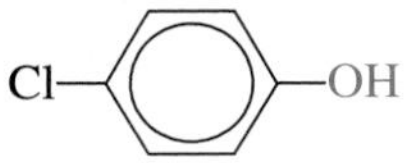	4-chlorophenol (*para*-chlorophenol)[b]

[a]The functional group is shown in color.
[b]Common name.
[c]Ar— stands for an aromatic (*aryl*) group such as the benzene ring.

Appendix D Data Tables

D.1 Thermodynamic Properties of Substances at 298.15 K

Substances are at 1 atm pressure.[a] For aqueous solutions solutes are at unit activity (≈1 M).

INORGANIC SUBSTANCES

	ΔH_f°, kJ/mol	ΔG_f°, kJ/mol	S°, $J \cdot mol^{-1} \cdot K^{-1}$
Aluminum			
Al(s)	0	0	28.3
$AlCl_3$(s)	−705.6	−630.1	109.3
Al_2Cl_6(g)	−1291	−1221	490
AlF_3(s)	−1504	−1425	66.48
Al_2O_3 (α, solid)	−1676	−1582	50.92
$Al(OH)_3$(s)	−1276	—	—
$Al_2(SO_4)_3$	−3441	−3100	239
Barium			
Ba(s)	0	0	62.3
$BaCO_3$(s)	−1216	−1138	112
$BaCl_2$(s)	−858.1	−810.4	123.7
BaF_2(s)	−1209	−1159	96.40
BaO(s)	−548.1	−520.4	72.09
$Ba(OH)_2$(s)	−946.0	−859.4	107
$Ba(OH)_2 \cdot 8H_2O$(s)	−3342	−2793	427
$BaSO_4$(s)	−1473	−1362	132
Beryllium			
Be(s)	0	0	9.54
$BeCl_2$(s)	−496.2	−449.5	75.81
BeF_2(s)	−1027	−979.5	53.35
BeO(s)	−608.4	−579.1	13.77
Bismuth			
Bi(s)	0	0	56.74

[a]The current IUPAC recommendation is that standard pressure be taken as 1 bar (10^5 Pa). The values given here for a standard pressure of 1 atm do not differ significantly from those at 1 bar. For example, for CO_2(g) the values of ΔH_f° and ΔG_f° are the same at 1 atm and 1 bar; the value of $S^\circ = 213.6\ J \cdot mol^{-1} \cdot K^{-1}$ at 1 atm and $213.8\ J \cdot mol^{-1} \cdot K^{-1}$ at 1 bar.

INORGANIC SUBSTANCES

	ΔH_f°, kJ/mol	ΔG_f°, kJ/mol	S°, J • mol^{-1} • K^{-1}
$BiCl_3(s)$	−379	−315	177
$Bi_2O_3(s)$	−573.9	−493.7	151
Boron			
B(s)	0	0	5.86
$BCl_3(l)$	−427.2	−387	206
$BF_3(g)$	−1137	−1120.3	254.0
$B_2H_6(g)$	36	86.6	232.0
$B_2O_3(s)$	−1273	−1194	53.97
Bromine			
Br(g)	111.9	82.43	174.9
$Br_2(g)$	30.91	3.14	245.4
$Br_2(l)$	0	0	152.2
BrCl(g)	14.6	−0.96	240.0
$BrF_3(g)$	−255.6	−229.5	292.4
$BrF_3(l)$	−300.8	−240.6	178.2
Cadmium			
Cd(s)	0	0	51.76
$CdCl_2(s)$	−391.5	−344.0	115.3
CdO(s)	−258	−228	54.8
Calcium			
Ca(s)	0	0	41.4
$CaCO_3(s)$	−1207	−1128	88.70
$CaCl_2(s)$	−795.8	−748.1	105
$CaF_2(s)$	−1220	−1167	68.87
$CaH_2(s)$	−186	−147	42
$Ca(NO_3)_2(s)$	−938.4	−743.2	193
CaO(s)	−635.1	−604.0	39.75
$Ca(OH)_2(s)$	−986.1	−898.6	83.39
$Ca_3(PO_4)_2(s)$	−4121	−3885	236
$CaSO_4(s)$	−1434	−1322	106.7
Carbon (See also the table of organic substances.)			
C(g)	716.7	671.3	158.0
C (diamond)	1.90	2.90	2.38
C (graphite)	0	0	5.74
$CCl_4(g)$	−102.9	−60.63	309.7
$CCl_4(l)$	−135.4	−65.27	216.2
$C_2N_2(g)$	308.9	297.2	242.3
CO(g)	−110.5	−137.2	197.6
$CO_2(g)$	−393.5	−394.4	213.6
$C_3O_2(g)$	−93.72	−109.8	276.4
$C_3O_2(l)$	−117.3	−105.0	181.1
$COCl_2(g)$	−220.9	−206.8	283.8
COS(g)	−138.4	−165.6	231.5
$CS_2(l)$	89.70	65.27	151.3
Chlorine			
Cl(g)	121.7	105.7	165.1
$Cl_2(g)$	0	0	223.0
$ClF_3(g)$	−163.2	−123.0	281.5
$ClO_2(g)$	103	120.3	256.8
$Cl_2O(g)$	80.33	97.49	267.9

INORGANIC SUBSTANCES

	ΔH_f°, kJ/mol	ΔG_f°, kJ/mol	S°, J • mol^{-1} • K^{-1}
Chromium			
Cr(s)	0	0	23.66
Cr_2O_3(s)	−1135	−1053	81.17
Cobalt			
Co(s)	0	0	30.0
CoO(s)	−237.9	−214.2	52.97
$Co(OH)_2$ (pink solid)	−539.7	−454.4	79
Copper			
Cu(s)	0	0	33.15
$CuCO_3$•$Cu(OH)_2$(s)	−1051	−893.7	186
CuO(s)	−157.3	−129.7	42.63
$Cu(OH)_2$(s)	−450.2	−373	108
$CuSO_4$•$5H_2O$(s)	−2279.6	−1880.1	300.4
Fluorine			
F(g)	78.99	61.92	158.7
F_2(g)	0	0	202.7
Helium			
He(g)	0	0	126.0
Hydrogen			
H(g)	218.0	203.3	114.6
H_2(g)	0	0	130.6
HBr(g)	−36.40	−53.43	198.6
HCl(g)	−92.31	−95.30	186.8
HCl(aq)	−167.2	−131.3	56.48
HCN(g)	135	125	201.7
HF(g)	−271.1	−273.2	173.7
HI(g)	26.48	1.72	206.5
HNO_3(l)	−173.2	−79.91	155.6
HNO_3(aq)	−207.4	−113.3	146.4
H_2O(g)	−241.8	−228.6	188.7
H_2O(l)	−285.8	−237.2	69.91
H_2O_2(g)	−136.1	−105.5	232.9
H_2O_2(l)	−187.8	−120.4	110
H_2S(g)	−20.63	−33.56	205.7
H_2SO_4(l)	−814.0	−690.1	156.9
H_2SO_4(aq)	−909.3	−744.6	20.08
Iodine			
I(g)	106.8	70.28	180.7
I_2(g)	62.44	19.36	260.6
I_2(s)	0	0	116.1
IBr(g)	40.84	3.72	258.7
ICl(g)	17.78	−5.44	247.4
ICl(l)	−23.89	−13.60	135.1
Iron			
Fe(s)	0	0	27.28
$FeCO_3$(s)	−740.6	−666.7	92.88
$FeCl_3$(s)	−399.5	−334.1	142.3
FeO(s)	−272	−251.5	60.75
Fe_2O_3(s)	−824.2	−742.2	87.40
Fe_3O_4(s)	−1118	−1015	146

INORGANIC SUBSTANCES

	ΔH_f°, kJ/mol	ΔG_f°, kJ/mol	S°, J • mol^{-1} • K^{-1}
$Fe(OH)_3$(s)	−823.0	−696.6	107
Lead			
Pb(s)	0	0	64.81
PbI_2(s)	−175.5	−173.6	174.8
PbO_2(s)	−277	−217.4	68.6
$PbSO_4$(s)	−919.9	−813.2	148.6
Lithium			
Li(s)	0	0	29.12
LiCl(s)	−408.6	−384.4	59.33
LiOH(s)	−484.9	−439.0	42.80
$LiNO_3$(s)	−483.1	−381.1	90.0
Magnesium			
Mg(s)	0	0	32.69
$MgCl_2$(s)	−641.3	−591.8	89.62
$MgCO_3$(s)	−1096	−1012	65.7
MgF_2(s)	−1124	−1071	57.24
MgO(s)	−601.7	−569.4	26.94
$Mg(OH)_2$(s)	−924.7	−833.9	63.18
$MgSO_4$(s)	−1285	−1171	91.6
Manganese			
Mn(s)	0	0	32.0
MnO_2(s)	−520	−465.2	53.05
Mercury			
Hg(g)	61.32	31.85	174.9
Hg(l)	0	0	76.02
HgO(s)	−90.83	−58.56	70.29
Nitrogen			
N(g)	472.7	455.6	153.2
N_2(g)	0	0	191.5
NF_3(g)	−124.7	−83.2	260.7
NH_3(g)	−46.11	−16.48	192.3
NH_3(aq)	−80.29	−26.57	111.3
NH_4Br(s)	−270.8	−175	113.0
NH_4Cl(s)	−314.4	−203.0	94.56
NH_4F(s)	−464.0	−348.8	71.96
NH_4HCO_3(s)	−849.4	−666.1	121
NH_4I(s)	−201.4	−113	117
NH_4NO_3(s)	−365.6	−184.0	151.1
NH_4NO_3(aq)	−339.9	−190.7	259.8
$(NH_4)_2SO_4$(s)	−1181	−901.9	220.1
N_2H_4(g)	95.40	159.3	238.4
N_2H_4(l)	50.63	149.2	121.2
NO(g)	90.25	86.57	210.6
N_2O(g)	82.05	104.2	219.7
NO_2(g)	33.18	51.30	240.0
N_2O_4(g)	9.16	97.82	304.2
N_2O_4(l)	−19.6	97.40	209.2
N_2O_5(g)	11.3	115.1	355.7
NOBr(g)	82.17	82.4	273.5
NOCl(g)	51.71	66.07	261.6

INORGANIC SUBSTANCES

	ΔH°_f, kJ/mol	ΔG°_f, kJ/mol	S°, $J \cdot mol^{-1} \cdot K^{-1}$
Oxygen			
O(g)	249.2	231.7	160.9
O_2(g)	0	0	205.0
O_3(g)	142.7	163.2	238.8
OF_2(g)	24.5	41.8	247.3
Phosphorus			
P (α, white)	0	0	41.1
P (red)	−17.6	−12.1	22.8
P_4(g)	58.9	24.5	279.9
PCl_3(g)	−287.0	−267.8	311.7
PCl_5(g)	−374.9	−305.0	364.5
PH_3(g)	5.4	13.4	210.1
P_4O_{10}(s)	−2984	−2698	228.9
Potassium			
K(g)	89.24	60.63	160.2
K(l)	2.28	0.26	71.46
K(s)	0	0	64.18
KBr(s)	−393.8	−380.7	95.90
KCN(s)	−113	−101.9	128.5
KCl(s)	−436.7	−409.2	82.59
$KClO_3$(s)	−397.7	−296.3	143
$KClO_4$(s)	−432.8	−303.2	151.0
KF(s)	−567.3	−537.8	66.57
KI(s)	−327.9	−324.9	106.3
KNO_3(s)	−494.6	−394.9	133.1
KOH(s)	−424.8	−379.1	78.87
KOH(aq)	−482.4	−440.5	91.63
K_2SO_4(s)	−1438	−1321	175.6
Silicon			
Si(s)	0	0	18.8
SiH_4(g)	34	56.9	204.5
Si_2H_6(g)	80.3	127	272.5
SiO_2 (quartz)	−910.9	−856.7	41.84
Silver			
Ag(s)	0	0	42.55
AgBr(s)	−100.4	−96.90	107
AgCl(s)	−127.1	−109.8	96.2
AgI(s)	−61.84	−66.19	115
$AgNO_3$(s)	−124.4	−33.5	140.9
Ag_2O(s)	−31.0	−11.2	121
Ag_2SO_4(s)	−715.9	−618.5	200.4
Sodium			
Na(g)	107.3	76.78	153.6
Na(l)	2.41	0.50	57.86
Na(s)	0	0	51.21
Na_2(g)	142.0	104.0	230.1
NaBr(s)	−361.1	−349.0	86.82
Na_2CO_3(s)	−1131	−1044	135.0
$NaHCO_3$(s)	−950.8	−851.0	102
NaCl(s)	−411.1	−384.0	72.13

INORGANIC SUBSTANCES

	ΔH_f°, kJ/mol	ΔG_f°, kJ/mol	S°, J • mol^{-1} • K^{-1}
NaCl(aq)	−407.3	−393.1	115.5
$NaClO_3$(s)	−365.8	−262.3	123
$NaClO_4$(s)	−383.3	−254.9	142.3
NaF(s)	−573.7	−543.5	51.46
NaH(s)	−56.27	−33.5	40.02
NaI(s)	−287.8	−286.1	98.53
$NaNO_3$(s)	−467.9	−367.1	116.5
$NaNO_3$(aq)	−447.4	−373.2	205.4
Na_2O_2(s)	−510.9	−447.7	94.98
NaOH(s)	−425.6	−379.5	64.48
NaOH(aq)	−469.2	−419.2	48.1
NaH_2PO_4(s)	−1537	−1386	127.5
Na_2HPO_4(s)	−1748	−1608	150.5
Na_3PO_4(s)	−1917	−1789	173.8
$NaHSO_4$(s)	−1125	−992.9	113
Na_2SO_4(s)	−1387	−1270	149.6
Na_2SO_4(aq)	−1390	−1268	138.1
$Na_2SO_4 \cdot 10H_2O$(s)	−4327	−3647	592.0
$Na_2S_2O_3$(s)	−1123	−1028	155
Sulfur			
S (rhombic)	0	0	31.8
S_8(g)	102.3	49.16	430.2
S_2Cl_2(g)	−18.4	−31.8	331.5
SF_6(g)	−1209	−1105	291.7
SO_2(g)	−296.8	−300.2	248.1
SO_3(g)	−395.7	−371.1	256.6
SO_2Cl_2(g)	−364.0	−320.0	311.8
SO_2Cl_2(l)	−394.1	−314	207
Tin			
Sn (white)	0	0	51.55
Sn (gray)	−2.1	0.1	44.14
$SnCl_4$(l)	−511.3	−440.2	259
SnO(s)	−286	−257	56.5
SnO_2(s)	−580.7	−519.7	52.3
Titanium			
Ti(s)	0	0	30.6
$TiCl_4$(g)	−763.2	−726.8	355
$TiCl_4$(l)	−804.2	−737.2	252.3
TiO_2(s)	−944.7	−889.5	50.33
Uranium			
U(s)	0	0	50.21
UF_6(g)	−2147	−2064	378
UF_6(s)	−2197	−2069	228
UO_2(s)	−1085	−1032	77.03
Zinc			
Zn(s)	0	0	41.6
ZnO(s)	−348.3	−318.3	43.64

ORGANIC SUBSTANCES

	Name	ΔH_f°, kJ/mol	ΔG_f°, kJ/mol	S°, J • mol^{-1} • K^{-1}
$CH_4(g)$	methane	−74.81	−50.75	186.2
$C_2H_2(g)$	acetylene	226.7	209.2	200.8
$C_2H_4(g)$	ethylene	52.26	68.12	219.4
$C_2H_6(g)$	ethane	−84.68	−32.89	229.5
$C_3H_8(g)$	propane	−103.8	−23.56	270.2
$C_4H_{10}(g)$	butane	−125.7	−17.15	310.1
$C_6H_6(g)$	benzene(g)	82.93	129.7	269.2
$C_6H_6(l)$	benzene(l)	48.99	124.4	173.3
$C_6H_{12}(g)$	cyclohexane(g)	−123.1	31.8	298.2
$C_6H_{12}(l)$	cyclohexane(l)	−156.2	26.7	204.3
$C_{10}H_8(g)$	naphthalene (g)	149	223.6	335.6
$C_{10}H_8(s)$	naphthalene(s)	75.3	201.0	166.9
$CH_2O(g)$	formaldehyde	−117.0	−110.0	218.7
$CH_3OH(g)$	methanol(g)	−200.7	−162.0	239.7
$CH_3OH(l)$	methanol(l)	−238.7	−166.4	126.8
$CH_3CHO(g)$	acetaldehyde(g)	−166.1	−133.4	246.4
$CH_3CHO(l)$	acetaldehyde(l)	−191.8	−128.3	160.4
$CH_3CH_2OH(g)$	ethanol(g)	−234.4	−167.9	282.6
$CH_3CH_2OH(l)$	ethanol(l)	−277.7	−174.9	160.7
$C_6H_5OH(s)$	phenol	−165.0	−50.42	144.0
$(CH_3)_2CO(g)$	acetone(g)	−216.6	−153.1	294.9
$(CH_3)_2CO(l)$	acetone(l)	−247.6	−155.7	200.4
$CH_3COOH(g)$	acetic acid(g)	−432.3	−374.0	282.5
$CH_3COOH(l)$	acetic acid(l)	−484.1	−389.9	159.8
$CH_3COOH(aq)$	acetic acid(aq)	−488.3	−396.6	178.7
$C_6H_5COOH(s)$	benzoic acid	−385.1	−245.3	167.6
$CH_3NH_2(g)$	methylamine	−23.0	32.3	242.6
$C_6H_5NH_2(g)$	aniline(g)	86.86	166.7	319.2
$C_6H_5NH_2(l)$	aniline(l)	31.6	149.1	191.3
$C_6H_{12}O_6(s)$	glucose	−1273.3	−910.4	212.1

D.2 Equilibrium Constants

A. Ionization Constants of Weak Acids at 25 °C

Name of Acid	Formula	K_a	Name of Acid	Formula	K_a
acetic	$HC_2H_3O_2$	1.8×10^{-5}	chloroacetic	$HC_2H_2ClO_2$	1.4×10^{-3}
acrylic	$HC_3H_3O_2$	5.5×10^{-5}	chlorous	$HClO_2$	1.1×10^{-2}
arsenic	H_3AsO_4	6.0×10^{-3}	citric	$H_3C_6H_5O_7$	7.4×10^{-4}
	$H_2AsO_4^-$	1.0×10^{-7}		$H_2C_6H_5O_7^-$	1.7×10^{-5}
	$HAsO_4^{2-}$	3.2×10^{-12}		$HC_6H_5O_7^{2-}$	4.0×10^{-7}
arsenous	H_3AsO_3	6.6×10^{-10}	cyanic	$HOCN$	3.5×10^{-4}
benzoic	$HC_7H_5O_2$	6.3×10^{-5}	dichloroacetic	$HC_2HCl_2O_2$	5.5×10^{-2}
bromoacetic	$HC_2H_2BrO_2$	1.3×10^{-3}	fluoroacetic	$HC_2H_2FO_2$	2.6×10^{-3}
butyric	$HC_4H_7O_2$	1.5×10^{-5}	formic	$HCHO_2$	1.8×10^{-4}
carbonic	H_2CO_3	4.4×10^{-7}	hydrazoic	HN_3	1.9×10^{-5}
	HCO_3^-	4.7×10^{-11}	hydrocyanic	HCN	6.2×10^{-10}

Name of Acid	Formula	K_a	Name of Acid	Formula	K_a
hydrofluoric	HF	6.6×10^{-4}	phosphoric	H_3PO_4	7.1×10^{-3}
hydrogen peroxide	H_2O_2	2.2×10^{-12}		$H_2PO_4^-$	6.3×10^{-8}
hydroselenic	H_2Se	1.3×10^{-4}		HPO_4^{2-}	4.2×10^{-13}
	HSe^-	1×10^{-11}	phosphorous	H_3PO_3	3.7×10^{-2}
hydrosulfuric	H_2S	1.0×10^{-7}		$H_2PO_3^-$	2.1×10^{-7}
	HS^-	1×10^{-19}	propionic	$HC_3H_5O_2$	1.3×10^{-5}
hydrotelluric	H_2Te	2.3×10^{-3}	pyrophosphoric	$H_4P_2O_7$	3.0×10^{-2}
	HTe^-	1.6×10^{-11}		$H_3P_2O_7^-$	4.4×10^{-3}
hypobromous	$HOBr$	2.5×10^{-9}		$H_2P_2O_7^{2-}$	2.5×10^{-7}
hypochlorous	$HOCl$	2.9×10^{-8}		$HP_2O_7^{3-}$	5.6×10^{-10}
hypoiodous	HOI	2.3×10^{-11}	selenic	H_2SeO_4	strong acid
hyponitrous	$HON{=}NOH$	8.9×10^{-8}		$HSeO_4^-$	2.2×10^{-2}
	$HON{=}NO^-$	4×10^{-12}	selenous	H_2SeO_3	2.3×10^{-3}
iodic	HIO_3	1.6×10^{-1}		$HSeO_3^-$	5.4×10^{-9}
iodoacetic	$HC_2H_2IO_2$	6.7×10^{-4}	succinic	$H_2C_4H_4O_4$	6.2×10^{-5}
malonic	$H_2C_3H_2O_4$	1.5×10^{-3}		$HC_4H_4O_4^-$	2.3×10^{-6}
	$HC_3H_2O_4^-$	2.0×10^{-6}	sulfuric	H_2SO_4	strong acid
nitrous	HNO_2	7.2×10^{-4}		HSO_4^-	1.1×10^{-2}
oxalic	$H_2C_2O_4$	5.4×10^{-2}	sulfurous	H_2SO_3	1.3×10^{-2}
	$HC_2O_4^-$	5.3×10^{-5}		HSO_3^-	6.2×10^{-8}
phenol	HOC_6H_5	1.0×10^{-10}	thiophenol	HSC_6H_5	3.2×10^{-7}
phenylacetic	$HC_8H_7O_2$	4.9×10^{-5}	trichloroacetic	$HC_2Cl_3O_2$	3.0×10^{-1}

B. Ionization Constants of Weak Bases at 25 °C

Name of Base	Formula	K_b	Name of Base	Formula	K_b
ammonia	NH_3	1.8×10^{-5}	isoquinoline	C_9H_7N	2.5×10^{-9}
aniline	$C_6H_5NH_2$	7.4×10^{-10}	methylamine	CH_3NH_2	4.2×10^{-4}
codeine	$C_{18}H_{21}O_3N$	8.9×10^{-7}	morphine	$C_{17}H_{19}O_3N$	7.4×10^{-7}
diethylamine	$(C_2H_5)_2NH$	6.9×10^{-4}	piperidine	$C_5H_{11}N$	1.3×10^{-3}
dimethylamine	$(CH_3)_2NH$	5.9×10^{-4}	pyridine	C_5H_5N	1.5×10^{-9}
ethylamine	$C_2H_5NH_2$	4.3×10^{-4}	quinoline	C_9H_7N	6.3×10^{-10}
hydrazine	NH_2NH_2	8.5×10^{-7}	triethanolamine	$C_6H_{15}O_3N$	5.8×10^{-7}
	$NH_2NH_3^+$	8.9×10^{-16}	triethylamine	$(C_2H_5)_3N$	5.2×10^{-4}
hydroxylamine	NH_2OH	9.1×10^{-9}	trimethylamine	$(CH_3)_3N$	6.3×10^{-5}

C. Solubility Product Constants[a]

Name of Solute	Formula	K_{sp}	Name of Solute	Formula	K_{sp}
aluminum hydroxide	$Al(OH)_3$	1.3×10^{-33}	barium sulfate	$BaSO_4$	1.1×10^{-10}
aluminum phosphate	$AlPO_4$	6.3×10^{-19}	barium sulfite	$BaSO_3$	8×10^{-7}
barium carbonate	$BaCO_3$	5.1×10^{-9}	barium thiosulfate	BaS_2O_3	1.6×10^{-5}
barium chromate	$BaCrO_4$	1.2×10^{-10}	bismuthyl chloride	$BiOCl$	1.8×10^{-31}
barium fluoride	BaF_2	1.0×10^{-6}	bismuthyl hydroxide	$BiOOH$	4×10^{-10}
barium hydroxide	$Ba(OH)_2$	5×10^{-3}	cadmium carbonate	$CdCO_3$	5.2×10^{-12}

[a] Data are at various temperatures around "room" temperature, from 18 to 25 °C.

Name of Solute	Formula	K_{sp}
cadmium hydroxide	$Cd(OH)_2$	2.5×10^{-14}
cadmium sulfide[b]	CdS	8×10^{-28}
calcium carbonate	$CaCO_3$	2.8×10^{-9}
calcium chromate	$CaCrO_4$	7.1×10^{-4}
calcium fluoride	CaF_2	5.3×10^{-9}
calcium hydrogen phosphate	$CaHPO_4$	1×10^{-7}
calcium hydroxide	$Ca(OH)_2$	5.5×10^{-6}
calcium oxalate	CaC_2O_4	4×10^{-9}
calcium phosphate	$Ca_3(PO_4)_2$	2.0×10^{-29}
calcium sulfate	$CaSO_4$	9.1×10^{-6}
calcium sulfite	$CaSO_3$	6.8×10^{-8}
chromium(II) hydroxide	$Cr(OH)_2$	2×10^{-16}
chromium(III) hydroxide	$Cr(OH)_3$	6.3×10^{-31}
cobalt(II) carbonate	$CoCO_3$	1.4×10^{-13}
cobalt(II) hydroxide	$Co(OH)_2$	1.6×10^{-15}
cobalt(III) hydroxide	$Co(OH)_3$	1.6×10^{-44}
copper(I) chloride	$CuCl$	1.2×10^{-6}
copper(I) cyanide	$CuCN$	3.2×10^{-20}
copper(I) iodide	CuI	1.1×10^{-12}
copper(II) arsenate	$Cu_3(AsO_4)_2$	7.6×10^{-36}
copper(II) carbonate	$CuCO_3$	1.4×10^{-10}
copper(II) chromate	$CuCrO_4$	3.6×10^{-6}
copper(II) ferrocyanide	$Cu_2[Fe(CN)_6]$	1.3×10^{-16}
copper(II) hydroxide	$Cu(OH)_2$	2.2×10^{-20}
copper(II) sulfide[b]	CuS	6×10^{-37}
iron(II) carbonate	$FeCO_3$	3.2×10^{-11}
iron(II) hydroxide	$Fe(OH)_2$	8.0×10^{-16}
iron(II) sulfide[b]	FeS	6×10^{-19}
iron(III) arsenate	$FeAsO_4$	5.7×10^{-21}
iron(III) ferrocyanide	$Fe_4[Fe(CN)_6]_3$	3.3×10^{-41}
iron(III) hydroxide	$Fe(OH)_3$	4×10^{-38}
iron(III) phosphate	$FePO_4$	1.3×10^{-22}
lead(II) arsenate	$Pb_3(AsO_4)_2$	4.0×10^{-36}
lead(II) azide	$Pb(N_3)_2$	2.5×10^{-9}
lead(II) bromide	$PbBr_2$	4.0×10^{-5}
lead(II) carbonate	$PbCO_3$	7.4×10^{-14}
lead(II) chloride	$PbCl_2$	1.6×10^{-5}
lead(II) chromate	$PbCrO_4$	2.8×10^{-13}
lead(II) fluoride	PbF_2	2.7×10^{-8}
lead(II) hydroxide	$Pb(OH)_2$	1.2×10^{-15}
lead(II) iodide	PbI_2	7.1×10^{-9}
lead(II) sulfate	$PbSO_4$	1.6×10^{-8}
lead(II) sulfide[b]	PbS	3×10^{-28}
lithium carbonate	Li_2CO_3	2.5×10^{-2}
lithium fluoride	LiF	3.8×10^{-3}
lithium phosphate	Li_3PO_4	3.2×10^{-9}
magnesium ammonium phosphate	$MgNH_4PO_4$	2.5×10^{-13}
magnesium carbonate	$MgCO_3$	3.5×10^{-8}
magnesium fluoride	MgF_2	3.7×10^{-8}
magnesium hydroxide	$Mg(OH)_2$	1.8×10^{-11}
magnesium phosphate	$Mg_3(PO_4)_2$	1×10^{-25}
manganese(II) carbonate	$MnCO_3$	1.8×10^{-11}
manganese(II) hydroxide	$Mn(OH)_2$	1.9×10^{-13}
manganese(II) sulfide[b]	MnS	3×10^{-14}
mercury(I) bromide	Hg_2Br_2	5.6×10^{-23}
mercury(I) chloride	Hg_2Cl_2	1.3×10^{-18}
mercury(I) iodide	Hg_2I_2	4.5×10^{-29}
mercury(II) sulfide[b]	HgS	2×10^{-53}
nickel(II) carbonate	$NiCO_3$	6.6×10^{-9}
nickel(II) hydroxide	$Ni(OH)_2$	2.0×10^{-15}
scandium fluoride	ScF_3	4.2×10^{-18}
scandium hydroxide	$Sc(OH)_3$	8.0×10^{-31}
silver arsenate	Ag_3AsO_4	1.0×10^{-22}
silver azide	AgN_3	2.8×10^{-9}
silver bromide	$AgBr$	5.0×10^{-13}
silver chloride	$AgCl$	1.8×10^{-10}
silver chromate	Ag_2CrO_4	2.4×10^{-12}
silver cyanide	$AgCN$	1.2×10^{-16}
silver iodate	$AgIO_3$	3.0×10^{-8}
silver iodide	AgI	8.5×10^{-17}
silver nitrite	$AgNO_2$	6.0×10^{-4}
silver sulfate	Ag_2SO_4	1.4×10^{-5}
silver sulfide[b]	Ag_2S	6×10^{-51}
silver sulfite	Ag_2SO_3	1.5×10^{-14}
silver thiocyanate	$AgSCN$	1.0×10^{-12}
strontium carbonate	$SrCO_3$	1.1×10^{-10}
strontium chromate	$SrCrO_4$	2.2×10^{-5}
strontium fluoride	SrF_2	2.5×10^{-9}
strontium sulfate	$SrSO_4$	3.2×10^{-7}
thallium(I) bromide	$TlBr$	3.4×10^{-6}
thallium(I) chloride	$TlCl$	1.7×10^{-4}
thallium(I) iodide	TlI	6.5×10^{-8}
thallium(III) hydroxide	$Tl(OH)_3$	6.3×10^{-46}
tin(II) hydroxide	$Sn(OH)_2$	1.4×10^{-28}
tin(II) sulfide[b]	SnS	1×10^{-26}
zinc carbonate	$ZnCO_3$	1.4×10^{-11}
zinc hydroxide	$Zn(OH)_2$	1.2×10^{-17}
zinc oxalate	ZnC_2O_4	2.7×10^{-8}
zinc phosphate	$Zn_3(PO_4)_2$	9.0×10^{-33}
zinc sulfide[b]	ZnS	2×10^{-25}

[b] For a solubility equilibrium of the type: $MS(s) + H_2O \rightleftharpoons M^{2+}(aq) + HS^-(aq) + OH^-(aq)$.

D. Complex-Ion Formation Constants[a]

Formula	K_f	Formula	K_f	Formula	K_f
$[Ag(CN)_2]^-$	5.6×10^{18}	$[Co(ox)_3]^{3-}$	10^{20}	$[HgI_4]^{2-}$	6.8×10^{29}
$[Ag(EDTA)]^{3-}$	2.1×10^{7}	$[Cr(EDTA)]^-$	10^{23}	$[Hg(ox)_2]^{2-}$	9.5×10^{6}
$[Ag(en)_2]^+$	5.0×10^{7}	$[Cr(OH)_4]^-$	8×10^{29}	$[Ni(CN)_4]^{2-}$	2×10^{31}
$[Ag(NH_3)_2]^+$	1.6×10^{7}	$[CuCl_3]^{2-}$	5×10^{5}	$[Ni(EDTA)]^{2-}$	3.6×10^{18}
$[Ag(SCN)_4]^{3-}$	1.2×10^{10}	$[Cu(CN)_4]^{3-}$	2.0×10^{30}	$[Ni(en)_3]^{2+}$	2.1×10^{18}
$[Ag(S_2O_3)_2]^{3-}$	1.7×10^{13}	$[Cu(EDTA)]^{2-}$	5×10^{18}	$[Ni(NH_3)_6]^{2+}$	5.5×10^{8}
$[Al(EDTA)]^-$	1.3×10^{16}	$[Cu(en)_2]^{2+}$	1×10^{20}	$[Ni(ox)_3]^{4-}$	3×10^{8}
$[Al(OH)_4]^-$	1.1×10^{33}	$[Cu(NH_3)_4]^{2+}$	1.1×10^{13}	$[PbCl_3]^-$	2.4×10^{1}
$[Al(ox)_3]^{3-}$	2×10^{16}	$[Cu(ox)_2]^{2-}$	3×10^{8}	$[Pb(EDTA)]^{2-}$	2×10^{18}
$[CdCl_4]^{2-}$	6.3×10^{2}	$[Fe(CN)_6]^{4-}$	10^{37}	$[PbI_4]^{2-}$	3.0×10^{4}
$[Cd(CN)_4]^{2-}$	6.0×10^{18}	$[Fe(EDTA)]^{2-}$	2.1×10^{14}	$[Pb(OH)_3]^-$	3.8×10^{14}
$[Cd(en)_3]^{2+}$	1.2×10^{12}	$[Fe(en)_3]^{2+}$	5.0×10^{9}	$[Pb(ox)_2]^{2-}$	3.5×10^{6}
$[Cd(NH_3)_4]^{2+}$	1.3×10^{7}	$[Fe(ox)_3]^{4-}$	1.7×10^{5}	$[Pb(S_2O_3)_3]^{4-}$	2.2×10^{6}
$[Co(EDTA)]^{2-}$	2.0×10^{16}	$[Fe(CN)_6]^{3-}$	10^{42}	$[PtCl_4]^{2-}$	1×10^{16}
$[Co(en)_3]^{2+}$	8.7×10^{13}	$[Fe(EDTA)]^-$	1.7×10^{24}	$[Pt(NH_3)_6]^{2+}$	2×10^{35}
$[Co(NH_3)_6]^{2+}$	1.3×10^{5}	$[Fe(ox)_3]^{3-}$	2×10^{20}	$[Zn(CN)_4]^{2-}$	1×10^{18}
$[Co(ox)_3]^{4-}$	5×10^{9}	$[Fe(SCN)]^{2+}$	8.9×10^{2}	$[Zn(EDTA)]^{2-}$	3×10^{16}
$[Co(SCN)_4]^{2-}$	1.0×10^{3}	$[HgCl_4]^{2-}$	1.2×10^{15}	$[Zn(en)_3]^{2+}$	1.3×10^{14}
$[Co(EDTA)]^-$	10^{36}	$[Hg(CN)_4]^{2-}$	3×10^{41}	$[Zn(NH_3)_4]^{2+}$	4.1×10^{8}
$[Co(en)_3]^{3+}$	4.9×10^{48}	$[Hg(EDTA)]^{2-}$	6.3×10^{21}	$[Zn(OH)_4]^{2-}$	4.6×10^{17}
$[Co(NH_3)_6]^{3+}$	4.5×10^{33}	$[Hg(en)_2]^{2+}$	2×10^{23}	$[Zn(ox)_3]^{4-}$	1.4×10^{8}

[a] The ligands referred to in this table are monodentate: Cl^-, CN^-, I^-, NH_3, OH^-, SCN^-, $S_2O_3^{2-}$; bidentate: ethylenediamine, en; oxalate ion, ox ($C_2O_4^{2-}$); tetradentate: ethylenediamine-tetraacetato ion, $EDTA^{4-}$.

D.3 Standard Electrode (Reduction) Potentials at 25 °C

Reduction Half-reaction	$E°$, V
$Li^+(aq) + e^- \longrightarrow Li(s)$	−3.040
$Rb^+(aq) + e^- \longrightarrow Rb(s)$	−2.924
$K^+(aq) + e^- \longrightarrow K(s)$	−2.924
$Cs^+(aq) + e^- \longrightarrow Cs(s)$	−2.923
$Ba^{2+}(aq) + 2e^- \longrightarrow Ba(s)$	−2.92
$Sr^{2+}(aq) + 2e^- \longrightarrow Sr(s)$	−2.89
$Ca^{2+}(aq) + 2e^- \longrightarrow Ca(s)$	−2.84
$Na^+(aq) + e^- \longrightarrow Na(s)$	−2.713
$La^{3+}(aq) + 3e^- \longrightarrow La(s)$	−2.38
$Mg^{2+}(aq) + 2e^- \longrightarrow Mg(s)$	−2.356
$Al^{3+}(aq) + 3e^- \longrightarrow Al(s)$	−1.676
$U^{3+}(aq) + 3e^- \longrightarrow U(s)$	−1.66
$Ti^{2+}(aq) + 2e^- \longrightarrow Ti(s)$	−1.63
$Mn^{2+}(aq) + 2e^- \longrightarrow Mn(s)$	−1.18
$Cr^{2+}(aq) + 2e^- \longrightarrow Cr(s)$	−0.90

Reduction Half-reaction	$E°$, V
$Zn^{2+}(aq) + 2\ e^- \longrightarrow Zn(s)$	−0.763
$2\ CO_2(g) + 2\ H^+(aq) + 2\ e^- \longrightarrow H_2C_2O_4(aq)$	−0.49
$Fe^{2+}(aq) + 2\ e^- \longrightarrow Fe(s)$	−0.440
$Cr^{3+}(aq) + e^- \longrightarrow Cr^{2+}(aq)$	−0.424
$Cd^{2+}(aq) + 2\ e^- \longrightarrow Cd(s)$	−0.403
$PbSO_4(s) + 2\ e^- \longrightarrow Pb(s) + SO_4^{2-}(aq)$	−0.356
$In^{3+}(aq) + 3\ e^- \longrightarrow In(s)$	−0.338
$Co^{2+}(aq) + 2\ e^- \longrightarrow Co(s)$	−0.277
$H_3PO_4(aq) + 2\ H^+(aq) + 2\ e^- \longrightarrow H_3PO_3(aq) + H_2O$	−0.276
$Ni^{2+}(aq) + 2\ e^- \longrightarrow Ni(s)$	−0.257
$V^{3+}(aq) + e^- \longrightarrow V^{2+}(aq)$	−0.255
$AgI(s) + e^- \longrightarrow Ag(s) + I^-(aq)$	−0.152
$Sn^{2+}(aq) + 2\ e^- \longrightarrow Sn(s)$	−0.137
$Pb^{2+}(aq) + 2\ e^- \longrightarrow Pb(s)$	−0.125
$2\ H^+(aq) + 2\ e^- \longrightarrow H_2(g)$	0
$AgBr(s) + e^- \longrightarrow Ag(s) + Br^-(aq)$	+0.071
$S(s) + 2\ H^+(aq) + 2\ e^- \longrightarrow H_2S(g)$	+0.14
$Sn^{4+}(aq) + 2\ e^- \longrightarrow Sn^{2+}(aq)$	+0.154
$Cu^{2+}(aq) + e^- \longrightarrow Cu^+(aq)$	+0.159
$SO_4^{2-}(aq) + 4\ H^+(aq) + 2\ e^- \longrightarrow 2\ H_2O + SO_2(g)$	+0.17
$AgCl(s) + e^- \longrightarrow Ag(s) + Cl^-(aq)$	+0.2223
$HAsO_2(aq) + 3\ H^+(aq) + 3\ e^- \longrightarrow As(s) + 2\ H_2O$	+0.240
$Hg_2Cl_2(s) + 2\ e^- \longrightarrow 2\ Hg(l) + 2\ Cl^-(aq)$	+0.2676
$PbO_2(s) + 2\ H^+(aq) + 2\ e^- \longrightarrow PbO(s) + H_2O$	+0.28
$Cu^{2+}(aq) + 2\ e^- \longrightarrow Cu(s)$	+0.337
$VO^{2+}(aq) + 2\ H^+(aq) + e^- \longrightarrow V^{3+}(aq) + H_2O$	+0.337
$[Fe(CN)_6]^{3-}(aq) + e^- \longrightarrow [Fe(CN)_6]^{4-}(aq)$	+0.361
$C_2N_2(g) + 2\ H^+(aq) + 2\ e^- \longrightarrow 2\ HCN(aq)$	+0.37
$H_2SO_3(aq) + 4\ H^+(aq) + 4\ e^- \longrightarrow S(s) + 3\ H_2O$	+0.449
$Cu^+(aq) + e^- \longrightarrow Cu(s)$	+0.520
$I_2(s) + 2\ e^- \longrightarrow 2\ I^-(aq)$	+0.535
$MnO_4^-(aq) + e^- \longrightarrow MnO_4^{2-}(aq)$	+0.56
$2\ HgCl_2(aq) + 2\ e^- \longrightarrow Hg_2Cl_2(s) + 2\ Cl^-(aq)$	+0.63
$O_2(g) + 2\ H^+(aq) + 2\ e^- \longrightarrow H_2O_2(aq)$	+0.695
$Fe^{3+}(aq) + e^- \longrightarrow Fe^{2+}(aq)$	+0.771
$Ag^+(aq) + e^- \longrightarrow Ag(s)$	+0.800
$Hg^{2+}(aq) + 2\ e^- \longrightarrow Hg(l)$	+0.854
$Cu^{2+}(aq) + I^-(aq) + e^- \longrightarrow CuI(s)$	+0.86
$NO_3^-(aq) + 4\ H^+(aq) + 3\ e^- \longrightarrow NO(g) + 2\ H_2O$	+0.956
$VO_2^+(aq) + 2\ H^+(aq) + e^- \longrightarrow VO^{2+}(aq) + H_2O$	+1.000
$[AuCl_4]^-(aq) + 3\ e^- \longrightarrow Au(s) + 4\ Cl^-(aq)$	+1.002
$NO_2(g) + 2\ H^+(aq) + 2\ e^- \longrightarrow NO(g) + H_2O$	+1.03
$Br_2(l) + 2\ e^- \longrightarrow 2\ Br^-(aq)$	+1.065
$NO_2(g) + H^+(aq) + e^- \longrightarrow HNO_2(aq)$	+1.07
$ClO_3^-(aq) + 2\ H^+(aq) + e^- \longrightarrow ClO_2(g) + H_2O$	+1.175
$ClO_4^-(aq) + 2\ H^+(aq) + 2\ e^- \longrightarrow ClO_3^-(aq) + H_2O$	+1.19
$2\ IO_3^-(aq) + 12\ H^+(aq) + 10\ e^- \longrightarrow I_2(s) + 6\ H_2O$	+1.20
$O_2(g) + 4\ H^+(aq) + 4\ e^- \longrightarrow 2\ H_2O$	+1.229

Reduction Half-reaction	$E°$, V
$MnO_2(s) + 4 H^+(aq) + 2 e^- \longrightarrow Mn^{2+}(aq) + 2 H_2O$	+1.23
$Cr_2O_7^{2-}(aq) + 14 H^+(aq) + 6 e^- \longrightarrow 2 Cr^{3+}(aq) + 7 H_2O$	+1.33
$Cl_2(g) + 2 e^- \longrightarrow 2 Cl^-(aq)$	+1.358
$Au^{3+}(aq) + 2 e^- \longrightarrow Au^+(aq)$	+1.36
$ClO_3^-(aq) + 6 H^+(aq) + 6 e^- \longrightarrow Cl^-(aq) + 3 H_2O$	+1.450
$PbO_2(s) + 4 H^+(aq) + 2 e^- \longrightarrow Pb^{2+}(aq) + 2 H_2O$	+1.455
$2 BrO_3^-(aq) + 12 H^+(aq) + 10 e^- \longrightarrow Br_2(l) + 6 H_2O$	+1.478
$MnO_4^-(aq) + 8 H^+(aq) + 5 e^- \longrightarrow Mn^{2+}(aq) + 4 H_2O$	+1.51
$Au^{3+}(aq) + 3 e^- \longrightarrow Au(s)$	+1.52
$PbO_2(s) + SO_4^{2-}(aq) + 4 H^+(aq) + 2 e^- \longrightarrow PbSO_4(s) + 2 H_2O$	+1.69
$MnO_4^-(aq) + 4 H^+(aq) + 3 e^- \longrightarrow MnO_2(s) + 2 H_2O$	+1.70
$H_2O_2(aq) + 2 H^+(aq) + 2 e^- \longrightarrow 2 H_2O$	+1.763
$Ag^{2+}(aq) + e^- \longrightarrow Ag^+(aq)$	+1.98
$S_2O_8^{2-}(aq) + 2 e^- \longrightarrow 2 SO_4^{2-}(aq)$	+2.01
$O_3(g) + 2 H^+(aq) + 2 e^- \longrightarrow O_2(g) + H_2O$	+2.075
$OF_2(g) + 2 H^+(aq) + 4 e^- \longrightarrow H_2O + 2 F^-(aq)$	+2.1
$F_2(g) + 2 e^- \longrightarrow 2 F^-(aq)$	+2.866
Basic Solution	
$Mg(OH)_2(s) + 2 e^- \longrightarrow Mg(s) + 2 OH^-(aq)$	−2.687
$Al(OH)_4^-(aq) + 3 e^- \longrightarrow Al(s) + 4 OH^-(aq)$	−2.310
$Sb(s) + 3 H_2O + 3 e^- \longrightarrow SbH_3(g) + 3 OH^-(aq)$	−1.338
$Zn(OH)_2(s) + 2 e^- \longrightarrow Zn(s) + 2 OH^-(aq)$	−1.246
$As(s) + 3 H_2O + 3 e^- \longrightarrow AsH_3(g) + 3 OH^-(aq)$	−1.21
$OCN^-(aq) + H_2O + 2 e^- \longrightarrow CN^-(aq) + 2 OH^-(aq)$	−0.97
$2 H_2O + 2 e^- \longrightarrow H_2(g) + 2 OH^-(aq)$	−0.828
$AsO_2^-(aq) + 2 H_2O + 3 e^- \longrightarrow As(s) + 4 OH^-(aq)$	−0.68
$AsO_4^{3-}(aq) + 2 H_2O + 2 e^- \longrightarrow AsO_2^-(aq) + 4 OH^-(aq)$	−0.67
$SO_3^{2-}(aq) + 3 H_2O + 4 e^- \longrightarrow S(s) + 6 OH^-(aq)$	−0.66
$HCHO(aq) + 2 H_2O + 2 e^- \longrightarrow CH_3OH(aq) + 2 OH^-(aq)$	−0.59
$HPbO_2^-(aq) + H_2O + 2 e^- \longrightarrow Pb(s) + 3 OH^-(aq)$	−0.54
$CrO_4^{2-}(aq) + 4 H_2O + 3 e^- \longrightarrow Cr(OH)_3(s) + 5 OH^-(aq)$	−0.11
$NO_3^-(aq) + H_2O + 2 e^- \longrightarrow NO_2^-(aq) + 2 OH^-(aq)$	+0.01
$Co(OH)_3(s) + e^- \longrightarrow Co(OH)_2(s) + OH^-(aq)$	+0.17
$Ag_2O(s) + H_2O + 2 e^- \longrightarrow 2 Ag(s) + 2 OH^-(aq)$	+0.342
$O_2(g) + 2 H_2O + 4 e^- \longrightarrow 4 OH^-(aq)$	+0.401
$2 IO^-(aq) + 2 H_2O + 2 e^- \longrightarrow I_2(s) + 4 OH^-(aq)$	+0.42
$2 BrO^-(aq) + 2 H_2O + 2 e^- \longrightarrow Br_2(l) + 4 OH^-(aq)$	+0.455
$Ni(OH)_3(s) + e^- \longrightarrow Ni(OH)_2(s) + OH^-(aq)$	+0.48
$BrO_3^-(aq) + 3 H_2O + 6 e^- \longrightarrow Br^-(aq) + 6 OH^-(aq)$	+0.584
$MnO_4^-(aq) + 2 H_2O + 3 e^- \longrightarrow MnO_2(s) + 4 OH^-(aq)$	+0.60
$2 AgO(s) + H_2O + 2 e^- \longrightarrow Ag_2O(s) + 2 OH^-(aq)$	+0.604
$ClO_3^-(aq) + 3 H_2O + 6 e^- \longrightarrow Cl^-(aq) + 6 OH^-(aq)$	+0.622
$BrO^-(aq) + H_2O + 2 e^- \longrightarrow Br^-(aq) + 2 OH^-(aq)$	+0.766
$HO_2^-(aq) + H_2O + 2 e^- \longrightarrow 3 OH^-(aq)$	+0.88
$ClO^-(aq) + H_2O + 2 e^- \longrightarrow Cl^-(aq) + 2 OH^-(aq)$	+0.890
$O_3(g) + H_2O + 2 e^- \longrightarrow O_2(g) + 2 OH^-(aq)$	+1.246

Appendix E
Glossary

Absolute humidity. *See* **humidity**.

Absolute zero is the lowest possible temperature—the temperature at which molecular motion ceases: −273.15 °C = 0 K.

An **acid** (1) a hydrogen-containing compound that, under appropriate conditions, can produce hydrogen ions, H^+ (Arrhenius theory); (2) a proton donor (Brønsted–Lowry theory); (3) an atom, ion, or molecule that can accept a pair of electrons to form a covalent bond (Lewis theory).

An **acid anhydride** is a nonmetal oxide whose reaction with water produces a ternary acid as its sole product. For example, SO_3 is the acid anhydride of H_2SO_4.

The **acid ionization constant (K_a)** is the equilibrium constant describing equilibrium in the reversible ionization of a weak acid.

Acid rain is rainfall that is more acidic than is water in equilibrium with atmospheric carbon dioxide. The excess acidity is produced by the action of air pollutants such as SO_3 and NO_2 on water.

The **accuracy** of a measurement refers to the closeness of the measured value to the "correct" or most probable value.

The **actinide** elements constitute the portion of the *f*-block of the periodic table in which the 5*f* subshell fills in the aufbau process. This series of elements is extracted from the main body of the table and printed at the bottom.

The **activated complex** is an aggregate of atoms in the transition state of a reaction formed by a favorable collision. (*See also* **transition state**.)

The **activation energy** of a reaction refers to the minimum total kinetic energy that molecules must bring into their collisions so that a chemical reaction may occur.

The **active site** on an enzyme is the region of the enzyme molecule to which the substrate attaches itself and where a chemical reaction occurs. (*See also* **enzyme** and **substrate**.)

The **activity series of the metals** is a listing of the metals in terms of their ability to displace one another from solutions of their ions or to displace H^+ as $H_2(g)$ from acidic solutions (*see* Figure 12.10).

The **actual yield** is the *measured* quantity of a desired product obtained in a chemical reaction. (*See also* **theoretical yield** and **percent yield**.)

Addition polymerization is a type of polymerization reaction in which monomers add to one another to produce a polymeric product that contains all the atoms of the starting monomers.

In an **addition reaction** substituent groups join to hydrocarbon molecules at points of unsaturation—double or triple bonds. This type of reaction is typical of alkenes and alkynes.

An **adduct** is a compound that results from the addition, through a coordinate covalent bond, of one structure to another. An example is the adduct of dimethylether, $(CH_3)_2O$, and borane, BH_3, in which the bond between the B and O atoms is through a lone pair of electrons on the O atom.

Adhesive forces are intermolecular forces between unlike molecules, for example, between those in a liquid and those in a surface over which the liquid is spread.

An **air pollutant** is a substance that is found in air in greater abundance than normally occurs naturally and that has some harmful effect(s) on the environment.

An **alcohol (ROH)** is an organic substance whose molecules contain the hydroxyl group, —OH, attached to an alkyl group.

Aldehydes are organic substances whose molecules have the functional group

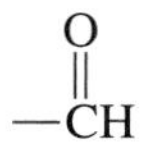

An **alkane** is a saturated hydrocarbon (carbon–hydrogen compound) having the general formula C_nH_{2n+2}; for example, CH_4, C_2H_6, C_3H_8, . . . (*See also* **saturated hydrocarbon**.)

An **alkene** is a hydrocarbon (carbon–hydrogen compound) whose molecules contain at least one carbon-to-carbon double bond.

An **alkyl group (R—)** is a substituent group in organic molecules that is derived from an alkane molecule by removal of one hydrogen atom.

An **alkyne** is a hydrocarbon (carbon–hydrogen compound) whose molecules contain at least one carbon-to-carbon triple bond.

Allotropes are two or more forms of an element that differ in their basic molecular structure. Diamond and graphite are alloptopes of carbon. Ordinary oxygen, O_2 (dioxygen), and ozone O_3 (trioxygen), are also allotropes.

An **alloy** is a metallic material consisting of two or more elements. Some alloys are homogeneous solid solutions; some are heterogeneous mixtures in which the separate phases are individual metals or, more commonly, solid solutions; some are intermetallic compounds.

An **alpha particle (α)** consists of two protons and two neutrons. It is identical to a doubly ionized helium ion, He^{2+}, and is emitted by the nuclei of certain radioactive atoms as they undergo decay. In the decaying nucleus, the atomic number decreases by two units and the mass number by four units.

An **amine** is an organic substance in which one or more H atoms of an ammonia molecule is replaced by a hydrocarbon residue.

The **ampere (A)** is the basic unit of electric current. One ampere is a current of one coulomb per second: 1 A = 1 C/s.

An **amphiprotic** substance can ionize either as a Brønsted–Lowry acid or base, depending on the acid–base properties of other species in the solution. For example, in the presence of CH_3COOH, an acid, H_2O acts as a base; in the presence of NH_3, a base, H_2O acts as an acid.

Amphoteric oxides and hydroxides are able to react with either acids or bases. The central element of the oxide or hydroxide appears in a cation in acidic solutions and in an anion in basic solutions. Some examples are Al_2O_3, $Al(OH)_3$, Cr_2O_3, $Cr(OH)_3$, ZnO, and $Zn(OH)_2$.

The angular momentum quantum number (*l*) is the second of three parameters that must be assigned a specific value to achieve a solution of Schrödinger's wave equation for the hydrogen atom: $l = 0, 1, 2, 3, \cdots, n - 1$. Its value establishes a particular sublevel or subshell within a principal energy level.

An **anion** is a negatively charged ion.

An **anode** is an electrode at which an oxidation half-reaction occurs. It is the negative electrode in a voltaic cell and the positive electrode in an electrolysis cell.

An **antibonding molecular orbital** places a high electron charge density (electron probability) away from the region between bonded atoms. (*See also* **bonding molecular orbital** and **molecular orbital**.)

An **anticodon** is a base triplet on a tRNA molecule that is complementary (that is, pairs up with) a base triplet on a segment of mRNA. The anticodon establishes the particular amino acid carried by the tRNA molecule. (*See also* **codon**.)

An **aqueous** solution is a solution in which water is the solvent.

Aromatic compounds are organic compounds with benzenelike structures. To describe their electronic structures (Lewis structures) one must use resonance theory.

The **atmosphere (atm)** is a unit used to measure gas pressure. It is equal to the pressure of a column of mercury having a height of exactly 760 mm. That is, 1 atm = 760 mmHg.

Atomic mass. *See* **atomic weight**.

An **atomic mass unit** is *exactly* one-twelfth the mass of an atom of carbon-12. The masses of the fundamental particles—electrons, protons, and neutrons—and of individual atoms are often expressed in these units.

The **atomic number (*Z*)** of an atom is the number of protons in the atomic nucleus.

An **atomic orbital** is a wave function for an electron corresponding to the assignment of specific values to the n, l, and m_l quantum numbers in a wave equation.

The **atomic radius** is a measure of the size of an atom based on the measurement of internuclear distances. (*See also* **covalent radius**, **ionic radius,** and **metallic radius.**)

The **atomic weight** of an element is the weighted average of the masses of the atoms of the naturally occurring isotopes of the element.

Atoms are the basic building blocks of matter. The number of types of atoms currently known is 111. A chemical element consists of a single type of atom and a chemical compound, of two or more different kinds of atoms.

The **aufbau principle** is a hypothetical process for building up an atom from the atom of preceding atomic number by adding a proton and the requisite number of neutrons to the nucleus and one electron to the appropriate atomic orbital.

Average bond energy is the average of the bond dissociation energies for a number of different molecular species containing the particular bond. (*See also* **bond dissociation energy**.)

Avogadro's law states that at a fixed temperature and pressure the volume of a gas is directly proportional to the amount of gas.

Avogadro's number (N_A) is the number of elementary units in a mole—6.0221367×10^{23} mol^{-1}.

A **band** is a collection of a large number of molecular orbitals, very closely spaced in energy, obtained by combining atomic orbitals of many atoms. Band theory is the principal theory used to describe bonding in metals and semiconductors.

A **barometer** is a device used to measure the pressure exerted by the atmosphere.

A **base** is (1) a compound that produces hydroxide ions, OH^-, in water solution (Arrhenius theory); (2) a proton acceptor (Brønsted–Lowry theory); (3) an atom, ion, or molecule that can donate a pair of electrons to form a covalent bond (Lewis theory).

The **base ionization constant (K_b)** is the equilibrium constant describing equilibrium in the reversible ionization of a weak base.

A **battery** is a device that stores chemical energy for later release as electricity as a result of a chemical reaction.

A **beta particle (β^-)** is identical to an electron and is emitted by the nuclei of certain radioactive atoms as they undergo decay. In the decaying nucleus, the atomic number increases by one unit and the mass number remains unchanged.

Beta particle (β^+). *See* **positron**.

A **bidentate** ligand has two points of attachment to a metal center in a complex. For example, $NH_2CH_2CH_2NH_2$ attaches through a lone pair of electrons on each N atom.

A **bimolecular** step is an elementary step in a reaction mechanism that involves the collision of two molecules.

Black-body radiation is electromagnetic radiation emitted by solids. The frequency of the radiation increases as the temperature of the solid is raised.

The **body-centered cubic (bcc)** crystal structure has as its unit cell a cube with a structural unit at each corner and one in the center of the cell. (*See also* **unit cell**.)

The **boiling point** of a liquid is the temperature at which the liquid boils—the temperature at which the vapor pressure of the liquid is equal to the prevailing atmospheric pressure.

The **bond dissociation energy (*D*)** of a particular covalent bond between two atoms is the quantity of energy required to break one mole of bonds of that type in a *gaseous* species.

A **bonding molecular orbital** places a high electron charge density (electron probability) in the region between two bonded atoms. (*See also* **antibonding molecular orbital** and **molecular orbital**.)

A **bonding pair** is a pair of electrons shared between two atoms in a molecule.

The **bond length** of a particular covalent bond is the distance between the nuclei of two atoms joined by that type of bond.

A **bond moment** describes the extent to which a separation of positive ($\delta+$) and negative ($\delta-$) charges exists in a covalent bond between two atoms.

Bond order is a term used to indicate the nature of a covalent bond—that is, whether *single* (bond order = 1), *double* (bond order = 2), or *triple* (bond order = 3). In molecular orbital theory it is one half the difference between the number of electrons in bonding molecular orbitals and in antibonding molecular orbitals.

Boyle's law states that for a given amount of gas at a constant temperature, the volume of a gas varies *inversely* with its pressure. That is, $V \propto 1/P$, or PV = constant.

A **buffer solution** is a solution containing a weak acid and its conjugate base, or a weak base and its conjugate acid. Small quantities of added acid are neutralized by one buffer component and small quantities of added base by the other. As a result the solution pH is maintained nearly constant.

A **buret** is a graduated, long glass tube constructed to deliver precise volumes of a liquid solution through a stopcock valve.

A **calorie (cal)** is the amount of energy needed to raise the temperature of 1 g of water by 1 °C (more precisely, from 14.5 to 15.5 °C). 1 cal = 4.184 J.

Calorimetry refers to the measurement of quantities of heat in a device called a **calorimeter**.

A **carbocation** is an intermediate species in certain reactions of organic compounds in which a positive charge is centered on a carbon atom in the species.

The **carbon cycle** refers to the sum of all the processes by which carbon atoms are cycled throughout Earth's solid crust, oceans, and atmosphere.

A **carboxylic acid** is an organic substance whose molecules contain the carboxyl group, —COOH.

A **catalyst** is a substance that increases the rate of a reaction without itself being consumed in the reaction. A catalyst changes a reaction mechanism to one with a lower activation energy.

A **cathode** is an electrode at which a reduction half-reaction occurs. It is the positive electrode in a voltaic cell and the negative electrode in an electrolysis cell.

A **cathode ray** is a beam of electrons that travels from the cathode to the anode when an electric discharge is passed through an evacuated tube.

A **cation** is a positively charged ion.

A **cell diagram** is a schematic representation of an electrochemical cell (see page 769).

Cell potential or **cell voltage** (E_{cell}) refers to the potential difference between the electrodes in a voltaic cell.

A **central atom** in a molecule or polyatomic ion is an atom bonded to two or more other atoms.

Charles's law states that the volume of a fixed amount of a gas at a constant pressure is *directly* proportional to its Kelvin temperature. That is, $V \propto T$, or V = constant $\times T$.

A **chelate** is a five- or six-membered ring structure(s) produced in a complex through the attachment of a polydentate ligand(s) to a metal center.

A **chemical bond** is a force that holds atoms together in molecules or ions in crystals.

A **chemical equation** is a description of a chemical reaction that uses symbols and formulas to represent the elements and compounds involved in the reaction. Numerical coefficients preceding each symbol or formula and indicating molar proportions may be needed to *balance* a chemical equation.

A **chemical formula** indicates the composition of a compound through symbols of the elements present and subscripts to indicate the relative numbers of atoms of each element.

A **chemical property** is displayed by a substance as it undergoes chemical change, that is, when it reacts with other substances to produce new substances with altered compositions.

Chemical shift is a term used in nuclear magnetic resonance (NMR) spectroscopy to indicate the location of an absorption peak relative to a standard. The magnitudes of the chemical shifts observed in an NMR spectrum can be used to determine structural features of a molecule.

A **chemical symbol** is a representation of an element made up of one or two letters derived from the English name of the element (or, sometimes, from the Latin name of the element or one of its compounds).

Chemistry is a study of the composition, structure, and properties of matter and of the changes that occur in matter.

The term **cis** is used to describe isomers in which two substituent groups are attached on the same side of a double bond in an organic molecule, or along the same edge of a square planar or octahedral complex ion. (*See also* **geometric isomerism**.)

A **codon** is a base triplet of mRNA that calls for a specific type of amino acid, carried by a tRNA molecule, to be added to a polypeptide chain in protein synthesis. (*See also* **anticodon**.)

Cohesive forces are intermolecular forces between like molecules.

A **colligative property** is a physical property of a solution that depends on the concentration of solute in the solution but not on the identity of the solute. The principal colligative properties are vapor pressure lowering, freezing point depression, boiling point elevation, and osmotic pressure.

A **colloid** is a dispersion in which the dispersed matter has one or more dimensions (length, width, or thickness) in the range from about 1 nm to 1000 nm. For a classification of different types of colloids, see Table 13.3, page 507.

The **common ion effect** refers to the ability of ions from a strong electrolyte to (a) suppress the ionization of a weak acid or weak base or (b) to reduce the solubility of a slightly soluble ionic compound.

A **complex** consists of a central atom, which is usually a metal ion, and attached groups called ligands.

A **complex ion** is a complex that carries a net electric charge, either positive (a complex cation) or negative (a complex anion).

Composition refers to the constituent elements in a sample of matter and their relative proportions. For example, the composition of water is 11.19% hydrogen and 88.81% oxygen, by mass.

A **compound** is a substance made up of atoms of two or more elements, with the different atoms joined in fixed proportions.

A **concentration cell** is a voltaic cell having identical electrodes in contact with solutions of different concentrations.

Condensation is the conversion of a gas (vapor) to a liquid.

In **condensation polymerization** monomers with at least two functional groups link together by eliminating small-molecule by-products.

A **conduction band** is a *partially* filled band of very closely spaced energy levels. Properties such as the electrical conductivity of metals can be explained in terms of electronic transitions between levels within the conduction band.

A **conjugate acid** is formed when a Brønsted–Lowry base accepts a proton. Every base has a conjugate acid.

A **conjugate base** is formed when a Brønsted–Lowry acid donates a proton. Every acid has a conjugate base.

A **conjugated system** refers to a molecular structure having a series of alternate single and double bonds. Substances having this feature can absorb UV and/or visible light.

Contributing structure. *See* **resonance structure**.

A **conversion factor** is a ratio in which the numerator and denominator are equivalent quantities but expressed in different units.

A **cooling curve** is a graph of temperature as a function of time obtained as a substance is cooled. Constant-temperature segments of the curve correspond to phase changes, for example, condensation and freezing.

A **coordinate covalent bond** is a linkage between two atoms in which one atom provides *both* of the electrons of the shared pair.

A **coordination compound** is a substance made up of one or more complexes. For example, $[Cu(NH_3)_4]SO_4$ is a coordination compound consisting of the complex cations $[Cu(NH_3)_4]^{2+}$ and the simple anions SO_4^{2-}

The **coordination number** of the metal center in a complex is the total number of points around a central atom at which bonding to ligands can occur.

Core electrons are electrons found in the inner electronic shells of atoms. (*See also* **valence electrons**.)

The **coulomb (C)** is the SI unit of electric charge. The electric charge on an electron, for example, is -1.602×10^{-19} C.

A **covalent bond** is a bond formed by a pair of electrons shared between atoms.

The **covalent radius** of an atom is one-half the distance between the nuclei of two like atoms joined into a molecule.

Critical mass is the minimum mass of a fissionable element that must be present to sustain a chain reaction. This is the mass required to produce an explosion of an atomic bomb.

The **critical point** refers to the condition at which the liquid and gaseous (vapor) states of a substance become identical. It is the highest temperature point on a vapor pressure curve.

The **critical pressure** of a substance is the pressure at its critical point.

The **critical temperature** of a substance is the temperature at its critical point.

A **crystal** is a structure having plane surfaces, sharp edges, and a regular geometric shape. The fundamental units—atoms, ions, or molecules—are assembled in a regular, repeating manner extending in three dimensions through the crystal.

Crystal field theory is a theory of bonding in complexes that focuses on the abilities of ligands to produce a splitting of a *d*-subshell energy level of the metal center in a complex.

Dalton's law of partial pressures states that in a mixture of gases each gas expands to fill the container and exerts its own pressure, called a partial pressure. The total pressure of the mixture is the sum of the partial pressures exerted by the separate gases.

Data are the facts of science obtained by careful observations and measurements made during experiments.

The ***d*-block** is the portion of the periodic table in which the $(n-1)d$ subshell (the *d* subshell of the next-to-outermost shell) fills in the aufbau process. The *d*-block comprises the B-group elements found in the main body of the periodic table.

The **debye (D)** is the unit used to express the dipole moments of polar molecules. One debye is equal to 3.34×10^{-30} C · m.

Degenerate is a term used to describe two or more orbitals that are at the same energy level.

Deionized water is water that has been freed of all ions through ion exchange processes. Specifically, cations are replaced by H^+ and anions by OH^-. The ions then combine to form H_2O.

Deliquescence is the condensation of water vapor on a solid followed by solution formation. The condensation continues until the vapor pressure of the solution equals the partial pressure of water vapor in the atmosphere.

The **density (*d*)** of a sample of matter is its mass per unit volume, that is, the mass of the sample divided by its volume.

Deoxyribonucleic acid (DNA) is a polymer of nucleotides. The nucleotides in DNA consist of the sugar deoxyribose, a phosphate ester, and a cyclic amine base. The possible bases in DNA are adenine, guanine, thymine, and cytosine.

A **diagonal relationship** refers to the similarity of a second-period element in one group of the periodic table with a third-period element of the next group. Specifically, resemblances in physical and chemical properties are found between Li and Mg, Be and Al, and B and Si.

Diamagnetic substances have all electrons paired and are weakly repelled by a magnetic field.

Diffusion is the process by which one substance mixes with one or more other substances as a result of the movement of molecules.

Dilution is a process of producing a more dilute solution from a more concentrated one by the addition of an appropriate quantity of solvent.

The **dipole moment (μ)** of a polar molecule is the product of the magnitude of the charges (δ) and the distance that separates them.

A **dispersion force** is an attractive force between an instantaneous dipole and an induced dipole.

A **disproportionation reaction** is an oxidation–reduction reaction in which the same substance is both oxidized and reduced.

Doping refers to the addition of trace amounts of certain elements to a semiconductor to change the semiconducting properties. (*See also* ***n*-type semiconductor** and ***p*-type semiconductor**.)

A **double bond** is a covalent linkage in which two atoms share *two* pairs of electrons between them.

Dynamic equilibrium occurs when two opposing processes occur at exactly the same rate, with the result that no net change occurs.

The **effective nuclear charge (Z_{eff})** acting on an electron in an atom is the actual nuclear charge less the screening effect of other electrons in the atom.

Effusion is a process in which a gas escapes from its container through a tiny hole (an orifice). (*See also* **Graham's law of effusion**.)

An **electrochemical cell** is a combination of two half-cells in which metal electrodes are joined by a wire and the solutions are brought into contact through a salt bridge or by other means. (*See also* **electrolytic cell**, **half-cell**, and **voltaic cell**.)

An **electrode** is a metal strip or carbon rod dipped into a solution or molten electrolyte to carry electricity to or from the liquid. (*See also* **anode** and **cathode**.)

Electrode potential is a property related to the electric charge density on an electrode in an oxidation–reduction equilibrium with ions in solution.

Electrolysis is the decomposition of compounds by passing electricity through an ionic solution or a molten salt. A nonspontaneous chemical change is produced.

An **electrolyte** is a compound that conducts electricity when molten or in a liquid solution.

An **electrolytic cell** is an electrochemical cell in which electricity is passed through a solution or a molten salt to produce a nonspontaneous chemical reaction.

The **electromagnetic spectrum** signifies the range of wavelengths and frequencies found for electromagnetic waves, extending from very long wavelength radio waves to the shortest gamma rays.

An **electromagnetic wave** originates in the vibrations of electrically charged objects and is propagated through oscillations of electric and magnetic fields.

Electromotive force (emf). *See* **cell potential**.

The **electron** is a particle carrying the fundamental unit of negative electric charge. Electrons have a mass of 0.0005486 u and an electric charge of -1.602×10^{-19} C and are found outside of the nuclei of atoms.

Electron affinity is the energy change that occurs when an electron is added to an atom in the gaseous state.

Electron capture (E.C.) is a type of radioactive decay in which a nucleus absorbs an electron from the first or second electronic shell. The atomic number of the nucleus decreases by one unit and its mass number is unchanged.

The **electron configuration** of an atom describes the distribution of electrons among atomic orbitals in the atom.

Electron dot structure. *See* **Lewis structure**.

Electron dot symbol. *See* **Lewis symbol**.

The **electronegativity** of an element is a measure of the tendency of its atoms in molecules to attract electrons to themselves.

An **electron group** is a collection of valence electrons localized in a region around a central atom that exerts repulsions on other groups of valence electrons. It may be a bonding pair of electrons in a single bond, two pairs of electrons in a double bond, three pairs of electrons in a triple bond, a lone pair of electrons, or even a lone unpaired electron.

The **electron spin quantum number (m_s)** is a fourth quantum number (in addition to the three required by the Schrödinger wave equation) necessary to characterize an electron in an orbital. The two possible values of the spin quantum number are $+\frac{1}{2}$ and $-\frac{1}{2}$.

An **element** is a substance composed of a single type of atom. Elements are the fundamental substances from which all material things are made.

An **elementary step** represents, at the molecular level, a single stage in the overall mechanism by which a chemical reaction occurs. (*See also* **unimolecular**, **bimolecular**, and **termolecular**.)

An **emission (line) spectrum** is a dispersion of electromagnetic radiation into a discrete set of wavelength components. These components can be rendered as images of a slit (lines) in light from a spectroscope. Line spectra are the result of light emission by excited gaseous atoms.

An **empirical formula** is the *simplest* formula describing the elements in a compound and the smallest integral ratio in which their atoms are united.

Enantiomers are pairs of mirror-image optical isomers that differ only in the direction in which they rotate the plane of polarized light. One isomer rotates the plane to the right, and the other rotates the plane to the same degree, but to the left.

The **end point** is the point in a titration at which an added indicator changes color. An indicator is chosen so that its end point matches the equivalence point of the reaction. (*See also* **equivalence point**.)

Endothermic describes a process or reaction in which thermal energy is converted to chemical energy. In an endothermic process a temperature decrease occurs in an *isolated* system, or, in a nonisolated system, heat is absorbed from the surroundings.

An **energy gap (E_g)** is the energy separation between a valence band and a conduction band that lies above it. In a semiconductor the gap is relatively small and in an insulator it is very large. (*See also* **valence band** and **conduction band**.)

An **energy level** is the state of an atom determined by the location of its electrons among the various principal shells and subshells.

Enthalpy (H) is a thermodynamic function defined as the sum of the internal energy and the pressure–volume product: $H = E + PV$.

The **enthalpy change (ΔH)** in a chemical reaction is equal to the heat of reaction at constant temperature and pressure, q_P.

An **enthalpy diagram** is a graphical representation of the change in enthalpy that occurs in a chemical reaction.

The **enthalpy (heat) of fusion** is the quantity of heat required to melt a given quantity of a solid.

The **enthalpy (heat) of sublimation** is the quantity of heat required to vaporize a given quantity of solid at a constant temperature. It is equal to the sum of the enthalpies of fusion and vaporization.

The **enthalpy (heat) of vaporization** is the quantity of heat required to vaporize a given quantity of liquid at a constant temperature.

Entropy (S) measures the degree of randomness or disorder in a system. The greater the degree of disorder in a system, the greater is its entropy.

Entropy change (ΔS) is the difference in entropy between two states of a system, as between the products and reactants of a chemical reaction.

Enzymes are high molecular weight proteins that catalyze reactions occuring in living organisms.

Equilibrium is a condition that is reached when two opposing processes occur at equal rates. As a result, the concentrations (or partial pressures) of the reacting species remain constant with time.

The **equilibrium constant (K_{eq})** is the form of the equilibrium constant needed in thermodynamic relationships, such as $\Delta G^\circ = -RT \ln K_{eq}$. In the K_{eq} expression species in solution are represented through their molarities and gases by their partial pressures in atm.

An **equilibrium constant expression** is a particular ratio of concentrations (or partial pressures) of products to reactants in a chemical reaction at equilibrium. The expression has a constant value that is independent of the manner in which equilibrium is reached. The value depends principally on temperature. (See page 609 for a statement on how the expression is formulated. *See also* K_c and K_p.)

The **equivalence point** of a titration is the point at which two reactants have been introduced into a reaction mixture in their stoichiometric proportions.

An **excited state** of an atom is one in which one or more electrons has been promoted to a higher energy level than in the ground state. (*See also* **ground state**.)

Exothermic describes a process or reaction in which chemical energy is converted to thermal energy. In an exothermic process a temperature increase occurs in an *isolated* system, or, in a nonisolated system, heat is given off to the surroundings.

The term **expanded octet** refers to a situation in which the central atom in a Lewis structure is able to accommodate more than the usual octet (8) of electrons in its valence shell. Expanded octets are encountered in molecules and polyatomic ions in which the central atom is a nonmetal of the third period or beyond.

An **experiment** is a carefully controlled procedure devised to test a hypothesis or a theory.

An **extensive property** is a physical property, such as mass or volume, that depends on the size or quantity of the sample of matter being considered.

A **face-centered cubic (fcc)** crystal structure has as its unit cell a cube with a structural unit at each of the corners and in the center of each face of the cube. (*See also* **unit cell**.)

The **Faraday constant (*F*)** is the electric charge, in coulombs, per mole of electrons—96,485 C/mol.

The ***f*-block** is the portion of the periodic table in which the $(n-2)f$ subshell (the *f* subshell of the second-from-outermost shell) fills in the aufbau process. The *f*-block consists of the two series of elements extracted from the main table (lanthanides and actinides) and printed at the bottom of the table.

Ferromagnetism is a magnetic effect much stronger than paramagnetism and associated with iron, cobalt, and nickel, and certain alloys. It requires that atoms be both paramagnetic and of the right size to be able to form magnetic domains.

The **first law of thermodynamics** states that the internal energy of an *isolated* system is constant, or if a system interacts with its surroundings by exchanging heat and/or work, the exchange must occur in such a way that no energy is created or destroyed. (*See also* **law of conservation of energy**.)

A **first-order reaction** has a rate equation in which the sum of the exponents, $m + n + \cdots = 1$.

Formal charge is a concept used in writing Lewis structures. It is the number of valence electrons in an isolated atom minus the number of electrons assigned to that atom in a Lewis structure.

The **formation constant (K_f)** describes equilibrium between a complex ion and the cation and ligands from which it is formed.

A **formula unit** is the simplest combination of atoms or ions consistent with the formula of a compound.

Formula weight is the mass of a formula unit relative to that of a carbon-12 atom; it is the sum of the weights of the atoms or ions represented by the formula.

Fractional crystallization is a method of purifying a solid by dissolving it in a suitable solvent and changing the solution temperature to a value where the solute solubility is lower (usually a lower temperature). Excess solute crystallizes as pure solid and soluble impurities remain in solution.

Fractional distillation is a method of separating the volatile components of a solution having different vapor pressure and boiling points. It involves large numbers of vaporizations and condensations occurring continuously in a distillation column.

Free energy (*G*) is a thermodynamic function used in establishing criteria for equilibrium and for spontaneous change. It is defined as $G = H - TS$, where H is the enthalpy; T, the Kelvin temperature; and S, the entropy of a system.

The **free energy change (ΔG)** is the difference in free energy between two states of a system, as between the free energies of the products and reactants of a chemical reaction. It is given by the equation $\Delta G = \Delta H - T\Delta S$. For a spontaneous change, $\Delta G < 0$, and for a nonspontaneous change, $\Delta G > 0$. For a system at equilibrium, $\Delta G = 0$.

A **free radical** is a highly reactive atom or molecular fragment characterized by having an unpaired electron(s). Free radicals are encountered as intermediates in some chemical reactions.

Freezing is the conversion of a liquid to a solid.

Freezing point is the temperature at which a liquid freezes—that is, the liquid comes into equilibrium with solid. For a pure substance the freezing point and melting point are the same.

The **frequency** of a wave is the number of cycles of the wave (the number of wavelengths) that pass through a point in a unit time.

A **functional group** is an atom or grouping of atoms attached to or within a hydrocarbon chain or ring that confers characteristic properties to the molecule as a whole.

Galvanic cell. *See* **voltaic cell**.

A **gamma ray (γ)** is a highly penetrating form of electromagnetic radiation emitted by the nuclei of certain radioactive atoms as they undergo decay.

Gas constant. *See* **universal gas constant**.

Genes are sections of the DNA molecule found in the chromosomes of cells; they are the basic units of heredity.

Geometric isomerism in organic compounds refers to the existence of isomers (cis, trans) that differ in the positions of attachment of substitutent groups at a double bond. In complexes the isomers differ in the positions of attachment of ligands to the central metal ion.

The **glass transition temperature (T_g)** is the temperature above which a polymer is tough and rubbery and below which it is glasslike—hard, stiff, and brittle.

Global warming refers to the anticipated increase in Earth's average temperature resulting from the injection of CO_2 and other infrared-absorbing gases into the atmosphere.

Graham's law of effusion states that the rates of effusion of gas molecules are inversely proportional to the square roots of their molar masses.

The **greenhouse effect** refers to the ability of $CO_2(g)$ and certain other gases to absorb and trap energy radiated by Earth's surface as infrared radiation.

The **ground state** of an atom is the atom at its lowest energy level. (*See also* **excited state**.)

Groups of the periodic table are the vertical columns of elements having similar properties.

A **half-cell** is a metal electrode partially immersed in a solution of ions that participate in an oxidation–reduction equilibrium at the electrode. (*See also* **electrochemical cell**.)

The **half-life** of a chemical reaction is the time required to consume one-half of the initial quantity of a reactant. For radioactive decay, it is the time in which one-half of the atoms of a radioisotope disintegrate.

A **half-reaction** is a portion of an oxidation–reduction reaction, representing either the oxidation process or the reduction process.

Hard water is groundwater containing significant concentrations of doubly-charged ions derived from natural sources, such as Ca^{2+}, Mg^{2+}, and Fe^{2+} and associated anions.

Heat (*q*) is an energy transfer into or out of a system caused by a difference in temperature between a system and its surroundings.

The **heat capacity** of a system is the quantity of heat required to change the temperature of the system by 1 °C (or 1 K). (*See also* **molar heat capacity** and **specific heat**.)

A **heating curve** is a graph of temperature as a function of time obtained by gradually heating a substance. Constant-temperature segments of the curve correspond to phase changes. (*See also* **cooling curve**.)

The **heat of reaction ($q_{reaction}$)** is the quantity of heat exchanged between a system and its surroundings when a chemical reaction occurs at a constant temperature and pressure.

The **Henderson–Hasselbalch** equation is used to relate the pH of a solution of a weak acid and its conjugate base to pK_a of the weak acid and the stoichiometric concentrations of the weak acid and of the conjugate base: pH = pK_a + log [conjugate base]/[weak acid].

Henry's law states that the solubility of a gas is directly proportional to the pressure maintained in the gas in equilibrium with the solution.

The **hertz (Hz)** is the unit used to describe the frequency of electromagnetic radiation: 1 Hz = 1 s^{-1}.

Hess's law states that the enthalpy change of a reaction is constant, whether the reaction is carried out directly in one step or indirectly through a number of steps.

A **heterogeneous mixture** is a mixture in which the composition and/or properties vary from one region to another within the mixture.

A **homogeneous mixture** is a mixture having the same composition and properties throughout the given mixture.

A **homologous series** is a series of organic compounds whose formulas and structures vary in a regular manner and whose properties are predictable based on this regularity.

Humidity is a general term referring to the water vapor content of air. **Absolute humidity** is the actual quantity of water vapor in air, expressed as g H_2O/m^3 air or as mmHg partial pressure of $H_2O(g)$. **Relative humidity** expresses the water vapor content as a percent of the maximum water vapor content possible.

Hund's rule states that electrons occupy atomic orbitals of identical energy singly before any pairing of electrons occurs. Furthermore, the electrons in the singly occupied orbitals have parallel spins.

Hybridization is a hypothetical process in which pure atomic orbitals are combined to produce a set of new orbitals called **hybrid orbitals** to describe covalent bonding by the valence bond method. (*See also* ***sp***, ***sp*²**, ***sp*³**, ***sp*³*d***, ***sp*³*d*²**.)

A **hydrate** is a compound that incorporates water molecules into its basic solid structure. For example, copper(II) sulfate pentahydrate has *five* H_2O molecules for every formula unit of copper(II) sulfate: $CuSO_4 \cdot 5H_2O$.

A **hydrocarbon** is a compound containing only hydrogen and carbon atoms.

In a **hydrogenation reaction**, $H_2(g)$ is a reactant and H atoms are added to C atoms at a carbon-to-carbon double or triple bond. For example, $CH_2{=}CH_2(g) + H_2(g) \longrightarrow CH_3CH_3$.

A **hydrogen bond** is a type of intermolecular force in which a hydrogen atom covalently bonded in one molecule is simultaneously attracted to a nonmetal atom in a neighboring molecule. In most cases, both the atom to which the hydrogen atom is bonded and the one to which it is attracted must be small atoms of high electronegativity—N, O, or F.

The **hydrologic cycle** is the series of natural processes by which water is recycled through the environment—Earth's solid crust, oceans and freshwater bodies, and the atmosphere.

In a general sense, **hydrolysis** is the reaction of a substance with water in which both the substance and the water molecules split apart. In a more limited sense, it is an acid-base reaction between an ion and water that usually results in the solution becoming either somewhat acidic or somewhat basic.

The **hydronium ion (H_3O^+)** is formed when a water molecule accepts a proton from an acid. For example, $HCl + H_2O \longrightarrow H_3O^+ + Cl^-$.

A **hypertonic solution** is a solution having an osmotic pressure greater than that of body fluids (blood, tears). A *hyper*tonic solution has a greater osmotic pressure than an *iso*tonic solution.

A **hypothesis** is a tentative explanation or prediction concerning some natural phenomenon.

A **hypotonic solution** is a solution having an osmotic pressure less than that of body fluids (blood, tears). A *hypo*tonic solution has a lower osmotic pressure than an *iso*tonic solution.

An **ideal gas** is a gas that strictly obeys the simple gas laws and the ideal gas law.

Ideal gas equation. *See* **ideal gas law**.

The **ideal gas law** states that the volume of a gas is directly proportional to the amount of a gas and its Kelvin temperature and inversely proportional to its pressure. Mathematically, it can be stated through the equation $PV = nRT$.

An **ideal solution** is one for which the heat of solution is zero and the volume of solution is the total of the volumes of the solution components. In general, the physical properties of an ideal solution (for example, density, vapor pressure) can be predicted from the properties of its components.

An **indicator** is a substance added to the reaction mixture in a titration that changes color at or near the equivalence point.

Industrial smog is polluted air associated with industrial activities. The principal pollutants are oxides of sulfur and particulate matter.

An **inert pair** refers to the ns^2 electrons in the valence shell of the posttransition elements of Groups 3A, 4A, and 5A. These electrons may remain in the valence shell following the loss of the np electrons, as in Tl^+, Sn^{2+}, Pb^{2+}, and Bi^{3+}.

The **initial rate of a reaction** is the rate of a reaction immediately after the reactants are brought together. The rate is generally expressed in terms of the rate of change with time of the concentration of one of the reactants or one of the products.

An **instantaneous rate of a reaction** is the rate of a reaction at some particular time in the course of a reaction. It is established through a tangent line to a concentration versus time graph at the time in question.

An **integrated rate equation** is an equation derived from the rate law for a reaction that expresses the concentration of a reactant as a function of time. (*See also* **rate law**.)

An **intensive property** is a property of a sample matter, such as temperature or density, that is independant of the quantity of matter being considered.

An **interhalogen compound** is a compound of two halogen elements (even three in a few cases). Some examples are ICl, ClF_3, and BrF_3.

An **intermediate** is a substance that is produced in one elementary step in a reaction mechanism and consumed in another. The intermediate does not appear in the chemical equation for the net reaction.

The **internal energy (*E*)** is the total amount of energy conatined in a thermodynamic system. The components of internal energy are energy associated with random molecular motion (thermal energy) and that associated with chemical bonds and intermolecular forces (chemical energy).

An **ion** is an electrically charged particle comprised of one or more atoms.

Ion exchange is the replacement in solution of ions carrying a single unit of charge for other ions, usually multiply charged. For example, Na^+ in a single cation resin replaces Ca^{2+}, Mg^{2+}, and Fe^{2+} in solution, and OH^- from an anion exchange resin replaces SO_4^{2-}.

Ionic bonds are attractive forces between positive and negative ions, holding them together in solid crystals.

The **ionic radius** is a measure of the size of a cation or anion based on the distance between the centers of ions in an ionic compound.

Ionization energy is the energy required to remove the least tightly bound electron from a ground-state atom (or ion) in the gaseous state.

The **ion product of water (K_w)** is the product of the concentration of hydronium ion, $[H_3O^+]$, and the concentration of hydroxide ion, $[OH^-]$, in pure water or a water solution. At 25 °C, its value is 1.0×10^{-14}.

Isoelectronic species (atoms, ions, molecules) have the same number of electrons in the same electron configuration.

Isomers are compounds having the same molecular formula but different structural formulas.

An **isotonic solution** is one that has the same osmotic pressure as body fluids (blood, tears).

Isotopes are atoms that have the same number of protons—the same atomic number—but different numbers of neutrons and, therefore, different mass numbers.

The **joule (J)** is the basic unit of energy in SI. It is the work done by a force of 1 newton (N) acting over a distance of 1 meter. That is, $1\ J = 1\ N \cdot m = 1 kg \cdot m^2 \cdot s^{-2}$.

K_c is the numerical value of an equilibrium constant expression in which molarities of products and reactants are used.

K_p is the numerical value of an equilibrium constant expression in which the partial pressures (usually in atm) of gaseous products and reactants are used.

A **kelvin (K)** is the SI base unit of temperature. An interval of one Kelvin on the Kelvin temperature scale is the same as one degree on the Celsius scale.

The **Kelvin scale** is an absolute temperature scale with its zero at −273.15 °C; its relationship to the Celsius scale is T (K) = t (°C) + 273.15.

Ketones are organic substances whose molecules have the *carbonyl* group (below) between two other C atoms.

The **kilogram (kg)** is the SI base unit of mass.

A **kilopascal (kPa)** is 1000 pascals (Pa). (*See also* **pascal**.)

Kinetic energy (KE) is energy of motion, given by the expression $KE = \frac{1}{2}mu^2$.

The **kinetic–molecular theory of gases** is a theory based on a small number of postulates concerning gas molecules from which simple gas laws, the ideal gas law, and equations dealing with temperature and molecular speeds can be derived.

The **lanthanide** elements constitute the portion of the *f*-block of the periodic table in which the 4*f* subshell fills in the aufbau process. This series of elements is extracted from the main body of the table and printed at the bottom.

Lanthanide contraction is a term used to describe the general downward trend in the radii of lanthanide atoms and ions with increasing atomic number. One of the important consequences of the lanthanide contraction is that atomic and ionic sizes are essentially unchanged between the fifth- and sixth-period members of a group of transition elements.

Lattice energy is the enthalpy change that accompanies the formation of one mole of an ionic solid from its gaseous ions.

The **law of combining volumes** states that, when gases measured at the same temperature and pressure are allowed to react, the volumes of gaseous reactants and products are in small whole-number ratios.

The **law of conservation of energy** states that in a physical or chemical change energy can neither be created nor destroyed.

The **law of conservation of mass** states that the total mass of the products of a reaction is always equal to the total mass of the reactants consumed.

The **law of definite proportions** states that a particular compound always contains its constituent elements in certain fixed proportions and in no other combinations.

The **law of multiple proportions** states that, if two elements form more than a single compound, the masses of one element that combine with a fixed mass of the second are in the ratio of small whole numbers.

Le Châtelier's principle is a statement that permits qualitative predictions about the effects produced by changes (amounts of reactants or products, reaction volume, temperature, . . .) imposed on a system at equilibrium. (See page 617 for a statement of the principle.)

Lewis acid. *See* **acid**.

Lewis base. *See* **base**.

A **Lewis structure** is a representation of covalent bonding through Lewis symbols, shared electron pairs, and lone-pair electrons.

A **Lewis symbol** is a representation of an element in which the chemical symbol stands for the core of the atom and dots placed around the symbol for its valence electrons.

A **ligand** is a species (atom, molecule, anion, or, rarely, cation) that is bonded to a metal center in a complex.

The **limiting reactant (reagent)** is the reactant that is completely consumed in a chemical reaction, thereby limiting the amounts of products formed.

Liquid crystal is the term used to describe a physical form of a substance that has the fluid properties of a liquid and the optical properties of a crystalline solid.

A **liter (L)** is a metric unit of volume equal to one cubic decimeter or 1000 cubic centimeters: $1\ L = 1\ dm^3 = 1000\ cm^3$.

Lone pairs are electron pairs assigned exclusively to one of the atoms in a Lewis structure. They are not shared and hence are not involved in the chemical bonding.

Macromolecules are giant molecules (polymers) having small molecules (monomers) as their building blocks.

The **magnetic quantum number (m_l)** is the last of three parameters that must be assigned a specific value to achieve a solution of Schrödinger's wave equation for the hydrogen atom: m_l is an integer between $-l$ and $+l$ (including 0). (*See also* **angular momentum quantum number** and **principal quantum number**.)

Main-group elements. *See* **representative elements**.

A **manometer** is a device used to measure a gas pressure. Generally a manometer must be used in combination with a barometer.

Mass is related to the force required to move an object or to change its velocity if it is already in motion. It is a measure of the quantity of matter in an object.

The **mass number (A)** is the sum of the number of protons and neutrons in the nucleus of an atom.

A **mass spectrometer** is a device that separates ions according to their mass-to-charge ratios.

Melting (fusion) is the conversion of a solid to a liquid.

The **melting point** of a solid is the temperature at which it melts, that is, the temperature at which it comes into equilibrium with the liquid phase. For a pure substance the melting point and the freezing point are the same.

A **meniscus** is the interface between a liquid and the air above it.

Metal carbonyls are complexes formed between certain transition metal atoms and carbon monoxide molecules, such as $Ni(CO)_4$ and $Fe(CO)_5$.

The **metallic radius** is one-half the distance between the nuclei of adjacent atoms in a solid metal.

A **metalloid** is an element that has the physical appearance of a metal but some nonmetallic properties as well. Metalloids are located along the stepped diagonal line in the periodic table.

Metals are elements having a distinctive set of properties: luster, good heat and electrical conductivity, malleability, and ductility. Metal atoms generally have small numbers of valence electrons and a tendency to form cations. Metals are found to the left of the stepped diagonal line in the periodic table. All *s*-block (except hydrogen and helium), *d*-block, and *f*-block elements are metals, as are a few in the *p*-block.

The **meter (m)** is the SI base unit of length.

The **method of initial rates** is an experimental method of establishing the rate law of a reaction. To establish the order of the reaction with respect to one of the reactants, the initial rates are compared for two different concentrations of that reactant, with the concentrations of all other reactants held constant. (*See also* **initial rate of reaction**, **rate law** and **order of a reaction**.)

A **millimeter of mercury (mmHg)** is a unit used to express gas pressure: 1 mmHg = 1/760 atm (exactly). (*See also* **atmosphere**.)

A **mineral** is a naturally occuring inorganic compound with a definite composition and a definite crystal structure. Some geologists consider certain organic substances, such as coal, to be minerals.

A **mixture** is a type of matter that has a composition and/or properties that may vary from one sample to another, or, in some cases, one part of the mixture to another.

The **molality (m)** of a solution is the amount of solute, in moles, per *kilogram of solvent* (not of solution).

Molar concentration. *See* **molarity**.

Molar heat capacity is the quantity of heat required to change the temperature of one mole of a substance by 1 °C (or 1 K); it is the heat capacity of one mole of substance.

The **molarity (M)** of a solution is the amount of solute, in moles, per liter of solution.

The **molar mass** of a substance is the mass of 1 mol of that substance. It is numerically equal to the atomic weight, molecular weight, or formula weight, and expressed as g/mol.

The **molar volume of a gas** refers to the volume occupied by 1 mol of gas at a fixed temperature and pressure; it is essentially independent of the identity of the gas. At standard temperature and pressure, the molar volume of an ideal gas is 22.4141 L.

A **mole (mol)** is an amount of substance that contains as many elementary units (atoms, molecules, formula units) as there are atoms in exactly 12 g of the isotope carbon-12.

Molecular geometry describes the geometrical figure formed when appropriate atomic nuclei in a molecular or polyatomic ion are joined by straight lines. Molecular geometry refers to the geometrical shape of a molecule or polyatomic ion.

A **molecular orbital** is a region in a molecule where there is a high electron charge density or a high probability of finding an electron(s). (*See also* **antibonding molecular orbital** and **bonding molecular orbital**.)

Molecular weight is the average mass of a molecule of a substance relative to that of a carbon-12 atom; it is the sum of the masses of the atoms represented in the molecular formula.

A **molecule** is a group of two or more atoms held together in a definite arrangement by forces called covalent bonds.

The **mole fraction (χ)** of a component in a homogeneous mixture (a solution) is the fraction of all the molecules in the mixture contributed by that component.

The **mole percent** of a component in a homogeneous mixture (a solution) is the percentage of all the molecules in the mixture contributed by that component.

A **monodentate** ligand attches to the metal center in a complex through one pair of electrons on a donor atom.

Monomers are small molecules that are capable of independent existence, but which under appropriate conditions can join together to form a giant molecule called a polymer. (*See also* **polymerization**.)

A **multidentate** ligand attaches to a metal center in a complex at more than one point. Ethylenediamine, $NH_2CH_2CH_2NH_2$, attaches at two points and the ethylenediaminetetraacetato ion, $[EDTA]^{4-}$, at six points.

A **multiple bond** is a covalent linkage in which two atoms share either *two* pairs (double bond) or *three* pairs (triple bond) of electrons between them.

The **Nernst equation** relates a cell voltage under nonstandard conditions, E_{cell}, to the standard cell potential, E°_{cell}, and the concentrations of reactants and products of a redox reaction. Its form at 25 °C is

$$E_{cell} = E^\circ_{cell} - (0.0592/n) \log Q$$

where n is the number of moles of electrons transferred in the redox reaction and Q is the reaction quotient.

A **net ionic equation** is an equation that represents the actual molecules or ions that participate in a chemical reaction, eliminating all nonparticipating species ("spectator" ions).

In a **network covalent solid**, covalent bonds extend throughout the crystalline solid. Intramolecular forces (covalent bonds) and intermolecular forces are indistinguishable.

A **neutralization** reaction is one in which an acid and a base react in such a manner that there is neither excess acid or base in the final solution. The products of the reaction are water and a salt.

The **neutron** is a fundamental particle of matter found in the nuclei of atoms. Neutrons have a mass of 1.0087 u and no electric charge.

A **newton (N)** is the basic unit of force in SI. It is the force required to give a 1 kg mass an acceleration of 1 m/s^2. That is, $1\ N = 1\ kg \cdot m \cdot s^{-2}$.

The **nitrogen cycle** refers to the totality of activities in which nitrogen atoms are recycled through the environment.

Nitrogen fixation refers to the conversion of atmospheric nitrogen into nitrogen compounds. This occurs naturally in the nitrogen cycle or artificially, as in the synthesis of ammonia. (*See also* **nitrogen cycle**.)

The **noble gases** are the elements in Group 8A of the periodic table. They have the valence-shell electron configuration ns^2np^6 (except helium, $1s^2$).

Nonbonding pairs. *See* **lone pairs**.

A **nonelectrolyte** is a substance that exists exclusively or almost exclusively in molecular form, whether in the pure state or in solution.

Nonmetals are elements that lack metallic properties. They are generally poor conductors of heat and electricity and brittle when in the solid state. Nonmetal atoms generally have larger numbers of valence electrons than do metals and a tendency to form anions. Nonmetal atoms are confined to the p-block of the periodic table (plus hydrogen).

In a **nonpolar bond** there is an equal sharing of the electrons between the bonded atoms. The electrons are not drawn any closer to one atom than to the other and so there is no charge separation.

A **nonspontaneous process** will not occur in a system left to itself. It can only be made to occur through intervention from outside the thermodynamic system.

The **normal boiling point** of a liquid is the temperature at which the liquid boils when the prevailing atmospheric pressure is 1 atm.

An ***n*-type semiconductor** is a semiconductor doped with donor atoms that can lose electrons to the conduction band. Electric current in this type of semiconductor is carried primarily by these donor electrons. (*See also* **doping**.)

Nuclear binding energy is the energy with which the nucleons are bound together into the nucleus of an atom. It is the energy equivalent of the mass lost in creating a nucleus from its individual protons and neutrons.

Nuclear fission is the splitting of a large unstable nucleus into two lighter fragments and two or more neutrons. Mass destroyed in this process is converted to an equivalent quantity of energy, which is evolved.

Nuclear fusion is the joining together or fusing of lighter nuclei into a heavier one. In the process some matter is converted to energy, which is released.

Nucleon is the general term for the nuclear particles protons and neutrons.

Nucleotides are the structural units that make up deoxyribonucleic acid (DNA) and ribonucleic acid (RNA). The nucleotides themselves consist of sugar, a phosphate ester group, and a cyclic amine base.

Nuclide is a term used to signify an atomic species having a particular atomic number and mass number, such as $^{12}_{6}C$. (*See also* **isotopes**.)

The **octet rule** states that most covalently bonded atoms represented in a Lewis structure have eight electrons in their outermost (valence) shells. In the formation of ionic compounds, the ions of the representative elements also tend to follow the octet rule.

An **orbital diagram** is a method of denoting an electron configuration in which parentheses or boxes are used to represent orbitals within subshells and arrows are used to represent electrons in the orbitals.

The **order of a reaction** is determined by the exponents of the concentration terms in the rate law for a reaction: Rate of reaction $= kA^mB^n \cdots$. The order of the reaction with respect to A is m; with respect to B it is n; and so on. The overall order of the reaction is $m + n + \cdots$.

An **ore** is a naturally occuring mineral containing a metal in a form and concentration that makes extraction of the metal feasible.

Osmosis is the net flow of a solvent through a semipermeable membrane, from pure solvent into a solution or from a solution of a lower concentration into one of a higher concentration.

The **osmotic pressure** of a solution is the pressure that must be applied to a solution to prevent the flow of solvent molecules into the solution when the solution and pure solvent are separated by a semipermeable membrane.

Overvoltage is the excess voltage above that calculated from $E°$ values required to bring about the electrode reactions on particular electrodes in an electrolysis.

Oxidation is a process in which the oxidation state of an element increases. It is the half-reaction of an oxidation–reduction reaction in which electrons are "lost."

The **oxidation state** or **oxidation number** of an element in a compound is a means of designating the number of electrons that its atoms have lost, gained, or shared in forming that compound.

An **oxidizing agent (oxidant)** is a substance that makes possible the oxidation that occurs in an oxidation–reduction reaction. The oxidizing agent itself is reduced.

The **ozone layer** is a band of the stratosphere, about 20 km thick and centered at an altitude of about 25 to 30 km, that has a much higher concentration of ozone than the rest of the atmosphere.

Paramagnetic substances have unpaired electrons and are attracted into an external magnetic field.

Partial pressure. *See* **Dalton's law of partial pressure**.

Particulate matter is an air pollutant consisting of solid and liquid particles of greater than molecular size but small enough to remain suspended in air.

Parts per billion (ppb) expresses the composition of a mixture as the number of parts of one component per billion parts of the mixture as a whole, usually on a mass basis for liquid solutions and a mole basis for gaseous mixtures.

Parts per million (ppm) expresses the composition of a mixture as the number of parts of one component per million parts of the mixture as a whole, usually on a mass basis for liquid solutions and a mole basis for gaseous mixtures.

Parts per trillion (ppt) expresses the composition of a mixture as the number of parts of one component per trillion parts of the mixture as a whole, usually on a mass basis for liquid solutions and a mole basis for gaseous mixtures.

A **pascal (Pa)** is the basic unit of pressure in SI. It is a pressure of 1 newton per square meter, $1\ N \cdot m^{-2}$.

The **Pauli exclusion principle** states that no two electrons in an atom may have all four quantum numbers alike. A consequence of this principle is that there may be no more than two electrons in an orbital, and that the two electrons must have opposing spins.

The ***p*-block** is the portion of the periodic table in which the *np* subshell (the *p* subshell of the outer shell) fills in the aufbau process. The *p*-block elements are all main-group or representative elements.

The **percent yield** is the ratio of the actual yield to the theoretical yield of a chemical reaction, expressed as a percentage.

The **periodic law** states that certain sets of physical and chemical properties recur at regular intervals (periodically) when the elements are arranged according to increasing atomic number.

A **periodic table** is a tabular arrangement of the elements according to increasing atomic number that places elements having similar properties into the same vertical columns. (Mendeleev's original periodic table was arranged according to atomic weights, not atomic numbers.)

Permanent hard water is hard water in which the predominant anions are other than HCO_3^-, for example Cl^- or SO_4^{2-}. (*See also* **hard water**.)

The **pH** is the negative of the logarithm of the hydronium ion concentration in a solution: $pH = -\log [H_3O^+]$.

A **phase diagram** is a pressure–temperature plot indicating the conditions under which a substance exists as a solid phase(s), a liquid, or a gas, or some combination of these in equilibrium.

Photochemical smog is air that is polluted with oxides of nitrogen and unburned hydrocarbons, together with ozone and several other components produced by the action of sunlight.

A **photoconductor** is a material whose electrical conductivity increases in the presence of light.

The **photoelectric effect** refers to the emission of electrons from the surface of certain materials when they are struck by light of the appropriate frequency.

A **photon** is a quantum of energy in the form of light. The energy of the photon is given by the expression $E = h\nu$.

A **photovoltaic (solar) cell** is a device that uses semiconductors to convert solar energy (light) into electricity.

A **physical property** is a property that can be observed and specified without reference to any other substance and that does not produce changes in composition. Some examples are color, density, hardness, and electrical conductivity.

A **pi (π) bond** forms by the overlap in a parallel or side-by-side fashion of *p* orbitals of the bonded atoms. A double bond consists of one σ and one π bond; a triple bond, of one σ and two π bonds.

pK_a is the negative of the logarithm of the ionization constant of an acid: $pK_a = -\log K_a$.

pK_b is the negative of the logarithm of the ionization constant of a base: $pK_b = -\log K_b$.

pK_w is the negative of the logarithm of the ion product of water: $pK_w = -\log K_w = -\log (1.0 \times 10^{-14}) = 14.00$ (at 25° C). (*See also* **ion product of water**.)

Planck's constant (*h*) is the numerical constant relating the energy of a photon of light and its frequency: $E = h\nu$. Its value is $6.6260755 \times 10^{-34}\ J \cdot s$.

A **plasticizer** is a substance that is physically mixed with a polymer and whose function is to lower the glass transition temperature (T_g) of the polymer. (*See also* **glass transition temperature**.)

pOH is the negative of the logarithm of the hydroxide concentration in an aqueous solution: $pOH = -\log [OH^-]$.

In a **polar bond** between two atoms, electrons are drawn closer to the more electronegative atom, creating a separation of charge. One end of the bond is thought of as having a small negative charge, δ–, and the other end, a small positive charge, δ+.

Polarizability is a measure of the ease with which electron charge density in an atom or molecule is distorted by an external electric field. It measures the ease with which a dipole can be induced in an atom or molecule.

In a **polar molecule**, depending on the electronegativities of bonded atoms and on the molecular geometry, a small separation of positive (δ+) and negative (δ–) charge exists.

A **polydentate** ligand attaches to a metal center in a complex at more than one point. Ethylendiamine, $NH_2CH_2CH_2NH_2$, attaches at two points and the ethylenediaminetetraacetato ion, $[EDTA]^{4-}$, at six points.

A **polymer** is a giant molecule formed by the combination of smaller molecules (monomers) in a repeating manner.

Polymerization is a type of reaction in which small repeating units (monomers) combine to form giant molecules (polymers).

A **polyprotic acid** has more than one ionizable H atom per molecule. The ionization of a polyprotic acid occurs in distinctive steps. For example, $H_2SO_4 + H_2O \longrightarrow H_3O^+ + HSO_4^-$, followed by $HSO_4^- + H_2O \longrightarrow H_3O^+ + SO_4^{2-}$.

A **positron** (β^+) is a positively charged particle having the same mass as a β^- particle. Sometimes called "positive" electrons, positrons are emitted by certain radioactive nuclei. In the decaying nucleus, the atomic number decreases by one unit and the mass number is unchanged.

The **potential difference**, measured in volts, is the difference in electric potential between two points in an electric circuit, for example, between the electrodes in an electrochemical cell.

Potential energy is energy due to position or arrangement. It is the energy associated with forces of attraction and repulsion between objects.

A **precipitation reaction** is a chemical reaction between ions in solution that produces an insoluble solid—a precipitate.

The **precision** of a set of measurements refers to the degree of reproducibility of the measurements.

Pressure (*P*) is a force per unit area—that is, $P = F/A$.

Principal level. *See* **principal shell**.

The **principal quantum number (*n*)** is the first of three quantum numbers that must be assigned a specific numerical value to achieve a solution to Schrödinger's wave equation for the hydrogen atom: $n = 1, 2, 3, \ldots$. Its value designates the main or principal energy level of an electron in an atom.

A **principal shell** refers to the collection of orbitals having the same principal quantum number.

The **products** are the substances that are produced in a chemical reaction. Their formulas appear on the right side of a chemical equation.

A **proton** is a particle carrying the fundamental unit of positive charge. Protons have a mass of 1.0073 u and are found in the nuclei of atoms.

A **proton acceptor** is a Brønsted–Lowry base. (*See also* **base**.)

A **proton donor** is a Brønsted–Lowry acid. (*See also* **acid**.)

Pseudohalogens are certain groupings of atoms that mimic the characteristics of a halogen atom. Examples include —CN and —OCN.

A ***p*-type semiconductor** is a semiconductor that has been doped with acceptor atoms. These atoms extract electrons from chemical bonds in the semiconductor, producing positive holes in the valence band. Electric current in these semiconductors is carried primarily by positive holes. (*See also* **doping**.)

Q_{ip}. *See* **reaction quotient**.

A **quantum** is the smallest quantity of energy that can be emitted or absorbed in a process, as given by the expression $E = h\nu$.

Quantum mechanics. *See* **wave mechanics**.

Quantum numbers are certain integral values assigned to three parameters in a wave equation to obtain acceptable solutions to the equation.

The **radioactive decay law** states that the rate of disintegration of a radioactive isotope, called the decay rate or activity, is directly proportional to the number of atoms present.

A **radioactive decay series** is a sequence of nuclear processes involving α and β^- emissions by which an initial long-lived radioactive nucleus is eventually converted to a stable nonradioactive nucleus, such as in the decay of $^{238}_{92}U$ to $^{206}_{82}Pb$.

A **radioactive tracer** is a radioisotope that can be used to follow a physical or chemical process through the ionizing radiation that it emits.

Radioactivity is the spontaneous emission of ionizing radiation by the atomic nuclei of certain isotopes.

Raoult's law states that the addition of a solute lowers the vapor pressure of the solvent, and that the fractional lowering of the vapor pressure is equal to the mole fraction of the solute.

The **rate constant (*k*)** of a reaction is a numerical constant that relates the rate of the reaction to the concentrations of the reactants. Rate constants are functions of temperature. (*See also* **rate law**.)

The **rate-determining step** in a reaction mechanism is the step (usually the slowest) that is crucial in establishing the rate of an overall reaction.

The **rate law (rate equation)** of a chemical reaction is an expression relating the rate of the reaction to the concentrations of the reactants: Rate of reaction $= k\text{A}^m\text{B}^n \cdots$.

The **reactants** are the starting materials or substances consumed in a chemical reaction. Their formulas appear on the left side of a chemical equation.

A **reaction mechanism** is a detailed representation of a chemical reaction consisting of a series of elementary steps. A plausible mechanism must be consistent with the stoichiometry and the rate law of the net reaction.

A **reaction profile** is a schematic representation of changes in energy during the course of a reaction. The profile identifies the energies of reactants, transition state(s), and products. Activation energies and enthalpies of reaction are shown in the reaction profile.

A **reaction quotient (*Q*)** has the same format as an equilibrium constant (*K*) but uses *initial* concentrations rather than equilibrium concentrations. The criterion for the precipitation of ions from solution is that $Q_{ip} \gg K_{sp}$

A **reducing agent (reductant)** is a substance that makes possible the reduction that occurs in an oxidation–reduction reaction. The reducing agent itself is oxidized.

Reduction is a process in which the oxidation state of an element decreases. It is the half-reaction of an oxidation–reduction reaction in which electrons are "gained."

Refining is the process of removing impurities from a metal by any of a variety of chemical or physical means.

Relative humidity. *See* **humidity**.

Representative elements are elements in which the subshell being filled in the aufbau process is either an *s* or *p* subshell of the principal shell of highest principal quantum number (the outermost shell). Representative elements are located in the *s*- and *p*-blocks of the periodic table.

Resonance is a term used to describe a situation in which several plausible Lewis structures can be written to represent a species but in which the true structure cannot be written. The plausible structures are called contributing structures or **resonance structures**, and the true structure, which is a composite of the contributing structures, is called the **resonance hybrid**.

The **resultant dipole moment** is the dipole moment of a molecule as a whole based on an assessment of bond moments and the molecular geometry. (*See also* **bond moment** and **dipole moment**.)

Reverse osmosis refers to the net flow of solvent through a semipermeable membrane in the opposite direction from that expected for osmosis. It is produced by applying a pressure to a solution that exceeds its osmotic pressure. (*See also* **osmosis**.)

Ribonucleic acid (RNA) is a polymer of nucleotides. The nucleotides in RNA consist of the sugar ribose, a phosphate ester, and a cyclic amine base. The possible bases in RNA are adenine, guanine, uracil, and cytosine.

The **root-mean-square speed (u_{rms})** of the molecules of a gas is the square root of the average of the squares of the molecular speeds. It is a key term in equations associated with the kinetic–molecular theory of gases.

A **salt** is an ionic compound in which hydrogen atoms of an acid are replaced by metal ions. Salts are produced in the reaction of an acid and a base.

A **saturated hydrocarbon** has molecules that contain the maximum number of hydrogen atoms for the carbon atoms present. All bonds in the molecules are single covalent bonds.

A **saturated solution** is one in which dynamic equilibrium exists between undissolved solute and a solution. The solution conatins the maximum amount of solute that can be dissolved in a particular quantity of solvent at the given temperature.

The ***s*-block** is the portion of the periodic table in which the *ns* subshell (the *s* subshell of the outer shell) fills in the aufbau process. The *s*-block elements are all main-group or representative elements.

A **scientific law** is a brief statement, sometimes in mathematical terms, used to summarize large amounts of scientific data.

The **second (s)** is the SI base unit of time.

One statement of the **second law of thermodynamics** is that all natural or spontaneous processes are accompanied by an increase in entropy of the universe. That is, $\Delta S_{univ} = \Delta S_{total} = \Delta S_{system} + \Delta S_{surroundings} > 0$.

A **second-order reaction** has a rate equation in which the sum of the exponents, $m + n + \cdots = 2$.

A **semiconductor** is a substance in which there is only a small energy gap between the valence and conduction band. The electrical conductivity of a semiconductor is not nearly as good as that of a metal, but still much better than that of an insulator. The electrical conductivity of a semiconductor increases with temperature—the reverse of the situation for metals.

A **semipermeable membrane** is a material that permits the flow of solvent molecues but severely restricts the flow of solute molecules of a solution.

In a measured, numerical value, **significant figures** include all digits known with certainty plus one digit of uncertain value. The quantity 12.011 ± 0.001, for example, is expressed to *five* significant figures when written as 12.011.

A **sigma (σ) bond** results from the end-to-end overlap of pure or hybridized atomic orbitals between the bonded atoms. A sigma bond exists along a line joining the nuclei of the bonded atoms.

A **single bond** is a covalent linkage in which two atoms share one pair of electrons between them.

The **skeletal structure** of a polyatomic species (molecule, ion) indicates the order in which atoms are attached to one another.

Slag is a term used in extractive metallurgy for a relatively low-melting-point product of the reaction of an acidic oxide and a basic oxide. For example, the slag calcium silicate, $CaSiO_3(l)$, is produced in the reaction of $SiO_2(s)$ and $CaO(s)$.

Soaps are salts of long-chain carboxylic acids called fatty acids (because they are derived from fats). Sodium and potassium salts are the soluble soaps used for cleaning. The soaps of multiply charged cations are insoluble.

Solar cell. *See* **photovoltaic cell**.

The **solubility** of a solute in a particular solvent refers to the concentration of the solute in a saturated solution.

A **solubility curve** is a graph of the solubility of a solute as a function of temperature.

The **solubility product constant** (K_{sp}) describes the equilibrium that exists between a slightly soluble ionic solute and its ions in a saturated aqueous solution.

The **solubility rules** are a set of generalizations used to describe classes of substances that are either soluble or insoluble in water.

A **solute** is a solution component that is dissolved in a solvent. A solution may have several solutes, which are generally present in lesser amounts than is the solvent.

A **solution** is a homogeneous mixture of two or more substances. The composition and properties are uniform throughout a solution.

The **solvent** is the solution component (usually present in greatest amount) in which one or more solutes are dissolved to form the solution.

sp describes a hybridization scheme in the valence bond method in which an *s* and one *p* orbital are combined into two *sp* hybrid orbitals oriented in a *linear* fashion.

sp^2 describes a hybridization scheme in the valence bond method in which an *s* and two *p* orbitals are combined into three sp^2 hybrid orbitals oriented in a *trigonal planar* fashion.

sp^3 describes a hybridization scheme in the valence bond method in which an *s* and three *p* orbitals are combined into four sp^3 hybrid orbitals oriented in a *tetrahedral* fashion.

sp^3d describes a hybridization scheme in the valence bond method in which an *s*, three *p*, and a *d* orbital are combined into five sp^3d hybrid orbitals oriented in a *trigonal bipyramidal* fashion.

sp^3d^2 describes a hybridization scheme in the valence bond method in which an *s*, three *p*, and two *d* orbitals are combined into six sp^3d^2 hybrid orbitals oriented in an *octahedral* fashion.

The **specific heat** of a substance is the quantity of heat required to raise the temperature of one gram of substance by 1 °C (or 1 K).

The **spectrochemical series** is a listing of ligands in terms of their abilities to produce a splitting of a *d*-subshell energy level in a complex. Ligands in the series are referred to as strong field, intermediate field, or weak field depending on the degree of splitting they produce. (*See also* **crystal field theory**.)

Spectrometer is the general name for an instrument that measures the extent to which electromagnetic radiation of particular wavelengths is absorbed by a sample.

Spectroscopy refers to various instrumental methods of chemical analysis and molecular structure determination that are based on the interaction of electromagnetic radiation and matter.

A **spontaneous process** is one that occurs in a system left to itself. Once started, no action from outside the system is required to keep the process going.

A **standard cell potential** or **standard cell voltage** (E°_{cell}) is the potential difference (in volts) between the electrodes in a voltaic cell when all species are present in their standard states. It can be obtained by adding together the standard potentials for the reduction and oxidation half-reactions of a net cell reaction.

The **standard conditions of temperature and pressure (STP)** for a gas are 273.15 K (0 °C) and 1 atm (760 mmHg).

The **standard electrode potential** (E°) measures the tendency of a *reduction* process to occur when all species in a half-cell are present in their standard states. This tendency is measured in volts, relative to an assigned value of zero for the standard hydrogen electrode.

The **standard enthalpy of formation** (ΔH°_f) of a substance is the enthalpy change that occurs in the formation of 1 mol of the substance in its standard state from the reference forms of its elements in their standard states. The reference forms of the elements are their most stable forms at the given temperature and 1 atm pressure.

The **standard enthalpy of reaction** (ΔH°) is the enthalpy change for a reaction in which all reactants and products are in their standard states.

The **standard free energy of formation** (ΔG°_f) of a substance is the free energy change that occurs in the formation of 1 mol of the substance in its standard state from the reference forms of its elements in their standard states. The reference forms of the elements are their most stable forms at the given temperature and 1 atm pressure.

The **standard free energy of reaction** (ΔG°) is the change in free energy for a reaction in which all reactants and products are in their standard states.

A **standard hydrogen electrode (SHE)** has hydrogen gas at 1 atm pressure and hydronium ion at unit activity (about 1 M) in oxidation–reduction equilibrium on an inert platinum electrode. The potential arbitrarily assigned to this electrode is *zero*.

The **standard state** of a solid or liquid substance is the pure element or compound at 1 atm pressure and at the temperature of interest. For a gaseous substance the standard state is the (hypothetical) pure gas behaving as an ideal gas at 1 atm pressure and the temperature of interest.

A **state function** is any property having a unique value when the state of a system is specified.

Stereisomerism is a type of isomerism in which the number and types of groups and their mode of attachment in a molecule or complex are the same in the isomers, but in which their spatial arrangements differ.

A **stoichiometric factor** is a conversion factor relating molar amounts of two species involved in a chemical reaction (that is, a reactant to a product, one reactant to another, and so on). The numerical values used in formulating the factor are the stoichiometric coefficients, the coefficients used to balance a chemical equation.

Stoichiometric proportions refer to relative amounts of reactants that are in the mole ratios corresponding to the coefficients in a balanced equation.

Stoichiometry refers to quantitative measurements and relationships involving substances and mixtures of chemical interest.

A **strong acid** is an acid that is essentially completely ionized in solution, that is, an acid that is a strong electrolyte. (*See also* **acid**.)

A **strong base** is a base that is essentially completely ionized in solution, that is, a base that is a strong electrolyte. (*See also* **base**.)

A **strong electrolyte** is a substance that exists exclusively or almost exclusively in ionic form in solution.

A **structural formula** is a chemical formula that shows how the atoms in a molecule are attached to one another.

Sublevel. *See* **subshell**.

Sublimation is the direct passage of molecules from the solid state to the vapor state.

A **sublimation curve** is a graph of the vapor pressure (sublimation pressure) of a solid as a function of temperature. It is analogous to the vapor pressure curve of a liquid.

Sublimation pressure is the pressure exerted by a vapor that is in equilibrium with the solid state. It is analogous to vapor pressure of a liquid.

A **subshell** is the collection of orbitals of a given type present in a principal shell. For example, the *three* 2*p* orbitals constitute the 2*p* subshell.

Subshell (sublevel) notation is a method of denoting an electron configuration that uses numbers to represent the principal shells and the letters *s*, *p*, *d*, and *f* for subshells. A superscript number following the letter indicates the number of electrons in the subshell.

A **substance** is a type of matter having a definite, or fixed, composition and fixed properties that do not vary from one sample to another.

In a **substitution reaction** a substituent group replaces a hydrogen atom in a hydrocarbon molecule. This type of reaction is characteristic of alkane and aromatic hydrocarbons.

The **substrate** in an enzyme-catalyzed reaction is the reactant species that attaches itself to the active site on an enzyme molecule and undergoes chemical reaction.

Supercooling is a condition in which a liquid is cooled below its freezing point without the appearance of any solid. A supercooled liquid usually does begin to freeze when a sufficient lowering of the temperature has occurred. The temperature of the freezing liquid then rises back to the normal freezing point.

A **supersaturated solution** contains more solute than is present in a saturated solution in equilibrium with undissolved solute. A supersaturated solution sometimes results when the temperature of an unsaturated solution is changed to a value where the solute solubility is smaller (usually this means cooling).

Surface tension is the amount of work required to extend a liquid surface, usually expressed in joules per square meter, $J \cdot m^{-2}$.

The **surroundings** refer to that part of the universe with which a system interacts by exchanging heat and/or work and/or matter.

A **system** is that part of the universe chosen for a thermochemical or thermodynamic study. (*See also* **surroundings**.)

Temperature is a physical property related to the kinetic energies of the atoms or molecules in a substance that indicates the direction of heat flow. Kinetic energy, as heat, is transferred from more energetic (higher temperature) to less energetic (lower temperature) atoms or molecules.

Temporary hard water is hard water that has hydrogen carbonate (bicarbonate) ion, HCO_3^- as its primary anion. (*See also* **hard water**.)

A **terminal atom** in a polyatomic species (molecule, ion) is bonded to just one other atom.

A **termolecular** step in a reaction mechanism involves the simultaneous collision of three molecules.

The **theoretical yield** is the *calculated* quantity of a product expected in a chemical reaction.

A **theory** is a framework for organizing scientific knowledge that provides explanations of observed natural phenomena and predictions that can be tested by further experiments.

Thermochemistry is the study of energy changes associated with chemical reactions or physical processes, especially energy changes that appear as heat.

Thermodynamics is the science dealing with the relationship between heat and motion (work) and with transformations of energy from one form to another. Thermochemistry is a subfield within thermodynamics.

A **thermoplastic polymer** is one that can be softened by heating and formed into desired shapes by applying pressure.

A **thermosetting polymer** becomes permanently hard at elevated temperatures and pressures.

The **third law of thermodynamics** states that the entropy of a pure, perfect crystal at 0 K is *zero*. This is the starting point for the experimental determination of absolute molar entropies.

Titration is a laboratory procedure in which two reactants in solution are made to react in their stoichiometric proportions.

A **titration curve** in a neutralization reaction is a graph of pH versus volume of titrant added from a buret.

A **torr** is a unit used to express gas pressure: 1 torr = 1 mmHg. (*See also* **millimeter of mercury**.)

The term **trans** is used to indicate geometric isomers in which two groups are attached to opposite sides of a double bond in an organic molecule, or at opposite corners of a square in a square planar complex, or above and below the central plane in an octahedral complex.

Transition elements are elements in which the subshell being filled in the aufbau process is in a principal shell of less than the highest quantum number (an inner shell). Transition elements are located in the *d*- and *f*-blocks of the periodic table.

A **transition state** is an intermediate state lying between the reactants and products of a chemical reaction. It is produced as a result of collisions between especially energetic molecules.

The **transuranium elements** are those with atomic number (Z) greater than 92. All the transuranium elements have now been produced through $Z = 111$.

A **triple bond** is a covalent linkage in which two atoms share *three* pairs of electrons between them.

A **triple point** is a particular temperature and pressure at which *three* phases of a pure substance are at equilibrium—solid, liquid, and vapor; or two solid phases and the liquid; or two solid phases and the vapor.

Trouton's rule is that the entropy of vaporization of a *nonpolar* liquid at its normal boiling point is approximately 87 $J \cdot K^{-1} \cdot mol^{-1}$. That is, $\Delta G^\circ_{vap} = \Delta H^\circ_{vap}/T \approx 87\ J \cdot K^{-1} \cdot mol^{-1}$.

The **Tyndall effect** is the scattering of light by colloidal particles, which makes a colloidal dispersion distinguishable from a true solution.

Heisenberg's **uncertainty principle** states that the product of the uncertainty in the position of an object and the uncertainty in its momentum (mass, m, times speed, u) cannot be less than $h/4\pi$. Thus, it is not possible to know with certainty both the position of a subatomic particle and details of its motion.

A **unimolecular** step in a reaction mechanism is one in which a single molecule undergoes rearrangement or decomposition.

The **unit cell** of a crystal structure is the simplest parallelpiped that can be used to generate the entire crystalline lattice through straight-line displacements in all three dimensions.

The **universal gas constant (R)** is the numerical constant required to relate pressure, volume, amount, and temperature of a gas in the ideal gas equation, $PV = nRT$. Its numerical value is $0.082057\ L \cdot atm \cdot mol^{-1} \cdot K^{-1}$ or $8.3145\ J \cdot mol^{-1} \cdot K^{-1}$.

An **unsaturated hydrocarbon** is a carbon–hydrogen compound having one or more multiple bonds (double, triple) between carbon atoms.

An **unsaturated solution** contains less of a solute in a given quantity of solution than is present in a saturated solution. It is a solution having a concentration less than the solubility limit.

A **valence band** is formed by combining atomic orbitals of the valence electrons of a large number of atoms into a set of molecular orbitals very closely spaced in energy. If the band is only partially filled with electrons it is also a conduction band. (*See also* **band** and **conduction band**.)

The **valence bond method** describes a covalent bond as a region of high electron charge density that results from the overlap of atomic orbitals between the bonded atoms and the sharing of a pair of electrons in the region of overlap.

Valence electrons are electrons with the highest principal quantum number. They are found in the outermost electronic shells of atoms. (*See also* **core electrons**.)

The **valence-shell electron-pair repulsion (VSEPR) theory** is an approach to describing the geometric shapes of molecules and polyatomic ions in terms of the geometrical distribution of electron groups in the valence shell(s) of central atom(s).

The **van't Hoff factor (i)** is a correction factor that must be incorporated in equations for colligative properties so that the equations may be applied to solutions of strong or weak electrolytes.

Vaporization or **evaporation** refers to the conversion of a liquid to a gas (vapor).

The **vapor pressure** of a liquid is the pressure exerted by the vapor in dynamic equilibrium with the liquid at a constant temperature.

A **vapor pressure curve** is a graph of the vapor pressure of a liquid as a function of temperature. The vapor pressure curve is the boundary between the liquid and vapor areas in a phase diagram.

Volt (V) is the unit used to measure electrode potentials and electrical potential differences.

A **voltaic cell** is an electrochemical cell that produces electricity through an oxidation–reduction reaction.

A **voltmeter** is an electrical instrument that measures a potential difference. (*See also* **potential difference**.)

Water cycle. *See* **hydrologic cycle**.

A **wave** is a progressive, repeating disturbance propagated from a point of origin to a more distant point.

The **wavelength** is the distance between any two identical points in consecutive cycles of a wave, for example, the distance between the peaks or crests of the wave.

Wave mechanics is the mathematical description of atomic structure based on the wave properties of subatomic properties.

The **wavenumber** ($\tilde{\nu}$) of radiation expresses the frequency of radiation as the number of cycles per cm of the wave: $\tilde{\nu} = 1/\lambda$

A **weak acid** is an acid that exists partly in ionic form and partly in molecular form in solution, that is, an acid that is a weak electrolyte. (*See also* **acid**.)

A **weak base** is a base that exists partly in ionic form and partly in molecular form in solution, that is, a base that is a weak electrolyte. (*See also* **base**.)

A **weak electrolyte** is a substance that is present partly in molecular form and partly in ionic form in its solutions.

Work is (1) the result of a force acting through a distance, for example, $1\ J = 1\ N \times 1\ m$, or (2) an energy transfer into or out of a thermodynamic system that can be expressed as the product of a force and a distance.

An **X-ray** is a type of electromagnetic radiation produced by the impact of cathode rays (electrons) on a solid, such as on a dense metal anode (a target) in a cathode-ray tube.

A **zero-order reaction** has a rate that is independent of the concentration of reactant(s). The sum of the exponents in its rate equation, $m + n + \cdots = 0$.

Appendix F
Answers to Selected Problems

Answers to In-Text Exercises and Selected Review Questions, Problems, and Additional Problems. *Note*: Some of your answers may differ slightly from those given here, depending on the number of steps used to solve a problem and whether intermediate results were rounded off.

Chapter 1

Exercises: 1.1 a. physical; **b.** chemical; **c.** physical. **1.2 a.** 7.42 ms; **b.** 5.41 μm; **c.** 1.19 mg; **d.** 5.98 km. **1.3 a.** 185 °F; **b.** 10.0 °F; **c.** 179 °C; **d.** −29.3 °C. **1.4 a.** 40 m^3; **b.** 0.196 cm^2; **c.** 91 cm^3; **d.** 11 m/s; **e.** 64.0 mi/h; **f.** 20.8 mi/gal. **1.5 a.** 100.5 m; **b.** 1.50×10^2 g; **c.** 415 g; **d.** 6.3 L. **1.6 a.** 0.0763 m; **b.** 85.6 g; **c.** 14.8 oz; **d.** 11.5 yd; **e.** 101 fl oz. **1.7 a.** 25.0 m/s; **b.** 1.53 km/h; **c.** 15.0 kg/h. **1.8 a.** 73.8 in.^{2}; **b.** 3.51 m^3; **c.** 1.11×10^4 kg/m^2. **1.9.** 13.6 g/mL. **1.10.** 5.44 mL. **1.11.** 3.16×10^4 mL. **Estimation Exercises: 1.1.** 70 kg. **1.2.** 20 kg. **Conceptual Exercises: 1.1.** The assumption that the hole is plugged was made implicitly in Conceptual Example 1.1 to obtain the density of 1.15 g/cm^3 and to conclude that the block would not float. **1.2.** 0.97 g/cm^3. The box is less dense than water and will float. **Review Questions: 13 a.** physical; **b.** chemical; **c.** chemical; **d.** chemical; **e.** physical. **14.** C, Cl, and Na are symbols representing elements. CO, $CaCl_2$, and KI are combinations of symbols for two elements (formulas) and represent compounds. **15.** Helium and salt are substances. Maple syrup and vinegar are mixtures. In this case, both are water solutions. **16.** Gasoline and white wine are homogeneous mixtures. Salad dressing and iced tea are heterogeneous mixtures. **17.** mass: kilogram (kg); length: meter (m); temperature: kelvin (K); time: second (s). **18.** area: m^2; volume: m^3; density: kg/m^3. **Problems: 23 a.** 4.54 ng; **b.** 3.76 km; **c.** 6.34 μg. **25 a.** 74.3 °F; **b.** 37.0 °C; **c.** 414 °F. **27 a.** 57.8 °C; **b.** −184 °F.

29 a. $5.00 \times 10^4\,\text{m}$; **b.** $546\,\text{mm} \times \dfrac{10^{-3}\,\text{m}}{\text{mm}} = 0.546\,\text{m}$; **c.** $98.5\,\text{kg} \times \dfrac{10^3\,\text{g}}{\text{kg}} = 9.85 \times 10^4\,\text{g}$; **d.** $47.9\,\text{mL} \times \dfrac{10^{-3}\,\text{L}}{\text{mL}} = 4.79 \times 10^{-2}\,\text{L}$;

e. $578\,\mu\text{s} \times \dfrac{10^{-6}\,\text{s}}{\mu\text{s}} \times \dfrac{\text{ms}}{10^{-3}\,\text{s}} = 0.578\,\text{ms}$;

f. $237\,\text{mm} \times \dfrac{10^{-3}\,\text{m}}{\text{mm}} \times \dfrac{\text{cm}}{10^{-2}\,\text{m}} = 23.7\,\text{cm}$.

31. (4) < (3) < (1) < (2). **33 a.** 4; **b.** 2; **c.** 5; **d.** 3; **e.** 4; **f.** 4. **35 a.** 2.800×10^3 m; **b.** 9.000×10^3 s; **c.** 9.0×10^{-4} cm; **d.** 2.000×10^1 s. **37 a.** 100.5 m; **b.** 153 g; **c.** 54.4 cm; **d.** 436 g; **e.** 111 mL; **f.** 2.4 cm. **39 a.** 40; **b.** 1.88; **c.** 4.80×10^3; **d.** 3.5×10^{-4}. **41.** 3.12 g/mL. **43.** 5.23 g/cm^3. **45.** 11.3 g/cm^3. **47.** 39.6 g. **49.** 1.16×10^5 cm^3; 7.73×10^4 cm. **51.** 4.2×10^{-3} g/cm^3. **53.** (2). **55.** The actual length is 0.071 m longer than what was measured. **57.** 7.92 in. **59.** $\dfrac{62.3\,\text{lb}}{\text{ft}^3}$. **61.** 1.07 oz. **63.** $1.3 \times 10^5 m^2$. **65.** Diameter = 9.30 cm. **67.** More water will overflow from the vessel of water on which the cork is floated (right).

Chapter 2

Exercises: 2.1 a. 0.612:1.142 = 0.536:1, not possible; **b.** 1.250: 1.142 = 1.095:1, not possible; **c.** 1.713:1.142 = 1.5:1 or 3:2, possible; **d.** 2.856:1.142 = 2.5:1 or 5:2, possible. **2.2.** 116. **2.3.** tetrasulfur dinitride. **2.4.** P_4O_{10}. **2.5 a.** AlF_3; **b.** K_2S; **c.** Ca_3N_2; **d.** Li_2O. **2.6 a.** calcium bromide; **b.** lithium sulfide; **c.** iron(II) bromide; **d.** copper(I) iodide. **2.7 a.** NH_4NO_3; **b.** iron(III) phosphate; **c.** NaClO; **d.** potassium hydrogen carbonate.

2.8. 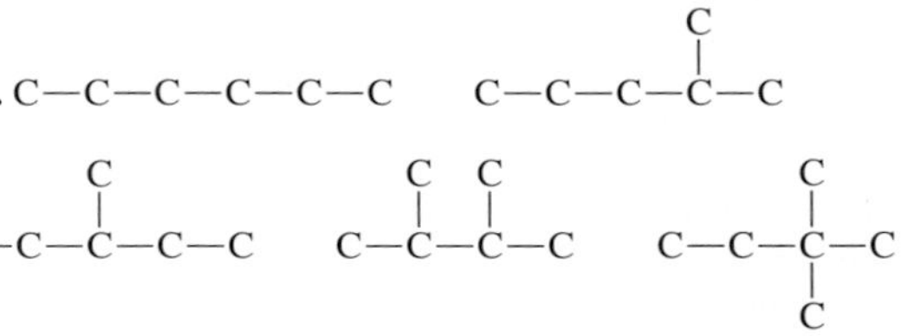

Estimation Exercise: 2.1. 129 °C. **Conceptual Exercise: 2.1 a.** No, the formulas are different: C_7H_{16} for the left structure and C_7H_{14} for the cyclic structure on the right; **b.** Yes, both hydrocarbons have the formula C_9H_{20}. One has methyl groups on the second and fourth carbon atoms, and the other, on the second and fifth carbon atoms. **Review Questions: 4.** Law of definite proportions. **5 a.** The total mass is 53.00 g. **b.** 11.00 g of carbon dioxide is formed. **c.** Law of conservation of mass and law of definite proportions. **6.** Compound B has an oxygen-to-sulfur mass ratio of 3:2. The law of multiple proportions. **7.** The law of conservation of mass. **17 a.** Different elements; **b.** isotopes; **c.** identical atoms; **d.** different elements; **e.** different elements. **18 a.** ${}^{8}_{5}B$; **b.** ${}^{14}_{6}C$; **c.** ${}^{235}_{92}U$; **d.** ${}^{60}_{27}Co$. **23 a.** Magnesium is an element. **b.** Hydroxyl is a prefix, meaning an OH group. **c.** Chloride is a negatively charged atom of chlorine, Cl^-. **d.** Ammonia is a compound, NH_3. **e.** Ammonium is an ion, NH_4^+. **f.** Ethane is a compound, CH_3CH_3. The only substances that could be found on a stockroom shelf are (a), (d), and (f). **29 a.** molecular formula; **b.** molecular formula; **c.** structural formula; **d.** molecular formula; **e.** structural formula. **Problems: 31.** Before reaction: 1.000 g Zn + 0.200 g S = 1.200 g; after reaction: 0.608 g ZnS + 0.592 g Zn = 1.200 g. **33.** Yes, each sample has the same percent carbon (62.5%) and percent hydrogen (4.19%). **35.** A and D are isotopes; B and C are isotopes. **37 a.** Ca, 20 e^-, 20 p^+; **b.** Na, 11 e^-, 11 p^+; **c.** F, 9 e^-, 9 p^+; **d.** Ar, 18 e^-, 18 p^+; **e.** Be, 4 e^-, 4 p^+. **39 a.** 30 p^+, 32 n; **b.** 94 p^+, 147 n; **c.** 43 p^+, 56 n; **d.** 42 p^+, 57 n. **41.** Element, Group, Period, Type: **a.** C, 4A, 2nd, nonmetal; **b.** Ca, 2A, 4th, metal; **c.** Cd, 2B, 5th, metal; **d.** Cl, 7A, 3rd, nonmetal; **e.** B, 3A, 2nd, nonmetal; **f.** Ba, 2A, 6th, metal; **g.** Bi, 5A, 6th, metal; **h.** Br, 7A, 4th, nonmetal. **43 a.** gallium; **b.** copper; **c.** iodine. **45 a.** He; **b.** O_2; **c.** Cl_2; **d.** P_4. **47 a.** N_2O; **b.** P_4S_3; **c.** PCl_5; **d.** SF_6. **49 a.** carbon disulfide; **b.** dinitrogen tetrasulfide; **c.** phosphorus trifluoride; **d.** disulfur decafluoride. **51 a.** sodium ion; **b.** magnesium ion; **c.** aluminum ion; **d.** chloride ion; **e.** oxide ion; **f.** nitride ion. **53 a.** iron(III) or ferric ion; **b.** copper(II) or cupric ion; **c.** silver ion. **55 a.** Br^-; **b.** Ca^{2+}; **c.** K^+; **d.** Fe^{2+}; **e.** Na^+. **57 a.** carbonate ion; **b.** hydrogen phosphate ion; **c.** permanganate ion; **d.** hydroxide ion. **59 a.** NH_4^+; **b.** HSO_4^-; **c.** CN^-; **d.** NO_2^-. **61 a.** sodium bromide; **b.** iron(III) chloride; **c.** lithium iodide; **d.** sodium oxide; **e.** potassium sulfide; **f.** copper(I) bromide; **g.** potassium chloride; **h.** magnesium bromide; **i.** calcium sulfide; **j.** iron(II) chloride; **k.** aluminum oxide. **63 a.** $MgSO_4$; **b.** $NaHCO_3$; **c.** KNO_3; **d.** $CaHPO_4$; **e.** $Ca(ClO_2)_2$; **f.** $CaCO_3$; **g.** $LiHSO_4$; **h.** $Mg(CN)_2$; **i.** KH_2PO_4; **j.** NaClO. **65 a.** sodium hydrogen sulfate; **b.** aluminum hydroxide; **c.** sodium carbonate; **d.** potassium hydrogen carbonate; **e.** ammonium nitrite. **67 a.** HCl; **b.** H_2SO_4; **c.** H_2CO_3; **d.** HCN; **e.** LiOH; **f.** $Mg(OH)_2$. **69 a.** sodium hydroxide; **b.** phosphoric acid; **c.** nitric acid; **d.** sulfurous acid; **e.** calcium hydroxide; **f.** hydrosulfuric acid.

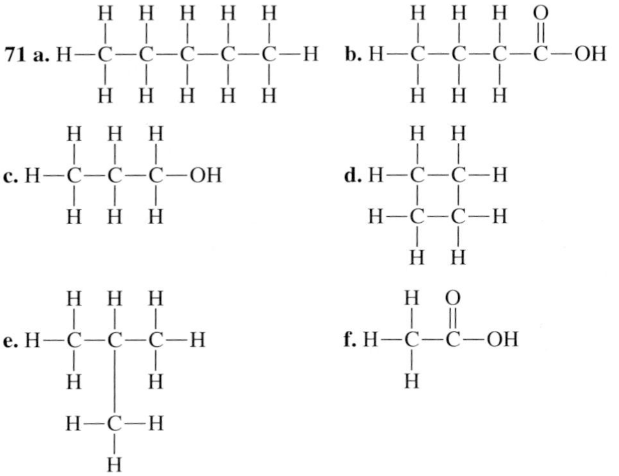

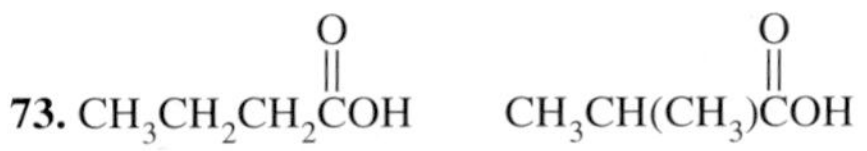

(butyric acid) (isobutyric acid)

75 a. straight-chain alkane hydrocarbon; **b.** alcohol; **c.** hydrocarbon (could be cyclic alkane); **d.** hydrocarbon; **e.** carboxylic acid; **f.** inorganic compound. **77.** b: 2,2,4-trimethylpentane. **79.** Before reaction: 124.8 g total; after reaction: 124.79 g total. The law of conservation of mass is confirmed. **81.** 0.422 g S burned.

83. $\frac{96.2\,g\,Hg}{3.8\,g\,O} = 25{:}1 \quad \frac{92.6\,g\,Hg}{7.4\,g\,O} = 12.5{:}1 \quad \frac{25}{12.5} = 2{:}1$

These ratios are in the proportions of small whole numbers. Thus, there are two compounds of different proportions. **85.** $^{40}_{20}Ca$. **87 a.** There is only one position where the OH group can be substituted on the ethanol molecule; ethanol does not have an isomer. **b.** The third carbon would have to form five bonds, but only four bonds are possible. **89 a.** No, one structure is simply a flipped-over version of the other. **b.** Yes, the structures have the same formula (C_9H_{20}), but the positions of the CH_3 group are different. **c.** No, the structures have different formulas. **d.** No, the structures and formulas are identical. **91 a.** isobutanol or 2-methyl-1-propanol; **b.** 3-methyl-2-pentanol; **c.** 2,2,4-trimethylpentane; **d.** 2,2-dimethyl-1-propanol.

Chapter 3

Exercises: 3.1 a. 243.97 u; **b.** 128.94 u; **c.** 257.95 u; **d.** 211.07 u. **3.2 a.** 294.94 u; **b.** 393.13 u; **c.** 150.82 u; **d.** 342.23 u. **3.3 a.** 0.0664 mol Fe; **b.** 7.76×10^{-4} mol H_3PO_4; **c.** 2.84×10^4 mol C_4H_{10}; **d.** 0.0102 mol K_2CrO_4. **3.4 a.** 1.00×10^3 g H_2O; **b.** 0.756 g $C_4H_{10}O$; **c.** 7.37×10^{-3} g C_2H_6; **d.** 134 g HNO_3. **3.5 a.** 9.27×10^{22} atoms; **b.** 2.15×10^{20} molecules; **c.** 1.92×10^{23} molecules; **d.** 1.70×10^{25} atoms. **3.6 a.** 6.65×10^{-24} g; **b.** 3.47×10^{-22} g; **c.** 2.55×10^{-22} g; **d.** 7.32×10^{-23} g. **3.7 a.** 21.20% N, 6.10% H, 24.27% S, 48.43% O; **b.** 46.65% N, 20.00% C, 26.64% O, 6.71% H; urea has the highest % N. **3.8.** 1.37×10^3 mg Na. **3.9.** $C_5H_{10}NO_2$. **3.10.** $C_7H_5N_3O_6$. **3.11.** C_2H_4; C_6H_{12}; C_5H_{10}. **3.12 a.** 80.24% C, 9.62% H, 10.14% O; **b.** $C_{21}H_{30}O_2$. **3.13 a.** $3\,Mg + B_2O_3 \longrightarrow 2\,B + 3\,MgO$; **b.** $3\,NO_2 + H_2O \longrightarrow 2\,HNO_3 + NO$; **c.** $3\,H_2 + Fe_2O_3 \longrightarrow 2\,Fe + 3\,H_2O$; **d.** $6\,CaO + P_4O_{10} \longrightarrow 2\,Ca_3(PO_4)_2$; **e.** $C_5H_{12} + 8\,O_2 \longrightarrow 5\,CO_2 + 6\,H_2O$; **f.** $2\,C_4H_{10} + 13\,O_2 \longrightarrow 8\,CO_2 + 10\,H_2O$. **3.14 a.** 1.59 mol CO_2; **b.** 305 mol H_2O **c.** 0.6060 mol CO_2. **3.15.** 21.4 g Mg. **3.16.** 0.967 g O_2. **3.17.** 774 mL. **3.18.** 4.77 g H_2S. **3.19.** 26.6 g. **3.20.** 345 g isopentyl alcohol. **3.21 a.** 9.00 M H_2SO_4; **b.** 1.26 M KI; **c.** 0.274 M HF; **d.** 0.0242 M HCl; **e.** 0.123 M $C_6H_{12}O_6$; **f.** 9.23 M C_2H_5OH. **3.22 a.** 673 g; **b.** 5.61 g; **c.** 0.0561 g; **d.** 4.63 g. **3.23.** 23.5 M HCOOH. **3.24.** 466 mL. **3.25.** 0.1724 M KOH. **3.26.** 375 mL $AgNO_3$. **Estimation Exercises: 3.1.** 4.0 g. **3.2.** 1.5×10^{23}. **3.3.** 6.7×10^{-23} g. **3.4.** NH_4NO_3. **3.5.** NH_3. **Conceptual Exercise: 3.1.** $Pb(s) + PbO_2(s) + 2\,H_2SO_4(aq) \longrightarrow 2\,PbSO_4(s) + 2\,H_2O(l)$. **Review Questions: 3.** 6.02×10^{23} O_2 molecules; 1.20×10^{24} O atoms. **4.** 6.02×10^{23} Ca^{2+} ions; 1.20×10^{24} Cl^- ions. **5.** molecular weight: 44.01 u; molar mass: 44.01 g/mol. **8 a.** HO; **b.** CH_2; **c.** C_5H_4; **d.** $C_6H_{16}O$. **Problems: 19 a.** 157.00 u; **b.** 97.994 u; **c.** 294.20 u; **d.** 666.46 u. **21 a.** 4.665×10^{-23} g/atom Si; **b.** 1.055×10^{-22} g/atom Cu; **c.** 1.709×10^{-22} g/atom Rh. **23 a.** 0.435 g MnO_2; **b.** 47.2 g CaH_2; **c.** 45.0 g $C_6H_{12}O_6$. **25 a.** 1.56 mol HNO_3; **b.** 2.85×10^{-2} mol CBr_4; **c.** 6.00×10^{-2} mol $FeSO_4$; **d.** 3.56×10^{-2} mol $Pb(NO_3)_2$. **27 a.** 64.35% Ba, 13.16% Si, 22.49% O; **b.** 58.53% C, 4.094% H,

11.38% N, 25.99% O; **c.** 16.61% Mg, 1.38% H, 16.41% C, 65.60% O; **d.** 4.71% Al, 41.85% Br, 50.28% O, 3.17% H. **29.** $C_6H_4Cl_2$. **31.** $CHCl_3$. **33.** empirical formula C_3H_3O, molecular formula $C_6H_6O_2$. **35.** 48.66% C, 8.214% H, 43.13% O. **37 a.** $Cl_2O_5 + H_2O \longrightarrow 2\ HClO_3$; **b.** $V_2O_5 + 2\ H_2 \longrightarrow V_2O_3 + 2\ H_2O$; **c.** $4\ Al + 3\ O_2 \longrightarrow 2\ Al_2O_3$; **d.** $2\ C_4H_{10} + 13\ O_2 \longrightarrow 8\ CO_2 + 10\ H_2O$; **e.** $Sn + 2\ NaOH \longrightarrow Na_2SnO_2 + H_2$; **f.** $PCl_5 + 4\ H_2O \longrightarrow H_3PO_4 + 5\ HCl$; **g.** $2\ CH_3OH + 3\ O_2 \longrightarrow 2\ CO_2 + 4\ H_2O$; **h.** $3\ Zn(OH)_2 + 2\ H_3PO_4 \longrightarrow Zn_3(PO_4)_2 + 6\ H_2O$. **39 a.** $2\ Mg(s) + O_2(g) \longrightarrow 2\ MgO(s)$; **b.** $NH_4NO_3(s) \longrightarrow N_2O(g) + 2\ H_2O(g)$; **c.** $CH_3CH(OH)CH_2CH_3(l) + 6\ O_2(g) \longrightarrow 4\ CO_2(g) + 5\ H_2O(l)$; **d.** $2\ Al(s) + 6\ HCl(aq) \longrightarrow 2\ AlCl_3(aq) + 3\ H_2(g)$. **41 a.** 1.6×10^{11} mol CO_2; **b.** 5.5×10^{11} mol O_2. **43 a.** 2.48×10^3 g NH_3; **b.** 193 g H_2. **45 a.** 932 g HNO_3; **b.** 2.04×10^3 g $C_7H_5N_3O_6$. **47.** 3.59×10^3 g HNO_3. **49 a.** 8.97×10^3 g CO_2; **b.** 422 mL $C_{14}H_{30}$. **51.** LiOH is limiting. 0.0750 mol Li_2CO_3. **53.** O_2 is limiting. 13.5 g CO_2. **55.** 0.727 g ZnS; 83.4%. **57 a.** 51.3 g NH_4HCO_3; **b.** 1.43×10^3 g $Zn(NO_3)_2$. **59 a.** 2.40 M HCl; **b.** 0.700 M Li_2CO_3. **61 a.** 0.907 M H_2SO_4; **b.** 1.95 M $C_6H_{12}O_6$. **63 a.** 80.0 g NaOH; **b.** 7.66 g $C_6H_{12}O_6$. **65.** 0.208 L. **67.** 2.05 L. **69.** 167 mL. **71.** 0.0520 M. **73.** 0.17 M. **75.** 14.9 M HNO_3. **77.** 0.2496 M HCl. **79.** 7.832×10^{-3} M $Ca(OH)_2$. **81.** 20.86 mL. **83.** Measure 100.0 mL of the 0.04000 M $AgNO_3$ solution in the 100.0-mL volumetric flask and add 1.0194 g of $AgNO_3$. **85 a.** 2.19 g $CaCO_3$; **b.** 4.76 g $Ca(C_3H_5O_3)_2$; **c.** 9.40 g $Ca(C_6H_{11}O_7)_2$; **d.** 3.63 g $Ca_3(C_6H_5O_7)_2$. **87.** 894 u. **89 a.** 90.49% C, 9.495% H; **b.** C_4H_5 empirical; **c.** C_8H_{10} molecular. **91.** Many compounds have the molecular formula $C_4H_8O_2$. Three of them are

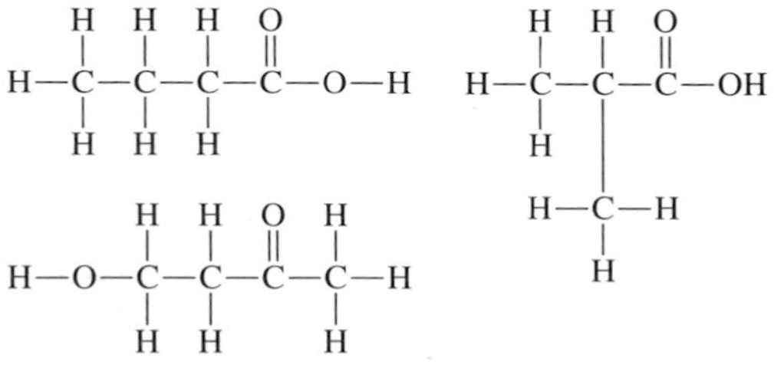

93. 12 g H_2. **95.** 5.59×10^6 metric tons. **97.** 24.0 g theoretical yield; 70.0% yield. **99.** 1.81 g. **101.** 31.6%.

Chapter 4

Exercises: 4.1 a. 722 mmHg; **b.** 737 torr; **c.** 760.7 torr; **d.** 1.01 atm. **4.2.** 6.50 m. **4.3.** 2.90 atm. **4.4.** 17.8 mL. **4.5.** 503 torr. **4.6.** 234 L. **4.7.** 9.82 kg. **4.8.** 671 °C. **4.9.** P/n = constant. If the number of molecules increases, the pressure must increase as more molecular collisions with the wall occur. **4.10.** 1.58 moles. **4.11.** −139 °C. **4.12.** 0.418 g. **4.13.** 74.3 g/mol. **4.14.** 126 g/mol. **4.15.** $C_4H_6O_2$. **4.16.** 1.25 g/L. **4.17.** 94 °C. **4.18.** 5.00 L CH_4. **4.19.** 4.16×10^4 L. **4.20.** 7.50 kg CaO **4.21.** $P_{O_2} = 0.209$ atm; $P_{Ar} = 0.00932$ atm; $P_{CO_2} = 0.0005$ atm; $P_{total} = 0.999$ atm. **4.22.** 12.90 atm. **4.23.** $P_{N_2} = 0.741$ atm; $P_{O_2} = 0.150$ atm; $P_{H_2O} = 0.060$ atm; $P_{Ar} = 0.009$ atm; $P_{CO_2} = 0.040$ atm. **4.24.** $\chi_{CH_4} = 0.664$; $\chi_{C_2H_6} = 0.264$; $\chi_{C_3H_8} = 0.057$; $\chi_{C_4H_{10}} = 0.015$. **4.25.** $V = 1.24$ L. **4.26.** 0.0197 g H_2; $m_{total} = 0.0235$ g. **4.27.** 0.502 g $KClO_3$. **4.28.** $rate_{N_2}/rate_{Ar} = 1.19$. **4.29.** 60 g/mol. **4.30.** 87.1 s. **Estimation Exercises: 4.1.** about 400 mmHg **4.2.** 30 °C. **4.3.** CH_4. **4.4.** 5000 K. **Conceptual Exercises: 4.1.** To achieve a 3.00 atm pressure with $P_{H_2} = 2.00$ atm, other gases must supply 1.00 atm pressure. The original He supplies only 0.50 atm. **Review Questions: 13.** $T_K = T_C + 273°$. **14 a.** volume decreases; **b.** volume decreases; **c.** volume increases. **15 a.** increase; **b.** increase; **c.** increase. **16 a.** temperature decreases; **b.** pressure decreases. **18 a.** container A; **b.** equal densities in containers A and B; **c.** container B. **19.** $T = 273$ K; $P = 1$ atm. If everyone picks the same standard conditions, it is much easier to compare the results of experiments. **25.** $\chi_A = \dfrac{n_A}{n_{total}}$; $\chi_A = \dfrac{P_A}{P_{total}}$. **Problems: 31.** 3.31×10^3 mmHg. **33 a.** 749 mmHg; **b.** 1.12 atm; **c.** 0.949 atm. **35 a.** 758 mmHg; **b.** 722 mmHg. **37 a.** 1.09 L; **b.** 304 mL. **c.** 2.60×10^3 torr; **d.** 0.392 atm. **39.** 1.4×10^9 ft³. **41.** 4.4 atm. **43.** 117 mL. **45.** 160 °C. **47.** a. **49.** 449 g Ne. **51.** 2.49×10^3 torr. **53.** 24.5 L. **55.** 0.489 m³. **57 a.** 22.3 L; **b.** 172 mL. **59 a.** 29.0 atm; **b.** 788 mmHg. **61 a.** 6.55 g Kr; **b.** 8.49 mg. **63.** 7.26×10^8 mol. **65.** 64.5 u. **67.** 153 u. **69 a.** 1.25 g/L; **b.** 1.03 g/L. **71.** 92 °C. **73.** 64.1 g/mol. **75.** 1.15 L SO_3. **77.** 2.48×10^5 L CO_2. **79.** 28.2 mg Mg. **81 a.** $\chi_{Ar} = 0.771$, $\chi_{Ne} = 0.225$, $\chi_{Kr} = 0.00435$; **b.** $\chi_{N_2} = 0.848$, $\chi_{O_2} = 0.146$, $\chi_{Ar} = 0.00603$. **83.** $P_{CO_2} = 3.6 \times 10^2$ torr; $P_{H_2} = 3.1 \times 10^2$ torr; $P_{N_2} = 1.4 \times 10^2$ torr; $P_{O_2} = 11$ torr; $P_{CH_4} = 2.5 \times 10^{-2}$ torr. **85.** 710 torr. **87.** 0.145 g $C_6H_{12}O_6$. **89.** b. **91 a.** 81 u; **b.** 26 u. **93.** $m = 3.14$ g He. **95.** These data are consistent with Avogadro's hypothesis. **97.** C_3H_6. **99.** 54.09 u. **101.** 4.1×10^4 L CO_2. **103.** 342 m/s. **105.** 3.1×10^2 s.

Chapter 5

Exercises: 5.1. −478 J. **5.2.** 142.7 kJ. **5.3.** −148.9 kJ. **5.4.** -1.17×10^3 kJ. **5.5.** 2.80×10^4 L CH_4. **5.6.** 11 J/°C. **5.7.** 6.67×10^4 cal = 66.7 kcal. **5.8.** 94 °C. **5.9.** 1.57×10^3 g Cu. **5.10.** 0.130 J • g⁻¹ • °C⁻¹. **5.11.** −56.2 kJ/mol H_2O. **5.12.** 393 kJ/mol C. **5.13.** 5.99 kJ/ °C. **5.14.** 52.3 kJ. **5.15.** −137 kJ. **5.16.** −44.2 kJ. **5.17.** 80.6 kJ/mol. **Estimation Exercises: 5.1.** The metal with the smallest specific heat will be raised to the highest temperature—silver. **5.2.** $\Delta H^\circ_{comb}(C_{10}H_{22}) \approx -6700$ kJ; $\Delta H^\circ_{comb}(C_{12}H_{26}) \approx -7900$ kJ. **Conceptual Exercises: 5.1.** The temperature increase in the 100-mL sample is twice as great as the temperature decrease in the 200-mL sample. The final temperature is 60 °C. **5.2.** The enthalpy of formation of CH_3OH is 39 kJ/mol greater than that of CH_3CH_2OH, but the enthalpies of the products of the combustion of CH_3CH_2OH are much lower than of the combustion of CH_3OH [by one mole each of $H_2O(l)$ and $CO_2(g)$]. CH_3CH_2OH has the greater negative heat of combustion (recall Figure 5.14). **Review Questions: 13.** endothermic; $\Delta H = 178$ kJ/mol $CaCO_3$. **15.** +46.11 kJ. **17.** Heat capacity is the amount of heat required to change the temperature of a system by one degree C. It is expressed in joules per degree C or joules per kelvin. **19.** A measured amount of substance of unknown specific heat is allowed to exchange a quantity of heat with a measured amount of a second substance of known specific heat, and the temperature changes in the two subsances are measured. The unknown specific heat is calculated with the equation: (mass × sp. ht. × ΔT)$_{unknown}$ = −(mass × sp. ht. × ΔT)$_{known}$. **21.** Reactants are sealed in a constant-volume "bomb," which is immersed in water in a calorimeter. Reaction is initiated, usually by electrical heating, and the change in temperature is measured. The heat of reaction is calculated from the temperature change and the heat capacity of the calorimeter (which is determined in a separate experiment). Bomb calorimetry is used mostly for combustion reactions. **23.** A substance is in its standard state when it is a pure material at one atm pressure. For a gas, it is the gas acting as a hypothetical ideal gas at 1 atm. **25.** $2\ Fe(s) + 3/2\ O_2(g) \longrightarrow Fe_2O_3(s)$. **27.** The relevant equation is $\Delta H° = \Sigma \nu_p \Delta H_f°(\text{products}) - \Sigma \nu_r \Delta H_f°(\text{reactants})$. Standard heats of reaction, $\Delta H°$, are calculated by substituting $\Delta H_f°$ values on the right side of the equation. To determine an unknown $\Delta H_f°$, a known $\Delta H°$ is substituted on the left side of the equation and known $\Delta H_f°$ values are substituted for all the terms on the right side, except for the unknown. **29.** The food Calorie is 1000 times as large as the ordinary calorie used in scientific work: 1 Cal = 1000 cal. **Problems: 31.** 130 J. **33.** 2.4×10^2 J. **35.** endothermic; 241 kJ/mol; $q_p = \Delta H$. **37.** $CaCO_3(s) \longrightarrow CaO(s) + CO_2(g)$; $\Delta H = 178$ kJ. **39 a.** −15 kJ; **b.** 23 kJ. **41.** −21.7 kJ. **43.** 1.45 kg CO. **45.** 4.1J/°C. **47 a.** 6.36 kJ; **b.** 16.2 kJ. **49.** 42.0 °C. **51.** $0.373\ J \cdot g^{-1} \cdot °C^{-1}$. **53.** 131 °C. **55.** $\Delta H = -53.7$ kJ. **57.** −27.4 kJ/g coal. The heat of reaction is ΔE, but the difference between ΔH and ΔE for this reaction is negligible because $\Delta n_g = 0$. **59.** 11.6 kJ/ °C. **61.** −24.0 kJ. **63.** −1370.9 kJ. **65 a.** −176.0 kJ; **b.** −146.3 kJ; **c.** −154.5 kJ; **d.** −11 kJ. **67.** −206.0 kJ /mol. **69.** −628 kJ /mol. **71 a.** C_8H_{18}; **b.** CH_4; **c.** CH_4; **d.** CH_4. **73.** 18.9 Cal; claim is verified. **75 a.** yes; **b.** yes; **c.** $\Delta E = 0$. **77.** $t_f = 27.8$ °C. **79.** H_2 produces no CO_2; CH_4, 4.94×10^{-2} g CO_2; CH_3OH, 6.06×10^{-2} g CO_2; C_8H_{16}, 6.46×10^{-2} g CO_2. **81.** 1.27×10^5 kJ. **83.** The liberated heat would raise the temperature to about 5000 °C, far above the melting point of iron (1530 °C). **85.** $\Delta H_{comb} = -5.64 \times 10^3$ kJ/mol thymol. **87.** The final temperature is approximately 60 °C above room temperature. **89.** $\Delta H° = 219.0$ kJ. **91.** Because $\Delta H_f°[H_2O(l)]$ is more negative than $\Delta H_f°[H_2O(g)]$, reaction (a) should liberate the greater quantity of heat, by 572 kJ/mol $C_{12}H_{26}$.

Chapter 6

Exercises: 6.1. 12.01 u. **6.2.** 20.18 u. **6.3.** 3.07×10^3 nm. **6.4.** 2.80×10^{11} Hz. **6.5.** 1.91×10^{-23} J/photon. **6.6.** 74.34 nm. **6.7.** -6.053×10^{-20} J. **6.8.** 4.086×10^{-19} J. **6.9.** 434.1 nm. **6.10.** 0.105 nm. **6.11 a.** not possible. For $l = 1$, m_l must be between +1 and −1; **b.** possible; **c.** possible; **d.** not possible. For $l = 2$, m_l must be between +2 and −2. **6.12 a.** three; **b.** sixteen; **c.** *f* subshell. **Estimation Exercises: 6.1.** ^{115}In. **6.2.** the infrared. **Conceptual Exercise: 6.1.** (a). **Review Questions: 22 a.** 3*d*; **b.** 2*s*; **c.** 4*p*; **d.** 4*f*. **23 a.** 3 sublevels; **b.** 2 sublevels; **c.** 4 sublevels. **Problems: 27.** 1.672×10^{-27} kg. **29.** All values are integral multiples of 1.6×10^{-19} C. **31 a.** 80:1; **b.** 9:1; **c.** 40:1. **33.** 151.9 u. **35.** 87.61 u. **37.** 1.5×10^4 s. **39.** 302 m. **41.** 0.0308 nm. **43.** $5.56 \times 10^{14}\ s^{-1}$. **45.** 4.04×10^{11} m; 2.70×10^{11} m. **47.** 2.47×10^{-19} J. **49.** 1.66×10^{-23} J/photon; 0.0100 kJ/mol. **51.** 427 nm. **53.** 246 nm, ultraviolet region. **55 a.** $4.568 \times 10^{14}\ s^{-1}$; **b.** $3.083 \times 10^{15}\ s^{-1}$. **57.** $E_1 = -2.179 \times 10^{-18}$ J; $E_2 = -5.448 \times 10^{-19}$ J; $E_3 = -2.421 \times 10^{-19}$ J; $E_4 = -1.362 \times 10^{-19}$ J; $E_5 = -8.716 \times 10^{-20}$ J; $E_6 = -6.053 \times 10^{-20}$ J; $E_7 = -4.447 \times 10^{-20}$ J; $E_\infty = 0$. **59.** 1.56×10^{-4} nm. **61.** 8.62×10^3 m/s. **63.** seven 4*f* orbitals. **65.** $n = 2$. **67.** If $n = 4$, $l = 0, 1, 2, 3, 4$. If $l = 3$, $m_l = -3, -2, -1, 0, 1, 2, 3$. **69 a.** permissible; **b.** not permissible because the value of l cannot be greater than $n - 1$; **c.** permissible; **d.** not permissible because *n* cannot have the value 0. **71 a.** For 3*s*, $n = 3$, $l = 0$, $m_l = 0$; **b.** for 5*f*, $n = 5$, $l = 3$, $m_l = -3, -2, -1, 0, 1, 2, 3$; **c.** 3*p*. **73.** 0.974 for ^{175}Lu; 0.026 for ^{176}Lu. **75.** 1.7×10^{-2} s/cycle; period and frequency are inversely related. **77.** Each photon that has more energy than the threshold energy will dislodge an electron. The energies of two photons do not add together to reach the threshold energy. Each photon interacts with the metal separately. **79.** c. **81.** 91.16 nm minimum; 121.6 nm maximum. **83.** $n = 4 \longrightarrow n = 2$. **85.** The sum of the energies of the transitions: $n = 3 \longrightarrow n = 2$ (1.634×10^{-18} J) and $n = 2 \longrightarrow n = 1$ (3.026×10^{-19} J) is the same as the energy of the transition: $n = 3 \longrightarrow n = 1$ (1.937×10^{-18} J). **87.** 2.41×10^{-34} m. **89.** The proton and electron must have the same momentum (mass times velocity) to have equal wavelengths. Because the electron has the smaller mass, it must have the greater velocity. **91.** 1.08×10^5 mol photons. **93.** Bohr meant that Einstein should accept some things as unknowable, that he should not assume that all natural phenomena were meant to be predictable.

Chapter 7

Exercises: 7.1 a. This notation indicates two electrons in the 1*s* subshell, two in the 2*s* subshell, six in the 2*p* subshell, two in the 3*s* subshell, six in the 3*p* subshell, ten in the 3*d* subshell, and two in the 4*s* subshell. This is the electron configuration of zinc ($Z = 30$); **b.** This orbital diagram has two electrons in the 1*s* subshell, two electrons in the 2*s* subshell, two electrons in each of the 2*p* orbitals, two electrons in the 3*s* subshell, and one electron in each of two 3*p* orbitals. Each pair of arrows represents two electrons with opposite spins. The electrons in the 3*p* orbitals have parallel spins. This element is silicon.

7.2. P: $1s^22s^22p^63s^23p^3$; $[Ne]3s^23p^3$; [Ne] [↑↓] (3*s*) [↑][↑][↑] (3*p*)

7.3. Ga: $4s^24p^1$; Te: $5s^25p^4$. **7.4 a.** $1s^22s^22p^63s^23p^63d^{10}4s^24p^2$; $[Ar]3d^{10}4s^24p^2$; **b.** $1s^22s^22p^63s^23p^63d^{10}4s^2$; $[Ar]3d^{10}4s^2$; **c.** $1s^22s^22p^63s^23p^63d^24s^2$; $[Ar]3d^24s^2$; **d.** $1s^22s^22p^63s^23p^63d^{10}4s^24p^64d^{10}5s^25p^5$; $[Kr]4d^{10}5s^25p^5$. **7.5 a.** K; **d.** Ga; **e.** S; **f.** Pb. **7.6 a.** F < N < Be; **b.** Be < Ca < Ba; **c.** F < Cl < S; **d.** Mg < Ca < K. **7.7.** $Y^{3+} < Sr^{2+} < Rb^+ < Br^- < Se^{2-}$. **7.8 a.** Be < N < F; **b.** Ba < Ca < Be; **c.** S < P < F; **d.** K < Ca < Mg. **7.9 a.** O; **b.** S; **c.** F. **Estimation Exercise: 7.1.** 400 kJ/mol. **Conceptual Exercise: 7.1 a.** $Z = 52$; **b.** $Z = 21$; **c.** $Z = 39$ (lowest) and $Z = 48$ (highest); **d.** N. **Review Questions: 3.** The orbital energies in (b) and (c) are identical for the H atom; those in (c) are also identical for other atoms. **9.** 3A elements, ns^2np^1; 5A elements, $ns^2np_x^1np_y^1np_z^1$. **10 a.** *ns*; **b.** *np*; **c.** $(n-1)d$; **d.** $(n-2)f$. **12 a.** 4; **b.** 8; **c.** 7; **d.** 3; **e.** 2. **15 a.** 2nd period, Group 8A; **b.** 3rd, 4A; **c.** 3rd, 1A; **d.** 2nd, 2A; **e.** 2nd, 5A; **f.** 3rd, 3A. **16.** 4A: ns^2np^2; 6A: ns^2np^4. **18.** *d* orbital, two electrons; *d* subshell, 10 electrons. **19 a.** 3*s*; **b.** 3*s*; **c.** 2*p*; **d.** 3*p*. **20 a.** Cs, $6s^1$; Se, $4s^24p^4$; In, $5s^25p^1$. **21 a.** Si: $1s^22s^22p^6$, core; $3s^23p^2$, valence; **b.** Rb: $1s^22s^22p^63s^23p^63d^{10}4s^24p^6$, core; $5s^1$, valence; **c.** Br: $1s^22s^22p^63s^23p^63d^{10}$, core; $4s^24p^5$, valence. **23 a.** diamagnetic; **b.** diamagnetic; **c.** paramagnetic; **d.** paramagnetic; **e.** diamagnetic; **f.** paramagnetic. **26 a.** 5; **b.** 32; **c.** 5; **d.** 2; **e.** 10. **27 a.** 3; **b.** 1; **c.** 0; **d.** 14; **e.** 1; **f.** 5; **g.** 32. **33 a.** Cl < S; **b.** Al < Mg; **c.** As < Ge; **d.** Ca < K. **37 a.** Sr < Ca < Mg; **b.** S < P < Cl; **c.** Sn < Ge < As; **d.** Se < Br < Cl. **Problems: 43 a.** Allowed. **b.** Not allowed. There are three electrons in one orbital and the 2*s* electrons have the same spin. **c.** Not allowed.

There are three electrons in one of the $2p$ orbitals. **d.** Not allowed. The unpaired electrons are not all the same spin. **e.** Not allowed. There should be one electron in each $3p$ orbital. **f.** Allowed. **45 a.** $3s$ subshell fills before any electron goes into $3p$ subshell. **b.** $4s$ subshell fills before $3d$ subshell. **c.** $3d$ subshell fills completely before $4p$ subshell. **47 a.** The subshell that fills after [Ar] is $4s$; $2d$ does not exist. **b.** The subshell that fills after [Ar]$4s^2$ is $3d$; $3f$ does not exist. **c.** The subshell that fills after [Kr]$4d^{10}5s^2$ is $5p$. **49 a.** $1s^22s^22p^63s^23p^6$; **b.** $1s^1$; **c.** $1s^22s^22p^6$; **d.** $1s^22s^2$; **e.** $1s^22s^22p^63s^23p^64s^1$; **f.** $1s^22s^22p^63s^23p^3$; **g.** $1s^22s^22p^63s^23p^64s^2$; **h.** $1s^22s^22p^63s^2$; **i.** $1s^22s^22p^63s^23p^63d^{10}4s^24p^5$. **51 a.** [Ar]$3d^{10}4s^24p^1$; **b.** [Kr]$4d^{10}5s^25p^4$; **c.** [Kr]$4d^{10}5s^25p^5$; **d.** [Xe]$6s^1$; **e.** [Kr]$4d^{10}5s^25p^3$; **f.** [Kr]$5s^2$.

53 a. $1s$ [↑↓] $2s$ [↑↓] $2p$ [↑|↑|↑]   **b.** $1s$ [↑↓] $2s$ [↑↓] $2p$ [↑| |]

c. $1s$ [↑↓] $2s$ [↑↓] $2p$ [↑↓|↑↓|↑↓] $3s$ [↑↓] $3p$ [↑|↑|]

d. $1s$ [↑↓] $2s$ [↑↓] $2p$ [↑↓|↑↓|↑↓] $3s$ [↑↓] $3p$ [↑↓|↑↓|↑↓] $4s$ [↑↓]

e. $1s$ [↑↓] $2s$ [↑↓] $2p$ [↑↓|↑↓|↑↓] $3s$ [↑↓] $3p$ [↑↓|↑↓|↑]

f. $1s$ [↑↓] $2s$ [↑↓] $2p$ [↑↓|↑↓|↑↓] $3s$ [↑↓] $3p$ [↑↓|↑↓|↑↓] $3d$ [↑| | | |] $4s$ [↑↓]

55. Hg is in the 6th period and Group 2B, and it is the last $5d$ transition element. [Xe] $4f$ [↑↓|↑↓|↑↓|↑↓|↑↓|↑↓|↑↓] $5d$ [↑↓|↑↓|↑↓|↑↓|↑↓] $6s$ [↑↓].

57 a. Ca < Sr < Rb; atomic radius increases from top to bottom in a group of elements and from right to left in a period of elements. **b.** C < Si < Al (for the same reason as in part a). **59 a.** K < Ca < Mg; a potassium atom is larger than a calcium atom, which is larger than a magnesium atom. **b.** Te < I < Br; a tellurium atom is larger than an iodine atom, which is larger than a bromine atom. **61 a.** Rb > Sr > Ca; rubidium is to the left of strontium in the fifth period, and strontium is below calcium in Group 2A. **b.** Al > Si > C; aluminum is to the left of silicon in the third period, and silicon is below carbon in Group 4A. **63.** The melting points are relatively low for the metals of Group 1A. They rise to a maximum with the Group 4A elements, and then decrease sharply for the nonmetals following Group 4A. It does not matter whether the temperatures plotted are in degrees Celsius or in kelvins because all points on the graph move up or down by the same number of degrees if a shift is made from one temperature scale to the other. **65.** [Rn]$5f^{14}6d^{10}7s^27p^2$. **67.** I_1(Cs) < I_1(B) < I_2(Sr) < I_2(In) < I_2(Xe) < I_3(Ca). A Cs atom is larger than a B atom and has a lower I_1. Second ionization energies of atoms are larger than their first ionization energies. I_2(Sr) should be larger than I_1(B), but it should be smaller than the other I_2 values, which increase from left to right in the fifth period. I_3(Ca) should be greatest of all because it represents removing an electron from the noble-gas electron configuration of Ca^{2+}. **69.** $Cl(g) + e^- \longrightarrow Cl^-(g)$; ΔH = EA = –349 kJ. For the overall process, ΔH = +242.8 kJ – 349 kJ = –106 kJ. The overall process is exothermic. **71.** $Na^+(g) + e^- \longrightarrow Na(g)$, ΔH = –496 kJ; $Cl^-(g) \longrightarrow e^- + Cl(g)$, ΔH = 349 kJ; total for the overall process: ΔH = –147 kJ. The overall process is exothermic.

73. $\dfrac{1\,\text{eV}}{\text{atom}} \times \dfrac{6.0221 \times 10^{23}\,\text{atoms}}{\text{mol}} \times \dfrac{1.6022 \times 10^{-19}\,\text{C}}{\text{electron}} \times \dfrac{1\,\text{J}}{1\,\text{V}\cdot\text{C}} \times \dfrac{1\,\text{kJ}}{10^3\,\text{J}} = 96.49\,\text{kJ/mol}$.

Chapter 8

Exercises: 8.1. 30.8 g H_2 from the reaction of Zn with an acid compared to 25.4 g H_2 from the spent uranium fuel. **8.2 a.** $2NaCl(l) \xrightarrow{\text{electrolysis}} 2Na(l) + Cl_2(g)$; $2Na(s) + H_2(g) \xrightarrow{\Delta} 2NaH(s)$; **b.** $2NaCl(aq) + 2H_2O(l) \xrightarrow{\text{electrolysis}} 2NaOH(aq) + Cl_2(g) + H_2(g)$; $2NaOH(aq) + Cl_2(g) \longrightarrow NaCl(aq) + NaOCl(aq) + H_2O(l)$. **8.3.** $3BaO(s) + 2Al(l) \xrightarrow{1800\ °C} 3Ba(g) + Al_2O_3(s)$.

Estimation Exercises: 8.1. CaH_2. **8.2.** $MgCl_2{\cdot}6H_2O$. **Conceptual Exercise: 8.1.** 71% MgO. **Review Questions: 1.** oxygen, hydrogen, hydrogen. **3.** sodium, calcium. **5.** sodium and potassium; magnesium and calcium. **6.** Group 1A: Li, Na, K, Rb, Cs, Fr; Group 2A: Ca, Sr, Ba, Ra. **15 a.** magnesium metal; **b.** sodium hydrogen carbonate (and sodium carbonate). **16 a.** calcium carbonate, $CaCO_3$; **b.** calcium oxide, CaO; **c.** calcium hydroxide, $Ca(OH)_2$; **d.** calcium sulfate dihydrate $CaSO_4{\cdot}2H_2O$; **e.** sodium sulfate decahydrate $Na_2SO_4{\cdot}10H_2O$. **17 a.** limestone, clay, and sand; **b.** limestone, sand, and sodium carbonate; **c.** slaked lime, sand, and water; **d.** gypsum. **Problems: 27 a.** lithium hydride; **b.** MgI_2; **c.** $Ba(OH)_2{\cdot}8H_2O$; **d.** calcium hydrogen carbonate. **29 a.** $Ca(s) + 2\ HCl(aq) \longrightarrow H_2(g) + CaCl_2(aq)$; **b.** $CaO(s) + 2\ HCl(aq) \longrightarrow CaCl_2(aq) + H_2O(l)$; **c.** $2\ NaF(s) + H_2SO_4(\text{concd aq}) \xrightarrow{\Delta} Na_2SO_4(s) + 2\ HF(g)$. **31 a.** $2\ Li(s) + Cl_2(g) \longrightarrow 2\ LiCl(s)$; **b.** $2\ K(s) + 2\ H_2O(l) \longrightarrow 2\ KOH(aq) + H_2(g)$; **c.** $2\ Cs(s) + Br_2(l) \longrightarrow 2\ CsBr(s)$; **d.** $K(s) + O_2(g) \longrightarrow KO_2(s)$. **33 a.** $MgCO_3(s) \xrightarrow{\Delta} MgO(s) + CO_2\ (g)$; **b.** $CaCl_2(l) \xrightarrow{\text{electrolysis}} Ca(s) + Cl_2(g)$; **c.** $Ca(s) + 2HCl(aq) \longrightarrow CaCl_2(aq) + H_2(g)$; **d.** $Ca(OH)_2(s) + H_2SO_4(aq) \longrightarrow CaSO_4(s) + 2\ H_2O(l)$. **35 a.** $Mg(OH)_2(s) + 2HCl(aq) \longrightarrow MgCl_2(aq) + 2\ H_2O(l)$; **b.** $CO_2(g) + 2\ KOH(aq) \longrightarrow K_2CO_3(aq) + H_2O(l)$; **c.** $2\ KCl(s) + H_2SO_4(aq) \longrightarrow K_2SO_4(s) + 2\ HCl(g)$. **37 a.** $2\ NaCl(s) + H_2SO_4(\text{concd aq}) \longrightarrow Na_2SO_4(s) + 2\ HCl(g)$; **b.** $Mg(HCO_3)_2(aq) \xrightarrow{\Delta} MgCO_3(s) + H_2O(l) + CO_2(g)$; $MgCO_3(s) \xrightarrow{\Delta} MgO(s) + CO_2(g)$. **39 a.** $2\ NH_4Cl(aq) + Ca(OH)_2(aq) \longrightarrow 2\ NH_3(g) + CaCl_2(aq) + 2\ H_2O(l)$; **b.** $CaCO_3(s) \xrightarrow{\Delta} CaO(s) + CO_2(g)$, followed by $CaO(s) + H_2O(l) \longrightarrow Ca(OH)_2(aq)$ (an alkaline solution). **41.** 6.50 L H_2; 183 g C_3H_8. **43.** 694 L seawater. The actual volume required is greater than 694 L because not every step in the overall process has a 100% yield. **45.** 267 m^3 CO_2. **47.** NH_3 does not soften permanent hard water. No carbonate ion is formed to yield precipitates of MCO_3, and the concentration of OH^- in the weakly basic solution is not high enough to cause the precipitation of insoluble $M(OH)_2$. **49.** 1.15×10^3 kg $Ca(OH)_2$. **51.** Some of the soap must be precipitated as a scum by hard water cations. Only after this occurs can additional soap be effective as a cleaning agent. **53 a.** Ca^{2+} + resin–2H ⟶ resin–Ca + $2H^+$; HCO_3^- + resin–OH ⟶ resin–HCO_3 + OH^-; $H^+ + OH^-$ ⟶ H_2O; **b.** Na^+ + resin–H ⟶ resin–Na + H^+; Cl^- + resin–OH ⟶ resin–Cl + OH^-; $H^+ + OH^-$ ⟶ H_2O; **c.** Na^+ + resin–H ⟶ resin–Na + H^+; $H^+ + OH^-$ ⟶ H_2O. **55 a.** 4 Li(s) + O_2(g) ⟶ 2 Li_2O(s); **b.** 6 Li(s) + N_2(g) ⟶ 2 Li_3N(s); **c.** Li_2CO_3(s) ⟶ Li_2O(s) + CO_2(g).

57. 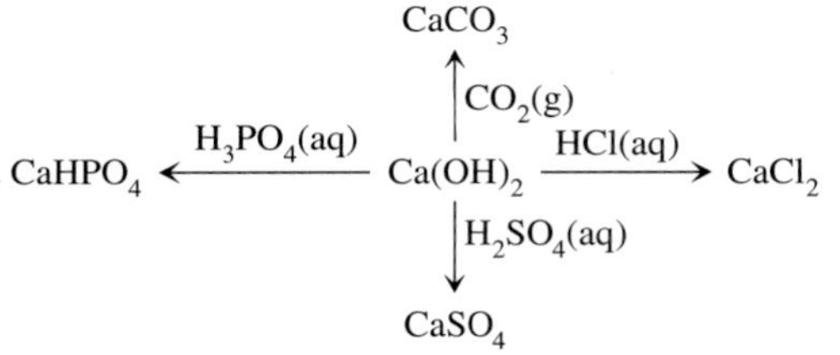

59 a. Ca^{2+}(aq) + Na_2CO_3(aq) ⟶ $CaCO_3$(s) + 2 Na^+(aq); **b.** 20.9 g Na_2CO_3. **61.** 9.23×10^{-3} M Na^+. **63 a.** 38.1 kg CaO; **b.** 54.4 kg NaOH. **65.** CaO(s) + CO_2(g) ⟶ $CaCO_3$(s) [or CaO(s) + H_2O ⟶ $Ca(OH)_2$(s), followed by $Ca(OH)_2$(s) + CO_2(g) ⟶ $CaCO_3$(s) + H_2O(g)]. **67 a.** H_2(l) has the greater heat of combustion on a per gram basis (–141.8 kJ/g H_2 compared to –46 kJ/g $C_{12}H_{26}$). **b.** $C_{12}H_{26}$(l) has the greater heat of combustion on a per mL basis [–34 kJ/mL $C_{12}H_{26}$(l) compared to –10.0 kJ/mL H_2(l)].

Chapter 9

Exercises: 9.1 a. :Ar: **b.** ·Ca· **c.** ·Br: **d.** :As· **e.** K· **f.** ·Se· **9.2.** 2 ·Al· + 3 ·O· ⟶ 2 $[Al]^{3+}$ + 3 $[:O:]^{2-}$ Al_2O_3 aluminum oxide. **9.3.** lattice energy: –863 kJ/mol LiCl(s). **9.4 a.** Ba < Ca < Be; **b.** Ga < Ge < Se; **c.** Te < S < Cl; **d.** Bi < P < S. **9.5.** C—S < C—H < C—Cl < C—O < C—Mg.

9.6. 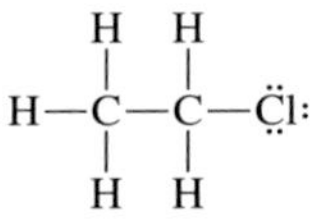 **9.7.** S=C=O **9.8.** 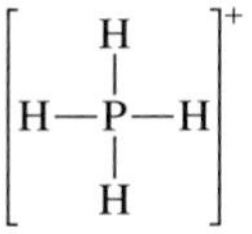

9.9. :O=N—Cl:

9.10. 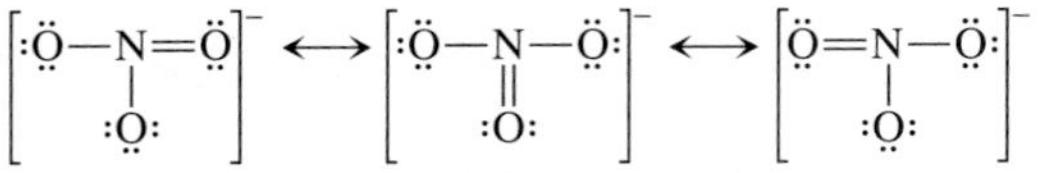

The resonance hybrid, which involves equal contributions from these three equivalent resonance structures, has N-to-O bonds with bond lengths and bond energies intermediate between single and double bonds. **9.11** 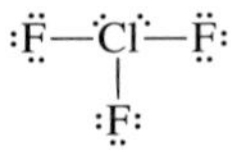 **Estimation Exercises:**

9.1 a. nitrogen-to-nitrogen bond length, 123 pm; **b.** oxygen-to-fluorine bond length, 144 pm. **9.2.** ΔH = –113 kJ. **Conceptual Exercises: 9.1 a.** Incorrect. The molecule ClO_2 has 19 valence electrons, but the Lewis structure shows 20—one too many. **b.** Correct; **c.** Incorrect. The structure shown has an incomplete octet on a N atom. The correct structure is :F—N=N—F: **Review Questions: 1.** Group 8A, the noble gases. **4 a.** Na· **b.** ·O· **c.** ·F: **d.** ·Al·

7 a. K^+, [Ar], $[K]^+$ **b.** S^{2-}, [Ne]$3s^23p^6$, $[:S:]^{2-}$ **c.** F^-, [He]$2s^22p^6$, $[:F:]^-$ **d.** Al^{3+}, [Ne], $[Al]^{3+}$ **8 a.** Mg^{2+}, [Ne], $[Mg]^{2+}$; **b.** Cl^-, [Ne]$3s^23p^6$, $[:Cl:]^-$ **c.** Li^+, [He], $[Li]^+$ **d.** N^{3-}, [He]$2s^22p^6$, $[:N:]^{3-}$

9 a. ·Ca· + 2 ·Br: ⟶ 2 $[:Br:]^-$ + Ca^{2+} **b.** ·Mg· + ·S· ⟶ $Mg^{2+}[:S:]^{2-}$ **10 a.** 2 ·Al· + 3 :S: ⟶ 2 $[Al]^{3+}$ + 3 $[:S:]^{2-}$ **b.** 3 ·Mg· + 2 :P· ⟶ 3 $[Mg]^{2+}$ + 2 $[:P:]^{3-}$

12. :I—I: The dash represents the bond pair of electrons and the pairs of dots, the lone pairs. **13.** $^{\delta+}$H—F:$^{\delta-}$ **14 a.** ionic; **b.** polar covalent; **c.** ionic; **d.** polar covalent; **e.** ionic; **f.** ionic; **g.** nonpolar covalent; **h.** nonpolar covalent; **i.** polar covalent. **16 a.** N; **b.** Cl; **c.** F; **d.** O. **17 a.** F; **b.** Br; **c.** Cl; **d.** N. **21 a.** 1; **b.** 4; **c.** 2; **d.** 1; **e.** 3; **f.** 1. **23.** 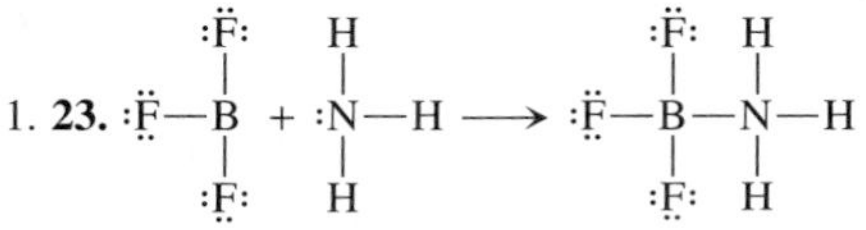

24. carbon, nitrogen, oxygen, and sulfur. **25.** carbon and nitrogen. **31.** (a), (b), and (c) are unsaturated. **33.** All the atoms of the reactants end up in a single product, such as CH_2=CH_2 + H_2 ⟶ CH_3CH_3. **37.** A delocalized bond extends over more than two atoms.

39.  **Problems. 41 b.** Sc^{3+}; **d.** Te^{2-}; **e.** Zr^{4+}. **43.** Sn: [Kr]$4d^{10}5s^25p^2$; Sn^{2+}: [Kr]$4d^{10}5s^2$; Sn^{4+}: [Kr]$4d^{10}$. **45 a.** K^+ + $[:I:]^-$ **b.** Ca^{2+} + $[:O:]^{2-}$; **c.** Ba^{2+} + 2$[:Cl:]^-$; **d.** 2Rb^+ + $[:S:]^{2-}$ **47.** From the Born–Haber cycle, ΔH_f° = –562 kJ/mol KI(s); from Appendix D, ΔH_f° = –567.3 kJ/mol KI(s). **49.** lattice energy, –669 kJ/mol KCl(s). **51 a.** 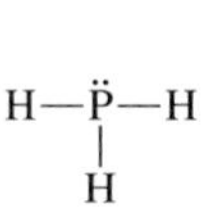 **b.** 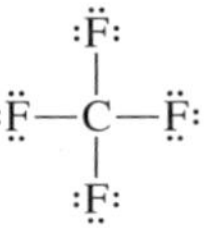 **53 a.** B < N < F; **b.** Ca < As < Br; **c.** Ga < C < O.

55 a. F—F < $\overset{\delta+}{Cl}$—$\overset{\delta-}{F}$ < $\overset{\delta+}{Br}$—$\overset{\delta-}{F}$ < $\overset{\delta+}{I}$—$\overset{\delta-}{F}$ < $\overset{\delta+}{H}$—$\overset{\delta-}{F}$;
b. H—H < $\overset{\delta+}{H}$—$\overset{\delta-}{I}$ < $\overset{\delta+}{H}$—$\overset{\delta-}{Br}$ < $\overset{\delta+}{H}$—$\overset{\delta-}{Cl}$ < $\overset{\delta+}{H}$—$\overset{\delta-}{F}$.

57 a. 233 pm; **b.** 149 pm. **59.** –535 kJ.

61 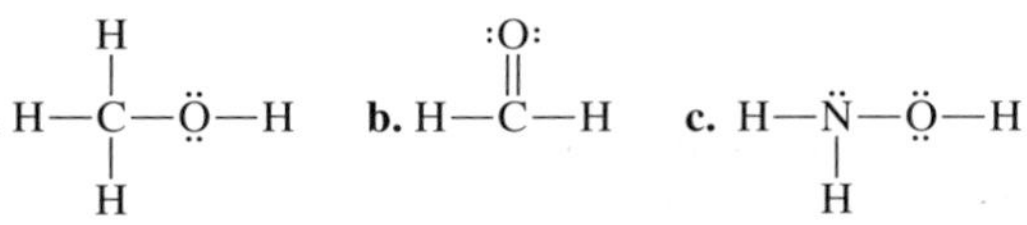

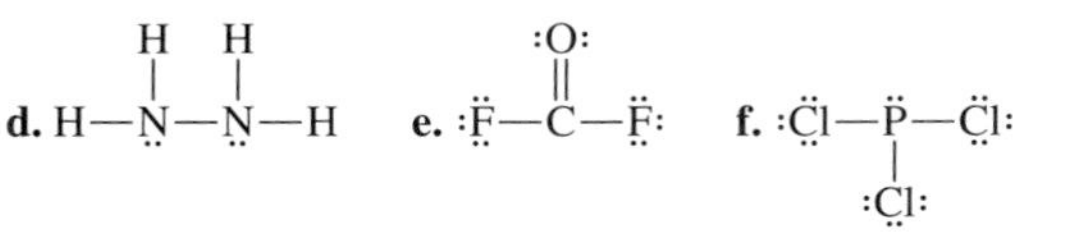

63. Formal charges: CNO^-, C(–2), N(+1), O(0); NCO^-, N(0), C(0), O(–1). The preferred structure is NCO^-.

65. 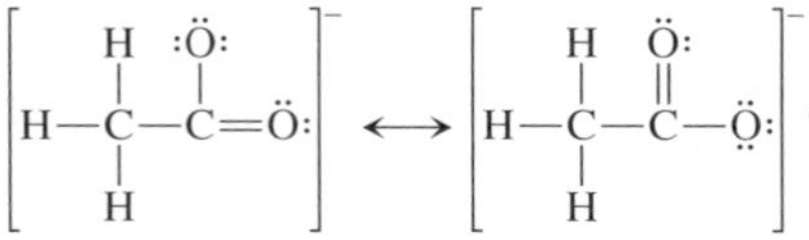 The resonance hybrid is a combination of these two structures, where the C-to-O bonds are intermediate between single and double bonds, and the C and O atoms share a pair of delocalized electrons.

67. 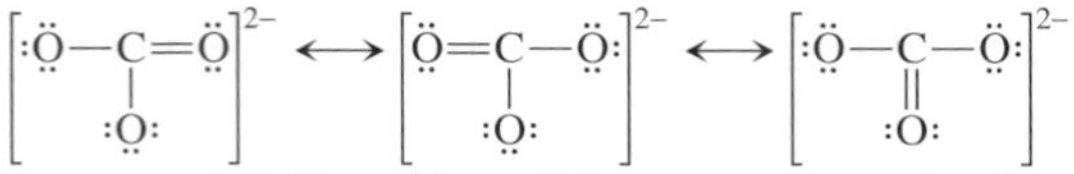

The resonance hybrid combines all three structures, so that the carbon and all three oxygen atoms share a pair of delocalized electrons.

69 a. 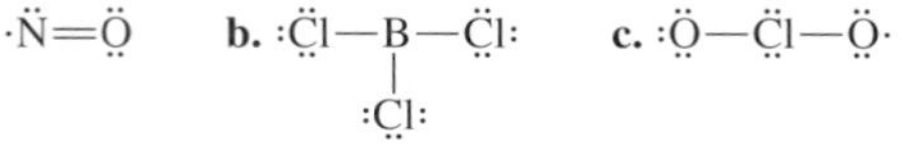

d. 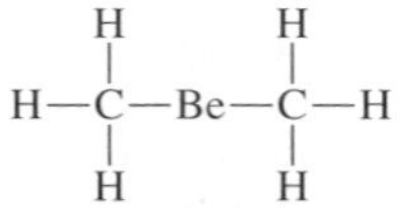

NO and ClO_2 are odd-electron molecules; BCl_3 and $Be(CH_3)_2$ have incomplete octets.

71 a. 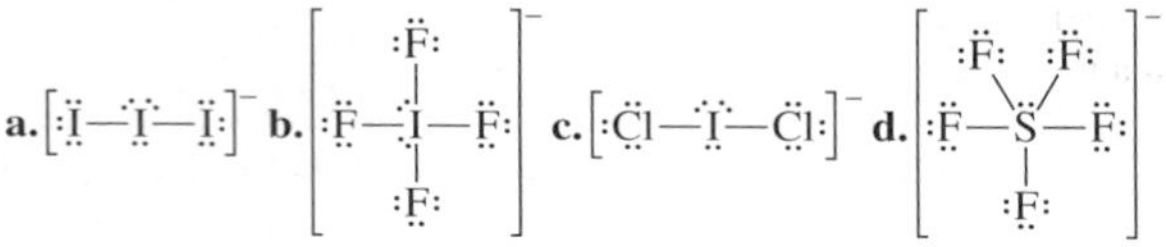

73 a. 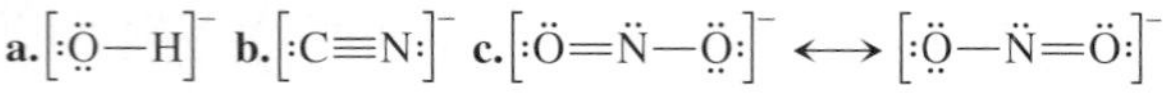

75 a. $H_2C{=}CH_2$; **b.** $H_2C{=}CHCH_2CH_3$; **c.** $HC{\equiv}CCH_3$; **d.** $H_3CC{\equiv}CCH_2CH_3$. **77 a.** $H_2C{=}CH_2 + H_2 \longrightarrow H_3C{-}CH_3$;

b. $H_2C{=}CHCH_3 + Br_2 \longrightarrow$ 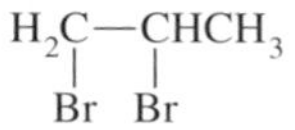

c. $HC{\equiv}CH + 2\,H_2 \longrightarrow H_3C{-}CH_3$.

79 a. $-CH_2CH_2CH_2CH_2CH_2CH_2CH_2CH_2-$

b. 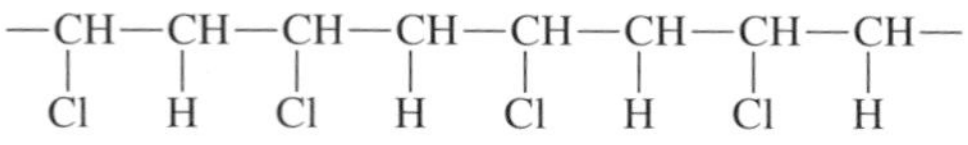

c. 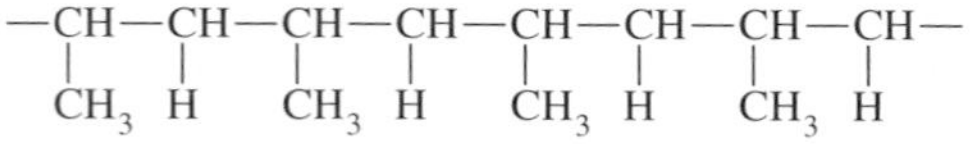

81.

W	X	Y	Z
7A	1A	2A	6A
:Ẅ̤:	X·	·Y·	:Ż:
1−	1+	2+	2−

83. 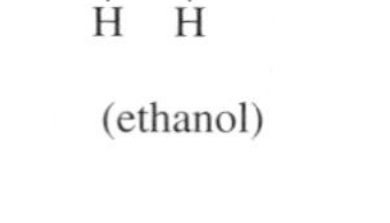 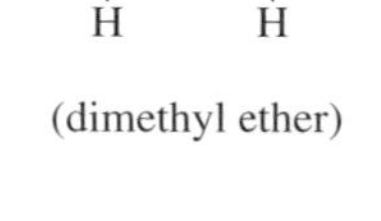

(ethanol) (dimethyl ether)

85. 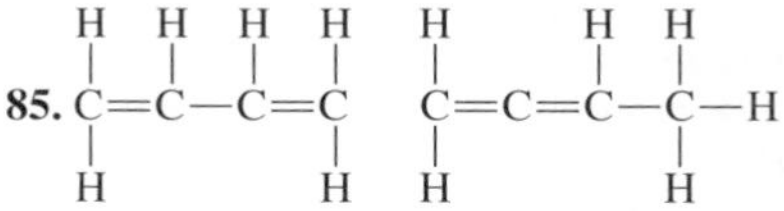

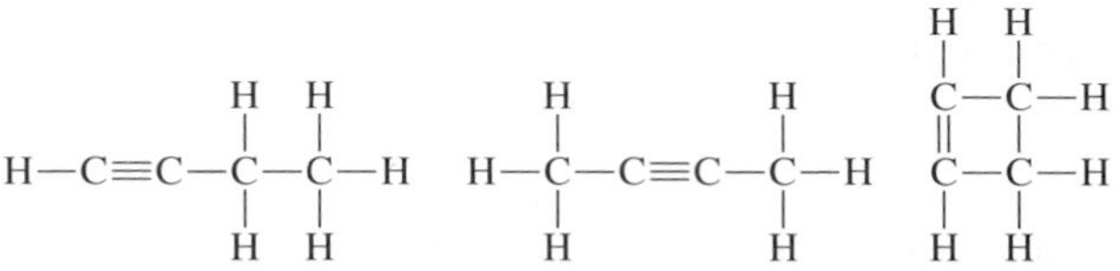

87. 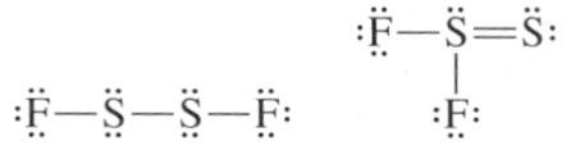

89. $Na^+(g) + Cl^-(g) \longrightarrow Na^+Cl^-(g)$, $\Delta H = -450$ kJ; $Na^+Cl^-(g) \longrightarrow NaCl(s)$, $\Delta H = -336$ kJ. **91.** $MgCl_2(s)$ is much more stable than MgCl(s) because $\Delta H_f^\circ[MgCl_2(s)] = -616$ kJ/mol compared to $\Delta H_f^\circ[MgCl(s)] = -15$ kJ/mol. **93.** 6.19×10^{18} polymer molecules.

Chapter 10

Exercises: 10.1. The five bonding pairs of electrons in $SbCl_5$ produce both a trigonal bipyramidal electron-group geometry and molecular geometry. **10.2.** The VSEPR notation for SF_4 is AX_4E. Electron-group repulsions are minimized if the lone pair of electrons is in the central plane with the S and two of the F atoms; the remaining two F atoms occupy positions leading to a "seesaw" shape. **10.3.** The ICl_4^- ion has the VSEPR notation AX_4E_2. The Cl atoms are coplanar with the central I atom, and the lone pair electrons are situated above and below the plane. **10.4.** The four electron groups distributed around the central S atom are all involved in bonding. The geometric shape corresponding to AX_4 is tetrahedral. **10.5.** The condensed structural formula is CH_3CH_2OH. All the electron pairs around the two C atoms are bonding pairs. The distribution of the bonds around these atoms is tetrahedral, and the bond angles are about 109°. The electron groups around the O atom are two bonding pairs and two lone pairs-that is, AX_2E_2. The C—O—H portion of the molecule is "bent," with a C—O—H bond angle of about 109°. **10.6 a.** nonpolar (trigonal planar); **b.** polar (bent structure); **c.** polar (linear, but with different electronegativities for Br and Cl); **d.** nonpolar (trigonal planar). **10.7.** In the I_3^- ion, five electron groups are distributed in trigonal bipyramidal fashion about the central I atom. Two of the five are bond pairs in axial positions and three are lone pairs in equatorial positions. A hybridization scheme for this linear ion is sp^3d. **Conceptual Exercises: 10.1.** The F—N—O bond angle in NOF (110°) is smaller than in NO_2F (118°) because of the strong repulsion of the bond pair electrons by the lone pair electrons of the central N atom in NOF. **10.2 a.** The distribution of the three H atoms and the O atom around the C atom of CH_3OH is tetrahedral, and the C—O—H bond is bent, with a bond angle of about 109°. **b.** The hybridization of both the C and O atoms is sp^3.

c. 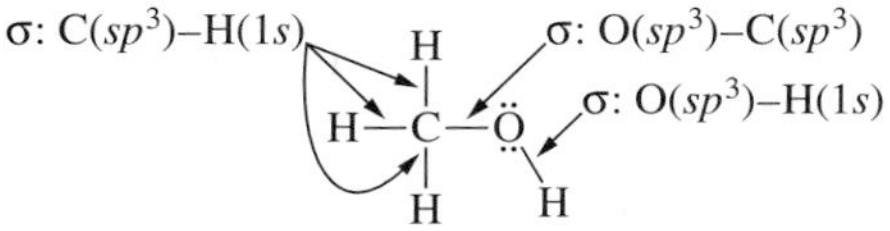

Review Questions: 6 a. AX_3E; **b.** AXE_3; **c.** AX_2E. **7 a.** 180°; **b.** 120°; **c.** 109.5°. **8.** Yes, if the molecule is diatomic—for example, HCl, which has the VSEPR notation AXE_3. **11.** No, a bond between two atoms of the same element is nonpolar, as is also a bond between two different atoms with the same electronegativity, such as C and S. **12.** No, a molecule with polar bonds may be nonpolar if the bond dipole moments cancel out, as in CCl_4. **13.** SO_2 has a bent shape and polar bonds; it is a polar molecule. The dipole moments in SO_3 cancel out because SO_3 has a trigonal planar shape. **18.** trigonal planar: sp^2; octahedral: sp^3d^2 (or d^2sp^3). **19 a.** tetrahedral: sp^3; **b.** linear: *sp*. **23.** Yes, all the bonds in an alkane molecule are single bonds and are thus σ bonds. **24.** No, an alkene hydrocarbon has one or more C-to-C double bonds, each consisting of a σ bond and a π bond. All C-to-C single bonds in the alkene hydrocarbon are σ bonds, as are all C-to-H bonds. **Problems: 29 a.** A linear molecule is improbable for H_2O, a triatomic molecule with a tetrahedral electron-group geometry around the O atom. **b.** A planar molecule is probable; SO_3 has trigonal planar electron-group geometry around the S atom. **c.** A planar molecule is not probable; PH_3 has a tetrahedral electron-group geometry. **31.** BF_3 has three electron groups, all bonding pairs; BF_3 is trigonal planar. ClF_3 has three bonding pairs and two lone pairs of electrons. The electron-group geometry is trigonal bipyramidal; the molecular geometry is T-shaped. **33 a.** pyramidal; **b.** tetrahedral; **c.** square planar; **d.** linear. **35 a.** The structure is "bent" at the O—O—H bonds, and the overall shape is a nonplanar "zigzag." **b.** The O—C—N portion is linear but the H—O—C portion is bent. **c.** The O—N—O portion is trigonal planar, but the O—H bond is bent out of the plane. **37 a.** sp^3; **b.** sp^3; **c.** *sp*; **d.** sp^2. **39 a.** A linear molecule; both C atoms are *sp* hybridized. **b.** The N—C—O portion is linear; the C atom is *sp* hybridized. The H—N—C portion is trigonal planar; the N atom is sp^2 hybridized. **c.** The NH_2—O portion is pyramidal; the N atom is sp^3 hybridized. The N—O—H portion is bent; the O atom is sp^3 hybridized. **d.** The CH_3—C portion is tetrahedral; the first C atom is sp^3 hybridized. The C—CO—O portion is trigonal planar; the second C atom is sp^2 hybridized. The C—O—H portion is bent; the central O atom is sp^3 hybridized.

41 a. 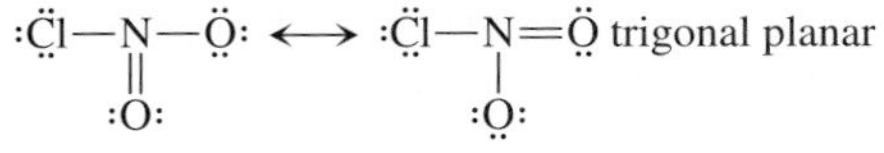

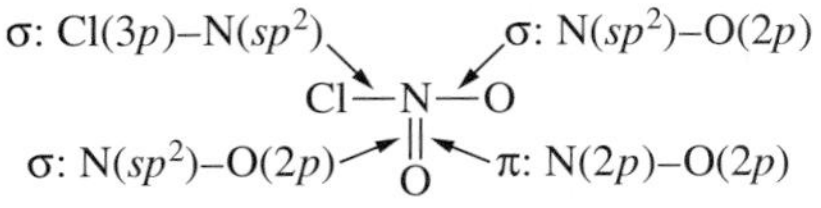

b. 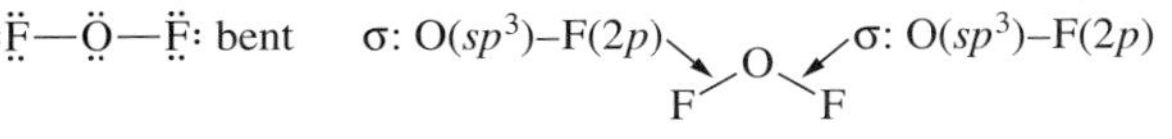

c. 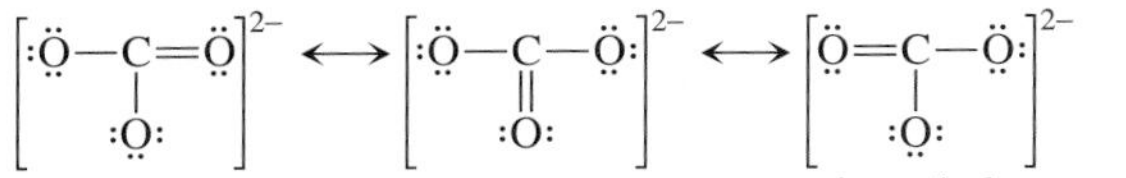

trigonal planar

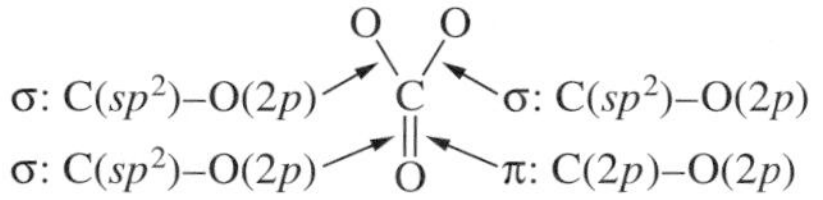

43. The electron-group geometry in the ICl_2^- ion is that of AX_2E_3, trigonal bipyramidal (the ion is linear). The hybridization scheme is sp^3d. The electron-group geometry in the ICl_2^- ion is that of AX_2E_2 (the ion is bent). The hybridization scheme is sp^3. **45.** The bonding scheme is

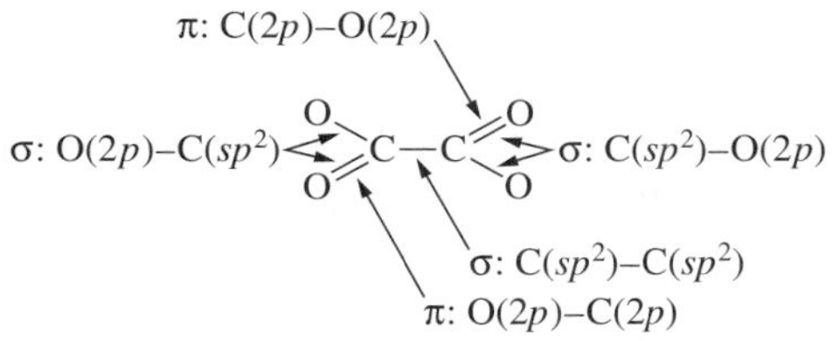

47. (a) and (c) can exist as cis and trans isomers. **49.** The molecular orbital occupancy in He_2 is $\sigma_{1s}^2\sigma_{1s}^{*2}$. The bond order, which is one-half the difference between the number of bonding and antibonding electrons, is zero. There can be no bond between the two atoms. **51.** No, although the statement is correct for diatomic molecules, for molecules with more than two atoms the dipole moments of different bonds may cancel each other out, resulting in a nonpolar molecule. **53.** seesaw shape. **55.** :N≡N: An *sp* hybridization scheme for the N atoms is consistent with this Lewis structure. Alternatively, the σ bond could be described through the overlap of p_x orbitals and the π bonds through the parallel overlap of p_y orbitals and of p_z orbitals. **57.** :O=C=C=C=O: According to VSEPR theory, the molecule should be linear, and this corresponds to *sp* hybridization of all three C atoms.

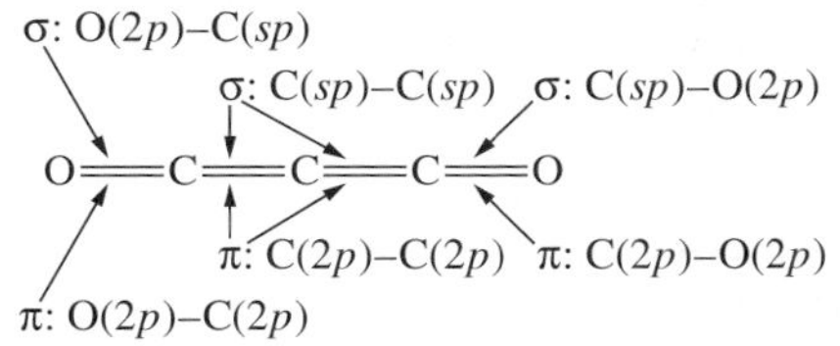

59. With one electron excited to a higher energy bonding orbital, the bond order = 1 and the electronically excited He_2 molecule could exist.

Chapter 11

Exercises: 11.1. ΔH_{vapn} = 34.0 kJ/mol. **11.2.** 5.33×10^{-3} g H_2O. **11.3.** IBr has the greater molecular mass and thus the greater London forces. The E.N. differences are small, and dipole–dipole attractions are not of great significance. IBr is a solid, and BrCl is a gas. **11.4.** NH_3, yes; CH_4, no; C_6H_5OH, yes; $CH_3C(=O)OH$, yes; H_2S, no; H_2O_2, yes. **11.5.** CsBr < KI < KCl < MgF_2. **11.6.** corner atoms: $\frac{1}{8} \times 8 = 1$; face atoms: $\frac{1}{2} \times 6 = 3$; 1 + 3 = 4 atoms per unit cell. **11.7** The Na^+ ion in the center of the unit cell is in contact with six face-centered Cl^- ions. Each Cl^- ion is in contact with two Na^+ ions at the centers of unit cells and four Na^+ ions on the edges of the unit cell. The coordination number is six, both for Na^+ and for Cl^-. A unit cell contains 1 center Na^+ and ($\frac{1}{4} \times 12$) edge Na^+, for a total of 4 Na^+. The unit cell also contains ($\frac{1}{8} \times 8$) corner Cl^- and ($\frac{1}{2} \times 6$) face-centered Cl^-, for a total of 4 Cl^-. The Na^+-to-Cl^- ratio is 4:4 = 1:1, the same ratio as indicated in the formula NaCl. **Estimation Exercises: 11.1.** H_2O, with a large value of ΔH_{vapn} and a low molar mass, requires the greatest amount of heat; CCl_4, with a relatively low value of ΔH_{vapn} and a large molar mass, requires the least. **11.2.** Approximate normal boiling points: CS_2, 47 °C; CH_3OH, 65 °C; CH_3CH_2OH, 78 °C; H_2O, 100 °C. **11.3.** Hg, with the smallest value of ΔH_{fusion} and the largest molar mass, requires the least heat; water, with an intermediate value of ΔH_{fusion} and by far the smallest molar mass, requires the most. **11.4.** ΔH_{vapn} = 34.3 kJ/mol. **Conceptual Exercises: 11.1.** Methane has a T_c = −82.4 °C. Room temperature is above T_c, so methane is present only as a gas. The pressure gauge method would work fine. **11.2.** If the H_2O were present only as a gas, its pressure would greatly exceed the vapor pressure at 30.0 °C. Some of the gas condenses to liquid, and the final condition is one of liquid and vapor in equilibrium. **Review Questions: 7.** Each area represents one phase. The curves represent two phases in equilibrium. The triple point represents three phases in equilibrium. **12.** Vapor pressure is affected by temperature (a), but not by the volume of liquid (b), the volume of vapor (c), or the area of contact between liquid and vapor (d), so long as both liquid and vapor are present. **18.** The polar substance will have the higher boiling point because of the presence of both dispersion *and* dipole–dipole forces. **19.** CH_4 has only weak dispersion forces as intermolecular forces. H_2O has dispersion forces, dipole–dipole forces, and hydrogen bonds to hold the molecules together and in the liquid state. **20.** Methane and ethane have only dispersion forces as intermolecular forces, whereas methanol and ethanol have dipole–dipole attractions and hydrogen bonds as well. **21.** Hexanoic acid has a higher molecular weight, but even more important are the dipole–dipole forces and hydrogen bonds that are present in hexanoic acid and absent in 2,2-dimethylbutane. **22.** C_6H_5COOH has a high molecular weight and dipole–dipole attractions, hydrogen bonds, and dispersion forces as intermolecular forces. It is likely to be the solid. $CH_3(CH_2)_8CH_3$ has a comparable molecular weight to C_6H_5COOH but has only dispersion forces; it is likely to be a liquid. CH_3OH has dipole–dipole forces and hydrogen bonds but a low molecular weight. The intermolecular forces are strong enough to keep it as a liquid at STP, but not a solid. $(CH_3CH_2)_2O$ has a higher molecular weight than CH_3OH but lacks hydrogen bonds; it should also be a liquid. **23.** BF_3 is the gas at STP. It has a moderately low molecular weight and is a nonpolar substance. NI_3 has a very high molecular weight and is a solid. PCl_3 has a relatively high molecular weight compared to BF_3, but not as high as NI_3. We might expect it to be a liquid. CH_3COOH has a molecular weight comparable to BF_3, but because of its polar character and ability to form hydrogen bonds, we expect it to be a liquid. **28.** The larger the ionic charge and the smaller the ionic size, the larger will be the lattice energy; and the larger the lattice energy, the higher will be the melting point. **Problems: 33.** ΔH_{total} = −48.5 kJ. **35 a.** about 430 mmHg; **b.** about 76 °C. **37.** The lowest temperature is the triple point (about 0 °C) temperature. The water must be placed in a vacuum chamber and the pressure reduced until the water boils and the temperature falls to the triple point value. (At this temperature the water will begin to freeze.) **39.** 1.29×10^{21} Hg atoms. **41.** P = 287 mmHg. **43.** The calculated pressure of $H_2O(g)$ (3.31 mmHg) is slightly less than the triple point pressure of water (4.58 mmHg). No vapor condenses to liquid or solid. The system is one of gaseous water only. **45.** 1.18×10^3 kJ. **47.** 2.77×10^3 kJ. **49.** 8.4 °C. **51.** The enthalpy change should be the same for the two-step process as for the direct sublimation. Enthalpy is a state function, so its value is independent of path. **53.** The vapor pressure curve extends from 3.4 mmHg at 2.0 °C (the triple point) to 145 atm at 380 °C (the critical point). The sublimation curve extends from 0 mmHg at the lowest temperature to 3.4 mmHg at 2.0 °C. The fusion curve is essentially a vertical straight line above the triple point. The area *above* the sublimation curve and to the *left* of the fusion curve represents the solid region. The liquid region lies *above* the vapor pressure curve and roughly between the temperatures of 2.0 °C and 380 °C. The area *below* the sublimation and vapor pressure curves, and at pressures below 145 atm at temperatures beyond 380 °C, represents the gaseous region. The region with pressures above 145 atm and temperatures above 380 °C may be referred to as the supercritical fluid region. **55.** $CO_2(s)$ melts to $CO_2(l)$ at a temperature somewhat above −56.7 °C (the triple point is at 5.1 atm and −56.7 °C). $CO_2(l)$ then converts to $CO_2(g)$ at a temperature that is probably below room temperature. (From Table 11.3, T_c = 304.2 K and P_c = 72.9 atm.) **57.** PF_3 is somewhat more polar than PI_3, but the molecular weight of PI_3 is much greater than that of PF_3. Dispersion forces predominate as intermolecular forces, and PF_3 should have the lower boiling point. **59.** 2,2-Dimethyl-1-butanol has a higher molecular weight and therefore stronger dispersion forces than hexane. Also, it can form hydrogen bonds, so its boiling point is higher than that of hexane. **61.** The molecular weights are comparable, but the compactness of the 3,3-dimethylpentane molecule and possibility of hydrogen bonding in 1-pentanol account for the higher melting point of 1-pentanol. **63.** $CH_4 < CH_3CH_3 < NH_3 < H_2O$. CH_4 and CH_3CH_3 are both nonpolar, but CH_4 has the lower molecular weight. Despite their lower molecular weights, both NH_3 and H_2O have higher boiling points than CH_3CH_3. The molecular weights of NH_3 and H_2O are comparable, but stronger hydrogen bonding in H_2O gives it a higher boiling point than NH_3. **65.** $CH_3OH < C_6H_5OH < NaOH < LiOH$. Both CH_3OH and C_6H_5OH involve hydrogen bonding, but the higher molecular weight of C_6H_5OH gives it the higher melting point. Both ionic compounds have higher melting points than the covalent ones, but the melting point of LiOH is greater than that of NaOH because the Li^+ ion is smaller than the Na^+ ion. **67.** Ethylene glycol can form more hydrogen bonds than isopropyl alcohol, so ethylene glycol has stronger intermolecular forces and a higher surface tension. **69.** In Figure 11.33, replace Na^+ by Mg^{2+} and Cl^- by O^{2-}. The length of the unit cell is twice the radius of a Mg^{2+} ion plus twice the radius of an O^{2-} ion: (2 × 65 pm) + (2 × 140 pm) = 410 pm. **71.** The surface tension of water must be reduced so that the water molecules will have less affinity for each other and a greater affinity for the surface to be wet. **73.** 96 mmHg. **75 a.** 152 mmHg; **b.** bp = 68.8 °C. **77.** The total pressure is the sum of the partial pressure of unreacted $O_2(g)$ and that of water vapor: 381 mmHg + 23.8 mmHg = 405 mmHg. **79.** 27 °C. **81.** If the CO_2 were present as a gas only, its pressure would be 73.0 atm. Because this pressure exceeds the vapor pressure of $CO_2(l)$ at 25 °C, some vapor condenses and the fire extinguisher contains both liquid and gaseous CO_2. **83.** $V = 6.59 \times 10^{-23}$ cm^3. **85.** simple cubic: 47.64% voids; fcc: 25.95% voids.

Chapter 12

Exercises: 12.1. 0.111 M glucose; $[C_6H_5O_7^{3-}] = 0.0112$ M; $[K^+] = 0.0201$ M; $[Cl^-] = 0.0800$ M; $[Na^+] = 0.0935$ M. **12.2 a.** $Ca(OH)_2(aq) + 2\,HCl(aq) \longrightarrow CaCl_2(aq) + 2\,H_2O(l)$; **b.** $Ca^{2+}(aq) + 2\,OH^-(aq) + 2\,H^+(aq) + 2\,Cl^-(aq) \longrightarrow Ca^{2+}(aq) + 2\,Cl^-(aq) + 2\,H_2O(l)$; **c.** $OH^-(aq) + H^+(aq) \longrightarrow H_2O(l)$. **12.3.** 12.6 mL. **12.4 a.** $Mg^{2+}(aq) + 2\,OH^-(aq) \longrightarrow Mg(OH)_2(s)$; **b.** $2\,Fe^{3+}(aq) + 3\,S^{2-}(aq) \longrightarrow Fe_2S_3(s)$; **c.** $CaCl_2$ and Na_2CO_3 are soluble, and $CaCO_3$ is insoluble; there is no reaction. **12.5 a.** Al: +3, O: −2; **b.** P: 0; **c.** Na: +1, Mn: +7, O: −2; **d.** Cl: +1, O: −2; **e.** H: +1, As: +5, O: −2; **f.** H: +1, Sb: +5, F: −1; **g.** Cs: +1, O: $-\frac{1}{2}$; **h.** C: −2, H: +1, F: −1; **i.** C: +2, H: +1, Cl: −1; **j.** C: 0, H: +1, O: −2. **12.6.** Yes, I_2 is oxidized and Cl_2 is reduced. **12.7.** $2\,MnO_4^- + 5\,C_2O_4^{2-} + 16\,H^+ \longrightarrow 2\,Mn^{2+} + 10\,CO_2 + 8\,H_2O$. **12.8.** $2\,CN^- + 5\,OCl^- + 2\,OH^- \longrightarrow 2\,CO_3^{2-} + N_2 + 5\,Cl^- + H_2O$. **12.9.** $2\,Fe(OH)_3 + 3\,OCl^- + 4\,OH^- \longrightarrow 2\,FeO_4^{2-} + 3\,Cl^- + 5\,H_2O$. **12.10.** 22.04 mL. **Conceptual Exercises: 12.1.** The CH_3NH_2 will cause a dimly lit bulb, as it is a weak base. HNO_3 is a strong acid and will cause a brightly lit bulb. The combination will produce a strong electrolyte, $CH_3NH_3^+NO_3^-$ (analogous to NH_4NO_3), and a brightly lit bulb. **12.2.** The acid will cause the $Fe(OH)_3$ to dissolve because the H^+ will react with the OH^- to form H_2O; $Fe(OH)_3(s) + 3\,H^+(aq) \longrightarrow Fe^{3+}(aq) + 3\,H_2O(l)$. **12.3.** $Cr_2O_7^{2-}(aq)$ is an oxidizing agent; it will react with a reducing agent. $HNO_3(aq)$ is also an oxidizing agent. There is no reaction between $Cr_2O_7^{2-}(aq)$ and $HNO_3(aq)$. HCl is a reducing agent (with Cl in the oxidation state −1). It is oxidized by $Cr_2O_7^{2-}(aq)$, probably to $Cl_2(g)$. **Review Questions: 1.** strong electrolytes: (b), (c), (e), (g); weak electrolytes: (d), (f); nonelectrolytes: (a), (h). **2.** strong acid: (e); weak acid: (d); strong base: (b), (h); weak base: (f); salt: (a), (c), (g). **3.** highest $[NO_3^-]$, (c); lowest $[NO_3^-]$, (b). **4.** $[Al^{3+}] > 0.0030$ M in (a) and (c). **5.** (d). **6.** 0.10 M NaCl. All the solutions are 0.10 M, but only NaCl is a strong electrolyte. **7.** 0.10 M H_2SO_4. Ionization is complete in the first step and partial in the second step, with the result that $[H^+] >$ 0.10 M. **10.** $Ca(OH)_2(s) + 2\,H^+(aq) \longrightarrow Ca^{2+}(aq) + 2\,H_2O(l)$. **11.** $CaCO_3(s) + 2\,CH_3COOH(aq) \longrightarrow Ca^{2+}(aq) + 2\,CH_3COO^-(aq) + H_2O(l) + CO_2(g)$. **12.** $HCO_3^-(aq) + HCOOH(aq) \longrightarrow HCOO^-(aq) + H_2O(l) + CO_2(g)$. **13.** The solution contains one or a combination of the ions Ag^+, Pb^{2+}, Hg_2^{2+}. **15.** Na_2CO_3. **19.** $Cu^{2+}(aq) + 2\,OH^-(aq) \longrightarrow Cu(OH)_2(s)$. **20.** $2\,Fe^{3+}(aq) + 3\,S^{2-}(aq) \longrightarrow Fe_2S_3(s)$. **21.** H is usually +1, O is usually −2. Hydrides have H as −1. Peroxides have O as −1, superoxides as $-\frac{1}{2}$. **25 a.** 0; **b.** +3; **c.** −2; **d.** +6; **e.** −3. **f.** +4; **g.** +4; **h.** +5; **i.** +2.5; **j.** −1. **26 a.** −3; **b.** +2; **c.** 0; **d.** −2; **e.** −2. **Problems: 29 a.** $[Li^+] = 0.0385$ M; **b.** $[Cl^-] =$ 0.070 M; **c.** $[Al^{3+}] = 0.0224$ M; **d.** $[Na^+] = 0.24$ M. **31.** $[NO_3^-] = 6.99 \times 10^{-3}$ M. **33.** $[Na^+] = 0.0844$ M, $[Cl^-] = 0.0554$ M, $[SO_4^{2-}] = 0.0145$ M. **35.** 67.5 mL. **37 a.** $HI(aq) \longrightarrow H^+(aq) + I^-(aq)$; **b.** $CH_3CH_2COOH(aq) \rightleftharpoons CH_3CH_2COO^-(aq) + H^+(aq)$; **c.** $HNO_2(aq) \rightleftharpoons H^+(aq) + NO_2^-(aq)$; **d.** $H_2PO_4^-(aq) \rightleftharpoons H^+(aq) + HPO_4^{2-}(aq)$. **39 a.** 46.8 mL; **b.** 11.9 mL; **c.** 23.1 mL. **41.** 0.8051 M. **43.** 3.43×10^4 L CO_2. **45.** $RbOH(aq) + HCl(aq) \longrightarrow RbCl(aq) + H_2O(l)$; $Rb^+(aq) + OH^-(aq) + H^+(aq) + Cl^-(aq) \longrightarrow Rb^+(aq) + Cl^-(aq) + H_2O(l)$; $OH^-(aq) + H^+(aq) \longrightarrow H_2O(l)$. **47 a.** $2\,I^-(aq) + Pb^{2+}(aq) \longrightarrow PbI_2(s)$; **b.** no reaction; **c.** $Cr^{3+}(aq) + 3\,OH^-(aq) \longrightarrow Cr(OH)_3(s)$; **d.** no reaction; **e.** $OH^- + H^+ \longrightarrow H_2O(l)$; **f.** $HSO_4^- + OH^- \longrightarrow H_2O(l) + SO_4^{2-}$. **49 a.** $MgO(s) + 2\,H^+(aq) \longrightarrow Mg^+(aq) + H_2O(l)$; **b.** $HCOOH(aq) + NH_3(aq) \longrightarrow NH_{4+}(aq) + HCOO^-(aq)$; **c.** no reaction; **d.** $Cu^{2+}(aq) + CO_3^{2-}(aq) \longrightarrow CuCO_3(s)$; **e.** no reaction. **51 a.** $ClO_2 + H_2O \longrightarrow ClO_3^- + 2\,H^+ + e^-$, oxidation; **b.** $MnO_4^- + 4\,H^+ + 3\,e^- \longrightarrow MnO_2 + 2\,H_2O$, reduction; **c.** $2\,BrO^- + 2\,H_2O + 2\,e^- \longrightarrow Br_2 + 4\,OH^-$, reduction; **d.** $SbH_3 + 3\,OH^- \longrightarrow Sb + 3\,H_2O + 3\,e^-$, oxidation. **53 a.** $3\,Ag + NO_3^- + 4\,H^+ \longrightarrow 3\,Ag^+ + NO(g) + 2\,H_2O$; **b.** $2\,MnO_4^- + 5\,H_2O_2 + 6\,H^+ \longrightarrow 5\,O_2 + 2\,Mn^{2+} + 8\,H_2O$; **c.** not possible; two oxidations. **55 a.** $3\,CN^- + BrO_3^- \longrightarrow 3\,OCN^- + Br^-$; **b.** $S_8 + 12\,OH^- \longrightarrow 2\,S_2O_3^{2-} + 4\,S^{2-} + 6\,H_2O$. **57 a.** $2\,NO + 5\,H_2 \longrightarrow 2\,NH_3 + 2\,H_2O$; **b.** $8\,Fe_2S_3 + 12\,O_2 + 24\,H_2O \longrightarrow 16\,Fe(OH)_3 + 3\,S_8$. **59 a.** $2\,MnO_4^- + 5\,C_2H_2O_4 + 6\,H^+ \longrightarrow 2\,Mn^{2+} + 10\,CO_2 + 8\,H_2O$; **b.** $2\,MnO_4^- + 3\,C_2H_4O + H_2O \longrightarrow 2\,MnO_2 + 3\,C_2H_4O_2 + 2\,OH^-$. **61 a.** oxidizing agents: NO_3^- (53a), MnO_4^- (53b), ClO_3^- (54a), O_2 (54c); reducing agents: Ag (53a), H_2O_2 (53b), Mn^{2+} (54a), S_8 (54c); no reaction: (53c, 54b). **63.** 0.1226 M Mn^{2+}. **65.** 9.26 g $Na_2C_2O_4$. **67.** NH_3 and $Ba(OH)_2$ are bases. CH_3CH_2COOH is a weak acid. HI is a strong acid, as is H_2SO_4 in its first ionization. Because H_2SO_4 ionizes further in a second step, it produces the highest $[H^+]$. **69.** The solution with 45.5 ppm K^+ has the higher molarity of K^+. **71.** Add 6.4 mL water. **73 a.** $H_2SO_4(aq)$: NaCl(s) dissolves and $BaCl_2(s)$ is converted to $BaSO_4(s)$. **b.** Water: $Na_2CO_3(s)$ dissolves and $MgCO_3(s)$ does not. **c.** HCl(aq): $KNO_3(s)$ dissolves and $AgNO_3(s)$ is converted to AgCl(s). **d.** HCl(aq): $CuCO_3$ reacts with HCl(aq), producing $Cu^{2+}(aq)$; $PbSO_4(s)$ does not react with HCl(aq). **e.** HCl(aq): $Mg(OH)_2(s)$ is neutralized, yielding $Mg^{2+}(aq)$; $BaSO_4(s)$ does not react with HCl(aq). **75.** 15.3 g NaCl.

77.

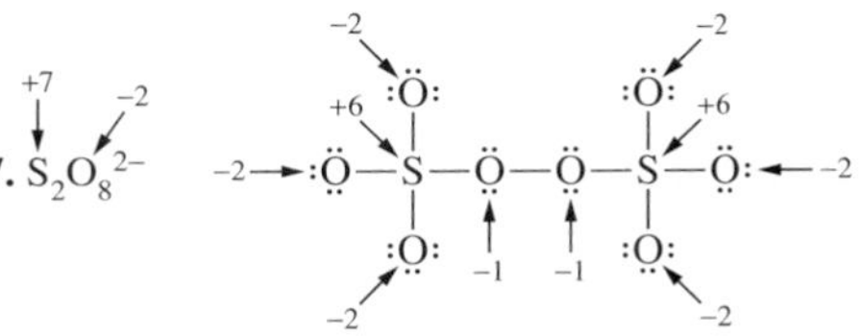

79. $2\,P_4 + 12\,H_2O \longrightarrow 5\,PH_3 + 3\,H_3PO_4$. **81 a.** $I_2 + 5\,H_5IO_6 \longrightarrow 7\,IO_3^- + 9\,H_2O + 7\,H^+$; **b.** $24\,SCl_2 + 64\,NH_3 \longrightarrow S_8 + 4\,S_4N_4 + 48\,NH_4Cl$; **c.** $2\,XeF_6 + 16\,OH^- \longrightarrow Xe + XeO_6^{4-} + 12\,F^- + O_2 + 8\,H_2O$; **d.** $S_4N_4 + 3\,H_2O + 6\,OH^- \longrightarrow S_2O_3^{2-} + 2\,SO_3^{2-} + 4\,NH_3$. **83.** $2\,C_{12}H_4Cl_6 + 23\,O_2 + 2\,H_2O \longrightarrow 24\,CO_2 + 12\,HCl$. **85.** 0.02401 M $KMnO_4$.

Chapter 13

Exercises: 13.1. 17.8% glucose by mass. **13.2.** 80 mL ethanol in 120 mL of water. The assumption is that the volumes will be exactly additive. **13.3 a.** 0.1 ppb; **b.** 100 ppt. **13.4.** 1.30 m. **13.5 a.** 8.50% CH_3OH by mass; **b.** 2.61 M; **c.** 4.97%. **13.6.** heptane < 1-octanol < pentanoic acid < acetic acid. **13.7.** 5.4×10^{-2} mg CO_2/100 g H_2O. **13.8.** 17.2 mmHg. **13.9.** Benzene has a lower molar mass than toluene. A solution with equal masses of benzene and toluene has a mole fraction of benzene that is greater than 50%. The vapor in equilibrium with this solution is the one richer in benzene. **13.10.** 17.6 g sucrose. **13.11.** $C_6H_6O_4$. **13.12.** molar mass = 6.87×10^4 g/mol. **Estimation Exercises: 13.1.** (a). This solution has 15.5 g solute per 100.0 g solution; the others have 15.5 g solute in 115.5 g solution. **13.2.** (b). Solution (b) has 25.0% methanol by mass. Solution (a) has 25.0 mL ethanol (25.0 mL × 0.789 g/mL) in 100.0 mL of solution (100.0 mL × 0.968 g/mL); this is only about 20 g solute per 100 g solution—20% ethanol by mass. **13.3.** (c). Solution (c) has a mole fraction of 0.10. Solution (a) contains 1 mol CH_3OH per 1000 g H_2O (55.5 mol H_2O), a mole fraction of CH_3OH much less than 0.10. Solution (b) has about 0.3 mol CH_3OH (10.0 g) in 90.0 g H_2O (5 mol); the mole fraction of CH_3OH is again less than 0.10. **13.4.** 0.022 m. **Conceptual Exercises: 13.1.** No, the net transfer of H_2O through the vapor will cease when the two solutions have attained equal concentrations. **13.2.** lowest freezing point: 0.008 M HCl; highest freezing point: 0.10 m $CO(NH_2)_2$. **Review Questions: 8.** 1 ppt < 1 ppb < 1 ppm < 1 mg/dL < 1%. **12 a.** $CHCl_3$ is slightly soluble in water. **b.** C_6H_5COOH is slightly soluble in water. **c.** $CH_3CHOHCH_2OH$ is highly soluble in water. **13 c.** Phenol is soluble both in water (because of hydrogen bonds as intermolecular forces) and in benzene (because of the structural similarities of C_6H_5OH and C_6H_6). **24.** Because NaCl produces two ions per formula unit, 0.16 M NaCl produces about the same number of particles in solution as 0.31 M glucose, making both solutions isotonic. **25 a.** The cells would shrink. **b.** The cells would swell. **26 a.** 0.10 M $NaHCO_3$; **b.** 1 M NaCl; **c.** 1 M $CaCl_2$; **d.** 3 M glucose. **27.** CH_3OH < CH_3COOH < NaCl < $MgBr_2$ < $Al_2(SO_4)_3$. **Problems: 29.** Pipet 40.0 mL of acetic acid into a 2.00-L volumetric flask. Fill with water. **31 a.** 3.96% by mass; **b.** 7.31% by mass; **c.** 8.21% by mass. **33 a.** 4.61% by volume; **b.** 0.811% by volume; **c.** 1.02% by volume. **35.** 1×10^2 mg/dL. **37 a.** 1 ppb benzene; **b.** 35 ppm NaCl; **c.** 1.3×10^{-4} M F^-. **39.** 1.25 m. **41.** 12.0 M; 30.6 m. **43.** 25% H_2SO_4 by mass. **45.** Water has a greater density and a smaller molar mass than methanol. Water as solute in methanol as solvent will yield the greater molality. **47.** $\chi = 0.0434$. **49.** 486 g. **51.** $\chi = 0.010$ has the greater mass percent urea. **53 a.** 55.8% by mass; **b.** 3.0 m; **c.** 0.062 g $Pb(NO_3)_2$ can be added per g solution. **55.** Exceptions are a few compounds that are more soluble at lower temperatures. **57 a.** 2.55 mg air; **b.** 1.97 mL. **59.** P_T = 10.3 mmHg; P_B = 60.8 mmHg; $\chi_T = 0.145$; $\chi_B = 0.855$. **61.** When the solute is nonvolatile, the vapor pressure of the solvent is lowered, and the boiling point is higher for the solution than for the pure solvent. When the solute is volatile, the vapor pressure of the solution may be greater than that of the pure solvent, in which case the boiling point is reduced. **63 a.** $t_f = -0.04$ °C; **b.** 0.328 m; **c.** 1.28 m. **65.** The 0.08 M is more concentrated and thus will have a higher boiling point. **67.** 265 g/mol. **69.** Since the cucumbers shrivel up, water leaves the cucumbers; the salt solution must have the higher osmotic pressure. **71.** $\pi = 9.94$ atm; the CH_3CH_2OH solution is hypertonic. **73.** Yes, HCl ionizes in water; $i \approx 2$, and $\Delta T_f \approx 2 \times 1.86 \times 0.01 \approx 0.04$ °C. HCl does not ionize in benzene, $i = 1$, and $\Delta T_f = 1 \times 5.12 \times 0.01 \approx 0.05$ °C. **75.** 1.2×10^2 g NaCl. **77.** (b) > (a) > (e) > (c) > (d). **79.** 6 m^2. **81 a.** 15.1% by volume; **b.** 12.2% by mass; **c.** 11.9% mass/volume; **d.** 7.25 mole percent. **83.** $CO_2(g)$ reacts with water to form $H_2CO_3(aq)$, which is neutralized by the NaOH(aq) to produce $Na_2CO_3(aq)$. The $CO_2(g)$ is highly soluble in NaOH(aq) because of this reaction. **85.** 1.9 mL air (STP)/100 g H_2O. **87.** Vitamin E has considerable hydrocarbon character in which the intermolecular forces are mostly dispersion forces; it should be insoluble in water but soluble in fats, where intermolecular forces are also of the dispersion type. Because of its OH and NH groups, vitamin B_2 is able to form hydrogen bonds with H_2O molecules and is, therefore, soluble in water. **89.** $\chi_{solute} = n_{solute}/(n_{solute} + n_{solvent})$. In a dilute solution, $n_{solvent} >> n_{solute}$, and $\chi_{solute} \approx n_{solute}/n_{solvent}$. Because $n_{solvent} \propto$ kg solvent, $\chi_{solute} \propto n_{solute}$/kg solvent $\propto$ molality. **91.** $T_f = -0.59$ °C; This is close to $T_f = -0.52$ °C for isotonic NaCl(aq). **93.** $C_6H_{12}O_3$.

Chapter 14

Exercises: 14.1. 6.74 mmHg. **14.2 a.** $NH_3(aq) + HNO_3(aq) \longrightarrow NH_4NO_3(aq)$; **b.** $2\ NH_3(aq) + H_2SO_4(aq) \longrightarrow (NH_4)_2SO_4(aq)$. **Estimation Exercises: 14.1.** The formula weights, and hence the mass percent N, are almost the same for $(NH_4)_2HPO_4$ and $(NH_4)_2SO_4$. Because the sum of the atomic weights of H and P is just slightly less than the atomic weight of S, $(NH_4)_2HPO_4$ has the smaller formula weight and a slightly greater mass percent N. **14.2.** 200 particles/cm^3. **Conceptual Exercise 14.1.** The warehouse would contain about 56 mol CO_2 in about 2.3×10^6 mol air—less than 35 ppm. **Review Questions: 10 a.** anhydrous ammonia; **b.** carbon monoxide, nitrogen monoxide, and carbon dioxide; **c.** methane, ozone, nitrous oxide, CFCs; **d.** radon; **e.** nitric acid or sulfuric acid; **f.** nitrogen, helium, and argon; **g.** helium and oxygen. **11 a.** magnesium nitride; **b.** N_2O_4; **c.** potassium peroxide; **d.** KO_2; **e.** NH_2CONH_2. **22 a.** NO is converted to NO_2, which initiates the other smog-forming reactions. **b.** Carbon monoxide is not involved in the formation of photochemical smog. **c.** Hydrocarbon vapors react with oxygen atoms to make free radicals, ozone, and PAN. **d.** Sulfur oxide is not important in photochemical smog. It is important in industrial smog. **Problems: 29.** argon, 9.34×10^3 ppm; neon, 18.18 ppm; helium, 5.24 ppm; krypton, 1.14 ppm; xenon, 0.09 ppm. **31.** 44.1%. **33.** 25 °C. The relative humidity—the ratio of the absolute humidity (18 mmHg) to the vapor pressure of water—will be greater, the lower the water vapor pressure. And the lower the temperature, the lower the vapor pressure of water. **35.** The absolute humidity of air above a kettle of boiling water is high—the air is nearly saturated in water vapor. The temperature must be reduced only slightly to reach the dew point. The absolute humidity in expired air is much less, and the temperature has to be lowered much further to reach the dew point. **37.** NH_4Cl < NO < N_2O < NH_3. **39 a.** $2\ C_8H_{18} + 25\ O_2 \longrightarrow 16\ CO_2 + 18\ H_2O$; **b.** $2\ CH_4 + 3\ O_2 \longrightarrow 2\ CO + 4\ H_2O$; **c.** $2\ CO + O_2 \xrightarrow{Pt} 2\ CO_2$.

41. Haber–Bosch process: $N_2(g) + 3H_2(g) \xrightarrow{\text{catalyst}} 2NH_3(g)$

followed by the Ostwald process: $4NH_3(g) + 5O_2(g) \xrightarrow{\text{Pt/Rh}}$

$4\ NO(g) + 6\ H_2O(g)$; $2\ NO(g) + O_2(g) \longrightarrow 2\ NO_2(g)$; $3\ NO_2(g) + H_2O(l) \longrightarrow 2\ HNO_3(aq) + NO(g)$ followed by $NH_3(aq) + HNO_3(aq) \longrightarrow NH_4NO_3(aq)$. **43.** 37.7 kg. **45 a.** $4\ Al + 3\ O_2(g) \longrightarrow 2\ Al_2O_3$; **b.** $2\ KClO_3 \longrightarrow 2\ KCl + 3\ O_2$; **c.** $2\ H_2O + 2\ Na_2O_2 \longrightarrow 4\ NaOH + O_2$. **47.** The pressure will increase because 3 mol $O_2(g)$ is produced for every 2 mol $CO_2(g)$ consumed. **49.** Combustion of a hydrocarbon is an oxidation–reduction reaction in which the oxidation state of carbon atoms can increase to either +2 (CO) or +4 (CO_2). The reaction of an acid with a metal carbonate is an acid– base reaction. The oxidation state of C is +4 in the metal carbonate and remains the same in CO_2. A reduction would be required to produce CO, but there is no accompanying oxidation. **51.** $2\ C_6H_{14} + 19\ O_2 \longrightarrow 12\ CO_2 + 14\ H_2O$; it is impossible to write a definitive equation for the incomplete combustion because one cannot know the ratio of CO to CO_2 produced at any one time. The ratio can even change during the reaction. **53 a.** 7.22 metric tons CH_4; **b.** 6.42 metric tons C_8H_{18}; **c.** 5.74 metric tons coal. **55 a.** $S(s) + O_2(g) \longrightarrow SO_2(g)$; **b.** $2\ ZnS(s) + 3\ O_2(g) \longrightarrow 2\ ZnO(s) + 2\ SO_2(g)$; **c.** $2\ SO_2(g) + O_2(g) \longrightarrow 2\ SO_3(g)$; **d.** $SO_3(g) + H_2O(l) \longrightarrow H_2SO_4(aq)$; **e.** $H_2SO_4(aq) + 2\ NH_3(aq) \longrightarrow (NH_4)_2SO_4(aq)$. **57.** ZnS, because it has a much higher % S by mass than the other materials. **59.** 600 times. **61.** Helium is formed from an alpha particle. There are many nuclear reactions that produce alpha particles. Argon is formed only when the potassium-40 isotope decays. **63.** 32.4%; $CaCl_2$ does take moisture out of the air, but only if the initial relative humidity is greater than 32.4%. **65.** % difference = 0.48%. **67.** No, the SO_2 level reached is 27.3 μg/m^3 air. **69.** 1.6% increase.

Chapter 15

Exercises: 15.1 a. $1.05 \times 10^{-5}\ M \cdot s^{-1}$; **b.** $3.15 \times 10^{-5}\ M \cdot s^{-1}$. **15.2 a.** $1.1 \times 10^{-3}\ M \cdot s^{-1}$; **b.** 0.287 M. **15.3.** rate = $113\ M \cdot s^{-1}$. **15.4.** t = 155 s. **15.5.** m_t = 7.83 g. **15.6.** 590 s. **15.7.** $t = 1.60 \times 10^5$ y. **15.8.** The claim is not authentic: the brandy is only about 12 y old. **15.9.** Use the equation that relates $t_{1/2}$ to $[A]_0$. For the first half-life period, $[A]_0 = 1/1.065 = 0.9390$ M, and $t_{1/2} = 4.63 \times 10^3$ min. For the second half-life period, $[A]_0 = \frac{1}{2} \times 0.9390$ M = 0.4695 M, and $t_{1/2} = 9.26 \times 10^3$ min. **15.10** 288 K (15 °C). **15.11 a.** $2\ NO + Cl_2 \longrightarrow 2\ NOCl$; **b.** $NOCl_2$; **c.** first step. **Estimation Exercises: 15.1 a.** $1.3 \times 10^{-3}\ M \cdot s^{-1}$; **b.** The instantaneous rate of reaction equals the average rate at about 260 s, and only at this one time during the reaction. **15.2.** 400 s. The answer lies between three (354 s) and four (472 s) half-lives. **15.3.** The time corresponds to 11.5 half-life periods. The activity should be between that for 12 half-life periods, 8.03×10^{10} atoms $\cdot\ s^{-1}$, and 11 half-life periods, 1.61×10^{11} atoms $\cdot\ s^{-1}$. **Conceptual Exercise: 15.1.** A graph of 1/[A] versus time, using points selected from Figure 15.8 (either Experiment 1 or 2), yields a straight line. This proves that the reaction is second order. **Review Questions: 3.** Only one O_2 molecule is formed for every two H_2O_2 molecules that react. The rate of formation of O_2 is only $\frac{1}{2}$ the rate of disappearance of H_2O_2. **8.** The reaction is second order. **12.** The isotope with a shorter half-life has a large decay constant, λ, and decays faster: $A = \lambda \cdot N$. **17 a.** Z; **b.** X; **c.** transition state (activated complex); **d.** activation energy for the forward reaction; **e.** activation energy for the reverse reaction. **30 a.** The rate doubles. **b.** The rate stays constant. **Problems: 33.** 0.2532 M. **35 a.** $3.1 \times 10^{-4}\ M \cdot s^{-1}$; **b.** $9.3 \times 10^{-4}\ M \cdot s^{-1}$. **37.** For the rate of reaction and the rate constant to have the same units, rate = k. This can occur only if a reaction is zero order overall. **39 a.** zero order in A, second order in B; **b.** second order overall; **c.** $k = 0.049\ M \cdot s^{-1}$. **41.** $3 \times 10^{-3}\ M \cdot s^{-1}$. **43.** 0.325 M. **45 a.** $5.87 \times 10^{-3}\ min^{-1}$; **b.** 118 min; **c.** 354 min. **47 a.** 134 mmHg; **b.** P_{total} = 1610 mmHg. **49.** Radioactive decay involves changes that occur within atomic nuclei. These are not chemical reactions and cannot be influenced by the factors that affect rates of chemical reactions, such as the presence of a catalyst. **51.** For less than a few hundred years old, there is not enough difference between the present activity and the activity of the object. After 50,000 years, there is so little activity left that it is almost impossible to get an accurate measurement. **53.** 2.44×10^4 y. **55.** More than 24 d but less than 32 d, and somewhat closer to 24 d: 10% = 0.10 is closer to $\frac{1}{8}$ (0.125) than to $\frac{1}{16}$ (0.0625). **57.** 0.0441 M. **59.** $[A]_t$ = 0.55 M. **61.** For a zero-order reaction the half-life gets longer as the initial concentration increases because the rate is constant and there are more molecules to react. For a second-order reaction, the rate of reaction increases as the square of the concentration of reactant. The half-life is inversely related to the concentration, and an increase in initial concentration means a shorter half-life. **63.** Increases in the average kinetic energies and collision frequencies of molecules amount to only a few percent for a 10 °C temperature increase. However, over this same temperature interval, the fraction of reactant molecules energetic enough to enter into a chemical reaction increases dramatically, perhaps doubling. **65.** The height of the peak representing the transition state is 171 kJ above the height of the reactants. The level of the products is 9.5 kJ below the level of the reactants. **67.** 165 kJ/mol. **69.** T_1 = 649 K. **71.** The rate equations for an elementary step and for an overall reaction are not likely to be the same unless (a) the reaction occurs in a single step or (b) the mechanism consists of a slow first step followed by a fast second step. If the rate-determining step is not the first step, the rate equation for the overall reaction is not likely to be the same as that of an elementary step. Also, the rate equations for some elementary steps have terms for reaction intermediates, and such terms cannot appear in the rate law for the overall reaction. **73 a.** $I + B \longrightarrow C + D$; **b.** Fast. If the second step were not fast, the rate equations for the overall reaction and the slow first step would not be the same. **75.** The reaction profile for the surface-catalyzed reaction is rather complex, but the highest energy point on the profile is still considerably below the transition state energy in the reaction profile of the noncatalyzed, homogeneous reaction, meaning that the surface-catalyzed reaction has the lower activation energy. **77.** The enzyme will be less active at 37 °C and even less active at 40 °C. **79.** An inhibitor may block the active site of the enzyme, or it may react with the enzyme to change the shape of the active site. **81.** 0 s, 35.3 mL; 60 s, 27.9 mL; 120 s, 22.6 mL; 180 s, 18.3 mL; 240 s, 14.9 mL; 300 s, 11.9 mL; 360 s, 9.44 mL; 420 s, 7.52 mL; 480 s, 6.08 mL; 540 s, 4.80 mL; 600 s, 3.76 mL. **83.** t = 337 min. **85 a.** 7.29 min; **b.** 5.7×10^{13} molecules/min. **87.** 0.036 dis/min. **89.** 8.12 g. **91.** A high activation energy. A reaction with a high E_a has a high value of $-E_a/R$, the slope of the graph of ln k versus $1/T$. A steep slope in this straight-line graph means that the rate of change ln k (and k) with $1/T$ (and T) is high.

Chapter 16

Exercises: 16.1. No, there would be a different value of $[COCl_2]$ for each set of values of [CO] and $[Cl_2]$. **16.2.** $K_c = 2.5 \times 10^{-3}$. **16.3.** $K_p = 4.6 \times 10^3$. **16.4.** $K_p = \dfrac{P_{CO}P_{H_2}}{P_{H_2O}}$ **16.5.** No, in order for the reaction to go to completion, we would expect a value of K_c much larger than 1.2×10^3. **16.6 a.** Equilibrium would shift to the right. The new equilibrium would have more H_2 and NH_3, and less N_2, than the original equilibrium. **b.** Equilibrium would shift to the left. There would be less N_2 and NH_3 but more H_2 in the new equilibrium than in the original equilibrium. **c.** Equilibrium

would shift to the right. The new equilibrium would have less N_2, H_2, and NH_3 than the original equilibrium. **16.7.** Changing the volume does not cause a change in the equilibrium amount of HI because there are the same number of moles of gas on both sides of the equation. **16.8.** At low temperatures. The formation of $SO_3(g)$, the forward reaction, is exothermic, and an exothermic reaction is favored at low temperatures. **16.9.** $K_p = 0.426$. **16.10.** $\chi_{NO} = 0.022$. **16.11.** 0.018 mol. **16.12.** 0.0176 mol H_2; 0.00756 mol I_2; 0.0849 mol HI. **16.13.** 0.658 atm. **Conceptual Exercises: 16.1 a.** Addition of H_2, a product, shifts equilibrium to the left. The new equilibrium has more CO(g), $H_2O(g)$, and $H_2(g)$, and less $CO_2(g)$, than the original equilibrium. **b.** Because both a reactant (H_2O) and a product (CO_2) are added to the equilibrium mixture, we cannot make a qualitative prediction of whether the equilibrium condition will shift to the left or to the right. **c.** Adding more H_2O favors the forward reaction, as does lowering the temperature (the forward reaction is exothermic). The new equilibrium will have more CO_2 and H_2 and less CO than the original equilibrium. The amount of H_2 is in doubt because we don't know if the combined effects of the two changes will consume more or less than the 1.00 mol H_2O added. **16.2.** With $P_{total} = 0.57$ atm, $P = 0.08$ atm. When $P = 0.08$ is substituted into the K_c expression, the value obtained, 0.019, is very close to $K_c = 0.023$. **Review Questions: 12 a.** $K_c = \dfrac{[CO]^2}{[CO_2]}$; **b.** $K_c = \dfrac{[HI]^2}{[H_2S]}$; **c.** $K_c = [O_2]$. **13 a.** $2\,H_2S(g) \rightleftharpoons 2\,H_2(g) + S_2(g)$, $K_c = \dfrac{[S_2][H_2]^2}{[H_2S]^2}$; **b.** $CS_2(g) + 4\,H_2(g) \rightleftharpoons CH_4(g) + 2\,H_2S(g)$, $K_c = \dfrac{[CH_4][H_2S]^2}{[CS_2][H_2]^4}$; **c.** $CO(g) + 3\,H_2(g) \longrightarrow CH_4(g) + H_2O(g)$, $K_c = \dfrac{[CH_4][H_2O]}{[CO][H_2]^3}$. **14 a.** $K_p = \dfrac{P_{CO_2}P_{H_2}}{P_{CO}P_{H_2O}}$; **b.** $K_p = \dfrac{P^2_{NH_3}}{P_{N_2}P^3_{H_2}}$; **c.** $K_p = P_{NH_3}P_{H_2S}$. **15 a.** $\frac{1}{2}\,N_2(g) + \frac{1}{2}\,O_2(g) \rightleftharpoons NO(g)$, $K_p = P_{NO}/(P_{N_2})^{1/2}(P_{O_2})^{1/2}$; **b.** $\frac{1}{2}N_2(g) + \frac{3}{2}H_2(g) \rightleftharpoons NH_3(g)$, $K_p = P_{NH_3}/(P_{N_2})^{1/2}(P_{H_2})^{3/2}$; **c.** $\frac{1}{2}\,N_2(g) + \frac{1}{2}\,O_2(g) + \frac{1}{2}\,Cl_2(g) \rightleftharpoons NOCl(g)$, $K_c = P_{NOCl}/[(P_{N_2})^{1/2} \times (P_{O_2})^{1/2}\,(P_{Cl_2})^{1/2}]$. **16 a.** $2\,CO(g) + 2\,NO(g) \rightleftharpoons N_2(g) + 2\,CO_2(g)$, $K_c = [N_2][CO_2]^2/[CO]^2[NO]^2$; **b.** $5\,O_2(g) + 4\,NH_3(g) \rightleftharpoons 4\,NO(g) + 6\,H_2O(g)$, $K_c = [NO]^4[H_2O]^6/[NH_3]^4[O_2]^5$; **c.** $2\,NaHCO_3(s) \rightleftharpoons Na_2CO_3(s) + H_2O(g) + CO_2(g)$, $K_c = [H_2O][CO_2]$.
17. $K_c\,(a) = \dfrac{1}{K_c\,(b)}$. **18.** K_c for the reaction $\frac{1}{2}\,N_2O_4 \rightleftharpoons NO_2$ is greater than that for the reaction $N_2O_4 \rightleftharpoons 2\,NO_2$. This is because the square root of a number smaller than 1 is *larger* than the number. That is, $\sqrt{0.113} = 0.336$. **20.** Adding HCl, adding O_2, increasing the total pressure on the gases, and lowering the temperature will all lead to a greater amount of $Cl_2(g)$ at equilibrium. **21 a.** Yes, there are fewer moles of gas on the right. **b.** No, there is the same number of moles on each side of the equation, so a pressure change will have no effect. **c.** Yes, there are fewer moles of gas on the right. **22 a.** The volume has no effect because there is the same number of moles of gas on each side of the equation. **b.** A larger volume will cause the reaction to go to the right to make more moles of gas. **23.** Reaction (a) is exothermic. It will not proceed as far in the forward direction at high temperature. The extent of dissociation decreases with increased temperature. Second reaction (b) is endothermic; the extent of dissociation increases with increased temperature. **Problems: 25 a.** 1.17×10^{-3}; **b.** 4.79×10^{-3}; **c.** 1.23×10^{3}. **27.** 3.59×10^{-3} M. **29.** 538. **31.** $[COCl_2] = 6.67 \times 10^{-3}$ M. **33.** $P_{total} = 6.9$ atm. **35.** $K_p = 1$. **37.** $Q_c < K_c$; the net reaction will go to the right. **39 a.** No, K_p would have to equal 1. **b.** Yes, it is possible. **c.** No, K_p would have to equal 1. **d.** Yes, it is possible. **41.** $K_c = 1.80 \times 10^{-6}$; $K_p = 6.76 \times 10^{-5}$. **43.** $K_c = 61.5$. **45.** $K_p = 6.66 \times 10^{-4}$. **47.** These dissociation reactions, because they require the breaking of bonds with no new bonds formed, are *endothermic*. The forward reaction is favored with increasing temperature—dissociation is more extensive when equilibrium is reached. **49 a.** An increase in pressure has *no effect* on the extent of the formation reaction because the number of moles of gaseous product equals that of gaseous reactants. **b.** The formation reaction occurs to a greater extent at higher pressures because the number of moles of gaseous product is less than the number of moles of gaseous reactants. **c.** No effect, for the same reason as (a). **d.** The formation reaction occurs to a greater extent with increased pressure, for the same reason as in (b). **51.** The density of ice is less than that of liquid water. An increased pressure on the ice favors the process in which the water molecules occupy a reduced volume—the liquid state. The ice melts. **53.** 12.5 g methylcyclopentane. **55.** 0.0436 mol CO. **57.** 0.131 mol COF_2. **59.** PCl_3 and $Cl_2 = 0.13$ mol; $PCl_5 = 0.067$ mol. **61 a.** 0.265 atm; **b.** 0.465 atm. **63.** 2.29 atm. **65.** $K_c = 3.9$. **67.** 0.534 atm. **69.** $K_c = 5.60$. **71.** 1.16 atm. **73.** $P_{SO_2} = 8.32 \times 10^{-3}$ atm.

Chapter 17

Exercises:

17.1 a. $HS^- + H_2O \rightleftharpoons H_2S + OH^-$
base(1) acid(2) acid(1) base(2)

b. $HNO_3 + H_2PO_4^- \rightleftharpoons H_3PO_4 + NO_3^-$
acid(1) base(2) acid(2) base(1)

17.2 a. H_2Te is the stronger acid. Because the Te atom is larger than the S atom, we expect the H—Te bond energy to be less than H—S bond energy, and the H—Te bond to be more easily broken than the H—S bond. **b.** $CH_3CH_2CH_2CHBrCOOH$ is the stronger acid because the Br atom on the second C atom has more electron-withdrawing power than the Cl atom on the fifth C atom. **17.3.** pH = 12.635. **17.4.** pH = 2.40. **17.5.** 10.10. **17.6.** $K_a = 6.3 \times 10^{-5}$; $pK_a = 4.20$. **17.7.** $[H_2PO_4^-] \approx [H_3O^+] = 0.06$ M; $[HPO_4^{2-}] \approx K_{a_2} = 6.3 \times 10^{-8}$ M. **17.8.** pH = 2.77. **17.9 a.** neutral; **b.** basic. **17.10.** pH = 5.27. **17.11.** 0.29 M. **17.12.** pH = 8.89. **17.13.** pH = 9.10. **17.14.** 3.22 g NH_4Cl. **17.15 a.** 2.90; **b.** 3.90; **c.** 10.10; **d.** 11.10. **17.16 a.** 4.97; **b.** 11.10. **Estimation Exercises: 17.1.** Because methylamine has both the higher molarity and the greater value of K_b, the 0.030 M methylamine has the greater $[OH^-]$. **17.2.** Ionization of 0.020 M H_2SO_4 is complete in the first step and partial in the second: 0.020 M < $[H_3O^+]$ < 0.040 M. Only response (b) fits this requirement: $[H_3O^+] = 0.025$ M. **17.3.** NH_4CN. K_b of CN^- is much greater than K_b of NO_2^- (K_a of HCN is much smaller than K_a of HNO_2). **Conceptual Exercises: 17.1.** The solution is basic. The reasoning is the same as in Conceptual Example 17.1, except it is based on OH^- instead of H_3O^+. **17.2.** The indicator color shows that the pH is in the range of about 4 to 5.5. Solutions (b) and (c) have pH values outside this range. The 1.00 M NH_4Cl would have a pH in this range because of hydrolysis of NH_4^+. The 1.00 M CH_3COOH–1.00 M CH_3COONa is a buffer with pH = 4.74 (pK_a of acetic acid). To distinguish between (a) and (d), add a small amount of either an acid or a base. The pH of the buffer solution (d) would not change, and that of the 1.00 M NH_4Cl (solution b) would. **17.3 a.** $K_b = 1.0 \times 10^{-5}$. **b.** From the graph, pH = 5. **Review Questions: 1.** Arrhenius: $HI \longrightarrow H^+ + I^-$; Brønsted–Lowry: $HI + H_2O \longrightarrow H_3O^+ + I^-$. **4.** acid: $HPO_4^{2-} + H_2O \rightleftharpoons H_3O^+ + PO_4^{3-}$; base: $HPO_4^{2-} + H_2O \rightleftharpoons OH^- + H_2PO_4^-$. **5 a.** $HClO_2 + H_2O \rightleftharpoons H_3O^+ + ClO_2^-$; **b.** $CH_3CH_2COOH + H_2O \rightleftharpoons H_3O^+ + CH_3CH_2COO^-$; **c.** $HCN + H_2O \rightleftharpoons H_3O^+ + CN^-$; **d.** $C_6H_5OH + H_2O \rightleftharpoons H_3O^+ + C_6H_5O^-$.

6 a. $K_a = \dfrac{[H_3O^+][ClO_2^-]}{[HClO_2]}$; **b.** $K_a = \dfrac{[H_3O^+][CH_3CH_2COO^-]}{[CH_3CH_2COOH]}$;

c. $K_a = \dfrac{[H_3O^+][CN^-]}{[HCN]}$; **d.** $K_a = \dfrac{[H_3O^+][C_6H_5O^-]}{[C_6H_5OH]}$. **7 a.**

$$\underset{\text{acid}(1)}{HOClO_2} + \underset{\text{base}(2)}{H_2O} \rightleftharpoons \underset{\text{acid}(2)}{H_3O^+} + \underset{\text{base}(1)}{OClO_2^-}$$; **b.**

$$\underset{\text{acid}(1)}{HSeO_4^-} + \underset{\text{base}(2)}{NH_3} \rightleftharpoons \underset{\text{acid}(2)}{NH_4^+} + \underset{\text{base}(1)}{SeO_4^{2-}}$$; **c.**

$$\underset{\text{acid}(1)}{HCO_3^-} + \underset{\text{base}(2)}{OH^-} \rightleftharpoons \underset{\text{base}(1)}{CO_3^{2-}} + \underset{\text{acid}(2)}{H_2O}$$;

d. $$\underset{\text{acid}(1)}{C_5H_5NH^+} + \underset{\text{base}(2)}{H_2O} \rightleftharpoons \underset{\text{base}(1)}{C_5H_5N} + \underset{\text{acid}(2).}{H_3O^+}$$

8. (a) benzoic acid < (b) formic acid < (c) hydrofluoric acid < (d) nitrous acid. **9.** (d) nitrite ion < (c) fluoride ion < (b) formate ion < (a) benzoate ion. **22.** pH less than 7: (a), a strong acid, and (c), a weak acid; pH greater than 7: (b) and (d), both weak bases. **23 a.** 0.05 M $KHSO_4$ has the lowest pH because of the acid anion, HSO_4^- ($K_{a_1} = 1.1 \times 10^{-2}$). Solution (b) is pH neutral, and solutions (c) and (d) are basic because of hydrolysis of the anions. **24.** Order of increasing pH: (f) < (a) < (d) < (g) < (c) < (b) < (e) < (h). **Problems: 25.** The sketch should show the transfer of a proton (the H atoms of the carboxylic acid group) from CH_3COOH to the O atom of H_2O, forming H_3O^+ and CH_3COO^-. In the reverse reaction, proton transfer is from H_3O^+ to CH_3COO^-, re-forming CH_3COOH and H_2O. For the ionization of HCl, proton transfer is from HCl to H_2O, forming H_3O^+ and Cl^-. Because HCl is a strong acid, there is no reverse reaction. **27.** Bicarbonate ion neutralizes excess stomach acid: $HCO_3^- + H_3O^+ \longrightarrow H_2CO_3 + H_2O \longrightarrow 2\,H_2O + CO_2(g)$. **29.** (d) < (b) < (a) < (c). **31.** CH_3COOH is a stronger acid than is H_2O. Because aniline is more able to accept a proton from CH_3COOH than from H_2O, it is a stronger base in $CH_3COOH(l)$ than it is in $H_2O(l)$. **33 a.** $1.4 \times 10^{-3} < K_a < 8.7 \times 10^{-3}$; **b.** $K_a \approx 3 \times 10^{-5}$. The Cl atom is far enough from the —COOH group to have little electron-withdrawing effect. The K_a value should be only slightly greater than those of acetic acid ($K_a = 1.8 \times 10^{-5}$) and pentanoic acid ($K_a = 1.5 \times 10^{-5}$). **35.** (e) < (a) < (c) < (b) < (d). **37.** $[OH^-] = 1.35 \times 10^{-3}$ M. **39.** 2.60; 11.51; 3.14; 10.34. **41.** 0.0062 M $Ba(OH)_2(aq)$. **43.** 0.44 g concd HCl. **45.** pH = 1.78. **47.** 3.3×10^{-2} M. **49.** $K_a = 3.6 \times 10^{-4}$. **51.** 0.0045 M H_2SO_4 has the lower pH. H_2SO_4 is a strong acid in its first ionization step; moreover, even K_{a_2} of H_2SO_4 is larger than K_{a_1} of H_3PO_4. **53.** The cola drink has $[H_3O^+] \approx 3.4 \times 10^{-3}$ and pH ≈ 2.5. **55.** $CH_3CH_2COO^- + H_2O \rightleftharpoons CH_3CH_2COOH + OH^-$; basic. **57.** $CO_3^{2-} + H_2O \rightleftharpoons HCO_3^- + OH^-$; $K_b = 2.1 \times 10^{-4}$. **59.** (c), NH_4I. Of the four solutions, (a) is pH neutral and (b) and (d) are basic because of hydrolysis of the anions. HN_4^+ produces an acidic solution by hydrolysis. **61.** pH = 10.22. **63.** 0.22 M CH_3COONa. **65.** H_3O^+ from HNO_3 (c) and $HCOO^-$ from $(HCOO)_2Ca$ (d) are common ions and suppress the ionization of HCOOH. NaCl (a) has no effect on the ionization equilibrium, and KOH (b) and Na_2CO_3 (e), by neutralizing HCOOH, stimulate its ionization. **67.** $[NH_4^+] = 1.8 \times 10^{-4}$ M. **69.** pH = 4.24. **71 a.** No, the HCl and NaOH would neutralize one another, leaving NaCl(aq) with either excess HCl or excess NaOH, not a buffer solution. **b.** No, the ratio $[CH_3COO^-]/[CH_3COOH] << 1$, much too small for the solution to act as a buffer. **73.** pH = 10.53. **75.** pH = 10.55. **77.** 0.88 g CH_3NH_3Cl. **79.** 8.22. **81.** yellow. **83.** A test with thymol blue should produce a blue color, and with alizarin yellow R, a yellow color. This places the pH between about 9.6 and 10.0. **85.** In contrast to a strong base–strong acid titration, in the titration of a weak base by a strong acid: (1) the initial pH is lower because the weak base is only partially ionized; (2) at the half-neutralization point, pH = pK_b, in a buffer solution in which the concentrations of the weak base and its conjugate acid are equal; (3) the pH < 7 at the equivalence because the cation of the weak base hydrolyzes; (4) the steep portion of the curve at the equivalence point is confined to a smaller pH range; (5) the choice of indicators is more limited. Only those with a color change in the pH range of about 3 to 6 will work (see Figure 17.16). **87.** No, the pH at the equivalence point of the weak base–strong acid titration is well below 7, whereas that of the weak acid–strong base titration is well above 7. **89.** 19.22 mL. **91.** $H_2NNH_2 + H_2O \rightleftharpoons H_2NNH_3^+ + OH^-$, $H_2NNH_3^+ + H_2O \rightleftharpoons {}^+H_3NNH_3^+ + OH^-$ The second H^+ is repelled by the charge on the $NH_3NH_2^+$ ion. This makes $K_{b_2} < K_{b_1}$ and pK_{b_2} > pK_{b_1}. **93 a.** 0.42%; **b.** 1.3%; **c.** 34%. **d.** The acidity does not increase. $[H_3O^+]$ *decreases* in the order (a) > (b) > (c). The percent ionization increases with dilution, but it is an increasing percentage of a *decreasing* total amount of acid. **95.** 0.20 mL. **97.** A solution may have $[H_3O^+] = 2 \times [OH^-]$ (corresponding to pH = 6.85), and a solution may have pH = 2 × pOH (corresponding to pH = 9.33), but obviously it cannot be the same solution. **99.** pH = 5.68. **101.** The portion of the titration curve at the equivalence point is limited to a narrow pH range and does not have a steep slope. It is difficult to determine the equivalence point with any precision. **103.** The solution at the first equivalence point in the titration of $H_3PO_4(aq)$ with NaOH(aq) is $NaH_2PO_4(aq)$. Its pH is represented by the midpoint of the first steep vertical rise in the titration curve. Adding small amounts of acid or base to this solution changes the pH significantly—it is not a buffer solution. A solution containing both $H_2PO_4^-$ and HPO_4^{2-} ions is an effective buffer solution. It is located about in the center of the second very slowly rising portion of the curve.

Chapter 18

Exercises: 18.1 a. $K_{sp} = [Mg^{2+}][OH^-]^2$; **b.** $K_{sp} = [Cu^{2+}]^3[AsO_4^{3-}]^2$. **18.2.** 7.1×10^{-8} M. **18.3.** $K_{sp} = 2.0 \times 10^{-11}$. **18.4.** 1.4×10^{-6} M. **18.5.** 1.4×10^{-5} M. **18.6.** A precipitate should form. **18.7.** Precipitation occurs. **18.8.** Yes, only 0.11% of the Ca^{2+} remains in solution. **18.9.** pH = 6.95. **18.10.** $[Ag^+] = 9.2 \times 10^{-15}$ M. **18.11.** No. **Estimation Exercises: 18.1.** The solutes are all of the same type, MX_2. Their molar solubilities parallel their K_{sp} values: CaF_2 (5.3×10^{-9}) < PbI_2 (7.1×10^{-9}) < MgF_2 (3.7×10^{-8}) < $PbCl_2$ (1.6×10^{-5}).

$[Ag(S_2O_3)_2]^{3-}$, AgI(s) is most soluble in 0.100 M NaCN. **Conceptual Exercises: 18.1.** $Mg(OH)_2(s) + 2\ NH_4^+(aq) \longrightarrow Mg^{2+}(aq) + 2\ H_2O(l) + 2\ NH_3(aq)$. **18.2.** No, NH_4^+ does not react with NH_3. The complex ion, $[Ag(NH_3)_2]^+$, is not destroyed, and the concentration of free Ag^+ remains too low for AgCl(s) to precipitate. **Review Questions: 1.** $K_{sp} = [Fe^{3+}][OH^-]^3$. **2.** $Zn_3(PO_4)_2(s) \rightleftharpoons 3\ Zn^{2+}(aq) + 2\ PO_4^{3-}(aq)$; $K_{sp} = 9.0 \times 10^{-33}$. **5.** $PbSO_4$ is more soluble because its K_{sp} value is larger and the two solutes are of the same type: MX. **7.** Yes. **11.** The solubility of $Fe(OH)_3(s)$ is increased by HCl(aq) and $CH_3COOH(aq)$, both acids. The solubility is lowered by $FeCl_3(aq)$ because of the common ion Fe^{3+}; it also lowered by NaOH(aq) and $NH_3(aq)$ because of the common ion OH^-. **13.** $CH_3COO^-(aq) + Ag^+(aq) \longrightarrow CH_3COOAg(s)$; $CH_3COOAg(s) + H_3O^+(aq) \longrightarrow Ag^+(aq) + CH_3COOH(aq) + H_2O(l)$. **15.** $Pb(NO_3)_2(aq)$, through the common ion Pb^{2+}, reduces the solubility of $PbCl_2(s)$; but because Pb^{2+} forms the complex ion $[PbCl_3]^-$, HCl(aq) increases the solubility of $PbCl_2(s)$. **16.** AgBr(s) dissolves to a greater extent in $Na_2S_2O_3(aq)$ than in $NH_3(aq)$ because the complex ion $[Ag(S_2O_3)_2]^{3-}$ is more stable than $[Ag(NH_3)_2]^+$. **18.** Cation Group 1 is separated from other groups by HCl(aq), and Group 2, by H_2S in 0.3 M HCl(aq). **19.** $PbCl_2(s)$ is separated from AgCl(s) and $Hg_2Cl_2(s)$ by hot water. AgCl(s) is separated from $Hg_2Cl_2(s)$ by $NH_3(aq)$. **Problems: 21 a.** $K_{sp} = [Fe^{3+}][OH^-]^3$; **b.** $K_{sp} = [Hg_2^{2+}][Cl^-]^2$; **c.** $K_{sp} = [Mg^{2+}][NH_4^+][PO_4^{3-}]$; **d.** $K_{sp} = [Li^+]^3[PO_4^{3-}]$. **23.** No, the molar solubility and K_{sp} cannot have the same value. The molar solubility must be raised to a power and generally multiplied by a factor to obtain K_{sp}. The molar solubility is larger than K_{sp} because the solubility is generally much smaller than 1 M, and raising such a number to a power (2, 3, 4, ...) produces a result that is smaller still. **25.** $K_{sp} = 1.6 \times 10^{-9}$. **27.** Ag_2CrO_4. **29.** $CaSO_4$. $CaSO_4(s)$ produces a greater $[Ca^{2+}]$ than $CaCO_3(s)$ because its K_{sp} is larger and both solutes are of the same type. In $CaSO_4$(satd aq), $[Ca^{2+}] = (9.1 \times 10^{-6})^{1/2} \approx 3 \times 10^{-3}$ M, and in CaF_2(satd aq), $[Ca^{2+}] = (\frac{1}{4} \times 5.3 \times 10^{-9})^{1/3} \approx 1 \times 10^{-3}$ M. **31.** 7.1 ppb Cu^{2+}. **33.** AgBr is more soluble in water because in each of the other solutions a common ion is present. **35.** 6.7×10^{-6} M. **37.** $[Pb^{2+}] = 4.4 \times 10^{-8}$ M; $[I^-] = 0.400$ M. **39.** $[CrO_4^{2-}] = 2.2 \times 10^{-6}$ M. **41.** Yes. **43 a.** $[SO_4^{2-}] = 3.0 \times 10^{-3}$ M; **b.** $[Pb^{2+}] = 5.3 \times 10^{-6}$ M. **45.** Precipitation is not complete; about 67% of the Mg^{2+} remains in solution. **47.** 4.0×10^{-5} M. **49.** Only $NaHSO_4$ will increase the molar solubility of $CaCO_3$. HSO_4^-, an acid, donates protons to CO_3^{2-}, a base, to form HCO_3^-. Further reaction produces H_2CO_3, which decomposes to H_2O and $CO_2(g)$ and stimulates further dissolving of $CaCO_3$. **51.** $CaCO_3(s) + 2\ H_3O^+(aq) \longrightarrow Ca^{2+}(aq) + 3\ H_2O(l) + CO_2(g)$; $CaCO_3(s) + 2\ CH_3COOH(aq) \longrightarrow Ca^{2+}(aq) + 2\ CH_3COO^-(aq) + H_2O(l) + CO_2(g)$. **53.** $K_3[Fe(CN)_6]$. **55.** Both H_3O^+ from HCl (d) and HSO_4^- from $NaHSO_4$ (b) can donate protons to NH_3 in the complex ion, causing $[Zn(NH_3)_4]^{2+}$ to dissociate and the concentration of free Zn^{2+} to increase. **57.** HNO_3 donates protons to NH_3, destroying the complex ion $[Ag(NH_3)_2]^+$ and allowing Ag^+ and Cl^- to precipitate. HNO_3 is not able to donate protons to Cl^- ions in $[AgCl_2]^-$ because Cl^- is too weak a base to accept them—HCl is a strong acid. **59.** $[NH_3] = 0.97$ M. **61 a.** $Zn^{2+}(aq) + 4\ OH^-(aq) \longrightarrow [Zn(OH)_4]^{2-}(aq)$; **b.** $Al_2O_3(s) + 6\ H_3O^+(aq) + 3\ H_2O(l) \longrightarrow 2\ [Al(H_2O)_6]^{3+}(aq)$; **c.** $Fe(OH)_3(s) + OH^-(aq) \longrightarrow$ N.R. **63.** $PbCl_2$ is soluble enough that $[Pb^{2+}]$ remaining in solution after the Group 1 precipitation is sufficiently high that K_{sp} of PbS is exceeded in Group 2. AgCl is so insoluble that $[Ag^+]$ remaining in solution after the Group 1 precipitation is not enough to yield a detectable precipitate of Ag_2S in Group 2. **65 a.** $Pb^{2+}(aq) + 2\ Cl^-(aq) \longrightarrow PbCl_2(s)$; **b.** $AgCl(s) + Hg_2Cl_2(s) + 4\ NH_3(aq) \longrightarrow [Ag(NH_3)_2]^+(aq) + Hg(l) + HgNH_2Cl(s) + NH_4^+(aq) + 2\ Cl^-(aq)$; **c.** $Al^{3+}(aq) + Fe^{3+}(aq) + 7\ OH^-(aq) \longrightarrow [Al(OH)_4]^-(aq) + Fe(OH)_3(s)$. **67.** Only Hg_2^{2+} is proven to be present, based on the gray color produced when the Group 1 precipitate is treated with $NH_3(aq)$. The presence of Pb^{2+} and Ag^+ remains uncertain. Treatment of the Group 1 precipitate with hot water and a subsequent test for Pb^{2+} was not performed, and the $NH_3(aq)$ was not tested for the presence of Ag^+. **69 a.** $BaSO_4$ is so insoluble that not enough dissolves to make it dangerous. **b.** 1.4 mg Ba^{2+}/L; **c.** $MgSO_4$ reduces the solubility of $BaSO_4$ through the common ion SO_4^{2-}. **71.** Yes. **73 a.** 4.5×10^{-6} M; **b.** yes. **75.** $PbCrO_4$ precipitates first. $[Pb^{2+}] = 2.8 \times 10^{-11}$ M at the time the first solid ($PbCrO_4$) appears. $[Pb^{2+}] = 1.6 \times 10^{-6}$ M at the time the second solid ($PbSO_4$) appears, and at this point $[CrO_4^{2-}] = 1.8 \times 10^{-7}$ M. Because practically all the CrO_4^{2-} has precipitated as $PbCrO_4$ before the first $PbSO_4$ appears, CrO_4^{2-} and SO_4^{2-} can be separated by fractional precipitation. **77.** 0.98 mol NH_3. **79.** *Not present:* Na^+, NH_4^+, and Cu^{2+}. All common compounds of Na^+ and NH_4^+ are soluble, and these cations would not form a precipitate in $NH_3(aq)$. Some compounds of Cu^{2+} are water soluble and some are water insoluble, but we expect the compounds to be colored and for $Cu^{2+}(aq)$ to have a characteristic blue color. *Possibly present:* Mg^{2+} and Ba^{2+}. Soluble magnesium compounds would yield a precipitate of $Mg(OH)_2(s)$ in $NH_3(aq)$. Some insoluble barium compounds are soluble in acids (e.g., $BaCO_3$), and $Ba^{2+}(aq)$ yields a white precipitate of $BaSO_4(s)$ when treated with $(NH_4)_2SO_4(aq)$.

Chapter 19

Exercises: 19.1 a. Spontaneous. The molecules in the wood (principally cellulose, a carbohydrate) would eventually oxidize to CO_2 and H_2O. The decay is greatly enhanced by the presence of microorganisms. **b.** Nonspontaneous. Stirring NaCl(aq) cannot supply the energy input required to dissociate NaCl into its elements. **c.** Indeterminate. $CaCO_3(s)$ should decompose on heating, but whether the decomposition is sufficient to produce $CO_2(g)$ at 1 atm at 650 °C, we cannot say. **19.2 a.** Decrease. Gaseous reactants yield a solid product. **b.** Increase. A solid reactant produces one solid product and a gas. **c.** No prediction. Two moles of gaseous reactants yield two moles of gaseous product. It is not immediately obvious whether entropy increases or decreases. **19.3.** $\Delta S° = -42.1$ J/K. **19.4 a.** Case 2: $\Delta H < 0$, $\Delta S < 0$; **b.** Case 3: $\Delta H > 0$, $\Delta S > 0$. **19.5 a.** $\Delta G° = -70.5$ kJ; **b.** $\Delta G° = -66.9$ kJ. **19.6.** $Mg(OH)_2(s) + 2\ H_3O^+(aq) \rightleftharpoons Mg^{2+}(aq) + 4\ H_2O(l)$; $K_{eq} = [Mg^{2+}]/[H_3O^+]^2$. **19.7.** $K_{eq} = 3.02 \times 10^{-21}$. **19.8.** $K_{eq} = 0.45$. **Estimation Exercise: 19.1.** 295 °C. **Conceptual Exercises: 19.1.** The temperature would be the same as the temperature at which the line $P = 1$ atm intersects the sublimation curve of $CO_2(s)$ in Figure 11.10: –78.5 °C. **19.2.** $\Delta S°_{298} = 0.15\ kJ \cdot K^{-1}$. The result is the same as that in Conceptual Example 19.2 because the same basic equations and same assumptions are used. **Review Questions: 3 a.** Spontaneous. Microorganisms are present in the milk that lead to its souring; no further intervention is needed. **b.** Nonspontaneous. The extraction of copper metal from copper ores requires a great deal of external intervention. **c.** Spontaneous. The corrosion of iron in moist air cannot be prevented from occurring without external intervention. **4 a.** Decrease. A liquid is converted to a solid. **b.** Increase. A solid is converted to a gas. **c.** Increase. A

a solid. **b.** Increase. A solid is converted to a gas. **c.** Increase. A liquid combines with oxygen gas to produce an even greater amount of gaseous products. **9.** NOF_3. It has a greater number of atoms than NO_2F, more vibrational modes, and a greater entropy. **11.** Low temperatures. At low temperatures, the ΔH term dominates in the Gibbs equation, and a $\Delta H < 0$ produces a $\Delta G < 0$. **12.** No, vaporization of water will occur spontaneously, but *not* to produce a vapor at 1 atm (which only occurs at 100 °C). **18 a.** $K_{eq} = (P_{NCl_3})^2/[(P_{N_2})(P_{Cl_2})^3]$; **b.** $K_{eq} = [Pb^{2+}][I^-]^2$; **c.** $K_{eq} = [H_3O^+][HSO_3^-]/(P_{SO_2})$; **d.** $K_{eq} = [OH^-]^2[CO_3^{2-}]$. **Problems: 21.** An arrangement in which all pennies are heads up is much too ordered to be likely. More likely is that about one-half the pennies will be "heads" and about one-half "tails." This is the highest entropy arrangement. **23 a.** Negative. Solids become more ordered as the temperature is lowered. **b.** Positive. The entropy of a gas increases when it is heated and its pressure is lowered. **c.** Negative. A large decrease in entropy occurs if a gas has its pressure increased, its temperature lowered, and is then converted to a solid. **25 a.** Decrease. A solid is more ordered than the liquid phase from which it is frozen. **b.** Indeterminate. The gases are all diatomic and the same number of moles of gas appear on each side of the equation. **c.** Increase. A liquid decomposes to produce a large amount of gas. **d.** Decrease. The conversion of a gas to an aqueous solution should produce a more ordered state. **27.** The value of ΔS is larger for (b). Sublimation of a solid produces much more disorder than conversion of one polymorphic form of a solid to another. **29.** Statement (b) is true. If the entropy of a system decreases, that of the surroundings must increase by an even greater amount, so that $\Delta S_{univ.} > 0$ and the process is spontaneous. **31.** A more accurate value of ΔS°_{298} can be obtained from tabulated S° values, as in Appendix D. The value obtained is 164.7 J • K^{-1}. **33.** Because the oxidation of Al(s) to Al_2O_3(s) is spontaneous, the environmental disintegration of aluminum is also spontaneous. This is true despite the fact that the oxidation occurs very slowly. **35.** In Table 19.1, cases 2 and 3 both represent situations in which the forward reaction is spontaneous at certain temperatures and nonspontaneous at others. When the forward reaction is spontaneous, the reverse reaction is nonspontaneous, and vice versa. **37.** The $T\Delta S$ line lies above the ΔH line and does not intersect it, so that $\Delta G < 0$ at all temperatures. **39 a.** Nonspontaneous at all temperatures because $\Delta H > 0$ and $\Delta S < 0$. **b.** Spontaneous at low temperatures (because of the large negative value of ΔH). Whether the reaction is spontaneous at very high temperatures is uncertain because the sign of ΔS is difficult to predict—three moles of gaseous reactants produce three moles of gaseous products. (The value of ΔS must be obtained from tabulated data.) **c.** Spontaneous at low temperatures but nonspontaneous at high temperatures because $\Delta H < 0$ and $\Delta S < 0$. **41.** The reaction should be nonspontaneous at all temperatures because $\Delta H > 0$ and $\Delta S < 0$. **43 a.** ΔG° = 22.9 kJ; **b.** ΔG° = −163 kJ. **45.** ΔH° = −1076.8 kJ; ΔS° = −56.5 J • K^{-1}; ΔG° = −1060.0 kJ (by the Gibbs equation) and −1060.1 kJ (from standard free energies of formation). **47.** 72 °C. **49.** ΔH°_{vapn} = 29.8 kJ/mol (by Trouton's rule), and 30.1 kJ/mol (from standard enthalpies of formation). **51 a.** $\Delta G = 0$ at equilibrium; **b.** K_p = 170. **53 a.** $K_{eq} = K_p = P_{H_2O} \times P_{SO_2}$; **b.** $K_{eq} = K_{sp} = [Mg^{2+}][OH^-]^2$; **c.** $K_{eq} = K_w/K_a = [CH_3COOH][OH^-]/[CH_3COO^-]$. **55 a.** $K_p = 7.2 \times 10^{24}$; **b.** $K_p = 1.3 \times 10^{-20}$. **57.** Pressure of naphthalene vapor = 0.063 mmHg. **59.** The vapor pressure of CCl_4 at 15 °C = 72.3 mmHg. **61.** 27 °C. **63.** 578 °C. **65.** 64 °C. **67.** $K_{eq} = 5.3 \times 10^9$. **69.** ΔG° = 9.8 kJ (estimate); 10.1 kJ (from standard free energies of formation). **71.** Energy cannot be created or destroyed. Only its form can change, and therefore energy remains constant. The entropy of the world constantly increases because all natural (spontaneous) processes produce an increase in entropy of the universe. **73.** ΔG = −9.34 kJ. **75.** The plot is a straight line with a slope of about 2.3×10^4 and a value of $\Delta H^\circ = -R \times$ slope ≈ −190 kJ/mol.

Chapter 20

Exercises: 20.1. 2 Al(s) + 3 Cu^{2+}(aq) ⟶ 2 Al^{3+}(aq) + 3 Cu(s). **20.2.** Net cell reaction: Zn(s) + Cl_2(g) ⟶ Zn^{2+}(aq) + 2 Cl^-(aq); Zn | Zn^{2+}(aq) || Cl^-(aq) | Cl_2(g), Pt. **20.3.** $E^\circ = E^\circ_{red}$ = 1.610 V. **20.4.** The reaction is not spontaneous in the forward direction; E°_{cell} = −0.434 V. **20.5.** $K_{eq} = 5 \times 10^{14}$. **20.6 a.** 1.053 V; **b.** 1.147 V; **c.** 1.048 V. **20.7 a.** Reduction: 2 H_2O + 2 e^- ⟶ 2 OH^-(aq) + H_2(g); oxidation: 2 I^-(aq) ⟶ I_2(s) + 2 e^-; net reaction: 2 I^-(aq) + 2 H_2O ⟶ I_2(s) + 2 OH^-(aq). **b.** Reduction: Cu^{2+}(aq) + 2 e^- ⟶ Cu(s); oxidation: 2 H_2O ⟶ O_2(g) + 4 H^+(aq) + 4 e^-; net reaction: 2 Cu^{2+}(aq) + 2 H_2O ⟶ 2 Cu(s) + 4 H^+(aq) + O_2(g). **20.9.** 22.5 min. **Estimation Exercises: 20.1 a.** Zn, Al; **b.** Co, Fe, Mg. **20.2.** Highest voltage (d); lowest voltage (c). **Conceptual Exercises: 20.1.** Oxidation of Zn to Zn^{2+} occurs on the zinc electrode. The reduction half-reaction is that of H^+(aq) in citric acid to H_2(g). In part this reduction occurs directly on the zinc electrode, but also some electrons pass through the electric measuring circuit to the copper electrode, where reduction of H^+(aq) also occurs. **20.2 a.** Anode: Ag(s), AgI(s) | I^-(1 M); cathode: Ag(s), AgCl(s) | Cl^-(1 M). **b.** Net reaction: AgCl(s) + I^-(1 M) ⟶ AgI(s) + Cl^-(1 M). **c.** E°_{cell} = 0.374 V. **20.3.** Cell reactions: (A) Zn^{2+}(1.0 M) + Cu(s) ⟶ Cu^{2+}(0.10 M) + Zn(s); (B) Zn(s) + Cu^{2+}(1.0 M) ⟶ Zn^{2+}(0.10 M) + Cu(s). Electric current will cease to flow when the concentrations of all the solutions become equal (0.55 M). **Review Questions: 3.** The anode is the electrode of an electrochemical cell where oxidation occurs; reduction occurs at the cathode. **6.** No, the basis of an electrochemical cell reaction must be an oxidation–reduction reaction, and a Brønsted–Lowry acid–base reaction does not involve changes in oxidation states. **9.** Standard electrode potentials are based on an *arbitrary* value of zero assigned to the half-reaction 2 H^+(1 M) + 2 e^- ⟶ H_2(g, 1 atm). Any reduction half-reaction that has a greater tendency to occur has a positive E° value, and any that has a lesser tendency, a negative E° value. **11.** E° is a standard electrode potential that describes the tendency for a particular reduction half-reaction to occur when reactants and products are in their standard states. E°_{cell} is a *difference* in E° values for two half-reactions, and its value depends on the specific half-reactions. **15.** Ag and Au. **18.** $\Delta G^\circ = -nFE^\circ_{cell} = -RT \ln K_{eq}$. **29.** Electroplating involves the reduction of metal cations to the free metal. The object to be electroplated is made the cathode, the electrode where reduction occurs. **30.** *Chlor–alkali* applies to the electrolysis of NaCl(aq), which produces Cl_2(g) ("chlor"), NaOH(aq) ("alkali"), and H_2(g). **Problems: 31.** E°_{red} = 0.796 V. **33.** E°_{red} = −0.17 V. **35.** E°_{red} = −1.13 V. **37 a.** Oxidation: Pb(s) ⟶ Pb^{2+}(aq) + 2 e^-; reduction: 2 H^+(aq) + 2 e^- ⟶ H_2(g, 1 atm); net reaction: Pb(s) + 2 H^+(aq) ⟶ Pb^{2+}(aq) + H_2(g); E°_{cell} = 0.125 V. **b.** Oxidation: 2 I^-(aq) ⟶ I_2(s) + 2 e^-; reduction: Cl_2(g) + 2 e^- ⟶ 2 Cl^-(aq); net reaction: 2 I^-(aq) + Cl_2(g) ⟶ I_2(s) + 2 Cl^-(aq); E°_{cell} = 0.823 V. **39 a.** Zn(s) ⟶ Zn^{2+}(aq) + 2 e^-; Ag^+(aq) + e^- ⟶ Ag(s); Zn(s) + 2 Ag^+(aq) ⟶ Zn^{2+}(aq) + 2 Ag(s); E°_{cell} = 1.563 V; Zn(s) | Zn^{2+}(aq) | | Ag^+(aq) | Ag(s). **b.** Fe^{2+}(aq) ⟶

Fe^{3+}(aq) + e^-; O_2(g) + 4 H^+(aq) + 4 e^- $\longrightarrow$ 2 H_2O; 4 Fe^{2+}(aq) + O_2(g) + 4 H^+(aq) $\longrightarrow$ 4 Fe^{3+}(aq) + 2 H_2O; E°_{cell} = 0.458 V; Pt | Fe^{2+}(aq), Fe^{3+}(aq) || H_2O, H^+(aq) | O_2(g), Pt. **41 a.** No, E°_{cell} = −0.140 V; **b.** Yes, E°_{cell} = 0.26 V. **43 a.** Yes, E°_{cell} = 1.273 V; **b.** No, E°_{cell} = −0.293 V; **c.** Yes, E°_{cell} = 0.26 V. **45 a.** Silver does not react with HCl(aq) because H^+(aq) is not a good enough oxidizing agent to oxidize Ag(s) to Ag^+(aq). E°_{cell} for the reaction is −0.800 V. **b.** Nitrate ion in acidic solution is a good enough oxidizing agent to oxidize Ag(s) to Ag^+(aq). E°_{cell} for the reaction is 0.156 V. **47.** 0.337 V < E°_{red} < 0.800 V. **49 a.** E°_{cell} = 2.476 V, ΔG° = −716.7 kJ; **b.** E°_{cell} = −0.03 V, ΔG° = 6 × 10^1 kJ. **51 a.** K_{eq} = [Fe^{3+}]/[Fe^{2+}][Ag^+] = 3.1; **b.** K_{eq} = [Mn^{2+}](P_{Cl_2})/[H^+]^4[Cl^-]^2 = 4 × 10^{-5}; **c.** K_{eq} = (P_{O_2})[Cl^-]^2/[OCl^-]^2 = 1 × 10^{33}. **53.** [Sn^{2+}] = 0.71 M; [Pb^{2+}] = 0.29 M. **55 a.** 0.691 V; **b.** 0.638 V; **c.** 0.566 V. **57.** pH = 1.82. **59.** 0.15 V. **61.** Anode: Zn(s) $\longrightarrow$ Zn^{2+}(aq) + 2 e^-; cathode: Cl_2(g) + 2 e^- $\longrightarrow$ 2 Cl^-(aq); net reaction: Zn(s) + Cl_2(g) $\longrightarrow$ Zn^{2+}(aq) + 2 Cl^-(aq); E°_{cell} = 2.121 V. **63.** Anode: Zn(s) + 2 OH^-(aq) $\longrightarrow$ ZnO(s) + H_2O + 2 e^-; cathode: Ag_2O(s) + H_2O + 2 e^- $\longrightarrow$ 2 Ag(s) + 2 OH^-(aq); net reaction: Zn(s) + Ag_2O(s) $\longrightarrow$ ZnO(s) + 2 Ag(s). **65.** Oxygen is the oxidizing agent required to oxidize Fe(s) to Fe^{2+} and then to Fe^{3+}. Water is a reactant in the reduction half-reaction, in which O_2(g) is reduced to OH^-(aq). Water is also a reactant in the conversion of $Fe(OH)_2$ to $Fe_2O_3 \cdot xH_2O$ ("rust"). The electrolyte completes the electrical circuit between the cathodic and anodic areas. **67.** The sacrificial anode, a more active metal, is consumed ("sacrificed") in place of the metal being protected. **69 a.** Ni(s, anode) $\longrightarrow$ Ni(s, cathode); **b.** Ni(s, anode) $\longrightarrow$ Ni(s, cathode); **c.** 2 Ni^{2+}(aq) + 2 H_2O $\longrightarrow$ 2 Ni(s, cathode) + 4 H^+(aq) + O_2(g). **71 a.** Ba(l) and Cl_2(g); **b.** H_2(g) and Br_2(l); **c.** H_2(g) and O_2(g). **73.** 14.3 g Ag. **75.** 7.59 × 10^4 C. **77.** Na(s) does not electrodeposit from $NaNO_3$(aq). Of the remaining solutions, $AgNO_3$(aq) yields the greatest number of moles of deposit. One mole of silver is formed for every mole of electrons (Ag^+ + e^- $\longrightarrow$ Ag), whereas only one-half mole of copper and zinc are formed (M^{2+} + 2 e^- $\longrightarrow$ M). Given that the atomic weight of Ag is greater than those of Cu and Zn, $AgNO_3$(aq) (solution c) yields the greatest mass of metal deposit. **79.** Sodium metal reacts with water, producing H_2(g) and NaOH(aq). Instead of just establishing a half-reaction electrode equilibrium, a piece of sodium enters into in a complete redox reaction with water. **81.** 52.9 mL O_2(g). **83.** E°_{cell} = −0.13 V, and the reaction should not occur with reactants and products in their standard states. However, because the acid used may have [H^+] > 1 M, and especially because Cl_2(g) is driven off by heating, the forward reaction is favored. **85.** E_{cell} = 0.032 V. **87.** E_{cell} = 0.269 V. **89.** As the cell operates, the concentrated solution becomes more dilute, the dilute solution becomes more concentrated, and the cell voltage decreases. The cell stops producing electricity when the concentrations in the two half-cell compartments have become equal. **91.** Anode: 2 H_2O $\longrightarrow$ 4 H^+(aq) + O_2(g) + 4 e^-; cathode: 2 H_2O + 2 e^- $\longrightarrow$ H_2(g) + 2 OH^-(aq); net reaction: 2 H_2O $\longrightarrow$ 2 H_2(g) + O_2(g). The H^+(aq) produced in the anode compartment and the OH^-(aq) in the cathode compartment are produced in exactly equal molar amounts, regardless of concentration of the Na_2SO_4, the electrolysis time, or the current used. When the two solutions are mixed, the result is simply Na_2SO_4(aq) at its characteristic pH (about 7).

Chapter 21

Exercise: 21.1. 2 BCl_3(g) + 3 H_2(g) $\longrightarrow$ 2 B(s) + 6 HCl(g). **Conceptual Exercise: 21.1.** The reaction has a negative E°_{cell} (−0.90 V) and a positive ΔG° (174 kJ), indicating that it is nonspontaneous, just as was concluded in Conceptual Example 21.1. The ΔG° values do not agree because the states of the reactants and products differ; for example, H_2SO_4(l) and Br_2(g) in Conceptual Example 21.1 and Br_2(g) and [H^+] = [SO_4^{2-}] = 1 M here. **Review Questions: 2.** boron and silicon. **3.** oxygen, silicon, and aluminum. **4.** aluminum (metal) and fluorine (nonmetal). **21 a.** silver azide; **b.** potassium thiocyanate; **c.** astatine oxide; **d.** telluric acid. **22 a.** H_2SeO_4; **b.** H_2Te; **c.** $Pb(N_3)_2$; **d.** AgAt. **23 a.** bauxite, Al_2O_3; boric oxide, B_2O_3; corundum, Al_2O_3; cyanogen, C_2N_2; hydrazine, N_2H_4; silica, SiO_2. **24 a.** apatite: Ca, P, O; borax: Na, B, O; corundum: Al, O; quartz: Si, O. **Problems: 25.** B_2O_3(s) + 3 Mg(s) $\longrightarrow$ 2 B(s) + 3 MgO(s). **27.** BH_3 is an electron-deficient structure that does not exist as a stable molecule (the stable species is diborane, B_2H_6). The possibility of resonance structures with B-to-F double bonds leads to a resonance hybrid for BF_3 that is a stable molecule. **29.** Al_2O_3(s) + 3 H_2SO_4(concd aq) + 15 H_2O(l) $\longrightarrow$ $Al_2(SO_4)_3 \cdot 18H_2O$(s). **31.** The fact that Al_2O_3 is amphoteric and Fe_2O_3 is not is the basis for purifying bauxite ore. Another key feature is the use of molten cryolite, Na_3AlF_6(l), as a solvent for Al_2O_3(s). Electrolysis can be conducted at a much lower temperature and in a better electrical conductor. **33.** $\Delta H^\circ \approx$ −852 kJ per mole of Fe(l) produced. This result is only approximate because it is based on data at 298 K, whereas the reaction occurs at a very high temperature. Also, the estimate assumes ΔH°_f = 0 for Fe(s), even though at the temperature of the reaction the stable form of iron is Fe(l). **35.** Aluminum could react with strongly acidic foods to produce H_2(g); the metal would become pitted. In a strongly basic medium (oven cleaner), aluminum could react to produce $[Al(OH)_4]^-$, and again the metal would become pitted. **37.** Na_2C_2(s) + 2 H_2O(l) $\longrightarrow$ 2 NaOH(aq) + C_2H_2(g).

39.

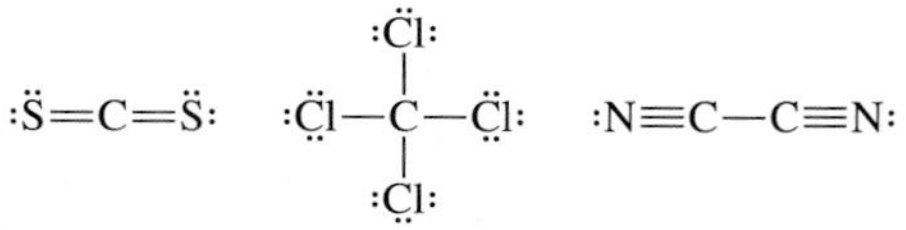

41. The geometric structure of the anion $Si_2O_7^{6-}$ is that of two SiO_4 tetrahedra sharing one O atom. The Lewis structure is

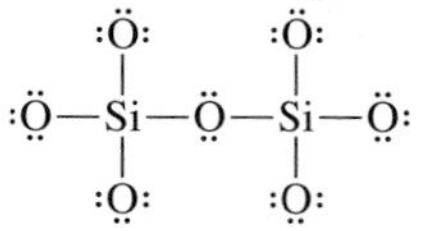

43. Each (OH) grouping has a total O.S. of −2 + 1 = −1; there are four such groups for a total of −4. The (Si_2O_5) grouping has a total O.S. of (2 × 4) − (5 × 2) = −2. The three Mg^{2+} ions have a total O.S. of +6. For the entire formula unit: total O.S. = +6 − 4 − 2 = 0, as required for a formula unit. **45 a.** $Si(CH_3)_4$; **b.** $SiCl_2(CH_3)_2$; **c.** $SiH(C_2H_5)_3$. **47.** Predictions are based on the values of E°_{cell} for the given oxidation half-reactions combined with the reduction half-reaction: PbO_2(s) + 4 H^+(aq) + 2 e^- $\longrightarrow$ Pb^{2+} + 2 H_2O; E° = 1.455 V. The results are **a.** yes; **b.** no; **c.** no; **d.** yes. **49 a.** Sn(s) + 2 HCl(aq) $\longrightarrow$ $SnCl_2$(aq) +

$H_2(g)$; **b.** $SnCl_2 + Cl_2(g) \longrightarrow SnCl_4$; **c.** $SnCl_4(aq) + 4\ NH_3(aq) + 2\ H_2O \longrightarrow SnO_2(s) + 4\ NH_4^+(aq) + 4\ Cl^-(aq)$. **51.** $(CH_3)_2NNH_2(l) + 4\ O_2(g) \longrightarrow 2\ CO_2(g) + 4\ H_2O(g) + N_2(g)$. **53 a.** $NH_2NH_2(aq) + 2\ HCl(aq) \longrightarrow NH_3NH_3^{2+}(aq) + 2\ Cl^-(aq)$; **b.** $3\ Cu(s) + 8\ H^+(aq) + 2\ NO_3^-(aq) \longrightarrow 3\ Cu^{2+}(aq) + 4\ H_2(g) + 2\ NO(g)$; **c.** $2\ NO(g) + O_2(g) \longrightarrow 2\ NO_2(g)$. **55.** reduction: $Fe_3^+(aq) + e^- \longrightarrow Fe^{2+}(aq)$; oxidation: $NH_2NH_3^+(aq) \longrightarrow N_2(g) + 5\ H^+ + 4\ e^-$; net reaction: $4\ Fe^{3+}(aq) + NH_2NH_3^+(aq) \longrightarrow 4\ Fe^{2+}(aq) + N_2(g) + 5\ H^+(aq)$; $E^\circ_{cell} = 1.00$ V. **57.** The principal allotropes of phosphorus are white P and red P, with white P being the more reactive. The molecular structure of white P consists of individual P_4 tetrahedra. In red P, the P_4 tetrahedra are joined into long chains. **59.** $P_4(s) + 3\ KOH(aq) + 3\ H_2O \longrightarrow 3\ KH_2PO_2(aq) + PH_3(g)$. **61.** reduction: $2\ SO_3^{2-} + 3\ H_2O + 4\ e^- \longrightarrow S_2O_3^{2-} + 6\ OH^-$; oxidation: $2\ S(s) + 6\ OH^- \longrightarrow S_2O_3^{2-} + 3\ H_2O + 4\ e^-$; net reaction: $SO_3^{2-}(aq) + S(s) \longrightarrow S_2O_3^{2-}(aq)$. **63.** Displacement of $Br_2(l)$ from $Br^-(aq)$ requires this oxidation half-reaction to occur: $2\ Br^-(aq) \longrightarrow Br_2(l) + 2\ e^-$; $E^\circ_{ox} = -1.065$ V. Only a reduction half-reaction with $E^\circ_{red} > 1.065$ V will work, and this must be $Cl_2(g) + 2\ e^- \longrightarrow 2\ Cl^-(aq)$, for which $E^\circ_{red} = 1.358$ V. I_2 is too poor an oxidizing agent to work, and I^-, Cl^-, and F^- can only be reducing agents, not oxidizing agents. **65.** $I^-(aq) + 3\ Cl_2(g) + 3\ H_2O \longrightarrow IO_3^-(aq) + 6\ H^+(aq) + 6\ Cl^-(aq)$; $Cl_2(g) + 2\ Br^-(aq) \longrightarrow Br_2(l) + 2\ Cl^-(aq)$. When the products of the reaction are treated with $CS_2(l)$, the Br_2 dissolves and the other products remain in the aqueous solution. **67.** ICl: interhalogen; NaCl: a halide salt (neither an interhalogen nor pseudohalogen); $(CN)_2$: pseudohalogen. **69 a.** $Cl_2(g) + H_2O(l) \longrightarrow HOCl(aq) + H^+(aq) + Cl^-(aq)$; **b.** $Cl_2(g) + 2\ NaOH(aq) \longrightarrow NaOCl(aq) + NaCl(aq) + H_2O(l)$; **c.** $3\ Cl_2(g) + 6\ NaOH(aq) \longrightarrow 5\ NaCl(aq) + NaClO_3(aq) + 3\ H_2O$. **71 a.** XeO_3: trigonal pyramidal (AXE_3); XeO_4: tetrahedral (AX_4). **73 a.** BrF_3: T-shaped (AX_3E_2); **b.** IF_5: square pyramidal (AX_5E). **75.** $Na_2B_4O_7 \cdot 10H_2O(s) + 6\ CaF_2(s) + 8\ H_2SO_4(\text{concd aq}) \longrightarrow 2\ NaHSO_4(s) + 6\ CaSO_4(s) + 4\ BF_3(g) + 17\ H_2O(l)$. **77.** $MnF_6^{2-} + 2\ SbF_5 \longrightarrow MnF_4 + 2\ SbF_6^-$; $MnF_4 \longrightarrow MnF_2 + F_2(g)$. **79.** Sulfur melts at 119 °C. Water is heated under pressure to produce steam at temperatures greater than 119 °C. The superheated steam melts the sulfur underground. Sulfur neither reacts with hot water nor dissolves in it, so the liquid sulfur can be brought to the surface in a pure condition. **81.** 1.4×10^4 L air. **83.** $Sn^{4+}(aq) + 2\ e^- \longrightarrow Sn^{2+}(aq)$, $E^\circ_{red} = 0.154$ V; $Sn(s) \longrightarrow Sn^{2+}(aq) + 2\ e^-$, $E^\circ_{ox} = 0.137$ V. For the reaction $Sn(s) + Sn^{4+}(aq) \longrightarrow 2\ Sn^{2+}(aq)$, $E^\circ_{cell} = 0.291$ V. The reduction of $Sn^{4+}(aq)$ by Sn(s) is a spontaneous reaction. Thus, all the tin ion in solution is $Sn^{2+}(aq)$ as long as some solid tin remains. **85.** length of unit cell: 335 pm; volume of unit cell: 3.76×10^{-23} cm^3; atoms per unit cell: one; density = 9.23 g/cm^3.

Chapter 22

Exercises: 22.1. At 800 °C: $TiO_2(s) + 2\ C(s) + 2\ Cl_2(g) \longrightarrow TiCl_4(g) + 2\ CO(g)$; at 1200 °C: $TiCl_4(g) + O_2(g) \longrightarrow TiO_2(s) + 2\ Cl_2(g)$. **22.2.** From Appendix D, $MnO_4^-(aq) + 2\ H_2O + 3\ e^- \longrightarrow MnO_2(s) + 4\ OH^-(aq)$; $E^\circ_{red} = 0.60$ V. $MnO_4^-(aq)$ in basic solution will oxidize any species for which $E^\circ_{ox} > -0.60$ V. This includes $Br^-(aq)$ to $BrO_3^-(aq)$, $Br_2(l)$ to $BrO^-(aq)$, Ag(s) to $Ag_2O(s)$, $NO_2^-(aq)$ to $NO_3^-(aq)$, S(s) to $SO_3^{2-}(aq)$, and so on. **Conceptual Exercise: 22.1.** $4\ FeCr_2O_4(s) + 16\ NaOH(l) + 7\ O_2(g) \longrightarrow 8\ Na_2CrO_4(s) + 4\ Fe(OH)_3(s) + 2\ H_2O(g)$. **Review Questions: 3.** cobalt. **4.** No, it is the representative element indium. The 4*d* subshell is filled and the 5*p* subshell is partially filled. **6.** copper, silver, and gold. **9.** Cr, Ni, Zn. **10.** V, Cr, Mn. **11.** $Sc(OH)_3$, $Cr(OH)_3$, $Zn(OH)_2$. **12.** $Fe(OH)_3$. **13 a.** iron coated with zinc; **b.** $FeCr_2O_4$; **c.** $Cu_2(OH)_2CO_3$. **14 a.** ferromanganese; **b.** brass; **c.** aqua regia. **17.** Au. **18.** Co and Ni atoms lose the two 4*s* electrons to form 2+ ions. An Fe atom can lose a third electron as well, a 3*d* electron. This leaves a half-filled 3*d* subshell, which is an especially stable electron configuration. **19 a.** scandium hydroxide; **b.** iron(II) silicate; **c.** sodium manganate; **d.** osmium pentacarbonyl. **20 a.** $BaCr_2O_7$; **b.** CrO_3; **c.** Hg_2Br_2; **d.** Na_3VO_4. **Problems: 25.** In both cases electrons are lost to produce the electron configuration of Ar. With calcium this means the two 4*s* electrons, and the ion Ca^{2+} is formed. With scandium, the 3*d* electron is lost as well, producing Sc^{3+}. **27.** (b). The transition elements are metals (EN < 2), but they are not as active as the alkali metals (some of which have EN < 1). **29 a.** The Ca atom is smaller than the K atom because it has a higher nuclear charge (+20 compared to +19) but the same number of electrons in its noble gas core (18), coupled with the fact that the two 4*s* electrons are not effective in screening one another. **b.** The Mn atom is smaller than the Ca atom because it has a higher nuclear charge (+25 compared to +20) and the same number of valence electrons in the same configuration (4*s*^{2}). **c.** The Mn and Fe atoms are about the same size because they have about the same nuclear charge (+25 and +26, respectively), the same number of valence electrons (4*s*^{2}), and inner shell electrons that are about equally effective in shielding the valence electrons from the nucleus.

31 a. Ti: [Ar] 3*d* [↑][↑][ ][ ][ ] 4*s* [↑↓]

b. Ag: [Kr] 4*d* [↑↓][↑↓][↑↓][↑↓][↑↓] 5*s* [↑]

c. Cr^{2+}: [Ar] 3*d* [↑][↑][↑][↑][ ] 4*s* [ ]

d. Mn^{2+}: [Ar] 3*d* [↑][↑][↑][↑][↑] 4*s* [ ]

33. oxidation state of V: +2 (V^{2+}); +4 (VO^{2+}); +5 (VO_2^+); +5 (VO_4^{3-}) **35 a.** $2\ Sc(s) + 6\ HCl(aq) \longrightarrow 2\ ScCl_3(aq) + 3\ H_2(g)$; **b.** $Sc(OH)_3(s) + 3\ HCl(aq) \longrightarrow ScCl_3(aq) + 3\ H_2O(l)$; **c.** $Sc(OH)_3(s) + 3\ Na^+(aq) + 3\ OH^-(aq) \longrightarrow 3\ Na^+(aq) + [Sc(OH)_6]^{3-}(aq)$;

d. $2\ ScCl_3(l) \xrightarrow{\text{electrolysis}} 2\ Sc(l) + 3\ Cl_2(g)$.

37 a. $Ba(s) + 2\ H^+(aq) + 2\ Cl^-(aq) \longrightarrow Ba^{2+}(aq) + 2\ Cl^-(aq) + H_2(g)$; followed by, $Ba^{2+}(aq) + 2\ Cl^-(aq) + 2\ K^+(aq) + CrO_4^{2-}(aq) \longrightarrow BaCrO_4(s) + 2\ K^+(aq) + 2\ Cl^-(aq)$. **b.** Combine the reduction half-reaction $MnO_4^-(aq) + 2\ H_2O + 3\ e^- \longrightarrow MnO_2(s) + 4\ OH^-(aq)$, $E^\circ_{red} = 0.60$ V with an oxidation half-reaction for which $E^\circ_{ox} > -0.60$ V. For example, $3\ NO_2^-(aq) + 2\ MnO_4^-(aq) + H_2O \longrightarrow 3\ NO_3^-(aq) + 2\ MnO_2(s) + 2\ OH^-(aq)$, $E^\circ_{cell} = 0.59$ V. **39.** $2\ Mn^{2+}(aq) + 5\ BiO_3^-$

$+ 14\ H^+ \longrightarrow 2\ MnO_4^-(aq) + 5\ Bi^{3+}(aq) + 7\ H_2O$. **41.** $[CrO_4{}^{2-}]$ in the equilibrium $Cr_2O_7{}^{2-}(aq) + H_2O \rightleftharpoons 2\ CrO_4{}^{2-}(aq) + 2\ H^+(aq)$ is large enough that K_{sp} of $PbCrO_4(s)$ is exceeded. **43.** Fe(s). The reducing agent is oxidized, and $E^\circ_{ox} = 0.440$ V for the half-reaction: $Fe(s) \longrightarrow Fe^{2+}(aq) + 2\ e^-$. For the oxidation of Co(s) to $Co^{2+}(aq)$, $E^\circ_{ox} = 0.277$ V, and for the oxidation of $Fe^{2+}(aq)$ to $Fe^{3+}(aq)$, $E^\circ_{ox} = -0.771$ V. $Co^{3+}(aq)$ is an oxidizing agent (reduced to Co^{2+}), not a reducing agent. **45.** $Fe_2O_3(s) + 3\ Cl_2(g) + 10\ OH^-(aq) \longrightarrow 2\ FeO_4{}^{2-}(aq) + 6\ Cl^-(aq) + 5\ H_2O$. **47 a.** $Cu(s) + SO_4{}^{2-}(aq) + 4\ H^+(aq) \longrightarrow Cu^{2+}(aq) + 2\ H_2O + SO_2(g)$; **b.** $3\ Cu(s) + 2\ NO_3^-(aq) + 8\ H^+(aq) \longrightarrow 3\ Cu^{2+}(aq) + 4\ H_2O + 2\ NO(g)$. **49.** For the reduction half-reaction $NO_3^-(aq) + 4\ H^+(aq) + 3\ e^- \longrightarrow 2\ NO(g) + 2\ H_2O$, $E^\circ_{red} = 0.96$ V. E°_{ox} for the oxidation of Ag(s) to $Ag^+(aq)$ is -0.800 V. Thus $HNO_3(aq)$ will react with Ag(s). E°_{ox} for the oxidation of Au(s) to $Au^{3+}(aq)$ is -1.52 V. As a consequence, $HNO_3(aq)$ will not react with Au(s). **51.** The Group 1B metals have higher melting points and densities than do the 1A metals. The Group 1B metals do not react with water or mineral acids, whereas the 1A metals do, to liberate $H_2(g)$. The Group 1B cations form many complex ions, whereas the Group 1A cations form very few. Almost all Group 1A compounds are water soluble, whereas many of the Group 1B compounds are not. **53 a.** $3\ Hg(l) + 2\ NO_3^-(aq) + 8\ H^+(aq) \longrightarrow 3\ Hg^{2+}(aq) + 4\ H_2O + 2\ NO(g)$; **b.** $ZnO(s) + 2\ CH_3COOH(aq) \longrightarrow Zn^{2+}(aq) + 2\ CH_3COO^-(aq) + H_2O(l)$. **55.** Commercially important elements are generally found in readily available mineral forms (ores) from which they can be extracted by straightforward chemical reactions. Some elements, even though abundant overall in Earth's crust, are too widely scattered in their mineral forms and/or too difficult to extract to make them commercially important. **57.** CO molecules have all electrons paired, but the Mn atom does not ($Z = 25$). The species $Mn(CO)_5$ has an unpaired electron. However, the dimer $Mn_2(CO)_{10}$ has all electrons paired. **59.** $2\ Cu(s) + O_2(g) + H_2O(l) + CO_2(g) \longrightarrow Cu_2(OH)_2CO_3(s)$. **61.** Ag^{2+} is very easily reduced to Ag^+ ($E^\circ_{red} = 1.98$ V), which means that Ag^{2+} is a powerful oxidizing agent. In the hypothetical compounds $AgBr_2$, $AgCl_2$, and AgS, the Ag^{2+} ion would oxidize the anions to the free elements. Thus, the compounds do not exist at all. Oxidation of O^{2-} and F^- to the free elements is much more difficult to accomplish, and so AgO and AgF_2 can be isolated. **63 a.** Sn^{2+}, V^{2+}, and Fe^{2+} are all capable; **b.** V^{2+} and Sn^{2+} are capable; **c.** only V^{2+} is capable. **65.** 2.70 g Cr. **67.** Let $[Cr_2O_7{}^{2-}] = x$ and $[CrO_4{}^{2-}] = 1 - 2x$, and substitute these values, together with $[H^+] = 1 \times 10^{-9}$, into the K_{eq} expression. $[Cr_2O_{72-}] \approx 3 \times 10^{-4}$, which is practically zero when compared to the total Cr(VI) concentration of 1 M.

Chapter 23

Exercises: 23.1 a. coordination number: 6, oxidation state: +3; **b.** coordination number: 6, oxidation state: +2. **23.2 a.** diamminesilver(I) ion; **b.** tetrachloroaurate(III) ion; **c.** pentaamminebromocobalt(III) bromide. **23.3 a.** $[Co(en)_3]^{3+}$; **b.** $[CrCl_4(NH_3)_2]^-$; **c.** $[PtCl_2(en)_2]SO_4$. **23.4 a.** not possible; **b.** yes, for example, $[Pt(NH_3)_4][CuCl_4]$ is a possible isomer. **23.5 a.** This is a strong-field, low-spin complex ion with the six 3*d* electrons paired in the lower energy set of three *d* orbitals. **b.** This is a weak-field complex ion with four of the eight 3*d* electrons paired in the lower energy set of two *d* orbitals and the remaining four distributed as one pair and 2 unpaired electrons in the upper energy set of three *d* orbitals. **Conceptual Exercise: 23.1.** There would be a single molecular form and no isomerism. **Review Questions: 8.** NH_3 is a Brønsted–Lowry base because it can accept a proton. It is a Lewis base because it can donate a pair of electrons to form a covalent bond. **11 a.** water; **b.** ammonia; **c.** ethylenediamine; **d.** oxalate ion. **12 a.** chloride ion; **b.** carbonate ion; **c.** nitrate ion; **d.** hydrogen sulfite ion. **14.** The metal atom is part of a complex anion. **15.** A nitro ligand is a nitrite ion in which N is the donor atom, whereas a nitrito ligand is a nitrite ion in which an O atom is the donor atom. **Problems: 21 a.** 6; **b.** 4; **c.** 4; **d.** 6. **23 a.** +2; **b.** +3; **c.** +3; **d.** +3. **25 a.** tetraamminecopper(II) ion; **b.** hexafluoroferrate(III) ion; **c.** tetraamminedichloroplatinum(IV) ion; **d.** tris(ethylenediamine)chromium(III) ion. **27 a.** $[Fe(H_2O)_6]^{3+}$; **b.** $[CrBrCl(NH_3)_4]^+$; **c.** $[Al(ox)_3]^{3-}$. **29 a.** potassium hexacyanochromate(II); **b.** potassium trioxalatochromate(III). **31 a.** $Na_2[Zn(OH)_4]$; **b.** $[Cr(en)_3]_2(SO_4)_3$; **c.** $K_2Na[Co(NO_2)_6]$. **33 a.** It is an anion and should have an *ate* ending: tetrahydroxozincate(II) ion. **b.** Because the complex ion is an anion, the metal should be named last and given an "ate" ending: hexafluoroferrate(III) ion. **35 a.** Listing the ligands in a different order does not make these structures isomers. Isomerism would only exist if a pair of ligands is cis in one structure and trans in the other, but this cannot be indicated by the formula alone. **b.** Yes, these are isomers. **c.** No, these are not isomers. The oxidation state of Fe is +3 in one ion and +2 in the other. **d.** Yes, these are isomers. They differ in the way in which the nitrite ion ligand is attached. **37.** The en ligand must attach through an adjacent pair of sites in this octahedral complex, and the other four sites are occupied by Cl^- ions. There is only one structure possible for this complex ion (cis-trans isomerism is not possible). **39.** Yes, two nonsuperimposable mirror-image structures can be seen by replacing the en groups above and below the central plane in Figure 23.8 with ox groups. **41 a.** Paramagnetic. The electron configuration of Mn^{2+} is $[Ar]3d^5$, and because of the odd number of electrons, there must be at least one unpaired electron present. If the complex ion is of the high-spin type, the number of unpaired electrons is five; for the low-spin type it is one. **b.** Diamagnetic. The electron configuration of Fe^{2+} is $[Ar]3d^6$. The complex ion should be low-spin because CN^- is a strong-field ligand. The six electrons should all be paired in the lower energy group of three 3*d* orbitals. **43.** $[Cr(H_2O)_6]^{3+}(aq)$ is violet and $[Cr(NH_3)_6]^{3+}(aq)$ is yellow. NH_3 is a strong-field ligand, and H_2O is a weaker field ligand. More energy is absorbed in promoting an electron from the lower to the higher energy group of *d* orbitals in the ammine complex than in the aqua complex. The light corresponding to the requisite energy is of high frequency and short wavelength: violet. The transmitted light is white light that is deficient in violet: yellow. Thus, $[Cr(NH_3)_6]^{3+}(aq)$ should be yellow. **45.** The trans isomer does not exhibit optical isomerism; the structure and its mirror image are superimposable. The cis isomer and its mirror image are optical isomers. (To see that the structures are not superimposable, replace the en group in the central plane of the complex ions in Figure 23.8 by Cl^- ions.) **47.** The $[PtCl_2(NH_3)_2]$ complex is square planar (Figure 23.5), whereas $[ZnCl_2(NH_3)_2]$ is tetrahedral (recall Figure 23.2). The tetrahedral structure does not display cis-trans isomerism. **49.** Cr^{3+} has the electron configuration $[Ar]3d^3$. The three 3*d* electrons will remain unpaired in the lower energy set of three 3*d* orbitals, regardless of whether the ligands L are strong field or weak field. **51.** $[Pt(NH_3)_4][PtCl_4]$. **53.** The electron configuration of Ni^{2+} is $[Ar]3d^8$. The complex ion $[Ni(CN)_4]^{2-}$ is a strong-field complex in which all the electrons are paired (diamagnetic). Refer to the *d*-level splitting diagrams of Figure 23.11. The assignment of eight electrons to the tetrahedral splitting diagram would leave two electrons unpaired. The assignment of eight electrons to the octahedral diagram would also leave two unpaired. The assignment of eight electrons to the diagram for a square planar complex would fill the four lowest energy *d* orbitals with electron pairs and leave the highest energy *d* orbital empty. The structure of the complex ion is square planar.

Credits

Chapter 1 Page 2/© Philippe Plailly/Science Photo Library/ Photo Researchers, Inc. Page 4 (T)/Carey B. Van Loon. Page 4 (BL)/© Professor C. M. Lang. Page 4 (BR)/© Rosenfeld Images. Page 5/© Civil Aviation Authority. Page 6 (Both)/Carey B. Van Loon. Page 11/Carey B. Van Loon. Page 15/© 1995 Paul Silverman/Fundamental Photographs. Page 22/© Frank Siteman/Stock, Boston.

Chapter 2 Page 32 (L)/© Professor C. M. Lang. Page 32 (BL)/ © 1995 Richard Megna/Fundamental Photographs. Page 32 (BR)/ © 1989 Paul Silverman/Fundamental Photographs. Page 33/© The Bettmann Archive. Page 38/© Science Photo Library/Photo Researchers, Inc. Page 40/© 1990 Richard Megna/Fundamental Photographs. Page 45/© 1995 Richard Megna/Fundamental Photographs. Page 49 (T)/© 1987 Paul Silverman/Fundamental Photographs. Page 49 (BL)/© 1992 Richard Megna/Fundamental Photographs. Page 49 (BR)/© 1992 Richard Megna/Fundamental Photographs. Page 58/ © 1993 Joseph P. Sinnot/Fundamental Photographs.

Chapter 3 Page 68/Carey B. Van Loon. Page 73/Carey B. Van Loon. Page 76/© 1995 Richard Megna/Fundamental Photographs. Page 92 (Both)/Carey B. Van Loon. Page 96 (All)/Carey B. Van Loon. Page 99 (TL)/© 1995 Richard Megna/Fundamental Photographs. Page 99 (TC)/© 1995 Richard Megna/Fundamental Photographs. Page 99 (TR)/© 1995 Richard Megna/Fundamental Photographs. Page 99 (BL)/Carey B. Van Loon. Page 99 (BC)/ Carey B. Van Loon. Page 99 (BR)/Carey B. Van Loon. Page 106/© 1995 Richard Megna/Fundamental Photographs.

Chapter 4 Page 108/© Philippe Plailly/Science Photo Library/ Photo Researchers, Inc. Page 110/© Ellis Herwig/Stock, Boston. Page 117/© Professor C. M. Lang. Page 123 (Both)/© 1991 Richard Megna/ Fundamental Photographs. Page 124/© Science Photo Library/ Photo Researchers, Inc. Page 125/Carey B. Van Loon. Page 132/© Archive/ Photo Researchers, Inc.

Chapter 5 Page 154/© 1992 Richard Megna/Fundamental Photographs. Page 156/© 1979 Bill W. Marsh/Photo Researchers, Inc. Page 157/© 1986 Richard Megna/Fundamental Photographs. Page 159 (Both)/© 1995 Richard Megna/Fundamental Photographs. Page 186/ Carey B. Van Loon.

Chapter 6 Page 194/© Norman Owen Tomalin/Bruce Coleman Inc. Page 196/© The Bettmann Archive. Page 197/Carey B. Van Loon. Page 198/© Professor C. M. Lang. Page 199/© Professor C. M. Lang. Page 200/© Science VU/Visuals Unlimited. Page 201/ © Professor C. M. Lang. Page 210/Carey B. Van Loon. Page 211/ © Department of Physics, Imperial College/Science Photo Library/ Photo Researchers, Inc. Page 212 (L)/Carey B. Van Loon. Page 212 (R)/© 1995 Paul Silverman/Fundamental Photographs. Page 213/ © 1993 Paul Silverman/Fundamental Photographs. Page 217/© Science Photo Library/Photo Researchers, Inc. Page 221/© Simon Fraser/Searle Pharmaceuticals/Science Photo Library/ Photo Researchers, Inc. Page 224 (T)/© UPI/Bettmann. Page 224 (B)/© The Bettmann Archive.

Chapter 7 Page 236/© Andrea Brizzi/The Stock Market.

Chapter 8 Page 274/Carey B. Van Loon. Page 276/© Royal Observatory, Edinburgh/AATB/Science Photo Library/Photo Researchers, Inc. Page 278 (Both)/Carey B. Van Loon. Page 281/ © UPI/Bettmann Newsphotos. Page 282/© 1993 Richard Megna/ Fundamental Photographs. Page 283 (All)/© 1995 Richard Megna/ Fundamental Photographs. Page 286/© Mark E. Gibson/Visuals Unlimited. Page 289/© James L. Amos/Peter Arnold, Inc. Page 291/© 1995 Richard Megna/Fundamental Photographs. Page 291 (All)/© 1995 Richard Megna/Fundamental Photographs. Page 292/© Albert Copley/Visuals Unlimited. Page 295/© 1991 Richard Megna/Fundamental Photographs. Page 296/© William E. Ferguson. Page 298/© Howard Sochurek/The Stock Market. Page 299/ © 1993 Richard Megna/Fundamental Photographs. Page Page 302 (Both)/© 1995 Richard Megna/Fundamental Photographs. Page 303/© Cabisco/Visuals Unlimited.

Chapter 9 Page 310/© E. R. Degginger/Bruce Coleman Inc. Page 312/Carey B. Van Loon. Page 313/© Albert Copley/Visuals Unlimited. Page 314/© University of Pennsylvania. Page 320/ © Science Photo Library/Photo Researchers, Inc. Page 344/ © 1992 Richard Megna/Fundamental Photographs. Page 345/ © 1995 Richard Megna/Fundamental Photographs. Page 346/ © Bob Masini/Phototake NYC.

Chapter 10 Page 356/© Clive Freeman, The Royal Institution/ Science Photo Library/Photo Researchers, Inc. Page 360/ Carey B. Van Loon. Page 362 (All)/Carey B. Van Loon. Page 363/Carey B. Van Loon. Page 368/ © Dan McCoy/Rainbow. Page 378 (Both)/Carey B. Van Loon. Page 388/© 1992 Richard Megna/Fundamental Photographs. Page 389/© Clive Freeman, The Royal Institution/Science Photo Library/Photo Researchers, Inc.

Chapter 11 Page 394/© C. Haagner/Bruce Coleman Inc. Page 402/Carey B. Van Loon. Page 405/© 1992 Richard Megna/Fundamental Photographs. Page 407/Carey B. Van Loon. Page 408/Carey B. Van Loon. Page 413/© Custom Medical Stock Photo. Page 419/ © Jim Zuckerman/Westlight. Page 423 (T)/© Herman Eisenbeiss. Page 423 (B)/Carey B. Van Loon. Page 424 (Both)/Carey B. Van Loon. Page 427/Carey B. Van Loon. Page 435/Carey B. Van Loon.

Chapter 12 Page 438 (All)/© 1995 Richard Megna/Fundamental Photographs. Page 441/Carey B. Van Loon. Page 442/Carey B. Van Loon. Page 444/Carey B. Van Loon. Page 446 (TL)/© 1995 Richard Megna/Fundamental Photographs. Page 446 (TC)/© 1995 Richard Megna/Fundamental Photographs. Page 446 (TR)/© 1995 Richard Megna/Fundamental Photographs. Page 446 (B)/Carey B. Van Loon. Page 447/© Clo Research. Page 449 (T)/© 1995 Richard Megna/Fundamental Photographs. Page 449 (B)/Carey B.

Van Loon. Page 452/© 1995 Richard Megna/Fundamental Photographs. Page 453 (Both)/© 1995 Richard Megna/Fundamental Photographs. Page 456 (All)/© 1995 Richard Megna/Fundamental Photographs. Page 462/© 1995 Richard Megna/Fundamental Photographs. Page 462/© 1988 Richard Megna/Fundamental Photographs. Page 463 (All)/Carey B. Van Loon. Page 465 (All)/ © 1995 Richard Megna/Fundamental Photographs. Page 467/ © Stephen Frisch/Stock, Boston. Page 468/Carey B. Van Loon.

Chapter 13 Page 476/© Veronika Burmeister/Visuals Unlimited. Page 479/Carey B. Van Loon. Page 481 (Both)/© 1995 Richard Megna/Fundamental Photographs. Page 485/© 1995 Richard Megna/Fundamental Photographs. Page 486/© Ken Graham/Bruce Coleman Inc. Page 490 (All)/© 1990 Richard Megna/Fundamental Photographs. Page 492/© Lee White/Westlight. Page 494/© 1993 Michael DeMocker/Visuals Unlimited. Page 500/© Jim Harrison/ Stock, Boston. Page 503/© Veronika Burmeister/Visuals Unlimited. Page 504/© Science Photo Library/Photo Researchers, Inc. Page 508 (T)/© Stephen Frisch/Stock, Boston. Page 508 (BL)/ Carey B. Van Loon. Page 508 (BR)/Carey B. Van Loon.

Chapter 14 Page 516/© 1987 Bruce Coleman Inc. Page 519/ © Norman Owen Tomalin/Bruce Coleman Inc. Page 520/© David Nunuk/© Science Photo Library/Photo Researchers, Inc. Page 521/ © 1992 Kristen Brochmann/Fundamental Photographs. Page 522 (T)/© D. Cavagnaro/Visuals Unlimited. Page 522 (B)/© 1992 Richard Megna/Fundamental Photographs. Page 524/© Ed Degginger/Bruce Coleman Inc. Page 526/© Ken Wagner/Phototake NYC. Page 527/© Science Photo Library/Photo Researchers, Inc. Page 528/© Science Photo Library/Photo Researchers, Inc. Page 530/© Science VU/Visuals Unlimited. Page 531/© Geoff Tompkinson/Science Photo Library/Photo Researchers, Inc. Page 532/ © Science VU/Visuals Unlimited. Page 534/© Peter S. Garra. Page 536/© NASA/Science Photo Library/Photo Researchers, Inc. Page 541 (T)/© M. Freeman/Bruce Coleman Inc. Page 541 (B)/© 1995 M. C. Chamberlain/DRK Photo. Page 543 (B)/© Dr. Gerald L. Fisher/Science Photo Library/ Photo Researchers, Inc. Page 543 (T)/© Marty Cordano/DRK Photo. Page 547/© Douglas Peebles/Westlight. Page 548/© Argonne National Laboratory Photo. Page 549/© 1991 Day Williams/Photo Researchers, Inc. Page 550/© Phototake NYC.

Chapter 15 Page 556/© Science VU/Visuals Unlimited. Page 558/© 1992 Richard Megna/Fundamental Photographs. Page 559/© Science VU/Visuals Unlimited. Page 594/Carey B. Van Loon.

Chapter 16 Page 608/© Clo Research. Page 619/© Wendell Metzen/Bruce Coleman Inc. Page 620/© Professor C.M. Lang.

Chapter 17 Page 644/© KAPIA-Frankfurt. Page 656/© 1995 Richard Megna/Fundamental Photographs. Page 657/© 1995 Richard Megna/Fundamental Photographs. Page 664/Carey B. Van Loon. Page 668/Carey B. Van Loon. Page 669 (All)/© 1995 Richard Megna/Fundamental Photographs. Page 673/Carey B. Van Loon. Page 679/©Dan McCoy/Rainbow. Page 681/Carey B. Van Loon. Page 682/© 1995 Richard Megna/Fundamental Photographs. Page 684 (All)/© 1995 Richard Megna/Fundamental Photographs.

Chapter 18 Page 706/Carey B. Van Loon. Page 709/© 1995 Richard Megna/Fundamental Photographs. Page 712/© 1995 Richard Megna/Fundamental Photographs. Page 713/Carey B. Van Loon. Page 715/© Max Listgarten/Visuals Unlimited. Page 716/ © 1995 Richard Megna/Fundamental Photographs. Page 718/ © 1995 Richard Megna/Fundamental Photographs. Page 720/ © 1995 Richard Megna/Fundamental Photographs. Page 722 (Both)/ © 1995 Richard Megna/Fundamental Photographs. Page 725/ Carey B. Van Loon. Page 727/© The Bettmann Archive.

Chapter 19 Page 734/© J. A. Strandberg/FDNY Forensic Unit. Page 737/Carey B. Van Loon. Page 739 (All)/© 1995 Kristen Brochmann/Fundamental Photographs. Page 741/Carey B. Van Loon. Page 743/© AP/Wide World Photos. Page 746/© Wesley Bocxe/Photo Researchers, Inc. Page 749/© Science Photo Library/Photo Researchers, Inc.

Chapter 20 Page 766/Carey B. Van Loon. Page 775/Carey B. Van Loon. Page 777 (Both)/Carey B. Van Loon. Page 778/Carey B. Van Loon. Page 781/© Science Photo Library/Photo Researchers, Inc. Page 785/Carey B. Van Loon. Page 791/NASA. Page 792/Courtesy of Alupower. Page 794/© Science VU-NSRDC/Visuals Unlimited. Page 800/© The Bettmann Archive. Page 803/© Science VU-AMAX/Visuals Unlimited.

Chapter 21 Page 810/Carey B. Van Loon. Page 814/© Stephen Frisch/Stock, Boston. Page 816 (T)/© UPI/Bettmann. Page 816 (B)/ © The Bettmann Archive. Page 817/© 1979 Ken Rogers/Westlight. Page 819/Courtesy of Hrand Djerahirdjian, S.A. Page 820/© 1993 Dorothy Littell Greco/Stock, Boston. Page 821/© Hank Morgan/ Science Source/Photo Researchers, Inc. Page 822/© Chuck O'Rear/ Westlight. Page 825/© 1995 Paul Silverman/Fundamental Photographs. Page 826 (T)/© Dr. Jeremy Burgess/Science Photo Library/Photo Researchers, Inc. Page 826 (B)/© 1992 Paul Silverman/ Fundamental Photographs. Page 827/© M. Angelo/Westlight. Page 830/NASA. Page 832/© Mark C. Burnett/Stock, Boston. Page 834 (T)/Carey B. Van Loon. Page 834 (B)/© Space Telescope Science Institute/NASA/Science Photo Library/Photo Researchers, Inc. Page 836/Carey B. Van Loon. Page 839/© Owen Franken/Stock, Boston. Page 842/© R. Marsh Starks/*Las Vegas Sun*.

Chapter 22 Page 856/© 1984 Rich Chisholm/The Stock Market. Page 857/© 1992 Richard Megna/Fundamental Photographs. Page 858 (T)/© N. H. (Dan) Cheatham/DRK Photo. Page 858 (BL)/ Carey B. Van Loon. Page 858 (BR)/Carey B. Van Loon. Page 859 (BL)/Carey B. Van Loon. Page 859 (BC)/Carey B. Van Loon. Page 859 (BR)/Carey B. Van Loon. Page 859 (T)/© 1995 Michael Dalton/Fundamental Photographs. Page 860 (All)/Carey B. Van Loon. Page 861/© Science VU/Visuals Unlimited. Page 867/© Gabe Palmer, Palmer/Kane Inc./The Stock Market. Page 868/© 1988 Michael Dalton/Fundamental Photographs. Page 869/© IBM Research/Peter Arnold, Inc.

Chapter 23 Page 880 (B)/© Science Photo Library/Photo Researchers, Inc. Page 880 (T)/Carey B. Van Loon. Page 894 (Both)/ Carey B. Van Loon. Page 895/Carey B. Van Loon. Page 896/Carey B. Van Loon.

Index

Page numbers followed by "f" refer to figures; page numbers followed by "n" refer to notes at the bottom of the page. Page numbers preceded by "A" refer to the Appendices.

B

C

F

T